Statistics for Social and Behavioral Sciences

Statistics for Social and Behavioral Sciences (SSBS) includes monographs and advanced textbooks relating to education, psychology, sociology, political science, public policy, and law.

More information about this series at http://www.springer.com/series/3463

Domingo Morales • María Dolores Esteban •
Agustín Pérez • Tomáš Hobza

A Course on Small Area Estimation and Mixed Models

Methods, Theory and Applications in R

Springer

Domingo Morales
Miguel Hernández University of Elche
Elche, Spain

María Dolores Esteban
Miguel Hernández University of Elche
Elche, Spain

Agustín Pérez
Miguel Hernández University of Elche
Elche, Spain

Tomáš Hobza
Czech Technical University in Prague
Prague, Czech Republic

ISSN 2199-7357 ISSN 2199-7365 (electronic)
Statistics for Social and Behavioral Sciences
ISBN 978-3-030-63756-9 ISBN 978-3-030-63757-6 (eBook)
https://doi.org/10.1007/978-3-030-63757-6

Mathematics Subject Classification: 62J12, 62P25, 62D05

This Springer imprint is published by the registered company Springer Nature Switzerland AG.
The registered company address is: Gewerbestrasse 11, 6330 Cham, Switzerland

Preface

Small area estimation (SAE) is a branch of statistical science that combines methods and tools of sampling and inference in finite populations, statistical models with random effects, and mathematical programming languages. Statisticians, both those from academic centers doing research in SAE and those from statistical offices who apply the SAE methodology to real data, need to be trained in the three disciplines to be competitive. SAE is basically a multidisciplinary branch of statistics, and this complicates the training of new researchers or statisticians specialized in its application. The university departments train researchers in statistical methodology, but there are still few doctoral programs specialized in SAE. The statistical offices have training programs, such as the "European Statistical Training Program (ESTP)" of EUROSTAT, where from time to time SAE courses are taught. Both in the doctorate courses and in the training programs, it is necessary to have books or manuals that facilitate the work of teaching and learning.

Currently, there are excellent books on SAE covering a broad spectrum of statistical inference procedures for SAE. These books try to cover most of the relevant theoretical developments. For this reason, some PhD students or applied statisticians have commented on the need to have a book with a different approach and that is complementary to the previous ones. The idea is to have a book that covers only some basic aspects of the theory and practice of SAE but that addresses those issues thinking that the reader is not an expert in sampling, statistical modeling, or programming languages.

This book aims to be useful to researchers and doctoral students. For this, the chapters expose the mathematical developments with plenty of details. In this way, the reader is provided with the follow-up and understanding of the usual reasoning in the research and development of new methodologies for SAE. The chapters of the book try to be self-contained, so that they can be read without having to necessarily read the previous chapters. The book wants to be useful to applied statisticians and statistical offices. For this, each chapter contains examples of application of SAE techniques to synthetic data. The data have been simulated by imitating the structure of labor force or living conditions surveys. Some of the socioeconomic indicators that are intended to be estimated are totals of unemployed people, unemployment rates, average annual net incomes, or poverty proportions and gaps. The examples are implemented in the R language, and the corresponding code is provided. For a better understanding, the programming is done in a simple language of didactic character. In this way, it is easier to identify the mathematical formulas and the methodology in the R code lines. As far as possible, functional programming is avoided. The contents of the book come from the subjects taught by the authors in SAE courses in universities and statistical offices and from the research they have done in recent years. This has conditioned the selection of contents, and it is also the reason why some very relevant aspects of the SAE theory are not covered by the book.

Chapter 1 gives some elementary comments on SAE and mixed models and presents the structure and description of the data files used in the examples with R. There are two files containing unit-level data and two files containing aggregated data at the domain level.

Chapters 2 and 3 introduce the most popular design-based estimators of domain means and totals and describe some resampling procedures for estimating their mean squared errors (MSEs). The book studies the direct estimators of Horvitz–Thompson and Hájeck type and the basic synthetic, post-stratified, and generalized regression estimators. It also treats the problem of adapting these estimators to complex design surveys, where inverse of inclusion probabilities (sampling weights) is corrected by non-response and calibrated to known population quantities for obtaining elevation factors.

As an alternative to the design-based approach, Chap. 4 gives an introduction to the prediction theory in finite populations and proves the general prediction theorem. This theorem gives the expression of the best linear unbiased predictor (BLUP) of a linear parameter and the corresponding prediction variance. Under the prediction theory, we think differently than under the theory of sampling in a finite and fixed population. For this reason, this chapter presents several examples in which the expression of the BLUP of a total is derived under linear models.

Chapter 5 presents some elements of linear models (LMs) in the framework of SAE problems. This chapter is devoted to the derivation of best linear unbiased predictors of domain linear parameters and presents some examples where the BLUP has an expression similar to that of some known estimator based on the distribution of the sample design.

Chapter 6 deals with linear mixed models (LMMs) and focusses on the methods and algorithms for estimating the model parameters of LMMs. Three methods for fitting LMMs are considered, namely the maximum likelihood (ML), the residual maximum likelihood (REML), and the Henderson 3 moment-based (H3) methods. The algorithms for implementing these methods are given, and two R functions (lmer from library lme4 and lme from library nlme) for fitting LMMs are described in examples of applications to the book data files.

Chapter 7 studies the basic unit-level model in SAE. This is the so-called nested error regression (NER) model. It gives the algorithms for calculating the ML, REML, and H3 estimators of the model parameters. The moments of the H3 estimator are also calculated. The chapter ends with a simulation experiment that compares the behavior of the three types of estimators.

Chapter 8 derives the empirical best linear unbiased predictor (EBLUP) of domain linear parameters under the NER model. It also introduces model-assisted estimators of domain means, which are constructed to be unbiased with respect to the design-based distribution. This chapter presents a design-based simulation study for comparing several SAE estimators. For this sake, an artificial population is generated and stratified random samples are drawn from the population.

Chapter 9 presents mathematical developments to approximate and estimate the MSEs of the BLUP and the EBLUP of a domain linear parameter. This chapter gives the mathematical derivations for the MSE of the BLUP with plenty of details. The derivation of the approximation of the MSE of the EBLUP and the properties of the corresponding estimator are only outlined. The chapter makes the particularizations of the obtained results to the NER model.

Chapter 10 addresses the problem of estimating domain nonlinear parameters under the NER model. It introduces the empirical best predictors (EBPs) for estimating additive parameters and treats the problem of estimating poverty indicators, such as poverty proportions and gaps. This chapter presents parametric bootstrap procedures for estimating the MSE of the EBP.

Chapters 11 and 12 deal with the two-fold nested error regression model. This model takes into account the variability between domains and between subdomains inside each domain. Chapter 11 introduces the model and develops the fitting methods. It also derives the EBLUPs of domain means and the corresponding MSE estimators. Chapter 12 contains the EBP theory for additive parameters with particularizations to the prediction of poverty proportions and gaps and of average income indicators.

In some SAE problems, we can intuitively expect that the slope parameters of some explanatory variable are not constant and therefore they should take different values in different domains. Chapter 13 studies the random regression coefficient (RRC) models, which give a practical solution to this problem by assuming that the regression parameters are random and therefore they give a more flexible way of modeling. Depending on the covariance structure of the random slopes, the chapter presents two RRC models and derives the EBLUPs of domain linear parameters with the corresponding MSE estimators.

Chapters 14 and 15 consider two different unit-level mixed models, namely the logit mixed model and the two-fold logit mixed model, which both belong to the class of generalized linear mixed models (GLMMs). These chapters adapt algorithms for fitting GLMMs to the considered logit mixed models. More concretely, the methods of simulated moments, EM and Laplace approximation, are considered. With regard to the estimation of domain proportions, EBPs are introduced. For calculating the EBPs, statisticians should employ an auxiliary census file, and this is a serious drawback in practice. This is why, the chapters also show how to compute the EBPs when the vector of auxiliary variables takes only a finite number of values. Both chapters give R codes with examples of applications to synthetic data.

SAE models can be formulated at the unit level or at the area level. The basic area-level model in SAE is the Fay–Herriot model. Chapter 16 studies this model, takes into account the problem of estimating the sampling error variances by using the generalized variance function approach, and introduces the EBLUPs of domain means with the corresponding MSE estimators. For estimating the model parameters, four types of estimators are considered. This is the only chapter that considers the Bayesian approach to SAE.

Chapters 17 and 18 generalizes the Fay–Herriot model by taking into account structures of temporal or spatial correlation. Chapter 17 studies two temporal area-level LMMs, the first one with independent time effects and the second one with AR(1)-correlated time effects. Chapter 18 deals with three spatial area-level LMMs. The first one does not include past data, while the other two models include spatial and temporal correlations.

Chapter 19 introduces a bivariate area-level model, gives algorithms to calculate the ML and REML estimators of model parameters, derives the EBLUPs of domain means, and approximates the matrix of MSEs. This is the only chapter devoted to multivariate LMMs.

Finally, Chaps. 20 and 21 deal with non-temporal and temporal area-level Poisson mixed models, respectively. These chapters present some fitting algorithms to estimate the model parameters, introduce the EBPs of functions of fixed and random effects, and give analytic and/or bootstrap procedures to estimate the corresponding MSEs.

This book does not cover many important methodologies of SAE. For example, it does not give predictors based on nonparametric, robust, or Bayesian approaches. The underlying philosophy is simple: to treat few topics in a depth and self-contained way.

This book is an introductory monograph for doctorate students and practitioners. It can be used as an auxiliary tool for SAE courses delivered in universities and statistical offices.

Elche, Spain — Domingo Morales
Elche, Spain — María Dolores Esteban
Elche, Spain — Agustín Pérez
Prague, Czech Republic — Tomáš Hobza
September 2020

Contents

1 Small Area Estimation ... 1
1.1 Introduction ... 1
1.2 Mixed Models ... 4
1.3 The Data Files ... 4
1.3.1 The LFS Data Files ... 5
1.3.2 The LCS Data Files ... 8
References ... 9

2 Design-Based Direct Estimation ... 13
2.1 Introduction ... 13
2.2 Survey Sampling Theory ... 14
2.3 Direct Estimator of the Total and the Mean ... 17
2.4 Estimator of the Ratio ... 21
2.5 Other Direct Estimators of the Mean and the Total ... 23
2.6 Bootstrap Resampling for Variance Estimation ... 26
2.7 Jackknife Resampling for Variance Estimation ... 27
2.7.1 Delete-One-Cluster Jackknife for Estimators of Domain Parameters ... 29
2.8 R Codes for Design-Based Direct Estimators ... 30
2.8.1 Horvitz–Thompson Direct Estimators of the Total and the Mean ... 30
2.8.2 Hájek Direct Estimator of the Mean and the Total ... 33
2.8.3 Jackknife Estimator of Variances ... 37
2.8.4 Functions for Calculating Direct Estimators ... 39
References ... 40

3 Design-Based Indirect Estimation ... 41
3.1 Introduction ... 41
3.2 Basic Synthetic Estimator ... 42
3.3 Post-Stratified Estimator ... 45
3.4 Sample Size Dependent Estimator ... 47
3.5 Generalized Regression Estimator ... 47

3.6 Estimators of Unemployment Rates ... 54
3.7 A Labor Force Survey ... 55
3.7.1 Weight Calibration and Benchmarking ... 58
3.7.2 Resampling Methods for the LFS ... 60
3.8 R Codes for Design-Based Indirect Estimators ... 62
3.8.1 Basic Synthetic Estimator of the Total ... 62
3.8.2 Post-stratified Estimator of the Total ... 66
3.8.3 Generalized Regression Estimator of the Mean ... 69
References ... 72

4 Prediction Theory ... 73
4.1 Introduction ... 73
4.2 The Predictive Approach ... 74
4.3 Prediction Theory Under the Linear Model ... 76
4.4 The General Prediction Theorem ... 80
4.5 BLUPs for Some Simple Models ... 85
4.6 R Codes for BLUPs ... 88
References ... 89

5 Linear Models ... 91
5.1 Introduction ... 91
5.2 Fixed Effects Linear Models ... 92
5.3 Linear Models with One Fixed Factor ... 97
5.4 BLUPs Based on Linear Models with Fixed Effects ... 101
5.4.1 Regression Synthetic Estimator ... 101
5.4.2 Estimators Without Domain Dependent Intercept ... 102
5.4.3 Estimators with Domain Dependent Intercept ... 104
5.5 R Codes for BLUPs ... 107
References ... 110

6 Linear Mixed Models ... 111
6.1 Introduction ... 111
6.2 Linear Mixed Models with Known Variances ... 112
6.2.1 Introduction ... 112
6.2.2 Least Squares Estimation of $\boldsymbol{\beta}$ ... 112
6.2.3 BLUP of a Linear Combination of Effects ... 113
6.3 Linear Mixed Models with Unknown Variances ... 117
6.4 Maximum Likelihood Estimation ... 119
6.4.1 Description of the Method ... 119
6.4.2 Maximum Likelihood Estimators for Alternative Parameters ... 122
6.5 Residual Maximum Likelihood Estimation ... 123
6.5.1 Description of the Method ... 123
6.5.2 REML Estimators for Alternative Parameters ... 127
6.5.3 Further REML Equations for Linear Mixed Models ... 128

6.6 Henderson 3 Estimation ... 137
6.6.1 Description of the Method ... 137
6.6.2 Moments of Henderson 3 Estimators ... 144
6.7 R Codes for Fitting Linear Mixed Models ... 148
6.7.1 Library lme4 ... 149
6.7.2 Library nlme ... 151
References ... 152

7 Nested Error Regression Models ... 155
7.1 Introduction ... 155
7.2 The NER Model ... 156
7.3 ML Estimators ... 159
7.4 ML Estimators for Alternative Parameters ... 163
7.5 REML Estimators ... 166
7.6 REML Estimators for Alternative Parameters ... 169
7.7 H3 Estimators ... 171
7.8 Moments of H3 Estimators ... 175
7.9 Simulation Experiment ... 181
7.10 R Codes ... 183
7.10.1 MLEs ... 183
7.10.2 Auxiliary Functions ... 185
References ... 187

8 EBLUPs Under Nested Error Regression Models ... 189
8.1 Introduction ... 189
8.2 The NER Model ... 190
8.3 BLUP of a Domain Mean ... 191
8.4 EBLUP of a Single Observation ... 192
8.5 Parametric Bootstrap Estimation of MSEs ... 194
8.6 Model-Assisted Estimation ... 195
8.7 Simulation Experiment ... 196
8.7.1 Artificial Population ... 197
8.7.2 Estimators and Performance Measures ... 198
8.7.3 Numerical Results and Conclusions ... 199
8.8 R Codes ... 201
8.8.1 EBLUPs for LFS Data ... 201
8.8.2 EBLUPs and MA Estimators for LCS Data ... 203
References ... 206

9 Mean Squared Error of EBLUPs ... 209
9.1 Introduction ... 209
9.2 The MSE of EBLUPs of Model Effects ... 210
9.2.1 All Model Parameters Are Known ... 211
9.2.2 Known Variances and Unknown Regression Parameters ... 212
9.2.3 All Model Parameters Are Unknown ... 214

9.3 The MSE of EBLUPs of Population Linear Parameters 219
9.4 Analytic Estimation of the MSE of EBLUPs 221
9.5 MSE Approximation in NER Models 223
9.6 MSE Estimation in NER Models 228
9.6.1 Henderson 3 Estimation of Variance Components 228
9.6.2 REML Estimation of Variance Components 229
9.6.3 ML Estimation of Variance Components 229
9.7 MSE Approximation in Linear Models with One Fixed Factor 231
9.8 Simulation Experiment 232
9.8.1 Samples 232
9.8.2 EBLUPs and MSEs 233
9.8.3 Algorithm 234
9.9 R Codes for MSEs 235
References 237

10 EBPs Under Nested Error Regression Models 239
10.1 Introduction 239
10.2 The Conditional Distribution of Normal Vectors 240
10.3 The Nested Error Regression Model 241
10.4 EBPs of Domain Means 245
10.5 EBPs of Additive Parameters 245
10.5.1 Poverty Proportion 249
10.5.2 Poverty Gap 250
10.5.3 Average Income 252
10.6 EBPs Under Subdomain-Level NER Models 253
10.6.1 Poverty Proportion 254
10.6.2 Poverty Gap 255
10.6.3 Average Income 255
10.7 ELL Predictors of Poverty Indicators 256
10.7.1 Poverty Proportion 257
10.7.2 Poverty Gap 257
10.7.3 Average Income 258
10.8 MSE of Empirical Best Predictors 259
10.8.1 Case 1 259
10.8.2 Case 2 260
10.8.3 Case 3 262
10.9 R Codes for EBPs 263
References 266

11 EBLUPs Under Two-Fold Nested Error Regression Models 267
11.1 Introduction 267
11.2 The Two-fold Nested Error Regression Model 268
11.3 The Model with Known Variance Components 269
11.4 REML Estimators for Alternative Parameters 271
11.4.1 Matrix Calculations 272

11.5 The Henderson 3 Method ... 274
11.5.1 Calculation of $\boldsymbol{M}_1$ ... 275
11.5.2 Calculation of $\boldsymbol{M}_2$ ... 277
11.5.3 Calculation of $\boldsymbol{M}_3$ ... 280
11.6 EBLUP of a Subdomain Mean ... 285
11.7 Mean Squared Error of the EBLUP of a Subdomain Mean ... 287
11.7.1 Calculation of $g_1(\boldsymbol{\theta})$ ... 287
11.7.2 Calculation of $g_2(\boldsymbol{\theta})$ ... 290
11.7.3 Calculation of $g_3(\boldsymbol{\theta})$ ... 291
11.7.4 Calculation of $g_4(\boldsymbol{\theta})$ ... 293
11.8 Simulation Experiments ... 294
11.8.1 Simulation 1 ... 294
11.8.2 Simulation 2 ... 295
11.9 R Codes for EBLUPs ... 297
References ... 299

12 EBPs Under Two-Fold Nested Error Regression Models ... 301
12.1 Introduction ... 301
12.2 Two-fold Nested Error Regression Models ... 302
12.2.1 The Population Model ... 302
12.2.2 The Sample Model ... 303
12.2.3 The Non-sample Model ... 303
12.2.4 The Inverse of the Variance Matrix ... 304
12.3 The Conditional Distribution of $\mathbf{y}_r$ given $\mathbf{y}_s$ ... 306
12.3.1 Conditional Mean Vector ... 306
12.3.2 Conditional Covariance Matrix ... 311
12.3.3 Conditional Variances ... 315
12.4 Monte Carlo EBP of an Additive Parameter ... 317
12.4.1 Introduction ... 317
12.4.2 Auxiliary Variables with Finite Number of Values ... 319
12.5 EBPs of Poverty Indicators ... 321
12.5.1 Poverty Proportion ... 321
12.5.2 Poverty Gap ... 322
12.6 EBPs of Average Income Indicators ... 323
12.7 Parametric Bootstrap MSE Estimator ... 324
12.8 R Codes for EBPs ... 325
References ... 330

13 Random Regression Coefficient Models ... 333
13.1 Introduction ... 333
13.2 The RRC Model with Covariance Parameters ... 334
13.2.1 The Model ... 334
13.2.2 REML Estimators ... 335
13.2.3 EBLUP of the Domain Mean ... 337

13.3 The RRC Model Without Covariance Parameters 339
13.3.1 The Model 339
13.3.2 REML Estimators 340
13.3.3 EBLUP of a Domain Mean 343
13.3.4 MSE of the EBLUP 344
13.4 R Codes for EBLUPs 349
References 351

14 EBPs Under Unit-Level Logit Mixed Models 353
14.1 Introduction 353
14.2 The Unit-Level Logit Mixed Model 354
14.3 MSM Algorithm 355
14.4 EM Algorithm 360
14.4.1 Introduction 360
14.4.2 EM Algorithm for the Logit Regression Model 362
14.5 ML-Laplace Approximation Algorithm 367
14.5.1 Introduction 367
14.5.2 The Laplace Approximation to the Likelihood 368
14.5.3 The AIC 372
14.6 Empirical Best Predictors 372
14.6.1 EBP of p_{dj} 373
14.6.2 EBP of μ_d and $\overline{\mu}_d$ 375
14.6.3 EBP of y_{dj} 377
14.6.4 EBP of $\overline{Y}_d$ 378
14.7 MSE of Empirical Best Predictors 380
14.7.1 Categorical Auxiliary Variables 380
14.7.2 Continuous Auxiliary Variables 382
14.8 R Codes for EBPs 386
References 389

15 EBPs Under Unit-Level Two-Fold Logit Mixed Models 391
15.1 Introduction 391
15.2 The Model 392
15.3 ML-Laplace Approximation Algorithm 393
15.3.1 The Laplace Approximation to the Likelihood 393
15.3.2 ML-Laplace Algorithm 397
15.3.3 Derivatives of $\boldsymbol{G}_d$ 398
15.3.4 AIC 402
15.4 Empirical Best Predictors 402
15.4.1 EBP of p_{dtj} 403
15.4.2 EBP of μ_{dt} and $\overline{\mu}_{dt}$ 405
15.4.3 EBP of y_{dtj} 407
15.4.4 EBP of $\overline{Y}_{dt}$ 408
15.5 MSE of Empirical Best Predictors 409
15.5.1 Bootstrap Estimation of the MSE of the EBP of μ_{dt} 409
15.5.2 Bootstrap Estimation of the MSE of the EBP of $\overline{Y}_{dt}$ 410

15.6 Simulation Experiment 411
15.7 R Codes for EBPs 412
References 416

16 Fay–Herriot Models 419
16.1 Introduction 419
16.2 BLUPs Under Area-Level Linear Mixed Models 421
16.3 The Area-Level Fay–Herriot Model 424
16.4 Sampling Error Variances 427
16.5 Estimation of Model Parameters 429
16.5.1 Prasad–Rao Estimator 429
16.5.2 Henderson 3 Estimator 429
16.5.3 Maximum Likelihood Method 432
16.5.4 Residual Maximum Likelihood Method 434
16.6 MSE of the EBLUP 437
16.6.1 Parametric Bootstrap 442
16.7 Bayesian Prediction 443
16.7.1 Unknown σ_u^2 447
16.8 Selection of Variables 451
16.8.1 Transformation of the Target Variable 451
16.8.2 Selection of Auxiliary Variables 453
16.9 R Codes for EBLUPs 454
References 458

17 Area-Level Temporal Linear Mixed Models 461
17.1 Introduction 461
17.2 Area-Level Model with Independent Time Effects 462
17.2.1 The Model 462
17.2.2 Residual Maximum Likelihood Estimation 463
17.2.3 EBLUP and Mean Squared Error 465
17.2.4 Simulations 470
17.3 Area-Level Model with Correlated Time Effects 472
17.3.1 The Model 472
17.3.2 Residual Maximum Likelihood Estimation 474
17.3.3 EBLUP and Mean Squared Error 476
17.3.4 Simulations 480
17.4 R Codes for EBLUPs 483
References 486

18 Area-Level Spatio-Temporal Linear Mixed Models 489
18.1 Introduction 489
18.2 Area-Level Spatial Linear Mixed Model 490
18.2.1 The Model 490
18.2.2 Fitting Methods Based on the Likelihood 491
18.2.3 Parametric Bootstrap Estimation of the MSE 494

18.3 Area-Level Spatio-Temporal Linear Mixed Model 1 495
18.3.1 The Model 495
18.3.2 Residual Maximum Likelihood Estimation 497
18.3.3 Simulations 498
18.4 Area-Level Spatio-Temporal Linear Mixed Model 2 501
18.4.1 The Model 501
18.4.2 Residual Maximum Likelihood Estimation 503
18.4.3 Simulations 504
18.5 R Codes for EBLUPs 505
References 506

19 Area-Level Bivariate Linear Mixed Models 509
19.1 Introduction 509
19.2 The Bivariate Fay–Herriot Model 510
19.3 Properties of the BLUPs 512
19.4 Maximum Likelihood Estimation 515
19.5 Residual Maximum Likelihood Estimation 518
19.6 The Matrix of Mean Squared Crossed Errors 522
19.7 Auxiliary Results 526
19.8 Simulations 531
19.9 R Codes for EBLUPs 534
19.9.1 Main Program 534
19.9.2 R Functions for the BFH Model 539
References 545

20 Area-Level Poisson Mixed Models 547
20.1 Introduction 547
20.2 The Model 548
20.3 MM Algorithm 549
20.4 EM Algorithm 552
20.5 ML-Laplace Approximation Algorithm 555
20.6 PQL Algorithm 560
20.7 Empirical Best Predictors 562
20.8 MSE of the EBP 564
20.8.1 Approximation of the MSE 565
20.8.2 Analytic Estimation of the MSE for MM Estimators 567
20.8.3 Bootstrap Estimation of the MSE 568
20.9 R Codes for EBPs 569
References 572

21 Area-Level Temporal Poisson Mixed Models 573
21.1 Introduction 573
21.2 The Model with Independent Time Effects 574
21.3 ML-Laplace Approximation Algorithm 576

21.4 Empirical Best Predictors 584
21.4.1 Bootstrap Estimation of the MSE 588
21.5 Simulation Experiment 588
21.6 R Codes for EBPs 590
References 593

A Some Useful Formulas 595

Index 597

Acronyms

AIC	Akaike information criterion
ANCOVA	Analysis of covariance
ANOVA	Analysis of variance
AR	Autoregressive
BFH	Bivariate Fay–Herriot
BLUE	Best linear unbiased estimator
BLUP	Best linear unbiased predictor
BP	Best predictor
EBLUE	Empirical best linear unbiased estimator
EBLUP	Empirical best linear unbiased predictor
EBP	Empirical best predictor
ELL	Elbers–Lanjouw–Lanjouw
EM	Expectation and maximization
ESTP	European statistical training program
EUROSTAT	European statistical office
FH	Fay–Herriot
GLM	Generalized linear model
GLMM	Generalized linear mixed model
GREG	Generalized regression
GVF	Generalized variance function
H3	Henderson 3
HB	Hierarchical Bayes
HT	Horvitz–Thompson
i.i.d.	Independent and identically distributed
LCS	Living conditions survey
LFS	Labor force survey
LGREG	Logistic generalized regression
LM	Linear model
LMM	Linear mixed model
MC	Monte Carlo
MCMC	Markov chain Monte Carlo

MFH	Multivariate Fay–Herriot
ML	Maximum likelihood
MLE	Maximum likelihood estimator
MM	Method of moments
MSE	Mean squared error
MSM	Method of simulated moments
NER	Nested error regression
p.d.f.	Probability density function
PQL	Penalized quasi-likelihood
REML	Residual maximum likelihood
RRC	Random regression coefficient
SAE	Small area estimation
SAR	Spatial or simultaneous autoregressive
SRS	Simple random sampling
SRSWOR	Simple random sampling without replacement
SRSWR	Simple random sampling with replacement
SSE	Sum of squared residuals

Chapter 1
Small Area Estimation

1.1 Introduction

Survey samples are designed and planned to produce parameter estimates of the population as a whole and of some specified subpopulations or domains. Sample sizes are selected in such a way that direct estimators, calculated using only the sample data from the corresponding domain, of parameters of planned domains achieve an a priori fixed precision. In practice, statisticians are often required to give estimates of domains that do not appear in the sampling design. Sample sizes in these unplanned domains may be too small for obtaining reliable direct estimates. To treat this problem, an easy solution would be to increase the sample sizes, but this may be too expensive or even impossible if we deal with data from the past surveys. A related approach would be to change the sampling design so that the unplanned domains become planned. Although this usually helps, it does not completely solve the problem. To employ indirect instead of direct estimators is thus a necessity. The use of models that include auxiliary variables and borrow strength from cross-sectional data or time and spatial correlation may produce more reliable estimates. Small area estimation (SAE) is a part of the statistical science that improves the efficiency of direct estimators by combining methodologies from survey sampling and finite population inference with statistical models.

SAE involves the estimation of parameters in small subsets (called small areas or domains) of an original population. A small area usually refers to a geographic territory, a demographic group, or a demographic group within a geographic region, where the sample size is small. The SAE setup often occurs if the sampling design is originally planned for estimating parameters of the whole population and not of its parts. In this context, the estimators of parameters have the desired precision for the whole population level but not at the small area level.

We can find a great number of works that describe in detail and exhaustively the existing theory of small area estimation. Among them, we can mention the review papers of Rao (1986, 1999, 2008), Ghosh and Rao (1994), Rao and Choudhry

D. Morales et al., *A Course on Small Area Estimation and Mixed Models*, Statistics for Social and Behavioral Sciences, https://doi.org/10.1007/978-3-030-63757-6_1

(1995), Pfeffermann (2002, 2013), Lahiri and Meza (2002), and Jiang and Lahiri (2006) and the monographs of Muckhopadhyay (1998), Rao (2003), Longford (2005), Rao and Molina (2015), and Pratesi (2016).

The direct estimators of domain parameters use only the data information of the considered domain. Horvitz and Thompson (1952) introduced a simple design-unbiased direct estimator of a domain mean as a sum of the sample values of the target variable multiplied by the sampling weights. By using weight calibration and non-response correction methods, many new weighted direct estimators can be found in the literature. The direct estimators do not use cross-sectional or temporal data. They are basically unbiased with respect to the sampling design distribution. However, they have large variances in small area estimation problems. Their estimated variances and coefficients of variation are usually greater than the ones of other more sophisticated estimators. The finding of these improved estimators is one of the main targets of SAE researchers.

SAE statistical techniques can be divided into three groups: design-based, model-assisted, and model-based methods. The three types of methodologies introduce and study estimators that are competitors of the direct estimators. The design-based approach to small area estimation looks for indirect estimators (basic synthetic, post-stratified, sample size dependent, and so on) with good properties with respect to the sampling design distribution. They employ auxiliary information from external data sources (from outside of the target domain), but they do not rely explicitly on models. For example, if the population sizes of domains crossed by sex–age groups are available from external data registers, then this information can be used for evaluating estimators that could provide better estimates than the direct ones. The design-based indirect estimators are optimized with respect to the sampling distribution. The design-based methods often use implicit models, although the bias and the variance of estimators are calculated with respect to the sampling design distribution; see e.g. Särndal et al. (1992), Lehtonen and Veijanen (2009), or Pratesi (2016). Some indirect estimators using implicit models are discussed in Rao (2003).

In the case of having auxiliary data related to the target variable, it is possible to obtain better accuracy for domain estimates by using explicit models, when compared to an estimation procedure not using auxiliary data. The model-assisted methodology considers the properties under the design-based distribution but employs explicit models to motivate the choice of estimators. Important examples of model-assisted estimators are the generalized regression (GREG) estimator, the calibration estimator introduced by Deville and Särndal (1992), and the LGREG estimator introduced by Lehtonen and Veijanen (1998). The GREG estimator uses a linear model as an assisting tool, estimates the domain mean of a continuous variable, and is constructed to be design-unbiased (or approximately so) irrespective of the fit of the model to data. The LGREG is a domain proportion estimator assisted by a logistic regression model. Different types of auxiliary data can be used in the model-assisted estimation. The GREG estimator uses auxiliary variables from survey files and their aggregated values from administrative registers. The LGREG estimator also uses a census file. They are estimators with a good balance between properties related to the design-based and the model-based distributions.

The model-based approach assumes that the data is generated by a true model, and therefore the inferences should be based on it. The use of explicit models in SAE gives an idea of how different sources of information are combined; see e.g. Fuller (1975), Fay and Herriot (1979), Holt et al. (1979), or Datta (2009). This approach can introduce estimators that may employ cross-sectional and temporal auxiliary information and that can take into account temporal and spatial correlation. The estimators have optimal properties with respect to the true model distribution. An important issue of this approach is the selection and the diagnostics of the selected model. A model with a good fit to data guarantees good model-based estimators with lower mean squared errors than the design-based estimators.

Based on the level of aggregation of the response variable, the small area models can be classified into two groups: (1) unit-level models and (2) area-level models. The basic SAE unit-level model is the nested error regression (NER) model. Battese et al. (1988) applied this model to the prediction of United States county crop areas using survey and satellite data. Since then, the empirical best linear unbiased predictors (EBLUPs) of domain means based on the NER model are being widely applied.

Many authors have further investigated the use of unit-level procedures for estimating nonlinear parameters that are not necessarily based on linear mixed models. Among the many contributions on this area, we cite, by way of example, some few papers related to poverty estimation. Molina and Rao (2010) derived empirical best predictors (EBPs) of nonlinear parameters based on the NER model, with applications to the estimation of poverty incidences and gaps. Hobza and Morales (2016) studied EBPs of poverty incidences based on unit-level logit mixed models. Tzavidis et al. (2008) and Marchetti et al. (2012) gave M-quantile estimators for poverty mapping. The SAE literature on unit-level model-based methods covers many other estimators and approaches not cited here.

The basic area-level model is the Fay–Herriot (FH) model. Fay and Herriot (1979) used an area-level linear mixed model to estimate the per capita income in small places of the United States of America. Several generalizations of the Fay–Herriot model have been applied to poverty estimation. For example, Esteban et al. (2012a,b), Marhuenda et al. (2013), and Morales et al. (2015) gave EBLUPs of Spanish poverty proportions based on temporal and spatio-temporal linear mixed models. Concerning generalized linear mixed models (GLMMs), López-Vizcaíno et al. (2013, 2015) and Boubeta et al. (2016, 2017) introduced EBPs of counts and proportions based on multinomial logit and Poisson area-level mixed models, respectively, with applications to Spanish data, and Chandra et al. (2017) gave small area predictors of counts under a non-stationary spatial GLMM model. The SAE literature on area-level model-based methods covers many other estimators employing nonparametric, robust, or Bayesian regression procedures, which are not described in this book.

1.2 Mixed Models

Linear models (LMs) assume that observations are drawn from the same population and are independent. Mixed models have a more complex multilevel or hierarchical structure. Observations in different levels or clusters are assumed to be independent, but observations within the same level or cluster are considered as dependent because they share common properties. For these data, we can speak about two sources of variation: between and within clusters. The possibility of modeling those sources of variation, commonly present in real data, gives a high flexibility, and therefore applicability, to mixed models.

Linear mixed models (LMMs) handle data in which observations are not independent. That is, LMMs correctly model correlated errors, whereas the procedures in the LM family usually do not. Mixed models are generalizations of LMs to better support the analysis of a dependent variable. Similarly, GLMMs are generalizations of LMMs for fitting non-normal target variables. In addition, they allow a linear prediction of a transformation of the expected value of the target variable.

Mixed models may incorporate random and hierarchical effects for treating repeated measures, spatial and temporal correlations, and SAE problems. Because of the flexibility and ability to combine different sources of information, mixed models are good tools for small area estimation. Usually, mixed models incorporate area random effects that explain the variability between areas not explained by the fixed effects.

This monograph studies unit-level and area-level linear mixed models as tools for construction of estimators of small area parameters with optimal properties. Among them, the NER and the FH models play a relevant role. It also describes some GLMMs for binomial and Poisson responses. The monograph assumes that the values of the auxiliary variables are known, so it does not consider models with random design or measurement errors.

Books dealing with LMMs and GLMMs include Searle et al. (1992), Longford (1995), McCullogh and Searle (2001), Goldstein (2003), Demidenko (2004), and Jiang (2007).

1.3 The Data Files

In all chapters of this monograph, the use of the described methodology is illustrated by practical applications, i.e. some R codes for fitting mixed models and calculating SAE estimators are given. At the website, https://github.com/small-area-estimation/sae-book, the reader can download all the R codes by chapter. This section describes the two data files employed in the examples throughout this book. The data files contain simulated data imitating the structure of a labor force survey and an income and living condition survey.

1.3.1 The LFS Data Files

The file `LFS20.txt` contains unit-level data from a labor force survey. The file stores the scores of 11 variables measured on 1050 individuals selected from a country with 20 regions. The variables are AREA (region), CLUSTER (primary sampling unit), AGE (in years), SEX (1 = men and 2 = women), WEIGHT (calibrated sample weight), EMPLOYED (1 = yes and 0 = no), UNEMPLOYED (1 = yes and 0 = no), INACTIVE (1 = yes and 0 = no), REGISTERED (1 if registered as an employment demandant, 0 otherwise) EDUCATION (1 = primary, 2 = secondary, and 3 = superior), and INCOME (annual net income in euros).

The following R code reads the data file `LFS20.txt` and calculates the number of columns, rows, areas, and sex categories.

```
dat <- read.table("LFS20.txt", header=TRUE, sep = "\t", dec = ".")

datcol <- ncol(dat); datcol           # number of columns (variables) in dat
n <- nrow(dat)                        # number of rows (cases) in dat
narea <- length(unique(dat$AREA))     # number of areas
nsex <- length(unique(dat$SEX))       # number of sex categories
output1 <- data.frame(area=dat$AREA, cluster=dat$CLUSTER, age=dat$AGE,
           sex=dat$SEX, weight=dat$WEIGHT, employed=dat$EMPLOYED,
           unemployed=dat$UNEMPLOYED, inactive=dat$INACTIVE,
           registered=dat$REGISTERED, education=dat$EDUCATION,
           income=dat$INCOME)
head(output1, 10)                     # some survey data
```

For AREA=1 and CLUSTER=1, Table 1.1 gives the scores of all the variables in the 10 first individuals.

The following code calculates the variable AGEG (age group), which takes the values 1 if AGE<25, 2 if 25≤AGE<54, and 3 if AGE≥54. The R function `table` can be used to calculate the sample sizes per AREA crossed by SEX and by SEX and AGEG.

```
summary(dat$AGE)
AGEG <- cut(dat$AGE, breaks=c(0,25,54,100), labels=c(1,2,3))
AGEG <- as.numeric(AGEG)            # variable age group
table(AGEG, dat$AREA, dat$SEX)      # sample sizes per age group, area and sex
table(dat$SEX, dat$AREA)            # sample sizes per sex and area
```

Table 1.1 Variables of `LFS20.txt` for the 10 first individuals

AGE	SEX	WEIGHT	EMPL	UNEMPL	INACT	REG	EDU	INCOME
23	2	154	0	1	0	1	2	25,883
55	1	190	1	0	0	0	1	47,249
49	2	174	1	0	0	0	1	34,052
41	1	231	1	0	0	0	2	48,149
40	2	231	1	0	0	0	3	56,230
38	2	196	1	0	0	0	2	41,829
39	1	266	1	0	0	0	2	34,071
37	2	266	1	0	0	0	2	42,344
29	1	325	1	0	0	0	1	48,192
28	2	325	0	0	1	0	1	42,320

Table 1.2 Sample sizes per AREA, SEX and AGEG

SEX	AGEG	1	2	3	4	5	6	7	8	9	10
1	1	6	1	2	7	2	6	5	5	10	4
1	2	20	7	15	15	15	17	9	12	35	8
1	3	3	10	7	5	5	7	9	6	18	7
2	1	3	8	3	7	8	3	5	5	10	6
2	2	16	6	8	16	16	7	14	15	33	12
2	3	12	5	12	5	4	3	6	5	19	4

SEX	AGEG	11	12	13	14	15	16	17	18	19	20
1	1	2	3	7	3	4	5	3	4	6	4
1	2	11	16	13	8	12	11	18	7	12	10
1	3	3	13	8	7	8	9	10	5	6	7
2	1	2	6	3	4	3	2	11	1	5	4
2	2	14	19	18	9	11	15	13	14	21	12
2	3	3	6	12	3	7	11	13	5	5	9

Table 1.3 Sample sizes per AREA and SEX

	1	2	3	4	5	6	7	8	9	10
1	29	18	24	27	22	30	23	23	63	19
2	31	19	23	28	28	13	25	25	62	22

	11	12	13	14	15	16	17	18	19	20
1	16	32	28	18	24	25	31	16	24	21
2	19	31	33	16	21	28	37	20	31	25

Table 1.2 gives the sample sizes per AREA (columns) and SEX and AGEG (rows). Table 1.3 gives the sample sizes per AREA (columns) and SEX (rows).

The file `Ndsa20.txt` contains the auxiliary aggregated data. The variables AREA, SEX, and AGEG are denoted here by area, sex, and age, respectively. The column N contains the population sizes per area, sex, and age. The columns reg, edu1, and edu2 contain the population totals of the auxiliary variables REGISTERED (level 1) and EDUCATION (levels 1, 2, and 3). The following R code reads the data file `Ndsa20.txt`.

```
dataux <- read.table("Ndsa20.txt", header=TRUE, sep = "\t", dec = ".")
# Population sizes by area, sex and age group
output2 <- data.frame(area=dataux$area, sex=dataux$sex, age=dataux$age,
           N=dataux$N, reg=dataux$reg, edu1=dataux$edu1, edu2=dataux$edu2,
           edu3=dataux$edu3)
head(output2, 10)
dim(dataux)     # file dimensions
```

Table 1.4 presents the first 10 registers of `Ndsa20.txt`. Table 1.5 gives the population sizes (values of N) per area (columns), sex, and age (rows).

The file `Nds20.txt` contains population sizes per area and sex. It also contains the corresponding totals of reg, edu1, edu2, and edu3. The data of this file are derived from the data in `Ndsa20.txt`. The following R code reads the data file `Nds20.txt`.

Table 1.4 Variables of `Ndsa20.txt` for the 10 first registers

area	sex	age	N	reg	edu1	edu2	edu3
1	1	1	1397	8	668	529	200
1	1	2	6097	12	2789	2196	1112
1	1	3	526	5	516	8	2
1	2	1	645	170	185	458	2
1	2	2	4199	7	659	3176	364
1	2	3	2941	16	2235	696	10
2	1	1	96	141	86	8	2
2	1	2	1769	222	447	1298	24
2	1	3	1711	6	1542	146	23
2	2	1	1701	232	661	1014	26

Table 1.5 Population sizes per area, sex, and age

sex	age	1	2	3	4	5	6	7	8	9	10
1	1	1397	96	301	1089	224	1149	829	775	1405	733
1	2	6097	1769	2952	2713	2410	3160	1848	2196	6641	1352
1	3	526	1711	1193	1005	618	1152	1346	845	2739	1171
2	1	645	1701	337	1256	1265	431	833	690	1420	901
2	2	4199	1245	1315	2852	2534	1050	2472	2504	5371	1964
2	3	2941	814	2068	806	411	569	1021	784	2971	540
sex	age	11	12	13	14	15	16	17	18	19	20
1	1	423	473	989	457	474	747	404	533	942	575
1	2	2188	3120	2520	1554	2165	2060	3135	1458	2592	2004
1	3	447	1876	1223	1243	1246	1455	1520	652	774	1143
2	1	384	823	324	481	560	370	1506	122	799	674
2	2	2446	3137	3116	1473	1998	2356	2298	2290	3898	2055
2	3	619	1006	1924	511	871	1668	1922	666	835	1346

Table 1.6 Population sizes per area and sex

sex	1	2	3	4	5	6	7	8	9	10
1	8020	3576	4446	4807	3252	5461	4023	3816	10,785	3256
2	7785	3760	3720	4914	4210	2050	4326	3978	9762	3405
sex	11	12	13	14	15	16	17	18	19	20
1	3058	5469	4732	3254	3885	4262	5059	2643	4308	3722
2	3449	4966	5364	2465	3429	4394	5726	3078	5532	4075

```
dataux <- read.table("Nds20.txt", header=TRUE, sep = "\t", dec = ".")
# Population sizes by area and sex
output3 <- data.frame(area=dataux$area, sex=dataux$sex, N=dataux$N,
           reg=dataux$reg, edu1=dataux$edu1, edu2=dataux$edu2,
           edu3=dataux$edu3)
head(output3, 10)
dim(dataux)            # file dimensions
```

Table 1.6 gives the population sizes per area (columns) and sex (rows).

1.3.2 The LCS Data Files

The file `datLCS.txt` contains unit-level data from a living conditions survey. The file stores the scores of 6 variables measured on 2512 individuals and selected from a country with 26 regions (domains). The variables are dom (domain), sex ($1=$ men and $2=$ women), house (household identification number), w (calibrated sampling weight), income (annual net equivalent income in euros), and lab (labor status, 0 if <16 years, $1=$ employed, $2=$ unemployed, and $3=$ inactive). The following R code reads the data file `datLCS.txt` and calculates the number of columns and rows. It also calculates the sample sizes per domain and sex.

```
dat <- read.table("datLCS.txt", header=TRUE, sep = "\t", dec = ",")
dat <- dat[order(dat$dom),]              # sort dat by dom
datcol <- ncol(dat)                      # number of columns (variables) in dat
n <- nrow(dat)                           # number of rows (cases) in dat
ndom <- length(unique(dat$dom))          # number of domains in dat
output4 <- data.frame(dom=dat$dom, sex=dat$sex, house=dat$house, w=dat$w,
               income=dat$income, lab=dat$lab)
head(output4, 10)                        # some survey data
table(dat$sex, dat$dom)                  # sample sizes per dom and sex
```

Table 1.7 presents the first 10 registers of dom $=3$. Table 1.8 gives the sample sizes per domain (columns) and sex (rows).

The file `auxLCS.txt` contains the auxiliary aggregated data. This file contains the scores of some variables that are aggregated by domains and sex categories. The variables are dom (domain), TOT (domain size), Mwork (domain mean of lab $=1$), Mnowork (domain mean of lab $=2$), and Minact (domain mean of lab $=3$). The

Table 1.7 Variables of `datLCS.txt` for the 10 first individuals

dom	sex	house	w	income	lab
3	2	3362	1939.533	7373.34	1
3	1	7442	2094.591	7330.40	2
3	2	7442	2094.591	7330.40	3
3	2	7443	2364.698	10021.20	3
3	1	7444	1610.814	4488.24	2
3	2	7444	1610.814	4488.24	0
3	1	7444	1610.814	4488.24	0
3	2	7444	1610.814	4488.24	1
3	1	7445	2018.978	13214.31	1
3	2	7445	2018.978	13214.31	1

Table 1.8 Sample sizes per domain and sex categories

	3	5	6	7	11	12	13	14	15	16	17	18	20
1	24	41	33	4	51	8	71	91	202	37	6	19	63
2	33	55	49	6	67	10	67	99	204	56	6	16	62

	21	22	23	24	25	27	28	29	30	31	32	33	34
1	26	7	15	34	40	35	27	34	70	28	147	65	37
2	23	6	25	31	39	47	30	35	65	30	146	67	23

Table 1.9 Variables of auxLCS.txt for the 10 first domains

dom	TOT	Mwork	Mnowork	Minact
3	82,001	0.3632226	0.127643	0.357414
5	251,866	0.3564652	0.155038	0.319212
6	190,653	0.3405221	0.158602	0.315825
7	24,699	0.3189367	0.121846	0.428671
11	154,625	0.3695341	0.163450	0.313515
12	90,315	0.3631095	0.096680	0.375870
13	223,742	0.3767652	0.123294	0.291349
14	346,216	0.3825718	0.146469	0.303392
15	779,492	0.371667	0.121645	0.354209
16	172,916	0.3008403	0.167355	0.296122

Table 1.10 Population sizes per domain

dom	3	5	6	7	11	12	13	14	15
size	82,001	251,866	190,653	24,699	154,625	90,315	223,742	346,216	779,492
dom	16	17	18	20	21	22	23	24	25
size	172,916	39,723	43,082	220,284	80,685	16,804	72,357	89,858	175,305
dom	27	28	29	30	31	32	33	34	
size	110,543	52,874	170,015	178,329	183,524	474,950	287,071	365,059	

following R code reads the data file auxLCS.txt and calculates the number of columns and rows. It also sorts the data by domains.

```
aux <- read.table("auxLCS.txt", header=TRUE, sep = "\t", dec = ",")
aux <- aux[order(aux$dom),]        # sort aux by dom
output5 <- data.frame(dom=aux$dom, TOT=aux$TOT, Mwork=aux$Mwork,
          Mnowork=aux$Mnowork, Minact=aux$Minact, ss=aux$ss)
head(output5,10)
dim(aux)          # file dimensions
```

Table 1.9 presents the aggregated data of the first 10 domains appearing in auxLCS.txt. Table 1.10 gives the population sizes (values of TOT) per domain.

References

Battese, G.E, Harter, R.M., Fuller, W.A.: An error-component model for prediction of county crop areas using survey and satellite data. J. Am. Stat. Assoc. **83**, 28–36 (1988)

Boubeta, M., Lombardía, M.J., Morales, D.: Empirical best prediction under area-level Poisson mixed models. TEST **25**, 548–569 (2016)

Boubeta, M., Lombardía, M.J., Morales, D.: Poisson mixed models for studying the poverty in small areas. Comput. Stat. Data Anal. **107**, 32–47 (2017)

Chandra, H., Salvati, N., Chambers, R.: Small area prediction of counts under a non-stationary spatial model. Spat. Stat. **20**, 30–56 (2017)

Datta, G.S.: Model-based approach to small area estimation. Handbook Stat. Samp. Surv. Infer. Anal. **29**(Part B), 251–288 (2009)

Demidenko, E.: Mixed Models, Theory and Applications. John Wiley, New York (2004)

Deville, J.C., Särndal, C.E.: Calibration estimators in survey sampling. J. Am. Stat. Assoc. **87**(418), 376–382 (1992)
Esteban, M.D., Morales, D., Pérez, A., Santamaría, L.: Two area-level time models for estimating small area poverty indicators. J. Indian Soc. Agricultural Stat. **66**(1), 75–89 (2012a)
Esteban, M.D., Morales, D., Pérez, A., Santamaría, L.: Small area estimation of poverty proportions under area-level time models. Comput. Stat. Data Anal. **56**, 2840–2855 (2012b)
Fay, R.E., Herriot, R.A.: Estimates of income for small places: an application of James-Stein procedures to census data. J. Am. Stat. Assoc. **74**, 269–277 (1979)
Fuller, W.A.: Regression analysis for sample survey. Sankhya **37**(3), 117–132 (1975)
Ghosh, M., Rao, J.N.K.: Small area estimation: an appraisal (with discussion). Stat. Sci. **9**, 55–93 (1994)
Goldstein, H.: Multilevel Statistical Models. Arnold, London (2003)
Hobza, T., Morales, D.: Empirical best prediction under unit-level logit mixed models. J. Off. Stat. **32**(3), 661–69 (2016)
Holt, D., Smith, T.M.F., Tomberlin, T.J.: A model-based approach to estimation for small subgroups of a population. J. Am. Stat. Assoc. **74**, 405–410 (1979)
Horvitz, D.G., Thompson, D.J.: A generalization of sampling without replacement from a finite universe. J. Am. Stat. Assoc. **47**, 663–685 (1952)
Jiang, J., Lahiri P.: Mixed model prediction and small area estimation. TEST **15**, 1–96 (2006)
Jiang, J.: Linear and Generalized Linear Mixed Models and Their Applications. Springer, Berlin (2007)
Lahiri, P., Meza, J.L.: Small-area estimation. In: El-Shaarawi, A.H., Piegorsch, W.W. (eds.) Encyclopedia of Environmetric, 4 p. 2010–2014. John Wiley & Sons, Chichester (2002)
Lehtonen, R., Veijanen, A.: Logistic generalized regression estimators. Surv. Methodol. **24**(1), 51–55 (1998)
Lehtonen, R., Veijanen, A.: Design-based methods of estimation for domains and small areas. In: Pfeffermann, D., Rao, C.R. (eds.) Sample Surveys – Inference and Analysis, vol. 29B, pp. 219–249. Elsevier, Amsterdam (2009)
Longford, N.T.: Random Coefficient Models. Clareton Press, London (1995)
Longford, N.T.: Missing Data and Small-Area Estimation: Modern Analytical Equipment for the Survey Statistician. Springer, Berlin (2005)
López-Vizcaíno, E., Lombardía, M.J., Morales, D.: Multinomial-based small area estimation of labour force indicators. Stat. Model. **13**(2), 153–178 (2013)
López-Vizcaíno, E., Lombardía, M.J. and Morales, D.: Small area estimation of labour force indicators under a multinomial model with correlated time and area effects. J. R. Stat. Assoc. Ser. A **178**(3), 535–565 (2015)
Marchetti, S., Tzavidis, N., Pratesi, M.: Non-parametric bootstrap mean squared error estimation for M-quantile estimators of small area averages, quantiles and poverty indicators. Comput. Stat. Data Anal. **56**, 2889–2902 (2012)
Marhuenda, Y., Molina, I., Morales, D.: Small area estimation with spatio-temporal Fay-Herriot models. Comput. Stat. Data Anal. **58**, 308–325 (2013)
McCullogh, C.E., Searle, S.R.: Generalized, Linear and Mixed Models. John Wiley, New York (2001)
Molina, I., Rao, J.N.K.: Small area estimation of poverty indicators. The Canadian Journal of Statistics **38**, 369–385 (2010)
Morales, D., Pagliarella, M.C., Salvatore, R.: Small area estimation of poverty indicators under partitioned area-level time models. Stat. Oper. Res. Trans. **39**(1), 19–34 (2015)
Muckhopadhyay, P.: Small Area Estimation in Survey Sampling. Narosa Publishing House, New Delhi (1998)
Pfeffermann, D.: Small area estimation – new developments and directions. Int. Stat. Rev. **70**, 125–143 (2002)
Pfeffermann, D.: New important developments in small area estimation. Stat. Sci. **28**, 40–68 (2013)
Pratesi, M.: Analysis of poverty data by small area estimation. Wiley Series in Survey Methodology. J. Wiley, New Jersey (2016)

Rao, J.N.K.: Synthetic estimators, SPREE and the best model based predictors. In: Proceedings of the Conference on Survey Methods in Agriculture, pp. 1–16. U. S. Department of Agriculture, Washington (1986)

Rao, J.N.K.: Some recent advances in model-based small area estimation. Surv. Methodol. **25**, 175–186 (1999)

Rao, J.N.K.: Small Area Estimation. John Wiley, Hoboken (2003)

Rao, J.N.K.: Some methods for small area estimation. Rivista Internazionale di Scienze Sociali. **4**, 387–406 (2008)

Rao, J.N.K., Choudhry, G.H.: Small area estimation: overview and empirical study. In: Cox, B., Chinnappa, C., Colledge, K. (eds.) Business Survey Methods, pp. 527–542. John Wiley, Hoboken (1995)

Rao, J.N.K., Molina, I.: Small Area Estimation, 2nd edn. Wiley, Hoboken (2015)

Särndal, C.E., Swensson, B., Wretman, J.: Model Assisted Survey Sampling. Springer, Berlin (1992)

Searle, S.R., Casella, G., McCullogh, C.E.: Variance Components. John Wiley, New York (1992)

Tzavidis, N., Salvati, N., Pratesi, M., Chambers, R.: M-quantile models with application to poverty mapping. Stat. Method. Appl. **17**, 393–411 (2008)

Chapter 2
Design-Based Direct Estimation

2.1 Introduction

Survey samples provide useful information about a population and avoid the need of carrying out the more expensive and time-consuming censuses. Sampling theory covers sampling designs and inference procedures for finite populations. If the population is partitioned in domains, the estimators of parameters of the global populations can be adapted and applied to estimate domain parameters. This can be done by treating the domains as independent new populations. This approach to small area estimation yields to design-based direct estimators.

The estimation of small area parameters, like domain means, totals, or ratios of a target variable, is an inference problem in finite populations. Historically, the first estimators of population parameters defined at the domain level were adaptations of the corresponding estimators defined for the global population. Direct estimators use only the data of the target variable in the domain of interest, and their properties are studied and optimized with respect to the probability distribution of the sample design. They do not use data from other domains or time periods. Since direct estimators are simple and intuitive, researchers use them as a benchmark to establish comparisons and to measure the efficiency gain obtained by using more sophisticated small area estimators.

This manuscript dedicates an initial chapter to introduce the basic concepts and tools of sampling and inference in finite populations. Inclusion probabilities and their inverses (sampling weights) play here a relevant role. For estimating means and totals, two types of direct estimators are considered. They were introduced by Horvitz and Thompson (1952) and Hájek (1971), respectively. For estimating ratios, plug-in estimators are employed. They are defined by substituting totals by their corresponding direct estimators.

The chapter gives a short introduction to the survey sampling theory and describes some properties of direct estimators, with special emphasis on estimators of means, totals, and ratios. For each estimator, the design-based expectation and

D. Morales et al., *A Course on Small Area Estimation and Mixed Models*, Statistics for Social and Behavioral Sciences, https://doi.org/10.1007/978-3-030-63757-6_2

variance are calculated and a direct estimator of the variance is given. In many practical cases, only first order inclusion probabilities are available, and therefore it is not possible to calculate unbiased direct estimators of variances. This is why the chapter also presents design-based resampling methods, like bootstrap and Jackknife, for variance estimation. The last section contains some examples giving R codes, including functions for calculating domain-level direct estimators.

2.2 Survey Sampling Theory

A *finite population* is a collection of different units, such as people, companies, households, hospitals, and so on. The *survey sampling theory* deals with the selection of samples (subsets of the population), the observation of characteristics of sampled units, and the use of the obtained data for doing inferences about the population.

Survey sampling is interested in a fixed population from which a part is observed. In other branches of statistics, observations are realizations of random variables, and the inferences are not referred to any actual population, but to a probability law on the random variables. The following example clarifies this point.

Example 2.1 An industry is interested in determining if the units of a production line fulfill some given specifications. By assuming the general approach to statistics, we can model the data (CORRECT $=0$ and DEFECTIVE $=1$) as realizations of independent and identically distributed Bernoulli variables with parameter θ. The statistical target is the estimation of the probability θ of making a defective unit. The problem becomes a finite population survey sampling problem if we are only interested in the units produced during a given day. In the last case, we are interested in estimating the proportion

$$p=\frac{\text{number of defective units produced during the day}}{\text{number of units produced during the day}}.$$

In survey sampling, there are two main approaches. The first one assumes that the data obey the probability distribution given by the random extraction of samples from the population. This is the design-based approach. In the second case, the scores of the target variable are assumed to be the realization of a random vector with distribution given by a statistical model. This is the model-based approach. The inference procedures are built and studied depending on the assumed probability distribution.

Under the design-based approach, the vector containing the values of a variable y in all the population units $(y_1,\ldots,y_N)$ is the basic parameter. A *probabilistic sampling plan (or design)* is a scheme for choosing the samples, such that each subset s of the population U has a known selection probability $p(s)$. Let us consider a population parameter T and its estimator $\widehat{T}$ based on s. The definitions of bias and variance of $\widehat{T}$ are based on $p(s)$, i.e.

$$\begin{aligned}&\text{BIAS:} && E_\pi[\widehat{T} - T] = \textstyle\sum_{s \subset U} p(s)[\widehat{T}(s) - T],\\ &\text{VARIANCE:} && \text{var}_\pi(\widehat{T}) = \textstyle\sum_{s \subset U} p(s)\big(\widehat{T}(s) - E_\pi[\widehat{T}]\big)^2.\end{aligned}$$

We use the notations E_π and var_π to emphasize the fact that we have expectations and variances with respect to the design-based probability distribution $p(s)$. Expectations and variances with respect to a model-based distribution are denoted by E_M and var_M.

In general, the calculation of $p(s)$ is not an easy task. Some simple cases are the simple random samplings with replacement (SRSWR) and without replacement (SRSWOR), i.e.

$$\begin{aligned}p(s) &= \tfrac{1}{N^n} \text{ for a SRSWR sample } s \text{ of size } n,\\ p(s) &= \tfrac{1}{\binom{N}{n}} \text{ for a SRSWOR sample } s \text{ of size } n.\end{aligned}$$

However, many calculations only require the inclusion probabilities π_i and π_{ij}, i.e.

$\pi_i = P(i \in s) = \sum_{s \in s(i)} p(s)$, where $s(i) = \{s \subset U : i \in s\}$ is the set of samples containing the unit i,

$\pi_{ij} = P(i \in s, j \in s) = \sum_{s \in s(i,j)} p(s)$, where $s(i, j) = \{s \subset U : i, j \in s\}$ is the set of samples containing the units i and j.

For example, under the SRSWOR, the inclusion probabilities are

$$\pi_i = n/N, \quad \pi_{ij} = \frac{n(n-1)}{N(N-1)} \quad \text{for } i, j \in U,\ i \neq j.$$

The following definition will be useful in some of the proofs.

Definition 2.1 The sampling design indicator functions are

$$\delta_i(s) = \begin{cases} 1 & \text{if the unit } i \text{ is in the sample } s \\ 0 & \text{otherwise} \end{cases} \overset{d}{=} \text{Bernoulli}(\pi_i).$$

It holds that

$$\begin{aligned}&(1)\ \textstyle\sum_{i=1}^N \delta_i(s) = n, && (2)\ P(\delta_i(s) = 1) = 1 - P(\delta_i(s) = 0) = \pi_i,\\ &(3)\ P(\delta_i(s) = 1, \delta_j(s) = 1) = \pi_{ij}, && (4)\ \pi_{ii} = \pi_i,\\ &(5)\ E_\pi[\delta_i(s)] = E_\pi[\delta_i^2(s)] = \pi_i, && (6)\ E_\pi[\delta_i(s)\delta_j(s)] = \pi_{ij},\\ &(7)\ \text{var}_\pi(\delta_i(s)) = \pi_i(1 - \pi_i), && (8)\ \text{cov}_\pi(\delta_i(s), \delta_j(s)) = \pi_{ij} - \pi_i\pi_j.\end{aligned}$$

In what follows, we simplify the notation and write δ_j instead of $\delta_j(s)$. Further, we consider only sampling without replacement, and we use the following notations:

- *Indexes:* s denotes a sample, and $d = 1, \ldots, D$, $j = 1, \ldots, N$, and $g = 1, \ldots, G$ denote domains (or small areas), units (or individuals), and groups, respectively.
- *Population and sample:* $U = \bigcup_{d=1}^{D} U_d$ for population and $s = \bigcup_{d=1}^{D} s_d$ for sample, where U_d and s_d are population and sample in domain d, respectively.
- *Sizes:* N for population and n for sample. When N and n have subindexes, they denote the corresponding size of the indexed set. For example, N_d is the population size of domain d.
- *Totals:* Y and X denote the population totals of variables y and x, respectively. If Y and X have subindexes, then they denote the corresponding totals of the indexed set.
- *Means:* $\overline{Y}$ and $\overline{X}$ denote the population means of variables y and x, respectively. If $\overline{Y}$ and $\overline{X}$ have subindexes, then they denote the corresponding means of the indexed set. For example, $\overline{Y}_d$ denotes the population mean of domain d.
- *Sampling weights:* w_j are the theoretical weights of the sampling design. They are the inverses of the inclusion probabilities, i.e. $w_j = 1/\pi_j$.

Example 2.2 For any individual j, interviewed at a labor force survey, some variables of interest are

$$y_j = \begin{cases} 1 \text{ if } j \text{ is unemployed}, \\ 0 \text{ otherwise}, \end{cases} \quad z_j = \begin{cases} 1 \text{ if } j \text{ is employed}, \\ 0 \text{ otherwise}, \end{cases} \quad t_j = \begin{cases} 1 \text{ if } j \text{ is inactive}, \\ 0 \text{ otherwise}. \end{cases}$$

Some target parameters are the totals of unemployed, employed, and inactive people and the unemployment rate, i.e.

$$Y_d = \sum_{j \in U_d} y_j, \quad Z_d = \sum_{j \in U_d} z_j, \quad T_d = \sum_{j \in U_d} t_j, \quad \text{and} \quad R_d = \frac{Y_d}{Y_d + Z_d} = \frac{\overline{Y}_d}{\overline{Y}_d + \overline{Z}_d},$$

where $\overline{Y}_d = Y_d/N_d$, $\overline{Z}_d = Z_d/N_d$, and N_d is the size of area d.

The following sections give estimators of the domain total and mean of a variable y, i.e.

$$Y_d = \sum_{j \in U_d} y_j, \quad \overline{Y}_d = \frac{1}{N_d} \sum_{j \in U_d} y_j.$$

Let us note that we assume that the units in U_d can be numbered, and in what follows, we sometimes use the notation

$$\sum_{j \in U_d} y_j = \sum_{j=1}^{N_d} y_j.$$

2.3 Direct Estimator of the Total and the Mean

Horvitz and Thompson (1952) proposed the following *direct* estimators of the total Y_d and the mean $\overline{Y}_d$ of domain d:

$$\hat{Y}_d^{dir1} = \sum_{j \in s_d} w_j y_j = \sum_{j \in s_d} \frac{1}{\pi_j} y_j, \quad \hat{\overline{Y}}_d^{dir1} = \frac{\hat{Y}_d^{dir1}}{N_d}, \tag{2.1}$$

where N_d is assumed to be known. Properties of these estimators are summarized in the following propositions.

Proposition 2.1 *If* $\pi_j > 0$, $\forall j \in U_d$, *then*

(a) $E_\pi\left[\hat{Y}_d^{dir1}\right] = Y_d$,

(b) $\text{var}_\pi\left(\hat{Y}_d^{dir1}\right) = \sum_{i \in U_d} \sum_{j \in U_d} (\pi_{ij} - \pi_i \pi_j) \frac{y_i}{\pi_i} \frac{y_j}{\pi_j}$, *and*

(c) $\widehat{\text{var}}_\pi\left(\hat{Y}_d^{dir1}\right)] = \sum_{i \in s_d} \sum_{j \in s_d} \frac{\pi_{ij} - \pi_i \pi_j}{\pi_{ij}} \frac{y_i}{\pi_i} \frac{y_j}{\pi_j}$ *is an unbiased estimator of* $\text{var}_\pi\left(\hat{Y}_d^{dir1}\right)$.

Proof We give two proofs of (a). The first one works directly with the probability distribution of samples s. Let $s_d(j)$ be the set of all samples such that $j \in s_d = s \cap U_d$. It holds that

$$E_\pi\left[\sum_{j \in s_d} \frac{y_j}{\pi_j}\right] = \sum_s p(s) \sum_{j \in s_d} \frac{y_j}{\pi_j} = \frac{y_1}{\pi_1} \sum_{s \in s_d(1)} p(s) + \frac{y_2}{\pi_2} \sum_{s \in s_d(2)} p(s)$$
$$+ \cdots + \frac{y_N}{\pi_N} \sum_{s \in s_d(N)} p(s) = \sum_{j \in U_d} \frac{y_j}{\pi_j} \pi_j = \sum_{j \in U_d} y_j = Y_d,$$

since $s_d(j) = \emptyset$ for $j \notin U_d$.

An alternative and more simpler proof is obtained by applying the indicator functions δ_j, i.e.

$$E_\pi\left[\hat{Y}_d^{dir1}\right] = E_\pi\left[\sum_{j \in s_d} \frac{y_j}{\pi_j}\right] = E_\pi\left[\sum_{i \in U_d} \frac{y_i}{\pi_i} \delta_i\right] = \sum_{i \in U_d} \frac{y_i}{\pi_i} E_\pi\left[\delta_i\right] = \sum_{i \in U_d} y_i = Y_d.$$

(b) By using the indicator functions, we get

$$\operatorname{var}_\pi\left(\hat{Y}_d^{dir1}\right) = \operatorname{var}_\pi\left(\sum_{j\in s_d}\frac{y_j}{\pi_j}\right) = \operatorname{var}_\pi\left(\sum_{i\in U_d}\frac{y_i}{\pi_i}\delta_i\right) = \sum_{i\in U_d}\sum_{j\in U_d}\frac{y_i}{\pi_i}\frac{y_j}{\pi_j}\operatorname{cov}_\pi\left(\delta_i,\delta_j\right)$$
$$= \sum_{i\in U_d}\sum_{j\in U_d}(\pi_{ij}-\pi_i\pi_j)\frac{y_i}{\pi_i}\frac{y_j}{\pi_j}.$$

(c) By using the indicator functions, we have

$$E_\pi\left[\widehat{\operatorname{var}}_\pi\left(\hat{Y}_d^{dir1}\right)\right] = \sum_{i\in U_d}\sum_{j\in U_d}\frac{\pi_{ij}-\pi_i\pi_j}{\pi_{ij}}\frac{y_j}{\pi_j}\frac{y_j}{\pi_j}E_\pi\left[\delta_i\delta_j\right]$$
$$= \sum_{i\in U_d}\sum_{j\in U_d}(\pi_{ij}-\pi_i\pi_j)\frac{y_i}{\pi_i}\frac{y_j}{\pi_j} = \operatorname{var}_\pi\left(\hat{Y}_d^{dir1}\right).$$

□

Corollary 2.1 *If $\pi_j > 0$, $\forall j \in U_d$, then*

(a) $E_\pi\left[\hat{\overline{Y}}_d^{dir1}\right] = \overline{Y}_d$,

(b) $\operatorname{var}_\pi\left(\hat{\overline{Y}}_d^{dir1}\right) = \frac{1}{N_d^2}\sum_{i\in U_d}\sum_{j\in U_d}(\pi_{ij}-\pi_i\pi_j)\frac{y_i}{\pi_i}\frac{y_j}{\pi_j}$, *and*

(c) $\widehat{\operatorname{var}}_\pi\left(\hat{\overline{Y}}_d^{dir1}\right) = \frac{1}{N_d^2}\sum_{i\in s_d}\sum_{j\in s_d}\frac{\pi_{ij}-\pi_i\pi_j}{\pi_{ij}}\frac{y_i}{\pi_i}\frac{y_j}{\pi_j}$ *is an unbiased estimator of* $\operatorname{var}_\pi\left(\hat{\overline{Y}}_d^{dir1}\right)$.

Let us consider now a simple random sampling design without replacement inside each domain (SRSWORD). This is to say, we consider a stratified random sampling design where the strata are the domains and the domain samples, $n_1, \ldots, n_D$, are fixed. For $i, j \in U_d$ we have

$$\pi_i = n_d/N_d, \quad \pi_{ii} = \pi_i = n_d/N_d, \quad \pi_{ij} = \frac{n_d(n_d-1)}{N_d(N_d-1)} \text{ if } i \neq j.$$

Proposition 2.2 *Under a SRSWORD design, the variance of the direct estimator of the total is*

$$\operatorname{var}_\pi\left(\hat{Y}_d^{dir1}\right) = \frac{(1-f_d)N_d^2}{n_d}S_{yd}^2, \quad S_{yd}^2 = \frac{1}{N_d-1}\sum_{i\in U_d}(y_i-\overline{Y}_d)^2, \quad f_d = \frac{n_d}{N_d}.$$

Proof It holds that

$$\mathrm{var}_\pi\left(\hat{Y}_d^{dir1}\right) = \sum_{i=1}^{N_d}(\pi_{ii} - \pi_i^2)\frac{y_i^2}{\pi_i^2} + \sum_{i=1}^{N_d}\sum_{\substack{j=1\\ i\neq j}}^{N_d}(\pi_{ij} - \pi_i\pi_j)\frac{y_i}{\pi_i}\frac{y_j}{\pi_j}$$

$$= \sum_{i=1}^{N_d}\frac{n_d}{N_d}\left(1 - \frac{n_d}{N_d}\right)\frac{N_d^2}{n_d^2}\,y_i^2 + \sum_{i=1}^{N_d}\sum_{\substack{j=1\\ i\neq j}}^{N_d}\left(\frac{n_d(n_d-1)}{N_d(N_d-1)} - \frac{n_d^2}{N_d^2}\right)\frac{N_d^2}{n_d^2}\,y_i y_j$$

$$= \sum_{i=1}^{N_d}\frac{N_d - n_d}{n_d}\,y_i^2 + \sum_{i=1}^{N_d}\sum_{\substack{j=1\\ i\neq j}}^{N_d}\frac{(n_d - N_d)}{(N_d-1)n_d}\,y_i y_j$$

$$= \frac{N_d - n_d}{n_d}\left[\sum_{i=1}^{N_d} y_i^2 - \frac{1}{N_d - 1}\sum_{i=1}^{N_d}\sum_{\substack{j=1\\ i\neq j}}^{N_d} y_i y_j\right] = \frac{N_d - n_d}{n_d}\left[\sum_{i=1}^{N_d} y_i^2\left(1 + \frac{1}{N_d-1}\right)\right.$$

$$\left. - \frac{1}{N_d - 1}\left(\sum_{i=1}^{N_d} y_i\right)^2\right] = \frac{(N_d - n_d)N_d}{n_d}\left[\frac{1}{N_d-1}\sum_{i=1}^{N_d} y_i^2 - \frac{Y_d^2}{N_d(N_d-1)}\right]$$

$$= \frac{(N_d - n_d)N_d}{n_d}\,S_{yd}^2 = \frac{(1 - f_d)N_d^2}{n_d}\,S_{yd}^2.$$

□

In sampling designs with $\pi_{ij} = \pi_i\pi_j,\ i \neq j$, and $\pi_{jj} = \pi_j$, it holds that

$$\mathrm{var}_\pi\left(\hat{Y}_d^{dir1}\right) = \sum_{j\in U_d}\frac{1-\pi_j}{\pi_j}\,y_j^2 = \sum_{j\in U_d}(w_j - 1)y_j^2, \tag{2.2}$$

$$\widehat{\mathrm{var}}_\pi\left(\hat{Y}_d^{dir1}\right) = \sum_{j\in s_d}\frac{1-\pi_j}{\pi_j^2}\,y_j^2 = \sum_{j\in s_d} w_j(w_j - 1)y_j^2. \tag{2.3}$$

For the estimator of the domain mean, we have

$$\mathrm{var}_\pi\left(\hat{\overline{Y}}_d^{dir1}\right) = \frac{1}{N_d^2}\sum_{j\in U_d}(w_j - 1)y_j^2, \quad \widehat{\mathrm{var}}_\pi\left(\hat{\overline{Y}}_d^{dir1}\right) = \frac{1}{N_d^2}\sum_{j\in s_d} w_j(w_j - 1)y_j^2. \tag{2.4}$$

The equalities $\pi_{ij} = \pi_i \pi_j,\ i \neq j$, hold for the Bernoulli sampling (BS) design and the SRSWR design. In sampling designs with $\pi_{ij} \approx \pi_i \pi_j$ if $i \neq j$ (i.e. under SRSWOR), the above formulas are approximations. If a SRSWORD design is employed, the approximation (2.2) is an upper bound of the variance of the estimator of the total, i.e.

$$\sum_{j \in U_d} \frac{1-\pi_j}{\pi_j} y_j^2 = \sum_{j=1}^{N_d} \frac{1-\frac{n_d}{N_d}}{n_d/N_d} y_j^2 = \sum_{j=1}^{N_d} \frac{N_d - n_d}{n_d} y_j^2 = \frac{(1-f_d)N_d^2}{n_d} \frac{1}{N_d} \sum_{j=1}^{N_d} y_j^2$$

$$= \frac{(1-f_d)N_d^2}{n_d} \left[\frac{N_d - 1}{N_d} S_{yd}^2 + \overline{Y}_d^2 \right] > \frac{(1-f_d)N_d^2}{n_d} S_{yd}^2 = \text{var}_\pi \left(\hat{Y}_d^{dir1} \right),$$

where the inequality holds if N_d is large enough and $\overline{Y}_d$ is not too close to zero. Särndal et al. (1992, p. 170), present the following formula for the covariance between two direct estimators:

$$\text{cov}_\pi(\hat{Y}_d^{dir1}, \hat{Z}_d^{dir1}) = \sum_{i \in U_d} \sum_{j \in U_d} \frac{\pi_{ij} - \pi_i \pi_j}{\pi_i \pi_j} y_i z_j.$$

An unbiased estimator of the covariance is

$$\widehat{\text{cov}}_\pi(\hat{Y}_d^{dir1}, \hat{Z}_d^{dir1}) = \sum_{i \in s_d} \sum_{j \in s_d} \frac{\pi_{ij} - \pi_i \pi_j}{\pi_{ij} \pi_i \pi_j} y_i z_j.$$

Remark 2.1 In sampling designs with $\pi_{ij} = \pi_i \pi_j,\ i \neq j$, and $\pi_{jj} = \pi_j$, we have

$$\text{cov}_\pi(\hat{Y}_d^{dir1}, \hat{Z}_d^{dir1}) = \sum_{j \in U_d} \frac{1-\pi_j}{\pi_j} y_j z_j,$$

$$\widehat{\text{cov}}_\pi(\hat{Y}_d^{dir1}, \hat{Z}_d^{dir1}) = \sum_{j \in s_d} \frac{1-\pi_j}{\pi_j^2} y_j z_j = \sum_{j \in s_d} w_j(w_j - 1)\, y_j z_j,$$

$$\text{cov}_\pi(\hat{\overline{Y}}_d^{dir1}, \hat{\overline{Z}}_d^{dir1}) = \frac{1}{N_d^2} \sum_{j \in U_d} \frac{1-\pi_j}{\pi_j} y_j z_j,$$

$$\widehat{\text{cov}}_\pi(\hat{\overline{Y}}_d^{dir1}, \hat{\overline{Z}}_d^{dir1}) = \frac{1}{N_d^2} \sum_{j \in s_d} \frac{1-\pi_j}{\pi_j^2} y_j z_j = \frac{1}{N_d^2} \sum_{j \in s_d} w_j(w_j - 1)\, y_j z_j.$$

Remark 2.2 For calculating $\widehat{Y}_d^{dir1}$, we need the sampling weights and the locations of sampled units. This is to say, we need the data $y_j,\ w_j,\ I_{U_d}(j),\ j \in s$, where $I_{U_d}(j)$ is the indicator function, i.e. $I_{U_d}(j) = 1$ if $j \in U_d$ and $I_{U_d}(j) = 0$ otherwise.

2.4 Estimator of the Ratio

In applications of statistical inference in finite populations, we often find situations where the target parameter is a ratio. Examples of ratio-type parameters are the unemployment rate or the domain mean when the population size in the denominator is unknown. This section gives some properties of estimators defined as a ratio of direct estimators of domain totals. Let us consider the domain ratio $R_d = Y_d/Z_d$, where $Y_d = \sum_{j\in U_d} y_j$ and $Z_d = \sum_{j\in U_d} z_j$, and the ratio estimator $\hat{R}_d = \hat{Y}_d^{dir1}/\hat{Z}_d^{dir1}$.

Proposition 2.3 *The standardized bias of $\hat{R}_d$ fulfills the inequality*

$$(B_\pi^{rel}[\hat{R}_d])^2 = \frac{(E_\pi[\hat{R}_d] - R_d)^2}{\text{var}_\pi(\hat{R}_d)} \leq \frac{\text{var}_\pi(\hat{Z}_d^{dir1})}{Z_d^2}.$$

Proof It holds that

$$\begin{aligned}
\text{cov}_\pi(\hat{R}_d, \hat{Z}_d^{dir1}) &= E_\pi[\hat{R}_d\hat{Z}_d^{dir1}] - E_\pi[\hat{R}_d]E_\pi[\hat{Z}_d^{dir1}] \\
&= E_\pi\left[\hat{Y}_d^{dir1}\right] - E_\pi[\hat{R}_d]E_\pi[\hat{Z}_d^{dir1}] \\
&= Y_d - E_\pi[\hat{R}_d]Z_d = -Z_d\left(E_\pi[\hat{R}_d] - R_d\right).
\end{aligned}$$

Therefore,

$$E_\pi[\hat{R}_d] - R_d = -\frac{\text{cov}_\pi(\hat{R}_d, \hat{Z}_d^{dir1})}{Z_d}.$$

By squaring both sides of the equality and using the symbol ρ_π for correlation with respect to the design-based probability, we obtain

$$\begin{aligned}
(E_\pi[\hat{R}_d] - R_d)^2 &= \frac{\left[\text{cov}_\pi(\hat{R}_d, \hat{Z}_d^{dir1})\right]^2}{Z_d^2} = \frac{\rho_\pi^2(\hat{R}_d, \hat{Z}_d^{dir1})\text{var}_\pi(\hat{R}_d)\text{var}_\pi(\hat{Z}_d^{dir1})}{Z_d^2} \\
&\leq \frac{\text{var}_\pi(\hat{R}_d)\text{var}_\pi(\hat{Z}_d^{dir1})}{Z_d^2},
\end{aligned}$$

which proves the stated result. □

Proposition 2.3 gives the following conclusion: if

$$B_\pi^{rel}[\hat{R}_d] = \frac{B_\pi[\hat{R}_d]}{(\text{var}_\pi(\hat{R}_d))^{1/2}} = \frac{E_\pi[\hat{R}_d] - R_d}{(\text{var}_\pi(\hat{R}_d))^{1/2}}$$

is the standardized bias of the ratio estimator $\hat{R}_d$, then

$$(B_\pi^{rel}[\hat{R}_d])^2 \leq \frac{\text{var}_\pi(\hat{Z}_d^{dir1})}{Z_d^2}.$$

Note that if the relative standard error (sampling error),

$$\frac{\sqrt{\text{var}_\pi(\hat{Z}_d^{dir1})}}{Z_d},$$

of the denominator of $\hat{R}_d$ tends to zero when the sample size increases, then the relative bias of $\hat{R}_d$ also tends to zero. This is an important property for building ratio estimators.

Proposition 2.4 *If $\hat{Y}_d^{dir1}$ and $\hat{Z}_d^{dir1}$ are consistent estimators of Y_d and Z_d, respectively, then*

(a) $\hat{R}_d$ is approximately unbiased.
(b) If n_d is large enough, an approximation to the variance of $\hat{R}_d$ is

$$\text{var}_\pi(\hat{R}_d) \approx \frac{1}{Z_d^2}\sum_{i\in U_d}\sum_{j\in U_d}(\pi_{ij}-\pi_i\pi_j)\frac{y_i - R_d z_i}{\pi_i}\frac{y_j - R_d z_j}{\pi_j}.$$

Proof The estimator $\hat{R}_d$ is a function of two variables, i.e.

$$\hat{R}_d = \frac{\hat{Y}_d^{dir1}}{\hat{Z}_d^{dir1}} = f(\hat{Y}_d^{dir1}, \hat{Z}_d^{dir1}).$$

As the partial derivatives of f are $\frac{\partial f}{\partial y} = \frac{1}{z}$ and $\frac{\partial f}{\partial z} = -\frac{y}{z^2}$, a first order Taylor series expansion of $f(\hat{Y}_d^{dir1}, \hat{Z}_d^{dir1})$ around (Y_d, Z_d) yields to

$$\hat{R}_d = f(\hat{Y}_d^{dir1}, \hat{Z}_d^{dir1}) \approx f(Y_d, Z_d) + \frac{\partial f(Y_d, Z_d)}{\partial y}(\hat{Y}_d^{dir1} - Y_d)$$
$$+ \frac{\partial f(Y_d, Z_d)}{\partial z}(\hat{Z}_d^{dir1} - Z_d) = R_d + \frac{1}{Z_d}(\hat{Y}_d^{dir1} - Y_d) - \frac{Y_d}{Z_d^2}(\hat{Z}_d^{dir1} - Z_d)$$
$$= R_d + \frac{1}{Z_d}(\hat{Y}_d^{dir1} - R_d\hat{Z}_d^{dir1}) = R_d + \frac{1}{Z_d}\sum_{j\in s_d}\frac{y_j - R_d z_j}{\pi_j}. \tag{2.5}$$

(a) By taking expectations in (2.5), we have

$$E_\pi[\hat{R}_d] \approx R_d + \frac{1}{Z_d}(Y_d - R_d Z_d) = R_d.$$

(b) By taking variances in (2.5) and using the sampling design indicator function δ_j, we get

$$\mathrm{var}_\pi(\hat{R}_d) \approx \frac{1}{Z_d^2} \sum_{i \in U_d} \sum_{j \in U_d} \frac{y_i - R_d z_i}{\pi_i} \frac{y_j - R_d z_j}{\pi_j} (\pi_{ij} - \pi_i \pi_j).$$

□

An estimator of the approximated variance of $\hat{R}_d$ is

$$\widehat{\mathrm{var}}_\pi(\hat{R}_d) = \frac{1}{(\hat{Z}_d^{dir1})^2} \sum_{i \in s_d} \sum_{j \in s_d} \frac{\pi_{ij} - \pi_i \pi_j}{\pi_{ij}} \frac{y_i - \hat{R}_d z_i}{\pi_i} \frac{y_j - \hat{R}_d z_j}{\pi_j}. \quad (2.6)$$

The estimator $\widehat{\mathrm{var}}_\pi(\hat{R}_d)$ is approximately unbiased if $E_\pi[\hat{R}_d] \approx R_d$ and $\mathrm{var}_\pi(\hat{Z}_d^{dir1}) \approx 0$. Otherwise, it is biased.

2.5 Other Direct Estimators of the Mean and the Total

Hájek (1971) proposed the following *direct* estimators of the domain mean and total:

$$\hat{\overline{Y}}_d^{dir2} = \frac{\hat{Y}_d^{dir1}}{\hat{N}_d} = \frac{\sum_{j \in s_d} w_j y_j}{\sum_{j \in s_d} w_j}, \quad \hat{Y}_d^{dir2} = N_d \hat{\overline{Y}}_d^{dir2}. \quad (2.7)$$

These estimators have the following properties.

Proposition 2.5 *If n_d is large enough and $\pi_j > 0\ \forall j \in U_d$, then*

(a) $E_\pi[\hat{\overline{Y}}_d^{dir2}] \approx \overline{Y}_d$ *and*
(b) $\mathrm{var}_\pi(\hat{\overline{Y}}_d^{dir2}) \approx \frac{1}{N_d^2} \sum_{i \in U_d} \sum_{j \in U_d} \frac{\pi_{ij} - \pi_i \pi_j}{\pi_i \pi_j} (y_i - \overline{Y}_d)(y_j - \overline{Y}_d).$

Proof Let $z_j = 1\ \forall j \in U_d$, then $Z_d = N_d$ and

$$R_d = \frac{Y_d}{Z_d} = \frac{Y_d}{N_d} = \overline{Y}_d.$$

The ratio estimator of R_d is

$$\hat{R}_d = \frac{\hat{Y}_d^{dir1}}{\hat{Z}_d^{dir1}} = \frac{\sum_{j \in s_d} w_j y_j}{\sum_{j \in s_d} w_j} = \hat{\overline{Y}}_d^{dir2}.$$

Since the Hájek estimator is consistent, the proof follows immediately from Proposition 2.4. □

An estimator of the approximated variance of $\overline{Y}_d^{dir2}$ is

$$\widehat{\text{var}}_\pi\left(\hat{\overline{Y}}_d^{dir2}\right) = \frac{1}{\hat{N}_d^2} \sum_{i \in s_d} \sum_{j \in s_d} \frac{\pi_{ij} - \pi_i \pi_j}{\pi_{ij} \pi_i \pi_j} \left(y_i - \hat{\overline{Y}}_d^{dir2}\right)\left(y_j - \hat{\overline{Y}}_d^{dir2}\right). \tag{2.8}$$

Corollary 2.2 *If n_d is large enough and $\pi_j > 0 \, \forall j \in U_d$, then*

(a) $E_\pi[\hat{Y}_d^{dir2}] \approx Y_d$ *and*

(b) $var_\pi\left(\hat{Y}_d^{dir2}\right) \approx \sum_{i \in U_d} \sum_{j \in U_d} \frac{\pi_{ij} - \pi_i \pi_j}{\pi_i \pi_j} (y_i - \overline{Y}_d)(y_j - \overline{Y}_d).$

An estimator of the approximated variance of $\hat{Y}_d^{dir2}$ is

$$\widehat{\text{var}}_\pi\left(\hat{Y}_d^{dir2}\right)] = \left(\frac{N_d}{\hat{N}_d}\right)^2 \sum_{i \in s_d} \sum_{j \in s_d} \frac{\pi_{ij} - \pi_i \pi_j}{\pi_{ij} \pi_i \pi_j} \left(y_i - \hat{\overline{Y}}_d^{dir2}\right)\left(y_j - \hat{\overline{Y}}_d^{dir2}\right). \tag{2.9}$$

Remark 2.3 In the case $\pi_{ij} = \pi_i \pi_j$, $i \neq j$, and $\pi_{jj} = \pi_j$, we get

$$\text{var}_\pi\left(\hat{\overline{Y}}_d^{dir2}\right) \approx \frac{1}{N_d^2} \sum_{j \in U_d} \frac{1 - \pi_j}{\pi_j} (y_j - \overline{Y}_d)^2,$$

$$\text{var}_\pi\left(\hat{Y}_d^{dir2}\right) \approx \sum_{j \in U_d} \frac{1 - \pi_j}{\pi_j} (y_j - \overline{Y}_d)^2,$$

$$\widehat{\text{var}}_\pi\left(\hat{\overline{Y}}_d^{dir2}\right) = \frac{1}{\hat{N}_d^2} \sum_{j \in s_d} \frac{1 - \pi_j}{\pi_j^2} \left(y_j - \hat{\overline{Y}}_d^{dir2}\right)^2$$

$$= \frac{1}{\hat{N}_d^2} \sum_{j \in s_d} w_j(w_j - 1)\left(y_j - \hat{\overline{Y}}_d^{dir2}\right)^2,$$

$$\widehat{\text{var}}_\pi\left(\hat{Y}_d^{dir2}\right) = \frac{N_d^2}{\hat{N}_d^2} \sum_{j \in s_d} w_j(w_j - 1)\left(y_j - \hat{\overline{Y}}_d^{dir2}\right)^2.$$

Estimators of the covariance between two direct estimators of domain means and totals, respectively, are

$$\widehat{\text{cov}}_\pi\left(\hat{\overline{Y}}_d^{dir2}, \hat{\overline{Z}}_d^{dir2}\right) = \frac{1}{\hat{N}_d^2} \sum_{j \in s_d} w_j(w_j - 1)\left(y_j - \hat{\overline{Y}}_d^{dir2}\right)\left(z_j - \hat{\overline{Z}}_d^{dir2}\right),$$

$$\widehat{\text{cov}}_\pi\left(\hat{Y}_d^{dir2}, \hat{Z}_d^{dir2}\right) = \frac{N_d^2}{\hat{N}_d^2} \sum_{j \in s_d} w_j(w_j - 1)\left(y_j - \hat{\overline{Y}}_d^{dir2}\right)\left(z_j - \hat{\overline{Z}}_d^{dir2}\right).$$

Under the SRSWORD design, it holds that

$$\hat{\overline{Y}}_d^{dir2} = \frac{\sum_{j \in s_d} \frac{N_d}{n_d} y_j}{\sum_{j \in s_d} \frac{N_d}{n_d}} = \frac{\frac{N_d}{n_d}}{N_d} \sum_{j \in s_d} y_j = \frac{1}{n_d} \sum_{j \in s_d} y_j = \bar{y}_{ds},$$

$$\widehat{\text{var}}_\pi(\hat{\overline{Y}}_d^{dir2}) = \frac{1}{N_d^2} \sum_{j \in s_d} \frac{N_d}{n_d} \frac{N_d - n_d}{n_d} (y_j - \bar{y}_{ds})^2 = \frac{1 - f_d}{n_d} \frac{1}{n_d} \sum_{j \in s_d} (y_j - \bar{y}_{ds})^2$$

$$\approx (1 - f_d) \frac{s_{yd}^2}{n_d},$$

where

$$s_{yd}^2 = \frac{1}{n_d - 1} \sum_{j \in s_d} (y_j - \bar{y}_{ds})^2.$$

As the direct estimator is approximately unbiased, the mean squared error and its estimator are

$$MSE(\hat{\overline{Y}}_d^{dir2}) \approx \text{var}_\pi\left(\hat{\overline{Y}}_d^{dir2}\right), \quad mse(\hat{\overline{Y}}_d^{dir2}) = \widehat{\text{var}}_\pi(\hat{\overline{Y}}_d^{dir2}).$$

For more details, see Särndal et al. (1992, pp. 185, 391), or Rao (2003, p. 12).

Although it is difficult to establish general conditions under which $\hat{\overline{Y}}_d^{dir2}$ is preferred to $\hat{\overline{Y}}_d^{dir1}$, Särndal et al. (1992, pp. 183–184), show some facts in favor of the first one.

1. By comparing the variances of both estimators, we have that $\hat{\overline{Y}}_d^{dir2}$ is preferred when the values of $y_j - \overline{Y}_d$ tend to be small. An extreme case is $y_j = c \; \forall j \in U_d$. In this case, it holds that

$$\overline{Y}_d = c, \qquad \hat{\overline{Y}}_d^{dir1} = c \frac{\sum_{j \in s_d} w_j}{N_d} = c \frac{\hat{N}_d}{N_d}, \qquad \hat{\overline{Y}}_d^{dir2} = c \frac{\hat{N}_d}{\hat{N}_d} = c.$$

   As $\text{var}_\pi\left(\hat{\overline{Y}}_d^{dir2}\right) = 0$, $\hat{\overline{Y}}_d^{dir2}$ is preferred to $\hat{\overline{Y}}_d^{dir1}$ if $\text{var}_\pi(\hat{N}_d) > 0$.
2. The estimator $\hat{\overline{Y}}_d^{dir2}$ behaves better than $\hat{\overline{Y}}_d^{dir1}$ when the sample size varies. If the sample size realization, $n_d = n_d(s)$, is larger than the average sample size, then the numerator and the denominator have many summands in $\hat{\overline{Y}}_d^{dir2}$. In the opposite case, the numerator and the denominator have few summands in $\hat{\overline{Y}}_d^{dir2}$. In this way, the ratio has some kind of stability. However, $\hat{\overline{Y}}_d^{dir1}$ does not present this stability because its denominator is a known constant.

In the case of the Bernoulli sampling where each individual is included in the sample independently with probability $\pi_j = \pi$, if $y_j = c \ \forall j \in U_d$, it holds that

$$\hat{\overline{Y}}_d^{dir1} = c \frac{n_d(s)}{\pi N_d}, \qquad \hat{\overline{Y}}_d^{dir2} = c.$$

Therefore, the variability of $\hat{\overline{Y}}_d^{dir1}$ is only ought to the variability of n_d for different samples s. In this case, $\text{var}_\pi(\hat{\overline{Y}}_d^{dir1}) > \text{var}_\pi(\hat{\overline{Y}}_d^{dir2}) = 0$.

3. Another situation where $\hat{\overline{Y}}_d^{dir2}$ is preferred to $\hat{\overline{Y}}_d^{dir1}$ is when the sample contains large values y_j of the target variable associated to small inclusion probabilities π_j. In this case, the value of the numerator of both estimators tends to be quite large. This fact is compensated by $\hat{\overline{Y}}_d^{dir2}$ because its denominator also tends to be large. This compensation produces stability. However, the denominator of $\hat{\overline{Y}}_d^{dir1}$ is constant and does not compensate the extreme values of the numerator.

Särndal et al. (1992, p. 184), give the following example that illustrates the above described situation. Let us consider a domain d with $N_d = 10$ units $y_1 = \ldots = y_9 = c$ e $y_{10} = 2c$. For estimating $\overline{Y}_d = 1.1c$, we draw a random sample of size $n_d = 1$ with inclusion probabilities $\pi_1 = \ldots = \pi_9 = 0.11$ and $\pi_{10} = 0.01$. Therefore, the unit 10 has the largest value of y and the smallest value of π. It holds that

$$\hat{\overline{Y}}_d^{dir2} = \begin{cases} c & \text{if } s = \{1\}, \ldots, \{9\}, \\ 2c & \text{if } s = \{10\}, \end{cases} \qquad \hat{\overline{Y}}_d^{dir1} = \begin{cases} \frac{c}{1.1} & \text{if } s = \{1\}, \ldots, \{9\}, \\ 20c & \text{if } s = \{10\}. \end{cases}$$

Obviously, with $\hat{\overline{Y}}_d^{dir2}$, we avoid the possibility of obtaining absurd estimates of $\overline{Y}_d = 1.1c$.

2.6 Bootstrap Resampling for Variance Estimation

In this section we present a basic bootstrap procedure for estimating the variance of an estimator.

Let us consider samples s drawn at random from a population U according to a given sampling design. Let $\hat{\theta}$ be the estimator of the population parameter θ. Särndal et al. (1992, p. 442), describe the following basic bootstrap procedure:

1. From the sample s, build an artificial population U^* mimicking U. This can be done by replicating each sample register as many times as the calibrated sample weight w_j (elevation factor).

2. Extract B independent bootstrap samples from U^* by using the same sampling design as the one used for obtaining s from U. For each bootstrap sample s_b, $b = 1, \ldots, B$, calculate the estimator $\hat{\theta}_b^*$ in the same form as $\hat{\theta}$ was calculated for s.
3. The observed distribution of $\hat{\theta}_1^*, \ldots, \hat{\theta}_B^*$ imitates the distribution of $\hat{\theta}$.
4. The bootstrap estimator of the *variance* of $\hat{\theta}$ is

$$\widehat{\mathrm{var}}_B(\hat{\theta}) = \frac{1}{B-1} \sum_{b=1}^{B} (\hat{\theta}_b^* - \hat{\theta}^*)^2, \qquad \text{where} \quad \hat{\theta}^* = \frac{1}{B} \sum_{b=1}^{B} \hat{\theta}_b^*.$$

5. The bootstrap estimator of the *mean squared error* of $\hat{\theta}$ is

$$mse_B(\hat{\theta}) = \frac{1}{B-1} \sum_{b=1}^{B} (\hat{\theta}_b^* - \hat{\theta})^2.$$

6. Given two population parameters θ and φ, with respective estimators $\hat{\theta}$ and $\hat{\varphi}$, the bootstrap estimators of the *covariance* and the *crossed mean squared error* of $\hat{\theta}$ and $\hat{\varphi}$ are

$$\widehat{\mathrm{cov}}_B(\hat{\theta}, \hat{\varphi}) = \frac{1}{B-1} \sum_{b=1}^{B} (\hat{\theta}_b^* - \hat{\theta}^*)(\hat{\varphi}_b^* - \hat{\varphi}^*),$$

$$mse_B(\hat{\theta}, \hat{\varphi}) = \frac{1}{B-1} \sum_{b=1}^{B} (\hat{\theta}_b^* - \hat{\theta})(\hat{\varphi}_b^* - \hat{\varphi}).$$

This bootstrap method has the disadvantage of requiring the construction of an artificial population for reproducing the original sampling design. In the case of complex sampling designs with strata and clusters, like the ones implemented in some labor force surveys, rebuilding the geographic structure of the population, within the bootstrap procedure, implies the construction of artificial populations with the same or similar cluster and strata sizes as the original one. In many cases, this is simply impossible to perform.

2.7 Jackknife Resampling for Variance Estimation

The jackknife method was developed by Quenouille (1949, 1956) as a technique for bias reduction in finite populations. Tukey (1958) suggested that jackknife could also be used for variance estimation, and Durbin (1959) applied this idea in infinite populations. The jackknife method is similar to the leave-one-out cross-validation procedure, and it can also be considered as a method for data partitioning. In what follows, the basic ideas for applying the Jackknife resampling are given. For more details, see Särndal et al. (1992, pp. 437–442).

Let s be a sample of n units drawn at random by a SRSWOR design. Let $\hat{\theta}$ be an estimator of the population parameter θ. The jackknife resampling procedure gives an estimator of $\mathrm{var}(\hat{\theta})$. The jackknife steps are

1. Partition at random the sample s in A groups of equal size $m = n/A$.
2. For each group a, $a = 1, \ldots, A$, build the subsample $s_{(a)}$ by eliminating from s the units of group a. Based on $s_{(a)}$, calculate the estimator $\hat{\theta}_{(a)}$ of θ in the same way as $\hat{\theta}$ was calculated for s.
3. The jackknife estimator of θ is $\hat{\theta}_J = \frac{1}{A}\sum_{a=1}^{A}\hat{\theta}_{(a)}$.
4. The jackknife variance estimator is $\mathrm{var}_{J1} = \frac{A-1}{A}\sum_{a=1}^{A}\left(\hat{\theta}_{(a)} - \hat{\theta}_J\right)^2$.

   In practice, var_{J1} is used as estimator of $\mathrm{var}(\hat{\theta})$ and $\mathrm{var}(\hat{\theta}_J)$. An alternative estimator is

$$\mathrm{var}_{J2} = \frac{A-1}{A}\sum_{a=1}^{A}\left(\hat{\theta}_{(a)} - \hat{\theta}\right)^2.$$

   It holds that $\mathrm{var}_{J2} \geq \mathrm{var}_{J1}$.
5. The jackknife bias estimator is $\mathrm{bias}_J = (A-1)(\hat{\theta}_J - \hat{\theta})$.

Remark 2.4 Särndal et al. (1992, pp. 437–442), introduce the jackknife estimator of the variance by using the pseudovalues

$$\hat{\theta}_a = A\hat{\theta} - (A-1)\hat{\theta}_{(a)}, \quad a = 1, \ldots, A.$$

They define the jackknife estimator of θ as bias-corrected estimator, i.e.

$$\hat{\theta}_{JK} = \frac{1}{A}\sum_{a=1}^{A}\hat{\theta}_a = A\hat{\theta} - (A-1)\hat{\theta}_J = \hat{\theta} - (A-1)\left(\hat{\theta}_J - \hat{\theta}\right) = \hat{\theta} - \mathrm{bias}_J\,.$$

Further, they give the variance estimator

$$\mathrm{var}_{JK1} = \frac{1}{A(A-1)}\sum_{a=1}^{A}\left(\hat{\theta}_a - \hat{\theta}_{JK}\right)^2,$$

which is equal to var_{J1}, because

$$\left(\hat{\theta}_a - \hat{\theta}_{JK}\right)^2 = \left\{\left[A\hat{\theta} - (A-1)\hat{\theta}_{(a)}\right] - \left[A\hat{\theta} - (A-1)\hat{\theta}_J\right]\right\}^2 = (A-1)^2\left(\hat{\theta}_{(a)} - \hat{\theta}_J\right)^2.$$

For applying the jackknife method, we have to fix a number of groups A. For having a variance estimator with a good accuracy, we could take as many groups as possible, i.e. $A = n$ and $m = 1$. On the other hand, because of the computational burden, we prefer working with few groups. The extreme cases are $A = 2$ and $m = n/2$. In practice, it is quite common to take a value of A between the extreme cases $A = n$ and $A = 2$.

Remark 2.5 If $\hat{\theta}_{(a)}$, $a = 1, \ldots, A$, were uncorrelated random variables with the same expectation, then var_{J1} should be unbiased for $\text{var}(\hat{\theta}_J)$. However, the $\hat{\theta}_{(a)}$'s are correlated, and therefore the unbiasedness property does not hold. The properties of the jackknife estimators of a general type parameter θ under a complex sampling design have not been studied in the literature. Under a SRS and linear target parameter, the jackknife variance estimator has, in general, a good behavior.

2.7.1 Delete-One-Cluster Jackknife for Estimators of Domain Parameters

The delete-one-cluster jackknife method (see e.g. Rao and Tausi 2004) generates jackknife samples by deleting a cluster each time. There are as many jackknife samples as clusters are in the sample. Consider the jackknife sample, $s^*_{(d_* c_*)}$, obtained by excluding the cluster c_* of the domain d_* from the sample s, and denote the corresponding domain d and cluster c subsample by $s^*_{dc(d_* c_*)}$. Let D_s be the number of domains in s, m_d be the number of clusters in s_d, $C = \sum_{d=1}^{D_s} m_d$, m_{d_*} be the number of clusters in d_*, and m_{Jd_*} be the number of clusters in the jackknife subsample $s^*_{(d_* c_*)}$. The jackknife weight of individual j, cluster c, and domain d in $s^*_{(d_* c_*)}$ is

$$w_{dcj(d_* c_*)} = w_{dcj} b_{dc(d_* c_*)}, \quad b_{dc(d_* c_*)} = \begin{cases} w_{d.}/w^*_{d.} & \text{if } d = d_*, c \neq c_*, \\ 1 & \text{if } d \neq d_*, \end{cases}$$

where $w_{d.} = \sum_{c=1}^{m_d} \sum_{j \in s_{dc}} w_{dcj}$ and $w^*_{d.} = \sum_{c=1, c \neq c_*}^{m_d} \sum_{j \in s^*_{dc(d_* c_*)}} w_{dcj}$. Note that the case $d = d_*$ and $c = c_*$ does not appear in the jackknife sample $s^*_{(d_* c_*)}$. The jackknife resampling method for estimating the variance of an estimator $\hat{\theta}$ of a population parameter θ is

1. By using the procedure described above, use sample s to draw jackknife samples $s^*_{(d_* c_*)}$, $d_* = 1, \ldots, D_s$, $c_* = 1, \ldots, m_{d_*}$. For every jackknife sample, calculate $\hat{\theta}^*_{(d_* c_*)}$ in the same way as $\hat{\theta}$ was calculated, but using the jackknife weights $w_{dcj(d_* c_*)}$.
2. The observed distribution of $\{\hat{\theta}^*_{(d_* c_*)} : d_* = 1, \ldots, D_s, c_* = 1, \ldots, m_{d_*}\}$ is expected to imitate the distribution of estimator $\hat{\theta}$.

3. The jackknife estimator of θ and bias($\hat{\theta}$) is

$$\hat{\theta}_J = \frac{1}{C}\sum_{d_*=1}^{D_s}\sum_{c_*=1}^{m_{d*}} \hat{\theta}^*_{(d_*c_*)}, \quad \text{bias}_J(\hat{\theta}) = \sum_{d_*=1}^{D_s}(m_{Jd_*}-1)\sum_{c_*=1}^{m_{d*}}\left(\hat{\theta}^*_{(d_*c_*)} - \hat{\theta}_J\right). \tag{2.10}$$

4. The design-based variance of $\hat{\theta}$ can be approximated by

$$\text{var}_J(\hat{\theta}) = \sum_{d_*=1}^{D_s}\frac{m_{Jd_*}-1}{m_{Jd_*}}\sum_{c_*=1}^{m_{d*}}(\hat{\theta}^*_{(d_*c_*)} - \hat{\theta}_J)^2. \tag{2.11}$$

2.8 R Codes for Design-Based Direct Estimators

This section presents some R codes illustrating the use of the studied estimators.

2.8.1 Horvitz–Thompson Direct Estimators of the Total and the Mean

We first read the auxiliary and sample data files and rename some variables.

```
# Auxiliary data
dataux <- read.table("Nds20.txt", header=TRUE, sep = "\t", dec = ".")
# Sort dataux by sex and area:
dataux <- dataux[order(dataux$sex, dataux$area),]
# Sample data
dat <- read.table("LFS20.txt", header=TRUE, sep = "\t", dec = ".")
# number of rows (cases) in dat:
n <- nrow(dat)
# Rename some variables
y1 <- dat$UNEMPLOYED; y2 <- dat$EMPLOYED
w <- dat$WEIGHT
area <- dat$AREA; sex <- dat$SEX
```

This section describes the following activities. For domains defined as AREA crossed by SEX, do:

A1. Estimate the totals of unemployed and employed people.
A2. Estimate the variances and the coefficients of variation.
A3. Repeat A1–A2 for means.
A4. Calculate the domain unemployment rates
A5. Estimate the variance of the unemployment rate estimator.
A6. Repeat A1–A5 for domains defined by AREA.

A1. For estimating the totals of unemployed and employed people by AREA and SEX, we apply formula (2.1), i.e.

$$\hat{Y}_d^{dir1} = \sum_{j \in s_d} w_j y_j.$$

The R code is

```
dir1.ds <- aggregate(w*data.frame(y1,y2), by=list(Area=area,Sex=sex), sum)
# Assign column names
names(dir1.ds) <- c("area", "sex", "y1tot", "y2tot")
```

A2. For estimating the variance of $\hat{Y}_d^{dir1}$, we apply the formula (2.4), i.e.

$$\widehat{\text{var}}_\pi\left(\hat{Y}_d^{dir1}\right) = \sum_{j \in s_d} w_j(w_j - 1)y_j^2.$$

The R code is

```
vardir1.ds <- aggregate(w*(w-1)*data.frame(y1^2,y2^2),
                        by=list(Area=area,Sex=sex), sum)
# Assign column names
names(vardir1.ds) <- c("area", "sex", "y1var", "y2var")
```

We build a table with direct estimates of totals, variances, and coefficients of variation.

```
# Add columns y1var and y2var
dir1.ds <- cbind(dir1.ds, vardir1.ds$y1var, vardir1.ds$y2var)
# CV for y1
y1cv <- 100*sqrt(vardir1.ds$y1var)/abs(dir1.ds$y1tot)
# CV for y2
y2cv <- 100*sqrt(vardir1.ds$y2var)/abs(dir1.ds$y2tot)
# Add columns y1cv and y2cv
dir1.ds <- cbind(dir1.ds, y1cv, y2cv)
# Change column names for dir1.ds
namesds <- c("area", "sex", "y1tot", "y2tot", "y1var", "y2var", "y1cv",
             "y2cv")
names(dir1.ds) <- namesds
```

A3. We calculate the estimators of the means and their variances by using the formulas (2.1) and (2.4), i.e.

$$\hat{\bar{Y}}_d^{dir1} = N_d^{-1}\hat{Y}_d^{dir1}, \quad \widehat{\text{var}}_\pi\left(\hat{\bar{Y}}_d^{dir1}\right) = N_d^{-2}\widehat{\text{var}}_\pi\left(\hat{Y}_d^{dir1}\right).$$

```
# Add column with population sizes
dir1.ds <- cbind(dir1.ds, dataux$N)
# Add columns with HT estimates of means
dir1.ds <- cbind(dir1.ds, dir1.ds$y1tot/dataux$N,
                 dir1.ds$y2tot/dataux$N)
# Variance estimates of HT estimator
dir1.ds <- cbind(dir1.ds, dir1.ds$y1var/dataux$N^2, dir1.ds$y2var/
                   dataux$N^2)
# Change column names for dir1.ds
names(dir1.ds) <- c(namesds, "Nds", "y1mean", "y2mean", "y1meanvar",
                    "y2meanvar")
```

A4. For estimating the unemployment rates (in %), we employ the ratio estimator

$$\hat{R}^{dir} = \frac{\hat{Y}_{1,d}^{dir1}}{\hat{Y}_{1,d}^{dir1} + \hat{Y}_{2,d}^{dir1}} 100,$$

where $\hat{Y}_{1,d}^{dir1}$ and $\hat{Y}_{2,d}^{dir1}$ are the direct estimators of the totals of unemployed and employed people, respectively. The R code is

```
# Include estimates of unemployment rates in table dir1.ds
dirrate.ds <- 100*dir1.ds$y1tot/(dir1.ds$y1tot + dir1.ds$y2tot)
dir1.ds <- cbind(dir1.ds, rate=dirrate.ds)
```

A5. For estimating the covariances $\widehat{\text{cov}}(\hat{Y}_{1,d}^{dir1}, \hat{Y}_{2,d}^{dir1})$, we apply the corresponding formula of Remark 2.1, i.e.

$$\widehat{\text{cov}}_\pi(\hat{Y}_{1,d}^{dir1}, \hat{Y}_{2,d}^{dir1}) = \sum_{j\in s_d} w_j(w_j - 1)y_{1,j}y_{2,j}.$$

The R code is

```
covardir1.ds <- aggregate(w*(w-1)*data.frame(y1*y2),
                          by=list(Area=area,Sex=sex), sum)
# Column names
names(covardir1.ds) <- c("area", "sex", "covar")
```

For estimating the variance of the unemployment rate estimator, we apply the formula (3.10) of Chap. 3, i.e.

$$\begin{aligned}\widehat{\text{var}}(\hat{R}_d) = &\frac{\hat{Y}_{2,d}^2}{(\hat{Y}_{1,d} + \hat{Y}_{2,d})^4}\widehat{\text{var}}(\hat{Y}_{1,d}) + \frac{\hat{Y}_{1,d}^2}{(\hat{Y}_{1,d} + \hat{Y}_{2,d})^4}\widehat{\text{var}}(\hat{Y}_{2,d}) \\ &- \frac{2\hat{Y}_{1,d}\hat{Y}_{2,d}}{(\hat{Y}_{1,d} + \hat{Y}_{2,d})^4}\widehat{\text{cov}}(\hat{Y}_{1,d}, \hat{Y}_{2,d}),\end{aligned}$$

where $\hat{Y}_{1,d} = \hat{Y}_{1,d}^{dir1}$ and $\hat{Y}_{2,d} = \hat{Y}_{2,d}^{dir1}$. The following R code calculates $\widehat{\text{var}}(\hat{R}_d)$

```
# Summands in formula of covariance estimator
s1.ds <- dir1.ds$y2tot^2*dir1.ds$y1var/(dir1.ds$y1tot+dir1.ds$y2tot)^4
s2.ds <- dir1.ds$y1tot^2*dir1.ds$y2var/(dir1.ds$y1tot+dir1.ds$y2tot)^4
s12.ds <- 2*dir1.ds$y1tot*dir1.ds$y2tot*covardir1.ds$covar/
          (dir1.ds$y1tot + dir1.ds$y2tot)^4
# Estimates of variances and coefficients of variation
dir1.ds$vrate <- 10^4*(s1.ds+s2.ds-s12.ds)
dir1.ds$cvrate <- 100*sqrt(dir1.ds$vrate)/abs(dir1.ds$rate)
```

Table 2.1 DIR1 estimates of labor status indicators for sex=1 (left) and sex=2 (right)

area	y1tot	y2tot	y1var	y2var	rate	y1tot	y2tot	y1var	y2var	rate
1	344	5422	117,992	1,548,184	5.97	452	3637	112,068	960,992	11.05
2	206	1782	42,230	433,104	10.36	222	1674	49,062	331,572	11.71
3	0	3452	0	676,846	0.00	165	1320	27,060	220,026	11.11
4	179	3388	31,862	613,772	5.02	187	2798	34,782	500,522	6.26
5	0	2549	0	421,576	0.00	137	2065	18,632	337,506	6.22
6	381	3658	72,380	695,074	9.43	200	735	39,800	108,008	21.39
7	137	2857	18,632	555,234	4.58	0	3121	0	606,322	0.00
8	188	2863	35,156	500,160	6.16	0	2625	0	452,400	0.00
9	600	6641	135,138	1,243,378	8.29	346	3124	64,512	514,402	9.97
10	156	1655	24,180	282,474	8.61	0	1313	0	233,774	0.00

The R code to save the results is

```
output1 <- data.frame(dir1.ds[,1:6], rate=round(dirrate.ds,2))
head(output1, 10)
```

A6. This activity is an exercise.

For the ten first areas, Table 2.1 presents some of the contents of the data frame dir1.ds. The columns y1tot and y2tot contain the direct estimates, $\hat{Y}_{1,d}^{dir1}$ and $\hat{Y}_{2,d}^{dir1}$, of totals of unemployed and employed people. The columns y1var, y2var, and rate give the variance estimates $\widehat{\text{var}}_\pi\left(\hat{Y}_{1,d}^{dir1}\right)$ and $\widehat{\text{var}}_\pi\left(\hat{Y}_{2d}^{dir1}\right)$ and the unemployment rates estimations $\hat{R}_d^{dir1} = \hat{Y}_{1,d}^{dir1}/(\hat{Y}_{1,d}^{dir1} + \hat{Y}_{2,d}^{dir1})$. The left (right) part of Table 2.1 contains the results for sex=1 (sex=2). In domains with null sample size, the dir1 estimator is not calculable, and we deliver the value of 0.

2.8.2 *Hájek Direct Estimator of the Mean and the Total*

This section describes the following activities. For domains defined as AREA crossed by SEX, do:

B1. Estimate the proportions of unemployed and employed people.
B2. Estimate the variances and the coefficients of variation.
B3. Repeat B1–B2 for totals.
B4. Estimate the unemployment rates.
B5. Estimate the variance of the unemployment rate estimator.
B6. Repeat B1–B5 for domains defined by AREA.

B1. By applying the formula (2.1), we calculate the estimator $\hat{Y}_d^{dir1}$ of the totals of unemployed and employed people by AREA and SEX. The R code is

```
dir <- aggregate(w*data.frame(1/w,1,y1,y2), by=list(Area=area,Sex=sex), sum)
# Column names
names(dir) <- c("area", "sex", "nds", "hatNds", "y1tot", "y2tot")
```

We calculate the direct estimates of means by AREA and SEX by applying the formula (2.7), i.e.

$$\hat{\overline{Y}}_d^{dir2} = \frac{\hat{Y}_d^{dir1}}{\hat{N}_d} = \frac{\sum_{j\in s_d} w_j y_j}{\sum_{j\in s_d} w_j}.$$

The R code is

```
dir2.ds <- data.frame(area=dir$area, sex=dir$sex, nds=dir$nds,
                      hatNds=dir$hatNds)
# Estimates of means of unemployed people
dir2.ds$y1mean <- dir$y1tot/dir$hatNds
# Estimates of means of employed people
dir2.ds$y2mean <- dir$y2tot/dir$hatNds
```

B2. For estimating the variance of $\hat{\overline{Y}}_d^{dir2}$, we apply the third formula of Remark 2.3, i.e.

$$\widehat{\mathrm{var}}_\pi(\hat{\overline{Y}}_d^{dir2}) = \frac{1}{\hat{N}_d^2}\sum_{j\in s_d} w_j(w_j-1)(y_j-\hat{\overline{Y}}_d^{dir2})^2.$$

The R code for the numerator is

```
# Define all the necessary objects
difference1 <- difference2 <- numerator1 <- numerator2 <- ww1 <- list()
for(d in 1:nrow(dir2.ds)){
  # Create a logic vector with the indexes of the corresponding domains
  condition <- paste(dat$AREA,dat$SEX,sep="")==paste(dir2.ds$area,
                         dir2.ds$sex,sep="")[d]
  # Calculate the difference between data and mean of each domain
  difference1[[d]] <- y1[condition]-dir2.ds$y1mean[d]
  difference2[[d]] <- y2[condition]-dir2.ds$y2mean[d]
  ww1[[d]] <- w[condition]*(w[condition]-1)
  numerator1[[d]] <- ww1[[d]]*difference1[[d]]^2
  numerator2[[d]] <- ww1[[d]]*difference2[[d]]^2
}
```

The following R code calculates $\widehat{\mathrm{var}}_\pi(\hat{\overline{Y}}_d^{dir2})$ by AREA and SEX:

```
dir2.ds$y1meanvar <- sapply(numerator1, sum)/dir2.ds$hatNds^2
dir2.ds$y2meanvar <- sapply(numerator2, sum)/dir2.ds$hatNds^2
```

We include in `dir2.ds` the estimated coefficients of variation $\mathrm{cv} = \mathrm{cv}(\hat{\overline{Y}}_d^{dir2})$.

```
# cv of y1-mean (in %)
dir2.ds$y1cv <- 100*sqrt(dir2.ds$y1meanvar)/abs(dir2.ds$y1mean)
# cv of y2-mean (in %)
dir2.ds$y2cv <- 100*sqrt(dir2.ds$y2meanvar)/abs(dir2.ds$y2mean)
```

B3. We repeat steps 1 and 2 for estimating the totals of unemployed and employed people. We use the estimators (2.7) and the fourth formula of Remark 2.3, i.e.

$$\hat{Y}_d^{dir2} = N_d \hat{\overline{Y}}_d^{dir2}, \quad \widehat{\text{var}}_\pi(\hat{Y}_d^{dir2}) = \frac{N_d^2}{\hat{N}_d^2} \sum_{j \in s_d} w_j(w_j - 1)(y_j - \hat{\overline{Y}}_d^{dir2})^2.$$

This is done with the R code

```
dir2.ds$y1tot <- dir2.ds$y1mean*dataux$N
dir2.ds$y2tot <- dir2.ds$y2mean*dataux$N
dir2.ds$y1totvar <- dir2.ds$y1meanvar*dataux$N^2
dir2.ds$y2totvar <- dir2.ds$y2meanvar*dataux$N^2
```

B4. The unemployment rate and its direct estimator are

$$R_d = \frac{Y_{1,d}}{Y_{1,d} + Y_{2,d}}, \quad \hat{R}_d = \frac{\hat{Y}_{1,d}^{dir2}}{\hat{Y}_{1,d}^{dir2} + \hat{Y}_{2,d}^{dir2}}.$$

The following R code estimates the unemployment rates (in %):

```
dir2.ds$rate <- 100*dir2.ds$y1tot/(dir2.ds$y1tot + dir2.ds$y2tot)
```

B5. For estimating the covariances $\widehat{\text{cov}}(\hat{Y}_{1,d}^{dir2}, \hat{Y}_{2,d}^{dir2})$, we apply the last formula of Remark 2.3, i.e.

$$\widehat{\text{cov}}_\pi(\hat{Y}_{1,d}^{dir2}, \hat{Y}_{2,d}^{dir2}) = \frac{N_d^2}{\hat{N}_d^2} \sum_{j \in s_d} w_j(w_j - 1)(y_{1,j} - \hat{\overline{Y}}_{1,d}^{dir2})(y_{2,j} - \hat{\overline{Y}}_{2,d}^{dir2}).$$

The R code is

```
ww1s1s2 <- mapply(ww1, mapply(difference1, difference2, FUN="*"),
                  FUN="*")
sumcovardir2 <- sapply(ww1s1s2, sum)
covardir2.ds <- sumcovardir2*dataux$N^2/dir2.ds$hatNds^2
```

For estimating the variance of the unemployment rate estimator, we apply the formula (3.10) of Chap. 3, i.e.

$$\begin{aligned}\widehat{\text{var}}(\hat{R}_d) = {} & \frac{\hat{Y}_{2,d}^2}{(\hat{Y}_{1,d} + \hat{Y}_{2,d})^4} \widehat{\text{var}}(\hat{Y}_{1,d}) + \frac{\hat{Y}_{1,d}^2}{(\hat{Y}_{1,d} + \hat{Y}_{2,d})^4} \widehat{\text{var}}(\hat{Y}_{2,d}) \\ & - \frac{2\hat{Y}_{1,d}\hat{Y}_{2,d}}{(\hat{Y}_{1,d} + \hat{Y}_{2,d})^4} \widehat{\text{cov}}(\hat{Y}_{1,d}, \hat{Y}_{2,d}),\end{aligned}$$

where $\hat{Y}_{1,d} = \hat{Y}_{1,d}^{dir2}$ and $\hat{Y}_{2,d} = \hat{Y}_{2,d}^{dir2}$. The following R code calculates $\widehat{\text{var}}(\hat{R}_d)$:

```
# Summands in formula of covariance estimator
s1.ds <- dir2.ds$y2tot^2*dir2.ds$y1totvar/(dir2.ds$y1tot+
         dir2.ds$y2tot)^4
s2.ds <- dir2.ds$y1tot^2*dir2.ds$y2totvar/(dir2.ds$y1tot+
         dir2.ds$y2tot)^4
s12.ds <- 2*dir2.ds$y1tot*dir2.ds$y2tot*covardir2.ds/
          (dir2.ds$y1tot+dir2.ds$y2tot)^4
# Estimates of variances and coefficients of variation
dir2.ds$vrate <- 10^4*(s1.ds+s2.ds-s12.ds)
dir2.ds$cvrate <- 100*sqrt(dir2.ds$vrate)/abs(dir2.ds$rate)
```

The R code to save the results is

```
output2 <- data.frame(dir2.ds[,1:2], round(dir2.ds[,11:14]),
                      rate=round(dir2.ds[,15],2))
head(output2, 10)
```

B6. This activity is an exercise.

For the ten first areas, Table 2.2 presents some of the contents of the data frame dir2.ds. The columns y1tot and y2tot contain the direct estimates, $\hat{Y}_{1,d}^{dir2}$ and $\hat{Y}_{2,d}^{dir2}$, of totals of unemployed and employed people. The columns y1var, y2var, and rate give the variance estimates $\widehat{\text{var}}_{\pi}\left(\hat{Y}_{1,d}^{dir2}\right)$ and $\widehat{\text{var}}_{\pi}\left(\hat{Y}_{2,d}^{dir2}\right)$ and the unemployment rates estimations $\hat{R}_d^{dir2} = \hat{Y}_{1,d}^{dir2}/(\hat{Y}_{1,d}^{dir2} + \hat{Y}_{2,d}^{dir2})$. The left (right) part of Table 2.2 contains the results for sex=1 (sex=2). In domains with null sample size, the dir2 estimator is not calculable, and we deliver the value of 0. By comparing the results presented in Tables 2.1 and 2.2, we conclude that dir2 estimators of totals have, in general, smaller variances than dir1 estimators. However, they both give the same estimates of unemployment ratios.

Comparing the results presented in Tables 2.1 and 2.2 one can observe that the Hájek type estimator dir2 has lower variance estimates than the Horvitz–Thompson estimator dir1, particularly in the columns denoted as y2var.

Table 2.2 dir2 estimates of labor status indicators for sex=1 (left) and sex=2 (right)

area	y1tot	y2tot	y1var	y2var	rate	y1tot	y2tot	y1var	y2var	rate
1	347	5470	114,455	610,953	5.97	453	3648	107,441	568,195	11.05
2	209	1809	41,081	192,151	10.36	225	1694	47,076	194,190	11.71
3	0	3521	0	122,182	0.00	165	1317	25,787	142,520	11.11
4	182	3436	31,534	173,090	5.02	189	2828	34,115	217,891	6.26
5	0	2456	0	84,070	0.00	137	2069	18,176	163,088	6.22
6	391	3758	70,745	213,647	9.43	194	712	33,309	71,319	21.39
7	138	2885	18,584	142,130	4.58	0	3071	0	150,426	0.00
8	189	2878	33,612	115,024	6.16	0	2648	0	139,145	0.00
9	595	6587	124,176	450,588	8.29	348	3142	62,643	350,470	9.97
10	159	1687	24,034	144,069	8.61	0	1289	0	133,244	0.00

2.8.3 *Jackknife Estimator of Variances*

This section describes the following activities. For domains defined by AREA, do:

C1. Estimate the totals of unemployed and employed people.
C2. Calculate direct estimators of variances and coefficients of variation.
C3. Calculate jackknife estimators of variances and coefficients of variation.

We first calculate some auxiliary parameters of the sample data file LFS20.txt.

```
# Number of domains
D <- length(unique(dat$AREA))
# Domain sample sizes
nd <- tapply(rep(1,n),INDEX=list(dat$AREA),FUN=sum)
# Clusters
nCLUSTER <- unique(dat$CLUSTER)
# Number of clusters
J <- length(unique(dat$CLUSTER))
md <- vector()
# Number of clusters by domains
for (d in 1:D)
    md[d] <- length(unique(dat$CLUSTER[dat$AREA==d]))
```

C1. By applying the formula (2.1), we calculate the direct estimates, dir1, of the totals of unemployed and employed people, i.e.

```
dir.d <- aggregate(w*data.frame(y1,y2), by=list(dat$AREA), sum)
# Assign column names
names(dir.d) <- c("area", "y1tot", "y2tot")
```

C2. By applying the formula (2.3), we calculate the direct estimators of the variances, i.e.

```
vardir.d <- aggregate(w*(w-1)*data.frame(y1^2,y2^2), by=list(dat$AREA), sum)
# Assign column names
names(vardir.d) <- c("area", "y1var", "y2var")
```

The direct estimators of the coefficients of variations are

```
cvdir1 <- round(100*sqrt(vardir.d$y1var)/abs(dir.d$y1tot),2) # CV for y1
cvdir2 <- round(100*sqrt(vardir.d$y2var)/abs(dir.d$y2tot),2) # CV for y2
```

C3. For calculating the jackknife estimators of the variances, we define the auxiliary arrays

```
jackdir1 <- jackdir2 <- matrix(0, nrow=D, ncol=J)
```

We run the following jackknife loop:

```
for (j in 1:J) {
  set <- subset(dat, dat$CLUSTER!=nCLUSTER[j], na.rm=TRUE)
  # Jackknife weights
  if (length(dat$AREA[dat$CLUSTER==j])>0) {
    domjack <- unique(dat$AREA[dat$CLUSTER==j])
    jfactor <- sum(dat$WEIGHT[dat$AREA==domjack])/
      sum(set$WEIGHT[set$AREA==domjack])
    set$WEIGHT[set$AREA==domjack] <- set$WEIGHT[set$AREA==domjack]*
      jfactor
  }
  # Direct estimators
  jdir.d <- aggregate(set$WEIGHT*data.frame(set$UNEMPLOYED,
                                  set$EMPLOYED), by=list(set$AREA), sum)
  # Assign column names
  names(jdir.d) <- c("area","y1tot","y2tot")
  jackdir1[,j] <- jdir.d$y1tot
  jackdir2[,j] <- jdir.d$y2tot
}
```

We calculate the jackknife means.

```
jmeandir1 <- rowMeans(jackdir1)
jmeandir2 <- rowMeans(jackdir2)
```

We apply the formulas of Sect. 2.7.1, for calculating the jackknife variances and coefficients of variation.

```
# Number of clusters by jackknife domain
md.J <- list()
for (d in 1:D){
  md.J[[d]] <- md
  md.J[[d]][d] <- md.J[[d]][d]-1
}
factor <- Map(f="/", lapply(md.J,1,FUN="-"), md.J)
# Jackknife variances
diff.cuad.1 <- (jackdir1-jmeandir1)^2
diff.cuad.2 <- (jackdir2-jmeandir2)^2
group <- rep(1:D, md)
jvardir1 <- jvardir2 <- vector()   # declare objects for indexing
for (d in 1:D) {
  jvardir1[d] <- sum(sapply(split(diff.cuad.1[d,],group), sum)*factor[[d]])
  jvardir2[d] <- sum(sapply(split(diff.cuad.2[d,],group), sum)*factor[[d]])
}
# Jackknife coefficients of variation
jcvdir1 <- round(100*sqrt(jvardir1)/jmeandir1,2)
jcvdir2 <- round(100*sqrt(jvardir2)/jmeandir2,2)
```

The R code to save the results is

```
output3 <- data.frame(nd, y1=dir.d$y1tot, v.y1=vardir.d$y1var,
                      vJ.y1=round(jvardir1), cv.y1=cvdir1, cvJ.y1=jcvdir1,
                      y2=dir.d$y2tot, v.y2=vardir.d$y2var,
                      vJ.y2=round(jvardir2), cv.y2=cvdir2, cvJ.y2=jcvdir2)
head(output3, 10)
```

Table 2.3 presents the results for the 10 first domains (AREA). The labels y_1 and y_2 denote the dir1 direct estimates of the totals of unemployed and employed people, respectively. The direct estimates of the variances of the direct estimators of totals are denoted by $\mathrm{v}(y_1)$ and $\mathrm{v}(y_2)$. The corresponding jackknife estimates are $\mathrm{v}_J(y_1)$ and $\mathrm{v}_J(y_2)$. The direct estimates of the coefficients of variation of the direct estimators of totals are denoted by $\mathrm{c}(y_1)$ and $\mathrm{c}(y_2)$. The corresponding jackknife estimates are $\mathrm{c}_J(y_1)$ and $\mathrm{c}_J(y_2)$. The direct and jackknife estimators of variances and coefficients of variation follow the same pattern. In any case, a finer analysis cannot be done because the data used is simulated and does not come from a real survey.

Table 2.3 dir1 estimates of unemployment (left) and employment (right) totals by area

d	n_d	y_1	$\mathrm{v}(y_1)$	$\mathrm{v}_J(y_1)$	$\mathrm{c}(y_1)$	$\mathrm{c}_J(y_1)$	y_2	$\mathrm{v}(y_2)$	$\mathrm{v}_J(y_2)$	$\mathrm{c}(y_2)$	$\mathrm{c}_J(y_2)$
1	60	796	230,060	329,637	60.26	72.19	9059	2,509,176	1,365,062	17.49	12.90
2	37	428	91,292	70,084	70.59	61.84	3456	764,676	674,173	25.30	23.76
3	47	165	27,060	26,103	99.70	97.87	4772	896,872	253,103	19.85	10.54
4	55	366	66,644	46,415	70.53	58.87	6186	1,114,294	313,081	17.06	9.05
5	50	137	18,632	17,774	99.63	97.30	4614	759,082	617,055	18.88	17.03
6	43	581	112,180	307,334	57.65	95.49	4393	803,082	50,480	20.40	5.11
7	48	137	18,632	17,338	99.63	96.15	5978	1,161,556	284,300	18.03	8.92
8	48	188	35,156	33,465	99.73	97.30	5488	952,560	198,549	17.78	8.12
9	125	946	199,650	242,903	47.23	52.09	9765	1,757,780	622,368	13.58	8.08
10	41	156	24,180	22,714	99.68	96.63	2968	516,248	491,492	24.21	23.62

2.8.4 Functions for Calculating Direct Estimators

The function `dir1` calculates the Horvitz–Thompson direct estimators of the mean and the total. The R code is

```
dir1 <- function(data, w, domain, Nd) {
  if(is.vector(data)){
    last <- length(domain) + 1
    Nd.hat <- aggregate(w, by=domain, sum)[,last]
    nd <- aggregate(rep(1, length(data)), by=domain, sum)[,last]
    tot <- aggregate(w*data, by=domain, sum)
    names(tot) <- c(names(domain), "tot")
    var.tot <- aggregate(w*(w-1)*data^2, by=domain, sum)[,last]
    if(missing(Nd)){
      return(cbind(tot, var.tot, Nd.hat, nd))
    }
    else{
      mean <- tot[,last]/Nd
      var.mean <- var.tot/Nd^2
      return(cbind(tot, var.tot, mean, var.mean, Nd.hat, Nd, nd))
    }
  }
  else{
    warning("Only a numeric or integer vector must be called as data",
            call. = FALSE)
  }
}
```

The function dir2 calculates the Hájek direct estimators of the mean and the total. The R code is

```
dir2 <- function(data, w, domain, Nd) {
  if(is.vector(data)){
    last <- length(domain) + 1
    Nd.hat <- aggregate(w, by=domain, sum)[,last]
    nd <- aggregate(rep(1, length(data)), by=domain, sum)[,last]
    Sum <- aggregate(w*data, by=domain, sum)
    mean <- Sum[,last]/Nd.hat
    dom <- as.numeric(Reduce("paste0", domain))
    if(length(domain)==1){
      domain.unique <- sort(unique(dom))
    }
    else{
      domain.unique <- as.numeric(Reduce("paste0", Sum[,1:length(domain)]))
    }
    difference <- list()
    for(d in 1:length(mean)){
      condition <- dom==domain.unique[d]
      difference[[d]] <- w[condition]*(w[condition]-1)*(data[condition]-mean[d])^2
    }
    var.mean <- unlist(lapply(difference, sum))/Nd.hat^2
    if(missing(Nd)){
      return(data.frame(Sum[,-last], mean, var.mean, Nd.hat, nd))
    }
    else{
      tot <- mean*Nd
      var.tot <- var.mean*Nd^2
      return(data.frame(Sum[,-last], tot, var.tot, mean, var.mean, Nd.hat, Nd, nd))
    }
  }
  else{
    warning("Only a numeric or integer vector must be called as data",
            call. = FALSE)
  }
}
```

The following R code illustrates the use of both functions, `dir1` and `dir2`, to the data set used in this chapter. We first read the sample data files and rename some variables.

```
# Auxiliary data
```

```
dataux <- read.table("Nds20.txt", header=TRUE, sep = "\t", dec = ".")
# Sort dataux by sex and area:
dataux <- dataux[order(dataux$sex, dataux$area),]
# Sample data
dat <- read.table("LFS20.txt", header=TRUE, sep = "\t", dec = ".")
# number of rows (cases) in dat:
n <- nrow(dat)
# Rename some  variables
y1 <- dat$UNEMPLOYED
w <- dat$WEIGHT
```

Note that data and w must be a vector R object and that domains must be introduced as a list R object. The following R code calculates the direct estimator for the totals and means of unemployed people:

```
# Horvitz-Thompson direct estimator for unemployed people
direct1 <- dir1(data=y1, w=dat$WEIGHT, domain=list(area=dat$AREA,
                sex=dat$SEX), Nd=dataux$N)
head(direct1, 10)
# Hajek direct estimator for unemployed people
direct2 <- dir2(data=y1, w=dat$WEIGHT, domain=list(area=dat$AREA,
                sex=dat$SEX), Nd=dataux$N)
head(direct2, 10)
```

References

Durbin, J.: A note on the application of Quenouille's method of bias reduction to the estimation of ratios. Biometrika **46**, 477–480 (1959)

Hájek, J.: Comment on "An Essay on the Logical Foundations of Survey Sampling, Part One". In: Godambe, V.P., Sprott, D.A. (eds.) The Foundations of Survey Sampling, pp. 236. Holt, Rinehart, and Winston, New York (1971)

Horvitz, D.G., Thompson, D.J.: A generalization of sampling without replacement from a finite universe. J. Am. Stat. Assoc. **47**, 663–685 (1952)

Quenouille, M.H.: Approximation tests of correlation in time series. J. R. Stat. Soc. Ser. B. **11**, 68–84 (1949)

Quenouille, M.H.: Notes on bias in estimation. Biometrika **34**, 353–360 (1956)

Rao, J.N.K.: Small Area Estimation. Wiley, New York (2003)

Rao, J.N.K., Tausi, M.: Estimating function jackknife variance estimators under stratified multi-stage sampling. Commun. Stat. Theory Methods **33**, 2087–2095 (2004)

Särndal, C.E., Swensson, B., Wretman, J.: Model Assisted Survey Sampling. Springer, Berlin (1992)

Tukey, J.: Bias and confidence in not quite large samples. Ann. Math. Stat. **29**, 614 (1958)

Chapter 3
Design-Based Indirect Estimation

3.1 Introduction

Direct estimators of domain parameters are calculated by using only data from the domain. In the context of small area estimation, domain sample sizes are small, and therefore direct estimators have large variances. Indirect estimators also employ data from outside the target domains to improve the quality of the estimates. This can be done by using models as auxiliary tools for constructing estimators. Nevertheless, the properties of these estimators are optimized with respect to the sampling distribution, so that they are called design-based indirect estimators. According to Särndal et al. (1992), this approach is called model-assisted.

In the case of having auxiliary data related to the target variable, it is possible to obtain better accuracy for domain estimates, when compared to an estimation procedure not using auxiliary data. For example, if the population sizes of domains crossed by sex–age groups are available from external data registers, then this information can be used for evaluating basic synthetic or post-stratified estimators that could provide better estimates than the direct ones.

Among the model-assisted estimators, the generalized regression estimator (GREG) plays a relevant role. The GREG estimator uses a linear model with fixed effects as an assisting tool, but it is constructed to be design-unbiased (or approximately so) irrespective of the fit of the model to data. Different types of auxiliary data can be used in model-assisted estimation. GREG estimators employ auxiliary variables from survey files and their aggregated values from administrative registers.

This chapter presents some design-based indirect small area estimators using different types of auxiliary information. More concretely, it describes the basic synthetic, the post-stratified, the sample size dependent, and the GREG estimators of domain totals and means. Further, the chapter presents an application to a labor force survey where the different steps in the construction of design-based indirect

D. Morales et al., *A Course on Small Area Estimation and Mixed Models*, Statistics for Social and Behavioral Sciences, https://doi.org/10.1007/978-3-030-63757-6_3

estimators are illustrated. Finally, some R codes for applying the introduced indirect estimators are given.

3.2 Basic Synthetic Estimator

Let us assume that the population of interest is stratified by a categorical variable in a small number of strata or groups. For example, a population of individuals could be stratified by sex–age groups with six categories (male and female for sex and ≤24, 25–53 and ≥54 years for age), and a population of households could be stratified by typology groups with four categories (only one adult, two adults without children, two adults with children, and other). Further, assume that the population is partitioned into a large number of domains. If the target variable y has a small variance inside the groups, so that the domain–group means within groups do not differ significantly, and the groups sizes are estimated with high precision, then a simple and efficient estimator is the basic synthetic estimator.

To introduce the basic synthetic estimator, we consider a population U partitioned in domains and groups, i.e. $U = \cup_{d=1}^{D} U_d$ with $U_{d_1} \cap U_{d_2} = \emptyset$ if $d_1 \neq d_2$ and $U = \cup_{g=1}^{G} U_g$ with $U_{g_1} \cap U_{g_2} = \emptyset$ if $g_1 \neq g_2$. Let $U_{dg} = U_d \cap U_g$, $d = 1, \ldots, D$, $g = 1, \ldots, G$. Let s, s_d, s_g, and s_{dg} be the sample and the corresponding subsamples. A direct estimator of $\overline{Y}_g = \frac{1}{N_g} \sum_{j \in U_g} y_j$ is

$$\hat{\overline{Y}}_g^{dir2} = \frac{1}{\hat{N}_g} \sum_{j \in s_g} w_j y_j, \quad \hat{N}_g = \sum_{j \in s_g} w_j.$$

The *basic synthetic* estimator of the total $Y_d = \sum_{j \in U_d} y_j$ is

$$\hat{Y}_d^{synth} = \sum_{g=1}^{G} N_{dg} \hat{\overline{Y}}_g^{dir2}, \tag{3.1}$$

where N_{dg} is the population size of U_{dg}. The basic synthetic estimator is a biased estimator of Y_d, see (Särndal et al. 1992, p. 410). Its approximated expectation and bias are

$$E_\pi[\hat{Y}_d^{synth}] \approx \sum_{g=1}^{G} N_{dg} \overline{Y}_g,$$

$$B_\pi[\hat{Y}_d^{synth}] = E_\pi[\hat{Y}_d^{synth}] - \sum_{j \in U_d} y_j \approx \sum_{g=1}^{G} N_{dg}(\overline{Y}_g - \overline{Y}_{dg}),$$

where $\overline{Y}_{dg} = \frac{1}{N_{dg}} \sum_{j \in U_{dg}} y_j$. The synthetic estimator is approximately unbiased if $\overline{Y}_{dg} = \overline{Y}_g$, $d = 1, \ldots, D$, that is, if the groups are internally homogeneous.

If conditions $\pi_{ij} = \pi_i \pi_j$ $(i \neq j)$ and $\pi_{jj} = \pi_j$ are fulfilled, we get (cf. Remark 2.3)

$$\text{var}_\pi\left(\hat{Y}_d^{synth}\right) = \sum_{g=1}^{G} N_{dg}^2 \, \text{var}_\pi\left(\hat{Y}_g^{dir2}\right) \approx \sum_{g=1}^{G} \frac{N_{dg}^2}{N_g^2} \sum_{j \in U_g} \frac{1-\pi_j}{\pi_j} (y_j - \overline{Y}_g)^2,$$

$$\widehat{\text{var}}_\pi\left(\hat{Y}_d^{synth}\right) = \sum_{g=1}^{G} N_{dg}^2 \widehat{\text{var}}_\pi\left(\hat{\overline{Y}}_g^{dir2}\right) = \sum_{g=1}^{G} \frac{N_{dg}^2}{\hat{N}_g^2} \sum_{j \in s_g} w_j (w_j - 1)\left(y_j - \hat{\overline{Y}}_g^{dir2}\right)^2.$$

The *basic synthetic* estimator of the mean is

$$\hat{\overline{Y}}_d^{synth} = \frac{1}{N_d} \sum_{g=1}^{G} N_{dg} \hat{\overline{Y}}_g^{dir2},$$

and it is a biased estimator of $\overline{Y}_d$. Its approximated expectation and bias are

$$E_\pi[\hat{\overline{Y}}_d^{synth}] \approx \frac{1}{N_d} \sum_{g=1}^{G} N_{dg} \overline{Y}_g,$$

$$B_\pi[\hat{\overline{Y}}_d^{synth}] = E_\pi[\hat{\overline{Y}}_d^{synth}] - \frac{1}{N_d} \sum_{j \in s_d} y_j \approx \frac{1}{N_d} \sum_{g=1}^{G} N_{dg} (\overline{Y}_g - \overline{Y}_{dg}).$$

The bias can be estimated in several ways. One option is the plug-in estimator

$$\widehat{B}_\pi^{(1)}[\hat{\overline{Y}}_d^{synth}] = \frac{1}{N_d} \sum_{g=1}^{G} N_{dg} (\hat{\overline{Y}}_g^{dir2} - \hat{\overline{Y}}_{dg}^{dir2}).$$

But in SAE problems, the sizes of the subsamples s_{dg} may be too small, which means that the direct estimators $\hat{\overline{Y}}_{dg}^{dir2}$ are not very precise. Other options are the estimators,

$$\widehat{B}_\pi^{(2)}[\hat{\overline{Y}}_d^{synth}] = \hat{\overline{Y}}_d^{synth} - \hat{\overline{Y}}_d^{dir2},$$

$$\widehat{B}_\pi^{(3)}[\hat{\overline{Y}}_d^{synth}] = \hat{\overline{Y}}_d^{synth2} - \hat{\overline{Y}}_d^{dir2}, \quad \hat{\overline{Y}}_d^{synth2} = \frac{1}{\hat{N}_d} \sum_{g=1}^{G} \hat{N}_{dg} \hat{\overline{Y}}_g^{dir2}.$$

In sampling designs with $\pi_{ij} = \pi_i\pi_j,\ i \neq j$, and $\pi_{jj} = \pi_j$, we get

$$\mathrm{var}_\pi(\hat{\overline{Y}}{}_d^{synth}) = \frac{1}{N_d^2}\sum_{g=1}^{G} N_{dg}^2\, \mathrm{var}_\pi(\hat{\overline{Y}}{}_g^{dir2}) \approx \frac{1}{N_d^2}\sum_{g=1}^{G}\frac{N_{dg}^2}{N_g^2}\sum_{j\in U_g}\frac{1-\pi_j}{\pi_j}(y_j - \overline{Y}_g)^2,$$

$$\widehat{\mathrm{var}}_\pi(\hat{\overline{Y}}{}_d^{synth}) = \frac{1}{N_d^2}\sum_{g=1}^{G} N_{dg}^2\,\widehat{\mathrm{var}}_\pi(\hat{\overline{Y}}{}_g^{dir2})$$

$$= \frac{1}{N_d^2}\sum_{g=1}^{G}\frac{N_{dg}^2}{\hat{N}_g^2}\sum_{j\in s_g} w_j(w_j - 1)(y_j - \hat{\overline{Y}}{}_g^{dir2})^2.$$

The mean squared error of the basic synthetic estimator is

$$MSE(\hat{\overline{Y}}{}_d^{synth}) = \mathrm{var}_\pi(\hat{\overline{Y}}{}_d^{synth}) + \left(B_\pi[\hat{\overline{Y}}{}_d^{synth}]\right)^2.$$

An upper biased estimator of $MSE(\hat{\overline{Y}}{}_d^{synth})$ is

$$mse(\hat{\overline{Y}}{}_d^{synth}) = \widehat{\mathrm{var}}_\pi(\hat{\overline{Y}}{}_d^{synth}) + (\hat{\overline{Y}}{}_d^{synth} - \hat{\overline{Y}}{}_d^{dir2})^2.$$

Given two basic synthetic estimators,

$$\hat{Y}_d^{synth} = \sum_{g=1}^{G} N_{dg}\hat{\overline{Y}}{}_g^{dir2}, \qquad \hat{Z}_d^{synth} = \sum_{g=1}^{G} N_{dg}\hat{\overline{Z}}{}_g^{dir2},$$

it holds that

$$\mathrm{cov}_\pi(\hat{Y}_d^{synth}, \hat{Z}_d^{synth}) = \sum_{g_1=1}^{G}\sum_{g_2=1}^{G} N_{dg_1}N_{dg_2}\mathrm{cov}_\pi(\hat{\overline{Y}}{}_{g_1}^{dir2}, \hat{\overline{Z}}{}_{g_2}^{dir2}).$$

By adapting the proof of Proposition 2.5 to two groups, g_1 and g_2, we get

$$\mathrm{cov}_\pi(\hat{\overline{Y}}{}_{g_1}^{dir2}, \hat{\overline{Z}}{}_{g_2}^{dir2}) = \frac{1}{N_{g_1}N_{g_2}}\sum_{i\in U_{g_1}}\sum_{j\in U_{g_2}}\frac{\pi_{ij} - \pi_i\pi_j}{\pi_i\pi_j}(y_i - \overline{Y}_{g_1})(z_j - \overline{Z}_{g_2}).$$

If $\pi_{ij} = \pi_i \pi_j$ $(i \neq j)$ and $g_1 \neq g_2$, it holds that $\text{cov}_\pi(\hat{\overline{Y}}_{g_1}^{dir2}, \hat{\overline{Z}}_{g_2}^{dir2}) = 0$. Similarly as in Remark 2.3, if $\pi_{jj} = \pi_j$, we have

$$\text{cov}_\pi(\hat{Y}_d^{synth}, \hat{Z}_d^{synth}) = \sum_{g=1}^{G} \frac{N_{dg}^2}{N_g^2} \sum_{j \in U_g} \frac{1-\pi_j}{\pi_j} (y_j - \overline{Y}_g)(z_j - \overline{Z}_g),$$

$$\widehat{\text{cov}}_\pi(\hat{Y}_d^{synth}, \hat{Z}_d^{synth}) = \sum_{g=1}^{G} \frac{N_{dg}^2}{\hat{N}_g^2} \sum_{j \in s_g} w_j(w_j - 1)\,(y_j - \hat{\overline{Y}}_g^{dir2})(z_j - \hat{\overline{Z}}_g^{dir2}).$$

The crossed mean squared error is

$$MSE(\hat{Y}_d^{synth}, \hat{Z}_d^{synth}) = \text{cov}_\pi(\hat{Y}_d^{synth}, \hat{Z}_d^{synth}) + B_\pi[\hat{Y}_d^{synth}]\, B_\pi[\hat{Z}_d^{synth}].$$

For calculating $\hat{\overline{Y}}_d^{synth}$, we need the following auxiliary information:

1. *Non-aggregated:* sampling weights w_j and group indicator $I_{U_g}(j)$, $j \in s$.
2. *Aggregated:* domain and domain–group population sizes N_d and N_{dg}, $d = 1, \ldots, D$, $g = 1, \ldots, G$.

The basic synthetic estimator is typically biased, but it has a small variance.

3.3 Post-Stratified Estimator

Let us consider a population U partitioned in domains and groups as described in Sect. 3.2. If the target variable y has a large variance inside the groups, then the domain–group means might differ significantly, and the basic synthetic estimator is not a good option. For these situations, the post-stratified estimator might be a good competitor.

The *post-stratified* estimator of the total is

$$\hat{Y}_d^{pst} = \sum_{g=1}^{G} N_{dg} \hat{\overline{Y}}_{dg}^{dir2}, \tag{3.2}$$

where

$$\hat{\overline{Y}}_{dg}^{dir2} = \frac{1}{\hat{N}_{dg}} \sum_{j \in s_{dg}} w_j y_j, \quad \hat{N}_{dg} = \sum_{j \in s_{dg}} w_j,$$

is the Hájek-type direct estimator of $\overline{Y}_{dg} = \frac{1}{N_{dg}} \sum_{j \in U_{dg}} y_j$.

The post-stratified estimator is an approximately unbiased estimator. If $\pi_{ij} = \pi_i\pi_j$ $(i \neq j)$ and $\pi_{jj} = \pi_j$, then

$$\text{var}_\pi\big(\hat{Y}_d^{pst}\big) = \sum_{g=1}^{G} N_{dg}^2 \text{var}_\pi\big(\hat{\overline{Y}}_{dg}^{dir2}\big) \approx \sum_{g=1}^{G}\sum_{j\in U_{dg}} \frac{1-\pi_j}{\pi_j}(y_j - \overline{Y}_{dg})^2,$$

$$\widehat{\text{var}}_\pi\big(\hat{Y}_d^{pst}\big) = \sum_{g=1}^{G} N_{dg}^2 \widehat{\text{var}}_\pi\big(\hat{\overline{Y}}_{dg}^{dir2}\big) = \sum_{g=1}^{G} \frac{N_{dg}^2}{\hat{N}_{dg}^2} \sum_{j\in s_{dg}} w_j(w_j-1)\big(y_j - \hat{\overline{Y}}_{dg}^{dir2}\big)^2.$$

The post-stratified estimator of the mean is

$$\hat{\overline{Y}}_d^{pst} = \frac{1}{N_d}\sum_{g=1}^{G} N_{dg}\hat{\overline{Y}}_{dg}^{dir2}.$$

This is an approximately unbiased estimator. If $\pi_{ij} = \pi_i\pi_j$ $(i \neq j)$ and $\pi_{jj} = \pi_j$, then

$$\text{var}_\pi\big(\hat{\overline{Y}}_d^{pst}\big) = \frac{1}{N_d^2}\sum_{g=1}^{G} N_{dg}^2 \text{var}_\pi\big(\hat{\overline{Y}}_{dg}^{dir2}\big) \approx \frac{1}{N_d^2}\sum_{g=1}^{G}\sum_{j\in U_{dg}} \frac{1-\pi_j}{\pi_j}(y_j - \overline{Y}_{dg})^2,$$

$$\widehat{\text{var}}_\pi\big(\hat{\overline{Y}}_d^{pst}\big) = \frac{1}{N_d^2}\sum_{g=1}^{G} N_{dg}^2 \widehat{\text{var}}_\pi\big(\hat{\overline{Y}}_{dg}^{dir2}\big)$$

$$= \frac{1}{N_d^2}\sum_{g=1}^{G} \frac{N_{dg}^2}{\hat{N}_{dg}^2} \sum_{j\in s_{dg}} w_j(w_j-1)\big(y_j - \hat{\overline{Y}}_{dg}^{dir2}\big)^2.$$

As the post-stratified estimator is approximately unbiased, the mean squared error can be approximated by the variance, i.e.

$$MSE\big(\hat{\overline{Y}}_d^{pst}\big) \approx \text{var}_\pi\big(\hat{\overline{Y}}_d^{pst}\big), \quad mse\big(\hat{\overline{Y}}_d^{pst}\big) = \widehat{\text{var}}_\pi\big(\hat{\overline{Y}}_d^{pst}\big).$$

Given two post-stratified estimators,

$$\hat{Y}_d^{pst} = \sum_{g=1}^{G} N_{dg}\hat{\overline{Y}}_{dg}^{dir2}, \qquad \hat{Z}_d^{pst} = \sum_{g=1}^{G} N_{dg}\hat{\overline{Z}}_{dg}^{dir2},$$

it holds that

$$\text{cov}_\pi\big(\hat{Y}_d^{pst}, \hat{Z}_d^{pst}\big) = \sum_{g_1=1}^{G}\sum_{g_2=1}^{G} N_{dg_1}N_{dg_2}\text{cov}_\pi\big(\hat{\overline{Y}}_{dg_1}^{dir2}, \hat{\overline{Z}}_{dg_2}^{dir2}\big).$$

If $\pi_{ij} = \pi_i \pi_j$ $(i \neq j)$, and applying $\pi_{jj} = \pi_j$, we get

$$\mathrm{cov}_\pi(\hat{Y}_d^{pst}, \hat{Z}_d^{pst}) = \sum_{g=1}^{G} \sum_{j \in U_{dg}} \frac{1-\pi_j}{\pi_j}(y_j - \overline{Y}_{dg})(z_j - \overline{Z}_{dg}),$$

$$\widehat{\mathrm{cov}}_\pi(\hat{Y}_d^{pst}, \hat{Z}_d^{pst}) = \sum_{g=1}^{G} \frac{N_{dg}^2}{\hat{N}_{dg}^2} \sum_{j \in s_{dg}} w_j(w_j - 1)\,(y_j - \hat{\overline{Y}}_{dg}^{dir2})(z_j - \hat{\overline{Z}}_{dg}^{dir2}).$$

For calculating $\hat{\overline{Y}}_d^{pst}$, we need the auxiliary information:

1. *Non-aggregated:* sampling weights w_j and domain–group indicator $I_{U_{dg}}(j),\ j \in s$.
2. *Aggregated:* domain and domain–group population sizes N_d and $N_{dg},\ d = 1, \ldots, D,\ g = 1, \ldots, G$.

The post-stratified estimator is typically unbiased, but it has a large variance.

3.4 Sample Size Dependent Estimator

Drew et al. (1982) proposed the sample size dependent estimator

$$\hat{Y}_d^{ssd} = \gamma_d \hat{Y}_d^{pst} + (1-\gamma_d)\hat{Y}_d^{synth},$$

where

$$\gamma_d = \begin{cases} 1 & \text{if } \hat{N}_d \geq \delta N_d \\ \frac{\hat{N}_d}{\delta N_d} & \text{otherwise.} \end{cases}$$

The constant δ controls the contribution of the synthetic component. Some possible values are $\delta = 1$ or $\delta = 2/3$ (as it was used in the Canadian labor force survey).

3.5 Generalized Regression Estimator

Särndal et al. (1992) presented the model-assisted approach to inference in finite populations. The model-assisted methodology considers the properties under the design-based distribution, but it employs a statistical model to motivate the choice of estimators. Under this approach, the generalized regression (GREG) estimators play a fundamental role. GREG estimation was introduced for domain estimation in Särndal (1981, 1984), Hidiroglou and Särndal (1985), and Särndal and Hidiroglou

(1989), and it was developed further (including computational tools) in Estevao et al. (1995). Lehtonen and Veijanen (2009, 2016) discussed the GREG estimators of means and proportions and presented empirical studies based on simulation experiments. This section uses this approach for introducing estimators of small area means and totals.

Let the subindexes d and j denote variables associated with the unit j of domain d. We consider p explanatory variables evaluated at the N population units, i.e. $\boldsymbol{x}_{dj} = (x_{dj,1}, \ldots, x_{dj,p})$, $d = 1, \ldots, D$, $j = 1, \ldots, N_d$. Let us define the population means and their direct estimators

$$\overline{\boldsymbol{X}}_d = \frac{1}{N_d} \sum_{j \in U_d} \boldsymbol{x}_{dj}, \quad \text{and} \quad \hat{\overline{\boldsymbol{X}}}_d^{dir2} = \frac{1}{\hat{N}_d} \sum_{j \in s_d} w_{dj} \boldsymbol{x}_{dj}, \quad \hat{N}_d = \sum_{j \in s_d} w_{dj}.$$

Further, let us consider the population model

$$\boldsymbol{y} = \boldsymbol{X}\boldsymbol{b} + \boldsymbol{e}, \quad E[\boldsymbol{e}] = \boldsymbol{0}, \quad \text{var}(\boldsymbol{e}) = \sigma^2 \boldsymbol{I}, \tag{3.3}$$

where $\boldsymbol{b} = \underset{1 \le k \le p}{\text{col}} (b_k)$ and

$$\boldsymbol{X} = \underset{1 \le d \le D}{\text{col}} (\underset{1 \le j \le N_d}{\text{col}} (\boldsymbol{x}_{dj})), \ \boldsymbol{y} = \underset{1 \le d \le D}{\text{col}} (\underset{1 \le j \le N_d}{\text{col}} (y_{dj})), \ \boldsymbol{e} = \underset{1 \le d \le D}{\text{col}} (\underset{1 \le j \le N_d}{\text{col}} (e_{dj})).$$

The population-based least squared approximation of $\boldsymbol{b}$ is

$$\boldsymbol{\beta} = (\boldsymbol{X}'\boldsymbol{X})^{-1}\boldsymbol{X}'\boldsymbol{y} = \Big(\sum_{d=1}^{D} \sum_{j \in U_d} \boldsymbol{x}'_{dj} \boldsymbol{x}_{dj} \Big)^{-1} \sum_{d=1}^{D} \sum_{j \in U_d} \boldsymbol{x}'_{dj} y_{dj}.$$

A direct least squared estimator of $\boldsymbol{\beta}$ is

$$\hat{\boldsymbol{\beta}} = \Big(\sum_{d=1}^{D} \sum_{j \in s_d} w_{dj} \boldsymbol{x}'_{dj} \boldsymbol{x}_{dj} \Big)^{-1} \sum_{d=1}^{D} \sum_{j \in s_d} w_{dj} \boldsymbol{x}'_{dj} y_{dj}.$$

The prediction of y_{dj} under model (3.3) is $\hat{y}_{dj} = \boldsymbol{x}_{dj}\hat{\boldsymbol{\beta}}$, and the residual is $\hat{e}_{dj} = y_{dj} - \hat{y}_{dj} = y_{dj} - \boldsymbol{x}_{dj}\hat{\boldsymbol{\beta}}$, $d = 1, \ldots, D$, $j = 1, \ldots, N_d$. The generalized regression (GREG) estimator of the total is constructed by summing up the predicted values and correcting by the residuals. The starting point is the equality

$$Y_d = \sum_{j \in U_d} y_{dj} = \sum_{j \in U_d} \hat{y}_{dj} + \sum_{j \in U_d} \hat{e}_{dj},$$

where we estimate the second summand with the dir2 estimator, i.e.

$$\hat{Y}_d^{greg} = \sum_{j\in U_d} \hat{y}_{dj} + \frac{N_d}{\hat{N}_d} \sum_{j\in s_d} w_{dj}\hat{e}_{dj}. \tag{3.4}$$

The first summand of (3.4) is

$$\sum_{j\in U_d} \hat{y}_{dj} = \sum_{j\in U_d} \boldsymbol{x}_{dj}\hat{\boldsymbol{\beta}} = \Big(\sum_{j\in U_d} \boldsymbol{x}_{dj}\Big)\hat{\boldsymbol{\beta}} = N_d\overline{\boldsymbol{X}}_d\hat{\boldsymbol{\beta}}.$$

The second summand of (3.4) is

$$\frac{N_d}{\hat{N}_d}\sum_{j\in s_d} w_{dj}\hat{e}_{dj} = \frac{N_d}{\hat{N}_d}\sum_{j\in s_d} w_{dj}(y_{dj} - \boldsymbol{x}_{dj}\hat{\boldsymbol{\beta}}) = \frac{N_d}{\hat{N}_d}\hat{Y}_d^{dir1} - \frac{N_d}{\hat{N}_d}\hat{\boldsymbol{X}}_d^{dir1}\hat{\boldsymbol{\beta}}$$

$$= N_d\hat{\overline{Y}}_d^{dir2} - N_d\hat{\overline{\boldsymbol{X}}}_d^{dir2}\hat{\boldsymbol{\beta}}.$$

Finally, the GREG estimator of the total is

$$\hat{Y}_d^{greg} = N_d\overline{\boldsymbol{X}}_d\hat{\boldsymbol{\beta}} + \Big(N_d\hat{\overline{Y}}_d^{dir2} - N_d\hat{\overline{\boldsymbol{X}}}_d^{dir2}\hat{\boldsymbol{\beta}}\Big) = N_d\hat{\overline{Y}}_d^{dir2} + N_d(\overline{\boldsymbol{X}}_d - \hat{\overline{\boldsymbol{X}}}_d^{dir2})\hat{\boldsymbol{\beta}}. \tag{3.5}$$

Note that the GREG estimator can be written in the form of a sum of some weights multiplied by the target variables, namely

$$\hat{Y}_d^{greg} = \sum_{\ell=1}^{D}\sum_{j\in s_\ell} g_{d,\ell j}w_{\ell j}y_{\ell j} \quad \text{and} \quad N_d\overline{\boldsymbol{X}}_d = \sum_{\ell=1}^{D}\sum_{j\in s_\ell} g_{d,\ell j}w_{\ell j}\boldsymbol{x}_{\ell j},$$

where

$$g_{d,\ell j} = \begin{cases} \frac{N_d}{\hat{N}_d} + N_d\big(\overline{\boldsymbol{X}}_d - \hat{\overline{\boldsymbol{X}}}_d^{dir2}\big)\Big(\sum_{d=1}^{D}\sum_{i\in s_d} w_{di}\boldsymbol{x}'_{di}\boldsymbol{x}_{di}\Big)^{-1}\boldsymbol{x}'_{\ell j} & \text{if } \ell = d,\ j\in s_\ell, \\ N_d\big(\overline{\boldsymbol{X}}_d - \hat{\overline{\boldsymbol{X}}}_d^{dir2}\big)\Big(\sum_{d=1}^{D}\sum_{i\in s_d} w_{di}\boldsymbol{x}'_{di}\boldsymbol{x}_{di}\Big)^{-1}\boldsymbol{x}'_{\ell j} & \text{if } \ell \neq d,\ j\in s_\ell. \end{cases} \tag{3.6}$$

If $\ell = d$, then we use the notation g_{dj}, i.e. $g_{dj} = g_{d,dj}$.
The GREG estimator of Y_d is approximately unbiased. The approximated variance of $\hat{Y}_d^{greg}$ is

$$\text{var}_\pi\big(\hat{Y}_d^{greg}\big) \approx \sum_{i\in U_d}\sum_{j\in U_d} \frac{\pi_{dij} - \pi_{di}\pi_{dj}}{\pi_{di}\pi_{dj}}(E_{di} - \overline{E}_d)(E_{dj} - \overline{E}_d),$$

where

$$\overline{E}_d = \frac{1}{N_d} \sum_{j \in U_d} E_{dj}, \quad E_{dj} = y_{dj} - \boldsymbol{x}_{dj} \boldsymbol{\beta}.$$

An estimator of $\mathrm{var}_\pi\left(\hat{Y}_d^{greg}\right)$ is

$$\widehat{\mathrm{var}}_\pi(\hat{Y}_d^{greg}) = \sum_{i \in s_d} \sum_{j \in s_d} \frac{\pi_{dij} - \pi_{di}\pi_{dj}}{\pi_{dij}\pi_{di}\pi_{dj}} g_{di} g_{dj} (y_{di} - \boldsymbol{x}_{di}\hat{\boldsymbol{\beta}})(y_{dj} - \boldsymbol{x}_{dj}\hat{\boldsymbol{\beta}}).$$

If $\pi_{dij} = \pi_{di}\pi_{dj}$ $(i \neq j)$ and $\pi_{djj} = \pi_{dj}$, we get

$$\mathrm{var}_\pi\left(\hat{Y}_d^{greg}\right) \approx \sum_{j \in U_d} \frac{1 - \pi_{dj}}{\pi_{dj}} (E_{dj} - \overline{E}_d)^2,$$

$$\widehat{\mathrm{var}}_\pi(\hat{Y}_d^{greg}) = \sum_{j \in s_d} w_{dj}(w_{dj} - 1) g_{dj}^2 (y_{dj} - \boldsymbol{x}_{dj}\hat{\boldsymbol{\beta}})^2.$$

The GREG estimator of the mean is

$$\hat{\overline{Y}}_d^{greg} = \hat{\overline{Y}}_d^{dir2} + (\overline{\boldsymbol{X}}_d - \hat{\overline{\boldsymbol{X}}}_d^{dir2})\hat{\boldsymbol{\beta}}.$$

Note that

$$\hat{\overline{Y}}_d^{greg} = \sum_{\ell=1}^{D} \sum_{j \in s_\ell} h_{d,\ell j} w_{\ell j} y_{\ell j}, \qquad \overline{\boldsymbol{X}}_d = \sum_{\ell=1}^{D} \sum_{j \in s_\ell} h_{d,\ell j} w_{\ell j} \boldsymbol{x}_{\ell j},$$

where $h_{d,\ell j} = g_{d,\ell j}/N_d$, i.e.

$$h_{d,\ell j} = \begin{cases} \hat{N}_d^{-1} + (\overline{\boldsymbol{X}}_d - \hat{\overline{\boldsymbol{X}}}_d^{dir2})\left(\sum_{d=1}^{D} \sum_{i \in s_d} w_{di} \boldsymbol{x}'_{di} \boldsymbol{x}_{di}\right)^{-1} \boldsymbol{x}'_{\ell j} & \text{if } \ell = d,\ j \in s_\ell, \\ (\overline{\boldsymbol{X}}_d - \hat{\overline{\boldsymbol{X}}}_d^{dir2})\left(\sum_{d=1}^{D} \sum_{i \in s_d} w_{di} \boldsymbol{x}'_{di} \boldsymbol{x}_{di}\right)^{-1} \boldsymbol{x}'_{\ell j} & \text{if } \ell \neq d,\ j \in s_\ell. \end{cases}$$

If $\ell = d$, then we use the notation h_{dj}, i.e. $h_{dj} = h_{d,dj}$.

The GREG estimator of $\overline{Y}_d$ is approximately unbiased. If $\pi_{dij} = \pi_{di}\pi_{dj}$ $(i \neq j)$, and applying $\pi_{djj} = \pi_{dj}$, we get

$$\mathrm{var}_\pi\left(\hat{\overline{Y}}_d^{greg}\right) \approx \frac{1}{N_d^2} \sum_{j \in U_d} \frac{1 - \pi_{dj}}{\pi_{dj}} (E_{dj} - \overline{E}_d)^2,$$

$$\widehat{\mathrm{var}}_\pi\left(\hat{\overline{Y}}_d^{greg}\right) = \frac{1}{N_d^2} \sum_{j \in s_d} w_{dj}(w_{dj} - 1) g_{dj}^2 (y_{dj} - \boldsymbol{x}_{dj}\hat{\boldsymbol{\beta}})^2.$$

As the GREG estimator of $\overline{Y}_d$ is approximately unbiased, the mean squared error can be approximated by the variance, i.e.

$$MSE\big(\widehat{\overline{Y}}_d^{greg}\big) \approx \text{var}_\pi\big(\widehat{\overline{Y}}_d^{greg}\big), \quad mse\big(\widehat{\overline{Y}}_d^{greg}\big) = \widehat{\text{var}}_\pi\big(\widehat{\overline{Y}}_d^{greg}\big).$$

Let us consider two GREG estimators of Y_d and Z_d that use the auxiliary variables X^y and X^z, respectively, i.e.

$$\hat{Y}_d^{greg} = \sum_{\ell=1}^{D} \sum_{j \in s_\ell} g_{d,\ell j}^{y} w_{\ell j} y_{\ell j}, \qquad \hat{Z}_d^{greg} = \sum_{\ell=1}^{D} \sum_{j \in s_\ell} g_{d,\ell j}^{z} w_{\ell j} z_{\ell j},$$

where

$$g_{d,\ell j}^{y} = \begin{cases} \frac{N_d}{\hat{N}_d} + N_d\big(\overline{X}_d^{y} - \widehat{\overline{X}}_d^{y,dir2}\big)\big(\sum_{d=1}^{D}\sum_{i \in s_d} w_{di} \boldsymbol{x}_{di}^{y\prime}\boldsymbol{x}_{di}^{y}\big)^{-1} \boldsymbol{x}_{\ell j}^{y\prime} & \text{if } \ell = d,\ j \in s_\ell, \\ N_d\big(\overline{X}_d^{y} - \widehat{\overline{X}}_d^{y,dir2}\big)\big(\sum_{d=1}^{D}\sum_{i \in s_d} w_{di} \boldsymbol{x}_{di}^{y\prime}\boldsymbol{x}_{di}^{y}\big)^{-1} \boldsymbol{x}_{\ell j}^{y\prime} & \text{if } \ell \neq d,\ j \in s_\ell, \end{cases} \tag{3.7}$$

and $g_{d,\ell j}^{z}$ is defined analogously. If $\pi_{dij} = \pi_{di}\pi_{dj}$ $(i \neq j)$ and $\pi_{djj} = \pi_{dj}$, we get

$$\text{cov}_\pi(\hat{Y}_d^{greg}, \hat{Z}_d^{greg}) \approx \sum_{j \in U_d} \frac{1-\pi_{dj}}{\pi_{dj}} (E_{dj}^{y} - \overline{E}_d^{y})(E_{dj}^{z} - \overline{E}_d^{z}),$$

$$\widehat{\text{cov}}_\pi(\hat{Y}_d^{greg}, \hat{Z}_d^{greg}) = \sum_{j \in s_d} w_{dj}(w_{dj}-1) g_{dj}^{y} g_{dj}^{z} (y_{dj} - \boldsymbol{x}_{dj}^{y}\hat{\boldsymbol{\beta}}^{y})(z_{dj} - \boldsymbol{x}_{dj}^{z}\hat{\boldsymbol{\beta}}^{z}),$$

where $E_{dj}^{y} = y_{dj} - \boldsymbol{x}_{dj}\boldsymbol{\beta}^{y}$, $E_{dj}^{z} = z_{dj} - \boldsymbol{x}_{dj}\boldsymbol{\beta}^{z}$, and

$$\overline{E}_d^{y} = \frac{1}{N_d}\sum_{j \in U_d} E_{dj}^{y}, \quad \boldsymbol{\beta}^{y} = \Big(\sum_{d=1}^{D}\sum_{j \in U_d} \boldsymbol{x}_{dj}^{y\prime}\boldsymbol{x}_{dj}^{y}\Big)^{-1} \sum_{d=1}^{D}\sum_{j \in U_d} \boldsymbol{x}_{dj}^{y\prime} y_{dj},$$

$$\overline{E}_d^{z} = \frac{1}{N_d}\sum_{j \in U_d} E_{dj}^{z}, \quad \boldsymbol{\beta}^{z} = \Big(\sum_{d=1}^{D}\sum_{j \in U_d} \boldsymbol{x}_{dj}^{z\prime}\boldsymbol{x}_{dj}^{z}\Big)^{-1} \sum_{d=1}^{D}\sum_{j \in U_d} \boldsymbol{x}_{dj}^{z\prime} z_{dj}.$$

For calculating $\widehat{\overline{Y}}_d^{greg}$ and estimating its variance, we need the auxiliary information:

1. *Non-aggregated:* sampling weights w_{dj} and explanatory variables $\boldsymbol{x}_{dj}$, $j \in s_d$, $d = 1, \ldots, D$, and domain indicator $I_{U_d}(j)$, $j \in s$.
2. *Aggregated:* population sizes N_d and population means $\overline{\boldsymbol{X}}_d$.

The GREG estimator is basically unbiased. Further, if the considered assisting linear model (3.3) has a good fit to data, then the GREG estimator has typically a small variance.

Example 3.1 This example shows that the post-stratified estimator of the domain total is a GREG estimator assisted by the population ANOVA model

$$y_{dgj} = a_{dg} + e_{dgj}, \quad d = 1, \ldots, D,\; g = 1, \ldots, G,\; j = 1, \ldots, N_{dg}, \tag{3.8}$$

where the e_{dgj}'s are independent errors with $E[e_{dgj}] = 0$ and $\text{var}(e_{dgj}) = \sigma^2$. In what follows, we derive the GREG estimator of the total Y_d assisted by this model. The population-based least squared approximation of a_{dg} is $\alpha_{dg} = \overline{Y}_{dg}$, and the least squared estimator of α_{dg} is $\hat{\alpha}_{dg} = \hat{\overline{Y}}_{dg}^{dir2}$. Model (3.8) can be written in the regression form

$$y_j = \boldsymbol{x}_j \boldsymbol{\beta} + e_j, \quad j = 1, \ldots, N,$$

where $\boldsymbol{\beta} = (a_{11}, \ldots, a_{1G}, \ldots, a_{D1}, \ldots, a_{DG})'$, and the vector of auxiliary variables is $\boldsymbol{x}_j = (x_{11,j}, \ldots, x_{1G,j}, \ldots, x_{D1,j}, \ldots, x_{DG,j})$, $x_{dg,j} = 1$ if $j \in U_{dg}$, $x_{dg,j} = 0$ otherwise. The row vector $\overline{\boldsymbol{X}}_d = \frac{1}{N_d}\sum_{j=1}^{N_d} \boldsymbol{x}_j$ can be written in the form $\overline{\boldsymbol{X}}_d = (\overline{x}_{11}, \ldots, \overline{x}_{1G}, \ldots, \overline{x}_{D1}, \ldots, \overline{x}_{DG})$, where the component is $\overline{x}_{\ell g} = 0$ if $\ell \neq d$ and $\overline{x}_{\ell g} = N_{dg}/N_d$ if $\ell = d,\; g = 1, \ldots, G$. The row vector $\hat{\overline{\boldsymbol{X}}}_d^{dir2}$ can be written in a similar form.

The GREG estimator (3.5) of Y_d is

$$\hat{Y}_d^{greg} == N_d \hat{\overline{Y}}_d^{dir2} + N_d(\overline{\boldsymbol{X}}_d - \hat{\overline{\boldsymbol{X}}}_d^{dir2})\hat{\boldsymbol{\beta}} = N_d \hat{\overline{Y}}_d^{dir2} + N_d \sum_{g=1}^{G}\Big(\frac{N_{dg}}{N_d} - \frac{\hat{N}_{dg}}{\hat{N}_d}\Big)\hat{\overline{Y}}_{dg}^{dir2}$$

$$= N_d \hat{\overline{Y}}_d^{dir2} + N_d \frac{1}{N_d}\sum_{g=1}^{G} N_{dg}\hat{\overline{Y}}_{dg}^{dir2} - N_d \frac{1}{\hat{N}_d}\sum_{g=1}^{G} \hat{Y}_{dg}^{dir1}$$

$$= N_d \hat{\overline{Y}}_d^{dir2} + \sum_{g=1}^{G} N_{dg}\hat{\overline{Y}}_{dg}^{dir2} - N_d \hat{\overline{Y}}_d^{dir2} = \hat{Y}_d^{pst}.$$

Example 3.2 This example shows that a bias-corrected basic synthetic estimator is a GREG estimator. Let us consider the population ANOVA model

$$y_{dgj} = a_g + e_{dgj}, \quad d = 1, \ldots, D,\; g = 1, \ldots, G,\; j = 1, \ldots, N_{dg}, \tag{3.9}$$

where the e_{dgj}'s are independent errors with $E[e_{dgj}] = 0$ and $\text{var}(e_{dgj}) = \sigma^2$. We derive the GREG estimator of the total Y_d assisted by this model. The population-based least squared approximation of a_g is $\alpha_g = \overline{Y}_g$, and the least squared estimator of α_g is $\hat{\alpha}_g = \hat{\overline{Y}}_g^{dir2}$. Model (3.9) can be written in the regression form:

$$y_j = \boldsymbol{x}_j\boldsymbol{\beta} + e_j, \quad j = 1, \ldots, N,$$

where $\boldsymbol{\beta} = (a_1, \ldots, a_G)'$ and $\boldsymbol{x}_j = (x_{1,j}, \ldots, x_{G,j})$, $x_{g,j} = 1$ if $j \in U_g$, $x_{g,j} = 0$ otherwise. The row vector $\overline{\boldsymbol{X}}_d = \frac{1}{N_d}\sum_{j=1}^{N_d}\boldsymbol{x}_j$ can be written in the form $\overline{\boldsymbol{X}}_d = (N_{d1}/N_d, \ldots, N_{dG}/N_d)$. The row vector $\hat{\overline{\boldsymbol{X}}}_d^{dir2}$ can be written in a similar form.

The GREG estimator (3.5) of Y_d is

$$\hat{Y}_d^{greg} = N_d\hat{\overline{Y}}_d^{dir2} + N_d(\overline{\boldsymbol{X}}_d - \hat{\overline{\boldsymbol{X}}}_d^{dir2})\hat{\boldsymbol{\beta}} = N_d\hat{\overline{Y}}_d^{dir2} + N_d\sum_{g=1}^{G}\left(\frac{N_{dg}}{N_d} - \frac{\hat{N}_{dg}}{\hat{N}_d}\right)\hat{\overline{Y}}_g^{dir2}$$

$$= N_d\hat{\overline{Y}}_d^{dir2} + \sum_{g=1}^{G} N_{dg}\hat{\overline{Y}}_g^{dir2} - N_d\sum_{g=1}^{G}\frac{\hat{N}_{dg}}{\hat{N}_d}\hat{\overline{Y}}_g^{dir2}.$$

As the second summand is the synthetic estimator of Y_d, we get

$$\hat{Y}_d^{greg} = N_d\hat{\overline{Y}}_d^{dir2} + \hat{Y}_d^{synth} - N_d\sum_{g=1}^{G}\frac{\hat{N}_{dg}}{\hat{N}_d}\hat{\overline{Y}}_g^{dir2}$$

$$= \hat{Y}_d^{synth} - N_d\Big(\frac{1}{\hat{N}_d}\sum_{g=1}^{G}\hat{N}_{dg}\hat{\overline{Y}}_g^{dir2} - \hat{\overline{Y}}_d^{dir2}\Big) = \hat{Y}_d^{synth} - \widehat{B}_\pi^{(3)}[\hat{Y}_d^{synth}].$$

The model-assisted approach can be applied with nonlinear assisting models. For example, Lehtonen and Veijanen (1998) considered the logistic regression population model

$$y_{dj} \sim \text{Bin}(1, p_{dj}), \quad \log\frac{p_{dj}}{1 - p_{dj}} = \boldsymbol{x}_{dj}\boldsymbol{b}, \quad d = 1, \ldots, D, \ j = 1, \ldots, N_d,$$

where $\boldsymbol{x}_{dj} = (x_{dj,1}, \ldots, x_{dj,p})$ and $\boldsymbol{b} = (b_1, \ldots, b_p)'$. The population log-likelihood is

$$\ell(\boldsymbol{b}) = \sum_{d=1}^{D}\sum_{j=1}^{N_d}\Big\{y_{dj}\log p_{dj} + (1 - y_{dj})\log(1 - p_{dj})\Big\}$$

$$= \sum_{d=1}^{D}\sum_{j=1}^{N_d}\Big\{y_{dj}\boldsymbol{x}_{dj}\boldsymbol{b} - \log\big(1 + \exp\{\boldsymbol{x}_{dj}\boldsymbol{b}\}\big)\Big\}.$$

Let $\boldsymbol{\beta}$ be the population maximum likelihood (ML) approximation of $\boldsymbol{b}$, i.e.

$$\boldsymbol{\beta} = \text{argmax}_{\boldsymbol{b}}\, \ell(\boldsymbol{b}).$$

Given a sample $s \subset U$ of size $n = n_1 + \ldots + n_D$, a direct estimator of $\ell(\boldsymbol{b})$, $\boldsymbol{b} \in R^p$ fixed, is

$$\hat{\ell}^{dir1}(\boldsymbol{b}) = \sum_{d=1}^{D}\sum_{j=1}^{n_d} w_{dj}\Big\{y_{dj}\boldsymbol{x}_{dj}\boldsymbol{b} - \log\big(1 + \exp\{\boldsymbol{x}_{dj}\boldsymbol{b}\}\big)\Big\}.$$

A weighted ML estimator of $\boldsymbol{\beta}$ is

$$\hat{\boldsymbol{\beta}} = \text{argmax}_{\boldsymbol{b}}\, \hat{\ell}^{dir1}(\boldsymbol{b}).$$

Under the logistic regression model, the predicted values are

$$\hat{y}_{dj} = \hat{p}_{dj} = \frac{\exp\{\boldsymbol{x}_{dj}\hat{\boldsymbol{\beta}}\}}{1 + \exp\{\boldsymbol{x}_{dj}\hat{\boldsymbol{\beta}}\}}, \quad d = 1, \ldots, D,\ j = 1, \ldots, N_d.$$

By applying the formula (3.4), a logistic regression-assisted estimator of Y_d is

$$\hat{Y}_d^{logit} = \sum_{j \in U_d} \hat{y}_{dj} + \frac{N_d}{\hat{N}_d} \sum_{j \in s_d} w_{dj}(y_{dj} - \hat{y}_{dj}).$$

For calculating $\hat{Y}_d^{logit}$, we need a census file containing the values of the auxiliary variables $\boldsymbol{x}_{dj}$ in all the units of the population.

3.6 Estimators of Unemployment Rates

Let Y_d and Z_d be the population totals of unemployed and employed people, respectively, in domain d. The unemployment rate is a ratio of totals, $R_d = Y_d/(Y_d + Z_d)$, which can be estimated by the plug-in estimator $\widehat{R}_d = \hat{Y}_d/(\hat{Y}_d + \hat{Z}_d)$. This section shows how to approximate its variance and bias by the Taylor linearization.

Let us consider the function $f(y, z) = y/(y + z)$, with partial derivatives

$$\frac{\partial f}{\partial y} = \frac{z}{(y + z)^2}, \quad \frac{\partial f}{\partial z} = \frac{-y}{(y + z)^2},$$

and first order Taylor series expansion

$$f(y,z) \approx f(y_0,z_0) + \frac{\partial f(y_0,z_0)}{\partial y}(y-y_0) + \frac{\partial f(y_0,z_0)}{\partial z}(z-z_0).$$

By doing the substitutions $y = \hat{Y}_d,\ z = \hat{Z}_d,\ y_0 = Y_d$, and $z_0 = Z_d$, we get

$$\widehat{R}_d \approx R_d + \frac{Z_d}{(Y_d+Z_d)^2}(\hat{Y}_d - Y_d) - \frac{Y_d}{(Y_d+Z_d)^2}(\hat{Z}_d - Z_d).$$

By taking expectations, we have

$$E_\pi[\widehat{R}_d] \approx R_d + \frac{Z_d}{(Y_d+Z_d)^2}B_\pi[\hat{Y}_d] - \frac{Y_d}{(Y_d+Z_d)^2}B_\pi[\hat{Z}_d].$$

For unbiased estimators $\hat{Y}_d$ and $\hat{Z}_d$, we obtain

$$E_\pi[\widehat{R}_d] \approx R_d.$$

For approximating the MSE (or the variance in the case of unbiased estimators $\hat{Y}_d$ and $\hat{Z}_d$), we square the term $\hat{R}_d - R_d$ taken from the first order Taylor expansion and take expectations. We get

$$MSE_\pi(\widehat{R}_d) \approx \frac{Z_d^2 MSE(\hat{Y}_d)}{(Y_d+Z_d)^4} + \frac{Y_d^2 MSE(\hat{Z}_d)}{(Y_d+Z_d)^4} - \frac{2Y_d Z_d cov(\hat{Y}_d,\hat{Z}_d)}{(Y_d+Z_d)^4},$$

which can be estimated by

$$mse(\widehat{R}_d) = \frac{\hat{Z}_d^2}{(\hat{Y}_d+\hat{Z}_d)^4} mse(\hat{Y}_d) + \frac{\hat{Y}_d^2}{(\hat{Y}_d+\hat{Z}_d)^4} mse(\hat{Z}_d) - \frac{2\hat{Y}_d\hat{Z}_d}{(\hat{Y}_d+\hat{Z}_d)^4}\widehat{cov}(\hat{Y}_d,\hat{Z}_d). \tag{3.10}$$

3.7 A Labor Force Survey

Let us consider a country with a nested geographical structure. The country is hierarchically partitioned into strata, domains, and clusters. Let us consider a labor force survey (LFS) with a two-stage sampling design and stratification in the first stage. The first stage units are the clusters (for example, census districts). Within each stratum, the clusters are selected without replacement and probabilities proportional to the number of households.

The second stage units are the dwellings. Inside each cluster, selected in the first stage, a fixed number of dwellings (for example, 20) are selected by SRSWOR. All the people living in the selected dwellings are interviewed. We use the following notation:

1. *Subindexes*: h for strata, a for clusters, v for dwellings, and j for individuals.
2. *Population sizes*: V_{ha} and V_h are the numbers of dwellings in cluster a of stratum h and of stratum h, respectively.
3. *Sample sizes*: m_h is the number of sampled clusters in stratum h, $h = 1, \dots, H$.
4. *Sample and subsamples*: the sample is $s = \cup_{h=1}^{H} s_h$, where $s_1, \dots, s_H$ are the stratum subsamples, and the stratum subsamples are $s_h = \cup_{a=1}^{m_h} s_{ha}$, where $s_{h1}, \dots, s_{hm_h}$ are the cluster subsamples in the m_h sampled clusters of stratum h.

We will approximate the probability of selecting the cluster a of stratum h in the first stage by assuming that the sampling was done with replacement. Let X_{ha} be the number of times that the cluster a of stratum h is in the sample. The probability of selecting the cluster a of stratum h is

$$P(cluster_{ha}) = 1 - P(X_{ha} = 0) = 1 - \left(1 - \frac{V_{ha}}{V_h}\right)^{m_h} = 1 - \sum_{k=0}^{m_h} \binom{m_h}{k} \left(-\frac{V_{ha}}{V_h}\right)^k$$
$$= m_h \frac{V_{ha}}{V_h} + o\left(\frac{V_{ha}}{V_h}\right) \approx m_h \frac{V_{ha}}{V_h}.$$

If m_h is small with respect to the number of all clusters in stratum h, the obtained approximation may also be used under the assumption of sampling without replacement, since the probability that one cluster is selected two or more times is very small. Further, if the cluster a of stratum h is selected in the first stage, then the probability of selecting the dwelling v of cluster a in stratum h by SRSWOR is

$$P(dwell_{hav} \,/\, cluster_{ha}) = \frac{\binom{1}{1}\binom{V_{ha}-1}{19}}{\binom{V_{ha}}{20}} = \frac{20}{V_{ha}}.$$

Therefore, the probability of selecting the dwelling v of cluster a in stratum h is

$$P(dwell_{hav}) = P(cluster_{ha}) P(dwell_{hav} \,/\, cluster_{ha}) \approx m_h \frac{V_{ha}}{V_h} \frac{20}{V_{ha}} = \frac{20\, m_h}{V_h}.$$

The inclusion probabilities of individual j of dwelling v and of dwelling v are the same. The selection probability of individual j of dwelling v of cluster a in stratum h ($j \in dwell_{hav}$) is

$$\pi_j = P(j \in s_h) = P(dwell_{hav}) = \frac{20\, m_h}{V_h} \triangleq \pi_h.$$

We have self-weighted samples for each stratum. By inverting the probabilities π_h, we obtain the theoretical sample weights, i.e.

$$w_j = \frac{1}{\pi_j} = \frac{V_h}{20\, m_h} = w_h, \qquad j \in s_h.$$

In the application to real data, the weights are corrected because of non-response. In what follows, w_j denotes the non-response corrected sample weights of individual j. The non-calibrated LFS estimator of the total Y of variable y is

$$\hat{Y}^{lfs*} = \sum_{h=1}^{H} \frac{N_h}{\hat{N}_h} \sum_{v \subset s_h} \sum_{j \in v} w_j y_j,$$

where

- N_h is the number of individuals living in dwellings of stratum h,
- $v \subset s_h$ denotes a dwelling which is included in the subsample of stratum h,
- $\hat{N}_h = \sum_{v \subset s_h} \sum_{j \in v} w_j = \sum_{j \in s_h} w_j = w_h \sum_{j \in s_h} 1 = w_h n_h$, and
- $n_h = \sum_{j \in s_h} 1$ is the number of sampled individuals in stratum h.

The non-calibrated LFS estimator is a ratio estimator that can be written in the alternative form

$$\hat{Y}^{lfs*} = \sum_{h=1}^{H} N_h \left(\frac{1}{\hat{N}_h} \sum_{j \in s_h} w_j y_j \right) = \sum_{h=1}^{H} N_h \hat{\bar{Y}}_h^{dir2}.$$

It is a post-stratified estimator, with corrected sample weights, where post-stratification groups are the strata. Another way of expressing $\hat{Y}^{lfs*}$ is

$$\hat{Y}^{lfs*} = \sum_{h=1}^{H} \sum_{j \in s_h} \frac{N_h\, w_h}{\hat{N}_h} y_j = \sum_{j \in s} w_j^b y_j,$$

where

$$w_j^b = w_j^b(s) = \frac{N_h\, w_h}{\hat{N}_h} = \frac{N_h}{n_h}, \qquad \text{if} \quad j \in s_h.$$

We remark that the non-calibrated weights w_j^b depend on the samples, while the theoretical weights do not. Therefore, the formulas of expectations and variances for post-stratified estimators are not applicable if we use w_j^b instead of w_j.

3.7.1 Weight Calibration and Benchmarking

Let us consider the non-calibrated LFS estimator

$$\hat{Y}^{lfs*} = \sum_{j \in s} w_j^b y_j$$

and K target variables with known totals

$$X_k = \sum_{j \in U} x_{jk}, \quad k = 1, \ldots, K.$$

The *calibration* problem is to find a new estimator

$$\hat{Y}_c = \sum_{j \in s} w_j^c y_j,$$

where the calibrated weights w_j^c are such that

1. they are close to the initial weights w_j^b, and
2. they fulfill the calibration equation

$$\sum_{j \in s} w_j^c x_{jk} = X_k, \quad k = 1, \ldots, K.$$

The problem is thus to find new weights w_j^c minimizing

$$\sum_{j \in s} w_j^b \phi(w_j^c / w_j^b), \qquad \text{subject to} \quad \sum_{j \in s} w_j^c x_{jk} = X_k, \; k = 1, \ldots, K, \tag{3.11}$$

where ϕ is a function that is monotonously decreasing at the left of $x = 1$, monotonously increasing at the right of $x = 1$, and such that $\phi(1) = 0$. Usual functions are $\phi(x) = (x-1)^2/2, \; x \in R$, or $\phi(x) = (x-1)\log x, \; x > 0$.

Proposition 3.1 *If $\phi(x) = (x-1)^2/2$, then the solution of the optimization problem (3.11) is*

$$w_j^c = w_j^b \Big(1 - \sum_{k=1}^{K} \lambda_k x_{jk}\Big), \quad j \in s,$$

where $\boldsymbol{\lambda} = (\lambda_1, \ldots, \lambda_K)' = \boldsymbol{A}^{-1}\boldsymbol{b}$, $\boldsymbol{A} = (\hat{X}_{\ell k}^{dir2})_{\ell,k=1,\ldots,K}$ and $\boldsymbol{b} = (\hat{X}_1^{dir2} - X_1, \ldots, \hat{X}_K^{dir2} - X_K)'$, under the assumption that the inverse matrix $\boldsymbol{A}^{-1}$ exists. The notations $\hat{X}_\ell^{dir2} = \sum_{j \in s} w_j^b x_{j\ell}$ and $\hat{X}_{\ell k}^{dir2} = \sum_{j \in s} w_j^b x_{j\ell} x_{jk}$ are used.

Proof For solving the optimization problem (3.11), the Lagrange function is

$$L = L(w_1^c, \ldots, w_n^c; \lambda_1, \ldots, \lambda_K) = \sum_{j \in s} w_j^b \phi(w_j^c / w_j^b) - \sum_{k=1}^{K} \lambda_k \Big(X_k - \sum_{j \in s} w_j^c x_{jk} \Big).$$

By taking derivatives and equating to zero, we have

$$0 = \frac{\partial L}{\partial w_j^c} = \phi'(w_j^c / w_j^b) + \sum_{k=1}^{K} \lambda_k x_{jk}, \quad j \in s,$$

$$0 = \frac{\partial L}{\partial \lambda_k} = \sum_{j \in s} w_j^c x_{jk} - X_k, \quad k = 1, \ldots, K.$$

For $\phi(x) = (x-1)^2/2$, $\phi'(x) = x - 1$, we have

$$0 = w_j^c - w_j^b + \sum_{k=1}^{K} \lambda_k w_j^b x_{jk}, \quad j \in s.$$

If we multiply by $x_{j\ell}$, we sum in $j \in s$, and we apply the equality restriction, we obtain

$$0 = X_\ell - \hat{X}_\ell^{dir2} + \sum_{k=1}^{K} \lambda_k \hat{X}_{\ell k}^{dir2}, \quad \ell = 1, \ldots, K.$$

We have obtained a linear system $\boldsymbol{A}\boldsymbol{\lambda} = \boldsymbol{b}$, with K equations and K unknowns, where $\boldsymbol{b} = (\hat{X}_1^{dir2} - X_1, \ldots, \hat{X}_K^{dir2} - X_K)'$, $\boldsymbol{A} = (\hat{X}_{\ell k}^{dir2})_{\ell,k=1,\ldots,K}$, and $\boldsymbol{\lambda} = (\lambda_1, \ldots, \lambda_K)'$. Pre-multiplying by $\boldsymbol{A}^{-1}$, we get $\boldsymbol{\lambda} = \boldsymbol{A}^{-1}\boldsymbol{b}$. Finally,

$$w_j^c = w_j^b \Big(1 - \sum_{k=1}^{K} \lambda_k x_{jk} \Big), \quad j \in s.$$

□

Remark 3.1 The calibrated weights depend on the sampling design, and they are functions of the sample, i.e. $w_j^c = w_j^c(s)$. Therefore, the calibrated weight of an individual in a selected sample depends on the rest of sampled individuals. However, the theoretical sample weights depend only on the sampling design. All the individuals in the population have a theoretical sample weight that remains unchanged when the samples are extracted.

The LFS estimator of the total Y of variable y is

$$\hat{Y}^{lfs} = \sum_{j \in s} w_j^c y_j,$$

where w_j^c are the calibrated weights. In the case of domains, the LFS estimator of Y_d is

$$\hat{Y}_d^{lfs} = \sum_{j \in s_d} w_j^c y_j.$$

For the population, $U = \cup_{d=1}^D U_d$ with $U_{d_1} \cap U_{d_2} = \emptyset$ if $d_1 \neq d_2$, it holds the benchmarking property

$$\hat{Y}^{lfs} = \sum_{d \in U} \hat{Y}_d^{lfs}.$$

The variance of $\hat{Y}^{lfs}$ can be estimated by a resampling procedure.

Let $\hat{Y}^{lfs}$ be the LFS estimator of the total Y. Let $\hat{Y}_1, \ldots, \hat{Y}_D$ be some estimators of the domain totals $Y_1, \ldots, Y_D$. In general, they will not fulfill the benchmarking property

$$\hat{Y}^{lfs} = \sum_{d=1}^{D} \hat{Y}_d.$$

We can force the benchmarking property by introducing the new estimators

$$\hat{Y}_d^c = \lambda_y \hat{Y}_d, \qquad \text{where} \quad \lambda_y = \frac{\hat{Y}^{lfs}}{\sum_{d=1}^D \hat{Y}_d}.$$

The benchmarked estimators fulfill the equality $\hat{Y}^{lfs} = \sum_{d=1}^D \hat{Y}_d^c$.

For estimating the variance and covariance of the benchmarked estimators, we can use the approximations

$$\widehat{\text{var}}_\pi(\hat{Y}_d^c) \approx \lambda_y^2 \widehat{\text{var}}_\pi(\hat{Y}_d), \qquad \widehat{\text{cov}}_\pi(\hat{Y}_d^c, \hat{Z}_d^c) \approx \lambda_y \lambda_z \widehat{\text{cov}}_\pi(\hat{Y}_d, \hat{Z}_d).$$

3.7.2 *Resampling Methods for the LFS*

The presented bootstrap does not require the construction of an artificial population. Let s be a LFS sample, which contains only registers of individuals. Let $s =$

$\cup_{h=1}^{H} s_h$, where $s_1, \ldots, s_H$ are the stratum subsamples. Let $s_h = \cup_{a=1}^{m_h} s_{ha}$, where $s_{h1}, \ldots, s_{hm_h}$ are the cluster subsamples in the m_h sampled clusters in stratum h.

The naive bootstrap tries to generate samples of the same size and design as the original sample, but it draws samples with replacement. The bootstrap samples are selected by using simple random sampling with replacement (SRSWR) in each stage, i.e.

1. Select a SRSWR sample of m_h cluster from the set of m_h clusters appearing in s.
2. Inside each selected cluster, extract a SRSWR sample of 20 dwellings from the 20 dwellings appearing in the cluster sample.
3. Include in the bootstrap sample all the registers from the selected dwellings.

The naive bootstrap produces samples that are self-weighted inside each stratum. Once the bootstrap samples are generated, the naive bootstrap MSE estimator is calculated in the same way as the basic bootstrap MSE estimator (cf. Sect. 2.6).

Rao and Tausi (2004) proposed a delete-one-cluster jackknife method for complex sampling designs. For calculating the delete-one-cluster jackknife variance estimator of $\hat{\theta}$, they generate jackknife samples by deleting a cluster each time. There are as many jackknife samples as clusters are in the corresponding SLFS sample. Consider the jackknife sample, $s_{(h_* c_*)}^*$, obtained by excluding the cluster c_* of stratum h_*. The jackknife weight of individual j, cluster c, and stratum h in the sample $s_{(h_* c_*)}^*$ is

$$w_{hcj(h_* c_*)} = w_{hcj} b_{hc(h_* c_*)}, \quad \text{where} \quad b_{hc(h_* c_*)} = \begin{cases} \frac{m_h}{m_h - 1} & \text{if } h = h_*, c \neq c_*, \\ 1 & \text{if } h \neq h_*, \end{cases}$$

and m_h is the number of clusters in the stratum h. Note that the case $h = h_*$ and $c = c_*$ does not appear in the jackknife sample $s_{(h_* c_*)}^*$. If H is the number of strata in the sample, the variance estimation is done as follows:

A. By using the procedure described above, use sample s to draw jackknife samples $s_{(h_* c_*)}^*$, $h_* = 1, \ldots, H$, $c_* = 1, \ldots, m_{h_*}$. For every jackknife sample, calculate $\hat{\theta}_{(h_* c_*)}^*$ in the same way as $\hat{\theta}$ was calculated. So, in each jackknife sample, the weights $w_{hcj(h_* c_*)}$ are adjusted by a calibration procedure to obtain calibrated weights $w_{hcj(h_* c_*)}^*$ in the same way as it was done with the SLFS sample. These calibrated weights $w_{hcj(h_* c_*)}^*$ are used to calculate $\hat{\theta}_{(h_* c_*)}^*$.
B. The observed distribution of $\{\hat{\theta}_{(h_* c_*)}^* : h_* = 1, \ldots, H, c_* = 1, \ldots, m_{h_*}\}$ is expected to imitate the distribution of estimator $\hat{\theta}$ in the SLFS sampling design.
C. The variance of $\hat{\theta}$ can be approximated by

$$\text{var}_J(\hat{\theta}) = \sum_{h_*=1}^{H} \frac{m_{h_*} - 1}{m_{h_*}} \sum_{c_*=1}^{m_{h_*}} (\hat{\theta}_{(h_* c_*)}^* - \hat{\theta}_J)^2,$$

where

$$\hat{\theta}_J = \frac{1}{C}\sum_{h_*=1}^{H}\sum_{c_*=1}^{m_{h*}} \hat{\theta}^*_{(h_*c_*)} \quad \text{and} \quad C = \sum_{h_*=1}^{H} m_{h_*}.$$

D. A jackknife estimator of the sampling error (coefficient of variation) in % of $\hat{\theta}$ is

$$\text{cv}_J(\hat{\theta}) = \frac{\sqrt{\text{var}_J(\hat{\theta})}}{\hat{\theta}} 100.$$

The jackknife method does not give good approximations to the design-based variances of small area estimators in domains containing few clusters. In those cases, the jackknife estimators tend to underestimate the variances of the small area estimators. For this sake, the following recommendations are given:

1. The jackknife estimator of the variance of a small area estimator is not reliable in domains with 1–5 clusters in the original sample. The domains in this situation could be marked with ** in tables of results.
2. The jackknife estimator of the variance of a small area estimator has a low reliability in domains with 5–10 clusters in the original sample. The domains in this situation could be marked with * in tables of results.
3. The jackknife method works acceptably well if the number of clusters in the domain is greater than or equal to 11.

3.8 R Codes for Design-Based Indirect Estimators

3.8.1 Basic Synthetic Estimator of the Total

For domains defined as AREA or AREA crossed by SEX, and using sex–age groups, this section gives R codes for calculating the basic synthetic estimator of totals of unemployed and employed people and unemployment rates. The following activities are considered:

A1. Calculate the estimated and true population sizes by area and sex–age group.
A2. Calculate the direct estimates of means in sex–age groups.
A3. Estimate the variance of direct estimators of means in sex–age groups.
A4. Calculate the synthetic estimators by AREA.
A5. Calculate the synthetic estimators by AREA crossed by SEX.

The following R code reads the data files:

```
# Auxiliary data
dataux <- read.table("Ndsa20.txt", header=TRUE, sep = "\t", dec = ".")
```

```
# Sort dataux by sex, area and age
dataux <- dataux[order(dataux$area, dataux$sex, dataux$age),]
# Survey data
dat <- read.table("LFS20.txt", header=TRUE, sep = "\t", dec = ".")
```

We build the age groups (Ga=1 if AGE<25, Ga=2 if 25≤AGE<54, and Ga=3 if 54≤AGE) at the unit level.

```
Ga <- cut(dat$AGE, breaks=c(0,25,54,max(dat$AGE)), labels=FALSE, right=TRUE)
```

We build the sex–age groups at the unit level.

```
Gsa <- paste0(dat$SEX, Ga)
gsa <- as.numeric(factor(Gsa, levels=c("11","12","13","21","22","23"),
                         labels=c(1,2,3,4,5,6)))
```

We fix the number of domains and groups.

```
D <- length(unique(dat$AREA))      # Number of areas
Ns <- length(unique(dat$SEX))      # Number of sex groups
Nga <- length(unique(Ga))          # Number of age groups
Ngsa <- length(unique(gsa))        # Number of sex--age groups
```

We rename some variables.

```
y1 <- dat$UNEMPLOYED; y2 <- dat$EMPLOYED
w <- dat$WEIGHT; area <- dat$AREA; sex <- dat$SEX
```

A1. We put the estimated population sizes by area and sex–age group in a matrix of dimension D×Ngsa.

```
hatNdg <- matrix(0, D, Ngsa)
for(i in 1:D) {
 for(j in 1:Ngsa) {
   hatNdg[i,j] <- sum(w[area==i & gsa==j])
 }
}
head(hatNdg)              # Some estimated sizes
```

We do the same arrangement for the true population sizes.

```
Ndg <- matrix(0, D, Ngsa)
for(i in 1:D) {
 for(j in 1:Ns) {
  for(k in 1:Nga) {
   Ndg[i,(j-1)*Nga+k] <- dataux$N[dataux$area==i & dataux$sex==j &
                                  dataux$age==k]
  }
 }
}
head(Ndg)                 # Some true sizes
```

A2. The direct estimates $\hat{\overline{Y}}_g^{dir2}$ of group means are defined in (2.7). The following R code calculates the direct estimates of means in sex–age groups:

```
# Mean of unemployed people
Y1bar.g <- aggregate(w*y1, by=list(gsa=gsa), sum)[,2]/colSums(hatNdg)
# Mean of employed people
Y2bar.g <- aggregate(w*y2, by=list(gsa=gsa), sum)[,2]/colSums(hatNdg)
```

A3. The estimator $\widehat{\text{var}}_\pi(\hat{\bar{Y}}_g^{dir2})$ is given in Remark 2.3, i.e.

$$\widehat{\text{var}}_\pi(\hat{\bar{Y}}_g^{dir2}) = \frac{1}{\hat{N}_g^2}\sum_{j\in s_g} w_j(w_j-1)(y_j-\hat{\bar{Y}}_g^{dir2})^2.$$

The following R code estimates the variance of direct estimators of means by sex–age groups:

```
varY1bar.g <- varY2bar.g <- vector() # Define variance vectors
den <- colSums(hatNdg)^2
for(i in 1:Ngsa) {
 varY1bar.g[i] <- sum(w[gsa==i]*(w[gsa==i]-1)*(y1[gsa==i]-Y1bar.g[i])^2)/
                  den[i]
 varY2bar.g[i] <- sum(w[gsa==i]*(w[gsa==i]-1)*(y2[gsa==i]-Y2bar.g[i])^2)/
                  den[i]
}
# Variance estimates in sex--age groups
varY1bar.g; varY2bar.g
```

We build a data.frame with the obtained results for sex–age groups.

```
gresults <- data.frame(g=1:Ngsa, Y1bar.g, Y2bar.g, varY1bar.g, varY2bar.g)
gresults
```

A4. The synthetic estimators of totals by area are defined in (3.1), i.e.

$$\hat{Y}_d^{synth} = \sum_{g=1}^{G} N_{dg}\hat{\bar{Y}}_g^{dir2}.$$

They can be calculated by using the following R code:

```
Synth.Y1 <- Ndg%*%Y1bar.g
Synth.Y2 <- Ndg%*%Y2bar.g
```

For estimating the variance of synthetic estimators of totals by area, we apply the formula

$$\widehat{\text{var}}_\pi(\hat{Y}_d^{synth}) = \sum_{g=1}^{G} N_{dg}^2\,\widehat{\text{var}}_\pi(\hat{\bar{Y}}_g^{dir2}).$$

The R code is

```
V.Synth.Y1 <- Ndg^2%*%varY1bar.g
V.Synth.Y2 <- Ndg^2%*%varY2bar.g
```

The unemployment rate (in %) by areas can be calculated as follows:

```
Synth.R.g <- Synth.Y1*100/(Synth.Y1 + Synth.Y2)
```

We build a data.frame with the obtained results.

```
dresults <- data.frame(d=1:D, Synth.Y1, Synth.Y2, V.Synth.Y1, V.Synth.Y2,
                       Synth.R.g)
```

A5. We first calculate the synthetic estimators of totals and their estimated variances by area for men.

```
Synth.Y1.gM <- Ndg[,1:3]%*%Y1bar.g[1:3]
Synth.Y2.gM <- Ndg[,1:3]%*%Y2bar.g[1:3]
V.Synth.Y1.gM <- Ndg[,1:3]^2%*%varY1bar.g[1:3]
V.Synth.Y2.gM <- Ndg[,1:3]^2%*%varY2bar.g[1:3]
```

The men unemployment rate (in %) by area can be calculated as follows:

```
Synth.R.gM <- Synth.Y1.gM*100/(Synth.Y1.gM + Synth.Y2.gM)
```

We build a data.frame with the obtained results.

```
SYNTH.M <- data.frame(area=1:D, y1tot=Synth.Y1.gM, y2tot=Synth.Y2.gM,
                      y1var=V.Synth.Y1.gM, y2var=V.Synth.Y2.gM,
                      rate=Synth.R.gM)
head(SYNTH.M)
```

We calculate the synthetic estimators of totals by area and their estimated variances for women.

```
Synth.Y1.gW <- Ndg[,4:6]%*%Y1bar.g[4:6]
Synth.Y2.gW <- Ndg[,4:6]%*%Y2bar.g[4:6]
V.Synth.Y1.gW <- Ndg[,4:6]^2%*%varY1bar.g[4:6]
V.Synth.Y2.gW <- Ndg[,4:6]^2%*%varY2bar.g[4:6]
```

The women unemployment rate (in %) by area can be calculated as follows:

```
Synth.R.gW <- Synth.Y1.gW*100/(Synth.Y1.gW + Synth.Y2.gW)
```

We build a data.frame with the obtained results.

```
SYNTH.W <- data.frame(area=1:D, y1tot=Synth.Y1.gW, y2tot=Synth.Y2.gW,
                     y1var=V.Synth.Y1.gW, y2var=V.Synth.Y2.gW,
                     rate=Synth.R.gW)
head(SYNTH.W)
```

The R code to save the results is

```
output1 <- rbind(SYNTH.M, SYNTH.W)
head(output1, 10)
```

For the ten first areas, Table 3.1 presents some of the outputs of the above R codes. The columns y1tot and y2tot contain the synthetic estimates, $\hat{Y}_{1d}^{synth}$ and $\hat{Y}_{2d}^{synth}$, of totals of unemployed and employed people. The columns y1var, y2var, and rate give the variance estimates $\widehat{\text{var}}_\pi\big(\hat{Y}_{1d}^{synth}\big)$ and $\widehat{\text{var}}_\pi\big(\hat{Y}_{2d}^{synth}\big)$ and the unemployment rate estimations $\hat{R}_d^{synth} = \hat{Y}_{1d}^{synth}/(\hat{Y}_{1d}^{synth} + \hat{Y}_{2d}^{synth})$. The left (right) part of Table 3.1 contains the results for sex=1 (sex=2).

The estimates of totals by domains provided by the synthetic estimator have a high degree of smoothing between domains. That is, the synthetic estimator approximates its estimates toward a central value to reduce its variance. As expected, the estimated variances of the synthetic estimator are smaller than the corresponding ones of the direct estimators, presented in Tables 2.1 and 2.2.

Table 3.1 SYNTH estimates of labor indicators for sex=1 (left) and sex=2 (right)

area	y1tot	y2tot	y1var	y2var	rate	y1tot	y2tot	y1var	y2var	rate
1	439	5959	9041	27,896	6.86	390	3190	5343	25,831	10.90
2	114	2039	745	5994	5.29	280	1467	3719	9234	16.03
3	185	2978	1739	7361	5.86	145	1247	781	6212	10.42
4	251	3115	2861	9275	7.45	346	2246	3874	11,932	13.34
5	145	2314	1111	4073	5.90	321	1994	3445	9946	13.87
6	280	3563	3513	11,656	7.28	126	863	505	1789	12.72
7	187	2362	1584	6616	7.33	274	1913	2379	8316	12.53
8	193	2477	1655	5622	7.22	259	1840	2149	7485	12.34
9	491	7078	10,739	42,227	6.48	558	4151	9994	40,130	11.86
10	152	1847	1087	4618	7.58	242	1557	1909	5799	13.45

3.8.2 Post-stratified Estimator of the Total

For domains defined as AREA or AREA crossed by SEX, and using sex–age groups, this section gives R codes for calculating the post-stratified estimator of totals of unemployed and employed people and unemployment rates. The following activities are considered:

B1. Calculate the estimated and true population sizes by area and sex–age group.
B2. Calculate the direct estimates of domain means in sex–age groups.
B3. Estimate the variance of direct estimators of means in sex–age groups.
B4. Calculate the post-stratified estimators by AREA.
B5. Calculate the post-stratified estimators by AREA crossed by SEX.

As the steps B1 and A1 are the same, we only present the steps B2–B5.

B2. The following R code calculates the direct estimates $\hat{\overline{Y}}_{dg}^{dir2}$ of domain–group means:

```
# Mean of unemployed people
Y1bar.dg <- aggregate(w*y1, by=list(area=area, gsa=gsa), sum)[,3]/hatNdg
# Mean of employed people
Y2bar.dg <- aggregate(w*y2, by=list(area=area, gsa=gsa), sum)[,3]/hatNdg
```

B3. We estimate the variances $\widehat{\text{var}}_\pi(\hat{\overline{Y}}_{dg}^{dir2})$ of direct estimators of means in domains crossed by sex–age groups.

```
varY1bar.dg <- varY2bar.dg <- matrix(0, D, Ngsa)
den <- hatNdg^2
for(i in 1:D) {
 for(j in 1:Ngsa) {
  varY1bar.dg[i,j] <- sum(w[area==i&gsa==j]*(w[area==i&gsa==j]-1)
                     *(y1[area==i&gsa==j]-Y1bar.dg[i,j])^2)/den[i,j]
  varY2bar.dg[i,j] <- sum(w[area==i&gsa==j]*(w[area==i&gsa==j]-1)
                     *(y2[area==i&gsa==j]-Y2bar.dg[i,j])^2)/den[i,j]
 }
}
```

B4. The post-stratified estimators of totals by area are defined in (3.2), i.e.

$$\hat{Y}_d^{pst} = \sum_{g=1}^{G} N_{dg} \hat{\bar{Y}}_{dg}^{dir2}.$$

They can be calculated by using the following R code:

```
# Post-stratified estimates of totals
Post.Y1 <- rowSums(Ndg*Y1bar.dg)
Post.Y2 <- rowSums(Ndg*Y2bar.dg)
```

For estimating the variance of post-stratified estimators of totals by area, we apply the formula

$$\widehat{\text{var}}_\pi(\hat{Y}_d^{pst}) = \sum_{g=1}^{G} N_{dg}^2 \widehat{\text{var}}_\pi(\hat{\bar{Y}}_{dg}^{dir2}).$$

The R code is

```
# Variance estimates in domains crossed by groups
V.Post.Y1 <- rowSums(Ndg^2*varY1bar.dg)
V.Post.Y2 <- rowSums(Ndg^2*varY2bar.dg)
```

The unemployment rate (in %) by areas can be calculated with the following R code:

```
Post.R.g <- Post.Y1*100/(Post.Y1 + Post.Y2)
```

We build a data.frame with the obtained results.

```
dresults <- data.frame(area=1:D, Post.Y1, Post.Y2, V.Post.Y1, V.Post.Y2,
                       Post.R.g)
```

B5. We calculate the post-stratified estimators of totals by area for men.

```
# Post-stratified estimates of totals for men
Post.Y1.gM <- rowSums(Ndg[,1:3]*Y1bar.dg[,1:3])
Post.Y2.gM <- rowSums(Ndg[,1:3]*Y2bar.dg[,1:3])
```

We calculate the estimates of the variances of the post-stratified estimators of totals for men.

```
# Variance estimates
V.Post.Y1.gM <- rowSums(Ndg[,1:3]^2*varY1bar.dg[,1:3])
V.Post.Y2.gM <- rowSums(Ndg[,1:3]^2*varY2bar.dg[,1:3])
```

The men unemployment rate (in %) by area can be calculated as follows:

```
Post.R.gM <- Post.Y1.gM*100/(Post.Y1.gM + Post.Y2.gM)
```

We build a data.frame with the obtained results.

```
Post.area.M <- data.frame(area=1:D, Post.Y1.gM, Post.Y2.gM, V.Post.Y1.gM,
                          V.Post.Y2.gM, Post.R.gM)
```

We calculate the post-stratified estimators of totals by area for women.

```
# Post-stratified estimates of totals women
Post.Y1.gW <- rowSums(Ndg[,4:6]*Y1bar.dg[,4:6])
Post.Y2.gW <- rowSums(Ndg[,4:6]*Y2bar.dg[,4:6])
```

We calculate the estimates of the variances of the post-stratified estimators of totals for women.

```
# Variance estimates
V.Post.Y1.gW <- rowSums(Ndg[,4:6]^2*varY1bar.dg[,4:6])
V.Post.Y2.gW <- rowSums(Ndg[,4:6]^2*varY2bar.dg[,4:6])
```

The women unemployment rate (in %) by area can be calculated as follows:

```
Post.R.gW <- Post.Y1.gW*100/(Post.Y1.gW + Post.Y2.gW)
```

We build a data.frame with the obtained results.

```
Post.area.W <- data.frame(area=1:D, Post.Y1.gW, Post.Y2.gW, V.Post.Y1.gW,
                          V.Post.Y2.gW, Post.R.gW)
```

The R code to save the results is

```
output2 <- cbind(Post.area.M, Post.area.W)
head(output2, 10)
```

For the ten first areas, Table 3.2 presents some of the outputs of the above R codes. The columns y1tot and y2tot contain the post-stratified estimates, $\hat{Y}_{1d}^{pst}$ and $\hat{Y}_{2d}^{pst}$, of totals of unemployed and employed people. The columns y1var, y2var, and rate give the variance estimates $\widehat{\text{var}}_\pi(\hat{Y}_{1d}^{pst})$ and $\widehat{\text{var}}_\pi(\hat{Y}_{2d}^{pst})$ and the unemployment rate estimations $\hat{R}_d^{pst} = \hat{Y}_{1d}^{pst}/(\hat{Y}_{1d}^{pst} + \hat{Y}_{2d}^{pst})$. The left (right) part of Table 3.2 contains the results for sex=1 (sex=2). In domains with null sample size, the post-stratified estimator is not calculable, and we deliver the value 0.

The estimates of totals by domains provided by the post-stratified estimator have a very low degree of smoothing between domains. That is, the post-stratified estimator takes into account the variability between domains and gives estimates that are highly dependent on the observed domain data. As expected, the estimated variances of the post-stratified estimator are greater than the corresponding ones of the synthetic estimator, presented in Table 3.1.

Table 3.2 PST estimates of labor indicators for sex=1 (left) and sex=2 (right)

area	y1tot	y2tot	y1var	y2var	rate	y1tot	y2tot	y1var	y2var	rate
1	354	5461	85,401	576,320	6.09	466	3626	25,853	402,117	11.38
2	212	1790	40,552	58,044	10.57	228	1731	44,606	165,552	11.63
3	0	3482	0	52,927	0.00	162	1298	22,953	108,106	11.09
4	177	3420	29,276	148,195	4.93	207	2825	34,731	217,697	6.82
5	0	2481	0	31,900	0.00	147	2080	19,085	141,567	6.59
6	405	3719	70,843	126,073	9.81	198	718	30,110	51,627	21.66
7	138	2902	17,138	86,419	4.53	0	3069	0	135,088	0.00
8	190	2895	32,788	84,223	6.16	0	2625	0	85,875	0.00
9	591	6606	117,286	372,943	8.22	344	3121	60,417	266,234	9.92
10	159	1693	20,940	106,462	8.61	0	1278	0	123,127	0.00

3.8.3 Generalized Regression Estimator of the Mean

For domains defined as AREA crossed by SEX, and using EDUCATION categories, this section gives R codes for calculating the generalized regression (GREG) estimator of average incomes. Read the LFS20 and the Nds20 auxiliary data and do the following activities:

C1. Fit a linear model to the survey data.
C2. Calculate the direct estimates of means and sizes by area–sex.
C3. Calculate the GREG estimates of the average income by area–sex.
C4. Estimate the variance of direct mean estimators by area–sex.
C5. Calculate the g-weights.
C6. Estimate the variance of the GREG estimators of means by area–sex.

The following R code reads the data files:

```
dataux <- read.table("Nds20.txt", header=TRUE, sep = "\t", dec = ".")
# Auxiliary data: population sizes by area and sex
dat <- read.table("LFS20.txt", header=TRUE, sep = "\t", dec = ".")
n <- nrow(dat)                          # Global sample size
narea <- length(unique(dataux$area))    # Number of areas
nsex <- length(unique(dataux$sex))      # Number of sex categories
```

We rename some variables and calculate the indicators of the EDUCATION categories.

```
y <- dat$INCOME; x1 <- dat$REGISTERED; x2 <- dat$EDUCATION;
w <- dat$WEIGH; area <- dat$AREA; sex <- dat$SEX
edu2 <- as.numeric(x2==2); edu3 <- as.numeric(x2==3)
```

C.1. We calculate the estimator of the regression parameter $\boldsymbol{\beta}$, i.e.

$$\hat{\boldsymbol{\beta}} = \Big(\sum_{d=1}^{D}\sum_{j\in s_d} w_{dj}\boldsymbol{x}'_{dj}\boldsymbol{x}_{dj}\Big)^{-1}\sum_{d=1}^{D}\sum_{j\in s_d} w_{dj}\boldsymbol{x}'_{dj}y_{dj}.$$

```
X <- cbind(1, x1, edu2, edu3)              # Matrix with auxiliary variables
p <- ncol(X)
Y <- matrix(y, nrow=n)                     # Vector with target variable
W <- diag(w)                               # Matrix with sampling weights
Q <- solve( crossprod(X,W)%*%X )
beta <- tcrossprod(Q,X) %*% crossprod(W,Y) # Estimator of beta parameter
```

Alternatively, we can calculate $\boldsymbol{\beta}$ by using the R function `lm`.

```
mod <- lm(y ~ x1 + edu2 + edu3, weights=w)
coef(mod)
```

C.2. The following R code builds a data frame with the direct estimates of sizes and means by area–sex:

```
# Estimated size
hatNds <- aggregate(w, by=list(sex=sex, area=area), sum)
# x1 means
dir.reg <- aggregate(w*x1, by=list(sex=sex, area=area), sum)[,3]/hatNds[,3]
# edu2 means
dir.edu2 <- aggregate(w*edu2, by=list(sex=sex, area=area), sum)[,3]/
                      hatNds[,3]
# edu3 means
dir.edu3 <- aggregate(w*edu3, by=list(sex=sex, area=area), sum)[,3]/
                      hatNds[,3]
# y means
```

```
dir.income <- aggregate(w*y, by=list(sex=sex, area=area), sum)[,3]/hatNds[,3]
hatdir <- data.frame(area=hatNds[,2], sex=hatNds[,1], N=hatNds[,3],
                     reg=dir.reg, edu2=dir.edu2, edu3=dir.edu3,
                     income=dir.income)
```

C.3. For estimating the GREG estimator of averages by area–sex, we apply the formula

$$\hat{\overline{Y}}_d^{greg} = \hat{\overline{Y}}_d^{dir2} + (\overline{\boldsymbol{X}}_d - \hat{\overline{\boldsymbol{X}}}_d^{dir2})\hat{\boldsymbol{\beta}}.$$

The following R code calculates the GREG estimates of the average incomes by area–sex:

```
# Direct estimates of y-mean
Ymean.dir <- as.matrix(hatdir$income)
# Direct estimates of X-mean
Xmean.dir <- as.matrix(cbind(1, hatdir[,4:6]))
# True X-means
Xmean <- cbind(1, dataux$reg/dataux$N, dataux$edu2/dataux$N,
               dataux$edu3/dataux$N)
# GREG estimates of y-means
Ymean.greg <- Ymean.dir + (Xmean-Xmean.dir)%*%beta
```

C.4. For estimating the variance of direct estimators of means by area–sex (by dir2 mode), we apply the following R code:
We estimate the variance by using `dir2` function described in Sect. 2.8.4.

```
dir2.est <- dir2(y, w, domain=list(sex=sex, area=area), Nd=dataux$N)
Yvar.dir <- dir2.est$var.mean
```

C5. We only need calculating the g-weights $g_{dj} = g_{d,dj}$, $\ell = d$, defined in (3.6), i.e.

$$g_{d,dj} = \frac{N_d}{\hat{N}_d} + N_d\big(\overline{\boldsymbol{X}}_d - \hat{\overline{\boldsymbol{X}}}_d^{dir2}\big)\Big(\sum_{d=1}^{D}\sum_{i\in s_d} w_{di}\boldsymbol{x}'_{di}\boldsymbol{x}_{di}\Big)^{-1}\boldsymbol{x}'_{dj}, \quad d = 1, \ldots, D,\ j \in s_d.$$

The following R code calculates the g-weights g_{dj}:

```
g <- vector()
for (k in 1:n) {
 condition <- dataux$area==dat$AREA[k]&dataux$sex==dat$SEX[k]
 Nk <- dataux$N[condition]
 regk <- dataux$reg[condition]/Nk
 edu2k <- dataux$edu2[condition]/Nk
 edu3k <- dataux$edu3[condition]/Nk
 Xk <- c(1, regk, edu2k, edu3k)
 condition2 <- hatdir$area==dat$AREA[k]&hatdir$sex==dat$SEX[k]
 Nhatk <- hatdir$N[condition2]
 reghatk <- hatdir$reg[condition2]
 edu2hatk <- hatdir$edu2[condition2]
 edu3hatk <- hatdir$edu3[condition2]
 Xdirk <- c(1, reghatk, edu2hatk ,edu3hatk)
 g[k] <- Nk/Nhatk + Nk*(Xk-Xdirk)%*%Q%*%X[k,]
}
```

C6. For estimating the variance of $\hat{\overline{Y}}_d^{greg}$, we apply the formula

$$\widehat{\text{var}}_{\pi}(\hat{\bar{Y}}_d^{greg}) = \frac{1}{N_d^2} \sum_{j \in s_d} w_{dj}(w_{dj} - 1) g_{dj}^2 (y_{dj} - \boldsymbol{x}_{dj} \hat{\boldsymbol{\beta}})^2.$$

The following R code calculates the estimates of the variance of the GREG estimators of means by area–sex:

```
yyds <- (y-fitted(mod))^2
# (alternatively fitted(mod) can be computed as X%*%beta)
vargreg.ds <- aggregate(w*(w-1)*g^2*yyds, by=list(sex, area), sum)
Yvar.greg <- vargreg.ds[,3]/dataux$N^2
```

We also calculate the estimated coefficients of variation.

```
cvdir <- round(100*sqrt(Yvar.dir)/as.vector(Ymean.dir),2)
cvgreg <- round(100*sqrt(Yvar.greg)/as.vector(Ymean.greg),2)
output3 <- data.frame(area=dataux$area, sex=dataux$sex, dir=Ymean.dir,
                      GREG=Ymean.greg, dirvar=Yvar.dir, GREGvar=Yvar.greg)
head(round(output3), 10)
```

For the ten first areas, Table 3.3 presents some of the outputs of the above R codes. The columns dir and GREG contain the direct (dir2) and GREG estimates of average annual incomes. The columns dirvar and GREGvar give the variance estimates. The left (right) part of Table 3.3 contains the results for sex=1 (sex=2). We recall that the target variables are continuous (income) and not dichotomic (employed or unemployed).

The estimates of average annual incomes by domains provided by the GREG estimator have a larger degree of smoothing between domains than the dir2 estimates. That is, the GREG estimator takes into account the variability between domains and the auxiliary information (at the unit level and at the area level) provided by the explanatory variables. As expected, the estimated variances of the GREG estimator are, in general, lower than the corresponding ones of the dir2 estimator.

Table 3.3 Estimates of average annual incomes for sex=1 (left) and sex=2 (right)

area	dir	GREG	dirvar	GREGvar	dir	GREG	dirvar	GREGvar
1	50,005	50,024	8,515,612	6,568,116	43,115	42,990	2,103,148	2,439,991
2	45,019	45,244	8,122,471	7,609,990	45,558	45,813	7,239,181	7,576,308
3	52,482	52,500	8,524,778	3,983,387	49,696	49,360	4,404,693	4,633,037
4	45,889	45,935	5,523,367	4,767,199	47,881	47,810	5,324,270	2,313,191
5	47,250	46,864	4,668,525	4,058,933	48,301	48,551	4,527,608	2,805,802
6	48,193	48,385	3,667,549	3,381,425	46,565	46,095	6,580,674	5,308,414
7	49,103	49,080	6,253,364	2,812,227	45,901	45,415	4,771,070	3,434,476
8	41,310	41,275	5,437,639	2,895,276	45,108	45,134	4,971,783	4,377,243
9	48,746	48,528	2,075,930	1,260,667	48,374	48,318	1,904,103	1,116,054
10	43,068	42,992	6,015,606	5,046,727	46,243	45,643	7,494,708	5,344,470

References

Drew, D., Singh, M.P., Choudhry, G.H.: Evaluation of small area estimation techniques for the Canadian labour force survey. Surv. Methodol. **8**, 17–47 (1982)

Estevao, V.M., Hidiroglou, M.A., Särndal, C.E.: Methodological principles for a generalized estimation system at Statistics Canada. J. Off. Stat. **11**, 181–204 (1995)

Hidiroglou, M.A., Särndal, C.E.: An empirical study of some regression estimators for small domains. Surv. Methodol. **11**, 65–77 (1985)

Lehtonen, R., Veijanen, A.: Logistic generalized regression estimators. Surv. Methodol. **24**(1), 51–55 (1998)

Lehtonen, R., Veijanen, A.: Design-based methods of estimation for domains and small areas. In: Pfeffermann, D., Rao, C.R. (eds.) Sample Surveys - Inference and Analysis, vol. 29B, pp. 219–249. Elsevier, Amsterdam (2009)

Lehtonen, R., Veijanen, A.: Model-assisted method for small area estimation of poverty indicators. In: M. Pratesi (ed.) Analysis of Poverty Data by Small Area Estimation, pp. 109–127. Wiley, Hoboken (2016)

Rao, J.N.K., Tausi, M.: Estimating function jackknife variance estimators under stratified multistage sampling. Commun. Stat. Theory Methods **33**, 2087–2095 (2004)

Särndal, C.E.: Frameworks for inference in survey sampling with applications to small area estimation and adjustment for nonresponse. Bull. Int. Stat. Inst. **49**, 494–513 (1981)

Särndal, C.E.: Design-consistent versus model-dependent estimation for small domains. J. Am. Stat. Assoc. **79**, 624–631 (1984)

Särndal, C.E., Hidiroglou, M.A.: Small domain estimation: a conditional analysis. J. Am. Stat. Assoc. **84**, 266–275 (1989)

Särndal, C.E., Swensson, B., Wretman, J.: Model Assisted Survey Sampling. Springer, Berlin (1992)

Chapter 4
Prediction Theory

4.1 Introduction

In the classical design-based inference theory, the population is fixed and the probability distribution of interest is determined by the random mechanism employed to extract the sample. As there are many ways of extracting a sample, the sampling design plays a relevant role and the finding of good estimators depends on it. On the other hand, the prediction theory treats the values of the target variable in all the units of the population as the realization of a random vector. The probability distribution of the population target vector is introduced by a statistical model and the inference procedures are optimized with respect to that distribution.

The prediction theory for finite populations relies on the so-called superpopulation models in which values of the target variables on population elements are considered as realizations of random variables having joint distributions. The model selection, fit, and diagnostic is the first and important step when applying a statistical methodology based on the prediction theory. Contrary to what happens under the sampling design theory, under the prediction theory there is no true model in applications to real data. There will only be useful models that adequately describe the behavior and relationships between the target variables and the auxiliary variables. The emphasis of this approach is thus on the analysis of data rather than on the design of samples. For more information about the prediction theory for finite populations, see e.g. the books of Cassel et al. (1977) Bolfarine and Zacks (1992) or Valliant et al. (2000).

This chapter gives a description of the prediction theory for finite populations. Section 4.1 introduces the basic notation and gives an illustrative example. Section 4.2 deals with the problem of predicting linear population parameters under a general linear model. Section 4.3 proves the general prediction theorem under a superpopulation linear model. Section 4.4 derives the best linear unbiased predictors of population totals under some linear models. It also shows that some of the

D. Morales et al., *A Course on Small Area Estimation and Mixed Models*, Statistics for Social and Behavioral Sciences, https://doi.org/10.1007/978-3-030-63757-6_4

obtained predictors are widely used estimators in the statistical inference of finite populations. Finally, Sect. 4.5 gives the corresponding R codes.

4.2 The Predictive Approach

Let N be the known number of units in a finite population and let y_j be the numeric value of the target variable measured at the population unit j. Without loss of generality, we write the population in the form $U = \{1, 2, \ldots, N\}$. A sample is a subset of U, i.e. $s \subset U$.

The *general problem* is to select sample $s \subset U$ of size n and to use the numerical values $y_1, \ldots, y_n$, associated to the units of s, for estimating

$$h(y_1, \ldots, y_N),$$

where the functional form of h is known (for example a linear function).

The *predictive approach* treats the numerical values $y_1, \ldots, y_N$ (y-values) as realizations of random variables $Y_1, \ldots, Y_N$. After observing the sample, estimating $h(y_1, \ldots, y_N)$ is in fact predicting the value of a function of the non-observed variables Y_j. The predictive approach models the relationships between the variables through the joint probability distribution of $(Y_1, \ldots, Y_N)$. The predictions are done with respect to that distribution (the model distribution).

We use the term "prediction" in the sense of "guessing," with statistical techniques, the values of the non-observed random variables Y. We do not use this term in the sense of "guessing" future values that might occur (like in time series).

Let $r \subset U$ be the set of non-sampled population units, i.e. $U = s \cup r$. The y-values in s are known, but the ones in r are not. The prediction effort is addressed to the y-values in r, or to a function of them. By using together what is known, for individuals in s, and what is predicted, for individuals in r, we can get predictions for the population U. The following example follows the same steps as the one appearing in Section 1.2 of Valliant et al. (2000).

Example 4.1 For a given region and time period, Table 4.1 shows data from hospitals. We want to estimate $T = \sum_{j=1}^{N} y_j$. We have a sample s of size $n = 10$, excluding the case 11. Note that $\sum_{j \in s} y_j$ is known and that

$$T = \sum_{j \in s} y_j + y_{11}.$$

Therefore, estimating T is equivalent to predicting y_{11}.

Let us assume that a simple regression model M holds, i.e.

$$E_M[Y_j] = \beta x_j, \; j = 1, \ldots, N, \qquad \mathrm{cov}_M(Y_j, Y_k) = \begin{cases} \sigma^2 x_j & \text{if } j = k, \\ 0 & \text{otherwise,} \end{cases}$$

Table 4.1 Data from hospitals

Hospital	N. of beds (x)	N. of discharges (y)
1	56	180
2	59	269
3	75	236
4	83	222
5	114	361
6	117	400
7	119	337
8	121	394
9	151	600
10	209	506
11	251	617
Total for $N = 11$	1355	4122
Totals for $n = 10$	1104	3505

where β is an unknown parameter that we have to estimate. The best linear unbiased estimator (BLUE) can be obtained by minimizing the weighted sum of squared errors

$$SWSE = \sum_{j \in s} \frac{1}{\sigma^2 x_j}(y_j - \beta x_j)^2.$$

By taking derivatives and equating to zero, we have

$$0 = \frac{\partial SWSE}{\partial \beta} = -\sum_{j \in s} \frac{2(y_j - \beta x_j)x_j}{\sigma^2 x_j} \iff \hat{\beta} = \frac{\sum_{j \in s} y_j}{\sum_{j \in s} x_j} = \frac{3505}{1104} = 3.175.$$

In a conventional regression analysis we should calculate a confidence interval for β or we should test the hypothesis $H_0 : \beta = 0$. In this case $\hat{\beta}$ is only an intermediate step for arriving to the final target: the estimation of T.

In hospital 11, we have $x_{11} = 251$, the corresponding prediction is $\hat{y}_{11} = \hat{\beta} x_{11} = 3.175 \cdot 251 = 796.88$ and

$$\hat{T} = \sum_{j \in s} y_j + \hat{y}_{11} = 3505 + 796.88 = 4301.88.$$

In this case the relative error is

$$\frac{|\hat{T} - T|}{T} 100\% = \frac{|4301.88 - 4122|}{4122} 100\% \approx 4.36\%,$$

which is rather moderate.

Example 4.1 can be extended to the case

$$T = \sum_{j \in s} y_j + \sum_{j \in r} y_j.$$

In this context, the estimator of T is

$$\hat{T} = \sum_{j \in s} y_j + \sum_{j \in r} \hat{\beta} x_j = \sum_{j \in s} y_j + \frac{\sum_{j \in s} y_j}{\sum_{j \in s} x_j} \sum_{j \in r} x_j = \left(\frac{\sum_{j \in s} y_j}{\sum_{j \in s} x_j} \right) \sum_{j \in s} x_j + \left(\frac{\sum_{j \in s} y_j}{\sum_{j \in s} x_j} \right) \sum_{j \in r} x_j$$

$$= \left(\frac{\sum_{j \in s} y_j}{\sum_{j \in s} x_j} \right) \sum_{j \in U} x_j = N \overline{y}_s \frac{\overline{x}}{\overline{x}_s} \triangleq \hat{T}_R \text{ (ratio estimator)},$$

where $\overline{y}_s = \frac{1}{n} \sum_{j \in s} y_j$, $\overline{x}_s = \frac{1}{n} \sum_{j \in s} x_j$ and $\overline{x} = \frac{1}{N} \sum_{j=1}^{N} x_j$. Section 4.3 extends these ideas to the prediction of linear parameters.

4.3 Prediction Theory Under the Linear Model

Let us consider a finite population $U = \{1, \ldots, N\}$. Let $\boldsymbol{y} = (y_1, \ldots, y_N)'$ be the vector containing the values of a variable Y in all the population units. The *target* is to estimate a linear combination of $y_1, \ldots, y_N$, $\boldsymbol{\gamma}'\boldsymbol{y}$, where $\boldsymbol{\gamma} = (\gamma_1, \ldots, \gamma_N)'$ is a vector containing N known constants. For example,

- If $\gamma_j = 1$, $j = 1 \ldots, N$, then $\boldsymbol{\gamma}'\boldsymbol{y} = \sum_{j=1}^{N} y_j$ is the population total,
- If $\gamma_j = \frac{1}{N}$, $j = 1 \ldots, N$, then $\boldsymbol{\gamma}'\boldsymbol{y} = \frac{1}{N} \sum_{j=1}^{N} y_j$ is the population mean.

Let us consider a sample $s \subset U$ of $n \leq N$ units. Let $r = U - s$ be the set of non-sampled units. Without loss of generality, we renumber the population units and we write $\boldsymbol{y} = (\boldsymbol{y}_s', \boldsymbol{y}_r')'$, where

- $\boldsymbol{y}_s$ is the vector of size n containing the values of Y in the observed units,
- $\boldsymbol{y}_r$ is the vector of size $N - n$ containing the values of Y in the non-observed units.

Similarly, we write $\boldsymbol{\gamma} = (\boldsymbol{\gamma}_s', \boldsymbol{\gamma}_r')'$ and $\boldsymbol{\gamma}'\boldsymbol{y} = \boldsymbol{\gamma}_s'\boldsymbol{y}_s + \boldsymbol{\gamma}_r'\boldsymbol{y}_r$. Note that the problem of estimating $\boldsymbol{\gamma}'\boldsymbol{y}$ is equivalent to the problem of predicting the value of the non-observed random variable $\boldsymbol{\gamma}_r'\boldsymbol{y}_r$.

Definition 4.1 A *linear estimator* of $\theta = \boldsymbol{\gamma}'\boldsymbol{y}$ is $\hat{\theta} = \boldsymbol{g}_s'\boldsymbol{y}_s$, where $\boldsymbol{g}_s = (g_1, \ldots, g_n)'$ is a vector of n coefficients.

Definition 4.2 The *estimation error* of the estimator $\hat{\theta} = \boldsymbol{g}_s'\boldsymbol{y}_s$ is $\hat{\theta} - \theta = \boldsymbol{g}_s'\boldsymbol{y}_s - \boldsymbol{\gamma}'\boldsymbol{y}$.

We can write the estimation error as a function of the observed and non-observed measurements, i.e.

$$\hat{\theta} - \theta = \boldsymbol{g}_s'\boldsymbol{y}_s - \boldsymbol{\gamma}'\boldsymbol{y} = (\boldsymbol{g}_s' - \boldsymbol{\gamma}_s')\boldsymbol{y}_s - \boldsymbol{\gamma}_r'\boldsymbol{y}_r = \boldsymbol{a}'\boldsymbol{y}_s - \boldsymbol{\gamma}_r'\boldsymbol{y}_r, \quad \text{with } \boldsymbol{a} = \boldsymbol{g}_s - \boldsymbol{\gamma}_s.$$

Note that

- The first component, $\boldsymbol{a}'\boldsymbol{y}_s$, depends only on the sampled units and its value can be calculated after observing the sample s.
- The second component depends on the non-sampled units and its value should be predicted.
- An "ideal" best estimator has the property $0 = \hat{\theta} - \theta = \boldsymbol{a}'\boldsymbol{y}_s - \boldsymbol{\gamma}_r'\boldsymbol{y}_r$. Therefore, using $\boldsymbol{g}_s'\boldsymbol{y}_s$ for estimating $\boldsymbol{\gamma}'\boldsymbol{y}$ is equivalent to using $\boldsymbol{a}'\boldsymbol{y}_s$ for predicting $\boldsymbol{\gamma}_r'\boldsymbol{y}_r$. This is to say, finding a good "$\boldsymbol{g}_s$" is equivalent to finding a good "$\boldsymbol{a}$."

In this section we study the prediction problem under the general linear model M:

$$E_M[\boldsymbol{y}] = \boldsymbol{X\beta}, \qquad \text{var}_M(\boldsymbol{y}) = \boldsymbol{V}, \tag{4.1}$$

where $\boldsymbol{X}_{N\times p}$ is the matrix of auxiliary variables, $\boldsymbol{\beta}_{p\times 1}$ is the vector of unknown regression parameters and $\boldsymbol{V}_{N\times N}$ is a known positive definite covariance matrix. We assume that the values of the auxiliary variables are known in all the population units, i.e. $\boldsymbol{X}_{N\times p}$ is known.

We sort the population units and we express the matrices X and V in the following block form:

$$\boldsymbol{X} = \begin{pmatrix} \boldsymbol{X}_s \\ \boldsymbol{X}_r \end{pmatrix}, \qquad \boldsymbol{V} = \begin{pmatrix} \boldsymbol{V}_s & \boldsymbol{V}_{sr} \\ \boldsymbol{V}_{rs} & \boldsymbol{V}_r \end{pmatrix},$$

where $\boldsymbol{X}_s$ is $n \times p$, $\boldsymbol{X}_r$ is $(N-n)\times p$, $\boldsymbol{V}_s$ is $n\times n$, $\boldsymbol{V}_r$ is $(N-n)\times(N-n)$, $\boldsymbol{V}_{sr}$ is $n \times (N-n)$, and $\boldsymbol{V}_{rs} = \boldsymbol{V}_{sr}'$. We further assume that $\boldsymbol{V}_s$ is positive definite and $\boldsymbol{X}_s$ has a full rank.

Definition 4.3 The estimator $\hat{\theta}$ is *unbiased* for θ under the model M if and only if $E_M[\hat{\theta} - \theta] = 0$. We can also say *predictively unbiased* or unbiased with respect to the model distribution.

Note that $E_M[\hat{\theta}] = \theta$ is not correct because θ is random.

Definition 4.4 The error variance (prediction variance) of $\hat{\theta}$ under M is $\text{var}_M(\hat{\theta} - \theta)$.

If $\hat{\theta}$ is predictively unbiased, then its error variance is equal to its mean squared error, i.e. $\text{var}_M(\hat{\theta} - \theta) = E_M[(\hat{\theta} - \theta)^2]$.

Example 4.2 (Ratio Estimator) We show that the ratio estimator of the total $T = \sum_{j=1}^{N} y_j$, $\hat{T}_R = N\overline{y}_s \frac{\overline{X}}{\overline{x}_s}$, is the best linear unbiased predictor (BLUP) if we work under the model

$$E_M[Y_j] = \beta x_j,\ j = 1, \ldots, N, \qquad \text{cov}_M(Y_j, Y_k) = \begin{cases} \sigma^2 x_j & \text{if } j = k, \\ 0 & \text{if } j \neq k. \end{cases}$$

Here, the best estimator means the estimator which minimizes the error variance. For a sample s, we have $T = \sum_{j\in s} y_j + \sum_{j\in r} y_j$. If we should know the value of the parameter β (which is unknown), then we could estimate T with $T^* = \sum_{j\in s} y_j + \beta \sum_{j\in r} x_j$, because $E_M[T^* - T] = 0$. On the other hand, every estimator $\hat{T}$ of T can be written as

$$\hat{T} = \sum_{j\in s} y_j + \left[\frac{\hat{T} - \sum_{j\in s} y_j}{\sum_{j\in r} x_j}\right] \sum_{j\in r} x_j,$$

so that $\hat{T}$ has the same form as T^* and $(\hat{T} - \sum_{j\in s} y_j)/\sum_{j\in r} x_j$ estimates β. We can write

$$\hat{T} = \sum_{j\in s} y_j + \hat{\beta} \sum_{j\in r} x_j \quad \text{and} \quad \hat{T} - T = \hat{\beta} \sum_{j\in r} x_j - \sum_{j\in r} y_j.$$

Therefore, $\hat{T}$ is predictively unbiased if

$$E_M\Big[\hat{\beta} \sum_{j\in r} x_j - \sum_{j\in r} y_j\Big] = (E_M[\hat{\beta}] - \beta) \sum_{j\in r} x_j = 0.$$

This is to say, $\hat{T}$ is predictively unbiased for T if and only if $\hat{\beta}$ is predictively unbiased for β.

The error variance of $\hat{T} = \sum_{j\in s} y_j + \hat{\beta} \sum_{j\in r} x_j$ is

$$\text{var}_M(\hat{T} - T) = \text{var}_M\Big(\hat{\beta} \sum_{j\in r} x_j - \sum_{j\in r} y_j\Big) = \Big(\sum_{j\in r} x_j\Big)^2 \text{var}_M(\hat{\beta}) + \text{var}_M\Big(\sum_{j\in r} y_j\Big).$$

For minimizing the error variance of $\hat{T}$ we have to minimize the variance of $\hat{\beta}$. Assume that we are restricted to linear unbiased estimators of β, i.e.

$$\hat{\beta} = \sum_{j\in s} a_j y_j, \quad E_M[\hat{\beta}] = \beta \sum_{j\in s} a_j x_j = \beta, \quad \sum_{j\in s} a_j x_j = 1.$$

The variance of $\hat{\beta}$ is

$$\mathrm{var}_M(\hat{\beta}) = \sigma^2 \sum_{j \in s} a_j^2 x_j.$$

We find the best linear unbiased estimator (BLUE) of β by applying the Lagrange multiplier method. The Lagrangian function is

$$L = \sigma^2 \sum_{j \in s} a_j^2 x_j + \lambda \left(\sum_{j \in s} a_j x_j - 1 \right).$$

By taking derivatives, we have

$$0 = \frac{\partial L}{\partial a_j} = 2\sigma^2 a_j x_j + \lambda x_j, \quad j \in s, \tag{4.2}$$

$$0 = \frac{\partial L}{\partial \lambda} = \sum_{j \in s} a_j x_j - 1, \tag{4.3}$$

and

$$0 = \sum_{j \in s} \frac{\partial L}{\partial a_j} = 2\sigma^2 \sum_{j \in s} a_j x_j + \lambda n \overline{x}_s = 2\sigma^2 + \lambda n \overline{x}_s,$$

which implies

$$\lambda = -\frac{2\sigma^2}{n\overline{x}_s}.$$

By substituting in (4.2), we get

$$a_j = -\lambda \frac{1}{2\sigma^2} = \frac{2\sigma^2}{n\overline{x}_s} \frac{1}{2\sigma^2} = \frac{1}{n\overline{x}_s}.$$

The BLUE of β is

$$\hat{\beta} = \sum_{j \in s} \frac{1}{n\overline{x}_s} y_j = \frac{\overline{y}_s}{\overline{x}_s}.$$

The BLUP of the total T, obtained from the BLUE of β, is

$$\begin{aligned}\hat{T} &= \sum_{j \in s} y_j + \hat{\beta} \sum_{j \in r} x_j = \sum_{j \in s} y_j + \frac{\overline{y}_s}{\overline{x}_s} \sum_{j \in r} x_j \\ &= \frac{n\overline{y}_s}{\overline{x}_s} \overline{x}_s + \frac{\overline{y}_s}{\overline{x}_s} \sum_{j \in r} x_j = N \frac{\overline{x}}{\overline{x}_s} \overline{y}_s = \hat{T}_R \text{ (ratio estimator).}\end{aligned}$$

The estimation error is

$$\hat{T}_R - T = \frac{N\bar{x}}{n\bar{x}_s}\sum_{j\in s} y_j - \sum_{j\in U} y_j = \left(\frac{N\bar{x}}{n\bar{x}_s} - 1\right)\sum_{j\in s} y_j - \sum_{j\in r} y_j$$
$$= \frac{(N-n)\bar{x}_r}{n\bar{x}_s}\sum_{j\in s} y_j - \sum_{j\in r} y_j,$$

where $\bar{x}_r = \frac{1}{N-n}\sum_{j\in r} x_j$. The error variance is

$$\mathrm{var}_M\left(\hat{T}_R - T\right) = \left(\frac{(N-n)\bar{x}_r}{n\bar{x}_s}\right)^2 n\bar{x}_s\sigma^2 + (N-n)\bar{x}_r\sigma^2$$
$$= \frac{(N-n)^2\bar{x}_r^2}{n\bar{x}_s}\sigma^2 + (N-n)\bar{x}_r\sigma^2 = (N-n)\bar{x}_r\sigma^2\left(\frac{(N-n)\bar{x}_r}{n\bar{x}_s} + 1\right)$$
$$= (N-n)\bar{x}_r\sigma^2\frac{(N-n)\bar{x}_r + n\bar{x}_s}{n\bar{x}_s} = \frac{(N-n)N}{n}\frac{\bar{x}_r\bar{x}}{\bar{x}_s}\sigma^2$$
$$= \frac{N^2}{n}(1-f)\frac{\bar{x}_r\bar{x}}{\bar{x}_s}\sigma^2, \quad \text{where } f = \frac{n}{N}.$$

4.4 The General Prediction Theorem

The following theorem gives the best linear unbiased predictor of a linear parameter under a superpopulation linear model (4.1). It also gives the corresponding prediction variance. For more details, see Chapter 2 of Valliant et al. (2000).

Theorem 4.1 *Among linear predictively unbiased estimators $\hat{\theta} = \boldsymbol{g}_s'\boldsymbol{y}_s$ of $\theta = \boldsymbol{\gamma}'\boldsymbol{y}$, the error variance is minimized by*

$$\hat{\theta}_{opt} = \boldsymbol{\gamma}_s'\boldsymbol{y}_s + \boldsymbol{\gamma}_r'\left[\boldsymbol{X}_r\hat{\boldsymbol{\beta}} + \boldsymbol{V}_{rs}\boldsymbol{V}_s^{-1}(\boldsymbol{y}_s - \boldsymbol{X}_s\hat{\boldsymbol{\beta}})\right], \tag{4.4}$$

where $\hat{\boldsymbol{\beta}} = (\boldsymbol{X}_s'\boldsymbol{V}_s^{-1}\boldsymbol{X}_s)^{-1}\boldsymbol{X}_s'\boldsymbol{V}_s^{-1}\boldsymbol{y}_s$. The error variance of $\hat{\theta}_{opt}$ is

$$\mathrm{var}_M(\hat{\theta}_{opt} - \theta) = \boldsymbol{\gamma}_r'(\boldsymbol{V}_r - \boldsymbol{V}_{rs}\boldsymbol{V}_s^{-1}\boldsymbol{V}_{sr})\boldsymbol{\gamma}_r$$
$$+ \boldsymbol{\gamma}_r'(\boldsymbol{X}_r - \boldsymbol{V}_{rs}\boldsymbol{V}_s^{-1}\boldsymbol{X}_s)(\boldsymbol{X}_s'\boldsymbol{V}_s^{-1}\boldsymbol{X}_s)^{-1}(\boldsymbol{X}_r - \boldsymbol{V}_{rs}\boldsymbol{V}_s^{-1}\boldsymbol{X}_s)'\boldsymbol{\gamma}_r.$$

Proof The error variance is

$$E_M\left[(g_s'y_s - \gamma'y)^2\right] = E_M\left[(a'y_s - \gamma_r'y_r)^2\right]$$
$$= \text{var}_M\left(a'y_s - \gamma_r'y_r\right) + \left(E_M\left[a'y_s - \gamma_r'y_r\right]\right)^2$$
$$= a'V_s a - 2a'V_{sr}\gamma_r + \gamma_r'V_r\gamma_r + \left[(a'X_s - \gamma_r'X_r)\beta\right]^2,$$

where $a = g_s - \gamma_s$. Since we are assuming that $\hat{\theta} = g_s'y_s$ is unbiased, then

$$E_M\left[g_s'y_s - \gamma'y\right] = E_M\left[a'y_s - \gamma_r'y_r\right] = (a'X_s - \gamma_r'X_r)\beta = 0,$$

i.e. the last term in the previous equation vanishes. The Lagrangian function for minimizing the error variance with respect to a is

$$L = L(a, \lambda) = a'V_s a - 2a'V_{sr}\gamma_r + 2(a'X_s - \gamma_r'X_r)\lambda.$$

By taking derivatives with respect to λ and equating to zero, we get

$$0 = \frac{\partial L}{\partial \lambda} = 2a'X_s - 2\gamma_r'X_r. \tag{4.5}$$

By taking derivatives with respect to a, we get

$$0 = \frac{\partial L}{\partial a} = 2V_s a - 2V_{sr}\gamma_r + 2X_s\lambda \tag{4.6}$$

and

$$a = V_s^{-1}(V_{sr}\gamma_r - X_s\lambda). \tag{4.7}$$

On the other hand, from (4.6) we have

$$X_s\lambda = V_{sr}\gamma_r - V_s a. \tag{4.8}$$

Pre-multiplying (4.8) by $X_s'V_s^{-1}$, and taking into account that $X_s'a = X_r'\gamma_r$, we get

$$X_s'V_s^{-1}X_s\lambda = X_s'V_s^{-1}V_{sr}\gamma_r - X_s'a = X_s'V_s^{-1}V_{sr}\gamma_r - X_r'\gamma_r = (X_s'V_s^{-1}V_{sr} - X_r')\gamma_r$$

and

$$\lambda = A_s^{-1}(X_s'V_s^{-1}V_{sr} - X_r')\gamma_r, \qquad \text{where} \quad A_s = X_s'V_s^{-1}X_s.$$

By substituting $\boldsymbol{\lambda}$ in (4.7) we obtain the optimal value of $\boldsymbol{a}$, i.e.

$$\boldsymbol{a}_{opt} = \boldsymbol{V}_s^{-1}\left[\boldsymbol{V}_{sr} + \boldsymbol{X}_s\boldsymbol{A}_s^{-1}(\boldsymbol{X}_r' - \boldsymbol{X}_s'\boldsymbol{V}_s^{-1}\boldsymbol{V}_{sr})\right]\boldsymbol{\gamma}_r.$$

The best linear unbiased estimator of $\boldsymbol{\gamma}_r'\boldsymbol{y}_r$ is

$$\begin{aligned}\boldsymbol{a}_{opt}'\boldsymbol{y}_s &= \boldsymbol{\gamma}_r'\left[\boldsymbol{V}_{sr}' + (\boldsymbol{X}_r - \boldsymbol{V}_{sr}'\boldsymbol{V}_s^{-1}\boldsymbol{X}_s)\boldsymbol{A}_s^{-1}\boldsymbol{X}_s'\right]\boldsymbol{V}_s^{-1}\boldsymbol{y}_s\\ &= \boldsymbol{\gamma}_r'\left[\boldsymbol{V}_{sr}'\boldsymbol{V}_s^{-1}\boldsymbol{y}_s + \boldsymbol{X}_r\boldsymbol{A}_s^{-1}\boldsymbol{X}_s'\boldsymbol{V}_s^{-1}\boldsymbol{y}_s - \boldsymbol{V}_{sr}'\boldsymbol{V}_s^{-1}\boldsymbol{X}_s\boldsymbol{A}_s^{-1}\boldsymbol{X}_s'\boldsymbol{V}_s^{-1}\boldsymbol{y}_s\right]\\ &= \boldsymbol{\gamma}_r'\left[\boldsymbol{X}_r\hat{\boldsymbol{\beta}} + \boldsymbol{V}_{rs}\boldsymbol{V}_s^{-1}(\boldsymbol{y}_s - \boldsymbol{X}_s\hat{\boldsymbol{\beta}})\right],\end{aligned}$$

where

$$\hat{\boldsymbol{\beta}} = \boldsymbol{A}_s^{-1}\boldsymbol{X}_s'\boldsymbol{V}_s^{-1}\boldsymbol{y}_s = (\boldsymbol{X}_s'\boldsymbol{V}_s^{-1}\boldsymbol{X}_s)^{-1}\boldsymbol{X}_s'\boldsymbol{V}_s^{-1}\boldsymbol{y}_s.$$

Therefore, we obtain

$$\hat{\theta}_{opt} = \boldsymbol{\gamma}_s'\boldsymbol{y}_s + \boldsymbol{a}_{opt}'\boldsymbol{y}_s = \boldsymbol{\gamma}_s'\boldsymbol{y}_s + \boldsymbol{\gamma}_r'\left[\boldsymbol{X}_r\hat{\boldsymbol{\beta}} + \boldsymbol{V}_{rs}\boldsymbol{V}_s^{-1}(\boldsymbol{y}_s - \boldsymbol{X}_s\hat{\boldsymbol{\beta}})\right].$$

Finally, the error variance is

$$\begin{aligned}V_M &= \text{var}_M(\hat{\theta}_{opt} - \theta) = \boldsymbol{a}_{opt}'\boldsymbol{V}_s\boldsymbol{a}_{opt} - 2\boldsymbol{a}_{opt}'\boldsymbol{V}_{sr}\boldsymbol{\gamma}_r + \boldsymbol{\gamma}_r'\boldsymbol{V}_r\boldsymbol{\gamma}_r\\ &= \boldsymbol{\gamma}_r'\left[\boldsymbol{V}_{rs} + (\boldsymbol{X}_r - \boldsymbol{V}_{rs}\boldsymbol{V}_s^{-1}\boldsymbol{X}_s)\boldsymbol{A}_s^{-1}\boldsymbol{X}_s'\right]\boldsymbol{V}_s^{-1}\left[\boldsymbol{V}_{sr} + \boldsymbol{X}_s\boldsymbol{A}_s^{-1}(\boldsymbol{X}_r' - \boldsymbol{X}_s'\boldsymbol{V}_s^{-1}\boldsymbol{V}_{sr})\right]\boldsymbol{\gamma}_r\\ &- 2\boldsymbol{\gamma}_r'\left[\boldsymbol{V}_{rs} + (\boldsymbol{X}_r - \boldsymbol{V}_{rs}\boldsymbol{V}_s^{-1}\boldsymbol{X}_s)\boldsymbol{A}_s^{-1}\boldsymbol{X}_s'\right]\boldsymbol{V}_s^{-1}\boldsymbol{V}_{sr}\boldsymbol{\gamma}_r + \boldsymbol{\gamma}_r'\boldsymbol{V}_r\boldsymbol{\gamma}_r\\ &= \boldsymbol{\gamma}_r'\left(\boldsymbol{V}_r - \boldsymbol{V}_{rs}\boldsymbol{V}_s^{-1}\boldsymbol{V}_{sr}\right)\boldsymbol{\gamma}_r + \boldsymbol{\gamma}_r'\left[\boldsymbol{V}_{rs}\boldsymbol{V}_s^{-1}\boldsymbol{X}_s\boldsymbol{A}_s^{-1}(\boldsymbol{X}_r' - \boldsymbol{X}_s'\boldsymbol{V}_s^{-1}\boldsymbol{V}_{sr})\right]\boldsymbol{\gamma}_r\\ &+ \boldsymbol{\gamma}_r'\left[(\boldsymbol{X}_r - \boldsymbol{V}_{rs}\boldsymbol{V}_s^{-1}\boldsymbol{X}_s)\boldsymbol{A}_s^{-1}\boldsymbol{X}_s'\boldsymbol{V}_s^{-1}\boldsymbol{X}_s\boldsymbol{A}_s^{-1}(\boldsymbol{X}_r' - \boldsymbol{X}_s'\boldsymbol{V}_s^{-1}\boldsymbol{V}_{sr})\right]\boldsymbol{\gamma}_r\\ &- \boldsymbol{\gamma}_r'\left[(\boldsymbol{X}_r - \boldsymbol{V}_{rs}\boldsymbol{V}_s^{-1}\boldsymbol{X}_s)\boldsymbol{A}_s^{-1}\boldsymbol{X}_s'\boldsymbol{V}_s^{-1}\boldsymbol{V}_{sr}\right]\boldsymbol{\gamma}_r\\ &= \boldsymbol{\gamma}_r'\left(\boldsymbol{V}_r - \boldsymbol{V}_{rs}\boldsymbol{V}_s^{-1}\boldsymbol{V}_{sr}\right)\boldsymbol{\gamma}_r + \boldsymbol{\gamma}_r'(\boldsymbol{X}_r - \boldsymbol{V}_{rs}\boldsymbol{V}_s^{-1}\boldsymbol{X}_s)\boldsymbol{A}_s^{-1}(\boldsymbol{X}_r - \boldsymbol{V}_{rs}\boldsymbol{V}_s^{-1}\boldsymbol{X}_s)'\boldsymbol{\gamma}_r.\end{aligned}$$

□

Corollary 4.1 *The estimator* $\hat{\boldsymbol{\beta}} = (\boldsymbol{X}_s'\boldsymbol{V}_s^{-1}\boldsymbol{X}_s)^{-1}\boldsymbol{X}_s'\boldsymbol{V}_s^{-1}\boldsymbol{y}_s$ *minimizes the weighted sum of squared residuals* $SSE = (\boldsymbol{y}_s - \boldsymbol{X}_s\boldsymbol{\beta})'\boldsymbol{V}_s^{-1}(\boldsymbol{y}_s - \boldsymbol{X}_s\boldsymbol{\beta})$.

Proof It holds that

$$SSE = \boldsymbol{y}_s'\boldsymbol{V}_s^{-1}\boldsymbol{y}_s + \boldsymbol{\beta}'\boldsymbol{X}_s'\boldsymbol{V}_s^{-1}\boldsymbol{X}_s\boldsymbol{\beta} - 2\boldsymbol{\beta}'\boldsymbol{X}_s'\boldsymbol{V}_s^{-1}\boldsymbol{y}_s.$$

Therefore, we have

$$0 = \frac{\partial SSE}{\partial \boldsymbol{\beta}} = 2\boldsymbol{X}_s'\boldsymbol{V}_s^{-1}\boldsymbol{X}_s\boldsymbol{\beta} - 2\boldsymbol{X}_s'\boldsymbol{V}_s^{-1}\boldsymbol{y}_s \quad \text{and} \quad \hat{\boldsymbol{\beta}} = (\boldsymbol{X}_s'\boldsymbol{V}_s^{-1}\boldsymbol{X}_s)^{-1}\boldsymbol{X}_s'\boldsymbol{V}_s^{-1}\boldsymbol{y}_s. \tag{4.9}$$

Since

$$\frac{\partial^2 SSE}{\partial \boldsymbol{\beta}^2} = 2\boldsymbol{X}_s'\boldsymbol{V}_s^{-1}\boldsymbol{X}_s$$

and we assume that $\boldsymbol{V}_s$ is symmetric and positive definite matrix, it follows that $\boldsymbol{V}_s^{-1}$ as well as $\boldsymbol{X}_s'\boldsymbol{V}_s^{-1}\boldsymbol{X}_s$ are positive definite matrices and thus $\hat{\boldsymbol{\beta}}$ is the point of minima of the function SSE. □

The equations $\boldsymbol{X}_s'\boldsymbol{V}_s^{-1}\boldsymbol{X}_s\hat{\boldsymbol{\beta}} = \boldsymbol{X}_s'\boldsymbol{V}_s^{-1}\boldsymbol{y}_s$, appearing in (4.9), are called *normal equations*. They are p equations with p unknowns $\beta_1, \ldots, \beta_p$, where $\boldsymbol{\beta} = (\beta_1, \ldots, \beta_p)'$. Further, the estimator $\hat{\boldsymbol{\beta}}$ of $\boldsymbol{\beta}$ is called *least squares estimator.*

Corollary 4.2 *If* $\boldsymbol{V}_{rs} = 0$, *then the BLUP and the error variance are*

$$\hat{\theta}_{opt} = \boldsymbol{\gamma}_s'\boldsymbol{y}_s + \boldsymbol{\gamma}_r'\boldsymbol{X}_r\hat{\boldsymbol{\beta}}, \quad \text{var}_M(\hat{\theta}_{opt} - \theta) = \boldsymbol{\gamma}_r'\left(\boldsymbol{V}_r + \boldsymbol{X}_r\boldsymbol{A}_s^{-1}\boldsymbol{X}_r'\right)\boldsymbol{\gamma}_r.$$

Therefore, it holds that $\hat{\theta}_{opt} = \boldsymbol{\gamma}_s'\boldsymbol{y}_s + \boldsymbol{\gamma}_r'\hat{\boldsymbol{y}}_r$, *where* $\hat{\boldsymbol{y}}_r = \boldsymbol{X}_r\hat{\boldsymbol{\beta}}$.

Proposition 4.1 *Among the linear predictively unbiased estimators* $\hat{\theta}$ *of* θ, *the variance is minimized by* $\hat{\theta}^* = \boldsymbol{\gamma}'\boldsymbol{X}\hat{\boldsymbol{\beta}}$, *where* $\hat{\boldsymbol{\beta}} = (\boldsymbol{X}_s'\boldsymbol{V}_s^{-1}\boldsymbol{X}_s)^{-1}\boldsymbol{X}_s'\boldsymbol{V}_s^{-1}\boldsymbol{y}_s$. *The estimator variance is*

$$\text{var}_M(\hat{\theta}^*) = \boldsymbol{\gamma}'\boldsymbol{X}\boldsymbol{A}_s^{-1}\boldsymbol{X}'\boldsymbol{\gamma}, \quad \textit{with} \quad \boldsymbol{A}_s = \boldsymbol{X}_s'\boldsymbol{V}_s^{-1}\boldsymbol{X}_s.$$

Proof Let $\hat{\theta} = \boldsymbol{g}_s'\boldsymbol{y}_s$ be a linear estimator of $\theta = \boldsymbol{\gamma}'\boldsymbol{y}$. The variance of $\hat{\theta}$ is $\text{var}_M(\hat{\theta}) = \boldsymbol{g}_s'\boldsymbol{V}_s\boldsymbol{g}_s$. As $\hat{\theta}$ is predictively unbiased, it holds that

$$0 = E_M\left[\hat{\theta} - \theta\right] = \boldsymbol{g}_s'\boldsymbol{X}_s\boldsymbol{\beta} - \boldsymbol{\gamma}'\boldsymbol{X}\boldsymbol{\beta} = (\boldsymbol{g}_s'\boldsymbol{X}_s - \boldsymbol{\gamma}'\boldsymbol{X})\boldsymbol{\beta}.$$

The Lagrangian function is

$$L = \boldsymbol{g}_s'\boldsymbol{V}_s\boldsymbol{g}_s + 2(\boldsymbol{g}_s'\boldsymbol{X}_s - \boldsymbol{\gamma}'\boldsymbol{X})\boldsymbol{\lambda}.$$

It holds that

$$0 = \frac{\partial L}{\partial \boldsymbol{g}_s} = 2\boldsymbol{V}_s\boldsymbol{g}_s + 2\boldsymbol{X}_s\boldsymbol{\lambda}. \tag{4.10}$$

From (4.10) we get $\boldsymbol{g}_s = -\boldsymbol{V}_s^{-1}\boldsymbol{X}_s\boldsymbol{\lambda}$. Pre-multiplying (4.10) by $\boldsymbol{X}_s'\boldsymbol{V}_s^{-1}$, and taking into account that $\boldsymbol{X}_s'\boldsymbol{g}_s = \boldsymbol{X}'\boldsymbol{\gamma}$, we obtain

$$0 = \boldsymbol{X}_s'\boldsymbol{V}_s^{-1}\boldsymbol{V}_s\boldsymbol{g}_s + \boldsymbol{X}_s'\boldsymbol{V}_s^{-1}\boldsymbol{X}_s\boldsymbol{\lambda} = \boldsymbol{X}'\boldsymbol{\gamma} + \boldsymbol{A}_s\boldsymbol{\lambda}.$$

Therefore, we have

$$\boldsymbol{\lambda} = -\boldsymbol{A}_s^{-1}\boldsymbol{X}'\boldsymbol{\gamma} \qquad \text{and} \quad \boldsymbol{g}_s^* = \boldsymbol{V}_s^{-1}\boldsymbol{X}_s\boldsymbol{A}_s^{-1}\boldsymbol{X}'\boldsymbol{\gamma}.$$

By substituting the expressions of $\hat{\theta}$ and $\text{var}_M(\hat{\theta})$, we get

$$\hat{\theta}^* = \boldsymbol{g}_s^{*\prime}\boldsymbol{y}_s = \boldsymbol{\gamma}'\boldsymbol{X}\boldsymbol{A}_s^{-1}\boldsymbol{X}_s'\boldsymbol{V}_s^{-1}\boldsymbol{y}_s = \boldsymbol{\gamma}'\boldsymbol{X}\hat{\boldsymbol{\beta}},$$

$$\text{var}_M(\hat{\theta}^*) = \boldsymbol{g}_s^{*\prime}\boldsymbol{V}_s\boldsymbol{g}_s^* = \boldsymbol{\gamma}'\boldsymbol{X}\boldsymbol{A}_s^{-1}\boldsymbol{X}_s'\boldsymbol{V}_s^{-1}\boldsymbol{V}_s\boldsymbol{V}_s^{-1}\boldsymbol{X}_s\boldsymbol{A}_s^{-1}\boldsymbol{X}'\boldsymbol{\gamma} = \boldsymbol{\gamma}'\boldsymbol{X}\boldsymbol{A}_s^{-1}\boldsymbol{X}'\boldsymbol{\gamma},$$

which completes the proof. □

The following remarks give some comments of interest about the best linear unbiased predictors.

Remark 4.1

1. The estimator $\hat{\theta}_{opt}$ can be expressed in the form

$$\hat{\theta}_{opt} = \boldsymbol{\gamma}_s'\boldsymbol{y}_s + \boldsymbol{\gamma}_r'\boldsymbol{X}_r\hat{\boldsymbol{\beta}} + \boldsymbol{\gamma}_r'\boldsymbol{V}_{rs}\boldsymbol{V}_s^{-1}(\boldsymbol{y}_s - \boldsymbol{X}_s\hat{\boldsymbol{\beta}}) = \boldsymbol{\gamma}_s'\boldsymbol{y}_s + \boldsymbol{\gamma}_r'\hat{\boldsymbol{y}}_r + \boldsymbol{\gamma}_r'\boldsymbol{V}_{rs}\boldsymbol{V}_s^{-1}\boldsymbol{e}_s.$$

   Therefore, $\hat{\theta}_{opt}$ uses the sampling units to reconstruct the "sample part" of the parameter $\theta = \boldsymbol{\gamma}'\boldsymbol{y}$. The term $\hat{\boldsymbol{y}}_r$ predicts the values of y in the non-sampled units and uses these predictions for reconstructing the non-sample part of the parameter. Finally, it uses the sample residuals $\boldsymbol{e}_s = \boldsymbol{y}_s - \boldsymbol{X}_s\hat{\boldsymbol{\beta}}$ for correcting the bias and obtaining the predictive unbiasedness.
2. The estimator $\hat{\theta}^* = \boldsymbol{\gamma}'\boldsymbol{X}\hat{\boldsymbol{\beta}} = \boldsymbol{\gamma}'\hat{\boldsymbol{y}}$ does not use explicitly the observed sample values and it reconstructs the parameter by using only the predictions of the y-values.

Remark 4.2 Let us assume that $\boldsymbol{V}_{rs} = \boldsymbol{0}$. If the target parameter is $y_j = \boldsymbol{\eta}'\boldsymbol{y}$, with $\boldsymbol{\eta} = (0, \ldots, 0, 1^{(j)}, 0, \ldots, 0)$, then the BLUP is

$$\hat{y}_j = \begin{cases} y_j & \text{if } j \in s, \\ \boldsymbol{x}_j\hat{\boldsymbol{\beta}} = \tilde{y}_j & \text{if } j \in r, \end{cases}$$

where $\boldsymbol{x}_j$ is the row j of matrix $\boldsymbol{X}$. For any other parameter $\theta = \boldsymbol{\gamma}'\boldsymbol{y}$, we have

$$\hat{\theta}_{opt} = \boldsymbol{\gamma}_s'\boldsymbol{y}_s + \boldsymbol{\gamma}_r'\boldsymbol{X}_r\hat{\boldsymbol{\beta}} = \sum_{j\in s}\gamma_j y_j + \sum_{j\in r}\gamma_j\boldsymbol{x}_j\hat{\boldsymbol{\beta}} = \sum_{j\in s}\gamma_j y_j + \sum_{j\in r}\gamma_j\tilde{y}_j,$$

$$\hat{\theta}^* = \boldsymbol{\gamma}'\boldsymbol{X}\hat{\boldsymbol{\beta}} = \sum_{j\in U}\gamma_j\boldsymbol{x}_j\hat{\boldsymbol{\beta}} = \sum_{j\in U}\gamma_j\tilde{y}_j.$$

The estimator $\hat{\theta}_{opt}$ is called *predictive* and the estimator $\hat{\theta}^*$ is called *projective*.

4.5 BLUPs for Some Simple Models

In some cases the BLUPs are classical estimators appearing in survey sampling methods for finite populations. In the following examples, we assume that

1. The *prediction target* is $T = \sum_{j=1}^{N} y_j$; this is to say, $T = \theta = \boldsymbol{\gamma}'\boldsymbol{y}$ with $\boldsymbol{\gamma} = (1,\ldots,1)'_{N\times 1}$.
2. The *notation* $e_j \sim (a,b)$ is used for indicating that e_j is a random error with $E[e_j] = a$ and $\text{var}(e_j) = b$. For vector and matrices, we use the notation

$$\boldsymbol{1}_n = \begin{pmatrix} 1 \\ \vdots \\ 1 \end{pmatrix}_{n\times 1}, \quad \boldsymbol{I}_n = \begin{pmatrix} 1 & \ldots & 0 \\ \vdots & \ddots & \vdots \\ 0 & \ldots & 1 \end{pmatrix}_{n\times n}, \quad \boldsymbol{1}_{n\times n} = \begin{pmatrix} 1 & \ldots & 1 \\ \vdots & \ddots & \vdots \\ \boldsymbol{1} & \ldots & 1 \end{pmatrix}_{n\times n}.$$

Example 4.3 (Expansive Estimator) Let us consider the model $y_j = \mu + e_j$, $j = 1,\ldots,N$, with uncorrelated random errors $e_j \sim (0,\sigma^2)$. In the framework of the general linear model, $E_M[\boldsymbol{y}] = \boldsymbol{X\beta}$, $\text{var}_M(\boldsymbol{y}) = \boldsymbol{V}$, we have that $\boldsymbol{\beta} = \mu$, $\boldsymbol{X} = \boldsymbol{1}_N$, $\boldsymbol{V} = \sigma^2\boldsymbol{I}_N$. Therefore, we have

$$\hat{\boldsymbol{\beta}} = (\boldsymbol{X}_s'\boldsymbol{V}_s^{-1}\boldsymbol{X}_s)^{-1}\boldsymbol{X}_s'\boldsymbol{V}_s^{-1}\boldsymbol{y}_s = \left(\sigma^{-2}\boldsymbol{1}_n'\boldsymbol{I}_n\boldsymbol{1}_n\right)^{-1}\boldsymbol{1}_n'\sigma^{-2}\boldsymbol{I}_n\boldsymbol{y}_s = \frac{1}{n}\sum_{j=1}^{n} y_j = \overline{y}_s.$$

The BLUP is

$$\hat{T} = \boldsymbol{\gamma}_s'\boldsymbol{y}_s + \boldsymbol{\gamma}_r'\boldsymbol{X}_r\hat{\boldsymbol{\beta}} = n\overline{y}_s + (N-n)\overline{y}_s = N\overline{y}_s.$$

The error variance is

$$\begin{aligned}
\text{var}_M(\hat{T} - T) &= \boldsymbol{\gamma}_r'(\boldsymbol{V}_r + \boldsymbol{X}_r(\boldsymbol{X}_s'\boldsymbol{V}_s^{-1}\boldsymbol{X}_s)^{-1}\boldsymbol{X}_r')\boldsymbol{\gamma}_r \\
&= \boldsymbol{1}_{N-n}'\left(\sigma^2\boldsymbol{I}_{N-n} + \boldsymbol{1}_{N-n}\frac{\sigma^2}{n}\boldsymbol{1}_{N-n}'\right)\boldsymbol{1}_{N-n} \\
&= \boldsymbol{1}_{N-n}'\sigma^2\left(\boldsymbol{I}_{N-n} + \frac{1}{n}\boldsymbol{1}_{(N-n)\times(N-n)}\right)\boldsymbol{1}_{N-n} = \sigma^2\left[(N-n) + \frac{1}{n}(N-n)^2\right] \\
&= \frac{\sigma^2(N-n)N}{n} = \frac{N^2(1-f)\sigma^2}{n}, \quad \text{where} \quad f = \frac{n}{N}.
\end{aligned}$$

Example 4.4 (Linear Regression Estimator) Let us consider the model $y_j = a + bx_j + e_j$, $j = 1,\ldots,N$, with uncorrelated random errors $e_j \sim (0,\sigma^2)$. Under the general linear model, $E_M[\boldsymbol{y}] = \boldsymbol{X\beta},\ \mathrm{var}_M(y) = \boldsymbol{V}$, we have

$$\boldsymbol{\beta} = \begin{pmatrix} a \\ b \end{pmatrix}, \quad \boldsymbol{X} = \begin{pmatrix} 1 & x_1 \\ \vdots & \vdots \\ 1 & x_N \end{pmatrix}, \quad \boldsymbol{V} = \sigma^2 \boldsymbol{I}_N.$$

Therefore, we have

$$\hat{\boldsymbol{\beta}} = \begin{pmatrix} \hat{a} \\ \hat{b} \end{pmatrix} = (\boldsymbol{X}_s' \boldsymbol{V}_s^{-1} \boldsymbol{X}_s)^{-1} \boldsymbol{X}_s' \boldsymbol{V}_s^{-1} \boldsymbol{y}_s = \begin{pmatrix} n & \sum_{j=1}^n x_j \\ \sum_{j=1}^n x_j & \sum_{j=1}^n x_j^2 \end{pmatrix}^{-1} \begin{pmatrix} \sum_{j=1}^n y_j \\ \sum_{j=1}^n x_j y_j \end{pmatrix}$$

$$= \frac{\begin{pmatrix} \sum_{j=1}^n x_j^2 & -\sum_{j=1}^n x_j \\ -\sum_{j=1}^n x_j & n \end{pmatrix} \begin{pmatrix} \sum_{j=1}^n y_j \\ \sum_{j=1}^n x_j y_j \end{pmatrix}}{n\left(\sum_{j=1}^n x_j^2\right) - \left(\sum_{j=1}^n x_j\right)^2} = \begin{pmatrix} \frac{\left(\sum_{j=1}^n y_j\right)\left(\sum_{j=1}^n x_j^2\right) - \left(\sum_{j=1}^n x_j\right)\left(\sum_{j=1}^n x_j y_j\right)}{n\left(\sum_{j=1}^n x_j^2\right) - \left(\sum_{j=1}^n x_j\right)^2} \\ \frac{n\left(\sum_{j=1}^n x_j y_j\right) - \left(\sum_{j=1}^n x_j\right)\left(\sum_{j=1}^n y_j\right)}{n\left(\sum_{j=1}^n x_j^2\right) - \left(\sum_{j=1}^n x_j\right)^2} \end{pmatrix}.$$

The estimators of b and a are

$$\hat{b} = \frac{\left(\frac{1}{n}\sum_{j=1}^n x_j y_j\right) - \left(\frac{1}{n}\sum_{j=1}^n x_j\right)\left(\frac{1}{n}\sum_{j=1}^n y_j\right)}{\left(\frac{1}{n}\sum_{j=1}^n x_j^2\right) - \left(\frac{1}{n}\sum_{j=1}^n x_j\right)^2} = \frac{\frac{1}{n}\sum_{j=1}^n (x_j - \overline{x}_s)(y_j - \overline{y}_s)}{\frac{1}{n}\sum_{j=1}^n (x_j - \overline{x}_s)^2},$$

$$\hat{a} = \frac{\left[\overline{y}_s\left(\frac{1}{n}\sum_{j=1}^n x_j^2\right) - \overline{y}_s \overline{x}_s^2\right] - \left[\overline{x}_s\left(\frac{1}{n}\sum_{j=1}^n x_j y_j\right) - \overline{y}_s \overline{x}_s^2\right]}{\left(\frac{1}{n}\sum_{j=1}^n x_j^2\right) - \overline{x}_s^2}$$

$$= \overline{y}_s - \overline{x}_s \frac{\left(\frac{1}{n}\sum_{j=1}^n x_j y_j\right) - \left(\frac{1}{n}\sum_{j=1}^n x_j\right)\left(\frac{1}{n}\sum_{j=1}^n y_j\right)}{\left(\frac{1}{n}\sum_{j=1}^n x_j^2\right) - \overline{x}_s^2} = \overline{y}_s - \overline{x}_s \hat{b}.$$

The BLUP is

$$\hat{T} = \boldsymbol{\gamma}_s' \boldsymbol{y}_s + \boldsymbol{\gamma}_r' \boldsymbol{X}_r \hat{\boldsymbol{\beta}} = n\overline{y}_s + \mathbf{1}_{N-n}' \begin{pmatrix} 1 & x_{n+1} \\ \vdots & \vdots \\ \mathbf{1} & x_N \end{pmatrix} \begin{pmatrix} \overline{y}_s - \overline{x}_s \hat{b} \\ \hat{b} \end{pmatrix}$$

$$= n\overline{y}_s + \left((N-n),\ \sum_{j=n+1}^N x_j\right) \begin{pmatrix} \overline{y}_s - \overline{x}_s \hat{b} \\ \hat{b} \end{pmatrix} = n\overline{y}_s + (N-n)\overline{y}_s - (N-n)\overline{x}_s \hat{b}$$

$$+ (N\overline{x} - n\overline{x}_s)\hat{b} = N\overline{y}_s + N(\overline{x} - \overline{x}_s)\hat{b} = N\left[\overline{y}_s + (\overline{x} - \overline{x}_s)\hat{b}\right].$$

The error variance is

$$V_M = \text{var}_M(\hat{T} - T) = \boldsymbol{\gamma}_r'(\boldsymbol{V}_r + \boldsymbol{X}_r(\boldsymbol{X}_s'\boldsymbol{V}_s^{-1}\boldsymbol{X}_s)^{-1}\boldsymbol{X}_r')\boldsymbol{\gamma}_r = \sigma^2\mathbf{1}_{N-n}'$$

$$\cdot \left(\boldsymbol{I}_{N-n} + \begin{pmatrix} 1 & x_{n+1} \\ \vdots & \vdots \\ \mathbf{1} & x_N \end{pmatrix} \frac{\begin{pmatrix} \sum_{j=1}^n x_j^2 & -\sum_{j=1}^n x_j \\ -\sum_{j=1}^n x_j & n \end{pmatrix}}{n\left(\sum_{j=1}^n x_j^2\right) - \left(\sum_{j=1}^n x_j\right)^2} \begin{pmatrix} 1 & \dots & 1 \\ x_{n+1} & \dots & x_N \end{pmatrix} \right) \mathbf{1}_{N-n}$$

$$= \sigma^2(N-n) + \sigma^2 \left(N-n, \sum_{j=n+1}^N x_j \right) \frac{\begin{pmatrix} \sum_{j=1}^n x_j^2 & -\sum_{j=1}^n x_j \\ -\sum_{j=1}^n x_j & n \end{pmatrix}}{n\sum_{j=1}^n (x_j - \overline{x}_s)^2} \begin{pmatrix} N-n \\ \sum_{j=n+1}^N x_j \end{pmatrix}$$

$$= \sigma^2(N-n) \left\{ 1 + \frac{(N-n)\sum_{j=1}^n x_j^2 - 2\sum_{j=1}^n x_j \sum_{j=n+1}^N x_j + \frac{n}{N-n}\left(\sum_{j=n+1}^N x_j\right)^2}{n\sum_{j=1}^n (x_j - \overline{x}_s)^2} \right\}$$

$$= \sigma^2(N-n)\left\{1 + \frac{A}{B}\right\}.$$

By taking into account that $n\sum_{j=1}^n x_j^2 - \left(\sum_{j=1}^n x_j\right)^2 = n\sum_{j=1}^n (x_j - \overline{x}_s)^2$ and that $f = n/N$, we get

$$A = (N-n)\sum_{j=1}^n (x_j - \overline{x}_s)^2 + \frac{N-n}{n}\left(\sum_{j=1}^n x_j\right)^2 - 2\sum_{j=1}^n x_j \left(\sum_{j=1}^N x_j - \sum_{j=1}^n x_j\right)$$

$$+ \frac{n}{N-n}\left(\sum_{j=1}^N x_j - \sum_{j=1}^n x_j\right)^2 = \frac{1}{n(N-n)}\left\{ n(N-n)^2\sum_{j=1}^n (x_j - \overline{x}_s)^2 + (N-n)^2\left(\sum_{j=1}^n x_j\right)^2 \right.$$

$$\left. - 2n(N-n)\left[\sum_{j=1}^N x_j \sum_{j=1}^n x_j - \left(\sum_{j=1}^n x_j\right)^2\right] + n^2\left[\left(\sum_{j=1}^N x_j\right)^2 - 2\sum_{j=1}^N x_j \sum_{j=1}^n x_j + \left(\sum_{j=1}^n x_j\right)^2\right]\right\}$$

$$= \frac{1}{n(N-n)}\left\{ n(N-n)^2\sum_{j=1}^n (x_j - \overline{x}_s)^2 + N^2\left(\sum_{j=1}^n x_j\right)^2 - 2nN\sum_{j=1}^N x_j \sum_{j=1}^n x_j + n^2\left(\sum_{j=1}^N x_j\right)^2 \right\}.$$

Therefore, we have

$$V_M = \sigma^2(N-n)\left\{ \frac{N}{n} + \frac{N^2\left(\sum_{j=1}^n x_j\right)^2 - 2nN\sum_{j=1}^N x_j \sum_{j=1}^n x_j + n^2\left(\sum_{j=1}^N x_j\right)^2}{n^2(N-n)\sum_{j=1}^n (x_j - \overline{x}_s)^2} \right\}$$

$$= \frac{\sigma^2(N-n)N}{n}\left\{ 1 + \frac{n^2N\overline{x}_s^2 - 2n^2N\overline{x}_s\overline{x} + n^2N\overline{x}^2}{n(N-n)\sum_{j=1}^n (x_j - \overline{x}_s)^2} \right\}$$

$$= \frac{N^2}{n}(1-f)\sigma^2\left[1+\frac{(\overline{x}_s-\overline{x})^2}{(1-f)\frac{1}{n}\sum_{j=1}^{n}(x_j-\overline{x}_s)^2}\right].$$

4.6 R Codes for BLUPs

This section gives R codes for calculating the expansive estimator and the linear regression estimator described in Examples 4.3 and 4.4, respectively. The target is estimating the population average of variable INCOME from the survey data file `LFS20.txt`. Let us note that there is a small difference with respect to Examples 4.3 and 4.4 since in the application we estimate population means instead of population totals. That means that the derived formulas for BLUP and error variance must be divided by N and N^2, respectively.

The following code reads the data file:

```
# Survey data
dat <- read.table("LFS20.txt", header=TRUE, sep = "\t", dec = ".")
# Sample size
n <- nrow(dat); n
# Rename variables
y <- dat$INCOME; x <- dat$REGISTERED
# Auxiliary data
dataux <- read.table("Nds20.txt", header=TRUE, sep = "\t", dec = ".")
```

Continuation of Example 4.3 (Expansive Estimator of the Average Income)

```
mod1 <- lm(y~1)                             # Assumed model
sigma12 <- as.numeric(anova(mod1)[3])       # Model error variance
beta1 <- as.numeric(mod1$coefficients)      # Regression parameter
Npop <- sum(dataux$N)                       # Population size
f <- n/Npop; f                              # Sampling fraction
Mincome1 <- beta1                           # Expansive estimator
Mincome1; mean(y)                           # Checking
varMincome1 <- (1-f)*sigma12/n              # Estimator error variance
```

Continuation of Example 4.4 (Linear Regression Estimator of the Average Income)

```
mod2 <- lm(y~x)                              # Assumed model
sigma22 <- anova(mod2)[2,3]                  # Model error variance
beta2 <- mod2$coefficients                   # Regression parameters
Npop <- sum(dataux$N)                        # Population size
f <- n/Npop; f                               # Sampling fraction
ymean <- mean(y); xmean <- mean(x)           # Sample means of y and x
Xmean <- sum(dataux$reg)/sum(dataux$N)       # Population mean of x
Mincome2 <- as.numeric(ymean+(Xmean-xmean)*  # Linear regression estimator
                    beta2[2])
xvar <- (n-1)*var(x)/n; xvar                 # Sample variance of x
varMincome2 <- ((1-f)*sigma22/n)*
            (1+((xmean-Xmean)^2/((1-f)*xvar))) # Estimator error variance
```

The R code to save the results is

```
model1 <- c(beta1, NA, sigma12, Mincome1, varMincome1)
model2 <- c(beta2, sigma22, Mincome2, varMincome2)
labels <- c("intercept", "beta1", "sigma2", "Mincome", "Mincome variance")
output <- data.frame(labels, model1, model2)
```

For model 1 (introduced in Example 4.3) and model 2 (introduced in Example 4.4), Table 4.2 presents the estimated intercept (intercept), covariate regression coeffi-

Table 4.2 Results of expansive estimator (model 1) and linear regression estimator (model 2) of the average income

	model 1	model 2
intercept	46,925.32	47,709.49
beta1		−9686.78
sigma2	157,023,019.44	150,178,380.43
Mincome	46,925.32	46,881.85
Mincome variance	148,684.02	142,241.59

cient (beta1), error variance (sigma2), average income (Mincome), and variance of average income estimator (Mincome variance).

For estimating the population average income, the linear regression estimator (derived under model 2) has lower estimated variance than the expansive estimator (derived under model 1).

References

Bolfarine, H., Zacks, S.: Prediction Theory for Finite Populations. Springer, Berlin (1992)

Cassel, C., Särndal, C.E., Wretman, J.: Foundations of Inference in Survey Sampling. Wiley, New York (1977)

Valliant, R., Dorfman, A.H., Royall, R.M.: Finite Population Sampling and Inference. A Prediction Approach. John Wiley, New York (2000)

Chapter 5
Linear Models

5.1 Introduction

Linear models are the simplest and most widely used statistical models. They are simple because they are easily applicable, they have good mathematical properties and they allow interpreting the relations between the target variable and the explanatory variables.

Fixed effect linear models assume that sample observations are independent and, therefore, they cannot take into account data correlation structures. This is why these models are not usually applied in small area estimation problems. However, there are situations where they are useful and convenient models. For example, if the number of domains D is small, then the variable indicating the domain where the observation lie is a categorical variable that can be introduced in the model as factor. The corresponding parameters $u_1, \ldots, u_{D-1}$ can be estimated precisely if the sample size n is big enough. Further, if the data are low correlated, then a fixed effect linear model including the domain variable as factor may be a competitive model for constructing predictors of small area linear parameters.

This chapter presents a short introduction to linear regression models with fixed effects. The considered models may contain continuous and categorical auxiliary variables, which are typically called covariates and factors, respectively. A proposition gives the explicit solutions of the likelihood equations and, therefore, it gives the maximum likelihood estimators of the model parameters. As a particular case, the linear model with one fixed factor is considered. This last model, with the domain indicator as factor, is sometimes used in small area estimation. It might be a good model when the number of domains is small. Based on this linear regression model, best linear unbiased predictors of domain-level linear parameters, like means and totals, are derived. Their model-based mean squared errors are also calculated.

The chapter also gives a collection of examples based on simple linear regression models. For each considered model, best linear unbiased predictors of domain-level

D. Morales et al., *A Course on Small Area Estimation and Mixed Models*, Statistics for Social and Behavioral Sciences, https://doi.org/10.1007/978-3-030-63757-6_5

means are obtained. The calculation is presented with plenty of details. The last section gives R codes for calculating the introduced small area predictors.

For a more in-depth study of linear models in statistics, the books Faraway (2005), Rencher and Schaalje (2008), Bingham et al. (2010), Montgomery et al. (2012) or Searle (2016) can be consulted.

5.2 Fixed Effects Linear Models

Let us consider a finite population U of N individuals. Let $\boldsymbol{X}$ be a $N \times p$ matrix containing the values of some quantitative auxiliary variables (covariates) and let $\boldsymbol{Z}$ be a $N \times q$ incidence matrix associated to some categorical variables (factors). For the vector of observations $\boldsymbol{y}_{N\times 1}$, we assume the model

$$\boldsymbol{y} = \boldsymbol{X}\boldsymbol{\beta} + \boldsymbol{Z}\boldsymbol{u} + \boldsymbol{e}, \tag{5.1}$$

where $\boldsymbol{\beta}_{p\times 1}$ and $\boldsymbol{u}_{q\times 1}$ are the vectors of parameters and $\boldsymbol{e}_{N\times 1}$ is the vector of random errors. We assume that the errors are independent and normally distributed with null mean vector and covariance matrix $\text{var}(\boldsymbol{e}) = E[\boldsymbol{e}\boldsymbol{e}'] = \sigma_e^2 \boldsymbol{W}^{-1}$, $\boldsymbol{W} = \text{diag}(w_1, \dots, w_N)$, $w_i > 0$, $i = 1, \dots, N$. The heteroscedasticity weights, w_i, are supposed to be known positive constants. The weights could also be functions of auxiliary variables or inverses of inclusion probabilities. Model (5.1) is also called *Analysis of Covariance* (ANCOVA) model.

Let $s \subset U$ be a sample of $n \leq N$ units and $r = U - s$ be the non-sampled part of the population. Without loss of generality, we assume that s contains the first n units of U. We further use the subindexes s and r to denote the sampled and non-sampled parts of vectors and matrices. Concerning the sample s, the submodel of (5.1) is

$$\boldsymbol{y}_s = \boldsymbol{X}_s\boldsymbol{\beta} + \boldsymbol{Z}_s\boldsymbol{u} + \boldsymbol{e}_s, \tag{5.2}$$

with $\text{var}(\boldsymbol{e}_s) = E[\boldsymbol{e}_s\boldsymbol{e}_s'] = \sigma_e^2 \boldsymbol{W}_s^{-1}$. We assume that $\boldsymbol{X}_s$ is a full rank matrix and that the columns of $\boldsymbol{X}_s$ are linearly independent from the columns of $\boldsymbol{Z}_s$. This is to say, the columns of $\boldsymbol{X}_s$ cannot be expressed as a linear combination of the columns of $\boldsymbol{Z}_s$. As we do not assume that the matrix of categorical variables for the sample units, $\boldsymbol{Z}_s$, is of full rank, then $\boldsymbol{Z}_s'\boldsymbol{W}_s\boldsymbol{Z}_s$ might not have inverse. However $\boldsymbol{X}_s'\boldsymbol{W}_s\boldsymbol{X}_s$ is invertible. The likelihood function of $\boldsymbol{y}_s$ is

$$f_{\boldsymbol{\beta},\boldsymbol{u},\sigma_e^2}(\boldsymbol{y}_s) = \left(2\pi\sigma_e^2\right)^{-n/2} |\boldsymbol{W}_s|^{1/2} \cdot \exp\left\{-\frac{1}{2\sigma_e^2}(\boldsymbol{y}_s - \boldsymbol{X}_s\boldsymbol{\beta} - \boldsymbol{Z}_s\boldsymbol{u})'\boldsymbol{W}_s(\boldsymbol{y}_s - \boldsymbol{X}_s\boldsymbol{\beta} - \boldsymbol{Z}_s\boldsymbol{u})\right\}.$$

The log-likelihood function, $\ell = \ell(\boldsymbol{\beta}, \boldsymbol{u}, \sigma_e^2) = \log f_{\boldsymbol{\beta},\boldsymbol{u},\sigma_e^2}(\boldsymbol{y}_s)$, is

$$\ell = \frac{1}{2}\log|\boldsymbol{W}_s| - \frac{n}{2}\log 2\pi - \frac{n}{2}\log\sigma_e^2 - \frac{1}{2\sigma_e^2}(\boldsymbol{y}_s - \boldsymbol{X}_s\boldsymbol{\beta} - \boldsymbol{Z}_s\boldsymbol{u})'\boldsymbol{W}_s(\boldsymbol{y}_s - \boldsymbol{X}_s\boldsymbol{\beta} - \boldsymbol{Z}_s\boldsymbol{u}).$$

By taking partial derivatives with respect to $\boldsymbol{\beta}$, $\boldsymbol{u}$ and σ_e^2 and equating to zero, we get

$$\begin{aligned}
\mathbf{0} &= \frac{\partial \ell}{\partial \boldsymbol{\beta}} = \frac{1}{\sigma_e^2}\boldsymbol{X}_s'\boldsymbol{W}_s(\boldsymbol{y}_s - \boldsymbol{X}_s\boldsymbol{\beta} - \boldsymbol{Z}_s\boldsymbol{u}),\\
\mathbf{0} &= \frac{\partial \ell}{\partial \boldsymbol{u}} = \frac{1}{\sigma_e^2}\boldsymbol{Z}_s'\boldsymbol{W}_s(\boldsymbol{y}_s - \boldsymbol{X}_s\boldsymbol{\beta} - \boldsymbol{Z}_s\boldsymbol{u}),\\
0 &= \frac{\partial \ell}{\partial \sigma_e^2} = -\frac{n}{2\sigma_e^2} + \frac{1}{2\sigma_e^4}(\boldsymbol{y}_s - \boldsymbol{X}_s\boldsymbol{\beta} - \boldsymbol{Z}_s\boldsymbol{u})'\boldsymbol{W}_s(\boldsymbol{y}_s - \boldsymbol{X}_s\boldsymbol{\beta} - \boldsymbol{Z}_s\boldsymbol{u}).
\end{aligned}$$

From the third equation, we obtain the maximum likelihood estimator of σ_e^2, i.e.

$$\hat{\sigma}_e^2 = \frac{1}{n}(\boldsymbol{y}_s - \boldsymbol{X}_s\hat{\boldsymbol{\beta}} - \boldsymbol{Z}_s\hat{\boldsymbol{u}})'\boldsymbol{W}_s(\boldsymbol{y}_s - \boldsymbol{X}_s\hat{\boldsymbol{\beta}} - \boldsymbol{Z}_s\hat{\boldsymbol{u}}).$$

The first two equations can be written in the matrix form

$$\begin{pmatrix} \boldsymbol{X}_s'\boldsymbol{W}_s\boldsymbol{X}_s & \boldsymbol{X}_s'\boldsymbol{W}_s\boldsymbol{Z}_s \\ \boldsymbol{Z}_s'\boldsymbol{W}_s\boldsymbol{X}_s & \boldsymbol{Z}_s'\boldsymbol{W}_s\boldsymbol{Z}_s \end{pmatrix}\begin{pmatrix} \boldsymbol{\beta} \\ \boldsymbol{u} \end{pmatrix} = \begin{pmatrix} \boldsymbol{X}_s'\boldsymbol{W}_s\boldsymbol{y}_s \\ \boldsymbol{Z}_s'\boldsymbol{W}_s\boldsymbol{y}_s \end{pmatrix}. \tag{5.3}$$

Let us define

$$\boldsymbol{G}_s = (\boldsymbol{Z}_s'\boldsymbol{W}_s\boldsymbol{Z}_s)^- \quad \text{and} \quad \boldsymbol{P}_s = \boldsymbol{W}_s - \boldsymbol{W}_s\boldsymbol{Z}_s\boldsymbol{G}_s\boldsymbol{Z}_s'\boldsymbol{W}_s,$$

where $\boldsymbol{A}^-$ denotes the generalized inverse of $\boldsymbol{A}$; this is to say, $\boldsymbol{A}^-$ is a matrix (not necessarily unique) such that $\boldsymbol{A}\boldsymbol{A}^-\boldsymbol{A} = \boldsymbol{A}$.

Proposition 5.1 *A solution of the likelihood equations (5.3) is*

$$\hat{\boldsymbol{\beta}} = (\boldsymbol{X}_s'\boldsymbol{P}_s\boldsymbol{X}_s)^{-1}\boldsymbol{X}_s'\boldsymbol{P}_s\boldsymbol{y}_s, \tag{5.4}$$

$$\hat{\boldsymbol{u}} = \boldsymbol{G}_s\boldsymbol{Z}_s'\boldsymbol{W}_s\boldsymbol{y}_s - \boldsymbol{G}_s\boldsymbol{Z}_s'\boldsymbol{W}_s\boldsymbol{X}_s\hat{\boldsymbol{\beta}}, \tag{5.5}$$

where $\hat{\boldsymbol{\beta}}$ is unique but $\hat{\boldsymbol{u}}$ may not be.

Proof We recall that $\mathrm{r}(\boldsymbol{Z}_s) \le q$. In the full rank case, $\mathrm{r}(\boldsymbol{Z}_s) = q$, the system of equations (5.3) has the unique solution

$$\begin{pmatrix} \hat{\boldsymbol{\beta}} \\ \hat{\boldsymbol{u}} \end{pmatrix} = \begin{pmatrix} \boldsymbol{X}_s'\boldsymbol{W}_s\boldsymbol{X}_s & \boldsymbol{X}_s'\boldsymbol{W}_s\boldsymbol{Z}_s \\ \boldsymbol{Z}_s'\boldsymbol{W}_s\boldsymbol{X}_s & \boldsymbol{Z}_s'\boldsymbol{W}_s\boldsymbol{Z}_s \end{pmatrix}^{-1}\begin{pmatrix} \boldsymbol{X}_s'\boldsymbol{W}_s\boldsymbol{y}_s \\ \boldsymbol{Z}_s'\boldsymbol{W}_s\boldsymbol{y}_s \end{pmatrix}. \tag{5.6}$$

In the incomplete rank case, $r(\boldsymbol{Z}_s) < q$, the system (5.3) has infinitely many solutions. If we change the symbol of inverse by the symbol of generalized inverse in (5.6), we obtain a set of possible solutions of (5.3). This is to say, we propose the solutions

$$\begin{pmatrix} \hat{\boldsymbol{\beta}} \\ \hat{\boldsymbol{u}} \end{pmatrix} = \begin{pmatrix} \boldsymbol{X}_s'\boldsymbol{W}_s\boldsymbol{X}_s & \boldsymbol{X}_s'\boldsymbol{W}_s\boldsymbol{Z}_s \\ \boldsymbol{Z}_s'\boldsymbol{W}_s\boldsymbol{X}_s & \boldsymbol{Z}_s'\boldsymbol{W}_s\boldsymbol{Z}_s \end{pmatrix}^{-} \begin{pmatrix} \boldsymbol{X}_s'\boldsymbol{W}_s\boldsymbol{y}_s \\ \boldsymbol{Z}_s'\boldsymbol{W}_s\boldsymbol{y}_s \end{pmatrix} = \begin{pmatrix} A^{11} & A^{12} \\ A^{21} & A^{22} \end{pmatrix} \begin{pmatrix} \boldsymbol{X}_s'\boldsymbol{W}_s\boldsymbol{y}_s \\ \boldsymbol{Z}_s'\boldsymbol{W}_s\boldsymbol{y}_s \end{pmatrix}.$$

By using the formula (A.1) from Appendix A which is valid in the present case for A_{22}^{-} in the place of A_{22}^{-1}, we calculate A^{11}, A^{12}, A^{21}, and A^{22}, i.e.

$$\begin{aligned}
A^{11} &= \left(A_{11} - A_{12}A_{22}^{-}A_{21}\right)^{-1} = \left(\boldsymbol{X}_s'\boldsymbol{W}_s\boldsymbol{X}_s - \boldsymbol{X}_s'\boldsymbol{W}_s\boldsymbol{Z}_s(\boldsymbol{Z}_s'\boldsymbol{W}_s\boldsymbol{Z}_s)^{-}\boldsymbol{Z}_s'\boldsymbol{W}_s\boldsymbol{X}_s\right)^{-1} \\
&= \left(\boldsymbol{X}_s'\left(\boldsymbol{W}_s - \boldsymbol{W}_s\boldsymbol{Z}_s(\boldsymbol{Z}_s'\boldsymbol{W}_s\boldsymbol{Z}_s)^{-}\boldsymbol{Z}_s'\boldsymbol{W}_s\right)\boldsymbol{X}_s\right)^{-1} = (\boldsymbol{X}_s'\boldsymbol{P}_s\boldsymbol{X}_s)^{-1}, \\
A^{12} &= -A^{11}A_{12}A_{22}^{-} = -(\boldsymbol{X}_s'\boldsymbol{P}_s\boldsymbol{X}_s)^{-1}\boldsymbol{X}_s'\boldsymbol{W}_s\boldsymbol{Z}_s(\boldsymbol{Z}_s'\boldsymbol{W}_s\boldsymbol{Z}_s)^{-} \\
&= -(\boldsymbol{X}_s'\boldsymbol{P}_s\boldsymbol{X}_s)^{-1}\boldsymbol{X}_s'\boldsymbol{W}_s\boldsymbol{Z}_s\boldsymbol{G}_s, \\
A^{21} &= (A^{12})' = -\boldsymbol{G}_s\boldsymbol{Z}_s'\boldsymbol{W}_s\boldsymbol{X}_s(\boldsymbol{X}_s'\boldsymbol{P}_s\boldsymbol{X}_s)^{-1}, \\
A^{22} &= A_{22}^{-} + A_{22}^{-}A_{21}A^{11}A_{12}A_{22}^{-} \\
&= (\boldsymbol{Z}_s'\boldsymbol{W}_s\boldsymbol{Z}_s)^{-} + (\boldsymbol{Z}_s'\boldsymbol{W}_s\boldsymbol{Z}_s)^{-}\boldsymbol{Z}_s'\boldsymbol{W}_s\boldsymbol{X}_s(\boldsymbol{X}_s'\boldsymbol{P}_s\boldsymbol{X}_s)^{-1}\boldsymbol{X}_s'\boldsymbol{W}_s\boldsymbol{Z}_s(\boldsymbol{Z}_s'\boldsymbol{W}_s\boldsymbol{Z}_s)^{-} \\
&= \boldsymbol{G}_s + \boldsymbol{G}_s\boldsymbol{Z}_s'\boldsymbol{W}_s\boldsymbol{X}_s(\boldsymbol{X}_s'\boldsymbol{P}_s\boldsymbol{X}_s)^{-1}\boldsymbol{X}_s'\boldsymbol{W}_s\boldsymbol{Z}_s\boldsymbol{G}_s.
\end{aligned}$$

Then, we have

$$\begin{aligned}
\hat{\boldsymbol{\beta}} &= A^{11}\boldsymbol{X}_s'\boldsymbol{W}_s\boldsymbol{y}_s + A^{12}\boldsymbol{Z}_s'\boldsymbol{W}_s\boldsymbol{y}_s \\
&= (\boldsymbol{X}_s'\boldsymbol{P}_s\boldsymbol{X}_s)^{-1}\boldsymbol{X}_s'\boldsymbol{W}_s\boldsymbol{y}_s - (\boldsymbol{X}_s'\boldsymbol{P}_s\boldsymbol{X}_s)^{-1}\boldsymbol{X}_s'\boldsymbol{W}_s\boldsymbol{Z}_s\boldsymbol{G}_s\boldsymbol{Z}_s'\boldsymbol{W}_s\boldsymbol{y}_s \\
&= (\boldsymbol{X}_s'\boldsymbol{P}_s\boldsymbol{X}_s)^{-1}\boldsymbol{X}_s'(\boldsymbol{W}_s - \boldsymbol{W}_s\boldsymbol{Z}_s\boldsymbol{G}_s\boldsymbol{Z}_s'\boldsymbol{W}_s)\boldsymbol{y}_s = (\boldsymbol{X}_s'\boldsymbol{P}_s\boldsymbol{X}_s)^{-1}\boldsymbol{X}_s'\boldsymbol{P}_s\boldsymbol{y}_s.
\end{aligned}$$

Note that $r(\boldsymbol{Z}_s\boldsymbol{G}_s\boldsymbol{Z}_s') \le q$ and $\boldsymbol{Z}_s\boldsymbol{G}_s\boldsymbol{Z}_s'$ is unique, because

$$\boldsymbol{Z}_s'\boldsymbol{W}_s\left[\boldsymbol{Z}_s\boldsymbol{G}_s\boldsymbol{Z}_s'\right]\boldsymbol{W}_s\boldsymbol{Z}_s = (\boldsymbol{Z}_s'\boldsymbol{W}_s\boldsymbol{Z}_s)(\boldsymbol{Z}_s'\boldsymbol{W}_s\boldsymbol{Z}_s)^{-}(\boldsymbol{Z}_s'\boldsymbol{W}_s\boldsymbol{Z}_s) = \boldsymbol{Z}_s'\boldsymbol{W}_s\boldsymbol{Z}_s$$

is unique. Therefore $\boldsymbol{P}_s$ is unique and $\hat{\boldsymbol{\beta}}$ is also unique. Further, it holds that

$$\begin{aligned}
\hat{\boldsymbol{u}} &= A^{21}\boldsymbol{X}_s'\boldsymbol{W}_s\boldsymbol{y}_s + A^{22}\boldsymbol{Z}_s'\boldsymbol{W}_s\boldsymbol{y}_s = -\boldsymbol{G}_s\boldsymbol{Z}_s'\boldsymbol{W}_s\boldsymbol{X}_s(\boldsymbol{X}_s'\boldsymbol{P}_s\boldsymbol{X}_s)^{-1}\boldsymbol{X}_s'\boldsymbol{W}_s\boldsymbol{y}_s \\
&+ \boldsymbol{G}_s\boldsymbol{Z}_s'\boldsymbol{W}_s\boldsymbol{y}_s + \boldsymbol{G}_s\boldsymbol{Z}_s'\boldsymbol{W}_s\boldsymbol{X}_s(\boldsymbol{X}_s'\boldsymbol{P}_s\boldsymbol{X}_s)^{-1}\boldsymbol{X}_s'\boldsymbol{W}_s\boldsymbol{Z}_s\boldsymbol{G}_s\boldsymbol{Z}_s'\boldsymbol{W}_s\boldsymbol{y}_s
\end{aligned}$$

$$= G_s Z_s' W_s y_s - G_s Z_s' W_s X_s \overbrace{(X_s' P_s X_s)^{-1} X_s' \underbrace{[W_s - W_s Z_s G_s Z_s' W_s]}_{P_s} y_s}^{\hat{\beta}}$$

$$= G_s Z_s' W_s y_s - G_s Z_s' W_s X_s \hat{\beta}.$$

Finally, we prove that $\hat{\beta}$ and $\hat{u}$ are a solution of the system of likelihood equations (5.3). Firstly, we check that $A = X_s' W_s X_s \hat{\beta} + X_s' W_s Z_s \hat{u}$ is equal to $X_s' W_s y_s$.

$$\begin{aligned} A &= X_s' W_s X_s \hat{\beta} + X_s' W_s Z_s \big(G_s Z_s' W_s y_s - G_s Z_s' W_s X_s \hat{\beta}\big) \\ &= X_s' W_s X_s \hat{\beta} + X_s' W_s Z_s G_s Z_s' W_s y_s - X_s' W_s Z_s G_s Z_s' W_s X_s \hat{\beta} \\ &= X_s' \big(W_s - W_s Z_s G_s Z_s' W_s\big) X_s \hat{\beta} + X_s' W_s Z_s G_s Z_s' W_s y_s \\ &= X_s' P_s X_s \big\{(X_s' P_s X_s)^{-1} X_s' P_s y_s\big\} + X_s' W_s Z_s G_s Z_s' W_s y_s \\ &= X_s' P_s y_s + X_s' W_s Z_s G_s Z_s' W_s y_s = X_s' \big(P_s + W_s Z_s G_s Z_s' W_s\big) y_s = X_s' W_s y_s. \end{aligned}$$

Secondly, we check that $B = Z_s' W_s X_s \hat{\beta} + Z_s' W_s Z_s \hat{u}$ is equal to $Z_s' W_s y_s$. Let us note that $Z_s' W_s Z_s G_s Z_s' = Z_s'$, because a post-multiplication by $W_s Z_s$ yields to

$$Z_s' W_s Z_s G_s Z_s' W_s Z_s = Z_s' W_s Z_s$$

and since G_s is the generalized inverse of $Z_s' W_s Z_s$ the left-hand side of the equation coincides with the right-hand side. Therefore, we get

$$\begin{aligned} B &= Z_s' W_s X_s \hat{\beta} + Z_s' W_s Z_s G_s Z_s' W_s \big(y_s - X_s \hat{\beta}\big) \\ &= Z_s' W_s X_s \hat{\beta} + Z_s' W_s \big(y_s - X_s \hat{\beta}\big) = Z_s' W_s y_s. \end{aligned}$$

□

If $\mathrm{r}(Z_s) = q$, an unbiased estimator of σ_e^2 is

$$\tilde{\sigma}_e^2 = \frac{1}{n - p - q} \big(y_s - X_s \hat{\beta} - Z_s \hat{u}\big)' W_s \big(y_s - X_s \hat{\beta} - Z_s \hat{u}\big),$$

which is unique because $Z_s \hat{u} = Z_s G_s Z_s' W_s y_s - Z_s G_s Z_s' W_s X_s \beta$ is unique, as $Z_s G_s Z_s'$ is unique.

Lemma 5.1 *If $Z_s' W_s Z_s$ is invertible, then*

$$G_s = (Z_s' W_s Z_s)^{-1}, \quad P_s W_s^{-1} P_s = P_s, \quad Z_s' P_s X_s = 0_{q\times p}, \quad X_s' P_s Z_s = 0_{p\times q}.$$

Proof It holds that

$$
\begin{aligned}
\boldsymbol{P}_s\boldsymbol{W}_s^{-1}\boldsymbol{P}_s &= (\boldsymbol{W}_s - \boldsymbol{W}_s\boldsymbol{Z}_s\boldsymbol{G}_s\boldsymbol{Z}_s'\boldsymbol{W}_s)\boldsymbol{W}_s^{-1}(\boldsymbol{W}_s - \boldsymbol{W}_s\boldsymbol{Z}_s\boldsymbol{G}_s\boldsymbol{Z}_s'\boldsymbol{W}_s)\\
&= \boldsymbol{W}_s - \boldsymbol{W}_s\boldsymbol{Z}_s\boldsymbol{G}_s\boldsymbol{Z}_s'\boldsymbol{W}_s - \boldsymbol{W}_s\boldsymbol{Z}_s\boldsymbol{G}_s\boldsymbol{Z}_s'\boldsymbol{W}_s + \boldsymbol{W}_s\boldsymbol{Z}_s\boldsymbol{G}_s\boldsymbol{Z}_s'\boldsymbol{W}_s\boldsymbol{Z}_s\boldsymbol{G}_s\boldsymbol{Z}_s'\boldsymbol{W}_s\\
&= \boldsymbol{W}_s - 2\boldsymbol{W}_s\boldsymbol{Z}_s\boldsymbol{G}_s\boldsymbol{Z}_s'\boldsymbol{W}_s + \boldsymbol{W}_s\boldsymbol{Z}_s\boldsymbol{G}_s\boldsymbol{G}_s^{-1}\boldsymbol{G}_s\boldsymbol{Z}_s'\boldsymbol{W}_s\\
&= \boldsymbol{W}_s - \boldsymbol{W}_s\boldsymbol{Z}_s\boldsymbol{G}_s\boldsymbol{Z}_s'\boldsymbol{W}_s = \boldsymbol{P}_s,\\
\boldsymbol{Z}_s'\boldsymbol{P}_s\boldsymbol{X}_s &= \boldsymbol{Z}_s'(\boldsymbol{W}_s - \boldsymbol{W}_s\boldsymbol{Z}_s\boldsymbol{G}_s\boldsymbol{Z}_s'\boldsymbol{W}_s)\boldsymbol{X}_s = \boldsymbol{Z}_s'\boldsymbol{W}_s\boldsymbol{X}_s - \boldsymbol{Z}_s'\boldsymbol{W}_s\boldsymbol{Z}_s\boldsymbol{G}_s\boldsymbol{Z}_s'\boldsymbol{W}_s\boldsymbol{X}_s\\
&= \boldsymbol{Z}_s'\boldsymbol{W}_s\boldsymbol{X}_s - \boldsymbol{G}_s^{-1}\boldsymbol{G}_s\boldsymbol{Z}_s'\boldsymbol{W}_s\boldsymbol{X}_s = \boldsymbol{Z}_s'\boldsymbol{W}_s\boldsymbol{X}_s - \boldsymbol{Z}_s'\boldsymbol{W}_s\boldsymbol{X}_s = \boldsymbol{0}.
\end{aligned}
$$

□

Proposition 5.2 *If $\boldsymbol{Z}_s'\boldsymbol{W}_s\boldsymbol{Z}_s$ is invertible, then*

$$
\begin{aligned}
\mathrm{var}(\hat{\boldsymbol{\beta}}) &= \sigma_e^2(\boldsymbol{X}_s'\boldsymbol{P}_s\boldsymbol{X}_s)^{-1}, \quad \mathrm{cov}(\hat{\boldsymbol{\beta}}, \hat{\boldsymbol{u}}) = -\sigma_e^2(\boldsymbol{X}_s'\boldsymbol{P}_s\boldsymbol{X}_s)^{-1}\boldsymbol{X}_s'\boldsymbol{W}_s\boldsymbol{Z}_s\boldsymbol{G}_s,\\
\mathrm{var}(\hat{\boldsymbol{u}}) &= \sigma_e^2\left\{\boldsymbol{G}_s + \boldsymbol{G}_s\boldsymbol{Z}_s'\boldsymbol{W}_s\boldsymbol{X}_s(\boldsymbol{X}_s'\boldsymbol{P}_s\boldsymbol{X}_s)^{-1}\boldsymbol{X}_s'\boldsymbol{W}_s\boldsymbol{Z}_s\boldsymbol{G}_s\right\}.
\end{aligned}
$$

Proof By applying Lemma 5.1, we obtain the variance matrix of $\hat{\boldsymbol{\beta}}$, i.e.

$$
\begin{aligned}
\mathrm{var}(\hat{\boldsymbol{\beta}}) &= (\boldsymbol{X}_s'\boldsymbol{P}_s\boldsymbol{X}_s)^{-1}\boldsymbol{X}_s'\boldsymbol{P}_s\mathrm{var}(\boldsymbol{y}_s)\boldsymbol{P}_s\boldsymbol{X}_s(\boldsymbol{X}_s'\boldsymbol{P}_s\boldsymbol{X}_s)^{-1}\\
&= \sigma_e^2(\boldsymbol{X}_s'\boldsymbol{P}_s\boldsymbol{X}_s)^{-1}\boldsymbol{X}_s'\boldsymbol{P}_s\boldsymbol{W}_s^{-1}\boldsymbol{P}_s\boldsymbol{X}_s(\boldsymbol{X}_s'\boldsymbol{P}_s\boldsymbol{X}_s)^{-1}\\
&= \sigma_e^2(\boldsymbol{X}_s'\boldsymbol{P}_s\boldsymbol{X}_s)^{-1}\boldsymbol{X}_s'\boldsymbol{P}_s\boldsymbol{X}_s(\boldsymbol{X}_s'\boldsymbol{P}_s\boldsymbol{X}_s)^{-1} = \sigma_e^2(\boldsymbol{X}_s'\boldsymbol{P}_s\boldsymbol{X}_s)^{-1}.
\end{aligned}
$$

The covariance matrix between $\hat{\boldsymbol{\beta}}$ and $\hat{\boldsymbol{u}}$ and the variance matrix of $\hat{\boldsymbol{u}}$ are

$$
\begin{aligned}
\mathrm{cov}(\hat{\boldsymbol{\beta}}, \hat{\boldsymbol{u}}) &= \mathrm{cov}(\hat{\boldsymbol{\beta}}, \boldsymbol{G}_s\boldsymbol{Z}_s'\boldsymbol{W}_s\boldsymbol{y}_s - \boldsymbol{G}_s\boldsymbol{Z}_s'\boldsymbol{W}_s\boldsymbol{X}_s\hat{\boldsymbol{\beta}})\\
&= \mathrm{cov}(\hat{\boldsymbol{\beta}}, \boldsymbol{y}_s)\boldsymbol{W}_s\boldsymbol{Z}_s\boldsymbol{G}_s - \mathrm{var}(\hat{\boldsymbol{\beta}})\boldsymbol{X}_s'\boldsymbol{W}_s\boldsymbol{Z}_s\boldsymbol{G}_s\\
&= \sigma_e^2(\boldsymbol{X}_s'\boldsymbol{P}_s\boldsymbol{X}_s)^{-1}\boldsymbol{X}_s'\boldsymbol{P}_s\boldsymbol{W}_s^{-1}\boldsymbol{W}_s\boldsymbol{Z}_s\boldsymbol{G}_s - \sigma_e^2(\boldsymbol{X}_s'\boldsymbol{P}_s\boldsymbol{X}_s)^{-1}\boldsymbol{X}_s'\boldsymbol{W}_s\boldsymbol{Z}_s\boldsymbol{G}_s\\
&= -\sigma_e^2(\boldsymbol{X}_s'\boldsymbol{P}_s\boldsymbol{X}_s)^{-1}\boldsymbol{X}_s'\boldsymbol{W}_s\boldsymbol{Z}_s\boldsymbol{G}_s,\\
\mathrm{var}(\hat{\boldsymbol{u}}) &= \mathrm{var}(\boldsymbol{G}_s\boldsymbol{Z}_s'\boldsymbol{W}_s\boldsymbol{y}_s - \boldsymbol{G}_s\boldsymbol{Z}_s'\boldsymbol{W}_s\boldsymbol{X}_s\hat{\boldsymbol{\beta}}) = \boldsymbol{G}_s\boldsymbol{Z}_s'\boldsymbol{W}_s\mathrm{var}(\boldsymbol{y}_s)\boldsymbol{W}_s\boldsymbol{Z}_s\boldsymbol{G}_s\\
&\quad - \boldsymbol{G}_s\boldsymbol{Z}_s'\boldsymbol{W}_s\mathrm{cov}(\boldsymbol{y}_s, \hat{\boldsymbol{\beta}})\boldsymbol{X}_s'\boldsymbol{W}_s\boldsymbol{Z}_s\boldsymbol{G}_s - \boldsymbol{G}_s\boldsymbol{Z}_s'\boldsymbol{W}_s\boldsymbol{X}_s\mathrm{cov}(\hat{\boldsymbol{\beta}}, \boldsymbol{y}_s)\boldsymbol{W}_s\boldsymbol{Z}_s\boldsymbol{G}_s\\
&\quad + \boldsymbol{G}_s\boldsymbol{Z}_s'\boldsymbol{W}_s\boldsymbol{X}_s\mathrm{var}(\hat{\boldsymbol{\beta}})\boldsymbol{X}_s'\boldsymbol{W}_s\boldsymbol{Z}_s\boldsymbol{G}_s\\
&= \sigma_e^2\boldsymbol{G}_s\boldsymbol{Z}_s'\boldsymbol{W}_s\boldsymbol{Z}_s\boldsymbol{G}_s - \sigma_e^2\boldsymbol{G}_s\boldsymbol{Z}_s'\boldsymbol{W}_s\{\boldsymbol{W}_s^{-1}\boldsymbol{P}_s\boldsymbol{X}_s(\boldsymbol{X}_s'\boldsymbol{P}_s\boldsymbol{X}_s)^{-1}\}\boldsymbol{X}_s'\boldsymbol{W}_s\boldsymbol{Z}_s\boldsymbol{G}_s\\
&\quad - \sigma_e^2\boldsymbol{G}_s\boldsymbol{Z}_s'\boldsymbol{W}_s\boldsymbol{X}_s\{(\boldsymbol{X}_s'\boldsymbol{P}_s\boldsymbol{X}_s)^{-1}\boldsymbol{X}_s'\boldsymbol{P}_s\boldsymbol{W}_s^{-1}\}\boldsymbol{W}_s\boldsymbol{Z}_s\boldsymbol{G}_s
\end{aligned}
$$

$$+ \sigma_e^2 G_s Z_s' W_s X_s (X_s' P_s X_s)^{-1} X_s' W_s Z_s G_s$$
$$= \sigma_e^2 \left\{ G_s + G_s Z_s' W_s X_s (X_s' P_s X_s)^{-1} X_s' W_s Z_s G_s \right\}.$$

□

The variance matrix of the $(p+q) \times 1$ vector $\hat{\boldsymbol{\theta}} = (\hat{\boldsymbol{\beta}}', \hat{\boldsymbol{u}}')'$ is

$$\mathrm{var}(\hat{\boldsymbol{\theta}}) = \sigma_e^2 \begin{bmatrix} (X_s' P_s X_s)^{-1} & -(X_s' P_s X_s)^{-1} X_s' W_s Z_s G_s \\ -G_s Z_s' W_s X_s (X_s' P_s X_s)^{-1} & G_s + G_s Z_s' W_s X_s (X_s' P_s X_s)^{-1} X_s' Z_s G_s \end{bmatrix}.$$

The best linear unbiased predictor (BLUP) of the population total is

$$\hat{Y} = \mathbf{1}_s' \boldsymbol{y}_s + \mathbf{1}_r' \hat{\boldsymbol{y}}_r = \mathbf{1}_s' \boldsymbol{y}_s + \mathbf{1}_r' X_r \hat{\boldsymbol{\beta}} + \mathbf{1}_r' Z_r \hat{\boldsymbol{u}}, \tag{5.7}$$

where $r = U - s$. The variance of the predictor (5.7), under the model (5.1), is

$$\mathrm{var}_M(\hat{Y}) = \sigma_e^2 \mathbf{1}_s' W_s^{-1} \mathbf{1}_s + \mathbf{1}_r' \, [X_r \; Z_r] \, \mathrm{var}(\hat{\boldsymbol{\theta}}) \begin{pmatrix} X_r' \\ Z_r' \end{pmatrix} \mathbf{1}_r.$$

5.3 Linear Models with One Fixed Factor

The linear model (5.1) is simpler in the case of having only one factor with D levels (domains). We assume the model

$$\boldsymbol{y} = \boldsymbol{X}\boldsymbol{\beta} + \boldsymbol{Z}\boldsymbol{u} + \boldsymbol{e}, \tag{5.8}$$

where $\boldsymbol{y} = \boldsymbol{y}_{N\times 1}$, $\boldsymbol{X} = \boldsymbol{X}_{N\times p}$ with $r(\boldsymbol{X}) = p$, $\boldsymbol{\beta} = \boldsymbol{\beta}_{p\times 1}$, $\boldsymbol{Z} = \boldsymbol{Z}_{N\times D} = \mathrm{diag}_{1\le d\le D} \mathbf{1}_{N_d}$, $\boldsymbol{u} = \boldsymbol{u}_{D\times 1}$, and $\boldsymbol{e} = \boldsymbol{e}_{N\times 1} \sim N(\boldsymbol{0}, \sigma_e^2 \boldsymbol{W}^{-1})$. Model (5.8) can alternatively be written in the form

$$y_{dj} = \boldsymbol{x}_{dj}\boldsymbol{\beta} + u_d + e_{dj} = \beta_1 x_{dj1} + \ldots + \beta_p x_{djp} + u_d + e_{dj}, \; d = 1, \ldots, D, \; j = 1, \ldots, N_d, \tag{5.9}$$

where N_d is the population size of domain d, y_{dj} is the value of the target variable at the unit j of domain d and $\boldsymbol{x}_{dj}$ is the row (d, j) of the matrix $\boldsymbol{X}$ containing quantitative auxiliary variables. For each domain d, the model (5.9) has an intercept u_d. The intercepts u_d are model parameters that vary with the domains. This is to say, each domain has, in general, a different intercept. However the slope parameters $\beta_1, \ldots, \beta_p$ are constant, i.e. they do not vary from one domain to another. We can write model (5.9) in the expanded matrix form

$$
\begin{pmatrix} y_{11} \\ \vdots \\ y_{1N_1} \\ \hline y_{21} \\ \vdots \\ y_{2N_2} \\ \hline \vdots \\ \hline y_{D1} \\ \vdots \\ y_{DN_D} \end{pmatrix}
=
\begin{pmatrix} x_{111} & \dots & x_{11p} \\ \vdots & \vdots & \vdots \\ x_{1N_11} & \dots & x_{1N_1p} \\ \hline x_{211} & \dots & x_{21p} \\ \vdots & \vdots & \vdots \\ x_{2N_21} & \dots & x_{2N_2p} \\ \hline \vdots & \vdots & \vdots \\ \hline x_{D11} & \dots & x_{D1p} \\ \vdots & \vdots & \vdots \\ x_{DN_D1} & \dots & x_{DN_Dp} \end{pmatrix}
\begin{pmatrix} \beta_1 \\ \beta_2 \\ \vdots \\ \beta_p \end{pmatrix}
+
\begin{pmatrix} 1\,0 \dots 0 \\ \vdots\ \vdots\ \ \ \vdots \\ 1\,0 \dots 0 \\ \hline 0\,1 \dots 0 \\ \vdots\ \vdots\ \ \ \vdots \\ 0\,1 \dots 0 \\ \hline \vdots\ \vdots\ \ \ \vdots \\ \hline 0\,0 \dots 1 \\ \vdots\ \vdots\ \ \ \vdots \\ 0\,0 \dots 1 \end{pmatrix}
\begin{pmatrix} u_1 \\ u_2 \\ \vdots \\ u_D \end{pmatrix}
+
\begin{pmatrix} e_{11} \\ \vdots \\ e_{1N_1} \\ \hline e_{21} \\ \vdots \\ e_{2N_2} \\ \hline \vdots \\ \hline e_{D1} \\ \vdots \\ e_{DN_D} \end{pmatrix},
$$

where $\mathrm{var}(e_{dj}) = \sigma_e^2 w_{dj}^{-1}$, $d = 1, \dots, D$, $j = 1, \dots, N_d$. Let s_d be the set of sampling units in the domain d, let $n, n_1, \dots, n_D$ and $N, N_1, \dots, N_D$ be the sample and the population sizes, respectively, and let us define $w_d = \sum_{j\in s_d} w_{dj}$. Then, we have

$$
\begin{aligned}
\boldsymbol{G}_s &= (\boldsymbol{Z}_s'\boldsymbol{W}_s\boldsymbol{Z}_s)^{-1} = \mathrm{diag}(1/w_1, \dots, 1/w_D), \\
\boldsymbol{P}_s &= \boldsymbol{W}_s - \boldsymbol{W}_s\boldsymbol{Z}_s\boldsymbol{G}_s\boldsymbol{Z}_s'\boldsymbol{W}_s = \boldsymbol{W}_s - \mathrm{diag}\,(\boldsymbol{A}_1, \dots, \boldsymbol{A}_D),
\end{aligned}
$$

where $\boldsymbol{A}_d = \frac{1}{w_d}\boldsymbol{w}_{n_d}\boldsymbol{w}_{n_d}'$ and $\boldsymbol{w}_{n_d} = (w_{d1}, \dots, w_{dn_d})'$. Further, for $\hat{\bar{X}}^w_{d\cdot i} = \frac{1}{w_d}\sum_{j\in s_d} w_{dj}x_{dji}$ and $\hat{\bar{Y}}^w_d = \frac{1}{w_d}\sum_{j\in s_d} w_{dj}y_{dj}$, we have

(a) $\boldsymbol{X}_s'\boldsymbol{P}_s\boldsymbol{X}_s = \boldsymbol{E}_{wxx}$, where $\boldsymbol{E}_{wxx} = \left(E_{wx_ix_k}\right)_{i,k=1,\dots,p}$ and

$$
E_{wx_ix_k} = \sum_{d=1}^{D}\sum_{j\in s_d} w_{dj}(x_{dji} - \hat{\bar{X}}^w_{d\cdot i})(x_{djk} - \hat{\bar{X}}^w_{d\cdot k}),
$$

(b) $\boldsymbol{X}_s'\boldsymbol{P}_s\boldsymbol{y}_s = \boldsymbol{E}_{wxy}$, where $\boldsymbol{E}_{wxy} = \left(E_{wx_1y}, \dots, E_{wx_py}\right)'$ and

$$
E_{wx_iy} = \sum_{d=1}^{D}\sum_{j\in s_d} w_{dj}(x_{dji} - \hat{\bar{X}}^w_{d\cdot i})(y_{dj} - \hat{\bar{Y}}^w_d),
$$

(c) $\boldsymbol{G}_s\boldsymbol{Z}_s'\boldsymbol{W}_s\boldsymbol{y}_s = \hat{\bar{\boldsymbol{y}}}^w_s$, where $\hat{\bar{\boldsymbol{y}}}^w_s = (\hat{\bar{Y}}^w_1, \dots, \hat{\bar{Y}}^w_D)'$,

(d) $G_s Z_s' W_s X_s = \overline{X}_s^{\wedge w}$, where $\overline{X}_s^{\wedge w}$ is a $D \times p$ matrix whose rows, $\overline{X}_d^{\wedge w}$, are the vectors of weighted means of the auxiliary variables, i.e. $\overline{X}_d^{\wedge w} = \frac{1}{w_d} \sum_{j \in s_d} w_{dj} \boldsymbol{x}_{dj} = \left(\overline{X}_{d\cdot 1}^{\wedge w}, \ldots, \overline{X}_{d\cdot p}^{\wedge w}\right)$, $d = 1, \ldots, D$.

Therefore, the maximum likelihood estimators of the regression parameters are

$$\hat{\boldsymbol{\beta}} = (X_s' P_s X_s)^{-1} X_s' P_s \boldsymbol{y}_s = E_{wxx}^{-1} E_{wxy},$$

$$\hat{\boldsymbol{u}} = G_s Z_s' W_s \boldsymbol{y}_s - G_s Z_s' W_s X_s \hat{\boldsymbol{\beta}} = \overline{\boldsymbol{y}}_s^{\wedge w} - \overline{X}_s^{\wedge w} \hat{\boldsymbol{\beta}},$$

with components $\hat{u}_d = \overline{Y}_d^{\wedge w} - \overline{X}_d^{\wedge w} \hat{\boldsymbol{\beta}}$, $d = 1, \ldots, D$. If matrix X_s has no intercept column of ones, then the ANOVA estimator of σ_e^2 is

$$\tilde{\sigma}_e^2 = \frac{1}{n - p - D} (\boldsymbol{y}_s - X_s \hat{\boldsymbol{\beta}} - Z_s \hat{\boldsymbol{u}})' W_s (\boldsymbol{y}_s - X_s \hat{\boldsymbol{\beta}} - Z_s \hat{\boldsymbol{u}}).$$

In the particular case $p = 2$, we have

$$\begin{pmatrix} \hat{\beta}_1 \\ \hat{\beta}_2 \end{pmatrix} = \begin{pmatrix} E_{wx_1x_1} & E_{wx_1x_2} \\ E_{wx_2x_1} & E_{wx_2x_2} \end{pmatrix}^{-1} \begin{pmatrix} E_{wx_1y} \\ E_{wx_2y} \end{pmatrix},$$

$$\hat{u}_d = \overline{Y}_d^{\wedge w} - \hat{\beta}_1 \overline{X}_{d\cdot 1}^{\wedge w} - \hat{\beta}_2 \overline{X}_{d\cdot 2}^{\wedge w}, \quad d = 1, \ldots, D,$$

$$\hat{y}_{dj} = \hat{\beta}_1 x_{dj1} + \hat{\beta}_2 x_{dj2} + \hat{u}_d, \quad d = 1, \ldots, D, \ j = 1, \ldots, N_d.$$

The domain mean of the model predictions is the projective estimator that is also called *regression estimator with domain auxiliary information*. The projective estimator is

$$\overline{Y}_d^{\wedge regd} = \frac{1}{N_d} \sum_{j=1}^{N_d} \hat{y}_{dj} = \overline{Y}_d^{\wedge w} + (\overline{X}_d - \overline{X}_d^{\wedge w}) \hat{\boldsymbol{\beta}}, \tag{5.10}$$

where $\overline{X}_d = \frac{1}{N_d} \sum_{j=1}^{N_d} \boldsymbol{x}_{dj}$ is the vector containing the population means of the auxiliary variables at domain d.

The BLUPs (predictive estimators) of the domain total Y_d and mean $\overline{Y}_d$ are

$$\hat{Y}_d^{blup} = \sum_{j \in s_d} y_{dj} + \sum_{j \notin s_d} \hat{y}_{dj} \quad \text{and} \quad \overline{Y}_d^{\wedge blup} = \frac{\hat{Y}_d^{blup}}{N_d}, \tag{5.11}$$

where s_d is the set of sampling units in the domain d.

For calculating $\overline{Y}_d^{\wedge blup}$ in (5.11), we assume that N_d and x_{dj}, $d = 1, \ldots, D$, $j = 1, \ldots, N_d$ are known. This is the case of having a census. Note that we calculate the estimators $\hat{\overline{Y}}_d^w$, $\hat{\boldsymbol{\beta}}$, and $\hat{\overline{\boldsymbol{X}}}_d^w$ from the sample, but we have to take the values of x_{dj} from a census file for individuals $j \notin s_d$ to be able to calculate the second sum in (5.11). Nevertheless, this problem can be overcome if we use the sum over the whole domain population for obtaining a simpler formula for $\hat{\overline{Y}}_d^{blup}$, i.e.

$$\begin{aligned}
\hat{\overline{Y}}_d^{blup} &= \frac{1}{N_d}\sum_{j=1}^{N_d}\hat{y}_{dj} + \frac{1}{N_d}\sum_{j\in s_d}(y_{dj}-\hat{y}_{dj})\\
&= \hat{\overline{Y}}_d^{regd} + \frac{n_d}{N_d}\frac{1}{n_d}\sum_{j\in s_d}\left(y_{dj}-\hat{\overline{Y}}_d^w-(\boldsymbol{x}_{dj}-\hat{\overline{\boldsymbol{X}}}_d^w)\hat{\boldsymbol{\beta}}\right)\\
&= (1-\frac{n_d}{N_d})\hat{\overline{Y}}_d^{regd} + \frac{n_d}{N_d}\left[\hat{\overline{Y}}_d^{regd} + \hat{\overline{Y}}_d - \hat{\overline{Y}}_d^w - (\hat{\overline{\boldsymbol{X}}}_d - \hat{\overline{\boldsymbol{X}}}_d^w)\hat{\boldsymbol{\beta}}\right]\\
&= (1-f_d)\hat{\overline{Y}}_d^{regd} + f_d\left[\hat{\overline{Y}}_d^w + (\overline{\boldsymbol{X}}_d-\hat{\overline{\boldsymbol{X}}}_d^w)\hat{\boldsymbol{\beta}} + \hat{\overline{Y}}_d - \hat{\overline{Y}}_d^w - (\hat{\overline{\boldsymbol{X}}}_d - \hat{\overline{\boldsymbol{X}}}_d^w)\hat{\boldsymbol{\beta}}\right]\\
&= (1-f_d)\hat{\overline{Y}}_d^{regd} + f_d\left[\hat{\overline{Y}}_d + (\overline{\boldsymbol{X}}_d-\hat{\overline{\boldsymbol{X}}}_d)\hat{\boldsymbol{\beta}}\right], \qquad (5.12)
\end{aligned}$$

where $f_d = n_d/N_d$, $\hat{\overline{Y}}_d = \frac{1}{n_d}\sum_{j\in s_d} y_{dj}$, and $\hat{\overline{\boldsymbol{X}}}_d = \frac{1}{n_d}\sum_{j\in s_d}\boldsymbol{x}_{dj}$ is a $1\times p$ vector. Thus, for calculating the BLUP we need only the sample data and the population means $\overline{\boldsymbol{X}}_d$.

Alternatively, we can calculate the predictive estimator of $\overline{Y}_d$, given in (5.12), by applying the formula

$$\hat{\overline{Y}}_d^{blup} = \boldsymbol{a}_s'\boldsymbol{y}_s + \boldsymbol{a}_r'(\boldsymbol{X}_r\hat{\boldsymbol{\beta}} + \boldsymbol{Z}_r\hat{\boldsymbol{u}}),$$

where $\boldsymbol{a}' = \frac{1}{N_d}\underset{1\le d\le D}{\text{col}'}(\delta_{\ell d}\boldsymbol{1}_{N_d}')$, $\boldsymbol{1}_{N_d} = \underset{1\le j\le N_d}{\text{col}}(1)$, $\delta_{\ell d}$ is the Kronecker delta and $\boldsymbol{a}_s, \boldsymbol{a}_r$ are the sample and non-sample parts of $\boldsymbol{a}$, respectively. Corollary 4.2 gives the error variance, i.e.

$$V_M = MSE(\hat{\overline{Y}}_d^{blup}) = \boldsymbol{a}_r'(\boldsymbol{V}_r + \boldsymbol{M}_r\boldsymbol{Q}_s\boldsymbol{M}_r')\boldsymbol{a}_r,$$

where

$$\boldsymbol{V}_r = \sigma_e^2\boldsymbol{W}_r^{-1},\ \ \boldsymbol{V}_s = \sigma_e^2\boldsymbol{W}_s^{-1},\ \ \boldsymbol{Q}_s = (\boldsymbol{M}_s'\boldsymbol{V}_s^{-1}\boldsymbol{M}_s)^{-1},\ \ \boldsymbol{M} = \begin{pmatrix}\boldsymbol{M}_s\\ \boldsymbol{M}_r\end{pmatrix} = \begin{pmatrix}\boldsymbol{X}_s & \boldsymbol{Z}_s\\ \boldsymbol{X}_r & \boldsymbol{Z}_r\end{pmatrix},$$

under the model (5.8). Therefore,

$$V_M = \sigma_e^2 \boldsymbol{a}_r' \boldsymbol{W}_r^{-1} \boldsymbol{a}_r + \boldsymbol{a}_r' \boldsymbol{M}_r (\boldsymbol{M}_s' \boldsymbol{V}_s^{-1} \boldsymbol{M}_s)^{-1} \boldsymbol{M}_r' \boldsymbol{a}_r$$
$$= \frac{\sigma_e^2}{N_d^2} \sum_{j \in U_d - s_d} w_{dj}^{-1} + \sigma_e^2 (1 - f_d)^2 \overline{\boldsymbol{M}}_d^* (\boldsymbol{M}_s' \boldsymbol{W}_s \boldsymbol{M}_s)^{-1} \overline{\boldsymbol{M}}_d^{*\prime},$$

where $\overline{\boldsymbol{M}}_d^* = \frac{1}{N_d - n_d} \sum_{j \in U_d - s_d} \boldsymbol{m}_{dj} = \frac{N_d}{N_d - n_d} \overline{\boldsymbol{M}}_d - \frac{n_d}{N_d - n_d} \overline{\boldsymbol{m}}_d$, where $\boldsymbol{m}_{dj}$ is the corresponding row of the matrix $\boldsymbol{M}$ and $\overline{\boldsymbol{M}}_d = \frac{1}{N_d} \sum_{j=1}^{N_d} \boldsymbol{m}_{dj}$, $\overline{\boldsymbol{m}}_d = \frac{1}{n_d} \sum_{j \in s_d} \boldsymbol{m}_{dj}$.

If $w_{dj} = 1$, $d = 1, \ldots, D$, $j = 1, \ldots, n_d$, then

$$MSE(\hat{\overline{Y}}_d^{blup}) = \sigma_e^2 \left[\frac{1 - f_d}{N_d} + (1 - f_d)^2 \overline{\boldsymbol{M}}_d^* (\boldsymbol{M}_s' \boldsymbol{W}_s \boldsymbol{M}_s)^{-1} \overline{\boldsymbol{M}}_d^{*\prime} \right].$$

Remark 5.1 In the case of complex sampling designs with inclusion probabilities π_{dj} (probability of including individual j of domain d in the sample), we could take as heteroscedasticity weights $w_{dj} = 1/\pi_{dj}$ in the model (5.8). In that case, we are assuming that $\text{var}_M(y_{dj}) = \pi_{dj} \sigma_e^2$; this is to say, the variability of Y under the model is directly proportional to the inclusion probabilities.

5.4 BLUPs Based on Linear Models with Fixed Effects

This section introduces some simple linear models with fixed effects and gives the corresponding projective and predictive (BLUP) estimators.

5.4.1 Regression Synthetic Estimator

The regression synthetic estimator is based on the model

$$y_{dj} = \beta x_{dj} + w_{dj}^{-1/2} e_{dj}, \; d = 1, \ldots, D, \; j = 1, \ldots, n_d, \qquad (5.13)$$

where the random errors e_{dj} are i.i.d. $N(0, \sigma^2)$, the parameters σ^2 and β are unknown and the heteroscedasticity weights w_{dj} are the sampling weights (known positive constants). The residual weighted sum of squares is

$$SSE = \sum_{d=1}^{D} \sum_{j \in s_d} w_{dj} \left(y_{dj} - \beta x_{dj} \right)^2 .$$

By taking partial derivatives with respect to β, we have

$$0 = -\frac{1}{2}\frac{\partial SSE}{\partial \beta} = \sum_{d=1}^{D}\sum_{j\in s_d} w_{dj}(y_{dj} - \beta x_{dj})x_{dj}$$

$$= \sum_{d=1}^{D}\sum_{j\in s_d} w_{dj}y_{dj}x_{dj} - \beta \sum_{d=1}^{D}\sum_{j\in s_d} w_{dj}x_{dj}^2,$$

and the weighted least squares estimator of β is

$$\hat{\beta} = \frac{\sum_{d=1}^{D}\sum_{j\in s_d} w_{dj}y_{dj}x_{dj}}{\sum_{d=1}^{D}\sum_{j\in s_d} w_{dj}x_{dj}^2}.$$

The predicted y-values are

$$\hat{y}_{dj} = \hat{\beta}x_{dj}, \quad d = 1, \ldots, D,\ j = 1, \ldots, n_d.$$

The population mean of the predicted y-values in the domain d is the *regression synthetic* (RSYNT) estimator

$$\hat{\overline{Y}}_d^{rsynt} = \frac{1}{N_d}\sum_{j=1}^{N_d} \hat{y}_{dj} = \overline{X}_d\hat{\beta}. \tag{5.14}$$

The BLUP under model (5.13) is

$$\hat{\overline{Y}}_d^{blup} = \frac{1}{N_d}\Big\{\sum_{j\in s_d} y_{dj} + \sum_{j\notin s_d} \hat{y}_{dj}\Big\}$$

$$= (1 - f_d)\hat{\overline{Y}}_d^{rsynt} + f_d\left\{\hat{\overline{Y}}_d + (\overline{X}_d - \hat{\overline{X}}_d)\hat{\beta}\right\}. \tag{5.15}$$

5.4.2 Estimators Without Domain Dependent Intercept

Let us consider the model

$$y_{dj} = \alpha + \beta x_{dj} + w_{dj}^{-1/2} e_{dj}, \ d = 1, \ldots, D,\ j = 1, \ldots, n_d, \tag{5.16}$$

where the random errors e_{dj} are i.i.d. $N(0, \sigma^2)$, the parameters σ^2, β, and α are unknown, and the heteroscedasticity weights w_{dj} are the sampling weights (known positive constants). The residual weighted sum of squares is

$$SSE = \sum_{d=1}^{D} \sum_{j \in s_d} w_{dj} \left(y_{dj} - \alpha - \beta x_{dj} \right)^2 .$$

By taking partial derivatives with respect to α, we have

$$0 = -\frac{1}{2} \frac{\partial SSE}{\partial \alpha} = \sum_{d=1}^{D} \sum_{j \in s_d} w_{dj} (y_{dj} - \alpha - \beta x_{dj}).$$

Therefore,

$$\alpha \sum_{d=1}^{D} \sum_{j \in s_d} w_{dj} = \sum_{d=1}^{D} \sum_{j \in s_d} w_{dj} y_{dj} - \beta \sum_{d=1}^{D} \sum_{j \in s_d} w_{dj} x_{dj}$$

and

$$\alpha = \hat{\overline{Y}}^{dir2} - \beta \hat{\overline{X}}^{dir2},$$

where

$$\hat{\overline{Y}}^{dir2} = \sum_{d=1}^{D} \sum_{j \in s_d} \omega_{dj} y_{dj}, \quad \hat{\overline{X}}^{dir2} = \sum_{d=1}^{D} \sum_{j \in s_d} \omega_{dj} x_{dj}, \quad \omega_{dj} = \frac{w_{dj}}{\sum_{d=1}^{D} \sum_{j \in s_d} w_{dj}}.$$

By taking partial derivatives with respect to β, we have

$$\begin{aligned} 0 &= -\frac{1}{2} \frac{1}{\sum_{d=1}^{D} \sum_{j \in s_d} w_{dj}} \frac{\partial SSE}{\partial \beta} = \sum_{d=1}^{D} \sum_{j \in s_d} \omega_{dj} (y_{dj} - \alpha - \beta x_{dj}) x_{dj} \\ &= \sum_{d=1}^{D} \sum_{j \in s_d} \omega_{dj} y_{dj} x_{dj} - \sum_{d=1}^{D} \sum_{j \in s_d} \omega_{dj} y_{dj} \sum_{d=1}^{D} \sum_{j \in s_d} \omega_{dj} x_{dj} \\ &+ \beta \left(\sum_{d=1}^{D} \sum_{j \in s_d} \omega_{dj} x_{dj} \right)^2 - \beta \sum_{d=1}^{D} \sum_{j \in s_d} \omega_{dj} x_{dj}^2, \end{aligned}$$

so the weighted least squares estimator of β is

$$\hat{\beta} = \frac{\sum_{d=1}^{D} \sum_{j \in s_d} w_{dj} \big(y_{dj} - \hat{\overline{Y}}^{dir2}\big)\big(x_{dj} - \hat{\overline{X}}^{dir2}\big)}{\sum_{d=1}^{D} \sum_{j \in s_d} w_{dj} \big(x_{dj} - \hat{\overline{X}}^{dir2}\big)^2}$$

and weighted least squares estimator of α is

$$\hat{\alpha} = \hat{\overline{Y}}^{dir2} - \hat{\beta}\hat{\overline{X}}^{dir2}.$$

The predicted y-values are

$$\hat{y}_{dj} = \hat{\alpha} + \hat{\beta} x_{dj} = \hat{\overline{Y}}^{dir2} - \hat{\beta}\hat{\overline{X}}^{dir2} + \hat{\beta} x_{dj} = \hat{\overline{Y}}^{dir2} + \hat{\beta}(x_{dj} - \hat{\overline{X}}^{dir2}).$$

The population mean of the predicted y-values in the domain d is the *projective estimator without domain dependent intercept*

$$\hat{\overline{Y}}_d^{reg} = \frac{1}{N_d} \sum_{j=1}^{N_d} \hat{y}_{dj} = \hat{\overline{Y}}^{dir2} + (\overline{X}_d - \hat{\overline{X}}^{dir2})\hat{\beta}. \tag{5.17}$$

The BLUP under model (5.16) is

$$\hat{\overline{Y}}_d^{blup} = \frac{1}{N_d}\Big\{\sum_{j \in s_d} y_{dj} + \sum_{j \notin s_d} \hat{y}_{dj}\Big\} = (1 - f_d)\hat{\overline{Y}}_d^{reg} + f_d \left\{\hat{\overline{Y}}_d + (\overline{X}_d - \hat{\overline{X}}_d)\hat{\beta}\right\}. \tag{5.18}$$

5.4.3 Estimators with Domain Dependent Intercept

Let us consider the model with one covariate

$$y_{dj} = \beta x_{dj} + u_d + w_{dj}^{-1/2} e_{dj}, \; d = 1, \ldots, D, \; j = 1, \ldots, n_d, \tag{5.19}$$

where the random errors e_{dj} are i.i.d. $N(0, \sigma^2)$, the parameters $\sigma^2, \; \beta, \; u_d, \; d = 1, \ldots, D$, are unknown and the heteroscedasticity weights w_{dj} are the sampling weights (known positive constants). It holds that

$$y_{dj} \sim N(\beta x_{dj} + u_d, \sigma^2 / w_{dj}), \quad d = 1, \ldots, D, \; j = 1, \ldots, n_d.$$

We can write model (5.19) in the form

$$w_{dj}^{1/2} y_{dj} = w_{dj}^{1/2}(\beta x_{dj} + u_d) + e_{dj}, \; d = 1, \dots, D, \; j = 1, \dots, n_d.$$

The residual weighted sum of squares is

$$SSE = \sum_{d=1}^{D} \sum_{j \in s_d} \left(w_{dj}^{1/2} y_{dj} - w_{dj}^{1/2}(\beta x_{dj} + u_d) \right)^2 = \sum_{d=1}^{D} \sum_{j \in s_d} w_{dj} \left(y_{dj} - (\beta x_{dj} + u_d) \right)^2.$$

By taking partial derivatives with respect to u_d, we have

$$0 = \frac{\partial SSE}{\partial u_d} = -2 \sum_{j \in s_d} w_{dj}(y_{dj} - \beta x_{dj} - u_d),$$

so that

$$u_d \sum_{j \in s_d} w_{dj} = \sum_{j \in s_d} w_{dj} y_{dj} - \beta \sum_{j \in s_d} w_{dj} x_{dj}$$

and

$$u_d = \hat{\overline{Y}}_d^{dir2} - \beta \hat{\overline{X}}_d^{dir2},$$

where

$$\hat{\overline{Y}}_d^{dir2} = \sum_{j \in s_d} \omega_{dj} y_{dj}, \quad \hat{\overline{X}}_d^{dir2} = \sum_{j \in s_d} \omega_{dj} x_{dj}, \quad \omega_{dj} = \frac{w_{dj}}{\sum_{j \in s_d} w_{dj}} = \frac{w_{dj}}{w_d}.$$

This is to say, $\hat{\overline{Y}}_d^{dir2}$ and $\hat{\overline{X}}_d^{dir2}$ are the direct estimators introduced in (2.7). By taking partial derivatives with respect to β, we have

$$\begin{aligned}
0 = -\frac{1}{2}\frac{\partial SSE}{\partial \beta} &= \sum_{d=1}^{D} \sum_{j \in s_d} w_{dj} \left(y_{dj} - \beta x_{dj} - u_d \right) x_{dj} \\
&= \sum_{d=1}^{D} \sum_{j \in s_d} w_{dj} y_{dj} x_{dj} - \beta \sum_{d=1}^{D} \sum_{j \in s_d} w_{dj} x_{dj}^2 - \sum_{d=1}^{D} \sum_{j \in s_d} w_{dj} (\hat{\overline{Y}}_d^{dir2} - \beta \hat{\overline{X}}_d^{dir2}) x_{dj} \\
&= \sum_{d=1}^{D} \sum_{j \in s_d} w_{dj} y_{dj} x_{dj} - \sum_{d=1}^{D} w_d \hat{\overline{Y}}_d^{dir2} \hat{\overline{X}}_d^{dir2} - \beta \Big\{ \sum_{d=1}^{D} \sum_{j \in s_d} w_{dj} x_{dj}^2 - \sum_{d=1}^{D} w_d (\hat{\overline{X}}_d^{dir2})^2 \Big\}.
\end{aligned}$$

The weighted least squares estimator of β is

$$\hat{\beta} = \frac{\sum_{d=1}^{D}\sum_{j\in s_d} w_{dj} y_{dj} x_{dj} - \sum_{d=1}^{D} w_d \hat{\overline{Y}}_d^{dir2} \hat{\overline{X}}_d^{dir2}}{\sum_{d=1}^{D}\sum_{j\in s_d} w_{dj} x_{dj}^2 - \sum_{d=1}^{D} w_d (\hat{\overline{X}}_d^{dir2})^2}$$

$$= \frac{\sum_{d=1}^{D}\sum_{j\in s_d} w_{dj} \big(y_{dj} - \hat{\overline{Y}}_d^{dir2}\big)\big(x_{dj} - \hat{\overline{X}}_d^{dir2}\big)}{\sum_{d=1}^{D}\sum_{j\in s_d} w_{dj} \big(x_{dj} - \hat{\overline{X}}_d^{dir2}\big)^2}$$

and the weighted least squares estimator of u_d is

$$\hat{u}_d = \hat{\overline{Y}}_d^{dir2} - \hat{\beta}\hat{\overline{X}}_d^{dir2}.$$

The predicted y-values are

$$\hat{y}_{dj} = \hat{u}_d + \hat{\beta} x_{dj} = \hat{\overline{Y}}_d^{dir2} + (x_{dj} - \hat{\overline{X}}_d^{dir2})\hat{\beta}, \quad d = 1, \ldots, D, \ j = 1, \ldots, n_d.$$

The population mean of the predicted y-values in the domain d is the *projective estimator with one covariate and domain dependent intercept*

$$\hat{\overline{Y}}_d^{regd} = \frac{1}{N_d}\sum_{j=1}^{N_d} \hat{y}_{dj} = \hat{\overline{Y}}_d^{dir2} + \hat{\beta}(\overline{X}_d - \hat{\overline{X}}_d^{dir2}). \tag{5.20}$$

The BLUP under model (5.19) is

$$\hat{Y}_d^{blup} = \sum_{j\in s_d} y_{dj} + \sum_{j\notin s_d} \hat{y}_{dj} \qquad \text{and} \qquad \hat{\overline{Y}}_d^{blup} = \frac{\hat{Y}_d^{blup}}{N_d},$$

where s_d is the subset of sampling units in the domain d. It holds that (cf. (5.12))

$$\hat{\overline{Y}}_d^{blup} = (1 - f_d)\hat{\overline{Y}}_d^{regd} + f_d\left\{\hat{\overline{Y}}_d + (\overline{X}_d - \hat{\overline{X}}_d)\hat{\beta}\right\}, \tag{5.21}$$

where $f_d = n_d/N_d$.

Remark 5.2 The GREG estimator (model-assisted) with one explanatory variable is (cf. (3.5))

$$\hat{\overline{Y}}_d^{greg} = \hat{\overline{Y}}_d^{dir2} + (\overline{X}_d - \hat{\overline{X}}_d^{dir2})\hat{\beta}.$$

If the model (5.16) with $w_{dj} = 1/\pi_{dj}$ is used as the assisting model, the parameter β is estimated by

$$\hat{\beta} = \frac{\sum_{d=1}^{D} \sum_{j \in s_d} w_{dj} \big(y_{dj} - \hat{\overline{Y}}^{dir2}\big)\big(x_{dj} - \hat{\overline{X}}^{dir2}\big)}{\sum_{d=1}^{D} \sum_{j \in s_d} w_{dj} \big(x_{dj} - \hat{\overline{X}}^{dir2}\big)^2}$$

and the GREG estimator is different from the projective estimators without and with domain dependent intercept.

If the model (5.19) is used as the assisting model with the corresponding estimate of β, then the GREG estimator coincides with the projective estimator with domain dependent intercept (5.20).

5.5 R Codes for BLUPs

This section gives R codes for calculating the projective and predictive (BLUP) estimators of domain means described in Sect. 5.4. The target is estimating the domain means of the variable INCOME from the survey data file `LFS20.txt`. The following code reads the data file:

```
dataux <- read.table("Nds20.txt", header=TRUE, sep = "\t", dec = ".")
dat <- read.table("LFS20.txt", header=TRUE, sep = "\t", dec = ".")
# Sample size
n <- nrow(dat); n
# Rename variables
y <- dat$INCOME; x <- dat$REGISTERED; w <- dat$WEIGHT; area <- dat$AREA
```

This section contains three examples. The first one assumes model (5.13), the second one assumes model (5.16), and the third one assumes model (5.19). The unique auxiliary variable for the three models is the dichotomic variable REGISTERED. Let us note that this variable is treated as the x variable in the following examples. As model (5.13) has a bad fit to data, their model-based estimator is less reliable than the corresponding ones based on models (5.16) and (5.19).

Example 5.1 (Regression Synthetic Estimators) The following R codes calculate the estimates of average incomes by area–sex by employing the projective and predictive estimators given in Sect. 5.4.1.

```
# Assumed model
mod1 <- lm(y~x-1, weights=w)
# Model error variance
sigma12 <- anova(mod1)[2,3]
# Regression parameter
beta1 <- as.numeric(mod1$coefficients)
# Population size and sampling fraction
Npop <- sum(dataux$N); f <- n/Npop
# Totals of x by area-sex
Xtot.ds <- tapply(dataux$reg,list(dataux$area,dataux$sex),mean)
# Sizes by area sex
Ntot.ds <- tapply(dataux$N,list(dataux$area,dataux$sex),mean)
# Means of x by area-sex
```

```
Xmean.ds <- Xtot.ds/Ntot.ds
# Sample means of y by area-sex
ymean.ds <- tapply(y,list(dat$AREA,dat$SEX),mean)
# Sample means of x by area-sex
xmean.ds <- tapply(x,list(dat$AREA,dat$SEX),mean)
# Regression synthetic estimators of y-means by area-sex
yreg1.ds <- beta1*Xmean.ds
# BLUP estimators of y-means by area-sex
yblup1.ds <- (1-f)*yreg1.ds+f*(ymean.ds+(Xmean.ds-xmean.ds)*beta1)
yblup1.ds
```

Example 5.2 (Projective Estimator Without Domain Dependent Intercept) The following R codes calculate the estimates of average incomes by area–sex by employing the projective and predictive estimators given in Sect. 5.4.2.

```
# Assumed model
mod2 <- lm(y~x,weights=w)
# Model error variance
sigma22 <- anova(mod2)[2,3]
# Regression parameters
beta2 <- as.numeric(mod2$coefficients)
# Regression parameter for x
beta2x <- beta2[2]; beta2x
# Population size and sampling fraction
Npop <- sum(dataux$N); f <- n/Npop
# Totals of x by area-sex
Xtot.ds <- tapply(dataux$reg,list(dataux$area,dataux$sex),mean)
# Sizes by area sex
Ntot.ds <- tapply(dataux$N,list(dataux$area,dataux$sex),mean)
# Means of x by area-sex
Xmean.ds <- Xtot.ds/Ntot.ds;
# Sample means of y by area-sex
ymean.ds <- tapply(y,list(dat$AREA,dat$SEX),mean)
# Sample means of x by area-sex
xmean.ds <- tapply(x,list(dat$AREA,dat$SEX),mean)
# Auxiliary terms for avoiding overflows
Ndir <- sum(w*10^(-3)); ydir <- sum(y*w*10^(-3)); xdir <- sum(x*w*10^(-3))
# Direct estimators of global means
ymeandir <- ydir/Ndir; xmeandir <- xdir/Ndir
# Projective estimator without domain auxiliary information
yreg2.ds <- ymeandir+(Xmean.ds-xmeandir)*beta2x; yreg.ds
# BLUP estimators of domain means
yblup2.ds <- (1-f)*yreg2.ds+f*(ymean.ds+(Xmean.ds-xmean.ds)*beta2x)
yblup2.ds
```

Example 5.3 (Projective Estimator with Domain Dependent Intercept) The following R codes calculate the estimates of average incomes by area–sex by employing the projective and predictive estimators given in Sect. 5.4.3.

```
# Assumed model
mod3 <- lm(y~as.factor(area)+x,weights=w)
# Model error variance
sigma32 <- anova(mod3)[3,3]
# Regression parameters
beta3 <- as.numeric(mod3$coefficients)
# Regression parameter for x
beta3x <- beta3[length(beta3)]
# Population size and sampling fraction
Npop <- sum(dataux$N); f <- n/Npop
# Totals of x by area-sex
Xtot.ds <- tapply(dataux$reg,list(dataux$area,dataux$sex),mean)
# Sizes by area sex
Ntot.ds <- tapply(dataux$N,list(dataux$area,dataux$sex),mean)
# Means of x by area-sex
Xmean.ds <- Xtot.ds/Ntot.ds
```

```
# Direct estimators of sizes
Ndir.ds <- tapply(w,list(dat$AREA,dat$SEX),sum)
# Direct estimators of y-totals
ydir.ds <- tapply(y*w,list(dat$AREA,dat$SEX),sum)
# Direct estimators of x-totals
xdir.ds <- tapply(x*w,list(dat$AREA,dat$SEX),sum)
# Direct estimators of y-means
ymeandir.ds <- ydir.ds/Ndir.ds
# Direct estimators of x-means
xmeandir.ds <- xdir.ds/Ndir.ds
# Projective estimator with domain auxiliary information
yreg3.ds <- ymeandir.ds+(Xmean.ds-xmeandir.ds)*beta3x; yregd.ds
# BLUP estimators of domain means
yblup3.ds <-  (1-f)*yreg3.ds+f*(ymean.ds+(Xmean.ds-xmean.ds)*beta3x)
yblup3.ds
```

For the ten first areas, Table 5.1 presents some of the outputs of the above R codes.

```
output <- data.frame(reg1=yreg1.ds[,1], blup1=yblup1.ds[,1],
                     reg2=yreg2.ds[,1], blup2=yblup2.ds[,1],
                     reg3=yreg3.ds[,1], blup3=yblup3.ds[,1],
                     reg1=yreg1.ds[,2], blup1=yblup1.ds[,2],
                     reg2=yreg2.ds[,2], blup2=yblup2.ds[,2],
                     reg3=yreg3.ds[,2], blup3=yblup3.ds[,2]
                     )
head(output, 10)
```

The columns reg and blup contain the projective and the BLUP estimates of average annual incomes under the three considered models. Subindexes 1, 2, and 3 refer to models of Sects. 5.4.1, 5.4.2, and 5.4.3, respectively. The left (right) part of Table 5.1 contains the results for sex=1 (sex=2).

Table 5.1 Estimates of average incomes for sex=1 (left) and sex=2 (right)

d	reg1	blup1	reg2	blup2	reg3	blup3	reg1	blup1	reg2	blup2	reg3	blup3
1	118	405	47,909	47,920	49,974	49,972	942	1180	47,693	47,665	43,065	43,063
2	3919	4147	46,913	46,895	44,950	44,944	2556	2804	47,270	47,257	45,483	45,480
3	265	564	47,871	47,894	52,412	52,409	276	556	47,868	47,872	49,624	49,617
4	1699	1954	47,495	47,484	45,819	45,819	1584	1848	47,525	47,522	47,848	47,844
5	549	815	47,796	47,787	47,105	47,099	3609	3868	46,994	46,997	48,219	48,215
6	2782	3047	47,211	47,218	48,178	48,179	7781	8018	45,900	45,903	46,379	46,380
7	208	485	47,886	47,887	49,048	49,042	255	520	47,874	47,863	45,834	45,835
8	2140	2372	47,379	47,347	41,242	41,246	1423	1671	47,567	47,551	45,043	45,041
9	2124	2387	47,383	47,389	48,713	48,711	5062	5314	46,613	46,625	48,335	48,337
10	4433	4661	46,778	46,759	42,997	43,000	2454	2708	47,297	47,287	46,147	46,144

References

Bingham, N.H., Fry, J.M.: Linear Models in Statistics. Springer, Berlin (2010)
Faraway, J.F.: Linear Models with R. Chapman & Hall/CRC, Boca Raton (2005)
Montgomery, D.C., Peck, E.A., Vining, G.G.: Introduction to Linear Regression Analysis. Wiley, New York (2012)
Rencher, A.C., Schaalje, G.B.: Linear Models in Statistics. Wiley, New York (2008)
Searle, S.R.: Linear Models. Wiley, New York (2016)

Chapter 6
Linear Mixed Models

6.1 Introduction

Linear models (LMs) are defined for independent random variables measured on units from the same population. Mixed models have a more complex multilevel or hierarchical structure. Variables are independent when they are observed in different levels or clusters, but within the same level or cluster they are considered as dependent because they share common properties. Clustered data have two sources of variation: between and within clusters. Mixed models describe these sources of variation, commonly present in real data, and give a high modeling flexibility and applicability.

Linear mixed models (LMMs) establish a linear relationship between a target variable and some explanatory variables. At the same time LMMs can handle data in which observations are not independent. LMMs might model correlated errors, whereas LMs usually do not. LMMs are generalizations of LMs which allow to better support the analysis of a dependent variable. These models can incorporate random effects, hierarchical effects, repeated measures, spatial and temporal correlations.

LMMs have a wide applicability in small area estimation (SAE), where the flexibility in combining different sources of information and explaining different sources of errors is of great help. Mixed models increase the effective information used in the estimation process by linking all observations of the sample, and at the same time they can allow for between-area variation. Concerning SAE, there are two main families of LMMs: area-level models and unit-level models. Fay and Herriot (1979) employed an area-level model in the USA to estimate per capita income for small areas. Battese, Harter and Fuller (1988) used a unit-level model to estimate county crop areas. These two pioneer applications of LMMs have been the source of further and fruitful research in the field of SAE.

This chapter gives an introduction to LMMs and to the most widely used fitting methods. Books dealing with LMMs include Searle et al. (1992), Longford (1995),

D. Morales et al., *A Course on Small Area Estimation and Mixed Models*, Statistics for Social and Behavioral Sciences, https://doi.org/10.1007/978-3-030-63757-6_6

McCullogh and Searle (2001), Goldstein (2003), Demidenko (2004) and Jiang (2007).

6.2 Linear Mixed Models with Known Variances

Let us start with models where the variance components are supposed to be known.

6.2.1 Introduction

Let us consider the model

$$y = X\beta + Zu + e, \quad (6.1)$$

where $y_{n\times 1}$ is the vector of observations, $\beta_{p\times 1}$ is the vector of fixed effects, $u_{q\times 1}$ is the vector of random effects, $X_{n\times p}$ and $Z_{n\times q}$ are the design matrices, and $e_{n\times 1}$ is the vector of model errors. Assume that sampling errors and random effects are independent and normally distributed with zero mean vectors and known variance matrices

$$\text{var}(u) = E[uu'] = V_u \quad \text{and} \quad \text{var}(e) = E[ee'] = V_e,$$

depending on a parameter σ containing the variance components. From (6.1) we obtain

$$V = \text{var}(y) = ZV_uZ' + V_e,$$

where V is assumed to be non-singular.

6.2.2 Least Squares Estimation of β

This section assumes that the variance components of model (6.1) are known. The random term is $Zu + e$, with variance $\text{var}(Zu + e) = ZV_uZ' + V_e = V$. We transform the model to have uncorrelated random terms and common variance equal to 1, i.e.

$$V^{-1/2}y = V^{-1/2}X\beta + V^{-1/2}(Zu + e).$$

Assuming that $\boldsymbol{y}^* = \boldsymbol{V}^{-1/2}\boldsymbol{y}$, $\boldsymbol{e}^* = \boldsymbol{V}^{-1/2}(\boldsymbol{Z}\boldsymbol{u}+\boldsymbol{e})$, and $\boldsymbol{X}^* = \boldsymbol{V}^{-1/2}\boldsymbol{X}$, the model is

$$\boldsymbol{y}^* = \boldsymbol{X}^*\boldsymbol{\beta} + \boldsymbol{e}^*,$$

with $\text{var}(\boldsymbol{e}^*) = \boldsymbol{V}^{-1/2}\text{var}(\boldsymbol{Z}\boldsymbol{u} + \boldsymbol{e})\boldsymbol{V}^{-1/2} = \boldsymbol{V}^{-1/2}\boldsymbol{V}\boldsymbol{V}^{-1/2} = \boldsymbol{I}_n$. Therefore, we can apply the ordinary least squares method, i.e.

$$\hat{\boldsymbol{\beta}} = \text{argmin}_{\boldsymbol{\beta}}\{\boldsymbol{e}^{*\prime}\boldsymbol{e}^*\}.$$

We observe that

$$\begin{aligned}\boldsymbol{e}^{*\prime}\boldsymbol{e}^* &= \left(\boldsymbol{V}^{-1/2}\boldsymbol{y} - \boldsymbol{V}^{-1/2}\boldsymbol{X}\boldsymbol{\beta}\right)'\left(\boldsymbol{V}^{-1/2}\boldsymbol{y} - \boldsymbol{V}^{-1/2}\boldsymbol{X}\boldsymbol{\beta}\right)\\ &= (\boldsymbol{y} - \boldsymbol{X}\boldsymbol{\beta})'\,\boldsymbol{V}^{-1}\,(\boldsymbol{y} - \boldsymbol{X}\boldsymbol{\beta}) = \boldsymbol{y}'\boldsymbol{V}^{-1}\boldsymbol{y} - 2\boldsymbol{\beta}'\boldsymbol{X}'\boldsymbol{V}^{-1}\boldsymbol{y} + \boldsymbol{\beta}'\boldsymbol{X}'\boldsymbol{V}^{-1}\boldsymbol{X}\boldsymbol{\beta}.\end{aligned}$$

By taking derivatives, we obtain

$$\frac{\partial \boldsymbol{e}^{*\prime}\boldsymbol{e}^*}{\partial \boldsymbol{\beta}} = -2\boldsymbol{X}'\boldsymbol{V}^{-1}\boldsymbol{y} + 2\boldsymbol{X}'\boldsymbol{V}^{-1}\boldsymbol{X}\boldsymbol{\beta}.$$

The normal equations are

$$\boldsymbol{X}'\boldsymbol{V}^{-1}\boldsymbol{X}\boldsymbol{\beta} = \boldsymbol{X}'\boldsymbol{V}^{-1}\boldsymbol{y} \tag{6.2}$$

and the solution is

$$\hat{\boldsymbol{\beta}} = (\boldsymbol{X}'\boldsymbol{V}^{-1}\boldsymbol{X})^{-1}\boldsymbol{X}'\boldsymbol{V}^{-1}\boldsymbol{y}, \tag{6.3}$$

when $\boldsymbol{X}'\boldsymbol{V}^{-1}\boldsymbol{X}$ and $\boldsymbol{V}$ are invertible. Under the assumed normality, $\hat{\boldsymbol{\beta}}$ is also the maximum likelihood estimator (MLE) of $\boldsymbol{\beta}$, i.e.

$$\hat{\boldsymbol{\beta}} = \underset{\boldsymbol{\beta}\in R^p}{\text{argmax}}\left\{-\frac{1}{2}(\boldsymbol{y} - \boldsymbol{X}\boldsymbol{\beta})'\boldsymbol{V}^{-1}(\boldsymbol{y} - \boldsymbol{X}\boldsymbol{\beta})\right\}.$$

6.2.3 BLUP of a Linear Combination of Effects

We consider the model (6.1) and we define the parameter of interest $\tau = \boldsymbol{a}_k'(\boldsymbol{X}_k\boldsymbol{\beta} + \boldsymbol{Z}_k\boldsymbol{u})$, where $\boldsymbol{a}_k$, $\boldsymbol{X}_k$, and $\boldsymbol{Z}_k$ are known vectors and matrices of sizes $k \times 1$, $k \times p$, and $k \times q$, respectively. Let $\hat{\tau} = \boldsymbol{g}'\boldsymbol{y} + g_0$ be a linear predictor of τ, where $\boldsymbol{g}_{n\times 1}$ and g_0 are such that

1. $\hat{\tau}$ is unbiased, i.e. $E[\tau] = \boldsymbol{a}_k'\boldsymbol{X}_k\boldsymbol{\beta}$ and $E[\hat{\tau}] = \boldsymbol{g}'\boldsymbol{X}\boldsymbol{\beta} + g_0$ are equal. Thus $g_0 = 0$ and $\boldsymbol{a}_k'\boldsymbol{X}_k = \boldsymbol{g}'\boldsymbol{X}$.
2. $\hat{\tau}$ minimizes the prediction error, i.e.

$$E[(\hat{\tau}-\tau)^2] = \text{var}(\hat{\tau}-\tau) = \text{var}(\boldsymbol{g}'\boldsymbol{y} - \boldsymbol{a}_k'\boldsymbol{X}_k\boldsymbol{\beta} - \boldsymbol{a}_k'\boldsymbol{Z}_k\boldsymbol{u}) = \text{var}(\boldsymbol{g}'\boldsymbol{y} - \boldsymbol{a}_k'\boldsymbol{Z}_k\boldsymbol{u})$$
$$= \boldsymbol{g}'\boldsymbol{V}\boldsymbol{g} + \boldsymbol{a}_k'\boldsymbol{Z}_k\boldsymbol{V}_u\boldsymbol{Z}_k'\boldsymbol{a}_k - 2\boldsymbol{g}'\boldsymbol{C}\boldsymbol{Z}_k'\boldsymbol{a}_k,$$

where $\boldsymbol{C} = \text{cov}(\boldsymbol{y}, \boldsymbol{u}) = \boldsymbol{Z}\boldsymbol{V}_u$.

The linear predictor $\hat{\tau}$ fulfilling the conditions (1) and (2) is called best linear unbiased predictor (BLUP) of τ.

Proposition 6.1 *Under the model (6.1), the BLUP of* $\tau = \boldsymbol{a}_k'(\boldsymbol{X}_k\boldsymbol{\beta} + \boldsymbol{Z}_k\boldsymbol{u})$ *is*

$$\hat{\tau} = \boldsymbol{a}_k'\left[\boldsymbol{X}_k\hat{\boldsymbol{\beta}} + \boldsymbol{Z}_k\boldsymbol{C}'\boldsymbol{V}^{-1}(\boldsymbol{y} - \boldsymbol{X}\hat{\boldsymbol{\beta}})\right], \quad \hat{\boldsymbol{\beta}} = (\boldsymbol{X}'\boldsymbol{V}^{-1}\boldsymbol{X})^{-1}\boldsymbol{X}'\boldsymbol{V}^{-1}\boldsymbol{y}. \tag{6.4}$$

Proof The problem to solve is

$$\text{minimize var}(\hat{\tau}-\tau), \qquad \text{restricted to } \boldsymbol{a}_k'\boldsymbol{X}_k = \boldsymbol{g}'\boldsymbol{X}.$$

Since $\boldsymbol{a}_k'\boldsymbol{Z}_k\boldsymbol{V}_u\boldsymbol{Z}_k'\boldsymbol{a}_k$ does not depend on $\boldsymbol{g}$, the Lagrangian function is

$$L(\boldsymbol{g}, \boldsymbol{\lambda}) = \boldsymbol{g}'\boldsymbol{V}\boldsymbol{g} - 2\boldsymbol{g}'\boldsymbol{C}\boldsymbol{Z}_k'\boldsymbol{a}_k + 2(\boldsymbol{g}'\boldsymbol{X} - \boldsymbol{a}_k'\boldsymbol{X}_k)\boldsymbol{\lambda}.$$

By taking partial derivatives with respect to $\boldsymbol{g}$ and $\boldsymbol{\lambda}$, we obtain

$$0 = \frac{\partial L(\boldsymbol{g}, \boldsymbol{\lambda})}{\partial \boldsymbol{g}} = 2\boldsymbol{V}\boldsymbol{g} - 2\boldsymbol{C}\boldsymbol{Z}_k'\boldsymbol{a}_k + 2\boldsymbol{X}\boldsymbol{\lambda} \iff \boldsymbol{V}\boldsymbol{g} + \boldsymbol{X}\boldsymbol{\lambda} = \boldsymbol{C}\boldsymbol{Z}_k'\boldsymbol{a}_k,$$
$$0 = \frac{\partial L(\boldsymbol{g}, \boldsymbol{\lambda})}{\partial \boldsymbol{\lambda}} = 2\boldsymbol{g}'\boldsymbol{X} - 2\boldsymbol{a}_k'\boldsymbol{X}_k \iff \boldsymbol{g}'\boldsymbol{X} = \boldsymbol{a}_k'\boldsymbol{X}_k.$$

In matrix form, the above equations are

$$\begin{pmatrix} \boldsymbol{V} & \boldsymbol{X} \\ \boldsymbol{X}' & \boldsymbol{0} \end{pmatrix}\begin{pmatrix} \boldsymbol{g} \\ \boldsymbol{\lambda} \end{pmatrix} = \begin{pmatrix} \boldsymbol{C}\boldsymbol{Z}_k'\boldsymbol{a}_k \\ \boldsymbol{X}_k'\boldsymbol{a}_k \end{pmatrix}.$$

If we apply the formula (cf. Appendix A)

$$\begin{pmatrix} \boldsymbol{A} & \boldsymbol{B} \\ \boldsymbol{B}' & \boldsymbol{C} \end{pmatrix}^{-1} = \begin{pmatrix} \boldsymbol{A}^{-1} & \boldsymbol{0} \\ \boldsymbol{0} & \boldsymbol{0} \end{pmatrix} + \begin{pmatrix} -\boldsymbol{A}^{-1}\boldsymbol{B} \\ \boldsymbol{I} \end{pmatrix}\left(\boldsymbol{C} - \boldsymbol{B}'\boldsymbol{A}^{-1}\boldsymbol{B}\right)^{-1}\left(-\boldsymbol{B}'\boldsymbol{A}^{-1}, \boldsymbol{I}\right),$$

with $A = V,\ B = X,\ C = 0$, then we obtain

$$\begin{pmatrix} V & X \\ X' & 0 \end{pmatrix}^{-1} = \begin{pmatrix} V^{-1} & 0 \\ 0 & 0 \end{pmatrix} - \begin{pmatrix} -V^{-1}X \\ I \end{pmatrix} \left(X'V^{-1}X\right)^{-1} \left(-X'V^{-1}, I\right)$$
$$= \begin{pmatrix} V^{-1} - V^{-1}X(X'V^{-1}X)^{-1}X'V^{-1} & V^{-1}X(X'V^{-1}X)^{-1} \\ (X'V^{-1}X)^{-1}X'V^{-1} & -(X'V^{-1}X)^{-1} \end{pmatrix}. \tag{6.5}$$

Therefore

$$\begin{pmatrix} g \\ \lambda \end{pmatrix} = \begin{pmatrix} V & X \\ X' & 0 \end{pmatrix}^{-1} \begin{pmatrix} CZ_k'a_k \\ X_k'a_k \end{pmatrix},$$

with

$$g = V^{-1}CZ_k'a_k - V^{-1}X(X'V^{-1}X)^{-1}X'V^{-1}CZ_k'a_k + V^{-1}X(X'V^{-1}X)^{-1}X_k'a_k.$$

The BLUP of τ is

$$\hat{\tau} = g'y = a_k'X_k\{(X'V^{-1}X)^{-1}X'V^{-1}y\} + a_k'Z_kC'V^{-1}y$$
$$- a_k'Z_kC'V^{-1}X\{(X'V^{-1}X)^{-1}X'V^{-1}y\} = a_k'\left[X_k\hat{\beta} + Z_kC'V^{-1}(y - X\hat{\beta})\right],$$

where

$$\hat{\beta} = (X'V^{-1}X)^{-1}X'V^{-1}y$$

is the least squares estimator of β. □

Let us note that in the sequel we will use the notation $E_{\beta}[u|y]$ to emphasize the dependence of the expectation on the parameter β.

Corollary 6.1 *Under the model (6.1), the BLUP of* u *is*

$$\hat{u} = V_uZ'V^{-1}\left(y - X\hat{\beta}\right) \tag{6.6}$$

and it holds $\hat{u} = E_{\hat{\beta}}[u|y]$.

Proof As $C = \text{cov}(y, u) = ZV_u$, by substituting $X_k = 0$, $a_k = 1_{(i)} = (0, \dots, 0, 1^{(i)}, 0, \dots, 0)'_{q\times 1}$, and $Z_k = I_{q\times q}$ in (6.4), we have $\hat{u}_i = 1'_{(i)}V_uZ'V^{-1}(y - X\hat{\beta})$, $i = 1, \dots, q$. This is to say, the BLUP of u is $\hat{u} = V_uZ'V^{-1}\left(y - X\hat{\beta}\right)$.

Second, we check that $\hat{u}$ is equal to $E_{\hat{\beta}}[u|y]$. Remind that if (u, y) has a multivariate normal distribution, then the distribution of u, given y, is also multivariate normal with the mean vector (see e.g. Theorem 2.2E in Rencher 1998)

$$E[\boldsymbol{u}|\boldsymbol{y}] = E[\boldsymbol{u}] + \text{cov}(\boldsymbol{u}, \boldsymbol{y})\,(\text{var}(\boldsymbol{y}))^{-1}\,(\boldsymbol{y} - E[\boldsymbol{y}]).$$

In the model (6.1), we have $E_{\boldsymbol{\beta}}[\boldsymbol{y}] = \boldsymbol{X\beta}$, $\text{var}(\boldsymbol{y}) = \boldsymbol{V}$, $E[\boldsymbol{u}] = 0$, and $\text{cov}(\boldsymbol{u}, \boldsymbol{y}) = \text{cov}(\boldsymbol{u}, \boldsymbol{X\beta} + \boldsymbol{Zu} + \boldsymbol{e}) = \text{var}(\boldsymbol{u})\boldsymbol{Z}' = \boldsymbol{V}_u\boldsymbol{Z}'$. Then

$$E_{\boldsymbol{\beta}}[\boldsymbol{u}|\boldsymbol{y}] = \boldsymbol{V}_u\boldsymbol{Z}'\boldsymbol{V}^{-1}(\boldsymbol{y} - \boldsymbol{X\beta}). \tag{6.7}$$

Therefore, $E_{\hat{\boldsymbol{\beta}}}[\boldsymbol{u}|\boldsymbol{y}]$ is equal to $\hat{\boldsymbol{u}}$ defined in (6.6). □

The BLUP (6.6) has the properties of being best in the sense that it minimizes $E[(\hat{\boldsymbol{u}}-\boldsymbol{u})'\boldsymbol{A}(\hat{\boldsymbol{u}}-\boldsymbol{u})]$ for any given positive definite matrix $\boldsymbol{A}$, it is linear with respect to $\boldsymbol{y}$ and it is unbiased, i.e. $E[\hat{\boldsymbol{u}}-\boldsymbol{u}] = \boldsymbol{0}$. See e.g. Section 7.2 of Searle et al. (1992).

As $\boldsymbol{e} \sim N(\boldsymbol{0}, \boldsymbol{V}_e)$ and $\boldsymbol{u} \sim N(\boldsymbol{0}, \boldsymbol{V}_u)$, the joint probability density of $\boldsymbol{y}$ and $\boldsymbol{u}$ is

$$\begin{aligned} f(\boldsymbol{y}, \boldsymbol{u}) &= f(\boldsymbol{y}|\,\boldsymbol{u})f(\boldsymbol{u}) \\ &= c\exp\left\{-\frac{1}{2}(\boldsymbol{y} - \boldsymbol{X\beta} - \boldsymbol{Zu})'\boldsymbol{V}_e^{-1}(\boldsymbol{y} - \boldsymbol{X\beta} - \boldsymbol{Zu})\right\}\exp\left\{-\frac{1}{2}\boldsymbol{u}'\boldsymbol{V}_u^{-1}\boldsymbol{u}\right\}, \end{aligned}$$

where c is a positive constant. Henderson et al. (1959) and Henderson (1975) gave a set of equations for obtaining simultaneously $\hat{\boldsymbol{\beta}}$ and $\hat{\boldsymbol{u}}$. The following result summarizes some of their mathematical derivations.

Proposition 6.2 *It holds that* $\hat{\boldsymbol{\beta}} = \tilde{\boldsymbol{\beta}}$ *and* $\hat{\boldsymbol{u}} = \tilde{\boldsymbol{u}}$, *where*

$$(\tilde{\boldsymbol{\beta}}, \tilde{\boldsymbol{u}}) = \textit{argmax}_{\boldsymbol{\beta},\boldsymbol{u}}\, f(\boldsymbol{y}, \boldsymbol{u}).$$

Proof The log-density is

$$\log f(\boldsymbol{y}, \boldsymbol{u}) = \log c - \frac{1}{2}(\boldsymbol{y} - \boldsymbol{X\beta} - \boldsymbol{Zu})'\boldsymbol{V}_e^{-1}(\boldsymbol{y} - \boldsymbol{X\beta} - \boldsymbol{Zu}) - \frac{1}{2}\boldsymbol{u}'\boldsymbol{V}_u^{-1}\boldsymbol{u}.$$

By taking derivatives with respect to $\boldsymbol{\beta}$ and equating to zero, we get

$$\boldsymbol{X}'\boldsymbol{V}_e^{-1}\boldsymbol{X\beta} + \boldsymbol{X}'\boldsymbol{V}_e^{-1}\boldsymbol{Zu} = \boldsymbol{X}'\boldsymbol{V}_e^{-1}\boldsymbol{y}. \tag{6.8}$$

By taking derivatives with respect to $\boldsymbol{u}$ and equating to zero, we get

$$\boldsymbol{Z}'\boldsymbol{V}_e^{-1}\boldsymbol{X\beta} + \left(\boldsymbol{Z}'\boldsymbol{V}_e^{-1}\boldsymbol{Z} + \boldsymbol{V}_u^{-1}\right)\boldsymbol{u} = \boldsymbol{Z}'\boldsymbol{V}_e^{-1}\boldsymbol{y}. \tag{6.9}$$

Let $\tilde{\boldsymbol{\beta}}$, $\tilde{\boldsymbol{u}}$ be the solution of the system of Eqs. (6.8) and (6.9). First, we check that $\tilde{\boldsymbol{\beta}}$ is equal to $\hat{\boldsymbol{\beta}}$ obtained from (6.3). From (6.9), we have

$$\tilde{\boldsymbol{u}} = \left(\boldsymbol{Z}'\boldsymbol{V}_e^{-1}\boldsymbol{Z} + \boldsymbol{V}_u^{-1}\right)^{-1}\boldsymbol{Z}'\boldsymbol{V}_e^{-1}\left(\boldsymbol{y} - \boldsymbol{X}\tilde{\boldsymbol{\beta}}\right). \tag{6.10}$$

By substituting $\tilde{\boldsymbol{u}}$ in (6.8), we obtain

$$\boldsymbol{X}'\boldsymbol{M}\boldsymbol{X}\tilde{\boldsymbol{\beta}} = \boldsymbol{X}'\boldsymbol{M}\boldsymbol{y}, \tag{6.11}$$

where $\boldsymbol{M} = \boldsymbol{V}_e^{-1} - \boldsymbol{V}_e^{-1}\boldsymbol{Z}\left(\boldsymbol{Z}'\boldsymbol{V}_e^{-1}\boldsymbol{Z} + \boldsymbol{V}_u^{-1}\right)^{-1}\boldsymbol{Z}'\boldsymbol{V}_e^{-1}$. Note that

$$\begin{aligned}
\boldsymbol{M}\boldsymbol{V} &= \left(\boldsymbol{V}_e^{-1} - \boldsymbol{V}_e^{-1}\boldsymbol{Z}\left(\boldsymbol{Z}'\boldsymbol{V}_e^{-1}\boldsymbol{Z} + \boldsymbol{V}_u^{-1}\right)^{-1}\boldsymbol{Z}'\boldsymbol{V}_e^{-1}\right)\left(\boldsymbol{Z}\boldsymbol{V}_u\boldsymbol{Z}' + \boldsymbol{V}_e\right)\\
&= \boldsymbol{V}_e^{-1}\boldsymbol{Z}\boldsymbol{V}_u\boldsymbol{Z}' + \boldsymbol{I} - \boldsymbol{V}_e^{-1}\boldsymbol{Z}\left(\boldsymbol{Z}'\boldsymbol{V}_e^{-1}\boldsymbol{Z} + \boldsymbol{V}_u^{-1}\right)^{-1}\left(\boldsymbol{Z}'\boldsymbol{V}_e^{-1}\boldsymbol{Z} + \boldsymbol{V}_u^{-1}\right)\boldsymbol{V}_u\boldsymbol{Z}'\\
&= \boldsymbol{V}_e^{-1}\boldsymbol{Z}\boldsymbol{V}_u\boldsymbol{Z}' + \boldsymbol{I} - \boldsymbol{V}_e^{-1}\boldsymbol{Z}\boldsymbol{V}_u\boldsymbol{Z}' = \boldsymbol{I},
\end{aligned}$$

so that $\boldsymbol{M} = \boldsymbol{V}^{-1}$ and (6.11) coincides with (6.2). This is to say, $\tilde{\boldsymbol{\beta}} = \hat{\boldsymbol{\beta}}$.

Second, we check that $\tilde{\boldsymbol{u}}$ defined in (6.10) is equal to $\hat{\boldsymbol{u}} = E_{\hat{\boldsymbol{\beta}}}[\boldsymbol{u}|\boldsymbol{y}]$. From (6.10) we get

$$\begin{aligned}
\tilde{\boldsymbol{u}} &= (\boldsymbol{Z}'\boldsymbol{V}_e^{-1}\boldsymbol{Z} + \boldsymbol{V}_u^{-1})^{-1}\boldsymbol{Z}'\boldsymbol{V}_e^{-1}(\boldsymbol{y} - \boldsymbol{X}\hat{\boldsymbol{\beta}})\\
&= \left(\boldsymbol{Z}'\boldsymbol{V}_e^{-1}\boldsymbol{Z} + \boldsymbol{V}_u^{-1}\right)^{-1}\boldsymbol{Z}'\boldsymbol{V}_e^{-1}\boldsymbol{V}\boldsymbol{V}^{-1}(\boldsymbol{y} - \boldsymbol{X}\hat{\boldsymbol{\beta}})\\
&= \left(\boldsymbol{Z}'\boldsymbol{V}_e^{-1}\boldsymbol{Z} + \boldsymbol{V}_u^{-1}\right)^{-1}\boldsymbol{Z}'\boldsymbol{V}_e^{-1}(\boldsymbol{Z}\boldsymbol{V}_u\boldsymbol{Z}' + \boldsymbol{V}_e)\boldsymbol{V}^{-1}(\boldsymbol{y} - \boldsymbol{X}\hat{\boldsymbol{\beta}})\\
&= \left(\boldsymbol{Z}'\boldsymbol{V}_e^{-1}\boldsymbol{Z} + \boldsymbol{V}_u^{-1}\right)^{-1}(\boldsymbol{Z}'\boldsymbol{V}_e^{-1}\boldsymbol{Z} + \boldsymbol{V}_u^{-1})\boldsymbol{V}_u\boldsymbol{Z}'\boldsymbol{V}^{-1}(\boldsymbol{y} - \boldsymbol{X}\hat{\boldsymbol{\beta}})\\
&= \boldsymbol{V}_u\boldsymbol{Z}'\boldsymbol{V}^{-1}(\boldsymbol{y} - \boldsymbol{X}\hat{\boldsymbol{\beta}}).
\end{aligned}$$

By combining this last result with (6.6), we obtain $\tilde{\boldsymbol{u}} = \hat{\boldsymbol{u}}$. □

The summary of this section is that the BLUE of $\boldsymbol{\beta}$ and the BLUP of $\boldsymbol{u}$ are

$$\hat{\boldsymbol{\beta}} = (\boldsymbol{X}'\boldsymbol{V}^{-1}\boldsymbol{X})^{-1}\boldsymbol{X}'\boldsymbol{V}^{-1}\boldsymbol{y}, \quad \hat{\boldsymbol{u}} = \boldsymbol{V}_u\boldsymbol{Z}'\boldsymbol{V}^{-1}\left(\boldsymbol{y} - \boldsymbol{X}\hat{\boldsymbol{\beta}}\right). \tag{6.12}$$

For more details see Searle (1971), 458–462, or chapter 7 of Searle et al. (1992).

6.3 Linear Mixed Models with Unknown Variances

Let us consider the linear mixed model

$$\boldsymbol{y} = \boldsymbol{X}\boldsymbol{\beta} + \boldsymbol{Z}_1\boldsymbol{u}_1 + \ldots + \boldsymbol{Z}_m\boldsymbol{u}_m + \boldsymbol{e}, \tag{6.13}$$

where $\boldsymbol{y} = (y_1, \ldots, y_n)'$ is the vector of sample observations, $\boldsymbol{\beta} = (\beta_1, \ldots, \beta_p)'$ is the vector of fixed effects, and $\boldsymbol{u}_i = (u_{i1}, \ldots, u_{iq_i})'$ is the vector containing the effects of the q_i levels of the i-th random factor, $i = 1, \ldots, m$. Finally, $\boldsymbol{e} = (e_1, \ldots, e_n)'$ is the vector of sampling errors and $\boldsymbol{X}, \boldsymbol{Z}_1, \ldots, \boldsymbol{Z}_m$ are design matrices with dimensions $n \times p$, $n \times q_1, \ldots, n \times q_m$, respectively.

The model (6.13) can be written in the form (6.1) if we define

$$\boldsymbol{Z} = (\boldsymbol{Z}_1, \ldots, \boldsymbol{Z}_m), \quad \boldsymbol{u} = (\boldsymbol{u}_1', \ldots, \boldsymbol{u}_m')' \text{ and } q = \sum_{i=1}^{m} q_i.$$

The following assumptions ensure that the model parameters are estimable.

(F1) $\boldsymbol{u}_1, \ldots, \boldsymbol{u}_m, \boldsymbol{e}$ are independent, and

$$\boldsymbol{e} \sim N_n(\boldsymbol{0}, \sigma_0^2 \boldsymbol{\Sigma}_e), \quad \boldsymbol{u}_i \sim N_{q_i}(\boldsymbol{0}, \sigma_i^2 \boldsymbol{\Sigma}_{u_i}),\ i = 1, \ldots, m,$$

with $\boldsymbol{\Sigma}_e$ and $\boldsymbol{\Sigma}_{u_i}$, $i = 1, \ldots, m$, known.

(F2) $\mathrm{r}(\boldsymbol{X}) = p$.

The assumption (F2) always holds if an adequate re-parametrization of the model is done. The next hypothesis states that the number of observations should be greater than the number of parameters.

(F3) $n \geq p + m + 1$.

If assumption (F4) on matrix ranks holds, then the fixed effects are not confused with the random effects of any factors.

(F4) $\mathrm{r}(\boldsymbol{X}, \boldsymbol{Z}_i) > p$, $i = 1, \ldots, m$.

Assumption (F5) ensures that random effects of a factor are not confused with random effects of other factors. Let $\boldsymbol{G}_0 = \boldsymbol{\Sigma}_e$ and $\boldsymbol{G}_i = \boldsymbol{Z}_i \boldsymbol{\Sigma}_{u_i} \boldsymbol{Z}_i'$, $i = 1, \ldots, m$.

(F5) $\boldsymbol{G}_0, \boldsymbol{G}_1, \ldots, \boldsymbol{G}_m$ are linearly independent, i.e.

$$\sum_{i=0}^{m} \alpha_i \boldsymbol{G}_i = \boldsymbol{0} \Longrightarrow \alpha_i = 0,\ i = 0, 1, \ldots, m.$$

Finally, assumption (F6) states that $\boldsymbol{Z}_i$, $i = 1, \ldots, m$, are standard design matrices. This assumption implies that $\boldsymbol{Z}_i'\boldsymbol{Z}_i$ is a $q_i \times q_i$ non-singular diagonal matrix, $\mathrm{r}(\boldsymbol{Z}_i) = q_i$ and $q_i \leq n$, $i = 1, \ldots, m$.

(F6) The matrix $\boldsymbol{Z}_i$ has only 0's and 1's, $i = 1, \ldots, m$. In each row there is exactly one element equal to 1, and in each column there is at least one element equal to 1.

Another consequence of the previous assumptions is that

$$\boldsymbol{y} \sim N_n(\boldsymbol{X\beta}, \boldsymbol{V}), \text{ with } \boldsymbol{V} = \sum_{i=0}^{m} \sigma_i^2 \boldsymbol{G}_i.$$

When necessary, we will emphasize the dependency of $\boldsymbol{V}$ on $\boldsymbol{\sigma} = (\sigma_0^2, \sigma_1^2, \ldots, \sigma_m^2)'$ by writing $\boldsymbol{V}(\boldsymbol{\sigma})$. Let $M = p+m+1$ and let $\boldsymbol{\theta}' = (\boldsymbol{\beta}', \boldsymbol{\sigma}')$ be the vector of unknown parameters. The parameter space is

$$\Theta = \left\{\boldsymbol{\theta}' = (\boldsymbol{\beta}', \boldsymbol{\sigma}') \in R^M : \ \boldsymbol{\beta} \in R^p, \ \sigma_0^2 > 0, \ \sigma_i^2 \geq 0, \ i = 1, \ldots, m\right\}.$$

6.4 Maximum Likelihood Estimation

In this section we derive maximum likelihood estimators of the parameters of the model defined in Sect. 6.3.

6.4.1 Description of the Method

Under the linear mixed model (6.13), the joint density function of $\boldsymbol{y}$, given $\boldsymbol{\theta}$, is

$$f_{\boldsymbol{\theta}}(\boldsymbol{y}) = (2\pi)^{-n/2}|\boldsymbol{V}|^{-1/2} \exp\left\{-\frac{1}{2}(\boldsymbol{y} - \boldsymbol{X}\boldsymbol{\beta})'\boldsymbol{V}^{-1}(\boldsymbol{y} - \boldsymbol{X}\boldsymbol{\beta})\right\}. \qquad (6.14)$$

The maximum likelihood (ML) estimator $\hat{\boldsymbol{\theta}} = (\hat{\beta}_1, \ldots, \hat{\beta}_p, \hat{\sigma}_0^2, \ldots, \hat{\sigma}_m^2)'$ of $\boldsymbol{\theta}$ satisfies

$$\hat{\boldsymbol{\theta}} = \text{argmax}_{\boldsymbol{\theta}\in\Theta} f_{\boldsymbol{\theta}}(\boldsymbol{y}) = \text{argmax}_{\boldsymbol{\theta}\in\Theta}\, l(\boldsymbol{\theta}),$$

where $f_{\boldsymbol{\theta}}(\boldsymbol{y})$ is defined in (6.14) and $l(\boldsymbol{\theta}) = \log f_{\boldsymbol{\theta}}(\boldsymbol{y})$is the log-likelihood. We denote the vector of scores as $\boldsymbol{S}(\boldsymbol{\theta}) = (\boldsymbol{S}_{\boldsymbol{\beta}}, S_{\sigma_0^2}, \ldots, S_{\sigma_m^2})'$, so that

$$\boldsymbol{S}(\boldsymbol{\theta}) = \frac{\partial l(\boldsymbol{\theta})}{\partial \boldsymbol{\theta}} = \left(\frac{\partial l(\boldsymbol{\theta})}{\partial \boldsymbol{\beta}}, \frac{\partial l(\boldsymbol{\theta})}{\partial \sigma_0^2}, \ldots, \frac{\partial l(\boldsymbol{\theta})}{\partial \sigma_m^2}\right)'.$$

If $\hat{\boldsymbol{\theta}}$ exists in the interior of Θ, then it is the solution of the likelihood equations which are obtained by equating to zero the components of the vector of scores. By deriving the log-likelihood with respect to the parameters we obtain the scores of model (6.13), i.e.

$$\boldsymbol{S}_{\boldsymbol{\beta}} = \boldsymbol{X}'\boldsymbol{V}^{-1}(\boldsymbol{y} - \boldsymbol{X}\boldsymbol{\beta}), \qquad (6.15)$$

$$S_{\sigma_i^2} = -\frac{1}{2}\frac{\partial \log|\boldsymbol{V}|}{\partial \sigma_i^2} - \frac{1}{2}(\boldsymbol{y} - \boldsymbol{X}\boldsymbol{\beta})'\frac{\partial \boldsymbol{V}^{-1}}{\partial \sigma_i^2}(\boldsymbol{y} - \boldsymbol{X}\boldsymbol{\beta}), \ \ i = 0, 1, \ldots, m.$$

We recall that (cf. Appendix A)

$$\frac{\partial \log |\boldsymbol{V}|}{\partial \sigma_i^2} = \text{tr}\left(\boldsymbol{V}^{-1}\frac{\partial \boldsymbol{V}}{\partial \sigma_i^2}\right), \quad \frac{\partial \boldsymbol{V}^{-1}}{\partial \sigma_i^2} = -\boldsymbol{V}^{-1}\frac{\partial \boldsymbol{V}}{\partial \sigma_i^2}\boldsymbol{V}^{-1}. \tag{6.16}$$

Since $\partial \boldsymbol{V}/\partial \sigma_i^2 = \boldsymbol{G}_i$, we have

$$S_{\sigma_i^2} = -\frac{1}{2}\text{tr}(\boldsymbol{V}^{-1}\boldsymbol{G}_i) + \frac{1}{2}(\boldsymbol{y} - \boldsymbol{X\beta})'\boldsymbol{V}^{-1}\boldsymbol{G}_i\boldsymbol{V}^{-1}(\boldsymbol{y} - \boldsymbol{X\beta}), \ i = 0, 1, \dots, m. \tag{6.17}$$

By equating (6.15) and (6.17) to zero, we obtain the likelihood equations

$$\boldsymbol{X}'\boldsymbol{V}^{-1}\boldsymbol{X\beta} = \boldsymbol{X}'\boldsymbol{V}^{-1}\boldsymbol{y},$$
$$\text{tr}(\boldsymbol{V}^{-1}\boldsymbol{G}_i) = (\boldsymbol{y} - \boldsymbol{X\beta})'\boldsymbol{V}^{-1}\boldsymbol{G}_i\boldsymbol{V}^{-1}(\boldsymbol{y} - \boldsymbol{X\beta}), \ i = 0, 1, \dots, m.$$

The likelihood equations have no explicit solutions. The Newton–Raphson or the Fisher-Scoring algorithms solve these equations iteratively, starting with an initial value $\boldsymbol{\theta}^{(0)}$. In each iteration, the Newton–Raphson method updates the estimator of $\boldsymbol{\theta}$ by using the formula

$$\boldsymbol{\theta}^{(r+1)} = \boldsymbol{\theta}^{(r)} - \boldsymbol{H}(\boldsymbol{\theta}^{(r)})^{-1}\boldsymbol{S}(\boldsymbol{\theta}^{(r)}),$$

where $\boldsymbol{S}(\boldsymbol{\theta}^{(r)})$ is the vector of scores and $\boldsymbol{H}(\boldsymbol{\theta}^{(r)})$ is the Hessian matrix of $l(\boldsymbol{\theta})$, both calculated at the value $\boldsymbol{\theta}^{(r)}$ obtained at the current iteration r. The elements of the Hessian matrix are obtained by taking second derivatives of the log-likelihood and by using (6.16). We apply $\partial \boldsymbol{V}/\partial \sigma_i^2 = \boldsymbol{G}_i$, $\partial \boldsymbol{V}^{-1}/\partial \sigma_i^2 = -\boldsymbol{V}^{-1}(\partial \boldsymbol{V}/\partial \sigma_i^2)\boldsymbol{V}^{-1}$ and the property that the derivative of the trace of a matrix is the trace of the derivative of the matrix. For $i, j = 0, 1, \dots, m$, we get

$$\frac{\partial^2 l(\boldsymbol{\theta})}{\partial \boldsymbol{\beta}\partial \boldsymbol{\beta}'} = -\boldsymbol{X}'\boldsymbol{V}^{-1}\boldsymbol{X}, \tag{6.18}$$

$$\frac{\partial^2 l(\boldsymbol{\theta})}{\partial \sigma_i^2 \partial \boldsymbol{\beta}} = \frac{\partial^2 l(\boldsymbol{\theta})}{\partial \boldsymbol{\beta}\partial \sigma_i^2} = -\boldsymbol{X}'\boldsymbol{V}^{-1}\boldsymbol{G}_i\boldsymbol{V}^{-1}(\boldsymbol{y} - \boldsymbol{X\beta}), \tag{6.19}$$

$$\begin{aligned}\frac{\partial^2 l(\boldsymbol{\theta})}{\partial \sigma_j^2 \partial \sigma_i^2} &= \frac{1}{2}\text{tr}(\boldsymbol{V}^{-1}\boldsymbol{G}_j\boldsymbol{V}^{-1}\boldsymbol{G}_i) \\ &\quad - \frac{1}{2}(\boldsymbol{y} - \boldsymbol{X\beta})'[\boldsymbol{V}^{-1}\boldsymbol{G}_j\boldsymbol{V}^{-1}\boldsymbol{G}_i\boldsymbol{V}^{-1} + \boldsymbol{V}^{-1}\boldsymbol{G}_i\boldsymbol{V}^{-1}\boldsymbol{G}_j\boldsymbol{V}^{-1}](\boldsymbol{y} - \boldsymbol{X\beta}) \\ &= \frac{1}{2}\text{tr}(\boldsymbol{V}^{-1}\boldsymbol{G}_j\boldsymbol{V}^{-1}\boldsymbol{G}_i) - (\boldsymbol{y} - \boldsymbol{X\beta})'\boldsymbol{V}^{-1}\boldsymbol{G}_j\boldsymbol{V}^{-1}\boldsymbol{G}_i\boldsymbol{V}^{-1}(\boldsymbol{y} - \boldsymbol{X\beta}).\end{aligned} \tag{6.20}$$

The Fisher-scoring algorithm for calculating the ML estimators of the model parameters replaces the Hessian matrix by its expectation with the sign changed, that is, the Fisher information matrix. The updating equation of the Fisher scoring is

$$\boldsymbol{\theta}^{(r+1)} = \boldsymbol{\theta}^{(r)} + \boldsymbol{F}(\boldsymbol{\theta}^{(r)})^{-1}\boldsymbol{S}(\boldsymbol{\theta}^{(r)}),$$

where $\boldsymbol{F}(\boldsymbol{\theta}^{(r)})$ is the Fisher information matrix defined by

$$\boldsymbol{F}(\boldsymbol{\theta}) = -E[\boldsymbol{H}(\boldsymbol{\theta})]$$

and evaluated at $\boldsymbol{\theta}^{(r)}$. Taking expectations in (6.18)–(6.20), changing the sign, and using the result (cf. Appendix A)

$$E[(\boldsymbol{y} - \boldsymbol{X\beta})'\boldsymbol{A}(\boldsymbol{y} - \boldsymbol{X\beta})] = \text{tr}\{\boldsymbol{AV}\},$$

valid for any non-random matrix $\boldsymbol{A}$, we get the elements of the Fisher information matrix

$$\boldsymbol{F}_{\beta\beta} = \boldsymbol{X}'\boldsymbol{V}^{-1}\boldsymbol{X}, \quad \boldsymbol{F}_{\sigma_i^2\beta} = \boldsymbol{F}_{\beta\sigma_i^2} = \boldsymbol{0}, \quad F_{\sigma_j^2\sigma_i^2} = \frac{1}{2}\text{tr}\{\boldsymbol{V}^{-1}\boldsymbol{G}_i\boldsymbol{V}^{-1}\boldsymbol{G}_j\}, \tag{6.21}$$

for $i, j = 0, 1, \dots, m$. Thus, the Fischer information matrix has the form

$$\boldsymbol{F}(\boldsymbol{\theta}) = \begin{pmatrix} \boldsymbol{F}_{\beta\beta} & \boldsymbol{0} & \boldsymbol{0} & \cdots & \boldsymbol{0} \\ \boldsymbol{0} & F_{\sigma_0^2\sigma_0^2} & F_{\sigma_0^2\sigma_1^2} & \cdots & F_{\sigma_0^2\sigma_m^2} \\ \boldsymbol{0} & F_{\sigma_1^2\sigma_0^2} & F_{\sigma_1^2\sigma_1^2} & \cdots & F_{\sigma_1^2\sigma_m^2} \\ \vdots & \vdots & \vdots & \ddots & \vdots \\ \boldsymbol{0} & F_{\sigma_m^2\sigma_0^2} & F_{\sigma_m^2\sigma_1^2} & \cdots & F_{\sigma_m^2\sigma_m^2} \end{pmatrix} = \begin{pmatrix} \boldsymbol{F}_1(\boldsymbol{\sigma}) & \boldsymbol{0} \\ \boldsymbol{0} & \boldsymbol{F}_2(\boldsymbol{\sigma}) \end{pmatrix}. \tag{6.22}$$

We can split the score vector in the same manner, i.e. $\boldsymbol{S}(\boldsymbol{\beta}, \boldsymbol{\sigma}) = (\boldsymbol{S}_1'(\boldsymbol{\beta}, \boldsymbol{\sigma}), \boldsymbol{S}_2'(\boldsymbol{\beta}, \boldsymbol{\sigma}))'$, where $\boldsymbol{S}_1(\boldsymbol{\beta}, \boldsymbol{\sigma}) = \boldsymbol{S}_\beta$ and $\boldsymbol{S}_2(\boldsymbol{\beta}, \boldsymbol{\sigma}) = (S_{\sigma_0^2}, \dots, S_{\sigma_m^2})'$ have the components calculated in (6.15) and (6.17), respectively. The block structure of matrix $\boldsymbol{F}(\boldsymbol{\theta})$ allows to separate the updating equation in two equations

$$\boldsymbol{\beta}^{(r+1)} = \boldsymbol{\beta}^{(r)} + \boldsymbol{F}_1^{-1}(\boldsymbol{\sigma}^{(r)})\boldsymbol{S}_1(\boldsymbol{\beta}^{(r)}, \boldsymbol{\sigma}^{(r)}), \quad \boldsymbol{\sigma}^{(r+1)} = \boldsymbol{\sigma}^{(r)} + \boldsymbol{F}_2^{-1}(\boldsymbol{\sigma}^{(r)})\boldsymbol{S}_2(\boldsymbol{\beta}^{(r)}, \boldsymbol{\sigma}^{(r)}),$$

with the additional simplification

$$\begin{aligned} \boldsymbol{\beta}^{(r+1)} &= \boldsymbol{\beta}^{(r)} + (\boldsymbol{X}'\boldsymbol{V}^{-1}(\boldsymbol{\sigma}^{(r)})\boldsymbol{X})^{-1}\boldsymbol{X}'\boldsymbol{V}^{-1}(\boldsymbol{\sigma}^{(r)})(\boldsymbol{y} - \boldsymbol{X}\boldsymbol{\beta}^{(r)}) \\ &= (\boldsymbol{X}'\boldsymbol{V}^{-1}(\boldsymbol{\sigma}^{(r)})\boldsymbol{X})^{-1}\boldsymbol{X}'\boldsymbol{V}^{-1}(\boldsymbol{\sigma}^{(r)})\boldsymbol{y}. \end{aligned}$$

Under regularity conditions (see e.g. Sections 1.3.1 and 1.8 of Jiang 2007), the ML estimator $\hat{\boldsymbol{\theta}}$ is consistent and asymptotically normal with asymptotic covariance matrix equal to $\boldsymbol{F}^{-1}(\boldsymbol{\theta})$. From (6.18) and (6.22), the asymptotic distributions of the ML estimators are

$$\hat{\boldsymbol{\beta}} \sim N_p\big(\boldsymbol{\beta}, (\boldsymbol{X}'\boldsymbol{V}^{-1}\boldsymbol{X})^{-1}\big), \quad \hat{\boldsymbol{\sigma}} \sim N_m\big(\boldsymbol{\sigma}, \boldsymbol{F}_2^{-1}(\boldsymbol{\sigma})\big).$$

6.4.2 *Maximum Likelihood Estimators for Alternative Parameters*

From computational reasons, it is sometimes useful to use the alternative parameters

$$\sigma^2 = \sigma_0^2, \quad \varphi_i = \sigma_i^2/\sigma_0^2, \quad i = 1, \ldots, m,$$

in the model (6.13). Let us define $\boldsymbol{\varphi}' = (\sigma^2, \varphi_1, \ldots, \varphi_m)$, $\boldsymbol{\theta}' = (\boldsymbol{\beta}', \boldsymbol{\varphi}')$, and $\boldsymbol{V} = \sigma^2\left(\boldsymbol{\Sigma}_e + \sum_{i=1}^m \varphi_i \boldsymbol{G}_i\right) = \sigma^2\boldsymbol{\Sigma}$. The joint density function of $\boldsymbol{y}$, given $\boldsymbol{\theta}$, is

$$f_{\boldsymbol{\theta}}(\boldsymbol{y}) = (2\pi)^{-n/2}(\sigma^2)^{-n/2}|\boldsymbol{\Sigma}|^{-1/2}\exp\left\{-\frac{1}{2\sigma^2}(\boldsymbol{y} - \boldsymbol{X\beta})'\boldsymbol{\Sigma}^{-1}(\boldsymbol{y} - \boldsymbol{X\beta})\right\}.$$

The log-likelihood function is

$$l(\boldsymbol{\theta}) = -\frac{n}{2}\log 2\pi - \frac{n}{2}\log\sigma^2 - \frac{1}{2}\log|\boldsymbol{\Sigma}| - \frac{1}{2\sigma^2}(\boldsymbol{y} - \boldsymbol{X\beta})'\boldsymbol{\Sigma}^{-1}(\boldsymbol{y} - \boldsymbol{X\beta}).$$

The components of the vector of scores are

$$S_{\boldsymbol{\beta}} = \frac{1}{\sigma^2}\boldsymbol{X}'\boldsymbol{\Sigma}^{-1}(\boldsymbol{y} - \boldsymbol{X\beta}), \; S_{\sigma^2} = -\frac{n}{2\sigma^2} + \frac{1}{2\sigma^4}(\boldsymbol{y} - \boldsymbol{X\beta})'\boldsymbol{\Sigma}^{-1}(\boldsymbol{y} - \boldsymbol{X\beta}), \tag{6.23}$$

$$S_{\varphi_i} = -\frac{1}{2}\text{tr}(\boldsymbol{\Sigma}^{-1}\boldsymbol{G}_i) + \frac{1}{2\sigma^2}(\boldsymbol{y} - \boldsymbol{X\beta})'\boldsymbol{\Sigma}^{-1}\boldsymbol{G}_i\boldsymbol{\Sigma}^{-1}(\boldsymbol{y} - \boldsymbol{X\beta}), \; i = 1, \ldots, m. \tag{6.24}$$

By putting $S_{\boldsymbol{\beta}} = \boldsymbol{0}$ and $S_{\sigma^2} = \boldsymbol{0}$ we obtain

$$\boldsymbol{\beta} = (\boldsymbol{X}'\boldsymbol{\Sigma}^{-1}\boldsymbol{X})^{-1}\boldsymbol{X}'\boldsymbol{\Sigma}^{-1}\boldsymbol{y} \quad \text{and} \quad \sigma^2 = \frac{1}{n}(\boldsymbol{y} - \boldsymbol{X\beta})'\boldsymbol{\Sigma}^{-1}(\boldsymbol{y} - \boldsymbol{X\beta}).$$

The second partial derivatives of the log-likelihood function are

$$
\begin{aligned}
&H_{\beta\beta} = -\tfrac{1}{\sigma^2}X'\Sigma^{-1}X, && H_{\beta\sigma^2} = -\tfrac{1}{\sigma^4}X'\Sigma^{-1}(y - X\beta),\\
&H_{\beta\varphi_i} = -\tfrac{1}{\sigma^2}X'\Sigma^{-1}G_i\Sigma^{-1}(y - X\beta), && H_{\sigma^2\sigma^2} = \tfrac{n}{2\sigma^4} - \tfrac{1}{\sigma^6}(y - X\beta)'\Sigma^{-1}(y - X\beta),\\
&H_{\sigma^2\varphi_i} = -\tfrac{1}{2\sigma^4}(y - X\beta)'\Sigma^{-1}G_i\Sigma^{-1}(y - X\beta),\\
&H_{\varphi_i\varphi_j} = \tfrac{1}{2}\text{tr}(\Sigma^{-1}G_j\Sigma^{-1}G_i) - \tfrac{1}{\sigma^2}(y - X\beta)'\Sigma^{-1}G_j\Sigma^{-1}G_i\Sigma^{-1}(y - X\beta).
\end{aligned}
$$

By taking expectations and changing the sign, we obtain the elements of the Fisher information matrix, i.e.

$$
\begin{aligned}
&F_{\beta\beta} = \tfrac{1}{\sigma^2}X'\Sigma^{-1}X, \quad F_{\beta\sigma^2} = 0, && F_{\beta\varphi_i} = 0,\\
&F_{\sigma^2\sigma^2} = \tfrac{n}{2\sigma^4}, \quad F_{\sigma^2\varphi_i} = \tfrac{1}{2\sigma^2}\,\text{tr}(\Sigma^{-1}G_i), && F_{\varphi_i\varphi_j} = \tfrac{1}{2}\,\text{tr}(\Sigma^{-1}G_j\Sigma^{-1}G_i).
\end{aligned}
\tag{6.25}
$$

Again, we can write

$$
F(\theta) = \begin{pmatrix} F_{\beta\beta} & 0 & 0 & \cdots & 0\\ 0 & F_{\sigma^2\sigma^2} & F_{\sigma^2\varphi_1} & \cdots & F_{\sigma^2\varphi_m}\\ 0 & F_{\varphi_1\sigma^2} & F_{\varphi_1\varphi_1} & \cdots & F_{\varphi_1\varphi_m}\\ \vdots & \vdots & \vdots & \ddots & \vdots\\ 0 & F_{\varphi_m\sigma^2} & F_{\varphi_m\varphi_1} & \cdots & F_{\varphi_m\varphi_m} \end{pmatrix} = \begin{pmatrix} F_1(\varphi) & 0\\ 0 & F_2(\varphi) \end{pmatrix}
\tag{6.26}
$$

and $S(\beta, \varphi) = (S_1'(\beta, \varphi), S_2'(\beta, \varphi))'$, where $S_1(\beta, \varphi) = S_\beta$ and $S_2(\beta, \varphi) = (S_{\sigma^2}, S_{\varphi_1}, \ldots, S_{\varphi_m})'$ have the components calculated in (6.23) and (6.24). The updating equations of the alternative Fisher-scoring algorithm are

$$
\beta^{(r+1)} = \beta^{(r)} + F_1^{-1}(\varphi^{(r)})S_1(\beta^{(r)}, \varphi^{(r)}), \quad \varphi^{(r+1)} = \varphi^{(r)} + F_2^{-1}(\varphi^{(r)})S_2(\beta^{(r)}, \varphi^{(r)}).
$$

6.5 Residual Maximum Likelihood Estimation

This section derives estimates of the parameters of model (6.13) by the method of residual maximum likelihood.

6.5.1 Description of the Method

Residual maximum likelihood estimation (REML) is introduced to reduce the bias of the maximum likelihood estimators of the variance components. For this sake, it transforms the vector y in two independent vectors $y_1^\star = K_1 y$ and $y_2^\star = K_2 y$, with the condition that the distribution of $\boldsymbol{y}_1^\star$ does not depend on the fixed regression parameter $\boldsymbol{\beta}$. Let $\boldsymbol{K}_1$ be a matrix such that $\boldsymbol{K}_1\boldsymbol{X} = \boldsymbol{0}$. Therefore

$$E[\boldsymbol{y}_1^\star] = E[\boldsymbol{K}_1\boldsymbol{y}] = E[\boldsymbol{K}_1(\boldsymbol{X}\boldsymbol{\beta} + \boldsymbol{Z}_1\boldsymbol{u}_1 + \ldots \boldsymbol{Z}_m\boldsymbol{u}_m + \boldsymbol{e})] = \boldsymbol{0}.$$

The vector $\boldsymbol{y}_2^\star$ is selected to be independent of $\boldsymbol{y}_1^\star$. Then, it satisfies

$$E[\boldsymbol{y}_1^\star \boldsymbol{y}_2^{\star\prime}] = \boldsymbol{K}_1 E[\boldsymbol{y}\boldsymbol{y}']\boldsymbol{K}_2' = \boldsymbol{K}_1\boldsymbol{V}\boldsymbol{K}_2' = \boldsymbol{0}.$$

Rows $\boldsymbol{k}'$ of matrix $\boldsymbol{K}_1$ are called contrasts, as they fulfill $\boldsymbol{k}'\boldsymbol{X} = \boldsymbol{0}$. The maximum number of contrasts linearly independent is $n - \mathrm{r}(\boldsymbol{X})$. We suppose that $\boldsymbol{X}$ has full rank p, so the matrix $\boldsymbol{K}_1$ can be selected such that its rank is $n - p$. Matrix $\boldsymbol{K}_2$ is selected with rank p.

To introduce matrix $\boldsymbol{K}_1$, we consider the model without random effects

$$\boldsymbol{y} = \boldsymbol{X}\boldsymbol{\beta} + \boldsymbol{\varepsilon}, \quad \text{with} \quad \boldsymbol{\varepsilon} \sim N(\boldsymbol{0}, \boldsymbol{\Sigma}_\varepsilon). \tag{6.27}$$

The maximum likelihood estimator of $\boldsymbol{\beta}$ in (6.27) is

$$\tilde{\boldsymbol{\beta}} = \left(\boldsymbol{X}'\boldsymbol{\Sigma}_\varepsilon^{-1}\boldsymbol{X}\right)^{-1} \boldsymbol{X}'\boldsymbol{\Sigma}_\varepsilon^{-1}\boldsymbol{y}.$$

We define the transformed vector (normalized residual)

$$\boldsymbol{y}_1^\star = \boldsymbol{\Sigma}_\varepsilon^{-1}(\boldsymbol{y} - \boldsymbol{X}\tilde{\boldsymbol{\beta}}) = \boldsymbol{\Sigma}_\varepsilon^{-1}\left(\boldsymbol{y} - \boldsymbol{X}(\boldsymbol{X}'\boldsymbol{\Sigma}_\varepsilon^{-1}\boldsymbol{X})^{-1}\boldsymbol{X}'\boldsymbol{\Sigma}_\varepsilon^{-1}\boldsymbol{y}\right) = \boldsymbol{K}_1\boldsymbol{y},$$

where $\boldsymbol{K}_1 = \boldsymbol{\Sigma}_\varepsilon^{-1} - \boldsymbol{\Sigma}_\varepsilon^{-1}\boldsymbol{X}(\boldsymbol{X}'\boldsymbol{\Sigma}_\varepsilon^{-1}\boldsymbol{X})^{-1}\boldsymbol{X}'\boldsymbol{\Sigma}_\varepsilon^{-1}$. Further we select $\boldsymbol{K}_2 = \boldsymbol{X}'\boldsymbol{V}^{-1}$.

Since $\boldsymbol{K}_1 = \boldsymbol{K}_1'$, it holds that

$$\begin{aligned}
E[\boldsymbol{y}_1^\star] &= E[\boldsymbol{K}_1\boldsymbol{y}] = \left(\boldsymbol{\Sigma}_\varepsilon^{-1} - \boldsymbol{\Sigma}_\varepsilon^{-1}\boldsymbol{X}(\boldsymbol{X}'\boldsymbol{\Sigma}_\varepsilon^{-1}\boldsymbol{X})^{-1}\boldsymbol{X}'\boldsymbol{\Sigma}_\varepsilon^{-1}\right)\boldsymbol{X}\boldsymbol{\beta} = \boldsymbol{0},\\
E[\boldsymbol{y}_2^\star] &= E[\boldsymbol{K}_2\boldsymbol{y}] = \boldsymbol{X}'\boldsymbol{V}^{-1}\boldsymbol{X}\boldsymbol{\beta},\\
\mathrm{var}(\boldsymbol{y}_1^\star) &= E[\boldsymbol{y}_1^\star\boldsymbol{y}_1^{\star\prime}] = \boldsymbol{K}_1\boldsymbol{V}\boldsymbol{K}_1,\\
\mathrm{var}(\boldsymbol{y}_2^\star) &= \boldsymbol{K}_2\boldsymbol{V}\boldsymbol{K}_2' = \boldsymbol{X}'\boldsymbol{V}^{-1}\boldsymbol{V}\boldsymbol{V}^{-1}\boldsymbol{X} = \boldsymbol{X}'\boldsymbol{V}^{-1}\boldsymbol{X},\\
E[\boldsymbol{y}_1^\star\boldsymbol{y}_2^{\star\prime}] &= \boldsymbol{K}_1E[\boldsymbol{y}\boldsymbol{y}']\boldsymbol{K}_2' = \boldsymbol{K}_1\boldsymbol{V}\boldsymbol{K}_2' = \boldsymbol{K}_1\boldsymbol{V}\boldsymbol{V}^{-1}\boldsymbol{X} = \boldsymbol{K}_1\boldsymbol{X} = \boldsymbol{0}.
\end{aligned}$$

As the maximum number of linearly independent columns in $\boldsymbol{K}_1$ is $n - \mathrm{r}(\boldsymbol{X})$, we can select $n - \mathrm{r}(\boldsymbol{X})$ of these columns for constructing a $n \times (n - \mathrm{r}(\boldsymbol{X}))$ sub-matrix $\boldsymbol{K}$ satisfying $\boldsymbol{K}'\boldsymbol{X} = \boldsymbol{0}$. We define the vectors $\boldsymbol{y}_1 = \boldsymbol{K}'\boldsymbol{y}$ and $\boldsymbol{y}_2 = \boldsymbol{y}_2^\star$. Since $\mathrm{r}(\boldsymbol{X}) = p$, we have that

$$\boldsymbol{y}_1 \sim N_{n-p}(\boldsymbol{0}, \boldsymbol{K}'\boldsymbol{V}\boldsymbol{K}), \quad \boldsymbol{y}_2 \sim N_p(\boldsymbol{X}'\boldsymbol{V}^{-1}\boldsymbol{X}\boldsymbol{\beta}, \boldsymbol{X}'\boldsymbol{V}^{-1}\boldsymbol{X})$$

and they are independent. We define $\boldsymbol{\sigma} = (\sigma_0^2, \sigma_1^2, \ldots, \sigma_m^2)'$ and $\boldsymbol{P} = \boldsymbol{K}(\boldsymbol{K}'\boldsymbol{V}\boldsymbol{K})^{-1}\boldsymbol{K}'$. The likelihood function of $\boldsymbol{y}_1$ is

$$l(\boldsymbol{\sigma}) = -\frac{1}{2}(n-p)\log 2\pi - \frac{1}{2}\log|\boldsymbol{K}'\boldsymbol{V}\boldsymbol{K}| - \frac{1}{2}\boldsymbol{y}_1'(\boldsymbol{K}'\boldsymbol{V}\boldsymbol{K})^{-1}\boldsymbol{y}_1,$$

where $\boldsymbol{V} = \sum_{i=0}^{m}\sigma_i^2\boldsymbol{G}_i$ and $\boldsymbol{y}_1 = \boldsymbol{K}'\boldsymbol{y}$. By taking partial derivatives with respect to σ_i^2 and using the fact that $\mathrm{tr}(\boldsymbol{A}\boldsymbol{B}) = \mathrm{tr}(\boldsymbol{B}\boldsymbol{A})$, we obtain the elements

$$\begin{aligned} S_{\sigma_i^2} &= \frac{\partial l(\boldsymbol{\sigma})}{\partial\sigma_i^2} = -\frac{1}{2}\frac{\partial}{\partial\sigma_i^2}\left\{\log|\boldsymbol{K}'\boldsymbol{V}\boldsymbol{K}|\right\} - \frac{1}{2}\frac{\partial}{\partial\sigma_i^2}\left\{\boldsymbol{y}'\boldsymbol{K}(\boldsymbol{K}'\boldsymbol{V}\boldsymbol{K})^{-1}\boldsymbol{K}'\boldsymbol{y}\right\} \\ &= -\frac{1}{2}\,\mathrm{tr}\left((\boldsymbol{K}'\boldsymbol{V}\boldsymbol{K})^{-1}\boldsymbol{K}'\boldsymbol{G}_i\boldsymbol{K}\right) + \frac{1}{2}\,\boldsymbol{y}'\boldsymbol{K}(\boldsymbol{K}'\boldsymbol{V}\boldsymbol{K})^{-1}(\boldsymbol{K}'\boldsymbol{G}_i\boldsymbol{K})(\boldsymbol{K}'\boldsymbol{V}\boldsymbol{K})^{-1}\boldsymbol{K}'\boldsymbol{y} \\ &= -\frac{1}{2}\,\mathrm{tr}(\boldsymbol{P}\boldsymbol{G}_i) + \frac{1}{2}\,\boldsymbol{y}'\boldsymbol{P}\boldsymbol{G}_i\boldsymbol{P}\boldsymbol{y} \end{aligned} \tag{6.28}$$

of the score vector $\boldsymbol{S}(\boldsymbol{\sigma}) = (S_{\sigma_0^2}, \ldots, S_{\sigma_m^2})'$. As

$$\frac{\partial\boldsymbol{P}}{\partial\sigma_j^2} = \frac{\partial[\boldsymbol{K}(\boldsymbol{K}'\boldsymbol{V}\boldsymbol{K})^{-1}\boldsymbol{K}']}{\partial\sigma_j^2} = -\boldsymbol{K}(\boldsymbol{K}'\boldsymbol{V}\boldsymbol{K})^{-1}\boldsymbol{K}'\boldsymbol{G}_j\boldsymbol{K}(\boldsymbol{K}'\boldsymbol{V}\boldsymbol{K})^{-1}\boldsymbol{K}' = -\boldsymbol{P}\boldsymbol{G}_j\boldsymbol{P},$$

the second order partial derivatives are

$$\frac{\partial l(\boldsymbol{\sigma})}{\partial\sigma_i^2\partial\sigma_j^2} = \frac{1}{2}\,\mathrm{tr}\left(\boldsymbol{P}\boldsymbol{G}_j\boldsymbol{P}\boldsymbol{G}_i\right) - \boldsymbol{y}'\boldsymbol{P}\boldsymbol{G}_j\boldsymbol{P}\boldsymbol{G}_i\boldsymbol{P}\boldsymbol{y}.$$

If we take expectations and change the sign, we obtain the Fisher information matrix. To calculate this matrix we use the relations $\boldsymbol{P}\boldsymbol{X} = \boldsymbol{0}$ and $\boldsymbol{P}\boldsymbol{V}\boldsymbol{P} = \boldsymbol{P}$, and the following result about expectations of quadratic forms (cf. Appendix A):

$$\text{If } E[\boldsymbol{y}] = \boldsymbol{\mu} \text{ and } \mathrm{var}(\boldsymbol{y}) = \boldsymbol{V}, \text{ then } E[\boldsymbol{y}'\boldsymbol{A}\boldsymbol{y}] = \mathrm{tr}(\boldsymbol{A}\boldsymbol{V}) + \boldsymbol{\mu}'\boldsymbol{A}\boldsymbol{\mu}. \tag{6.29}$$

The elements of the Fisher information matrix $\boldsymbol{F}(\boldsymbol{\sigma}) = (F_{\sigma_i^2\sigma_j^2})_{i,j=0}^{m}$ are

$$\begin{aligned} F_{\sigma_i^2\sigma_j^2} &= -E\left[\frac{\partial l(\boldsymbol{\sigma})}{\partial\sigma_i^2\partial\sigma_j^2}\right] = -\frac{1}{2}\,\mathrm{tr}(\boldsymbol{P}\boldsymbol{G}_j\boldsymbol{P}\boldsymbol{G}_i) + \mathrm{tr}(\boldsymbol{P}\boldsymbol{G}_j\boldsymbol{P}\boldsymbol{G}_i\boldsymbol{P}\boldsymbol{V}) + \boldsymbol{\beta}'\boldsymbol{X}'\boldsymbol{P}\boldsymbol{G}_j\boldsymbol{P}\boldsymbol{G}_i\boldsymbol{P}\boldsymbol{X}\boldsymbol{\beta} \\ &= -\frac{1}{2}\,\mathrm{tr}(\boldsymbol{P}\boldsymbol{G}_j\boldsymbol{P}\boldsymbol{G}_i) + \mathrm{tr}(\boldsymbol{G}_j\boldsymbol{P}\boldsymbol{G}_i\boldsymbol{P}\boldsymbol{V}\boldsymbol{P}) = \frac{1}{2}\,\mathrm{tr}(\boldsymbol{P}\boldsymbol{G}_j\boldsymbol{P}\boldsymbol{G}_i). \end{aligned} \tag{6.30}$$

To calculate the REML estimators, the Fisher-scoring method uses the updating formula

$$\boldsymbol{\sigma}^{(r+1)} = \boldsymbol{\sigma}^{(r)} + \boldsymbol{F}^{-1}(\boldsymbol{\sigma}^{(r)})\boldsymbol{S}(\boldsymbol{\sigma}^{(r)}),$$

where $\boldsymbol{F}(\boldsymbol{\sigma}^{(r)})$ is the Fisher information matrix calculated at $\boldsymbol{\sigma}^{(r)}$. We observe that $\boldsymbol{F}(\boldsymbol{\sigma})$ is a $(m+1) \times (m+1)$ matrix; however the Fisher information matrix needed to calculate maximum likelihood estimators, $\boldsymbol{F}(\boldsymbol{\theta})$, is $(p+m+1) \times (p+m+1)$.

The Fisher-scoring algorithm gives the estimate of $\boldsymbol{\sigma}$. If we plug that estimate in the likelihood function of $\boldsymbol{y}_2$, we consider it as a constant, and we maximize on $\boldsymbol{\beta}$, we get the REML estimators of $\boldsymbol{\beta}$. The likelihood function of $\boldsymbol{y}_2$ is

$$\begin{aligned} l(\boldsymbol{\beta}) &= -\frac{p}{2}\log 2\pi - \frac{1}{2}\log|\boldsymbol{X}'\boldsymbol{V}^{-1}\boldsymbol{X}| \\ &\quad - \frac{1}{2}(\boldsymbol{y}_2 - \boldsymbol{X}'\boldsymbol{V}^{-1}\boldsymbol{X}\boldsymbol{\beta})'\left(\boldsymbol{X}'\boldsymbol{V}^{-1}\boldsymbol{X}\right)^{-1}(\boldsymbol{y}_2 - \boldsymbol{X}'\boldsymbol{V}^{-1}\boldsymbol{X}\boldsymbol{\beta}). \end{aligned}$$

By taking partial derivatives with respect to $\boldsymbol{\beta}$, and equating to zero, we obtain

$$0 = \frac{\partial l(\boldsymbol{\beta})}{\partial \boldsymbol{\beta}} = \boldsymbol{X}'\boldsymbol{V}^{-1}\boldsymbol{X}\left(\boldsymbol{X}'\boldsymbol{V}^{-1}\boldsymbol{X}\right)^{-1}(\boldsymbol{y}_2 - \boldsymbol{X}'\boldsymbol{V}^{-1}\boldsymbol{X}\boldsymbol{\beta}) = \boldsymbol{X}'\boldsymbol{V}^{-1}(\boldsymbol{y} - \boldsymbol{X}\boldsymbol{\beta}).$$

Therefore

$$\hat{\boldsymbol{\beta}}_{REML} = (\boldsymbol{X}'\hat{\boldsymbol{V}}^{-1}\boldsymbol{X})^{-1}\boldsymbol{X}'\hat{\boldsymbol{V}}^{-1}\boldsymbol{y},$$

where $\hat{V} = \sum_{i=0}^{m}\hat{\sigma}_i^2\boldsymbol{G}_i$ and $\hat{\sigma}_0^2, \hat{\sigma}_1^2, \ldots, \hat{\sigma}_m^2$ are the REML estimators of $\sigma_0^2, \sigma_1^2, \ldots, \sigma_m^2$.

By taking again derivatives with respect to $\boldsymbol{\beta}$, we get

$$\boldsymbol{F}_{\boldsymbol{\beta}\boldsymbol{\beta}} = -E\left[\partial^2 l(\boldsymbol{\beta})/\partial\boldsymbol{\beta}\partial\boldsymbol{\beta}'\right] = \boldsymbol{X}'\boldsymbol{V}^{-1}\boldsymbol{X},$$

that is the same value of $\boldsymbol{F}_{\boldsymbol{\beta}\boldsymbol{\beta}}$ as obtained in the maximum likelihood procedure.

Theorem 6.1 states that the REML method does not depend on the selected matrix $\boldsymbol{K}$ (with $\boldsymbol{K}'\boldsymbol{X} = \boldsymbol{0}$).

Theorem 6.1 *Let $\boldsymbol{K}'$ be a full rank $(n-p) \times n$ matrix. Let $\boldsymbol{V}$ be a symmetric and positive definite $n \times n$ matrix. Let $\boldsymbol{X}$ be a $n \times p$ matrix with rank p. If $\boldsymbol{K}'\boldsymbol{X} = \boldsymbol{0}$, then*

$$\boldsymbol{K}(\boldsymbol{K}'\boldsymbol{V}\boldsymbol{K})^{-1}\boldsymbol{K}' = \boldsymbol{P}, \quad \textit{with} \quad \boldsymbol{P} = \boldsymbol{V}^{-1} - \boldsymbol{V}^{-1}\boldsymbol{X}(\boldsymbol{X}'\boldsymbol{V}^{-1}\boldsymbol{X})^{-1}\boldsymbol{X}'\boldsymbol{V}^{-1}.$$

Proof The matrices $\boldsymbol{K}(\boldsymbol{K}'\boldsymbol{K})^{-1}\boldsymbol{K}'$ and $\boldsymbol{X}(\boldsymbol{X}'\boldsymbol{X})^{-1}\boldsymbol{X}'$ are symmetric and idempotent. Furthermore, it holds that $\boldsymbol{K}(\boldsymbol{K}'\boldsymbol{K})^{-1}\boldsymbol{K}'\boldsymbol{X} = \boldsymbol{0}$ and $\boldsymbol{X}(\boldsymbol{X}'\boldsymbol{X})^{-1}\boldsymbol{X}'\boldsymbol{K} = \boldsymbol{0}$. The matrix

$$\boldsymbol{T} = \boldsymbol{I} - \boldsymbol{X}(\boldsymbol{X}'\boldsymbol{X})^{-1}\boldsymbol{X}' - \boldsymbol{K}(\boldsymbol{K}'\boldsymbol{K})^{-1}\boldsymbol{K}'$$

is symmetric and idempotent. Therefore

$$\begin{aligned}\operatorname{tr}(\boldsymbol{T}\boldsymbol{T}') = \operatorname{tr}(\boldsymbol{T}^2) = \operatorname{tr}(\boldsymbol{T}) &= \operatorname{tr}(\boldsymbol{I}) - \operatorname{tr}(\boldsymbol{X}(\boldsymbol{X}'\boldsymbol{X})^{-1}\boldsymbol{X}') - \operatorname{tr}(\boldsymbol{K}(\boldsymbol{K}'\boldsymbol{K})^{-1}\boldsymbol{K}') \\ &= n - \operatorname{r}(\boldsymbol{X}) - \operatorname{r}(\boldsymbol{K}) = n - p - (n-p) = 0.\end{aligned}$$

As $\boldsymbol{T}$ is idempotent with $\operatorname{tr}(\boldsymbol{T}\boldsymbol{T}') = 0$, then $\operatorname{r}(\boldsymbol{T}) = 0$ and $\boldsymbol{T} = \boldsymbol{0}$. Therefore, we have

$$\boldsymbol{I} - \boldsymbol{X}(\boldsymbol{X}'\boldsymbol{X})^{-1}\boldsymbol{X}' = \boldsymbol{K}(\boldsymbol{K}'\boldsymbol{K})^{-1}\boldsymbol{K}'. \tag{6.31}$$

As $\boldsymbol{V}$ is symmetric and positive definite, there exists $\boldsymbol{V}^{1/2}$ symmetric and positive definite such that $\boldsymbol{V} = \boldsymbol{V}^{1/2}\boldsymbol{V}^{1/2}$. It holds

$$\left(\boldsymbol{V}^{1/2}\boldsymbol{K}\right)' \boldsymbol{V}^{-1/2}\boldsymbol{X} = \boldsymbol{0}.$$

The result (6.31) is also valid if we replace $\boldsymbol{K}$ and $\boldsymbol{X}$ by $\boldsymbol{V}^{1/2}\boldsymbol{K}$ and $\boldsymbol{V}^{-1/2}\boldsymbol{X}$, respectively. We now obtain

$$\boldsymbol{I} - \boldsymbol{V}^{-1/2}\boldsymbol{X}(\boldsymbol{X}'\boldsymbol{V}^{-1}\boldsymbol{X})^{-1}\boldsymbol{X}'\boldsymbol{V}^{-1/2} = \boldsymbol{V}^{1/2}\boldsymbol{K}(\boldsymbol{K}'\boldsymbol{V}\boldsymbol{K})^{-1}\boldsymbol{K}'\boldsymbol{V}^{1/2},$$

or equivalently

$$\boldsymbol{P} = \boldsymbol{V}^{-1} - \boldsymbol{V}^{-1}\boldsymbol{X}(\boldsymbol{X}'\boldsymbol{V}^{-1}\boldsymbol{X})^{-1}\boldsymbol{X}'\boldsymbol{V}^{-1} = \boldsymbol{K}(\boldsymbol{K}'\boldsymbol{V}\boldsymbol{K})^{-1}\boldsymbol{K}',$$

and the stated result is obtained. □

Under regularity conditions (see e.g. Sections 1.3.2 and 1.8 of Jiang 2007), the REML estimators of $\boldsymbol{\beta}$ and $\hat{\boldsymbol{\sigma}}$ are consistent and asymptotically normal, i.e.

$$\hat{\boldsymbol{\beta}} \sim N_p\big(\boldsymbol{\beta}, (\boldsymbol{X}'\boldsymbol{V}^{-1}\boldsymbol{X})^{-1}\big), \quad \hat{\boldsymbol{\sigma}} \sim N_{m+1}\big(\boldsymbol{\sigma}, \boldsymbol{F}^{-1}(\boldsymbol{\sigma})\big).$$

6.5.2 REML Estimators for Alternative Parameters

In the model (6.13), we consider the parameters

$$\sigma^2 = \sigma_0^2, \quad \varphi_i = \sigma_i^2/\sigma_0^2, \quad i = 1, \ldots, m.$$

Let us denote $\boldsymbol{\varphi}' = (\sigma^2, \varphi_1, \ldots, \varphi_m)$, $\boldsymbol{\theta}' = (\boldsymbol{\beta}', \boldsymbol{\varphi}')$, and $\boldsymbol{V} = \sigma^2 \left(\boldsymbol{\Sigma}_e + \sum_{i=1}^{m} \varphi_i \, \boldsymbol{G}_i\right) = \sigma^2 \boldsymbol{\Sigma}$. For the REML method, the log-likelihood associated to this parametrization is

$$l(\boldsymbol{\varphi}) = -\frac{1}{2}(n-p) \log 2\pi - \frac{1}{2}(n-p) \log \sigma^2 - \frac{1}{2} \log |\boldsymbol{K}'\boldsymbol{\Sigma}\boldsymbol{K}| - \frac{1}{2\sigma^2} \boldsymbol{y}'\boldsymbol{P}\boldsymbol{y}, \qquad (6.32)$$

where $\boldsymbol{P} = \boldsymbol{K}(\boldsymbol{K}'\boldsymbol{\Sigma}\boldsymbol{K})^{-1}\boldsymbol{K}' = \boldsymbol{\Sigma}^{-1} - \boldsymbol{\Sigma}^{-1}\boldsymbol{X}(\boldsymbol{X}'\boldsymbol{\Sigma}^{-1}\boldsymbol{X})^{-1}\boldsymbol{X}'\boldsymbol{\Sigma}^{-1}$. The components of the vector of scores are

$$S_{\sigma^2} = -\frac{n-p}{2\sigma^2} + \frac{1}{2\sigma^4} \boldsymbol{y}'\boldsymbol{P}\boldsymbol{y}, \quad S_{\varphi_i} = -\frac{1}{2}\mathrm{tr}(\boldsymbol{P}\boldsymbol{G}_i) + \frac{1}{2\sigma^2} \boldsymbol{y}'\boldsymbol{P}\boldsymbol{G}_i\boldsymbol{P}\boldsymbol{y}, \quad i = 1, \ldots, m.$$

Second partial derivatives of the log-likelihood are

$$H_{\sigma^2\sigma^2} = \frac{n-p}{2\sigma^4} - \frac{1}{\sigma^6}\boldsymbol{y}'\boldsymbol{P}\boldsymbol{y}, \quad H_{\sigma^2\varphi_i} = -\frac{1}{2\sigma^4}\boldsymbol{y}'\boldsymbol{P}\boldsymbol{G}_i\boldsymbol{P}\boldsymbol{y},$$
$$H_{\varphi_i\varphi_j} = \frac{1}{2}\mathrm{tr}(\boldsymbol{P}\boldsymbol{G}_j\boldsymbol{P}\boldsymbol{G}_i) - \frac{1}{\sigma^2}\boldsymbol{y}'\boldsymbol{P}\boldsymbol{G}_j\boldsymbol{P}\boldsymbol{G}_i\boldsymbol{P}\boldsymbol{y}.$$

By taking expectations, changing the sign, and applying $\boldsymbol{P}\boldsymbol{X} = \boldsymbol{0}$ and $\boldsymbol{P}\boldsymbol{\Sigma}\boldsymbol{P} = \boldsymbol{P}$, we obtain the elements of the Fisher information matrix

$$F_{\sigma^2\sigma^2} = -\frac{n-p}{2\sigma^4} + \frac{1}{\sigma^4}\mathrm{tr}(\boldsymbol{P}\boldsymbol{\Sigma}) = \frac{n-p}{2\sigma^4}, \qquad (6.33)$$
$$F_{\sigma^2\varphi_i} = \frac{1}{2\sigma^2}\mathrm{tr}(\boldsymbol{P}\boldsymbol{G}_i), \quad F_{\varphi_i\varphi_j} = \frac{1}{2}\mathrm{tr}(\boldsymbol{P}\boldsymbol{G}_j\boldsymbol{P}\boldsymbol{G}_i).$$

The Fisher-scoring updating formula for $\boldsymbol{\varphi}$ is

$$\boldsymbol{\varphi}^{(r+1)} = \boldsymbol{\varphi}^{(r)} + \boldsymbol{F}^{-1}(\boldsymbol{\varphi}^{(r)})\boldsymbol{S}(\boldsymbol{\varphi}^{(r)}),$$

where the components of the Fisher information matrix $\boldsymbol{F}$ and the score vector $\boldsymbol{S}$ were defined above. Moreover, from equation $S_{\sigma^2} = 0$, we get

$$\hat{\sigma}^2 = \frac{1}{n-p}\boldsymbol{y}'\boldsymbol{P}\boldsymbol{y}. \qquad (6.34)$$

6.5.3 Further REML Equations for Linear Mixed Models

This section gives the REML equations for linear mixed models. Some of the mathematical derivations are based on the works of Cressie and Lahiri (1993) and Harville (1974).

Let us consider the model

$$y = X\beta + Zu + e,$$

where X is $n \times p$ with $\mathrm{r}(X) = p$, β is $p \times 1$, $Z = [Z_1, \ldots, Z_m]$ is $n \times q$, Z_j is $n \times q_j$, $u = [u_1', \ldots, u_m']' \sim N(0, \sigma^2\Sigma_u)$ with $\Sigma_u = \mathrm{diag}(\varphi_1\Sigma_{u_1}, \ldots, \varphi_m\Sigma_{u_m})$ and e is $n \times 1$ with $e \sim N_n(0, \sigma^2\Sigma_e)$, and $\Sigma_e = \mathrm{diag}(D_1, \ldots, D_n)$. The distribution of y is

$$y \sim N(X\beta, \sigma^2\Sigma), \quad \Sigma = \Sigma_e + Z\Sigma_u Z'.$$

Let us define the non-singular symmetric and idempotent $n \times n$ matrix $S = I - X(X'X)^{-1}X'$. We apply the spectral (Jordan) decomposition theorem, then there exists a matrix A of dimension $n \times (n-p)$, such that

$$S = AA', \quad A'A = I. \tag{6.35}$$

It holds that $SA = A$, $A'SA = A'A = I$, and

$$I = A'SA = A'A - A'X(X'X)^{-1}X'A = I - A'X(X'X)^{-1}X'A \quad \Longleftrightarrow \quad A'X = 0.$$

The following propositions give some auxiliary results.

Proposition 6.3 *Let $G = \Sigma^{-1}X(X'\Sigma^{-1}X)^{-1}$ be an $n \times p$ matrix with rank p. Let A be the matrix given in (6.35). The determinant of the $n \times n$ matrix $[A, G]$ is*

$$|[A, G]| = |X'X|^{-1/2}. \tag{6.36}$$

Proof By taking into account that (cf. Appendix A)

$$\begin{vmatrix} A_{11} & A_{12} \\ A_{21} & A_{22} \end{vmatrix} = |A_{11}| \left|A_{22} - A_{21}A_{11}^{-1}A_{12}\right| \quad \text{and} \quad |A'A| = |A|^2,$$

we get

$$|[A, G]| = \left(|[A, G]'[A, G]|\right)^{1/2} = \left|\begin{bmatrix} A' \\ G' \end{bmatrix}[A, G]\right|^{1/2} = \begin{vmatrix} A'A & A'G \\ G'A & G'G \end{vmatrix}^{1/2}$$
$$= |A'A|^{1/2}|G'G - G'A(A'A)^{-1}A'G|^{1/2} = |G'G - G'AA'G|^{1/2},$$

because $A'A = I$. On the other hand

$$G'AA'G = G'SG = G'G - G'X(X'X)^{-1}X'G.$$

Therefore, we have

$$\begin{aligned}\|[A, G]\| &= |G'G - G'G + G'X(X'X)^{-1}X'G|^{1/2}\\ &= |(X'\Sigma^{-1}X)^{-1}X'\Sigma^{-1}X(X'X)^{-1}X'\Sigma^{-1}X(X'\Sigma^{-1}X)^{-1}|^{1/2}\\ &= |X'X|^{-1/2}.\end{aligned}$$

□

Proposition 6.4 *Let A be the matrix defined in (6.35). The vector $\omega = A'y$ is independent of the least squares estimator of the regression parameters*

$$\hat{\beta} = G'y = (X'\Sigma^{-1}X)^{-1}X'\Sigma^{-1}y.$$

Proof As ω and $\hat{\beta}$ are jointly normal distributed, it is sufficient to prove that they are uncorrelated. As $A'X = 0$, then

$$\begin{aligned}\mathrm{cov}(\omega, \hat{\beta}) &= \mathrm{cov}(A'y, G'y) = \sigma^2 A'\Sigma G \\ &= \sigma^2 A'\Sigma\Sigma^{-1}X(X'\Sigma^{-1}X)^{-1} = \sigma^2 A'X(X'\Sigma^{-1}X)^{-1} = 0.\end{aligned} \tag{6.37}$$

□

Proposition 6.5 *It holds that*

$$(y - X\beta)'\Sigma^{-1}(y - X\beta) = (y - X\hat{\beta})'\Sigma^{-1}(y - X\hat{\beta}) + (\hat{\beta} - \beta)'X'\Sigma^{-1}X(\hat{\beta} - \beta). \tag{6.38}$$

Proof By adding and subtracting $X\hat{\beta}$, we get

$$\begin{aligned}&(y - X\beta)'\Sigma^{-1}(y - X\beta) = (y - X\hat{\beta} + X\hat{\beta} - X\beta)'\Sigma^{-1}(y - X\hat{\beta} + X\hat{\beta} - X\beta)\\ &= (y - X\hat{\beta})'\Sigma^{-1}(y - X\hat{\beta}) + 2(y - X\hat{\beta})'\Sigma^{-1}X(\hat{\beta} - \beta) + (\hat{\beta} - \beta)'X'\Sigma^{-1}X(\hat{\beta} - \beta).\end{aligned}$$

Finally

$$\begin{aligned}(y - X\hat{\beta})'\Sigma^{-1}X(\hat{\beta} - \beta) &= y'[I - X(X'\Sigma^{-1}X)^{-1}X'\Sigma^{-1}]'\Sigma^{-1}X(\hat{\beta} - \beta)\\ &= y'[\Sigma^{-1}X - \Sigma^{-1}X(X'\Sigma^{-1}X)^{-1}X'\Sigma^{-1}X](\hat{\beta} - \beta) = 0.\end{aligned}$$

□

Proposition 6.6 *The joint density of ω and $\hat{\beta}$, evaluated at $\omega = A'y$ and $\hat{\beta} = G'y$, is*

$$\begin{aligned}f_{\omega,\hat{\beta}}(A'y, G'y|\varphi, \beta) &= |X'X|^{1/2} f_y(y|\varphi, \beta) = (2\pi\sigma^2)^{-n/2}|X'X|^{1/2}|\Sigma|^{-1/2}\\ &\cdot \exp\left\{\frac{-1}{2\sigma^2}(y - X\beta)'\Sigma^{-1}(y - X\beta)\right\}.\end{aligned}$$

Proof The vector

$$z = \begin{pmatrix} \omega \\ \hat{\beta} \end{pmatrix} = \begin{pmatrix} A'y \\ G'y \end{pmatrix} = [A, G]'y$$

is a linear transformation of y. Therefore z is normally distributed. The mean vector of z is

$$E(z) = \begin{pmatrix} A'X\beta \\ G'X\beta \end{pmatrix} = \begin{pmatrix} 0 \\ \beta \end{pmatrix},$$

because $G'X\beta = (X'\Sigma^{-1}X)^{-1}X'\Sigma^{-1}X\beta = \beta$. The covariance matrix of z is

$$\mathrm{var}(z) = \sigma^2 \begin{pmatrix} A' \\ G' \end{pmatrix} \Sigma\,[A, G] = \sigma^2 \begin{pmatrix} A'\Sigma A & A'\Sigma G \\ G'\Sigma A & G'\Sigma G \end{pmatrix}.$$

From (6.37), we have $A'\Sigma G = 0$ and

$$G'\Sigma G = (X'\Sigma^{-1}X)^{-1}X'\Sigma^{-1}\Sigma\Sigma^{-1}X(X'\Sigma^{-1}X)^{-1} = (X'\Sigma^{-1}X)^{-1},$$

so that

$$\mathrm{var}(z) = \sigma^2 \begin{pmatrix} A'\Sigma A & 0 \\ 0 & (X'\Sigma^{-1}X)^{-1} \end{pmatrix}.$$

From (6.36), we have that $|[A, G]| = |X'X|^{-1/2} \neq 0$ and $\mathrm{r}([A, G]) = n$. As $|B| = |B'|$, the determinant of var(z) is

$$\begin{aligned}|\mathrm{var}(z)| &= (\sigma^2)^n \left| \begin{pmatrix} A' \\ G' \end{pmatrix} \Sigma\,[A, G] \right| = (\sigma^2)^n \left| \begin{pmatrix} A' \\ G' \end{pmatrix} \right| |\Sigma||[A, G]| \\ &= (\sigma^2)^n |[A, G]|^2 |\Sigma| = (\sigma^2)^n |X'X|^{-1} |\Sigma|.\end{aligned}$$

The joint density of ω and $\hat{\beta}$ is

$$\begin{aligned}f_z(z|\varphi, \beta) &= (2\pi)^{-n/2}|\mathrm{var}(z)|^{-1/2} \exp\left\{-\frac{1}{2}(z - E(z))'\mathrm{var}(z)^{-1}(z - E(z))\right\} \\ &= \eta \exp\left\{\frac{-1}{2\sigma^2}(y'A, y'G - \beta')\begin{pmatrix} (A'\Sigma A)^{-1} & 0 \\ 0 & X'\Sigma^{-1}X \end{pmatrix}\begin{pmatrix} A'y \\ G'y - \beta \end{pmatrix}\right\} \\ &= \eta \exp\left\{\frac{-1}{2\sigma^2}[y'A(A'\Sigma A)^{-1}A'y + (y'G - \beta')X'\Sigma^{-1}X(G'y - \beta)]\right\} \\ &= \eta \exp\left\{\frac{-1}{2\sigma^2}[y'A(A'\Sigma A)^{-1}A'y + (\hat{\beta} - \beta)'X'\Sigma^{-1}X(\hat{\beta} - \beta)]\right\}, \quad (6.39)\end{aligned}$$

where $\eta = (2\pi\sigma^2)^{-n/2}|\boldsymbol{X}'\boldsymbol{X}|^{1/2}|\boldsymbol{\Sigma}|^{-1/2}$. As $\boldsymbol{A}'\boldsymbol{X} = \boldsymbol{0}$, $\mathrm{r}(\boldsymbol{A}') = n - p$ and $\boldsymbol{\Sigma}$ is positive definite, by applying Theorem 6.1, we get

$$\boldsymbol{A}(\boldsymbol{A}'\boldsymbol{\Sigma}\boldsymbol{A})^{-1}\boldsymbol{A}' = \boldsymbol{\Sigma}^{-1} - \boldsymbol{\Sigma}^{-1}\boldsymbol{X}(\boldsymbol{X}'\boldsymbol{\Sigma}^{-1}\boldsymbol{X})^{-1}\boldsymbol{X}'\boldsymbol{\Sigma}^{-1}.$$

Further, it holds that

$$\boldsymbol{\Sigma}^{-1} - \boldsymbol{\Sigma}^{-1}\boldsymbol{X}(\boldsymbol{X}'\boldsymbol{\Sigma}^{-1}\boldsymbol{X})^{-1}\boldsymbol{X}'\boldsymbol{\Sigma}^{-1} = [\boldsymbol{I} - \boldsymbol{X}\boldsymbol{G}']'\boldsymbol{\Sigma}^{-1}[\boldsymbol{I} - \boldsymbol{X}\boldsymbol{G}'].$$

Therefore

$$\begin{aligned}\boldsymbol{y}'\boldsymbol{A}(\boldsymbol{A}'\boldsymbol{\Sigma}\boldsymbol{A})^{-1}\boldsymbol{A}'\boldsymbol{y} &= \boldsymbol{y}'[\boldsymbol{I} - \boldsymbol{X}\boldsymbol{G}']'\boldsymbol{\Sigma}^{-1}[\boldsymbol{I} - \boldsymbol{X}\boldsymbol{G}']\boldsymbol{y}\\ &= \left[(\boldsymbol{I} - \boldsymbol{X}\boldsymbol{G}')\boldsymbol{y}\right]'\boldsymbol{\Sigma}^{-1}(\boldsymbol{I} - \boldsymbol{X}\boldsymbol{G}')\boldsymbol{y} = (\boldsymbol{y} - \boldsymbol{X}\hat{\boldsymbol{\beta}})'\boldsymbol{\Sigma}^{-1}(\boldsymbol{y} - \boldsymbol{X}\hat{\boldsymbol{\beta}}).\end{aligned}$$

By applying (6.38), we get

$$\begin{aligned}&\boldsymbol{y}'\boldsymbol{A}(\boldsymbol{A}'\boldsymbol{\Sigma}\boldsymbol{A})^{-1}\boldsymbol{A}'\boldsymbol{y} + (\hat{\boldsymbol{\beta}} - \boldsymbol{\beta})'\boldsymbol{X}'\boldsymbol{\Sigma}^{-1}\boldsymbol{X}(\hat{\boldsymbol{\beta}} - \boldsymbol{\beta})\\ &= (\boldsymbol{y} - \boldsymbol{X}\hat{\boldsymbol{\beta}})'\boldsymbol{\Sigma}^{-1}(\boldsymbol{y} - \boldsymbol{X}\hat{\boldsymbol{\beta}}) + (\hat{\boldsymbol{\beta}} - \boldsymbol{\beta})'\boldsymbol{X}'\boldsymbol{\Sigma}^{-1}\boldsymbol{X}(\hat{\boldsymbol{\beta}} - \boldsymbol{\beta}) = (\boldsymbol{y} - \boldsymbol{X}\boldsymbol{\beta})'\boldsymbol{\Sigma}^{-1}(\boldsymbol{y} - \boldsymbol{X}\boldsymbol{\beta}).\end{aligned}$$

By substituting in (6.39), we finally obtain the joint probability density function of $\boldsymbol{\omega}$ and $\hat{\boldsymbol{\beta}}$, evaluated at $\boldsymbol{\omega} = \boldsymbol{A}'\boldsymbol{y}$ and $\hat{\boldsymbol{\beta}} = \boldsymbol{G}'\boldsymbol{y}$, i.e.

$$\begin{aligned}f_{\boldsymbol{\omega},\hat{\boldsymbol{\beta}}}(\boldsymbol{A}'\boldsymbol{y}, \boldsymbol{G}'\boldsymbol{y}|\boldsymbol{\varphi}, \boldsymbol{\beta}) &= (2\pi\sigma^2)^{-n/2}|\boldsymbol{X}'\boldsymbol{X}|^{1/2}|\boldsymbol{\Sigma}|^{-1/2}\\ &\quad \cdot \exp\left\{\frac{-1}{2\sigma^2}(\boldsymbol{y} - \boldsymbol{X}\boldsymbol{\beta})'\boldsymbol{\Sigma}^{-1}(\boldsymbol{y} - \boldsymbol{X}\boldsymbol{\beta})\right\}\\ &= |\boldsymbol{X}'\boldsymbol{X}|^{1/2} f_{\boldsymbol{y}}(\boldsymbol{y}|\boldsymbol{\varphi}, \boldsymbol{\beta}).\end{aligned}$$

□

Proposition 6.7 *The density of* $\boldsymbol{\omega}$, *evaluated at* $\boldsymbol{A}'\boldsymbol{y}$, *is*

$$f_{\boldsymbol{\omega}}(\boldsymbol{A}'\boldsymbol{y}|\boldsymbol{\varphi}) = (2\pi\sigma^2)^{-(n-p)/2}|\boldsymbol{X}'\boldsymbol{X}|^{1/2}|\boldsymbol{\Sigma}|^{-1/2}|\boldsymbol{X}'\boldsymbol{\Sigma}^{-1}\boldsymbol{X}|^{-1/2}\exp\left\{\frac{-1}{2\sigma^2}\boldsymbol{y}'\boldsymbol{P}\boldsymbol{y}\right\},$$

where

$$\begin{aligned}\boldsymbol{P} &= \boldsymbol{\Sigma}^{-1} - \boldsymbol{\Sigma}^{-1}\boldsymbol{X}(\boldsymbol{X}'\boldsymbol{\Sigma}^{-1}\boldsymbol{X})^{-1}\boldsymbol{X}'\boldsymbol{\Sigma}^{-1}\\ &= [\boldsymbol{I} - \boldsymbol{X}(\boldsymbol{X}'\boldsymbol{\Sigma}^{-1}\boldsymbol{X})^{-1}\boldsymbol{X}'\boldsymbol{\Sigma}^{-1}]'\boldsymbol{\Sigma}^{-1}[\boldsymbol{I} - \boldsymbol{X}(\boldsymbol{X}'\boldsymbol{\Sigma}^{-1}\boldsymbol{X})^{-1}\boldsymbol{X}'\boldsymbol{\Sigma}^{-1}].\end{aligned}$$

Proof As

$$\int_{R^p} f_{\hat{\beta}}(\hat{\beta}|\varphi, \beta)\, d\hat{\beta} = 1 \quad \text{and} \quad f_{\hat{\beta}}(\hat{\beta}|\varphi, \beta) = f_{\hat{\beta}}(\beta|\varphi, \hat{\beta}),$$

it holds that

$$\begin{aligned} f_\omega(A'y|\varphi) &= f_\omega(A'y|\varphi) \int_{R^p} f_{\hat{\beta}}(\hat{\beta}|\varphi, \beta)\, d\hat{\beta} \\ &= f_\omega(A'y|\varphi) \int_{R^p} f_{\hat{\beta}}(\hat{\beta}|\varphi, \beta)\, d\beta \\ &= \int_{R^p} f_\omega(A'y|\varphi) f_{\hat{\beta}}(\hat{\beta}|\varphi, \beta)\, d\beta = \int_{R^p} f_{\omega,\hat{\beta}}(A'y, G'y|\varphi, \beta)\, d\beta \\ &= (2\pi\sigma^2)^{-n/2} |X'X|^{1/2} |\Sigma|^{-1/2} \int_{R^p} \exp\left\{\frac{-1}{2\sigma^2}(y - X\beta)'\Sigma^{-1}(y - X\beta)\right\} d\beta. \end{aligned}$$

The density $f_\omega(A'y|\varphi)$ can be written in the form

$$\begin{aligned} f_\omega(A'y|\varphi) &= (2\pi\sigma^2)^{-(n-p)/2} |X'X|^{1/2} |\Sigma|^{-1/2} |X'\Sigma^{-1}X|^{-1/2} \exp\left\{\frac{-1}{2\sigma^2} y'\Sigma^{-1}y\right\} \\ &\cdot \int_{R^p} \exp\left\{\frac{1}{\sigma^2} y'\Sigma^{-1}X\beta\right\} (2\pi\sigma^2)^{-p/2} |X'\Sigma^{-1}X|^{1/2} \exp\left\{\frac{-1}{2\sigma^2}\beta'X'\Sigma^{-1}X\beta\right\} d\beta. \end{aligned} \tag{6.40}$$

Let us remind that the moment generation function of $Y \sim N_n(\mu, V)$ is

$$M_Y(a) = E[\exp\{a'Y\}] = \exp\{-a'\mu + \frac{1}{2}a'Va\}.$$

By taking $Y = \beta,\ \mu = 0,\ V = \sigma^2(X'\Sigma^{-1}X)^{-1}$, and $a = \frac{1}{\sigma^2}X'\Sigma^{-1}y$ in (6.40), we get

$$\begin{aligned} f_\omega(A'y|\varphi) &= \xi \exp\left\{\frac{-1}{2\sigma^2} y'\Sigma^{-1}y\right\} \exp\left\{\frac{1}{2\sigma^2} y'\Sigma^{-1}X(X'\Sigma^{-1}X)^{-1}X'\Sigma^{-1}y\right\} \\ &= \xi \exp\left\{\frac{-1}{2\sigma^2} y'[\Sigma^{-1} - \Sigma^{-1}X(X'\Sigma^{-1}X)^{-1}X'\Sigma^{-1}]y\right\} \\ &= \xi \exp\left\{\frac{-1}{2\sigma^2} y'Py\right\}, \end{aligned}$$

where $\xi = (2\pi\sigma^2)^{-(n-p)/2} |X'X|^{1/2} |\Sigma|^{-1/2} |X'\Sigma^{-1}X|^{-1/2}$. □

The log-likelihood of $\boldsymbol{\omega}$ is

$$\ell(\boldsymbol{\varphi}; \boldsymbol{\omega}) = -\frac{n-p}{2}\log 2\pi - \frac{n-p}{2}\log\sigma^2 + \frac{1}{2}\log|\boldsymbol{X}'\boldsymbol{X}| - \frac{1}{2}\log|\boldsymbol{\Sigma}| - \frac{1}{2}\log|\boldsymbol{X}'\boldsymbol{\Sigma}^{-1}\boldsymbol{X}| - \frac{1}{2\sigma^2}\boldsymbol{y}'\boldsymbol{P}\boldsymbol{y}. \tag{6.41}$$

We remind that $\boldsymbol{\Sigma} = \left(\boldsymbol{\Sigma}_e + \sum_{i=1}^m \varphi_i \boldsymbol{G}_i\right)$, where $\boldsymbol{G}_i = \boldsymbol{Z}_i\boldsymbol{\Sigma}_{u_i}\boldsymbol{Z}_i'$ and thus, $\frac{\partial\boldsymbol{\Sigma}}{\partial\varphi_i} = \boldsymbol{G}_i$. By taking derivatives of $\ell(\boldsymbol{\sigma}; \boldsymbol{\omega})$ with respect to φ_i, we get

$$\begin{aligned}
\frac{\partial\ell(\boldsymbol{\varphi}; \boldsymbol{\omega})}{\partial\varphi_i} &= -\frac{1}{2}\operatorname{tr}(\boldsymbol{\Sigma}^{-1}\boldsymbol{G}_i) + \frac{1}{2}\operatorname{tr}\{(\boldsymbol{X}'\boldsymbol{\Sigma}^{-1}\boldsymbol{X})^{-1}\boldsymbol{X}'\boldsymbol{\Sigma}^{-1}\boldsymbol{G}_i\boldsymbol{\Sigma}^{-1}\boldsymbol{X}\} - \frac{1}{2\sigma^2}\boldsymbol{y}'\frac{\partial\boldsymbol{P}}{\partial\varphi_i}\boldsymbol{y} \\
&= -\frac{1}{2}\operatorname{tr}(\boldsymbol{\Sigma}^{-1}\boldsymbol{G}_i) + \frac{1}{2}\operatorname{tr}\{[\boldsymbol{\Sigma}^{-1}\boldsymbol{X}(\boldsymbol{X}'\boldsymbol{\Sigma}^{-1}\boldsymbol{X})^{-1}\boldsymbol{X}'\boldsymbol{\Sigma}^{-1}]\boldsymbol{G}_i\} - \frac{1}{2\sigma^2}\boldsymbol{y}'\frac{\partial\boldsymbol{P}}{\partial\varphi_i}\boldsymbol{y} \\
&= -\frac{1}{2}\operatorname{tr}(\boldsymbol{\Sigma}^{-1}\boldsymbol{G}_i) + \frac{1}{2}\operatorname{tr}([\boldsymbol{\Sigma}^{-1} - \boldsymbol{P}]\boldsymbol{G}_i) - \frac{1}{2\sigma^2}\boldsymbol{y}'\frac{\partial\boldsymbol{P}}{\partial\varphi_i}\boldsymbol{y} \\
&= -\frac{1}{2}\operatorname{tr}(\boldsymbol{P}\boldsymbol{G}_i) - \frac{1}{2\sigma^2}\boldsymbol{y}'\frac{\partial\boldsymbol{P}}{\partial\varphi_i}\boldsymbol{y}.
\end{aligned} \tag{6.42}$$

The REML equations for $\varphi_1, \ldots, \varphi_m$ are

$$\sigma^2\operatorname{tr}(\boldsymbol{P}\boldsymbol{G}_i) + \boldsymbol{y}'\frac{\partial\boldsymbol{P}}{\partial\varphi_i}\boldsymbol{y} = 0, \quad i = 1, \ldots, m.$$

The derivative of $\boldsymbol{P} = \boldsymbol{\Sigma}^{-1} - \boldsymbol{\Sigma}^{-1}\boldsymbol{X}(\boldsymbol{X}'\boldsymbol{\Sigma}^{-1}\boldsymbol{X})^{-1}\boldsymbol{X}'\boldsymbol{\Sigma}^{-1}$ with respect to φ_i is

$$\begin{aligned}
\frac{\partial\boldsymbol{P}}{\partial\varphi_i} &= -(\boldsymbol{\Sigma}^{-1}\boldsymbol{G}_i\boldsymbol{\Sigma}^{-1}) + (\boldsymbol{\Sigma}^{-1}\boldsymbol{G}_i\boldsymbol{\Sigma}^{-1})\boldsymbol{X}(\boldsymbol{X}'\boldsymbol{\Sigma}^{-1}\boldsymbol{X})^{-1}\boldsymbol{X}'\boldsymbol{\Sigma}^{-1} \\
&\quad - \boldsymbol{\Sigma}^{-1}\boldsymbol{X}(\boldsymbol{X}'\boldsymbol{\Sigma}^{-1}\boldsymbol{X})^{-1}\boldsymbol{X}'(\boldsymbol{\Sigma}^{-1}\boldsymbol{G}_i\boldsymbol{\Sigma}^{-1})\boldsymbol{X}(\boldsymbol{X}'\boldsymbol{\Sigma}^{-1}\boldsymbol{X})^{-1}\boldsymbol{X}'\boldsymbol{\Sigma}^{-1} \\
&\quad + \boldsymbol{\Sigma}^{-1}\boldsymbol{X}(\boldsymbol{X}'\boldsymbol{\Sigma}^{-1}\boldsymbol{X})^{-1}\boldsymbol{X}'(\boldsymbol{\Sigma}^{-1}\boldsymbol{G}_i\boldsymbol{\Sigma}^{-1}) \\
&= -(\boldsymbol{\Sigma}^{-1}\boldsymbol{G}_i\boldsymbol{\Sigma}^{-1}) + (\boldsymbol{\Sigma}^{-1}\boldsymbol{G}_i\boldsymbol{\Sigma}^{-1})\boldsymbol{X}\boldsymbol{G}' \\
&\quad - \boldsymbol{G}\boldsymbol{X}'(\boldsymbol{\Sigma}^{-1}\boldsymbol{G}_i\boldsymbol{\Sigma}^{-1})\boldsymbol{X}\boldsymbol{G}' + \boldsymbol{G}\boldsymbol{X}'(\boldsymbol{\Sigma}^{-1}\boldsymbol{G}_i\boldsymbol{\Sigma}^{-1}) \\
&= -(\boldsymbol{I} - \boldsymbol{G}\boldsymbol{X}')\boldsymbol{\Sigma}^{-1}\boldsymbol{G}_i\boldsymbol{\Sigma}^{-1}(\boldsymbol{I} - \boldsymbol{G}\boldsymbol{X}')' = -\boldsymbol{P}\boldsymbol{G}_i\boldsymbol{P},
\end{aligned} \tag{6.43}$$

where $\boldsymbol{G} = \boldsymbol{\Sigma}^{-1}\boldsymbol{X}(\boldsymbol{X}'\boldsymbol{\Sigma}^{-1}\boldsymbol{X})^{-1}$. As $\boldsymbol{y} - \boldsymbol{X}\hat{\boldsymbol{\beta}} = (\boldsymbol{I} - \boldsymbol{G}\boldsymbol{X}')'\boldsymbol{y}$, we have

$$\begin{aligned}
\boldsymbol{y}'\frac{\partial\boldsymbol{P}}{\partial\varphi_i}\boldsymbol{y} &= -(\boldsymbol{y} - \boldsymbol{X}\hat{\boldsymbol{\beta}})'\boldsymbol{\Sigma}^{-1}\boldsymbol{G}_i\boldsymbol{\Sigma}^{-1}(\boldsymbol{y} - \boldsymbol{X}\hat{\boldsymbol{\beta}}) \\
&= -(\boldsymbol{y} - \boldsymbol{X}\hat{\boldsymbol{\beta}})'\boldsymbol{\Sigma}^{-1}\boldsymbol{Z}_i\boldsymbol{\Sigma}_{u_i}\boldsymbol{Z}_i'\boldsymbol{\Sigma}^{-1}(\boldsymbol{y} - \boldsymbol{X}\hat{\boldsymbol{\beta}}).
\end{aligned} \tag{6.44}$$

The REML equations for $\varphi_1, \ldots, \varphi_m$ are

$$\sigma^2 \operatorname{tr}(\boldsymbol{P}\boldsymbol{G}_i) = (\boldsymbol{y} - \boldsymbol{X}\hat{\boldsymbol{\beta}})'\boldsymbol{\Sigma}^{-1}\boldsymbol{Z}_i\boldsymbol{\Sigma}_{u_i}\boldsymbol{Z}_i'\boldsymbol{\Sigma}^{-1}(\boldsymbol{y} - \boldsymbol{X}\hat{\boldsymbol{\beta}}), \quad i = 1, \ldots, m. \tag{6.45}$$

Let us define

$$\hat{\boldsymbol{u}} = \boldsymbol{\Sigma}_u\boldsymbol{Z}'\boldsymbol{\Sigma}^{-1}(\boldsymbol{y} - \boldsymbol{X}\hat{\boldsymbol{\beta}}) = (\hat{\boldsymbol{u}}_1, \ldots, \hat{\boldsymbol{u}}_m)',$$

where

$$\hat{\boldsymbol{u}}_i = \varphi_i\boldsymbol{\Sigma}_{u_i}\boldsymbol{Z}_i'\boldsymbol{\Sigma}^{-1}(\boldsymbol{y} - \boldsymbol{X}\hat{\boldsymbol{\beta}}), \quad i = 1, \ldots, m.$$

Using this notation the REML equations can be written in the form

$$\sigma^2 \operatorname{tr}(\boldsymbol{P}\boldsymbol{G}_i) = \frac{1}{\varphi_i^2}\hat{\boldsymbol{u}}_i'\boldsymbol{\Sigma}_{u_i}^{-1}\hat{\boldsymbol{u}}_i, \quad i = 1, \ldots, m, \tag{6.46}$$

or equivalently

$$\varphi_i = \frac{\hat{\boldsymbol{u}}_i'\boldsymbol{\Sigma}_{u_i}^{-1}\hat{\boldsymbol{u}}_i}{\varphi_i\sigma^2 \operatorname{tr}(\boldsymbol{P}\boldsymbol{G}_i)}, \quad i = 1, \ldots, m. \tag{6.47}$$

The REML equation for σ^2 is (obtained by deriving (6.41) with respect to σ^2)

$$\sigma^2 = \frac{1}{n-p}\boldsymbol{y}'\boldsymbol{P}\boldsymbol{y}. \tag{6.48}$$

By using (6.47) and (6.48), we obtain the following algorithm.

1. Set initial values of σ^2 and φ_i, $i = 1, \ldots, m$.
2. Update their values by using (6.47) and (6.48) until convergence.
3. Output: $\hat{\sigma}^2$ and $\hat{\varphi}_i$, $i = 1, \ldots, m$.
4. Calculate $\hat{\boldsymbol{\beta}} = (\boldsymbol{X}'\hat{\boldsymbol{\Sigma}}^{-1}\boldsymbol{X})^{-1}\boldsymbol{X}'\hat{\boldsymbol{\Sigma}}^{-1}\boldsymbol{y}$, where $\hat{\boldsymbol{\Sigma}} = \left(\boldsymbol{\Sigma}_e + \sum_{i=1}^m \hat{\varphi}_i\boldsymbol{G}_i\right)$.

The particularization of Eqs. (6.46) to the case $m = 1$ can be obtained as follows. If $m = 1$, we have that $\boldsymbol{Z} = \boldsymbol{Z}_1$ and $\boldsymbol{\Sigma}_u = \varphi_1\boldsymbol{\Sigma}_{u_1}$ are $n \times q_1$ and $q_1 \times q_1$ matrices, respectively. The trace $\operatorname{tr}(\boldsymbol{P}\boldsymbol{G}_1)$ can be written in the form

$$\operatorname{tr}(\boldsymbol{P}\boldsymbol{G}_1) = \operatorname{tr}(\boldsymbol{P}\boldsymbol{Z}\boldsymbol{\Sigma}_{u_1}\boldsymbol{Z}') = \operatorname{tr}(\boldsymbol{\Sigma}_{u_1}^{-1}\boldsymbol{\Sigma}_{u_1}\boldsymbol{Z}'\boldsymbol{P}\boldsymbol{Z}\boldsymbol{\Sigma}_{u_1}). \tag{6.49}$$

We have that

$$\boldsymbol{\Sigma}_{u_1}\boldsymbol{Z}'\boldsymbol{P}\boldsymbol{Z}\boldsymbol{\Sigma}_{u_1} = \boldsymbol{\Sigma}_{u_1}\boldsymbol{Z}'\boldsymbol{\Sigma}^{-1}\boldsymbol{Z}\boldsymbol{\Sigma}_{u_1} - \boldsymbol{\Sigma}_{u_1}\boldsymbol{Z}'\boldsymbol{\Sigma}^{-1}\boldsymbol{X}(\boldsymbol{X}'\boldsymbol{\Sigma}^{-1}\boldsymbol{X})^{-1}\boldsymbol{X}'\boldsymbol{\Sigma}^{-1}\boldsymbol{Z}\boldsymbol{\Sigma}_{u_1}. \tag{6.50}$$

By applying the formula $(A+BCD)^{-1} = A^{-1} - A^{-1}B(C^{-1}+DA^{-1}B)^{-1}DA^{-1}$, we get

$$\boldsymbol{T} = (\boldsymbol{\Sigma}_u^{-1} + \boldsymbol{Z}'\boldsymbol{\Sigma}_e^{-1}\boldsymbol{Z})^{-1} = \boldsymbol{\Sigma}_u - \boldsymbol{\Sigma}_u\boldsymbol{Z}'\boldsymbol{\Sigma}^{-1}\boldsymbol{Z}\boldsymbol{\Sigma}_u. \tag{6.51}$$

Pre and post-multiplying by $\boldsymbol{\Sigma}_u^{-1}$ in (6.51), we get

$$\boldsymbol{Z}'\boldsymbol{\Sigma}^{-1}\boldsymbol{Z} = \frac{1}{\varphi_1}\boldsymbol{\Sigma}_{u_1}^{-1} - \frac{1}{\varphi_1^2}\boldsymbol{\Sigma}_{u_1}^{-1}\boldsymbol{T}\boldsymbol{\Sigma}_{u_1}^{-1}. \tag{6.52}$$

Post-multiplying by $\boldsymbol{Z}'\boldsymbol{\Sigma}_e^{-1}\varphi_1^{-1}$ in (6.51) and doing some transformations, we get

$$\begin{aligned} \boldsymbol{T}\boldsymbol{Z}'\boldsymbol{\Sigma}_e^{-1}\varphi_1^{-1} &= \boldsymbol{\Sigma}_u\boldsymbol{Z}'\boldsymbol{\Sigma}_e^{-1}\varphi_1^{-1} - \boldsymbol{\Sigma}_u\boldsymbol{Z}'\boldsymbol{\Sigma}^{-1}\boldsymbol{Z}\boldsymbol{\Sigma}_u\boldsymbol{Z}'\boldsymbol{\Sigma}_e^{-1}\varphi_1^{-1} \\ &= \varphi_1^{-1}\boldsymbol{\Sigma}_u\boldsymbol{Z}'\boldsymbol{\Sigma}^{-1}(\boldsymbol{\Sigma} - \boldsymbol{Z}\boldsymbol{\Sigma}_u\boldsymbol{Z}')\boldsymbol{\Sigma}_e^{-1} = \varphi^{-1}\boldsymbol{\Sigma}_u\boldsymbol{Z}'\boldsymbol{\Sigma}^{-1} = \boldsymbol{\Sigma}_{u_1}\boldsymbol{Z}'\boldsymbol{\Sigma}^{-1}. \end{aligned} \tag{6.53}$$

By replacing (6.52) and (6.53) in (6.50), we get

$$\begin{aligned} \boldsymbol{\Sigma}_{u_1}\boldsymbol{Z}'\boldsymbol{P}\boldsymbol{Z}\boldsymbol{\Sigma}_{u_1} &= \frac{1}{\varphi_1}\boldsymbol{\Sigma}_{u_1} - \frac{1}{\varphi_1^2}\boldsymbol{T} - \frac{1}{\varphi_1^2}\boldsymbol{T}\boldsymbol{Z}'\boldsymbol{\Sigma}_e^{-1}\boldsymbol{X}(\boldsymbol{X}'\boldsymbol{\Sigma}^{-1}\boldsymbol{X})^{-1}\boldsymbol{X}'\boldsymbol{\Sigma}_e^{-1}\boldsymbol{Z}\boldsymbol{T} \\ &= \frac{1}{\varphi_1}\left[\boldsymbol{\Sigma}_{u_1} - \frac{1}{\varphi_1}(\boldsymbol{T} + \boldsymbol{T}\boldsymbol{Z}'\boldsymbol{\Sigma}_e^{-1}\boldsymbol{X}(\boldsymbol{X}'\boldsymbol{\Sigma}^{-1}\boldsymbol{X})^{-1}\boldsymbol{X}'\boldsymbol{\Sigma}_e^{-1}\boldsymbol{Z}\boldsymbol{T})\right]. \end{aligned}$$

Let us consider the matrix

$$\boldsymbol{T}^* = \boldsymbol{T} + \boldsymbol{T}\boldsymbol{Z}'\boldsymbol{\Sigma}_e^{-1}\boldsymbol{X}(\boldsymbol{X}'\boldsymbol{\Sigma}^{-1}\boldsymbol{X})^{-1}\boldsymbol{X}'\boldsymbol{\Sigma}_e^{-1}\boldsymbol{Z}\boldsymbol{T}.$$

Then, we have

$$\boldsymbol{\Sigma}_{u_1}\boldsymbol{Z}'\boldsymbol{P}\boldsymbol{Z}\boldsymbol{\Sigma}_{u_1} = \frac{1}{\varphi_1}\left[\boldsymbol{\Sigma}_{u_1} - \frac{1}{\varphi_1}\boldsymbol{T}^*\right] \tag{6.54}$$

and by replacing this expression in (6.49), we get

$$\mathrm{tr}(\boldsymbol{P}\boldsymbol{G}_1) = \frac{1}{\varphi_1}\left[q_1 - \frac{1}{\varphi_1}\mathrm{tr}(\boldsymbol{\Sigma}_{u_1}^{-1}\boldsymbol{T}^*)\right]. \tag{6.55}$$

The particularization of Eqs. (6.46) to the case $m = 1$ is the REML equation for φ_1, i.e.

$$\sigma^2\left[q_1 - \frac{1}{\varphi_1}\mathrm{tr}(\boldsymbol{\Sigma}_{u_1}^{-1}\boldsymbol{T}^*)\right] = \frac{1}{\varphi_1}\hat{\boldsymbol{u}}_1'\boldsymbol{\Sigma}_{u_1}^{-1}\hat{\boldsymbol{u}}_1,$$

or equivalently

$$\varphi_1 = \frac{\hat{\boldsymbol{u}}_1' \boldsymbol{\Sigma}_{u_1}^{-1} \hat{\boldsymbol{u}}_1}{\sigma^2 \left[q_1 - \frac{1}{\varphi_1} \mathrm{tr}(\boldsymbol{\Sigma}_{u_1}^{-1} \boldsymbol{T}^*) \right]}. \tag{6.56}$$

6.6 Henderson 3 Estimation

This sections presents Henderson 3 method for estimating parameters of the model (6.13).

6.6.1 *Description of the Method*

The maximum likelihood method gives at the same time the estimates of the model coefficients $\boldsymbol{\beta}$ and the variance components $\sigma_0^2, \ldots, \sigma_m^2$. This section presents the method of fitting constants for estimating the variance components. This approach estimates the regression parameter $\boldsymbol{\beta}$ by the least squares method and predicts the random effects by using the BLUP formulas, but replacing the true variances by their obtained estimates. The predictor of $\boldsymbol{u}$ is called EBLUP (empirical BLUP). The method of fitting constants is also known as Henderson 3 method, since its introduction by Henderson (1953). We write the general linear mixed model, $\boldsymbol{y} = \mathbb{X}\boldsymbol{\beta} + \boldsymbol{e}$, in the form

$$\boldsymbol{y} = \boldsymbol{X}_1 \boldsymbol{\beta}_1 + \boldsymbol{X}_2 \boldsymbol{\beta}_2 + \boldsymbol{e}, \tag{6.57}$$

where $\mathbb{X} = (\boldsymbol{X}_1, \boldsymbol{X}_2)$, $\boldsymbol{\beta} = (\boldsymbol{\beta}_1', \boldsymbol{\beta}_2')'$, $\boldsymbol{e} \sim N(\boldsymbol{0}, \sigma_0^2 \boldsymbol{W}^{-1})$ and $\boldsymbol{W}$ is a known symmetric and positive definite matrix. We assume that $\mathbb{X}'\boldsymbol{W}\mathbb{X}$ and $\boldsymbol{X}_1'\boldsymbol{W}\boldsymbol{X}_1$ are invertible. The partition simply divides $\boldsymbol{\beta}$ in two groups of effects $\boldsymbol{\beta}_1$ and $\boldsymbol{\beta}_2$, without taking into account if they represent fixed or random effects. This issue will be considered later.

We apply the transformation

$$\boldsymbol{W}^{1/2}\boldsymbol{y} = \boldsymbol{W}^{1/2}\boldsymbol{X}_1\boldsymbol{\beta}_1 + \boldsymbol{W}^{1/2}\boldsymbol{X}_2\boldsymbol{\beta}_2 + \boldsymbol{W}^{1/2}\boldsymbol{e}$$

and we denote $\boldsymbol{y}^* = \boldsymbol{W}^{1/2}\boldsymbol{y}$, $\boldsymbol{X}_1^* = \boldsymbol{W}^{1/2}\boldsymbol{X}_1$, $\boldsymbol{X}_2^* = \boldsymbol{W}^{1/2}\boldsymbol{X}_2$, and $\boldsymbol{e}^* = \boldsymbol{W}^{1/2}\boldsymbol{e}$. The new model is

$$\boldsymbol{y}^* = \boldsymbol{X}_1^*\boldsymbol{\beta}_1 + \boldsymbol{X}_2^*\boldsymbol{\beta}_2 + \boldsymbol{e}^*, \tag{6.58}$$

with $\boldsymbol{e}^* \sim N(\boldsymbol{0}, \sigma_0^2 \boldsymbol{I}_n)$.

If we fit the model (6.58) under the assumption that $\boldsymbol{\beta}_1$ and $\boldsymbol{\beta}_2$ are fixed effects, the total sum of squares is

$$SST = \boldsymbol{y}^{*\prime}\boldsymbol{y}^* = \boldsymbol{y}'\boldsymbol{W}\boldsymbol{y}. \tag{6.59}$$

The residual sum of squares is

$$SSE(\boldsymbol{\beta}_1, \boldsymbol{\beta}_2) = \boldsymbol{y}^{*\prime}\boldsymbol{M}^*\boldsymbol{y}^* = \boldsymbol{y}'\boldsymbol{M}\boldsymbol{y}, \tag{6.60}$$

where $\boldsymbol{M}^* = [\boldsymbol{I}_n - \mathbb{X}^*(\mathbb{X}^{*\prime}\mathbb{X}^*)^{-1}\mathbb{X}^{*\prime}]'[\boldsymbol{I}_n - \mathbb{X}^*(\mathbb{X}^{*\prime}\mathbb{X}^*)^{-1}\mathbb{X}^{*\prime}]$ and so

$$\boldsymbol{M} = [\boldsymbol{I}_n - \mathbb{X}(\mathbb{X}'\boldsymbol{W}\mathbb{X})^{-1}\mathbb{X}'\boldsymbol{W}]'\boldsymbol{W}[\boldsymbol{I}_n - \mathbb{X}(\mathbb{X}'\boldsymbol{W}\mathbb{X})^{-1}\mathbb{X}'\boldsymbol{W}].$$

The reduction in the total sum of squares (regression sum of squares) is

$$SSR(\boldsymbol{\beta}_1, \boldsymbol{\beta}_2) = SST - SSE(\boldsymbol{\beta}_1, \boldsymbol{\beta}_2) = \boldsymbol{y}'\boldsymbol{Q}\boldsymbol{y},$$

where $\boldsymbol{Q} = \boldsymbol{W}\mathbb{X}(\mathbb{X}'\boldsymbol{W}\mathbb{X})^{-1}\mathbb{X}'\boldsymbol{W}$.

If we fit the submodel

$$\boldsymbol{y}^* = \boldsymbol{X}_1^*\boldsymbol{\beta}_1 + \boldsymbol{e}^*,$$

under the assumption that $\boldsymbol{\beta}_1$ is a fixed effect, the residual sum of squares is

$$SSE(\boldsymbol{\beta}_1) = \boldsymbol{y}'\boldsymbol{M}_1\boldsymbol{y}, \tag{6.61}$$

where $\boldsymbol{M}_1 = [\boldsymbol{I}_n - \boldsymbol{X}_1(\boldsymbol{X}_1'\boldsymbol{W}\boldsymbol{X}_1)^{-1}\boldsymbol{X}_1'\boldsymbol{W}]'\boldsymbol{W}[\boldsymbol{I}_n - \boldsymbol{X}_1(\boldsymbol{X}_1'\boldsymbol{W}\boldsymbol{X}_1)^{-1}\boldsymbol{X}_1'\boldsymbol{W}]$. The reduction in the sum of squares (regression sum of squares) is

$$SSR(\boldsymbol{\beta}_1) = SST - SSE(\boldsymbol{\beta}_1) = \boldsymbol{y}'\boldsymbol{Q}_1\boldsymbol{y},$$

where $\boldsymbol{Q}_1 = \boldsymbol{W}\boldsymbol{X}_1(\boldsymbol{X}_1'\boldsymbol{W}\boldsymbol{X}_1)^{-1}\boldsymbol{X}_1'\boldsymbol{W}$. The reduction of the total sum of squares because of the introduction of $\boldsymbol{X}_2$ in the model, with respect to the model that only had $\boldsymbol{X}_1$, is

$$SSR(\boldsymbol{\beta}_2|\boldsymbol{\beta}_1) = SSR(\boldsymbol{\beta}_1, \boldsymbol{\beta}_2) - SSR(\boldsymbol{\beta}_1) = SSE(\boldsymbol{\beta}_1) - SSE(\boldsymbol{\beta}_1, \boldsymbol{\beta}_2).$$

To introduce the Henderson 3 method, we first calculate the expectation of $SSR(\boldsymbol{\beta}_2|\boldsymbol{\beta}_1)$ and $SSR(\boldsymbol{\beta}_1, \boldsymbol{\beta}_2)$. In a second step, we modify these statistics to make them unbiased. Note that all the considered sums of squares are quadratic functions of $\boldsymbol{y}$, so that we will apply (6.29) systematically. For a general linear model $\boldsymbol{y} = \mathbb{X}\boldsymbol{\beta} + \boldsymbol{e}$, where $\boldsymbol{e} \sim N(\boldsymbol{0}, \sigma_0^2\boldsymbol{W}^{-1})$ and $\boldsymbol{\beta}$ may contain fixed or random effects, we have $E[\boldsymbol{y}] = \mathbb{X}E[\boldsymbol{\beta}]$ and $\text{var}(\boldsymbol{y}) = \mathbb{X}\text{var}(\boldsymbol{\beta})\mathbb{X}' + \sigma_0^2\boldsymbol{W}^{-1}$ (under the assumption that the possibly random part of $\boldsymbol{\beta}$ and the vector $\boldsymbol{e}$ are independent).

From (6.29), we obtain for any symmetric matrix $\boldsymbol{Q}$

$$
\begin{aligned}
E[\boldsymbol{y}'\boldsymbol{Q}\boldsymbol{y}] &= \mathrm{tr}\big(\boldsymbol{Q}\big[\mathbb{X}\mathrm{var}(\boldsymbol{\beta})\mathbb{X}' + \sigma_0^2\boldsymbol{W}^{-1}\big]\big) + E[\boldsymbol{\beta}]'\mathbb{X}'\boldsymbol{Q}\mathbb{X}E[\boldsymbol{\beta}] \\
&= \mathrm{tr}\big(\boldsymbol{Q}\mathbb{X}\mathrm{var}(\boldsymbol{\beta})\mathbb{X}'\big) + \sigma_0^2\mathrm{tr}\big(\boldsymbol{Q}\boldsymbol{W}^{-1}\big) + \mathrm{tr}\big(\boldsymbol{Q}\mathbb{X}E[\boldsymbol{\beta}]E[\boldsymbol{\beta}]'\mathbb{X}'\big) \\
&= \mathrm{tr}\big(\boldsymbol{Q}\mathbb{X}E[\boldsymbol{\beta}\boldsymbol{\beta}']\mathbb{X}'\big) + \sigma_0^2\mathrm{tr}\big(\boldsymbol{Q}\boldsymbol{W}^{-1}\big) = \mathrm{tr}\big(\mathbb{X}'\boldsymbol{Q}\mathbb{X}E[\boldsymbol{\beta}\boldsymbol{\beta}']\big) + \sigma_0^2\mathrm{tr}\big(\boldsymbol{Q}\boldsymbol{W}^{-1}\big).
\end{aligned}
$$

The expectation of the total sum of squares appearing in (6.59) is

$$
E[SST] = E[\boldsymbol{y}'\boldsymbol{W}\boldsymbol{y}] = \mathrm{tr}\left(\mathbb{X}'\boldsymbol{W}\mathbb{X}E[\boldsymbol{\beta}\boldsymbol{\beta}']\right) + \sigma_0^2\mathrm{tr}\left(\boldsymbol{I}_n\right) = \mathrm{tr}\left(\mathbb{X}'\boldsymbol{W}\mathbb{X}E[\boldsymbol{\beta}\boldsymbol{\beta}']\right) + n\sigma_0^2. \tag{6.62}
$$

The expectation of the sum of residual squares in (6.60) is

$$
E[SSE(\boldsymbol{\beta}_1, \boldsymbol{\beta}_2)] = E[\boldsymbol{y}'\boldsymbol{M}\boldsymbol{y}] = \mathrm{tr}\big(\mathbb{X}'\boldsymbol{M}\mathbb{X}E[\boldsymbol{\beta}\boldsymbol{\beta}']\big) + \sigma_0^2\mathrm{tr}\big(\boldsymbol{M}\boldsymbol{W}^{-1}\big).
$$

This expression can be simplified if we take into account that

$$
\begin{aligned}
\mathbb{X}'\boldsymbol{M}\mathbb{X} &= \mathbb{X}'[\boldsymbol{I}_n - \mathbb{X}(\mathbb{X}'\boldsymbol{W}\mathbb{X})^{-1}\mathbb{X}'\boldsymbol{W}]'\boldsymbol{W}[\boldsymbol{I}_n - \mathbb{X}(\mathbb{X}'\boldsymbol{W}\mathbb{X})^{-1}\mathbb{X}'\boldsymbol{W}]\mathbb{X} = \mathbb{X}'\boldsymbol{W}\mathbb{X} \\
&\quad - 2\mathbb{X}'\boldsymbol{W}\mathbb{X}(\mathbb{X}'\boldsymbol{W}\mathbb{X})^{-1}\mathbb{X}'\boldsymbol{W}\mathbb{X} + \mathbb{X}'\boldsymbol{W}\mathbb{X}(\mathbb{X}'\boldsymbol{W}\mathbb{X})^{-1}\mathbb{X}'\boldsymbol{W}\mathbb{X}(\mathbb{X}'\boldsymbol{W}\mathbb{X})^{-1}\mathbb{X}'\boldsymbol{W}\mathbb{X} \\
&= \boldsymbol{0}
\end{aligned}
$$

and

$$
\begin{aligned}
\boldsymbol{M}\boldsymbol{W}^{-1} &= [\boldsymbol{I}_n - \mathbb{X}(\mathbb{X}'\boldsymbol{W}\mathbb{X})^{-1}\mathbb{X}'\boldsymbol{W}]'\boldsymbol{W}[\boldsymbol{I}_n - \mathbb{X}(\mathbb{X}'\boldsymbol{W}\mathbb{X})^{-1}\mathbb{X}'\boldsymbol{W}]\boldsymbol{W}^{-1} \\
&= [\boldsymbol{I}_n - \mathbb{X}(\mathbb{X}'\boldsymbol{W}\mathbb{X})^{-1}\mathbb{X}'\boldsymbol{W}]'[\boldsymbol{I}_n - \boldsymbol{W}\mathbb{X}(\mathbb{X}'\boldsymbol{W}\mathbb{X})^{-1}\mathbb{X}'] \\
&= \boldsymbol{I}_n - 2\boldsymbol{W}\mathbb{X}(\mathbb{X}'\boldsymbol{W}\mathbb{X})^{-1}\mathbb{X}' + \boldsymbol{W}\mathbb{X}(\mathbb{X}'\boldsymbol{W}\mathbb{X})^{-1}\mathbb{X}'\boldsymbol{W}\mathbb{X}(\mathbb{X}'\boldsymbol{W}\mathbb{X})^{-1}\mathbb{X}' \\
&= \boldsymbol{I}_n - \boldsymbol{W}\mathbb{X}(\mathbb{X}'\boldsymbol{W}\mathbb{X})^{-1}\mathbb{X}'.
\end{aligned}
$$

Since $\mathbb{X}'\boldsymbol{W}\mathbb{X}(\mathbb{X}'\boldsymbol{W}\mathbb{X})^{-1}$ is equal to the identity matrix, we obtain that

$$
\begin{aligned}
\mathrm{tr}\big(\boldsymbol{M}\boldsymbol{W}^{-1}\big) &= \mathrm{tr}\big(\boldsymbol{I}_n - \boldsymbol{W}\mathbb{X}(\mathbb{X}'\boldsymbol{W}\mathbb{X})^{-1}\mathbb{X}'\big) = n - \mathrm{tr}\big(\mathbb{X}'\boldsymbol{W}\mathbb{X}(\mathbb{X}'\boldsymbol{W}\mathbb{X})^{-1}\big) \\
&= n - p = n - \mathrm{r}\big(\mathbb{X}\big),
\end{aligned}
$$

where $\mathrm{r}(\mathbb{X})$ denotes the rank of $\mathbb{X}$. This result can also be proved in the case $\mathrm{r}(\mathbb{X}) < p$. Therefore

$$
E[SSE(\boldsymbol{\beta}_1, \boldsymbol{\beta}_2)] = \sigma_0^2\big[n - \mathrm{r}(\mathbb{X})\big] \tag{6.63}
$$

and also with (6.62) and (6.63) we obtain that

$$E[SSR(\boldsymbol{\beta}_1, \boldsymbol{\beta}_2)] = E[SST] - E[SSE(\boldsymbol{\beta}_1, \boldsymbol{\beta}_2)] = \mathrm{tr}\big(\mathbb{X}'\boldsymbol{W}\mathbb{X}E[\boldsymbol{\beta}\boldsymbol{\beta}']\big) + \sigma_0^2 \mathrm{r}(\mathbb{X}).$$

From the model (6.57), it follows that

$$\mathbb{X}'\boldsymbol{W}\mathbb{X} = \begin{pmatrix} \boldsymbol{X}_1' \\ \boldsymbol{X}_2' \end{pmatrix} \boldsymbol{W}\,(\boldsymbol{X}_1,\ \boldsymbol{X}_2) = \begin{pmatrix} \boldsymbol{X}_1'\boldsymbol{W}\boldsymbol{X}_1 & \boldsymbol{X}_1'\boldsymbol{W}\boldsymbol{X}_2 \\ \boldsymbol{X}_2'\boldsymbol{W}\boldsymbol{X}_1 & \boldsymbol{X}_2'\boldsymbol{W}\boldsymbol{X}_2 \end{pmatrix}.$$

Consequently

$$E[SSR(\boldsymbol{\beta}_1, \boldsymbol{\beta}_2)] = \mathrm{tr}\left\{ \begin{pmatrix} \boldsymbol{X}_1'\boldsymbol{W}\boldsymbol{X}_1 & \boldsymbol{X}_1'\boldsymbol{W}\boldsymbol{X}_2 \\ \boldsymbol{X}_2'\boldsymbol{W}\boldsymbol{X}_1 & \boldsymbol{X}_2'\boldsymbol{W}\boldsymbol{X}_2 \end{pmatrix} E[\boldsymbol{\beta}\boldsymbol{\beta}'] \right\} + \sigma_0^2 \mathrm{r}(\mathbb{X}). \tag{6.64}$$

The following calculations are done under the assumption that model (6.57) holds with $\boldsymbol{\beta}_1$ and $\boldsymbol{\beta}_2$ possibly random. From (6.61) and (6.29), we obtain

$$\begin{aligned} E[SSE(\boldsymbol{\beta}_1)] &= \mathrm{tr}\left(\mathbb{X}'\boldsymbol{M}_1\mathbb{X}E[\boldsymbol{\beta}\boldsymbol{\beta}']\right) + \sigma_0^2 \mathrm{tr}\left(\boldsymbol{M}_1\boldsymbol{W}^{-1}\right) \\ &= \mathrm{tr}\left(\mathbb{X}'\boldsymbol{M}_1\mathbb{X}E[\boldsymbol{\beta}\boldsymbol{\beta}']\right) + \sigma_0^2\,[n - \mathrm{r}\,(\boldsymbol{X}_1)]\,. \end{aligned} \tag{6.65}$$

From (6.62) and (6.65), we have that

$$E[SSR(\boldsymbol{\beta}_1)] = E[SST] - E[SSE(\boldsymbol{\beta}_1)] = \mathrm{tr}\left(\mathbb{X}'\boldsymbol{Q}_1\mathbb{X}E[\boldsymbol{\beta}\boldsymbol{\beta}']\right) + \sigma_0^2 \mathrm{r}\,(\boldsymbol{X}_1)\,,$$

where $\boldsymbol{Q}_1 = \boldsymbol{W} - \boldsymbol{M}_1 = \boldsymbol{W}\boldsymbol{X}_1(\boldsymbol{X}_1'\boldsymbol{W}\boldsymbol{X}_1)^{-1}\boldsymbol{X}_1'\boldsymbol{W}$. If $\boldsymbol{X}_1'\boldsymbol{W}\boldsymbol{X}_1$ is invertible, then we get

$$\begin{aligned} \mathbb{X}'\boldsymbol{Q}_1\mathbb{X} &= \begin{pmatrix} \boldsymbol{X}_1' \\ \boldsymbol{X}_2' \end{pmatrix} \boldsymbol{W}\boldsymbol{X}_1(\boldsymbol{X}_1'\boldsymbol{W}\boldsymbol{X}_1)^{-1}\boldsymbol{X}_1'\boldsymbol{W}\,(\boldsymbol{X}_1\ \boldsymbol{X}_2) \\ &= \begin{pmatrix} \boldsymbol{X}_1'\boldsymbol{W}\boldsymbol{X}_1 & \boldsymbol{X}_1'\boldsymbol{W}\boldsymbol{X}_1(\boldsymbol{X}_1'\boldsymbol{W}\boldsymbol{X}_1)^{-1}\boldsymbol{X}_1'\boldsymbol{W}\boldsymbol{X}_2 \\ \boldsymbol{X}_2'\boldsymbol{W}\boldsymbol{X}_1(\boldsymbol{X}_1'\boldsymbol{W}\boldsymbol{X}_1)^{-1}\boldsymbol{X}_1'\boldsymbol{W}\boldsymbol{X}_1 & \boldsymbol{X}_2'\boldsymbol{W}\boldsymbol{X}_1(\boldsymbol{X}_1'\boldsymbol{W}\boldsymbol{X}_1)^{-1}\boldsymbol{X}_1'\boldsymbol{W}\boldsymbol{X}_2 \end{pmatrix} \\ &= \begin{pmatrix} \boldsymbol{X}_1'\boldsymbol{W}\boldsymbol{X}_1 & \boldsymbol{X}_1'\boldsymbol{W}\boldsymbol{X}_2 \\ \boldsymbol{X}_2'\boldsymbol{W}\boldsymbol{X}_1 & \boldsymbol{X}_2'\boldsymbol{W}\boldsymbol{X}_1(\boldsymbol{X}_1'\boldsymbol{W}\boldsymbol{X}_1)^{-1}\boldsymbol{X}_1'\boldsymbol{W}\boldsymbol{X}_2 \end{pmatrix} \end{aligned}$$

and

$$E[SSR(\boldsymbol{\beta}_1)] = \mathrm{tr}\left\{ \begin{pmatrix} \boldsymbol{X}_1'\boldsymbol{W}\boldsymbol{X}_1 & \boldsymbol{X}_1'\boldsymbol{W}\boldsymbol{X}_2 \\ \boldsymbol{X}_2'\boldsymbol{W}\boldsymbol{X}_1 & \boldsymbol{X}_2'\boldsymbol{W}\boldsymbol{X}_1(\boldsymbol{X}_1'\boldsymbol{W}\boldsymbol{X}_1)^{-1}\boldsymbol{X}_1'\boldsymbol{W}\boldsymbol{X}_2 \end{pmatrix} E[\boldsymbol{\beta}\boldsymbol{\beta}'] \right\} + \sigma_0^2 \mathrm{r}(\boldsymbol{X}_1). \tag{6.66}$$

Applying (6.64) and (6.66), we obtain

$$E[SSR(\boldsymbol{\beta}_2|\boldsymbol{\beta}_1)] = E[SSR(\boldsymbol{\beta}_1, \boldsymbol{\beta}_2)] - E[SSR(\boldsymbol{\beta}_1)] \tag{6.67}$$

$$= \operatorname{tr}\left\{\begin{pmatrix} \mathbf{0} & \mathbf{0} \\ \mathbf{0} & \boldsymbol{X}_2'\boldsymbol{W}[\boldsymbol{W}^{-1} - \boldsymbol{X}_1(\boldsymbol{X}_1'\boldsymbol{W}\boldsymbol{X}_1)^{-1}\boldsymbol{X}_1']\boldsymbol{W}\boldsymbol{X}_2 \end{pmatrix} E[\boldsymbol{\beta}\boldsymbol{\beta}']\right\} + \sigma_0^2[\mathrm{r}(\mathbb{X}) - \mathrm{r}(\boldsymbol{X}_1)]$$

$$= \operatorname{tr}\left\{\boldsymbol{X}_2'\boldsymbol{W}[\boldsymbol{W}^{-1} - \boldsymbol{X}_1(\boldsymbol{X}_1'\boldsymbol{W}\boldsymbol{X}_1)^{-1}\boldsymbol{X}_1']\boldsymbol{W}\boldsymbol{X}_2 E[\boldsymbol{\beta}_2\boldsymbol{\beta}_2']\right\} + \sigma_0^2[\mathrm{r}(\mathbb{X}) - \mathrm{r}(\boldsymbol{X}_1)].$$

We observe that $E\left[SSR(\boldsymbol{\beta}_2|\,\boldsymbol{\beta}_1)\right]$ is simply a function of $E\left[\boldsymbol{\beta}_2\boldsymbol{\beta}_2'\right]$ and σ_0^2. It does not depend on $E\left[\boldsymbol{\beta}_1\boldsymbol{\beta}_1'\right]$ and $E\left[\boldsymbol{\beta}_1\boldsymbol{\beta}_2'\right]$. We also observe that (6.67) was obtained without doing assumptions about the form of $E\left[\boldsymbol{\beta}\boldsymbol{\beta}'\right]$. Therefore, (6.67) holds for any structure of the covariance matrix of $\boldsymbol{\beta}$.

Remark 6.1 The formula for $E[SSR(\boldsymbol{\beta}_2|\boldsymbol{\beta}_1)]$ can be obtained in a more easy and direct way if we use the equality

$$SSR(\boldsymbol{\beta}_2|\boldsymbol{\beta}_1) = SSE(\boldsymbol{\beta}_1) - SSE(\boldsymbol{\beta}_1, \boldsymbol{\beta}_2).$$

As we derived that

$$E[SSE(\boldsymbol{\beta}_1, \boldsymbol{\beta}_2)] = \sigma_0^2\,[n - \mathrm{r}(\mathbb{X})]$$

and

$$E[SSE(\boldsymbol{\beta}_1)] = \operatorname{tr}\left(\mathbb{X}'\boldsymbol{M}_1\mathbb{X}E[\boldsymbol{\beta}\boldsymbol{\beta}']\right) + \sigma_0^2\,[n - \mathrm{r}\,(\boldsymbol{X}_1)]\,,$$

we have

$$E(SSR(\boldsymbol{\beta}_2|\boldsymbol{\beta}_1)) = \operatorname{tr}\left(\mathbb{X}'\boldsymbol{M}_1\mathbb{X}E[\boldsymbol{\beta}\boldsymbol{\beta}']\right) + \sigma_0^2\,[\mathrm{r}\,(\mathbb{X}) - \mathrm{r}\,(\boldsymbol{X}_1)]\;.$$

Since

$$\mathbb{X}'\boldsymbol{M}_1\mathbb{X} = \begin{pmatrix}\boldsymbol{X}_1' \\ \boldsymbol{X}_2'\end{pmatrix}\left[\boldsymbol{W} - \boldsymbol{W}\boldsymbol{X}_1(\boldsymbol{X}_1'\boldsymbol{W}\boldsymbol{X}_1)^{-1}\boldsymbol{X}_1'\boldsymbol{W}\right](\boldsymbol{X}_1\;\boldsymbol{X}_2)$$

$$= \begin{pmatrix} \mathbf{0} & \mathbf{0} \\ \mathbf{0} & \boldsymbol{X}_2'\boldsymbol{W}\boldsymbol{X}_2 - \boldsymbol{X}_2'\boldsymbol{W}\boldsymbol{X}_1(\boldsymbol{X}_1'\boldsymbol{W}\boldsymbol{X}_1)^{-1}\boldsymbol{X}_1'\boldsymbol{W}\boldsymbol{X}_2 \end{pmatrix},$$

we finally get

$$E(SSR(\boldsymbol{\beta}_2|\boldsymbol{\beta}_1)) = \operatorname{tr}\left\{\boldsymbol{X}_2'\boldsymbol{W}[\boldsymbol{W}^{-1} - \boldsymbol{X}_1(\boldsymbol{X}_1'\boldsymbol{W}\boldsymbol{X}_1)^{-1}\boldsymbol{X}_1']\boldsymbol{W}\boldsymbol{X}_2 E[\boldsymbol{\beta}_2\boldsymbol{\beta}_2']\right\}$$
$$+ \sigma_0^2[\mathrm{r}(\mathbb{X}) - \mathrm{r}(\boldsymbol{X}_1)].$$

Let us consider again the model (6.13)

$$y = X\beta + Z_1 u_1 + \ldots + Z_m u_m + e,$$

with $e \sim N_n(0, \sigma_0^2 W^{-1})$ and $u_i \sim N_{q_i}(0, \sigma_i^2 I_{q_i}),\ i = 1, \ldots, m$. We define

$$\beta^{(i)} = (\beta', u_1', \ldots, u_{i-1}')' \quad \text{and} \quad u^{(i)} = (u_i', \ldots, u_m')'.$$

For $i = 1, \ldots, m$, we consider the case

$$\begin{aligned} X_1 = X_1^{(i)} &= (X, Z_1, \ldots, Z_{i-1}),\ \beta_1 = \beta^{(i)}, \\ X_2 = X_2^{(i)} &= (Z_i, \ldots, Z_m), \qquad \beta_2 = u^{(i)} \end{aligned}$$

and we define

$$M_i = W - W X_1^{(i)} (X_1^{(i)\prime} W X_1^{(i)})^{-1} X_1^{(i)\prime} W, \tag{6.68}$$

$$L_{i,k} = Z_k' W [W^{-1} - X_1^{(i)} (X_1^{(i)\prime} W X_1^{(i)})^{-1} X_1^{(i)\prime}] W Z_k.$$

We further use the notation

$$\begin{aligned} \mathbb{X} = X_1^{(m+1)} &= (X, Z) = (X, Z_1, \ldots, Z_m), \\ M = M_{m+1} &= W - W\mathbb{X}(\mathbb{X}' W \mathbb{X})^{-1} \mathbb{X}' W. \end{aligned}$$

Then, (6.63) and (6.67) become

$$E[SSE(\beta^{(i)}, u^{(i)})] = E[SSE(\beta, u)] = \sigma_0^2 [n - \mathrm{r}(X, Z)], \tag{6.69}$$

and

$$E[SSR(u^{(i)} | \beta^{(i)})] = \sum_{k=i}^{m} \mathrm{tr}\left(L_{i,k}\right) \sigma_k^2 + \sigma_0^2 [\mathrm{r}(X, Z) - \mathrm{r}(X, Z_1, \ldots, Z_{i-1})]. \tag{6.70}$$

From (6.69) and (6.70), and applying the method of moments, we get the following linear and triangular system of equations:

$$\begin{aligned} SSE(\beta, u) &= \sigma_0^2 [n - \mathrm{r}(X, Z_1, \ldots, Z_m)], \\ SSR(u^{(m)} | \beta^{(m)}) &= \sigma_0^2 [\mathrm{r}(X, Z) - \mathrm{r}(X, Z_1, \ldots, Z_{m-1})] + \sigma_m^2 \mathrm{tr}\left(L_{m,m}\right), \\ SSR(u^{(m-1)} | \beta^{(m-1)}) &= \sigma_0^2 [\mathrm{r}(X, Z) - \mathrm{r}(X, Z_1, \ldots, Z_{m-2})] \\ &\quad + \sigma_m^2 \mathrm{tr}\left(L_{m-1,m}\right) + \sigma_{m-1}^2 \mathrm{tr}\left(L_{m-1,m-1}\right), \end{aligned}$$

$$\vdots$$

$$SSR(\boldsymbol{u}^{(1)}|\boldsymbol{\beta}^{(1)}) = \sigma_0^2[\mathrm{r}(\boldsymbol{X}, \boldsymbol{Z}) - \mathrm{r}(\boldsymbol{X})] + \sum_{k=1}^{m} \sigma_i^2 \mathrm{tr}\left(\boldsymbol{L}_{1,k}\right).$$

From the first equation, we obtain an unbiased estimator of σ_0^2,

$$\hat{\sigma}_0^2 = \frac{SSE(\boldsymbol{\beta}, \boldsymbol{u})}{n - \mathrm{r}(\boldsymbol{X}, \boldsymbol{Z})} = MSE(\boldsymbol{\beta}, \boldsymbol{u}). \tag{6.71}$$

From the second equation, we get an unbiased estimator of σ_m^2,

$$\hat{\sigma}_m^2 = \frac{SSR(\boldsymbol{u}^{(m)}|\boldsymbol{\beta}^{(m)}) - \hat{\sigma}_0^2[\mathrm{r}(\boldsymbol{X}, \boldsymbol{Z}) - \mathrm{r}(\boldsymbol{X}, \boldsymbol{Z}_1, \ldots, \boldsymbol{Z}_{m-1})]}{\mathrm{tr}\left(\boldsymbol{L}_{m,m}\right)}. \tag{6.72}$$

From the third equation, we get an unbiased estimator of σ_{m-1}^2,

$$\begin{aligned}\hat{\sigma}_{m-1}^2 = & \frac{1}{\mathrm{tr}(\boldsymbol{L}_{m-1,m-1})}\Big(SSR(\boldsymbol{u}^{(m-1)}|\boldsymbol{\beta}^{(m-1)}) \\ & - \hat{\sigma}_0^2\big[\mathrm{r}(\boldsymbol{X}, \boldsymbol{Z}) - \mathrm{r}(\boldsymbol{X}, \boldsymbol{Z}_1, \ldots, \boldsymbol{Z}_{m-2})\big] - \hat{\sigma}_m^2 \mathrm{tr}(\boldsymbol{L}_{m-1,m})\Big),\end{aligned}$$

and so on. As

$$SSR(\boldsymbol{u}^{(r)}|\,\boldsymbol{\beta}^{(r)}) = SSE(\boldsymbol{\beta}^{(r)}) - SSE(\boldsymbol{\beta}^{(r)}, \boldsymbol{u}^{(r)}) = SSE(\boldsymbol{\beta}^{(r)}) - SSE(\boldsymbol{\beta}, \boldsymbol{u}),$$

then the previous formula can be expressed as a function of residual sum of squares. That is,

$$\begin{aligned}
\hat{\sigma}_0^2 &= \frac{\boldsymbol{y}'\boldsymbol{M}_{m+1}\boldsymbol{y}}{n - \mathrm{r}(\boldsymbol{X}_1^{(m+1)})}, \\
\hat{\sigma}_m^2 &= \frac{\boldsymbol{y}'\boldsymbol{M}_m\boldsymbol{y} - \boldsymbol{y}'\boldsymbol{M}_{m+1}\boldsymbol{y} - \hat{\sigma}_0^2\left[\mathrm{r}(\boldsymbol{X}_1^{(m+1)}) - \mathrm{r}(\boldsymbol{X}_1^{(m)})\right]}{\mathrm{tr}(\boldsymbol{L}_{m,m})}, \\
\cdots & \quad \cdots \\
\hat{\sigma}_i^2 &= \frac{\boldsymbol{y}'\boldsymbol{M}_i\boldsymbol{y} - \boldsymbol{y}'\boldsymbol{M}_{m+1}\boldsymbol{y} - \hat{\sigma}_0^2\left[\mathrm{r}(\boldsymbol{X}_1^{(m+1)}) - \mathrm{r}(\boldsymbol{X}_1^{(i)})\right] - \sum_{k=i+1}^{m} \hat{\sigma}_k^2 \mathrm{tr}(\boldsymbol{L}_{i,k})}{\mathrm{tr}(\boldsymbol{L}_{i,i})}, \\
\cdots & \quad \cdots \\
\hat{\sigma}_1^2 &= \frac{\boldsymbol{y}'\boldsymbol{M}_1\boldsymbol{y} - \boldsymbol{y}'\boldsymbol{M}_{m+1}\boldsymbol{y} - \hat{\sigma}_0^2\left[\mathrm{r}(\boldsymbol{X}_1^{(m+1)}) - \mathrm{r}(\boldsymbol{X}_1^{(1)})\right] - \sum_{k=2}^{m} \hat{\sigma}_k^2 \mathrm{tr}(\boldsymbol{L}_{1,k})}{\mathrm{tr}(\boldsymbol{L}_{1,1})}.
\end{aligned}$$

For more details see Searle et al. (1992), 202–208, or Searle (1971), 443–445. If we replace the variance components $\sigma_0^2, \sigma_1^2, \ldots, \sigma_m^2$ by their estimators $\hat{\sigma}_0^2, \hat{\sigma}_1^2, \ldots, \hat{\sigma}_m^2$ in (6.3) and (6.6), we obtain the estimator of $\boldsymbol{\beta}$ and the predictors $\boldsymbol{u}_1, \ldots, \boldsymbol{u}_m$.

Remark 6.2 If we use the alternative parametrization $(\sigma^2, \varphi_1, \ldots, \varphi_m)$, then the system of equations is not linear any more. Consequently, by solving the transformed system one does not obtain unbiased estimators.

6.6.2 Moments of Henderson 3 Estimators

Let us write the general linear mixed model

$$\boldsymbol{y} = \mathbb{X}\boldsymbol{\beta} + \boldsymbol{e}, \tag{6.73}$$

in the form

$$\boldsymbol{y} = \boldsymbol{X}_1\boldsymbol{\beta}_1 + \boldsymbol{X}_2\boldsymbol{\beta}_2 + \boldsymbol{e}, \tag{6.74}$$

where $\boldsymbol{e} \sim N_n(\boldsymbol{0}, \sigma_0^2\boldsymbol{W}^{-1})$ and $\mathbb{X}'\mathbb{X},\ \boldsymbol{X}_1'\boldsymbol{X}_1$ are invertible. We have done a partition of $\boldsymbol{\beta} = (\boldsymbol{\beta}_1', \boldsymbol{\beta}_2')'$ in two groups of effects. In principle, $\boldsymbol{\beta}_1$ and $\boldsymbol{\beta}_2$ can be random with $E[\boldsymbol{\beta}_2] = 0$ and $\boldsymbol{e},\ \boldsymbol{\beta}_1$ and $\boldsymbol{\beta}_2$ are independent. In the model (6.73), we have that

$$E[\boldsymbol{y}] = \mathbb{X}E[\boldsymbol{\beta}], \quad \operatorname{var}(\boldsymbol{y}) = \mathbb{X}\operatorname{var}(\boldsymbol{\beta})\mathbb{X}' + \sigma_0^2\boldsymbol{W}^{-1}.$$

For calculating the expectation and the variance of a quadratic form under the model (6.73), we use the formulas

$$\begin{aligned}
E[\boldsymbol{y}'\boldsymbol{Q}\boldsymbol{y}] &= \operatorname{tr}\{\boldsymbol{Q}\operatorname{var}(\boldsymbol{y})\} + E[\boldsymbol{y}]'\boldsymbol{Q}E[\boldsymbol{y}],\\
\operatorname{var}(\boldsymbol{y}'\boldsymbol{Q}\boldsymbol{y}) &= 2\operatorname{tr}\left\{(\boldsymbol{Q}\operatorname{var}(\boldsymbol{y}))^2\right\} + 4E[\boldsymbol{y}]'\boldsymbol{Q}\operatorname{var}(\boldsymbol{y})\,\boldsymbol{Q}E[\boldsymbol{y}],\\
\operatorname{cov}(\boldsymbol{y}'\boldsymbol{A}\boldsymbol{y},\ \boldsymbol{y}'\boldsymbol{B}\boldsymbol{y})] &= 2\operatorname{tr}\{\boldsymbol{A}\operatorname{var}(\boldsymbol{y})\boldsymbol{B}\operatorname{var}(\boldsymbol{y})\} + 4E[\boldsymbol{y}]'\boldsymbol{A}\operatorname{var}(\boldsymbol{y})\,\boldsymbol{B}E[\boldsymbol{y}],
\end{aligned}$$

where the last two formulas are valid if the vector $\boldsymbol{y}$ has multivariate normal distribution. By particularizing these three formulas to the model (6.73), we get

$$\begin{aligned}
E[\boldsymbol{y}'\boldsymbol{Q}\boldsymbol{y}] &= \operatorname{tr}\left\{\boldsymbol{Q}\,(\mathbb{X}\operatorname{var}(\boldsymbol{\beta})\mathbb{X}' + \sigma_0^2\boldsymbol{W}^{-1})\right\} + E[\boldsymbol{\beta}]'\mathbb{X}'\boldsymbol{Q}\mathbb{X}E[\boldsymbol{\beta}],\\
\operatorname{var}(\boldsymbol{y}'\boldsymbol{Q}\boldsymbol{y}) &= 2\operatorname{tr}\left\{(\boldsymbol{Q}\,(\mathbb{X}\operatorname{var}(\boldsymbol{\beta})\mathbb{X}' + \sigma_0^2\boldsymbol{W}^{-1}))^2\right\}\\
&\quad + 4E[\boldsymbol{\beta}]'\mathbb{X}'\boldsymbol{Q}\,(\mathbb{X}\operatorname{var}(\boldsymbol{\beta})\mathbb{X}' + \sigma_0^2\boldsymbol{W}^{-1})\,\boldsymbol{Q}\mathbb{X}E[\boldsymbol{\beta}],
\end{aligned}$$

$$\operatorname{cov}(\boldsymbol{y}'\boldsymbol{A}\boldsymbol{y},\ \boldsymbol{y}'\boldsymbol{B}\boldsymbol{y}) = 2\operatorname{tr}\left\{\boldsymbol{A}(\mathbb{X}\operatorname{var}(\boldsymbol{\beta})\mathbb{X}' + \sigma_0^2\boldsymbol{W}^{-1})\boldsymbol{B}(\mathbb{X}\operatorname{var}(\boldsymbol{\beta})\mathbb{X}' + \sigma_0^2\boldsymbol{W}^{-1})\right\}$$
$$+\ 4E[\boldsymbol{\beta}]'\mathbb{X}'\boldsymbol{A}\,(\mathbb{X}\operatorname{var}(\boldsymbol{\beta})\mathbb{X}' + \sigma_0^2\boldsymbol{W}^{-1})\,\boldsymbol{B}\mathbb{X}E[\boldsymbol{\beta}].$$

First, we consider the residual sum of squares of the model (6.74) with $\boldsymbol{\beta}_1,\ \boldsymbol{\beta}_2$ fixed, i.e.

$$SSE(\boldsymbol{\beta}_1, \boldsymbol{\beta}_2) = \boldsymbol{y}'\boldsymbol{M}\boldsymbol{y},$$

where $\boldsymbol{M} = \boldsymbol{W} - \boldsymbol{W}\mathbb{X}(\mathbb{X}'\boldsymbol{W}\mathbb{X})^{-1}\mathbb{X}'\boldsymbol{W}$. We know that $\boldsymbol{M} = \boldsymbol{M}',\ \boldsymbol{M}^2 = \boldsymbol{M}$,

$$(\boldsymbol{M}\boldsymbol{W}^{-1})^2 = \boldsymbol{M}\boldsymbol{W}^{-1},\ \boldsymbol{M}\boldsymbol{W}^{-1}\boldsymbol{M} = \boldsymbol{M},\ \boldsymbol{M}\mathbb{X} = \boldsymbol{0} \text{ and } \operatorname{tr}\{\boldsymbol{M}\boldsymbol{W}^{-1}\} = n - \operatorname{r}(\mathbb{X}).$$

We do the following calculations under the assumption that model (6.74) holds with $\boldsymbol{\beta}_1$ and $\boldsymbol{\beta}_2$ possibly random:

$$E[SSE(\boldsymbol{\beta}_1, \boldsymbol{\beta}_2)] = E[\boldsymbol{y}'\boldsymbol{M}\boldsymbol{y}] = \operatorname{tr}\left\{\boldsymbol{M}(\mathbb{X}\operatorname{var}(\boldsymbol{\beta})\mathbb{X}' + \sigma_0^2\boldsymbol{W}^{-1})\right\}$$
$$+ \operatorname{tr}\left\{\boldsymbol{M}\mathbb{X}E[\boldsymbol{\beta}]E[\boldsymbol{\beta}]'\mathbb{X}'\right\} = \sigma_0^2\operatorname{tr}(\boldsymbol{M}\boldsymbol{W}^{-1}) = \sigma_0^2\,(n - \operatorname{r}(\mathbb{X})),$$
$$\operatorname{var}[SSE(\boldsymbol{\beta}_1, \boldsymbol{\beta}_2)] = \operatorname{var}[\boldsymbol{y}'\boldsymbol{M}\boldsymbol{y}] = 2\operatorname{tr}\left\{\left[\boldsymbol{M}(\mathbb{X}\operatorname{var}(\boldsymbol{\beta})\mathbb{X}' + \sigma_0^2\boldsymbol{W}^{-1})\right]^2\right\}$$
$$+ 4\operatorname{tr}\left\{\boldsymbol{M}(\mathbb{X}\operatorname{var}(\boldsymbol{\beta})\mathbb{X}' + \sigma_0^2\boldsymbol{W}^{-1})\boldsymbol{M}\mathbb{X}E[\boldsymbol{\beta}]E[\boldsymbol{\beta}]'\mathbb{X}'\right\}$$
$$= 2\sigma_0^4\operatorname{tr}\{(\boldsymbol{M}\boldsymbol{W}^{-1})^2\} = 2\sigma_0^4\operatorname{tr}(\boldsymbol{M}\boldsymbol{W}^{-1})$$
$$= 2\sigma_0^4\,(n - \operatorname{r}(\mathbb{X}))\,. \tag{6.75}$$

Second, we consider the residual sum of squares of the model $\boldsymbol{y} = \boldsymbol{X}_1\boldsymbol{\beta}_1 + \boldsymbol{e}$, with $\boldsymbol{\beta}_1$ fixed, i.e.

$$SSE(\boldsymbol{\beta}_1) = \boldsymbol{y}'\boldsymbol{M}_1\boldsymbol{y},$$

where $\boldsymbol{M}_1 = \boldsymbol{W} - \boldsymbol{W}\boldsymbol{X}_1(\boldsymbol{X}_1'\boldsymbol{W}\boldsymbol{X}_1)^{-1}\boldsymbol{X}_1'\boldsymbol{W}$. We know that $\boldsymbol{M}_1 = \boldsymbol{M}_1',\ \boldsymbol{M}_1^2 = \boldsymbol{M}_1$,

$$(\boldsymbol{M}_1\boldsymbol{W}^{-1})^2 = \boldsymbol{M}_1\boldsymbol{W}^{-1},\ \boldsymbol{M}\boldsymbol{W}^{-1}\boldsymbol{M} = \boldsymbol{M},\ \boldsymbol{M}_1\boldsymbol{X}_1 = \boldsymbol{0} \text{ and } \operatorname{tr}(\boldsymbol{M}_1\boldsymbol{W}^{-1}) = n - \operatorname{r}(\boldsymbol{X}_1).$$

We do the following calculations under the assumption that model (6.74) holds with $\boldsymbol{\beta}_1$ and $\boldsymbol{\beta}_2$ possibly random:

$$E[SSE(\boldsymbol{\beta}_1)] = E[\boldsymbol{y}'\boldsymbol{M}_1\boldsymbol{y}] = \operatorname{tr}\left\{\boldsymbol{M}_1(\mathbb{X}\operatorname{var}(\boldsymbol{\beta})\mathbb{X}' + \sigma_0^2\boldsymbol{W}^{-1})\right\}$$
$$+ \operatorname{tr}\left\{\boldsymbol{M}_1\mathbb{X}E[\boldsymbol{\beta}]E[\boldsymbol{\beta}]'\mathbb{X}'\right\}.$$

Note that $\mathbb{X} = (\boldsymbol{X}_1,\ \boldsymbol{X}_2)$ and, by assuming that $\boldsymbol{\beta}_1$ and $\boldsymbol{\beta}_2$ are uncorrelated, we get

$$
\begin{aligned}
\boldsymbol{M}_1\mathbb{X}\mathrm{var}(\boldsymbol{\beta})\mathbb{X}' &= (\boldsymbol{0},\ \boldsymbol{M}_1\boldsymbol{X}_2)\mathrm{var}(\boldsymbol{\beta})\begin{pmatrix}\boldsymbol{X}_1'\\ \boldsymbol{X}_2'\end{pmatrix} = \boldsymbol{M}_1\boldsymbol{X}_2\mathrm{var}(\boldsymbol{\beta}_2)\boldsymbol{X}_2',\\
\boldsymbol{M}_1\mathbb{X}E[\boldsymbol{\beta}]E[\boldsymbol{\beta}]'\mathbb{X}' &= (\boldsymbol{0},\ \boldsymbol{M}_1\boldsymbol{X}_2)E[\boldsymbol{\beta}]E[\boldsymbol{\beta}]'\begin{pmatrix}\boldsymbol{X}_1'\\ \boldsymbol{X}_2'\end{pmatrix}\\
&= \boldsymbol{M}_1\boldsymbol{X}_2E[\boldsymbol{\beta}_2]E[\boldsymbol{\beta}_1]'\boldsymbol{X}_1' + \boldsymbol{M}_1\boldsymbol{X}_2E[\boldsymbol{\beta}_2]E[\boldsymbol{\beta}_2]'\boldsymbol{X}_2' = 0,
\end{aligned}
$$

since $E[\boldsymbol{\beta}_2] = \boldsymbol{0}$. Further, we have

$$
\begin{aligned}
E[SSE(\boldsymbol{\beta}_1)] &= \mathrm{tr}\left\{\boldsymbol{M}_1\boldsymbol{X}_2E[\boldsymbol{\beta}_2\boldsymbol{\beta}_2']\boldsymbol{X}_2'\right\} + \mathrm{tr}\left\{\sigma_0^2\boldsymbol{M}_1\boldsymbol{W}^{-1}\right\}\\
&= \mathrm{tr}\left\{\boldsymbol{X}_2'\boldsymbol{M}_1\boldsymbol{X}_2E[\boldsymbol{\beta}_2\boldsymbol{\beta}_2']\right\} + \sigma_0^2(n - \mathrm{r}(\boldsymbol{X}_1)),\\
\mathrm{var}(SSE(\boldsymbol{\beta}_1)) &= 2\mathrm{tr}\left\{\left[\boldsymbol{M}_1(\mathbb{X}\mathrm{var}(\boldsymbol{\beta})\mathbb{X}' + \sigma_0^2\boldsymbol{W}^{-1})\right]^2\right\}\\
&\quad + 4\mathrm{tr}\left\{\boldsymbol{M}_1(\mathbb{X}\mathrm{var}(\boldsymbol{\beta})\mathbb{X}' + \sigma_0^2\boldsymbol{W}^{-1})\boldsymbol{M}_1\mathbb{X}E[\boldsymbol{\beta}]E[\boldsymbol{\beta}]'\mathbb{X}'\right\}\\
&= 2\mathrm{tr}\left\{\left[\boldsymbol{M}_1(\boldsymbol{X}_2\mathrm{var}(\boldsymbol{\beta}_2)\boldsymbol{X}_2' + \sigma_0^2\boldsymbol{W}^{-1})\right]^2\right\} \qquad (6.76)\\
&= 2\mathrm{tr}\left\{\boldsymbol{M}_1(\boldsymbol{X}_2\mathrm{var}(\boldsymbol{\beta}_2)\boldsymbol{X}_2' + \sigma_0^2\boldsymbol{W}^{-1})\boldsymbol{M}_1(\boldsymbol{X}_2\mathrm{var}(\boldsymbol{\beta}_2)\boldsymbol{X}_2' + \sigma_0^2\boldsymbol{W}^{-1})\right\}
\end{aligned}
$$

and

$$
\begin{aligned}
&\mathrm{cov}(SSE(\boldsymbol{\beta}_1, \boldsymbol{\beta}_2), SSE(\boldsymbol{\beta}_1)) = \mathrm{cov}(\boldsymbol{y}'\boldsymbol{M}\boldsymbol{y}, \boldsymbol{y}'\boldsymbol{M}_1\boldsymbol{y})\\
&\quad = 2\mathrm{tr}\left\{\boldsymbol{M}(\mathbb{X}\mathrm{var}(\boldsymbol{\beta})\mathbb{X}' + \sigma_0^2\boldsymbol{W}^{-1})\boldsymbol{M}_1(\mathbb{X}\mathrm{var}(\boldsymbol{\beta})\mathbb{X}' + \sigma_0^2\boldsymbol{W}^{-1})\right\}\\
&\quad + 4\mathrm{tr}\left\{\boldsymbol{M}(\mathbb{X}\mathrm{var}(\boldsymbol{\beta})\mathbb{X}' + \sigma_0^2\boldsymbol{W}^{-1})\boldsymbol{M}_1\mathbb{X}E[\boldsymbol{\beta}]E[\boldsymbol{\beta}]'\mathbb{X}'\right\}\\
&\quad = 2\mathrm{tr}\left\{\sigma_0^2\boldsymbol{M}\boldsymbol{W}^{-1}\boldsymbol{M}_1(\boldsymbol{X}_2\mathrm{var}(\boldsymbol{\beta}_2)\boldsymbol{X}_2' + \sigma_0^2\boldsymbol{W}^{-1})\right\}. \qquad (6.77)
\end{aligned}
$$

Remark 6.3 Let us consider the linear mixed model

$$
\boldsymbol{y} = \boldsymbol{X}\boldsymbol{\beta} + \boldsymbol{Z}\boldsymbol{u} + \boldsymbol{e},
$$

where $\boldsymbol{\beta}$ is vector of fixed regression parameters, random vectors $\boldsymbol{u}$ and $\boldsymbol{e}$ are independent, and

$$
\boldsymbol{e} \sim N_n(\boldsymbol{0}, \sigma_0^2\boldsymbol{W}^{-1}), \quad \boldsymbol{u} \sim N_q(\boldsymbol{0}, \sigma_1^2\boldsymbol{\Sigma}_u),
$$

with $\boldsymbol{W}$ and $\boldsymbol{\Sigma}_u$ known. That means we consider the case, where

$$m = 1, \quad \boldsymbol{X}_1 = \boldsymbol{X}_1^{(1)} = \boldsymbol{X}, \quad \boldsymbol{X}_2 = \boldsymbol{Z}_1 = \boldsymbol{Z}, \quad \mathbb{X} = (\boldsymbol{X}, \boldsymbol{Z}), \quad \boldsymbol{\beta}_1 = \boldsymbol{\beta}, \quad \boldsymbol{\beta}_2 = \boldsymbol{u}$$

and

$$\boldsymbol{M} = \boldsymbol{W} - \boldsymbol{W}\mathbb{X}(\mathbb{X}'\boldsymbol{W}\mathbb{X})^{-1}\mathbb{X}'\boldsymbol{W}, \quad \boldsymbol{M}_1 = \boldsymbol{W} - \boldsymbol{W}\boldsymbol{X}(\boldsymbol{X}'\boldsymbol{W}\boldsymbol{X})^{-1}\boldsymbol{X}'\boldsymbol{W}.$$

By the same steps as presented on page 143 we get the following Henderson 3 estimates of the variance parameters:

$$\hat{\sigma}_0^2 = \frac{\boldsymbol{y}'\boldsymbol{M}\boldsymbol{y}}{n - \mathrm{r}(\boldsymbol{X}, \boldsymbol{Z})}, \qquad \hat{\sigma}_1^2 = \frac{\boldsymbol{y}'\boldsymbol{M}_1\boldsymbol{y} - \boldsymbol{y}'\boldsymbol{M}\boldsymbol{y} - \hat{\sigma}_0^2\,[\mathrm{r}(\boldsymbol{X}, \boldsymbol{Z}) - \mathrm{r}(\boldsymbol{X})]}{\mathrm{tr}(\boldsymbol{L}_{1,1})},$$

which can be written in the form

$$\hat{\sigma}_0^2 = \frac{1}{n - \mathrm{r}(\boldsymbol{X}, \boldsymbol{Z})}\boldsymbol{y}'\boldsymbol{M}\boldsymbol{y} = a_0\,\boldsymbol{y}'\boldsymbol{M}\boldsymbol{y} = a_0\,SSE(\boldsymbol{\beta}, \boldsymbol{u})$$

and

$$\begin{aligned}\hat{\sigma}_1^2 &= \frac{1}{\mathrm{tr}(\boldsymbol{L}_{1,1})}\boldsymbol{y}'\boldsymbol{M}_1\boldsymbol{y} - \frac{1}{\mathrm{tr}(\boldsymbol{L}_{1,1})}\left(1 + \frac{\mathrm{r}(\boldsymbol{X}, \boldsymbol{Z}) - \mathrm{r}(\boldsymbol{X})}{n - \mathrm{r}(\boldsymbol{X}, \boldsymbol{Z})}\right)\boldsymbol{y}'\boldsymbol{M}\boldsymbol{y} \\ &= a_{1,1}\boldsymbol{y}'\boldsymbol{M}_1\boldsymbol{y} + a_{1,2}\boldsymbol{y}'\boldsymbol{M}\boldsymbol{y} = a_{1,1}\,SSE(\boldsymbol{\beta}) + a_{1,2}\,SSE(\boldsymbol{\beta}, \boldsymbol{u}).\end{aligned}$$

Thus, for their variances and covariance it holds,

$$\begin{aligned}\mathrm{var}(\hat{\sigma}_0^2) &= a_0^2\,\mathrm{var}(SSE(\boldsymbol{\beta}, \boldsymbol{u})) = a_0^2\,2\sigma_0^4\,(n - \mathrm{r}(\boldsymbol{X}, \boldsymbol{Z})) = \frac{2\sigma_0^4}{n - \mathrm{r}(\boldsymbol{X}, \boldsymbol{Z})}, \\ \mathrm{var}(\hat{\sigma}_1^2) &= a_{1,1}^2\,\mathrm{var}(SSE(\boldsymbol{\beta})) + a_{1,2}^2\,\mathrm{var}(SSE(\boldsymbol{\beta}, \boldsymbol{u})) \\ &\quad + 2a_{1,1}a_{1,2}\,\mathrm{cov}(SSE(\boldsymbol{\beta}, \boldsymbol{u}), SSE(\boldsymbol{\beta}))\end{aligned}$$

and

$$\mathrm{cov}(\hat{\sigma}_0^2, \hat{\sigma}_1^2) = a_0a_{1,1}\mathrm{cov}(SSE(\boldsymbol{\beta}, \boldsymbol{u}), SSE(\boldsymbol{\beta})) + a_0a_{1,2}\mathrm{var}(SSE(\boldsymbol{\beta}, \boldsymbol{u})),$$

where (with respect to (6.75), (6.76), and (6.77))

$$\begin{aligned}\mathrm{var}(SSE(\boldsymbol{\beta}, \boldsymbol{u})) &= 2\sigma_0^4\,(n - \mathrm{r}(\boldsymbol{X}, \boldsymbol{Z}))\,, \\ \mathrm{var}(SSE(\boldsymbol{\beta})) &= 2\mathrm{tr}([\boldsymbol{M}_1\boldsymbol{V}]^2), \\ \mathrm{cov}(SSE(\boldsymbol{\beta}, \boldsymbol{u}), SSE(\boldsymbol{\beta})) &= 2\mathrm{tr}(\sigma_0^2\boldsymbol{M}\boldsymbol{W}^{-1}\boldsymbol{M}_1\boldsymbol{V}),\end{aligned}$$

with $\boldsymbol{V} = \sigma_1^2\boldsymbol{Z}\boldsymbol{\Sigma}_u\boldsymbol{Z}' + \sigma_0^2\boldsymbol{W}^{-1}$.

Let us return again the general model (6.13),

$$\boldsymbol{y} = \boldsymbol{X\beta} + \boldsymbol{Z}_1\boldsymbol{u}_1 + \ldots + \boldsymbol{Z}_m\boldsymbol{u}_m + \boldsymbol{e},$$

and the notation introduced around the formula (6.68). As the Henderson 3 estimators have the general expression

$$\begin{aligned}
\hat{\sigma}_0^2 &= a_{0,m+1}\boldsymbol{y}'\boldsymbol{M}_{m+1}\boldsymbol{y},\\
\hat{\sigma}_m^2 &= a_{m,m+1}\boldsymbol{y}'\boldsymbol{M}_{m+1}\boldsymbol{y} + a_{m,m}\boldsymbol{y}'\boldsymbol{M}_m\boldsymbol{y},\\
\ldots &\quad \ldots\\
\hat{\sigma}_i^2 &= a_{i,m+1}\boldsymbol{y}'\boldsymbol{M}_{m+1}\boldsymbol{y} + \ldots + a_{i,i}\boldsymbol{y}'\boldsymbol{M}_i\boldsymbol{y},\\
\ldots &\quad \ldots\\
\hat{\sigma}_1^2 &= a_{1,m+1}\boldsymbol{y}'\boldsymbol{M}_{m+1}\boldsymbol{y} + \ldots + a_{1,1}\boldsymbol{y}'\boldsymbol{M}_1\boldsymbol{y},
\end{aligned}$$

their variances and covariances can be calculated in the following way:

$$\begin{aligned}
\text{var}(\hat{\sigma}_0^2) &= a_{0,m+1}^2\text{var}(\boldsymbol{y}'\boldsymbol{M}_{m+1}\boldsymbol{y}),\\
\text{cov}(\hat{\sigma}_0^2, \hat{\sigma}_i^2) &= a_{0,m+1}\sum_{r=i}^{m+1} a_{i,r}\text{cov}(\boldsymbol{y}'\boldsymbol{M}_{m+1}\boldsymbol{y}, \boldsymbol{y}'\boldsymbol{M}_r\boldsymbol{y}), \quad i = 1, \ldots, m,\\
\text{cov}(\hat{\sigma}_i^2, \hat{\sigma}_j^2) &= \sum_{r=i}^{m+1}\sum_{s=j}^{m+1} a_{i,r}a_{j,s}\text{cov}(\boldsymbol{y}'\boldsymbol{M}_r\boldsymbol{y}, \boldsymbol{y}'\boldsymbol{M}_s\boldsymbol{y}), \quad i, j = 1, \ldots, m,
\end{aligned}$$

where

$$\begin{aligned}
\text{cov}(\boldsymbol{y}'\boldsymbol{M}_r\boldsymbol{y}, \boldsymbol{y}'\boldsymbol{M}_s\boldsymbol{y}) = {}& 2\text{tr}\left\{\boldsymbol{M}_r(\mathbb{X}\text{var}(\boldsymbol{\beta})\mathbb{X}' + \sigma_0^2\boldsymbol{W}^{-1})\boldsymbol{M}_s(\mathbb{X}\text{var}(\boldsymbol{\beta})\mathbb{X}' + \sigma_0^2\boldsymbol{W}^{-1})\right\}\\
&+ 4\text{tr}\left\{\boldsymbol{M}_r(\mathbb{X}\text{var}(\boldsymbol{\beta})\mathbb{X}' + \sigma_0^2\boldsymbol{W}^{-1})\boldsymbol{M}_s\mathbb{X}E[\boldsymbol{\beta}]E[\boldsymbol{\beta}]'\mathbb{X}'\right\}.
\end{aligned}$$

6.7 R Codes for Fitting Linear Mixed Models

This section gives R codes for fitting linear mixed models to data from the survey data file `LFS20.txt`. The target variable y is INCOME and the auxiliary variables x_1, x_2 and x_3 are AGE, EDUCATION2 and EDUCATION3, respectively. The following code reads the data file and installs some R packages.

```
dat <- read.table("LFS20.txt", header=TRUE, sep = "\t", dec = ".")
Ga <- cut(dat$AGE, breaks=c(0,25,54,1000), labels=c(1,2,3), right=TRUE)
dat$ageG <- as.numeric(Ga)                       # Age groups
dat$EDUCATION <- as.factor(dat$EDUCATION)   # EDUCATION as factor
```

```
# Install and load libraries
if(!require(lme4)){
  install.packages("lme4")
  library(lme4)
}
if(!require(nlme)){
  install.packages("nlme")
  library(nlme)
}
```

Let us first consider the linear model LM1, i.e.

$$y_j = \beta_0 + \beta_1 x_{1,j} + \beta_2 x_{2,j} + \beta_3 x_{3,j} + e_j, \quad j = 1, \ldots, n,$$

where the e_j's are i.i.d. $N(0, \sigma_e^2)$. The following R code fits the model LM1 to the data.

```
lm.1 <- lm(formula=INCOME~AGE+EDUCATION, data=dat)
summary(lm.1)
```

Table 6.1 gives the estimates of the regression parameters. The residual standard error is $\hat{\sigma}_e = 10390$.

6.7.1 Library lme4

The linear mixed model LMM1 is the nested error regression model

$$y_{dj} = \beta_0 + \beta_1 x_{1,dj} + \beta_2 x_{2,dj} + \beta_3 x_{3,dj} + u_d + e_{dj},$$

$d = 1, \ldots, D$, $j = 1, \ldots, n_d$, where the u_d's are i.i.d. $N(0, \sigma_u^2)$, the e_{dj}'s are i.i.d. $N(0, \sigma_e^2)$, and they are all mutually independent. The variable AREA, taking the values $d = 1, \ldots, D$, defines the random effect u_d. By using the function `lmer` from the library `lme4`, the following R code fits the model LMM1 to the data.

```
lmm.1 <- lmer(formula=INCOME~AGE+EDUCATION+(1|AREA), data=dat, REML=TRUE)
summary(lmm.1)                              # Summary of the fitting procedure
anova(lmm.1)                                # Analysis of Variance Table
fixef(lmm.1)                                # Regression parameters
var <- as.data.frame(VarCorr(lmm.1))        # Variance parameters
sigmau <- var$sdcor[1]; sigmau              # Random effect standard deviation
sigmae <- var$sdcor[2]; sigmae              # Residual standard deviation
ranef(lmm.1)                                # Modes of the random effects
lmm.1.fit <- fitted(lmm.1)                  # Predicted values
```

Table 6.1 Regression parameters of model LM1

Variable	Estimate	Std. error	t-value	Pr(> $\|t\|$)
Intercept	38,121.19	1229.72	31.000	<2e-16
AGE	27.13	20.60	1.317	0.188
EDUCATION2	9566.88	796.58	12.010	<2e-16
EDUCATION3	19,978.16	974.13	20.509	< 2e-16

The REML estimates of the regression parameters are $\beta_0 = 38120.90$, $\beta_1 = 26.79$, $\beta_2 = 9572.10$, and $\beta_3 = 19983.27$. The estimated standard deviations are $\hat{\sigma}_u = 615.6$ and $\hat{\sigma}_e = 10376.7$. By fixing the argument REML=FALSE, the obtained ML estimates are $\beta_0 = 38120.24$, $\beta_1 = 26.89$, $\beta_2 = 9570.95$, $\beta_3 = 19981.26$, $\hat{\sigma}_u = 500.3$, and $\hat{\sigma}_e = 10362.0$.

The model LMM2 is a nested error regression model with covariable random slope, i.e.

$$y_{dj} = \beta_0 + \beta_1 x_{1,dj} + \beta_2 x_{2,dj} + \beta_3 x_{3,dj} + v_d x_{1,dj} + u_d + e_{dj},$$

$d = 1, \ldots, D$, $j = 1, \ldots, n_d$, where the u_d's are i.i.d. $N(0, \sigma_u^2)$, the v_d's are i.i.d. $N(0, \sigma_v^2)$, the e_{dj}'s are i.i.d. $N(0, \sigma_e^2)$, and they are all mutually independent. The variable AREA, taking the values $d = 1, \ldots, D$, defines the random effects u_d and v_d. By using the function `lmer` from the library `lme4`, the following R code fits the model LMM2 to the data.

```
lmm.2 <- lmer(formula=INCOME~AGE+EDUCATION+(AGE|AREA), data=dat, REML=FALSE)
summary(lmm.2)                # Summary of the fitting procedure
```

The ML estimates of the regression parameters are $\beta_0 = 38120.24$, $\beta_1 = 26.88$, $\beta_2 = 9571.02$, and $\beta_3 = 19981.36$. The estimated standard deviations are $\hat{\sigma}_u = 509.3$, $\hat{\sigma}_v = 0.06$, and $\hat{\sigma}_e = 10360$.

The nested error regression model LMM3, with random effects at the subdomain level, is

$$y_{dtj} = \beta_0 + \beta_1 x_{1,dtj} + \beta_2 x_{2,dtj} + \beta_3 x_{3,dtj} + u_{dt} + e_{dtj},$$
$$d = 1, \ldots, D,\ t = 1, \ldots, T,\ j = 1, \ldots, n_{dt},$$

where the u_{dt}'s are i.i.d. $N(0, \sigma_u^2)$, the e_{dtj}'s are i.i.d. $N(0, \sigma_e^2)$, and they are all mutually independent. The random effect, u_{dt}, is associated to the variable AREA crossed by age group. By using the function `lmer` from the library `lme4`, the following R code fits the model LMM3 to the data.

```
lmm.3 <- lmer(formula=INCOME~AGE+EDUCATION+(1|AREA:ageG), data=dat,
              REML=TRUE)
summary(lmm.3)                  # Summary of the fitting procedure
```

The REML estimates of the regression parameters are $\beta_0 = 38088.56$, $\beta_1 = 27.53$, $\beta_2 = 9586.55$, and $\beta_3 = 19967.86$. The estimated standard deviations are $\hat{\sigma}_u = 737.4$ and $\hat{\sigma}_e = 10368.3$.

The two-fold nested error regression model LMM4 is

$$y_{dtj} = \beta_0 + \beta_1 x_{1,dtj} + \beta_2 x_{2,dtj} + \beta_3 x_{3,dtj} + u_{1,d} + u_{2,dt} + e_{dtj},$$
$$d = 1, \ldots, D,\ t = 1, \ldots, T,\ j = 1, \ldots, n_{dt},$$

where the $u_{1,d}$'s are i.i.d. $N(0, \sigma_1^2)$, the $u_{2,dt}$'s are i.i.d. $N(0, \sigma_2^2)$, the e_{dtj}'s are i.i.d. $N(0, \sigma_0^2)$, and they are all mutually independent. The indexes d, t, and j are for AREA, age group, and individual, respectively. By using the function `lmer` from the library `lme4`, the following R code fits the model LMM4 to the data.

```
lmm.4 <- lmer(formula=INCOME~AGE+EDUCATION+(1|AREA/ageG), data=dat,
         REML=TRUE)
summary(lmm.4)       # Summary of the fitting procedure
anova(lmm.4)         # Analysis of Variance Table
fixef(lmm.4)         # Regression parameters
var <- as.data.frame(VarCorr(lmm.4)); var    # Variance parameters
sigmau2 <- var$sdcor[1]; sigmau2             # u_{2,dt} standard deviation
sigmau1 <- var$sdcor[2]; sigmau1             # u_{1,d} standard deviation
sigmae <- var$sdcor[3]; sigmae               # Residual standard deviation
```

The REML estimates of the regression parameters are $\beta_0 = 38099.47$, $\beta_1 = 27.22$, $\beta_2 = 9582.65$, and $\beta_3 = 19974.69$. The estimated standard deviations are $\hat{\sigma}_1 = 451.36$, $\hat{\sigma}_2 = 594.98$, and $\hat{\sigma}_e = 10367.89$.

6.7.2 Library nlme

The function `lme` of the R library `nlme` also fits the models LMM1-LMM4 to the sample data. The following R codes fit model LMM1 to data.

```
LMM.1 <- lme(INCOME~AGE+EDUCATION, random=~1|AREA, data=dat, method="REML")
summary(LMM.1)
```

Table 6.2 gives the estimates of the regression parameters. The estimated standard deviations are $\hat{\sigma}_u = 615.59$ and $\hat{\sigma}_e = 10376.7$.

The following R codes fit model LMM2 to data.

```
LMM.2 <- lme(INCOME~AGE+EDUCATION, random=~AGE|AREA, data=dat, method="ML")
summary(LMM.2)
```

Table 6.3 gives the estimates of the regression parameters. The estimated standard deviations are $\hat{\sigma}_u = 500.23$, $\hat{\sigma}_v = 0.03$, and $\hat{\sigma}_e = 10362.03$.

The following R codes fit model LMM3 to data.

```
inter <- interaction(dat$AREA, dat$ageG)
LMM.3 <- lme(INCOME~AGE+EDUCATION, random=~1|inter, data=dat, method="REML")
summary(LMM.3)
```

Table 6.4 gives the estimates of the regression parameters. The estimated standard deviations are $\hat{\sigma}_u = 737.29$ and $\hat{\sigma}_e = 10368.26$.

Table 6.2 Regression parameters of model LMM1

Variable	Estimate	Std. error	DF	t-value	p-value
Intercept	38,120.90	1239.96	1027	30.74	0.0000
AGE	26.79	20.62	1027	1.30	0.1943
EDUCATION2	9572.10	798.03	1027	11.99	0.0000
EDUCATION3	19,983.27	979.15	1027	20.41	0.0000

Table 6.3 Regression parameters of model LMM2

Variable	Estimate	Std. error	DF	t-value	p-value
Intercept	38,120.24	1236.58	1027	30.83	0.0000
AGE	26.89	20.62	1027	1.30	0.1925
EDUCATION2	9570.95	797.58	1027	12.00	0.0000
EDUCATION3	19,981.26	977.58	1027	20.44	0.0000

Table 6.4 Regression parameters of model LMM3

Variable	Estimate	Std. error	DF	t-value	p-value
Intercept	38,088.57	1248.74	987	30.50	0.0000
AGE	27.53	21.06	987	1.31	0.1915
EDUCATION2	9586.54	797.67	987	12.02	0.0000
EDUCATION3	19,967.86	977.74	987	20.42	0.0000

Table 6.5 Regression parameters of model LMM4

Variable	Estimate	Std.Error	DF	t-value	p-value
Intercept	38,099.49	1247.71	987	30.54	0.0000
AGE	27.22	20.92	987	1.30	0.1935
EDUCATION2	9582.64	798.07	987	12.01	0.0000
EDUCATION3	19,974.71	979.17	987	20.40	0.0000

The following R codes fit model LMM4 to data.

```
LMM.4 <- lme(INCOME~AGE+EDUCATION, random=~1|AREA/ageG, data=dat,
       method="REML")
summary(LMM.4)
```

Table 6.5 gives the estimates of the regression parameters. The estimated standard deviations are $\hat{\sigma}_1 = 452.13$, $\hat{\sigma}_2 = 594.70$, and $\hat{\sigma}_e = 10367.88$.

References

Battese, G.E, Harter, R.M., Fuller, W.A.: An error-component model for prediction of county crop areas using survey and satellite data. J. Am. Stat. Assoc. **83**, 28–36 (1988)

Cressie, N., Lahiri, S.N.: The asymptotic distribution of RMLE estimators. J. Multivar. Anal. **45**, 217–233 (1993)

Demidenko E.: Mixed Models, Theory and Applications. John Wiley, New York (2004)

Fay, R.E., Herriot, R.A.: Estimates of income for small places: an application of James-Stein procedures to census data. J. Am. Stat. Assoc. **74**, 269–277 (1979)

Goldstein, H.: Multilevel Statistical Models. Arnold, London (2003)

Harville, D.A.: Bayesian inference for variance components using only error contrasts. Biometrika **61**(2), 383–386 (1974)

Henderson, C.R.: Estimation of variance and covariance components. Biometrics **9**, 226–252 (1953)

Henderson, C.R.: Best linear unbiased estimation and prediction under selection model. Biometrics **31**, 423–427 (1975)

Henderson C.R., Kempthorne O., Searle S.R., Von Krosigk C.M.: The Estimation of environmental and genetic trends from records subject to culling. Biometrics **15**, 192–218 (1959)

Jiang, J.: Linear and Generalized Linear Mixed Models and Their Applications. Springer, Berlin (2007)

Longford, N.T.: Random Coefficient Models. Clareton Press, London (1995)

McCullogh, C.E., Searle, S.R.: Generalized, Linear and Mixed Models. John Wiley, New York (2001)

Rencher, A.C.: Multivariate Statistical Inference and Applications. John Wiley, New York (1998)

Searle, S.R.: Linear Models. John Wiley & Sons, Hoboken (1971)

Searle, S.R., Casella, G., McCullogh, C.E.: Variance Components. John Wiley, New York (1992)

Chapter 7
Nested Error Regression Models

7.1 Introduction

The linear mixed model (6.1) is more simple in the particular case of having only one random effect with D levels. In that case we have a unit-level linear mixed model with one random factor that is known as *nested error regression model* (NER), or random intercept model, in the statistical literature.

Battese and Fuller (1981), Harter (1983), and Battese et al. (1988) applied the NER model for the first time to small area estimation (SAE) problems. They predicted areas under corn and soybeans for 12 counties in north-central Iowa, based on the 1978 June Enumerative Survey and satellite data. Since then, the NER model is considered as the basic unit-level model in SAE. With regard to applications to real data, the empirical best predictors based on NER models have been employed to predict domain parameters.

This chapter deals with the estimation of the regression and variance components' parameters of the NER model. It describes three fitting methods. Namely, maximum likelihood (ML), residual maximum likelihood (REML), and Henderson 3 (H3). These three types of estimators of model parameters are widely used under linear mixed models. The respective sections give, with plenty of details, the derivations of the algorithm formulas required for programming the three fitting algorithms.

Section 7.2 introduces the NER model. Sections 7.3 and 7.4 give algorithms for calculating ML estimators. The ML estimators maximize the log-likelihood of the vector $\boldsymbol{y}$ containing the values of the dependent variable in all the sample units.

Sections 7.5 and 7.6 deal with the REML estimators. The REML estimators maximize the log-likelihood of a linear transformation of $\boldsymbol{y}$. For solving the system of log-likelihood equations, Fisher-scoring algorithms are provided under two different parametrizations.

The H3 estimators are moment-based estimators that are obtained as the solution of a system of two linear equations. Section 7.7 derives the expression of the H3

D. Morales et al., *A Course on Small Area Estimation and Mixed Models*, Statistics for Social and Behavioral Sciences, https://doi.org/10.1007/978-3-030-63757-6_7

estimators and calculates their first and second moments. Section 7.9 presents a simulation experiment for empirically investigating the behavior of the ML, REML, and H3 estimators of the parameters of a NER model.

Finally, Sect. 7.10 gives R codes for calculating the ML estimators of the parameters of the NER model.

7.2 The NER Model

The NER model is

$$y_{dj} = \boldsymbol{x}_{dj}\boldsymbol{\beta} + u_d + e_{dj}, \quad d = 1, \ldots, D,\, j = 1, \ldots, n_d, \tag{7.1}$$

where y_{dj} is the target variable measured in the sampled unit j of the domain d, $\boldsymbol{x}_{dj}$ is a $1 \times p$ vector containing the values of the auxiliary variables, $\boldsymbol{\beta} = (\beta_1, \ldots, \beta_p)'$ is the vector of regression parameters, $u_d \sim N(0, \sigma_u^2)$ is a domain random effect, $e_{dj} \sim N(0, w_{dj}^{-1}\sigma_e^2)$ is a model error, and $w_{11} > 0, \ldots, w_{Dn_D} > 0$ are known heteroscedasticity weights. Model (7.1) assumes that the u_d's and the e_{dj}'s are all mutually independent. The variance of y_{dj} is $\text{var}(y_{dj}) = \sigma_u^2 + w_{dj}^{-1}\sigma_e^2$.

The model (7.1) can be interpreted as a random intercept model when we take $x_{dj1} = 1$ and $\beta_1 = \alpha$. The variable $\alpha_d = \alpha + u_d$ is the random ordinate at the origin (intercept) in domain d.

At the domain level, the NER model is

$$\boldsymbol{y}_d = \boldsymbol{X}_d\boldsymbol{\beta} + \boldsymbol{Z}_d u_d + \boldsymbol{e}_d, \quad d = 1, \ldots, D, \tag{7.2}$$

where $\boldsymbol{y}_d = \underset{1\le j\le n_d}{\text{col}}(y_{dj})$, $\boldsymbol{X}_d = \underset{1\le j\le n_d}{\text{col}}(\boldsymbol{x}_{dj})$, $\boldsymbol{Z}_d = \boldsymbol{1}_{n_d}$, $u_d \sim N(0, \sigma_u^2)$ $\boldsymbol{e}_d = \underset{1\le j\le n_d}{\text{col}}(e_{dj}) \sim N(\boldsymbol{0}_{n_d}, \sigma_e^2\boldsymbol{W}_d^{-1})$, $\boldsymbol{W}_d = \underset{1\le j\le n_d}{\text{diag}}(w_{dj})$, and $\boldsymbol{0}_{n_d}$ and $\boldsymbol{1}_{n_d}$ are the $n_d \times 1$ vectors with all components equal to zero and one, respectively. Model (7.2) assumes that the u_d's and the $\boldsymbol{e}_d$'s are all mutually independent. The variance matrix of $\boldsymbol{y}_d$ is

$$\boldsymbol{V}_d = \text{var}(\boldsymbol{y}_d) = \sigma_u^2\boldsymbol{1}_{n_d}\boldsymbol{1}'_{n_d} + \sigma_e^2\boldsymbol{W}_d^{-1}, \quad d = 1, \ldots, D. \tag{7.3}$$

In the full matrix form, the NER model is

$$\boldsymbol{y} = \boldsymbol{X}\boldsymbol{\beta} + \boldsymbol{Z}\boldsymbol{u} + \boldsymbol{e}, \tag{7.4}$$

where $\boldsymbol{y} = \underset{1\le d\le D}{\text{col}}(\boldsymbol{y}_d)$, $\boldsymbol{X} = \underset{1\le d\le D}{\text{col}}(\boldsymbol{X}_d)$, $\boldsymbol{Z} = \underset{1\le d\le D}{\text{diag}}(\boldsymbol{Z}_d)$, $\boldsymbol{u} = \underset{1\le d\le D}{\text{col}}(u_d) \sim N(\boldsymbol{0}_D, \boldsymbol{V}_u)$, $\boldsymbol{V}_u = \sigma_u^2\boldsymbol{I}_D$, $\boldsymbol{e} = \underset{1\le d\le D}{\text{col}}(\boldsymbol{e}_d) \sim N(\boldsymbol{0}_n, \boldsymbol{V}_e)$, $\boldsymbol{V}_e = \sigma_e^2\boldsymbol{W}^{-1}$, $\boldsymbol{W} =$

$\underset{1\le d\le D}{\text{diag}}(\boldsymbol{W}_d)$, and $\boldsymbol{I}_D$ is the $D\times D$ identity matrix. Model (7.4) assumes that the vectors $\boldsymbol{u}$ and $\boldsymbol{e}$ are independent. The variance matrix of $\boldsymbol{y}$ is

$$\boldsymbol{V}=\text{var}(\boldsymbol{y})=\boldsymbol{Z}\boldsymbol{V}_u\boldsymbol{Z}'+\boldsymbol{V}_e=\text{diag}(\boldsymbol{V}_1,\ldots,\boldsymbol{V}_D).$$

Under model (7.4), it holds that $\boldsymbol{Z}'\boldsymbol{Z}=\text{diag}(n_1,\ldots,n_D)$, $\text{tr}(\boldsymbol{Z}'\boldsymbol{Z})=\sum_{d=1}^{D}n_d=n$, and $\text{r}(\boldsymbol{Z})=D$. If the matrix $\boldsymbol{X}$ has a full rank and the columns of $\boldsymbol{X}$ are linearly independent from the columns of $\boldsymbol{Z}$, then we have $\text{r}[\boldsymbol{X},\boldsymbol{Z}]=p+D$. Further, if $x_{dj1}=1$, for $d=1,\ldots,D$, $j=1,\ldots,n_d$, then $\text{r}[\boldsymbol{X},\boldsymbol{Z}]=p+D-1$, because the columns of $\boldsymbol{Z}$ sum up to $\boldsymbol{1}_n$. We define

$$\boldsymbol{w}_{n_d}=\boldsymbol{W}_d\boldsymbol{1}_{n_d}=(w_{d1},\ldots,w_{dn_d})'_{n_d\times 1},\ w_d=\boldsymbol{1}'_{n_d}\boldsymbol{w}_{n_d}=\sum_{j=1}^{n_d}w_{dj},\ \gamma_d^w=\frac{\sigma_u^2}{\sigma_u^2+\frac{\sigma_e^2}{w_d}}. \tag{7.5}$$

For calculating $\boldsymbol{V}_d^{-1}$, we use the matrix inversion formula

$$(\boldsymbol{A}+\boldsymbol{a}\boldsymbol{b}')^{-1}=\boldsymbol{A}^{-1}-\frac{\boldsymbol{A}^{-1}\boldsymbol{a}\boldsymbol{b}'\boldsymbol{A}^{-1}}{1+\boldsymbol{b}'\boldsymbol{A}^{-1}\boldsymbol{a}},$$

with $\boldsymbol{A}=\sigma_e^2\boldsymbol{W}_d^{-1}=\sigma_e^2\text{diag}\left(w_{d1}^{-1},\ldots,w_{dn_d}^{-1}\right)$ and $\boldsymbol{a}=\boldsymbol{b}=\sigma_u\boldsymbol{1}_{n_d}=(\sigma_u,\ldots,\sigma_u)'$. Then $\boldsymbol{A}^{-1}=\sigma_e^{-2}\boldsymbol{W}_d$ and

$$\begin{aligned}
1+\boldsymbol{b}'\boldsymbol{A}^{-1}\boldsymbol{a}&=1+\frac{\sigma_u^2}{\sigma_e^2}w_d,\\
\boldsymbol{A}^{-1}\boldsymbol{a}\boldsymbol{b}'\boldsymbol{A}^{-1}&=\sigma_e^{-4}\sigma_u^2\boldsymbol{W}_d\boldsymbol{1}_{n_d}\boldsymbol{1}'_{n_d}\boldsymbol{W}_d=\sigma_e^{-4}\sigma_u^2\boldsymbol{w}_{n_d}\boldsymbol{w}'_{n_d},\\
\frac{\boldsymbol{A}^{-1}\boldsymbol{a}\boldsymbol{b}'\boldsymbol{A}^{-1}}{1+\boldsymbol{b}'\boldsymbol{A}^{-1}\boldsymbol{a}}&=\frac{\sigma_e^{-4}\sigma_u^2\boldsymbol{w}_{n_d}\boldsymbol{w}'_{n_d}}{1+\frac{\sigma_u^2}{\sigma_e^2}w_d}=\sigma_e^{-2}w_d^{-1}\gamma_d^w\boldsymbol{w}_{n_d}\boldsymbol{w}'_{n_d}.
\end{aligned}$$

Therefore, we have

$$\boldsymbol{V}_d^{-1}=\frac{1}{\sigma_e^2}\left(\boldsymbol{W}_d-\frac{\gamma_d^w}{w_d}\boldsymbol{w}_{n_d}\boldsymbol{w}'_{n_d}\right). \tag{7.6}$$

For general linear mixed models fulfilling Eq. (7.4), the best linear unbiased estimator (BLUE) of $\boldsymbol{\beta}$ and the best linear unbiased predictor (BLUP) of $\boldsymbol{u}$ are given in (6.12). Under the NER model, if $\sigma_e^2>0$ and $\sigma_u^2>0$ are known, the BLUE of $\boldsymbol{\beta}$ and BLUP of $\boldsymbol{u}$ are

$$\tilde{\boldsymbol{\beta}}=(\boldsymbol{X}'\boldsymbol{V}^{-1}\boldsymbol{X})^{-1}\boldsymbol{X}'\boldsymbol{V}^{-1}\boldsymbol{y}\quad\text{and}\quad\tilde{\boldsymbol{u}}=\boldsymbol{V}_u\boldsymbol{Z}'\boldsymbol{V}^{-1}(\boldsymbol{y}-\boldsymbol{X}\tilde{\boldsymbol{\beta}}). \tag{7.7}$$

A computationally efficient formula for calculating $\tilde{\boldsymbol{\beta}}$ is

$$\tilde{\boldsymbol{\beta}} = \Big(\sum_{d=1}^{D} \boldsymbol{X}_d' \boldsymbol{V}_d^{-1} \boldsymbol{X}_d\Big)^{-1} \Big(\sum_{d=1}^{D} \boldsymbol{X}_d' \boldsymbol{V}_d^{-1} \boldsymbol{y}_d\Big).$$

A simpler expression for $\tilde{\boldsymbol{u}}$ can be derived. By applying (7.6) to the product $\boldsymbol{V}_u \boldsymbol{Z}' \boldsymbol{V}^{-1}$, we get

$$\begin{aligned}\boldsymbol{V}_u \boldsymbol{Z}' \boldsymbol{V}^{-1} &= \sigma_u^2 \underset{1\le d\le D}{\text{diag}} (\boldsymbol{1}_{n_d}') \underset{1\le d\le D}{\text{diag}} \left(\frac{1}{\sigma_e^2}\Big[\boldsymbol{W}_d - \frac{\gamma_d^w}{w_d} \boldsymbol{w}_{n_d} \boldsymbol{w}_{n_d}'\Big]\right) \\ &= \underset{1\le d\le D}{\text{diag}} \left(\frac{\sigma_u^2}{\sigma_e^2}\Big[\boldsymbol{w}_{n_d}' - \frac{\gamma_d^w}{w_d} w_d \boldsymbol{w}_{n_d}'\Big]\right) = \underset{1\le d\le D}{\text{diag}} \left(\frac{\sigma_u^2}{\sigma_e^2}[1-\gamma_d^w]\boldsymbol{w}_{n_d}'\right).\end{aligned}$$

By taking into account that $(\sigma_u^2/\sigma_e^2)(1-\gamma_d^w) = \gamma_d^w/w_d$, we have

$$\boldsymbol{V}_u \boldsymbol{Z}' \boldsymbol{V}^{-1} = \text{diag}\left(\frac{\gamma_1^w}{w_1} \boldsymbol{w}_{n_1}', \ldots, \frac{\gamma_D^w}{w_D} \boldsymbol{w}_{n_D}'\right)_{D\times n}.$$

By substituting $\boldsymbol{V}_u \boldsymbol{Z}' \boldsymbol{V}^{-1}$ in (7.7), we obtain the intuitive formula

$$\tilde{u}_d = \gamma_d^w \Big(\hat{\overline{Y}}_d^w - \hat{\overline{\boldsymbol{X}}}_d^w \tilde{\boldsymbol{\beta}}\Big), \quad d = 1, \ldots, D, \tag{7.8}$$

where $\hat{\overline{Y}}_d^w = \frac{1}{w_d} \sum_{j=1}^{n_d} w_{dj} y_{dj}$ and $\hat{\overline{\boldsymbol{X}}}_d^w = \frac{1}{w_d} \sum_{j=1}^{n_d} w_{dj} \boldsymbol{x}_{dj}$.

The empirical BLUE (EBLUE) of $\boldsymbol{\beta}$ and the empirical BLUP (EBLUP) of $\boldsymbol{u}$ are obtained by substituting the variances σ_u^2 and σ_e^2 by consistent estimators $\hat{\sigma}_u^2$ and $\hat{\sigma}_e^2$, respectively. They are

$$\hat{\boldsymbol{\beta}} = (\boldsymbol{X}'\hat{\boldsymbol{V}}^{-1}\boldsymbol{X})^{-1}\boldsymbol{X}'\hat{\boldsymbol{V}}^{-1}\boldsymbol{y} \quad \text{and} \quad \hat{\boldsymbol{u}} = \hat{\boldsymbol{V}}_u \boldsymbol{Z}'\hat{\boldsymbol{V}}^{-1}(\boldsymbol{y} - \boldsymbol{X}\hat{\boldsymbol{\beta}}), \tag{7.9}$$

where

$$\hat{\boldsymbol{V}} = \boldsymbol{Z}\hat{\boldsymbol{V}}_u\boldsymbol{Z}' + \hat{\boldsymbol{V}}_e, \quad \hat{\boldsymbol{V}}_u = \hat{\sigma}_u^2 \boldsymbol{I}_D, \quad \hat{\boldsymbol{V}}_e = \hat{\sigma}_e^2 \boldsymbol{W}^{-1}.$$

The following sections introduce three procedures (ML, REML, and H3) for estimating the variance components, σ_u^2 and σ_e^2, of the NER model.

7.3 ML Estimators

The ML approach calculates simultaneously the estimators of $\boldsymbol{\beta}$, σ_u^2, and σ_e^2 and later substitutes their values in (7.8) for obtaining the predictor $\hat{\boldsymbol{u}}$. Section 6.4 describes the two algorithms (Fisher scoring and Newton–Raphson) for calculating the ML estimators of the regression coefficients and of the variance components in the more general model (6.1).

Because the expressions (6.21) of the Fisher-scoring algorithm are simpler than the expressions (6.18)–(6.20) of the Newton–Raphson algorithm, this section gives plenty of details about the formulas needed for applying the Fisher-scoring algorithm in the particular model (7.1) with only one random factor.

Let $\boldsymbol{\theta} = (\boldsymbol{\beta}', \sigma_u^2, \sigma_e^2)'$. The log-likelihood of the NER model (7.4) is

$$l(\boldsymbol{\theta}) = -\frac{n}{2}\log(2\pi) - \frac{1}{2}\log|\boldsymbol{V}| - \frac{1}{2}(\boldsymbol{y} - \boldsymbol{X}\boldsymbol{\beta})'\boldsymbol{V}^{-1}(\boldsymbol{y} - \boldsymbol{X}\boldsymbol{\beta}). \tag{7.10}$$

Let $S' = (S_{\boldsymbol{\beta}}', S_{\sigma_u^2}, S_{\sigma_e^2})$ be the vector of scores, i.e. $S_{\boldsymbol{\beta}} = \partial l(\boldsymbol{\theta})/\partial\boldsymbol{\beta}$, $S_{\sigma_u^2} = \partial l(\boldsymbol{\theta})/\partial\sigma_u^2$ and $S_{\sigma_e^2} = \partial l(\boldsymbol{\theta})/\partial\sigma_e^2$. Under the model (7.1), it is not difficult to check that the scores (6.15) and (6.17) can be expressed as sums of domain-based components, i.e.

$$S_{\boldsymbol{\beta}} = \sum_{d=1}^{D} \boldsymbol{X}_d{}'\boldsymbol{V}_d^{-1}(\boldsymbol{y}_d - \boldsymbol{X}_d\boldsymbol{\beta}), \tag{7.11}$$

$$S_{\sigma_u^2} = -\frac{1}{2}\sum_{d=1}^{D}\operatorname{tr}\left(\boldsymbol{V}_d^{-1}\boldsymbol{J}_{n_d}\right) + \frac{1}{2}\sum_{d=1}^{D}(\boldsymbol{y}_d - \boldsymbol{X}_d\boldsymbol{\beta})'\,\boldsymbol{V}_d^{-1}\boldsymbol{J}_{n_d}\boldsymbol{V}_d^{-1}\left(\boldsymbol{y}_d - \boldsymbol{X}_d\boldsymbol{\beta}\right),$$

$$S_{\sigma_e^2} = -\frac{1}{2}\sum_{d=1}^{D}\operatorname{tr}\left(\boldsymbol{V}_d^{-1}\boldsymbol{W}_d^{-1}\right) + \frac{1}{2}\sum_{d=1}^{D}(\boldsymbol{y}_d - \boldsymbol{X}_d\boldsymbol{\beta})'\,\boldsymbol{V}_d^{-1}\boldsymbol{W}_d^{-1}\boldsymbol{V}_d^{-1}\left(\boldsymbol{y}_d - \boldsymbol{X}_d\boldsymbol{\beta}\right),$$

where $\boldsymbol{J}_{n_d} = \boldsymbol{1}_{n_d}\boldsymbol{1}_{n_d}'$, $\boldsymbol{V}_d$, $\boldsymbol{W}_d$ and $\boldsymbol{1}_{n_d}$ are defined around (7.3). The elements of the Fisher information matrix are

$$F_{\boldsymbol{\beta}\boldsymbol{\beta}} = \sum_{d=1}^{D}\boldsymbol{X}_d'\boldsymbol{V}_d^{-1}\boldsymbol{X}_d, \quad F_{\sigma_u^2\sigma_u^2} = \frac{1}{2}\sum_{d=1}^{D}\operatorname{tr}\left((\boldsymbol{V}_d^{-1}\boldsymbol{J}_{n_d})^2\right), \tag{7.12}$$

$$F_{\sigma_u^2\sigma_e^2} = \frac{1}{2}\sum_{d=1}^{D}\operatorname{tr}\left(\boldsymbol{V}_d^{-1}\boldsymbol{W}_d^{-1}\boldsymbol{V}_d^{-1}\boldsymbol{J}_{n_d}\right), \quad F_{\sigma_e^2\sigma_e^2} = \frac{1}{2}\sum_{d=1}^{D}\operatorname{tr}\left((\boldsymbol{V}_d^{-1}\boldsymbol{W}_d^{-1})^2\right).$$

By using the formulas (7.5) and (7.6), about $\boldsymbol{w}_{n_d}$, w_d, γ_d^w and $\boldsymbol{V}_d^{-1}$, and doing some straightforward algebra, we get

$$S_{\beta} = \frac{1}{\sigma_e^2}\sum_{d=1}^{D}\left(\boldsymbol{X}_d'\boldsymbol{W}_d\boldsymbol{\varsigma}_d - \frac{\gamma_d^w}{w_d}\boldsymbol{X}_d'\boldsymbol{w}_{n_d}\boldsymbol{w}_{n_d}'\boldsymbol{\varsigma}_d\right),$$

$$S_{\sigma_u^2} = -\frac{1}{2}\sum_{d=1}^{D}\frac{1-\gamma_d^w}{\sigma_e^2}w_d + \frac{1}{2}\sum_{d=1}^{D}\left(\frac{1-\gamma_d^w}{\sigma_e^2}\right)^2\boldsymbol{\varsigma}_d'\boldsymbol{w}_{n_d}\boldsymbol{w}_{n_d}'\boldsymbol{\varsigma}_d,$$

$$S_{\sigma_e^2} = -\frac{1}{2}\sum_{d=1}^{D}\frac{n_d-\gamma_d^w}{\sigma_e^2} + \frac{1}{2(\sigma_e^2)^2}\sum_{d=1}^{D}\left[\boldsymbol{\varsigma}_d'\boldsymbol{W}_d\boldsymbol{\varsigma}_d + \frac{\gamma_d^w(\gamma_d^w-2)}{w_d}\boldsymbol{\varsigma}_d'\boldsymbol{w}_{n_d}\boldsymbol{w}_{n_d}'\boldsymbol{\varsigma}_d\right],$$

where $\boldsymbol{\varsigma}_d = \boldsymbol{y}_d - \boldsymbol{X}_d\boldsymbol{\beta}$, $d = 1,\ldots,D$. The elements of the Fisher information matrix are

$$F_{\beta\beta} = \tfrac{1}{\sigma_e^2}\textstyle\sum_{d=1}^{D}\left(\boldsymbol{X}_d'\boldsymbol{W}_d\boldsymbol{X}_d - \tfrac{\gamma_d^w}{w_d}\boldsymbol{X}_d'\boldsymbol{w}_{n_d}\boldsymbol{w}_{n_d}'\boldsymbol{X}_d\right),\quad F_{\sigma_u^2\sigma_u^2} = \tfrac{1}{2}\textstyle\sum_{d=1}^{D}\left(\tfrac{1-\gamma_d^w}{\sigma_e^2}\right)^2 w_d^2,$$

$$F_{\sigma_u^2\sigma_e^2} = \tfrac{1}{2}\textstyle\sum_{d=1}^{D}\left(\tfrac{1-\gamma_d^w}{\sigma_e^2}\right)^2 w_d,\quad F_{\sigma_e^2\sigma_e^2} = \tfrac{1}{2(\sigma_e^2)^2}\textstyle\sum_{d=1}^{D}\left[n_d + \gamma_d^w(\gamma_d^w-2)\right].$$

Note that the Fisher-scoring algorithm requires some initial values (seeds) to start with updating cycles. A possible seed for $\boldsymbol{\beta}$ is the ML estimator under the model $\boldsymbol{y} = \boldsymbol{X}\boldsymbol{\beta} + \boldsymbol{e}$, obtained by removing the vector $\boldsymbol{u}$ of random effects from model (7.4). This is to say,

$$\boldsymbol{\beta}_0 = (\boldsymbol{X}'\boldsymbol{W}\boldsymbol{X})^{-1}\boldsymbol{X}'\boldsymbol{W}\boldsymbol{y}. \tag{7.13}$$

A possible seed for σ_u^2 is

$$\sigma_{u,0}^2 = \frac{1}{D}\sum_{d=1}^{D}\left(\hat{\overline{Y}}_d^w - \hat{\overline{\boldsymbol{X}}}_d^w\boldsymbol{\beta}_0\right)^2,$$

which, for computational convenience, can be written in the form

$$\sigma_{u,0}^2 = \frac{1}{D}\left[\sum_{d=1}^{D}\frac{1}{w_d^2}\boldsymbol{y}_d'\boldsymbol{w}_{n_d}\boldsymbol{w}_{n_d}'\boldsymbol{y}_d - 2\boldsymbol{\beta}_0'\sum_{d=1}^{D}\frac{1}{w_d^2}\boldsymbol{X}_d'\boldsymbol{w}_{n_d}\boldsymbol{w}_{n_d}'\boldsymbol{y}_d\right.$$
$$\left. + \boldsymbol{\beta}_0'\left(\sum_{d=1}^{D}\frac{1}{w_d^2}\boldsymbol{X}_d'\boldsymbol{w}_{n_d}\boldsymbol{w}_{n_d}'\boldsymbol{X}_d\right)\boldsymbol{\beta}_0\right]. \tag{7.14}$$

Finally, a possible seed for σ_e^2 is the residual mean of squares under the model $\boldsymbol{y} = \boldsymbol{X}\boldsymbol{\beta} + \boldsymbol{e}$, i.e.

$$\sigma_{e,0}^2 = \frac{(\boldsymbol{y} - \boldsymbol{X}\boldsymbol{\beta}_0)'\boldsymbol{W}(\boldsymbol{y} - \boldsymbol{X}\boldsymbol{\beta}_0)}{n - \mathrm{r}(\boldsymbol{X})} = \frac{1}{n - \mathrm{r}(\boldsymbol{X})}\Bigg[\sum_{d=1}^{D} \boldsymbol{y}_d'\boldsymbol{W}_d\boldsymbol{y}_d - 2\boldsymbol{\beta}_0'\sum_{d=1}^{D}\boldsymbol{X}_d'\boldsymbol{W}_d\boldsymbol{y}_d$$
$$+ \boldsymbol{\beta}_0'\Big(\sum_{d=1}^{D}\boldsymbol{X}_d'\boldsymbol{W}_d\boldsymbol{X}_d\Big)\boldsymbol{\beta}_0\Bigg]. \tag{7.15}$$

Fisher-Scoring Algorithm

The following steps implement the Fisher-scoring algorithm.

1 Initial constants and calculations.

1.1 Set the maximal number of iterations, the desired algorithm tolerance and $i = 0$.

1.2 For $d = 1, \ldots, D$, calculate n_d, $\boldsymbol{w}_{n_d}$, w_d, $\boldsymbol{W}_d$ and

$$\boldsymbol{X}_d'\boldsymbol{w}_{n_d}\boldsymbol{w}_{n_d}'\boldsymbol{X}_d,\ \boldsymbol{y}_d'\boldsymbol{w}_{n_d}\boldsymbol{w}_{n_d}'\boldsymbol{y}_d,\ \boldsymbol{X}_d'\boldsymbol{w}_{n_d}\boldsymbol{w}_{n_d}'\boldsymbol{y}_d,\ \boldsymbol{X}_d'\boldsymbol{W}_d\boldsymbol{X}_d,\ \boldsymbol{y}_d'\boldsymbol{W}_d\boldsymbol{y}_d,\ \boldsymbol{X}_d'\boldsymbol{W}_d\boldsymbol{y}_d.$$

2 Initial values for estimators.

Calculate the initial values of $\boldsymbol{\beta}_0$, $\sigma_{u,0}^2$ and $\sigma_{e,0}^2$ by applying (7.13), (7.14) and (7.15).

3 Iteration $i + 1$.

Let $\boldsymbol{\theta}_i = (\boldsymbol{\beta}_i', \sigma_{u,i}^2, \sigma_{e,i}^2)$ be the vector of estimators obtained after the iteration i.

3.1 For $d = 1, \ldots, D$, do

$$\gamma_{d,i}^w = \frac{\sigma_{u,i}^2}{\sigma_{u,i}^2 + \sigma_{e,i}^2/w_d}, \qquad \boldsymbol{\varsigma}_{d,i} = \boldsymbol{y}_d - \boldsymbol{X}_d\boldsymbol{\beta}_i.$$

3.2 Calculate the score vector

$$S_{\boldsymbol{\beta}_i} = \sum_{d=1}^{D}\frac{1}{\sigma_{e,i}^2}\left(\boldsymbol{X}_d{}'\boldsymbol{W}_d\boldsymbol{\varsigma}_{d,i} - \frac{\gamma_{d,i}^w}{w_d}\boldsymbol{X}_d'\boldsymbol{w}_{n_d}\boldsymbol{w}_{n_d}'\boldsymbol{\varsigma}_{d,i}\right),$$

$$S_{\sigma_{u,i}^2} = -\sum_{d=1}^{D}\frac{1-\gamma_{d,i}^w}{2\sigma_{e,i}^2}w_d + \frac{1}{2}\sum_{d=1}^{D}\left(\frac{1-\gamma_{d,i}^w}{\sigma_{e,i}^2}\right)^2\boldsymbol{\varsigma}_{d,i}'\boldsymbol{w}_{n_d}\boldsymbol{w}_{n_d}'\boldsymbol{\varsigma}_{d,i},$$

$$S_{\sigma^2_{e,i}} = -\sum_{d=1}^{D} \frac{1}{2} \frac{n_d - \gamma^w_{d,i}}{\sigma^2_{e,i}}$$

$$+ \frac{1}{2(\sigma^2_{e,i})^2} \sum_{d=1}^{D} \left[\boldsymbol{\varsigma}'_{d,i} \boldsymbol{W}_d \boldsymbol{\varsigma}_{d,i} + \frac{\gamma^w_{d,i}(\gamma^w_{d,i} - 2)}{w_d} \boldsymbol{\varsigma}'_{d,i} \boldsymbol{w}_{n_d} \boldsymbol{w}'_{n_d} \boldsymbol{\varsigma}_{d,i} \right],$$

and the Fisher information matrix

$$F_{\boldsymbol{\beta}_i \boldsymbol{\beta}_i} = \sum_{d=1}^{D} \frac{1}{\sigma^2_{e,i}} \left(\boldsymbol{X}_d' \boldsymbol{W}_d \boldsymbol{X}_d - \frac{\gamma^w_{d,i}}{w_d} \boldsymbol{X}'_d \boldsymbol{w}_{n_d} \boldsymbol{w}'_{n_d} \boldsymbol{X}_d \right),$$

$$F_{\sigma^2_{u,i}\sigma^2_{u,i}} = \sum_{d=1}^{D} \frac{1}{2} \left(\frac{1 - \gamma^w_{d,i}}{\sigma^2_{e,i}} \right)^2 w_d^2, \quad F_{\sigma^2_{u,i}\sigma^2_{e,i}} = \sum_{d=1}^{D} \frac{1}{2} \left(\frac{1 - \gamma^w_{d,i}}{\sigma^2_{e,i}} \right)^2 w_d,$$

$$F_{\sigma^2_{e,i}\sigma^2_{e,i}} = \sum_{d=1}^{D} \frac{1}{2(\sigma^2_{e,i})^2} \left[n_d + \gamma^w_{d,i}(\gamma^w_{d,i} - 2) \right].$$

3.3 Update the estimator values

$$\boldsymbol{\beta}_{i+1} = \boldsymbol{\beta}_i + \left(F_{\boldsymbol{\beta}_i \boldsymbol{\beta}_i}\right)^{-1} S_{\boldsymbol{\beta}_i},$$

$$\begin{pmatrix} \sigma^2_{u,i+1} \\ \sigma^2_{e,i+1} \end{pmatrix} = \begin{pmatrix} \sigma^2_{u,i} \\ \sigma^2_{e,i} \end{pmatrix} + \begin{pmatrix} F_{\sigma^2_{u,i}\sigma^2_{u,i}} & F_{\sigma^2_{u,i}\sigma^2_{e,i}} \\ F_{\sigma^2_{u,i}\sigma^2_{e,i}} & F_{\sigma^2_{e,i}\sigma^2_{e,i}} \end{pmatrix}^{-1} \begin{pmatrix} S_{\sigma^2_{u,i}} \\ S_{\sigma^2_{e,i}} \end{pmatrix}.$$

3.4 Check the stopping rules.

The relative tolerance of one component of an estimator $\hat{\boldsymbol{\theta}}_{i+1} = (\hat{\theta}_{1,i+1}, \ldots, \hat{\theta}_{p+2,i+1})$ at the iteration $i+1$ is

$$RT_j = \frac{|\hat{\theta}_{j,i+1} - \hat{\theta}_{j,i}|}{|\hat{\theta}_{j,i+1}|}, \qquad j = 1, \ldots, p+2.$$

If at least one of the following conditions hold:

- The number $i+1$ of iterations is greater than the maximum number of iterations given at Step 1,
- For every component of the vector of parameter estimates, the relative tolerance is lower than the one given in Step 1,

then STOP. Otherwise, do $i \leftarrow i+1$ and go to Step 3.

4 Final estimators.

If i^* is the last iteration, then the ML estimators are

$$\hat{\boldsymbol{\beta}} = \boldsymbol{\beta}_{i^*}, \quad \hat{\sigma}_u^2 = \max\left\{\sigma_{u,i^*}^2, 0\right\}, \quad \hat{\sigma}_e^2 = \max\left\{\sigma_{e,i^*}^2, 0\right\}.$$

7.4 ML Estimators for Alternative Parameters

Section 6.4 describes the algorithms (Fisher scoring and Newton–Raphson) for calculating the ML estimates under model (6.1) with the alternative parametrization

$$\sigma^2 = \sigma_e^2, \quad \varphi = \frac{\sigma_u^2}{\sigma_e^2}.$$

This section gives the Fisher-scoring algorithm for the particular model (7.1) with only one random factor. Under the model (7.1), we have the following simplifications:

$$\boldsymbol{V} = \sigma_e^2 \boldsymbol{\Sigma} = \sigma_e^2 (\boldsymbol{W}^{-1} + \varphi \boldsymbol{Z} \boldsymbol{I}_D \boldsymbol{Z}') = \sigma_e^2 \boldsymbol{W}^{-1} + \sigma_u^2 \boldsymbol{Z}\boldsymbol{Z}' = \text{diag}(\boldsymbol{V}_1, \ldots, \boldsymbol{V}_D),$$

$$\boldsymbol{V}_d = \sigma_e^2 \boldsymbol{W}_d^{-1} + \sigma_u^2 \mathbf{1}_{n_d} \mathbf{1}_{n_d}' = \sigma_e^2 (\boldsymbol{W}_d^{-1} + \varphi \mathbf{1}_{n_d} \mathbf{1}_{n_d}') = \sigma_e^2 \boldsymbol{\Sigma}_d, , \quad d = 1, \ldots, D,$$

$$\boldsymbol{\Sigma}_d = \boldsymbol{W}_d^{-1} + \varphi \mathbf{1}_{n_d} \mathbf{1}_{n_d}', \quad d = 1, \ldots, D.$$

We calculate $\boldsymbol{\Sigma}^{-1} = \text{diag}(\boldsymbol{\Sigma}_1^{-1}, \ldots, \boldsymbol{\Sigma}_D^{-1})$ instead of calculating $\boldsymbol{V}^{-1}$. We apply the matrix property that was applied for calculating $\boldsymbol{V}_d^{-1}$ in (7.6), and we use the same definitions as in (7.5), i.e.

$$(\boldsymbol{A} + \boldsymbol{u}\boldsymbol{v}')^{-1} = \boldsymbol{A}^{-1} - \frac{\boldsymbol{A}^{-1}\boldsymbol{u}\boldsymbol{v}'\boldsymbol{A}^{-1}}{1 + \boldsymbol{v}'\boldsymbol{A}^{-1}\boldsymbol{u}}.$$

In this case $\boldsymbol{A} = \boldsymbol{W}_d^{-1}$, $\boldsymbol{u} = \varphi \mathbf{1}_{n_d}$, and $\boldsymbol{v}' = \mathbf{1}_{n_d}'$, then $\boldsymbol{A}^{-1} = \boldsymbol{W}_d$ and

$$1 + \boldsymbol{v}'\boldsymbol{A}^{-1}\boldsymbol{u} = 1 + \varphi \mathbf{1}_{n_d}' \boldsymbol{W}_d \mathbf{1}_{n_d} = 1 + \varphi w_d,$$

$$\boldsymbol{A}^{-1}\boldsymbol{u}\boldsymbol{v}'\boldsymbol{A}^{-1} = \varphi \boldsymbol{W}_d \mathbf{1}_{n_d} \mathbf{1}_{n_d}' \boldsymbol{W}_d = \varphi \boldsymbol{w}_{n_d} \boldsymbol{w}_{n_d}',$$

$$\frac{\boldsymbol{A}^{-1}\boldsymbol{u}\boldsymbol{v}'\boldsymbol{A}^{-1}}{1 + \boldsymbol{v}'\boldsymbol{A}^{-1}\boldsymbol{u}} = \frac{\varphi}{1 + \varphi w_d} \boldsymbol{w}_{n_d} \boldsymbol{w}_{n_d}' = \frac{\gamma_d^w}{w_d} \boldsymbol{w}_{n_d} \boldsymbol{w}_{n_d}'.$$

Therefore,

$$\boldsymbol{\Sigma}_d^{-1} = \boldsymbol{W}_d - \frac{\gamma_d^w}{w_d} \boldsymbol{w}_{n_d} \boldsymbol{w}_{n_d}'. \tag{7.16}$$

In what follows, we simplify the notation, and we write σ^2 instead of σ_e^2.

Let $S' = (S'_{\boldsymbol{\beta}}, S_{\sigma^2}, S_{\varphi})$ be the vector of scores, $S_{\boldsymbol{\beta}} = \partial l(\boldsymbol{\theta})/\partial \boldsymbol{\beta}$, $S_{\sigma^2} = \partial l(\boldsymbol{\theta})/\partial \sigma^2$, and $S_{\varphi} = \partial l(\boldsymbol{\theta})/\partial \varphi$. Under model (7.1), the scores are sums of domain components, i.e. (c.f. (6.23), (6.24))

$$S_{\boldsymbol{\beta}} = \frac{1}{\sigma^2}\sum_{d=1}^{D} \boldsymbol{X}_d{}'\boldsymbol{\Sigma}_d^{-1}(\boldsymbol{y}_d - \boldsymbol{X}_d\boldsymbol{\beta}),$$

$$S_{\sigma^2} = -\frac{n}{2\sigma^2} + \frac{1}{2\sigma^4}\sum_{d=1}^{D} (\boldsymbol{y}_d - \boldsymbol{X}_d\boldsymbol{\beta})'\,\boldsymbol{\Sigma}_d^{-1}\,(\boldsymbol{y}_d - \boldsymbol{X}_d\boldsymbol{\beta}),$$

$$S_{\varphi} = -\frac{1}{2}\sum_{d=1}^{D} \operatorname{tr}\left(\boldsymbol{\Sigma}_d^{-1}\boldsymbol{1}_{n_d}\boldsymbol{1}'_{n_d}\right) + \frac{1}{2\sigma^2}\sum_{d=1}^{D} (\boldsymbol{y}_d - \boldsymbol{X}_d\boldsymbol{\beta})'\,\boldsymbol{\Sigma}_d^{-1}\boldsymbol{1}_{n_d}\boldsymbol{1}'_{n_d}\boldsymbol{\Sigma}_d^{-1}\,(\boldsymbol{y}_d - \boldsymbol{X}_d\boldsymbol{\beta}).$$

The elements of the Fisher information matrix are (c.f. (6.25))

$$\begin{array}{ll} F_{\boldsymbol{\beta}\boldsymbol{\beta}} = \frac{1}{2\sigma^2}\sum_{d=1}^{D} \boldsymbol{X}'_d\boldsymbol{\Sigma}_d^{-1}\boldsymbol{X}_d, & F_{\boldsymbol{\beta}\sigma^2} = 0, \\ F_{\boldsymbol{\beta}\varphi} = 0, & F_{\sigma^2\sigma^2} = \frac{n}{2\sigma^4} \\ F_{\sigma^2\varphi} = \frac{1}{2\sigma^2}\sum_{d=1}^{D} \operatorname{tr}\left(\boldsymbol{\Sigma}_d^{-1}\boldsymbol{1}_{n_d}\boldsymbol{1}'_{n_d}\right), & F_{\varphi\varphi} = \frac{1}{2}\sum_{d=1}^{D} \operatorname{tr}\left(\boldsymbol{\Sigma}_d^{-1}\boldsymbol{1}_{n_d}\boldsymbol{1}'_{n_d}\boldsymbol{\Sigma}_d^{-1}\boldsymbol{1}_{n_d}\boldsymbol{1}'_{n_d}\right). \end{array}$$

By applying the formulas (7.5) and (7.16), about $\boldsymbol{w}_{n_d}$, w_d, γ_d^w, and $\boldsymbol{\Sigma}_d^{-1}$, and doing some calculations, we get

$$S_{\boldsymbol{\beta}} = \frac{1}{\sigma^2}\sum_{d=1}^{D}\left(\boldsymbol{X}_d{}'\boldsymbol{W}_d\boldsymbol{\varsigma}_d - \frac{\gamma_d^w}{w_d}\boldsymbol{X}'_d\boldsymbol{w}_{n_d}\boldsymbol{w}'_{n_d}\boldsymbol{\varsigma}_d\right),$$

$$S_{\sigma^2} = -\frac{n}{2\sigma^2} + \frac{1}{2(\sigma^2)^2}\sum_{d=1}^{D}\left(\boldsymbol{\varsigma}'_d\boldsymbol{W}_d\boldsymbol{\varsigma}_d - \frac{\gamma_d^w}{w_d}\boldsymbol{\varsigma}'_d\boldsymbol{w}_{n_d}\boldsymbol{w}'_{n_d}\boldsymbol{\varsigma}_d\right),$$

$$S_{\varphi} = -\frac{1}{2}\sum_{d=1}^{D}(1-\gamma_d^w)w_d + \frac{1}{2\sigma^2}\sum_{d=1}^{D}(1-\gamma_d^w)^2\boldsymbol{\varsigma}'_d\boldsymbol{w}_{n_d}\boldsymbol{w}'_{n_d}\boldsymbol{\varsigma}_d,$$

where $\boldsymbol{\varsigma}_d = \boldsymbol{y}_d - \boldsymbol{X}_d\boldsymbol{\beta}$, $d = 1, \ldots, D$. The elements of the Fisher information matrix are

$$F_{\boldsymbol{\beta}\boldsymbol{\beta}} = \frac{1}{\sigma^2}\sum_{d=1}^{D}\left(\boldsymbol{X}_d{}'\boldsymbol{W}_d\boldsymbol{X}_d - \frac{\gamma_d^w}{w_d}\boldsymbol{X}'_d\boldsymbol{w}_{n_d}\boldsymbol{w}'_{n_d}\boldsymbol{X}_d\right), \quad F_{\sigma^2\sigma^2} = \frac{n}{2(\sigma^2)^2},$$

$$F_{\sigma^2\varphi} = \frac{1}{2\sigma^2}\sum_{d=1}^{D}\left(1-\gamma_d^w\right)w_d, \quad F_{\varphi\varphi} = \frac{1}{2}\sum_{d=1}^{D}\left((1-\gamma_d^w)w_d\right)^2.$$

Note that the Fisher-scoring algorithm requires some initial values (seeds) to start with updating cycles. A possible seed for $\boldsymbol{\beta}$ is its ML estimator $\boldsymbol{\beta}_0$ under the model (7.1), but treating the random effects as fixed, which is defined in (7.13). For σ^2, we propose to take the residual mean of squares under the model without random effects, i.e. $\sigma_0^2 = \sigma_{e,0}^2$, where $\sigma_{e,0}^2$ is defined in (7.15). Finally, a possible seed for φ is

$$\varphi_0 = \frac{\sigma_{u,0}^2}{\sigma_0^2}, \tag{7.17}$$

where $\sigma_{u,0}^2$ is expressed in (7.14). By using the above expressions, we can implement the Fisher-scoring algorithm with the following steps.

Fisher-Scoring Algorithm for Alternative Parameters

1 Initial constants and calculations.

1.1 Set the maximal number of iterations, the desired algorithm tolerance, and $i = 0$.

1.2 For $d = 1, \ldots, D$, calculate n_d, $\boldsymbol{w}_{n_d}$, w_d, $\boldsymbol{W}_d$, and

$$\boldsymbol{X}_d'\boldsymbol{w}_{n_d}\boldsymbol{w}_{n_d}'\boldsymbol{X}_d,\ \boldsymbol{y}_d'\boldsymbol{w}_{n_d}\boldsymbol{w}_{n_d}'\boldsymbol{y}_d,\ \boldsymbol{X}_d'\boldsymbol{w}_{n_d}\boldsymbol{w}_{n_d}'\boldsymbol{y}_d,\ \boldsymbol{X}_d'\boldsymbol{W}_d\boldsymbol{X}_d,\ \boldsymbol{y}_d'\boldsymbol{W}_d\boldsymbol{y}_d,\ \boldsymbol{X}_d'\boldsymbol{W}_d\boldsymbol{y}_d.$$

2 Initial values for estimators.

Calculate the initial values of $\boldsymbol{\beta}_0$, σ_0^2, and φ_0 by using (7.13), (7.15), and (7.17).

3 Iteration $i + 1$.

Let $\boldsymbol{\theta}_i = (\boldsymbol{\beta}_i', \sigma_i^2, \varphi_i)$ be the vector of estimators obtained after the iteration i.

3.1 For $d = 1, \ldots, D$, do

$$\gamma_{d,i}^w = \frac{\varphi_i\sigma_i^2}{\varphi_i\sigma_i^2 + \sigma_i^2/w_d}, \qquad \boldsymbol{\varsigma}_{d,i} = \boldsymbol{y}_d - \boldsymbol{X}_d\boldsymbol{\beta}_i.$$

3.2 Calculate the score vector

$$S_{\boldsymbol{\beta}_i} = \sum_{d=1}^D \frac{1}{\sigma_i^2}\Big(\boldsymbol{X}_d'\boldsymbol{W}_d\boldsymbol{\varsigma}_{d,i} - \frac{\gamma_{d,i}^w}{w_d}\boldsymbol{X}_d'\boldsymbol{w}_{n_d}\boldsymbol{w}_{n_d}'\boldsymbol{\varsigma}_{d,i}\Big),$$

$$S_{\sigma_i^2} = -\frac{n}{2\sigma_i^2} + \frac{1}{2(\sigma_i^2)^2}\sum_{d=1}^D\Big(\boldsymbol{\varsigma}_{d,i}'\boldsymbol{W}_d\boldsymbol{\varsigma}_{d,i} - \frac{\gamma_{d,i}^w}{w_d}\boldsymbol{\varsigma}_{d,i}'\boldsymbol{w}_{n_d}\boldsymbol{w}_{n_d}'\boldsymbol{\varsigma}_{d,i}\Big),$$

$$S_{\varphi_i} = -\sum_{d=1}^D \frac{1}{2}(1-\gamma_{d,i}^w)w_d + \frac{1}{2\sigma_i^2}\sum_{d=1}^D(1-\gamma_{d,i}^w)^2\boldsymbol{\varsigma}_{d,i}'\boldsymbol{w}_{n_d}\boldsymbol{w}_{n_d}'\boldsymbol{\varsigma}_{d,i},$$

and the elements of the Fisher information matrix

$$F_{\beta_i\beta_i} = \sum_{d=1}^{D} \frac{1}{\sigma_i^2}\Big(X_d{}'W_dX_d - \frac{\gamma_{d,i}^w}{w_d} X_d' w_{nd} w_{nd}' X_d\Big), \; F_{\sigma_i^2\sigma_i^2} = \sum_{d=1}^{D} \frac{n}{2(\sigma_i^2)^2},$$

$$F_{\sigma_i^2\varphi_i} = \sum_{d=1}^{D} \frac{1}{2\sigma_i^2}(1-\gamma_{d,i}^w)w_d, \; F_{\varphi_i\varphi_i} = \sum_{d=1}^{D} \frac{1}{2}\left((1-\gamma_{d,i}^w)w_d\right)^2.$$

3.3 Update the estimator values

$$\beta_{i+1} = \beta_i + \left(F_{\beta_i\beta_i}\right)^{-1} S_{\beta_i},$$

$$\begin{pmatrix} \sigma_{i+1}^2 \\ \varphi_{i+1} \end{pmatrix} = \begin{pmatrix} \sigma_i^2 \\ \varphi_i \end{pmatrix} + \begin{pmatrix} F_{\sigma_i^2\sigma_i^2} & F_{\sigma_i^2\varphi_i} \\ F_{\sigma_i^2\varphi_i} & F_{\varphi_i\varphi_i} \end{pmatrix}^{-1} \begin{pmatrix} S_{\sigma_i^2} \\ S_{\varphi_i} \end{pmatrix}.$$

3.4 Check the stopping rules.
The relative tolerance of one component of an estimator $\hat{\theta}_{i+1} = (\hat{\theta}_{1,i+1}, \ldots, \hat{\theta}_{p+2,i+1})$ at the iteration $i+1$ is

$$RT_j = \frac{|\hat{\theta}_{j,i+1} - \hat{\theta}_{j,i}|}{|\hat{\theta}_{j,i+1}|}, \qquad j = 1, \ldots, p+2.$$

If at least one of the following conditions hold:

- The number $i+1$ of iterations is greater than the maximum number of iterations given at Step 1,
- For every component of the vector of parameter estimates, the relative tolerance is lower than the one given in Step 1,

then STOP. Otherwise, do $i \leftarrow i+1$ and go to Step 3.

4 Final estimators.
If i^* is the last iteration, then the ML estimators are

$$\hat{\beta} = \beta_{i^*}, \quad \hat{\sigma}^2 = \max\left\{\sigma_{i^*}^2, 0\right\}, \quad \hat{\varphi} = \max\{\varphi_{i^*}, 0\}.$$

7.5 REML Estimators

The REML method estimates separately the regression parameters β and the variances σ_u^2 and σ_e^2. Section 6.5 gives the Fisher-scoring algorithm for calculating the REML estimators of the variance components. The estimator β is an explicit function of the data and the variance components' estimates. Therefore, the REML method is computationally more efficient than the ML method. This section gives the particularization of the Fisher-scoring algorithm to model (7.1) with only one random factor.

We define $\boldsymbol{P} = \boldsymbol{K}(\boldsymbol{K}\boldsymbol{V}\boldsymbol{K})^{-1}\boldsymbol{K} = \boldsymbol{V}^{-1} - \boldsymbol{V}^{-1}\boldsymbol{X}(\boldsymbol{X}'\boldsymbol{V}^{-1}\boldsymbol{X})^{-1}\boldsymbol{X}'\boldsymbol{V}^{-1}$ and $\boldsymbol{\sigma} = (\sigma_u^2, \sigma_e^2)'$, where $\boldsymbol{K} = \sigma_e^{-2}(\boldsymbol{W} - \boldsymbol{W}\boldsymbol{X}(\boldsymbol{X}'\boldsymbol{W}\boldsymbol{X})^{-1}\boldsymbol{X}'\boldsymbol{W})$ (particular case where $\boldsymbol{V}_e = \sigma_e^2\boldsymbol{W}^{-1}$) and $\boldsymbol{V} = \sigma_e^2\boldsymbol{W}^{-1} + \sigma_u^2\boldsymbol{Z}\boldsymbol{Z}'$. By adapting the development of Sect. 6.5.1 to the NER model, we get the REML log-likelihood

$$l(\boldsymbol{\sigma}) = -\frac{n-p}{2}\log 2\pi + \frac{1}{2}\log|\boldsymbol{X}'\boldsymbol{X}| - \frac{1}{2}\log|\boldsymbol{V}| - \frac{1}{2}\log|\boldsymbol{X}'\boldsymbol{V}^{-1}\boldsymbol{X}| - \frac{1}{2}\boldsymbol{y}'\boldsymbol{P}\boldsymbol{y}.$$

Let $S' = (S_{\sigma_u^2}, S_{\sigma_e^2})$ be the vector of scores, where $S_{\sigma_u^2} = \partial l(\boldsymbol{\sigma})/\partial\sigma_u^2$ and $S_{\sigma_e^2} = \partial l(\boldsymbol{\sigma})/\partial\sigma_e^2$ can be written in the form (cf. 6.28)

$$S_{\sigma_u^2} = -\frac{1}{2}\operatorname{tr}(\boldsymbol{P}\boldsymbol{Z}\boldsymbol{Z}') + \frac{1}{2}\boldsymbol{y}'\boldsymbol{P}\boldsymbol{Z}\boldsymbol{Z}'\boldsymbol{P}\boldsymbol{y}, \quad S_{\sigma_e^2} = -\frac{1}{2}\operatorname{tr}(\boldsymbol{P}\boldsymbol{W}^{-1}) + \frac{1}{2}\boldsymbol{y}'\boldsymbol{P}\boldsymbol{W}^{-1}\boldsymbol{P}\boldsymbol{y}.$$

By taking second partial derivatives and changing the sign and taking expectations, we get the components of the Fisher information matrix, i.e. (cf. 6.30)

$$F_{\sigma_u^2\sigma_u^2} = \frac{1}{2}\operatorname{tr}(\boldsymbol{P}\boldsymbol{Z}\boldsymbol{Z}'\boldsymbol{P}\boldsymbol{Z}\boldsymbol{Z}'), \quad F_{\sigma_e^2\sigma_e^2} = \frac{1}{2}\operatorname{tr}(\boldsymbol{P}\boldsymbol{W}^{-1}\boldsymbol{P}\boldsymbol{W}^{-1}),$$
$$F_{\sigma_u^2\sigma_e^2} = F_{\sigma_e^2\sigma_u^2} = \frac{1}{2}\operatorname{tr}(\boldsymbol{P}\boldsymbol{Z}\boldsymbol{Z}'\boldsymbol{P}\boldsymbol{W}^{-1}). \tag{7.18}$$

By applying the formula $\boldsymbol{P} = \boldsymbol{V}^{-1} - \boldsymbol{V}^{-1}\boldsymbol{X}\boldsymbol{Q}\boldsymbol{X}'\boldsymbol{V}^{-1}$ with $\boldsymbol{Q} = (\boldsymbol{X}'\boldsymbol{V}^{-1}\boldsymbol{X})^{-1}$ and doing some calculations, we get the scores

$$\begin{aligned}
S_{\sigma_u^2} &= -\frac{1}{2}\sum_{d=1}^{D}\boldsymbol{1}_{n_d}'\boldsymbol{V}_d^{-1}\boldsymbol{1}_{n_d} + \frac{1}{2}\sum_{d=1}^{D}\boldsymbol{1}_{n_d}'\boldsymbol{V}_d^{-1}\boldsymbol{X}_d\boldsymbol{Q}\boldsymbol{X}_d'\boldsymbol{V}_d^{-1}\boldsymbol{1}_{n_d} \\
&+ \frac{1}{2}\sum_{d=1}^{D}\boldsymbol{y}_d'\boldsymbol{V}_d^{-1}\boldsymbol{1}_{n_d}\boldsymbol{1}_{n_d}'\boldsymbol{V}_d^{-1}\boldsymbol{y}_d - \left(\sum_{d=1}^{D}\boldsymbol{y}_d'\boldsymbol{V}_d^{-1}\boldsymbol{1}_{n_d}\boldsymbol{1}_{n_d}'\boldsymbol{V}_d^{-1}\boldsymbol{X}_d\right)\boldsymbol{Q}\left(\sum_{d=1}^{D}\boldsymbol{X}_d'\boldsymbol{V}_d^{-1}\boldsymbol{y}_d\right) \\
&+ \frac{1}{2}\left(\sum_{d=1}^{D}\boldsymbol{y}_d'\boldsymbol{V}_d^{-1}\boldsymbol{X}_d\right)\boldsymbol{Q}\left(\sum_{=1}^{D}\boldsymbol{X}_d'\boldsymbol{V}_d^{-1}\boldsymbol{1}_{n_d}\boldsymbol{1}_{n_d}'\boldsymbol{V}_d^{-1}\boldsymbol{X}_d\right)\boldsymbol{Q}\left(\sum_{d=1}^{D}\boldsymbol{X}_d'\boldsymbol{V}_d^{-1}\boldsymbol{y}_d\right), \\
S_{\sigma_e^2} &= -\frac{1}{2}\sum_{d=1}^{D}\operatorname{tr}\left(\boldsymbol{V}_d^{-1}\boldsymbol{W}_d^{-1}\right) + \frac{1}{2}\sum_{d=1}^{D}\operatorname{tr}\left(\boldsymbol{V}_d^{-1}\boldsymbol{X}_d\boldsymbol{Q}\boldsymbol{X}_d'\boldsymbol{V}_d^{-1}\boldsymbol{W}_d^{-1}\right) \\
&+ \frac{1}{2}\sum_{d=1}^{D}\boldsymbol{y}_d'\boldsymbol{V}_d^{-1}\boldsymbol{W}_d^{-1}\boldsymbol{V}_d^{-1}\boldsymbol{y}_d - \left(\sum_{d=1}^{D}\boldsymbol{y}_d'\boldsymbol{V}_d^{-1}\boldsymbol{W}_d^{-1}\boldsymbol{V}_d^{-1}\boldsymbol{X}_d\right)\boldsymbol{Q}\left(\sum_{d=1}^{D}\boldsymbol{X}_d'\boldsymbol{V}_d^{-1}\boldsymbol{y}_d\right) \\
&+ \frac{1}{2}\left(\sum_{d=1}^{D}\boldsymbol{y}_d'\boldsymbol{V}_d^{-1}\boldsymbol{X}_d\right)\boldsymbol{Q}\left(\sum_{d=1}^{D}\boldsymbol{X}_d'\boldsymbol{V}_d^{-1}\boldsymbol{W}_d^{-1}\boldsymbol{V}_d^{-1}\boldsymbol{X}_d\right)\boldsymbol{Q}\left(\sum_{d=1}^{D}\boldsymbol{X}_d'\boldsymbol{V}_d^{-1}\boldsymbol{y}_d\right),
\end{aligned}$$

and the components of the Fisher information matrix

$$F_{\sigma_u^2\sigma_u^2} = \frac{1}{2}\sum_{d=1}^{D} \mathbf{1}'_{n_d}\mathbf{V}_d^{-1}\mathbf{1}_{n_d}\mathbf{1}'_{n_d}\mathbf{V}_d^{-1}\mathbf{1}_{n_d} - \sum_{d=1}^{D} \mathbf{1}'_{n_d}\mathbf{V}_d^{-1}\mathbf{1}_{n_d}\mathbf{1}'_{n_d}\mathbf{V}_d^{-1}\mathbf{X}_d\mathbf{Q}\mathbf{X}'_d\mathbf{V}_d^{-1}\mathbf{1}_{n_d}$$

$$+ \frac{1}{2}\sum_{d_1=1}^{D}\sum_{d_2=1}^{D} \mathbf{1}'_{n_{d_1}}\mathbf{V}_{d_1}^{-1}\mathbf{X}_{d_1}\mathbf{Q}\mathbf{X}'_{d_2}\mathbf{V}_{d_2}^{-1}\mathbf{1}_{n_{d_2}}\mathbf{1}'_{n_{d_2}}\mathbf{V}_{d_2}^{-1}\mathbf{X}_{d_2}\mathbf{Q}\mathbf{X}'_{d_1}\mathbf{V}_{d_1}^{-1}\mathbf{1}_{n_{d_1}},$$

$$F_{\sigma_u^2\sigma_e^2} = \frac{1}{2}\sum_{d=1}^{D} \mathbf{1}'_{n_d}\mathbf{V}_d^{-1}\mathbf{W}_d^{-1}\mathbf{V}_d^{-1}\mathbf{1}_{n_d} - \sum_{d=1}^{D} \mathbf{1}'_{n_d}\mathbf{V}_d^{-1}\mathbf{W}_d^{-1}\mathbf{V}_d^{-1}\mathbf{X}_d\mathbf{Q}\mathbf{X}'_d\mathbf{V}_d^{-1}\mathbf{1}_{n_d}$$

$$+ \frac{1}{2}\sum_{d_1=1}^{D}\sum_{d_2=1}^{D} \mathbf{1}'_{n_{d_1}}\mathbf{V}_{d_1}^{-1}\mathbf{X}_{d_1}\mathbf{Q}\mathbf{X}'_{d_2}\mathbf{V}_{d_2}^{-1}\mathbf{W}_{d_2}^{-1}\mathbf{V}_{d_2}^{-1}\mathbf{X}_{d_2}\mathbf{Q}\mathbf{X}'_{d_1}\mathbf{V}_{d_1}^{-1}\mathbf{1}_{n_{d_1}},$$

$$F_{\sigma_e^2\sigma_e^2} = \frac{1}{2}\sum_{d=1}^{D} \operatorname{tr}\left(\mathbf{V}_d^{-1}\mathbf{W}_d^{-1}\mathbf{V}_d^{-1}\mathbf{W}_d^{-1}\right) - \sum_{d=1}^{D} \operatorname{tr}\left(\mathbf{V}_d^{-1}\mathbf{W}_d^{-1}\mathbf{V}_d^{-1}\mathbf{X}_d\mathbf{Q}\mathbf{X}'_d\mathbf{V}_d^{-1}\mathbf{W}_d^{-1}\right)$$

$$+ \frac{1}{2}\sum_{d_1=1}^{D}\sum_{d_2=1}^{D} \operatorname{tr}\left(\mathbf{V}_{d_1}^{-1}\mathbf{X}_{d_1}\mathbf{Q}\mathbf{X}'_{d_2}\mathbf{V}_{d_2}^{-1}\mathbf{W}_{d_2}^{-1}\mathbf{V}_{d_2}^{-1}\mathbf{X}_{d_2}\mathbf{Q}\mathbf{X}'_{d_1}\mathbf{V}_{d_1}^{-1}\mathbf{W}_{d_1}^{-1}\right).$$

The updating formula for calculating the REML estimates of the variance components is

$$\boldsymbol{\sigma}_{k+1} = \boldsymbol{\sigma}_k + \boldsymbol{F}(\boldsymbol{\sigma}_k)^{-1}\boldsymbol{S}(\boldsymbol{\sigma}_k),$$

where $\boldsymbol{F}(\boldsymbol{\sigma}_k)$ is the Fisher information matrix evaluated at $\boldsymbol{\sigma}_k$. Note that the Fisher-scoring algorithm requires initial values. As initial values for σ_u^2, σ_e^2 we can take estimators $\sigma_{u,0}^2$ and $\sigma_{e,0}^2$ defined in (7.14) and (7.15), respectively. By using these seeds, we can implement the Fisher-scoring algorithm for calculating σ_u^2 and σ_e^2. Later, we calculate the REML estimator of $\boldsymbol{\beta}$ by applying the explicit formula

$$\hat{\boldsymbol{\beta}}_{REML} = (\mathbf{X}'\hat{\mathbf{V}}^{-1}\mathbf{X})^{-1}\mathbf{X}'\hat{\mathbf{V}}^{-1}\mathbf{y} = \Big(\sum_{d=1}^{D}\mathbf{X}'_d\hat{\mathbf{V}}_d^{-1}\mathbf{X}_d\Big)^{-1}\Big(\sum_{d=1}^{D}\mathbf{X}'_d\hat{\mathbf{V}}_d^{-1}\mathbf{y}_d\Big),$$

where $\hat{\mathbf{V}} = \hat{\sigma}_e^2\mathbf{W}^{-1} + \hat{\sigma}_u^2\mathbf{Z}\mathbf{Z}'$, $\hat{\mathbf{V}}_d^{-1} = \frac{1}{\hat{\sigma}_e^2}(\mathbf{W}_d - \frac{\hat{\gamma}_d^w}{w_d}\mathbf{w}_{n_d}\mathbf{w}'_{n_d})$, $\hat{\gamma}_d^w = \frac{\hat{\sigma}_u^2}{\hat{\sigma}_u^2 + \frac{\hat{\sigma}_e^2}{w_d}}$, and $\hat{\sigma}_e^2, \hat{\sigma}_u^2$ are the REML estimators of σ_e^2 and σ_u^2, respectively.

7.6 REML Estimators for Alternative Parameters

Section 6.5.2 describes the Fisher-scoring algorithm for calculating the REML estimator of the variance components under the model (6.1) with alternative parametrization $\sigma^2 = \sigma_e^2,\ \varphi = \frac{\sigma_u^2}{\sigma_e^2}$. This section gives the particularization of the Fisher-scoring algorithm to model (7.1) with only one random factor. Under the model with only one random factor, it holds that

$$V = \sigma_e^2\boldsymbol{\Sigma} = \sigma_e^2(W^{-1} + \varphi Z I_D Z') = \sigma_e^2 W^{-1} + \sigma_u^2 ZZ' = \operatorname{diag}(V_1, \dots, V_D),$$

with $V_d = \sigma_e^2 W_d^{-1} + \sigma_u^2 \mathbf{1}_{n_d}\mathbf{1}'_{n_d} = \sigma_e^2(W_d^{-1} + \varphi \mathbf{1}_{n_d}\mathbf{1}'_{n_d}) = \sigma_e^2 \boldsymbol{\Sigma}_d,\ d = 1, \dots, D$. In this case we have to calculate $\boldsymbol{\Sigma}^{-1} = \operatorname{diag}(\boldsymbol{\Sigma}_1^{-1}, \dots, \boldsymbol{\Sigma}_D^{-1})$, where $\boldsymbol{\Sigma}_d = W_d^{-1} + \varphi \mathbf{1}_{n_d}\mathbf{1}'_{n_d},\ d = 1, \dots, D$, and $\boldsymbol{\Sigma}_d^{-1} = W_d - \frac{\gamma_d^w}{w_d} w_{n_d} w'_{n_d}$ is given in (7.16).

Let us define $\boldsymbol{\sigma}' = (\sigma^2, \varphi),\ \boldsymbol{\theta}' = (\boldsymbol{\beta}', \boldsymbol{\sigma}')$ and

$$P = K(K\boldsymbol{\Sigma}K)^{-1}K = \boldsymbol{\Sigma}^{-1} - \boldsymbol{\Sigma}^{-1}X(X'\boldsymbol{\Sigma}^{-1}X)^{-1}X'\boldsymbol{\Sigma}^{-1}.$$

Under the alternative parametrization, the scores are (c.f. Sect. 6.5.2)

$$S_{\sigma^2} = -\frac{n-p}{2\sigma^2} + \frac{1}{2\sigma^4} y'Py, \quad S_\varphi = -\frac{1}{2}\operatorname{tr}(PZZ') + \frac{1}{2\sigma^2} y'PZZ'Py.$$

By taking partial derivatives, changing the sign, taking expectations, and applying $PX = \mathbf{0}$ and $P\boldsymbol{\Sigma}P = \mathbf{0}$, we obtain the components of the Fisher information matrix, i.e.

$$F_{\sigma^2\sigma^2} = -\frac{n-p}{2\sigma^4} + \frac{1}{\sigma^4}\operatorname{tr}(P\boldsymbol{\Sigma}) = \frac{n-p}{2\sigma^4},$$

$$F_{\sigma^2\varphi} = \frac{1}{2\sigma^2}\operatorname{tr}(PZZ'), \quad F_{\varphi\varphi} = \frac{1}{2}\operatorname{tr}(PZZ'PZZ').$$

Let us define $Q = (X'\boldsymbol{\Sigma}^{-1}X)^{-1}$. Then we have

$$S_{\sigma^2} = -\frac{n-p}{2\sigma^2} + \frac{1}{2\sigma^4} y'Py,$$

where

$$y'Py = \sum_{d=1}^{D} y'_d\boldsymbol{\Sigma}_d^{-1}y_d - \Big(\sum_{d=1}^{D} y'_d\boldsymbol{\Sigma}_d^{-1}X_d\Big)Q\Big(\sum_{d=1}^{D} X'_d\boldsymbol{\Sigma}_d^{-1}y_d\Big).$$

From the equation $S_{\sigma^2} = \mathbf{0}$, we get (c.f. (6.48))

$$\hat{\sigma}^2 = \frac{1}{n-p} \boldsymbol{y}'\boldsymbol{P}\boldsymbol{y} = \frac{1}{n-p} \left\{ \sum_{d=1}^{D} \boldsymbol{y}_d' \boldsymbol{\Sigma}_d^{-1} \boldsymbol{y}_d - \Big(\sum_{d=1}^{D} \boldsymbol{y}_d' \boldsymbol{\Sigma}_d^{-1} \boldsymbol{X}_d\Big) \boldsymbol{Q} \Big(\sum_{d=1}^{D} \boldsymbol{X}_d' \boldsymbol{\Sigma}_d^{-1} \boldsymbol{y}_d\Big) \right\}. \tag{7.19}$$

Therefore, we can implement the Fisher-scoring algorithm with updating formula for φ and the updating formula (7.19) for σ^2. The updating formula for φ is

$$\varphi_{i+1} = \varphi_i + F(\varphi_i)^{-1} S(\varphi_i),$$

where

$$\begin{aligned} S(\varphi) &= -\frac{1}{2} \sum_{d=1}^{D} \mathbf{1}_{n_d}' \boldsymbol{\Sigma}_d^{-1} \mathbf{1}_{n_d} + \frac{1}{2} \sum_{d=1}^{D} \mathbf{1}_{n_d}' \boldsymbol{\Sigma}_d^{-1} \boldsymbol{X}_d \boldsymbol{Q} \boldsymbol{X}_d' \boldsymbol{\Sigma}_d^{-1} \mathbf{1}_{n_d} \\ &+ \frac{1}{2\sigma^2} \sum_{d=1}^{D} \boldsymbol{y}_d' \boldsymbol{\Sigma}_d^{-1} \mathbf{1}_{n_d} \mathbf{1}_{n_d}' \boldsymbol{\Sigma}_d^{-1} \boldsymbol{y}_d \\ &- \frac{1}{\sigma^2} \Big(\sum_{d=1}^{D} \boldsymbol{y}_d' \boldsymbol{\Sigma}_d^{-1} \mathbf{1}_{n_d} \mathbf{1}_{n_d}' \boldsymbol{\Sigma}_d^{-1} \boldsymbol{X}_d\Big) \boldsymbol{Q} \Big(\sum_{d=1}^{D} \boldsymbol{X}_d' \boldsymbol{\Sigma}_d^{-1} \boldsymbol{y}_d\Big) \\ &+ \frac{1}{2\sigma^2} \Big(\sum_{d=1}^{D} \boldsymbol{y}_d' \boldsymbol{\Sigma}_d^{-1} \boldsymbol{X}_d\Big) \boldsymbol{Q} \Big(\sum_{d=1}^{D} \boldsymbol{X}_d' \boldsymbol{\Sigma}_d^{-1} \mathbf{1}_{n_d} \mathbf{1}_{n_d}' \boldsymbol{\Sigma}_d^{-1} \boldsymbol{X}_d\Big) \boldsymbol{Q} \Big(\sum_{d=1}^{D} \boldsymbol{X}_d' \boldsymbol{\Sigma}_d^{-1} \boldsymbol{y}_d\Big), \\ F(\varphi) &= \frac{1}{2} \sum_{d=1}^{D} \mathbf{1}_{n_d}' \boldsymbol{\Sigma}_d^{-1} \mathbf{1}_{n_d} \mathbf{1}_{n_d}' \boldsymbol{\Sigma}_d^{-1} \mathbf{1}_{n_d} - \sum_{d=1}^{D} \mathbf{1}_{n_d}' \boldsymbol{\Sigma}_d^{-1} \mathbf{1}_{n_d} \mathbf{1}_{n_d}' \boldsymbol{\Sigma}_d^{-1} \boldsymbol{X}_d \boldsymbol{Q} \boldsymbol{X}_d' \boldsymbol{\Sigma}_d^{-1} \mathbf{1}_{n_d} \\ &+ \frac{1}{2} \sum_{d_1=1}^{D} \sum_{d_2=1}^{D} \mathbf{1}_{n_{d_1}}' \boldsymbol{\Sigma}_{d_1}^{-1} \boldsymbol{X}_{d_1} \boldsymbol{Q} \boldsymbol{X}_{d_2}' \boldsymbol{\Sigma}_{d_2}^{-1} \mathbf{1}_{n_{d_2}} \mathbf{1}_{n_{d_2}}' \boldsymbol{\Sigma}_{d_2}^{-1} \boldsymbol{X}_{d_2} \boldsymbol{Q} \boldsymbol{X}_{d_1}' \boldsymbol{\Sigma}_{d_1}^{-1} \mathbf{1}_{n_{d_1}}. \end{aligned}$$

As initial value for σ^2 we can use $\sigma_0^2 = \sigma_{e,0}^2$ defined in (7.15). A possible seed for φ is

$$\varphi_0 = \frac{\sigma_{u,0}^2}{\sigma_0^2}, \tag{7.20}$$

where $\sigma_{u,0}^2$ is expressed in (7.14). Finally, we calculate the REML estimator of $\boldsymbol{\beta}$ by applying the explicit formula

$$\hat{\boldsymbol{\beta}}_{REML} = \left(\boldsymbol{X}'\hat{\boldsymbol{\Sigma}}^{-1}\boldsymbol{X}\right)^{-1} \boldsymbol{X}'\hat{\boldsymbol{\Sigma}}^{-1}\boldsymbol{y} = \Big(\sum_{d=1}^{D} \boldsymbol{X}_d' \hat{\boldsymbol{\Sigma}}_d^{-1} \boldsymbol{X}_d\Big)^{-1} \Big(\sum_{d=1}^{D} \boldsymbol{X}_d' \hat{\boldsymbol{\Sigma}}_d^{-1} \boldsymbol{y}_d\Big),$$

where $\hat{\boldsymbol{\Sigma}} = \boldsymbol{W}^{-1} + \hat{\varphi}\boldsymbol{Z}\boldsymbol{Z}'$ and $\hat{\varphi}$ is the REML estimator of φ obtained as the output of the Fisher-scoring algorithm.

7.7 H3 Estimators

The H3 method unbiasedly estimates σ_e^2 and σ_u^2 by using quadratic forms and makes substitutions in (7.7) for obtaining an estimator of $\boldsymbol{\beta}$ and a predictor of $\boldsymbol{u}$. By using the H3 method, the estimators of the variance components are (cf. (6.71) and (6.72))

$$\hat{\sigma}_e^2 = \frac{SSE(\boldsymbol{\beta}, \boldsymbol{u})}{n - \mathrm{r}(\boldsymbol{X}, \boldsymbol{Z})} \tag{7.21}$$

and

$$\hat{\sigma}_u^2 = \frac{SSR(\boldsymbol{u}|\boldsymbol{\beta}) - \hat{\sigma}_e^2[\mathrm{r}(\boldsymbol{X}, \boldsymbol{Z}) - \mathrm{r}(\boldsymbol{X})]}{\mathrm{tr}\left(\boldsymbol{L}_{1,1}\right)} \tag{7.22}$$

$$= \frac{SSE(\boldsymbol{\beta}) - SSE(\boldsymbol{\beta}, \boldsymbol{u}) - \hat{\sigma}_e^2[\mathrm{r}(\boldsymbol{X}, \boldsymbol{Z}) - \mathrm{r}(\boldsymbol{X})]}{\mathrm{tr}\left(\boldsymbol{Z}'\boldsymbol{W}\left[\boldsymbol{W}^{-1} - \boldsymbol{X}(\boldsymbol{X}'\boldsymbol{W}\boldsymbol{X})^{-1}\boldsymbol{X}'\right]\boldsymbol{W}\boldsymbol{Z}\right)}$$

$$= \frac{SSE(\boldsymbol{\beta}) - \hat{\sigma}_e^2[n - \mathrm{r}(\boldsymbol{X})]}{\mathrm{tr}\left(\boldsymbol{Z}'\boldsymbol{W}\left[\boldsymbol{W}^{-1} - \boldsymbol{X}(\boldsymbol{X}'\boldsymbol{W}\boldsymbol{X})^{-1}\boldsymbol{X}'\right]\boldsymbol{W}\boldsymbol{Z}\right)}, \tag{7.23}$$

where $SSE(\boldsymbol{\beta}, \boldsymbol{u})$ and $SSE(\boldsymbol{\beta})$ are defined below. These estimators were explicitly calculated by Prasad and Rao (1990). In what follows, we give alternative developed expressions of formulas (7.21) and (7.23).

Calculation of $\hat{\sigma}_e^2$

The quadratic form $SSE(\boldsymbol{\beta}, \boldsymbol{u})$ is the residual sum of squares of the ANCOVA model (5.8), with one fixed factor and full rank, i.e.

$$\boldsymbol{y} = \boldsymbol{X}\boldsymbol{\beta} + \boldsymbol{Z}\boldsymbol{u} + \boldsymbol{e},$$

where $\boldsymbol{y} = \boldsymbol{y}_{n\times 1}$, $\boldsymbol{\beta} = \boldsymbol{\beta}_{p\times 1}$, $\boldsymbol{X} = \boldsymbol{X}_{n\times p}$, with $\mathrm{rank}(\boldsymbol{X}) = p$, $\boldsymbol{Z} = \boldsymbol{Z}_{n\times D} = \underset{1\le d\le D}{\mathrm{diag}}\,(\boldsymbol{1}_{n_d})$, $\boldsymbol{e} \sim N_n(\boldsymbol{0}, \sigma^2\boldsymbol{W}^{-1})$, and $\boldsymbol{u} = \boldsymbol{u}_{D\times 1}$ is a vector of parameters. The ML estimators of $\boldsymbol{\beta}$ and $\boldsymbol{u}$ are given in (5.4) and (5.5), respectively. They are

$$\hat{\boldsymbol{\beta}}^e = (\boldsymbol{X}'\boldsymbol{P}\boldsymbol{X})^{-1}\boldsymbol{X}'\boldsymbol{P}\boldsymbol{y}, \quad \hat{\boldsymbol{u}}^e = \boldsymbol{G}\boldsymbol{Z}'\boldsymbol{W}(\boldsymbol{y} - \boldsymbol{X}\hat{\boldsymbol{\beta}}^e),$$

where $\boldsymbol{G} = (\boldsymbol{Z}'\boldsymbol{W}\boldsymbol{Z})^{-1} = \text{diag}_{D\times D}\left(\frac{1}{w_1}, \ldots, \frac{1}{w_D}\right)$ and $\boldsymbol{P} = \boldsymbol{W} - \boldsymbol{W}\boldsymbol{Z}\boldsymbol{G}\boldsymbol{Z}'\boldsymbol{W}$. The ANCOVA estimator of σ_e^2 is

$$\hat{\sigma}_e^2 = \frac{SSE(\boldsymbol{\beta}, \boldsymbol{u})}{n - \text{r}(\boldsymbol{X}, \boldsymbol{Z})} = \frac{(\boldsymbol{y} - \hat{\boldsymbol{y}}^e)'\boldsymbol{W}(\boldsymbol{y} - \hat{\boldsymbol{y}}^e)}{n - D - p},$$

where

$$\hat{\boldsymbol{y}}^e = \boldsymbol{X}\hat{\boldsymbol{\beta}}^e + \boldsymbol{Z}\hat{\boldsymbol{u}}^e, \quad \hat{\boldsymbol{\beta}}^e = (\boldsymbol{X}'\boldsymbol{P}\boldsymbol{X})^{-1}\boldsymbol{X}'\boldsymbol{P}\boldsymbol{y}, \quad \hat{\boldsymbol{u}}^e = \boldsymbol{G}\boldsymbol{Z}'\boldsymbol{W}(\boldsymbol{y} - \boldsymbol{X}\hat{\boldsymbol{\beta}}^e).$$

Remark 7.1 Note that

$$SSE(\boldsymbol{\beta}, \boldsymbol{u}) = \sum_{d=1}^{D}\sum_{j=1}^{n_d} w_{dj}\hat{\varepsilon}_{dj}^2, \quad \hat{\varepsilon}_{dj} = y_{dj} - \hat{y}_{dj},$$

where $\hat{y}_{dj}$ and $\hat{\varepsilon}_{dj}$ are the predicted values and the residuals of the ANCOVA model

$$y_{dj} = \boldsymbol{x}_{dj}\boldsymbol{\beta} + u_d + \varepsilon_{dj}, \quad \varepsilon_{dj} \overset{\text{i.i.d.}}{\sim} N(0, \sigma_e^2 w_{dj}^{-1}), \quad d = 1, \ldots, D, \; j = 1, \ldots, n_d.$$

The estimator of u_d under the ANCOVA model is

$$\hat{u}_d = \hat{\overline{Y}}_d^w - \hat{\overline{\boldsymbol{X}}}_d^w\hat{\boldsymbol{\beta}}, \quad d = 1, \ldots, D.$$

Then, the residuals can be written in the form

$$\hat{\varepsilon}_{dj} = y_{dj} - \hat{y}_{dj} = (y_{dj} - \hat{\overline{Y}}_d^w) - (\boldsymbol{x}_{dj} - \hat{\overline{\boldsymbol{X}}}_d^w)\hat{\boldsymbol{\beta}},$$

and they are the residuals obtained when fitting the regression model

$$(y_{dj} - \hat{\overline{Y}}_d^w) = (\boldsymbol{x}_{dj} - \hat{\overline{\boldsymbol{X}}}_d^w)\boldsymbol{\beta} + \varepsilon_{dj}, \quad \varepsilon_{dj} \sim \text{ i.i.d. } N(0, \sigma_e^2 w_{dj}^{-1}).$$

The ANCOVA estimator of σ_e^2 can be written in the form

$$\hat{\sigma}_e^2 = \frac{SSE(\boldsymbol{\beta}, \boldsymbol{u})}{n - \text{r}(\boldsymbol{X}, \boldsymbol{Z})} = \frac{\sum_{d=1}^{D}\sum_{j=1}^{n_d} w_{dj}\hat{\varepsilon}_{dj}^2}{n - D - p}.$$

Calculation of $\hat{\sigma}_u^2$

The quadratic form $SSE(\boldsymbol{\beta})$ is the residual sum of squares of the regression model

$$\boldsymbol{y} = \boldsymbol{X}\boldsymbol{\beta} + \boldsymbol{e}, \tag{7.24}$$

where $y = y_{n\times 1}$, $\beta = \beta_{p\times 1}$, $X = X_{n\times p}$ with $r(X) = p$ and $e \sim N_n(0, \sigma^2 W^{-1})$. It holds that

$$SSE(\beta) = (y - \hat{y}^u)' W (y - \hat{y}^u), \quad \text{where } \hat{y}^u = X\hat{\beta}^u, \quad \hat{\beta}^u = (X'WX)^{-1}X'Wy.$$

The estimator of σ_u^2 is

$$\hat{\sigma}_u^2 = \frac{SSE(\beta) - \hat{\sigma}_e^2[n-p]}{\text{tr}\left(Z'W\left[W^{-1} - X(X'WX)^{-1}X'\right]WZ\right)}.$$

The estimator $\hat{\sigma}_u^2$ can take negative values, but $Pr(\hat{\sigma}_u^2 \leq 0)$ tends to 0 as $D \to \infty$. If $\hat{\sigma}_u^2$ is negative, then we equate it to zero and we define

$$\tilde{\sigma}_u^2 = \max\left\{\hat{\sigma}_u^2, 0\right\}. \tag{7.25}$$

Remark 7.2 Note that

$$SSE(\beta) = \sum_{d=1}^{D}\sum_{j=1}^{n_d} w_{dj}\hat{e}_{dj}^2,$$

where $\hat{e}_{dj}$ are the residuals of the regression model

$$y_{dj} = x_{dj}\beta + e_{dj}, \qquad e_{dj} \overset{\text{i.i.d.}}{\sim} N(0, \sigma_e^2 w_{dj}^{-1}).$$

If we assume $r(X) = p$ and $w_{dj} = 1$, $d = 1, \ldots, D$, $j = 1, \ldots, n_d$, then we get

$$\hat{\sigma}_u^2 = \frac{SSE(\beta) - \hat{\sigma}_e^2[n - r(X)]}{\text{tr}\left(Z'\left[I - X(X'X)^{-1}X'\right]Z\right)} = \frac{\sum_{d=1}^{D}\sum_{j=1}^{n_d}\hat{e}_{dj}^2 - (n-p)\hat{\sigma}_e^2}{n - \text{tr}\left((X'X)^{-1}\sum_{d=1}^{D} n_d^2 \hat{\bar{X}}_d' \hat{\bar{X}}_d\right)}. \tag{7.26}$$

Calculation of $\hat{\beta}$

The H3 estimator of β is

$$\hat{\beta} = (X'\hat{V}^{-1}X)^{-1}X'\hat{V}^{-1}y,$$

where

$$\hat{V}^{-1} = \text{diag}(\hat{V}_1^{-1}, \ldots, \hat{V}_D^{-1}), \quad \hat{V}_d^{-1} = \frac{1}{\hat{\sigma}_e^2}\left(W_d - \frac{\hat{\gamma}_d^w}{w_d} w_{n_d} w_{n_d}'\right), \quad \hat{\gamma}_d^w = \frac{\hat{\sigma}_u^2}{\hat{\sigma}_u^2 + \frac{\hat{\sigma}_e^2}{w_d}}.$$

The Case of One Explanatory Variable
Let us consider the model

$$y_{dj} = \beta x_{dj} + u_d + e_{dj}, \; d = 1, \ldots, D, \; j = 1, \ldots, n_d, \tag{7.27}$$

where u_d and e_{dj} are independent domain-level and unit-level random variables with distributions $N(0, \sigma_u^2)$ and $N(0, w_{dj}^{-1}\sigma_e^2)$, respectively. Let us define

$$\overline{Y}_d^{\,w} = \frac{1}{w_d}\sum_{j=1}^{n_d} w_{dj}y_{dj}, \quad \overline{X}_d^{\,w} = \frac{1}{w_d}\sum_{j=1}^{n_d} w_{dj}x_{dj}, \quad w_d = \sum_{j=1}^{n_d} w_{dj}, \quad w = \sum_{d=1}^{D}\sum_{j=1}^{n_d} w_{dj}.$$

By using the H3 method, the estimator of β is

$$\hat{\beta} = \frac{\sum_{d=1}^{D}\sum_{j=1}^{n_d} w_{dj}y_{dj}x_{dj} - \sum_{d=1}^{D}\hat{\gamma}_d^w w_d \overline{Y}_d^{\,w}\overline{X}_d^{\,w}}{\sum_{d=1}^{D}\sum_{j=1}^{n_d} w_{dj}x_{dj}^2 - \sum_{d=1}^{D}\hat{\gamma}_d^w w_d \left(\overline{X}_d^{\,w}\right)^2}, \quad \hat{\gamma}_d^w = \frac{\hat{\sigma}_u^2}{\hat{\sigma}_u^2 + \frac{\hat{\sigma}_e^2}{w_d}},$$

the estimator of σ_e^2 is

$$\hat{\sigma}_e^2 = \frac{1}{n - D - 1}\sum_{d=1}^{D}\sum_{j=1}^{n_d} w_{dj}\left(y_{dj} - \overline{Y}_d^{\,w} - \hat{b}_e(x_{dj} - \overline{X}_d^{\,w})\right)^2,$$

$$\hat{b}_e = \frac{\sum_{d=1}^{D}\sum_{j=1}^{n_d} w_{dj}(y_{dj} - \overline{Y}_d^{\,w})(x_{dj} - \overline{X}_d^{\,w})}{\sum_{d=1}^{D}\sum_{j=1}^{n_d} w_{dj}(x_{dj} - \overline{X}_d^{\,w})^2},$$

and the estimator of σ_u^2 is

$$\hat{\sigma}_u^2 = \max\left\{\frac{1}{n^*}\left[\sum_{d=1}^{D}\sum_{j=1}^{n_d} w_{dj}(y_{dj} - \hat{b}_u x_{dj})^2 - (n-1)\hat{\sigma}_e^2\right], 0\right\},$$

$$\hat{b}_u = \frac{\sum_{d=1}^{D}\sum_{j=1}^{n_d} w_{dj}y_{dj}x_{dj}}{\sum_{d=1}^{D}\sum_{j=1}^{n_d} w_{dj}x_{dj}^2}, \quad n^* = w - \frac{\sum_{d=1}^{D}\left(\sum_{j=1}^{n_d} w_{dj}x_{dj}\right)^2}{\sum_{d=1}^{D}\sum_{j=1}^{n_d} w_{dj}X_{dj}^2}.$$

Algorithm
The algorithm that implements the H3 method has the following steps:

1 Initial calculations.

1.1 For $d = 1, \ldots, D$, calculate n_d, $\boldsymbol{w}_{n_d}$, w_d, $\boldsymbol{W}_d$.
1.2 Calculate $\boldsymbol{W}$, $\boldsymbol{G}$, $\boldsymbol{P}$, $\hat{\boldsymbol{\beta}}^e$, $\hat{\boldsymbol{u}}^e$, $\hat{\sigma}_e^2$, $\hat{\boldsymbol{\beta}}^u$, $\hat{\sigma}_u^2$.

1.3 For $d = 1, \ldots, D$, calculate $\hat{\gamma}_d^w$, $\hat{V}_d^{-1}$.

1.4 Calculate $\hat{V}^{-1}$, $\hat{\boldsymbol{\beta}}$.

2 Output: $\hat{\boldsymbol{\beta}}$, $\hat{\sigma}_e^2$, $\hat{\sigma}_u^2$.

Note that we can easily calculate $\hat{\sigma}_e^2$ and $\hat{\sigma}_u^2$ by using standard statistical software for linear models.

7.8 Moments of H3 Estimators

This section particularizes the results of Sect. 6.6.2 to the NER model defined in (7.4). This is to say, we assume that the vectors and matrices of model (6.74) are $\boldsymbol{\beta}_1 = \boldsymbol{\beta}_{p\times 1}$, $\boldsymbol{\beta}_2 = \boldsymbol{u}_{D\times 1}$, $\boldsymbol{e} = \boldsymbol{e}_{n\times 1}$, $\boldsymbol{X}_1 = \boldsymbol{X}_{n\times p}$, and $\boldsymbol{X}_2 = \boldsymbol{Z}_{n\times D}$, where $\boldsymbol{u} \sim N_D(\boldsymbol{0}, \sigma_u^2 \boldsymbol{I}_D)$ is independent of $\boldsymbol{e} \sim N_n(\boldsymbol{0}, \sigma_e^2 \boldsymbol{W}^{-1})$ and $\boldsymbol{W} = \text{diag}(w_{11}, \ldots, w_{Dn_D})$. We further define $\boldsymbol{w}_{n_d} = (w_{d1}, \ldots, w_{dn_d})'_{n_d\times 1}$, $w_d = \sum_{j=1}^{n_d} w_{dj}$ and $w = \sum_{d=1}^{D} w_d$. Then

$$\boldsymbol{Z}_d'\boldsymbol{W}_d\boldsymbol{X}_d = \boldsymbol{1}_{n_d}'\boldsymbol{W}_d\boldsymbol{X}_d = w_d\hat{\overline{\boldsymbol{X}}}_d^{w}, \quad \boldsymbol{Z}_d'\boldsymbol{W}_d\boldsymbol{Z}_d = w_d,$$

$$\boldsymbol{Z}'\boldsymbol{W}\boldsymbol{X} = \underset{1\le d\le D}{\text{col}}\left(w_d\hat{\overline{\boldsymbol{X}}}_d^{w}\right), \quad \boldsymbol{Z}'\boldsymbol{W}\boldsymbol{Z} = \underset{1\le d\le D}{\text{diag}}\left(w_d\right), \quad \text{tr}(\boldsymbol{Z}'\boldsymbol{W}\boldsymbol{Z}) = w,$$

$\text{r}(\boldsymbol{Z}) = D$ and $\text{r}(\mathbb{X}) = \text{r}(\boldsymbol{X}, \boldsymbol{Z}) = D + p - \lambda$, where $\lambda = 1$ if the model has an intercept and $\lambda = 0$ otherwise. In this section, all the calculations are done under the assumption $\lambda = 0$ so that the inverse of the matrix $\mathbb{X}'\boldsymbol{W}\mathbb{X}$ exists. If $\lambda = 1$, then $\mathbb{X}'\boldsymbol{W}\mathbb{X}$ is not invertible and similar calculation could be done by using generalized inverse matrices. In order to derive the moments of the H3 estimators, we first calculate some matrix traces under the model (7.4).

Trace of $\boldsymbol{C}_1 = \boldsymbol{M}\boldsymbol{W}^{-1}\boldsymbol{M}_1\boldsymbol{W}^{-1}$
We remind that $\boldsymbol{M} = \boldsymbol{W} - \boldsymbol{W}\mathbb{X}(\mathbb{X}'\boldsymbol{W}\mathbb{X})^{-1}\mathbb{X}'\boldsymbol{W}$ and $\boldsymbol{M}_1 = \boldsymbol{W} - \boldsymbol{W}\boldsymbol{X}(\boldsymbol{X}'\boldsymbol{W}\boldsymbol{X})^{-1}\boldsymbol{X}'\boldsymbol{W}$. We have that

$$\begin{aligned}
\boldsymbol{M}\boldsymbol{W}^{-1}\boldsymbol{M}_1\boldsymbol{W}^{-1} &= (\boldsymbol{I} - \boldsymbol{W}\mathbb{X}(\mathbb{X}'\boldsymbol{W}\mathbb{X})^{-1}\mathbb{X}')(\boldsymbol{I} - \boldsymbol{W}\boldsymbol{X}(\boldsymbol{X}'\boldsymbol{W}\boldsymbol{X})^{-1}\boldsymbol{X}') \\
&\triangleq (\boldsymbol{I} - \boldsymbol{H})(\boldsymbol{I} - \boldsymbol{H}_1) = \boldsymbol{I} - \boldsymbol{H} - \boldsymbol{H}_1 + \boldsymbol{H}\boldsymbol{H}_1, \\
\text{tr}(\boldsymbol{M}\boldsymbol{W}^{-1}\boldsymbol{M}_1\boldsymbol{W}^{-1}) &= \text{tr}(\boldsymbol{I} - \boldsymbol{H} - \boldsymbol{H}_1 + \boldsymbol{H}\boldsymbol{H}_1) = n - r(\boldsymbol{X}, \boldsymbol{Z}) - r(\boldsymbol{X}) + \text{tr}(\boldsymbol{H}\boldsymbol{H}_1), \\
\text{tr}(\boldsymbol{H}\boldsymbol{H}_1) &= \text{tr}\left(\boldsymbol{W}\mathbb{X}(\mathbb{X}'\boldsymbol{W}\mathbb{X})^{-1}\mathbb{X}'\boldsymbol{W}\boldsymbol{X}(\boldsymbol{X}'\boldsymbol{W}\boldsymbol{X})^{-1}\boldsymbol{X}'\right) \\
&= \text{tr}\left([(\mathbb{X}'\boldsymbol{W}\mathbb{X})^{-1}\mathbb{X}'\boldsymbol{W}\boldsymbol{X}][(\boldsymbol{X}'\boldsymbol{W}\boldsymbol{X})^{-1}\boldsymbol{X}'\boldsymbol{W}\mathbb{X}]\right).
\end{aligned}$$

On the one hand, $\boldsymbol{K}_1 \triangleq (\mathbb{X}'\boldsymbol{W}\mathbb{X})^{-1}\mathbb{X}'\boldsymbol{W}\boldsymbol{X}$ gives the weighted least squares estimators of the linear regressions of columns from matrix $\boldsymbol{X}$ on the set of explanatory variables defined by the columns of matrix $\mathbb{X}$. The p equations of the cited regression models are

$$x_{dj}^{(r)} = \beta_1^{(r)} x_{dj}^{(1)} + \ldots + \beta_p^{(r)} x_{dj}^{(r)} + u_d^{(r)} + e_{dj}^{(r)},\ r = 1, \ldots, p,\ d = 1, \ldots, D,\ j = 1, \ldots, n_D.$$

The least squares estimators are $u_d^{(r)} = 0$, $\beta_r^{(r)} = 1$ and $\beta_s^{(r)} = 0$, $s = 1, \ldots, p$, $s \neq r$. Let $\boldsymbol{x}^{(1)}, \ldots, \boldsymbol{x}^{(p)}$ be the columns of $\boldsymbol{X}$, then

$$\boldsymbol{K}_1 = (\mathbb{X}'\boldsymbol{W}\mathbb{X})^{-1}\mathbb{X}'\boldsymbol{W}\boldsymbol{X} = (\mathbb{X}'\boldsymbol{W}\mathbb{X})^{-1}\mathbb{X}'\boldsymbol{W}(\boldsymbol{x}^{(1)}, \ldots, \boldsymbol{x}^{(p)}) = \begin{pmatrix} \boldsymbol{I}_p \\ \boldsymbol{0}_{(D-\lambda)\times p} \end{pmatrix}_{(p+D-\lambda)\times p}.$$

On the other hand, $\boldsymbol{K}_2 \triangleq (\boldsymbol{X}'\boldsymbol{W}\boldsymbol{X})^{-1}\boldsymbol{X}'\boldsymbol{W}\mathbb{X}$ gives the least squares estimators of the linear regressions of the columns of $\mathbb{X}$ on the set of explanatory variables defined by the columns of $\boldsymbol{X}$. The $p + D - \lambda$ equations of the cited regression models are

$$\begin{aligned} x_{dj}^{(r)} &= 1 \cdot x_{dj}^{(r)}, \quad r = 1, \ldots, p, \\ \xi_{dj}^{(1)} &= \beta_1^{(1)} x_{dj}^{(1)} + \ldots + \beta_p^{(1)} x_{dj}^{(p)} + e_{dj}^{(1)}, \\ &\vdots \\ \xi_{dj}^{(D-\lambda)} &= \beta_1^{(D-\lambda)} x_{dj}^{(1)} + \ldots + \beta_p^{(D-\lambda)} x_{dj}^{(p)} + e_{dj}^{(D-\lambda)}, \end{aligned}$$

where $\xi_{dj}^{(i)} = 1$ if $d = i$ and $\xi_{dj}^{(i)} = 0$ otherwise. Therefore, we have

$$\boldsymbol{K}_2 = (\boldsymbol{X}'\boldsymbol{W}\boldsymbol{X})^{-1}\boldsymbol{X}'\boldsymbol{W}\left(\boldsymbol{x}^{(1)}, \ldots, \boldsymbol{x}^{(p)}, \xi^{(1)}, \ldots, \xi^{(D-\lambda)}\right) = \left(\begin{array}{ccc|ccc} 1 & \ldots & 0 & \hat{\beta}_1^{(1)} & \ldots & \hat{\beta}_1^{(D-\lambda)} \\ \vdots & \ddots & \vdots & \vdots & \ddots & \vdots \\ 0 & \ldots & 1 & \hat{\beta}_p^{(1)} & \ldots & \hat{\beta}_p^{(D-\lambda)} \end{array}\right)$$

$$= \left(\boldsymbol{I}_p, \boldsymbol{B}_{p\times(D-\lambda)}\right)_{p\times(p+D-\lambda)}.$$

By putting together the two previous results, we get

$$\boldsymbol{K}_1\boldsymbol{K}_2 = \begin{pmatrix} \boldsymbol{I}_p \\ \boldsymbol{0}_{(D-\lambda)\times p} \end{pmatrix} \left(\boldsymbol{I}_p, \boldsymbol{B}_{p\times(D-\lambda)}\right) = \begin{pmatrix} \boldsymbol{I}_p & \boldsymbol{B}_{p\times(D-\lambda)} \\ \boldsymbol{0}_{(D-\lambda)\times p} & \boldsymbol{0}_{(D-\lambda)\times(D-\lambda)} \end{pmatrix}.$$

Finally, $\mathrm{tr}(\boldsymbol{H}\boldsymbol{H}_1) = \mathrm{tr}(\boldsymbol{K}\boldsymbol{K}_1) = p$ and

$$\mathrm{tr}(\boldsymbol{C}_1) = n - r(\mathbb{X}) - r(\boldsymbol{X}) + \mathrm{tr}(\boldsymbol{H}\boldsymbol{H}_1) = n - (p + D - \lambda) - p + p = n - p - D + \lambda.$$

Trace of $C_2 = Z'MW^{-1}M_1Z$
We have that

$$\begin{aligned}\mathrm{tr}(C_2) &= \mathrm{tr}(Z'WZ) - \mathrm{tr}(Z'W\mathbb{X}(\mathbb{X}'W\mathbb{X})^{-1}\mathbb{X}'WZ)\\ &\quad - \mathrm{tr}(Z'WX(X'WX)^{-1}X'WZ)\\ &\quad + \mathrm{tr}(Z'W\mathbb{X}(\mathbb{X}'W\mathbb{X})^{-1}\mathbb{X}'WX(X'WX)^{-1}X'WZ)\\ &= \mathrm{tr}(Z'WZ) - \mathrm{tr}(Z'HWZ) - \mathrm{tr}(Z'H_1WZ) + \mathrm{tr}(Z'HH_1WZ).\end{aligned}$$

We calculate separately each of the previous traces. First, we have

$$\mathrm{tr}(Z'WZ) = w.$$

Second, for calculating $\mathrm{tr}(Z'HWZ) = \mathrm{tr}(Z'W\mathbb{X}(\mathbb{X}'W\mathbb{X})^{-1}\mathbb{X}'WZ)$, we consider

$$M_2 = W - WZ(Z'WZ)^{-1}Z'W \quad \text{and} \quad \mathbb{X}'W\mathbb{X} = \begin{pmatrix} X'WX & X'WZ \\ Z'WX & Z'WZ \end{pmatrix}.$$

By applying the inversion formula for block matrices (A.1) to $(\mathbb{X}'W\mathbb{X})$, i.e.

$$A^{11} = \left(A_{11} - A_{12}A_{22}^{-1}A_{21}\right)^{-1}, \quad A^{12} = (A^{21})' = -A^{11}A_{12}A_{22}^{-1}$$

and

$$A^{22} = A_{22}^{-1} + A_{22}^{-1}A_{21}A^{11}A_{12}A_{22}^{-1},$$

we obtain the blocks of matrix $(\mathbb{X}'W\mathbb{X})^{-1}$. They are

$$\begin{aligned}(\mathbb{X}'W\mathbb{X})^{11} &= (X'M_2X)^{-1}, \quad (\mathbb{X}'W\mathbb{X})^{12} = -(X'M_2X)^{-1}X'WZ(Z'WZ)^{-1},\\ (\mathbb{X}'W\mathbb{X})^{21} &= -(Z'WZ)^{-1}Z'WX(X'M_2X)^{-1},\\ (\mathbb{X}'W\mathbb{X})^{22} &= (Z'WZ)^{-1} + (Z'WZ)^{-1}Z'WX(X'M_2X)^{-1}X'WZ(Z'WZ)^{-1}.\end{aligned}$$

Pre-multiplying by $Z'W\mathbb{X}$, we obtain

$$Z'W\mathbb{X}(\mathbb{X}'W\mathbb{X})^{-1} = Z'W\,(X,\ Z)\,(\mathbb{X}'W\mathbb{X})^{-1} = \left(A_{(D-\lambda)\times p},\ B_{(D-\lambda)\times(D-\lambda)}\right),$$

where

$$\begin{aligned}A &= Z'WX(X'M_2X)^{-1} - (Z'WZ)(Z'WZ)^{-1}Z'WX(X'M_2X)^{-1} = 0_{(D-\lambda)\times p},\\ B &= -Z'WX(X'M_2X)^{-1}X'WZ(Z'WZ)^{-1} + (Z'WZ)(Z'WZ)^{-1}\\ &\quad + (Z'WZ)(Z'WZ)^{-1}Z'WX(X'M_2X)^{-1}X'WZ(Z'WZ)^{-1} = I_{(D-\lambda)\times(D-\lambda)}.\end{aligned}$$

Therefore,

$$\boldsymbol{Z}'\boldsymbol{W}\mathbb{X}(\mathbb{X}'\boldsymbol{W}\mathbb{X})^{-1}\mathbb{X}'\boldsymbol{W}\boldsymbol{Z} = (\boldsymbol{0}_{(D-\lambda)\times p}, \boldsymbol{I}_{(D-\lambda)\times(D-\lambda)}) \begin{pmatrix} \boldsymbol{X}' \\ \boldsymbol{Z}' \end{pmatrix} \boldsymbol{W}\boldsymbol{Z} = \boldsymbol{Z}'\boldsymbol{W}\boldsymbol{Z},$$

$$\text{tr}(\boldsymbol{Z}'\boldsymbol{W}\mathbb{X}(\mathbb{X}'\boldsymbol{W}\mathbb{X})^{-1}\mathbb{X}'\boldsymbol{W}\boldsymbol{Z}) = \text{tr}(\boldsymbol{Z}'\boldsymbol{W}\boldsymbol{Z}) = w.$$

Third, as $\boldsymbol{Z}'\boldsymbol{H} = \boldsymbol{Z}'\boldsymbol{W}\mathbb{X}(\mathbb{X}'\boldsymbol{W}\mathbb{X})^{-1}\mathbb{X}' = \boldsymbol{Z}'$, we get

$$\text{tr}(\boldsymbol{Z}'\boldsymbol{H}\boldsymbol{H}_1\boldsymbol{W}\boldsymbol{Z}) = \text{tr}(\boldsymbol{Z}'\boldsymbol{H}_1\boldsymbol{W}\boldsymbol{Z}).$$

Finally,

$$\text{tr}(\boldsymbol{C}_2) = \text{tr}(\boldsymbol{Z}'\boldsymbol{M}\boldsymbol{W}^{-1}\boldsymbol{M}_1\boldsymbol{Z}) = w - w - \text{tr}(\boldsymbol{Z}'\boldsymbol{H}_1\boldsymbol{W}\boldsymbol{Z}) + \text{tr}(\boldsymbol{Z}'\boldsymbol{H}_1\boldsymbol{W}\boldsymbol{Z}) = 0.$$

Trace of $\boldsymbol{L}_{1,1} = \boldsymbol{Z}'\boldsymbol{M}_1\boldsymbol{Z}$
We observe that

$$(\boldsymbol{X}'\boldsymbol{W}\boldsymbol{Z})(\boldsymbol{X}'\boldsymbol{W}\boldsymbol{Z}) = \underset{1\le d\le D}{\text{col}'}\left(w_d\hat{\overline{\boldsymbol{X}}}_d^{w\prime}\right) \underset{1\le d\le D}{\text{col}}\left(w_d\hat{\overline{\boldsymbol{X}}}_d^{w}\right) = \sum_{d=1}^{D} w_d^2 \hat{\overline{\boldsymbol{X}}}_d^{w\prime} \hat{\overline{\boldsymbol{X}}}_d^{w}.$$

The trace of $\boldsymbol{L}_{1,1}$ is

$$\begin{aligned} n_* = \text{tr}(\boldsymbol{L}_{1,1}) &= \text{tr}(\boldsymbol{Z}'\boldsymbol{M}_1\boldsymbol{Z}) = \text{tr}(\boldsymbol{Z}'(\boldsymbol{W} - \boldsymbol{W}\boldsymbol{X}(\boldsymbol{X}'\boldsymbol{W}\boldsymbol{X})^{-1}\boldsymbol{X}'\boldsymbol{W})\boldsymbol{Z}) \\ &= \text{tr}(\boldsymbol{Z}'\boldsymbol{W}\boldsymbol{Z}) - \text{tr}(\boldsymbol{Z}'\boldsymbol{W}\boldsymbol{X}(\boldsymbol{X}'\boldsymbol{W}\boldsymbol{X})^{-1}\boldsymbol{X}'\boldsymbol{W}\boldsymbol{Z}) \\ &= w - \text{tr}\big((\boldsymbol{X}'\boldsymbol{W}\boldsymbol{X})^{-1}(\boldsymbol{X}'\boldsymbol{W}\boldsymbol{Z})(\boldsymbol{Z}'\boldsymbol{W}\boldsymbol{X})\big). \end{aligned}$$

Therefore, we have

$$n_* = w - \text{tr}\left((\boldsymbol{X}'\boldsymbol{W}\boldsymbol{X})^{-1}\sum_{d=1}^{D} w_d^2 \hat{\overline{\boldsymbol{X}}}_d^{w\prime} \hat{\overline{\boldsymbol{X}}}_d^{w}\right), \tag{7.28}$$

where $\hat{\overline{\boldsymbol{X}}}_d^{w} = \frac{1}{w_d}\sum_{j=1}^{n_d} w_{dj}\boldsymbol{x}_{dj}$ has dimension $1 \times p$, and it is the weighted mean of the rows (d, j), $j = 1, \ldots, n_d$, from the matrix $\boldsymbol{X}$.

Trace of $\boldsymbol{L}_{1,1}^2 = \boldsymbol{Z}'\boldsymbol{M}_1\boldsymbol{Z}\boldsymbol{Z}'\boldsymbol{M}_1\boldsymbol{Z}$
The trace of $\boldsymbol{L}_{1,1}^2$ is

$$\begin{aligned} n_{**} = \text{tr}(\boldsymbol{L}_{1,1}^2) &= \text{tr}\big(\boldsymbol{Z}'(\boldsymbol{W} - \boldsymbol{W}\boldsymbol{X}(\boldsymbol{X}'\boldsymbol{W}\boldsymbol{X})^{-1}\boldsymbol{X}'\boldsymbol{W})\boldsymbol{Z}\boldsymbol{Z}'(\boldsymbol{W} - \boldsymbol{W}\boldsymbol{X}(\boldsymbol{X}'\boldsymbol{W}\boldsymbol{X})^{-1}\boldsymbol{X}'\boldsymbol{W})\boldsymbol{Z}\big) \\ &= \text{tr}(\boldsymbol{Z}'\boldsymbol{W}\boldsymbol{Z}\boldsymbol{Z}'\boldsymbol{W}\boldsymbol{Z}) \\ &- \text{tr}(\boldsymbol{Z}'\boldsymbol{W}\boldsymbol{Z}\boldsymbol{Z}'\boldsymbol{W}\boldsymbol{X}(\boldsymbol{X}'\boldsymbol{W}\boldsymbol{X})^{-1}\boldsymbol{X}'\boldsymbol{W}\boldsymbol{Z}) - \text{tr}(\boldsymbol{Z}'\boldsymbol{W}\boldsymbol{X}(\boldsymbol{X}'\boldsymbol{W}\boldsymbol{X})^{-1}\boldsymbol{X}'\boldsymbol{W}\boldsymbol{Z}\boldsymbol{Z}'\boldsymbol{W}\boldsymbol{Z}) \end{aligned}$$

$$+ \operatorname{tr}(\boldsymbol{Z}'\boldsymbol{W}\boldsymbol{X}(\boldsymbol{X}'\boldsymbol{W}\boldsymbol{X})^{-1}\boldsymbol{X}'\boldsymbol{W}\boldsymbol{Z}\boldsymbol{Z}'\boldsymbol{W}\boldsymbol{X}(\boldsymbol{X}'\boldsymbol{W}\boldsymbol{X})^{-1}\boldsymbol{X}'\boldsymbol{W}\boldsymbol{Z}).$$

We observe that

$$\begin{aligned}(\boldsymbol{X}'\boldsymbol{W}\boldsymbol{Z})(\boldsymbol{Z}'\boldsymbol{W}\boldsymbol{Z})(\boldsymbol{Z}'\boldsymbol{W}\boldsymbol{X}) &= \underset{1\le d\le D}{\operatorname{col}'}\Big(w_d\hat{\bar{\boldsymbol{X}}}_d^{w\prime}\Big)\underset{1\le d\le D}{\operatorname{diag}}\big(w_d\big)\underset{1\le d\le D}{\operatorname{col}}\Big(w_d\hat{\bar{\boldsymbol{X}}}_d^{w}\Big)\\ &= \sum_{d=1}^{D} w_d^3\hat{\bar{\boldsymbol{X}}}_d^{w\prime}\hat{\bar{\boldsymbol{X}}}_d^{w}.\end{aligned}$$

Therefore, we have

$$\begin{aligned}n_{**} &= \operatorname{tr}((\boldsymbol{Z}'\boldsymbol{W}\boldsymbol{Z})(\boldsymbol{Z}'\boldsymbol{W}\boldsymbol{Z}))\\ &- 2\operatorname{tr}((\boldsymbol{X}'\boldsymbol{W}\boldsymbol{X})^{-1}(\boldsymbol{X}'\boldsymbol{W}\boldsymbol{Z})(\boldsymbol{Z}'\boldsymbol{W}\boldsymbol{Z})(\boldsymbol{Z}'\boldsymbol{W}\boldsymbol{X}))\\ &+ \operatorname{tr}\big(\big[(\boldsymbol{X}'\boldsymbol{W}\boldsymbol{X})^{-1}(\boldsymbol{X}'\boldsymbol{W}\boldsymbol{Z})(\boldsymbol{Z}'\boldsymbol{W}\boldsymbol{X})\big]\big[(\boldsymbol{X}'\boldsymbol{W}\boldsymbol{X})^{-1}(\boldsymbol{X}'\boldsymbol{W}\boldsymbol{Z})(\boldsymbol{Z}'\boldsymbol{W}\boldsymbol{X})\big]\big)\\ &= \sum_{d=1}^{D} w_d^2 - 2\operatorname{tr}\Big((\boldsymbol{X}'\boldsymbol{W}\boldsymbol{X})^{-1}\sum_{d=1}^{D} w_d^3\hat{\bar{\boldsymbol{X}}}_d^{w\prime}\hat{\bar{\boldsymbol{X}}}_d^{w}\Big)\\ &+ \operatorname{tr}\Big(\Big[\boldsymbol{X}'\boldsymbol{W}\boldsymbol{X})^{-1}\sum_{d=1}^{D} w_d^2\hat{\bar{\boldsymbol{X}}}_d^{w\prime}\hat{\bar{\boldsymbol{X}}}_d^{w}\Big]^2\Big).\end{aligned} \tag{7.29}$$

Variances and Covariances of H3 Estimators

The H3 estimators of the variance components under the model (7.4) are

$$\hat{\sigma}_e^2 = \frac{SSE(\boldsymbol{\beta}, \boldsymbol{u})}{n - \operatorname{r}(\boldsymbol{X}, \boldsymbol{Z})}, \quad \hat{\sigma}_u^2 = \frac{SSE(\boldsymbol{\beta})[n - \operatorname{r}(\boldsymbol{X}, \boldsymbol{Z})] - SSE(\boldsymbol{\beta}, \boldsymbol{u})[n - \operatorname{r}(\boldsymbol{X})]}{[n - \operatorname{r}(\boldsymbol{X}, \boldsymbol{Z})]\operatorname{tr}(\boldsymbol{L}_{1,1})}.$$

General formulas for variances and covariances of these estimators were derived in Remark 6.3. For the variance of $\hat{\sigma}_e^2$, it holds

$$\operatorname{var}\left(\hat{\sigma}_e^2\right) = \operatorname{var}\left(\frac{SSE(\boldsymbol{\beta}, \boldsymbol{u})}{n - D - p + \lambda}\right) = \frac{2\sigma_e^4(n - D - p + \lambda)}{(n - D - p + \lambda)^2} = \frac{2\sigma_e^4}{n - D - p + \lambda}.$$

The covariance of $\hat{\sigma}_e^2$ and $\hat{\sigma}_u^2$ is

$$\operatorname{cov}(\hat{\sigma}_e^2, \hat{\sigma}_u^2) = \frac{\operatorname{cov}(SSE(\boldsymbol{\beta}, \boldsymbol{u}), SSE(\boldsymbol{\beta}))}{[n - \operatorname{r}(\boldsymbol{X}, \boldsymbol{Z})]\operatorname{tr}(\boldsymbol{L}_{1,1})} - \frac{[n - \operatorname{r}(\boldsymbol{X})]\operatorname{var}(SSE(\boldsymbol{\beta}, \boldsymbol{u}))}{[n - \operatorname{r}(\boldsymbol{X}, \boldsymbol{Z})]^2\operatorname{tr}(\boldsymbol{L}_{1,1})}. \tag{7.30}$$

It holds (cf. Remark 6.3)

$$\mathrm{var}(SSE(\boldsymbol{\beta}, \boldsymbol{u})) = 2\sigma_e^4\,(n - \mathrm{r}(\boldsymbol{X}, \boldsymbol{Z}))\,,$$
$$\mathrm{cov}(SSE(\boldsymbol{\beta}, \boldsymbol{u}), SSE(\boldsymbol{\beta})) = 2\sigma_e^2 \mathrm{tr}(\boldsymbol{M}\boldsymbol{W}^{-1}\boldsymbol{M}_1\boldsymbol{V}),$$

with $\boldsymbol{V} = \sigma_u^2\boldsymbol{Z}\boldsymbol{Z}' + \sigma_e^2\boldsymbol{W}^{-1}$ and thus

$$\begin{aligned}\mathrm{cov}(SSE(\boldsymbol{\beta}, \boldsymbol{u}), SSE(\boldsymbol{\beta})) &= 2\sigma_e^2\sigma_u^2\mathrm{tr}(\boldsymbol{M}\boldsymbol{W}^{-1}\boldsymbol{M}_1\boldsymbol{Z}\boldsymbol{Z}') + 2\sigma_e^4\mathrm{tr}(\boldsymbol{M}\boldsymbol{W}^{-1}\boldsymbol{M}_1\boldsymbol{W}^{-1})\\ &= 2\sigma_e^2\sigma_u^2\mathrm{tr}(\boldsymbol{C}_2) + 2\sigma_e^4\mathrm{tr}(\boldsymbol{C}_1) = 2\sigma_e^4(n - p - D + \lambda).\end{aligned}$$

Substituting into (7.30) we finally obtain

$$\begin{aligned}\mathrm{cov}(\hat{\sigma}_e^2, \hat{\sigma}_u^2) &= \frac{2\sigma_e^4}{\mathrm{tr}(\boldsymbol{L}_{1,1})} - \frac{2\sigma_e^4[n - \mathrm{r}(\boldsymbol{X})]}{[n - \mathrm{r}(\boldsymbol{X}, \boldsymbol{Z})]\mathrm{tr}(\boldsymbol{L}_{1,1})}\\ &= \frac{2\sigma_e^4(\lambda - D)}{\mathrm{tr}(\boldsymbol{L}_{1,1})(n - D - p + \lambda)} = \frac{-2\sigma_e^4(D - \lambda)}{n_*(n - D - p + \lambda)}.\end{aligned}$$

The variance of $\hat{\sigma}_u^2$ is (cf. Remark 6.3)

$$\begin{aligned}\mathrm{var}(\hat{\sigma}_u^2) = &\frac{\mathrm{var}(SSE(\boldsymbol{\beta}))}{[\mathrm{tr}(\boldsymbol{L}_{1,1})]^2} + \frac{[n - \mathrm{r}(\boldsymbol{X})]^2\mathrm{var}(SSE(\boldsymbol{\beta}, \boldsymbol{u}))}{[\mathrm{tr}(\boldsymbol{L}_{1,1})]^2[n - \mathrm{r}(\boldsymbol{X}, \boldsymbol{Z})]^2}\\ &- \frac{2[n - \mathrm{r}(\boldsymbol{X})]\mathrm{cov}(SSE(\boldsymbol{\beta}, \boldsymbol{u}), SSE(\boldsymbol{\beta}))}{[n - \mathrm{r}(\boldsymbol{X}, \boldsymbol{Z})][\mathrm{tr}(\boldsymbol{L}_{1,1})]^2}.\end{aligned}$$

For the variance of $SSE(\boldsymbol{\beta})$, it holds

$$\begin{aligned}&\mathrm{var}(SSE(\boldsymbol{\beta})) = 2\mathrm{tr}([\boldsymbol{M}_1\boldsymbol{V}]^2)\\ &= 2\mathrm{tr}\left(\boldsymbol{M}_1(\sigma_u^2\boldsymbol{Z}\boldsymbol{Z}' + \sigma_e^2\boldsymbol{W}^{-1})\boldsymbol{M}_1(\sigma_u^2\boldsymbol{Z}\boldsymbol{Z}' + \sigma_e^2\boldsymbol{W}^{-1})\right)\\ &= 2\sigma_u^4\mathrm{tr}(\boldsymbol{M}_1\boldsymbol{Z}\boldsymbol{Z}'\boldsymbol{M}_1\boldsymbol{Z}\boldsymbol{Z}') + 4\sigma_u^2\sigma_e^2\mathrm{tr}(\boldsymbol{M}_1\boldsymbol{Z}\boldsymbol{Z}'\boldsymbol{M}_1\boldsymbol{W}^{-1}) + 2\sigma_e^4\mathrm{tr}(\boldsymbol{M}_1\boldsymbol{W}^{-1}\boldsymbol{M}_1\boldsymbol{W}^{-1})\\ &= 2\sigma_u^4\mathrm{tr}(\boldsymbol{L}_{1,1}^2) + 4\sigma_u^2\sigma_e^2\mathrm{tr}(\boldsymbol{L}_{1,1}) + 2\sigma_e^4\mathrm{tr}(\boldsymbol{M}_1\boldsymbol{W}^{-1}),\end{aligned}$$

since $\boldsymbol{M}_1\boldsymbol{W}^{-1}\boldsymbol{M}_1 = \boldsymbol{M}_1$. Further, it holds $\mathrm{tr}(\boldsymbol{M}_1\boldsymbol{W}^{-1}) = n - p$, and as $\mathrm{r}(\boldsymbol{X}) = p$ and $\mathrm{r}(\boldsymbol{X}, \boldsymbol{Z}) = p + D - \lambda$, we get

$$\begin{aligned}\mathrm{var}(\hat{\sigma}_u^2) = &\frac{2\sigma_u^4\mathrm{tr}(\boldsymbol{L}_{1,1}^2) + 4\sigma_e^2\sigma_u^2\mathrm{tr}(\boldsymbol{L}_{1,1}) + 2(n - p)\sigma_e^4}{[\mathrm{tr}(\boldsymbol{L}_{1,1})]^2}\\ &+ \frac{2\sigma_e^4(n - p)^2}{[\mathrm{tr}(\boldsymbol{L}_{1,1})]^2(n - p - D + \lambda)} - \frac{4(n - p)\sigma_e^4}{[\mathrm{tr}(\boldsymbol{L}_{1,1})]^2}\end{aligned}$$

$$= \frac{2}{[\mathrm{tr}(\boldsymbol{L}_{1,1})]^2}\Big\{\mathrm{tr}(\boldsymbol{L}_{1,1}^2)\sigma_u^4 + 2\mathrm{tr}(\boldsymbol{L}_{1,1})\sigma_u^2\sigma_e^2$$

$$+ \frac{(n-p)(n-p-D+\lambda) + (n-p)^2 - 2(n-p)(n-p-D+\lambda)}{n-p-D+\lambda}\sigma_e^4\Big\}$$

$$= \frac{2}{[\mathrm{tr}(\boldsymbol{L}_{1,1})]^2}\left\{\frac{(n-p)(D-\lambda)}{n-p-D+\lambda}\sigma_e^4 + 2\mathrm{tr}(\boldsymbol{L}_{1,1})\sigma_u^2\sigma_e^2 + \mathrm{tr}(\boldsymbol{L}_{1,1}^2)\sigma_u^4\right\}$$

$$= \frac{2}{n_*^2}\left\{\frac{(n-p)(D-\lambda)}{n-p-D+\lambda}\sigma_e^4 + 2n_*\sigma_u^2\sigma_e^2 + n_{**}\sigma_u^4\right\},$$

where $n_* = \mathrm{tr}(\boldsymbol{L}_{1,1})$ and $n_{**} = \mathrm{tr}(\boldsymbol{L}_{1,1}^2)$. In the particular case $\boldsymbol{W} = \boldsymbol{I}$, we have $w_d = n_d$, $d = 1, \ldots, D$, and $w = n$. Therefore, $n_* = \mathrm{tr}\{\boldsymbol{M}_1\boldsymbol{Z}\boldsymbol{Z}'\} = n - \mathrm{tr}\left\{(\boldsymbol{X}'\boldsymbol{X})^{-1}\sum_{d=1}^{D} n_d^2 \hat{\overline{\boldsymbol{X}}}_d' \hat{\overline{\boldsymbol{X}}}_d\right\}$ and $n_{**} = \mathrm{tr}\{(\boldsymbol{M}_1\boldsymbol{Z}\boldsymbol{Z}')^2\}$, where $\boldsymbol{M}_1 = \boldsymbol{I}_n - \boldsymbol{X}(\boldsymbol{X}'\boldsymbol{X})^{-1}\boldsymbol{X}'$.

7.9 Simulation Experiment

This section presents a simulation experiment that compares the behavior of the fitting methods H3, ML, and REML. We simulate samples in the following way.

`Simulation of the explanatory variable.` For $d = 1, \ldots, D$, $j = 1, \ldots, n_d$, generate

$$x_{dj} \sim U(a_d, b_d); \qquad \text{i.e. } x_{dj} = (b_d - a_d)U_{dj} + a_d, \text{ with } U_{dj} \sim U(0, 1).$$

Take $D = 30$, $n_1 = \ldots = n_D$, $a_d = 1$, $b_d = 1 + d$, $d = 1, \ldots, D$.

`Weights:` For $d = 1, \ldots, D$, $j = 1, \ldots, n_d$, put $w_{dj} = 1/x_{dj}^{\ell}$, $\ell = 0, 1$.

`Simulation of random effects and errors:` For $d = 1, \ldots, D$, $j = 1, \ldots, n_d$, generate

$$u_d \sim N(0, \sigma_u^2), \quad e_{dj} \sim N(0, \sigma_e^2), \quad \text{with } \sigma_u^2 = 1, \ \sigma_e^2 = 1.$$

`Simulation of the target variable:` For $d = 1, \ldots, D$, $j = 1, \ldots, n_d$, calculate

$$y_{dj} = \beta x_{dj} + u_d + w_{dj}^{-1/2} e_{dj}, \quad \text{with } \beta = 1.$$

The steps of the simulation are

1. Repeat $K = 10^4$ times ($k = 1, \ldots, K$).

   1.1. Generate a random sample of size $n = \sum_{d=1}^{D} n_d$.

1.2. Calculate $\hat{\beta}_k$, $\hat{\sigma}^2_{ek}$, and $\hat{\sigma}^2_{uk}$ by using the H3, ML, and REML methods.

2. Output: For $\hat{\theta}_k \in \{\hat{\beta}_k, \hat{\sigma}^2_{ek}, \hat{\sigma}^2_{uk}\}$ and $\theta \in \{\beta, \sigma^2_e, \sigma^2_u\}$, calculate the empirical mean squared errors and empirical biases

$$EMSE(\hat{\theta}) = \frac{10^5}{K}\sum_{k=1}^{K}(\hat{\theta}_k - \theta)^2, \quad EBIAS(\hat{\theta}) = \frac{10^5}{K}\sum_{k=1}^{K}(\hat{\theta}_k - \theta).$$

Tables 7.1 and 7.2 present the simulation results for the case $\ell = 0$.

Concerning $EMSE$, the estimators ML and REML behave better than the estimators H3, particularly for the parameter β. Between ML and REML, we do not find significative differences. Nevertheless, ML estimators seem to have slightly lower $EMSE$s than REML estimators.

Concerning $EBIAS$, the REML estimators behave better than the ML and H3 estimators. We do not find significative differences between REML and H3 methods when estimating σ^2_e. However, the REML estimators have lower bias than the ML

Table 7.1 *EMSE* of $\hat{\beta}$, $\hat{\sigma}^2_e$, and $\hat{\sigma}^2_u$ for $\ell = 0$

	n	150	300	450	600	750	900	1050	1200	1350	1500
	n_d	5	10	15	20	25	30	35	40	45	50
H3	β	243	105	76	92	62	73	71	76	72	68
	σ^2_e	1668	758	459	349	284	229	194	170	156	135
	σ^2_u	9809	8173	7799	7410	7445	7481	7385	7099	7260	7216
ML	β	23	11	7	6	4	4	3	3	3	2
	σ^2_e	1659	756	458	349	283	229	194	170	155	134
	σ^2_u	9712	8122	7691	7314	7345	7351	7232	6997	7160	7117
REML	β	23	11	7	6	4	4	3	3	3	2
	σ^2_e	1664	758	459	349	284	229	194	170	155	135
	σ^2_u	9765	8147	7708	7319	7354	7356	7238	6999	7161	7118

Table 7.2 *EBIAS* of $\hat{\beta}$, $\hat{\sigma}^2_e$, and $\hat{\sigma}^2_u$ for $\ell = 0$

	n	150	300	450	600	750	900	1050	1200	1350	1500
	n_d	5	10	15	20	25	30	35	40	45	50
H3	β	−4444	−2684	−2127	−2452	−1799	−1856	−1870	−1939	−1885	−1735
	σ^2_e	−233	−22	9	−104	40	−38	61	−19	0	−1
	σ^2_u	−1055	−758	−12	−355	173	42	169	−89	−210	−119
ML	β	5	2	−11	2	4	4	4	−1	−1	−1
	σ^2_e	−723	−275	−180	−253	−79	−140	−26	−97	−70	−64
	σ^2_u	−1850	−1433	−467	−693	−218	−196	−49	−265	−367	−243
REML	β	5	2	−11	2	4	4	4	−1	−1	−1
	σ^2_e	−234	−21	9	−104	39	−37	62	−19	0	−1
	σ^2_u	−389	−373	209	−201	260	152	306	5	−121	−27

and H3 method when estimating σ_u^2. Finally, the H3 method has greater bias than the ML and the REML when estimating $\boldsymbol{\beta}$. In this last case, we do not find differences between the ML and the REML methods.

7.10 R Codes

7.10.1 MLEs

This section gives R codes for calculating the ML estimators of the parameters of the NER model (7.1) and presents applications to data from the file `LFS20.txt`. The target variable y is INCOME and the auxiliary variables x_1 and x_2 are EDUCATION2 and EDUCATION3, respectively. The following code reads the data file and makes some calculations.

```
dat <- read.table("LFS20.txt", header=TRUE, sep = "\t", dec = ".")
dat$EDUCATION <- as.factor(dat$EDUCATION)  # EDUCATION as factor
y <- 10^(-4)*dat$INCOME
age <- dat$ageG
edu2 <- as.numeric(dat$EDUCATION==2)
edu3 <- as.numeric(dat$EDUCATION==3)
n <- length(dat$AREA)                      # Global sample size
domain <- sort(unique(dat$AREA))           # Domains
D <- length(domain)                        # Number of domains
# Domain sizes
nd <- tapply(rep(1, n), list(dat$AREA), sum, simplify = FALSE)
yd <- Xd <- Vd <- list()
for (d in 1:D) {
  condition <- dat$AREA==domain[d]
  # Auxiliary variables
  Xd[[d]] <- cbind(1, edu2[condition],edu3[condition])
  # Target variable
  yd[[d]] <- y[condition]
}
```

We first apply the function `lmer` from the R library `lme4`.

```
if(!require(lme4)){
  install.packages("lme4")
  library(lme4)
}
lmm.1 <- lmer(formula=y~EDUCATION+(1|AREA), data=dat, REML=FALSE)
summary(lmm.1)
```

Before running the remaining R codes, it is necessary to upload the functions of Sect. 7.10.2. The following procedures need initial values (seeds) for starting the programmed algorithms. We calculate the seeds for the model parameters by using the formulas (7.13), (7.14), and (7.15). Section 7.10.2 contains the R function `seed`.

```
start <- seed(Xd, yd, n, nd)
```

The function `FisherScoring` applies the Fisher-scoring algorithm described in Sect. 7.3. This function already has R codes for calculating the seeds (7.13), (7.14), and (7.15).

```
ML.fit <- FisherScoring(Xd, yd, n, nd, n.iter=200)
```

The R function `nlm` also calculates the MLE estimators of the parameters of a NER model by minimizing the log-likelihood function.

```
ML.nlm <- nlm(log_lik, p=start, Xd=Xd, yd=yd, n=n, nd=nd, hessian=TRUE)
```

Similarly, we can apply the function `optimx` of the R package `OPTIMX`.

```
if(!require(optimx)){
  install.packages("optimx")
  library(optimx)
}
ML.optimx <- optimx(par=start, log_lik, Xd=Xd, yd=yd, n=n, nd=nd,
                    method="BFGS")
```

The function `nleqslv` of the R package `nleqslv` can be applied to solve a system of nonlinear equations. The following codes solve the system of log-likelihood Eqs. (7.11) of the NER model and, therefore, calculate the ML estimators of the model parameters.

```
if(!require(nleqslv)){
  install.packages("nleqslv")
  library(nleqslv)
}
ML.nleqslv <- nleqslv(start, scores, Xd=Xd, yd=yd, n=n, nd=nd, jacobian=TRUE)
```

The functions `dfsane` and `BBsolve` of the R package `BB` also solve the system of log-likelihood Eqs. (7.11). The function `BBsolve` is especially useful in problems where the algorithms in `dfsane` are likely to experience difficulties in convergence.

```
if(!require(BB)){
  install.packages("BB")
  library(BB)
}
ML.dfsane <- dfsane(start, scores, Xd=Xd, yd=yd, n=n, nd=nd, method=1)
# Alternative for dfsane
ML.BBsolve <- BBsolve(start, scores, Xd=Xd, yd=yd, n=n, nd=nd, method=1)
```

Summary of results

```
var <- as.data.frame(VarCorr(lmm.1))
output <- data.frame(seeds=start,
                     lmer=c(lmm.1@beta, var$vcov[1], var$vcov[2]),
                     FisherScoring=ML.fit[[5]],
                     nlm=ML.nlm$estimate,
                     optimx=as.numeric(ML.optimx[1,1:5]),
                     nleqslv=ML.nleqslv$x, dfsane=ML.dfsane$par)
round(output, 5)
```

Table 7.3 presents the ML estimates of the model parameters when they are calculated by the considered R functions. The first row contains the numerical values of the seeds (7.13), (7.14), and (7.15).

Table 7.3 ML estimates of model parameters

Method	$\hat{\beta}_0$	$\hat{\beta}_1$	$\hat{\beta}_2$	$\hat{\sigma}_u^2$	$\hat{\sigma}_e^2$
seeds	3.95957	0.90851	1.95432	0.02422	1.08110
lmer	3.95807	0.90926	1.95475	0.00280	1.07521
FisherScoring	3.95807	0.90926	1.95475	0.00280	1.07521
nlm	3.95807	0.90926	1.95476	0.00280	1.07521
optimx	3.95812	0.90927	1.95464	0.00282	1.07514
nleqslv	3.95807	0.90926	1.95475	0.00279	1.07521
dfsane	3.95807	0.90926	1.95475	0.00279	1.07521

7.10.2 Auxiliary Functions

The function seed calculates initial estimates of the parameters of a NER model by using the formulas (7.13), (7.14), and (7.15).

```
seed <- function(Xd, yd, n, nd){
  XtX <- Reduce(Map(Xd, Xd, f=crossprod), f="+")
  Xty <- Reduce(Map(Xd, yd, f=crossprod), f="+")
  beta0 <- solve(XtX)%*%Xty
  sigmau2.0 <- mean((sapply(yd, mean) - sapply(lapply(Xd, colMeans), beta0,
               FUN=crossprod))^2)
  yxbeta <- Map(yd, lapply(Xd, beta0, FUN="%*%"), f="-")
  sigmae2.0 <- sum(mapply(yxbeta, yxbeta, FUN=crossprod))/(n-ncol(Xd[[1]]))
  return(c(beta0, sigmau2.0, sigmae2.0))
}
```

The function log_lik calculates the log-likelihood function (7.10) of the NER model, with negative sign, for applying minimization algorithms.

```
log_lik <- function (parameters, Xd, yd, n, nd) {
  beta <- parameters[1:3]
  sigmau <- parameters[4]
  sigmae <- parameters[5]
  sum1 <- n/2*(log(2*pi))
  Vd <- lapply(mapply(mapply(sigmau/sigmae,nd,nd, FUN="matrix"),
        lapply(nd, FUN=diag),FUN="+"), sigmae, FUN="*")
  Ad <- mapply(mapply(sigmau/sigmae,nd,nd, FUN="matrix"),
        lapply(nd, FUN=diag),FUN="+")
  Ad.det <- lapply(Ad, FUN=det)
  sum2 <- 0.5*sum(mapply(lapply(nd, log(sigmae), FUN="*"),
          lapply(Ad.det, FUN=log), FUN="+"))
  Vd.inv <- lapply(Vd,solve)
  yxbeta <- mapply(yd, lapply(Xd, beta, FUN="%*%"), FUN="-")
  yxbetat <- mapply(mapply(yd, lapply(Xd, beta, FUN="%*%"), FUN="-"),
             FUN=t)
  yxbetat.v.yxbeta <- mapply(mapply(yxbetat, Vd.inv, FUN="%*%"), yxbeta,
                      FUN="%*%")
  sum3 <- 0.5*sum(mapply(mapply(yxbetat, Vd.inv, FUN="%*%"), yxbeta,
          FUN="%*%"))
  l <- sum1 + sum2 + sum3
  return(l)
}
```

The function scores applies the formula (7.11) for calculating the first partial derivatives of the log-likelihood function of the NER model. Under the default option small=TRUE, the function scores divides the β-scores by D and the variance scores by n. This option is recommended for applying R functions that solve systems of nonlinear equations.

```
scores <- function(parameters, Xd, yd, n, nd, small=TRUE){
  beta <- parameters[1:3]
  sigmau <- parameters[4]
  sigmae <- parameters[5]
  Vd <- lapply(mapply(mapply(sigmau/sigmae,nd,nd, FUN="matrix"),
        lapply(nd, FUN=diag),FUN="+"), sigmae, FUN="*")
  Vd.inv <- lapply(Vd,solve)
  S.beta <- apply(mapply(lapply(Xd, FUN=t), mapply(Vd.inv, mapply(yd,
            lapply(Xd, beta, FUN="%*%"),FUN="-"), FUN="%*%"),
            FUN="%*%"), 1, sum)
  yxbeta <- mapply(yd, lapply(Xd, beta, FUN="%*%"), FUN="-")
  yxbetat <- mapply(mapply(yd, lapply(Xd, beta, FUN="%*%"), FUN="-"), FUN=t)
  S.sigmau1 <- Reduce(lapply(lapply(mapply(Vd.inv, mapply(1, nd, nd,
               FUN="matrix"), FUN="%*%"), FUN=diag), FUN=sum), f=sum)
  S.sigmau2 <- sum(mapply(mapply(yxbetat, mapply(mapply(Vd.inv,
               mapply(1, nd, nd, FUN="matrix"), FUN="%*%"), Vd.inv,
               FUN="%*%"), FUN="%*%"), yxbeta, FUN="%*%"))
  S.sigmau <- -0.5*S.sigmau1 + 0.5*S.sigmau2
  S.sigmae1 <- Reduce(lapply(lapply(Vd.inv, diag), sum), f=sum)
  S.sigmae2 <- sum(mapply(yxbetat, mapply(Vd.inv, mapply(Vd.inv, yxbeta,
               FUN="%*%"), FUN="%*%"), FUN="%*%"))
  S.sigmae <- -0.5*S.sigmae1 + 0.5*S.sigmae2
  S <- c(S.beta, S.sigmau, S.sigmae)
  if(small==TRUE)
    S <- c(S.beta/D, S.sigmau/n, S.sigmae/n)
  return(S)
}
```

The function `Inf.Fisher` applies the formula (7.12) for calculating the Fisher information matrix.

```
Inf.Fisher <- function(parameters, Xd, yd, n, nd){
  beta <- parameters[1:3]
  sigmau <- parameters[4]
  sigmae <- parameters[5]
  Vd <- lapply(mapply(mapply(sigmau/sigmae, nd, nd, FUN="matrix"), lapply(nd,
        FUN=diag),FUN="+"), sigmae, FUN="*")
  Vd.inv <- lapply(Vd, solve)
  yxbeta <- mapply(yd, lapply(Xd, beta, FUN="%*%"), FUN="-")
  yxbetat <- mapply(mapply(yd, lapply(Xd, beta, FUN="%*%"), FUN="-"), FUN=t)
  F.beta.beta <- Reduce(Map(lapply(Xd, FUN=t), mapply(Vd.inv,  Xd,
                 FUN="%*%"), f="%*%"), f="+")
  Vd.inv.Jnd <- mapply(Vd.inv, mapply(1,nd,nd, FUN="matrix"), FUN="%*%")
  F.sigmau.sigmau <- 0.5*sum(sapply(lapply(Map(Vd.inv.Jnd, Vd.inv.Jnd,
                     f="%*%"), FUN=diag),FUN=sum))
  F.sigmau.sigmae <- 0.5*sum(sapply(lapply(Map(Vd.inv, Vd.inv.Jnd, f="%*%"),
                     FUN=diag), FUN=sum))
  F.sigmae.sigmae <- 0.5*sum(sapply(lapply(Map(Vd.inv, Vd.inv, f="%*%"),
                     FUN=diag), FUN=sum))
  F.sigma <- matrix(c(F.sigmau.sigmau, F.sigmau.sigmae, F.sigmau.sigmae,
             F.sigmae.sigmae), ncol=2)
  return(list(F.beta.beta, F.sigma))
}
```

The function `FisherScoring` runs the Fisher-scoring algorithm described in Sect. 7.3. This algorithm calculates the ML estimates of the parameters of a NER model.

```
FisherScoring <- function(Xd, yd, n, nd, n.iter=100){
  parameters <- list()
  parameters[[1]] <- seed(Xd, yd, n, nd)
```

```
    p <- ncol(Xd[[1]])
    iter <- 1
    while(iter < n.iter){
      S <- scores(parameters[[iter]], Xd, yd, n, nd, small=FALSE)
      F <- Inf.Fisher(parameters[[iter]], Xd, yd, n, nd)
      diff.beta <- solve(F[[1]])%*%S[1:p]
      diff.sigma <- solve(F[[2]])%*%S[(p+1):(p+2)]
      diff <- rbind(diff.beta, diff.sigma)
      parameters[[iter+1]] <- parameters[[iter]] + diff
      RT <- abs(diff)/abs(parameters[[iter+1]])
      out <- RT>0.01
      if(sum(out)==0)
        break
      iter <- iter + 1
    }
    return(parameters)
  }
```

References

Battese, G.E., Fuller, W.A.: Prediction of county crop areas using survey and satellite data. In: Proceedings of the Section on Survey Research Methods, pp. 500–505. American Statistical Association, Washington (1981)

Battese, G.E, Harter, R.M., Fuller, W.A.: An error-component model for prediction of county crop areas using survey and satellite data. J. Am. Stat. Asso. **83**, 28–36 (1988)

Harter, R.M.: Small area estimation using nested-error models and auxiliary data. Ph.D. Thesis. Iowa State University (1983)

Prasad, N.G.N., Rao, J.N.K.: The estimation of the mean squared error of small-area estimators. J. Am. Stat. Asso. **85**, 163–171 (1990)

Chapter 8
EBLUPs Under Nested Error Regression Models

8.1 Introduction

Many social and economic indicators are linear combinations of the values that a target variable takes in all units of a finite population. These indicators have the form of weighted sums with known positive (or null) weights. Examples of linear indicators are a domain mean, a domain total, or a given observed value of the target variable.

Linear parameter predictors based on linear models are required to meet some good properties, such as being unbiased with minimum variance. Under linear models with fixed effects, the best linear unbiased predictors (BLUP) fulfill these conditions. Under linear models with random effects, BLUPs are not calculable because the variance component parameters are unknown in practice. The empirical best linear unbiased predictors (EBLUP) are plug-in approximations of the BLUPs that are obtained by substituting the unknown variance component parameters by consistent estimators. In that way, they inherit the good properties of BLUPs asymptotically. The aim of this chapter is to introduce EBLUPs of linear parameters under the nested error regression (NER) model.

This chapter introduces the NER model in vectorial form and derives the expressions of the BLUP and the EBLUP of a single observation and of a domain mean. This is done in Sects. 8.2, 8.3, and 8.4. The rest of the chapter is organized as follows. Section 8.5 describes the parametric bootstrap approach given by González-Manteiga et al. (2008a) for estimating the mean squared error of the EBLUPs of linear parameters. By using the predictions and residuals of a NER model, Sect. 8.6 introduces a model-assisted estimator of the domain mean of a continuous variable. Section 8.7 contains a comparative simulation experiment and shows the corresponding results. Finally, Sect. 8.8 presents an application to a labor force survey data file and gives the R code for calculating EBLUPs of domain means.

D. Morales et al., *A Course on Small Area Estimation and Mixed Models*, Statistics for Social and Behavioral Sciences, https://doi.org/10.1007/978-3-030-63757-6_8

8.2 The NER Model

Let U be a finite population partitioned in domains, i.e. $U = \cup_{d=1}^{D} U_d$ and $U_{d_1} \cap U_{d_2} = \emptyset$ if $d_1 \neq d_2$. Let N and N_d be the sizes of U and U_d, so that $N = \sum_{d=1}^{D} N_d$. We assume that the population target vector $\boldsymbol{y} = \boldsymbol{y}_{N\times 1}$ follows the NER model

$$\boldsymbol{y} = \boldsymbol{X}\boldsymbol{\beta} + \boldsymbol{Z}\boldsymbol{u} + \boldsymbol{e}, \tag{8.1}$$

where $\boldsymbol{X} = \boldsymbol{X}_{N\times p}$, $\boldsymbol{Z} = \boldsymbol{Z}_{N\times D} = \text{diag}\left(\boldsymbol{1}_{N_d}, d = 1, \ldots, D\right)$, $\boldsymbol{\beta} = \boldsymbol{\beta}_{p\times 1}$, $\boldsymbol{1}'_a = (1, \ldots, 1)_{1\times a}$, $\boldsymbol{u} = \boldsymbol{u}_{D\times 1} \sim N(\boldsymbol{0}, \sigma_u^2 \boldsymbol{I}_D)$ is independent of $\boldsymbol{e} = \boldsymbol{e}_{N\times 1} \sim N(\boldsymbol{0}, \sigma_e^2 \boldsymbol{W}^{-1})$, $\boldsymbol{I}_D = \text{diag}(1, \ldots, 1)_{D\times D}$ and $\boldsymbol{W} = \text{diag}(w_{11}, \ldots, w_{DN_D})_{N\times N}$ is a diagonal matrix containing known heteroscedastic weights $w_{11} > 0, \ldots, w_{DN_D} > 0$. The model (8.1) can alternatively be written as

$$y_{dj} = \boldsymbol{x}_{dj}\boldsymbol{\beta} + u_d + e_{dj}, \quad d = 1, \ldots, D,\ j = 1, \ldots, N_d, \tag{8.2}$$

where y_{dj} is the target variable measured at unit j of domain d, $\boldsymbol{x}_{dj}$ is the corresponding row of the matrix $\boldsymbol{X}$ containing the auxiliary variables, u_d is the domain random effect, and e_{dj} is the random error.

Let $s \subset U$ be a sample of $n \leq N$ units and let $r = U - s$ be the set of non-sampled units. The domain subsets of s and r are denoted by s_d and r_d, respectively. The subindexes s and r in vectors or matrices are used to denote their sampled and their non-sampled parts. Without loss of generality, we renumber the population units and, for example, we write

$$\boldsymbol{Z} = \begin{pmatrix} \boldsymbol{Z}_s \\ \boldsymbol{Z}_r \end{pmatrix}, \quad \boldsymbol{X} = \begin{pmatrix} \boldsymbol{X}_s \\ \boldsymbol{X}_r \end{pmatrix}, \quad \boldsymbol{y} = \begin{pmatrix} \boldsymbol{y}_s \\ \boldsymbol{y}_r \end{pmatrix}, \quad \boldsymbol{V} = \text{var}(\boldsymbol{y}) = \begin{pmatrix} \boldsymbol{V}_s & \boldsymbol{V}_{sr} \\ \boldsymbol{V}_{rs} & \boldsymbol{V}_r \end{pmatrix}.$$

For a sample s of size n with n_d units in domain d, $n = \sum_{d=1}^{D} n_d$, we have that

$$\boldsymbol{V}_s = \text{var}(\boldsymbol{y}_s) = \sigma_u^2 \boldsymbol{Z}_s \boldsymbol{Z}'_s + \sigma_e^2 \boldsymbol{W}_s^{-1} = \text{diag}(\boldsymbol{V}_{s1}, \ldots, \boldsymbol{V}_{sD})_{n\times n},$$

where $\boldsymbol{Z}_s$ and $\boldsymbol{W}_s$ are the sample parts of $\boldsymbol{Z}$ and $\boldsymbol{W}$, respectively, $\boldsymbol{V}_{sd} = \sigma_e^2 \boldsymbol{W}_{sd}^{-1} + \sigma_u^2 \boldsymbol{1}_{n_d}\boldsymbol{1}'_{n_d}$ and $\boldsymbol{W}_{sd} = \text{diag}(w_{d1}, \ldots, w_{dn_d})_{n_d\times n_d}$. Similarly, we have

$$\begin{aligned} \boldsymbol{V}_{rs} &= E\left[(\boldsymbol{y}_r - \boldsymbol{X}_r\boldsymbol{\beta})(\boldsymbol{y}_s - \boldsymbol{X}_s\boldsymbol{\beta})'\right] = E\left[(\boldsymbol{Z}_r\boldsymbol{u} + \boldsymbol{e}_r)(\boldsymbol{Z}_s\boldsymbol{u} + \boldsymbol{e}_s)'\right] \\ &= \boldsymbol{Z}_r E\left[\boldsymbol{u}\boldsymbol{u}'\right]\boldsymbol{Z}'_s = \boldsymbol{Z}_r \sigma_u^2 \boldsymbol{I}_D \boldsymbol{Z}'_s = \sigma_u^2 \boldsymbol{Z}_r \boldsymbol{Z}'_s. \end{aligned}$$

By substituting $\boldsymbol{V}_s$ and $\boldsymbol{V}_{rs}$ in (4.4), we obtain the best linear unbiased predictor (BLUP) of a linear parameter $\boldsymbol{\gamma}'\boldsymbol{y}$, where $\boldsymbol{\gamma}' = (\gamma_1, \ldots, \gamma_N)$ is a known vector.

8.3 BLUP of a Domain Mean

This section presents the particularization of Theorem 4.1 to model (8.2) and calculates the BLUP of a domain mean. The target parameter is

$$\theta = \overline{Y}_d = \frac{1}{N_d}\sum_{j=1}^{N_d} y_{dj} = \boldsymbol{\gamma}'\boldsymbol{y},$$

where $\boldsymbol{\gamma}' = \frac{1}{N_d}\left(\mathbf{0}'_{N_1},\ldots,\mathbf{0}'_{N_{d-1}},\mathbf{1}'_{N_d},\mathbf{0}'_{N_{d+1}},\ldots,\mathbf{0}'_{N_D}\right)$ and $\mathbf{0}'_a = (0,\ldots,0)_{1\times a}$. For calculating the inverse of $\boldsymbol{V}_{sd} = \sigma_e^2\boldsymbol{W}_{sd}^{-1} + \sigma_u^2\mathbf{1}_{n_d}\mathbf{1}'_{n_d}$, we apply the formula

$$(\boldsymbol{A}+\boldsymbol{a}\boldsymbol{b}')^{-1} = \boldsymbol{A}^{-1} - \frac{\boldsymbol{A}^{-1}\boldsymbol{a}\boldsymbol{b}'\boldsymbol{A}^{-1}}{1+\boldsymbol{b}'\boldsymbol{A}^{-1}\boldsymbol{a}},$$

with $\boldsymbol{A} = \sigma_e^2\text{diag}(w_{d1}^{-1},\ldots,w_{dn_d}^{-1})$ and $\boldsymbol{a} = \boldsymbol{b} = (\sigma_u,\ldots\sigma_u)'$. In a similar way to (7.6), we get

$$\boldsymbol{V}_{sd}^{-1} = \frac{1}{\sigma_e^2}\Big(\boldsymbol{W}_{sd} - \frac{\gamma_d^w}{w_d}\boldsymbol{w}_{n_d}\boldsymbol{w}'_{n_d}\Big),$$

$$\gamma_d^w = \frac{\sigma_u^2}{\sigma_u^2+\frac{\sigma_e^2}{w_d}},\qquad \boldsymbol{w}'_{n_d} = (w_{d1},\ldots,w_{dn_d}),\qquad w_d = \sum_{j=1}^{n_d} w_{dj}.$$

The inverse of $\boldsymbol{V}_s$ is $\boldsymbol{V}_s^{-1} = \text{diag}(\boldsymbol{V}_{s1}^{-1},\ldots,\boldsymbol{V}_{sD}^{-1})$. In what follows we present some auxiliary calculations for obtaining the BLUP $\hat{\theta}_{opt}$ of $\overline{Y}_d$. It holds

$$\begin{aligned}
\boldsymbol{\gamma}'_r\boldsymbol{V}_{rs} &= \frac{\sigma_u^2}{N_d}\left(\mathbf{0}'_{N_1-n_1},\ldots,\mathbf{0}'_{N_{d-1}-n_{d-1}},\mathbf{1}'_{N_d-n_d},\mathbf{0}'_{N_{d+1}-n_{d+1}}\ldots,\mathbf{0}'_{N_D-n_D}\right)\\
&\quad\cdot \underset{1\le d\le D}{\text{diag}}\,(\mathbf{1}_{N_d-n_d})\,\underset{1\le d\le D}{\text{diag}}\,(\mathbf{1}'_{n_d})\\
&= \sigma_u^2\frac{N_d-n_d}{N_d}\left(\mathbf{0}'_{n_1},\ldots,\mathbf{0}'_{n_{d-1}},\mathbf{1}'_{n_d},\mathbf{0}'_{n_{d+1}}\ldots,\mathbf{0}'_{n_D}\right)\\
&= \sigma_u^2(1-f_d)\left(\mathbf{0}'_{n_1},\ldots,\mathbf{0}'_{n_{d-1}},\mathbf{1}'_{n_d},\mathbf{0}'_{n_{d+1}}\ldots,\mathbf{0}'_{n_D}\right),\\
\boldsymbol{\gamma}'_r\boldsymbol{V}_{rs}\boldsymbol{V}_s^{-1}\boldsymbol{y}_s &= \sigma_u^2(1-f_d)\left(\mathbf{0}'_{n_1},\ldots,\mathbf{0}'_{n_{d-1}},\mathbf{1}'_{n_d},\mathbf{0}'_{n_{d+1}}\ldots,\mathbf{0}'_{n_D}\right)\underset{1\le d\le D}{\text{diag}}\,(\boldsymbol{V}_{sd}^{-1})\boldsymbol{y}_s\\
&= \sigma_u^2(1-f_d)\frac{1}{\sigma_e^2}\mathbf{1}'_{n_d}\Big(\boldsymbol{W}_{sd} - \frac{\gamma_d^w}{w_d}\boldsymbol{w}_{nd}\boldsymbol{w}'_{nd}\Big)\boldsymbol{y}_{sd}
\end{aligned}$$

$$= \sigma_u^2 \frac{1}{\sigma_e^2}(1-f_d)\Big(\sum_{j=1}^{n_d} w_{dj} y_{dj} - \frac{\gamma_d^w}{w_d} w_d \sum_{j=1}^{n_d} w_{dj} y_{dj}\Big)$$

$$= \sigma_u^2 \frac{1}{\sigma_e^2}(1-f_d) w_d \hat{\overline{Y}}_d^w (1-\gamma_d^w) = (1-f_d)\gamma_d^w \hat{\overline{Y}}_d^w,$$

$$\boldsymbol{\gamma}_r' \boldsymbol{V}_{rs} \boldsymbol{V}_s^{-1} \boldsymbol{X}_s = (1-f_d)\gamma_d^w \hat{\overline{\boldsymbol{X}}}_d^w, \ \boldsymbol{\gamma}_r' \boldsymbol{X}_r = \frac{1}{N_d} \sum_{j \in r_d} \boldsymbol{x}_{dj} = \overline{\boldsymbol{X}}_d - f_d \hat{\overline{\boldsymbol{X}}}_d, \ \boldsymbol{\gamma}_s' \boldsymbol{y}_s = f_d \hat{\overline{Y}}_d,$$

where $f_d = \frac{n_d}{N_d}$, $\hat{\overline{Y}}_d^w = \frac{1}{w_d}\sum_{j=1}^{n_d} w_{dj} y_{dj}$, $\hat{\overline{Y}}_d = \frac{1}{n_d}\sum_{j=1}^{n_d} y_{dj}$, $\overline{\boldsymbol{X}}_d = \frac{1}{N_d}\sum_{j=1}^{N_d} \boldsymbol{x}_{dj}$, $\hat{\overline{\boldsymbol{X}}}_d = \frac{1}{n_d}\sum_{j=1}^{n_d} \boldsymbol{x}_{dj}$, and $\hat{\overline{\boldsymbol{X}}}_d^w = \frac{1}{w_d}\sum_{j=1}^{n_d} w_{dj} \boldsymbol{x}_{dj}$.

By substituting the above derived terms in (4.4) and by doing some straightforward calculations, we get the BLUP $\hat{\theta}_{opt} = \hat{\overline{Y}}_d^{blup}$ of $\overline{Y}_d$, i.e.

$$\begin{aligned}
\hat{\overline{Y}}_d^{blup} &= \gamma_s' \boldsymbol{y}_s + \boldsymbol{\gamma}_r'\big[\boldsymbol{X}_r \hat{\boldsymbol{\beta}} + \boldsymbol{V}_{rs}\boldsymbol{V}_s^{-1}(\boldsymbol{y}_s - \boldsymbol{X}_s \hat{\boldsymbol{\beta}})\big] \\
&= f_d \hat{\overline{Y}}_d + \overline{\boldsymbol{X}}_d \hat{\boldsymbol{\beta}} - f_d \hat{\overline{\boldsymbol{X}}}_d \hat{\boldsymbol{\beta}} + (1-f_d)\gamma_d^w \hat{\overline{Y}}_d^w - (1-f_d)\gamma_d^w \hat{\overline{\boldsymbol{X}}}_d^w \hat{\boldsymbol{\beta}} \\
&= (1-f_d)\big[\overline{\boldsymbol{X}}_d \hat{\boldsymbol{\beta}} + \gamma_d^w(\hat{\overline{Y}}_d^w - \hat{\overline{\boldsymbol{X}}}_d^w \hat{\boldsymbol{\beta}})\big] + f_d\big[\hat{\overline{Y}}_d + (\overline{\boldsymbol{X}}_d - \hat{\overline{\boldsymbol{X}}}_d)\hat{\boldsymbol{\beta}}\big]. \quad (8.3)
\end{aligned}$$

Note that σ_u^2 and σ_e^2 are unknown and therefore $\hat{\overline{Y}}_d^{blup}$ is not calculable. If we substitute σ_u^2 and σ_e^2 by convenient estimators, we obtain the EBLUP $\hat{\overline{Y}}_d^{eblup}$ of $\overline{Y}_d$.

8.4 EBLUP of a Single Observation

Let the target parameter be $\theta = y_{dj} = \boldsymbol{\gamma}'\boldsymbol{y}$, where $\boldsymbol{\gamma}'$ is a $1 \times N$ vector having all the components equal to zero except the one in the position $j + \sum_{k=1}^{d-1} N_k$ that is equal to one. We particularize the BLUP formula $\hat{\theta}_{opt} = \boldsymbol{\gamma}_s' \boldsymbol{y}_s + \boldsymbol{\gamma}_r'\Big[\boldsymbol{X}_r \hat{\boldsymbol{\beta}} + \boldsymbol{V}_{rs}\boldsymbol{V}_s^{-1}(\boldsymbol{y}_s - \boldsymbol{X}_s \hat{\boldsymbol{\beta}})\Big]$ to the target parameter y_{dj}.

Case 1: $j \in r_d$, so that $\boldsymbol{\gamma}_s' = (0, \ldots, 0)_{1\times n}$ and $\boldsymbol{\gamma}_r' = (0, \ldots, 0, 1, 0, \ldots, 0)_{1\times(N-n)}$. We have

$$\boldsymbol{\gamma}_r' \boldsymbol{V}_{rs} = \sigma_u^2 (0, \ldots, 0, 1, 0, \ldots, 0) \underset{1\le d\le D}{\text{diag}}\big(\boldsymbol{1}_{N_d-n_d}\big) \underset{1\le d\le D}{\text{diag}}\big(\boldsymbol{1}_{n_d}'\big)$$

$$= \sigma_u^2 \big(\boldsymbol{0}_{n_1}', \ldots, \boldsymbol{0}_{n_{d-1}}', \boldsymbol{1}_{n_d}', \boldsymbol{0}_{n_{d+1}}' \ldots, \boldsymbol{0}_{n_D}'\big),$$

$$\boldsymbol{\gamma}_r' \boldsymbol{V}_{rs} \boldsymbol{V}_s^{-1} = \frac{\sigma_u^2}{\sigma_e^2}\big(\boldsymbol{0}_{n_1}', \ldots, \boldsymbol{0}_{n_{d-1}}', \boldsymbol{1}_{n_d}', \boldsymbol{0}_{n_{d+1}}' \ldots, \boldsymbol{0}_{n_D}'\big) \underset{1\le d\le D}{\text{diag}}\Big(\boldsymbol{W}_{sd} - \frac{\gamma_d^w}{w_d}\boldsymbol{w}_{nd}\boldsymbol{w}_{nd}'\Big)$$

$$= \frac{\sigma_u^2}{\sigma_e^2}(1-\gamma_d^w)\big(\mathbf{0}'_{n_1},\ldots,\mathbf{0}'_{n_{d-1}},\boldsymbol{w}'_{nd},\mathbf{0}'_{n_{d+1}}\ldots,\mathbf{0}'_{n_D}\big),$$

$$\boldsymbol{\gamma}'_r\boldsymbol{V}_{rs}\boldsymbol{V}_s^{-1}\boldsymbol{y}_s = \frac{\sigma_u^2}{\sigma_e^2}(1-\gamma_d^w)\big(\mathbf{0}'_{n_1},\ldots,\mathbf{0}'_{n_{d-1}},\boldsymbol{w}'_{nd},\mathbf{0}'_{n_{d+1}}\ldots,\mathbf{0}'_{n_D}\big)\underset{1\le d\le D}{\mathrm{col}}(\boldsymbol{y}_{sd})$$

$$= \frac{\sigma_u^2}{\sigma_e^2}(1-\gamma_d^w)\sum_{j=1}^{n_d} w_{dj}y_{dj} = \gamma_d^w\hat{\overline{Y}}{}_d^w.$$

Therefore,

$$\hat{\theta}_{opt} = \hat{y}_{dj}^{blup} = \boldsymbol{x}_{dj}\hat{\boldsymbol{\beta}} + \gamma_d^w\big(\hat{\overline{Y}}{}_d^w - \hat{\overline{\boldsymbol{X}}}{}_d^w\hat{\boldsymbol{\beta}}\big).$$

Case 2: $j \in s_d$, so that $\boldsymbol{\gamma}'_s = (0,\ldots,0,1,0,\ldots,0)_{1\times n}$ and $\boldsymbol{\gamma}'_r = (0,\ldots,0)_{1\times(N-n)}$. We get

$$\hat{\theta}_{opt} = \hat{y}_{dj}^{blup} = y_{dj}.$$

In summary, the BLUP of y_{dj} is

$$\hat{y}_{dj}^{blup} = \begin{cases} \boldsymbol{x}_{dj}\hat{\boldsymbol{\beta}} + \gamma_d^w(\hat{\overline{Y}}{}_d^w - \hat{\overline{\boldsymbol{X}}}{}_d^w\hat{\boldsymbol{\beta}}) & \text{if } j \in r_d,\\ y_{dj} & \text{if } j \in s_d. \end{cases}$$

If we substitute σ_u^2 and σ_e^2 by convenient estimators in the expression of $\hat{y}_{dj}^{blup}$, we obtain the EBLUP $\hat{y}_{dj}^{eblup}$ of y_{dj}, i.e.

$$\hat{y}_{dj}^{eblup} = \begin{cases} \boldsymbol{x}_{dj}\hat{\boldsymbol{\beta}} + \hat{\gamma}_d^w(\hat{\overline{Y}}{}_d^w - \hat{\overline{\boldsymbol{X}}}{}_d^w\hat{\boldsymbol{\beta}}) & \text{if } j \in r_d,\\ y_{dj} & \text{if } j \in s_d, \end{cases} \quad \text{where } \hat{\gamma}_d^w = \frac{w_d\hat{\sigma}_u^2}{w_d\hat{\sigma}_u^2+\hat{\sigma}_e^2}.$$

Remark 8.1 For estimating the mean of domain d, $\overline{Y}_d = \frac{1}{N_d}\sum_{j=1}^{N_d} y_{dj}$, we can use the *predictive* estimator $\hat{\overline{Y}}{}_d^{(1)} = \frac{1}{N_d}\sum_{j=1}^{N_d}\hat{y}_{dj}^{eblup}$. Note that

$$\hat{\overline{Y}}{}_d^{(1)} = \frac{1}{N_d}\sum_{j\in s_d} y_{dj} + \frac{1}{N_d}\sum_{j\in r_d}\Big\{\boldsymbol{x}_{dj}\hat{\boldsymbol{\beta}} + \hat{\gamma}_d^w\big(\hat{\overline{Y}}{}_d^w - \hat{\overline{\boldsymbol{X}}}{}_d^w\hat{\boldsymbol{\beta}}\big)\Big\}$$

$$= (1-f_d)\Big[\overline{\boldsymbol{X}}_d\hat{\boldsymbol{\beta}} + \hat{\gamma}_d^w\big(\hat{\overline{Y}}{}_d^w - \hat{\overline{\boldsymbol{X}}}{}_d^w\hat{\boldsymbol{\beta}}\big)\Big] + f_d\big[\hat{\overline{Y}}_d + (\overline{\boldsymbol{X}}_d - \hat{\overline{\boldsymbol{X}}}_d)\hat{\boldsymbol{\beta}}\big],$$

which is the EBLUP, i.e. $\hat{\overline{Y}}{}_d^{(1)} = \hat{\overline{Y}}{}_d^{eblup}$. On the other hand, the *projective* estimator is

$$\hat{\overline{Y}}{}_d^{(2)} = \frac{1}{N_d}\sum_{j=1}^{N_d}\Big\{\boldsymbol{x}_{dj}\hat{\boldsymbol{\beta}} + \hat{\gamma}_d^w\big(\hat{\overline{Y}}{}_d^w - \hat{\overline{\boldsymbol{X}}}{}_d^w\hat{\boldsymbol{\beta}}\big)\Big\} = (1-\gamma_d^w)\overline{\boldsymbol{X}}_d\hat{\boldsymbol{\beta}} + \hat{\gamma}_d^w\big(\hat{\overline{Y}}{}_d^w + (\overline{\boldsymbol{X}}_d - \hat{\overline{\boldsymbol{X}}}{}_d^w)\hat{\boldsymbol{\beta}}\big).$$

We recall that the w_{dj}'s are heteroscedastic weights and not sampling weights, so that $\hat{\overline{Y}}_d^{w}$ and $\hat{\overline{X}}_d^{w}$ differ from $\hat{\overline{Y}}_d^{dir2}$ and $\hat{\overline{X}}_d^{dir2}$, respectively.

8.5 Parametric Bootstrap Estimation of MSEs

Some resampling methods for estimating the mean squared error (MSE) of empirical predictors can be found in the literature. The jackknife methodology proposed by Jiang et al. (2002) provides estimators with bias of order $O(D^{-3/2})$. Pfefferman and Tiller (2005) proposed a parametric and a nonparametric bootstrap method for estimating the same quantity under state-space models. Hall and Maiti (2006a,b) introduced parametric and matched-moment double-bootstrap algorithms. This section presents the parametric bootstrap approach given by González-Manteiga et al. (2008a).

We assume that the values of the target variable in the units of a finite population are realizations of random variables following a NER model. Under this framework, González-Manteiga et al. (2008a) proposed a bootstrap procedure for estimating MSEs of linear parameters in finite populations, when the values of the auxiliary variables are available for all units in the population. The knowledge of the x-values for the entire population, although a limitation, is one of the sources of strength of the method. It is worthwhile to emphasize that the finiteness of the population is taken into account in the proposed bootstrap method, so that bootstrap populations are built and bootstrap samples are drawn with the same subset indexes as the original sample. We remind that, under the prediction theory, the source of randomness is in the realization of the population and not in the sample extraction.

Let $\boldsymbol{y}$ be the population vector that contains the values of the target variable measured in the units of a finite population U. Let $\boldsymbol{X}$ be the matrix of auxiliary variables. Assume that $\boldsymbol{y}$ is generated by the superpopulation NER model (8.1), i.e.

$$\xi: \ \boldsymbol{y} = \boldsymbol{X}\boldsymbol{\beta} + \boldsymbol{Z}\boldsymbol{u} + \boldsymbol{e}.$$

Let s be a sample (subset) extracted from U and let $\boldsymbol{y}_s$ and $\boldsymbol{X}_s$ be the sampled parts of $\boldsymbol{y}$ and $\boldsymbol{X}$, respectively. Let $\eta = \eta(\boldsymbol{y})$ be a linear parameter of $\boldsymbol{y}$ and denote the EBLUP of η by $\hat{\eta} = \hat{\eta}(\boldsymbol{y}_s, \boldsymbol{X}_s)$. An example of linear parameter is the domain mean $\eta = \overline{Y}_d$. The following steps give a parametric bootstrap procedure for estimating the MSE of the EBLUP $\hat{\eta}$ of η.

Step 1. From s, calculate estimators $\hat{\sigma}_u^2$, $\hat{\sigma}_e^2$ and $\hat{\boldsymbol{\beta}}$ of σ_u^2, σ_e^2 and $\boldsymbol{\beta}$, respectively.

Step 2. Generate the vector of random effects and of random errors with multivariate normal distributions, i.e. $\boldsymbol{u}^* \sim N_D(\boldsymbol{0}_D, \hat{\sigma}_u^2 \boldsymbol{I}_D)$ and $\boldsymbol{e}^* \sim N_N(\boldsymbol{0}_N, \hat{\sigma}_e^2 \boldsymbol{W}^{-1})$.

Step 3. With the known incidence matrices $\boldsymbol{X}$ and $\boldsymbol{Z}$ and heteroscedasticity matrix $\boldsymbol{W}$, construct the bootstrap superpopulation model

$$\xi^*: \ \boldsymbol{y}^* = \boldsymbol{X}\hat{\boldsymbol{\beta}} + \boldsymbol{Z}\boldsymbol{u}^* + \boldsymbol{e}^*, \tag{8.4}$$

where $\boldsymbol{y}^*$ is a bootstrap population vector.

For this vector, $\eta^* = \eta(\boldsymbol{y}^*)$ is defined by analogy to $\eta = \eta(\boldsymbol{y})$, but as function of the bootstrap vector $\boldsymbol{y}^*$. Let $\hat{\eta}^*$ be the EBLUP of η^* constructed from $(\boldsymbol{y}_s^*, \boldsymbol{X}_s)$ in the same way as $\hat{\eta}$ was obtained from $(\boldsymbol{y}_s, \boldsymbol{X}_s)$. Under model ξ^*, given the initial sample s, the mean squared error of $\hat{\eta}^*$ is denoted by $MSE_*(\hat{\eta}^*)$. Thus, for estimating the MSE of $\hat{\eta}$, the parametric bootstrap estimator of the mean squared error is $\text{MSE}_*(\hat{\eta}) = MSE_*(\hat{\eta}^*)$. In practice, this estimator is approximated via Monte Carlo in the following way.

Step 4. Given the bootstrap superpopulation model ξ^*, generate independent and identically distributed bootstrap populations vectors $\boldsymbol{y}^{*(b)}$, $b = 1, \ldots, B$, of size N,

$$\boldsymbol{y}^{*(b)} = \boldsymbol{X}\hat{\boldsymbol{\beta}} + \boldsymbol{Z}\boldsymbol{u}^{*(b)} + \boldsymbol{e}^{*(b)},$$

and calculate bootstrap population parameters $\eta^{*(b)} = \eta(\boldsymbol{y}^{*(b)})$, $b = 1, \ldots, B$.

Step 5. From each vector $\boldsymbol{y}^{*(b)}$ generated in previous step, take the subvector $\boldsymbol{y}_s^{*(b)}$ with the same subset of indexes $s \subset U$ as in the initial sample.

Step 6. Calculate the bootstrap EBLUP $\hat{\eta}^{*(b)} = \hat{\eta}^*(\boldsymbol{y}_s^{*(b)}, \boldsymbol{X}_s)$ of $\eta^{*(b)}$. The Monte Carlo approximation of the bootstrap estimator $MSE_*(\hat{\eta}^*)$ is given by

$$mse_*(\hat{\eta}^*) = B^{-1} \sum_{b=1}^{B} (\hat{\eta}^{*(b)} - \eta^{*(b)})^2. \tag{8.5}$$

8.6 Model-Assisted Estimation

The design-based approach relies on the sampling distribution. Let $s \subset U$ be a random sample of fixed size n drawn according to a specified sampling design $\pi(s)$ and let $s_1, \ldots, s_D$ be the corresponding domain subsamples of sizes $n_1, \ldots, n_D$. The first and second order inclusion probabilities are the probabilities of obtaining the unit j and the units j and k of domain d, respectively, while sampling from the population according to the sampling design. They are $\pi_{dj} = \sum_{s:j \in s_d} \pi(s)$ and $\pi_{djk} = \sum_{s:j,k \in s_d} \pi(s)$, respectively.

Horvitz and Thompson (1952) introduced a design-unbiased estimator of $\overline{Y}_d$. The Horvitz–Thompson (HT) estimator is

$$\hat{\overline{Y}}_d^{dir1} = N_d^{-1} \sum_{j \in s_d} \pi_{dj}^{-1} y_{dj}.$$

The HT estimator possesses properties of a good estimator (is admissible within the class of all unbiased estimators) but makes no use of auxiliary information. Hájek (1971) proposed the ratio estimator

$$\hat{\overline{Y}}_d^{dir2} = \frac{1}{\hat{N}_d^{dir}} \sum_{j\in s_d} \pi_{dj}^{-1} y_{dj}, \quad \hat{N}_d^{dir} = \sum_{j\in s_d} \pi_{dj}^{-1},$$

and Särndal et al. (1992, p. 182) gave several reasons for regarding the Hájek estimator as usually better than the HT estimator. However, it is common to assume that there exist some auxiliary variables (to be obtained from census results, administrative files, etc.) that can be essential for efficient estimation of domain means. The model-assisted estimators incorporate the auxiliary information by using models and preserving good design-based properties. We follow Cassel et al. (1977) and introduce a design-unbiased estimator of the domain mean $\overline{Y}_d$ assisted by a NER model. The estimator is obtained by summing up the predicted values $\hat{y}_{dj}^{eblup} = \boldsymbol{x}_{dj}\hat{\boldsymbol{\beta}} + \hat{u}_d$, $d = 1, \ldots, D$, $j = 1, \ldots, N_d$, and adjusting by residuals, i.e.

$$\begin{aligned} \hat{\overline{Y}}_d^{ma} &= \frac{1}{N_d}\Big\{ \sum_{j\in U_d} (\boldsymbol{x}_{dj}\hat{\boldsymbol{\beta}} + \hat{u}_d) + \sum_{j\in s_d} \pi_{dj}^{-1}(y_{dj} - \boldsymbol{x}_{dj}\hat{\boldsymbol{\beta}} - \hat{u}_d)\Big\} \\ &= \overline{\boldsymbol{X}}_d\hat{\boldsymbol{\beta}} + \hat{u}_d + \frac{\hat{N}_d^{dir}}{N_d}\big(\hat{\overline{Y}}_d^{dir2} - \hat{\overline{\boldsymbol{X}}}_d^{dir2}\hat{\boldsymbol{\beta}} - \hat{u}_d\big), \quad \hat{\overline{\boldsymbol{X}}}_d^{dir2} = \frac{1}{\hat{N}_d^{dir}} \sum_{j\in s_d} \pi_{dj}^{-1}\boldsymbol{x}_{dj}. \end{aligned}$$

Morales et al. (2018) applied the model-assisted estimator $\hat{\overline{Y}}_d^{ma}$ to the estimation of poverty proportions in counties of the region of Valencia (East of Spain). They employed data from the Spanish living condition survey of 2013.

8.7 Simulation Experiment

This section summarizes the design-based Monte Carlo simulation experiment carried out by Santamaría et al. (2004). The experiment was designed for investigating the behavior of classical estimators and BLUPs. First, we briefly describe the artificial population that Santamaría et al. (2004) generated for simulating a stratified random sampling design. Second, we introduce the estimators and the performance measures. Third, we describe the simulation experiment. Finally, we present the numerical results and we give some conclusions.

8.7.1 Artificial Population

The artificial population is a data file with 11 variables and 300,000 records. Each record represents a household of an imaginary country. The file is generated with the purpose of simulating surveys on income and living conditions. The data file is described in Table 8.1.

The target variable y is generated from the model

$$y_{hd_6gj} = u_{d_6} + A_h + B_g + \beta_1 x_{1hd_6gj} + \beta_2 x_{2hd_6gj} + e_{hd_6gj}, \tag{8.6}$$

where $h = 1, \dots, 6,\ d_6 = 1, \dots, 256,\ g = 1, \dots, 4$, and $j = 1, \dots, N_{hd_6g}$ denote stratum, zone, socioeconomic group, and household, respectively. Further, u_{d_6} and e_{hd_6gj} are the zone random effects and the random errors that are independent with distributions $N(0, \sigma_u^2)$ and $N(0, \sigma_e^2)$, respectively. The model parameters are $\beta_1 = 500,\ \beta_2 = 25,\ \sigma_u^2 = 1000,\ \sigma_e^2 = 750,\ A_h = 4000 + 300h,\ h = 1, \dots, 6$, and $B_g = 5000 + 500g,\ g = 1, \dots, 4$.

The target population parameters are

$$\overline{Y}_{d_\ell} = \frac{1}{N_{d_\ell}} \sum_{h=1}^{6} \sum_{g=1}^{4} \sum_{j=1}^{N_{hd_\ell g}} y_{hd_\ell gj}, \quad \ell = 1, \dots, 6,$$

where $d_\ell = 1, \dots, D_\ell$ is the index of the ℓ-th geographic aggregation level, $\ell = 1, \dots, 6$.

Table 8.1 Description of the artificial population

Variable	Description	Values
Geographical characteristics		
dom01	Region (d_1)	1–8
dom02	Community (d_2)	01–16
dom03	Province (d_3)	01–32
dom04	County (d_4)	01–64
dom05	District (d_5)	01–128
dom06	Zone (d_6)	01–256
strat	Stratum (h)	1–6
Household characteristics		
x_1	Number of members	01–30
x_2	Dwelling area in m^2	0000–9999
x_3	Socioeconomic condition group (g)	1–4
y	Annual net monetary income	00000–99999

8.7.2 Estimators and Performance Measures

Let us denote the sampling weights by $\omega_{dj} = \pi_{dj}^{-1}$, $d = 1, \ldots, D$, $j = 1, \ldots, n_d$. The simulation considers the following estimators:

Estimator 1 Direct estimator.

$$\hat{\overline{Y}}_d^{dir2} = \frac{\sum_{j\in s_d} \omega_{dj} y_j}{\hat{N}_d^{dir}}, \quad \text{with } \hat{N}_d^{dir} = \sum_{j\in s_d} \omega_{dj}.$$

Estimator 2 Post-stratified estimator.

$$\hat{\overline{Y}}_d^{psst} = \frac{1}{N_d}\sum_{g=1}^{G} N_{dg}\hat{\overline{Y}}_{dg}^{dir2} = \frac{1}{N_d}\sum_{g=1}^{G} N_{dg}\left(\frac{\sum_{j\in s_{dg}} \omega_{dj} y_j}{\sum_{j\in s_{dg}} \omega_{dj}}\right).$$

Estimator 3 Basic synthetic estimator.

$$\hat{\overline{Y}}_d^{synt} = \frac{1}{N_d}\sum_{g=1}^{G} N_{dg}\hat{\overline{Y}}_{g}^{dir2} = \frac{1}{N_d}\sum_{g=1}^{G} N_{dg}\left(\frac{\sum_{j\in s_{g}} \omega_{dj} y_j}{\sum_{j\in s_{g}} \omega_{dj}}\right).$$

Estimator 4 Sampling size dependent composite estimator.

$$\hat{\overline{Y}}_d^{ssd} = \gamma_d \hat{\overline{Y}}_d^{dir2} + (1-\gamma_d)\hat{\overline{Y}}_d^{synt}, \quad \text{where} \quad \gamma_d = \begin{cases} 1 & \text{if } \hat{N}_d^{dir2} \ge N_d, \\ \frac{\hat{N}_d^{dir2}}{N_d} & \text{otherwise.} \end{cases}$$

Estimator 5 Projective estimator based on the fixed effect linear model

$$y_{dj} = u_d + \beta x_{1dj} + e_{dj}, \; d = 1, \ldots, D, \; j = 1, \ldots, n_d,$$

where the random errors e_{dj} are independent with distributions $N(0, \omega_{dj}^{-1}\sigma_e^2)$. Note that sampling weights are here the model heteroscedasticity weights. The estimator is

$$\hat{\overline{Y}}_d^{regd} = \hat{\overline{Y}}_d^{w} + \hat{\beta}(\overline{X}_{1d} - \hat{\overline{X}}_{1d}^{dir2}), \quad \hat{\beta} = \frac{\sum_{d=1}^{D}\sum_{j=1}^{n_d} \omega_{dj}\big(y_{dj} - \hat{\overline{Y}}_d^{dir2}\big)\big(x_{1dj} - \hat{\overline{X}}_{1d}^{dir2}\big)}{\sum_{d=1}^{D}\sum_{j=1}^{n_d} \omega_{dj}\big(x_{1dj} - \hat{\overline{X}}_{1d}^{dir2}\big)^2}.$$

Estimator 6 EBLUP based on the linear mixed model

$$y_{dj} = u_d + \beta x_{1dj} + e_{dj}, \; d = 1, \ldots, D, \; j = 1, \ldots, n_d, \tag{8.7}$$

where the u_d's and the e_{dj}'s are independent with distributions $N(0, \sigma_u^2)$ and $N(0, \omega_{dj}^{-1}\sigma_e^2)$, respectively. The simulation estimates the variance components by the Henderson 3 method.

Estimator 7 EBLUP based on model (8.7), but the model parameters are estimated by the ML method with the Fisher-scoring algorithm.

Estimator 8 EBLUP based on the nested error regression model

$$y_{dj} = u_d + \beta_1 x_{1dj} + \beta_2 x_{2dj} + e_{dj}, \ d = 1, \dots, D, \ j = 1, \dots, n_d,$$

where the u_d's and the e_{dj}'s are independent with distributions $N(0, \sigma_u^2)$ and $N(0, \omega_{dj}^{-1}\sigma_e^2)$, respectively. The simulation estimates the model parameters by the ML method with the Fisher-scoring algorithm.

Estimator 9 EBLUP based on the nested error regression model

$$y_{dgj} = u_d + \gamma_g + \beta_1 x_{1dgj} + \beta_2 x_{2dgj} + e_{dgj},$$

where the u_d's and the e_{dj}'s are independent with distributions $N(0, \sigma_u^2)$ and $N(0, \omega_{dj}^{-1}\sigma_e^2)$, respectively. The simulation estimates the model parameters by the ML method with the Fisher-scoring algorithm. The parameters are γ_g, $g = 1, \dots, G-1$ ($\gamma_G = 0$), for socioeconomic groups, and β_1, β_2 for the covariates x_1 and x_2.

For comparing the estimators of $\overline{Y}_d$, the simulation extracts K samples from the artificial population. Let $\hat{\overline{Y}}_d^{(k)}$ be the estimator of $\overline{Y}_d$ in the sample k. The performance measures are

1. The average of the relative percentages of bias across the domains, i.e.

$$BIAS = \frac{1}{D}\sum_{d=1}^{D} |BIAS_d|, \quad \text{with} \quad BIAS_d = \frac{100}{K}\sum_{k=1}^{K} \frac{\hat{\overline{Y}}_d^{(k)}}{\overline{Y}_d}.$$

2. The average of the relative percentages of root mean squared error across the domains, i.e.

$$RMSE = \frac{1}{D}\sum_{d=1}^{D} RMSE_d, \quad \text{with} \quad RMSE_d = \frac{100}{\overline{Y}_d}\sqrt{\frac{1}{K}\sum_{k=1}^{K}\left(\hat{\overline{Y}}_d^{(k)} - \overline{Y}_d\right)^2}.$$

8.7.3 Numerical Results and Conclusions

The simulation extracts $K = 10{,}000$ samples of size 600. The sampling design extracts independent simple random samples with replacement within each stratum.

Table 8.2 Sizes and weights of strata

Stratum	1	2	3	4	5	6
N_h	49,828	52,717	48,051	48,865	48,831	51,708
n_h	100	105	96	98	98	103
w_h	498.28	502.07	500.53	498.62	498.28	502.02

Table 8.3 Relative percentages of bias $BIAS$

Estimator	dom01	dom02	dom03	dom04	dom05	dom06	Mean
1	99.998	100.005	100.026	100.001	100.001	100.145	100.029
2	99.905	99.007	96.326	90.943	77.544	55.690	86.569
3	100.012	100.037	100.027	100.143	100.337	100.704	100.210
4	100.012	100.037	100.026	100.020	100.062	100.219	100.063
5	100.002	99.999	100.002	100.005	100.004	100.142	100.026
6	100.000	99.994	99.992	99.972	98.890	90.194	98.174
7	99.984	99.975	99.938	99.869	98.740	90.172	98.113
8	99.977	99.960	99.885	99.601	99.681	99.827	99.822
9	100.001	100.026	100.020	100.046	100.112	100.264	100.078

Table 8.4 Relative percentages of root mean squared errors $RMSE$

Estimator	dom01	dom02	dom03	dom04	dom05	dom06	Mean
9	0.913	1.240	1.642	2.220	2.977	3.842	2.139
8	1.067	1.503	2.132	2.538	3.406	4.449	2.516
5	1.107	1.558	2.211	3.074	4.246	5.087	2.880
4	1.801	2.148	2.693	3.267	4.319	5.575	3.300
3	1.745	2.101	2.659	3.324	4.572	6.065	3.411
1	1.320	1.873	2.697	3.805	5.420	6.687	3.634
7	1.110	1.561	2.205	3.214	9.860	27.697	7.608
6	1.107	1.558	2.211	3.218	9.868	28.056	7.669
2	1.505	3.219	6.502	13.961	31.099	53.112	18.233

Table 8.2 presents the sampling weights (elevation factors). The sampling weights of all units of stratum h are $\omega_h = N_h/n_h$, where n_h and N_h are the sample and population sizes, respectively. Note that the sampling weights appear in the definition of the models for estimators 5–9.

In the first simulation experiment, the geographical variable dom01 defines the domains. Nevertheless, the simulations are repeated for domains defined by dom02–dom06. Each replication extracts a sample of size 600 with the cited stratified random sampling design and calculates the estimators and compares them with the target parameter $\overline{Y}_d$. After $K = 10\,000$ replications, the simulation calculates the performance measures.

Table 8.3 presents the values of $BIAS$. Table 8.4 presents the values of $RMSE$. The first column of Tables 8.3 and 8.4 gives the estimator, columns 2–7 give the performance measure for the different types of domain, and column 8 gives the

Table 8.5 Average domain sample sizes

	dom01	dom02	dom03	dom04	dom05	dom06
D	8	16	32	64	128	256
$600/D$	75	37.5	18.75	9.375	4.6875	2.34375

average of columns 2–7. Table 8.4 sorts the rows in accordance with the average performance value appearing in column 8.

To clarify the effect of the sample size ($n = 600$) and the number of domains (D) in the analysis of the simulation results, Table 8.5 presents the ratios n/D for dom01–dom06. Note that sample sizes for dom01–dom03 are basically large and sample sizes for dom04–dom06 are small. Therefore, the most interesting results correspond to domains dom04–dom06.

The estimator 1 is a direct estimator that only uses the domain sample data of the target variable. The estimator 2 should be used only in the case where the ratio n/D is large, i.e. for domains dom01 and dom02. Note that the samples s_{dg} often remain empty (or almost empty) when the sample size is small. The estimator 3 is quite stable with respect to n/D. As the variability of y cannot be explained only with the socioeconomic groups g, the estimator 3 has moderate results. The estimator 4, that is a composition of estimators 1 and 3, can be recommended. It is a good alternative to the case where it is not easy to find a model that properly fits to the data. The estimator 5 is preferable to estimators 6 and 7. Note that the estimator 5 is based on a model with fixed effects in the domains; however, the estimators 6 and 7 are based on the same model but with random effects. The simulation suggests that the estimators based on linear models with random effects in the domains are more sensitive to type of fitting than those based on the models with fixed effects. It is interesting that ML Fisher-scoring algorithm is slightly preferable to Henderson 3 method. The estimators 8 and 9 are based on models that are quite close to the true model. Thus they present the best performance.

For model-based estimators, the best numerical results are obtained for those models with a good fit to the data. A bad model produces a bad estimator. Moreover, if no good model is available, it might be preferable to use models with fixed effects. Otherwise, random effects models tend to perform better.

8.8 R Codes

8.8.1 EBLUPs for LFS Data

This section gives R codes for calculating the EBLUPs of domain means by using data from the survey data file `LFS20.txt`. It employs functions of the R package `sae` described by Molina and Marhuenda (2015). The domains of interest are the areas crossed by sex. The target variable y is INCOME and the auxiliary variables x_1, x_2, and x_3 are REGISTERED, EDUCATION2, and EDUCATION3, respectively. The following code loads the R package sae. It also reads the data file and the file with the aggregated auxiliary variables.

```
if(!require(sae)){
  install.packages("sae")
  library(sae)
}
dataux <- read.table("Nds20.txt", header=TRUE, sep="\t", dec=".")
dat <- read.table("LFS20.txt", header=TRUE, sep="\t", dec=".")
dat$AREASEX <- paste(dat$AREA, dat$SEX, sep="s") # Domains
n <- dim(dat)[1]                           # Global sample size
dat <- dat[order(dat$AREA, dat$SEX),]      # Sort dat by area and sex
narea <- length(unique(dataux$area))       # Number of areas
nsex <- length(unique(dataux$sex))         # Number of sex categories
```

We rename some variables.

```
y <- dat$INCOME; x1 <- dat$REGISTERED; x2 <- dat$EDUCATION
area <- dat$AREA; sex<-dat$SEX; AREASEX <- dat$AREASEX
edu2 <- as.numeric(x2==2)       # Category 2 of EDUCATION
edu3 <- as.numeric(x2==3)       # Category 3 of EDUCATION
```

We construct a data frame with the domain means of the auxiliary variables. We further build a data frame with domain sizes. In both cases, the first column contains the domain index.

```
areasex <- paste(dataux$area, dataux$sex, sep="s")
Xmean <- data.frame(areasex, reg=dataux$reg/dataux$N,
                    edu2=dataux$edu2/dataux$N,
                    edu3=dataux$edu3/dataux$N)
Popn <- data.frame(areasex, N=dataux$N)
```

We fit a NER model and we calculate the EBLUPs of domain-sex average incomes.

```
result <- eblupBHF(y~x1+edu2+edu3, dom=dat$AREASEX, meanxpop=Xmean,
                   popnsize=Popn, method="ML")
result$eblup[1:10,]             # Domains, EBLUPs and sample sizes
eblup <- result$eblup[,2]       # EBLUPS
# Additional results
result$fit$summary
# Regression parameters of the fitted NER models
result$fit$fixed
result$fit$random               # Domain random effects
result$fit$errorvar             # Error variance
result$fit$refvar               # Random effect variance
result$fit$loglike              # Log-likelihood estimated value
```

Table 8.6 presents the estimated model parameters and p-values. The AREASEX random intercept and the residual standard deviations are $\sigma_u = 482.93$ and $\sigma_e = 9875.89$, respectively.

Table 8.6 Estimated parameters of NER model

Parameter	Estimate	Std. error	t-value	p-value
Intercept	40,172.36	494.15	81.30	0.00
Registered	−11,643.58	1123.83	−10.36	0.00
edu2	9703.05	675.88	14.36	0.00
edu3	20,079.15	874.53	22.96	0.00

Table 8.7 EBLUPS and MSEs for sex = 1 (left) and sex = 2 (right)

Area	eblup	mse	cv	eblup	mse	cv
1	46,895.50	379,261.39	1.31	45,978.28	406,093.26	1.39
2	43,265.75	396,859.37	1.46	44,687.09	404,172.20	1.42
3	49,815.35	427,493.97	1.31	46,747.19	421,587.33	1.39
4	45,119.70	378,843.49	1.36	48,690.60	317,123.54	1.16
5	46,578.48	445,548.41	1.43	47,340.73	337,781.17	1.23
6	45,995.08	417,151.94	1.40	44,114.57	429,116.39	1.48
7	47,124.39	398,708.53	1.34	47,143.04	378,006.77	1.30
8	43,046.64	458,818.78	1.57	45,347.07	397,512.16	1.39
9	48,025.49	375,496.08	1.28	46,523.04	348,388.16	1.27
10	45,944.37	363,180.12	1.31	44,549.06	380,222.06	1.38

The following code calculates the parametric bootstrap MSEs of the EBLUPs of domain-sex average incomes.

```
set.seed(123)
msey <- pbmseBHF(y~x1+edu2+edu3, dom=dat$AREASEX, meanxpop=Xmean,
                 popnsize=Popn, B=500, method="ML")
msey$mse                                       # Domains and MSEs
mseEBLUP <- msey$mse[,2]                       # MSEs
cvEBLUP <- round(100*sqrt(mseEBLUP)/eblup,2)   # Coefficients of variation
```

Summary of results

```
output1 <- data.frame(area=result$eblup[1:10,1], eblup=eblup, mse=msey$mse[,2],
                      cv=cvEBLUP)
output1
```

As the bootstrap procedure has had too many warning messages, the MSE calculations might not be fully reliable. Some caution should be taken when publishing the MSE results. For the ten first areas, Table 8.7 gives the EBLUPs, the MSEs, and the coefficients of variations for sex=1 (left) and sex=2 (right).

8.8.2 EBLUPs and MA Estimators for LCS Data

This section gives R codes for calculating the direct estimator, the model-assisted estimator, and the EBLUP of domain means by using data from the survey data file `datLCS.txt`. It employs functions of the R package `sae` described by Molina and Marhuenda (2015). The domain variable is denoted by dom. The target variable y is income, the sampling weight is w and the dichotomic auxiliary variables, x_1 and x_2, are employed and unemployed. The R function `direct` defined in Sect. 2.8.4 is needed.

The following code loads the package `sae` and reads the sample file and the file containing the aggregated auxiliary variables.

```
if(!require(sae)){
  install.packages("sae")
  library(sae)
}
aux <- read.table("auxLCS.txt", header=TRUE, sep = "\t", dec = ",")
dat <- read.table("datLCS.txt", header=TRUE, sep = "\t", dec = ",")
aux <- aux[order(aux$dom),]                    # sort the file by domains
D <- nrow(aux)                                 # number of domains
```

The variable `lab` gives the labor situation, with values 0: $\leq$ 15 years, 1: working, 2: not working, 3: inactive. We define some variables.

```
income <- dat$income; w <- dat$w; dom <- dat$dom
work <- as.numeric(dat$lab==1)
nowork <- as.numeric(dat$lab==2)
inact <- as.numeric(dat$lab==3)
```

We calculate the direct Hájek estimators of domain average incomes.

```
income.dir <- dir2(data=income, w, domain=list(dom=dom))
diry <- income.dir$mean; hatNd <- income.dir$Nd.hat; nd <- income.dir$nd
```

We calculate the direct estimators of proportions of employed and unemployed people.

```
work.dir <- dir2(data=work, w, domain=list(dom=dom))
dirwork <- work.dir$mean
nowork.dir <- dir2(data=nowork, w, domain=list(dom=dom))
dirnowork <- nowork.dir$mean
```

We calculate domain sizes and x-means.

```
Nd <- aux$TOT                                   # domain sizes
round(hatNd[1:10],0)                            # estimated domain sizes
# Dataframes needed for applying the function eblupBHF of the library sae
Xmean <- data.frame(dom=aux$dom, work=aux$Mwork, nowork=aux$Mnowork)
PopN <- data.frame(dom=aux$dom, Nd)
```

We fit a NER model and we calculate the EBLUPs of income means by domains. This is done by applying the R function `eblupBHF`.

```
EB <- eblupBHF(income~work+nowork, dom=dom, meanxpop=Xmean, popnsize=PopN,
               method="REML", data=dat)
anova <- EB$fit$summary                         # ANOVA Table
2*pnorm(abs(anova$coefficients[,3]), low=F)     # p-values
beta <- EB$fit$fixed                            # betas
sigmae2 <- EB$fit$errorvar                      # Error variance
sigmau2 <- EB$fit$refvar                        # Random effect variance
head(EB$fit$random, 10)                         # Random effects
Residuals <- EB$fit$residuals                   # Residuals
EBLUP <- EB$eblup[order(EB$eblup$domain),]      # Sort EB$EBLUP by dom
# Inserts EB$fit$random in EBLUP
EBLUP$random <- EB$fit$random[,1]
eblup <- EBLUP$eblup                            # EBLUPs sorted by domains
# Random effects sorted by domains
ud <- EBLUP$random
```

Table 8.8 presents the estimated model parameters and p-values. The domain random intercept and the residual standard deviations are $\sigma_u = 2140.91$ and $\sigma_e = 8801.08$, respectively.

Table 8.8 Estimated parameters of NER model

Parameter	Estimate	Std. error	t-value	p-value
Intercept	13,226.38	508.55	26.008	0.00
Work	3551.72	389.75	9.113	0.00
No work	−2188.06	519.68	−4.210	0.00

We calculate the parametric bootstrap MSE of the EBLUP. For this sake, we apply the R function `pbmseBHF`.

```
set.seed(123)
MSEeblup <- pbmseBHF(income~work+nowork, dom=dom, meanxpop=Xmean,
                     popnsize=PopN, B=500, method="REML", data=dat)
# Sort MSEeblup$mse by dom
EBMSE <- MSEeblup$mse[order(MSEeblup$mse$domain),]
ebmse <- EBMSE$mse                              # Model-based MSEs of EBLUPS
mCVeb <- round(100*sqrt(ebmse)/eblup,2)         # Model-based CVs of EBLUPS
```

We calculate the model-assisted counterpart of the EBLUP.

```
s1 <- beta[1] + beta[2]*Xmean$work + beta[3]*Xmean$nowork + ud; s1
s2 <- hatNd*(diry-beta[1]-beta[2]*dirwork-beta[3]*dirnowork-ud)/Nd; s2
MA <- s1+s2
```

We give a summary of the obtained results

```
output2 <- data.frame(dom=aux$dom, Nd, hatNd=round(hatNd,0), nd,
                      DIR=round(diry,0), MA=round(MA,0), EB=round(eblup,0),
                      mMSEeb=round(ebmse,0), mCVeb)
head(output2, 10)
```

For the ten first areas, Table 8.9 gives the direct (DIR), model-assisted (MA), and EBLUP (EB) estimators of domain income means. It also gives the parametric bootstrap estimators of the MSEs and the CVs of the EBLUPs.

Table 8.9 shows that the estimated MSEs of the EBLUPs tend to be smaller in those domains with larger sample sizes.

Table 8.9 DIR, MA and EB estimators of income domain means

d	N_d	$\hat{N}_d$	n_d	DIR	MA	EB	mMSEeb	mCVeb
3	82, 001	123, 590	57	8361	7639	10,662	1, 042, 121	9.57
5	251, 866	203, 416	96	13,334	13,511	16,666	696, 889	5.01
6	190, 653	168, 144	82	15,869	15,677	15,572	812, 825	5.79
7	24, 699	14, 369	10	13,245	13,857	13,964	2, 626, 551	11.61
11	154, 625	154, 801	118	11,662	11,787	12,579	616, 950	6.24
12	90, 315	35, 275	18	16,785	15,838	15,688	2, 347, 230	9.77
13	223, 742	202, 894	138	16,057	15,803	14,455	489, 786	4.84
14	346, 216	404, 616	190	13,370	13,560	14,477	415, 288	4.45
15	779, 492	839, 092	406	15,211	15,353	16,001	174, 299	2.61
16	172, 916	237, 995	93	14,531	13,420	16,640	687, 910	4.98
17	39, 723	20, 118	12	18,238	17,371	15,584	2, 876, 743	10.88
18	43, 082	50, 511	35	21,590	21,946	17,596	1, 437, 648	6.81
20	220, 284	236, 153	125	17,224	17,328	15,970	607, 043	4.88
21	80, 685	108, 153	49	11,340	11,730	12,293	1, 209, 800	8.95
22	16, 804	23, 063	13	10,174	9148	12,351	2, 632, 350	13.14
23	72, 357	68, 308	40	10,109	10,973	11,779	1, 443, 794	10.2
24	89, 858	126, 016	65	13,600	13,366	13,920	874, 731	6.72
25	175, 305	124, 844	79	12,238	12,468	12,539	753, 276	6.92
27	110, 543	152, 125	82	11,738	11,261	12,576	749, 037	6.88
28	52, 874	98, 278	57	12,471	11,892	13,021	1, 176, 872	8.33
29	170, 015	139, 770	69	15,825	15,021	15,153	834, 316	6.03
30	178, 329	274, 619	135	10,872	10,922	12,857	495, 003	5.47
31	183, 524	115, 214	58	12,405	12,991	13,591	847, 049	6.77
32	474, 950	502, 865	293	16,821	16,918	16,791	257, 352	3.02
33	287, 071	335, 968	132	9478	9590	11,034	573, 620	6.86
34	365, 059	170, 242	60	15,640	15,110	14,464	1, 062, 634	7.13

References

Cassel, C., Särndal, C.E., Wretman, J.: Foundations of Inference in Survey Sampling. Wiley, New York (1977)

González-Manteiga, W., Lombardía, M. J., Molina, I., Morales, D., Santamaría, L.: Bootstrap mean squared error of small-area EBLUP. J. Stat. Comput. Simul., **78**, 443–462. (2008a).

Hájek, J.: Comment on "An essay on the logical foundations of survey sampling, part one". In: Godambe, V.P., Sprott, D.A. (eds.) The Foundations of Survey Sampling, 236 pp. Holt, Rinehart, and Winston (1971)

Hall, P., Maiti, T.: Nonparametric estimation of mean-squared prediction error in nested-error regression models. Ann. Stat. **34**(4), 1733–1750 (2006a)

Hall, P., Maiti, T.: On parametric bootstrap methods for small-area prediction. J. R. Stat. Soc. B. **68**, 221–238 (2006b)

Horvitz, D.G., Thompson, D.J.: A generalization of sampling without replacement from a finite universe. J. Am. Stat. Assoc. **47**, 663–685 (1952)

Jiang, J., Lahiri, P., Wan, S.: A unified jackknife theory for empirical best prediction with M-estimation. Ann. Stat. **30**, 1782–1810 (2002)

Molina, I., Marhuenda, Y.: sae: An R package for small area estimation. R J. **7**, 81–98 (2015)
Morales, D., Rueda, M.M., Esteban, M.D.: Model-assisted estimation of small area poverty measures: an application within the Valencia Region in Spain. Soc. Indic. Res. **138**(3), 873–900 (2018)
Pfefferman, D., Tiller, R.: Bootstrap approximation to prediction MSE for state-space models with estimated parameters. J. Time Ser. Anal. **26**, 893–916 (2005)
Santamaría, L., Morales, D., Molina, I.: A comparative study of small area estimators of average household incomes. SORT Stat. Oper. Res. Trans. **28**(2), 215–229 (2004)
Särndal, C.E., Swensson, B., Wretman, J.: Model Assisted Survey Sampling. Springer, New York (1992)

Chapter 9
Mean Squared Error of EBLUPs

9.1 Introduction

The General Prediction Theorem 4.1 gives the mean squared error (MSE) of the best linear unbiased predictor (BLUP) of a population linear parameter. The BLUP depends on some unknown parameters, typically, variances and/or correlations. When those parameters are replaced by suitable estimators, then the resulting predictor is the empirical BLUP (EBLUP). However, the exact MSE of the EBLUP has not been analytically derived. For this reason, some approximations have appeared in the literature.

The first simplification of the MSE was obtained by Kackar and Harville (1981), assuming normality of the errors and the random effects. In the second paper Kackar and Harville (1984) gave an approximation of the mentioned MSE and proposed an estimator based on it.

Prasad and Rao (1990) gave a new approximation for models with block-diagonal covariance matrices. Under certain regularity assumptions for the model and the estimators of variance components, they showed that when the number of blocks D tends to infinity, their approximation is of order $o(D^{-1})$. They proposed an estimator of the MSE and gave the specific expressions of this estimator for Fay–Herriot, nested error, and random coefficient models. The conditions imposed on the estimators of the variance components are satisfied by estimators obtained by the fitting constant method, also called Henderson method 3 (see Searle et al. 1992), but they are not fulfilled by the maximum likelihood (ML) estimators.

Datta and Lahiri (2000) provided the analog MSE estimator for general models with block-diagonal covariance matrices, when variance components are estimated by the ML or the residual maximum likelihood (REML) methods. Das et al. (2004) studied the approximation of the MSE for a wider class of models, including ANOVA and longitudinal random effects models, when ML or REML estimators of the variance components are employed.

D. Morales et al., *A Course on Small Area Estimation and Mixed Models*, Statistics for Social and Behavioral Sciences, https://doi.org/10.1007/978-3-030-63757-6_9

Resampling represents a solution when estimators cannot be obtained by traditional methods. But even when such estimators are available, by using resampling techniques like bootstrap or jackknife, it is possible to obtain more accurate alternatives. Concerning the estimation of the mean squared error of empirical predictors, some of these techniques can already be found in the literature. The jackknife methodology proposed by Jiang et al. (2002) succeeds to provide estimators with bias of order $O(D^{-3/2})$. Pfefferman and Tiller (2005) proposed parametric and nonparametric bootstrap methods for estimating the same quantity under state-space models. Hall and Maiti (2006a,b) introduced parametric and matched-moment, double-bootstrap algorithms. González-Manteiga et al. (2008a) applied a parametric bootstrap procedure to linear mixed models. The approach of the last paper is described more in detail in Sect. 8.5.

This chapter treats the problem of approximating and estimating the MSE of the EBLUP under linear mixed models. Section 9.2 deals with the calculation of the MSE of the BLUP and the EBLUP of a linear function of fixed and random effects. Section 9.3 approximates the MSE of the EBLUP of a population linear parameter. Section 9.4 gives an analytic estimator of the MSE. Sections 9.5 and 9.6 particularize the results of Sects. 9.3 and 9.4 to the nested error regression model. Section 9.7 gives similar results for a linear model with no random effects. Section 9.8 presents some simulation results. Finally, Sect. 9.9 gives R codes for calculating the analytic estimators of the MSEs of the EBLUPs under the NER model.

9.2 The MSE of EBLUPs of Model Effects

Let $U = \{1, \ldots, N\}$ be a finite population. Let $\boldsymbol{y} = (y_1, \ldots, y_N)'$ be the vector containing the values of a target variable in the units of U. We assume the model

$$\boldsymbol{y} = \boldsymbol{X\beta} + \boldsymbol{Zu} + \boldsymbol{e}, \tag{9.1}$$

where $\boldsymbol{\beta}_{p\times 1}$ is the vector of fixed effects, $\boldsymbol{u}_{q\times 1}$ is the vector of random effects, $\boldsymbol{X}_{N\times p}$ and $\boldsymbol{Z}_{N\times q}$ are known incidence matrices, and $\boldsymbol{e}_{N\times 1}$ is the vector of random errors. We assume that the random effects and the random errors are independent and normally distributed with zero mean vectors and covariance matrices,

$$\text{var}(\boldsymbol{u}) = E[\boldsymbol{uu}'] = \boldsymbol{V}_u \quad \text{and} \quad \text{var}(\boldsymbol{e}) = E[\boldsymbol{ee}'] = \boldsymbol{V}_e,$$

which are functions (with continuous first partial derivatives) of the parameters $\boldsymbol{\theta} = (\theta_0, \theta_1, \ldots, \theta_m)$ called *variance components*. We assume that $\boldsymbol{V} = \text{var}(\boldsymbol{y})$, $\boldsymbol{V}_u$, and $\boldsymbol{V}_e$ are non-singular. From (9.1), we have

$$\boldsymbol{V} = \boldsymbol{ZV}_u\boldsymbol{Z}' + \boldsymbol{V}_e.$$

Let $s \subset U$ be a sample of $n \leq N$ units. Let $r = U - s$ be the set of units that are not in the sample. Without loss of generality, we can reorder the population, and we can write $\boldsymbol{y} = (\boldsymbol{y}_s', \boldsymbol{y}_r')'$, where $\boldsymbol{y}_s$ is the vector of n observed units and $\boldsymbol{y}_r$ is the vector of $N - n$ non-observed units. In what follows, the index s denotes the observed part of the model (9.1), and the index r denotes the non-observed part.

Let $\boldsymbol{a} = (\boldsymbol{a}_s', \boldsymbol{a}_r')'$ be a $N \times 1$ vector of known constants. We are interested in estimating $\eta = \boldsymbol{a}'\boldsymbol{y} = \boldsymbol{a}_s'\boldsymbol{y}_s + \boldsymbol{a}_r'\boldsymbol{y}_r$. We define $\tau = \boldsymbol{a}_r'(\boldsymbol{X}_r\boldsymbol{\beta} + \boldsymbol{Z}_r\boldsymbol{u})$, so that $\boldsymbol{a}_r'\boldsymbol{y}_r = \tau + \boldsymbol{a}_r'\boldsymbol{e}_r$.

This section calculates the MSE of the BLUP and the EBLUP of τ. We consider the following cases: (1) $\boldsymbol{\beta}$, and $\theta_0, \theta_1, \ldots, \theta_m$ are known, (2) $\theta_0, \theta_1, \ldots, \theta_m$ are known, but $\boldsymbol{\beta}$ is unknown, and (3) all model parameters are unknown.

9.2.1 All Model Parameters Are Known

This section assumes that $\boldsymbol{\beta}$ and $\theta_0, \theta_1, \ldots, \theta_m$ are known. As the BLUP of τ is (cf. Proposition 6.1)

$$\tilde{\tau} = \boldsymbol{a}_r'(\boldsymbol{X}_r\boldsymbol{\beta} + \boldsymbol{Z}_r\tilde{\boldsymbol{u}}), \quad \text{with} \quad \tilde{\boldsymbol{u}} = \boldsymbol{C}_s'\boldsymbol{V}_s^{-1}(\boldsymbol{y}_s - \boldsymbol{X}_s\boldsymbol{\beta})$$

and $\boldsymbol{C}_s = \text{cov}(\boldsymbol{y}_s, \boldsymbol{u}) = \boldsymbol{Z}_s\boldsymbol{V}_u$, the prediction error is $\tilde{\tau} - \tau = \boldsymbol{a}_r'\boldsymbol{Z}_r(\tilde{\boldsymbol{u}} - \boldsymbol{u})$. The MSE is

$$MSE(\tilde{\tau}) = E[(\tilde{\tau} - \tau)^2] = \text{var}(\tilde{\tau} - \tau) = \boldsymbol{a}_r'\boldsymbol{Z}_r\text{var}(\tilde{\boldsymbol{u}} - \boldsymbol{u})\boldsymbol{Z}_r'\boldsymbol{a}_r.$$

On the other hand, we have

$$\begin{aligned}\text{var}(\tilde{\boldsymbol{u}} - \boldsymbol{u}) &= \text{var}(\tilde{\boldsymbol{u}}) + \text{var}(\boldsymbol{u}) - 2\text{cov}(\tilde{\boldsymbol{u}}, \boldsymbol{u}) \\ &= \boldsymbol{C}_s'\boldsymbol{V}_s^{-1}\boldsymbol{V}_s\boldsymbol{V}_s^{-1}\boldsymbol{C}_s + \boldsymbol{V}_u - 2\boldsymbol{C}_s'\boldsymbol{V}_s^{-1}\boldsymbol{C}_s = \boldsymbol{V}_u - \boldsymbol{V}_u\boldsymbol{Z}_s'\boldsymbol{V}_s^{-1}\boldsymbol{Z}_s\boldsymbol{V}_u.\end{aligned}$$

We know that $\boldsymbol{V}_s^{-1} = (\boldsymbol{V}_{es} + \boldsymbol{Z}_s\boldsymbol{V}_u\boldsymbol{Z}_s')^{-1}$. By using the matrix inversion formula

$$(A + BCD)^{-1} = A^{-1} - A^{-1}B(C^{-1} + DA^{-1}B)^{-1}DA^{-1}, \tag{9.2}$$

we get

$$\boldsymbol{V}_s^{-1} = \boldsymbol{V}_{es}^{-1} - \boldsymbol{V}_{es}^{-1}\boldsymbol{Z}_s(\boldsymbol{V}_u^{-1} + \boldsymbol{Z}_s'\boldsymbol{V}_{es}^{-1}\boldsymbol{Z}_s)^{-1}\boldsymbol{Z}_s'\boldsymbol{V}_{es}^{-1}.$$

Let $\boldsymbol{T}_s = (\boldsymbol{V}_u^{-1} + \boldsymbol{Z}_s'\boldsymbol{V}_{es}^{-1}\boldsymbol{Z}_s)^{-1}$, and then we can write $\boldsymbol{V}_s^{-1}$ as a function of $\boldsymbol{T}_s$, i.e.

$$\boldsymbol{V}_s^{-1} = \boldsymbol{V}_{es}^{-1} - \boldsymbol{V}_{es}^{-1}\boldsymbol{Z}_s\boldsymbol{T}_s\boldsymbol{Z}_s'\boldsymbol{V}_{es}^{-1}.$$

In the same way, by applying the formula (9.2) to $\boldsymbol{T}_s$, we obtain

$$\boldsymbol{T}_s = \boldsymbol{V}_u - \boldsymbol{V}_u\boldsymbol{Z}_s'(\boldsymbol{V}_{es} + \boldsymbol{Z}_s\boldsymbol{V}_u\boldsymbol{Z}_s')^{-1}\boldsymbol{Z}_s\boldsymbol{V}_u = \boldsymbol{V}_u - \boldsymbol{V}_u\boldsymbol{Z}_s'\boldsymbol{V}_s^{-1}\boldsymbol{Z}_s\boldsymbol{V}_u.$$

Therefore, we have

$$\mathrm{var}(\tilde{\boldsymbol{u}} - \boldsymbol{u}) = \boldsymbol{T}_s$$

and

$$MSE(\tilde{\tau}) = \boldsymbol{a}_r'\boldsymbol{Z}_r\boldsymbol{T}_s\boldsymbol{Z}_r'\boldsymbol{a}_r \triangleq g_1(\boldsymbol{\theta}).$$

9.2.2 Known Variances and Unknown Regression Parameters

This section assumes that $\theta_0, \theta_1, \ldots, \theta_m$ are known, but $\boldsymbol{\beta}$ is unknown. Let us define $\boldsymbol{Q}_s = (\boldsymbol{X}_s'\boldsymbol{V}_s^{-1}\boldsymbol{X}_s)^{-1}$ and $\boldsymbol{C}_s = \mathrm{cov}(\boldsymbol{y}_s, \boldsymbol{u}) = \boldsymbol{Z}_s\boldsymbol{V}_u$, and then the BLUP of τ is (cf. Proposition 6.1)

$$\hat{\tau}_{blup} = \boldsymbol{a}_r'(\boldsymbol{X}_r\hat{\boldsymbol{\beta}} + \boldsymbol{Z}_r\hat{\boldsymbol{u}}),$$

where

$$\hat{\boldsymbol{u}} = \boldsymbol{C}_s'\boldsymbol{V}_s^{-1}(\boldsymbol{y}_s - \boldsymbol{X}_s\hat{\boldsymbol{\beta}}) \quad \text{and} \quad \hat{\boldsymbol{\beta}} = (\boldsymbol{X}_s'\boldsymbol{V}_s^{-1}\boldsymbol{X}_s)^{-1}\boldsymbol{X}_s'\boldsymbol{V}_s^{-1}\boldsymbol{y}_s = \boldsymbol{Q}_s\boldsymbol{X}_s'\boldsymbol{V}_s^{-1}\boldsymbol{y}_s.$$

Therefore,

$$\hat{\tau}_{blup} - \tau = \boldsymbol{a}_r'\boldsymbol{X}_r(\hat{\boldsymbol{\beta}} - \boldsymbol{\beta}) + \boldsymbol{a}_r'\boldsymbol{Z}_r(\hat{\boldsymbol{u}} - \boldsymbol{u})$$

and

$$\begin{aligned}(\hat{\tau}_{blup} - \tau)^2 &= \left[\boldsymbol{a}_r'\boldsymbol{X}_r(\hat{\boldsymbol{\beta}} - \boldsymbol{\beta}) + \boldsymbol{a}_r'\boldsymbol{Z}_r(\hat{\boldsymbol{u}} - \boldsymbol{u})\right]\left[(\hat{\boldsymbol{\beta}} - \boldsymbol{\beta})'\boldsymbol{X}_r'\boldsymbol{a}_r + (\hat{\boldsymbol{u}} - \boldsymbol{u})'\boldsymbol{Z}_r'\boldsymbol{a}_r\right]\\
&= \boldsymbol{a}_r'\boldsymbol{X}_r(\hat{\boldsymbol{\beta}} - \boldsymbol{\beta})(\hat{\boldsymbol{\beta}} - \boldsymbol{\beta})'\boldsymbol{X}_r'\boldsymbol{a}_r + \boldsymbol{a}_r'\boldsymbol{X}_r(\hat{\boldsymbol{\beta}} - \boldsymbol{\beta})(\hat{\boldsymbol{u}} - \boldsymbol{u})'\boldsymbol{Z}_r'\boldsymbol{a}_r\\
&\quad + \boldsymbol{a}_r'\boldsymbol{Z}_r(\hat{\boldsymbol{u}} - \boldsymbol{u})(\hat{\boldsymbol{\beta}} - \boldsymbol{\beta})'\boldsymbol{X}_r'\boldsymbol{a}_r + \boldsymbol{a}_r'\boldsymbol{Z}_r(\hat{\boldsymbol{u}} - \boldsymbol{u})(\hat{\boldsymbol{u}} - \boldsymbol{u})'\boldsymbol{Z}_r'\boldsymbol{a}_r.\end{aligned}$$

In matrix form, we have

$$(\hat{\tau}_{blup} - \tau)^2 = \left[\boldsymbol{a}_r'\boldsymbol{X}_r,\ \boldsymbol{a}_r'\boldsymbol{Z}_r\right]\begin{bmatrix}\hat{\boldsymbol{\beta}} - \boldsymbol{\beta}\\ \hat{\boldsymbol{u}} - \boldsymbol{u}\end{bmatrix}\left[(\hat{\boldsymbol{\beta}} - \boldsymbol{\beta})',\ (\hat{\boldsymbol{u}} - \boldsymbol{u})'\right]\begin{bmatrix}\boldsymbol{X}_r'\boldsymbol{a}_r\\ \boldsymbol{Z}_r'\boldsymbol{a}_r\end{bmatrix}.$$

Therefore,

$$
\begin{aligned}
MSE(\hat{\tau}_{blup}) &= E\left[(\hat{\tau}_{blup} - \tau)^2\right] \\
&= \left[\boldsymbol{a}_r' \boldsymbol{X}_r,\ \boldsymbol{a}_r' \boldsymbol{Z}_r\right] E\left[\begin{bmatrix} \hat{\boldsymbol{\beta}} - \boldsymbol{\beta} \\ \hat{\boldsymbol{u}} - \boldsymbol{u} \end{bmatrix} \left[(\hat{\boldsymbol{\beta}} - \boldsymbol{\beta})',\ (\hat{\boldsymbol{u}} - \boldsymbol{u})'\right]\right] \begin{bmatrix} \boldsymbol{X}_r' \boldsymbol{a}_r \\ \boldsymbol{Z}_r' \boldsymbol{a}_r \end{bmatrix}.
\end{aligned}
$$

We calculate each block matrix appearing in the formula of $MSE(\hat{\tau}_{blup})$.

The matrix $\boldsymbol{R}_{11} \triangleq E\left[(\hat{\boldsymbol{\beta}} - \boldsymbol{\beta})(\hat{\boldsymbol{\beta}} - \boldsymbol{\beta})'\right]$ is

$$
\boldsymbol{R}_{11} = \operatorname{var}(\hat{\boldsymbol{\beta}}) = \boldsymbol{Q}_s \boldsymbol{X}_s' \boldsymbol{V}_s^{-1} \boldsymbol{V}_s \boldsymbol{V}_s^{-1} \boldsymbol{X}_s \boldsymbol{Q}_s = \boldsymbol{Q}_s \boldsymbol{X}_s' \boldsymbol{V}_s^{-1} \boldsymbol{X}_s \boldsymbol{Q}_s = \boldsymbol{Q}_s \boldsymbol{Q}_s^{-1} \boldsymbol{Q}_s = \boldsymbol{Q}_s.
$$

The matrix $\boldsymbol{R}_{12} \triangleq E\left[(\hat{\boldsymbol{\beta}} - \boldsymbol{\beta})(\hat{\boldsymbol{u}} - \boldsymbol{u})'\right]$ is

$$
\begin{aligned}
\boldsymbol{R}_{12} &= \operatorname{cov}(\hat{\boldsymbol{\beta}}, \hat{\boldsymbol{u}} - \boldsymbol{u}) = \operatorname{cov}(\hat{\boldsymbol{\beta}}, \hat{\boldsymbol{u}}) - \operatorname{cov}(\hat{\boldsymbol{\beta}}, \boldsymbol{u}) \\
&= \operatorname{cov}\big(\boldsymbol{Q}_s \boldsymbol{X}_s' \boldsymbol{V}_s^{-1} \boldsymbol{y}_s, \boldsymbol{C}_s' \boldsymbol{V}_s^{-1} (\boldsymbol{I} - \boldsymbol{X}_s \boldsymbol{Q}_s \boldsymbol{X}_s' \boldsymbol{V}_s^{-1}) \boldsymbol{y}_s\big) - \operatorname{cov}\big(\boldsymbol{Q}_s \boldsymbol{X}_s' \boldsymbol{V}_s^{-1} \boldsymbol{y}_s, \boldsymbol{u}\big) \\
&= \boldsymbol{Q}_s \boldsymbol{X}_s' \boldsymbol{V}_s^{-1} \boldsymbol{V}_s (\boldsymbol{I} - \boldsymbol{V}_s^{-1} \boldsymbol{X}_s \boldsymbol{Q}_s \boldsymbol{X}_s') \boldsymbol{V}_s^{-1} \boldsymbol{C}_s - \boldsymbol{Q}_s \boldsymbol{X}_s' \boldsymbol{V}_s^{-1} \boldsymbol{C}_s \\
&= -\boldsymbol{Q}_s (\boldsymbol{X}_s' \boldsymbol{V}_s^{-1} \boldsymbol{X}_s) \boldsymbol{Q}_s \boldsymbol{X}_s' \boldsymbol{V}_s^{-1} \boldsymbol{C}_s = -\boldsymbol{Q}_s \boldsymbol{X}_s' \boldsymbol{V}_s^{-1} \boldsymbol{Z}_s \boldsymbol{V}_u.
\end{aligned}
$$

Finally, the matrix $\boldsymbol{R}_{22} \triangleq E\left[(\hat{\boldsymbol{u}} - \boldsymbol{u})(\hat{\boldsymbol{u}} - \boldsymbol{u})'\right]$ is

$$
\begin{aligned}
\boldsymbol{R}_{22} &= \operatorname{cov}(\hat{\boldsymbol{u}} - \boldsymbol{u}, \hat{\boldsymbol{u}} - \boldsymbol{u}) = \operatorname{var}(\hat{\boldsymbol{u}}) - \operatorname{cov}(\hat{\boldsymbol{u}}, \boldsymbol{u}) - \operatorname{cov}(\boldsymbol{u}, \hat{\boldsymbol{u}}) + \operatorname{var}(\boldsymbol{u}) \\
&= \boldsymbol{C}_s' \boldsymbol{V}_s^{-1} (\boldsymbol{I} - \boldsymbol{X}_s \boldsymbol{Q}_s \boldsymbol{X}_s' \boldsymbol{V}_s^{-1}) \boldsymbol{V}_s (\boldsymbol{I} - \boldsymbol{V}_s^{-1} \boldsymbol{X}_s \boldsymbol{Q}_s \boldsymbol{X}_s') \boldsymbol{V}_s^{-1} \boldsymbol{C}_s \\
&- \boldsymbol{C}_s' \boldsymbol{V}_s^{-1} (\boldsymbol{I} - \boldsymbol{X}_s \boldsymbol{Q}_s \boldsymbol{X}_s' \boldsymbol{V}_s^{-1}) \boldsymbol{C}_s - \boldsymbol{C}_s' (\boldsymbol{I} - \boldsymbol{V}_s^{-1} \boldsymbol{X}_s \boldsymbol{Q}_s \boldsymbol{X}_s') \boldsymbol{V}_s^{-1} \boldsymbol{C}_s + \boldsymbol{V}_u \\
&= \boldsymbol{C}_s' \boldsymbol{V}_s^{-1} (\boldsymbol{I} - \boldsymbol{X}_s \boldsymbol{Q}_s \boldsymbol{X}_s' \boldsymbol{V}_s^{-1}) \boldsymbol{V}_s \boldsymbol{V}_s^{-1} \boldsymbol{C}_s \\
&- \boldsymbol{C}_s' \boldsymbol{V}_s^{-1} (\boldsymbol{I} - \boldsymbol{X}_s \boldsymbol{Q}_s \boldsymbol{X}_s' \boldsymbol{V}_s^{-1}) \boldsymbol{V}_s \boldsymbol{V}_s^{-1} \boldsymbol{X}_s \boldsymbol{Q}_s \boldsymbol{X}_s' \boldsymbol{V}_s^{-1} \boldsymbol{C}_s \\
&- \boldsymbol{C}_s' \boldsymbol{V}_s^{-1} (\boldsymbol{I} - \boldsymbol{X}_s \boldsymbol{Q}_s \boldsymbol{X}_s' \boldsymbol{V}_s^{-1}) \boldsymbol{C}_s - \boldsymbol{C}_s' (\boldsymbol{I} - \boldsymbol{V}_s^{-1} \boldsymbol{X}_s \boldsymbol{Q}_s \boldsymbol{X}_s') \boldsymbol{V}_s^{-1} \boldsymbol{C}_s + \boldsymbol{V}_u \\
&= -\boldsymbol{C}_s' \boldsymbol{V}_s^{-1} \boldsymbol{X}_s \boldsymbol{Q}_s \boldsymbol{X}_s' \boldsymbol{V}_s^{-1} \boldsymbol{C}_s + \boldsymbol{C}_s' \boldsymbol{V}_s^{-1} \boldsymbol{X}_s \boldsymbol{Q}_s (\boldsymbol{X}_s' \boldsymbol{V}_s^{-1} \boldsymbol{X}_s) \boldsymbol{Q}_s \boldsymbol{X}_s' \boldsymbol{V}_s^{-1} \boldsymbol{C}_s \\
&- \boldsymbol{C}_s' \boldsymbol{V}_s^{-1} \boldsymbol{C}_s + \boldsymbol{C}_s' \boldsymbol{V}_s^{-1} \boldsymbol{X}_s \boldsymbol{Q}_s \boldsymbol{X}_s' \boldsymbol{V}_s^{-1} \boldsymbol{C}_s + \boldsymbol{V}_u \\
&= \boldsymbol{V}_u - \boldsymbol{V}_u \boldsymbol{Z}_s' \boldsymbol{V}_s^{-1} \boldsymbol{Z}_s \boldsymbol{V}_u + (\boldsymbol{V}_u \boldsymbol{Z}_s' \boldsymbol{V}_s^{-1}) \boldsymbol{X}_s \boldsymbol{Q}_s \boldsymbol{X}_s' (\boldsymbol{V}_s^{-1} \boldsymbol{Z}_s \boldsymbol{V}_u).
\end{aligned}
$$

For obtaining a more synthetic formula of $\boldsymbol{R}_{22}$, we apply the equalities

$$
\begin{aligned}
\boldsymbol{V}_s^{-1} &= (\boldsymbol{V}_{es} + \boldsymbol{Z}_s \boldsymbol{V}_u \boldsymbol{Z}_s')^{-1} = \boldsymbol{V}_{es}^{-1} - \boldsymbol{V}_{es}^{-1} \boldsymbol{Z}_s \boldsymbol{T}_s \boldsymbol{Z}_s' \boldsymbol{V}_{es}^{-1}, \\
\boldsymbol{T}_s &= (\boldsymbol{V}_u^{-1} + \boldsymbol{Z}_s' \boldsymbol{V}_{es}^{-1} \boldsymbol{Z}_s)^{-1} = \boldsymbol{V}_u - \boldsymbol{V}_u \boldsymbol{Z}_s' \boldsymbol{V}_s^{-1} \boldsymbol{Z}_s \boldsymbol{V}_u.
\end{aligned}
$$

We obtain

$$\boldsymbol{Z}_s'\boldsymbol{V}_{es}^{-1}\boldsymbol{Z}_s = \boldsymbol{T}_s^{-1} - \boldsymbol{V}_u^{-1},$$

$$\boldsymbol{Z}_s'\boldsymbol{V}_s^{-1} = \boldsymbol{Z}_s'\boldsymbol{V}_{es}^{-1} - \boldsymbol{Z}_s'\boldsymbol{V}_{es}^{-1}\boldsymbol{Z}_s\boldsymbol{T}_s\boldsymbol{Z}_s'\boldsymbol{V}_{es}^{-1}$$

$$= \boldsymbol{Z}_s'\boldsymbol{V}_{es}^{-1} - \boldsymbol{Z}_s'\boldsymbol{V}_{es}^{-1} + \boldsymbol{V}_u^{-1}\boldsymbol{T}_s\boldsymbol{Z}_s'\boldsymbol{V}_{es}^{-1} = \boldsymbol{V}_u^{-1}\boldsymbol{T}_s\boldsymbol{Z}_s'\boldsymbol{V}_{es}^{-1},$$

$$\boldsymbol{V}_u\boldsymbol{Z}_s'\boldsymbol{V}_s^{-1} = \boldsymbol{T}_s\boldsymbol{Z}_s'\boldsymbol{V}_{es}^{-1}.$$

Finally, we obtain

$$\boldsymbol{R}_{12} = E\left[(\hat{\boldsymbol{\beta}} - \boldsymbol{\beta})(\hat{\boldsymbol{u}} - \boldsymbol{u})'\right] = -\boldsymbol{Q}_s\boldsymbol{X}_s'\boldsymbol{V}_{es}^{-1}\boldsymbol{Z}_s\boldsymbol{T}_s,$$

$$\boldsymbol{R}_{22} = E\left[(\hat{\boldsymbol{u}} - \boldsymbol{u})(\hat{\boldsymbol{u}} - \boldsymbol{u})'\right] = \boldsymbol{T}_s + \boldsymbol{T}_s\boldsymbol{Z}_s'\boldsymbol{V}_{es}^{-1}\boldsymbol{X}_s\boldsymbol{Q}_s\boldsymbol{X}_s'\boldsymbol{V}_{es}^{-1}\boldsymbol{Z}_s\boldsymbol{T}_s.$$

Coming back to the calculation of $MSE(\hat{\tau}_{blup})$, we have

$$MSE(\hat{\tau}_{blup}) = \left[\boldsymbol{a}_r'\boldsymbol{X}_r, \boldsymbol{a}_r'\boldsymbol{Z}_r\right] E\begin{bmatrix}(\hat{\boldsymbol{\beta}} - \boldsymbol{\beta})(\hat{\boldsymbol{\beta}} - \boldsymbol{\beta})' & (\hat{\boldsymbol{\beta}} - \boldsymbol{\beta})(\hat{\boldsymbol{u}} - \boldsymbol{u})' \\ (\hat{\boldsymbol{u}} - \boldsymbol{u})(\hat{\boldsymbol{\beta}} - \boldsymbol{\beta})' & (\hat{\boldsymbol{u}} - \boldsymbol{u})(\hat{\boldsymbol{u}} - \boldsymbol{u})'\end{bmatrix}\begin{bmatrix}\boldsymbol{X}_r'\boldsymbol{a}_r \\ \boldsymbol{Z}_r'\boldsymbol{a}_r\end{bmatrix}$$

$$= \boldsymbol{a}_r'\boldsymbol{X}_r\boldsymbol{R}_{11}\boldsymbol{X}_r'\boldsymbol{a}_r + \boldsymbol{a}_r'\boldsymbol{X}_r\boldsymbol{R}_{12}\boldsymbol{Z}_r'\boldsymbol{a}_r + \boldsymbol{a}_r'\boldsymbol{Z}_r\boldsymbol{R}_{12}'\boldsymbol{X}_r'\boldsymbol{a}_r$$

$$+ \boldsymbol{a}_r'\boldsymbol{Z}_r\boldsymbol{R}_{22}\boldsymbol{Z}_r'\boldsymbol{a}_r = \boldsymbol{a}_r'\boldsymbol{X}_r\boldsymbol{Q}_s\boldsymbol{X}_r'\boldsymbol{a}_r - \boldsymbol{a}_r'\boldsymbol{X}_r\boldsymbol{Q}_s\boldsymbol{X}_s'\boldsymbol{V}_{es}^{-1}\boldsymbol{Z}_s\boldsymbol{T}_s\boldsymbol{Z}_r'\boldsymbol{a}_r$$

$$- \boldsymbol{a}_r'\boldsymbol{Z}_r\boldsymbol{T}_s\boldsymbol{Z}_s'\boldsymbol{V}_{es}^{-1}\boldsymbol{X}_s\boldsymbol{Q}_s\boldsymbol{X}_r'\boldsymbol{a}_r + \boldsymbol{a}_r'\boldsymbol{Z}_r\boldsymbol{T}_s\boldsymbol{Z}_r'\boldsymbol{a}_r$$

$$+ \boldsymbol{a}_r'\boldsymbol{Z}_r\boldsymbol{T}_s\boldsymbol{Z}_s'\boldsymbol{V}_{es}^{-1}\boldsymbol{X}_s\boldsymbol{Q}_s\boldsymbol{X}_s'\boldsymbol{V}_{es}^{-1}\boldsymbol{Z}_s\boldsymbol{T}_s\boldsymbol{Z}_r'\boldsymbol{a}_r = \boldsymbol{a}_r'\boldsymbol{Z}_r\boldsymbol{T}_s\boldsymbol{Z}_r'\boldsymbol{a}_r$$

$$+ [\boldsymbol{a}_r'\boldsymbol{X}_r - \boldsymbol{a}_r'\boldsymbol{Z}_r\boldsymbol{T}_s\boldsymbol{Z}_s'\boldsymbol{V}_{es}^{-1}\boldsymbol{X}_s]\boldsymbol{Q}_s[\boldsymbol{X}_r'\boldsymbol{a}_r - \boldsymbol{X}_s'\boldsymbol{V}_{es}^{-1}\boldsymbol{Z}_s\boldsymbol{T}_s\boldsymbol{Z}_r'\boldsymbol{a}_r].$$

Given the vector of variance components $\boldsymbol{\theta} = (\theta_0, \theta_1, \ldots, \theta_m)$, we get

$$MSE(\hat{\tau}_{blup}) = g_1(\boldsymbol{\theta}) + g_2(\boldsymbol{\theta}),$$

where

$$g_1(\boldsymbol{\theta}) = \boldsymbol{a}_r'\boldsymbol{Z}_r\boldsymbol{T}_s\boldsymbol{Z}_r'\boldsymbol{a}_r,$$

$$g_2(\boldsymbol{\theta}) = [\boldsymbol{a}_r'\boldsymbol{X}_r - \boldsymbol{a}_r'\boldsymbol{Z}_r\boldsymbol{T}_s\boldsymbol{Z}_s'\boldsymbol{V}_{es}^{-1}\boldsymbol{X}_s]\boldsymbol{Q}_s[\boldsymbol{X}_r'\boldsymbol{a}_r - \boldsymbol{X}_s'\boldsymbol{V}_{es}^{-1}\boldsymbol{Z}_s\boldsymbol{T}_s\boldsymbol{Z}_r'\boldsymbol{a}_r].$$

9.2.3 All Model Parameters Are Unknown

If the variance components $\boldsymbol{\theta} = (\theta_0, \theta_1, \ldots, \theta_m)$ are known, the BLUP of τ is $\hat{\tau}_{blup} = \hat{\tau}(\boldsymbol{\theta})$. If $\boldsymbol{\theta}$ is unknown, then we substitute $\boldsymbol{\theta}$ by a convenient estimator, and we obtain the EBLUP of τ, i.e.

$$\hat{\tau}_{eblup} = \hat{\tau}(\hat{\boldsymbol{\theta}}).$$

The mean squared error of $\hat{\tau}_{eblup}$ is

$$\begin{aligned} MSE(\hat{\tau}_{eblup}) &= E\left[(\hat{\tau}_{eblup} - \hat{\tau}_{blup} + \hat{\tau}_{blup} - \tau)^2\right] \\ &= MSE(\hat{\tau}_{blup}) + E\left[(\hat{\tau}_{eblup} - \hat{\tau}_{blup})^2\right] \\ &+ 2E\left[(\hat{\tau}_{eblup} - \hat{\tau}_{blup})(\hat{\tau}_{blup} - \tau)\right]. \end{aligned}$$

A function (possibly vectorial) of the vector of observations $s(\boldsymbol{y}_s)$ is *even* if for every $\boldsymbol{y}_s$ the equality $s(-\boldsymbol{y}_s) = s(\boldsymbol{y}_s)$ holds. A function $s(\boldsymbol{y}_s)$ is *translation invariant* if $s(\boldsymbol{y}_s + \boldsymbol{X}_s\boldsymbol{\beta}) = s(\boldsymbol{y}_s)$ for every $\boldsymbol{y}_s \in R^n$, $\boldsymbol{\beta} \in R^p$. Kackar and Harville (1981) proved that Henderson 3, ML, and REML estimators of $\boldsymbol{\theta}$ are even and translation invariant functions of $\boldsymbol{y}_s$. They also proved that if $E[\hat{\tau}(\boldsymbol{\theta})]$ is finite and the estimator $\hat{\boldsymbol{\theta}}$ is an even and translation invariant function of $\boldsymbol{y}_s$, then $\hat{\tau}_{eblup} = \hat{\tau}(\hat{\boldsymbol{\theta}})$ is unbiased ($E[\hat{\tau}(\hat{\boldsymbol{\theta}}) - \tau] = 0$). Kackar and Harville (1984) proved that if $\hat{\boldsymbol{\theta}}$ is an even and translation invariant function of $\boldsymbol{y}_s$, then

$$E\left[(\hat{\tau}_{eblup} - \hat{\tau}_{blup})(\hat{\tau}_{blup} - \tau)\right] = 0. \tag{9.3}$$

This section assumes that (9.3) holds. Therefore, we have the equality

$$MSE(\hat{\tau}_{eblup}) = MSE(\hat{\tau}_{blup}) + E\left[(\hat{\tau}_{eblup} - \hat{\tau}_{blup})^2\right]. \tag{9.4}$$

In what follows, we look for an approximation of the term

$$E\left[(\hat{\tau}_{eblup} - \hat{\tau}_{blup})^2\right].$$

Let us define the vector $\boldsymbol{d}(\boldsymbol{\theta}) = (d_0(\boldsymbol{\theta}), d_1(\boldsymbol{\theta}), \ldots, d_m(\boldsymbol{\theta}))'$, where

$$d_j(\boldsymbol{\theta}) = \frac{\partial \hat{\tau}(\boldsymbol{\theta})}{\partial \theta_j}, \quad j = 0, 1, \ldots, m.$$

The first order Taylor expansion of $\hat{\tau}(\hat{\boldsymbol{\theta}})$ around $\boldsymbol{\theta}$ is

$$\hat{\tau}(\hat{\boldsymbol{\theta}}) \approx \hat{\tau}(\boldsymbol{\theta}) + \sum_{j=0}^{m} d_j(\boldsymbol{\theta})(\hat{\theta}_j - \theta_j).$$

In other words, we have

$$\hat{\tau}_{eblup} \approx \hat{\tau}_{blup} + \sum_{j=0}^{m} d_j(\boldsymbol{\theta})(\hat{\theta}_j - \theta_j) = \hat{\tau}_{blup} + \boldsymbol{d}'(\boldsymbol{\theta})(\hat{\boldsymbol{\theta}} - \boldsymbol{\theta}).$$

Let us assume that $\hat{\boldsymbol{\theta}}$ is asymptotically unbiased, i.e.

$$E[\hat{\theta}_j - \theta_j] \underset{n\to\infty}{\longrightarrow} 0, \quad j = 0, 1, \ldots, m.$$

Then, we have

$$\begin{aligned} E\big[(\hat{\tau}_{eblup} - \hat{\tau}_{blup})^2\big] &\approx E\big[(\boldsymbol{d}'(\boldsymbol{\theta})(\hat{\boldsymbol{\theta}} - \boldsymbol{\theta}))^2\big] \\ &= \sum_{i=0}^{m}\sum_{j=0}^{m} E\big[d_i(\boldsymbol{\theta})(\hat{\theta}_i - \theta_i)d_j(\boldsymbol{\theta})(\hat{\theta}_j - \theta_j)\big]. \end{aligned} \tag{9.5}$$

As $\boldsymbol{d}(\boldsymbol{\theta})$ is a random vector, the summand (i, j) in (9.5) is

$$E\Big[d_i(\boldsymbol{\theta})d_j(\boldsymbol{\theta})(\hat{\theta}_i - \theta_i)(\hat{\theta}_j - \theta_j)\Big] = E_{\hat{\boldsymbol{\theta}}}\Big[(\hat{\theta}_i - \theta_i)(\hat{\theta}_j - \theta_j)E_{\boldsymbol{d}}\Big[d_i(\boldsymbol{\theta})d_j(\boldsymbol{\theta})\,|\,\hat{\boldsymbol{\theta}}\Big]\Big].$$

Lemma 9.1 (see the proof below) states that

$$E\left[d_j(\boldsymbol{\theta})\right] = 0, \quad j = 0, 1, \ldots, m,$$

and therefore

$$\mathrm{cov}\big(d_i(\boldsymbol{\theta}), d_j(\boldsymbol{\theta})\big) = E\big[d_i(\boldsymbol{\theta})d_j(\boldsymbol{\theta})\big], \quad i, j = 0, 1, \ldots, m.$$

In the case of calculating $\hat{\boldsymbol{\theta}}$ with a data set that is different and independent from the data that is used to calculate $\hat{\tau}_{blup} = \hat{\tau}(\boldsymbol{\theta})$, we have

$$E\big[d_i(\boldsymbol{\theta})d_j(\boldsymbol{\theta})\,|\,\hat{\boldsymbol{\theta}}\big] = E\big[d_i(\boldsymbol{\theta})d_j(\boldsymbol{\theta})\big], \quad \mathrm{cov}\big(d_i(\boldsymbol{\theta}), d_j(\boldsymbol{\theta})\,|\,\hat{\boldsymbol{\theta}}\big) = \mathrm{cov}\big(d_i(\boldsymbol{\theta}), d_j(\boldsymbol{\theta})\big),$$

and consequently

$$\begin{aligned} E\big[d_i(\boldsymbol{\theta})d_j(\boldsymbol{\theta})(\hat{\theta}_i - \theta_i)(\hat{\theta}_j - \theta_j)\big] &= \mathrm{cov}\left(d_i(\boldsymbol{\theta}), d_j(\boldsymbol{\theta})\right) E_{\hat{\boldsymbol{\theta}}}\big[(\hat{\theta}_i - \theta_i)(\hat{\theta}_j - \theta_j)\big] \\ &\approx \mathrm{cov}\left(d_i(\boldsymbol{\theta}), d_j(\boldsymbol{\theta})\right) \mathrm{cov}(\hat{\theta}_i, \hat{\theta}_j). \end{aligned}$$

Therefore, the second summand of (9.4) is

$$E\left[(\hat{\tau}_{eblup} - \hat{\tau}_{blup})^2\right] \approx \sum_{i=0}^{m}\sum_{j=0}^{m} \operatorname{cov}\left(d_i(\boldsymbol{\theta}), d_j(\boldsymbol{\theta})\right) \operatorname{cov}(\hat{\theta}_i, \hat{\theta}_j).$$

Let $\boldsymbol{G}(\boldsymbol{\theta})$ and $\boldsymbol{B}(\boldsymbol{\theta})$ be the covariance matrices of $\boldsymbol{d}(\boldsymbol{\theta})$ and $\hat{\boldsymbol{\theta}}$, respectively. Then,

$$\sum_{j=0}^{m} \operatorname{cov}\left(d_i(\boldsymbol{\theta}), d_j(\boldsymbol{\theta})\right) \operatorname{cov}(\hat{\theta}_i, \hat{\theta}_j)$$

is the element i of the main diagonal of $\boldsymbol{G}(\boldsymbol{\theta})\boldsymbol{B}(\boldsymbol{\theta})$ and

$$E\left[(\hat{\tau}_{eblup} - \hat{\tau}_{blup})^2\right] \approx \operatorname{tr}\left\{\boldsymbol{G}(\boldsymbol{\theta})\boldsymbol{B}(\boldsymbol{\theta})\right\}.$$

In the case of calculating $\hat{\boldsymbol{\theta}}$ and $\hat{\tau}_{blup} = \hat{\tau}(\boldsymbol{\theta})$ from the same data set, Kackar and Harville (1984) proposed the approximation

$$E\left[(\hat{\tau}_{eblup} - \hat{\tau}_{blup})^2\right] \approx \operatorname{tr}\left\{\boldsymbol{G}(\boldsymbol{\theta})\boldsymbol{B}(\boldsymbol{\theta})\right\}.$$

Therefore, we can approximate the MSE of $\hat{\tau}_{eblup}$ by

$$MSE(\hat{\tau}_{eblup}) \approx MSE(\hat{\tau}_{blup}) + \operatorname{tr}\left\{\boldsymbol{G}(\boldsymbol{\theta})\boldsymbol{B}(\boldsymbol{\theta})\right\}.$$

Prasad and Rao (1990) gave the following approximation:

$$\operatorname{tr}\left\{\boldsymbol{G}(\boldsymbol{\theta})\boldsymbol{B}(\boldsymbol{\theta})\right\} \approx \operatorname{tr}\left\{(\nabla \boldsymbol{b}')\boldsymbol{V}_s(\nabla \boldsymbol{b}')' E\left[(\hat{\boldsymbol{\theta}} - \boldsymbol{\theta})(\hat{\boldsymbol{\theta}} - \boldsymbol{\theta})'\right]\right\}, \tag{9.6}$$

where $\boldsymbol{b}' = (b_1, \ldots, b_n) = \boldsymbol{a}_r'\boldsymbol{Z}_r\boldsymbol{V}_u\boldsymbol{Z}_s'\boldsymbol{V}_s^{-1}$,

$$\frac{\partial \boldsymbol{b}'}{\partial \theta_j} = \left(\frac{\partial b_1}{\partial \theta_j}, \ldots, \frac{\partial b_n}{\partial \theta_j}\right) \quad \text{and} \quad \nabla \boldsymbol{b}' = \begin{pmatrix} \frac{\partial \boldsymbol{b}'}{\partial \theta_0} \\ \frac{\partial \boldsymbol{b}'}{\partial \theta_1} \\ \vdots \\ \frac{\partial \boldsymbol{b}'}{\partial \theta_m} \end{pmatrix} = \begin{pmatrix} \frac{\partial b_1}{\partial \theta_0} & \cdots & \frac{\partial b_n}{\partial \theta_0} \\ \frac{\partial b_1}{\partial \theta_1} & \cdots & \frac{\partial b_n}{\partial \theta_1} \\ \vdots & \cdots & \vdots \\ \frac{\partial b_1}{\partial \theta_m} & \cdots & \frac{\partial b_n}{\partial \theta_m} \end{pmatrix}_{(m+1)\times n}.$$

If the vector of variance components $\boldsymbol{\theta} = (\theta_0, \theta_1, \ldots, \theta_m)$ is unknown, we finally have

$$MSE(\hat{\tau}_{eblup}) = g_1(\boldsymbol{\theta}) + g_2(\boldsymbol{\theta}) + g_3(\boldsymbol{\theta}), \tag{9.7}$$

where

$$g_1(\boldsymbol{\theta}) = \boldsymbol{a}_r'\boldsymbol{Z}_r\boldsymbol{T}_s\boldsymbol{Z}_r'\boldsymbol{a}_r.$$
$$g_2(\boldsymbol{\theta}) = [\boldsymbol{a}_r'\boldsymbol{X}_r - \boldsymbol{a}_r'\boldsymbol{Z}_r\boldsymbol{T}_s\boldsymbol{Z}_s'\boldsymbol{V}_{es}^{-1}\boldsymbol{X}_s]\boldsymbol{Q}_s[\boldsymbol{X}_r'\boldsymbol{a}_r - \boldsymbol{X}_s'\boldsymbol{V}_{es}^{-1}\boldsymbol{Z}_s\boldsymbol{T}_s\boldsymbol{Z}_r'\boldsymbol{a}_r],$$
$$g_3(\boldsymbol{\theta}) \approx \text{tr}\left\{(\nabla\boldsymbol{b}')\boldsymbol{V}_s(\nabla\boldsymbol{b}')'E\big[(\hat{\boldsymbol{\theta}}-\boldsymbol{\theta})(\hat{\boldsymbol{\theta}}-\boldsymbol{\theta})'\big]\right\},$$

and $\boldsymbol{T}_s = \boldsymbol{V}_u - \boldsymbol{V}_u\boldsymbol{Z}_s'\boldsymbol{V}_s^{-1}\boldsymbol{Z}_s\boldsymbol{V}_u,\ \boldsymbol{Q}_s = (\boldsymbol{X}_s'\boldsymbol{V}_s^{-1}\boldsymbol{X}_s)^{-1}$.

Lemma 9.1 *It holds that*

$$E\left[\boldsymbol{d}(\boldsymbol{\theta})\right] = \boldsymbol{0}.$$

Proof We calculate the derivatives

$$d_j(\boldsymbol{\theta}) = \frac{\partial\hat{\tau}(\boldsymbol{\theta})}{\partial\theta_j}, \qquad j = 0, 1, \ldots, m.$$

We use the following expression of $\hat{\tau}(\boldsymbol{\theta})$:

$$\begin{aligned}
\hat{\tau}(\boldsymbol{\theta}) &= \boldsymbol{a}_r'\boldsymbol{X}_r\hat{\boldsymbol{\beta}} + \boldsymbol{a}_r'\boldsymbol{Z}_r\hat{\boldsymbol{u}} = \boldsymbol{a}_r'\boldsymbol{X}_r\boldsymbol{Q}_s\boldsymbol{X}_s'\boldsymbol{V}_s^{-1}\boldsymbol{y}_s + \boldsymbol{a}_r'\boldsymbol{Z}_r\boldsymbol{V}_u\boldsymbol{Z}_s'\boldsymbol{V}_s^{-1}(\boldsymbol{y}_s - \boldsymbol{X}_s\hat{\boldsymbol{\beta}})\\
&= \boldsymbol{a}_r'\boldsymbol{X}_r\boldsymbol{Q}_s\boldsymbol{X}_s'\boldsymbol{V}_s^{-1}\boldsymbol{y}_s + \boldsymbol{a}_r'\boldsymbol{Z}_r\boldsymbol{V}_u\boldsymbol{Z}_s'\boldsymbol{V}_s^{-1}\boldsymbol{y}_s - \boldsymbol{a}_r'\boldsymbol{Z}_r\boldsymbol{V}_u\boldsymbol{Z}_s'\boldsymbol{V}_s^{-1}\boldsymbol{X}_s\boldsymbol{Q}_s\boldsymbol{X}_s'\boldsymbol{V}_s^{-1}\boldsymbol{y}_s\\
&= \boldsymbol{a}_r'\boldsymbol{X}_r\boldsymbol{G}\boldsymbol{y}_s + \boldsymbol{b}'\boldsymbol{y}_s - \boldsymbol{b}'\boldsymbol{X}_s\boldsymbol{G}\boldsymbol{y}_s,
\end{aligned}$$

where $\boldsymbol{G} = \boldsymbol{Q}_s\boldsymbol{X}_s'\boldsymbol{V}_s^{-1}$ and $\boldsymbol{b}' = \boldsymbol{a}_r'\boldsymbol{Z}_r\boldsymbol{V}_u\boldsymbol{Z}_s'\boldsymbol{V}_s^{-1}$. Therefore,

$$d_j(\boldsymbol{\theta}) = \boldsymbol{a}_r'\boldsymbol{X}_r\frac{\partial\boldsymbol{G}}{\partial\theta_j}\boldsymbol{y}_s + \frac{\partial\boldsymbol{b}'}{\partial\theta_j}\boldsymbol{y}_s - \frac{\partial\boldsymbol{b}'}{\partial\theta_j}\boldsymbol{X}_s\boldsymbol{G}\boldsymbol{y}_s - \boldsymbol{b}'\boldsymbol{X}_s\frac{\partial\boldsymbol{G}}{\partial\theta_j}\boldsymbol{y}_s.$$

We define

$$\boldsymbol{P}(\boldsymbol{\theta}) = \boldsymbol{V}_s^{-1} - \boldsymbol{V}_s^{-1}\boldsymbol{X}_s\boldsymbol{Q}_s\boldsymbol{X}_s'\boldsymbol{V}_s^{-1},$$

and we calculate the partial derivatives $d_j(\boldsymbol{\theta})$ by applying the formula $\frac{\partial\boldsymbol{A}^{-1}}{\partial\theta_j} = -\boldsymbol{A}^{-1}\frac{\partial\boldsymbol{A}}{\partial\theta_j}\boldsymbol{A}^{-1}$. For the first summand of $d_j(\boldsymbol{\theta})$, we have

$$\begin{aligned}
\frac{\partial\boldsymbol{G}}{\partial\theta_j} &= -\boldsymbol{Q}_s\frac{\partial\boldsymbol{Q}_s^{-1}}{\partial\theta_j}\boldsymbol{Q}_s\boldsymbol{X}_s'\boldsymbol{V}_s^{-1} - \boldsymbol{Q}_s\boldsymbol{X}_s'\boldsymbol{V}_s^{-1}\frac{\partial\boldsymbol{V}_s}{\partial\theta_j}\boldsymbol{V}_s^{-1}\\
&= -\boldsymbol{Q}_s\boldsymbol{X}_s'\frac{\partial\boldsymbol{V}_s^{-1}}{\partial\theta_j}\boldsymbol{X}_s\boldsymbol{Q}_s\boldsymbol{X}_s'\boldsymbol{V}_s^{-1} - \boldsymbol{Q}_s\boldsymbol{X}_s'\boldsymbol{V}_s^{-1}\frac{\partial\boldsymbol{V}_s}{\partial\theta_j}\boldsymbol{V}_s^{-1}\\
&= \boldsymbol{Q}_s\boldsymbol{X}_s'\boldsymbol{V}_s^{-1}\frac{\partial\boldsymbol{V}_s}{\partial\theta_j}\boldsymbol{V}_s^{-1}\boldsymbol{X}_s\boldsymbol{Q}_s\boldsymbol{X}_s'\boldsymbol{V}_s^{-1} - \boldsymbol{Q}_s\boldsymbol{X}_s'\boldsymbol{V}_s^{-1}\frac{\partial\boldsymbol{V}_s}{\partial\theta_j}\boldsymbol{V}_s^{-1} = -\boldsymbol{Q}_s\boldsymbol{X}_s'\boldsymbol{V}_s^{-1}\frac{\partial\boldsymbol{V}_s}{\partial\theta_j}\boldsymbol{P}(\boldsymbol{\theta}).
\end{aligned}$$

For the second and third summands, we have

$$\frac{\partial \boldsymbol{b}'}{\partial \theta_j} = \boldsymbol{a}_r'\boldsymbol{Z}_r\frac{\partial \boldsymbol{V}_u}{\partial \theta_j}\boldsymbol{Z}_s'\boldsymbol{V}_s^{-1} - \boldsymbol{a}_r'\boldsymbol{Z}_r\boldsymbol{V}_u\boldsymbol{Z}_s'\boldsymbol{V}_s^{-1}\frac{\partial \boldsymbol{V}_s}{\partial \theta_j}\boldsymbol{V}_s^{-1} \triangleq \boldsymbol{C}_1 - \boldsymbol{C}_2,$$

$$\frac{\partial \boldsymbol{b}'}{\partial \theta_j}\boldsymbol{X}_s\boldsymbol{G} = \boldsymbol{a}_r'\boldsymbol{Z}_r\frac{\partial \boldsymbol{V}_u}{\partial \theta_j}\boldsymbol{Z}_s'\boldsymbol{V}_s^{-1}\boldsymbol{X}_s\boldsymbol{Q}_s\boldsymbol{X}_s'\boldsymbol{V}_s^{-1}$$
$$- \boldsymbol{a}_r'\boldsymbol{Z}_r\boldsymbol{V}_u\boldsymbol{Z}_s'\boldsymbol{V}_s^{-1}\frac{\partial \boldsymbol{V}_s}{\partial \theta_j}\boldsymbol{V}_s^{-1}\boldsymbol{X}_s\boldsymbol{Q}_s\boldsymbol{X}_s'\boldsymbol{V}_s^{-1} \triangleq \boldsymbol{D}_1 - \boldsymbol{D}_2,$$

$$\frac{\partial \boldsymbol{b}'}{\partial \theta_j} - \frac{\partial \boldsymbol{b}'}{\partial \theta_j}\boldsymbol{X}_s\boldsymbol{G} = (\boldsymbol{C}_1 - \boldsymbol{D}_1) + (\boldsymbol{D}_2 - \boldsymbol{C}_2)$$
$$= \boldsymbol{a}_r'\boldsymbol{Z}_r\frac{\partial \boldsymbol{V}_u}{\partial \theta_j}\boldsymbol{Z}_s'\boldsymbol{P}(\boldsymbol{\theta}) - \boldsymbol{a}_r'\boldsymbol{Z}_r\boldsymbol{V}_u\boldsymbol{Z}_s'\boldsymbol{V}_s^{-1}\frac{\partial \boldsymbol{V}_s}{\partial \theta_j}\boldsymbol{P}(\boldsymbol{\theta}).$$

Finally, for the fourth summand of $d_j(\boldsymbol{\theta})$, we have

$$\boldsymbol{b}'\boldsymbol{X}_s\frac{\partial \boldsymbol{G}}{\partial \theta_j} = -\boldsymbol{a}_r'\boldsymbol{Z}_r\boldsymbol{V}_u\boldsymbol{Z}_s'\boldsymbol{V}_s^{-1}\boldsymbol{X}_s\boldsymbol{Q}_s\boldsymbol{X}_s'\boldsymbol{V}_s^{-1}\frac{\partial \boldsymbol{V}_s}{\partial \theta_j}\boldsymbol{P}(\boldsymbol{\theta}).$$

Putting all the summands together, we get

$$d_j(\boldsymbol{\theta}) = \left\{-\boldsymbol{a}'\boldsymbol{X}_r\boldsymbol{Q}_s\boldsymbol{X}_s'\boldsymbol{V}_s^{-1} - \boldsymbol{a}_r'\boldsymbol{Z}_r\boldsymbol{V}_u\boldsymbol{Z}_s'\boldsymbol{V}_s^{-1} + \boldsymbol{a}_r'\boldsymbol{Z}_r\boldsymbol{V}_u\boldsymbol{Z}_s'\boldsymbol{V}_s^{-1}\boldsymbol{X}_s\boldsymbol{Q}_s\boldsymbol{X}_s'\boldsymbol{V}_s^{-1}\right\}$$
$$\cdot \frac{\partial \boldsymbol{V}_s}{\partial \theta_j}\boldsymbol{P}(\boldsymbol{\theta})\boldsymbol{y}_s + \boldsymbol{a}_r'\boldsymbol{Z}_r\frac{\partial \boldsymbol{V}_u}{\partial \theta_j}\boldsymbol{Z}_s'\boldsymbol{P}(\boldsymbol{\theta})\boldsymbol{y}_s \triangleq \Delta_j(\boldsymbol{\theta})\boldsymbol{P}(\boldsymbol{\theta})\boldsymbol{y}_s$$

and

$$E\left[d_j(\boldsymbol{\theta})\right] = \Delta_j(\boldsymbol{\theta})\boldsymbol{P}(\boldsymbol{\theta})\boldsymbol{X}_s\boldsymbol{\beta}, \qquad j = 0, 1, \ldots, m.$$

Further, we have

$$\boldsymbol{P}(\boldsymbol{\theta})\boldsymbol{X}_s = \boldsymbol{V}_s^{-1}\boldsymbol{X}_s - \boldsymbol{V}_s^{-1}\boldsymbol{X}_s\boldsymbol{Q}_s\boldsymbol{X}_s'\boldsymbol{V}_s^{-1}\boldsymbol{X}_s = \boldsymbol{0},$$

because $\boldsymbol{Q}_s = (\boldsymbol{X}_s'\boldsymbol{V}_s^{-1}\boldsymbol{X}_s)^{-1}$. Finally, we obtain $E\left[\boldsymbol{d}(\boldsymbol{\theta})\right] = \boldsymbol{0}$. □

9.3 The MSE of EBLUPs of Population Linear Parameters

Let us assume that the aim is to predict the population linear parameter

$$\eta = \boldsymbol{a}_s'\boldsymbol{y}_s + \boldsymbol{a}_r'\boldsymbol{y}_r = \boldsymbol{a}_s'\boldsymbol{y}_s + \boldsymbol{a}_r'(\boldsymbol{X}_r\boldsymbol{\beta} + \boldsymbol{Z}_r\boldsymbol{u} + \boldsymbol{e}_r) = \boldsymbol{a}_s'\boldsymbol{y}_s + \tau + \boldsymbol{a}_r'\boldsymbol{e}_r$$

with the predictor

$$\hat{\eta} = \boldsymbol{a}'_s \boldsymbol{y}_s + \hat{\tau}_{eblup}.$$

The mean squared error is

$$\begin{aligned}
MSE(\hat{\eta}) &= E\left[(\hat{\eta} - \eta)^2\right] = E\left[(\hat{\tau}_{eblup} - \tau - \boldsymbol{a}'_r \boldsymbol{e}_r)^2\right] \\
&= E\left[(\hat{\tau}_{eblup} - \tau)^2\right] + E\left[(\boldsymbol{a}'_r \boldsymbol{e}_r)^2\right] - 2E\left[(\hat{\tau}_{eblup} - \tau)\boldsymbol{a}'_r \boldsymbol{e}_r\right] \\
&= MSE(\hat{\tau}_{eblup}) + E\left[\boldsymbol{a}'_r \boldsymbol{e}_r \boldsymbol{e}'_r \boldsymbol{a}_r\right] \\
&- 2E\left[\boldsymbol{a}'_r \left(\boldsymbol{X}_r(\hat{\boldsymbol{\beta}} - \boldsymbol{\beta}) + \boldsymbol{Z}_r(\hat{\boldsymbol{u}} - \boldsymbol{u})\right) \boldsymbol{e}'_r \boldsymbol{a}_r\right] \\
&= MSE(\hat{\tau}_{eblup}) + \boldsymbol{a}'_r \boldsymbol{V}_{er} \boldsymbol{a}_r - 2\boldsymbol{a}'_r \boldsymbol{X}_r E\left[(\hat{\boldsymbol{\beta}} - \boldsymbol{\beta})\boldsymbol{e}'_r\right] \boldsymbol{a}_r \\
&- 2\boldsymbol{a}'_r \boldsymbol{Z}_r E\left[(\hat{\boldsymbol{u}} - \boldsymbol{u})\boldsymbol{e}'_r\right] \boldsymbol{a}_r.
\end{aligned}$$

We note that

$$\begin{aligned}
E\left[(\hat{\boldsymbol{\beta}} - \boldsymbol{\beta})\boldsymbol{e}'_r\right] &= \text{cov}(\hat{\boldsymbol{\beta}}, \boldsymbol{e}_r) = \text{cov}(\boldsymbol{Q}_s \boldsymbol{X}'_s \boldsymbol{V}_s^{-1} \boldsymbol{y}_s, \boldsymbol{e}_r) = 0, \\
E\left[(\hat{\boldsymbol{u}} - \boldsymbol{u})\boldsymbol{e}'_r\right] &= \text{cov}(\hat{\boldsymbol{u}} - \boldsymbol{u}, \boldsymbol{e}_r) = \text{cov}(\hat{\boldsymbol{u}}, \boldsymbol{e}_r) - \text{cov}(\boldsymbol{u}, \boldsymbol{e}_r) \\
&= \text{cov}(\boldsymbol{V}_u \boldsymbol{Z}'_s \boldsymbol{V}_s^{-1}(\boldsymbol{I} - \boldsymbol{X}_s \boldsymbol{Q}_s \boldsymbol{X}'_s \boldsymbol{V}_s^{-1})\boldsymbol{y}_s, \boldsymbol{e}_r) = 0.
\end{aligned}$$

Finally, we obtain

$$\begin{aligned}
MSE(\hat{\eta}) &= MSE(\hat{\tau}_{eblup}) + \boldsymbol{a}'_r \boldsymbol{V}_{er} \boldsymbol{a}_r \\
&= MSE(\hat{\tau}_{blup}) + E\left[(\hat{\tau}_{eblup} - \hat{\tau}_{blup})^2\right] + \boldsymbol{a}'_r \boldsymbol{V}_{er} \boldsymbol{a}_r \\
&\approx g_1(\boldsymbol{\theta}) + g_2(\boldsymbol{\theta}) + g_3(\boldsymbol{\theta}) + g_4(\boldsymbol{\theta}),
\end{aligned}$$

where

$$\begin{aligned}
g_1(\boldsymbol{\theta}) &= \boldsymbol{a}'_r \boldsymbol{Z}_r \boldsymbol{T}_s \boldsymbol{Z}'_r \boldsymbol{a}_r, \\
g_2(\boldsymbol{\theta}) &= [\boldsymbol{a}'_r \boldsymbol{X}_r - \boldsymbol{a}'_r \boldsymbol{Z}_r \boldsymbol{T}_s \boldsymbol{Z}'_s \boldsymbol{V}_{es}^{-1} \boldsymbol{X}_s] \boldsymbol{Q}_s [\boldsymbol{X}'_r \boldsymbol{a}_r - \boldsymbol{X}'_s \boldsymbol{V}_{es}^{-1} \boldsymbol{Z}_s \boldsymbol{T}_s \boldsymbol{Z}'_r \boldsymbol{a}_r], \\
g_3(\boldsymbol{\theta}) &\approx \text{tr}\left\{(\nabla \boldsymbol{b}')\boldsymbol{V}_s(\nabla \boldsymbol{b}')' E\left[(\hat{\boldsymbol{\theta}} - \boldsymbol{\theta})(\hat{\boldsymbol{\theta}} - \boldsymbol{\theta})'\right]\right\}, \\
g_4(\boldsymbol{\theta}) &= \boldsymbol{a}'_r \boldsymbol{V}_{er} \boldsymbol{a}_r.
\end{aligned}$$

9.4 Analytic Estimation of the MSE of EBLUPs

We need an estimator of $MSE(\hat{\eta})$ for measuring the variability of $\hat{\eta}$ in applications. Let us first consider the problem of estimating the MSE of $\hat{\tau} = \boldsymbol{a}_r'(\boldsymbol{X}_r\hat{\boldsymbol{\beta}} + \boldsymbol{Z}_r\hat{\boldsymbol{u}})$. A simple procedure for estimating $MSE(\hat{\tau})$ is substituting $\boldsymbol{\theta}$ by $\hat{\boldsymbol{\theta}}$ in the expression of the MSE. In this case, we obtain the plug-in estimator

$$mse_1(\hat{\tau}_{eblup}) = g_1(\hat{\boldsymbol{\theta}}) + g_2(\hat{\boldsymbol{\theta}}) + g_3(\hat{\boldsymbol{\theta}}). \tag{9.8}$$

If we use a consistent estimator $\hat{\boldsymbol{\theta}}$ of $\boldsymbol{\theta}$, it holds that $E[g_2(\hat{\boldsymbol{\theta}})] \cong g_2(\boldsymbol{\theta})$ and $E[g_3(\hat{\boldsymbol{\theta}})] \cong g_3(\boldsymbol{\theta})$. However, g_1 does not fulfill the property of being approximately unbiased.

For approximating the bias of $g_1(\hat{\boldsymbol{\theta}})$, we expand $g_1(\hat{\boldsymbol{\theta}})$ in Taylor series around $\boldsymbol{\theta}$. We obtain

$$g_1(\hat{\boldsymbol{\theta}}) \approx g_1(\boldsymbol{\theta}) + (\hat{\boldsymbol{\theta}} - \boldsymbol{\theta})'\nabla g_1(\boldsymbol{\theta}) + \frac{1}{2}(\hat{\boldsymbol{\theta}} - \boldsymbol{\theta})'\nabla^2 g_1(\boldsymbol{\theta})(\hat{\boldsymbol{\theta}} - \boldsymbol{\theta}) \triangleq g_1(\boldsymbol{\theta}) + \Delta_1 + \Delta_2,$$

where $\nabla g_1(\boldsymbol{\theta})$ is the vector of first partial derivatives of $g_1(\boldsymbol{\theta})$ with respect to $\boldsymbol{\theta}$ and $\nabla^2 g_1(\boldsymbol{\theta})$ is the matrix of second partial derivatives of $g_1(\boldsymbol{\theta})$ with respect to $\boldsymbol{\theta}$. If $\hat{\boldsymbol{\theta}}$ is an unbiased estimator of $\boldsymbol{\theta}$, then $E[\Delta_1] = 0$. In general, if the term $E[\Delta_1] \approx \boldsymbol{b}_{\hat{\boldsymbol{\theta}}}'(\boldsymbol{\theta})\nabla g_1(\boldsymbol{\theta})$, where $\boldsymbol{b}_{\hat{\boldsymbol{\theta}}}(\boldsymbol{\theta})$ is an approximation to the bias of $E[\hat{\boldsymbol{\theta}} - \boldsymbol{\theta}]$, is of inferior order than $E[\Delta_2]$, then a possible approximation of $E[g_1(\hat{\boldsymbol{\theta}})]$ is

$$E[g_1(\hat{\boldsymbol{\theta}})] \approx g_1(\boldsymbol{\theta}) + \frac{1}{2}\,\mathrm{tr}\left(\nabla^2 g_1(\boldsymbol{\theta})\overline{\boldsymbol{V}}[\hat{\boldsymbol{\theta}}]\right), \tag{9.9}$$

where $\overline{\boldsymbol{V}}[\hat{\boldsymbol{\theta}}]$ is the asymptotic covariance matrix of $\hat{\boldsymbol{\theta}}$. Further, if the covariance matrix $\boldsymbol{V}$ has a linear structure in $\boldsymbol{\theta}$, Prasad and Rao (1990) proved that (9.9) takes the simpler form

$$E[g_1(\hat{\boldsymbol{\theta}})] \approx g_1(\boldsymbol{\theta}) - g_3(\boldsymbol{\theta}). \tag{9.10}$$

From (9.8) and (9.10), the bias of $mse_1(\hat{\tau}_{eblup})$ is

$$\begin{aligned} E[mse_1(\hat{\tau}_{eblup})] - MSE(\hat{\tau}_{eblup}) &\approx (g_1(\boldsymbol{\theta}) - g_3(\boldsymbol{\theta}) + g_2(\boldsymbol{\theta}) + g_3(\boldsymbol{\theta})) \\ &\quad - (g_1(\boldsymbol{\theta}) + g_2(\boldsymbol{\theta}) + g_3(\boldsymbol{\theta})) = -g_3(\boldsymbol{\theta}). \end{aligned}$$

Therefore, we can estimate $MSE(\hat{\tau}_{eblup})$ with

$$mse(\hat{\tau}_{eblup}) = g_1(\hat{\boldsymbol{\theta}}) + g_2(\hat{\boldsymbol{\theta}}) + 2g_3(\hat{\boldsymbol{\theta}}). \tag{9.11}$$

Let us note that in the calculated bias $-g_3(\boldsymbol{\theta})$ of $mse_1(\hat{\tau}_{eblup})$, the unknown parameter $\boldsymbol{\theta}$ was again substituted by its estimate $\hat{\boldsymbol{\theta}}$.

The formula (9.11) is valid when we estimate $\boldsymbol{\theta}$ by the Henderson 3 or the REML method, because they give unbiased or almost unbiased estimators. However, for the ML estimator, it holds that $E[\Delta_1] \approx \boldsymbol{b}'_{\hat{\theta}}(\boldsymbol{\theta})\nabla g_1(\boldsymbol{\theta}) \neq 0$. In the last case, we can estimate $MSE(\hat{\tau}_{eblup})$ with

$$mse(\hat{\tau}_{eblup}) = g_1(\hat{\boldsymbol{\theta}}) + g_2(\hat{\boldsymbol{\theta}}) + 2g_3(\hat{\boldsymbol{\theta}}) - \boldsymbol{b}'_{\hat{\theta}}(\hat{\boldsymbol{\theta}})\nabla g_1(\hat{\boldsymbol{\theta}}). \tag{9.12}$$

We can calculate the term $\boldsymbol{b}_{\hat{\theta}}(\boldsymbol{\theta})$ more easily in the case that $\boldsymbol{V}$ is block-diagonal, i.e.

$$\boldsymbol{V} = \text{diag}(\boldsymbol{V}_1, \ldots, \boldsymbol{V}_D),$$

with

$$\boldsymbol{V}_d = \boldsymbol{Z}_d\boldsymbol{V}_{ud}\boldsymbol{Z}'_d + \boldsymbol{V}_{ed}, \quad d = 1, \ldots, D,$$

so the components of model (9.1) take the form $\boldsymbol{y} = (\boldsymbol{y}'_1, \ldots, \boldsymbol{y}'_D)'$, $\boldsymbol{X} = (\boldsymbol{X}'_1, \ldots, \boldsymbol{X}'_D)'$, $\boldsymbol{Z} = \text{diag}(\boldsymbol{Z}_1, \ldots, \boldsymbol{Z}_D)$, $\boldsymbol{u} = (\boldsymbol{u}'_1, \ldots, \boldsymbol{u}'_D)'$, and $\boldsymbol{e} = (\boldsymbol{e}'_1, \ldots, \boldsymbol{e}'_D)'$, where $\boldsymbol{X}_d$ is $N_d \times p$, $\boldsymbol{Z}_d$ is $N_d \times q_d$, $\boldsymbol{y}_d$ is $N_d \times 1$, $N = \sum_{d=1}^{D} N_d$, and $q = \sum_{d=1}^{D} q_d$. This kind of models can be decomposed in D submodels, i.e.

$$\boldsymbol{y}_d = \boldsymbol{X}_d\boldsymbol{\beta} + \boldsymbol{Z}_d\boldsymbol{u}_d + \boldsymbol{e}_d, \quad d = 1, \ldots, D. \tag{9.13}$$

Under the model (9.13), if $\hat{\boldsymbol{\theta}}$ is the ML estimator of $\boldsymbol{\theta}$ based on a sample s, an approximation to the bias is (see formula 6.3.13 in Rao (2003))

$$\boldsymbol{b}_{\hat{\theta}}(\boldsymbol{\theta}) = \frac{1}{2D}\left\{\mathcal{I}^{-1}(\boldsymbol{\theta}) \underset{0\leq j\leq m}{\text{col}}\left(\text{tr}\left[\sum_{d=1}^{D}(\boldsymbol{X}'_{sd}\boldsymbol{V}^{-1}_{sd}\boldsymbol{X}_{sd})^{-1}\Big(\sum_{d=1}^{D}\boldsymbol{X}'_{sd}\boldsymbol{V}^{(j)}_{sd}\boldsymbol{X}_{sd}\Big)\right]\right)\right\},$$

where $\text{col}_{0\leq j\leq m}(a_j)$ is a column vector with components a_j, $j = 0, \ldots, m$, and

$$\boldsymbol{V}^{(j)}_{sd} = \frac{\partial \boldsymbol{V}^{-1}_{sd}}{\partial \theta_j} = -\boldsymbol{V}^{-1}_{sd}\frac{\partial \boldsymbol{V}_{sd}}{\partial \theta_j}\boldsymbol{V}^{-1}_{sd} \quad \mathcal{I}_{jk}(\boldsymbol{\theta}) = \frac{1}{2}\sum_{d=1}^{D}\text{tr}\left[\left(\boldsymbol{V}^{-1}_{sd}\frac{\partial \boldsymbol{V}_{sd}}{\partial \theta_j}\right)\left(\boldsymbol{V}^{-1}_{sd}\frac{\partial \boldsymbol{V}_{sd}}{\partial \theta_k}\right)\right].$$

Prasad and Rao (1990) obtained the MSE estimator (9.11) for three linear mixed models with block-diagonal covariance matrices. From their results, Harville and Jeske (1992) proposed (9.11) for a general linear mixed model (9.1) under the assumption that $E[\hat{\boldsymbol{\theta}} - \boldsymbol{\theta}] = 0$. Das et al. (2001) gave rigorous proofs of approximations (9.11) and (9.12) for REML and ML estimators. Lahiri and Rao (1995) studied the robustness of the given approximations.

When predicting η by $\hat{\eta} = \boldsymbol{a}_s'\boldsymbol{y}_s + \hat{\tau}_{eblup}$, we use the MSE estimator

$$mse_1(\hat{\eta}) = g_1(\hat{\boldsymbol{\theta}}) + g_2(\hat{\boldsymbol{\theta}}) + 2g_3(\hat{\boldsymbol{\theta}}) + g_4(\hat{\boldsymbol{\theta}}),$$

if $\hat{\boldsymbol{\theta}}$ is approximately unbiased. Otherwise, we use the alternative estimator

$$mse_1(\hat{\eta}) = g_1(\hat{\boldsymbol{\theta}}) + g_2(\hat{\boldsymbol{\theta}}) + 2g_3(\hat{\boldsymbol{\theta}}) + g_4(\hat{\boldsymbol{\theta}}) - \boldsymbol{b}_{\hat{\theta}}'(\hat{\boldsymbol{\theta}})\nabla g_1(\hat{\boldsymbol{\theta}}).$$

9.5 MSE Approximation in NER Models

In this section we particularize the derived formulas for MSE approximation to the case of nested error regression model. Let us consider the linear mixed model

$$\boldsymbol{y} = \boldsymbol{X}\boldsymbol{\beta} + \boldsymbol{Z}\boldsymbol{u} + \boldsymbol{e}, \tag{9.14}$$

where $\boldsymbol{y} = \boldsymbol{y}_{N\times 1}$, $\boldsymbol{X} = \boldsymbol{X}_{N\times p}$, $\boldsymbol{\beta} = \boldsymbol{\beta}_{p\times 1}$, $\boldsymbol{Z} = \boldsymbol{Z}_{N\times D} = \text{diag}(\boldsymbol{1}_{N_1}, \ldots, \boldsymbol{1}_{N_D})$, $\boldsymbol{u} = \boldsymbol{u}_{D\times 1} \sim N(\boldsymbol{0}, \sigma_u^2 \boldsymbol{I}_D)$ is independent of $\boldsymbol{e} = \boldsymbol{e}_{N\times 1} \sim N(\boldsymbol{0}, \sigma_e^2 \boldsymbol{W}^{-1})$, $\boldsymbol{W} = \text{diag}_{1\le d\le D}(\boldsymbol{W}_d)$, $\boldsymbol{W}_d = \text{diag}_{1\le j\le N_D}(w_{dj})$, $\boldsymbol{I}_D = \text{diag}_{1\le d\le D}(1)$, and $\boldsymbol{1}_m = \text{col}_{1\le j\le m}(1)$. Alternatively, we can write this model in the form

$$y_{dj} = \boldsymbol{x}_{dj}\boldsymbol{\beta} + u_d + e_{dj}, \quad d = 1, \ldots, D,\ j = 1, \ldots, N_d, \tag{9.15}$$

where y_{dj} is the target variable measured at sample unit j of domain d and $\boldsymbol{x}_{dj}$ is the row (d, j) of matrix $\boldsymbol{X}$ containing the corresponding auxiliary variables.

Let us assume that the target parameter η is the domain mean $\overline{Y}_d = \frac{1}{N_d}\sum_{j=1}^{N_d} y_{dj}$, i.e.

$$\eta = \boldsymbol{a}'\boldsymbol{y}, \quad \text{with} \quad \boldsymbol{a}' = \frac{1}{N_d}(\boldsymbol{0}_{N_1}', \ldots, \boldsymbol{0}_{N_{d-1}}', \boldsymbol{1}_{N_d}', \boldsymbol{0}_{N_{d+1}}', \ldots, \boldsymbol{0}_{N_D}').$$

The EBLUP of $\overline{Y}_d$ based on a sample s with domain sample sizes n_d is (cf. (8.3))

$$\hat{\overline{Y}}_d^{eblup} = (1 - f_d)\left[\overline{\boldsymbol{X}}_d\hat{\boldsymbol{\beta}} + \hat{\gamma}_d^w(\hat{\overline{Y}}_d^w - \hat{\overline{\boldsymbol{X}}}_d^w\hat{\boldsymbol{\beta}})\right] + f_d\left[\hat{\overline{Y}}_d + (\overline{\boldsymbol{X}}_d - \hat{\overline{\boldsymbol{X}}}_d)\hat{\boldsymbol{\beta}}\right],$$

where $w_d = \sum_{j=1}^{n_d} w_{dj}$, $f_d = \frac{n_d}{N_d}$, $\hat{\gamma}_d^w = \hat{\sigma}_u^2/(\hat{\sigma}_u^2 + \hat{\sigma}_e^2/w_d)$,

$$\hat{\overline{Y}}_d^w = \frac{1}{w_d}\sum_{j=1}^{n_d} w_{dj}y_{dj},\ \hat{\overline{\boldsymbol{X}}}_d^w = \frac{1}{w_d}\sum_{j=1}^{n_d} w_{dj}\boldsymbol{x}_{dj},\ \hat{\overline{\boldsymbol{X}}}_d = \frac{1}{n_d}\sum_{j=1}^{n_d}\boldsymbol{x}_{dj},\ \overline{\boldsymbol{X}}_d = \frac{1}{N_d}\sum_{j=1}^{N_d}\boldsymbol{x}_{dj}.$$

This section calculates the MSE of the EBLUP of $\overline{Y}_d$. First, we calculate $g_i(\boldsymbol{\theta})$, $i = 1, 2, 3, 4$, with $\boldsymbol{\theta} = (\sigma_e^2, \sigma_u^2)$. For calculating $\boldsymbol{T}_s$, we use the formula

$$\boldsymbol{T}_s = \boldsymbol{V}_u - \boldsymbol{V}_u \boldsymbol{Z}_s' \boldsymbol{V}_s^{-1} \boldsymbol{Z}_s \boldsymbol{V}_u,$$

where $\boldsymbol{V}_u = \sigma_u^2 \boldsymbol{I}_D$, $\boldsymbol{V}_e = \sigma_e^2 \boldsymbol{W}^{-1}$, $\boldsymbol{Z}_s = \underset{1\leq d\leq D}{\text{diag}}(\boldsymbol{1}_{n_d})$, $\boldsymbol{V}_s^{-1} = \text{diag}(\boldsymbol{V}_{s1}^{-1}, \ldots, \boldsymbol{V}_{sD}^{-1})$, $\boldsymbol{V}_{sd}^{-1} = \frac{1}{\sigma_e^2}\left(\boldsymbol{W}_{sd} - \frac{\gamma_d^w}{w_d} \boldsymbol{w}_{n_d} \boldsymbol{w}_{n_d}'\right)$, and $\boldsymbol{w}_{n_d}' = (w_{d1}, \ldots, w_{dn_d})$. By substitution and using the fact that $\sigma_u^2/\sigma_e^2(1-\gamma_d^w) = \gamma_d^w/w_d$, we obtain

$$\begin{aligned}
\boldsymbol{V}_u \boldsymbol{Z}_s' \boldsymbol{V}_s^{-1} \boldsymbol{Z}_s \boldsymbol{V}_u &= \frac{\sigma_u^4}{\sigma_e^2} \underset{1\leq d\leq D}{\text{diag}}\left(\boldsymbol{1}_{n_d}'\right) \underset{1\leq d\leq D}{\text{diag}}\left(\boldsymbol{W}_{sd} - \frac{\gamma_d^w}{w_d} \boldsymbol{w}_{n_d} \boldsymbol{w}_{n_d}'\right) \underset{1\leq d\leq D}{\text{diag}}\left(\boldsymbol{1}_{n_d}\right) \\
&= \sigma_u^2 \underset{1\leq d\leq D}{\text{diag}}\left(\frac{\gamma_d^w}{w_d} \boldsymbol{w}_{n_d}'\right) \underset{1\leq d\leq D}{\text{diag}}\left(\boldsymbol{1}_{n_d}\right) = \sigma_u^2 \underset{1\leq d\leq D}{\text{diag}}\left(\gamma_d^w\right), \\
\boldsymbol{T}_s &= \sigma_u^2 \underset{1\leq d\leq D}{\text{diag}}\left(1 - \gamma_d^w\right).
\end{aligned}$$

As $\boldsymbol{1}_{N_d-n_d}' \boldsymbol{1}_{N_d-n_d} \boldsymbol{1}_{N_d-n_d}' \boldsymbol{1}_{N_d-n_d} = (N_d - n_d)^2$, the term g_1 is

$$\begin{aligned}
g_1(\boldsymbol{\theta}) &= \boldsymbol{a}_r' \boldsymbol{Z}_r \boldsymbol{T}_s \boldsymbol{Z}_r' \boldsymbol{a}_r = \frac{\sigma_u^2}{N_d^2} (\boldsymbol{0}', \ldots, \boldsymbol{0}', \boldsymbol{1}_{N_d-n_d}', \boldsymbol{0}', \ldots, \boldsymbol{0}') \underset{1\leq d\leq D}{\text{diag}}\left(\boldsymbol{1}_{N_d-n_d}\right) \\
&\cdot \underset{1\leq d\leq D}{\text{diag}}\left(1 - \gamma_d^w\right) \underset{1\leq d\leq D}{\text{diag}}\left(\boldsymbol{1}_{N_d-n_d}'\right) (\boldsymbol{0}', \ldots, \boldsymbol{0}', \boldsymbol{1}_{N_d-n_d}', \boldsymbol{0}' \ldots, \boldsymbol{0}')' \\
&= \frac{\sigma_u^2}{N_d^2} (\boldsymbol{0}', \ldots, \boldsymbol{0}', \boldsymbol{1}_{N_d-n_d}', \boldsymbol{0}' \ldots, \boldsymbol{0}') \underset{1\leq d\leq D}{\text{diag}}\left((1-\gamma_d^w) \boldsymbol{1}_{N_d-n_d} \boldsymbol{1}_{N_d-n_d}'\right) \\
&\cdot (\boldsymbol{0}', \ldots, \boldsymbol{0}', \boldsymbol{1}_{N_d-n_d}', \boldsymbol{0}' \ldots, \boldsymbol{0}')' = \frac{\sigma_u^2}{N_d^2} (1-\gamma_d^w)(N_d - n_d)^2 \\
&= (1-f_d)^2 (1-\gamma_d^w) \sigma_u^2 \approx (1-\gamma_d^w)\sigma_u^2 \quad \text{if } n_d \ll N_d.
\end{aligned}$$

The term g_2 is

$$g_2(\boldsymbol{\theta}) = [\boldsymbol{a}_r' \boldsymbol{X}_r - \boldsymbol{a}_r' \boldsymbol{Z}_r \boldsymbol{T}_s \boldsymbol{Z}_s' \boldsymbol{V}_{es}^{-1} \boldsymbol{X}_s] \boldsymbol{Q}_s [\boldsymbol{X}_r' \boldsymbol{a}_r - \boldsymbol{X}_s' \boldsymbol{V}_{es}^{-1} \boldsymbol{Z}_s \boldsymbol{T}_s \boldsymbol{Z}_r' \boldsymbol{a}_r],$$

where $\boldsymbol{Q}_s = (\boldsymbol{X}_s' \boldsymbol{V}_s^{-1} \boldsymbol{X}_s)^{-1}$. Let us define

$$\overline{\boldsymbol{X}}_d^* = \frac{1}{N_d - n_d} \sum_{j=n_d+1}^{N_d} \boldsymbol{x}_{dj}.$$

We obtain

$$\begin{aligned}
\boldsymbol{a}_r'\boldsymbol{X}_r &= \frac{1}{N_d}(\mathbf{0}',\ldots,\mathbf{0}',\mathbf{1}_{N_d-n_d}',\mathbf{0}'\ldots,\mathbf{0}')\boldsymbol{X}_r = (1-f_d)\overline{\boldsymbol{X}}_d^*,\\
\boldsymbol{Z}_r\boldsymbol{T}_s\boldsymbol{Z}_s' &= \sigma_u^2 \underset{1\le d\le D}{\operatorname{diag}}\left(\mathbf{1}_{N_d-n_d}\right)\underset{1\le d\le D}{\operatorname{diag}}\left(1-\gamma_d^w\right)\underset{1\le d\le D}{\operatorname{diag}}\left(\mathbf{1}_{n_d}'\right)\\
&= \sigma_u^2 \underset{1\le d\le D}{\operatorname{diag}}\left((1-\gamma_d^w)\mathbf{1}_{N_d-n_d}\mathbf{1}_{n_d}'\right)
\end{aligned}$$

and

$$\begin{aligned}
\boldsymbol{a}_r'\boldsymbol{Z}_r\boldsymbol{T}_s\boldsymbol{Z}_s'\boldsymbol{V}_{es}^{-1}\boldsymbol{X}_s &= \frac{1}{N_d}\frac{\sigma_u^2}{\sigma_e^2}(\mathbf{0}',\ldots,\mathbf{0}',\mathbf{1}_{N_d-n_d}',\mathbf{0}',\ldots,\mathbf{0}')\\
&\quad\cdot \underset{1\le d\le D}{\operatorname{diag}}\left((1-\gamma_d^w)\mathbf{1}_{N_d-n_d}\mathbf{1}_{n_d}'\right)\boldsymbol{W}_s\boldsymbol{X}_s\\
&= \frac{1}{N_d}\frac{\sigma_u^2}{\sigma_e^2}(1-\gamma_d^w)(N_d-n_d)\,(\mathbf{0}',\ldots,\mathbf{0}',\boldsymbol{w}_{n_d}',\mathbf{0}'\ldots,\mathbf{0}')\boldsymbol{X}_s\\
&= (1-f_d)\gamma_d^w\frac{1}{w_d}(\mathbf{0}',\ldots,\mathbf{0}',\boldsymbol{w}_{n_d}',\mathbf{0}'\ldots,\mathbf{0}')\boldsymbol{X}_s = (1-f_d)\gamma_d^w\hat{\overline{\boldsymbol{X}}}_d^w.
\end{aligned}$$

Finally, we have

$$\begin{aligned}
g_2(\boldsymbol{\theta}) &= (1-f_d)^2(\overline{\boldsymbol{X}}_d^* - \gamma_d^w\hat{\overline{\boldsymbol{X}}}_d^w)(\boldsymbol{X}_s'\boldsymbol{V}_s^{-1}\boldsymbol{X}_s)^{-1}(\overline{\boldsymbol{X}}_d^* - \gamma_d^w\hat{\overline{\boldsymbol{X}}}_d^w)'\\
&\approx (\overline{\boldsymbol{X}}_d - \gamma_d^w\hat{\overline{\boldsymbol{X}}}_d^w)(\boldsymbol{X}_s'\boldsymbol{V}_s^{-1}\boldsymbol{X}_s)^{-1}(\overline{\boldsymbol{X}}_d - \gamma_d^w\hat{\overline{\boldsymbol{X}}}_d^w)' \quad \text{if } n_d \ll N_d.
\end{aligned}$$

The term g_3 is

$$g_3(\boldsymbol{\theta}) \approx \operatorname{tr}\left\{(\nabla\boldsymbol{b}')\boldsymbol{V}_s(\nabla\boldsymbol{b}')'E\left[(\hat{\boldsymbol{\theta}}-\boldsymbol{\theta})(\hat{\boldsymbol{\theta}}-\boldsymbol{\theta})'\right]\right\},$$

where

$$\begin{aligned}
\boldsymbol{b}' = \boldsymbol{a}_r'\boldsymbol{Z}_r\boldsymbol{V}_u\boldsymbol{Z}_s'\boldsymbol{V}_s^{-1} &= \frac{1}{N_d}\frac{\sigma_u^2}{\sigma_e^2}(\mathbf{0}',\ldots,\mathbf{0}',\mathbf{1}_{N_d-n_d}',\mathbf{0}'\ldots,\mathbf{0}')\\
&\quad\cdot \underset{1\le d\le D}{\operatorname{diag}}\left(\mathbf{1}_{N_d-n_d}\right)\underset{1\le d\le D}{\operatorname{diag}}\left(\mathbf{1}_{n_d}'\right)\underset{1\le d\le D}{\operatorname{diag}}\left(\boldsymbol{W}_{sd}-\frac{\gamma_d^w}{w_d}\boldsymbol{w}_{n_d}\boldsymbol{w}_{n_d}'\right)\\
&= \frac{1}{N_d}(\mathbf{0}',\ldots,\mathbf{0}',\mathbf{1}_{N_d-n_d}',\mathbf{0}'\ldots,\mathbf{0}')\underset{1\le d\le D}{\operatorname{diag}}\left(\mathbf{1}_{N_d-n_d}\right)\underset{1\le d\le D}{\operatorname{diag}}\left(\frac{\gamma_d^w}{w_d}\boldsymbol{w}_{n_d}'\right)\\
&= \frac{1}{N_d}(\mathbf{0}',\ldots,\mathbf{0}',\mathbf{1}_{N_d-n_d}',\mathbf{0}'\ldots,\mathbf{0}')\underset{1\le d\le D}{\operatorname{diag}}\left(\frac{\gamma_d^w}{w_d}\mathbf{1}_{N_d-n_d}\boldsymbol{w}_{n_d}'\right)\\
&= \left(\mathbf{0}',\ldots,\mathbf{0}',\frac{\gamma_d^w}{w_d}\frac{N_d-n_d}{N_d}\boldsymbol{w}_{n_d}',\mathbf{0}',\ldots,\mathbf{0}'\right)
\end{aligned}$$

and

$$(\nabla \boldsymbol{b}') = \begin{pmatrix} \boldsymbol{0}', \ldots, \boldsymbol{0}', (1-f_d)\frac{\partial \gamma_d^w}{\partial \sigma_e^2}\frac{1}{w_d}\boldsymbol{w}'_{n_d}, \boldsymbol{0}' \ldots, \boldsymbol{0}' \\ \boldsymbol{0}', \ldots, \boldsymbol{0}', (1-f_d)\frac{\partial \gamma_d^w}{\partial \sigma_u^2}\frac{1}{w_d}\boldsymbol{w}'_{n_d}, \boldsymbol{0}' \ldots, \boldsymbol{0}' \end{pmatrix},$$

$$(\nabla \boldsymbol{b}')\boldsymbol{V}_s(\nabla \boldsymbol{b}')' = (1-f_d)^2 \frac{1}{w_d^2}\boldsymbol{w}'_{n_d}\boldsymbol{V}_{sd}\boldsymbol{w}_{n_d} \begin{pmatrix} \partial \gamma_d^w/\partial \sigma_e^2 \\ \partial \gamma_d^w/\partial \sigma_u^2 \end{pmatrix} \left(\partial \gamma_d^w/\partial \sigma_e^2,\ \partial \gamma_d^w/\partial \sigma_u^2\right).$$

$$\begin{aligned} \boldsymbol{w}'_{n_d}\boldsymbol{V}_{sd}\boldsymbol{w}_{n_d} &= \boldsymbol{w}'_{n_d}\left(\sigma_e^2 \boldsymbol{W}_{sd}^{-1} + \sigma_u^2 \boldsymbol{1}_{n_d}\boldsymbol{1}'_{n_d}\right)\boldsymbol{w}_{n_d} = \sigma_e^2 w_d + \sigma_u^2 w_d^2 \\ &= w_d^2\left(\sigma_u^2 + \sigma_e^2/w_d\right). \end{aligned}$$

Therefore, we get

$$\begin{aligned} g_3(\boldsymbol{\theta}) &\approx \operatorname{tr}\left\{(\nabla \boldsymbol{b}')\boldsymbol{V}_s(\nabla \boldsymbol{b}')' E\left[(\hat{\boldsymbol{\theta}}-\boldsymbol{\theta})(\hat{\boldsymbol{\theta}}-\boldsymbol{\theta})'\right]\right\} \\ &= \operatorname{tr}\left\{(1-f_d)^2\left(\sigma_u^2+\frac{\sigma_e^2}{w_d}\right)\begin{pmatrix} \left(\frac{\partial \gamma_d^w}{\partial \sigma_e^2}\right)^2 & \frac{\partial \gamma_d^w}{\partial \sigma_e^2}\frac{\partial \gamma_d^w}{\partial \sigma_u^2} \\ \frac{\partial \gamma_d^w}{\partial \sigma_e^2}\frac{\partial \gamma_d^w}{\partial \sigma_u^2} & \left(\frac{\partial \gamma_d^w}{\partial \sigma_u^2}\right)^2 \end{pmatrix}\begin{pmatrix} \operatorname{var}(\hat{\sigma}_e^2) & \operatorname{cov}(\hat{\sigma}_e^2, \hat{\sigma}_u^2) \\ \operatorname{cov}(\hat{\sigma}_e^2, \hat{\sigma}_u^2) & \operatorname{var}(\hat{\sigma}_u^2) \end{pmatrix}\right\} \\ &= (1-f_d)^2\left(\sigma_u^2+\frac{\sigma_e^2}{w_d}\right) \\ &\quad \cdot \left\{\left(\frac{\partial \gamma_d^w}{\partial \sigma_e^2}\right)^2 \operatorname{var}(\hat{\sigma}_e^2) + 2\frac{\partial \gamma_d^w}{\partial \sigma_e^2}\frac{\partial \gamma_d^w}{\partial \sigma_u^2}\operatorname{cov}(\hat{\sigma}_e^2, \hat{\sigma}_u^2) + \left(\frac{\partial \gamma_d^w}{\partial \sigma_u^2}\right)^2 \operatorname{var}(\hat{\sigma}_u^2)\right\}. \end{aligned}$$

By taking into account that

$$\gamma_d^w = \frac{\sigma_u^2}{\sigma_u^2+\sigma_e^2/w_d}, \quad \frac{\partial \gamma_d^w}{\partial \sigma_e^2} = \frac{-\sigma_u^2/w_d}{(\sigma_u^2+\sigma_e^2/w_d)^2} \quad \text{and} \quad \frac{\partial \gamma_d^w}{\partial \sigma_u^2} = \frac{\sigma_e^2/w_d}{(\sigma_u^2+\sigma_e^2/w_d)^2},$$

we obtain

$$\begin{aligned} g_3(\boldsymbol{\theta}) &= \frac{(1-f_d)^2}{w_d^2}\left(\sigma_u^2+\frac{\sigma_e^2}{w_d}\right)^{-3}\left\{\sigma_u^4 \operatorname{var}(\hat{\sigma}_e^2) - 2\sigma_u^2\sigma_e^2\operatorname{cov}(\hat{\sigma}_e^2, \hat{\sigma}_u^2) + \sigma_e^4 \operatorname{var}(\hat{\sigma}_u^2)\right\} \\ &\approx \left(\sigma_u^2+\frac{\sigma_e^2}{w_d}\right)^{-3} w_d^{-2}\operatorname{var}\left(\sigma_u^2\hat{\sigma}_e^2 - \sigma_e^2\hat{\sigma}_u^2\right) \quad \text{if } n_d \ll N_d. \end{aligned}$$

For g_4, we have

$$g_4(\boldsymbol{\theta}) = \boldsymbol{a}_r' \boldsymbol{V}_{er} \boldsymbol{a}_r = \frac{\sigma_e^2}{N_d^2}(\boldsymbol{0}', \ldots, \boldsymbol{1}_{N_d-n_d}', \ldots, \boldsymbol{0}')\boldsymbol{W}_r^{-1}(\boldsymbol{0}', \ldots, \boldsymbol{1}_{N_d-n_d}', \ldots, \boldsymbol{0}')'$$

$$= \frac{\sigma_e^2}{N_d^2}\boldsymbol{1}_{N_d-n_d}' \boldsymbol{W}_{rd}^{-1} \boldsymbol{1}_{N_d-n_d}.$$

If $w_{dj} = 1,\ d = 1, \ldots, D,\ j = 1, \ldots, N_d$, and N_d is sufficiently large, then

$$g_4(\boldsymbol{\theta}) = \frac{\sigma_e^2(N_d - n_d)}{N_d^2} = \frac{1 - f_d}{N_d}\sigma_e^2 \approx 0.$$

In summary, under the NER model (9.15), the MSE approximation is

$$MSE(\hat{\overline{Y}}_d^{eblup}) \approx g_1(\sigma_e^2, \sigma_u^2) + g_2(\sigma_e^2, \sigma_u^2) + g_3(\sigma_e^2, \sigma_u^2) + g_4(\sigma_e^2, \sigma_u^2),$$

where

$$g_1(\sigma_e^2, \sigma_u^2) = (1 - f_d)^2(1 - \gamma_d^w)\sigma_u^2,$$

$$g_2(\sigma_e^2, \sigma_u^2) = (1 - f_d)^2(\overline{\boldsymbol{X}}_d^* - \gamma_d^w \hat{\overline{\boldsymbol{X}}}_d^w)(\boldsymbol{X}_s' \boldsymbol{V}_s^{-1} \boldsymbol{X}_s)^{-1}(\overline{\boldsymbol{X}}_d^* - \gamma_d^w \hat{\overline{\boldsymbol{X}}}_d^w)',$$

$$g_3(\sigma_e^2, \sigma_u^2) \approx \frac{(1 - f_d)^2}{w_d^2}\left(\sigma_u^2 + \frac{\sigma_e^2}{w_d}\right)^{-3} \cdot \left\{\sigma_u^4 \overline{\text{var}}(\hat{\sigma}_e^2) - 2\sigma_u^2\sigma_e^2 \overline{\text{cov}}(\hat{\sigma}_e^2, \hat{\sigma}_u^2) + \sigma_e^4 \overline{\text{var}}(\hat{\sigma}_u^2)\right\},$$

$$g_4(\sigma_e^2, \sigma_u^2) = \frac{\sigma_e^2}{N_d^2}\boldsymbol{1}_{N_d-n_d}' \boldsymbol{W}_{rd}^{-1} \boldsymbol{1}_{N_d-n_d},$$

and $\overline{\text{var}}(\cdot)$ and $\overline{\text{cov}}(\cdot, \cdot)$ denote asymptotic variance and covariance, respectively.

If, moreover, N_d is large, $n_d \ll N_d$ and $w_{dj} = 1$ $(d = 1, \ldots, D,\ j = 1, \ldots, N_d)$, we obtain simplified formulas

$$MSE(\hat{\overline{Y}}_d^{eblup}) \approx g_1(\sigma_e^2, \sigma_u^2) + g_2(\sigma_e^2, \sigma_u^2) + g_3(\sigma_e^2, \sigma_u^2),$$

where

$$g_1(\sigma_e^2, \sigma_u^2) \approx (1 - \gamma_d^w)\sigma_u^2,$$

$$g_2(\sigma_e^2, \sigma_u^2) \approx (\overline{\boldsymbol{X}}_d - \gamma_d^w \hat{\overline{\boldsymbol{X}}}_d)(\boldsymbol{X}_s' \boldsymbol{V}_s^{-1} \boldsymbol{X}_s)^{-1}(\overline{\boldsymbol{X}}_d - \gamma_d^w \hat{\overline{\boldsymbol{X}}}_d)',$$

$$g_3(\sigma_e^2, \sigma_u^2) \approx \left(\sigma_u^2 + \frac{\sigma_e^2}{n_d}\right)^{-3} \frac{1}{n_d^2}\left\{\sigma_u^4 \overline{\text{var}}(\hat{\sigma}_e^2) - 2\sigma_u^2\sigma_e^2 \overline{\text{cov}}(\hat{\sigma}_e^2, \hat{\sigma}_u^2) + \sigma_e^4 \overline{\text{var}}(\hat{\sigma}_u^2)\right\}.$$

9.6 MSE Estimation in NER Models

This section considers three cases depending on the employed estimators of σ_e^2 and σ_u^2: Henderson 3, REML, or ML.

9.6.1 Henderson 3 Estimation of Variance Components

Let $\hat{\sigma}_e^2$ and $\hat{\sigma}_u^2$ be the Henderson 3 estimators of σ_e^2 and σ_u^2. The MSE estimator is

$$mse(\hat{\overline{Y}}_d^{eblup}) = g_1(\hat{\sigma}_e^2, \hat{\sigma}_u^2) + g_2(\hat{\sigma}_e^2, \hat{\sigma}_u^2) + 2g_3(\hat{\sigma}_e^2, \hat{\sigma}_u^2) + g_4(\hat{\sigma}_e^2, \hat{\sigma}_u^2).$$

The calculation of g_3 requires calculating the asymptotic variance and covariance of $\hat{\sigma}_e^2$ and $\hat{\sigma}_u^2$. The asymptotic variance of $\hat{\sigma}_e^2$ is

$$\overline{\text{var}}(\hat{\sigma}_e^2) = 2(n - D - p + \lambda)^{-1}\sigma_e^4,$$

where $\lambda = 1$ if the model (9.1) has intercept (auxiliary variable $X_0 \equiv 1$), and $\lambda = 0$ otherwise. The asymptotic variance of $\hat{\sigma}_u^2$ is

$$\overline{\text{var}}(\hat{\sigma}_u^2) = 2n_*^{-2}\Big[(n - D - p + \lambda)^{-1}(D - \lambda)(n - p)\sigma_e^4 + 2n_*\sigma_e^2\sigma_u^2 + n_{**}\sigma_u^4\Big],$$

where n_* and n_{**} are given in (7.28) and (7.29), respectively, i.e.

$$n_* = \sum_{d=1}^{D} w_d - \text{tr}\Big[(\boldsymbol{X}_s'\boldsymbol{W}_s\boldsymbol{X}_s)^{-1}\sum_{d=1}^{D} w_d^2(\hat{\overline{\boldsymbol{X}}}_d^w)'\hat{\overline{\boldsymbol{X}}}_d^w\Big],$$

$$n_{**} = \sum_{d=1}^{D} w_d^2 - 2\,\text{tr}\Big[(\boldsymbol{X}_s'\boldsymbol{W}_s\boldsymbol{X}_s)^{-1}\sum_{d=1}^{D} w_d^3(\hat{\overline{\boldsymbol{X}}}_d^w)'\hat{\overline{\boldsymbol{X}}}_d^w\Big] \qquad (9.16)$$

$$+\, \text{tr}\Bigg[\Big((\boldsymbol{X}_s'\boldsymbol{W}_s\boldsymbol{X}_s)^{-1}\sum_{d=1}^{D} w_d^2(\hat{\overline{\boldsymbol{X}}}_d^w)'\hat{\overline{\boldsymbol{X}}}_d^w\Big)^2\Bigg].$$

The asymptotic covariance of $\hat{\sigma}_e^2$ and $\hat{\sigma}_u^2$ is

$$\overline{\text{cov}}(\hat{\sigma}_e^2, \hat{\sigma}_u^2) = -2n_*^{-1}(n - D - p + \lambda)^{-1}(D - \lambda)\sigma_e^4.$$

9.6.2 REML Estimation of Variance Components

Let $\hat{\sigma}_e^2$ and $\hat{\sigma}_u^2$ be the REML estimators of σ_e^2 and σ_u^2. The MSE estimator is again

$$mse(\hat{\overline{Y}}_d^{eblup}) = g_1(\hat{\sigma}_e^2, \hat{\sigma}_u^2) + g_2(\hat{\sigma}_e^2, \hat{\sigma}_u^2) + 2g_3(\hat{\sigma}_e^2, \hat{\sigma}_u^2) + g_4(\hat{\sigma}_e^2, \hat{\sigma}_u^2).$$

The calculation of g_3 requires calculating the asymptotic variance and covariance of $\hat{\sigma}_e^2$ and $\hat{\sigma}_u^2$. The asymptotic covariance matrix of $(\hat{\sigma}_e^2, \hat{\sigma}_u^2)$ is the inverse of the Fisher information matrix, i.e.

$$\overline{\text{var}}(\hat{\sigma}_e^2, \hat{\sigma}_u^2) = \mathcal{I}^{-1}(\sigma_e^2, \sigma_u^2).$$

The elements of the Fisher information matrix are given in formulas (7.18) of Sect. 7.5.

9.6.3 ML Estimation of Variance Components

The model (9.14) is a particular case of (9.13) and can be written in the form

$$\boldsymbol{y}_d = \boldsymbol{X}_d\boldsymbol{\beta} + \boldsymbol{1}_{N_d}u_d + \boldsymbol{e}_d, \quad d = 1, \dots, D,$$

where $\boldsymbol{X}_d$ is $N_d \times p$, $\boldsymbol{\beta}$ is $p \times 1$, $\boldsymbol{1}_{N_d}$ is $N_d \times 1$, u_d is 1×1, and $\boldsymbol{e}_d$ is $N_d \times 1$. With this notation, the MSE estimator is

$$\begin{aligned} mse(\hat{\overline{Y}}_d^{eblup}) &= g_1(\hat{\sigma}_e^2, \hat{\sigma}_u^2) + g_2(\hat{\sigma}_e^2, \hat{\sigma}_u^2) + 2g_3(\hat{\sigma}_e^2, \hat{\sigma}_u^2) + g_4(\hat{\sigma}_e^2, \hat{\sigma}_u^2) \\ &\quad - \boldsymbol{b}_{\hat{\boldsymbol{\theta}}}'(\hat{\sigma}_e^2, \hat{\sigma}_u^2)\nabla g_1(\hat{\sigma}_e^2, \hat{\sigma}_u^2), \end{aligned}$$

where $\hat{\sigma}_e^2$ and $\hat{\sigma}_u^2$ are the ML estimators of σ_e^2 and σ_u^2, respectively. Concerning the bias correction term, we recall that

$$g_1(\sigma_e^2, \sigma_u^2) = (1-f_d)^2(1-\gamma_d^w)\sigma_u^2 = (1-f_d)^2\Big(1-\frac{w_d\sigma_u^2}{w_d\sigma_u^2+\sigma_e^2}\Big)\sigma_u^2 = \frac{(1-f_d)^2\sigma_u^2\sigma_e^2}{w_d\sigma_u^2+\sigma_e^2}.$$

The first derivatives of g_1 are

$$\frac{\partial g_1}{\partial \sigma_e^2} = \frac{(1-f_d)^2}{(w_d\sigma_u^2+\sigma_e^2)^2}\{\sigma_u^2(w_d\sigma_u^2+\sigma_e^2) - \sigma_u^2\sigma_e^2\} = \frac{(1-f_d)^2 w_d\sigma_u^4}{(w_d\sigma_u^2+\sigma_e^2)^2}$$

and

$$\frac{\partial g_1}{\partial \sigma_u^2} = \frac{(1-f_d)^2}{(w_d\sigma_u^2+\sigma_e^2)^2}\{\sigma_e^2(w_d\sigma_u^2+\sigma_e^2) - w_d\sigma_u^2\sigma_e^2\} = \frac{(1-f_d)^2\sigma_e^4}{(w_d\sigma_u^2+\sigma_e^2)^2}.$$

Therefore, we have

$$\nabla g_1(\sigma_e^2, \sigma_u^2) = \frac{(1-f_d)^2}{(\sigma_e^2+w_d\sigma_u^2)^2}(w_d\sigma_u^4, \sigma_e^4)'.$$

The first partial derivatives of $\gamma_d^w = w_d\sigma_u^2/(w_d\sigma_u^2+\sigma_e^2)$ are

$$\frac{\partial \gamma_d^w}{\partial \sigma_e^2} = -\frac{w_d\sigma_u^2}{(w_d\sigma_u^2+\sigma_e^2)^2}, \quad \frac{\partial \gamma_d^w}{\partial \sigma_u^2} = \frac{w_d(w_d\sigma_u^2+\sigma_e^2) - w_d^2\sigma_u^2}{(w_d\sigma_u^2+\sigma_e^2)^2} = \frac{w_d\sigma_e^2}{(w_d\sigma_u^2+\sigma_e^2)^2}.$$

The first partial derivatives of γ_d^w/σ_e^2 are

$$\frac{\partial}{\partial \sigma_e^2}\left\{\frac{\gamma_d^w}{\sigma_e^2}\right\} = \frac{\frac{\partial \gamma_d^w}{\partial \sigma_e^2}\sigma_e^2 - \gamma_d^w}{\sigma_e^4} = \frac{-\frac{w_d\sigma_u^2}{(w_d\sigma_u^2+\sigma_e^2)^2}\sigma_e^2 - \frac{w_d\sigma_u^2}{w_d\sigma_u^2+\sigma_e^2}}{\sigma_e^4}$$

$$= -\frac{w_d\sigma_u^2\sigma_e^2 + w_d\sigma_u^2(w_d\sigma_u^2+\sigma_e^2)}{\sigma_e^4(w_d\sigma_u^2+\sigma_e^2)^2},$$

$$\frac{\partial}{\partial \sigma_u^2}\left\{\frac{\gamma_d^w}{\sigma_e^2}\right\} = \frac{1}{\sigma_e^2}\frac{\partial \gamma_d^w}{\partial \sigma_u^2} = \frac{w_d\sigma_e^2}{\sigma_e^2(w_d\sigma_u^2+\sigma_e^2)^2} = \frac{w_d}{(w_d\sigma_u^2+\sigma_e^2)^2}.$$

The first partial derivatives of

$$\boldsymbol{V}_{sd}^{-1} = \frac{1}{\sigma_e^2}\boldsymbol{W}_{sd} - \frac{1}{w_d}\frac{\gamma_d^w}{\sigma_e^2}\boldsymbol{w}_{n_d}\boldsymbol{w}_{n_d}'$$

are

$$\frac{\partial \boldsymbol{V}_{sd}^{-1}}{\partial \sigma_e^2} = -\frac{1}{\sigma_e^4}\boldsymbol{W}_{sd} + \frac{1}{w_d}\frac{w_d\sigma_u^2(2\sigma_e^2+w_d\sigma_u^2)}{\sigma_e^4(w_d\sigma_u^2+\sigma_e^2)^2}\boldsymbol{w}_{n_d}\boldsymbol{w}_{n_d}',$$

$$\frac{\partial \boldsymbol{V}_{sd}^{-1}}{\partial \sigma_u^2} = -\frac{1}{w_d}\frac{w_d}{(w_d\sigma_u^2+\sigma_e^2)^2}\boldsymbol{w}_{n_d}\boldsymbol{w}_{n_d}'.$$

We finally obtain (cf. page 222)

$$b_{\hat{\theta}}(\sigma_e^2, \sigma_u^2) = \frac{1}{2D} \mathcal{I}^{-1}(\sigma_e^2, \sigma_u^2)$$
$$\cdot \begin{pmatrix} \operatorname{tr}\left\{ \sum_{d=1}^{D} (X'_{sd} V_{sd}^{-1} X_{sd})^{-1} \left(\sum_{d=1}^{D} X'_{sd} \frac{\partial V_{sd}^{-1}}{\partial \sigma_e^2} X_{sd} \right) \right\} \\ \operatorname{tr}\left\{ \sum_{d=1}^{D} (X'_{sd} V_{sd}^{-1} X_{sd})^{-1} \left(\sum_{d=1}^{D} X'_{sd} \frac{\partial V_{sd}^{-1}}{\partial \sigma_u^2} X_{sd} \right) \right\} \end{pmatrix},$$

where

$$X'_{sd} \frac{\partial V_{sd}^{-1}}{\partial \sigma_e^2} X_{sd} = -\frac{1}{\sigma_e^4} \sum_{j=1}^{n_d} w_{dj} x'_{dj} x_{dj} + \frac{\sigma_u^2 (2\sigma_e^2 + w_{d}\sigma_u^2)}{\sigma_e^4 (\sigma_e^2 + w_{d}\sigma_u^2)^2} w_d^2 \overline{\hat{X}}_d^{w\prime} \overline{\hat{X}}_d^{w},$$

$$X'_{sd} \frac{\partial V_{sd}^{-1}}{\partial \sigma_u^2} X_{sd} = -(\sigma_e^2 + w_{d}\sigma_u^2)^{-2} w_d^2 \overline{\hat{X}}_d^{w\prime} \overline{\hat{X}}_d^{w},$$

$$X'_{sd} V_{sd}^{-1} X_{sd} = \frac{1}{\sigma_e^2} \left[\sum_{j=1}^{n_d} w_{dj} x'_{dj} x_{dj} - \gamma_d w_d \overline{\hat{X}}_d^{w\prime} \overline{\hat{X}}_d^{w} \right],$$

and $\mathcal{I}(\sigma_e^2, \sigma_u^2)$ is the 2×2 Fisher information matrix, whose elements are given in formulas (7.12) of Sect. 7.3.

9.7 MSE Approximation in Linear Models with One Fixed Factor

Let us consider the model (9.15), where the u_d's are unknown parameters (and not random variables). In this case, $V_s = \sigma_e^2 W_s^{-1}$, and the EBLUP is BLUP because it does not depend on the estimator of the unique variance σ_e^2. The BLUP of the mean of the target variable at domain d is given in (5.12). Let us define $a'_r = (\mathbf{0}', \dots, \mathbf{1}'_{N_d - n_d}, \dots, \mathbf{0}')$,

$$M = \begin{pmatrix} M_s \\ M_r \end{pmatrix} = \begin{pmatrix} X_s & Z_s \\ X_r & Z_r \end{pmatrix},$$

and $\overline{M}_d^* = \frac{1}{N_d - n_d} \sum_{j \in n_d + 1}^{N_d} m_{dj} = \frac{N_d}{N_d - n_d} \overline{M}_d - \frac{n_d}{N_d - n_d} \overline{m}_d$, where m_{dj} is the corresponding row of the matrix M and $\overline{M}_d = \frac{1}{N_d} \sum_{j=1}^{N_d} m_{dj}$, $\overline{m}_d = \frac{1}{n_d} \sum_{j=1}^{n_d} m_{dj}$. By applying Corollary 4.2, we obtain the MSE

$$MSE(\hat{\overline{Y}}_d^{blup}) = h_2(\sigma_e^2) + h_4(\sigma_e^2),$$

where

$$h_2(\sigma_e^2) = \sigma_e^2 \boldsymbol{a}_r' \boldsymbol{M}_r (\boldsymbol{M}_s' \boldsymbol{W}_s \boldsymbol{M}_s)^{-1} \boldsymbol{M}_r' \boldsymbol{a}_r = (1 - f_d)^2 \sigma_e^2 \overline{\boldsymbol{M}}_d^* (\boldsymbol{M}_s' \boldsymbol{W}_s \boldsymbol{M}_s)^{-1} \overline{\boldsymbol{M}}_d^{*\prime}$$
$$\approx \sigma_e^2 \overline{\boldsymbol{M}}_d (\boldsymbol{M}_s' \boldsymbol{W}_s \boldsymbol{M}_s)^{-1} \overline{\boldsymbol{M}}_d' \quad \text{if } n_d \ll N_d,$$
$$h_4(\sigma_e^2) = \sigma_e^2 \boldsymbol{a}_r' \boldsymbol{W}_r^{-1} \boldsymbol{a}_r = \frac{\sigma_e^2}{N_d^2} \sum_{j=n_d+1}^{N_d} w_{dj}^{-1} = \frac{1 - f_d}{N_d} \sigma_e^2 \approx 0,$$

if $w_{dj} = 1$ for all d, j, and N_d is large.

9.8 Simulation Experiment

This section presents simulation experiment studying behavior of the MSE approximation for EBLUPs under a simple model.

9.8.1 Samples

The sample sizes are selected to be $n_d \equiv 5$. For $d = 1, \ldots, D$, $j = 1, \ldots, n_d$, the regressors are generated as $x_{dj} \sim U(a_d, b_d)$, i.e.

$$x_{dj} = (b_d - a_d) U_{dj} + a_d, \quad U_{dj} \sim U(0, 1), \quad a_d = 1,\ b_d = 1 + d, \quad d = 1, \ldots, D.$$

The heteroscedastic weights are $w_{dj} = 1/x_{dj}^{\ell}$, $\ell = 0, 1/2, 1$, $d = 1, \ldots, D$, $j = 1, \ldots, n_d$.

For $d = 1, \ldots, D$, $j = 1, \ldots, n_d$, the random domain effects and the random errors are generated in the following way:

$$u_d \sim N(0, \sigma_u^2), \quad e_{dj} \sim N(0, \sigma_e^2), \quad \text{with } (\sigma_u^2, \sigma_e^2) = (1, 1),$$

and the target variables are calculated as

$$y_{dj} = \beta x_{dj} + u_d + w_{dj}^{-1/2} e_{dj}, \quad \text{with } \beta = 1.$$

9.8.2 EBLUPs and MSEs

We consider the sample model

$$y_s = X_s\beta + Z_s u + e_s, \qquad (y_{sd} = X_{sd}\beta + u_d 1_{n_d} + e_{sd},\ d = 1, \ldots, D).$$

We calculate $\hat{\beta}$, $\hat{\sigma}_e^2$, and $\hat{\sigma}_u^2$ by the Henderson 3 method. We apply the projective EBLUP

$$\hat{\overline{Y}}_d^{eblup} = \overline{X}_d\hat{\beta} + \hat{\gamma}_d^w(\hat{\overline{Y}}_d^w - \hat{\overline{X}}_d^w\hat{\beta}),$$

where $w_d = \sum_{j=1}^{n_d} w_{dj}$ and

$$\hat{\overline{Y}}_d^w = \frac{1}{w_d}\sum_{j=1}^{n_d} w_{dj}y_{dj}, \overline{X}_d = \frac{1}{N_d}\sum_{j=1}^{N_d} x_{dj}, \hat{\overline{X}}_d^w = \frac{1}{w_d}\sum_{j=1}^{n_d} w_{dj}x_{dj}, \hat{\gamma}_d^w = \frac{\hat{\sigma}_u^2}{\hat{\sigma}_u^2 + \frac{\hat{\sigma}_e^2}{w_d}}.$$

Let us define $W_{sd} = \text{diag}\left(w_{d1}, \ldots, w_{dn_d}\right)_{n_d\times n_d}$, $w_{n_d}' = (w_{d1}, \ldots, w_{dn_d})_{1\times n_d}$ and

$$\hat{V}_{sd}^{-1} = \frac{1}{\hat{\sigma}_e^2}\left(W_{sd} - \frac{\hat{\gamma}_d^w}{w_d}w_{n_d}w_{n_d}'\right)_{n_d\times n_d}.$$

The estimator of the MSE of the EBLUP is

$$mse(\hat{\overline{Y}}_d^{eblup}) = g_1(\hat{\sigma}_e^2, \hat{\sigma}_u^2) + g_2(\hat{\sigma}_e^2, \hat{\sigma}_u^2) + 2g_3(\hat{\sigma}_e^2, \hat{\sigma}_u^2).$$

The functions g_1, g_2, and g_3 are

$$g_1(\hat{\sigma}_e^2, \hat{\sigma}_u^2) \approx (1 - \hat{\gamma}_d^w)\hat{\sigma}_u^2.$$

$$g_2(\hat{\sigma}_e^2, \hat{\sigma}_u^2) \approx (\overline{X}_d - \hat{\gamma}_d^w\hat{\overline{X}}_d^w)\Big(\sum_{d=1}^{D} X_{sd}'\hat{V}_{sd}^{-1}X_{sd}\Big)^{-1}(\overline{X}_d - \hat{\gamma}_d^w\hat{\overline{X}}_d^w)',$$

$$g_3(\hat{\sigma}_e^2, \hat{\sigma}_u^2) \approx \left(\hat{\sigma}_u^2 + \frac{\hat{\sigma}_e^2}{w_d}\right)^{-3}\frac{1}{w_d^2}\left\{\hat{\sigma}_u^4\hat{\text{var}}(\hat{\sigma}_e^2) - 2\hat{\sigma}_u^2\hat{\sigma}_e^2\hat{\text{cov}}(\hat{\sigma}_e^2, \hat{\sigma}_u^2) + \hat{\sigma}_e^4\hat{\text{var}}(\hat{\sigma}_u^2)\right\},$$

where

$$\begin{aligned}
\hat{\text{var}}(\hat{\sigma}_e^2) &= 2(n - D - p)^{-1}\hat{\sigma}_e^4,\\
\hat{\text{var}}(\hat{\sigma}_u^2) &= 2n_*^{-2}\left[D(n-p)(n-D-p)^{-1}\hat{\sigma}_e^4 + 2n_*\hat{\sigma}_e^2\hat{\sigma}_u^2 + n_{**}\hat{\sigma}_u^4\right],\\
\hat{\text{cov}}(\hat{\sigma}_e^2, \hat{\sigma}_u^2) &= -2n_*^{-1}(n - D - p)^{-1}D\hat{\sigma}_e^4,
\end{aligned}$$

n_* and n_{**} are given in (9.16) and p is the number of covariables ($p = 1$ in the simulation experiment).

9.8.3 Algorithm

Take $D = 30$ and $N_d = 1000$. The simulation algorithm has the following steps:

1. Repeat I times ($i = 1, \dots, I$).

   1.1. *Generation of the population* $U_i = \cup_{d=1}^{D} U_{id}$.

   - For $d = 1, \dots, D$, $j = 1, \dots, N_d$, generate $u_d^{(i)} \sim N(0, \sigma_u^2)$, $e_{dj}^{(i)} \sim N(0, \sigma_e^2)$.
   - For $d = 1, \dots, D$, $j = 1, \dots, N_d$, generate $x_{dj}^{(i)} \sim U(a_d, b_d)$.
   - For $d = 1, \dots, D$, $j = 1, \dots, N_d$, generate $y_{dj}^{(i)} = \beta x_{dj}^{(i)} + u_d^{(i)} + w_{dj}^{-1/2} e_{dj}^{(i)}$.
   - For $d = 1, \dots, D$, calculate $\overline{Y}_d^{(i)}$, $\overline{X}_d^{(i)}$.

   1.2. *Extraction of sample* s_i.
   For $d = 1, \dots, D$, extract a random sample of size n_d from the population U_{id}.

   1.3. Calculate the H3 estimates $\hat{\beta}_i$, $\hat{\sigma}_{ei}^2$, and $\hat{\sigma}_{ui}^2$.

   1.4. For $d = 1, \dots, D$, calculate $\hat{\overline{Y}}_d^{eblup,i}$,

$$mse_{di} = g_1(\hat{\sigma}_{ei}^2, \hat{\sigma}_{ui}^2) + g_2(\hat{\sigma}_{ei}^2, \hat{\sigma}_{ui}^2) + 2g_3(\hat{\sigma}_{ei}^2, \hat{\sigma}_{ui}^2)$$

$$\xi_{di} = I\left(\overline{Y}_d^{(i)} \in \left[\hat{\overline{Y}}_d^{eblup,i} - 1.96\sqrt{mse_{di}},\ \hat{\overline{Y}}_d^{eblup,i} + 1.96\sqrt{mse_{di}}\right]\right),$$

2. Output: For $d = 1, \dots, D$, calculate

$$EMSE_d = \frac{1}{I}\sum_{i=1}^{I}(\hat{\overline{Y}}_d^{eblup,i} - \overline{Y}_d^{(i)})^2, \quad mse_d = \frac{1}{I}\sum_{i=1}^{I} mse_{di}, \quad C_d = \frac{1}{I}\sum_{i=1}^{I}\xi_{di}.$$

Table 9.1 presents the MSEs of the parameter estimators for $\sigma_e^2 = \sigma_u^2 = 1$, $I = 1000$, and $n_d = 5$. For the 10 first domains, Table 9.2 presents the empirical MSEs of the EBLUPs ($EMSE_d$), the corresponding MSE estimates (mse_d), and the coverture of the 95% confidence intervals of the $\overline{Y}_d$'s (C_d).

Table 9.1 MSEs for $I = 1000$

	$\ell = 0$	$\ell = 0.5$	$\ell = 1$
$MSE(\hat{\beta})$	0.0026	0.0017	0.0016
$MSE(\hat{\sigma}_{e^2})$	0.0166	0.0164	0.0163
$MSE(\hat{\sigma}_{u^2})$	0.0990	0.1564	0.2864

Table 9.2 Results for $\sigma_e^2 = \sigma_u^2 = 1$, $I = 1000$, and $n_d = 5$

	$\ell = 0$			$\ell = 0.5$		
d	$EMSE_d$	mse_d	C_d	$EMSE_d$	mse_d	C_d
1	0.1594	0.1692	0.958	0.1887	0.2009	0.956
2	0.1664	0.1692	0.950	0.2208	0.2222	0.944
3	0.1696	0.1692	0.948	0.2405	0.2391	0.945
4	0.1685	0.1693	0.940	0.2548	0.2554	0.947
5	0.1730	0.1693	0.938	0.2720	0.2676	0.945
6	0.1631	0.1693	0.946	0.2667	0.2789	0.939
7	0.1728	0.1694	0.944	0.2951	0.2902	0.940
8	0.1704	0.1695	0.943	0.3089	0.2991	0.946
9	0.1561	0.1695	0.963	0.2874	0.3108	0.962
10	0.1834	0.1696	0.941	0.3442	0.3162	0.949
Mean	0.1874	0.1704	0.936	0.3845	0.3495	0.936

9.9 R Codes for MSEs

This section gives R codes for calculating the analytic estimators of the MSEs of the EBLUPs under the NER model. The R codes of this section are a continuation of the ones given in Sect. 8.8.2. First, run the R codes of Sect. 8.8.2. Second, run the codes appearing below.

Do some preliminary matrix calculations.

```
gammad <- nd*sigmau2/(nd*sigmau2+sigmae2)
udom <- sort(unique(dom))
Q <- QQ <- QQQ <- matrix(0,3,3); Vd.inv <- unod <- Xd <- list()
for (d in 1:D) {
 unod[[d]] <- matrix(1,nd[d])
 condition <- dom==udom[[d]]
 # Matrix Xd
 Xd[[d]] <- cbind(1, work[condition], nowork[condition])
 # Identity matrix
 Id <- diag(1,nd[d])
 # Inverse of Vd matrix
 Vd.inv[[d]] <- (Id-(gammad[d]/nd[d])*unod[[d]]%*%t(unod[[d]]))/sigmae2
 Q <- Q + t(Xd[[d]])%*%Vd.inv[[d]]%*%Xd[[d]]
 QQ <- QQ + t(Xd[[d]])%*%Vd.inv[[d]]%*%unod[[d]]%*%t(unod[[d]])%*%
           Vd.inv[[d]]%*%Xd[[d]]
 QQQ <- QQQ + t(Xd[[d]])%*%Vd.inv[[d]]%*%Vd.inv[[d]]%*%Xd[[d]]
}
Q <- solve(Q)
```

The following codes calculate the REML Fisher information matrix defined in Sect. 7.5. For calculating the element $F_{\sigma_u^2\sigma_u^2}$, run the code

```
Fuu1 <- Fuu2 <- Fuu3 <- 0
for (d in 1:D) {
 oneVone <- t(unod[[d]])%*%Vd.inv[[d]]%*%unod[[d]]
 Fuu1 <- Fuu1 + oneVone^2
 Fuu2 <- Fuu2 +
         oneVone*t(unod[[d]])%*%Vd.inv[[d]]%*%Xd[[d]]%*%Q%*%t(Xd[[d]])%*%
         Vd.inv[[d]]%*%unod[[d]]
 Fuu3 <- Fuu3 +
         t(unod[[d]])%*%Vd.inv[[d]]%*%Xd[[d]]%*%Q%*%QQ%*%Q%*%t(Xd[[d]])%*%
         Vd.inv[[d]]%*%unod[[d]]
}
Fuu <- 0.5*Fuu1 - Fuu2 + 0.5*Fuu3; Fuu
```

Calculate the element $F_{\sigma_u^2\sigma_e^2}$.

```
Fue1 <- Fue2 <- Fue3 <- 0
for (d in 1:D) {
 Fue1 <- Fue1 + t(unod[[d]])%*%Vd.inv[[d]]%*%Vd.inv[[d]]%*%unod[[d]]
 Fue2 <- Fue2 + t(unod[[d]])%*%Vd.inv[[d]]%*%Vd.inv[[d]]%*%Xd[[d]]%*%Q%*%
                t(Xd[[d]])%*%Vd.inv[[d]]%*% unod[[d]]
 Fue3 <- Fue3 + t(unod[[d]])%*%Vd.inv[[d]]%*%Xd[[d]]%*%Q%*%QQQ%*%Q%*%
                t(Xd[[d]])%*%Vd.inv[[d]]%*%unod[[d]]
}
Fue <- 0.5*Fue1 - Fue2 + 0.5*Fue3; Fue
```

Calculate the element $F_{\sigma_e^2\sigma_e^2}$.

```
Fee1 <- Fee2 <- Fee3 <- 0
for (d in 1:D) {
 Fee1 <- Fee1 + sum(diag(Vd.inv[[d]]%*%Vd.inv[[d]]))
 Fee2 <- Fee2 + sum(diag(Vd.inv[[d]]%*%Vd.inv[[d]]%*%Xd[[d]]%*%Q%*%
                t(Xd[[d]])%*%Vd.inv[[d]]))
 Fee3 <- Fee3 + sum(diag(Vd.inv[[d]]%*%Xd[[d]]%*%Q%*%QQQ%*%Q%*%
                t(Xd[[d]])%*%Vd.inv[[d]]))
}
Fee <- 0.5*Fee1 - Fee2 + 0.5*Fee3; Fee
```

Calculate the REML Fisher information matrix.

```
F <- matrix(c(Fuu, Fue, Fue, Fee), 2, 2); F
F.inv <- solve(F); F.inv
```

Calculate g_1.

```
g1 <- (1-gammad)*sigmau2; g1
```

Calculate g_2.

```
g2 <- vector()
for (d in 1:D) {
 xd.bar <- c(1, Xmean$work[d], Xmean$nowork[d])
 xd.dird <- c(1, dirwork[d], dirnowork[d])
 ad <- xd.bar - gammad[d]*xd.dird
 g2[d] <- t(ad)%*%Q%*%ad
}
g2
```

Table 9.3 MSE estimations

d	3	5	6	7	11	12	13	14	15
mse_B	1,042,121	696,889	812,825	2,626,551	616,950	2,347,230	489,786	415,288	174,299
mse_g	1,110,668	716,467	820,023	3,151,846	595,666	2,421,292	516,222	384,865	187,246
$R_{B/g}$	−6.17	−2.73	−0.88	−16.67	3.57	−3.06	−5.12	7.91	−6.91
d	16	17	18	20	21	22	23	24	25
mse_B	687,910	2,876,743	1,437,648	607,043	1,209,800	2,632,350	1,443,794	874,731	753,276
mse_g	735,225	2,934,141	1,605,401	564,633	1,256,542	2,838,051	1,460,025	998,611	847,011
$R_{B/g}$	−6.44	−1.96	−10.45	7.51	−3.72	−7.25	−1.11	−12.41	−11.07
d	27	28	29	30	31	32	33	34	
mse_B	749,037	1,176,872	834,316	495,003	847,049	257,352	573,620	1,062,634	
mse_g	819,962	1,110,847	952,852	530,319	1,094,738	254,306	539,273	1,065,668	
$R_{B/g}$	−8.65	5.94	−12.44	−6.66	−22.63	1.20	6.37	−0.28	

Calculate g_3.

```
g31 <- g32 <- g33 <- g3 <- vector()
for (d in 1:D) {
  gfix <- nd[d]/((nd[d]*sigmau2+sigmae2)^3)
  g31[d] <- sigmau2^2*F.inv[2,2]*gfix
  g32[d] <- -2*sigmau2*sigmae2*F.inv[1,2]*gfix
  g33[d] <- sigmae2^2*F.inv[1,1]*gfix
}
g3 <- g31 + g32 + g33; g3
```

Calculate g_4 and $mse = g_1 + g_2 + 2g_3 + g_4$.

```
g4 <- sigmae2*(aux$TOT-nd)/(aux$TOT^2); g4
mseg <- g1 + g2 + 2*g3 + g4; mseg
```

Table 9.3 presents the bootstrap (mse_B) and the analytic (mse_g) estimations of the MSEs of the EBLUPs. It also presents the relative difference in percentage, i.e. $R_{B/g} = 100\frac{mse_B - mse_g}{mse_g}$.

```
R.Bg <- round(100*(ebmse-mseg)/mseg,2); R.Bg
```

Give a summary of the obtained results.

```
output <- data.frame(mseB=round(ebmse,0), mseg=round(mseg,0), R.Bg)
output
```

We remind that the bootstrap estimators are calculated with $B = 500$ replicates.

References

Das, K., Jiang, J., Rao, J.N.K.: Mean squared error of empirical predictor. Technical Report, School of Mathematics and Statistics, Carleton University, Ottawa, Canada. (2001)

Das, K., Jiang, J., Rao, J.N.K.: Mean squared error of empirical predictor. Ann. Stat. **32**, 818–840 (2004)

Datta, G.S., Lahiri, P.: A unified measure of uncertainty of estimated best linear unbiased predictors in small area estimation problems. Stat. Sin. **10**, 613–627 (2000)

González-Manteiga, W., Lombardía, M.J., Molina, I., Morales, D., Santamaría, L.: Bootstrap mean squared error of small-area EBLUP. J. Stat. Comput. Simul. **78**, 443–462 (2008a)

Hall, P., Maiti, T.: Nonparametric estimation of mean-squared prediction error in nested-error regression models. Ann. Stat. **34**(4), 1733–1750 (2006a)

Hall, P., Maiti, T.: On parametric bootstrap methods for small-area prediction. J. R. Stat. Soc. B **68**, 221–238 (2006b)

Harville, D.A., Jeske, D.R.: Mean squared error of estimation or prediction under General Linear Model. J. Am. Stat. Assoc. **87**, 724–731 (1992)

Jiang, J., Lahiri, P., Wan, S.: A unified jackknife theory for empirical best prediction with M-estimation. Ann. Stat. **30**, 1782–1810 (2002)

Kackar, R.N., Harville, D.A.: Unbiasedness of two-stage estimation and prediction procedures for mixed linear models. Commun. Stat. Theory Methods **10**, 1249–1261 (1981)

Kackar, R.N., Harville, D.A.: Approximations for standard errors of estimators of fixed and random effects in mixed linear models. J. Am. Stat. Assoc. **79**, 853–862 (1984)

Lahiri, P., Rao, J.N.K.: Robust estimation of mean squared error of small area estimators. J. Am. Stat. Assoc. **90**, 758–766 (1995)

Pfefferman, D., Tiller, R.: Bootstrap approximation to prediction MSE for state-space models with estimated parameters. J. Time Ser. Anal. **26**, 893–916 (2005)

Prasad, N.G.N., Rao, J.N.K.: The estimation of the mean squared error of small-area estimators. J. Am. Stat. Assoc. **85**, 163–171 (1990)

Rao, J.N.K.: Small Area Estimation. John Wiley (2003)

Searle, S.R., Casella, G., McCullogh, C.E.: Variance Components. Wiley, New York (1992)

Chapter 10
EBPs Under Nested Error Regression Models

10.1 Introduction

The estimation of domain totals and means has been widely treated in the previous chapters via design-based estimators or empirical best linear unbiased predictors (EBLUPs). These parameters are linear combinations of a target population vector $\boldsymbol{y}$. In practice, there are many socioeconomic indicators that are not linear. Therefore, statisticians need further unit-level model-based procedures for estimating specific types of nonlinear parameters.

For the estimation of complex domain parameters such as certain poverty indicators, Molina and Rao (2010) proposed the empirical best predictors (EBPs), based on the assumption that a one-to-one transformation of the target variable follows the nested error regression (NER) model of Battese et al. (1988). Under that model, the EBP method gives approximately the "best" predictor in the sense of having minimum error variance in the class of unbiased predictors.

Elbers et al. (2003) introduced the ELL method by assuming that $\boldsymbol{y}$ follows a NER model, where the random effects are included for subdomains of the domains of estimation. They further derived predictors of additive parameters based on the marginal distribution of the assumed NER model. Since many household surveys use a two-stage sampling with primary sampling units (clusters) nested within the domains of estimation, the ELL method uses random effects for these clusters rather than for the domains of interest as in the EBP method.

This chapter derives empirical best predictors of additive parameters based on NER models and pays special attention to the family of poverty indicators given by Foster et al. (1984). There are other currently available model-based approaches for estimating small area poverty indicators that are not treated in this chapter. Without being exhaustive, we cite here some procedures based on unit-level linear mixed models. Molina et al. (2014) uses a Bayesian version of the NER model to derive hierarchical Bayes (HB) predictors for poverty mapping in Spanish provinces by gender. Guadarrama et al. (2016) compared several methods to predict

D. Morales et al., *A Course on Small Area Estimation and Mixed Models*, Statistics for Social and Behavioral Sciences, https://doi.org/10.1007/978-3-030-63757-6_10

poverty indicators, including EBP, ELL, and HB approaches, under NER models. Marhuenda et al. (2017) gave several procedures for estimating poverty indicators under two-fold NER models. The selection of the transformation was included in the estimation process by Rojas-Perilla et al. (2020). For the case where normality cannot be assumed, this methodology was extended by Graf et al. (2019) and Diallo and Rao (2018) to unit-level mixed models with skewed distributions.

On the other hand, by using other modeling strategies, we cite some proposals for the estimation of poverty proportions. Farrell et al. (1997) gives bootstrap adjustments for empirical Bayes interval estimates of small area proportions. Malec et al. (1997) gave some small area inference methods for binary variables. Chambers et al. (2012) introduce an M-quantile regression approach for binary data. Hobza and Morales (2016) and Hobza et al. (2020) introduced predictors based on unit-level logit and generalized mixed models, respectively. Based on temporal and spatio-temporal area-level models, Esteban et al. (2012a,b), Marhuenda et al. (2013), and Morales et al. (2015) gave EBLUPs of poverty proportions.

This chapter is organized as follows. Section 10.2 gives the conditional distribution of normal vectors. Under the domain-level NER model, Sect. 10.3 calculates the distribution of the non-observed target vector y_{dr} conditioned to the observed vector y_{ds}. Based on the conditional distribution, Sects. 10.4 and 10.5 derive the empirical best predictors (EBPs) of means and additive parameters. Section 10.6 gives the corresponding EBPs based on subdomain-level NER models. Section 10.7 introduces the ELL prediction methodology. Section 10.8 presents parametric bootstrap algorithms for estimation of the mean squared error of EBPs. Section 10.9 gives some R codes for calculating EBPs.

10.2 The Conditional Distribution of Normal Vectors

Consider a vector $\boldsymbol{y} = (y_1, \ldots, y_N)'$ containing the values that a random variable takes in the N units of a finite population. Let $\boldsymbol{y}_s$ be the sub-vector of $\boldsymbol{y}$ corresponding to sample elements, and let $\boldsymbol{y}_r$ be the sub-vector of $\boldsymbol{y}$ corresponding to the out-of-sample elements; that is, $\boldsymbol{y} = (\boldsymbol{y}_s', \boldsymbol{y}_r')'$. The inference target is to predict the value of a real-valued function $\delta = h(\boldsymbol{y})$ of the random vector $\boldsymbol{y}$ using the sample data $\boldsymbol{y}_s$.

The best predictor of δ is $\hat{\delta}^B = E_{\boldsymbol{y}_r}\big[\delta|\boldsymbol{y}_s\big]$, where the expectation is taken with respect to the conditional distribution of $\boldsymbol{y}_r$ given $\boldsymbol{y}_s$. If $\boldsymbol{y}$ is normally distributed with mean vector $\boldsymbol{\mu} = (\boldsymbol{\mu}_s', \boldsymbol{\mu}_r')'$ and covariance matrix

$$\boldsymbol{V} = \begin{pmatrix} \boldsymbol{V}_s & \boldsymbol{V}_{sr} \\ \boldsymbol{V}_{rs} & \boldsymbol{V}_r \end{pmatrix},$$

then the distribution of $\boldsymbol{y}_r$, given $\boldsymbol{y}_s$, is also multivariate normal, i.e.

$$\boldsymbol{y}_r|\boldsymbol{y}_s \sim N(\boldsymbol{\mu}_{r|s}, \boldsymbol{V}_{r|s}), \tag{10.1}$$

with mean vector and covariance matrix (see e.g. Theorem 2.2E in Rencher (1998))

$$\boldsymbol{\mu}_{r|s} = \boldsymbol{\mu}_r + \boldsymbol{V}_{rs}\boldsymbol{V}_s^{-1}(\boldsymbol{y}_s - \boldsymbol{\mu}_s), \quad \boldsymbol{V}_{r|s} = \boldsymbol{V}_r - \boldsymbol{V}_{rs}\boldsymbol{V}_s^{-1}\boldsymbol{V}_{sr}.$$

10.3 The Nested Error Regression Model

Let us assume that $\boldsymbol{y}$ follows the NER superpopulation model

$$y_{dj} = \boldsymbol{x}_{dj}\boldsymbol{\beta} + u_d + e_{dj}, \quad d = 1, \ldots, D, \; j = 1, \ldots, N_d, \tag{10.2}$$

where $\boldsymbol{x}_{dj}$ is a row vector containing the values of p auxiliary variables and $\boldsymbol{\beta}$ is the column vector of regression parameters. The NER model assumes that the random effects u_d and the errors e_{dj} are mutually independent, $u_d \sim N(0, \sigma_u^2)$, and $e_{dj} \sim N(0, w_{dj}^{-1}\sigma_e^2)$ with known heteroscedasticity weights $w_{dj} > 0$. Let us define $\boldsymbol{e}_d = \underset{1\le j\le N_d}{\text{col}}(e_{dj})$ and $\boldsymbol{1}_a = \underset{1\le j\le a}{\text{col}}(1)$, where the operator $\underset{1\le j\le a}{\text{col}}$ defines a column vector of dimension a. Then, the model (10.2) can be written in the vector form

$$\boldsymbol{y}_d = \boldsymbol{X}_d\boldsymbol{\beta} + \boldsymbol{1}_{N_d}u_d + \boldsymbol{e}_d, \quad d = 1, \ldots, D.$$

The vectors $\boldsymbol{e}_d$ are independent with $\boldsymbol{e}_d \sim N(\boldsymbol{0}_{N_d}, \sigma_e^2\boldsymbol{W}_d^{-1})$, where $\boldsymbol{0}_a = \underset{1\le j\le a}{\text{col}}(0)$, $\boldsymbol{W}_d = \underset{1\le j\le N_d}{\text{diag}}(w_{dj})$, and the operator $\underset{1\le j\le a}{\text{diag}}$ defines an $a \times a$ diagonal matrix. The vectors $\boldsymbol{y}_d$ are independent with $\boldsymbol{y}_d \sim N(\boldsymbol{\mu}_d, \boldsymbol{V}_d)$, $\boldsymbol{\mu}_d = \boldsymbol{X}_d\boldsymbol{\beta}$, $\boldsymbol{V}_d = \sigma_u^2\boldsymbol{1}_{N_d}\boldsymbol{1}_{N_d}' + \sigma_e^2\boldsymbol{W}_d^{-1}$, and $\boldsymbol{X}_d = \underset{1\le j\le N_d}{\text{col}}(\boldsymbol{x}_{dj})$.

In practice, inference is carried out based on a sample, $s = \cup_{d=1}^D s_d$, drawn from the finite population U with sample size n_d in domain d. The non-sampled part of the population is denoted by $r = U - s = \cup_{d=1}^D r_d$. Let $\boldsymbol{y}_{ds}$ be the sub-vector of $\boldsymbol{y}_d$ corresponding to sample elements and $\boldsymbol{y}_{dr}$ the sub-vector of $\boldsymbol{y}_d$ containing the out-of-sample elements. Without loss of generality, we can write $\boldsymbol{y}_d = \text{col}(\boldsymbol{y}_{ds}, \boldsymbol{y}_{dr}) = (\boldsymbol{y}_{ds}', \boldsymbol{y}_{dr}')'$. We also define the corresponding decompositions $\boldsymbol{X}_d = \text{col}(\boldsymbol{X}_{ds}, \boldsymbol{X}_{dr})$ and

$$\boldsymbol{V}_d = \begin{pmatrix} \boldsymbol{V}_{ds} & \boldsymbol{V}_{dsr} \\ \boldsymbol{V}_{drs} & \boldsymbol{V}_{dr} \end{pmatrix}.$$

The sample sub-vectors $\boldsymbol{y}_{ds}$ follow the marginal models derived from the population model (10.2), i.e.

$$\boldsymbol{y}_{ds} = \boldsymbol{X}_{ds}\boldsymbol{\beta} + \boldsymbol{1}_{n_d}u_d + \boldsymbol{e}_{ds}, \quad d = 1, \ldots, D. \tag{10.3}$$

The vectors $\boldsymbol{e}_{ds}$ are independent with $\boldsymbol{e}_{ds} \sim N(\boldsymbol{0}_{n_d}, \sigma_e^2 \boldsymbol{W}_{ds}^{-1})$, where $\boldsymbol{W}_{ds} = \underset{1\leq j\leq n_d}{\text{diag}} (w_{dj})$. The vectors $\boldsymbol{y}_{ds}$ are independent with $\boldsymbol{y}_{ds} \sim N(\boldsymbol{\mu}_{ds}, \boldsymbol{V}_{ds})$, $\boldsymbol{\mu}_{ds} = \boldsymbol{X}_{ds}\boldsymbol{\beta}$, $\boldsymbol{V}_{ds} = \sigma_u^2 \boldsymbol{1}_{n_d}\boldsymbol{1}_{n_d}' + \sigma_e^2 \boldsymbol{W}_{ds}^{-1}$.

If $\sigma_e^2 > 0$ and $\sigma_u^2 > 0$ are known, the best linear unbiased estimator (BLUE) of $\boldsymbol{\beta}$ and the best linear unbiased predictor (BLUP) of u_d, $d = 1, \dots, D$, are given in (7.7). They are

$$\tilde{\boldsymbol{\beta}} = \Big(\sum_{d=1}^{D} \boldsymbol{X}_{ds}' \boldsymbol{V}_{ds}^{-1} \boldsymbol{X}_{ds}\Big)^{-1} \sum_{d=1}^{D} \boldsymbol{X}_{ds}' \boldsymbol{V}_{ds}^{-1} \boldsymbol{y}_{ds}, \; \tilde{u}_d = \sigma_u^2 \boldsymbol{1}_{n_d}' \boldsymbol{V}_{ds}^{-1} \left(\boldsymbol{y}_{ds} - \boldsymbol{X}_{ds}\tilde{\boldsymbol{\beta}}\right), \tag{10.4}$$

where the inverse of $\boldsymbol{V}_{ds}$ was obtained in (7.6), i.e.

$$\boldsymbol{V}_{ds}^{-1} = \frac{1}{\sigma_e^2}\left(\boldsymbol{W}_{ds} - \frac{\gamma_d^w}{w_d} \boldsymbol{w}_{n_d} \boldsymbol{w}_{n_d}'\right), \quad \gamma_d^w = \frac{w_d \sigma_u^2}{w_d \sigma_u^2 + \sigma_e^2},$$

with $\boldsymbol{w}_{n_d} = \underset{1\leq j\leq n_d}{\text{col}} (w_{dj})$ and $w_d = \boldsymbol{1}_{n_d}' \boldsymbol{w}_{n_d} = \sum_{j=1}^{n_d} w_{dj}$. If we substitute σ_e^2 and σ_u^2 by consistent estimators $\hat{\sigma}_e^2$ and $\hat{\sigma}_u^2$ in (10.4) and we define $\hat{\boldsymbol{V}}_{ds} = \hat{\sigma}_u^2 \boldsymbol{1}_{n_d}\boldsymbol{1}_{n_d}' + \hat{\sigma}_e^2 \boldsymbol{W}_{ds}^{-1}$, then we obtain the empirical BLUE (EBLUE) of $\boldsymbol{\beta}$ and the empirical BLUP (EBLUP) of u_d in the form

$$\hat{\boldsymbol{\beta}} = \Big(\sum_{d=1}^{D} \boldsymbol{X}_{ds}' \hat{\boldsymbol{V}}_{ds}^{-1} \boldsymbol{X}_{ds}\Big)^{-1} \sum_{d=1}^{D} \boldsymbol{X}_{ds}' \hat{\boldsymbol{V}}_{ds}^{-1} \boldsymbol{y}_{ds}, \; \hat{u}_d = \hat{\sigma}_u^2 \boldsymbol{1}_{n_d}' \hat{\boldsymbol{V}}_{ds}^{-1} (\boldsymbol{y}_{ds} - \boldsymbol{X}_{ds}\hat{\boldsymbol{\beta}}). \tag{10.5}$$

The non-sampled sub-vectors $\boldsymbol{y}_{dr}$ follow the marginal models derived from the population model (10.2), i.e.

$$\boldsymbol{y}_{dr} = \boldsymbol{X}_{dr}\boldsymbol{\beta} + \boldsymbol{1}_{N_d-n_d} u_d + \boldsymbol{e}_{dr}, \quad d = 1, \dots, D. \tag{10.6}$$

The vectors $\boldsymbol{e}_{dr}$ are independent with $\boldsymbol{e}_{dr} \sim N(\boldsymbol{0}_{N_d-n_d}, \sigma_e^2 \boldsymbol{W}_{dr}^{-1})$, where $\boldsymbol{W}_{dr} = \underset{n_d+1\leq j\leq N_d}{\text{diag}} (w_{dj})$. The vectors $\boldsymbol{y}_{dr}$ are independent and normally distributed with $\boldsymbol{y}_{dr} \sim N(\boldsymbol{\mu}_{dr}, \boldsymbol{V}_{dr})$, $\boldsymbol{\mu}_{dr} = \boldsymbol{X}_{dr}\boldsymbol{\beta}$, and $\boldsymbol{V}_{dr} = \sigma_u^2 \boldsymbol{1}_{N_d-n_d}\boldsymbol{1}_{N_d-n_d}' + \sigma_e^2 \boldsymbol{W}_{dr}^{-1}$. The covariance matrix between $\boldsymbol{y}_{dr}$ and $\boldsymbol{y}_{ds}$ is

$$\begin{aligned}\boldsymbol{V}_{drs} &= \text{cov}\big(\boldsymbol{y}_{dr}, \boldsymbol{y}_{ds}\big) = \text{cov}\big(\boldsymbol{X}_{dr}\boldsymbol{\beta} + \boldsymbol{1}_{N_d-n_d} u_d + \boldsymbol{e}_{dr}, \boldsymbol{X}_{ds}\boldsymbol{\beta} + \boldsymbol{1}_{n_d} u_d + \boldsymbol{e}_{ds}\big) \\ &= \boldsymbol{1}_{N_d-n_d} \text{var}(u_d) \boldsymbol{1}_{n_d}' = \sigma_u^2 \boldsymbol{1}_{N_d-n_d} \boldsymbol{1}_{n_d}'.\end{aligned}$$

The distribution of $\boldsymbol{y}_{dr}$, given the sample data $\boldsymbol{y}_s$, is

$$\boldsymbol{y}_{dr}|\boldsymbol{y}_s \sim \boldsymbol{y}_{dr}|\boldsymbol{y}_{ds} \sim N(\boldsymbol{\mu}_{dr|s}, \boldsymbol{V}_{dr|s}), \tag{10.7}$$

where the conditional mean vector is

$$\boldsymbol{\mu}_{dr|s} = \boldsymbol{\mu}_{dr} + \boldsymbol{V}_{drs}\boldsymbol{V}_{ds}^{-1}(\boldsymbol{y}_{ds} - \boldsymbol{\mu}_{ds}) = \boldsymbol{X}_{dr}\boldsymbol{\beta} + \sigma_u^2 \mathbf{1}_{N_d-n_d}\mathbf{1}'_{n_d}\boldsymbol{V}_{ds}^{-1}(\boldsymbol{y}_{ds} - \boldsymbol{X}_{ds}\boldsymbol{\beta})$$

and the conditional covariance matrix is

$$\begin{aligned}
\boldsymbol{V}_{dr|s} &= \boldsymbol{V}_{dr} - \boldsymbol{V}_{drs}\boldsymbol{V}_{ds}^{-1}\boldsymbol{V}_{dsr} = \sigma_u^2 \mathbf{1}_{N_d-n_d}\mathbf{1}'_{N_d-n_d} + \sigma_e^2 \mathbf{W}_{dr}^{-1} \\
&\quad - \sigma_u^2 \mathbf{1}_{N_d-n_d}\mathbf{1}'_{n_d}\frac{1}{\sigma_e^2}\Big(\mathbf{W}_{ds} - \frac{\gamma_d^w}{w_d}\boldsymbol{w}_{n_d}\boldsymbol{w}'_{n_d}\Big)\sigma_u^2 \mathbf{1}_{n_d}\mathbf{1}'_{N_d-n_d} \\
&= \sigma_u^2 \mathbf{1}_{N_d-n_d}\mathbf{1}'_{N_d-n_d} + \sigma_e^2 \mathbf{W}_{dr}^{-1} - \frac{\sigma_u^4}{\sigma_e^2}\Big(w_d - \frac{\gamma_d^w}{w_d}w_d^2\Big)\mathbf{1}_{N_d-n_d}\mathbf{1}'_{N_d-n_d} \\
&= \sigma_u^2\Big(1 - \frac{\sigma_u^2}{\sigma_e^2}w_d(1-\gamma_d^w)\Big)\mathbf{1}_{N_d-n_d}\mathbf{1}'_{N_d-n_d} + \sigma_e^2 \mathbf{W}_{dr}^{-1}.
\end{aligned}$$

It holds that

$$1-\frac{\sigma_u^2}{\sigma_e^2}w_d(1-\gamma_d^w) = 1-\frac{\sigma_u^2}{\sigma_e^2}w_d\frac{\sigma_e^2}{w_d\sigma_u^2+\sigma_e^2} = 1-\frac{w_d\sigma_u^2}{w_d\sigma_u^2+\sigma_e^2} = \frac{\sigma_e^2}{w_d\sigma_u^2+\sigma_e^2} = 1-\gamma_d^w.$$

Therefore, the conditional covariance matrix is

$$\boldsymbol{V}_{dr|s} = \sigma_u^2(1-\gamma_d^w)\mathbf{1}_{N_d-n_d}\mathbf{1}'_{N_d-n_d} + \sigma_e^2 \mathbf{W}_{dr}^{-1}.$$

If $n_d \neq 0$ and $j \in U_d - s_d$, the conditional mean is

$$\begin{aligned}
\mu_{dj|s} &= \boldsymbol{x}_{dj}\boldsymbol{\beta} + \frac{\sigma_u^2}{\sigma_e^2}\mathbf{1}'_{n_d}\Big(\mathbf{W}_{ds} - \frac{\gamma_d^w}{w_d}\boldsymbol{w}_{n_d}\boldsymbol{w}'_{n_d}\Big)(\boldsymbol{y}_{ds} - \boldsymbol{X}_{ds}\boldsymbol{\beta}) \\
&= \boldsymbol{x}_{dj}\boldsymbol{\beta} + \frac{\sigma_u^2}{\sigma_e^2}w_d(1-\gamma_d^w)(\hat{\bar{Y}}_d^w - \hat{\bar{\boldsymbol{X}}}_d^w\boldsymbol{\beta}) = \boldsymbol{x}_{dj}\boldsymbol{\beta} + \gamma_d^w(\hat{\bar{Y}}_d^w - \hat{\bar{\boldsymbol{X}}}_d^w\boldsymbol{\beta}),
\end{aligned}$$

where $\hat{\bar{Y}}_d^w = w_d^{-1}\sum_{j=1}^{n_d} w_{dj}y_{dj}$ and $\hat{\bar{\boldsymbol{X}}}_d^w = w_d^{-1}\sum_{j=1}^{n_d} w_{dj}\boldsymbol{x}_{dj}$. For any $j \in U_d - s_d$, it thus holds that

$$\mu_{dj|s} = \begin{cases} \boldsymbol{x}_{dj}\boldsymbol{\beta} + \gamma_d^w(\hat{\bar{Y}}_d^w - \hat{\bar{\boldsymbol{X}}}_d^w\boldsymbol{\beta}) & \text{if } n_d \neq 0, \\ \boldsymbol{x}_{dj}\boldsymbol{\beta} & \text{if } n_d = 0, \end{cases} \qquad \gamma_d^w = \frac{w_d\sigma_u^2}{w_d\sigma_u^2+\sigma_e^2}.$$

For any $j \in U_d - s_d$, the conditional variance is

$$v_{dj|s} = \begin{cases} \sigma_u^2(1-\gamma_d^w) + w_{dj}^{-1}\sigma_e^2 & \text{if } n_d \neq 0, \\ \sigma_u^2 + w_{dj}^{-1}\sigma_e^2 & \text{if } n_d = 0. \end{cases}$$

The conditional distribution of $\mathbf{y}_r$, given $\mathbf{y}_s$, plays an important role in the calculations of the best predictors (BPs) of population parameters $\delta = \delta(\mathbf{y})$. Assume that the model parameters $\boldsymbol{\theta} = (\boldsymbol{\beta}', \sigma_u^2, \sigma_e^2)'$ are known. Under the NER model, the BP is an unbiased predictor $\hat{\delta} = \hat{\delta}(\mathbf{y}_s)$ of δ that minimizes the MSE. To obtain the BP of δ, we have to solve the problem

$$\min_{\hat{\delta}} E\big[(\hat{\delta} - \delta)^2\big] \quad \text{subject to} \quad E[\hat{\delta} - \delta] = 0,$$

where the expectation is taken with respect to the probability density function of $\mathbf{y}$.

Proposition 10.1 *The BP of* δ *is* $\hat{\delta}^{bp} = E_{\mathbf{y}_r}[\delta | \mathbf{y}_s]$.

Proof Let $\hat{\delta}$ be a predictor of δ. Note that $\hat{\delta}$ is a function of $\mathbf{y}_s$ and δ is a function of $\mathbf{y}$. In the proof we simplify the notation and we do not use the subindexes in the expectations, i.e. we use the notation $E_{\mathbf{y}_r}[\delta|\mathbf{y}_s] = E[\delta|\mathbf{y}_s]$. From the formulas, it is clear with respect to which random vector the expectations are taken. It holds

$$\begin{aligned}
E[(\hat{\delta} - \delta)^2] &= E\Big[E[(\hat{\delta} - \delta)^2 | \mathbf{y}_s]\Big] \\
&= E\Big[E[\{(\hat{\delta} - E[\delta|\mathbf{y}_s]) + (E[\delta|\mathbf{y}_s] - \delta)\}^2 | \mathbf{y}_s]\Big] \\
&= E\Big[E[(\hat{\delta} - E[\delta|\mathbf{y}_s])^2 | \mathbf{y}_s]\Big] + E\Big[E[(E[\delta|\mathbf{y}_s] - \delta)^2 | \mathbf{y}_s]\Big] \\
&+ 2E\Big[(\hat{\delta} - E[\delta|\mathbf{y}_s])E\big[E[\delta|\mathbf{y}_s] - \delta | \mathbf{y}_s\big]\Big].
\end{aligned}$$

The last summand is zero because

$$E\big[E[\delta|\mathbf{y}_s] - \delta | \mathbf{y}_s\big] = E[\delta|\mathbf{y}_s] - E[\delta|\mathbf{y}_s] = 0.$$

Since the second summand does not depend on $\hat{\delta}$, by taking $\hat{\delta} = E[\delta|\mathbf{y}_s]$, the first summand is zero and $E\big[(\hat{\delta} - \delta)^2\big]$ is minimized. Further, it holds that

$$\begin{aligned}
E\big[\hat{\delta} - \delta\big] = E\big[E[\delta|\mathbf{y}_s] - \delta\big] &= E\Big[E\big[E[\delta|\mathbf{y}_s] - \delta | \mathbf{y}_s\big]\Big] \\
&= E\Big[E[\delta|\mathbf{y}_s] - E[\delta|\mathbf{y}_s]\Big] = 0.
\end{aligned}$$

This is to say, $\hat{\delta} = E[\delta|\mathbf{y}_s]$ is an unbiased predictor of δ. □

For domain parameters, we substitute $\mathbf{y}$ and $\mathbf{y}_s$ by $\mathbf{y}_d$ and $\mathbf{y}_{ds}$, respectively. Further, if the parameter of interest takes the linear form $\delta_d = \frac{1}{N_d}\sum_{j=1}^{N_d} \delta_{dj}$, then the BP is

$$\hat{\delta}_d^{bp} = E[\delta_d|\mathbf{y}_s] = \frac{1}{N_d}\sum_{j=1}^{N_d} E[\delta_{dj}|\mathbf{y}_s] = \frac{1}{N_d}\sum_{j=1}^{N_d} \hat{\delta}_{dj}^{bp}.$$

10.4 EBPs of Domain Means

This section derives the EBPs of $\overline{Y}_d$, where $\overline{Y}_d = \frac{1}{N_d}\sum_{j=1}^{N_d} y_{dj}$. Let $\hat{\boldsymbol{\beta}}$, $\hat{\sigma}_u^2$, and $\hat{\sigma}_e^2$ be consistent estimators of the model parameters $\boldsymbol{\beta}$, σ_u^2, and σ_e^2, respectively. Under the conditioned distribution (10.7), the predicted values are

$$\hat{\boldsymbol{y}}_{ds}^{ebp} = \boldsymbol{y}_{ds}, \quad \hat{\boldsymbol{y}}_{dr}^{ebp} = \hat{\boldsymbol{\mu}}_{dr|s} = \boldsymbol{X}_{dr}\hat{\boldsymbol{\beta}} + \hat{\sigma}_u^2 \mathbf{1}_{N_d-n_d}\mathbf{1}_{n_d}' \hat{\boldsymbol{V}}_{ds}^{-1}(\boldsymbol{y}_{ds} - \boldsymbol{X}_{ds}\hat{\boldsymbol{\beta}}),$$

or equivalently

$$\hat{y}_{dj}^{ebp} = \begin{cases} y_{dj} & \text{if } j \in s_d, \\ \boldsymbol{x}_{dj}\hat{\boldsymbol{\beta}} + \hat{u}_d & \text{if } j \in r_d = U_d - s_d, \end{cases}$$

where the predicted random effects, see (10.5), are

$$\hat{u}_d = \hat{\sigma}_u^2 \mathbf{1}_{n_d}' \hat{\boldsymbol{V}}_{ds}^{-1}(\boldsymbol{y}_{ds} - \boldsymbol{X}_{ds}\hat{\boldsymbol{\beta}}) = \hat{\gamma}_d^w(\hat{\overline{Y}}_d^w - \hat{\overline{\boldsymbol{X}}}_d^w\hat{\boldsymbol{\beta}}), \ \hat{\gamma}_d^w = \frac{w_d\hat{\sigma}_u^2}{w_d\hat{\sigma}_u^2 + \hat{\sigma}_e^2}, \ d = 1, \dots, D.$$

The EBP or predictive estimator of $\overline{Y}_d$ is

$$\begin{aligned}
\hat{\overline{Y}}_d^{ebp} &= \frac{1}{N_d}\sum_{j=1}^{N_d} \hat{y}_{dj}^{ebp} = \frac{1}{N_d}\sum_{j\in s_d} y_{dj} + \frac{1}{N_d}\sum_{j\in r_d}\{\boldsymbol{x}_{dj}\hat{\boldsymbol{\beta}} + \hat{u}_d\} \\
&= f_d\hat{\overline{Y}}_d + \frac{1}{N_d}\sum_{j\in U_d}\{\boldsymbol{x}_{dj}\hat{\boldsymbol{\beta}} + \hat{u}_d\} - f_d\frac{1}{n_d}\sum_{j\in s_d}\{\boldsymbol{x}_{dj}\hat{\boldsymbol{\beta}} + \hat{u}_d\} \\
&= (1 - f_d)\big[\overline{\boldsymbol{X}}_d\hat{\boldsymbol{\beta}} + \hat{u}_d\big] + f_d\big[\hat{\overline{Y}}_d + (\overline{\boldsymbol{X}}_d - \hat{\overline{\boldsymbol{X}}}_d)\hat{\boldsymbol{\beta}}\big],
\end{aligned}$$

where $f_d = \frac{n_d}{N_d}$ is the domain sample fraction, $\hat{\overline{Y}}_d = \frac{1}{n_d}\sum_{j\in s_d} y_{dj}$, $\hat{\overline{\boldsymbol{X}}}_d = \frac{1}{n_d}\sum_{j\in s_d} \boldsymbol{x}_{dj}$, and $\overline{\boldsymbol{X}}_d = \frac{1}{N_d}\sum_{j\in U_d} \boldsymbol{x}_{dj}$.

Let us note that the derived formulas for EBP of domain means coincide with the formulas for EBLUPs of domain means derived in Sect. 8.3 by using Theorem 4.1.

10.5 EBPs of Additive Parameters

This section considers additive domain parameters that can be written in the form

$$\delta_d = \frac{1}{N_d}\sum_{j=1}^{N_d} h(y_{dj}), \tag{10.8}$$

where h is a known measurable function. A class of additive parameters is the FGT family of poverty indicators given by Foster et al. (1984) and presented with more details in (12.28). We consider three additive parameters based on income variables that belong to the FGT family.

Let z_{dj} be the annual net income of individual j from domain d, $d = 1, \ldots, D$, $j = 1, \ldots, N_d$, and z be a poverty threshold. Let $I(z_{dj} < z)$ be the indicator function defined by $I(z_{dj} < z) = 1$ if $z_{dj} < z$ and $I(z_{dj} < z) = 0$ otherwise. The poverty proportions, the poverty gaps, and the average incomes, at the domain level, are

$$P_d = \frac{1}{N_d}\sum_{j=1}^{N_d} p_{dj}, \quad G_d = \frac{1}{N_d}\sum_{j=1}^{N_d} g_{dj}, \quad \overline{Z}_d = \frac{1}{N_d}\sum_{j=1}^{N_d} z_{dj}, \qquad d = 1, \ldots, D, \tag{10.9}$$

where $p_{dj} = I(z_{dj} < z)$, $g_{dj} = \left(\frac{z - z_{dj}}{z}\right) I(z_{dj} < z)$, $d = 1, \ldots, D$, $j = 1, \ldots, N_d$. If $y_{dj} = T(z_{dj})$, where $T : R \mapsto R$ is a one-to-one increasing function, then the three socioeconomic indicators defined in (10.9) are small area additive parameters, as they can be written in the form (10.8). For example, the poverty proportions can be written as

$$P_d = \frac{1}{N_d}\sum_{j=1}^{N_d} I\left(T^{-1}(y_{dj}) < z\right) = \frac{1}{N_d}\sum_{j=1}^{N_d} I\left(y_{dj} < T(z)\right).$$

The best predictor (BP) of δ_d is the conditional expectation

$$\hat{\delta}_d^{bp} = E_{y_r}\left[\frac{1}{N_d}\sum_{j=1}^{N_d} h(y_{dj})\bigg|\boldsymbol{y}_s\right] = \frac{1}{N_d}\left\{\sum_{j\in s_d} h(y_{dj}) + \sum_{j\in r_d} E_{y_r}\left[h(y_{dj})|\boldsymbol{y}_s\right]\right\}.$$

The conditional distribution (10.7) depends on the vector $\boldsymbol{\theta} = (\boldsymbol{\beta}', \sigma_u^2, \sigma_e^2)'$ of unknown model parameters, which must be estimated, that is,

$$E_{y_r}\left[h(y_{dj})|\boldsymbol{y}_s\right] = E_{y_r}\left[h(y_{dj})|\boldsymbol{y}_s; \boldsymbol{\theta}\right].$$

Let $\hat{\boldsymbol{\theta}} = (\hat{\boldsymbol{\beta}}', \hat{\sigma}_u^2, \hat{\sigma}_e^2)'$ be an estimator based on sample data $\boldsymbol{y}_s$. The EBP of δ_d is

$$\hat{\delta}_d^{ebp} = \frac{1}{N_d}\left\{\sum_{j\in s_d} h(y_{dj}) + \sum_{j\in r_d} E_{y_r}\left[h(y_{dj})|\boldsymbol{y}_s; \hat{\boldsymbol{\theta}}\right]\right\}. \tag{10.10}$$

For a general function h, the EBP expectation might be not calculable analytically. When this happens, the following Monte Carlo procedure can be applied:

(a) Estimate the unknown parameter $\boldsymbol{\theta} = (\boldsymbol{\beta}', \sigma_u^2, \sigma_e^2)'$ using the sample data $\boldsymbol{y}_s$.

(b) Using the estimate $\hat{\boldsymbol{\theta}} = (\hat{\boldsymbol{\beta}}', \hat{\sigma}_u^2, \hat{\sigma}_e^2)'$ obtained in (a), draw L copies of each non-sample variable y_{dj} as

$$y_{dj}^{(\ell)} \sim N(\hat{\mu}_{dj|s}, \hat{v}_{dj|s}), \quad j \in r_d, \; d = 1, \ldots, D, \; \ell = 1, \ldots, L,$$

where

$$\hat{\mu}_{dj|s} = \boldsymbol{x}_{dj}\hat{\boldsymbol{\beta}} + \hat{\gamma}_d^w(\hat{\overline{Y}}_d^w - \hat{\overline{\boldsymbol{X}}}_d^w \hat{\boldsymbol{\beta}}), \; \hat{v}_{dj|s} = \hat{\sigma}_u^2(1 - \hat{\gamma}_d^w) + w_{dj}^{-1}\hat{\sigma}_e^2 \quad \text{if} \quad n_d \neq 0;$$

$$\hat{\mu}_{dj|s} = \boldsymbol{x}_{dj}\hat{\boldsymbol{\beta}}, \qquad \hat{v}_{dj|s} = \hat{\sigma}_u^2 + w_{dj}^{-1}\hat{\sigma}_e^2 \quad \text{if} \quad n_d = 0,$$

and $\hat{\gamma}_d^w = \frac{w_{d}\hat{\sigma}_u^2}{w_{d}\hat{\sigma}_u^2 + \hat{\sigma}_e^2}$, $\hat{\overline{Y}}_d^w = w_{d}^{-1}\sum_{j=1}^{n_d} w_{dj}y_{dj}$, and $\hat{\overline{\boldsymbol{X}}}_d^w = w_{d}^{-1}\sum_{j=1}^{n_d} w_{dj}\boldsymbol{x}_{dj}$.

(c) Approximate the EBP expectation by the Monte Carlo method, i.e.

$$E_{\boldsymbol{y}_r}\left[h(y_{dj})|\boldsymbol{y}_s; \hat{\boldsymbol{\theta}}\right] \approx \frac{1}{L}\sum_{\ell=1}^{L} h(y_{dj}^{(\ell)}), \quad j \in r_d, \; d = 1, \ldots, D.$$

The Monte Carlo approximation of the EBP of δ_d is

$$\hat{\delta}_d^{ebp} \approx \frac{1}{L}\sum_{\ell=1}^{L} \delta_d^{(\ell)}, \quad \delta_d^{(\ell)} = \frac{1}{N_d}\left\{\sum_{j \in s_d} h(y_{dj}) + \sum_{j \in r_d} h(y_{dj}^{(\ell)})\right\}. \tag{10.11}$$

Remark 10.1 The EBP of a domain additive parameter is calculated by applying the formula (10.11) or formula (10.10) if the expectation can be calculated analytically. For applying one of these formulas, we need a census file containing:

(A) the values of the auxiliary variables in all the units of the population,
(B) the same values of the auxiliary variables for the sample units as in the sample file, and
(C) the variable that identifies the sample units in the sample data file, in order to separate the sampled from the non-sampled parts of the population.

If a census file, fulfilling conditions (A), (B), and (C), is available, then the EBP based on the selected NER model with continuous auxiliary variables is a very competitive estimator. However, updated census files are rarely available in practice. This is to say, in many cases the values of the auxiliary variables are not available for all the population units. If in addition some of the variables are continuous, the EBP method is not applicable.

An important particular case, where the EBP method is applicable, is when the model (10.2) is homoscedastic and the number of values of the vector of auxiliary variables is finite. More concretely, suppose that the auxiliary variables are categorical such that $\boldsymbol{x}_{dj} \in \{\boldsymbol{x}_1, \ldots, \boldsymbol{x}_K\}$ and that $w_{dj} = 1$, $d = 1, \ldots, D$, $j = 1, \ldots, N_d$. In this case, the population $U = \cup_{d=1}^{D} U_d$ can be partitioned into domain-class subsets $U_{dk} = \{j \in U_d : \boldsymbol{x}_{dj} = \boldsymbol{x}_k\}$ such that $U_d = \cup_{k=1}^{K} U_{dk}$ and the NER model (10.2) can be written in the form

$$y_{dkj} = \boldsymbol{x}_k \boldsymbol{\beta} + u_d + e_{dkj}, \quad d = 1, \ldots, D, \; k = 1, \ldots, K, \; j = 1, \ldots, N_{dk},$$

where the size N_{dk} of U_{dk} is supposed to be available from external data sources (aggregated auxiliary information). Let us define $n_{dk} = \#\{j \in s_d : \boldsymbol{x}_{dj} = \boldsymbol{x}_k\}$,

$$\begin{array}{lll} \hat{\mu}_{dk|s} = \boldsymbol{x}_k \hat{\boldsymbol{\beta}} + \hat{\gamma}_d(\hat{\overline{Y}}_d - \hat{\overline{\boldsymbol{X}}}_d \hat{\boldsymbol{\beta}}), & \hat{v}_{d|s} = \hat{\sigma}_u^2(1 - \hat{\gamma}_d) + \hat{\sigma}_e^2 & \text{if} \quad n_d \neq 0; \\ \hat{\mu}_{dk|s} = \boldsymbol{x}_k \hat{\boldsymbol{\beta}}, & \hat{v}_{d|s} = \hat{\sigma}_u^2 + \hat{\sigma}_e^2 & \text{if} \quad n_d = 0, \end{array}$$

where

$$\hat{\gamma}_d = \frac{n_d \hat{\sigma}_u^2}{n_d \hat{\sigma}_u^2 + \hat{\sigma}_e^2}, \quad \hat{\overline{Y}}_d = \frac{1}{n_d} \sum_{k=1}^{K} \sum_{j=1}^{n_{dk}} y_{dkj}, \quad \hat{\overline{\boldsymbol{X}}}_d = \frac{1}{n_d} \sum_{k=1}^{K} n_{dk} \boldsymbol{x}_k.$$

Then we can calculate $\delta_d^{(\ell)}$, defined in (10.11), as

$$\delta_d^{(\ell)} = \frac{1}{N_d} \left\{ \sum_{k=1}^{K} \sum_{j=1}^{n_{dk}} h(y_{dkj}) + \sum_{k=1}^{K} \sum_{j=n_{dk}+1}^{N_{dk}} h(y_{dkj}^{(\ell)}) \right\}, \tag{10.12}$$

where $y_{dkj}^{(\ell)} \sim N(\hat{\mu}_{dk|s}, \hat{v}_{d|s})$, $k = 1, \ldots, K, d = 1, \ldots, D$, $\ell = 1, \ldots, L$.

Remark 10.2 If the selected NER model contains continuous auxiliary variables and there is no available census file, it is possible to use a design-based approximation to $\delta_d^{(\ell)}$ when $n_d > 0$. This is to say,

$$\hat{\delta}_d^{(\ell)} = \frac{1}{N_d} \left\{ \sum_{j \in s_d} \left(h(y_{dj}) - h(y_{dj}^{(\ell)})\right) + \sum_{j \in s_d} \omega_{dj} h(y_{dj}^{(\ell)}) \right\}, \tag{10.13}$$

where ω_{dj}'s are calibrated sampling weights and

$$y_{dj}^{(\ell)} \sim N(\hat{\mu}_{dj|s}, \hat{v}_{dj|s}), \quad j \in s_d, \; d = 1, \ldots, D, \; \ell = 1, \ldots, L,$$

with $\hat{\mu}_{dj|s} = \boldsymbol{x}_{dj} \hat{\boldsymbol{\beta}} + \hat{\gamma}_d^w(\hat{\overline{Y}}_d^w - \hat{\overline{\boldsymbol{X}}}_d^w \hat{\boldsymbol{\beta}})$, $\hat{v}_{dj|s} = \hat{\sigma}_u^2(1 - \hat{\gamma}_d^w) + w_{dj}^{-1} \hat{\sigma}_e^2$, and

$$\hat{\gamma}_d^w = \frac{w_d \hat{\sigma}_u^2}{w_d \hat{\sigma}_u^2 + \hat{\sigma}_e^2}, \quad \hat{\overline{Y}}_d^w = w_d^{-1} \sum_{j=1}^{n_d} w_{dj} y_{dj}, \quad \hat{\overline{\boldsymbol{X}}}_d^w = w_d^{-1} \sum_{j=1}^{n_d} w_{dj} \boldsymbol{x}_{dj}.$$

Let us note that the second sum in the formula (10.13) is a dir1 estimator (Horvitz–Thompson) of $h_{tot} = \sum_{j\in U_d} h(y_{dj}^{(\ell)})$. If we subtract $\sum_{j\in s_d} h(y_{dj}^{(\ell)})$ from h_{tot}, we obtain a design-based approximation to $\sum_{j\in r_d} h(y_{dj}^{(\ell)})$ in (10.11). As the summand $\sum_{j\in s_d} h(y_{dj})$ appears in (10.11) and (10.13), then $\hat{\delta}_d^{(\ell)}$ is a design-based approximation of $\delta_d^{(\ell)}$. However, there are two main concerns for using the approximation (10.13):

1. It generates $y_{dj}^{(\ell)}$ for $j \in s_d$, but the normal distribution parameters, $\hat{\mu}_{dj|s}$ and $\hat{v}_{dj|s}$, are only applicable for $j \in r_d$.
2. It is a kind of model-based direct estimator with no optimal properties under design-based or model-based distributions.

In the next subsections we particularize the derived formula for the EBP of a general additive parameter δ_d to three specific cases, namely the poverty proportion, the poverty gap, and the average income.

10.5.1 Poverty Proportion

Let z_{dj} be an income variable measured on individual j from domain d, and let z be the poverty line. The domain poverty proportion is

$$P_d = \frac{1}{N_d}\sum_{j=1}^{N_d} I(z_{dj} < z),$$

where $I(z_{dj} < z) = 1$ if $z_{dj} < z$ and $I(z_{dj} < z) = 0$ otherwise. Assume that a one-to-one increasing transformation of the income variable for each unit, $y_{dj} = T(z_{dj})$, follows the NER model (10.2). Then, the domain poverty proportion is the additive parameter

$$P_d = \frac{1}{N_d}\sum_{j=1}^{N_d} h_0(y_{dj}), \quad h_0(y_{dj}) = I(y_{dj} < T(z)).$$

The EBP of P_d is

$$\hat{p}_d^{ebp} = \frac{1}{N_d}\left\{\sum_{j\in s_d} h_0(y_{dj}) + \sum_{j\in r_d} \hat{p}_{dj}^{ebp}\right\}, \tag{10.14}$$

where

$$\hat{p}_{dj}^{ebp} = E_{y_r}\big[I(y_{dj} < T(z))|\boldsymbol{y}_s;\hat{\boldsymbol{\theta}}\big] = P_{y_r}\big(y_{dj} < T(z)|\boldsymbol{y}_s;\hat{\boldsymbol{\theta}}\big) = P\big(N(0,1) < \hat{\alpha}_{dj}\big)$$
$$= \Phi(\hat{\alpha}_{dj}), \quad \hat{\alpha}_{dj} = \hat{v}_{dj|s}^{-1/2}(T(z) - \hat{\mu}_{dj|s}),$$

$\hat{v}_{dj|s}$ and $\hat{\mu}_{dj|s}$ are plug-in estimators of the conditional variance $v_{dj|s}$ and mean $\mu_{dj|s}$, respectively, and $\Phi(\cdot)$ is the cumulative distribution function of a standard normal random variable.

10.5.2 Poverty Gap

Let z_{dj} be the income of individual j from domain d, and let z be the poverty line. The domain poverty gap is

$$G_d = \frac{1}{N_d} \sum_{j=1}^{N_d} \frac{z - z_{dj}}{z} I(z_{dj} < z),$$

where $I(z_{dj} < z) = 1$ if $z_{dj} < z$ and $I(z_{dj} < z) = 0$ otherwise. Assume that a one-to-one increasing transformation of the income variable for each unit, $y_{dj} = T(z_{dj})$, follows the nested error regression model (10.2). The domain poverty gap is the additive parameter

$$G_d = \frac{1}{N_d} \sum_{j=1}^{N_d} h_1(y_{dj}), \quad h_1(y_{dj}) = \frac{z - T^{-1}(y_{dj})}{z} I(T^{-1}(y_{dj}) < z).$$

The EBP of G_d is

$$\hat{g}_d^{ebp} = \frac{1}{N_d} \left\{ \sum_{j \in s_d} h_1(y_{dj}) + \sum_{j \in r_d} \hat{g}_{dj}^{ebp} \right\}, \tag{10.15}$$

where $\hat{g}_{dj}^{ebp} \triangleq E_{\boldsymbol{y}_r}[h_1(y_{dj})|\boldsymbol{y}_s; \hat{\boldsymbol{\theta}}]$, which can be expressed as

$$\begin{aligned}
\hat{g}_{dj}^{ebp} &= E_{\boldsymbol{y}_r}\left[\frac{z - T^{-1}(y_{dj})}{z} I\big(y_{dj} < T(z)\big) \Big| \boldsymbol{y}_s; \hat{\boldsymbol{\theta}} \right] \\
&= E_{\boldsymbol{y}_r}\left[I\big(y_{dj} < T(z)\big) \Big| \boldsymbol{y}_s; \hat{\boldsymbol{\theta}} \right] - \frac{1}{z} E_{\boldsymbol{y}_r}\left[T^{-1}(y_{dj}) I\big(y_{dj} < T(z)\big) \Big| \boldsymbol{y}_s; \hat{\boldsymbol{\theta}} \right] \\
&\triangleq S_1 - \frac{1}{z} S_2.
\end{aligned}$$

We have proved that the first summand is

$$S_1 = E_{\boldsymbol{y}_r}\left[I\big(y_{dj} < T(z)\big) \Big| \boldsymbol{y}_s; \hat{\boldsymbol{\theta}} \right] = \Phi(\hat{\alpha}_{dj}), \quad \hat{\alpha}_{dj} = \hat{v}_{dj|s}^{-1/2}(T(z) - \hat{\mu}_{dj|s}).$$

For calculating S_2, we simplify the notation, i.e. we denote $y_{dj} = y$, $\hat{\mu}_{dj|s} = \mu$, $\hat{v}_{dj|s}^{1/2} = \sigma$ and $\hat{\alpha}_{dj} = \alpha$ and in the integral below we do the following change of variables:

$$x = \frac{y-\mu}{\sigma}, \quad y = \sigma x + \mu, \quad dy = \sigma dx, \quad y = T(z) \Leftrightarrow x = \frac{T(z)-\mu}{\sigma} = \alpha.$$

We further assume that $y = T(z) = \log(z+c)$ or $z = T^{-1}(y) = e^y - c$. Then, the term S_2 is

$$\begin{aligned}
S_2 &= E_{y_r}\left[T^{-1}(y_{dj})I\big(y_{dj} < T(z)\big)\Big|\boldsymbol{y}_s; \hat{\boldsymbol{\theta}}\right] = \int_{-\infty}^{T(z)} T^{-1}(y) f_{N(\mu,\sigma^2)}(y)\, dy \\
&= \int_{-\infty}^{\alpha} T^{-1}(\sigma x + \mu) f_{N(\mu,\sigma^2)}(\sigma x + \mu)\, \sigma dx \\
&= \int_{-\infty}^{\alpha} \left(e^{\mu}\exp\{\sigma x\} - c\right)\frac{1}{\sqrt{2\pi}\,\sigma}\exp\Big\{-\frac{1}{2}x^2\Big\}\,\sigma dx \\
&= -c\Phi(\alpha) + e^{\mu}\int_{-\infty}^{\alpha}\frac{1}{\sqrt{2\pi}}\exp\Big\{-\frac{1}{2}(x^2 - 2\sigma x + \sigma^2)\Big\}\exp\Big\{\frac{1}{2}\sigma^2\Big\}\, dx.
\end{aligned}$$

Applying another change of variables

$$u = x - \sigma, \quad x = u + \sigma, \quad dx = du, \quad x = \alpha \Leftrightarrow u = \alpha - \sigma,$$

we obtain

$$\begin{aligned}
S_2 &= -c\Phi(\alpha) + \exp\Big\{\frac{1}{2}\sigma^2 + \mu\Big\}\int_{-\infty}^{\alpha} f_{N(\sigma,1)}(x)\, dx \\
&= -c\Phi(\alpha) + \exp\Big\{\frac{1}{2}\sigma^2 + \mu\Big\}\int_{-\infty}^{\alpha-\sigma} f_{N(0,1)}(u)\, du \\
&= -c\Phi(\alpha) + \exp\Big\{\frac{1}{2}\sigma^2 + \mu\Big\}\Phi(\alpha - \sigma).
\end{aligned}$$

Finally, we get

$$\begin{aligned}
\hat{g}_{dj}^{ebp} &= \Phi(\hat{\alpha}_{dj}) - \frac{1}{z}\left(\exp\Big\{\frac{1}{2}\,\hat{v}_{dj|s} + \hat{\mu}_{dj|s}\Big\}\Phi(\hat{\alpha}_{dj} - \hat{v}_{dj|s}^{1/2}) - c\Phi(\hat{\alpha}_{dj})\right) \\
&= \frac{z+c}{z}\,\Phi(\hat{\alpha}_{dj}) - \frac{1}{z}\exp\Big\{\frac{1}{2}\,\hat{v}_{dj|s} + \hat{\mu}_{dj|s}\Big\}\Phi\big(\hat{\alpha}_{dj} - \hat{v}_{dj|s}^{1/2}\big).
\end{aligned}$$

10.5.3 Average Income

Let z_{dj} be an income variable for individual j from domain d. The domain average income is

$$\overline{Z}_d = \frac{1}{N_d}\sum_{j=1}^{N_d} z_{dj}.$$

For estimating the average income in small areas, we assume that a one-to-one transformation of the income variable for each unit, $y_{dj} = T(z_{dj})$, follows the NER model (10.2). Using the inverse transformation, we express z_{dj} in terms of the model response variables y_{dj} as

$$z_{dj} = T^{-1}(y_{dj}) \triangleq h(y_{dj}),$$

which means that $\overline{Z}_d$ is an additive parameter. Therefore, the EBP of $\overline{Z}_d$ is

$$\hat{\overline{Z}}_d^{ebp} = \frac{1}{N_d}\Big\{\sum_{j\in s_d} z_{dj} + \sum_{j\in r_d} \hat{z}_{dj}^{ebp}\Big\}, \tag{10.16}$$

where

$$\hat{z}_{dj}^{ebp} \triangleq E_{y_r}\big[z_{dj}|\boldsymbol{y}_s, \hat{\boldsymbol{\theta}}\big] = E_{y_r}\big[h(y_{dj})|\boldsymbol{y}_s; \hat{\boldsymbol{\theta}}\big] = E_{y_r}\big[T^{-1}(y_{dj})|\boldsymbol{y}_s; \hat{\boldsymbol{\theta}}\big].$$

Let us now assume that $y = T(z) = \log(z + c)$ or $z = T^{-1}(y) = e^y - c$. For calculating $\hat{z}_{dj}^{ebp}$, we simplify the notation, i.e. we denote $y_{dj} = y$, $\hat{\mu}_{dj|s} = \mu$, and $\hat{v}_{dj|s}^{1/2} = \sigma$ and we do a change of variables

$$x = \frac{y-\mu}{\sigma}, \quad y = \sigma x + \mu, \quad dy = \sigma dx.$$

It holds that

$$\begin{aligned}
\hat{z}_{dj}^{ebp} &= \int_{-\infty}^{\infty} T^{-1}(y) f_{N(\mu,\sigma^2)}(y)\,dy = \int_{-\infty}^{\infty} T^{-1}(\sigma x + \mu) f_{N(\mu,\sigma^2)}(\sigma x + \mu)\,\sigma dx \\
&= \int_{-\infty}^{\infty} (e^{\mu}\exp\{\sigma x\} - c)\frac{1}{\sqrt{2\pi}\,\sigma}\exp\Big\{-\frac{1}{2}x^2\Big\}\sigma dx \\
&= -c + e^{\mu}\int_{-\infty}^{\infty}\frac{1}{\sqrt{2\pi}}\exp\Big\{-\frac{1}{2}(x^2 - 2\sigma x + \sigma^2)\Big\}\exp\Big\{\frac{1}{2}\sigma^2\Big\}dx \\
&= -c + \exp\Big\{\frac{1}{2}\sigma^2 + \mu\Big\}\int_{-\infty}^{\infty} f_{N(\sigma,1)}(x)\,dx = -c + \exp\Big\{\frac{1}{2}\sigma^2 + \mu\Big\}.
\end{aligned}$$

Therefore,

$$\hat{z}_{dj}^{ebp} = -c + \exp\left\{\frac{1}{2}\,\hat{v}_{dj|s} + \hat{\mu}_{dj|s}\right\}.$$

10.6 EBPs Under Subdomain-Level NER Models

This section assumes that the population U of size N is partitioned into D domains $U_1, \ldots, U_D$, of sizes $N_1, \ldots, N_D$, and that each domain U_d is partitioned into M_d subdomains $U_{d1}, \ldots, U_{dM_d}$, of sizes $N_{d1}, \ldots, N_{dM_d}$, respectively, $d = 1, \ldots, D$. The components y_{dtj} of the target population vector $\boldsymbol{y}$ have three subindexes denoting domain d, subdomain t, and unit j, respectively.

Instead of assuming the domain-level NER model defined in (10.2), this section assumes that $\boldsymbol{y}$ follows the subdomain-level NER model

$$y_{dtj} = \boldsymbol{x}_{dtj}\boldsymbol{\beta} + u_{2,dt} + w_{dtj}^{-1/2} e_{dtj}, \quad d = 1, \ldots, D, t = 1, \ldots, M_d, j = 1, \ldots, N_{dt}, \tag{10.17}$$

where $\boldsymbol{x}_{dtj}$ is a row vector containing p auxiliary variables, w_{dtj} is a known heteroscedasticity weight, and the random effects $u_{2,dt} \sim N(0, \sigma_2^2)$ and errors $e_{dtj} \sim N(0, \sigma_0^2)$ are all mutually independent. We note that in this section we use notation of the random effects and variance components, which is coherent with the following chapters dealing with two-fold nested error regression models. Let us define $w_{dt.} = \sum_{j=1}^{n_{dt}} w_{dtj}$, $\boldsymbol{V}_{dts} = \sigma_2^2 \boldsymbol{1}_{n_{dt}} \boldsymbol{1}_{n_{dt}}' + \sigma_0^2 \boldsymbol{W}_{dts}^{-1}$, and

$$\boldsymbol{V}_{dts}^{-1} = \frac{1}{\sigma_0^2}\left(\boldsymbol{W}_{dts} - \frac{\gamma_{dt}^{w}}{w_{dt.}} \boldsymbol{w}_{n_{dt}} \boldsymbol{w}_{n_{dt}}'\right), \quad \gamma_{dt}^{w} = \frac{w_{dt.}\sigma_2^2}{w_{dt.}\sigma_2^2 + \sigma_0^2},$$

where $\boldsymbol{W}_{dts} = \underset{1\le j\le n_{dt}}{\text{diag}}\,(w_{dtj})$, $\boldsymbol{w}_{n_{dt}} = \underset{1\le j\le n_{dt}}{\text{col}}\,(w_{dtj})$, and n_{dt} is the sample size in domain d and subdomain t.

By adapting the notation to subdomains and doing similar calculations as in Sect. 10.3, we obtain the conditional distribution of $\boldsymbol{y}_{dtr}$ given $\boldsymbol{y}_{dts}$, which is

$$\boldsymbol{y}_{dtr}|\boldsymbol{y}_{dts} \sim N(\boldsymbol{\mu}_{dtr|s}^{(2)}, \boldsymbol{V}_{dtr|s}^{(2)}),$$

where the conditional mean vector and covariance matrix are

$$\boldsymbol{\mu}_{dtr|s}^{(2)} = \boldsymbol{X}_{dtr}\boldsymbol{\beta} + \sigma_2^2 \boldsymbol{1}_{N_{dt}-n_{dt}} \boldsymbol{1}_{n_{dt}}' \boldsymbol{V}_{dts}^{-1}(\boldsymbol{y}_{dts} - \boldsymbol{X}_{dts}\boldsymbol{\beta}),$$

$$\boldsymbol{V}_{dtr|s}^{(2)} = \sigma_2^2(1 - \gamma_{dt}^{w}) \boldsymbol{1}_{N_{dt}-n_{dt}} \boldsymbol{1}_{N_{dt}-n_{dt}}' + \sigma_0^2 \boldsymbol{W}_{dtr}^{-1}.$$

For any $j \in U_{dt} - s_{dt}$, the conditional mean is

$$\mu^{(2)}_{dtj|s} = \begin{cases} \boldsymbol{x}_{dtj}\boldsymbol{\beta} + \gamma^w_{dt}(\hat{\overline{Y}}^w_{dt} - \hat{\overline{\boldsymbol{X}}}^w_{dt}\boldsymbol{\beta}) & \text{if } n_{dt} \neq 0, \\ \boldsymbol{x}_{dtj}\boldsymbol{\beta} & \text{if } n_{dt} = 0, \end{cases}$$

where $\hat{\overline{Y}}^w_{dt} = w^{-1}_{dt.} \sum_{j=1}^{n_{dt}} w_{dtj} y_{dtj}$ and $\hat{\overline{\boldsymbol{X}}}^w_{dt} = w^{-1}_{dt.} \sum_{j=1}^{n_{dt}} w_{dtj} \boldsymbol{x}_{dtj}$. For any $j \in U_{dt} - s_{dt}$, the conditional variance is

$$v^{(2)}_{dtj|s} = \begin{cases} \sigma_2^2(1 - \gamma^w_{dt}) + w^{-1}_{dtj}\sigma_0^2 & \text{if } n_{dt} \neq 0, \\ \sigma_2^2 + w^{-1}_{dtj}\sigma_0^2 & \text{if } n_{dt} = 0. \end{cases}$$

Under the subdomain-level NER model (10.17), the following sections give the expressions of the EBPs of domain and subdomain poverty proportions, poverty gaps, and average incomes.

10.6.1 Poverty Proportion

Let z_{dtj} be an income variable of individual j from domain d and subdomain t, and let z be the poverty line. Assume that a one-to-one increasing transformation of the income variable for each unit, $y_{dtj} = T(z_{dtj})$, follows the subdomain-level NER model (10.17). The domain and subdomain poverty proportions are the additive parameters

$$P_d = \frac{1}{N_d} \sum_{t=1}^{M_d} \sum_{j=1}^{N_{dt}} h_0(y_{dtj}), \; P_{dt} = \frac{1}{N_{dt}} \sum_{j=1}^{N_{dt}} h_0(y_{dtj}), \quad h_0(y_{dtj}) = I(y_{dtj} < T(z)).$$

The EBPs of P_d and P_{dt} are

$$\hat{p}^{ebp2}_d = \frac{1}{N_d}\Big\{ \sum_{t=1}^{M_d} \sum_{j \in s_{dt}} h_0(y_{dtj}) + \sum_{t=1}^{M_d} \sum_{j \in r_{dt}} \hat{p}^{ebp2}_{dtj} \Big\},$$

$$\hat{p}^{ebp2}_{dt} = \frac{1}{N_{dt}}\Big\{ \sum_{j \in s_{dt}} h_0(y_{dtj}) + \sum_{j \in r_{dt}} \hat{p}^{ebp2}_{dtj} \Big\},$$

where s_{dt} and r_{dt} denote the sets of sampled and non-sampled units in domain d and subdomain t, respectively,

$$\begin{aligned} \hat{p}^{ebp2}_{dtj} &= E_{\boldsymbol{y}_r}\Big[h_0(y_{dtj}) | \boldsymbol{y}_s; \hat{\boldsymbol{\theta}} \Big] = E_{\boldsymbol{y}_r}\big[I(y_{dtj} < T(z)) | \boldsymbol{y}_s; \hat{\boldsymbol{\theta}} \big] \\ &= P_{\boldsymbol{y}_r}\big(y_{dtj} < T(z) | \boldsymbol{y}_s; \hat{\boldsymbol{\theta}} \big) = P\big(N(0,1) < \hat{\alpha}^{(2)}_{dtj} \big) = \Phi(\hat{\alpha}^{(2)}_{dtj}), \end{aligned}$$

$$\hat{\alpha}^{(2)}_{dtj} = \frac{T(z) - \hat{\mu}^{(2)}_{dtj|s}}{\hat{v}^{(2)1/2}_{dtj|s}},$$

$\hat{v}^{(2)}_{dtj|s}$ and $\hat{\mu}^{(2)}_{dtj|s}$ are plug-in estimators of the conditional variance $v^{(2)}_{dtj|s}$ and mean $\mu^{(2)}_{dtj|s}$, respectively, and $\Phi(\cdot)$ is the cumulative distribution function of a standard normal random variable.

10.6.2 Poverty Gap

If a one-to-one increasing transformation $y_{dtj} = T(z_{dtj})$ follows the subdomain-level NER model (10.17), the domain and subdomain poverty gaps are the additive parameters

$$G_d = \frac{1}{N_d}\sum_{t=1}^{M_d}\sum_{j=1}^{N_{dt}} h_1(y_{dtj}), \quad G_{dt} = \frac{1}{N_{dt}}\sum_{j=1}^{N_{dt}} h_1(y_{dtj}),$$

where

$$h_1(y_{dtj}) = \frac{z - T^{-1}(y_{dtj})}{z} I(T^{-1}(y_{dtj}) < z).$$

If $y = T(z) = \log(z + c)$, then the EBPs of G_d and G_{dt} are

$$\hat{g}_d^{ebp2} = \frac{1}{N_d}\Big\{\sum_{t=1}^{M_d}\sum_{j\in s_{dt}} h_1(y_{dtj}) + \sum_{t=1}^{M_d}\sum_{j\in r_{dt}} \hat{g}_{dtj}^{ebp2}\Big\},$$

$$\hat{g}_{dt}^{ebp2} = \frac{1}{N_{dt}}\Big\{\sum_{j\in s_{dt}} h_1(y_{dtj}) + \sum_{j\in r_{dt}} \hat{g}_{dtj}^{ebp2}\Big\},$$

where

$$\begin{aligned}\hat{g}_{dtj}^{ebp2} &= E_{\boldsymbol{y}_r}\left[h_1(y_{dtj})|\boldsymbol{y}_s;\hat{\boldsymbol{\theta}}\right] \\ &= \frac{z+c}{z}\Phi(\hat{\alpha}^{(2)}_{dtj}) - \frac{1}{z}\exp\Big\{\frac{1}{2}\hat{v}^{(2)}_{dtj|s} + \hat{\mu}^{(2)}_{dtj|s}\Big\}\Phi\big(\hat{\alpha}^{(2)}_{dtj} - \hat{v}^{(2)1/2}_{dtj|s}\big),\end{aligned}$$

and $\hat{\alpha}^{(2)}_{dtj}$, $\hat{\mu}^{(2)}_{dtj|s}$, and $\hat{v}^{(2)}_{dtj|s}$ are given above.

10.6.3 Average Income

If a one-to-one increasing transformation $y_{dtj} = T(z_{dtj})$ follows the subdomain-level NER model (10.17), the domain and subdomain average incomes are the additive parameters

$$\overline{Z}_d = \frac{1}{N_d}\sum_{t=1}^{M_d}\sum_{j=1}^{N_{dt}} h(y_{dtj}),\quad \overline{Z}_{dt} = \frac{1}{N_{dt}}\sum_{j=1}^{N_{dt}} h(y_{dtj}),\quad h(y_{dtj}) = T^{-1}(y_{dtj}) = z_{dtj}.$$

If $y = T(z) = \log(z + c)$, then the EBPs of $\overline{Z}_d$ and $\overline{Z}_{dt}$ are

$$\hat{\overline{Z}}_d^{ebp2} = \frac{1}{N_d}\Big\{\sum_{t=1}^{M_d}\sum_{j\in s_{dt}} z_{dtj} + \sum_{t=1}^{M_d}\sum_{j\in r_{dt}} \hat{z}_{dtj}^{ebp2}\Big\},\quad \hat{\overline{Z}}_{dt}^{ebp2} = \frac{1}{N_{dt}}\Big\{\sum_{j\in s_{dt}} z_{dtj} + \sum_{j\in r_{dt}} \hat{z}_{dtj}^{ebp2}\Big\},$$

where

$$\hat{z}_{dtj}^{ebp2} = -c + \exp\Big\{\frac{1}{2}\,\hat{v}_{dtj|s}^{(2)} + \hat{\mu}_{dtj|s}^{(2)}\Big\}$$

and $\hat{\mu}_{dtj|s}^{(2)}$ and $\hat{v}_{dtj|s}^{(2)}$ are defined above.

10.7 ELL Predictors of Poverty Indicators

The ELL approach assumes that the subdomain-level NER model (10.17) is correct. Under this model, the marginal mean and variance of y_{dtj} are

$$\mu_{dtj}^{(2)} = E(y_{dtj}) = \boldsymbol{x}_{dtj}\boldsymbol{\beta},\quad v_{dtj}^{(2)} = \text{var}(y_{dtj}) = \sigma_2^2 + \sigma_0^2/w_{dtj}. \tag{10.18}$$

The domain and subdomain additive parameters takes the form

$$\delta_d = \frac{1}{N_d}\sum_{t=1}^{M_d}\sum_{j=1}^{N_{dt}} h(y_{dtj}),\quad \delta_{dt} = \frac{1}{N_{dt}}\sum_{j=1}^{N_{dt}} h(y_{dtj}).$$

The ELL predictors of δ_d and δ_{dt} are

$$\hat{\delta}_d^{ELL} = \frac{1}{N_d}\sum_{t=1}^{M_d}\sum_{j=1}^{N_{dt}} \hat{\delta}_{dtj}^{ELL},\quad \hat{\delta}_{dt}^{ELL} = \frac{1}{N_{dt}}\sum_{j=1}^{N_{dt}} \hat{\delta}_{dtj}^{ELL},\quad \hat{\delta}_{dtj}^{ELL} = E[h(y_{dtj}); \hat{\boldsymbol{\theta}}], \tag{10.19}$$

which is in practice approximated by a bootstrap procedure (Monte Carlo method with estimated model parameters). From the formula, one can see that the ELL predictor is defined by expectation calculated with respect to the marginal distribution of y_{dtj} and not the conditional one, which was the case of EBP. Because of (10.18), not containing a random effect prediction, the ELL predictor is essentially synthetic.

The ELL predictors of domain and subdomain additive parameters are calculated by applying the formula (10.19). For applying this formula, we need a census file containing the values of the auxiliary variables in all the units of the population, and the same variable for identifying the units in the sample and census files, in order to separate the sampled from the non-sampled parts of the population. This is to say, it requires the same auxiliary information as the EBPs.

The following sections give the expressions of the ELL predictors for the poverty proportion, the poverty gap, and the average income.

10.7.1 Poverty Proportion

For the poverty proportion, we have $h_0(y_{dtj}) = I(T^{-1}(y_{dtj}) < z) = I\big(y_{dtj} < T(z)\big)$. The ELL predictors of the poverty proportions P_d and P_{dt} are

$$\hat{p}_d^{ELL} = \frac{1}{N_d}\sum_{t=1}^{M_d}\sum_{j=1}^{N_{dt}} \hat{p}_{dtj}^{ELL}, \quad \hat{p}_{dt}^{ELL} = \frac{1}{N_{dt}}\sum_{j=1}^{N_{dt}} \hat{p}_{dtj}^{ELL}, \quad \hat{p}_{dtj}^{ELL} = E[h_0(y_{dtj}); \hat{\boldsymbol{\theta}}].$$

If T is monotone nondecreasing, we get

$$\begin{aligned}\hat{p}_{dtj}^{ELL} &= E\left[I\big(y_{dtj} < T(z)\big); \hat{\boldsymbol{\theta}}\right] = P\big(y_{dtj} < T(z); \hat{\boldsymbol{\theta}}\big) = P\big(N(0,1) < \hat{\alpha}_{dtj}^{ELL}\big)\\ &= \Phi(\hat{\alpha}_{dtj}^{ELL}),\end{aligned}$$

where Φ denotes the cumulative distribution function of a standard normal random variable,

$$\hat{\alpha}_{dtj}^{ELL} = \frac{T(z) - \hat{\mu}_{dtj}^{(2)}}{\big(\hat{v}_{dtj}^{(2)}\big)^{1/2}} \tag{10.20}$$

and $\hat{\mu}_{dtj}^{(2)}$ and $\hat{v}_{dtj}^{(2)}$ are obtained from (10.18) by substituting $\boldsymbol{\beta}$, σ_0^2 and σ_2^2 for consistent estimators.

10.7.2 Poverty Gap

For the poverty gap, we have

$$h_1(y_{dtj}) = \frac{z - T^{-1}(y_{dtj})}{z} I(T^{-1}(y_{dtj}) < z).$$

The ELL predictors of the poverty gaps G_d and G_{dt} are

$$\hat{g}_d^{ELL} = \frac{1}{N_d}\sum_{t=1}^{M_d}\sum_{j=1}^{N_{dt}} \hat{g}_{dtj}^{ELL}, \quad \hat{g}_{dt}^{ELL} = \frac{1}{N_{dt}}\sum_{j=1}^{N_{dt}} \hat{g}_{dtj}^{ELL}, \quad \hat{g}_{dtj}^{ELL} = E[h_1(y_{dtj}); \hat{\boldsymbol{\theta}}].$$

If $y = T(z) = \log(z + c)$ or $z = T^{-1}(y) = e^y - c$, we obtain

$$\begin{aligned}
\hat{g}_{dtj}^{ELL} &= E\left[h_1(y_{dtj}); \hat{\boldsymbol{\theta}}\right] = E\left[\frac{z - T^{-1}(y_{dtj})}{z} I\big(y_{dtj} < T(z)\big); \hat{\boldsymbol{\theta}}\right] \\
&= E\left[I\big(y_{dtj} < T(z)\big); \hat{\boldsymbol{\theta}}\right] - \frac{1}{z} E\left[T^{-1}(y_{dtj}) I\big(y_{dtj} < T(z)\big); \hat{\boldsymbol{\theta}}\right] \\
&= \Phi(\hat{\alpha}_{dtj}^{ELL}) - \frac{1}{z} E\left[T^{-1}(y_{dtj}) I\big(y_{dtj} < T(z)\big); \hat{\boldsymbol{\theta}}\right].
\end{aligned}$$

The last step is due to the equality $E[I\big(y_{dtj} < T(z)\big); \boldsymbol{\theta}] = \Phi(\hat{\alpha}_{dtj}^{ELL})$, which was proved when deriving the ELL predictor of the poverty proportion. Similarly as in Sect. 10.5.2, we get

$$E\left[T^{-1}(y_{dtj}) I\big(y_{dtj} < T(z)\big); \hat{\boldsymbol{\theta}}\right] = -c\Phi(\hat{\alpha}_{dtj}^{ELL}) + \exp\left\{\frac{1}{2}\hat{\upsilon}_{dtj}^{(2)} + \hat{\mu}_{dtj}^{(2)}\right\}\Phi(\hat{\alpha}_{dtj}^{ELL} - \hat{\upsilon}_{dtj}^{(2)1/2}). \tag{10.21}$$

Therefore,

$$\begin{aligned}
\hat{g}_{dtj}^{ELL} &= \Phi(\hat{\alpha}_{dtj}^{ELL}) - \frac{1}{z}\left(\exp\left\{\frac{1}{2}\hat{\upsilon}_{dtj}^{(2)} + \hat{\mu}_{dtj}^{(2)}\right\}\Phi(\hat{\alpha}_{dtj}^{ELL} - \hat{\upsilon}_{dtj}^{(2)1/2}) - c\Phi(\hat{\alpha}_{dtj}^{ELL})\right) \\
&= \frac{z + c}{z}\Phi(\hat{\alpha}_{dtj}^{ELL}) - \frac{1}{z}\exp\left\{\frac{1}{2}\hat{\upsilon}_{dtj}^{(2)} + \hat{\mu}_{dtj}^{(2)}\right\}\Phi(\hat{\alpha}_{dtj}^{ELL} - \hat{\upsilon}_{dtj}^{(2)1/2}),
\end{aligned}$$

where $\hat{\alpha}_{dtj}^{ELL}$ is given in (10.20) and $\hat{\mu}_{dtj}^{(2)}$ and $\hat{\upsilon}_{dtj}^{(2)}$ are obtained from (10.18) by substituting $\boldsymbol{\beta}$, σ_0^2, and σ_2^2 for consistent estimators.

10.7.3 Average Income

For the average income, we have $h(y_{dtj}) = T^{-1}(y_{dtj}) = z_{dtj}$. The ELL predictors of the average incomes $\overline{Z}_d$ and $\overline{Z}_{dt}$ are

$$\hat{\overline{Z}}_d^{ELL} = \frac{1}{N_d}\sum_{t=1}^{M_d}\sum_{j=1}^{N_{dt}} \hat{z}_{dtj}^{ELL}, \quad \hat{\overline{Z}}_{dt}^{ELL} = \frac{1}{N_{dt}}\sum_{j=1}^{N_{dt}} \hat{z}_{dtj}^{ELL}, \quad \hat{z}_{dtj}^{ELL} = E[h(y_{dtj}); \hat{\boldsymbol{\theta}}].$$

If $y = T(z) = \log(z + c)$ or $z = T^{-1}(y) = e^y - c$, we obtain

$$\hat{z}_{dtj}^{ELL} = -c + \exp\left\{\frac{1}{2}\,\hat{v}_{dtj}^{(2)} + \hat{\mu}_{dtj}^{(2)}\right\},$$

where $\hat{\mu}_{dtj}^{(2)}$ and $\hat{v}_{dtj}^{(2)}$ are obtained from (10.18) by substituting $\boldsymbol{\beta}$, σ_0^2, and σ_2^2 for consistent estimators.

10.8 MSE of Empirical Best Predictors

This section introduces parametric bootstrap procedures to estimate the mean squared error (MSE) of EBPs of poverty proportions. The introduced methods can be easily adapted to other domain additive parameters. We assume that the domain-level NER model is homoscedastic, i.e. $w_{dj} = 1,\ d = 1, \ldots, D,\ j = 1, \ldots, N_d$. We consider three cases:

1. A census file fulfilling properties (A), (B), and (C), given in Remark 10.1, is available.
2. A census file fulfilling properties (A), but not (B) and (C), is available.
3. A census file is not available, but the auxiliary variables are categorical such that $\boldsymbol{x}_{dj} \in \{\boldsymbol{x}_1, \ldots, \boldsymbol{x}_K\}$ and the sizes N_{dk} of $U_{dk} = \{j \in U_d : \boldsymbol{x}_{dj} = \boldsymbol{x}_k\}$, $d = 1, \ldots, D,\ k = 1, \ldots, K$, are available.

10.8.1 Case 1

González-Manteiga et al. (2008a) introduced a parametric bootstrap procedure for estimating the MSEs of EBLUPs that was later adapted by Molina and Rao (2010) to estimate the MSEs of EBPs under the assumptions of Case 1. Their bootstrap procedure for estimating $MSE(\hat{p}_d^{ebp})$ has the following steps:

1. Fit a NER model to the sample and calculate $\hat{\boldsymbol{\theta}} = (\hat{\boldsymbol{\beta}}', \hat{\sigma}_u^2, \hat{\sigma}_e^2)'$.
2. Repeat B times ($b = 1, \ldots, B$):

   (a) *Bootstrap population*: Generate $u_d^{*(b)} \sim N(0, \hat{\sigma}_u^2)$, $e_{dj}^{*(b)} \sim N(0, \hat{\sigma}_e^2)$ $d = 1, \ldots, D$ $j = 1, \ldots, N_d$. Generate the bootstrap target variables

   $$y_{dj}^{*(b)} = \boldsymbol{x}_{dj}\hat{\boldsymbol{\beta}} + u_d^{*(b)} + e_{dj}^{*(b)}, \quad d = 1, \ldots, D, j = 1, \ldots, N_d,$$

   and calculate the bootstrap population quantities

   $$P_d^{*(b)} = \frac{1}{N_d}\sum_{j \in U_d} h_0(y_{dj}^{*(b)}), \quad d = 1, \ldots, D.$$

(b) *Bootstrap sample*: The bootstrap sample has the same units as the real data sample. It is not extracted at random. This is to say, $s_d^{*(b)} = s_d, \; b = 1, \ldots, B$. The model is on the population, and therefore the source of randomness comes from the generation of the population.

(c) *Bootstrap model*: Fit the same NER model as in Step 1 to the bootstrap sample $(y_{dj}^{*(b)}, \boldsymbol{x}_{dj}), \; d = 1, \ldots, D, \; j \in s_d$. Calculate the estimator of the vector of model parameters $\hat{\boldsymbol{\theta}}^{*(b)} = (\hat{\boldsymbol{\beta}}^{*(b)\prime}, \hat{\sigma}_u^{2*(b)}, \hat{\sigma}_e^{2*(b)})'$. Calculate the EBPs of the poverty proportions $P_d^{*(b)}$ of the bootstrap population, i.e.

$$\hat{p}_d^{ebp*(b)} = \frac{1}{N_d}\Big\{\sum_{j \in s_d} h_0(y_{dj}^{*(b)}) + \sum_{j \in r_d} \hat{p}_{dj}^{ebp*(b)}\Big\},$$

where

$$\hat{p}_{dj}^{ebp*(b)} = P\big(N(0,1) < \hat{\alpha}_{dj}^{*(b)}\big) = \Phi(\hat{\alpha}_{dj}^{*(b)}), \quad \hat{\alpha}_{dj}^{*(b)} = \hat{v}_{d|s}^{-1/2*(b)}(T(z) - \hat{\mu}_{dj|s}^{*(b)}),$$

$$\mu_{dj|s}^{*(b)} = \begin{cases} \boldsymbol{x}_{dj}\hat{\boldsymbol{\beta}}^{*(b)} + \hat{\gamma}_d^{*(b)}(\hat{\overline{Y}}_d^{*(b)} - \hat{\overline{\boldsymbol{X}}}_d\hat{\boldsymbol{\beta}}^{*(b)}) & \text{if } n_d \neq 0, \\ \boldsymbol{x}_{dj}\hat{\boldsymbol{\beta}}^{*(b)} & \text{if } n_d = 0, \end{cases}$$

$$\hat{\gamma}_d^{*(b)} = \frac{n_d\hat{\sigma}_u^{2*(b)}}{n_d\hat{\sigma}_u^{2*(b)} + \hat{\sigma}_e^{2*(b)}}, \quad \hat{\overline{\boldsymbol{X}}}_d = \frac{1}{n_d}\sum_{j=1}^{n_d}\boldsymbol{x}_{dj}, \quad \hat{\overline{Y}}_d^{*(b)} = \frac{1}{n_d}\sum_{j=1}^{n_d} y_{dj}^{*(b)},$$

$$\hat{v}_{d|s}^{*(b)} = \begin{cases} \hat{\sigma}_u^{2*(b)}(1 - \hat{\gamma}_d^{*(b)}) + \hat{\sigma}_e^{2*(b)} & \text{if } n_d \neq 0, \\ \hat{\sigma}_u^{2*(b)} + \hat{\sigma}_e^{2*(b)} & \text{if } n_d = 0. \end{cases}$$

3. Output: $mse^*(\hat{p}_d^{ebp}) = \frac{1}{B}\sum_{b=1}^{B}(\hat{p}_d^{ebp*(b)} - P_d^{*(b)})^2$.

10.8.2 Case 2

We assume that the census file C fulfills Property A, but not Properties B and C. As we cannot identify the sampled units in the census file C, we construct an enlarged file G by inserting the n_d registers of the sample file S within each domain. The new registers will contain the sampled values of the auxiliary variables, i.e. $\boldsymbol{x}_{dj}, \; d = 1, \ldots, D, \; j \in s_d$. We also add a new column with a sample identifier taking the value one in the inserted registers and zero otherwise. The new enlarged census file G fulfills Properties A, B, and C with domain sizes $N_d + n_d, \; d = 1, \ldots, D$. The approximated EBP (ABP) of P_d is

$$\hat{p}_d^{abp} = \frac{1}{N_d + n_d}\Big\{\sum_{j \in s_d} y_{dj} + \sum_{j \in U_d} \hat{p}_{dj}^{ebp}\Big\}, \tag{10.22}$$

where the first sum is calculated by using the target data of the sample file S and the second sum is calculated by using data from the census file C. Note that the set r_d of the enlarged census file G is equal to the set U_d of the original census file C.

By using the file G, a bootstrap procedure for estimating $MSE(\hat{p}_d^{abp})$ has the following steps:

1. Fit a NER model to the sample and calculate $\hat{\boldsymbol{\theta}} = (\hat{\boldsymbol{\beta}}', \hat{\sigma}_u^2, \hat{\sigma}_e^2)'$.
2. Repeat B times ($b = 1, \ldots, B$):

   (a) *Bootstrap population*: Generate $u_d^{*(b)} \sim N(0, \hat{\sigma}_u^2)$, $e_{dj}^{*(b)} \sim N(0, \hat{\sigma}_e^2)$ $d = 1, \ldots, D$ $j = 1, \ldots, N_d + n_d$. Generate the bootstrap target variables

   $$y_{dj}^{*(b)} = \boldsymbol{x}_{dj}\hat{\boldsymbol{\beta}} + u_d^{*(b)} + e_{dj}^{*(b)}, \quad d = 1, \ldots, D, \, j = 1, \ldots, N_d + n_d,$$

   where $\boldsymbol{x}_{dj}$ is taken from S if $j \leq n_d$ and from C if $j > n_d$. Calculate the bootstrap population quantities

   $$P_d^{*(b)} = \frac{1}{N_d + n_d}\Big(\sum_{j \in s_d} h_0(y_{dj}^{*(b)}) + \sum_{j \in U_d} h_0(y_{dj}^{*(b)})\Big), \quad d = 1, \ldots, D,$$

   where $\sum_{j \in s_d}$ and $\sum_{j \in U_d}$ denote domain sums in S and C, respectively.

   (b) *Bootstrap sample*: The bootstrap sample, extracted from G, has the same units as the real data sample. It is not extracted at random, i.e. $s_d^{*(b)} = s_d$, $b = 1, \ldots, B$.

   (c) *Bootstrap model*: Fit the same NER model as in Step 1 to the bootstrap sample $(y_{dj}^{*(b)}, \boldsymbol{x}_{dj})$, $d = 1, \ldots, D$, $j \in s_d$. Calculate the estimator of the vector of model parameters $\hat{\boldsymbol{\theta}}^{*(b)} = (\hat{\boldsymbol{\beta}}^{*(b)\prime}, \hat{\sigma}_u^{2*(b)}, \hat{\sigma}_e^{2*(b)})'$. Calculate the EBPs of the poverty proportions $P_d^{*(b)}$ of the bootstrap population, i.e.

   $$\hat{p}_d^{abp*(b)} = \frac{1}{N_d + n_d}\Big\{\sum_{j \in s_d} h_0(y_{dj}^{*(b)}) + \sum_{j \in U_d} \hat{p}_{dj}^{ebp*(b)}\Big\},$$

   where

   $$\hat{p}_{dj}^{ebp*(b)} = P\big(N(0,1) < \hat{\alpha}_{dj}^{*(b)}\big) = \Phi(\hat{\alpha}_{dj}^{*(b)}), \quad \hat{\alpha}_{dj}^{*(b)} = \hat{v}_{d|s}^{-1/2*(b)}(T(z) - \hat{\mu}_{dj|s}^{*(b)}),$$

   $$\mu_{dj|s}^{*(b)} = \begin{cases} \boldsymbol{x}_{dj}\hat{\boldsymbol{\beta}}^{*(b)} + \hat{\gamma}_d^{*(b)}(\hat{\overline{Y}}_d^{*(b)} - \hat{\overline{\boldsymbol{X}}}_d\hat{\boldsymbol{\beta}}^{*(b)}) & \text{if } n_d \neq 0, \\ \boldsymbol{x}_{dj}\hat{\boldsymbol{\beta}}^{*(b)} & \text{if } n_d = 0, \end{cases}$$

   $$\hat{\gamma}_d^{*(b)} = \frac{n_d\hat{\sigma}_u^{2*(b)}}{n_d\hat{\sigma}_u^{2*(b)} + \hat{\sigma}_e^{2*(b)}}, \quad \hat{\overline{\boldsymbol{X}}}_d = \frac{1}{n_d}\sum_{j=1}^{n_d}\boldsymbol{x}_{dj}, \quad \hat{\overline{Y}}_d^{*(b)} = \frac{1}{n_d}\sum_{j=1}^{n_d} y_{dj}^{*(b)},$$

   $$\hat{v}_{d|s}^{*(b)} = \begin{cases} \hat{\sigma}_u^{2*(b)}(1 - \hat{\gamma}_d^{*(b)}) + \hat{\sigma}_e^{2*(b)} & \text{if } n_d \neq 0, \\ \hat{\sigma}_u^{2*(b)} + \hat{\sigma}_e^{2*(b)} & \text{if } n_d = 0. \end{cases}$$

3. Output: $mse^*(\hat{p}_d^{abp}) = \frac{1}{B}\sum_{b=1}^{B}\big(\hat{p}_d^{abp*(b)} - P_d^{*(b)}\big)^2$.

10.8.3 Case 3

Let us denote the size of the covariate class $\boldsymbol{x}_k$ crossed by domain d by $N_{dk} = \#\{j \in U_d : \boldsymbol{x}_{dj} = \boldsymbol{x}_k\}$. Similarly, the size of the covariate class $\boldsymbol{x}_k$ crossed by r_d is $N_{dk,r} = \#\{j \in r_d : \boldsymbol{x}_{dj} = \boldsymbol{x}_k\}$. A bootstrap procedure for estimating $MSE(\hat{p}_d^{ebp})$ has the following steps:

1. Fit a NER model to the sample and calculate $\hat{\boldsymbol{\theta}} = (\hat{\boldsymbol{\beta}}', \hat{\sigma}_u^2, \hat{\sigma}_e^2)'$.
2. Repeat B times ($b = 1, \ldots, B$):
   (a) *Bootstrap population*: Generate $u_d^{*(b)} \sim N(0, \hat{\sigma}_u^2)$, $e_{dkj}^{*(b)} \sim N(0, \hat{\sigma}_e^2)$, $d = 1, \ldots, D$, $k = 1, \ldots, K$, $j = 1, \ldots, N_{dk}$. Generate the bootstrap target variables

$$y_{dkj}^{*(b)} = \boldsymbol{x}_k\hat{\boldsymbol{\beta}} + u_d^{*(b)} + e_{dkj}^{*(b)}, \quad d = 1, \ldots, D,\ k = 1, \ldots, K,\ j = 1, \ldots, N_{dk},$$

   and calculate the bootstrap population quantities

$$P_d^{*(b)} = \frac{1}{N_d}\sum_{k=1}^{K}\sum_{j=1}^{N_{dk}} h_0\left(y_{dkj}^{*(b)}\right), \quad d = 1, \ldots, D.$$

   (b) *Bootstrap sample*: The bootstrap sample has the same units as the real data sample. It is not extracted at random, i.e. $s_d^{*(b)} = s_d$, $b = 1, \ldots, B$.
   (c) *Bootstrap model*: Fit the same NER model as in Step 1 to the bootstrap sample $(y_{dkj}^{*(b)}, \boldsymbol{x}_k)$, $d = 1, \ldots, D$, $k = 1, \ldots, K$, $j = 1, \ldots, n_{dk}$. Calculate the estimator of the vector of model parameters $\hat{\boldsymbol{\theta}}^{*(b)} = (\hat{\boldsymbol{\beta}}^{*(b)\prime}, \hat{\sigma}_u^{2*(b)}, \hat{\sigma}_e^{2*(b)})'$. Calculate the EBPs of the poverty proportions $P_d^{*(b)}$ of the bootstrap population, i.e.

$$\hat{p}_d^{ebp*(b)} = \frac{1}{N_d}\sum_{k=1}^{K}\left\{\sum_{j=1}^{n_{dk}} h_0\left(y_{dkj}^{*(b)}\right) + (N_{dk} - n_{dk})\hat{p}_{dk}^{ebp*(b)}\right\},$$

   where

$$\hat{p}_{dk}^{ebp*(b)} = P\big(N(0,1) < \hat{\alpha}_{dk}^{*(b)}\big) = \Phi(\hat{\alpha}_{dk}^{*(b)}),\ \hat{\alpha}_{dk}^{*(b)} = \hat{v}_{d|s}^{-1/2*(b)}(T(z) - \hat{\mu}_{dk|s}^{*(b)}),$$

$$\mu_{dk|s}^{*(b)} = \begin{cases} \boldsymbol{x}_k\hat{\boldsymbol{\beta}}^{*(b)} + \hat{\gamma}_d^{*(b)}(\hat{\overline{Y}}_d^{*(b)} - \hat{\overline{\boldsymbol{X}}}_d\hat{\boldsymbol{\beta}}^{*(b)}) & \text{if } n_d \neq 0, \\ \boldsymbol{x}_k\hat{\boldsymbol{\beta}}^{*(b)} & \text{if } n_d = 0, \end{cases}$$

$$\hat{\gamma}_d^{*(b)} = \frac{n_d \hat{\sigma}_u^{2*(b)}}{n_d \hat{\sigma}_u^{2*(b)} + \hat{\sigma}_e^{2*(b)}}, \ \hat{\bar{X}}_d = \frac{1}{n_d} \sum_{k=1}^{K} n_{dk} \boldsymbol{x}_k, \ \hat{\bar{Y}}_d^{*(b)} = \frac{1}{n_d} \sum_{k=1}^{K} \sum_{j=1}^{n_{dk}} y_{dkj}^{*(b)},$$

$$\hat{v}_{d|s}^{*(b)} = \begin{cases} \hat{\sigma}_u^{2*(b)}(1 - \hat{\gamma}_d^{*(b)}) + \hat{\sigma}_e^{2*(b)} & \text{if } n_d \neq 0, \\ \hat{\sigma}_u^{2*(b)} + \hat{\sigma}_e^{2*(b)} & \text{if } n_d = 0. \end{cases}$$

3. Output: $mse^*(\hat{p}_d^{ebp}) = \frac{1}{B} \sum_{b=1}^{B} \left(\hat{p}_d^{ebp*(b)} - P_d^{*(b)}\right)^2$.

10.9 R Codes for EBPs

This section gives R codes for calculating EBPs and direct estimators of poverty proportions, poverty gaps, and average incomes by domains under a NER model with all heteroscedasticity weights equal to one. The domain variable is denoted by dom. The target variable y is income, the sampling weight is w, and the dichotomic auxiliary variables x_1 and x_2 are employed and unemployed.

The following code loads the package `lme4` and reads the auxiliary data file:

```
if(!require(lme4)){
  install.packages("lme4")
  library(lme4)
}
aux <- read.table("auxLCS.txt", header=TRUE, sep = "\t", dec = ",")
# Number of employees by domain
aux$Twork <- round(aux$TOT*aux$Mwork,0)
# Number of unemployees by domain
aux$Tnowork <- round(aux$TOT*aux$Mnowork,0)
# Rest of labour cases by domain
aux$Tres <- aux$TOT - aux$Twork - aux$Tnowork
# Sort aux by dom
aux <- aux[order(aux$dom),]
```

The following code reads the sample file and defines some variables:

```
dat <- read.table("datLCS.txt", header=TRUE, sep = "\t", dec = ",")
z0 <- 7280                          # poverty threshold.
income <- dat$income                # income variable
dom <- dat$dom; w <- dat$w          # sampling weight and domain
poor <- as.numeric(income<z0)       # poverty variable: 1 if yes (income<z0),
                                    # 0 otherwise
gap <- (z0-income)*poor/z0          # gap variable
work <- as.numeric(dat$lab==1)      # employed
nowork <- as.numeric(dat$lab==2)    # unemployed
inact <- as.numeric(dat$lab==3)     # inactive
one <- rep(1, nrow(dat))            # variable one
```

We calculate sample sizes by domain and labor status.

```
# Sample sizes by domains
nd <- tapply(X=one, INDEX=dom, FUN=sum)
# Sample sizes by domains and employment category
ndworking <- tapply(X=one, INDEX=list(dom,work), FUN=sum, default=0)
ndwork <- ndworking[,2]
# Sample sizes by domains and unemployment category
ndnoworking <- tapply(X=one, INDEX=list(dom,nowork), FUN=sum, default=0)
ndnowork <- ndnoworking[,2]
# By domains and innactive or <16 categories
ndres <- nd - ndwork - ndnowork
```

We calculate some direct estimators and population sizes by domains.

```
# Sizes
hatNd <- tapply(X=w, INDEX=dom, FUN=sum)
# Totals of employed people
hatNwork <- tapply(X=work*w, INDEX=dom, FUN=sum)
# Totals of unemployed people
hatNnowork <- tapply(X=nowork*w, INDEX=dom, FUN=sum)
# Totals of innactive or <16 people
hatNdres <- hatNd - hatNwork - hatNnowork
dirptot <- tapply(X=poor*w, INDEX=dom, FUN=sum)
# Poverty proportions
dirp <- dirptot/hatNd
dirgtot <- tapply(X=gap*w, INDEX=dom, FUN=sum)
# Poverty gaps
dirg <- dirgtot/hatNd
diritot <- tapply(X=income*w, INDEX=dom, FUN=sum)
# Average incomes
diri <- diritot/hatNd
```

We apply the log transformation, we fit a nested error regression model, and we calculate the components appearing in the formula of the EBPs of poverty proportions by domains.

```
# Log transformation
c <- 10
y <- log(income+c)
y0 <- log(z0+c)
# Fitting the model
lmm <- lmer(formula=y ~ work+nowork+(1|dom), data = dat, REML = TRUE)
# Beta
beta <- fixef(lmm); beta
bwork <- beta[1] + beta[2]                # x1=1, x2=0
bnowork <- beta[1] + beta[3]              # x1=0, x2=1
bres <- beta[1]                           # x1=0, x2=0
# u (sorted by dom)
ud <- ranef(lmm)$dom; ud; dim(ud)
# mu
muwork <- bwork + ud
munowork <- bnowork + ud
mures <- bres + ud
# sigma
var <- as.data.frame(VarCorr(lmm))
sigmau2 <- var$vcov[1]
sigmae2 <- var$vcov[2]
# gammad
gammad <- sigmau2*nd/(sigmau2*nd+sigmae2)
# vd|s
vd <- sigmau2*(1-gammad) + sigmae2; vd
# alphad
alphadwork <- vd^(-1/2)*(y0-muwork)
alphadnowork <- vd^(-1/2)*(y0-munowork)
alphadres <- vd^(-1/2)*(y0-mures)
```

We calculate the EBPs of poverty proportions by domains.

```
# Normal CDF
nor1work <- pnorm(alphadwork[,1], mean=0, sd=1)
nor1nowork <- pnorm(alphadnowork[,1], mean=0, sd=1)
nor1res <- pnorm(alphadres[,1], mean=0, sd=1)
# Poverty sample totals
totp <- tapply(X=poor, INDEX=dom, FUN=sum)
# Poverty proportion EBPs
ebpp <- (totp + (aux$Twork-ndwork)*nor1work + (aux$Tnowork-ndnowork)*
        nor1nowork + (aux$Tres-ndres)*nor1res)/aux$TOT
```

We calculate the EBPs of poverty gaps by domains.

```
# Normal CDF
nor2work <- pnorm(alphadwork[,1]-vd^(1/2), mean=0, sd=1)
nor2nowork <- pnorm(alphadnowork[,1]-vd^(1/2), mean=0, sd=1)
nor2res <- pnorm(alphadres[,1]-vd^(1/2), mean=0, sd=1)
# Exponential terms
expwork <- exp(vd/2 + muwork)
expnowork <- exp(vd/2 + munowork)
expres <- exp(vd/2 + mures)
# Poverty gap summands
gapwork <- ((z0+c)/z0)*nor1work - expwork*nor2work/z0
gapnowork <- ((z0+c)/z0)*nor1nowork - expnowork*nor2nowork/z0
gapres <- ((z0+c)/z0)*nor1res - expres*nor2res/z0
# Poverty gap sample totals
totg <- tapply(X=gap, INDEX=dom, FUN=sum)
# Poverty gap EBPs
ebpg <- (totg+(aux$Twork-ndwork)*gapwork+(aux$Tnowork-ndnowork)*gapnowork +
        (aux$Tres-ndres)*gapres)/aux$TOT
```

We calculate the EBPs of average incomes by domains.

```
# Income summands
incomework <- expwork-c
incomenowork <- expnowork-c
incomeres <- expres-c
# Income sample totals
toti <- tapply(X=income, INDEX=dom, FUN=sum); toti
# Average income EBPs
ebpi <- (toti + (aux$Twork-ndwork)*incomework + (aux$Tnowork-ndnowork)*
        incomenowork + (aux$Tres-ndres)*incomeres)/aux$TOT
```

We summarize results in a table.

```
output <- data.frame(dom=aux$dom, nd, dirp=round(dirp,5), ebpp=round(ebpp,5),
               dirg=round(dirg,5), ebpg=round(ebpg,5), diri=round(diri,0),
               ebpi=round(ebpi,0))
output
```

For the ten first domains, Table 10.1 presents the direct and the EBP estimates of poverty proportions, poverty gaps, and average incomes.

Table 10.1 Estimates of additive parameters by domains

d	n_d	$\hat{p}_d^{dir}$	$\hat{p}_d^{ebp}$	$\hat{g}_d^{dir}$	$\hat{g}_d^{ebp}$	$\hat{\bar{Z}}_d^{dir}$	$\hat{\bar{Z}}_d^{ebp}$
3	57	0.4612	0.4574	0.2822	0.1770	8361	10,450
5	96	0.3629	0.2521	0.1336	0.0817	13,334	15,944
6	82	0.1805	0.2306	0.0652	0.0732	15,869	16,799
7	10	0.0000	0.2514	0.0000	0.0818	13,245	16,114
11	118	0.1668	0.3582	0.0940	0.1278	11,662	12,680
12	18	0.1770	0.2913	0.1651	0.0981	16,785	14,634
13	138	0.0253	0.2215	0.0058	0.0694	16,057	17,101
14	190	0.1553	0.2813	0.0749	0.0936	13,370	14,882
15	406	0.1068	0.2016	0.0517	0.0618	15,211	17,959
16	93	0.0743	0.2202	0.0245	0.0690	14,531	17,215

References

Battese, G.E., Harter, R.M., Fuller, W.A.: An error-component model for prediction of county crop areas using survey and satellite data. J. Am. Stat. Assoc. **83**, 28–36 (1988)

Chambers R., Salvati, N., Tzavidis, N.: M-Quantile regression for binary data with application to small area estimation. Centre for Statistical and Survey Methodology, University of Wollongong. (Working Paper 12–12, 2012, 24). Available at http://ro.uow.edu.au/cssmwp/101. Accessed January 2016

Diallo, M.S., Rao, J.N.K.: Small area estimation of complex parameters under unit-level models with skew-normal errors. Scand. J. Stat. **45**, 1092–1116 (2018)

Elbers, C., Lanjouw, J.O., Lanjouw, P.: Micro-level estimation of poverty and inequality. Econometrica **71**, 355–364 (2003)

Esteban, M.D., Morales, D., Pérez, A., Santamaría, L.: Two area-level time models for estimating small area poverty indicators. J. Indian Soc. Agric. Stat. **66**(1), 75–89 (2012a)

Esteban, M.D., Morales, D., Pérez, A., Santamaría, L.: Small area estimation of poverty proportions under area-level time models. Comput. Stat. Data Anal. **56**, 2840–2855 (2012b)

Farrell, P., MacGibbon, B., Tomberlin, T.: Bootstrap adjustments for empirical Bayes interval estimates of small area proportions. Can. J. Stat. **25**, 75–89 (1997)

Foster, J., Greer, J., Thorbecke, E.: A class of decomposable poverty measures. Econometrica **52**, 761–766 (1984)

González-Manteiga, W., Lombardía, M.J., Molina, I., Morales, D., Santamaría, L.: Bootstrap mean squared error of small-area EBLUP. J. Stat. Comput. Simul.. **78**, 443–462 (2008a)

Graf, M., Marín, J.M., Molina, I.: A generalized mixed model for skewed distributions applied to small area estimation. Test **28**, 565–597 (2019)

Guadarrama, M., Molina, I., Rao, J.N.K.: A comparison of small area estimation methods for poverty mapping. Stat. Transition New Ser. **17**(1), 41–66 (2016)

Hobza, T., Morales, D.: Empirical best prediction under unit-level logit mixed models. J. Off. Stat. **32**(3), 661–69 (2016)

Hobza, T., Marhuenda, Y., Morales, D.: Small area estimation of additive parameters under unit-level generalized linear mixed models. SORT Stat. Oper. Res. Trans. **44**(1), 3–38 (2020)

Malec, D., Sedransk, J., Moriarity, C., LeClere, F.: Small area inference for binary variables in the National Health Interview Survey. J. Am. Stat. Assoc. **92**, 815–826 (1997)

Marhuenda, Y., Molina, I., Morales, D.: Small area estimation with spatio-temporal Fay-Herriot models. Comput. Stat. Data Anal. **58**, 308–325 (2013)

Marhuenda, Y., Molina, I., Morales, D., Rao, J.N.K.: Poverty mapping in small areas under a two-fold nested error regression model. J. R. Stat. Soc. A. **180**(4), 1111–1136 (2017)

Molina, I., Rao, J.N.K.: Small area estimation of poverty indicators. Can. J. Stat. **38**, 369–385 (2010)

Molina I., Nandram, B., Rao, J.N.K.: Small area estimation of general parameters with application to poverty indicators: a hierarchical Bayes approach. Ann. Appl. Stat. **8**, 852–885 (2014)

Morales, D., Pagliarella, M.C., Salvatore, R.: Small area estimation of poverty indicators under partitioned area-level time models. SORT-Stat. Oper. Res. Trans. **39**(1), 19–34 (2015)

Rencher, A.C.: Multivariate Statistical Inference and Applications. Wiley, New York (1998)

Rojas-Perilla, N., Pannier, S., Schmid, T., Tzavidis, N.: Data-driven transformations in small area estimation. J. R. Stat. Soc. A. **183**, 121–148 (2020)

Chapter 11
EBLUPs Under Two-Fold Nested Error Regression Models

11.1 Introduction

Nested error regression models are linear mixed models where the regression parameter is constant but the intercept is random with realizations on the domains. Searle et al. (1992) provided a detailed description of linear mixed models. In the Small Area Estimation setup, the one-fold nested error regression model was first employed by Battese et al. (1988) to estimate areas under corn and soya beans. Datta and Gosh (1991) and Pfeffermann and Barnard (1991) used the two-fold nested error regression model for the special case of cluster-specific covariates. Stukel and Rao (1999) proposed to use empirical best linear unbiased predictors (EBLUP) of small area means under a general two-fold nested error regression model. These last authors used the Henderson 3 moment estimators to fit the model and derived an estimator of the mean squared error (MSE) of the EBLUP, correct to second order term. Morales and Santamaría (2019) applied this model over time to give estimates of average incomes in Spanish provinces.

This chapter introduces the Henderson 3 (H3), maximum likelihood (ML), and residual maximum likelihood (REML) estimators of the two-fold nested error regression model parameters. Because of efficiency and computational reasons we recommend the use of REML estimators. The chapter gives the mathematical derivations of the EBLUPs of population linear parameters and describes the MSE estimator given by Stukel and Rao (1999), conveniently adapted to REML estimators.

Some simulation experiments empirically investigate the behavior of the REML Fisher-scoring algorithm, the EBLUPs of domains means, and the MSE estimators for different sets of sample sizes. The simulations report tables with empirical biases and mean squared errors as performance measures.

The last section gives R codes for calculating the EBLUPs of domain means under the two-fold nested error regression model by using data from the survey data file LFS20.txt.

D. Morales et al., *A Course on Small Area Estimation and Mixed Models*, Statistics for Social and Behavioral Sciences, https://doi.org/10.1007/978-3-030-63757-6_11

11.2 The Two-fold Nested Error Regression Model

Let us consider a linear mixed model with two nested random effects. The first random effect has D levels. For each level d of the first random effect, the second one has m_d levels. More concretely, the two-fold nested error regression (NER2) model is

$$y_{dtj} = \boldsymbol{x}_{dtj}\boldsymbol{\beta} + u_{1,d} + u_{2,dt} + w_{dtj}^{-1/2} e_{dtj}, \quad d = 1, \ldots, D, \; t = 1, \ldots, m_d, \; j = 1, \ldots, n_{dt}, \tag{11.1}$$

where y_{dtj} is the target variable and $\boldsymbol{x}_{dtj}$ is the $1 \times p$ row vector of auxiliary variables measured at the unit j, subdomain t and domain d, $\boldsymbol{\beta} = (\beta_1, \ldots, \beta_p)'$ is the $p \times 1$ vector of regression parameters, $u_{1,d} \sim N(0, \sigma_1^2)$ is the random effect of domain d, $u_{2,dt} \sim N(0, \sigma_2^2)$ is the random effect of subdomain t of domain d, $e_{dtj} \sim N(0, \sigma_0^2)$ is a model error, and $w_{dtj} > 0$ is a known heteroscedasticity weight. Model (11.1) assumes that the $u_{1,d}$'s, the $u_{2,dt}$'s and the e_{dtj}'s are all mutually independent. The variance of y_{dtj} is $\text{var}(y_{dtj}) = \sigma_1^2 + \sigma_2^2 + w_{dtj}^{-1}\sigma_0^2$.

Let $\boldsymbol{I}_a$ be the $a \times a$ identity matrix and let $\boldsymbol{1}_a$ and $\boldsymbol{0}_a$ be the $a \times 1$ column vectors with all the components equal to one and zero, respectively. Let us define $\boldsymbol{y}_{dt} = \underset{1\le j\le n_{dt}}{\text{col}}(y_{dtj})$, $\boldsymbol{X}_{dt} = \underset{1\le j\le n_{dt}}{\text{col}}(\boldsymbol{x}_{dtj})$, $\boldsymbol{W}_{dt} = \underset{1\le j\le n_{dt}}{\text{diag}}(w_{dtj})$, $\boldsymbol{e}_{dt} = \underset{1\le j\le n_{dt}}{\text{col}}(e_{dtj})$. At the subdomain level, model (11.1) is

$$\boldsymbol{y}_{dt} = \boldsymbol{X}_{dt}\boldsymbol{\beta} + u_{1,d} + u_{2,dt} + \boldsymbol{W}_{dt}^{-1/2}\boldsymbol{e}_{dt}, \quad d = 1, \ldots, D, \; t = 1, \ldots, m_d,$$

where $u_{1,d}$, $u_{2,dt}$, $\boldsymbol{e}_{dt} \sim N(\boldsymbol{0}_{n_{dt}}, \sigma_0^2\boldsymbol{I}_{n_{dt}})$, $d = 1, \ldots, D$, $t = 1, \ldots, m_d$, are mutually independent.

Let us define $n_d = \sum_{t=1}^{m_d} n_{dt}$, $\boldsymbol{y}_d = \underset{1\le t\le m_d}{\text{col}}(\boldsymbol{y}_{dt})$, $\boldsymbol{X}_d = \underset{1\le t\le m_d}{\text{col}}(\boldsymbol{X}_{dt})$, $\boldsymbol{u}_{2,d} = \underset{1\le t\le m_d}{\text{col}}(u_{2,dt})$ $\boldsymbol{W}_d = \underset{1\le t\le m_d}{\text{diag}}(\boldsymbol{W}_{dt})$, $\boldsymbol{e}_d = \underset{1\le t\le m_d}{\text{col}}(\boldsymbol{e}_{dt})$. At the domain level, model (11.1) is

$$\boldsymbol{y}_d = \boldsymbol{X}_d\boldsymbol{\beta} + u_{1,d} + \boldsymbol{u}_{2,d} + \boldsymbol{W}_d^{-1/2}\boldsymbol{e}_d, \quad d = 1, \ldots, D,$$

where $u_{1,d}$, $\boldsymbol{u}_{2,d} \sim N(\boldsymbol{0}_{m_d}, \sigma_2^2\boldsymbol{I}_{m_d})$, $\boldsymbol{e}_d \sim N(\boldsymbol{0}_{n_d}, \sigma_0^2\boldsymbol{I}_{n_d})$, $d = 1, \ldots, D$, are mutually independent.

Let us define $n = \sum_{d=1}^{D} n_d$, $m = \sum_{d=1}^{D} m_d$, $\boldsymbol{y} = \underset{1\le d\le D}{\text{col}}(\boldsymbol{y}_d)$, $\boldsymbol{X} = \underset{1\le d\le D}{\text{col}}(\boldsymbol{X}_d)$, $\boldsymbol{u}_1 = \underset{1\le d\le D}{\text{col}}(u_{1,d})$ $\boldsymbol{u}_2 = \underset{1\le d\le D}{\text{col}}(\boldsymbol{u}_{2,d})$ $\boldsymbol{W} = \underset{1\le d\le D}{\text{diag}}(\boldsymbol{W}_d)$, $\boldsymbol{e} = \underset{1\le d\le D}{\text{col}}(\boldsymbol{e}_d)$. Let us also define the $n \times D$ and $n \times m$ incidence matrices $\boldsymbol{Z}_1 =$

$\underset{1 \le d \le D}{\text{diag}}(\mathbf{1}_{n_d})$ and $Z_2 = \underset{1 \le d \le D}{\text{diag}}\left(\underset{1 \le t \le m_d}{\text{diag}}(\mathbf{1}_{n_{dt}})\right)$. In matrix form, model (11.1) is

$$y = X\beta + Z_1 u_1 + Z_2 u_2 + W^{-1/2} e, \tag{11.2}$$

where $u_1 \sim N(\mathbf{0}, \sigma_1^2 I_D)$, $u_2 \sim N(\mathbf{0}, \sigma_2^2 I_m)$, and $e \sim N(\mathbf{0}, \sigma_0^2 I_n)$ are mutually independent. By defining $Z = (Z_1, Z_2)$ and $u = (u_1', u_2')'$, model (11.2) can be expressed in the standard form of linear mixed models, i.e.

$$y = X\beta + Zu + W^{-1/2} e.$$

The NER2 model was used by Stukel and Rao (1999) for the estimation of linear parameters, such as domain means.

In what follows, we employ the alternative variance parameters

$$\sigma^2 = \sigma_0^2, \quad \varphi_1 = \frac{\sigma_1^2}{\sigma_0^2}, \quad \varphi_2 = \frac{\sigma_2^2}{\sigma_0^2}.$$

11.3 The Model with Known Variance Components

Let $\theta = (\sigma^2, \varphi_1, \varphi_2)$ be the vector of variance components, with $\sigma^2 > 0$, $\varphi_1 > 0$, and $\varphi_2 > 0$. Under model (11.2), we have $\text{var}(u_1) = \sigma^2 \varphi_1 I_D$, $\text{var}(u_2) = \sigma^2 \varphi_2 I_m$, $\text{var}(e) = \sigma^2 I_n$. Further, the covariance matrices of u and y are $V_u = \text{var}(u) = \sigma^2 \text{diag}(\varphi_1 I_D, \varphi_2 I_m)$ and

$$V = \text{var}(y) = Z_1 \text{var}(u_1) Z_1{}' + Z_2 \text{var}(u_2) Z_2{}' + \sigma^2 W^{-1} = \sigma^2 \Sigma = \sigma^2 \underset{1 \le d \le D}{\text{diag}}(\Sigma_d),$$

where $V = V(\theta)$, $\Sigma = \Sigma(\varphi_1, \varphi_1)$ and

$$\Sigma_d = \varphi_1 \mathbf{1}_{n_d} \mathbf{1}_{n_d}' + L_d, \quad L_d = \varphi_2 \underset{1 \le t \le m_d}{\text{diag}}(\mathbf{1}_{n_{dt}})\, I_{m_d} \underset{1 \le t \le m_d}{\text{diag}}(\mathbf{1}_{n_{dt}}') + W_d^{-1}, \quad d = 1, \ldots, D.$$

If θ is known, then the BLUE of $\beta = (\beta_1, \ldots, \beta_p)'$ and the BLUP of u, see (6.12), are

$$\tilde{\beta} = (X'V^{-1}X)^{-1} X'V^{-1} y, \quad \tilde{u} = V_u Z' V^{-1} \left(y - X\tilde{\beta}\right). \tag{11.3}$$

A programmable expression of $\tilde{\boldsymbol{\beta}}$ is

$$\tilde{\boldsymbol{\beta}} = \Big(\sum_{d=1}^{D} \boldsymbol{X}_d' \boldsymbol{\Sigma}_d^{-1} \boldsymbol{X}_d\Big)^{-1} \Big(\sum_{d=1}^{D} \boldsymbol{X}_d' \boldsymbol{\Sigma}_d^{-1} \boldsymbol{y}_d\Big).$$

The components of the BLUP of $\boldsymbol{u}$ are

$$\begin{aligned}
\tilde{\boldsymbol{u}} &= \sigma^2 \begin{pmatrix} \varphi_1 \boldsymbol{I}_D & \boldsymbol{0} \\ \boldsymbol{0} & \varphi_2 \boldsymbol{I}_m \end{pmatrix} \begin{bmatrix} \boldsymbol{Z}_1' \\ \boldsymbol{Z}_2' \end{bmatrix} \underset{1\le d\le D}{\text{diag}} \Big(\frac{1}{\sigma^2}\boldsymbol{\Sigma}_d^{-1}\Big) \underset{1\le d\le D}{\text{col}} (\boldsymbol{y}_d - \boldsymbol{X}_d \tilde{\boldsymbol{\beta}}) \\
&= \begin{bmatrix} \varphi_1 \underset{1\le d\le D}{\text{diag}} (\boldsymbol{1}_{n_d}') \underset{1\le d\le D}{\text{diag}} \big(\boldsymbol{\Sigma}_d^{-1}\big) \underset{1\le d\le D}{\text{col}} (\boldsymbol{y}_d - \boldsymbol{X}_d \tilde{\boldsymbol{\beta}}) \\ \varphi_2 \underset{1\le d\le D}{\text{diag}} \big(\underset{1\le t\le m_d}{\text{diag}} (\boldsymbol{1}_{n_{dt}}')\big) \underset{1\le d\le D}{\text{diag}} \big(\boldsymbol{\Sigma}_d^{-1}\big) \underset{1\le d\le D}{\text{col}} (\boldsymbol{y}_d - \boldsymbol{X}_d \tilde{\boldsymbol{\beta}}) \end{bmatrix} \\
&= \begin{bmatrix} \varphi_1 \underset{1\le d\le D}{\text{col}} \Big(\boldsymbol{1}_{n_d}' \boldsymbol{\Sigma}_d^{-1} (\boldsymbol{y}_d - \boldsymbol{X}_d \tilde{\boldsymbol{\beta}})\Big) \\ \varphi_2 \underset{1\le d\le D}{\text{col}} \Big(\underset{1\le t\le m_d}{\text{diag}} (\boldsymbol{1}_{n_{dt}}') \boldsymbol{\Sigma}_d^{-1} (\boldsymbol{y}_d - \boldsymbol{X}_d \tilde{\boldsymbol{\beta}})\Big) \end{bmatrix}.
\end{aligned}$$

To prevent the computer from consuming excessive time with the matrix calculations of (11.3), we calculate the expressions of some inverse matrices. For calculating $\boldsymbol{L}_d^{-1}$ we apply the formula

$$(A + CBD)^{-1} = A^{-1} - A^{-1}C(B^{-1} + DA^{-1}C)^{-1}DA^{-1},$$

with $A = \boldsymbol{W}_d^{-1}$, $C = \varphi_2 \underset{1\le t\le m_d}{\text{diag}} (\boldsymbol{1}_{n_{dt}})$, $B = \boldsymbol{I}_{m_d}$ and $D = \underset{1\le t\le m_d}{\text{diag}} (\boldsymbol{1}_{n_{dt}}')$. We obtain

$$\boldsymbol{L}_d^{-1} = \boldsymbol{W}_d - \varphi_2 \boldsymbol{W}_d \underset{1\le t\le m_d}{\text{diag}} (\boldsymbol{1}_{n_{dt}}) \Big[\boldsymbol{I}_{m_d} + \varphi_2 \underset{1\le t\le m_d}{\text{diag}} (\boldsymbol{1}_{n_{dt}}') \boldsymbol{W}_d \underset{1\le t\le m_d}{\text{diag}} (\boldsymbol{1}_{n_{dt}}) \Big]^{-1} \cdot \underset{1\le t\le m_d}{\text{diag}} (\boldsymbol{1}_{n_{dt}}') \boldsymbol{W}_d.$$

For calculating $\boldsymbol{\Sigma}_d^{-1}$, we use the formula

$$(\boldsymbol{A} + \boldsymbol{u}\boldsymbol{v}')^{-1} = \boldsymbol{A}^{-1} - \frac{\boldsymbol{A}^{-1}\boldsymbol{u}\boldsymbol{v}'\boldsymbol{A}^{-1}}{1 + \boldsymbol{v}'\boldsymbol{A}^{-1}\boldsymbol{u}}$$

with $A = \boldsymbol{L}_d$, $\boldsymbol{u} = \varphi_1 \boldsymbol{1}_{n_d}$, $\boldsymbol{v}' = \boldsymbol{1}_{n_d}'$. We obtain

$$\boldsymbol{\Sigma}_d^{-1} = \boldsymbol{L}_d^{-1} - \frac{\varphi_1}{1 + \varphi_1 \boldsymbol{1}_{n_d}' \boldsymbol{L}_d^{-1} \boldsymbol{1}_{n_d}} \boldsymbol{L}_d^{-1} \boldsymbol{1}_{n_d} \boldsymbol{1}_{n_d}' \boldsymbol{L}_d^{-1}. \tag{11.4}$$

11.4 REML Estimators for Alternative Parameters

The REML log-likelihood of $\boldsymbol{\theta} = (\sigma^2, \varphi_1, \varphi_2)$ is (cf. (6.32))

$$l_{reml}(\boldsymbol{\theta}) = -\frac{1}{2}(n-p)\log 2\pi - \frac{1}{2}(n-p)\log\sigma^2 - \frac{1}{2}\log|\boldsymbol{K}'\boldsymbol{\Sigma}\boldsymbol{K}| - \frac{1}{2\sigma^2}\,\boldsymbol{y}'\boldsymbol{P}\boldsymbol{y},$$

where $\boldsymbol{K} = \boldsymbol{W} - \boldsymbol{W}\boldsymbol{X}(\boldsymbol{X}'\boldsymbol{W}\boldsymbol{X})^{-1}\boldsymbol{X}'\boldsymbol{W}$ and

$$\boldsymbol{P} = \boldsymbol{K}(\boldsymbol{K}'\boldsymbol{\Sigma}\boldsymbol{K})^{-1}\boldsymbol{K}' = \boldsymbol{\Sigma}^{-1} - \boldsymbol{\Sigma}^{-1}\boldsymbol{X}(\boldsymbol{X}'\boldsymbol{\Sigma}^{-1}\boldsymbol{X})^{-1}\boldsymbol{X}'\boldsymbol{\Sigma}^{-1}.$$

By taking partial derivatives with respect to σ^2, φ_1, and φ_2, we calculate the components of the score vector $\boldsymbol{S}(\boldsymbol{\theta})$. As it was shown in Sect. 6.5, we get

$$\begin{aligned}
S_{\sigma^2} &= -\frac{n-p}{2\sigma^2} + \frac{1}{2\sigma^4}\,\boldsymbol{y}'\boldsymbol{P}\boldsymbol{y},\\
S_{\varphi_1} &= -\frac{1}{2}\,\mathrm{tr}(\boldsymbol{P}\boldsymbol{Z}_1\boldsymbol{Z}_1') + \frac{1}{2\sigma^2}\,\boldsymbol{y}'\boldsymbol{P}\boldsymbol{Z}_1\boldsymbol{Z}_1'\boldsymbol{P}\boldsymbol{y},\\
S_{\varphi_2} &= -\frac{1}{2}\,\mathrm{tr}(\boldsymbol{P}\boldsymbol{Z}_2\boldsymbol{Z}_2') + \frac{1}{2\sigma^2}\,\boldsymbol{y}'\boldsymbol{P}\boldsymbol{Z}_2\boldsymbol{Z}_2'\boldsymbol{P}\boldsymbol{y}.
\end{aligned}$$

The second partial derivatives are

$$\begin{aligned}
H_{\sigma^2\sigma^2} &= \frac{n-p}{2\sigma^4} - \frac{1}{\sigma^6}\boldsymbol{y}'\boldsymbol{P}\boldsymbol{y},\\
H_{\sigma^2\varphi_1} &= -\frac{1}{2\sigma^4}\boldsymbol{y}'\boldsymbol{P}\boldsymbol{Z}_1\boldsymbol{Z}_1'\boldsymbol{P}\boldsymbol{y},\ H_{\sigma^2\varphi_2} = -\frac{1}{2\sigma^4}\boldsymbol{y}'\boldsymbol{P}\boldsymbol{Z}_2\boldsymbol{Z}_2'\boldsymbol{P}\boldsymbol{y},\\
H_{\varphi_1\varphi_1} &= \frac{1}{2}\,\mathrm{tr}(\boldsymbol{P}\boldsymbol{Z}_1\boldsymbol{Z}_1'\boldsymbol{P}\boldsymbol{Z}_1\boldsymbol{Z}_1') - \frac{1}{\sigma^2}\boldsymbol{y}'\boldsymbol{P}\boldsymbol{Z}_1\boldsymbol{Z}_1'\boldsymbol{P}\boldsymbol{Z}_1\boldsymbol{Z}_1'\boldsymbol{P}\boldsymbol{y},\\
H_{\varphi_1\varphi_2} &= \frac{1}{2}\,\mathrm{tr}(\boldsymbol{P}\boldsymbol{Z}_1\boldsymbol{Z}_1'\boldsymbol{P}\boldsymbol{Z}_2\boldsymbol{Z}_2') - \frac{1}{\sigma^2}\boldsymbol{y}'\boldsymbol{P}\boldsymbol{Z}_1\boldsymbol{Z}_1'\boldsymbol{P}\boldsymbol{Z}_2\boldsymbol{Z}_2'\boldsymbol{P}\boldsymbol{y},\\
H_{\varphi_2\varphi_2} &= \frac{1}{2}\,\mathrm{tr}(\boldsymbol{P}\boldsymbol{Z}_2\boldsymbol{Z}_2'\boldsymbol{P}\boldsymbol{Z}_2\boldsymbol{Z}_2') - \frac{1}{\sigma^2}\boldsymbol{y}'\boldsymbol{P}\boldsymbol{Z}_2\boldsymbol{Z}_2'\boldsymbol{P}\boldsymbol{Z}_2\boldsymbol{Z}_2'\boldsymbol{P}\boldsymbol{y}.
\end{aligned}$$

By taking expectations, changing the sign and taking into account that $\boldsymbol{P}\boldsymbol{X} = \boldsymbol{0}$ and $\boldsymbol{P}\boldsymbol{\Sigma}\boldsymbol{P} = \boldsymbol{P}$, we calculate the components of the Fisher information matrix $\boldsymbol{F}(\boldsymbol{\theta})$. As it was shown in Sect. 6.5, we get

$$\begin{aligned}
F_{\sigma^2\sigma^2} &= -\frac{n-p}{2\sigma^4} + \frac{1}{\sigma^4}\,\mathrm{tr}(\boldsymbol{P}\boldsymbol{\Sigma}) = \frac{n-p}{2\sigma^4}, \quad F_{\sigma^2\varphi_1} = \frac{1}{2\sigma^2}\,\mathrm{tr}(\boldsymbol{P}\boldsymbol{Z}_1\boldsymbol{Z}_1'),\\
F_{\sigma^2\varphi_2} &= \frac{1}{2\sigma^2}\,\mathrm{tr}(\boldsymbol{P}\boldsymbol{Z}_2\boldsymbol{Z}_2'), \quad F_{\varphi_1\varphi_1} = \frac{1}{2}\,\mathrm{tr}(\boldsymbol{P}\boldsymbol{Z}_1\boldsymbol{Z}_1'\boldsymbol{P}\boldsymbol{Z}_1\boldsymbol{Z}_1'),\\
F_{\varphi_1\varphi_2} &= \frac{1}{2}\,\mathrm{tr}(\boldsymbol{P}\boldsymbol{Z}_1\boldsymbol{Z}_1'\boldsymbol{P}\boldsymbol{Z}_2\boldsymbol{Z}_2'), \quad F_{\varphi_2\varphi_2} = \frac{1}{2}\,\mathrm{tr}(\boldsymbol{P}\boldsymbol{Z}_2\boldsymbol{Z}_2'\boldsymbol{P}\boldsymbol{Z}_2\boldsymbol{Z}_2').
\end{aligned}$$

The Fisher-scoring updating equation is

$$\boldsymbol{\theta}^{(i+1)} = \boldsymbol{\theta}^{(i)} + \boldsymbol{F}^{-1}(\boldsymbol{\theta}^{(i)})\boldsymbol{S}(\boldsymbol{\theta}^{(i)}).$$

As seeds, we can take $\sigma^{2(0)} = \sigma_0^{2(0)}$, $\varphi_1^{(0)} = \sigma_1^{2(0)}/\sigma_0^{2(0)}$, and $\varphi_2^{(0)} = \sigma_2^{2(0)}/\sigma_0^{2(0)}$, where $\sigma_0^{2(0)}$, $\sigma_1^{2(0)}$, and $\sigma_2^{2(0)}$ are the Henderson 3 estimates given in Sect. 11.5. The REML estimator of $\boldsymbol{\beta}$ is

$$\hat{\boldsymbol{\beta}} = (\boldsymbol{X}'\hat{\boldsymbol{V}}^{-1}\boldsymbol{X})^{-1}\boldsymbol{X}'\hat{\boldsymbol{V}}^{-1}\boldsymbol{y}, \quad \hat{\boldsymbol{V}} = \boldsymbol{V}(\hat{\boldsymbol{\theta}}).$$

The asymptotic distributions of the REML estimators of $\boldsymbol{\theta}$ and $\boldsymbol{\beta}$ are

$$\hat{\boldsymbol{\theta}} \sim N_3(\boldsymbol{\theta}, \boldsymbol{F}^{-1}(\boldsymbol{\theta})), \quad \hat{\boldsymbol{\beta}} \sim N_p(\boldsymbol{\beta}, (\boldsymbol{X}'\boldsymbol{V}^{-1}(\boldsymbol{\theta})\boldsymbol{X})^{-1}).$$

Asymptotic confidence intervals of level $1-\alpha$ for θ_a and β_j are

$$\hat{\theta}_a \pm z_{1-\alpha/2}\, \nu_{aa}^{1/2}, \ a = 1, 2, 3; \quad \hat{\beta}_j \pm z_{1-\alpha/2}\, q_{jj}^{1/2}, \ j = 1, \ldots, p,$$

where $\hat{\boldsymbol{\theta}} = \boldsymbol{\theta}^{\kappa}$, $\boldsymbol{F}^{-1}(\boldsymbol{\theta}^{\kappa}) = (\nu_{ab})_{a,b=1,2,3}$, $(\boldsymbol{X}'\boldsymbol{V}^{-1}(\boldsymbol{\theta}^{\kappa})\boldsymbol{X})^{-1} = (q_{ij})_{i,j=1,\ldots,p}$, κ is the final iteration of the fitting algorithm, and z_α is the α-quantile of the $N(0,1)$ distribution. If we observe $\hat{\beta}_j = \beta_{obs}$, the p-value for testing $H_0 : \beta_j = 0$ is

$$p = 2P_{H_0}(\hat{\beta}_j > |\beta_{obs}|) = 2P(N(0,1) > |\beta_{obs}|/\sqrt{q_{jj}}).$$

11.4.1 Matrix Calculations

In what follows, we give some matrix calculations for implementing the REML Fisher-scoring algorithm without using $n \times n$ matrices. We define

$$\boldsymbol{Q} = (\boldsymbol{X}'\boldsymbol{\Sigma}^{-1}\boldsymbol{X})^{-1} = \Big(\sum_{d=1}^{D} \boldsymbol{X}_d'\boldsymbol{\Sigma}_d^{-1}\boldsymbol{X}_d\Big)^{-1},$$

so that

$$\boldsymbol{P} = \boldsymbol{\Sigma}^{-1} - \boldsymbol{\Sigma}^{-1}\boldsymbol{X}\boldsymbol{Q}\boldsymbol{X}'\boldsymbol{\Sigma}^{-1} = \underset{1\le d\le D}{\text{diag}}\,(\boldsymbol{\Sigma}_d^{-1}) - \underset{1\le d\le D}{\text{col}}\,(\boldsymbol{\Sigma}_d^{-1}\boldsymbol{X}_d)\boldsymbol{Q}\,\underset{1\le d\le D}{\text{col}'}(\boldsymbol{X}_d'\boldsymbol{\Sigma}_d^{-1}).$$

The components of the REML score vector are

$$S_{\sigma^2} = -\frac{n-p}{2\sigma^2} + \frac{1}{2\sigma^4}\sum_{d=1}^{D} \boldsymbol{y}_d'\boldsymbol{\Sigma}_d^{-1}\boldsymbol{y}_d - \frac{1}{2\sigma^4}\Big(\sum_{d=1}^{D} \boldsymbol{y}_d'\boldsymbol{\Sigma}_d^{-1}\boldsymbol{X}_d\Big)\boldsymbol{Q}\Big(\sum_{d=1}^{D} \boldsymbol{X}_d'\boldsymbol{\Sigma}_d^{-1}\boldsymbol{y}_d\Big)$$

$$S_{\varphi_1} = -\frac{1}{2}\,\mathrm{tr}\{\boldsymbol{Z}_1'\boldsymbol{P}\boldsymbol{Z}_1\} + \frac{1}{2\sigma^2}\,\boldsymbol{y}'\boldsymbol{P}\boldsymbol{Z}_1\boldsymbol{Z}_1'\boldsymbol{P}\boldsymbol{y}$$

$$= -\frac{1}{2}\sum_{d=1}^{D} \boldsymbol{1}_{n_d}'[\boldsymbol{\Sigma}_d^{-1} - \boldsymbol{\Sigma}_d^{-1}\boldsymbol{X}_d\boldsymbol{Q}\boldsymbol{X}_d'\boldsymbol{\Sigma}_d^{-1}]\boldsymbol{1}_{n_d} + \frac{1}{2\sigma^2}\sum_{d=1}^{D} \boldsymbol{y}_d'\boldsymbol{\Sigma}_d^{-1}\boldsymbol{1}_{n_d}\boldsymbol{1}_{n_d}'\boldsymbol{\Sigma}_d^{-1}\boldsymbol{y}_d$$

$$- \frac{1}{\sigma^2}\Big(\sum_{d=1}^{D} \boldsymbol{y}_d'\boldsymbol{\Sigma}_d^{-1}\boldsymbol{1}_{n_d}\boldsymbol{1}_{n_d}'\boldsymbol{\Sigma}_d^{-1}\boldsymbol{X}_d\Big)\boldsymbol{Q}\Big(\sum_{d=1}^{D} \boldsymbol{X}_d'\boldsymbol{\Sigma}_d^{-1}\boldsymbol{y}_d\Big)$$

$$+ \frac{1}{2\sigma^2}\Big(\sum_{d=1}^{D} \boldsymbol{y}_d'\boldsymbol{\Sigma}_d^{-1}\boldsymbol{X}_d\Big)\boldsymbol{Q}\Big(\sum_{d=1}^{D} \boldsymbol{X}_d'\boldsymbol{\Sigma}_d^{-1}\boldsymbol{1}_{n_d}\boldsymbol{1}_{n_d}'\boldsymbol{\Sigma}_d^{-1}\boldsymbol{X}_d\Big)\boldsymbol{Q}\Big(\sum_{d=1}^{D} \boldsymbol{X}_d'\boldsymbol{\Sigma}_d^{-1}\boldsymbol{y}_d\Big),$$

$$S_{\varphi_2} = -\frac{1}{2}\,\mathrm{tr}\{\boldsymbol{Z}_2'\boldsymbol{P}\boldsymbol{Z}_2\} + \frac{1}{2\sigma^2}\,\boldsymbol{y}'\boldsymbol{P}\boldsymbol{Z}_2\boldsymbol{Z}_2'\boldsymbol{P}\boldsymbol{y}$$

$$= -\frac{1}{2}\sum_{d=1}^{D} \mathrm{tr}\left\{\underset{1\le t\le m_d}{\mathrm{diag}}\,(\boldsymbol{1}_{n_{dt}}')[\boldsymbol{\Sigma}_d^{-1} - \boldsymbol{\Sigma}_d^{-1}\boldsymbol{X}_d\boldsymbol{Q}\boldsymbol{X}_d'\boldsymbol{\Sigma}_d^{-1}]\,\underset{1\le t\le m_d}{\mathrm{diag}}\,(\boldsymbol{1}_{n_{dt}})\right\}$$

$$+ \frac{1}{2\sigma^2}\sum_{d=1}^{D} \boldsymbol{y}_d'\boldsymbol{\Sigma}_d^{-1}\,\underset{1\le t\le m_d}{\mathrm{diag}}\,(\boldsymbol{1}_{n_{dt}})\,\underset{1\le t\le m_d}{\mathrm{diag}}\,(\boldsymbol{1}_{n_{dt}}')\boldsymbol{\Sigma}_d^{-1}\boldsymbol{y}_d$$

$$- \frac{1}{\sigma^2}\Big(\sum_{d=1}^{D} \boldsymbol{y}_d'\boldsymbol{\Sigma}_d^{-1}\,\underset{1\le t\le m_d}{\mathrm{diag}}\,(\boldsymbol{1}_{n_{dt}})\,\underset{1\le t\le m_d}{\mathrm{diag}}\,(\boldsymbol{1}_{n_{dt}}')\boldsymbol{\Sigma}_d^{-1}\boldsymbol{X}_d\Big)\boldsymbol{Q}\Big(\sum_{d=1}^{D} \boldsymbol{X}_d'\boldsymbol{\Sigma}_d^{-1}\boldsymbol{y}_d\Big)$$

$$+ \frac{1}{2\sigma^2}\Big(\sum_{d=1}^{D} \boldsymbol{y}_d'\boldsymbol{\Sigma}_d^{-1}\boldsymbol{X}_d\Big)\boldsymbol{Q}\Big(\sum_{d=1}^{D} \boldsymbol{X}_d'\boldsymbol{\Sigma}_d^{-1}\,\underset{1\le t\le m_d}{\mathrm{diag}}\,(\boldsymbol{1}_{n_{dt}})\,\underset{1\le t\le m_d}{\mathrm{diag}}\,(\boldsymbol{1}_{n_{dt}}')\boldsymbol{\Sigma}_d^{-1}\boldsymbol{X}_d\Big)$$

$$\cdot\ \boldsymbol{Q}\Big(\sum_{d=1}^{D} \boldsymbol{X}_d'\boldsymbol{\Sigma}_d^{-1}\boldsymbol{y}_d\Big).$$

The components of the REML Fisher information matrix are

$$F_{\sigma^2\sigma^2} = \frac{n-p}{2\sigma^4}$$

$$F_{\sigma^2\varphi_1} = \frac{1}{2\sigma^2}\mathrm{tr}\{\boldsymbol{Z}_1'\boldsymbol{P}\boldsymbol{Z}_1\} = \frac{1}{2\sigma^2}\sum_{d=1}^{D} \boldsymbol{1}_{n_d}'\left[\boldsymbol{\Sigma}_d^{-1} - \boldsymbol{\Sigma}_d^{-1}\boldsymbol{X}_d\boldsymbol{Q}\boldsymbol{X}_d'\boldsymbol{\Sigma}_d^{-1}\right]\boldsymbol{1}_{n_d},$$

$$F_{\sigma^2\varphi_2} = \frac{1}{2\sigma^2}\mathrm{tr}\{\mathbf{Z}_2'\mathbf{P}\mathbf{Z}_2\} = \frac{1}{2\sigma^2}\sum_{d=1}^{D}\mathrm{tr}\{\underset{1\le t\le m_d}{\mathrm{diag}}(\mathbf{1}_{n_{dt}}')\left[\mathbf{\Sigma}_d^{-1} - \mathbf{\Sigma}_d^{-1}\mathbf{X}_d\mathbf{Q}\mathbf{X}_d'\mathbf{\Sigma}_d^{-1}\right]\underset{1\le t\le m_d}{\mathrm{diag}}(\mathbf{1}_{n_{dt}})\},$$

$$F_{\varphi_1\varphi_1} = \frac{1}{2}\mathrm{tr}\{\mathbf{Z}_1'\mathbf{P}\mathbf{Z}_1\mathbf{Z}_1'\mathbf{P}\mathbf{Z}_1\} = \frac{1}{2}\sum_{d=1}^{D}(\mathbf{1}_{n_d}'\mathbf{\Sigma}_d^{-1}\mathbf{1}_{n_d})^2 - \sum_{d=1}^{D}\mathbf{1}_{n_d}'\mathbf{\Sigma}_d^{-1}\mathbf{1}_{n_d}\mathbf{1}_{n_d}'\mathbf{\Sigma}_d^{-1}\mathbf{X}_d\mathbf{Q}\mathbf{X}_d'\mathbf{\Sigma}_d^{-1}\mathbf{1}_{n_d}$$

$$+\frac{1}{2}\sum_{d=1}^{D}\mathbf{1}_{n_d}'\mathbf{\Sigma}_d^{-1}\mathbf{X}_d\mathbf{Q}\Big(\sum_{d=1}^{D}\mathbf{X}_d'\mathbf{\Sigma}_d^{-1}\mathbf{1}_{n_d}\mathbf{1}_{n_d}'\mathbf{\Sigma}_d^{-1}\mathbf{X}_d\Big)\mathbf{Q}\mathbf{X}_d'\mathbf{\Sigma}_d^{-1}\mathbf{1}_{n_d},$$

$$F_{\varphi_1\varphi_2} = \frac{1}{2}\mathrm{tr}\{\mathbf{Z}_2'\mathbf{P}\mathbf{Z}_1\mathbf{Z}_1'\mathbf{P}\mathbf{Z}_2\} = \frac{1}{2}\sum_{d=1}^{D}\mathrm{tr}\Big\{\underset{1\le t\le m_d}{\mathrm{diag}}(\mathbf{1}_{n_{dt}}')\mathbf{\Sigma}_d^{-1}\mathbf{1}_{n_d}\mathbf{1}_{n_d}'\mathbf{\Sigma}_d^{-1}\underset{1\le t\le m_d}{\mathrm{diag}}(\mathbf{1}_{n_{dt}})\Big\}$$

$$-\sum_{d=1}^{D}\mathrm{tr}\Big\{\underset{1\le t\le m_d}{\mathrm{diag}}(\mathbf{1}_{n_{dt}}')\mathbf{\Sigma}_d^{-1}\mathbf{X}_d\mathbf{Q}\mathbf{X}_d'\mathbf{\Sigma}_d^{-1}\mathbf{1}_{n_d}\mathbf{1}_{n_d}'\mathbf{\Sigma}_d^{-1}\underset{1\le t\le m_d}{\mathrm{diag}}(\mathbf{1}_{n_{dt}})\Big\}$$

$$+\frac{1}{2}\sum_{d=1}^{D}\mathrm{tr}\Big\{\underset{1\le t\le m_d}{\mathrm{diag}}(\mathbf{1}_{n_{dt}}')\mathbf{\Sigma}_d^{-1}\mathbf{X}_d\mathbf{Q}\Big(\sum_{d=1}^{D}\mathbf{X}_d'\mathbf{\Sigma}_d^{-1}\mathbf{1}_{n_d}\mathbf{1}_{n_d}'\mathbf{\Sigma}_d^{-1}\mathbf{X}_d\Big)\mathbf{Q}\mathbf{X}_d'\mathbf{\Sigma}_d^{-1}\underset{1\le t\le m_d}{\mathrm{diag}}(\mathbf{1}_{n_{dt}})\Big\},$$

$$F_{\varphi_2\varphi_2} = \frac{1}{2}\mathrm{tr}\{\mathbf{Z}_2'\mathbf{P}\mathbf{Z}_2\mathbf{Z}_2'\mathbf{P}\mathbf{Z}_2\}$$

$$= \frac{1}{2}\sum_{d=1}^{D}\mathrm{tr}\Big\{\underset{1\le t\le m_d}{\mathrm{diag}}(\mathbf{1}_{n_{dt}}')\mathbf{\Sigma}_d^{-1}\underset{1\le t\le m_d}{\mathrm{diag}}(\mathbf{1}_{n_{dt}})\underset{1\le t\le m_d}{\mathrm{diag}}(\mathbf{1}_{n_{dt}}')\mathbf{\Sigma}_d^{-1}\underset{1\le t\le m_d}{\mathrm{diag}}(\mathbf{1}_{n_{dt}})\Big\}$$

$$-\sum_{d=1}^{D}\mathrm{tr}\Big\{\underset{1\le t\le m_d}{\mathrm{diag}}(\mathbf{1}_{n_{dt}}')\mathbf{\Sigma}_d^{-1}\mathbf{X}_d\mathbf{Q}\mathbf{X}_d'\mathbf{\Sigma}_d^{-1}\underset{1\le t\le m_d}{\mathrm{diag}}(\mathbf{1}_{n_{dt}})\underset{1\le t\le m_d}{\mathrm{diag}}(\mathbf{1}_{n_{dt}}')\mathbf{\Sigma}_d^{-1}\underset{1\le t\le m_d}{\mathrm{diag}}(\mathbf{1}_{n_{dt}})\Big\}$$

$$+\frac{1}{2}\sum_{d=1}^{D}\mathrm{tr}\Big\{\underset{1\le t\le m_d}{\mathrm{diag}}(\mathbf{1}_{n_{dt}}')\mathbf{\Sigma}_d^{-1}\mathbf{X}_d\mathbf{Q}\Big(\sum_{d=1}^{D}\mathbf{X}_d'\mathbf{\Sigma}_d^{-1}\underset{1\le t\le m_d}{\mathrm{diag}}(\mathbf{1}_{n_{dt}})\underset{1\le t\le m_d}{\mathrm{diag}}(\mathbf{1}_{n_{dt}}')\mathbf{\Sigma}_d^{-1}\mathbf{X}_d\Big)$$

$$\cdot\ \mathbf{Q}\mathbf{X}_d'\mathbf{\Sigma}_d^{-1}\underset{1\le t\le m_d}{\mathrm{diag}}(\mathbf{1}_{n_{dt}})\Big\}.$$

11.5 The Henderson 3 Method

This section presents the fitting constant method for estimating the variance components. This method was introduced by Henderson (1953) and it is also known as Henderson 3 (H3). For applying the H3 method, we treat the effects $\boldsymbol{u}_1$ and $\boldsymbol{u}_2$ as fixed and we fit the model

$$\boldsymbol{y} = \boldsymbol{X}\boldsymbol{\beta} + \boldsymbol{Z}_1\boldsymbol{u}_1 + \boldsymbol{Z}_2\boldsymbol{u}_2 + \boldsymbol{W}^{-1/2}\boldsymbol{e}, \tag{11.5}$$

by using the least squares estimation method. For having a full rank model (11.5), we assume that the matrix $\boldsymbol{X}$ does not contain a column of ones and we equate to zero the parameters of the last subdomain within each domain. This is to say, we set $u_{2,dm_d} = 0$, $d = 1, \ldots, D$. This is equivalent to deleting the columns $\sum_{j=1}^{d} m_j$, $d = 1, \ldots, D$, of the matrix $Z_2 = \underset{1 \le d \le D}{\text{diag}} \left(\underset{1 \le t \le m_d}{\text{diag}} (\mathbf{1}_{n_{dt}}) \right)$. We employ the incidence matrices

$$Z_1 = \underset{1 \le d \le D}{\text{diag}} (\mathbf{1}_{n_d})_{n \times D} \quad \text{and} \quad \widetilde{Z}_2 = \underset{1 \le d \le D}{\text{diag}} \left(\text{col} \left\{ \underset{1 \le t \le m_d - 1}{\text{diag}} (\mathbf{1}_{n_{dt}}), \mathbf{0}_{n_{dm_d} \times (m_d - 1)} \right\} \right).$$

The H3 estimators are (cf. page 143 in Sect. 6.4)

$$\hat{\sigma}_0^2 = \frac{y' M_3 y}{n - \text{r}(X^{(3)})} = \frac{y' M_3 y}{n - p - m},$$

$$\hat{\sigma}_2^2 = \frac{y' M_2 y - y' M_3 y - \hat{\sigma}_0^2 [\text{r}(X^{(3)}) - \text{r}(X^{(2)})]}{\text{tr}(L_{2,2})} = \frac{y' M_2 y - y' M_3 y - (m - D) \hat{\sigma}_0^2}{\text{tr}(L_{2,2})},$$

$$\hat{\sigma}_1^2 = \frac{y' M_1 y - y' M_3 y - \hat{\sigma}_0^2 \left[\text{r}(X^{(3)}) - \text{r}(X^{(1)}) \right] - \hat{\sigma}_2^2 \text{tr}(L_{1,2})}{\text{tr}(L_{1,1})}$$

$$= \frac{y' M_1 y - y' M_3 y - m \hat{\sigma}_0^2 - \hat{\sigma}_2^2 \text{tr}(L_{1,2})}{\text{tr}(L_{1,1})},$$

where

$$X^{(1)} = X, \qquad X^{(2)} = (X, Z_1), \qquad X^{(3)} = (X, Z_1, \widetilde{Z}_2),$$

$$M_1 = W - W X^{(1)} (X^{(1)\prime} W X^{(1)})^{-1} X^{(1)\prime} W, \quad L_{1,1} = Z_1' M_1 Z_1,$$

$$M_2 = W - W X^{(2)} (X^{(2)\prime} W X^{(2)})^{-1} X^{(2)\prime} W, \quad L_{2,2} = \widetilde{Z}_2' M_2 \widetilde{Z}_2,$$

$$M_3 = W - W X^{(3)} (X^{(3)\prime} W X^{(3)})^{-1} X^{(3)\prime} W, \quad L_{1,2} = \widetilde{Z}_2' M_1 \widetilde{Z}_2.$$

The H3 estimators can be calculated by programming the above formulas and inverting matrices of order $p + m$. As m is usually large, this approach is not computationally efficient. This is why we give below some more efficient formulas for calculating the H3 estimators.

11.5.1 Calculation of M_1

We define $C = (X^{(1)\prime} W X^{(1)})^{-1} = \left(\sum_{d=1}^{D} \sum_{t=1}^{m_d} X_{dt}' W_{dt} X_{dt} \right)^{-1}$. We have

$$y' M_1 y = y' W y - y' W X C X' W y$$

$$= \sum_{d=1}^{D} y_d' W_d y_d - \left(\sum_{d=1}^{D} y_d' W_d X_d \right) C \left(\sum_{d=1}^{D} y_d' W_d X_d \right)'$$

$$= \sum_{d=1}^{D}\sum_{t=1}^{m_d} \boldsymbol{y}'_{dt}\boldsymbol{W}_{dt}\boldsymbol{y}_{dt} - \left(\sum_{d=1}^{D}\sum_{t=1}^{m_d} \boldsymbol{y}'_{dt}\boldsymbol{W}_{dt}\boldsymbol{X}_{dt}\right)\boldsymbol{C}\left(\sum_{d=1}^{D}\sum_{t=1}^{m_d} \boldsymbol{y}'_{dt}\boldsymbol{W}_{dt}\boldsymbol{X}_{dt}\right)',$$

$$\boldsymbol{L}_{1,1} = \boldsymbol{Z}'_1\boldsymbol{M}_1\boldsymbol{Z}_1 = \underset{1\le d\le D}{\text{diag}}(\boldsymbol{1}'_{n_d}\boldsymbol{W}_d\boldsymbol{1}_{n_d}) - \underset{1\le d\le D}{\text{col}}(\boldsymbol{1}'_{n_d}\boldsymbol{W}_d\boldsymbol{X}_d)\boldsymbol{C}\underset{1\le d\le D}{\text{col}'}(\boldsymbol{X}'_d\boldsymbol{W}_d\boldsymbol{1}_{n_d}),$$

$$\text{tr}(\boldsymbol{L}_{1,1}) = w_{...} - \sum_{d=1}^{D}\text{tr}\left\{\boldsymbol{1}'_{n_d}\boldsymbol{W}_d\boldsymbol{X}_d\boldsymbol{C}\boldsymbol{X}'_d\boldsymbol{W}_d\boldsymbol{1}_{n_d}\right\}$$

$$= w_{...} - \sum_{d=1}^{D}\text{tr}\left\{(\sum_{t=1}^{m_d}\boldsymbol{w}'_{n_{dt}}\boldsymbol{X}_{dt})\boldsymbol{C}(\sum_{t=1}^{m_d}\boldsymbol{w}'_{n_{dt}}\boldsymbol{X}_{dt})'\right\},$$

where

$$w_{...} = \sum_{d=1}^{D}\sum_{t=1}^{m_d}\sum_{j=1}^{n_{dt}} w_{dtj}$$

and $\boldsymbol{w}'_{n_d} = \boldsymbol{1}'_{n_d}\boldsymbol{W}_d$ and $\boldsymbol{w}'_{n_{dt}} = \boldsymbol{1}'_{n_{dt}}\boldsymbol{W}_{dt}$.

For the matrix $\boldsymbol{L}_{1,2}$ it holds

$$\boldsymbol{L}_{1,2} = \widetilde{\boldsymbol{Z}}'_2\boldsymbol{M}_1\widetilde{\boldsymbol{Z}}_2 = \widetilde{\boldsymbol{Z}}'_2\boldsymbol{W}\widetilde{\boldsymbol{Z}}_2 - \widetilde{\boldsymbol{Z}}'_2\boldsymbol{W}\,\boldsymbol{X}\boldsymbol{C}\,\boldsymbol{X}'\boldsymbol{W}\widetilde{\boldsymbol{Z}}_2$$
$$= \widetilde{\boldsymbol{Z}}'_2\boldsymbol{W}\widetilde{\boldsymbol{Z}}_2 - (\widetilde{\boldsymbol{Z}}'_2\boldsymbol{W}\boldsymbol{X})\boldsymbol{C}(\widetilde{\boldsymbol{Z}}'_2\boldsymbol{W}\boldsymbol{X})',$$

where

$$\widetilde{\boldsymbol{Z}}'_2\boldsymbol{W}\widetilde{\boldsymbol{Z}}_2 = \underset{1\le d\le D}{\text{diag}}\left\{\left[\underset{1\le t\le m_d-1}{\text{diag}}(\boldsymbol{1}'_{n_{dt}}), \boldsymbol{0}'\right]\begin{pmatrix}\underset{1\le t\le m_d-1}{\text{diag}}(\boldsymbol{W}_{dt}) & \boldsymbol{0}\\ \boldsymbol{0} & \boldsymbol{W}_{dm_d}\end{pmatrix}\begin{bmatrix}\underset{1\le t\le m_d-1}{\text{diag}}(\boldsymbol{1}_{n_{dt}})\\ \boldsymbol{0}\end{bmatrix}\right\}$$
$$= \underset{1\le d\le D}{\text{diag}}\left(\underset{1\le t\le m_d-1}{\text{diag}}(\boldsymbol{1}'_{n_{dt}}\boldsymbol{W}_{dt}\boldsymbol{1}_{n_{dt}})\right) = \underset{1\le d\le D}{\text{diag}}\left(\underset{1\le t\le m_d-1}{\text{diag}}(w_{dt.})\right),$$

$$\widetilde{\boldsymbol{Z}}'_2\boldsymbol{W}\boldsymbol{X} = \underset{1\le d\le D}{\text{col}}\left\{\left[\underset{1\le t\le m_d-1}{\text{diag}}(\boldsymbol{1}'_{n_{dt}}), \boldsymbol{0}'\right]\begin{pmatrix}\underset{1\le t\le m_d-1}{\text{diag}}(\boldsymbol{W}_{dt}) & \boldsymbol{0}\\ \boldsymbol{0} & \boldsymbol{W}_{dm_d}\end{pmatrix}\begin{bmatrix}\underset{1\le t\le m_d-1}{\text{col}}(\boldsymbol{X}_{dt})\\ \boldsymbol{X}_{dm_d}\end{bmatrix}\right\}$$
$$= \underset{1\le d\le D}{\text{col}}\left(\underset{1\le t\le m_d-1}{\text{col}}(\boldsymbol{1}'_{n_{dt}}\boldsymbol{W}_{dt}\boldsymbol{X}_{dt})\right)$$

and $w_{dt.} = \sum_{j=1}^{n_{dt}} w_{dtj}$. The trace of $\boldsymbol{L}_{1,2}$ is

$$\text{tr}(\boldsymbol{L}_{1,2}) = \text{tr}(\widetilde{\boldsymbol{Z}}'_2\boldsymbol{W}\widetilde{\boldsymbol{Z}}_2) - \text{tr}((\widetilde{\boldsymbol{Z}}'_2\boldsymbol{W}\boldsymbol{X})\boldsymbol{C}(\widetilde{\boldsymbol{Z}}'_2\boldsymbol{W}\boldsymbol{X})') = \left(w_{...} - \sum_{d=1}^{D} w_{dm_d.}\right) - t_C,$$

where

$$t_C = \sum_{d=1}^{D} \operatorname{tr}\left(\underset{1 \le t \le m_d - 1}{\operatorname{col}} (\mathbf{1}'_{n_{dt}} \boldsymbol{W}_{dt} \boldsymbol{X}_{dt}) \boldsymbol{C} \underset{1 \le t \le m_d - 1}{\operatorname{col}'} (\boldsymbol{X}'_{dt} \boldsymbol{W}_{dt} \mathbf{1}_{n_{dt}}) \right)$$

$$= \sum_{d=1}^{D} \sum_{t=1}^{m_d - 1} \operatorname{tr}\left(\mathbf{1}'_{n_{dt}} \boldsymbol{W}_{dt} \boldsymbol{X}_{dt} \boldsymbol{C} \boldsymbol{X}'_{dt} \boldsymbol{W}_{dt} \mathbf{1}_{n_{dt}}\right) = \sum_{d=1}^{D} \sum_{t=1}^{m_d - 1} \boldsymbol{w}'_{n_{dt}} \boldsymbol{X}_{dt} \boldsymbol{C} \boldsymbol{X}'_{dt} \boldsymbol{w}_{n_{dt}}.$$

11.5.2 Calculation of M_2

We define $\boldsymbol{G}_1 = (\boldsymbol{Z}'_1 \boldsymbol{W} \boldsymbol{Z}_1)^{-1} = \underset{1 \le d \le D}{\operatorname{diag}} (w_{d..}^{-1})$, $\boldsymbol{P}_1 = \boldsymbol{W} - \boldsymbol{W} \boldsymbol{Z}_1 \boldsymbol{G}_1 \boldsymbol{Z}'_1 \boldsymbol{W}$, and

$$\boldsymbol{B} = \left(\boldsymbol{X}^{(2)\prime} \boldsymbol{W} \boldsymbol{X}^{(2)}\right)^{-1} = \begin{pmatrix} \boldsymbol{X}' \boldsymbol{W} \boldsymbol{X} & \boldsymbol{X}' \boldsymbol{W} \boldsymbol{Z}_1 \\ \boldsymbol{Z}'_1 \boldsymbol{W} \boldsymbol{X} & \boldsymbol{Z}'_1 \boldsymbol{W} \boldsymbol{Z}_1 \end{pmatrix}^{-1} = \begin{pmatrix} \boldsymbol{B}^{11} & \boldsymbol{B}^{12} \\ \boldsymbol{B}^{21} & \boldsymbol{B}^{22} \end{pmatrix},$$

where $w_{d..} = \sum_{t=1}^{m_d} \sum_{j=1}^{n_{dt}} w_{dtj}$. Using the formula (A.1), we obtain

$$\boldsymbol{B}^{11} = (\boldsymbol{X}' \boldsymbol{P}_1 \boldsymbol{X})^{-1},$$

$$\boldsymbol{B}^{12} = -\boldsymbol{B}^{11} \boldsymbol{X}' \boldsymbol{W} \boldsymbol{Z}_1 \boldsymbol{G}_1 = -\boldsymbol{B}^{11} \underset{1 \le d \le D}{\operatorname{col}'} (w_{d..}^{-1} \boldsymbol{X}'_d \boldsymbol{W}_d \mathbf{1}_{n_d}), \quad \boldsymbol{B}^{21} = \boldsymbol{B}^{12\prime},$$

$$\boldsymbol{B}^{22} = \boldsymbol{G}_1 + \boldsymbol{G}_1 \boldsymbol{Z}'_1 \boldsymbol{W} \boldsymbol{X} \boldsymbol{B}^{11} \boldsymbol{X}' \boldsymbol{W} \boldsymbol{Z}_1 \boldsymbol{G}_1$$

$$= \underset{1 \le d \le D}{\operatorname{diag}} (w_{d..}^{-1}) + \underset{1 \le d \le D}{\operatorname{col}} (w_{d..}^{-1} \boldsymbol{w}'_{n_d} \boldsymbol{X}_d) \boldsymbol{B}^{11} \underset{1 \le d \le D}{\operatorname{col}'} (w_{d..}^{-1} \boldsymbol{X}'_d \boldsymbol{w}_{n_d})$$

$$= \underset{1 \le d \le D}{\operatorname{diag}} (w_{d..}^{-1}) + \underset{1 \le d \le D}{\operatorname{col}} (w_{d..}^{-1} \sum_{t=1}^{m_d} \boldsymbol{w}'_{n_{dt}} \boldsymbol{X}_{dt}) \boldsymbol{B}^{11} \Big[\underset{1 \le d \le D}{\operatorname{col}} (w_{d..}^{-1} \sum_{t=1}^{m_d} \boldsymbol{w}'_{n_{dt}} \boldsymbol{X}_{dt}) \Big]',$$

where

$$\boldsymbol{X}' \boldsymbol{P}_1 \boldsymbol{X} = \boldsymbol{X}' \boldsymbol{W} \boldsymbol{X} - \boldsymbol{X}' \boldsymbol{W} \boldsymbol{Z}_1 \boldsymbol{G}_1 \boldsymbol{Z}'_1 \boldsymbol{W} \boldsymbol{X}$$

$$= \sum_{d=1}^{D} \boldsymbol{X}'_d \boldsymbol{W}_d \boldsymbol{X}_d - \sum_{d=1}^{D} \boldsymbol{X}'_d \boldsymbol{W}_d \mathbf{1}_{n_d} w_{d..}^{-1} \mathbf{1}'_{n_d} \boldsymbol{W}_d \boldsymbol{X}_d$$

$$= \sum_{d=1}^{D} \sum_{t=1}^{m_d} \boldsymbol{X}'_{dt} \boldsymbol{W}_{dt} \boldsymbol{X}_{dt} - \sum_{d=1}^{D} w_{d..}^{-1} \left(\sum_{t=1}^{m_d} \boldsymbol{X}'_{dt} \boldsymbol{W}_{dt} \mathbf{1}_{n_{dt}} \right) \left(\sum_{t=1}^{m_d} \boldsymbol{X}'_{dt} \boldsymbol{W}_{dt} \mathbf{1}_{n_{dt}} \right)'.$$

The quadratic form is

$$\begin{aligned}\boldsymbol{y}'\boldsymbol{M}_2\boldsymbol{y} &= \boldsymbol{y}'\boldsymbol{W}\boldsymbol{y} - \boldsymbol{y}'\boldsymbol{W}[\boldsymbol{X},\boldsymbol{Z}_1]\boldsymbol{B}[\boldsymbol{X},\boldsymbol{Z}_1]'\boldsymbol{W}\boldsymbol{y}\\ &= \boldsymbol{y}'\boldsymbol{W}\boldsymbol{y} - \Big[\boldsymbol{y}'\boldsymbol{W}\boldsymbol{X}\boldsymbol{B}^{11}\boldsymbol{X}'\boldsymbol{W}\boldsymbol{y} + \boldsymbol{y}'\boldsymbol{W}\boldsymbol{Z}_1\boldsymbol{B}^{22}\boldsymbol{Z}_1'\boldsymbol{W}\boldsymbol{y} + 2\boldsymbol{y}'\boldsymbol{W}\boldsymbol{X}\boldsymbol{B}^{12}\boldsymbol{Z}_1'\boldsymbol{W}\boldsymbol{y}\Big].\end{aligned}$$

Therefore, we have

$$\begin{aligned}\boldsymbol{y}'\boldsymbol{M}_2\boldsymbol{y} =& \sum_{d=1}^{D}\sum_{t=1}^{m_d}\boldsymbol{y}_{dt}'\boldsymbol{W}_{dt}\boldsymbol{y}_{dt} - \left(\sum_{d=1}^{D}\sum_{t=1}^{m_d}\boldsymbol{y}_{dt}'\boldsymbol{W}_{dt}\boldsymbol{X}_{dt}\right)\boldsymbol{B}^{11}\left(\sum_{d=1}^{D}\sum_{t=1}^{m_d}\boldsymbol{y}_{dt}'\boldsymbol{W}_{dt}\boldsymbol{X}_{dt}\right)'\\ &- \left[\underset{1\le d\le D}{\text{col}'}\left(\sum_{t=1}^{m_d}\boldsymbol{y}_{dt}'\boldsymbol{w}_{n_{dt}}\right)\right]\boldsymbol{B}^{22}\left[\underset{1\le d\le D}{\text{col}'}\left(\sum_{t=1}^{m_d}\boldsymbol{y}_{dt}'\boldsymbol{w}_{n_{dt}}\right)\right]'\\ &- 2\left(\sum_{d=1}^{D}\sum_{t=1}^{m_d}\boldsymbol{y}_{dt}'\boldsymbol{W}_{dt}\boldsymbol{X}_{dt}\right)\boldsymbol{B}^{12}\left[\underset{1\le d\le D}{\text{col}'}\left(\sum_{t=1}^{m_d}\boldsymbol{y}_{dt}'\boldsymbol{w}_{n_{dt}}\right)\right]'.\end{aligned}$$

The matrix $\boldsymbol{L}_{2,2}$ is

$$\begin{aligned}\boldsymbol{L}_{2,2} &= \widetilde{\boldsymbol{Z}}_2'\boldsymbol{M}_2\widetilde{\boldsymbol{Z}}_2 = \widetilde{\boldsymbol{Z}}_2'\boldsymbol{W}\widetilde{\boldsymbol{Z}}_2 - \widetilde{\boldsymbol{Z}}_2'\boldsymbol{W}[\boldsymbol{X},\boldsymbol{Z}_1]\boldsymbol{B}[\boldsymbol{X},\boldsymbol{Z}_1]'\boldsymbol{W}\widetilde{\boldsymbol{Z}}_2\\ &= \widetilde{\boldsymbol{Z}}_2'\boldsymbol{W}\widetilde{\boldsymbol{Z}}_2 - \widetilde{\boldsymbol{Z}}_2'\boldsymbol{W}\Big[\boldsymbol{X}\boldsymbol{B}^{11}\boldsymbol{X}' + \boldsymbol{Z}_1\boldsymbol{B}^{22}\boldsymbol{Z}_1' + \boldsymbol{X}\boldsymbol{B}^{12}\boldsymbol{Z}_1' + \boldsymbol{Z}_1\boldsymbol{B}^{21}\boldsymbol{X}'\Big]\boldsymbol{W}\widetilde{\boldsymbol{Z}}_2\\ &= (\widetilde{\boldsymbol{Z}}_2'\boldsymbol{W}\widetilde{\boldsymbol{Z}}_2) - (\widetilde{\boldsymbol{Z}}_2'\boldsymbol{W}\boldsymbol{X})\boldsymbol{B}^{11}(\widetilde{\boldsymbol{Z}}_2'\boldsymbol{W}\boldsymbol{X})' - (\widetilde{\boldsymbol{Z}}_2'\boldsymbol{W}\boldsymbol{Z}_1)\boldsymbol{B}^{22}(\widetilde{\boldsymbol{Z}}_2'\boldsymbol{W}\boldsymbol{Z}_1)'\\ &- (\widetilde{\boldsymbol{Z}}_2'\boldsymbol{W}\boldsymbol{X})\boldsymbol{B}^{12}(\widetilde{\boldsymbol{Z}}_2'\boldsymbol{W}\boldsymbol{Z}_1)' - (\widetilde{\boldsymbol{Z}}_2'\boldsymbol{W}\boldsymbol{Z}_1)\boldsymbol{B}^{21}(\widetilde{\boldsymbol{Z}}_2'\boldsymbol{W}\boldsymbol{X})'.\end{aligned}$$

The components of $\boldsymbol{L}_{2,2}$ are

$$\widetilde{\boldsymbol{Z}}_2'\boldsymbol{W}\widetilde{\boldsymbol{Z}}_2 = \underset{1\le d\le D}{\text{diag}}\Big(\underset{1\le t\le m_d-1}{\text{diag}}(w_{dt\cdot})\Big),$$

$$\begin{aligned}\widetilde{\boldsymbol{Z}}_2'\boldsymbol{W}\boldsymbol{Z}_1 &= \underset{1\le d\le D}{\text{diag}}\left\{\Big[\underset{1\le t\le m_d-1}{\text{diag}}(\boldsymbol{1}_{n_{dt}}'),\boldsymbol{0}'\Big]\begin{pmatrix}\underset{1\le t\le m_d-1}{\text{diag}}(\boldsymbol{W}_{dt}) & \boldsymbol{0}\\ \boldsymbol{0} & \boldsymbol{W}_{dm_d}\end{pmatrix}\begin{bmatrix}\underset{1\le t\le m_d-1}{\text{col}}(\boldsymbol{1}_{n_{dt}})\\ \boldsymbol{1}_{n_{dm_d}}\end{bmatrix}\right\}\\ &= \underset{1\le d\le D}{\text{diag}}\Big(\underset{1\le t\le m_d-1}{\text{col}}(\boldsymbol{1}_{n_{dt}}'\boldsymbol{W}_{dt}\boldsymbol{1}_{n_{dt}})\Big) = \underset{1\le d\le D}{\text{diag}}\Big(\underset{1\le t\le m_d-1}{\text{col}}(w_{dt\cdot})\Big),\end{aligned}$$

$$\widetilde{\boldsymbol{Z}}_2'\boldsymbol{W}\boldsymbol{X} = \underset{1\le d\le D}{\text{col}}\Big(\underset{1\le t\le m_d-1}{\text{col}}(\boldsymbol{1}_{n_{dt}}'\boldsymbol{W}_{dt}\boldsymbol{X}_{dt})\Big).$$

Finally, the trace of $\boldsymbol{L}_{2,2}$ is

$$\begin{aligned}\text{tr}(\boldsymbol{L}_{2,2}) &= \text{tr}(\widetilde{\boldsymbol{Z}}_2'\boldsymbol{W}\widetilde{\boldsymbol{Z}}_2) - \text{tr}((\widetilde{\boldsymbol{Z}}_2'\boldsymbol{W}\boldsymbol{X})\boldsymbol{B}^{11}(\widetilde{\boldsymbol{Z}}_2'\boldsymbol{W}\boldsymbol{X})') - \text{tr}((\widetilde{\boldsymbol{Z}}_2'\boldsymbol{W}\boldsymbol{Z}_1)\boldsymbol{B}^{22}(\widetilde{\boldsymbol{Z}}_2'\boldsymbol{W}\boldsymbol{Z}_1)')\\ &- 2\text{tr}((\widetilde{\boldsymbol{Z}}_2'\boldsymbol{W}\boldsymbol{X})\boldsymbol{B}^{12}(\widetilde{\boldsymbol{Z}}_2'\boldsymbol{W}\boldsymbol{Z}_1)') = \left(w_{\cdots} - \sum_{d=1}^{D}w_{dm_d\cdot}\right) - t_{11} - t_{22} - 2t_{12},\end{aligned}$$

where

$$t_{11} = \sum_{d=1}^{D} \operatorname{tr}\Big(\underset{1\le t\le m_d-1}{\operatorname{col}} (\mathbf{1}'_{n_{dt}} \boldsymbol{W}_{dt} \boldsymbol{X}_{dt}) \boldsymbol{B}^{11} \underset{1\le t\le m_d-1}{\operatorname{col}'} (\boldsymbol{X}'_{dt} \boldsymbol{W}_{dt} \mathbf{1}_{n_{dt}}) \Big)$$

$$= \sum_{d=1}^{D} \sum_{t=1}^{m_d-1} \operatorname{tr}\Big(\mathbf{1}'_{n_{dt}} \boldsymbol{W}_{dt} \boldsymbol{X}_{dt} \boldsymbol{B}^{11} \boldsymbol{X}'_{dt} \boldsymbol{W}_{dt} \mathbf{1}_{n_{dt}} \Big) = \sum_{d=1}^{D} \sum_{t=1}^{m_d-1} \boldsymbol{w}'_{n_{dt}} \boldsymbol{X}_{dt} \boldsymbol{B}^{11} \boldsymbol{X}'_{dt} \boldsymbol{w}_{n_{dt}},$$

$$t_{22} = \operatorname{tr}\Big(\underset{1\le d\le D}{\operatorname{diag}} \big(\underset{1\le t\le m_d-1}{\operatorname{col}} (w_{dt.}) \big)$$

$$\cdot \Big[\underset{1\le d\le D}{\operatorname{diag}} (w_{d..}^{-1}) + \underset{1\le d\le D}{\operatorname{col}} (w_{d..}^{-1} \boldsymbol{w}'_{n_d} \boldsymbol{X}_d) \boldsymbol{B}^{11} \underset{1\le d\le D}{\operatorname{col}'} (w_{d..}^{-1} \boldsymbol{X}'_d \boldsymbol{w}_{n_d}) \Big]$$

$$\cdot \underset{1\le d\le D}{\operatorname{diag}} \big(\underset{1\le t\le m_d-1}{\operatorname{col}'} (w_{dt.}) \big) \Big) = \sum_{d=1}^{D} w_{d..}^{-1} \operatorname{tr}\Big(\underset{1\le t\le m_d-1}{\operatorname{col}} (w_{dt.}) \underset{1\le t\le m_d-1}{\operatorname{col}'} (w_{dt.}) \Big)$$

$$+ \operatorname{tr}\Big(\underset{1\le d\le D}{\operatorname{col}} (w_{d..}^{-1} \underset{1\le t\le m_d-1}{\operatorname{col}} (w_{dt.}) \boldsymbol{w}'_{n_d} \boldsymbol{X}_d) \boldsymbol{B}^{11} \underset{1\le d\le D}{\operatorname{col}'} (w_{d..}^{-1} \boldsymbol{X}'_d \boldsymbol{w}_{n_d} \underset{1\le t\le m_d-1}{\operatorname{col}'} (w_{dt.})) \Big)$$

$$= \sum_{d=1}^{D} w_{d..}^{-1} \sum_{t=1}^{m_d-1} w_{dt.}^2 + \sum_{d=1}^{D} w_{d..}^{-2} \operatorname{tr}\Big(\underset{1\le t\le m_d-1}{\operatorname{col}} (w_{dt.}) \boldsymbol{w}'_{n_d} \boldsymbol{X}_d \boldsymbol{B}^{11} \boldsymbol{X}'_d \boldsymbol{w}_{n_d} \underset{1\le t\le m_d-1}{\operatorname{col}'} (w_{dt.}) \Big)$$

$$= \sum_{d=1}^{D} w_{d..}^{-1} \sum_{t=1}^{m_d-1} w_{dt.}^2 + \sum_{d=1}^{D} w_{d..}^{-2} (\boldsymbol{w}'_{n_d} \boldsymbol{X}_d \boldsymbol{B}^{11} \boldsymbol{X}'_d \boldsymbol{w}_{n_d}) \sum_{t=1}^{m_d-1} w_{dt.}^2$$

$$= \sum_{d=1}^{D} w_{d..}^{-1} \left\{ \left[1 + w_{d..}^{-1} \Big(\sum_{t=1}^{m_d} \boldsymbol{w}'_{n_{dt}} \boldsymbol{X}_{dt} \Big) \boldsymbol{B}^{11} \Big(\sum_{t=1}^{m_d} \boldsymbol{w}'_{n_{dt}} \boldsymbol{X}_{dt} \Big)' \right] \sum_{t=1}^{m_d-1} w_{dt.}^2 \right\}$$

and

$$t_{12} = -\operatorname{tr}\Big(\underset{1\le d\le D}{\operatorname{col}} \big(\underset{1\le t\le m_d-1}{\operatorname{col}} (\mathbf{1}'_{n_{dt}} \boldsymbol{W}_{dt} \boldsymbol{X}_{dt}) \big) \boldsymbol{B}^{11}$$

$$\cdot \underset{1\le d\le D}{\operatorname{col}'} (w_{d..}^{-1} \boldsymbol{X}'_d \boldsymbol{W}_d \mathbf{1}_{n_d}) \underset{1\le d\le D}{\operatorname{diag}} \big(\underset{1\le t\le m_d-1}{\operatorname{col}'} (w_{dt.}) \big) \Big)$$

$$= -\operatorname{tr}\Big(\underset{1\le d\le D}{\operatorname{col}} \big(\underset{1\le t\le m_d-1}{\operatorname{col}} (\mathbf{1}'_{n_{dt}} \boldsymbol{W}_{dt} \boldsymbol{X}_{dt}) \big) \boldsymbol{B}^{11} \underset{1\le d\le D}{\operatorname{col}'} (w_{d..}^{-1} \boldsymbol{X}'_d \boldsymbol{W}_d \mathbf{1}_{n_d} \underset{1\le t\le m_d-1}{\operatorname{col}'} (w_{dt.})) \Big)$$

$$= -\sum_{d=1}^{D} w_{d..}^{-1} \operatorname{tr}\Big(\underset{1\le t\le m_d-1}{\operatorname{col}} (\mathbf{1}'_{n_{dt}} \boldsymbol{W}_{dt} \boldsymbol{X}_{dt}) \boldsymbol{B}^{11} \boldsymbol{X}'_d \boldsymbol{W}_d \mathbf{1}_{n_d} \underset{1\le t\le m_d-1}{\operatorname{col}'} (w_{dt.}) \Big)$$

$$= -\sum_{d=1}^{D} w_{d..}^{-1} \sum_{t=1}^{m_d-1} \mathbf{1}'_{n_{dt}} \boldsymbol{W}_{dt} \boldsymbol{X}_{dt} \boldsymbol{B}^{11} \boldsymbol{X}'_d \boldsymbol{W}_d \mathbf{1}_{n_d} w_{dt}.$$

$$= -\sum_{d=1}^{D} w_{d..}^{-1} \Big(\sum_{t=1}^{m_d-1} w_{dt.} \mathbf{1}'_{n_{dt}} \boldsymbol{W}_{dt} \boldsymbol{X}_{dt} \Big) \boldsymbol{B}^{11} \sum_{t=1}^{m_d} \boldsymbol{X}'_{dt} \boldsymbol{w}_{n_{dt}}.$$

11.5.3 Calculation of M_3

This section gives a computationally efficient formula of $\boldsymbol{y}'\boldsymbol{M}_3\boldsymbol{y}$, where

$$\boldsymbol{M}_3 = \boldsymbol{W} - \boldsymbol{W}\boldsymbol{X}^{(3)}(\boldsymbol{X}^{(3)\prime}\boldsymbol{W}\boldsymbol{X}^{(3)})^{-1}\boldsymbol{X}^{(3)\prime}\boldsymbol{W}, \;\; \boldsymbol{X}^{(3)} = (\boldsymbol{X}, \boldsymbol{Z}_1, \widetilde{\boldsymbol{Z}}_2), \;\; \boldsymbol{Z} = (\boldsymbol{Z}_1, \widetilde{\boldsymbol{Z}}_2).$$

The first step is calculating the inverse of $\boldsymbol{X}^{(3)\prime}\boldsymbol{W}\boldsymbol{X}^{(3)}$. We have

$$\boldsymbol{A} = (\boldsymbol{X}^{(3)\prime}\boldsymbol{W}\boldsymbol{X}^{(3)})^{-1} = \begin{pmatrix} \boldsymbol{X}'\boldsymbol{W}\boldsymbol{X} & \boldsymbol{X}'\boldsymbol{W}\boldsymbol{Z} \\ \boldsymbol{Z}'\boldsymbol{W}\boldsymbol{X} & \boldsymbol{Z}'\boldsymbol{W}\boldsymbol{Z} \end{pmatrix}^{-1} = \begin{pmatrix} \boldsymbol{D}^{11} & \boldsymbol{D}^{12} \\ \boldsymbol{D}^{21} & \boldsymbol{D}^{22} \end{pmatrix},$$

where $\boldsymbol{D}^{11} = (\boldsymbol{X}'\boldsymbol{P}\boldsymbol{X})^{-1}$, $\boldsymbol{D}^{12} = -\boldsymbol{D}^{11}\boldsymbol{X}'\boldsymbol{W}\boldsymbol{Z}\boldsymbol{G}$, $\boldsymbol{D}^{21} = (\boldsymbol{D}^{12})'$, $\boldsymbol{D}^{22} = \boldsymbol{G} + \boldsymbol{G}\boldsymbol{Z}'\boldsymbol{W}\boldsymbol{X}\boldsymbol{D}^{11}\boldsymbol{X}'\boldsymbol{W}\boldsymbol{Z}\boldsymbol{G}$, $\boldsymbol{G} = (\boldsymbol{Z}'\boldsymbol{W}\boldsymbol{Z})^{-1}$, and $\boldsymbol{P} = \boldsymbol{W} - \boldsymbol{W}\boldsymbol{Z}\boldsymbol{G}\boldsymbol{Z}'\boldsymbol{W}$.

The components of matrix $\boldsymbol{G}^{-1} = \boldsymbol{Z}'\boldsymbol{W}\boldsymbol{Z}$ are

$$\boldsymbol{G}^{11} = \boldsymbol{Z}_1'\boldsymbol{W}\boldsymbol{Z}_1 = \underset{1\le d\le D}{\text{diag}}\,(\mathbf{1}'_{n_d})\,\underset{1\le d\le D}{\text{diag}}\,(\boldsymbol{W}_d)\,\underset{1\le d\le D}{\text{diag}}\,(\mathbf{1}_{n_d}) = \underset{1\le d\le D}{\text{diag}}\,(\mathbf{1}'_{n_d}\boldsymbol{W}_d\mathbf{1}_{n_d})$$
$$= \underset{1\le d\le D}{\text{diag}}\,(w_{d..}),$$

$$\boldsymbol{G}^{12} = \boldsymbol{Z}_1'\boldsymbol{W}\widetilde{\boldsymbol{Z}}_2 = \underset{1\le d\le D}{\text{diag}}\left(\underset{1\le t\le m_d-1}{\text{col}'}\,(w_{dt.})\right), \quad \boldsymbol{G}^{21} = (\boldsymbol{G}^{12})',$$

$$\boldsymbol{G}^{22} = \widetilde{\boldsymbol{Z}}_2'\boldsymbol{W}\widetilde{\boldsymbol{Z}}_2 = \underset{1\le d\le D}{\text{diag}}\left(\underset{1\le t\le m_d-1}{\text{diag}}\,(w_{dt.})\right).$$

The components of matrix $\boldsymbol{G}$ are

$$G_{11} = [\boldsymbol{G}^{11} - \boldsymbol{G}^{12}(\boldsymbol{G}^{22})^{-1}\boldsymbol{G}^{21}]^{-1} = \Big[\underset{1\le d\le D}{\text{diag}}\,(w_{d..})$$
$$- \underset{1\le d\le D}{\text{diag}}\Big(\underset{1\le t\le m_d-1}{\text{col}'}(w_{dt.})\Big)\,\underset{1\le d\le D}{\text{diag}}\Big(\underset{1\le t\le m_d-1}{\text{diag}}(w_{dt.}^{-1})\Big)\,\underset{1\le d\le D}{\text{diag}}\Big(\underset{1\le t\le m_d-1}{\text{col}}(w_{dt.})\Big)\Big]^{-1}$$

$$= \left[\underset{1\le d\le D}{\text{diag}} \left(w_{d..} - \sum_{t=1}^{m_d-1} w_{dt.} \right) \right]^{-1} = \underset{1\le d\le D}{\text{diag}} \, (w_{dm_d.}^{-1}),$$

$$\begin{aligned}
\boldsymbol{G}_{12} &= -\boldsymbol{G}_{11}\boldsymbol{G}^{12}(\boldsymbol{G}^{22})^{-1} \\
&= -\underset{1\le d\le D}{\text{diag}} \, (w_{dm_d.}^{-1}) \, \underset{1\le d\le D}{\text{diag}} \Big(\underset{1\le t\le m_d-1}{\text{col}'} (w_{dt.}) \Big) \, \underset{1\le d\le D}{\text{diag}} \Big(\underset{1\le t\le m_d-1}{\text{diag}} \, (w_{dt.}^{-1}) \Big) \\
&= -\underset{1\le d\le D}{\text{diag}} \Big(w_{dm_d.}^{-1} \underset{1\le t\le m_d-1}{\text{col}'} (w_{dt.}) \underset{1\le t\le m_d-1}{\text{diag}} \, (w_{dt.}^{-1}) \Big) = -\underset{1\le d\le D}{\text{diag}} \, (w_{dm_d.}^{-1} \boldsymbol{1}'_{m_d-1}), \\
\boldsymbol{G}_{22} &= (\boldsymbol{G}^{22})^{-1} + (\boldsymbol{G}^{22})^{-1}\boldsymbol{G}^{21}\boldsymbol{G}_{11}\boldsymbol{G}^{12}(\boldsymbol{G}^{22})^{-1} = \underset{1\le d\le D}{\text{diag}} \Big(\underset{1\le t\le m_d-1}{\text{diag}} \, (w_{dt.}^{-1}) \Big) \\
&+ \underset{1\le d\le D}{\text{diag}} \Big(\underset{1\le t\le m_d-1}{\text{diag}} \, (w_{dt.}^{-1}) \underset{1\le t\le m_d-1}{\text{col}} (w_{dt.}) \, w_{dm_d.}^{-1} \underset{1\le t\le m_d-1}{\text{col}'} (w_{dt.}) \underset{1\le t\le m_d-1}{\text{diag}} \, (w_{dt.}^{-1}) \Big) \\
&= \underset{1\le d\le D}{\text{diag}} \Big(\underset{1\le t\le m_d-1}{\text{diag}} \, (w_{dt.}^{-1}) \Big) + \underset{1\le d\le D}{\text{diag}} \Big(w_{dm_d.}^{-1} \boldsymbol{1}_{m_d-1} \boldsymbol{1}'_{m_d-1} \Big).
\end{aligned}$$

For obtaining a computationally efficient formula for the matrix $\boldsymbol{P} = \boldsymbol{W} - \boldsymbol{W}\boldsymbol{Z}\boldsymbol{G}\boldsymbol{Z}'\boldsymbol{W}$ we do some calculations. First,

$$\begin{aligned}
\boldsymbol{W}\boldsymbol{Z}\boldsymbol{G}\boldsymbol{Z}'\boldsymbol{W} &= \underset{1\le d\le D}{\text{diag}} \, (\boldsymbol{W}_d)[\boldsymbol{Z}_1, \widetilde{\boldsymbol{Z}}_2] \begin{pmatrix} \boldsymbol{G}_{11} & \boldsymbol{G}_{12} \\ \boldsymbol{G}_{21} & \boldsymbol{G}_{22} \end{pmatrix} \begin{bmatrix} \boldsymbol{Z}'_1 \\ \widetilde{\boldsymbol{Z}}'_2 \end{bmatrix} \underset{1\le d\le D}{\text{diag}} \, (\boldsymbol{W}_d) \\
&= \underset{1\le d\le D}{\text{diag}} \, (\boldsymbol{W}_d) \big[\boldsymbol{Z}_1\boldsymbol{G}_{11}\boldsymbol{Z}'_1 + \boldsymbol{Z}_1\boldsymbol{G}_{12}\widetilde{\boldsymbol{Z}}'_2 + \widetilde{\boldsymbol{Z}}_2\boldsymbol{G}_{21}\boldsymbol{Z}'_1 + \widetilde{\boldsymbol{Z}}_2\boldsymbol{G}_{22}\widetilde{\boldsymbol{Z}}'_2 \big] \\
&\cdot \underset{1\le d\le D}{\text{diag}} \, (\boldsymbol{W}_d) = \boldsymbol{Z}_{11} + \boldsymbol{Z}_{12} + \boldsymbol{Z}_{21} + \boldsymbol{Z}_{22}.
\end{aligned}$$

Further, we have

$$\begin{aligned}
\boldsymbol{Z}_{11} &= \underset{1\le d\le D}{\text{diag}} \, (\boldsymbol{W}_d \boldsymbol{1}_{n_d} w_{dm_d.}^{-1} \boldsymbol{1}'_{n_d} \boldsymbol{W}_d) = \underset{1\le d\le D}{\text{diag}} \, (w_{dm_d.}^{-1} \boldsymbol{w}_{n_d} \boldsymbol{w}'_{n_d}), \\
\boldsymbol{Z}_{12} &= -\underset{1\le d\le D}{\text{diag}} \Big(\boldsymbol{W}_d \boldsymbol{1}_{n_d} w_{dm_d.}^{-1} \boldsymbol{1}'_{m_d-1} \Big[\underset{1\le t\le m_d-1}{\text{diag}} \, (\boldsymbol{1}'_{n_{dt}}), \boldsymbol{0} \Big] \boldsymbol{W}_d \Big) \\
&= -\underset{1\le d\le D}{\text{diag}} \Big(w_{dm_d.}^{-1} \boldsymbol{w}_{n_d} \boldsymbol{1}'_{m_d-1} \Big[\underset{1\le t\le m_d-1}{\text{diag}} \, (\boldsymbol{w}'_{n_{dt}}), \boldsymbol{0} \Big] \Big) \\
&= -\underset{1\le d\le D}{\text{diag}} \Big(w_{dm_d.}^{-1} \boldsymbol{w}_{n_d} \Big[\underset{1\le t\le m_d-1}{\text{col}'} (\boldsymbol{w}'_{n_{dt}}), \boldsymbol{0} \Big] \Big) \\
&= -\underset{1\le d\le D}{\text{diag}} \Big(w_{dm_d.}^{-1} \Big[\underset{1\le t\le m_d}{\text{col}} (\boldsymbol{w}_{n_{dt}}) \underset{1\le t\le m_d-1}{\text{col}'} (\boldsymbol{w}'_{n_{dt}}), \boldsymbol{0} \Big] \Big), \\
\boldsymbol{Z}_{22} &= \underset{1\le d\le D}{\text{diag}} \Big(\boldsymbol{W}_d \Big[\underset{1\le t\le m_d-1}{\text{diag}} \, (\boldsymbol{1}'_{n_{dt}}), \boldsymbol{0} \Big]' \underset{1\le t\le m_d-1}{\text{diag}} \, (w_{dt.}^{-1}) \Big[\underset{1\le t\le m_d-1}{\text{diag}} \, (\boldsymbol{1}'_{n_{dt}}), \boldsymbol{0} \Big] \boldsymbol{W}_d \Big)
\end{aligned}$$

$$+ \operatorname*{diag}_{1\le d\le D}\Big(\boldsymbol{W}_d\Big[\operatorname*{diag}_{1\le t\le m_d-1}(\boldsymbol{1}'_{n_{dt}}),\boldsymbol{0}\Big]' w_{dm_d.}^{-1}\boldsymbol{1}_{m_d-1}\boldsymbol{1}'_{m_d-1}\Big[\operatorname*{diag}_{1\le t\le m_d-1}(\boldsymbol{1}'_{n_{dt}}),\boldsymbol{0}\Big]\boldsymbol{W}_d\Big)$$

$$= \operatorname*{diag}_{1\le d\le D}\Big(\Big[\operatorname*{diag}_{1\le t\le m_d-1}(\boldsymbol{w}'_{n_{dt}}),\boldsymbol{0}\Big]'\operatorname*{diag}_{1\le t\le m_d-1}(w_{dt.}^{-1})\Big[\operatorname*{diag}_{1\le t\le m_d-1}(\boldsymbol{w}'_{n_{dt}}),\boldsymbol{0}\Big]\Big)$$

$$+ \operatorname*{diag}_{1\le d\le D}\Big(w_{dm_d.}^{-1}\Big[\operatorname*{diag}_{1\le t\le m_d-1}(\boldsymbol{w}'_{n_{dt}}),\boldsymbol{0}\Big]'\boldsymbol{1}_{m_d-1}\boldsymbol{1}'_{m_d-1}\Big[\operatorname*{diag}_{1\le t\le m_d-1}(\boldsymbol{w}'_{n_{dt}}),\boldsymbol{0}\Big]\Big)$$

$$= \operatorname*{diag}_{1\le d\le D}\Big(\operatorname{diag}\Big[\operatorname*{diag}_{1\le t\le m_d-1}(w_{dt.}^{-1}\boldsymbol{w}_{n_{dt}}\boldsymbol{w}'_{n_{dt}}),\boldsymbol{0}\Big]\Big)$$

$$+ \operatorname*{diag}_{1\le d\le D}\Big(w_{dm_d.}^{-1}\Big[\operatorname*{col}'_{1\le t\le m_d-1}(\boldsymbol{w}'_{n_{dt}}),\boldsymbol{0}\Big]'\Big[\operatorname*{col}'_{1\le t\le m_d-1}(\boldsymbol{w}'_{n_{dt}}),\boldsymbol{0}\Big]\Big).$$

The multivariate quadratic form $\boldsymbol{X}'\boldsymbol{P}\boldsymbol{X}$ can be decomposed in summands, i.e.

$$\boldsymbol{X}'\boldsymbol{P}\boldsymbol{X} = \boldsymbol{X}'\boldsymbol{W}\boldsymbol{X} - \boldsymbol{X}'\boldsymbol{Z}_{11}\boldsymbol{X} - 2\boldsymbol{X}'\boldsymbol{Z}_{12}\boldsymbol{X} - \boldsymbol{X}'\boldsymbol{Z}_{22}\boldsymbol{X},$$

where

$$\boldsymbol{X}'\boldsymbol{W}\boldsymbol{X} = \operatorname*{col}'_{1\le d\le D}(\boldsymbol{X}'_d)\operatorname*{diag}_{1\le d\le D}(\boldsymbol{W}_d)\operatorname*{col}_{1\le d\le D}(\boldsymbol{X}_d) = \sum_{d=1}^{D}\sum_{t=1}^{m_d}\boldsymbol{X}'_{dt}\boldsymbol{W}_{dt}\boldsymbol{X}_{dt},$$

$$\boldsymbol{X}'\boldsymbol{Z}_{11}\boldsymbol{X} = \operatorname*{col}'_{1\le d\le D}(\boldsymbol{X}'_d)\operatorname*{diag}_{1\le d\le D}(w_{dm_d.}^{-1}\boldsymbol{w}_{n_d}\boldsymbol{w}'_{n_d})\operatorname*{col}_{1\le d\le D}(\boldsymbol{X}_d)$$

$$= \sum_{d=1}^{D} w_{dm_d.}^{-1}\Big(\sum_{t=1}^{m_d}\boldsymbol{X}'_{dt}\boldsymbol{w}_{n_{dt}}\Big)\Big(\sum_{t=1}^{m_d}\boldsymbol{X}'_{dt}\boldsymbol{w}_{n_{dt}}\Big)',$$

$$\boldsymbol{X}'\boldsymbol{Z}_{12}\boldsymbol{X} = -\operatorname*{col}'_{1\le d\le D}(\boldsymbol{X}'_d)\operatorname*{diag}_{1\le d\le D}\Big(w_{dm_d.}^{-1}\Big[\operatorname*{col}_{1\le t\le m_d}(\boldsymbol{w}_{n_{dt}})\operatorname*{col}'_{1\le t\le m_d-1}(\boldsymbol{w}'_{n_{dt}}),\boldsymbol{0}\Big]\Big)$$

$$\cdot \operatorname*{col}_{1\le d\le D}(\boldsymbol{X}_d) = -\sum_{d=1}^{D} w_{dm_d.}^{-1}\Big[\Big(\sum_{t=1}^{m_d}\boldsymbol{X}'_{dt}\boldsymbol{w}_{n_{dt}}\Big)\Big(\sum_{t=1}^{m_d-1}\boldsymbol{X}'_{dt}\boldsymbol{w}_{n_{dt}}\Big)',\boldsymbol{0}\Big],$$

$$\boldsymbol{X}'\boldsymbol{Z}_{22}\boldsymbol{X} = \operatorname*{col}'_{1\le d\le D}(\boldsymbol{X}'_d)\operatorname*{diag}_{1\le d\le D}\Big(\operatorname{diag}\Big[\operatorname*{diag}_{1\le t\le m_d-1}(w_{dt.}^{-1}\boldsymbol{w}_{n_{dt}}\boldsymbol{w}'_{n_{dt}}),\boldsymbol{0}\Big]\Big)\operatorname*{col}_{1\le d\le D}(\boldsymbol{X}_d)$$

$$+ \operatorname*{col}'_{1\le d\le D}(\boldsymbol{X}'_d)\operatorname*{diag}_{1\le d\le D}\Big(w_{dm_d.}^{-1}\Big[\operatorname*{col}_{1\le t\le m_d-1}(\boldsymbol{w}_{n_{dt}}),\boldsymbol{0}\Big]'\Big[\operatorname*{col}'_{1\le t\le m_d-1}(\boldsymbol{w}'_{n_{dt}}),\boldsymbol{0}\Big]\Big)\operatorname*{col}_{1\le d\le D}(\boldsymbol{X}_d)$$

$$= \sum_{d=1}^{D}\sum_{t=1}^{m_d-1} w_{dt.}^{-1}\boldsymbol{X}'_{dt}\boldsymbol{w}_{n_{dt}}\boldsymbol{w}'_{n_{dt}}\boldsymbol{X}_{dt} + \sum_{d=1}^{D} w_{dm_d.}^{-1}\Big(\sum_{t=1}^{m_d-1}\boldsymbol{X}'_{dt}\boldsymbol{w}_{n_{dt}}\Big)\Big(\sum_{t=1}^{m_d-1}\boldsymbol{X}'_{dt}\boldsymbol{w}_{n_{dt}}\Big)'.$$

We do some calculations for obtaining a computationally efficient formula for the matrix $\boldsymbol{G}\boldsymbol{Z}'\boldsymbol{W}\boldsymbol{X}$.

$$\boldsymbol{G}\boldsymbol{Z}'\boldsymbol{W}\boldsymbol{X} = \begin{pmatrix} \boldsymbol{G}_{11} & \boldsymbol{G}_{12} \\ \boldsymbol{G}_{21} & \boldsymbol{G}_{22} \end{pmatrix} \begin{pmatrix} \boldsymbol{Z}_1' \\ \widetilde{\boldsymbol{Z}}_2' \end{pmatrix} \boldsymbol{W}\boldsymbol{X} = \begin{pmatrix} \boldsymbol{R}_{11} + \boldsymbol{R}_{12} \\ \boldsymbol{R}_{21} + \boldsymbol{R}_{22} \end{pmatrix},$$

where

$$\boldsymbol{R}_{11} = \boldsymbol{G}_{11}\boldsymbol{Z}_1'\boldsymbol{W}\boldsymbol{X} = \underset{1\le d\le D}{\text{diag}}\,(w_{dm_d.}^{-1})\,\underset{1\le d\le D}{\text{diag}}\,(\boldsymbol{1}_{n_d}')\,\underset{1\le d\le D}{\text{col}}(\boldsymbol{W}_d\boldsymbol{X}_d)$$
$$= \underset{1\le d\le D}{\text{col}}\Big(w_{dm_d.}^{-1}\sum_{t=1}^{m_d}\boldsymbol{w}_{n_{dt}}'\boldsymbol{X}_{dt}\Big),$$

$$\boldsymbol{R}_{12} = \boldsymbol{G}_{12}\widetilde{\boldsymbol{Z}}_2'\boldsymbol{W}\boldsymbol{X}$$
$$= -\underset{1\le d\le D}{\text{diag}}\,(w_{dm_d.}^{-1}\boldsymbol{1}_{m_d-1}')\,\underset{1\le d\le D}{\text{diag}}\Big(\Big[\underset{1\le t\le m_d-1}{\text{diag}}(\boldsymbol{1}_{n_{dt}}'),\boldsymbol{0}\Big]\Big)\underset{1\le d\le D}{\text{col}}(\boldsymbol{W}_d\boldsymbol{X}_d)$$
$$=-\underset{1\le d\le D}{\text{col}}\Big(w_{dm_d.}^{-1}\boldsymbol{1}_{m_d-1}'\underset{1\le t\le m_d-1}{\text{col}}(\boldsymbol{w}_{n_{dt}}'\boldsymbol{X}_{dt})\Big)=-\underset{1\le d\le D}{\text{col}}\Big(w_{dm_d.}^{-1}\sum_{t=1}^{m_d-1}\boldsymbol{w}_{n_{dt}}'\boldsymbol{X}_{dt}\Big),$$

$$\boldsymbol{R}_{21} = \boldsymbol{G}_{21}\boldsymbol{Z}_1'\boldsymbol{W}\boldsymbol{X} = -\underset{1\le d\le D}{\text{diag}}\,(w_{dm_d.}^{-1}\boldsymbol{1}_{m_d-1})\,\underset{1\le d\le D}{\text{diag}}\,(\boldsymbol{1}_{n_d}')\,\underset{1\le d\le D}{\text{col}}(\boldsymbol{W}_d\boldsymbol{X}_d)$$
$$= -\underset{1\le d\le D}{\text{col}}(w_{dm_d.}^{-1}\boldsymbol{1}_{m_d-1}\boldsymbol{w}_{n_d}'\boldsymbol{X}_d) = -\underset{1\le d\le D}{\text{col}}\Big(\underset{1\le t\le m_d-1}{\text{col}}\Big(w_{dm_d.}^{-1}\sum_{t=1}^{m_d}\boldsymbol{w}_{n_{dt}}'\boldsymbol{X}_{dt}\Big)\Big),$$

$$\boldsymbol{R}_{22} = \boldsymbol{G}_{22}\widetilde{\boldsymbol{Z}}_2'\boldsymbol{W}\boldsymbol{X} = \Big[\underset{1\le d\le D}{\text{diag}}\Big(\underset{1\le t\le m_d-1}{\text{diag}}(w_{dt.}^{-1})\Big) + \underset{1\le d\le D}{\text{diag}}\Big(w_{dm_d.}^{-1}\boldsymbol{1}_{m_d-1}\boldsymbol{1}_{m_d-1}'\Big)\Big]$$
$$\cdot\ \underset{1\le d\le D}{\text{diag}}\Big(\Big[\underset{1\le t\le m_d-1}{\text{diag}}(\boldsymbol{1}_{n_{dt}}'),\boldsymbol{0}\Big]\Big)\underset{1\le d\le D}{\text{col}}(\boldsymbol{W}_d\boldsymbol{X}_d)$$
$$= \underset{1\le d\le D}{\text{col}}\Big(\underset{1\le t\le m_d-1}{\text{col}}(w_{dt.}^{-1}\boldsymbol{w}_{n_{dt}}'\boldsymbol{X}_{dt})\Big) + \underset{1\le d\le D}{\text{col}}\Big(\underset{1\le t\le m_d-1}{\text{col}}\Big(w_{dm_d.}^{-1}\sum_{t=1}^{m_d-1}\boldsymbol{w}_{n_{dt}}'\boldsymbol{X}_{dt}\Big)\Big).$$

The components of matrix

$$\boldsymbol{A} = (\boldsymbol{X}^{(3)\prime}\boldsymbol{W}\boldsymbol{X}^{(3)})^{-1} = \begin{pmatrix} \boldsymbol{D}^{11} & \boldsymbol{D}^{12} \\ \boldsymbol{D}^{21} & \boldsymbol{D}^{22} \end{pmatrix} = \begin{pmatrix} \boldsymbol{D}^{11} & \boldsymbol{A}^{12} & \boldsymbol{A}^{13} \\ \boldsymbol{A}^{21} & \boldsymbol{A}^{22} & \boldsymbol{A}^{23} \\ \boldsymbol{A}^{31} & \boldsymbol{A}^{32} & \boldsymbol{A}^{33} \end{pmatrix}$$

can be calculated as follows:

$$\begin{aligned}
\boldsymbol{D}^{11} &= (\boldsymbol{X}'\boldsymbol{P}\boldsymbol{X})^{-1} = (\boldsymbol{X}'\boldsymbol{W}\boldsymbol{X} - \boldsymbol{X}'\boldsymbol{Z}_{11}\boldsymbol{X} - 2\boldsymbol{X}'\boldsymbol{Z}_{12}\boldsymbol{X} - \boldsymbol{X}'\boldsymbol{Z}_{22}\boldsymbol{X})^{-1},\\
\boldsymbol{D}^{21} &= -\boldsymbol{G}\boldsymbol{Z}'\boldsymbol{W}\boldsymbol{X}\boldsymbol{D}^{11} = [\boldsymbol{A}^{21}, \boldsymbol{A}^{31}]', \quad \boldsymbol{D}^{12} = (\boldsymbol{D}^{21})' = [(\boldsymbol{A}^{21})', (\boldsymbol{A}^{13})']\\
&= [\boldsymbol{A}^{12}, \boldsymbol{A}^{13}],\\
\boldsymbol{A}^{21} &= -[\boldsymbol{G}_{11}\boldsymbol{Z}_1' + \boldsymbol{G}_{12}\widetilde{\boldsymbol{Z}}_2']\boldsymbol{W}\boldsymbol{X}\boldsymbol{D}^{11} = -[\boldsymbol{R}_{11} + \boldsymbol{R}_{12}]\boldsymbol{D}^{11},\\
\boldsymbol{A}^{31} &= -[\boldsymbol{G}_{21}\boldsymbol{Z}_1' + \boldsymbol{G}_{22}\widetilde{\boldsymbol{Z}}_2']\boldsymbol{W}\boldsymbol{X}\boldsymbol{D}^{11} = -[\boldsymbol{R}_{21} + \boldsymbol{R}_{22}]\boldsymbol{D}^{11}.
\end{aligned}$$

The matrix $\boldsymbol{D}^{22}$ can be decomposed as

$$\boldsymbol{D}^{22} = \boldsymbol{G} + \boldsymbol{G}\boldsymbol{Z}'\boldsymbol{W}\boldsymbol{X}\boldsymbol{D}^{11}\boldsymbol{X}'\boldsymbol{W}\boldsymbol{Z}\boldsymbol{G} = \begin{pmatrix} \boldsymbol{A}^{22} & \boldsymbol{A}^{23} \\ \boldsymbol{A}^{31} & \boldsymbol{A}^{33} \end{pmatrix},$$

where

$$\begin{aligned}
\boldsymbol{A}^{22} &= \boldsymbol{G}_{11} + [\boldsymbol{R}_{11} + \boldsymbol{R}_{12}]\boldsymbol{D}^{11}[\boldsymbol{R}_{11} + \boldsymbol{R}_{12}]',\\
\boldsymbol{A}^{23} &= \boldsymbol{G}_{12} + [\boldsymbol{R}_{11} + \boldsymbol{R}_{12}]\boldsymbol{D}^{11}[\boldsymbol{R}_{21} + \boldsymbol{R}_{22}]',\\
\boldsymbol{A}^{32} &= \boldsymbol{G}_{21} + [\boldsymbol{R}_{21} + \boldsymbol{R}_{22}]\boldsymbol{D}^{11}[\boldsymbol{R}_{11} + \boldsymbol{R}_{12}]',\\
\boldsymbol{A}^{33} &= \boldsymbol{G}_{22} + [\boldsymbol{R}_{21} + \boldsymbol{R}_{22}]\boldsymbol{D}^{11}[\boldsymbol{R}_{21} + \boldsymbol{R}_{22}]'.
\end{aligned}$$

Finally, the quadratic form $\boldsymbol{y}'\boldsymbol{M}_3\boldsymbol{y}$ is

$$\begin{aligned}
&\boldsymbol{y}'\boldsymbol{M}_3\boldsymbol{y} = \boldsymbol{y}'\boldsymbol{W}\boldsymbol{y} - \boldsymbol{y}'\boldsymbol{W}[\boldsymbol{X}, \boldsymbol{Z}_1, \widetilde{\boldsymbol{Z}}_2]\boldsymbol{A}[\boldsymbol{X}', \boldsymbol{Z}_1', \widetilde{\boldsymbol{Z}}_2']'\boldsymbol{W}\boldsymbol{y} = \boldsymbol{y}'\boldsymbol{W}\boldsymbol{y} - \boldsymbol{y}'\boldsymbol{W}\\
&\cdot [\boldsymbol{X}\boldsymbol{D}^{11}\boldsymbol{X}' + \boldsymbol{Z}_1\boldsymbol{A}^{22}\boldsymbol{Z}_1' + \widetilde{\boldsymbol{Z}}_2\boldsymbol{A}^{33}\widetilde{\boldsymbol{Z}}_2' + 2\boldsymbol{X}\boldsymbol{A}^{12}\boldsymbol{Z}_1' + 2\boldsymbol{X}\boldsymbol{A}^{13}\widetilde{\boldsymbol{Z}}_2' + 2\boldsymbol{Z}_1\boldsymbol{A}^{23}\widetilde{\boldsymbol{Z}}_2']\boldsymbol{W}\boldsymbol{y}.
\end{aligned}$$

Therefore, we have

$$\begin{aligned}
\boldsymbol{y}'\boldsymbol{M}_3\boldsymbol{y} = &\sum_{d=1}^{D}\sum_{t=1}^{m_d} \boldsymbol{y}_{dt}'\boldsymbol{W}_{dt}\boldsymbol{y}_{dt} - \Bigg\{ \Big(\sum_{d=1}^{D}\sum_{t=1}^{m_d} \boldsymbol{y}_{dt}'\boldsymbol{W}_{dt}\boldsymbol{X}_{dt}\Big)\boldsymbol{D}^{11}\Big(\sum_{d=1}^{D}\sum_{t=1}^{m_d} \boldsymbol{y}_{dt}'\boldsymbol{W}_{dt}\boldsymbol{X}_{dt}\Big)'\\
&+ \underset{1\le d\le D}{\text{col}'}\Big(\sum_{t=1}^{m_d} \boldsymbol{y}_{dt}'\boldsymbol{w}_{n_{dt}}\Big)\boldsymbol{A}^{22}\Big[\underset{1\le d\le D}{\text{col}'}\Big(\sum_{t=1}^{m_d} \boldsymbol{y}_{dt}'\boldsymbol{w}_{n_{dt}}\Big)\Big]'\\
&+ \underset{1\le d\le D}{\text{col}'}\Big(\underset{1\le t\le m_d-1}{\text{col}'}(\boldsymbol{y}_{dt}'\boldsymbol{w}_{n_{dt}})\Big)\boldsymbol{A}^{33}\Big[\underset{1\le d\le D}{\text{col}'}\Big(\underset{1\le t\le m_d-1}{\text{col}'}(\boldsymbol{y}_{dt}'\boldsymbol{w}_{n_{dt}})\Big)\Big]'\\
&+ 2\Big(\sum_{d=1}^{D}\sum_{t=1}^{m_d} \boldsymbol{y}_{dt}'\boldsymbol{W}_{dt}\boldsymbol{X}_{dt}\Big)\boldsymbol{A}^{12}\Big[\underset{1\le d\le D}{\text{col}'}\Big(\sum_{t=1}^{m_d} \boldsymbol{y}_{dt}'\boldsymbol{w}_{n_{dt}}\Big)\Big]'
\end{aligned}$$

$$+ 2\Big(\sum_{d=1}^{D}\sum_{t=1}^{m_d} \boldsymbol{y}'_{dt}\boldsymbol{W}_{dt}\boldsymbol{X}_{dt}\Big)\boldsymbol{A}^{13}\Big[\underset{1\le d\le D}{\text{col}'}\big(\underset{1\le t\le m_d-1}{\text{col}'}(\boldsymbol{y}'_{dt}\boldsymbol{w}_{n_{dt}})\big)\Big]'$$

$$+ 2\underset{1\le d\le D}{\text{col}'}\Big(\sum_{t=1}^{m_d} \boldsymbol{y}'_{dt}\boldsymbol{w}_{n_{dt}}\Big)\boldsymbol{A}^{23}\Big[\underset{1\le d\le D}{\text{col}'}\big(\underset{1\le t\le m_d-1}{\text{col}'}(\boldsymbol{y}'_{dt}\boldsymbol{w}_{n_{dt}})\big)\Big]'\Bigg\}.$$

11.6 EBLUP of a Subdomain Mean

Let U be a finite population partitioned in domains and subdomains, i.e. $U = \cup_{d=1}^{D} U_d$, $U_{d_1} \cap U_{d_2} = \emptyset$ if $d_1 \neq d_2$, and $U_d = \cup_{t=1}^{M_d} U_{dt}$, $U_{dt_1} \cap U_{dt_2} = \emptyset$ if $t_1 \neq t_2$. Let N, N_d, and N_{dt} be the sizes of U, U_d, and U_{dt}, so that $N = \sum_{d=1}^{D} N_d$ and $N_d = \sum_{t=1}^{M_d} N_{dt}$. We assume that the population target vector $\boldsymbol{y} = \boldsymbol{y}_{N\times 1}$ follows the NER2 model (11.2) with the obvious size changes. This is to say, with N, N_d, N_{dt}, and M_d in the place of n, n_d, n_{dt}, and m_d, respectively.

Let $s \subset U$ be a sample of $n \le N$ units and let $r = U - s$ be the set of non-sampled units. The domain and subdomain subsets of s and r are denoted by s_d, r_d, s_{dt}, and r_{dt}, respectively. The subindexes s and r in vectors or matrices are used to denote their sampled and the non-sampled parts. Without loss of generality, we renumber the population units and we write

$$\boldsymbol{Z}_1 = \begin{pmatrix} \boldsymbol{Z}_{1s} \\ \boldsymbol{Z}_{1r} \end{pmatrix}, \quad \boldsymbol{Z}_2 = \begin{pmatrix} \boldsymbol{Z}_{2s} \\ \boldsymbol{Z}_{2r} \end{pmatrix}, \quad \boldsymbol{X} = \begin{pmatrix} \boldsymbol{X}_s \\ \boldsymbol{X}_r \end{pmatrix}, \quad \boldsymbol{y} = \begin{pmatrix} \boldsymbol{y}_s \\ \boldsymbol{y}_r \end{pmatrix}, \quad \boldsymbol{e} = \begin{pmatrix} \boldsymbol{e}_s \\ \boldsymbol{e}_r \end{pmatrix},$$

and

$$\boldsymbol{V} = \text{var}(\boldsymbol{y}) = \begin{pmatrix} \boldsymbol{V}_s & \boldsymbol{V}_{sr} \\ \boldsymbol{V}_{rs} & \boldsymbol{V}_r \end{pmatrix}.$$

The EBLUP of the linear parameter $\eta = \boldsymbol{a}'\boldsymbol{y} = \boldsymbol{a}'_s\boldsymbol{y}_s + \boldsymbol{a}'_r\boldsymbol{y}_r$ is (cf. Theorem 4.1)

$$\hat{\eta} = \boldsymbol{a}'_s\boldsymbol{y}_s + \boldsymbol{a}'_r\Big[\boldsymbol{X}_r\hat{\boldsymbol{\beta}} + \hat{\boldsymbol{V}}_{rs}\hat{\boldsymbol{V}}_s^{-1}(\boldsymbol{y}_s - \boldsymbol{X}_s\hat{\boldsymbol{\beta}})\Big].$$

As $\boldsymbol{V}_{ers} = \text{cov}(\boldsymbol{e}_r, \boldsymbol{e}_s) = \boldsymbol{0}$, $\boldsymbol{V}_{rs} = \boldsymbol{Z}_r\boldsymbol{V}_u\boldsymbol{Z}'_s + \boldsymbol{V}_{ers} = \boldsymbol{Z}_r\boldsymbol{V}_u\boldsymbol{Z}'_s$, and $\hat{\boldsymbol{u}} = \hat{\boldsymbol{V}}_u\boldsymbol{Z}'_s\hat{\boldsymbol{V}}_s^{-1}(\boldsymbol{y}_s - \boldsymbol{X}_s\hat{\boldsymbol{\beta}})$ (cf. Corollary 6.1) with

$$\boldsymbol{Z}_s = [\boldsymbol{Z}_{1s}, \boldsymbol{Z}_{2s}], \quad \boldsymbol{Z}_r = [\boldsymbol{Z}_{1r}, \boldsymbol{Z}_{2r}], \quad \hat{\boldsymbol{u}} = \begin{pmatrix} \hat{\boldsymbol{u}}_1 \\ \hat{\boldsymbol{u}}_2 \end{pmatrix},$$

we have

$$\hat{\eta} = \boldsymbol{a}_s'\boldsymbol{y}_s + \boldsymbol{a}_r'\left[\boldsymbol{X}_r\hat{\boldsymbol{\beta}} + \boldsymbol{Z}_r\hat{\boldsymbol{V}}_u\boldsymbol{Z}_s'\hat{\boldsymbol{V}}_s^{-1}(\boldsymbol{y}_s - \boldsymbol{X}_s\hat{\boldsymbol{\beta}})\right] = \boldsymbol{a}_s'\boldsymbol{y}_s + \boldsymbol{a}_r'\left[\boldsymbol{X}_r\hat{\boldsymbol{\beta}} + \boldsymbol{Z}_r\hat{\boldsymbol{u}}\right]$$

$$= \boldsymbol{a}'\left[\boldsymbol{X}\hat{\boldsymbol{\beta}} + \boldsymbol{Z}_1\hat{\boldsymbol{u}}_1 + \boldsymbol{Z}_2\hat{\boldsymbol{u}}_2\right] + \boldsymbol{a}_s'\left[\boldsymbol{y}_s - \boldsymbol{X}_s\hat{\boldsymbol{\beta}} - \boldsymbol{Z}_{s1}\hat{\boldsymbol{u}}_1 - \boldsymbol{Z}_{s2}\hat{\boldsymbol{u}}_2\right].$$

In the notation of model (11.1), $\overline{Y}_{dt} = \frac{1}{N_{dt}}\sum_{j=1}^{N_{dt}} y_{dtj}$ can be written as the linear parameter $\eta = \boldsymbol{a}'\boldsymbol{y}$, where

$$\boldsymbol{a}' = \frac{1}{N_{dt}}\Big((\boldsymbol{0}_{N_1}', \ldots, \boldsymbol{0}_{N_{d-1}}', \underset{1\le k\le M_d}{\text{col}'}\,[\delta_{tk}\boldsymbol{1}_{N_{dk}}'], \boldsymbol{0}_{N_{d+1}}', \ldots, \boldsymbol{0}_{N_D}'\Big)$$

$$= \frac{1}{N_{dt}}\underset{1\le\ell\le D}{\text{col}'}\Big(\delta_{d\ell}\underset{1\le k\le M_\ell}{\text{col}'}\,[\delta_{tk}\boldsymbol{1}_{N_{\ell k}}']\Big)$$

and $\delta_{ab} = 1$ if $a = b$ and $\delta_{ab} = 0$ if $a \ne b$. It holds that $\boldsymbol{a}'\boldsymbol{X} = \overline{\boldsymbol{X}}_{dt}$,

$$\boldsymbol{a}'\boldsymbol{Z}_1 = \frac{1}{N_{dt}}\underset{1\le\ell\le D}{\text{col}'}\{\delta_{d\ell}\underset{1\le k\le M_\ell}{\text{col}'}\,[\delta_{tk}\boldsymbol{1}_{N_{\ell k}}']\}\underset{1\le\ell\le D}{\text{diag}}\,(\boldsymbol{1}_{N_\ell}) = \underset{1\le\ell\le D}{\text{col}'}\{\delta_{d\ell}\} = \overline{\boldsymbol{Z}}_{1,dt},$$

$$\boldsymbol{a}'\boldsymbol{Z}_2 = \frac{1}{N_{dt}}\underset{1\le\ell\le D}{\text{col}'}\{\delta_{d\ell}\underset{1\le k\le M_\ell}{\text{col}'}\,[\delta_{tk}\boldsymbol{1}_{N_{\ell k}}']\}\underset{1\le\ell\le D}{\text{diag}}\,(\underset{1\le k\le M_\ell}{\text{diag}}\,(\boldsymbol{1}_{N_{\ell k}}))$$

$$= \underset{1\le\ell\le D}{\text{col}'}\{\underset{1\le k\le M_\ell}{\text{col}'}\{\delta_{d\ell}\delta_{tk}\}\} = \overline{\boldsymbol{Z}}_{2,dt}.$$

The EBLUP of $\overline{Y}_{dt}$, when $n_{dt} > 0$, is

$$\hat{\overline{Y}}_{dt}^{eblup} = \overline{\boldsymbol{X}}_{dt}\hat{\boldsymbol{\beta}} + \overline{\boldsymbol{Z}}_{1,dt}\hat{\boldsymbol{u}}_1 + \overline{\boldsymbol{Z}}_{2,dt}\hat{\boldsymbol{u}}_2 + f_{dt}\left[\overline{\boldsymbol{y}}_{s,dt} - \overline{\boldsymbol{X}}_{s,dt}\hat{\boldsymbol{\beta}} - \overline{\boldsymbol{Z}}_{1,dt}\hat{\boldsymbol{u}}_1 - \overline{\boldsymbol{Z}}_{2,dt}\hat{\boldsymbol{u}}_2\right]$$

$$= \overline{\boldsymbol{X}}_{dt}\hat{\boldsymbol{\beta}} + \hat{u}_{1,d} + \hat{u}_{2,dt} + f_{dt}\left[\overline{\boldsymbol{y}}_{s,dt} - \overline{\boldsymbol{X}}_{s,dt}\hat{\boldsymbol{\beta}} - \hat{u}_{1,d} - \hat{u}_{2,dt}\right],$$

where $\overline{\boldsymbol{y}}_{s,dt} = \frac{1}{n_{dt}}\sum_{j=1}^{n_{dt}} y_{dtj}$, $\overline{\boldsymbol{X}}_{s,dt} = \frac{1}{n_{dt}}\sum_{j=1}^{n_{dt}} \boldsymbol{x}_{dtj}$, and $f_{dt} = \frac{n_{dt}}{N_{dt}}$. The EBLUP of $\overline{Y}_{dt}$, when $n_{dt} = 0$, is the synthetic part

$$\hat{\overline{Y}}_{dt}^{eblup} = \overline{\boldsymbol{X}}_{dt}\hat{\boldsymbol{\beta}} + \overline{\boldsymbol{Z}}_{1,dt}\hat{\boldsymbol{u}}_1 + \overline{\boldsymbol{Z}}_{2,dt}\hat{\boldsymbol{u}}_2 = \overline{\boldsymbol{X}}_{dt}\hat{\boldsymbol{\beta}} + \hat{u}_{1,d} + \hat{u}_{2,dt}.$$

11.7 Mean Squared Error of the EBLUP of a Subdomain Mean

Let $\boldsymbol{\theta} = (\sigma^2, \varphi_1, \varphi_2)$ be the vector of variance components. Let us assume that $m_d = M_d$, $d = 1, \dots, D$, i.e. all the subdomains are in the sample. Section 9.3 gives an approximation of the MSE of the EBLUP of $\eta = \boldsymbol{a}'\boldsymbol{y} = \overline{Y}_{dt}$ under a superpopulation linear mixed model. The approximation is

$$
\begin{aligned}
MSE(\hat{\overline{Y}}_{dt}^{eblup}) &= g_1(\boldsymbol{\theta}) + g_2(\boldsymbol{\theta}) + g_3(\boldsymbol{\theta}) + g_4(\boldsymbol{\theta}),\\
g_1(\boldsymbol{\theta}) &= \boldsymbol{a}_r'\boldsymbol{Z}_r\boldsymbol{T}_s\boldsymbol{Z}_r'\boldsymbol{a}_r,\\
g_2(\boldsymbol{\theta}) &= [\boldsymbol{a}_r'\boldsymbol{X}_r - \boldsymbol{a}_r'\boldsymbol{Z}_r\boldsymbol{T}_s\boldsymbol{Z}_s'\boldsymbol{V}_{es}^{-1}\boldsymbol{X}_s]\boldsymbol{Q}_s[\boldsymbol{X}_r'\boldsymbol{a}_r - \boldsymbol{X}_s'\boldsymbol{V}_{es}^{-1}\boldsymbol{Z}_s\boldsymbol{T}_s\boldsymbol{Z}_r'\boldsymbol{a}_r],\\
g_3(\boldsymbol{\theta}) &\approx \operatorname{tr}\left\{(\nabla \boldsymbol{b}')\boldsymbol{V}_s(\nabla \boldsymbol{b}')'E\left[(\hat{\boldsymbol{\theta}} - \boldsymbol{\theta})(\hat{\boldsymbol{\theta}} - \boldsymbol{\theta})'\right]\right\},\\
g_4(\boldsymbol{\theta}) &= \boldsymbol{a}_r'\boldsymbol{V}_{er}\boldsymbol{a}_r.
\end{aligned}
$$

The MSE estimator is

$$
mse(\hat{\overline{Y}}_{dt}^{eblup}) = g_1(\hat{\boldsymbol{\theta}}) + g_2(\hat{\boldsymbol{\theta}}) + 2g_3(\hat{\boldsymbol{\theta}}) + g_4(\hat{\boldsymbol{\theta}}),
$$

under the assumption that $\hat{\boldsymbol{\theta}}$ are REML or H3 estimates.

11.7.1 Calculation of $g_1(\boldsymbol{\theta})$

The elements of the formula $g_1(\boldsymbol{\theta}) = \boldsymbol{a}_r'\boldsymbol{Z}_r\boldsymbol{T}_s\boldsymbol{Z}_r'\boldsymbol{a}_r$ are

$$
\boldsymbol{a}_r' = \frac{1}{N_{dt}}\big(\boldsymbol{0}_{N_1-n_1}', \dots, \boldsymbol{0}_{N_{d-1}-n_{d-1}}', \underset{1\le k\le m_d}{\operatorname{col}'}[\delta_{tk}\boldsymbol{1}_{N_{dk}-n_{dk}}'], \boldsymbol{0}_{N_{d+1}-n_{d+1}}', \dots, \boldsymbol{0}_{N_D-n_D}'\big),
$$

$$
\boldsymbol{Z}_r = [\boldsymbol{Z}_{1r}, \boldsymbol{Z}_{2r}], \quad \boldsymbol{T}_s = \boldsymbol{V}_u - \boldsymbol{V}_u\boldsymbol{Z}_s'\boldsymbol{V}_s^{-1}\boldsymbol{Z}_s\boldsymbol{V}_u = \begin{pmatrix} \boldsymbol{T}_{11s} & \boldsymbol{T}_{12s}\\ \boldsymbol{T}_{21s} & \boldsymbol{T}_{22s}\end{pmatrix},
$$

$$
\boldsymbol{V}_u = \begin{pmatrix} \sigma_1^2\boldsymbol{I}_D & \boldsymbol{0}\\ \boldsymbol{0} & \sigma_2^2\boldsymbol{I}_m\end{pmatrix}, \quad \boldsymbol{Z}_s = [\boldsymbol{Z}_{1s}, \boldsymbol{Z}_{2s}], \quad \boldsymbol{V}_s^{-1} = \underset{1\le d\le D}{\operatorname{diag}}(\boldsymbol{V}_{ds}^{-1}).
$$

It holds that

$$\boldsymbol{V}_u \boldsymbol{Z}_s' \boldsymbol{V}_s^{-1} \boldsymbol{Z}_s \boldsymbol{V}_u = \begin{pmatrix} \sigma_1^2 \boldsymbol{Z}_{1s}' \\ \sigma_2^2 \boldsymbol{Z}_{2s}' \end{pmatrix} \underset{1\le d\le D}{\text{diag}} (\boldsymbol{V}_{ds}^{-1}) \left[\sigma_1^2 \boldsymbol{Z}_{1s}, \sigma_2^2 \boldsymbol{Z}_{2s}\right]$$

$$= \begin{pmatrix} \sigma_1^4 \boldsymbol{Z}_{1s}' \underset{1\le d\le D}{\text{diag}} (\boldsymbol{V}_{ds}^{-1}) \boldsymbol{Z}_{1s} & \sigma_1^2 \sigma_2^2 \boldsymbol{Z}_{1s}' \underset{1\le d\le D}{\text{diag}} (\boldsymbol{V}_{ds}^{-1}) \boldsymbol{Z}_{2s} \\ \sigma_1^2 \sigma_2^2 \boldsymbol{Z}_{2s}' \underset{1\le d\le D}{\text{diag}} (\boldsymbol{V}_{ds}^{-1}) \boldsymbol{Z}_{1s} & \sigma_2^4 \boldsymbol{Z}_{2s}' \underset{1\le d\le D}{\text{diag}} (\boldsymbol{V}_{ds}^{-1}) \boldsymbol{Z}_{2s} \end{pmatrix},$$

where

$$\boldsymbol{Z}_{1s}' \underset{1\le d\le D}{\text{diag}} (\boldsymbol{V}_{ds}^{-1}) \boldsymbol{Z}_{1s} = \underset{1\le d\le D}{\text{diag}} (\boldsymbol{1}_{n_d}') \underset{1\le d\le D}{\text{diag}} (\boldsymbol{V}_{ds}^{-1}) \underset{1\le d\le D}{\text{diag}} (\boldsymbol{1}_{n_d})$$

$$= \underset{1\le d\le D}{\text{diag}} (\boldsymbol{1}_{nd}' \boldsymbol{V}_{ds}^{-1} \boldsymbol{1}_{nd}),$$

$$\boldsymbol{Z}_{1s}' \underset{1\le d\le D}{\text{diag}} (\boldsymbol{V}_{ds}^{-1}) \boldsymbol{Z}_{2s} = \underset{1\le d\le D}{\text{diag}} (\boldsymbol{1}_{n_d}') \underset{1\le d\le D}{\text{diag}} (\boldsymbol{V}_{ds}^{-1}) \underset{1\le d\le D}{\text{diag}} \{ \underset{1\le k\le m_d}{\text{diag}} (\boldsymbol{1}_{n_{dk}})\}$$

$$= \underset{1\le d\le D}{\text{diag}} \{\boldsymbol{1}_{nd}' \boldsymbol{V}_{ds}^{-1} \underset{1\le k\le m_d}{\text{diag}} (\boldsymbol{1}_{n_{dk}})\},$$

$$\boldsymbol{Z}_{2s}' \underset{1\le d\le D}{\text{diag}} (\boldsymbol{V}_{ds}^{-1}) \boldsymbol{Z}_{2s} = \underset{1\le d\le D}{\text{diag}} \{ \underset{1\le k\le m_d}{\text{diag}} (\boldsymbol{1}_{n_{dk}}') \boldsymbol{V}_{ds}^{-1} \underset{1\le k\le m_d}{\text{diag}} (\boldsymbol{1}_{n_{dk}})\}.$$

The blocks of matrix $\boldsymbol{T}_s$ are

$$\boldsymbol{T}_{11s} = \sigma_1^2 \underset{1\le d\le D}{\text{diag}} (1 - \sigma_1^2 \boldsymbol{1}_{n_d}' \boldsymbol{V}_{ds}^{-1} \boldsymbol{1}_{n_d}),$$

$$\boldsymbol{T}_{12s} = -\sigma_1^2 \sigma_2^2 \underset{1\le d\le D}{\text{diag}} \{\boldsymbol{1}_{n_d}' \boldsymbol{V}_{ds}^{-1} \underset{1\le k\le m_d}{\text{diag}} (\boldsymbol{1}_{n_{dk}})\}, \quad \boldsymbol{T}_{21s} = (\boldsymbol{T}_{12s})',$$

$$\boldsymbol{T}_{22s} = \sigma_2^2 \underset{1\le d\le D}{\text{diag}} \{ \boldsymbol{I}_{m_d} - \sigma_2^2 \underset{1\le k\le m_d}{\text{diag}} (\boldsymbol{1}_{n_{dk}}') \boldsymbol{V}_{ds}^{-1} \underset{1\le k\le m_d}{\text{diag}} (\boldsymbol{1}_{n_{dk}}) \}.$$

In what follows, we calculate the product $\boldsymbol{Z}_r \boldsymbol{T}_s \boldsymbol{Z}_r'$.

$$\boldsymbol{Z}_r \boldsymbol{T}_s \boldsymbol{Z}_r' = [\boldsymbol{Z}_{1r}, \boldsymbol{Z}_{2r}] \boldsymbol{T}_s \left[\boldsymbol{Z}_{1r}', \boldsymbol{Z}_{2r}'\right]'$$

$$= \boldsymbol{Z}_{1r} \boldsymbol{T}_{11s} \boldsymbol{Z}_{1r}' + \boldsymbol{Z}_{1r} \boldsymbol{T}_{12s} \boldsymbol{Z}_{2r}' + \boldsymbol{Z}_{2r} \boldsymbol{T}_{21s} \boldsymbol{Z}_{1r}' + \boldsymbol{Z}_{2r} \boldsymbol{T}_{22s} \boldsymbol{Z}_{2r}'.$$

It holds that

$$\boldsymbol{M}_{11}^{rr} = \boldsymbol{Z}_{1r} \boldsymbol{T}_{11s} \boldsymbol{Z}_{1r}' = \sigma_1^2 \underset{1\le d\le D}{\text{diag}} (\boldsymbol{1}_{N_d - n_d}) \underset{1\le d\le D}{\text{diag}} (1 - \sigma_1^2 \boldsymbol{1}_{n_d}' \boldsymbol{V}_{ds}^{-1} \boldsymbol{1}_{n_d}) \underset{1\le d\le D}{\text{diag}} (\boldsymbol{1}_{N_d - n_d}')$$

$$= \sigma_1^2 \underset{1\le d\le D}{\text{diag}} (\boldsymbol{1}_{N_d - n_d} [1 - \sigma_1^2 \boldsymbol{1}_{n_d}' \boldsymbol{V}_{ds}^{-1} \boldsymbol{1}_{n_d}] \boldsymbol{1}_{N_d - n_d}'),$$

$$
\begin{aligned}
\boldsymbol{M}_{12}^{rr} = \boldsymbol{Z}_{1r}\boldsymbol{T}_{12s}\boldsymbol{Z}_{2r}' &= -\sigma_1^2\sigma_2^2 \underset{1\le d\le D}{\text{diag}}\,(\mathbf{1}_{N_d-n_d})\underset{1\le d\le D}{\text{diag}}\,\{\mathbf{1}_{n_d}'\boldsymbol{V}_{ds}^{-1}\underset{1\le k\le m_d}{\text{diag}}\,(\mathbf{1}_{n_{dk}})\}\\
&\cdot \underset{1\le d\le D}{\text{diag}}\,\{\underset{1\le k\le m_d}{\text{diag}}\,(\mathbf{1}_{N_{dk}-n_{dk}}')\}\\
&= -\sigma_1^2\sigma_2^2 \underset{1\le d\le D}{\text{diag}}\,\{\mathbf{1}_{N_d-n_d}\mathbf{1}_{n_d}'\boldsymbol{V}_{ds}^{-1}\underset{1\le k\le m_d}{\text{diag}}\,(\mathbf{1}_{n_{dk}})\underset{1\le k\le m_d}{\text{diag}}\,(\mathbf{1}_{N_{dk}-n_{dk}}')\},\\
\boldsymbol{M}_{22}^{rr} = \boldsymbol{Z}_{2r}\boldsymbol{T}_{22s}\boldsymbol{Z}_{2r}' &= \sigma_2^2 \underset{1\le d\le D}{\text{diag}}\,\{\underset{1\le k\le m_d}{\text{diag}}\,(\mathbf{1}_{N_{dk}-n_{dk}})\}\\
&\cdot \underset{1\le d\le D}{\text{diag}}\,\{\boldsymbol{I}_{m_d}-\sigma_2^2\underset{1\le k\le m_d}{\text{diag}}\,(\mathbf{1}_{n_{dk}}')\boldsymbol{V}_{ds}^{-1}\underset{1\le k\le m_d}{\text{diag}}\,(\mathbf{1}_{n_{dk}})\}\underset{1\le d\le D}{\text{diag}}\,\{\underset{1\le k\le m_d}{\text{diag}}\,(\mathbf{1}_{N_{dk}-n_{dk}}')\}\\
&= \sigma_2^2 \underset{1\le d\le D}{\text{diag}}\,\{\underset{1\le k\le m_d}{\text{diag}}\,(\mathbf{1}_{N_{dk}-n_{dk}})\underset{1\le k\le m_d}{\text{diag}}\,(\mathbf{1}_{N_{dk}-n_{dk}}')\}\\
&- \sigma_2^4 \underset{1\le d\le D}{\text{diag}}\,\{\underset{1\le k\le m_d}{\text{diag}}\,(\mathbf{1}_{N_{dk}-n_{dk}})\underset{1\le k\le m_d}{\text{diag}}\,(\mathbf{1}_{n_{dk}}')\boldsymbol{V}_{ds}^{-1}\underset{1\le k\le m_d}{\text{diag}}\,(\mathbf{1}_{n_{dk}})\\
&\cdot \underset{1\le k\le m_d}{\text{diag}}\,(\mathbf{1}_{N_{dk}-n_{dk}}')\}
\end{aligned}
$$

and $\boldsymbol{M}_{21}^{rr} = (\boldsymbol{M}_{12}^{rr})'$. As

$$
\boldsymbol{a}_r' = \frac{1}{N_{dt}}\underset{1\le\ell\le D}{\text{col}'}\left[\delta_{d\ell}\underset{1\le k\le m_\ell}{\text{col}'}\,[\delta_{tk}\mathbf{1}_{N_{\ell k}-n_{\ell k}}']\right],\quad f_{dt} = \frac{n_{dt}}{N_{dt}}\quad\text{and}\quad \boldsymbol{V}_{ds} = \sigma^2\boldsymbol{\Sigma}_{ds},
$$

we get

$$
\begin{aligned}
\boldsymbol{a}_r'\boldsymbol{M}_{11}^{rr}\boldsymbol{a}_r &= \sigma_1^2\boldsymbol{a}_r'\underset{1\le d\le D}{\text{diag}}\,\big(\mathbf{1}_{N_d-n_d}[1-\sigma_1^2\mathbf{1}_{n_d}'\boldsymbol{V}_{ds}^{-1}\mathbf{1}_{n_d}]\mathbf{1}_{N_d-n_d}'\big)\boldsymbol{a}_r\\
&= \sigma_1^2(1-f_{dt})^2[1-\sigma_1^2\mathbf{1}_{n_d}'\boldsymbol{V}_{ds}^{-1}\mathbf{1}_{n_d}] = \sigma^2\varphi_1(1-f_{dt})^2[1-\varphi_1\mathbf{1}_{n_d}'\boldsymbol{\Sigma}_{ds}^{-1}\mathbf{1}_{n_d}],\\
\boldsymbol{a}_r'\boldsymbol{M}_{12}^{rr}\boldsymbol{a}_r &= -\sigma_1^2\sigma_2^2\boldsymbol{a}_r'\underset{1\le d\le D}{\text{diag}}\,\{\mathbf{1}_{N_d-n_d}\mathbf{1}_{n_d}'\boldsymbol{V}_{ds}^{-1}\underset{1\le k\le m_d}{\text{diag}}\,(\mathbf{1}_{n_{dk}})\underset{1\le k\le m_d}{\text{diag}}\,(\mathbf{1}_{N_{dk}-n_{dk}}')\}\boldsymbol{a}_r\\
&= -\sigma^2\varphi_1\varphi_2(1-f_{dt})\mathbf{1}_{n_d}'\boldsymbol{\Sigma}_{ds}^{-1}\underset{1\le k\le m_d}{\text{diag}}\,(\mathbf{1}_{n_{dk}})\underset{1\le k\le m_d}{\text{col}}\,[\delta_{tk}(1-f_{dk})],\\
\boldsymbol{a}_r'\boldsymbol{M}_{22}^{rr}\boldsymbol{a}_r &= \sigma_2^2\boldsymbol{a}_r'\underset{1\le d\le D}{\text{diag}}\,\{\underset{1\le k\le m_d}{\text{diag}}\,(\mathbf{1}_{N_{dk}-n_{dk}})\underset{1\le k\le m_d}{\text{diag}}\,(\mathbf{1}_{N_{dk}-n_{dk}}')\}\boldsymbol{a}_r\\
&- \sigma_2^4\boldsymbol{a}_r'\underset{1\le d\le D}{\text{diag}}\,\{\underset{1\le k\le m_d}{\text{diag}}\,(\mathbf{1}_{N_{dk}-n_{dk}})\underset{1\le k\le m_d}{\text{diag}}\,(\mathbf{1}_{n_{dk}}')\boldsymbol{V}_{ds}^{-1}\underset{1\le k\le m_d}{\text{diag}}\,(\mathbf{1}_{n_{dk}})\\
&\cdot \underset{1\le k\le m_d}{\text{diag}}\,(\mathbf{1}_{N_{dk}-n_{dk}}')\}\boldsymbol{a}_r,
\end{aligned}
$$

so that

$$\boldsymbol{a}_r'\boldsymbol{M}_{22}^{rr}\boldsymbol{a}_r = \sigma^2\varphi_2(1-f_{dt})^2 - \sigma^2\varphi_2^2 \underset{1\le k\le m_d}{\text{col}'} [(1-f_{dk})\delta_{tk}]$$
$$\cdot \underset{1\le k\le m_d}{\text{diag}} (\boldsymbol{1}_{n_{dk}}')\boldsymbol{\Sigma}_{ds}^{-1} \underset{1\le k\le m_d}{\text{diag}} (\boldsymbol{1}_{n_{dk}}) \underset{1\le k\le m_d}{\text{col}} [(1-f_{dk})\delta_{tk}].$$

Finally, we have

$$g_1(\boldsymbol{\theta}) = \boldsymbol{a}_r'\boldsymbol{Z}_r\boldsymbol{T}_s\boldsymbol{Z}_r'\boldsymbol{a}_r = \boldsymbol{a}_r'\boldsymbol{M}_{11}^{rr}\boldsymbol{a}_r + 2\boldsymbol{a}_r'\boldsymbol{M}_{12}^{rr}\boldsymbol{a}_r + \boldsymbol{a}_r'\boldsymbol{M}_{22}^{rr}\boldsymbol{a}_r.$$

11.7.2 Calculation of $g_2(\boldsymbol{\theta})$

The expression of $g_2(\theta)$ is

$$\begin{aligned} g_2(\boldsymbol{\theta}) &= [\boldsymbol{a}_r'\boldsymbol{X}_r - \boldsymbol{a}_r'\boldsymbol{Z}_r\boldsymbol{T}_s\boldsymbol{Z}_s'\boldsymbol{V}_{es}^{-1}\boldsymbol{X}_s]\boldsymbol{Q}_s[\boldsymbol{X}_r'\boldsymbol{a}_r - \boldsymbol{X}_s'\boldsymbol{V}_{es}^{-1}\boldsymbol{Z}_s\boldsymbol{T}_s\boldsymbol{Z}_r'\boldsymbol{a}_r] \\ &= [\boldsymbol{a}_{21}' - \boldsymbol{a}_{22}']\boldsymbol{Q}_s[\boldsymbol{a}_{21} - \boldsymbol{a}_{22}], \end{aligned}$$

where $\boldsymbol{Q}_s = (\boldsymbol{X}_s'\boldsymbol{V}^{-1}\boldsymbol{X}_s)^{-1} = \sigma^2\left(\sum_{d=1}^D \boldsymbol{X}_{ds}'\boldsymbol{\Sigma}_{ds}^{-1}\boldsymbol{X}_{ds}\right)^{-1}$ and $\boldsymbol{V}_{es}^{-1} = \sigma^{-2}\boldsymbol{W}_s$. On the one hand, we have

$$\boldsymbol{a}_{21}' = \boldsymbol{a}_r'\boldsymbol{X}_r = \frac{1}{N_{dt}}\boldsymbol{1}_{N_{dt}-n_{dt}}'\boldsymbol{X}_{dt,r} = \frac{1}{N_{dt}}\sum_{j\in r_{dt}}\boldsymbol{x}_{dtj} = (1-f_{dt})\overline{\boldsymbol{X}}_{dt}^*,$$

where

$$\overline{\boldsymbol{X}}_{dt}^* = \frac{1}{N_{dt}-n_{dt}}\sum_{j\in r_{dt}}\boldsymbol{x}_{dtj}.$$

On the other hand, we get

$$\begin{aligned} \boldsymbol{a}_{22}' &= \boldsymbol{a}_r'\boldsymbol{Z}_r\boldsymbol{T}_s\boldsymbol{Z}_s'\boldsymbol{V}_{es}^{-1}\boldsymbol{X}_s = \sigma^{-2}\boldsymbol{a}_r'(\boldsymbol{M}_{11}^{rs} + \boldsymbol{M}_{12}^{rs} + \boldsymbol{M}_{21}^{rs} + \boldsymbol{M}_{22}^{rs})\boldsymbol{W}_s\boldsymbol{X}_s \\ &= \boldsymbol{G}_{11} + \boldsymbol{G}_{12} + \boldsymbol{G}_{21} + \boldsymbol{G}_{22}, \end{aligned}$$

where

$$\begin{aligned} \boldsymbol{M}_{11}^{rs} &= \boldsymbol{Z}_{1r}\boldsymbol{T}_{11s}\boldsymbol{Z}_{1s}', \quad \boldsymbol{M}_{12}^{rs} = \boldsymbol{Z}_{1r}\boldsymbol{T}_{12s}\boldsymbol{Z}_{2s}' \\ \boldsymbol{M}_{21}^{rs} &= \boldsymbol{Z}_{2r}\boldsymbol{T}_{21s}\boldsymbol{Z}_{1s}' = (\boldsymbol{M}_{12}^{sr})', \quad \boldsymbol{M}_{22}^{rs} = \boldsymbol{Z}_{2r}\boldsymbol{T}_{22s}\boldsymbol{Z}_{2s}'. \end{aligned}$$

Let $\boldsymbol{w}'_{n_{dk}} = (w_{dk1}, \ldots, w_{dkn_{dk}})$. We obtain

$$\boldsymbol{G}_{11} = \sigma^{-2}\boldsymbol{a}'_r\boldsymbol{M}^{rs}_{11}\boldsymbol{W}_s\boldsymbol{X}_s = \frac{\sigma_1^2}{\sigma^2 N_{dt}} \underset{1\le k\le m_d}{\text{col}'}[\delta_{tk}\boldsymbol{1}'_{N_{dk}-n_{dk}}]\boldsymbol{1}_{N_d-n_d}[1-\sigma_1^2\boldsymbol{1}'_{n_d}\boldsymbol{V}^{-1}_{ds}\boldsymbol{1}_{n_d}]$$

$$\cdot\ \boldsymbol{1}'_{n_d}\boldsymbol{W}_{ds}\boldsymbol{X}_{ds} = \varphi_1(1-f_{dt})[1-\varphi_1\boldsymbol{1}'_{n_d}\boldsymbol{\Sigma}^{-1}_{ds}\boldsymbol{1}_{n_d}]\sum_{k=1}^{m_d}\boldsymbol{w}'_{n_{dk}}\boldsymbol{X}_{dk,s},$$

$$\boldsymbol{G}_{12} = \sigma^{-2}\boldsymbol{a}'_r\boldsymbol{M}^{rs}_{12}\boldsymbol{W}_s\boldsymbol{X}_s = -\frac{\sigma_1^2\sigma_2^2}{\sigma^2 N_{dt}}$$

$$\cdot\ \underset{1\le k\le m_d}{\text{col}'}[\delta_{tk}\boldsymbol{1}'_{N_{dk}-n_{dk}}]\boldsymbol{1}_{N_d-n_d}\boldsymbol{1}'_{n_d}\boldsymbol{V}^{-1}_{ds}\underset{1\le k\le m_d}{\text{diag}}(\boldsymbol{1}_{n_{dk}})\underset{1\le k\le m_d}{\text{diag}}(\boldsymbol{1}'_{n_{dk}})\boldsymbol{W}_{ds}\boldsymbol{X}_{ds}$$

$$= -\varphi_1\varphi_2(1-f_{dt})\boldsymbol{1}'_{n_d}\boldsymbol{\Sigma}^{-1}_{ds}\underset{1\le k\le m_d}{\text{diag}}(\boldsymbol{1}_{n_{dk}})\underset{1\le k\le m_d}{\text{col}}(\boldsymbol{w}'_{n_{dk}}\boldsymbol{X}_{dk,s}),$$

$$\boldsymbol{G}_{21} = \sigma^{-2}\boldsymbol{a}'_r\boldsymbol{M}^{rs}_{21}\boldsymbol{W}_s\boldsymbol{X}_s = -\frac{\sigma_1^2\sigma_2^2}{\sigma^2 N_{dt}}$$

$$\cdot\ \underset{1\le k\le m_d}{\text{col}'}[\delta_{tk}\boldsymbol{1}'_{N_{dk}-n_{dk}}]\underset{1\le k\le m_d}{\text{diag}}(\boldsymbol{1}_{N_{dk}-n_{dk}})\underset{1\le k\le m_d}{\text{diag}}(\boldsymbol{1}'_{n_{dk}})\boldsymbol{V}^{-1}_{ds}\boldsymbol{1}_{n_d}\boldsymbol{1}'_{n_d}\boldsymbol{W}_{ds}\boldsymbol{X}_{ds}$$

$$= -\varphi_1\varphi_2(1-f_{dt})\underset{1\le k\le m_d}{\text{col}'}[\delta_{tk}]\underset{1\le k\le m_d}{\text{diag}}(\boldsymbol{1}'_{n_{dk}})\boldsymbol{\Sigma}^{-1}_{ds}\boldsymbol{1}_{n_d}\sum_{k=1}^{m_d}\boldsymbol{w}'_{n_{dk}}\boldsymbol{X}_{dk,s},$$

$$\boldsymbol{G}_{22} = \sigma^{-2}\boldsymbol{a}'_r\boldsymbol{M}^{rs}_{22}\boldsymbol{W}_s\boldsymbol{X}_s$$

$$= \frac{\sigma_2^2}{\sigma^2 N_{dt}}\underset{1\le k\le m_d}{\text{col}'}[\delta_{tk}\boldsymbol{1}'_{N_{dk}-n_{dk}}]\{\underset{1\le k\le m_d}{\text{diag}}(\boldsymbol{1}_{N_{dk}-n_{dk}})\underset{1\le k\le m_d}{\text{diag}}(\boldsymbol{1}'_{n_{dk}})$$

$$-\ \sigma_2^2\underset{1\le k\le m_d}{\text{diag}}(\boldsymbol{1}_{N_{dk}-n_{dk}})\underset{1\le k\le m_d}{\text{diag}}(\boldsymbol{1}'_{n_{dk}})\boldsymbol{V}^{-1}_{ds}\underset{1\le k\le m_d}{\text{diag}}(\boldsymbol{1}_{n_{dk}})\underset{1\le k\le m_d}{\text{diag}}(\boldsymbol{1}'_{n_{dk}})\}\boldsymbol{W}_{ds}\boldsymbol{X}_{ds}$$

$$= \varphi_2(1-f_{dt})\underset{1\le k\le m_d}{\text{col}'}[\delta_{tk}]\Big[\boldsymbol{I}_{m_d}-\varphi_2\underset{1\le k\le m_d}{\text{diag}}(\boldsymbol{1}'_{n_{dk}})\boldsymbol{\Sigma}^{-1}_{ds}\underset{1\le k\le m_d}{\text{diag}}(\boldsymbol{1}_{n_{dk}})\Big]$$

$$\cdot\ \underset{1\le k\le m_d}{\text{col}}(\boldsymbol{w}'_{n_{dk}}\boldsymbol{X}_{dk,s}).$$

11.7.3 Calculation of $g_3(\boldsymbol{\theta})$

The formula of $g_3(\theta)$ is

$$g_3(\boldsymbol{\theta}) \approx \text{tr}\left\{(\nabla\boldsymbol{b}')\boldsymbol{V}_s(\nabla\boldsymbol{b}')'E\left[(\hat{\boldsymbol{\theta}}-\boldsymbol{\theta})(\hat{\boldsymbol{\theta}}-\boldsymbol{\theta})'\right]\right\},$$

where

$$\begin{aligned}
\boldsymbol{b}' &= \boldsymbol{a}_r'\boldsymbol{Z}_r\boldsymbol{V}_u\boldsymbol{Z}_s'\boldsymbol{V}_s^{-1} = \boldsymbol{a}_r'[\boldsymbol{Z}_{1r}, \boldsymbol{Z}_{2r}]\,\text{diag}\left(\sigma_1^2\boldsymbol{I}_D, \sigma_2^2\boldsymbol{I}_M\right)[\boldsymbol{Z}_{1s}', \boldsymbol{Z}_{2s}']'\boldsymbol{V}_s^{-1} \\
&= \boldsymbol{a}_r'\left[\sigma_1^2\boldsymbol{Z}_{1r}\boldsymbol{Z}_{1s}' + \sigma_2^2\boldsymbol{Z}_{2r}\boldsymbol{Z}_{2s}'\right]\boldsymbol{V}_s^{-1} = \sigma_1^2\boldsymbol{a}_r'\boldsymbol{Z}_{1r}\boldsymbol{Z}_{1s}'\boldsymbol{V}_s^{-1} + \sigma_2^2\boldsymbol{a}_r'\boldsymbol{Z}_{2r}\boldsymbol{Z}_{2s}'\boldsymbol{V}_s^{-1} \\
&= \boldsymbol{b}_1' + \boldsymbol{b}_2' = \underset{1\le\ell\le D}{\text{col}'}[\delta_{d\ell}\boldsymbol{b}_{1\ell}'] + \underset{1\le\ell\le D}{\text{col}'}[\delta_{d\ell}\boldsymbol{b}_{2\ell}'],
\end{aligned}$$

$$\boldsymbol{b}_{1d}' = \frac{\sigma^2\varphi_1}{N_{dt}}\underset{1\le k\le m_d}{\text{col}'}[\delta_{tk}\boldsymbol{1}_{N_{dk}-n_{dk}}']\boldsymbol{1}_{N_d-n_d}\boldsymbol{1}_{n_d}'\boldsymbol{V}_{ds}^{-1} = \varphi_1(1-f_{dt})\boldsymbol{1}_{n_d}'\boldsymbol{\Sigma}_{ds}^{-1},$$

$$\begin{aligned}
\boldsymbol{b}_{2d}' &= \frac{\sigma^2\varphi_2}{N_{dt}}\underset{1\le k\le m_d}{\text{col}'}[\delta_{tk}\boldsymbol{1}_{N_{dk}-n_{dk}}']\underset{1\le k\le m_d}{\text{diag}}(\boldsymbol{1}_{N_{dk}-n_{dk}})\underset{1\le k\le m_d}{\text{diag}}(\boldsymbol{1}_{n_{dk}}')\boldsymbol{V}_{ds}^{-1}, \\
&= \varphi_2(1-f_{dt})\underset{1\le k\le m_d}{\text{col}'}[\delta_{tk}]\underset{1\le k\le m_d}{\text{diag}}(\boldsymbol{1}_{n_{dk}}')\boldsymbol{\Sigma}_{ds}^{-1}.
\end{aligned}$$

For $\boldsymbol{A}_{ds} = \boldsymbol{I}_{m_d} + \varphi_2\underset{1\le k\le m_d}{\text{diag}}(\boldsymbol{1}_{n_{dk}}')\boldsymbol{W}_{ds}\underset{1\le k\le m_d}{\text{diag}}(\boldsymbol{1}_{n_{dk}})$, we have (cf. (11.4))

$$\begin{aligned}
\boldsymbol{\Sigma}_{ds}^{-1} &= \boldsymbol{L}_{ds}^{-1} - \frac{\varphi_1}{1+\varphi_1\boldsymbol{1}_{n_d}'\boldsymbol{L}_{ds}^{-1}\boldsymbol{1}_{n_d}}\boldsymbol{L}_{ds}^{-1}\boldsymbol{1}_{n_d}\boldsymbol{1}_{n_d}'\boldsymbol{L}_{ds}^{-1}, \\
\boldsymbol{L}_{ds}^{-1} &= \boldsymbol{W}_{ds} - \varphi_2\boldsymbol{W}_{ds}\underset{1\le k\le m_d}{\text{diag}}(\boldsymbol{1}_{n_{dk}})\boldsymbol{A}_{ds}^{-1}\underset{1\le k\le m_d}{\text{diag}}(\boldsymbol{1}_{n_{dk}}')\boldsymbol{W}_{ds}.
\end{aligned}$$

By applying the formula $\frac{\partial \boldsymbol{A}^{-1}}{\partial\gamma} = -\boldsymbol{A}^{-1}\frac{\partial\boldsymbol{A}}{\partial\gamma}\boldsymbol{A}^{-1}$, we obtain the partial derivatives of $\boldsymbol{L}_{ds}^{-1}$.

$$\begin{aligned}
&\frac{\partial\boldsymbol{L}_{ds}^{-1}}{\partial\sigma^2} = \boldsymbol{0}_{n_d\times n_d}, \quad \frac{\partial\boldsymbol{L}_{ds}^{-1}}{\partial\varphi_1} = \boldsymbol{0}_{n_d\times n_d}, \\
&\frac{\partial\boldsymbol{L}_{ds}^{-1}}{\partial\varphi_2} = -\boldsymbol{W}_{ds}\underset{1\le k\le m_d}{\text{diag}}(\boldsymbol{1}_{n_{dk}})\boldsymbol{A}_{ds}^{-1}\underset{1\le k\le m_d}{\text{diag}}(\boldsymbol{1}_{n_{dk}}')\boldsymbol{W}_{ds} + \varphi_2\boldsymbol{W}_{ds}\underset{1\le k\le m_d}{\text{diag}}(\boldsymbol{1}_{n_{dk}})\boldsymbol{A}_{ds}^{-1} \\
&\quad\cdot \underset{1\le k\le m_d}{\text{diag}}(\boldsymbol{1}_{n_{dk}}')\boldsymbol{W}_{ds}\underset{1\le k\le m_d}{\text{diag}}(\boldsymbol{1}_{n_{dk}})\boldsymbol{A}_{ds}^{-1}\underset{1\le k\le m_d}{\text{diag}}(\boldsymbol{1}_{n_{dk}}')\boldsymbol{W}_{ds}.
\end{aligned}$$

The partial derivatives of $\boldsymbol{\Sigma}_{ds}^{-1}$ are

$$\begin{aligned}
&\frac{\partial\boldsymbol{\Sigma}_{ds}^{-1}}{\partial\sigma^2} = \boldsymbol{0}_{n_d\times n_d}, \quad \frac{\partial\boldsymbol{\Sigma}_{ds}^{-1}}{\partial\varphi_1} = \frac{1}{[1+\varphi_1\boldsymbol{1}_{n_d}'\boldsymbol{L}_{ds}^{-1}\boldsymbol{1}_{n_d}]^2}\boldsymbol{L}_{ds}^{-1}\boldsymbol{1}_{n_d}\boldsymbol{1}_{n_d}'\boldsymbol{L}_{ds}^{-1}, \\
&\frac{\partial\boldsymbol{\Sigma}_{ds}^{-1}}{\partial\varphi_2} = \frac{\partial\boldsymbol{L}_{ds}^{-1}}{\partial\varphi_2} + \frac{\varphi_1^2\boldsymbol{1}_{n_d}'\frac{\partial\boldsymbol{L}_{ds}^{-1}}{\partial\varphi_2}\boldsymbol{1}_{n_d}}{[1+\varphi_1\boldsymbol{1}_{n_d}'\boldsymbol{L}_{ds}^{-1}\boldsymbol{1}_{n_d}]^2}\boldsymbol{L}_{ds}^{-1}\boldsymbol{1}_{n_d}\boldsymbol{1}_{n_d}'\boldsymbol{L}_{ds}^{-1}
\end{aligned}$$

$$-\frac{\varphi_1}{1+\varphi_1 \mathbf{1}'_{n_d} \boldsymbol{L}_{ds}^{-1} \mathbf{1}_{n_d}} \left[\frac{\partial \boldsymbol{L}_{ds}^{-1}}{\partial \varphi_2} \mathbf{1}_{n_d} \mathbf{1}'_{n_d} \boldsymbol{L}_{ds}^{-1} + \boldsymbol{L}_{ds}^{-1} \mathbf{1}_{n_d} \mathbf{1}'_{n_d} \frac{\partial \boldsymbol{L}_{ds}^{-1}}{\partial \varphi_2} \right].$$

Let $\boldsymbol{\theta} = (\theta_1, \theta_2, \theta_3) = (\sigma^2, \varphi_1, \varphi_2)$. The partial derivatives of $\boldsymbol{b}'_{1d}$ and $\boldsymbol{b}'_{2d}$ are

$$\frac{\partial \boldsymbol{b}'_{1d}}{\partial \sigma^2} = \mathbf{0}_{1\times n_d} \quad \frac{\partial \boldsymbol{b}'_{1d}}{\partial \varphi_1} = (1-f_{dt}) \mathbf{1}'_{n_d} \left[\boldsymbol{\Sigma}_{ds}^{-1} + \varphi_1 \frac{\partial \boldsymbol{\Sigma}_{ds}^{-1}}{\partial \varphi_1} \right],$$

$$\frac{\partial \boldsymbol{b}'_{1d}}{\partial \varphi_2} = \varphi_1 (1-f_{dt}) \mathbf{1}'_{n_d} \frac{\partial \boldsymbol{\Sigma}_{ds}^{-1}}{\partial \varphi_2}, \quad \frac{\partial \boldsymbol{b}'_{2d}}{\partial \sigma^2} = \mathbf{0}_{1\times n_d},$$

$$\frac{\partial \boldsymbol{b}'_{2d}}{\partial \varphi_1} = \varphi_2 (1-f_{dt}) \underset{1\le k\le m_d}{\text{col}'} [\delta_{tk}] \underset{1\le k\le m_d}{\text{diag}} (\mathbf{1}'_{n_{dk}}) \frac{\partial \boldsymbol{\Sigma}_{ds}^{-1}}{\partial \varphi_1},$$

$$\frac{\partial \boldsymbol{b}'_{2d}}{\partial \varphi_2} = (1-f_{dt}) \underset{1\le k\le m_d}{\text{col}'} [\delta_{tk}] \underset{1\le k\le m_d}{\text{diag}} (\mathbf{1}'_{n_{dk}}) \left[\boldsymbol{\Sigma}_{ds}^{-1} + \varphi_2 \frac{\partial \boldsymbol{\Sigma}_{ds}^{-1}}{\partial \varphi_2} \right].$$

Let $\boldsymbol{Q} = (q_{ab})_{a,b=1,\ldots,3}$, where

$$q_{ab} = \left(\frac{\partial \boldsymbol{b}'_{1d}}{\partial \theta_a} + \frac{\partial \boldsymbol{b}'_{2d}}{\partial \theta_a} \right) \sigma^2 \boldsymbol{\Sigma}_{ds} \left(\frac{\partial \boldsymbol{b}'_{1d}}{\partial \theta_b} + \frac{\partial \boldsymbol{b}'_{2d}}{\partial \theta_b} \right)', \quad a, b = 1, 2, 3.$$

Let $F_{ab} = F_{\theta_a,\theta_b}$ be the components of the REML Fisher information matrix. It holds that

$$g_3(\boldsymbol{\theta}) \approx \text{tr} \left\{ \boldsymbol{Q} E \left[(\hat{\boldsymbol{\theta}} - \boldsymbol{\theta})(\hat{\boldsymbol{\theta}} - \boldsymbol{\theta})' \right] \right\}$$

$$\approx \text{tr} \left\{ \begin{pmatrix} q_{11} & q_{12} & q_{13} \\ q_{21} & q_{22} & q_{23} \\ q_{31} & q_{32} & q_{33} \end{pmatrix} \begin{pmatrix} F_{\sigma^2\sigma^2} & F_{\sigma^2\varphi_1} & F_{\sigma^2\varphi_2} \\ F_{\varphi_1\sigma^2} & F_{\varphi_1\varphi_1} & F_{\varphi_1\varphi_2} \\ F_{\varphi_2\sigma^2} & F_{\varphi_2\varphi_1} & F_{\varphi_2\varphi_2} \end{pmatrix}^{-1} \right\}.$$

11.7.4 Calculation of $g_4(\boldsymbol{\theta})$

We have that $g_4(\boldsymbol{\theta}) = \boldsymbol{a}'_r \boldsymbol{V}_{er} \boldsymbol{a}_r$, where

$$\boldsymbol{a}'_r = \frac{1}{N_{dt}} \underset{1\le \ell\le D}{\text{col}'} \left[\delta_{d\ell} \underset{1\le k\le m_\ell}{\text{col}'} [\delta_{tk} \mathbf{1}'_{N_{\ell k}-n_{\ell k}}] \right], \quad \boldsymbol{V}_{er}^{-1} = \sigma^{-2} \boldsymbol{W}_r = \sigma^{-2} \underset{1\le d\le D}{\text{diag}} (\boldsymbol{W}_{dr}).$$

Therefore, we get

$$
\begin{aligned}
g_4(\boldsymbol{\theta}) &= \frac{1}{N_{dt}} \underset{1\le \ell \le D}{\text{col}'} \left[\delta_{d\ell} \underset{1\le k \le m_\ell}{\text{col}'} \left[\delta_{tk} \mathbf{1}'_{N_{\ell k}-n_{\ell k}} \right] \right] \sigma^2 \underset{1\le d \le D}{\text{diag}} \, (\mathbf{W}_{dr}^{-1}) \\
&\cdot \frac{1}{N_{dt}} \underset{1\le \ell \le D}{\text{col}} \left[\delta_{d\ell} \underset{1\le k \le m_\ell}{\text{col}} \left[\delta_{tk} \mathbf{1}_{N_{\ell k}-n_{\ell k}} \right] \right] \\
&= \frac{\sigma^2}{N_{dt}^2} \mathbf{1}'_{N_{dt}-n_{dt}} \underset{j \in r_{dt}}{\text{diag}}(w_{dtj}^{-1}) \mathbf{1}_{N_{dt}-n_{dt}} = \frac{\sigma^2}{N_{dt}^2} \sum_{j \in r_{dt}} \frac{1}{w_{dtj}}.
\end{aligned}
$$

11.8 Simulation Experiments

This section presents two simulation experiments. Simulation 1 investigates the REML Fisher-scoring fitting method. Simulation 2 studies the behavior of the EBLUPs of subdomain means when the NER2 model is fitted by the REML method. The explanatory and target variables are simulated as follows:

Explanatory variable: For $d = 1, \dots, D$, $t = 1, \dots, m_d$, $j = 1, \dots, n_{dt}$, generate $x_{dtj} = (b_{dt} - a_{dt})U_{dtj} + a_{dt}$, with $U_{dtj} = \frac{j}{n_{dt}+1}$, $j = 1, \dots, n_{dt}$. Take $a_{dt} = 1$, $b_{dt} = 1 + \frac{1}{m_d}(m_d(d-1)+t)$, $d = 1, \dots, D$, $t = 1, \dots, m_d$.

Random effects and errors: For $d = 1, \dots, D$, $t = 1, \dots, m_d$, $j = 1, \dots, n_{dt}$, generate $u_{1,d} \sim N(0, \sigma_1^2)$, $u_{2,dt} \sim N(0, \sigma_2^2)$, $e_{dtj} \sim N(0, \sigma_0^2)$, with $\sigma_1^2 = \sigma_2^2 = \sigma_0^2 = 1$.

Target variable: For $d = 1, \dots, D$, $t = 1, \dots, m_d$, $j = 1, \dots, n_{dt}$, generate

$$
y_{dtj} = \beta x_{dtj} + u_{1,d} + u_{2,dt} + e_{dtj}, \quad \text{with } \beta = 1.
$$

11.8.1 Simulation 1

For empirically investigating the REML fitting method, we implement a Fisher-scoring algorithm with two stopping rules: (1) the number of iterations is greater than a fixed number itermax (we take itermax = 500), or (2) the absolute difference between the values of all the estimators, in two consecutive iterations, is lower than a tolerance ϵ (we take $\epsilon = 0.00001$).

The steps of the simulation experiment are:

1. Simulate the explanatory variables.
2. Repeat $I = 1000$ times ($i = 1, \dots, I$)

Table 11.1 EMSE and BIAS ($\times 10^3$) of $\hat{\beta}$, $\hat{\sigma}_0^2$, $\hat{\sigma}_1^2$, and $\hat{\sigma}_2^2$

	n	450	600	750	900	1050	1200	1350	1500
	n_d	15	20	25	30	35	40	45	50
	n_{dt}	3	4	5	6	7	8	9	10
$EMSE$	β	0.126	0.087	0.068	0.054	0.046	0.040	0.034	0.030
	σ_0^2	6.583	4.386	3.403	2.605	2.187	1.879	1.661	1.479
	σ_1^2	110.26	106.36	106.56	105.14	102.23	101.38	98.96	99.23
	σ_2^2	30.90	26.26	23.87	22.86	21.75	21.49	20.40	20.26
$BIAS$	β	0.069	−0.069	−0.036	−0.003	0.082	−0.083	0.065	0.009
	σ_0^2	−0.311	1.011	0.328	−0.179	−0.184	−0.239	−0.087	−0.385
	σ_1^2	−8.586	−0.432	0.562	5.623	−1.291	2.186	−4.623	1.709
	σ_2^2	−0.542	−2.486	1.926	0.680	−0.288	0.336	−1.626	0.947

2.1. Generate the values of the target variable with total sample size $n = \sum_{d=1}^{D}\sum_{t=1}^{m_d} n_{dt}$.
2.2. Calculate $\hat{\beta}^{(i)}$, $\hat{\sigma}_0^{2(i)}$, $\hat{\sigma}_1^{2(i)}$, and $\hat{\sigma}_2^{2(i)}$ by using the REML method.

3. The empirical mean squared errors and biases of $\hat{\eta}^{(i)} \in \{\hat{\beta}^{(i)}, \hat{\sigma}_0^{2(i)}, \hat{\sigma}_1^{2(i)}, \hat{\sigma}_2^{2(i)}\}$ are

$$EMSE(\hat{\eta}) = \frac{1}{I}\sum_{i=1}^{I}(\hat{\eta}^{(i)} - \eta)^2, \quad BIAS(\hat{\eta}) = \frac{1}{I}\sum_{i=1}^{I}(\hat{\eta}^{(i)} - \eta).$$

The simulation experiment is carried out with a constant number of levels and sublevels of the random effects, but with different values of the sample sizes. We take $D = 30$, $m_d = 5$, $d = 1, \ldots, D$. We run the simulation experiment for the eight different groups of sample sizes appearing in Table 11.1, where are presented also the results of Simulation 1. As expected, $EMSE$ decreases as sample size increases.

11.8.2 Simulation 2

This section presents a simulation experiment designed to study the behavior of the EBLUPs of subdomain means when the NER2 model is fitted by the REML method. For simulating the samples and calculating the performance measures, the simulation steps are:

1. Generate the deterministic elements of the population. This is to say, calculate x_{dtj}, $d = 1, \ldots, D$, $t = 1, \ldots, m_d$, $j = 1, \ldots, N_{dt}$, in the same way as in Simulation 1 with n_{dt} replaced by N_{dt}.

2. Repeat $I = 1000$ times $(i = 1, \dots, I)$
   (a) Generation of the random elements of the population
      - Random effects and errors: For $d = 1, \dots, D$, $t = 1, \dots, m_d$, $j = 1, \dots, N_{dt}$, generate $u_{1,d}^{(i)} \sim N(0, \sigma_1^2)$, $u_{2,dt}^{(i)} \sim N(0, \sigma_2^2)$, $e_{dtj}^{(i)} \sim N(0, \sigma_0^2)$, with $\sigma_1^2 = 1$, $\sigma_2^2 = 1$, $\sigma_0^2 = 1$.
      - Target variable: For $d = 1, \dots, D$, $t = 1, \dots, m_d$, $j = 1, \dots, N_{dt}$, generate

$$y_{dtj}^{(i)} = \beta x_{dtj} + u_{1,d}^{(i)} + u_{2,dt}^{(i)} + e_{dtj}^{(i)}, \quad \text{with } \beta = 1.$$

   (b) Extraction of samples. For $d = 1, \dots, D$, $t = 1, \dots, m_d$, select the n_{dt} units of the level (d, t) in the positions $\left[\frac{N_{dt}}{1+n_{dt}}\right] j$, $j = 1, \dots, n_{dt}$.
   (c) Calculate $\hat{\beta}^{(i)}$, $\hat{\sigma}_0^{2(i)}$, $\hat{\sigma}_1^{2(i)}$, and $\hat{\sigma}_2^{2(i)}$ by using the REML fitting method.
   (d) For $d = 1, \dots, D$, $t = 1, \dots, m_d$ calculate $\hat{\overline{Y}}_{dt}^{eblup,(i)}$ and $\overline{Y}_{dt}^{(i)} = \frac{1}{N_{dt}} \sum_{j=1}^{N_{dt}} y_{dtj}^{(i)}$.
3. Output: For $d = 1, \dots, D$, $t = 1, \dots, m_d$, calculate

$$EMSE_{dt} = \frac{1}{I} \sum_{i=1}^{I} \left(\hat{\overline{Y}}_{dt}^{eblup,(i)} - \overline{Y}_{dt}^{(i)} \right)^2, \quad BIAS_{dt} = \frac{1}{I} \sum_{i=1}^{I} \left(\hat{\overline{Y}}_{dt}^{eblup,(i)} - \overline{Y}_{dt}^{(i)} \right),$$

$$EMSE = \frac{1}{m} \sum_{d=1}^{D} \sum_{t=1}^{m_d} EMSE_{dt}, \quad BIAS = \frac{1}{m} \sum_{d=1}^{D} \sum_{t=1}^{m_d} BIAS_{dt}.$$

Table 11.2 presents the results of Simulation 2. As in Simulation 1, the MSE decreases as sample size increases.

Table 11.2 EMSE and BIAS of $\hat{\overline{Y}}_{dt}^{eblup}$

N	4500	6000	7500	9000	10,500	12,000	13,500	15,000
N_d	150	200	250	300	350	400	450	500
N_{dt}	30	40	50	60	70	80	90	100
n_{dt}	3	4	5	6	7	8	9	10
$BIAS$	0.00043	−0.00004	0.00122	−0.00010	0.00008	−0.00034	0.00001	0.00007
MSE	0.25191	0.19521	0.15969	0.13521	0.11699	0.10357	0.09280	0.08401

11.9 R Codes for EBLUPs

This section gives R codes for calculating the EBLUPs of domain means under the NER2 model by using data from the survey data file `LFS20.txt`. The domains of interest are the areas crossed by sex–age groups. The target variable is INCOME and the auxiliary variables are EDUCATION2 and EDUCATION3.

The following code reads the file with the aggregated auxiliary variables and sorts the file by area, sex, and age groups.

```
aux <- read.table("Ndsa20.txt", header=TRUE, sep = "\t", dec = ".")
# sort auxiliary data by area (ascending), sex (ascending), age (ascending)
aux <- aux[order(aux$sex, aux$age, aux$area), ]
```

The following code reads the survey data file and calculates new variables. The age groups are: 1 if AGE < 25, 2 if $25 \leq$ AGE < 54, and 3 if AGE ≥ 54.

```
dat <- read.table("LFS20.txt", header=TRUE, sep = "\t", dec = ".")
ns <- nrow(dat)     # global sample size
one <- rep(1,ns)    # auxiliary 1-variable
# Age groups
ageG <- as.numeric(cut(dat$AGE, breaks=c(0,25,54,max(dat$AGE)),
                   labels=c(1,2,3), right=TRUE))
# EDUCATION categories
edu2 <- as.numeric(dat$EDUCATION==2)
edu3 <- as.numeric(dat$EDUCATION==3)
```

We calculate direct estimates of means and sizes by area and sex–age groups, by using `dir2` function described in Sect. 2.8.4.

```
dir.income <- dir2(data=dat$INCOME, w=dat$WEIGHT, domain=list(area=dat$AREA,
                   age=ageG, sex=dat$SEX))
dir <- dir.income$mean
n <- dir.income$nd
py <- dir2(data=dat$INCOME, w=one, domain=list(area=dat$AREA, age=ageG,
           sex=dat$SEX))$mean
pedu2 <- dir2(data=edu2, w=one, domain=list(area=dat$AREA, age=ageG,
              sex=dat$SEX))$mean
pedu3 <- dir2(data=edu3, w=one, domain=list(area=dat$AREA, age=ageG,
              sex=dat$SEX))$mean
```

We install and/or load some R packages.

```
if(!require(Matrix)){
  install.packages("Matrix")
  library(Matrix)
}
if(!require(lme4)){
  install.packages("lme4")
  library(lme4)
}
```

We fit a NER2 model with the R library `lme4`.

```
# Define new variables
edu <- as.factor(dat$EDUCATION)
sexage <- paste(dat$SEX, ageG, sep="")
# Apply lmer function
lmm <- lmer(formula=INCOME ~ edu + (1|AREA/sexage), data=dat, REML=TRUE)
# Summary of the fitting procedure
summary(lmm)
# Analysis of Variance Table
anova <- anova(lmm)
# Regression parameters
```

Table 11.3 Estimated parameters of NER2 model

Parameter	Estimate	Std. error	z-value	p-value
Intercept	39,526.9	534.3	73.98	0.00
edu2	9203.5	716.5	12.85	0.00
edu3	19,539.5	926.8	21.08	0.00

```
beta <- fixef(lmm); beta
# Variance parameters
var <- as.data.frame(VarCorr(lmm))
# Standard deviation of u2dt
sigmau2 <- var$sdcor[1]
# Standard deviation of u1d
sigmau1 <- var$sdcor[2]
# Standard deviation of edtj
sigmae <- var$sdcor[3]
# Modes of the random effects
ranef(lmm)
# Modes of subdomain random effects
udt <- ranef(lmm)[[1]]
# Modes of domain random effects
ud <- ranef(lmm)[[2]]
# Predicted values
ypred <- fitted(lmm)
# p values
p.values <- 2*pnorm(abs(coef(summary(lmm))[,3]), low=F); p.values
```

Table 11.3 gives the estimates of the regression parameters (Estimate), the standard error (Std. Error), the test statistics (z-value), and the p-value (p-value) for testing if the regression parameter is equal to zero.

The estimated standard deviations of $u_{1,d}$, $u_{2,dt}$, and e_{dtj} are $\sigma_1 = 0.205532$, $\sigma_2 = 1512.037$, and $\sigma_0 = 10285.84$.

We calculate the EBLUPs of the average incomes by areas and sex–age groups.

```
xbeta <- beta[1] + beta[2]*(aux$edu2/aux$N) + beta[3]*(aux$edu3/aux$N)
zu <- ud[,1] + udt[,1]
e1 <- xbeta + zu
fdt <- n/aux$N
xsbeta <- beta[1] + beta[2]*pedu2 + beta[3]*pedu3
e2 <- fdt*(py-xsbeta)
eblup <- e1+e2
# Summary of results
output <- data.frame(aux[1:4], n, dir, eblup)
head(output, 10)
```

For the ten first areas and the first sex–age group, Table 11.4 gives the population and sample sizes (N and n) and the direct (dir) and EBLUP (eblup) estimates of average incomes. The EBLUP estimates are smoother than the direct ones.

Table 11.4 Estimates of average incomes

area	sex	age	N	n	dir	eblup
1	1	1	1397	6	43,494.80	45,830.45
2	1	1	96	1	35,814.00	40,583.71
3	1	1	301	2	48,138.33	42,610.45
4	1	1	1089	7	40,216.09	41,453.50
5	1	1	224	2	45,255.90	48,067.34
6	1	1	1149	6	55,943.13	51,760.30
7	1	1	829	5	52,833.80	47,679.64
8	1	1	775	5	45,500.28	47,081.01
9	1	1	1405	10	49,696.52	48,162.38
10	1	1	733	4	37,998.28	46,207.23

References

Battese, G.E, Harter, R.M., Fuller, W.A.: An error-component model for prediction of county crop areas using survey and satellite data. J. Am. Stat. Assoc. **83**, 28–36 (1988)

Datta, G.S., Gosh, M.: Bayesian prediction in linear models: applications to small area estimation. Ann. Stat. **19**, 1748–1770 (1991)

Henderson, C.R.: Estimation of variance and covariance components. Biometrics **9**, 226–252 (1953)

Morales, D., Santamaría, L.: Small area estimation under unit-level temporal linear mixed models J. Stat. Comput. Simul. **89**(9), 1592–1620 (2019)

Pfeffermann, D., Barnard, C.: Some new estimators for small area means with applications to the assessment of farmland values. J. Bus. Econ. Stat. **9**, 73–84 (1991)

Searle, S.R., Casella, G., McCullogh, C.E.: Variance Components. Wiley, New York (1992)

Stukel, D.M., Rao, J.N.K.: On small-area estimation under two-fold nested error regression models. J. Stat. Plan. Inference **78**, 131–147 (1999)

Chapter 12
EBPs Under Two-Fold Nested Error Regression Models

12.1 Introduction

For the estimation of complex domain parameters such as certain poverty indicators, Molina and Rao (2010) proposed the empirical best (EB) method, based on assuming that a one-to-one transformation of the target variable follows the unit-level nested error model of Battese et al. (1988) with random effects for the domains of interest. Under that model, EB method gives approximately the "best" estimator in the sense of being unbiased with minimum variance error.

When the target population is naturally divided in subpopulations at two nested aggregation levels (e.g. in provinces and counties within provinces), or when the sampling design has two stages, as it is usual in many household surveys, it is reasonable to assume a two-fold nested error regression model including random effects at the two levels of aggregation, domains and subdomains. Marhuenda et al. (2017) developed the EB method for predicting additive parameters under the two-fold nested error regression model.

This chapter describes the EB methodology given by Marhuenda et al. (2017) for predicting additive parameters and provides analytical expressions for the EB predictors (EBP) of poverty proportions, poverty gaps, and average incomes. It gives Monte Carlo algorithms for approximating the EB predictors of more complex domain or subdomain parameters. The case of using only categorical explanatory variables is also treated, because it does not require the use of an auxiliary census data file. The obtained EB estimates of subdomain parameters have the good property of being consistent with the corresponding domain estimate.

For estimating the error variances of the EBPs, a parametric bootstrap procedure is given. The EBP methodology is illustrated with an application to the survey data file LFS20.txt. The given R codes calculate EBPs of poverty proportions, poverty gaps, and average incomes by areas and age groups.

D. Morales et al., *A Course on Small Area Estimation and Mixed Models*, Statistics for Social and Behavioral Sciences, https://doi.org/10.1007/978-3-030-63757-6_12

12.2 Two-fold Nested Error Regression Models

This section considers vectors $\boldsymbol{y} = (y_1, \ldots, y_N)'$ containing the values of a target random variable associated with N units of a finite population. Let $\boldsymbol{y}_s$ be the sub-vector of $\boldsymbol{y}$ corresponding to sample elements and $\boldsymbol{y}_r$ the sub-vector of $\boldsymbol{y}$ corresponding to the out-of-sample elements; that is, $\boldsymbol{y} = (\boldsymbol{y}_s', \boldsymbol{y}_r')'$. The inference problem is to predict the value of a real-valued function $\delta = h(\boldsymbol{y})$ of the random vector $\boldsymbol{y}$ using the sample data $\boldsymbol{y}_s$. The best predictor of δ is given by $\hat{\delta}^{bp} = E_{\boldsymbol{y}_r}(\delta|\boldsymbol{y}_s)$, where the expectation is taken with respect to the conditional distribution of $\boldsymbol{y}_r$ given $\boldsymbol{y}_s$.

The population of interest is hierarchically divided in domains and subdomains. More concretely, let U be a population of size N partitioned into D domains or areas $U_1, \ldots, U_D$ of sizes $N_1, \ldots, N_D$, respectively. Additionally, each domain U_d is partitioned into M_d subdomains $U_{d1}, \ldots, U_{dM_d}$, of sizes $N_{d1}, \ldots, N_{dM_d}$, respectively, $d = 1, \ldots, D$. The components of vector $\boldsymbol{y}$ are referenced with three subindexes. This is to say, y_{dtj} denotes the value that the study variable takes on the sample unit j of subdomain t and domain d.

12.2.1 The Population Model

At the population level, the two-fold nested error regression model is

$$y_{dtj} = \boldsymbol{x}_{dtj}\boldsymbol{\beta} + u_{1,d} + u_{2,dt} + w_{dtj}^{-1/2} e_{dtj},\ d = 1, \ldots, D, t = 1, \ldots, M_d, j = 1, \ldots, N_{dt}, \tag{12.1}$$

where $\boldsymbol{x}_{dtj}$ is a row vector containing p auxiliary variables, $w_{dtj} > 0$ is a known heteroscedasticity weight, and the random effects and errors are all mutually independent and such that $u_{1,d} \sim N(0, \sigma_1^2)$, $u_{2,dt} \sim N(0, \sigma_2^2)$, and $e_{dtj} \sim N(0, \sigma_0^2)$.

The population model (12.1) can be written in the matrix form as (without taking into account reordering with respect to sampled and non-sampled elements)

$$\boldsymbol{y} = \boldsymbol{X}\boldsymbol{\beta} + \boldsymbol{Z}_1\boldsymbol{u}_1 + \boldsymbol{Z}_2\boldsymbol{u}_2 + \boldsymbol{W}^{-1/2}\boldsymbol{e}, \tag{12.2}$$

where $\boldsymbol{u}_1 = \boldsymbol{u}_{1,D\times 1} \sim N(\boldsymbol{0}, \sigma_1^2 \boldsymbol{I}_D)$, $\boldsymbol{u}_2 = \boldsymbol{u}_{2,M\times 1} \sim N(\boldsymbol{0}, \sigma_2^2 \boldsymbol{I}_M)$, and $\boldsymbol{e} = \boldsymbol{e}_{N\times 1} \sim N(\boldsymbol{0}, \sigma_0^2 \boldsymbol{I}_N)$ are independent, $\boldsymbol{y}=\boldsymbol{y}_{N\times 1} = \underset{1\le d\le D}{\text{col}}\,(\underset{1\le t\le M_d}{\text{col}}\,(\underset{1\le j\le N_{dt}}{\text{col}}\,(y_{dtj})))$, $\boldsymbol{\beta} = \boldsymbol{\beta}_{p\times 1}$, $\boldsymbol{X} = \boldsymbol{X}_{N\times p} = \underset{1\le d\le D}{\text{col}}\,(\underset{1\le t\le M_d}{\text{col}}\,(\underset{1\le j\le N_{dt}}{\text{col}}\,(\boldsymbol{x}_{dtj})))$ with rank$(\boldsymbol{X}) = p$, $M = \sum_{d=1}^D M_d$, $N = \sum_{d=1}^D N_d$, $N_d = \sum_{t=1}^{M_d} N_{dt}$, $\boldsymbol{Z}_1 = \underset{1\le d\le D}{\text{diag}}\,(\boldsymbol{1}_{N_d})_{N\times D}$, $\boldsymbol{Z}_2 = \underset{1\le d\le D}{\text{diag}}\,(\underset{1\le t\le M_d}{\text{diag}}\,(\boldsymbol{1}_{N_{dt}}))_{N\times M}$, $\boldsymbol{I}_a$ is the $a \times a$ identity matrix, $\boldsymbol{1}_a$ is the $a \times 1$

vector with all its elements equal to 1, $\boldsymbol{W} = \underset{1\leq d\leq D}{\text{diag}}(\boldsymbol{W}_d)$, $\boldsymbol{W}_d = \underset{1\leq t\leq M_d}{\text{diag}}(\boldsymbol{W}_{dt})$, $\boldsymbol{W}_{dt} = \underset{1\leq j\leq N_{dt}}{\text{diag}}(w_{dtj})_{N_{dt}\times N_{dt}}$ with known heteroscedasticity weights $w_{dtj} > 0$, $d = 1,\ldots,D$, $t = 1,\ldots,M_d$, $j = 1,\ldots,N_{dt}$.

Without loss of generality we can reorder the population so that the target vector takes the form $\boldsymbol{y} = (\boldsymbol{y}_s', \boldsymbol{y}_r')'$. We describe the two corresponding sub-models more in detail.

12.2.2 The Sample Model

In practice, inference is carried out based on a sample drawn from the population. We assume that a sample s_d of size n_d is drawn from domain U_d, $d = 1,\ldots,D$. Let s_{dt} be the subsample from subdomain U_{dt}, $t = 1,\ldots,M_d$. We allow the existence of subdomains with no observations in the sample. Without loss of generality, we assume that these are the last $M_d - m_d$ subdomains; that is, $s_{dt} = \emptyset$, for $m_d + 1 \leq t \leq M_d$ whereas $s_{dt} \neq \emptyset$, for $1 \leq t \leq m_d$. The sample sub-vector $\boldsymbol{y}_s$ follows the marginal model derived from the population model (12.1), i.e.

$$y_{dtj} = \boldsymbol{x}_{dtj}\boldsymbol{\beta} + u_{1,d} + u_{2,dt} + w_{dtj}^{-1/2} e_{dtj},\ d = 1,\ldots,D, t = 1,\ldots,m_d, j = 1,\ldots,n_{dt}, \tag{12.3}$$

where we change M, M_d and N, N_d, and N_{dt} by the sample counterparts m, m_d and n, n_d, and n_{dt}, respectively. In matrix notation, the model is

$$\boldsymbol{y}_s = \boldsymbol{X}_s\boldsymbol{\beta} + \boldsymbol{Z}_{1s}\boldsymbol{u}_1 + \boldsymbol{Z}_{2s}\boldsymbol{u}_{2s} + \boldsymbol{W}_s^{-1/2}\boldsymbol{e}_s, \tag{12.4}$$

where $\boldsymbol{u}_1 = \boldsymbol{u}_{1,D\times 1} \sim N(\boldsymbol{0}, \sigma_1^2\boldsymbol{I}_D)$, $\boldsymbol{u}_{2s} = \boldsymbol{u}_{2s,m\times 1} \sim N(\boldsymbol{0}, \sigma_2^2\boldsymbol{I}_m)$, and $\boldsymbol{e}_s = \boldsymbol{e}_{s,n\times 1} \sim N(\boldsymbol{0}, \sigma_0^2\boldsymbol{I}_n)$ are independent, $\boldsymbol{y}_s = \boldsymbol{y}_{s,n\times 1}$, $\boldsymbol{\beta} = \boldsymbol{\beta}_{p\times 1}$, $\boldsymbol{X}_s = \boldsymbol{X}_{s,n\times p} = \underset{1\leq d\leq D}{\text{col}}(\underset{1\leq t\leq m_d}{\text{col}}(\underset{1\leq j\leq n_{dt}}{\text{col}}(\boldsymbol{x}_{dtj})))$ with $\text{rank}(\boldsymbol{X}_s) = p$, $\boldsymbol{Z}_{1s} = \underset{1\leq d\leq D}{\text{diag}}(\boldsymbol{1}_{n_d})_{n\times D}$, $\boldsymbol{Z}_{2s} = \underset{1\leq d\leq D}{\text{diag}}(\underset{1\leq t\leq m_d}{\text{diag}}(\boldsymbol{1}_{n_{dt}}))_{n\times m}$, $m = \sum_{d=1}^D m_d$, $n = \sum_{d=1}^D n_d$, $n_d = \sum_{t=1}^{m_d} n_{dt}$, $\boldsymbol{W}_s = \underset{1\leq d\leq D}{\text{diag}}(\boldsymbol{W}_{ds})$, $\boldsymbol{W}_{ds} = \underset{1\leq t\leq m_d}{\text{diag}}(\boldsymbol{W}_{dts})$, $\boldsymbol{W}_{dts} = \underset{1\leq j\leq n_{dt}}{\text{diag}}(w_{dtj})_{n_{dt}\times n_{dt}}$ with known heteroscedasticity weights $w_{dtj} > 0$, $d = 1,\ldots,D$, $t = 1,\ldots,m_d$, $j = 1,\ldots,n_{dt}$.

12.2.3 The Non-sample Model

Let r be the subset of units not appearing in the sample s. The corresponding sub-vector $\boldsymbol{y}_r$ follows the model (12.1), with the immediate modifications. For $d = 1,\ldots,D$, $t = 1,\ldots,M_d$, $j = n_{dt}+1,\ldots,N_{dt}$, the non-sample model is

$$y_{dtj} = \boldsymbol{x}_{dtj}\boldsymbol{\beta} + u_{1,d} + u_{2,dt} + w_{dtj}^{-1/2} e_{dtj},$$

where we change N, N_d and N_{dt} by $N-n$, N_d-n_d, and $N_{dt}-n_{dt}$, respectively. In matrix notation, the model is

$$\boldsymbol{y}_r = \boldsymbol{X}_r\boldsymbol{\beta} + \boldsymbol{Z}_{1r}\boldsymbol{u}_1 + \boldsymbol{Z}_{2r}\boldsymbol{u}_{2r} + \boldsymbol{W}_r^{-1/2}\boldsymbol{e}_r, \qquad (12.5)$$

where $\boldsymbol{u}_1 = \boldsymbol{u}_{1,D\times 1} \sim N(\boldsymbol{0}, \sigma_1^2\boldsymbol{I}_D)$, $\boldsymbol{u}_{2r} = \boldsymbol{u}_{2r,M\times 1} \sim N(\boldsymbol{0}, \sigma_2^2\boldsymbol{I}_M)$, and $\boldsymbol{e}_r = \boldsymbol{e}_{r,(N-n)\times 1} \sim N(\boldsymbol{0}, \sigma_0^2\boldsymbol{I}_{N-n})$ are independent, $\boldsymbol{y}_r = \boldsymbol{y}_{r,(N-n)\times 1}$, $\boldsymbol{\beta} = \boldsymbol{\beta}_{p\times 1}$, $\boldsymbol{X}_r = \boldsymbol{X}_{r,(N-n)\times p} = \underset{1\le d\le D}{\text{col}}\,(\underset{1\le t\le M_d}{\text{col}}\,(\underset{n_{dt}+1\le j\le N_{dt}}{\text{col}}\,(\boldsymbol{x}_{dtj})))$ with $\text{rank}(\boldsymbol{X}_r) = p$, $\boldsymbol{Z}_{1r} = \underset{1\le d\le D}{\text{diag}}\,(\boldsymbol{1}_{N_d-n_d})_{(N-n)\times D}$, $\boldsymbol{Z}_{2r} = \underset{1\le d\le D}{\text{diag}}\,(\underset{1\le t\le M_d}{\text{diag}}\,(\boldsymbol{1}_{N_{dt}-n_{dt}}))_{(N-n)\times M}$, $\boldsymbol{W}_r = \underset{1\le d\le D}{\text{diag}}\,(\boldsymbol{W}_{dr})$, $\boldsymbol{W}_{dr} = \underset{1\le t\le M_d}{\text{diag}}\,(\boldsymbol{W}_{dtr})$, $\boldsymbol{W}_{dtr} = \underset{n_{dt}+1\le j\le N_{dt}}{\text{diag}}\,(w_{dtj})_{(N_{dt}-n_{dt})\times(N_{dt}-n_{dt})}$ with known $w_{dtj} > 0$, $d = 1,\ldots,D$, $t = 1,\ldots,M_d$, $j = n_{dt}+1,\ldots,N_{dt}$, and $n_{dt} = 0$ if $t > m_d$.

12.2.4 The Inverse of the Variance Matrix

Let $\boldsymbol{V}_s$ denote the covariance matrix of the sample vector $\boldsymbol{y}_s$. Direct calculation of $\boldsymbol{V}_s^{-1}$ is not computationally efficient because it requires the inversion of the $n\times n$ matrix $\boldsymbol{V}_s$. This is why we apply the inversion formula (cf. Appendix A)

$$(\boldsymbol{A}+\boldsymbol{u}\boldsymbol{v}')^{-1} = \boldsymbol{A}^{-1} - \frac{\boldsymbol{A}^{-1}\boldsymbol{u}\boldsymbol{v}'\boldsymbol{A}^{-1}}{1+\boldsymbol{v}'\boldsymbol{A}^{-1}\boldsymbol{u}} \qquad (12.6)$$

for deriving an expression for $\boldsymbol{V}_s^{-1}$. Note that variance of $\boldsymbol{y}_s$ is

$$\boldsymbol{V}_s = \text{var}(\boldsymbol{y}_s) = \boldsymbol{Z}_{1s}\text{var}(\boldsymbol{u}_1)\boldsymbol{Z}_{1s}' + \boldsymbol{Z}_{2s}\text{var}(\boldsymbol{u}_{2s})\boldsymbol{Z}_{2s}' + \sigma_0^2\boldsymbol{W}_s^{-1} = \text{diag}(\boldsymbol{V}_{1s},\ldots,\boldsymbol{V}_{Ds}), \qquad (12.7)$$

where

$$\boldsymbol{V}_{ds} = \sigma_1^2\boldsymbol{1}_{n_d}\boldsymbol{1}_{n_d}' + \sigma_2^2 \underset{1\le t\le m_d}{\text{diag}}\,(\boldsymbol{1}_{n_{dt}}\boldsymbol{1}_{n_{dt}}') + \sigma_0^2\boldsymbol{W}_{ds}^{-1} = \sigma_1^2\boldsymbol{1}_{n_d}\boldsymbol{1}_{n_d}' + \boldsymbol{R}_{ds},\ d = 1,\ldots,D,$$

$$\boldsymbol{R}_{ds} = \underset{1\le t\le m_d}{\text{diag}}\,(\sigma_2^2\boldsymbol{1}_{n_{dt}}\boldsymbol{1}_{n_{dt}}' + \sigma_0^2\boldsymbol{W}_{dts}^{-1}) = \underset{1\le t\le m_d}{\text{diag}}\,(\boldsymbol{R}_{dts}),\quad d = 1,\ldots,D.$$

For $d = 1,\ldots,D$, $t = 1,\ldots,m_d$, we introduce the notation $\boldsymbol{w}_{n_{dt}} = \boldsymbol{W}_{dts}\boldsymbol{1}_{n_{dt}} = (w_{dt1},\ldots,w_{dtn_{dt}})_{n_{dt}\times 1}'$, $w_{dt\cdot} = \boldsymbol{1}_{n_{dt}}'\boldsymbol{w}_{n_{dt}} = \sum_{j=1}^{n_{dt}} w_{dtj}$ and

$$\gamma_{dt} = \frac{\sigma_2^2}{\sigma_2^2 + \frac{\sigma_0^2}{w_{dt\cdot}}},\quad \varphi_d = \frac{\sigma_1^2}{\sigma_0^2 + \sigma_1^2\sum_{\ell=1}^{m_d}(1-\gamma_{d\ell})w_{d\ell\cdot}}. \qquad (12.8)$$

For calculating $\boldsymbol{V}_s^{-1} = \text{diag}(\boldsymbol{V}_{1s}^{-1}, \ldots, \boldsymbol{V}_{Ds}^{-1})$ it is necessary to obtain $\boldsymbol{R}_{ds}^{-1}$. Here we use twice the formula (12.6), first to calculate $\boldsymbol{R}_{dts}^{-1}$ and second to obtain $\boldsymbol{V}_{ds}^{-1}$. For calculating $\boldsymbol{R}_{dts}^{-1} = (\sigma_2^2 \boldsymbol{1}_{n_{dt}} \boldsymbol{1}'_{n_{dt}} + \sigma_0^2 \boldsymbol{W}_{dts}^{-1})^{-1}$, we put $\boldsymbol{A} = \sigma_0^2 \boldsymbol{W}_{dts}^{-1}$, $\boldsymbol{u} = \sigma_2^2 \boldsymbol{1}_{n_{dt}}$, $\boldsymbol{v}' = \boldsymbol{1}'_{n_{dt}}$ and we get

$$\boldsymbol{R}_{dts}^{-1} = \frac{1}{\sigma_0^2}\boldsymbol{W}_{dts} - \frac{\sigma_2^2}{\sigma_0^4} \frac{\boldsymbol{W}_{dts}\boldsymbol{1}_{n_{dt}}\boldsymbol{1}'_{n_{dt}}\boldsymbol{W}_{dts}}{1 + \frac{\sigma_2^2}{\sigma_0^2}\boldsymbol{1}'_{n_{dt}}\boldsymbol{W}_{dts}\boldsymbol{1}_{n_{dt}}} = \frac{1}{\sigma_0^2}\left(\boldsymbol{W}_{dts} - \frac{\sigma_2^2 \boldsymbol{w}_{n_{dt}} \boldsymbol{w}'_{n_{dt}}}{\sigma_0^2 (1 + \frac{\sigma_2^2}{\sigma_0^2} w_{dt\cdot})}\right)$$

$$= \frac{1}{\sigma_0^2}\left(\boldsymbol{W}_{dts} - \frac{\gamma_{dt}}{w_{dt\cdot}} \boldsymbol{w}_{n_{dt}} \boldsymbol{w}'_{n_{dt}}\right), \quad d = 1, \ldots, D, \ t = 1, \ldots, m_d.$$

If we define $\boldsymbol{B}_{dts} = \boldsymbol{W}_{dts} - \frac{\gamma_{dt}}{w_{dt\cdot}} \boldsymbol{w}_{n_{dt}} \boldsymbol{w}'_{n_{dt}}$, we have

$$\boldsymbol{R}_{ds}^{-1} = \frac{1}{\sigma_0^2} \underset{1 \le t \le m_d}{\text{diag}} (\boldsymbol{B}_{dts}).$$

For calculating $\boldsymbol{V}_{ds}^{-1} = (\sigma_1^2 \boldsymbol{1}_{n_d} \boldsymbol{1}'_{n_d} + \boldsymbol{R}_{ds})^{-1}$, we put $\boldsymbol{A} = \boldsymbol{R}_{ds}$, $\boldsymbol{u} = \sigma_1^2 \boldsymbol{1}_{n_d}$, $\boldsymbol{v}' = \boldsymbol{1}'_{n_d}$ in (12.6) and we obtain

$$\boldsymbol{V}_{ds}^{-1} = \frac{1}{\sigma_0^2} \underset{1 \le t \le m_d}{\text{diag}} (\boldsymbol{B}_{dts}) \tag{12.9}$$

$$- \frac{\sigma_1^2}{\sigma_0^4} \frac{\underset{1 \le t \le m_d}{\text{col}} \left[\boldsymbol{w}_{n_{dt}} - \frac{\gamma_{dt}}{w_{dt\cdot}} w_{dt\cdot} \boldsymbol{w}_{n_{dt}}\right] \underset{1 \le t \le m_d}{\text{col}'} \left[\boldsymbol{w}'_{n_{dt}} - \frac{\gamma_{dt}}{w_{dt\cdot}} w_{dt\cdot} \boldsymbol{w}'_{n_{dt}}\right]}{1 + \frac{\sigma_1^2}{\sigma_0^2} (\sum_{\ell=1}^{m_d} w_{d\ell\cdot} - \sum_{\ell=1}^{m_d} \gamma_{d\ell} w_{d\ell\cdot})}$$

$$= \frac{1}{\sigma_0^2} \Big[\underset{1 \le t \le m_d}{\text{diag}} (\boldsymbol{B}_{dts}) - \varphi_d \underset{1 \le t \le m_d}{\text{col}} \left[(1 - \gamma_{dt}) \boldsymbol{w}_{n_{dt}}\right] \underset{1 \le t \le m_d}{\text{col}'} \left[(1 - \gamma_{dt}) \boldsymbol{w}'_{n_{dt}}\right] \Big].$$

By applying (12.9), no matrix inversions are needed in programs. We also give an alternative formula for $\boldsymbol{V}_{ds}^{-1}$ requiring the inversion of a $m_d \times m_d$ matrix. Let us note that $\boldsymbol{V}_{ds} = \sigma_1^2 \boldsymbol{1}_{n_d} \boldsymbol{1}'_{n_d} + \boldsymbol{R}_{ds}$, where

$$\boldsymbol{R}_{ds} = \sigma_2^2 \underset{1 \le t \le m_d}{\text{diag}} (\boldsymbol{1}_{n_{dt}}) \boldsymbol{I}_{m_d} \underset{1 \le t \le m_d}{\text{diag}} (\boldsymbol{1}'_{n_{dt}}) + \sigma_0^2 \boldsymbol{W}_{ds}^{-1}.$$

For calculating $\boldsymbol{R}_{ds}^{-1}$, we apply the formula (cf. Appendix A)

$$(A + CBD)^{-1} = A^{-1} - A^{-1}C(B^{-1} + DA^{-1}C)^{-1}DA^{-1},$$

with $A = \sigma_0^2 \boldsymbol{W}_{ds}^{-1}$, $C = \sigma_2^2 \underset{1\le t\le m_d}{\text{diag}} (\mathbf{1}_{n_{dt}})$, $B = \boldsymbol{I}_{m_d}$, and $D = \underset{1\le t\le m_d}{\text{diag}} (\mathbf{1}'_{n_{dt}})$. We obtain

$$\boldsymbol{R}_{ds}^{-1} = \sigma_0^{-2}\boldsymbol{W}_{ds} - \sigma_0^{-2}\sigma_2^2\boldsymbol{W}_{ds} \underset{1\le t\le m_d}{\text{diag}} (\mathbf{1}_{n_{dt}})$$
$$\cdot \left[\boldsymbol{I}_{m_d} + \sigma_0^{-2}\sigma_2^2 \underset{1\le t\le m_d}{\text{diag}} (\mathbf{1}'_{n_{dt}})\boldsymbol{W}_{ds} \underset{1\le t\le m_d}{\text{diag}} (\mathbf{1}_{n_{dt}})\right]^{-1} \sigma_0^{-2} \underset{1\le t\le m_d}{\text{diag}} (\mathbf{1}'_{n_{dt}})\boldsymbol{W}_{ds}.$$

For calculating $\boldsymbol{V}_{ds}^{-1}$, we use the formula (12.6) with $A = \boldsymbol{R}_{ds}$, $\boldsymbol{u} = \sigma_1^2\mathbf{1}_{n_d}$, $\boldsymbol{v}' = \mathbf{1}'_{n_d}$. Finally, we obtain

$$\boldsymbol{V}_{ds}^{-1} = \boldsymbol{R}_{ds}^{-1} - \frac{\sigma_1^2}{1+\sigma_1^2\mathbf{1}'_{n_d}\boldsymbol{R}_{ds}^{-1}\mathbf{1}_{n_d}}\boldsymbol{R}_{ds}^{-1}\mathbf{1}_{n_d}\mathbf{1}'_{n_d}\boldsymbol{R}_{ds}^{-1}. \tag{12.10}$$

12.3 The Conditional Distribution of $\mathbf{y}_r$ given $\mathbf{y}_s$

Due to the normality assumptions of the population model (12.1), the vector $\boldsymbol{y} = (\boldsymbol{y}'_s, \boldsymbol{y}'_r)'$ is normally distributed with mean vector $\boldsymbol{\mu} = (\boldsymbol{\mu}'_s, \boldsymbol{\mu}'_r)'$ and covariance matrix

$$\boldsymbol{V} = \begin{pmatrix} \boldsymbol{V}_s & \boldsymbol{V}_{sr} \\ \boldsymbol{V}_{rs} & \boldsymbol{V}_r \end{pmatrix},$$

where $\boldsymbol{V}_s = \text{var}(\boldsymbol{y}_s)$, $\boldsymbol{V}_r = \text{var}(\boldsymbol{y}_r)$, $\boldsymbol{V}_{rs} = \text{cov}(\boldsymbol{y}_r, \boldsymbol{y}_s)$, and $\boldsymbol{V}_{sr} = \boldsymbol{V}'_{rs}$. Thus, the conditional distribution of $\boldsymbol{y}_r|\boldsymbol{y}_s$ is

$$\boldsymbol{y}_r|\boldsymbol{y}_s \sim N(\boldsymbol{\mu}_{r|s}, \boldsymbol{V}_{r|s}),$$

where the conditional mean vector and covariance matrix are (see e.g. Theorem 2.2E in Rencher (1998))

$$\boldsymbol{\mu}_{r|s} = \boldsymbol{X}_r\boldsymbol{\beta} + \boldsymbol{V}_{rs}\boldsymbol{V}_s^{-1}(\boldsymbol{y}_s - \boldsymbol{X}_s\boldsymbol{\beta}), \quad \boldsymbol{V}_{r|s} = \boldsymbol{V}_r - \boldsymbol{V}_{rs}\boldsymbol{V}_s^{-1}\boldsymbol{V}_{sr}. \tag{12.11}$$

12.3.1 Conditional Mean Vector

This section derives a programmable formula for the conditional mean $\boldsymbol{\mu}_{r|s}$. We know that $\boldsymbol{V}_s^{-1} = \underset{1\le d\le D}{\text{diag}} (\boldsymbol{V}_{ds}^{-1})$, where $\boldsymbol{V}_{ds}^{-1}$ is given in (12.9) or alternatively in (12.10). Since the population vector $\boldsymbol{y}$ follows the model (12.1), it holds that

$$\boldsymbol{y} - E[\boldsymbol{y}] = (\boldsymbol{X}\boldsymbol{\beta} + \boldsymbol{Z}_1\boldsymbol{u}_1 + \boldsymbol{Z}_2\boldsymbol{u}_2 + \boldsymbol{e}) - \boldsymbol{X}\boldsymbol{\beta} = \boldsymbol{Z}_1\boldsymbol{u}_1 + \boldsymbol{Z}_2\boldsymbol{u}_2 + \boldsymbol{e}.$$

Therefore, the covariance $\boldsymbol{V}_{rs} = \text{cov}(\boldsymbol{y}_r, \boldsymbol{y}_s)$ is

$$\begin{aligned}\boldsymbol{V}_{rs} &= E[(\boldsymbol{Z}_{1r}\boldsymbol{u}_1 + \boldsymbol{Z}_{2r}\boldsymbol{u}_{2r} + \boldsymbol{e}_r)(\boldsymbol{Z}_{1s}\boldsymbol{u}_1 + \boldsymbol{Z}_{2s}\boldsymbol{u}_{2s} + \boldsymbol{e}_s)'] \\ &= \boldsymbol{Z}_{1r}\sigma_1^2\boldsymbol{I}_D\boldsymbol{Z}_{1s}' + \boldsymbol{Z}_{2r}E[\boldsymbol{u}_{2r}\boldsymbol{u}_{2s}']\boldsymbol{Z}_{2s}'.\end{aligned}$$

Let us now calculate the $M \times m$ matrix $E[\boldsymbol{u}_{2r}\boldsymbol{u}_{2s}']$. As $E[u_{2,d_1t_1}u_{2,d_2t_2}] = 0$ if $d_1 \neq d_2$ or $t_1 \neq t_2$, we get

$$\begin{aligned}E[\boldsymbol{u}_{2r}\boldsymbol{u}_{2s}'] &= E\Big[\underset{1\le d\le D}{\text{col}}\big(\underset{1\le t\le M_d}{\text{col}}(u_{2,dt})\big)\underset{1\le d\le D}{\text{col}'}\big(\underset{1\le t\le m_d}{\text{col}'}(u_{2,dt})\big)\Big] \\ &= \underset{1\le d\le D}{\text{diag}}\Big(E\Big[\text{col}\big\{\underset{1\le t\le m_d}{\text{col}}(u_{2,dt}), \underset{m_d+1\le t\le M_d}{\text{col}}(u_{2,dt})\big\}\underset{1\le t\le m_d}{\text{col}'}(u_{2,dt})\Big]\Big) \\ &= \sigma_2^2 \underset{1\le d\le D}{\text{diag}}\left(\text{col}\{\boldsymbol{I}_{m_d}, \boldsymbol{0}_{M_d-m_d\times m_d}\}\right).\end{aligned}$$

As $\boldsymbol{Z}_{2r} = \underset{1\le d\le D}{\text{diag}}\big(\text{diag}\{\underset{1\le t\le m_d}{\text{diag}}(\boldsymbol{1}_{N_{dt}-n_{dt}}), \underset{m_d+1\le t\le M_d}{\text{diag}}(\boldsymbol{1}_{N_{dt}})\}\big)$, we get

$$\begin{aligned}&\boldsymbol{Z}_{2r}E[\boldsymbol{u}_{2r}\boldsymbol{u}_{2s}']\boldsymbol{Z}_{2s}' \\ &= \sigma_2^2 \underset{1\le d\le D}{\text{diag}}\Big(\underset{1\le t\le M_d}{\text{diag}}(\boldsymbol{1}_{N_{dt}-n_{dt}})\text{col}\{\boldsymbol{I}_{m_d}, \boldsymbol{0}_{M_d-m_d\times m_d}\}\underset{1\le t\le m_d}{\text{diag}}(\boldsymbol{1}_{n_{dt}}')\Big) \\ &= \sigma_2^2 \underset{1\le d\le D}{\text{diag}}\Big(\text{col}\{\underset{1\le t\le m_d}{\text{diag}}(\boldsymbol{1}_{N_{dt}-n_{dt}}), \boldsymbol{0}_{N_d-N_{ds}\times m_d}\}\underset{1\le t\le m_d}{\text{diag}}(\boldsymbol{1}_{n_{dt}}')\Big) \\ &= \sigma_2^2 \underset{1\le d\le D}{\text{diag}}\Big(\text{col}\{\underset{1\le t\le m_d}{\text{diag}}(\boldsymbol{1}_{N_{dt}-n_{dt}}\boldsymbol{1}_{n_{dt}}'), \boldsymbol{0}_{N_d-N_{ds}\times n_d}\}\Big),\end{aligned}$$

where $N_{ds} = \sum_{t=1}^{m_d} N_{dt}$. We have obtained that $\boldsymbol{V}_{rs} = \underset{1\le d\le D}{\text{diag}}(\boldsymbol{V}_{drs})$, with

$$\boldsymbol{V}_{drs} = \sigma_1^2\boldsymbol{1}_{N_d-n_d}\boldsymbol{1}_{n_d}' + \text{col}\Big\{\sigma_2^2 \underset{1\le t\le m_d}{\text{diag}}(\boldsymbol{1}_{N_{dt}-n_{dt}}\boldsymbol{1}_{n_{dt}}'), \boldsymbol{0}_{(N_d-N_{ds})\times n_d}\Big\} = A + B. \tag{12.12}$$

Moreover, formula (12.9) can be written in the form

$$\begin{aligned}\boldsymbol{V}_{ds}^{-1} &= \frac{1}{\sigma_0^2}\underset{1\le t\le m_d}{\text{diag}}(\boldsymbol{B}_{dts}) \\ &\quad - \frac{1}{\sigma_0^2}\varphi_d \underset{1\le t\le m_d}{\text{col}}\big[(1-\gamma_{dt})\boldsymbol{w}_{n_{dt}}\big]\underset{1\le t\le m_d}{\text{col}'}\big[(1-\gamma_{dt})\boldsymbol{w}_{n_{dt}}'\big] = C - D,\end{aligned}$$

where $\boldsymbol{B}_{dts} = \boldsymbol{W}_{dts} - \frac{\gamma_{dt}}{w_{dt.}}\boldsymbol{w}_{n_{dt}}\boldsymbol{w}'_{n_{dt}}$. Then $\boldsymbol{V}_{rs}\boldsymbol{V}_s^{-1} = \underset{1\le d\le D}{\text{diag}}\,(\boldsymbol{V}_{drs}\boldsymbol{V}_{ds}^{-1})$, where

$$\boldsymbol{V}_{drs}\boldsymbol{V}_{ds}^{-1} = (A+B)(C-D) = (AC-AD)+(BC-BD).$$

The intermediate calculations are

$$\begin{aligned}
AC &= \sigma_1^2\boldsymbol{1}_{N_d-n_d}\underset{1\le t\le m_d}{\text{col}'}(\boldsymbol{1}'_{n_{dt}})\frac{1}{\sigma_0^2}\underset{1\le t\le m_d}{\text{diag}}\Big(\boldsymbol{W}_{dts}-\frac{\gamma_{dt}}{w_{dt.}}\boldsymbol{w}_{n_{dt}}\boldsymbol{w}'_{n_{dt}}\Big)\\
&= \frac{\sigma_1^2}{\sigma_0^2}\boldsymbol{1}_{N_d-n_d}\underset{1\le t\le m_d}{\text{col}'}\Big[\boldsymbol{1}'_{n_{dt}}\Big(\boldsymbol{W}_{dts}-\frac{\gamma_{dt}}{w_{dt.}}\boldsymbol{w}_{n_{dt}}\boldsymbol{w}'_{n_{dt}}\Big)\Big]\\
&= \frac{\sigma_1^2}{\sigma_0^2}\boldsymbol{1}_{N_d-n_d}\underset{1\le t\le m_d}{\text{col}'}\Big[\boldsymbol{w}'_{n_{dt}}-\gamma_{dt}\boldsymbol{w}'_{n_{dt}}\Big] = \frac{\sigma_1^2}{\sigma_0^2}\boldsymbol{1}_{N_d-n_d}\underset{1\le t\le m_d}{\text{col}'}\big[(1-\gamma_{dt})\boldsymbol{w}'_{n_{dt}}\big],\\
AD &= \sigma_1^2\boldsymbol{1}_{N_d-n_d}\underset{1\le t\le m_d}{\text{col}'}(\boldsymbol{1}'_{n_{dt}})\frac{\varphi_d}{\sigma_0^2}\underset{1\le t\le m_d}{\text{col}}\big[(1-\gamma_{dt})\boldsymbol{w}_{n_{dt}}\big]\underset{1\le t\le m_d}{\text{col}'}\big[(1-\gamma_{dt})\boldsymbol{w}'_{n_{dt}}\big]\\
&= \frac{\sigma_1^2}{\sigma_0^2}\varphi_d\boldsymbol{1}_{N_d-n_d}\Big(\sum_{\ell=1}^{m_d}(1-\gamma_{d\ell})\boldsymbol{1}'_{n_{d\ell}}\boldsymbol{w}_{n_{d\ell}}\Big)\underset{1\le t\le m_d}{\text{col}'}\big[(1-\gamma_{dt})\boldsymbol{w}'_{n_{dt}}\big]\\
&= \frac{\sigma_1^2}{\sigma_0^2}\varphi_d\Big(\sum_{\ell=1}^{m_d}w_{d\ell.}(1-\gamma_{d\ell})\Big)\boldsymbol{1}_{N_d-n_d}\underset{1\le t\le m_d}{\text{col}'}\big[(1-\gamma_{dt})\boldsymbol{w}'_{n_{dt}}\big],
\end{aligned}$$

$$\begin{aligned}
BC &= \text{col}\Big\{\sigma_2^2\underset{1\le t\le m_d}{\text{diag}}(\boldsymbol{1}_{N_{dt}-n_{dt}}\boldsymbol{1}'_{n_{dt}}),\ \boldsymbol{0}_{(N_d-N_{ds})\times n_d}\Big\}\\
&\quad\cdot\frac{1}{\sigma_0^2}\underset{1\le t\le m_d}{\text{diag}}\Big(\boldsymbol{W}_{dts}-\frac{\gamma_{dt}}{w_{dt.}}\boldsymbol{w}_{n_{dt}}\boldsymbol{w}'_{n_{dt}}\Big)\\
&= \frac{\sigma_2^2}{\sigma_0^2}\text{col}\Big\{\underset{1\le t\le m_d}{\text{diag}}\Big(\boldsymbol{1}_{N_{dt}-n_{dt}}\boldsymbol{w}'_{n_{dt}}-\frac{\gamma_{dt}}{w_{dt.}}\boldsymbol{1}_{N_{dt}-n_{dt}}w_{dt.}\boldsymbol{w}'_{n_{dt}}\Big),\ \boldsymbol{0}_{(N_d-N_{ds})\times n_d}\Big\}\\
&= \frac{\sigma_2^2}{\sigma_0^2}\text{col}\Big\{\underset{1\le t\le m_d}{\text{diag}}\big((1-\gamma_{dt})\boldsymbol{1}_{N_{dt}-n_{dt}}\boldsymbol{w}'_{n_{dt}}\big),\ \boldsymbol{0}_{(N_d-N_{ds})\times n_d}\Big\},\\
BD &= \text{col}\Big\{\sigma_2^2\underset{1\le t\le m_d}{\text{diag}}(\boldsymbol{1}_{N_{dt}-n_{dt}}\boldsymbol{1}'_{n_{dt}}),\ \boldsymbol{0}_{(N_d-N_{ds})\times n_d}\Big\}\\
&\quad\cdot\frac{\varphi_d}{\sigma_0^2}\underset{1\le t\le m_d}{\text{col}}\big[(1-\gamma_{dt})\boldsymbol{w}_{n_{dt}}\big]\underset{1\le t\le m_d}{\text{col}'}\big[(1-\gamma_{dt})\boldsymbol{w}'_{n_{dt}}\big]\\
&= \frac{\varphi_d}{\sigma_0^2}\text{col}\Big\{\sigma_2^2\underset{1\le t\le m_d}{\text{diag}}(\boldsymbol{1}_{N_{dt}-n_{dt}}\boldsymbol{1}'_{n_{dt}})\underset{1\le t\le m_d}{\text{col}}\,[(1-\gamma_{dt})\boldsymbol{w}_{n_{dt}}],\ \boldsymbol{0}_{(N_d-N_{ds})\times 1}\Big\}
\end{aligned}$$

$$\cdot \underset{1\le t\le m_d}{\text{col}'}[(1-\gamma_{dt})\boldsymbol{w}'_{n_{dt}}] = \frac{\sigma_2^2\varphi_d}{\sigma_0^2}$$

$$\cdot \text{col}\left\{\underset{1\le t\le m_d}{\text{col}}[\mathbf{1}_{N_{dt}-n_{dt}}(1-\gamma_{dt})w_{dt.}],\ \mathbf{0}_{(N_d-N_{ds})\times 1}\right\}\underset{1\le t\le m_d}{\text{col}'}[(1-\gamma_{dt})\boldsymbol{w}'_{n_{dt}}].$$

Note that

$$\frac{\sigma_1^2}{\sigma_0^2}\left[1-\varphi_d\sum_{\ell=1}^{m_d}w_{d\ell.}(1-\gamma_{d\ell})\right] = \frac{\sigma_1^2}{\sigma_0^2}\left[1-\frac{\sigma_1^2\sum_{\ell=1}^{m_d}w_{d\ell.}(1-\gamma_{d\ell})}{\sigma_0^2+\sigma_1^2\sum_{\ell=1}^{m_d}(1-\gamma_{d\ell})w_{d\ell.}}\right]$$

$$= \frac{\sigma_1^2}{\sigma_0^2+\sigma_1^2\sum_{\ell=1}^{m_d}(1-\gamma_{d\ell})w_{d\ell.}} = \varphi_d.$$

Therefore,

$$AC-AD = \frac{\sigma_1^2}{\sigma_0^2}\mathbf{1}_{N_d-n_d}\underset{1\le t\le m_d}{\text{col}'}\left[(1-\gamma_{dt})\boldsymbol{w}'_{n_{dt}}\right]$$

$$-\frac{\sigma_1^2}{\sigma_0^2}\varphi_d\left(\sum_{\ell=1}^{m_d}w_{d\ell.}(1-\gamma_{d\ell})\right)\mathbf{1}_{N_d-n_d}\underset{1\le t\le m_d}{\text{col}'}[(1-\gamma_{dt})\boldsymbol{w}'_{n_{dt}}]$$

$$=\frac{\sigma_1^2}{\sigma_0^2}\left[1-\varphi_d\sum_{\ell=1}^{m_d}w_{d\ell.}(1-\gamma_{d\ell})\right]\mathbf{1}_{N_d-n_d}\underset{1\le t\le m_d}{\text{col}'}[(1-\gamma_{dt})\boldsymbol{w}'_{n_{dt}}]$$

$$=\varphi_d\mathbf{1}_{N_d-n_d}\underset{1\le t\le m_d}{\text{col}'}[(1-\gamma_{dt})\boldsymbol{w}'_{n_{dt}}]$$

and

$$\boldsymbol{V}_{drs}\boldsymbol{V}_{ds}^{-1} = (AC-AD)+BC-BD = \varphi_d\,\mathbf{1}_{N_d-n_d}\underset{1\le t\le m_d}{\text{col}'}[(1-\gamma_{dt})\boldsymbol{w}'_{n_{dt}}]$$

$$+\frac{\sigma_2^2}{\sigma_0^2}\text{col}\left\{\underset{1\le t\le m_d}{\text{diag}}\left((1-\gamma_{dt})\mathbf{1}_{N_{dt}-n_{dt}}\boldsymbol{w}'_{n_{dt}}\right),\ \mathbf{0}_{(N_d-N_{ds})\times n_d}\right\}$$

$$-\frac{\sigma_2^2\varphi_d}{\sigma_0^2}\text{col}\left\{\underset{1\le t\le m_d}{\text{col}}[w_{dt.}(1-\gamma_{dt})\mathbf{1}_{N_{dt}-n_{dt}}],\ \mathbf{0}_{(N_d-N_{ds})\times 1}\right\}$$

$$\cdot\underset{1\le t\le m_d}{\text{col}'}[(1-\gamma_{dt})\boldsymbol{w}'_{n_{dt}}].$$

The conditional mean vector is $\boldsymbol{\mu}_{r|s} = \underset{1\le d\le D}{\text{col}}(\boldsymbol{\mu}_{dr|s})$, where

$$\boldsymbol{\mu}_{dr|s} = \boldsymbol{X}_{dr}\boldsymbol{\beta}+\boldsymbol{V}_{drs}\boldsymbol{V}_{ds}^{-1}(\boldsymbol{y}_{ds}-\boldsymbol{X}_{ds}\boldsymbol{\beta}) = \underset{1\le t\le M_d}{\text{col}}(\boldsymbol{\mu}_{dtr|s})$$

$$= \underset{1\le t\le M_d}{\text{col}}[\boldsymbol{X}_{dtr}\boldsymbol{\beta}]+\boldsymbol{V}_{drs}\boldsymbol{V}_{ds}^{-1}\underset{1\le t\le m_d}{\text{col}}[\boldsymbol{y}_{dts}-\boldsymbol{X}_{dts}\boldsymbol{\beta}]$$

and, correspondingly to the notation of this chapter, $\boldsymbol{y}_{dts} = \underset{1\leq j\leq n_{dt}}{\text{col}}(y_{dtj})$, $\boldsymbol{y}_{ds} = \underset{1\leq t\leq m_d}{\text{col}}(\boldsymbol{y}_{dts})$, $\boldsymbol{X}_{dts} = \underset{1\leq j\leq n_{dt}}{\text{col}}(\boldsymbol{x}_{dtj})$, $\boldsymbol{X}_{ds} = \underset{1\leq t\leq m_d}{\text{col}}(\boldsymbol{X}_{dts})$, $\boldsymbol{X}_{dtr} = \underset{n_{dt}+1\leq j\leq N_{dt}}{\text{col}}(\boldsymbol{x}_{dtj})$, and $\boldsymbol{X}_{dr} = \underset{1\leq t\leq M_d}{\text{col}}(\boldsymbol{X}_{dtr})$. By doing some algebra, we get

$$\begin{aligned}\boldsymbol{\mu}_{dr|s} &= \underset{1\leq t\leq M_d}{\text{col}}[\boldsymbol{X}_{dtr}\boldsymbol{\beta}] + \varphi_d \mathbf{1}_{N_d-n_d}\sum_{\ell=1}^{m_d}(1-\gamma_{d\ell})\boldsymbol{w}'_{n_{d\ell}}(\boldsymbol{y}_{d\ell s}-\boldsymbol{X}_{d\ell s}\boldsymbol{\beta}) \\ &+ \frac{\sigma_2^2}{\sigma_0^2}\text{col}\left\{\underset{1\leq t\leq m_d}{\text{col}}\left[(1-\gamma_{dt})\mathbf{1}_{N_{dt}-n_{dt}}\boldsymbol{w}'_{n_{dt}}(\boldsymbol{y}_{dts}-\boldsymbol{X}_{dts}\boldsymbol{\beta})\right], \mathbf{0}_{(N_d-N_{ds})\times 1}\right\} \\ &- \frac{\sigma_2^2}{\sigma_0^2}\varphi_d\,\text{col}\left\{\underset{1\leq t\leq m_d}{\text{col}}[w_{dt.}(1-\gamma_{dt})\mathbf{1}_{N_{dt}-n_{dt}}], \mathbf{0}_{(N_d-N_{ds})\times 1}\right\} \\ &\cdot \sum_{\ell=1}^{m_d}(1-\gamma_{d\ell})\boldsymbol{w}'_{n_{d\ell}}(\boldsymbol{y}_{d\ell s}-\boldsymbol{X}_{d\ell s}\boldsymbol{\beta}).\end{aligned}$$

Note that

$$1-\gamma_{dt} = 1-\frac{\sigma_2^2}{\sigma_0^2}w_{dt.}(1-\gamma_{dt}), \quad \gamma_{dt} = w_{dt.}(1-\gamma_{dt})\frac{\sigma_2^2}{\sigma_0^2}$$

and let us denote by $\bar{y}_{d\ell s} = w_{d\ell.}^{-1}\sum_{j\in s_{d\ell}} w_{d\ell j}y_{d\ell j}$ and $\bar{\boldsymbol{x}}_{d\ell s} = w_{d\ell.}^{-1}\sum_{j\in s_{d\ell}} w_{d\ell j}\boldsymbol{x}_{d\ell j}$ the weighted sample means of the response and auxiliary variables in subdomain ℓ from domain d. For $1\leq t\leq m_d$, the conditional mean vector is

$$\begin{aligned}\boldsymbol{\mu}_{dtr|s} &= \boldsymbol{X}_{dtr}\boldsymbol{\beta} + \mathbf{1}_{N_{dt}-n_{dt}}\varphi_d\sum_{\ell=1}^{m_d}(1-\gamma_{d\ell})w_{d\ell.}(\bar{y}_{d\ell s}-\bar{\boldsymbol{x}}_{d\ell s}\boldsymbol{\beta}) \\ &+ \mathbf{1}_{N_{dt}-n_{dt}}\frac{\sigma_2^2}{\sigma_0^2}(1-\gamma_{dt})w_{dt.}(\bar{y}_{dts}-\bar{\boldsymbol{x}}_{dts}\boldsymbol{\beta}) \\ &- \mathbf{1}_{N_{dt}-n_{dt}}\frac{\sigma_2^2}{\sigma_0^2}\varphi_d w_{dt.}(1-\gamma_{dt})\sum_{\ell=1}^{m_d}(1-\gamma_{d\ell})w_{d\ell.}(\bar{y}_{d\ell s}-\bar{\boldsymbol{x}}_{d\ell s}\boldsymbol{\beta}) \\ &= \boldsymbol{X}_{dtr}\boldsymbol{\beta}+\mathbf{1}_{N_{dt}-n_{dt}}\varphi_d\sum_{\ell=1}^{m_d}(1-\gamma_{d\ell})w_{d\ell.}(\bar{y}_{d\ell s}-\bar{\boldsymbol{x}}_{d\ell s}\boldsymbol{\beta})\left[1-\frac{\sigma_2^2}{\sigma_0^2}w_{dt.}(1-\gamma_{dt})\right] \\ &+ \mathbf{1}_{N_{dt}-n_{dt}}w_{dt.}(1-\gamma_{dt})\frac{\sigma_2^2}{\sigma_0^2}(\bar{y}_{dts}-\bar{\boldsymbol{x}}_{dts}\boldsymbol{\beta}) \\ &= \boldsymbol{X}_{dtr}\boldsymbol{\beta}+\mathbf{1}_{N_{dt}-n_{dt}}w_{dt.}(1-\gamma_{dt})\frac{\sigma_2^2}{\sigma_0^2}\end{aligned}$$

$$\cdot \left\{ \bar{y}_{dts} - \bar{\boldsymbol{x}}_{dts}\boldsymbol{\beta} + \frac{\varphi_d \sigma_0^2}{w_{dt.}\sigma_2^2} \sum_{\ell=1}^{m_d} (1-\gamma_{d\ell}) w_{d\ell.} (\bar{y}_{d\ell s} - \bar{\boldsymbol{x}}_{d\ell s}\boldsymbol{\beta}) \right\}$$

$$= \boldsymbol{X}_{dtr}\boldsymbol{\beta} + \mathbf{1}_{N_{dt}-n_{dt}} \gamma_{dt} \left\{ \bar{y}_{dts} - \bar{\boldsymbol{x}}_{dts}\boldsymbol{\beta} + \frac{\sigma_0^4}{\sigma_2^4} \frac{\varphi_d}{w_{dt.}} \sum_{\ell=1}^{m_d} \gamma_{d\ell} (\bar{y}_{d\ell s} - \bar{\boldsymbol{x}}_{d\ell s}\boldsymbol{\beta}) \right\}.$$

For $m_d + 1 \le t \le M_d$, the conditional mean vector is

$$\boldsymbol{\mu}_{dtr|s} = \boldsymbol{X}_{dtr}\boldsymbol{\beta} + \mathbf{1}_{N_{dt}} \varphi_d \sum_{\ell=1}^{m_d} (1-\gamma_{d\ell}) w_{d\ell.} (\bar{y}_{d\ell s} - \bar{\boldsymbol{x}}_{d\ell s}\boldsymbol{\beta})$$

$$= \boldsymbol{X}_{dtr}\boldsymbol{\beta} + \mathbf{1}_{N_{dt}} \frac{\sigma_0^2}{\sigma_2^2} \varphi_d \sum_{\ell=1}^{m_d} \gamma_{d\ell} (\bar{y}_{d\ell s} - \bar{\boldsymbol{x}}_{d\ell s}\boldsymbol{\beta}).$$

Finally, if $m_d = 0$ or equivalently if $n_d = 0$, the conditional mean vector is the marginal mean vector, i.e.

$$\boldsymbol{\mu}_{dtr|s} = \boldsymbol{\mu}_{dt} = \boldsymbol{X}_{dt}\boldsymbol{\beta}, \quad t = 1, \ldots, M_d.$$

In summary, for $m_d > 0$ and $j \in r_{dt} = \{n_{dt}+1, \ldots, N_{dt}\}$, we have

$$\mu_{dtj|s} = \boldsymbol{x}_{dtj}\boldsymbol{\beta} + \gamma_{dt} \left\{ \bar{y}_{dts} - \bar{\boldsymbol{x}}_{dts}\boldsymbol{\beta} + \frac{\sigma_0^4}{\sigma_2^4} \frac{\varphi_d}{w_{dt.}} \sum_{\ell=1}^{m_d} \gamma_{d\ell} (\bar{y}_{d\ell s} - \bar{\boldsymbol{x}}_{d\ell s}\boldsymbol{\beta}) \right\}, 1 \le t \le m_d,$$

$$\mu_{dtj|s} = \boldsymbol{x}_{dtj}\boldsymbol{\beta} + \frac{\sigma_0^2}{\sigma_2^2} \varphi_d \sum_{\ell=1}^{m_d} \gamma_{d\ell} (\bar{y}_{d\ell s} - \bar{\boldsymbol{x}}_{d\ell s}\boldsymbol{\beta}), \quad m_d + 1 \le t \le M_d.$$

For $m_d = 0$ and $j \in U_{dt}$, we have $\mu_{dtj|s} = \boldsymbol{x}_{dtj}\boldsymbol{\beta}$.

12.3.2 Conditional Covariance Matrix

This section derives a programmable formula for the conditional covariance matrix $\boldsymbol{V}_{r|s}$. By (12.11), since $\boldsymbol{V}_r$, $\boldsymbol{V}_{rs}$, and $\boldsymbol{V}_s^{-1}$ are all block-diagonal, it holds that $\boldsymbol{V}_{r|s} = \underset{1\le d\le D}{\text{diag}} (\boldsymbol{V}_{dr|s})$, where

$$\boldsymbol{V}_{dr|s} = \boldsymbol{V}_{dr} - \boldsymbol{V}_{drs} \boldsymbol{V}_{ds}^{-1} \boldsymbol{V}_{dsr}.$$

Under the non-sample model (12.5), we have

$$\boldsymbol{V}_{dr} = \sigma_0^2 \boldsymbol{W}_{dr}^{-1} + \sigma_1^2 \mathbf{1}_{N_d-n_d} \mathbf{1}'_{N_d-n_d} + \sigma_2^2 \underset{1\le t\le M_d}{\text{diag}} (\mathbf{1}_{N_{dt}-n_{dt}} \mathbf{1}'_{N_{dt}-n_{dt}}). \qquad (12.13)$$

Moreover, using the expression of $\boldsymbol{V}_{drs}$ given in (12.12), we have

$$\begin{aligned}\boldsymbol{V}_{drs}\boldsymbol{V}_{ds}^{-1}\boldsymbol{V}_{dsr} &= (A+B)\boldsymbol{V}_{ds}^{-1}(A+B)'\\ &= A\boldsymbol{V}_{ds}^{-1}A' + A\boldsymbol{V}_{ds}^{-1}B' + B\boldsymbol{V}_{ds}^{-1}A' + B\boldsymbol{V}_{ds}^{-1}B'\\ &= \boldsymbol{L}_{d1} + \boldsymbol{L}_{d2} + \boldsymbol{L}_{d3} + \boldsymbol{L}_{d4}. \end{aligned} \tag{12.14}$$

The first term on the right hand side of (12.14) is

$$\begin{aligned}\boldsymbol{L}_{d1} &= A\boldsymbol{V}_{ds}^{-1}A = \sigma_1^4 \boldsymbol{1}_{N_d-n_d}\boldsymbol{1}'_{n_d}\boldsymbol{V}_{ds}^{-1}\boldsymbol{1}_{n_d}\boldsymbol{1}'_{N_d-n_d}\\ &= \sigma_1^4(\boldsymbol{1}'_{n_d}\boldsymbol{V}_{ds}^{-1}\boldsymbol{1}_{n_d})\boldsymbol{1}_{N_d-n_d}\boldsymbol{1}'_{N_d-n_d}, \end{aligned} \tag{12.15}$$

where

$$\begin{aligned}\boldsymbol{1}'_{n_d}\boldsymbol{V}_{ds}^{-1}\boldsymbol{1}_{n_d} &= \frac{1}{\sigma_0^2}\sum_{t=1}^{m_d}\boldsymbol{1}'_{n_{dt}}\boldsymbol{B}_{dts}\boldsymbol{1}_{n_{dt}}\\ &- \frac{\varphi_d}{\sigma_0^2}\boldsymbol{1}'_{n_d}\underset{1\le t\le m_d}{\text{col}}((1-\gamma_{dt})\boldsymbol{w}_{n_{dt}})\underset{1\le t\le m_d}{\text{col}'}((1-\gamma_{dt})\boldsymbol{w}'_{n_{dt}})\boldsymbol{1}_{n_d}. \end{aligned} \tag{12.16}$$

Note that

$$\boldsymbol{B}_{dts} = \boldsymbol{W}_{dts} - \frac{\gamma_{dt}}{w_{dt.}}\boldsymbol{w}_{n_{dt}}\boldsymbol{w}'_{n_{dt}},$$

$$\underset{1\le t\le m_d}{\text{col}'}((1-\gamma_{dt})\boldsymbol{w}'_{n_{dt}})\boldsymbol{1}_{n_d} = \sum_{t=1}^{m_d} w_{dt.}(1-\gamma_{dt}), \tag{12.17}$$

$$\begin{aligned}\boldsymbol{1}'_{n_{dt}}\boldsymbol{B}_{dts}\boldsymbol{1}_{n_{dt}} &= \boldsymbol{1}'_{n_{dt}}\left(\boldsymbol{W}_{dts} - \frac{\gamma_{dt}}{w_{dt.}}\boldsymbol{w}_{n_{dt}}\boldsymbol{w}'_{n_{dt}}\right)\boldsymbol{1}_{n_{dt}}\\ &= w_{dt.} - \frac{\gamma_{dt}}{w_{dt.}}w_{dt.}^2 = w_{dt.}(1-\gamma_{dt}). \end{aligned} \tag{12.18}$$

Replacing (12.17) and (12.18) in (12.16), we get

$$\begin{aligned}\boldsymbol{1}'_{n_d}\boldsymbol{V}_{ds}^{-1}\boldsymbol{1}_{n_d} &= \frac{1}{\sigma_0^2}\left[\sum_{t=1}^{m_d} w_{dt.}(1-\gamma_{dt}) - \varphi_d\left\{\sum_{t=1}^{m_d} w_{dt.}(1-\gamma_{dt})\right\}^2\right]\\ &= \frac{1}{\sigma_0^2}\sum_{t=1}^{m_d} w_{dt.}(1-\gamma_{dt})\left\{1-\varphi_d\sum_{t=1}^{m_d} w_{dt.}(1-\gamma_{dt})\right\}\\ &= \frac{\varphi_d}{\sigma_1^2}\sum_{t=1}^{m_d} w_{dt.}(1-\gamma_{dt}). \end{aligned} \tag{12.19}$$

Replacing (12.19) in (12.15), we finally obtain

$$\boldsymbol{L}_{d1} = \sigma_1^2 \varphi_d \sum_{t=1}^{m_d} w_{dt.}(1-\gamma_{dt}) \mathbf{1}_{N_d-n_d} \mathbf{1}'_{N_d-n_d}.$$

The second term on the right hand side of (12.14) is

$$\boldsymbol{L}_{d2} = \sigma_1^2 \mathbf{1}_{N_d-n_d} \mathbf{1}'_{n_d} \boldsymbol{V}_{ds}^{-1} \text{col}' \left\{ \sigma_2^2 \underset{1\le t\le m_d}{\text{diag}} (\mathbf{1}_{n_{dt}} \mathbf{1}'_{N_{dt}-n_{dt}}), \mathbf{0}_{n_d\times(N_d-N_{ds})} \right\}$$

$$= \sigma_1^2\sigma_2^2 \underset{1\le t\le M_d}{\text{diag}} (\mathbf{1}_{N_{dt}-n_{dt}}) \mathbf{1}_{M_d} \mathbf{1}'_{n_d} \boldsymbol{V}_{ds}^{-1} \text{col}' \left\{ \underset{1\le t\le m_d}{\text{diag}} (\mathbf{1}_{n_{dt}}), \mathbf{0}_{n_d\times(M_d-m_d)} \right\}$$

$$\cdot \underset{1\le t\le M_d}{\text{diag}} (\mathbf{1}'_{N_{dt}-n_{dt}}) = \sigma_1^2\sigma_2^2 \underset{1\le t\le M_d}{\text{diag}} (\mathbf{1}_{N_{dt}-n_{dt}}) \mathbf{1}_{M_d} \underset{1\le t\le m_d}{\text{col}'} (\mathbf{1}'_{n_{dt}})$$

$$\cdot \frac{1}{\sigma_0^2} \left[\underset{1\le t\le m_d}{\text{diag}} (\boldsymbol{B}_{dts}) - \varphi_d \underset{1\le t\le m_d}{\text{col}} \left[(1-\gamma_{dt}) \boldsymbol{w}_{n_{dt}}\right] \underset{1\le t\le m_d}{\text{col}'} \left[(1-\gamma_{dt}) \boldsymbol{w}'_{n_{dt}}\right] \right]$$

$$\cdot \text{col}' \left\{ \underset{1\le t\le m_d}{\text{diag}} (\mathbf{1}_{n_{dt}}), \mathbf{0}_{n_d\times(M_d-m_d)} \right\} \underset{1\le t\le M_d}{\text{diag}} (\mathbf{1}'_{N_{dt}-n_{dt}})$$

$$= \frac{\sigma_1^2\sigma_2^2}{\sigma_0^2} \underset{1\le t\le M_d}{\text{diag}} (\mathbf{1}_{N_{dt}-n_{dt}}) \mathbf{1}_{M_d} \left[\text{col}' \left\{ \underset{1\le t\le m_d}{\text{col}'} (\mathbf{1}'_{n_{dt}} \boldsymbol{B}_{dts} \mathbf{1}_{n_{dt}}), \mathbf{0}_{1\times(M_d-m_d)} \right\} \right.$$

$$\left. - \text{col}' \left\{ \varphi_d \Big(\sum_{\ell=1}^{m_d} w_{d\ell.}(1-\gamma_{d\ell})\Big) \underset{1\le t\le m_d}{\text{col}'} (w_{dt.}(1-\gamma_{dt})), \mathbf{0}_{1\times(M_d-m_d)} \right\} \right]$$

$$\cdot \underset{1\le t\le M_d}{\text{diag}} (\mathbf{1}'_{N_{dt}-n_{dt}}).$$

Using (12.18), we obtain

$$\boldsymbol{L}_{d2} = \underset{1\le t\le M_d}{\text{diag}} (\mathbf{1}_{N_{dt}-n_{dt}}) \mathbf{1}_{M_d} \sigma_2^2 \varphi_d \, \text{col}' \left\{ \underset{1\le t\le m_d}{\text{col}'} (w_{dt.}(1-\gamma_{dt})), \mathbf{0}_{1\times(M_d-m_d)} \right\}$$

$$\cdot \underset{1\le t\le M_d}{\text{diag}} (\mathbf{1}'_{N_{dt}-n_{dt}}).$$

The third term satisfies $\boldsymbol{L}_{d3} = \boldsymbol{L}'_{d2}$. The last term on the right hand side of (12.14) is

$$\boldsymbol{L}_{d4} = \sigma_2^4 \text{col} \left\{ \underset{1\le t\le m_d}{\text{diag}} (\mathbf{1}_{N_{dt}-n_{dt}} \mathbf{1}'_{n_{dt}}), \mathbf{0}_{(N_d-N_{ds})\times n_d} \right\} \boldsymbol{V}_{ds}^{-1}$$

$$\cdot \text{col}' \left\{ \underset{1\le t\le m_d}{\text{diag}} (\mathbf{1}_{n_{dt}} \mathbf{1}'_{N_{dt}-n_{dt}}), \mathbf{0}_{n_d\times(N_d-N_{ds})} \right\}$$

$$= \frac{\sigma_2^4}{\sigma_0^2} \operatorname*{diag}_{1\le t\le M_d} (\mathbf{1}_{N_{dt}-n_{dt}}) \operatorname{diag}\Bigg\{ \operatorname*{diag}_{1\le t\le m_d} (\mathbf{1}'_{n_{dt}} \boldsymbol{B}_{dts} \mathbf{1}_{n_{dt}})$$

$$- \varphi_d \operatorname*{col}_{1\le t\le m_d} (w_{dt.}(1-\gamma_{dt})) \operatorname*{col'}_{1\le t\le m_d} (w_{dt.}(1-\gamma_{dt})), \mathbf{0}_{(M_d-m_d)\times(M_d-m_d)} \Bigg\}$$

$$\cdot \operatorname*{diag}_{1\le t\le M_d} (\mathbf{1}'_{N_{dt}-n_{dt}}).$$

Using (12.18) again, we obtain

$$\boldsymbol{L}_{d4} = \frac{\sigma_2^4}{\sigma_0^2} \operatorname*{diag}_{1\le t\le M_d} (\mathbf{1}_{N_{dt}-n_{dt}}) \operatorname{diag}\Bigg\{ \operatorname*{diag}_{1\le t\le m_d} (w_{dt.}(1-\gamma_{dt}))$$

$$- \varphi_d \operatorname*{col}_{1\le t\le m_d} (w_{dt.}(1-\gamma_{dt})) \operatorname*{col'}_{1\le t\le m_d} (w_{dt.}(1-\gamma_{dt})), \mathbf{0}_{(M_d-m_d)\times(M_d-m_d)} \Bigg\}$$

$$\cdot \operatorname*{diag}_{1\le t\le M_d} (\mathbf{1}'_{N_{dt}-n_{dt}}).$$

Summarizing, we have obtained

$$\boldsymbol{L}_{d1} = \sigma_1^2 \varphi_d \sum_{t=1}^{m_d} w_{dt.}(1-\gamma_{dt}) \mathbf{1}_{N_d-n_d} \mathbf{1}'_{N_d-n_d},$$

$$\boldsymbol{L}_{d2} = \operatorname*{diag}_{1\le t\le M_d} (\mathbf{1}_{N_{dt}-n_{dt}}) \boldsymbol{V}_{2d} \operatorname*{diag}_{1\le t\le M_d} (\mathbf{1}'_{N_{dt}-n_{dt}}), \quad \boldsymbol{L}_{d3} = \boldsymbol{L}'_{d2},$$

$$\boldsymbol{L}_{d4} = \operatorname*{diag}_{1\le t\le M_d} (\mathbf{1}_{N_{dt}-n_{dt}}) \boldsymbol{V}_{4d} \operatorname*{diag}_{1\le t\le M_d} (\mathbf{1}'_{N_{dt}-n_{dt}}),$$

where

$$\boldsymbol{V}_{2d} = \mathbf{1}_{M_d} \sigma_2^2 \varphi_d \operatorname{col}' \left\{ \operatorname*{col'}_{1\le t\le m_d} (w_{dt.}(1-\gamma_{dt})), \mathbf{0}_{1\times(M_d-m_d)} \right\},$$

$$\boldsymbol{V}_{4d} = \frac{\sigma_2^4}{\sigma_0^2} \operatorname{diag}\Bigg\{ \operatorname*{diag}_{1\le t\le m_d} (w_{dt.}(1-\gamma_{dt}))$$

$$- \varphi_d \operatorname*{col}_{1\le t\le m_d} (w_{dt.}(1-\gamma_{dt})) \operatorname*{col'}_{1\le t\le m_d} (w_{dt.}(1-\gamma_{dt})), \mathbf{0}_{(M_d-m_d)\times(M_d-m_d)} \Bigg\}.$$

Recalling (12.13), we obtain

$$\boldsymbol{V}_{dr|s} = \boldsymbol{V}_{dr} - \boldsymbol{V}_{drs} \boldsymbol{V}_{ds}^{-1} \boldsymbol{V}_{dsr} \tag{12.20}$$

$$= \sigma_0^2 \boldsymbol{W}_{dr}^{-1} + \sigma_1^2 \left\{ 1 - \varphi_d \sum_{t=1}^{m_d} w_{dt.}(1-\gamma_{dt}) \right\} \mathbf{1}_{N_d-n_d} \mathbf{1}'_{N_d-n_d}$$

$$
\begin{aligned}
&+ \underset{1\le t\le M_d}{\text{diag}}\,(\mathbf{1}_{N_{dt}-n_{dt}})\left(\sigma_2^2 \boldsymbol{I}_{M_d} - \boldsymbol{V}_{2d} - \boldsymbol{V}_{2d}' - \boldsymbol{V}_{4d}\right)\underset{1\le t\le M_d}{\text{diag}}\,(\mathbf{1}_{N_{dt}-n_{dt}}')\\
&= \sigma_0^2 \boldsymbol{W}_{dr}^{-1} + \sigma_0^2 \varphi_d \mathbf{1}_{N_d-n_d}\mathbf{1}_{N_d-n_d}' + \underset{1\le t\le M_d}{\text{diag}}\,(\mathbf{1}_{N_{dt}-n_{dt}})\boldsymbol{T}_{dr|s}\underset{1\le t\le M_d}{\text{diag}}\,(\mathbf{1}_{N_{dt}-n_{dt}}'),
\end{aligned}
$$

where

$$
\boldsymbol{T}_{dr|s} = \sigma_2^2 \boldsymbol{I}_{M_d} - (\boldsymbol{V}_{2d} + \boldsymbol{V}_{2d}') - \boldsymbol{V}_{4d}.
$$

12.3.3 Conditional Variances

For any subdomain $t = 1, \ldots, M_d$ and unit $j = n_{dt} + 1, \ldots, N_{dt}$, let $v_{dtj|s}$ be the corresponding diagonal element in the matrix $\boldsymbol{V}_{dr|s}$. The diagonal elements can be written in the form

$$
v_{dtj|s} = \boldsymbol{a}_{dtj}' \boldsymbol{V}_{dr|s} \boldsymbol{a}_{dtj}, \quad \text{where } \boldsymbol{a}_{dtj}' = \underset{1\le \ell\le M_d}{\text{col}'}\,(\delta_{t\ell} \underset{n_{d\ell}+1\le i\le N_{d\ell}}{\text{col}'}\,(\delta_{ij}))
$$

and δ_{ij} is the delta of Kronecker, i.e. $\delta_{ij} = 1$ if $i = j$ and $\delta_{ij} = 0$ if $i \neq j$. Thus, $\boldsymbol{a}_{dtj}'$ is a $1 \times (N_d - n_d)$ vector with a 1 in the position (t, j) and with 0's in the remaining positions. Replacing the expression of $\boldsymbol{V}_{dr|s}$ given in (12.20), we have

$$
\begin{aligned}
v_{dtj|s} = \boldsymbol{a}_{dtj}' \Big[&\sigma_0^2 \boldsymbol{W}_{dr}^{-1} + \sigma_0^2 \varphi_d \mathbf{1}_{N_d-n_d}\mathbf{1}_{N_d-n_d}'\\
&+ \underset{1\le \ell\le M_d}{\text{diag}}\,(\mathbf{1}_{N_{d\ell}-n_{d\ell}})\boldsymbol{T}_{dr|s}\underset{1\le \ell\le M_d}{\text{diag}}\,(\mathbf{1}_{N_{d\ell}-n_{d\ell}}')\Big]\boldsymbol{a}_{dtj}.
\end{aligned}
$$

By defining $\boldsymbol{a}_{dt}' = \underset{1\le \ell\le M_d}{\text{col}'}\,(\delta_{t\ell})$, it holds

$$
\begin{aligned}
&\boldsymbol{a}_{dtj}' \boldsymbol{W}_{dr}^{-1} \boldsymbol{a}_{dtj} = w_{dtj}^{-1}, \quad \boldsymbol{a}_{dtj}' \mathbf{1}_{N_d-n_d}\mathbf{1}_{N_d-n_d}' \boldsymbol{a}_{dtj} = 1,\\
&\boldsymbol{a}_{dtj}' \underset{1\le \ell\le M_d}{\text{diag}}\,(\mathbf{1}_{N_{d\ell}-n_{d\ell}})\boldsymbol{T}_{dr|s}\underset{1\le \ell\le M_d}{\text{diag}}\,(\mathbf{1}_{N_{d\ell}-n_{d\ell}}')\boldsymbol{a}_{dtj} = \boldsymbol{a}_{dt}' \boldsymbol{T}_{dr|s} \boldsymbol{a}_{dt}.
\end{aligned}
$$

Consequently, we get

$$
v_{dtj|s} = \sigma_0^2 (w_{dtj}^{-1} + \varphi_d) + \boldsymbol{a}_{dt}' \boldsymbol{T}_{dr|s} \boldsymbol{a}_{dt}. \tag{12.21}
$$

Further, we have

$$
\begin{aligned}
\boldsymbol{a}_{dt}' \boldsymbol{T}_{dr|s} \boldsymbol{a}_{dt} &= \boldsymbol{a}_{dt}' \left(\sigma_2^2 \boldsymbol{I}_{M_d} - \boldsymbol{V}_{2d} - \boldsymbol{V}_{2d}' - \boldsymbol{V}_{4d}\right) \boldsymbol{a}_{dt}\\
&= \sigma_2^2\, \boldsymbol{a}_{dt}' \boldsymbol{I}_{M_d} \boldsymbol{a}_{dt} - 2\boldsymbol{a}_{dt}' \boldsymbol{V}_{2d} \boldsymbol{a}_{dt} - \boldsymbol{a}_{dt}' \boldsymbol{V}_{4d} \boldsymbol{a}_{dt}.
\end{aligned} \tag{12.22}
$$

Moreover, for any $M_d \times M_d$ matrix $\boldsymbol{A}$, the product $\boldsymbol{a}'_{dt}\boldsymbol{A}\boldsymbol{a}_{dt}$ gives the t-th diagonal element of $\boldsymbol{A}$. Then $\boldsymbol{a}'_{dt}\boldsymbol{I}_{M_d}\boldsymbol{a}_{dt} = 1$ for any t and

$$\boldsymbol{a}'_{dt}\boldsymbol{V}_{2d}\boldsymbol{a}_{dt} = \boldsymbol{a}'_{dt}\boldsymbol{V}'_{2d}\boldsymbol{a}_{dt} = \boldsymbol{a}'_{dt}\boldsymbol{V}'_{4d}\boldsymbol{a}_{dt} = 0 \quad \text{for} \quad m_d + 1 \le t \le M_d.$$

On the other hand, by defining $b_{dt} = w_{dt.}(1-\gamma_{dt})$, we get for $1 \le t \le m_d$

$$\begin{aligned}
\boldsymbol{a}'_{dt}\boldsymbol{V}_{2d}\boldsymbol{a}_{dt} &= \sigma_2^2\,\varphi_d\,\boldsymbol{a}'_{dt}\,\boldsymbol{1}_{M_d}\,\text{col}'\left\{\underset{1\le t\le m_d}{\text{col}'}(w_{dt.}(1-\gamma_{dt})),\ \boldsymbol{0}_{1\times(M_d-m_d)}\right\}\boldsymbol{a}_{dt}\\
&= \sigma_2^2\,\varphi_d\,b_{dt}, \qquad (12.23)\\
\boldsymbol{a}'_{dt}\boldsymbol{V}_{4d}\boldsymbol{a}_{dt} &= \frac{\sigma_2^4}{\sigma_0^2}\,\boldsymbol{a}'_{dt}\text{diag}\Big\{\underset{1\le t\le m_d}{\text{diag}}(w_{dt.}(1-\gamma_{dt})) - \varphi_d\underset{1\le t\le m_d}{\text{col}}(w_{dt.}(1-\gamma_{dt}))\\
&\quad\cdot \underset{1\le t\le m_d}{\text{col}'}(w_{dt.}(1-\gamma_{dt})),\boldsymbol{0}_{(M_d-m_d)\times(M_d-m_d)}\Big\}\boldsymbol{a}_{dt}\\
&= \frac{\sigma_2^4}{\sigma_0^2}\left[w_{dt.}(1-\gamma_{dt}) - \varphi_d\,w_{dt.}^2(1-\gamma_{dt})^2\right]\\
&= \frac{\sigma_2^4}{\sigma_0^2}\,b_{dt}(1-\varphi_d\,b_{dt}). \qquad (12.24)
\end{aligned}$$

Substituting (12.23) and (12.24) in (12.22), and recalling that $b_{dt} = w_{dt.}(1-\gamma_{dt}) = \gamma_{dt}\frac{\sigma_0^2}{\sigma_2^2}$, we obtain

$$\begin{aligned}
\boldsymbol{a}'_{dt}\boldsymbol{T}_{dr|s}\boldsymbol{a}_{dt} &= \sigma_2^2\left\{1 - 2\varphi_d\,b_{dt} - \frac{\sigma_2^2}{\sigma_0^2}b_{dt}(1-\varphi_d\,b_{dt})\right\}\\
&= \sigma_2^2\left\{1 - 2\varphi_d\gamma_{dt}\frac{\sigma_0^2}{\sigma_2^2} - \gamma_{dt}\left(1-\varphi_d\,\gamma_{dt}\frac{\sigma_0^2}{\sigma_2^2}\right)\right\}\\
&= \sigma_2^2 - 2\varphi_d\gamma_{dt}\sigma_0^2 - \gamma_{dt}\sigma_2^2 + \gamma_{dt}^2\varphi_d\sigma_0^2\\
&= \sigma_0^2\varphi_d\gamma_{dt}(\gamma_{dt}-2) + \sigma_2^2(1-\gamma_{dt}).
\end{aligned}$$

Replacing this expression in (12.21), we finally obtain for $1 \le t \le m_d$ and $n_{dt}+1 \le j \le N_{dt}$ that

$$\begin{aligned}
v_{dtj|s} &= \sigma_0^2(w_{dtj}^{-1} + \varphi_d) + \sigma_0^2\varphi_d\gamma_{dt}(\gamma_{dt}-2) + \sigma_2^2(1-\gamma_{dt})\\
&= \sigma_0^2\left[w_{dtj}^{-1} + \varphi_d\left\{1+\gamma_{dt}(\gamma_{dt}-2)\right\}\right] + \sigma_2^2(1-\gamma_{dt}).
\end{aligned}$$

For $m_d + 1 \leq t \leq M_d$, we have

$$v_{dtj|s} = \sigma_0^2(w_{dtj}^{-1} + \varphi_d) + \sigma_2^2, \quad j = 1, \ldots, N_{dt}.$$

If $m_d = 0$ or equivalently if $n_d = 0$, the conditional variance is the marginal variance, i.e.

$$v_{dtj|s} = v_{dtj} = w_{dtj}^{-1}\sigma_0^2 + \sigma_1^2 + \sigma_2^2, \quad t = 1, \ldots, M_d, \; j = 1, \ldots, N_{dt}.$$

12.4 Monte Carlo EBP of an Additive Parameter

12.4.1 Introduction

The target of this section is to estimate a small area additive parameter of the form

$$\delta_d = \frac{1}{N_d}\sum_{t=1}^{M_d}\sum_{j=1}^{N_{dt}} h(y_{dtj}),$$

where h is a known measurable function. The best predictor of δ_d is

$$\begin{aligned}\hat{\delta}_d^{bp} &= E_{\boldsymbol{y}_r}\Big[\frac{1}{N_d}\sum_{t=1}^{M_d}\sum_{j=1}^{N_{dt}} h(y_{dtj})|\boldsymbol{y}_s\Big] \\ &= \frac{1}{N_d}\sum_{t=1}^{M_d}\Big\{\sum_{j\in s_{dt}} h(y_{dtj}) + \sum_{j\in U_{dt}-s_{dt}} E_{\boldsymbol{y}_r}\big[h(y_{dtj})|\boldsymbol{y}_s\big]\Big\}.\end{aligned}$$

The conditional distribution of $\boldsymbol{y}_r|\boldsymbol{y}_s$ depends on the vector $\boldsymbol{\theta} = (\boldsymbol{\beta}', \sigma_0^2, \sigma_1^2, \sigma_2^2)'$ of unknown model parameters, which must be estimated, that is,

$$E_{\boldsymbol{y}_r}\big[h(y_{dtj})|\boldsymbol{y}_s\big] = E_{\boldsymbol{y}_r}\big[h(y_{dtj})|\boldsymbol{y}_s; \boldsymbol{\theta}\big].$$

Let $\hat{\boldsymbol{\theta}} = (\hat{\boldsymbol{\beta}}', \hat{\sigma}_0^2, \hat{\sigma}_1^2, \hat{\sigma}_2^2)'$ be an estimator based on sample data $\boldsymbol{y}_s$. The result of replacing $\boldsymbol{\theta}$ by $\hat{\boldsymbol{\theta}}$ in the formula of the best predictor is called empirical best predictor, that is,

$$\hat{\delta}_d^{ebp} = \frac{1}{N_d}\sum_{t=1}^{M_d}\Big\{\sum_{j\in s_{dt}} h(y_{dtj}) + \sum_{j\in U_{dt}-s_{dt}} E_{\boldsymbol{y}_r}\big[h(y_{dtj})|\boldsymbol{y}_s; \hat{\boldsymbol{\theta}}\big]\Big\}.$$

For a general function $h(\cdot)$, the expected value above might be not tractable analytically. When this occurs, the following Monte Carlo procedure can be applied.

(a) Estimate the unknown parameter $\boldsymbol{\theta} = (\boldsymbol{\beta}', \sigma_0^2, \sigma_1^2, \sigma_2^2)'$ using sample data $\boldsymbol{y}_s$.
(b) Replacing $\boldsymbol{\theta} = (\boldsymbol{\beta}', \sigma_0^2, \sigma_1^2, \sigma_2^2)'$ by the estimate $\hat{\boldsymbol{\theta}} = (\hat{\boldsymbol{\beta}}', \hat\sigma_0^2, \hat\sigma_1^2, \hat\sigma_2^2)'$, obtained in (a), draw L copies of each non-sample variable y_{dtj}. That means, for $d = 1,\dots,D$, $t = 1,\dots,M_d$, $j \in U_{dt} - s_{dt}$, $\ell = 1,\dots,L$, generate $y_{dtj}^{(\ell)} \sim N(\hat\mu_{dtj|s}, \hat v_{dtj|s})$ with $\hat\mu_{dtj|s}$ and $\hat v_{dtj|s}$ obtained by replacing $\boldsymbol{\theta}$ by $\hat{\boldsymbol{\theta}}$ in the formulas for $\mu_{dtj|s}$ and $v_{dtj|s}$, where

$$\mu_{dtj|s}=\boldsymbol{x}_{dtj}\boldsymbol{\beta}+\gamma_{dt}\left\{\bar y_{dts}-\bar{\boldsymbol{x}}_{dts}\boldsymbol{\beta}+\frac{\sigma_0^4}{\sigma_2^4}\frac{\varphi_d}{w_{dt.}}\sum_{\ell=1}^{m_d}\gamma_{d\ell}(\bar y_{d\ell s}-\bar{\boldsymbol{x}}_{d\ell s}\boldsymbol{\beta})\right\},\ 1\le t\le m_d,$$

$$\mu_{dtj|s} = \boldsymbol{x}_{dtj}\boldsymbol{\beta} + \frac{\sigma_0^2}{\sigma_2^2}\varphi_d\sum_{\ell=1}^{m_d}\gamma_{d\ell}\,(\bar y_{d\ell s} - \bar{\boldsymbol{x}}_{d\ell s}\boldsymbol{\beta}), \quad \text{if } m_d + 1 \le t \le M_d,$$

$$\mu_{dtj|s} = \boldsymbol{x}_{dtj}\boldsymbol{\beta}, \quad \text{if } m_d = 0,$$

$$v_{dtj|s} = \begin{cases} \sigma_0^2\left[w_{dtj}^{-1} + \varphi_d\{1+\gamma_{dt}(\gamma_{dt}-2)\}\right] + \sigma_2^2(1-\gamma_{dt}), & 1 \le t \le m_d, \\ \sigma_0^2(w_{dtj}^{-1} + \varphi_d) + \sigma_2^2, & m_d + 1 \le t \le M_d, \\ w_{dtj}^{-1}\sigma_0^2 + \sigma_1^2 + \sigma_2^2, & m_d = 0, \end{cases}$$

$\bar y_{d\ell s} = w_{d\ell.}^{-1}\sum_{j\in s_{d\ell}} w_{d\ell j}y_{d\ell j}$, $\bar{\boldsymbol{x}}_{d\ell s} = w_{d\ell.}^{-1}\sum_{j\in s_{d\ell}} w_{d\ell j}\boldsymbol{x}_{d\ell j}$ and

$$\varphi_d = \frac{\sigma_1^2}{\sigma_0^2 + \sigma_1^2\sum_{\ell=1}^{m_d}(1-\gamma_{d\ell})w_{d\ell.}}, \qquad \gamma_{dt} = \frac{\sigma_2^2}{\sigma_2^2 + \sigma_0^2/w_{dt.}}.$$

(c) For $d = 1,\dots,D$, $t = 1,\dots,M_d$, $j \in U_{dt} - s_{dt}$, the Monte Carlo approximation of the expected value is

$$E_{\boldsymbol{y}_r}\left[h(y_{dtj})|\boldsymbol{y}_s;\hat{\boldsymbol{\theta}}\right] \approx \frac{1}{L}\sum_{\ell=1}^{L} h(y_{dtj}^{(\ell)})$$

and the Monte Carlo approximation of the EBP of the additive domain parameter δ_d is

$$\hat\delta_d^{ebp} \approx \frac{1}{N_d}\sum_{t=1}^{M_d}\Big\{\sum_{j\in s_{dt}} h(y_{dtj}) + \sum_{j\in U_{dt}-s_{dt}}\Big(\frac{1}{L}\sum_{\ell=1}^{L} h(y_{dtj}^{(\ell)})\Big)\Big\} = \frac{1}{L}\sum_{\ell=1}^{L}\delta_d^{(\ell)},$$

where

$$\delta_d^{(\ell)} = \frac{1}{N_d}\sum_{t=1}^{M_d}\Big\{\sum_{j\in s_{dt}} h(y_{dtj}) + \sum_{j\in U_{dt}-s_{dt}} h(y_{dtj}^{(\ell)})\Big\}.$$

If we were interested in estimating the subdomain additive parameter

$$\delta_{dt} = \frac{1}{N_{dt}}\sum_{j=1}^{N_{dt}} h(y_{dtj}),$$

the corresponding Monte Carlo approximation of the EBP is

$$\hat{\delta}_{dt}^{ebp} \approx \frac{1}{N_{dt}}\Big\{\sum_{j\in s_{dt}} h(y_{dtj}) + \sum_{j\in U_{dt}-s_{dt}}\Big(\frac{1}{L}\sum_{\ell=1}^{L} h(y_{dtj}^{(\ell)})\Big)\Big\} = \frac{1}{L}\sum_{\ell=1}^{L}\delta_{dt}^{(\ell)},$$

where

$$\delta_{dt}^{(\ell)} = \frac{1}{N_{dt}}\Big\{\sum_{j\in s_{dt}} h(y_{dtj}) + \sum_{j\in U_{dt}-s_{dt}} h(y_{dtj}^{(\ell)})\Big\}.$$

Remark 12.1 If the selected two-fold NER model contains continuous auxiliary variables and there is no available census file, it is possible to use a design-based approximation to $\delta_{dt}^{(\ell)}$ when $n_{dt} > 0$. This approximation is

$$\delta_{dt}^{(\ell)} = \frac{1}{N_{dt}}\Big\{\sum_{j\in s_{dt}} \big(h(y_{dtj}) - h(y_{dtj}^{(\ell)})\big) + \sum_{j\in s_{dt}} \omega_{dtj} h(y_{dtj}^{(\ell)})\Big\}, \tag{12.25}$$

where ω_{dtj}'s are the calibrated sample weights and

$$y_{dtj}^{(\ell)} \sim N(\hat{\mu}_{dtj|s}, \hat{v}_{dtj|s}), \quad j \in s_{dt},\ t = 1,\dots,m_d,\ d = 1,\dots,D,\ \ell = 1,\dots,L,$$

with $\hat{\mu}_{dtj|s}$ and $\hat{v}_{dtj|s}$ obtained by replacing $\boldsymbol{\theta}$ by $\hat{\boldsymbol{\theta}}$ in the formulas for $\mu_{dtj|s}$ and $v_{dtj|s}$ given above. Let us note that the second sum in the formula (12.25) stands in fact for summation over the whole population, this is why the sum of $h(y_{dj}^{(\ell)})$ over the sample is subtracted. For more details concerning use of the described approximation we refer to Remark 10.2.

12.4.2 Auxiliary Variables with Finite Number of Values

In many practical cases the values of the auxiliary variables are not available for all the population units. If in addition some of the variables are continuous, the EBP method is not applicable. An important particular case, where this method is applicable, is when the model is homoscedastic and the number of values of the vector of auxiliary variables is finite. More concretely, suppose that the covariates are categorical such that $\boldsymbol{x}_{dtj} \in \{\boldsymbol{x}_1, \ldots, \boldsymbol{x}_K\}$ and that $w_{dtj} = 1$ for all d, t, j. Then, we can calculate $\delta_d^{(\ell)}$ as

$$\delta_d^{(\ell)} = \frac{1}{N_d}\Big\{\sum_{t=1}^{m_d}\sum_{k=1}^{K}\sum_{j=1}^{n_{dtk}} h(y_{dtkj}) + \sum_{t=1}^{M_d}\sum_{k=1}^{K}\sum_{j=n_{dtk}+1}^{N_{dtk}} h(y_{dtkj}^{(\ell)})\Big\}, \tag{12.26}$$

where $N_{dtk} = \#\{j \in U_{dt} : \boldsymbol{x}_{dtj} = \boldsymbol{x}_k\}$ is supposed to be available from external data sources (aggregated auxiliary information), $n_{dtk} = \#\{j \in s_{dt} : \boldsymbol{x}_{dtj} = \boldsymbol{x}_k\}$, $y_{dtkj}^{(\ell)} \sim N(\hat{\mu}_{dtk|s}, \hat{v}_{dt|s})$, $k = 1, \ldots, K$, $t = 1, \ldots, M_d$, $d = 1, \ldots, D$, $\ell = 1, \ldots, L$. Further,

$$\hat{\mu}_{dtk|s} = \begin{cases} \boldsymbol{x}_k\hat{\boldsymbol{\beta}} + \hat{\gamma}_{dt}\big\{\bar{y}_{dts} - \bar{\boldsymbol{x}}_{dts}\hat{\boldsymbol{\beta}} + \frac{\hat{\sigma}_0^4}{\hat{\sigma}_2^4}\frac{\hat{\varphi}_d}{n_{dt}}\sum_{\ell=1}^{m_d}\hat{\gamma}_{d\ell}(\bar{y}_{d\ell s} - \bar{\boldsymbol{x}}_{d\ell s}\hat{\boldsymbol{\beta}})\big\}, & 1 \le t \le m_d, \\ \boldsymbol{x}_k\hat{\boldsymbol{\beta}} + \frac{\hat{\sigma}_0^2}{\hat{\sigma}_2^2}\hat{\varphi}_d\sum_{\ell=1}^{m_d}\hat{\gamma}_{d\ell}(\bar{y}_{d\ell s} - \bar{\boldsymbol{x}}_{d\ell s}\hat{\boldsymbol{\beta}}), & m_d + 1 \le t \le M_d, \\ \boldsymbol{x}_k\hat{\boldsymbol{\beta}}, \quad m_d = 0, & \end{cases}$$

$$\hat{v}_{dt|s} = \begin{cases} \hat{\sigma}_0^2\big[1 + \hat{\varphi}_d\{1 + \hat{\gamma}_{dt}(\hat{\gamma}_{dt} - 2)\}\big] + \hat{\sigma}_2^2(1 - \hat{\gamma}_{dt}), & 1 \le t \le m_d, \\ \hat{\sigma}_0^2(1 + \hat{\varphi}_d) + \hat{\sigma}_2^2, & m_d + 1 \le t \le M_d, \\ \hat{\sigma}_0^2 + \hat{\sigma}_1^2 + \hat{\sigma}_2^2, & m_d = 0, \end{cases}$$

where

$$\bar{y}_{dts} = \frac{1}{n_{dt}}\sum_{k=1}^{K}\sum_{j=1}^{n_{dtk}} y_{dtkj}, \qquad \bar{\boldsymbol{x}}_{dts} = \frac{1}{n_{dt}}\sum_{k=1}^{K} n_{dtk}\boldsymbol{x}_k$$

and

$$\hat{\varphi}_d = \frac{\hat{\sigma}_1^2}{\hat{\sigma}_0^2 + \hat{\sigma}_1^2\sum_{\ell=1}^{m_d}(1 - \hat{\gamma}_{d\ell})n_{d\ell}}, \qquad \hat{\gamma}_{dt} = \frac{n_{dt}\hat{\sigma}_2^2}{n_{dt}\hat{\sigma}_2^2 + \hat{\sigma}_0^2}.$$

Similarly, we can calculate $\delta_{dt}^{(\ell)}$ as

$$\delta_{dt}^{(\ell)} = \frac{1}{N_{dt}}\Big\{\sum_{k=1}^{K}\sum_{j=1}^{n_{dtk}} h(y_{dtkj}) + \sum_{k=1}^{K}\sum_{j=n_{dtk}+1}^{N_{dtk}} h(y_{dtkj}^{(\ell)})\Big\}. \tag{12.27}$$

12.5 EBPs of Poverty Indicators

Let z_{dtj} be a welfare variable (e.g. income or expenditure) for individual j from subdomain t within domain d and let z be the poverty line. For a given power $\alpha \geq 0$, the FGT poverty indicator of order α (Foster et al. 1984) for subdomain t within domain d is defined as

$$F_{\alpha,dt} = \frac{1}{N_{dt}} \sum_{j=1}^{N_{dt}} F_{\alpha,dtj}, \quad F_{\alpha,dtj} = \left(\frac{z - z_{dtj}}{z}\right)^{\alpha} I(z_{dtj} < z), \tag{12.28}$$

where $I(z_{dtj} < z) = 1$ if $z_{dtj} < z$ and $I(z_{dtj} < z) = 0$ otherwise. For $\alpha = 0$, we obtain the subdomain poverty proportion, which measures the proportion of people in the subdomain whose welfare is below the poverty line z. For $\alpha = 1$, we obtain the subdomain poverty gap, measuring the degree of poverty of the people in that subdomain. These indicators are defined analogously for domains.

For estimating these indicators in domains or subdomains, we assume that a one-to-one transformation of the welfare variable for each unit, $y_{dtj} = T(z_{dtj})$, follows the two-fold nested error model (12.1). Then, using the inverse transformation $z_{dtj} = T^{-1}(y_{dtj})$, we can express $F_{\alpha,dtj}$ in terms of the model response variables y_{dtj} as

$$F_{\alpha,dtj} = \left(\frac{z - T^{-1}(y_{dtj})}{z}\right)^{\alpha} I(T^{-1}(y_{dtj}) < z) = h_{\alpha}(y_{dtj}), \tag{12.29}$$

which means that $F_{\alpha,dt}$ is an additive parameter. Therefore, the best predictor of $F_{\alpha,dt}$ is

$$\hat{F}_{\alpha,dt}^{bp} = \frac{1}{N_{dt}} \left\{ \sum_{j \in s_{dt}} F_{\alpha,dtj} + \sum_{j \in U_{dt} - s_{dt}} \hat{F}_{\alpha,dtj}^{bp} \right\},$$

where $\hat{F}_{\alpha,dtj}^{bp} = E_{\mathbf{y}_r}\left[F_{\alpha,dtj} | \mathbf{y}_s, \boldsymbol{\theta}\right]$. For $\alpha = 0, 1$ and for certain transformations T, the expectation $\hat{F}_{\alpha,dtj}^{bp}$ can be calculated analytically, avoiding Monte Carlo simulation.

12.5.1 Poverty Proportion

For the poverty proportion, $\alpha = 0$, we have $h_0(y_{dtj}) = I(T^{-1}(y_{dtj}) < z) = I\big(y_{dtj} < T(z)\big)$. If T is a nondecreasing monotonous function, we obtain

$$\begin{aligned} \hat{F}_{0,dtj}^{bp} &= E_{\mathbf{y}_r}\left[F_{0,dtj} | \mathbf{y}_s; \boldsymbol{\theta}\right] = E_{\mathbf{y}_r}\left[h_0(y_{dtj}) | \mathbf{y}_s; \boldsymbol{\theta}\right] = E_{\mathbf{y}_r}\left[I\big(y_{dtj} < T(z)\big) | \mathbf{y}_s; \boldsymbol{\theta}\right] \\ &= P_{\mathbf{y}_r}\big(y_{dtj} < T(z) | \mathbf{y}_s; \boldsymbol{\theta}\big) = P\big(N(0,1) < \alpha_{dtj}\big) = \Phi(\alpha_{dtj}), \end{aligned}$$

where

$$\alpha_{dtj} = \frac{T(z) - \mu_{dtj|s}}{v_{dtj|s}^{1/2}}$$

and Φ denotes the cumulative distribution function of a standard normal random variable.

12.5.2 Poverty Gap

For the poverty gap, $\alpha = 1$, we have $h_1(y_{dtj}) = \frac{z - T^{-1}(y_{dtj})}{z} I(T^{-1}(y_{dtj}) < z)$. In this section we assume that $y = T(z) = \log(z + c)$ or $z = T^{-1}(y) = e^y - c$. First, we obtain

$$\begin{aligned}
\hat{F}_{1,dtj}^{bp} &= E_{y_r}\left[F_{1,dtj}|\boldsymbol{y}_s;\boldsymbol{\theta}\right] = E_{y_r}\left[h_1(y_{dtj})|\boldsymbol{y}_s;\boldsymbol{\theta}\right] \\
&= E_{y_r}\left[\frac{z - T^{-1}(y_{dtj})}{z} I\big(y_{dtj} < T(z)\big)\Big|\boldsymbol{y}_s;\boldsymbol{\theta}\right] \\
&= E_{y_r}\left[I\big(y_{dtj} < T(z)\big)\Big|\boldsymbol{y}_s;\boldsymbol{\theta}\right] - \frac{1}{z} E_{y_r}\left[T^{-1}(y_{dtj}) I\big(y_{dtj} < T(z)\big)\Big|\boldsymbol{y}_s;\boldsymbol{\theta}\right] \\
&= S_1 - \frac{1}{z} S_2.
\end{aligned}$$

We have proved that the first summand is

$$S_1 = E_{y_r}\left[I\big(y_{dtj} < T(z)\big)\Big|\boldsymbol{y}_s;\boldsymbol{\theta}\right] = \Phi(\alpha_{dtj}).$$

For calculating S_2, we simplify the notation, i.e. $y_{dtj} = y$, $\mu_{dtj|s} = \mu$, $v_{dtj|s}^{1/2} = \sigma$, and $\alpha_{dtj} = \alpha$ and in the integrals below we apply the following changes of variables:

$$x = \frac{y - \mu}{\sigma}, \quad y = \sigma x + \mu, \quad dy = \sigma dx, \quad y = T(z) \Leftrightarrow x = \frac{T(z) - \mu}{\sigma} = \alpha,$$

$$u = x - \sigma, \quad x = u + \sigma, \quad dx = du, \quad x = \alpha \Leftrightarrow u = \alpha - \sigma.$$

It holds that

$$\begin{aligned}
S_2 &= E_{y_r}\left[T^{-1}(y_{dtj}) I\big(y_{dtj} < T(z)\big)\Big|\boldsymbol{y}_s;\boldsymbol{\theta}\right] = \int_{-\infty}^{T(z)} T^{-1}(y) f_{N(\mu,\sigma^2)}(y)\, dy \\
&= \int_{-\infty}^{\alpha} T^{-1}(\sigma x + \mu) f_{N(\mu,\sigma^2)}(\sigma x + \mu)\, \sigma dx
\end{aligned}$$

$$
\begin{aligned}
&= \int_{-\infty}^{\alpha} \left(e^{\mu} \exp\{\sigma x\} - c\right) \frac{1}{\sqrt{2\pi}\,\sigma} \exp\left\{-\frac{1}{2}x^2\right\} \sigma dx \\
&= -c\Phi(\alpha) + e^{\mu} \int_{-\infty}^{\alpha} \frac{1}{\sqrt{2\pi}} \exp\left\{-\frac{1}{2}(x^2 - 2\sigma x + \sigma^2)\right\} \exp\left\{\frac{1}{2}\sigma^2\right\} dx \\
&= -c\Phi(\alpha) + \exp\left\{\frac{1}{2}\sigma^2 + \mu\right\} \int_{-\infty}^{\alpha} f_{N(\sigma,1)}(x)\, dx \\
&= -c\Phi(\alpha) + \exp\left\{\frac{1}{2}\sigma^2 + \mu\right\} \int_{-\infty}^{\alpha-\sigma} f_{N(0,1)}(u)\, du.
\end{aligned}
$$

Therefore

$$
S_2 = -c\Phi(\alpha) + \exp\left\{\frac{1}{2}\sigma^2 + \mu\right\}\Phi(\alpha - \sigma)
$$

and

$$
\begin{aligned}
\hat{F}_{1,dtj}^{bp} &= \Phi(\alpha_{dtj}) - \frac{1}{z}\left(\exp\left\{\frac{1}{2}v_{dtj|s} + \mu_{dtj|s}\right\}\Phi(\alpha_{dtj} - v_{dtj|s}^{1/2}) - c\Phi(\alpha_{dtj})\right) \\
&= \frac{z+c}{z}\Phi(\alpha_{dtj}) - \frac{1}{z}\exp\left\{\frac{1}{2}v_{dtj|s} + \mu_{dtj|s}\right\}\Phi\left(\alpha_{dtj} - v_{dtj|s}^{1/2}\right).
\end{aligned}
$$

12.6 EBPs of Average Income Indicators

Let z_{dtj} be a welfare variable (e.g. income or expenditure) for individual j from subdomain t within domain d. The average income of subdomain t within domain d is

$$
\bar{z}_{dt} = \frac{1}{N_{dt}} \sum_{j=1}^{N_{dt}} z_{dtj}.
$$

This indicator is defined analogously for domains. For estimating the average incomes in domains or subdomains, we assume that a one-to-one transformation of the welfare variable for each unit, $y_{dtj} = T(z_{dtj})$, follows the two-fold nested error model (12.1). Then, using the inverse transformation we can express z_{dtj} in terms of the model response variables y_{dtj} as

$$
z_{dtj} = T^{-1}(y_{dtj}) = h(y_{dtj}),
$$

which means that $\overline{z}_{dt}$ is an additive parameter. Therefore, the best predictor of $\overline{z}_{dt}$ is

$$\hat{\overline{z}}_{dt}^{bp} = \frac{1}{N_{dt}} \left(\sum_{j \in s_{dt}} z_{dtj} + \sum_{j \in U_{dt} - s_{dt}} \hat{z}_{dtj}^{bp} \right),$$

where

$$\hat{z}_{dtj}^{bp} = E_{y_r}\left[z_{dtj} | \boldsymbol{y}_s, \boldsymbol{\theta}\right] = E_{y_r}\left[h(y_{dtj}) | \boldsymbol{y}_s; \boldsymbol{\theta}\right] = E_{y_r}[T^{-1}(y_{dtj}) | \boldsymbol{y}_s; \boldsymbol{\theta}].$$

Let us now assume that $y = T(z) = \log(z + c)$ or $z = T^{-1}(y) = e^y - c$. For calculating $\hat{z}_{dtj}^{bp}$, we simplify the notation, i.e. $y_{dtj} = y$, $\mu_{dtj|s} = \mu$, $v_{dtj|s}^{1/2} = \sigma$ and in the integral below we do the following change of variables:

$$x = \frac{y - \mu}{\sigma}, \quad y = \sigma x + \mu, \quad dy = \sigma dx, \quad y = T(z) \Leftrightarrow x = \frac{T(z) - \mu}{\sigma} = \alpha.$$

It holds that

$$\begin{aligned}
\hat{z}_{dtj}^{bp} &= \int_{-\infty}^{\infty} T^{-1}(y) f_{N(\mu,\sigma^2)}(y)\, dy = \int_{-\infty}^{\infty} T^{-1}(\sigma x + \mu) f_{N(\mu,\sigma^2)}(\sigma x + \mu)\, \sigma dx \\
&= \int_{-\infty}^{\infty} \left(e^{\mu} \exp\{\sigma x\} - c\right) \frac{1}{\sqrt{2\pi}\,\sigma} \exp\left\{ -\frac{1}{2} x^2 \right\} \sigma dx \\
&= -c + e^{\mu} \int_{-\infty}^{\infty} \frac{1}{\sqrt{2\pi}} \exp\left\{ -\frac{1}{2}(x^2 - 2\sigma x + \sigma^2) \right\} \exp\left\{ \frac{1}{2}\sigma^2 \right\} dx \\
&= -c + \exp\left\{ \frac{1}{2}\sigma^2 + \mu \right\} \int_{-\infty}^{\infty} f_{N(\sigma,1)}(x)\, dx = -c + \exp\left\{ \frac{1}{2}\sigma^2 + \mu \right\}.
\end{aligned}$$

Therefore

$$\hat{z}_{dtj}^{bp} = -c + \exp\left\{ \frac{1}{2} v_{dtj|s} + \mu_{dtj|s} \right\}.$$

12.7 Parametric Bootstrap MSE Estimator

Analytical approximations to the MSE of empirical best predictors are difficult to derive in the case of complex parameters such as the FGT poverty measures. We therefore present a parametric bootstrap MSE estimator by following the bootstrap method for finite populations of González-Manteiga et al. (2008a). This bootstrap method can be readily applied to other complex parameters not necessarily of the additive form as the FGT measures. Steps for implementing this method are given now.

1. Fit the model (12.1) to sample data $(\boldsymbol{y}_s, \boldsymbol{X}_s)$ and obtain an estimator $\hat{\boldsymbol{\theta}} = (\hat{\boldsymbol{\beta}}', \hat{\sigma}_0^2, \hat{\sigma}_1^2, \hat{\sigma}_2^2)'$ of $\boldsymbol{\theta} = (\boldsymbol{\beta}', \sigma_0^2, \sigma_1^2, \sigma_2^2)'$.
2. Repeat B times $(b = 1, \dots, B)$:

   (a) For $d = 1, \dots, D$, $t = 1, \dots, M_d$, $j = 1, \dots, N_{dt}$, generate independently $u_{1,d}^* \sim N(0, \hat{\sigma}_1^2)$, $u_{2,dt}^* \sim N(0, \hat{\sigma}_2^2)$ and $e_{dtj}^* \sim N(0, \hat{\sigma}_0^2)$.
   (b) For $d = 1, \dots, D$, $t = 1, \dots, M_d$, $j = 1, \dots, N_{dt}$, generate independently the bootstrap population

$$y_{dtj}^{*(b)} = \boldsymbol{x}_{dtj}\hat{\boldsymbol{\beta}} + u_{1,d}^* + u_{2,dt}^* + w_{dtj}^{-1/2} e_{dtj}^*$$

   and calculate the bootstrap population parameters

$$\delta_d^{*(b)} = \frac{1}{N_d} \sum_{t=1}^{M_d} \sum_{j=1}^{N_{dt}} h(y_{dtj}^{*(b)}), \quad \delta_{dt}^{*(b)} = \frac{1}{N_{dt}} \sum_{j=1}^{N_{dt}} h(y_{dtj}^{*(b)}).$$

   (c) From the bootstrap population generated in Step (b), take the sample with the same indices $s \subset U$ as the initial sample, and calculate the bootstrap EBPs, $\hat{\delta}_d^{ebp*(b)}$ and $\hat{\delta}_{dt}^{ebp*(b)}$, as described in Sect. 12.4 using the bootstrap sample data $\boldsymbol{y}_s^{*(b)}$ and the known values $\boldsymbol{x}_{dtj}$.

3. Output: the bootstrap estimators of $MSE(\hat{\delta}_d^{ebp})$ and $MSE(\hat{\delta}_{dt}^{ebp})$

$$mse^*(\hat{\delta}_d^{ebp}) = \frac{1}{B} \sum_{b=1}^{B} \left(\hat{\delta}_d^{ebp*(b)} - \delta_d^{*(b)} \right)^2,$$

$$mse^*(\hat{\delta}_{dt}^{ebp}) = \frac{1}{B} \sum_{b=1}^{B} \left(\hat{\delta}_{dt}^{ebp*(b)} - \delta_{dt}^{*(b)} \right)^2.$$

Remark 12.2 The described bootstrap estimator is applicable if a census file fulfilling properties (A), (B), and (C), given in Remark 10.1, is available. Nevertheless, it can be easily modified to the cases that only a census file fulfilling property (A) is available or no census file is available but the auxiliary variables are categorical and the sizes of the population classes are known. This modification can be done in the same manner as described in Sect. 10.8 for the NER model.

12.8 R Codes for EBPs

This section gives R codes for fitting the NER2 model to the survey data file `LFS20.txt`. We employ the R package `lme4`. The domains are defined by the variable AREA and the subdomains are obtained by crossing this variable with the age groups. The age groups (ageG) are defined by ageG = 1 if AGE < 25, ageG = 2 if $25 \leq$ AGE < 54, and ageG = 3 if AGE ≥ 54. The first target parameters are the proportions of poor people by domains or subdomains. The second and the third target parameters are the domain and subdomain poverty gaps and average incomes, respectively. As auxiliary variables, we take the dichotomic variables defining the three categories of the variable EDUCATION (primary or less, secondary and superior). The categories are named edu1, edu2, and edu3, respectively. The following R code reads the unit-level data files, load some R packages and defines some variables.

```
if(!require(Matrix)){
  install.packages("Matrix")
  library(Matrix)
}
if(!require(lme4)){
  install.packages("lme4")
  library(lme4)
}
dat <- read.table("LFS20.txt", header=TRUE, sep = "\t", dec = ".")
z0 <- 36500                                     # poverty threshold
ns <- nrow(dat)                                 # global sample size
poor <- as.numeric(dat$INCOME<z0)               # variable poor
gap <- (z0-dat$INCOME)*poor/z0                  # variable gap
one <- rep(1,nrow(dat))                         # variable one
Ga <- cut(dat$AGE, breaks=c(0,25,54,max(dat$AGE)), labels=c(1,2,3),
          right=TRUE)
ageG <- as.numeric(Ga)                          # age group
edu2 <- as.numeric(dat$EDUCATION==2)            # secondary education
edu3 <- as.numeric(dat$EDUCATION==3)            # superior education
y <- log(dat$INCOME)                            # variable y=log(income)
```

The following code reads the auxiliary data file and renames some variables.

```
aux <- read.table("Ndsa20.txt", header=TRUE, sep = "\t", dec = ".")
# Sort aux by sex, age and area
aux <- aux[order(aux$sex, aux$age, aux$area), ]
```

We calculate sample sizes, counts, and means by subdomains.

```
# Sizes
ndt <- tapply(X=one, INDEX=list(dat$AREA,ageG), FUN=sum)
# Sample counts of edu3
ndtedu3 <- tapply(edu3, list(dat$AREA,ageG), sum)
# Sample counts of edu2
ndtedu2 <- tapply(edu2, list(dat$AREA,ageG), sum)
# Sample counts of edu1
ndtedu1 <- ndt - ndtedu3 - ndtedu2
# Sample counts of poor people
ndtpoor <- tapply(poor, list(dat$AREA,ageG), sum)
# Sample poverty proportion
mdtpoor <- ndtpoor/ndt
# Sample sum of gap variable
ndtgap <- tapply(gap, list(dat$AREA,ageG), sum)
# Sample poverty gap
mdtgap <- ndtgap/ndt
# Sample sum of log-income
ndty <- tapply(y, list(dat$AREA,ageG), sum)
# Sample log-income mean
mdty <- ndty/ndt
```

We calculate direct estimators of sizes, poverty proportions and gaps and income means by subdomains, by using dir2 function described in Sect. 2.8.4.

```
dir.poor <- dir2(data=poor, w=dat$WEIGHT, domain=list(area=dat$AREA,
                 ageG=ageG))
dir.gap <- dir2(data=gap, w=dat$WEIGHT, domain=list(area=dat$AREA,
                 ageG=ageG))
dir.income <- dir2(data=dat$INCOME, w=dat$WEIGHT, domain=list(area=dat$AREA,
                 ageG=ageG))
hatNdt <- dir.poor$Nd.hat    # sizes
dirp <- dir.poor$mean        # poverty proportions
dirg <- dir.gap$mean         # poverty gaps
diri <- dir.income$mean      # income means
```

We fit a two-fold nested error regression model (NER2) to the variable $y = \log(\text{INCOME})$ under the assumption $w_{dtj} = 1$ for all d, t, j. We apply R function lmer with the REML fitting method.

```
lmm <- lmer(formula=y ~ edu3 + edu2 + (1|AREA/ageG), data=dat, REML=TRUE)
summary(lmm)                          # summary of the fitting procedure
anova(lmm)                            # analysis of variance table
beta <- fixef(lmm)                    # regression parameters
bedu3 <- beta[1] + beta[2]            # beta for x1=1, x2=1, x3=0 (edu3)
bedu2 <- beta[1] + beta[3]            # beta for x1=1, x2=0, x3=1 (edu2)
bedu1 <- beta[1]                      # beta for x1=1, x2=0, x3=0 (edu0)
var <- as.data.frame(VarCorr(lmm))    # variance parameters
sigmau2 <- var$sdcor[1]               # standard deviation of u_{2,dt}
sigmau1 <- var$sdcor[2]               # standard deviation of u_{1,d}
sigmae <- var$sdcor[3]                # residual standard deviation
ranef(lmm)                            # modes of the random effects
ypred <- fitted(lmm)                  # predictions
residuals <- resid(lmm)               # residuals
p.values <- 2*pnorm(abs(coef(summary(lmm))[,3]), low=F)      # p values
```

Table 12.1 gives the estimates of the regression parameters. The standard deviations of $u_{1,d}$, $u_{2,dt}$, and e_{dtj} are $\sigma_1 = 0.02114$, $\sigma_2 = 0.02321$, and $\sigma_0 = 0.27282$.

Figure 12.1 (left) plots a dispersion graph of model residuals. The residuals are situated symmetrically around zero. Figure 12.1 (right) plots an histogram of the model residuals that shows some lack of normality.

We first calculate the means and the variances of unobserved values of the income variable conditioned to observed ones. The following R code calculates γ_{dt}, ϕ_d, and $v_{dtj|s}$.

```
# Calculation of gammadt, gammadt, by subdomains
gammadt <- sigmau2^2*ndt/(sigmau2^2*ndt+sigmae^2)
# Calculation of deltad
gammad <- apply((1-gammadt)*ndt, 1, sum)
# phid by domains
phid <- sigmau1^2/(sigmae^2+sigmau1^2*gammad)
# Calculation of the conditioned variances, vdt, by subdomains
vdt <- sigmae^2*(1+phid*(1+gammadt*(gammadt-2))) + sigmau2^2*(1-gammadt)
```

Table 12.1 Estimated parameters of NER2 model

Parameter	Estimate	Std. error	z-value	p-value
Intercept	10.5406	0.01475	714.8	0.00
edu3	0.4282	0.02454	17.4	0.00
edu2	0.2275	0.01888	12.0	0.00

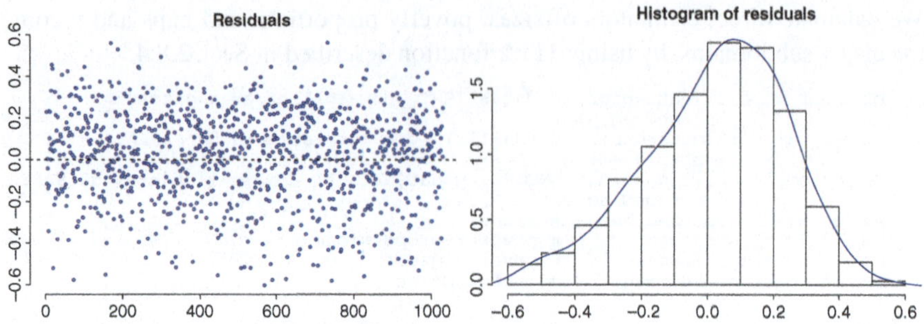

Fig. 12.1 Dispersion graph and histogram of residuals

The following R code calculates μ_{dtj} and α_{dtj}. We recall that for each area d and age group t, there are only three different values of μ_{dtj} corresponding to edu3, edu2, and edu1, respectively. The same happens for α_{dtj}.

```
# Preliminary calculations
mdtxbeta <- (bedu3*ndtedu3 + bedu2*ndtedu2 + bedu1*ndtedu1)/ndt
gammayxd <- apply(gammadt*(mdty-mdtxbeta), 1, sum)
uudt <- gammadt*(mdty-mdtxbeta+(sigmae^2/sigmau2^2)^2*(phid/ndt)*gammayxd)
# Calculation of the conditioned means
muedu3 <- bedu3 + uudt; muedu2 <- bedu2 + uudt; muedu1 <- bedu1 + uudt
# alphadt
y0 <- log(z0)
alphadedu3 <- vdt^(-1/2)*(y0-muedu3)
alphadedu2 <- vdt^(-1/2)*(y0-muedu2)
alphadedu1 <- vdt^(-1/2)*(y0-muedu1)
```

The following R code calculates $\Phi(\alpha_{dtj})$ and the EBPs of the poverty proportions by subdomains.

```
# Normal CDF
noredu3 <- pnorm(alphadedu3, mean=0, sd=1)
noredu2 <- pnorm(alphadedu2, mean=0, sd=1)
noredu1 <- pnorm(alphadedu1, mean=0, sd=1)
# Populations sizes by subdomains
Ndt <- tapply(X=aux$N, INDEX=list(aux$area,aux$age), FUN=sum)   # global
Nedu3 <- tapply(aux$edu3, list(aux$area,aux$age), sum)           # edu3
Nedu2 <- tapply(aux$edu2, list(aux$area,aux$age), sum)           # edu2
Nedu1 <- Ndt - Nedu3 - Nedu2                                     # edu1
# Poverty proportion EBPs by subdomains
ebpptot <- ndtpoor + (Nedu3-ndtedu3)*noredu3 + (Nedu2-ndtedu2)*noredu2 +
           (Nedu1-ndtedu1)*noredu1
ebp.poor <- ebpptot/Ndt; ebp.poor
```

The following R code calculates the poverty gaps by subdomains.

```
gap3 <- noredu3 - exp(vdt/2+muedu3)*pnorm(alphadedu3-vdt^{1/2})/z0
gap2 <- noredu2 - exp(vdt/2+muedu2)*pnorm(alphadedu2-vdt^{1/2})/z0
gap1 <- noredu1 - exp(vdt/2+muedu1)*pnorm(alphadedu1-vdt^{1/2})/z0
# Poverty gap EBPs by subdomains
ebpgtot <- ndtgap + (Nedu3-ndtedu3)*gap3 + (Nedu2-ndtedu2)*gap2 +
           (Nedu1-ndtedu1)*gap1
ebp.gap <- ebpgtot/Ndt; ebp.gap
```

The following R code calculates the income means by subdomains.

```
inc3 <- exp(vdt/2+muedu3)
inc2 <- exp(vdt/2+muedu2)
```

```
inc1 <- exp(vdt/2+muedu1)
# EBPs of income means by subdomains
ebpitot <- ndty + (Nedu3-ndtedu3)*inc3 + (Nedu2-ndtedu2)*inc2 +
                   (Nedu1-ndtedu1)*inc1
ebpi <- ebpitot/Ndt; ebpi
```

Summary of results for poverty proportions

```
output1 <- data.frame(n1=ndt[,1], n2=ndt[,2], n3=ndt[,3],
             ebpp1=round(ebp.poor[,1],4), ebpp2=round(ebp.poor[,2],4),
             ebpp3=round(ebp.poor[,3],4), dirp1=round(subset(dir.poor,
             dom.ageG=="1")$mean,4), dirp2=round(subset(dir.poor,
             dom.ageG=="2")$mean,4), dirp3=round(subset(dir.poor,
             dom.ageG=="3")$mean,4))
head(output1, 10)
```

Summary of results for poverty gaps

```
output2 <- data.frame(n1=ndt[,1], n2=ndt[,2], n3=ndt[,3],
             ebpg1=round(ebp.gap[,1],4), ebpg2=round(ebp.gap[,2],4),
             ebpg3=round(ebp.gap[,3],4), dirg1=round(subset(dir.gap,
             dom.ageG=="1")$mean,4), dirg2=round(subset(dir.gap,
             dom.ageG=="2")$mean,4), dirg3=round(subset(dir.gap,
             dom.ageG=="3")$mean,4))
head(output2, 10)
```

Summary of results for income means

```
output3 <- data.frame(n1=ndt[,1], n2=ndt[,2], n3=ndt[,3],
             ebpi1=round(ebpi[,1],0), ebpi2=round(ebpi[,2],0),
             ebpi3=round(ebpi[,3],0), diri1=round(subset(dir.income,
             dom.ageG=="1")$mean,0), diri2=round(subset(dir.income,
             dom.ageG=="2")$mean,0), diri3=round(subset(dir.income,
             dom.ageG=="3")$mean,0))
head(output3, 10)
```

Table 12.2 (left) gives the sample sizes by subdomains (AREA crossed by ageG). We note that sample sizes are very small. The columns labeled by ebpp1, ebpp2, and ebpp3 contain the EBPs of poverty proportions by areas and age groups 1, 2, and 3, respectively. The columns labeled by dirp1, dirp2, and dirp3 contain the direct estimates of poverty proportions by areas and age groups 1, 2, and 3, respectively.

Table 12.3 gives the EBPs of poverty gap by subdomains. The columns labeled by ebpg1, ebpg2, and ebpg3 contain the EBPs of poverty gaps by areas and age groups 1, 2, and 3, respectively. The columns labeled by dirg1, dirg2, and dirg3

Table 12.2 EBPs and direct estimates of subdomain poverty proportions

d	n_1	n_2	n_3	ebpp1	ebpp2	ebpp3	dirp1	dirp2	dirp3
1	9	36	15	0.2813	0.2372	0.4019	0.4229	0.1098	0.2662
2	9	13	15	0.2737	0.2278	0.4192	0.2049	0.2576	0.3303
3	5	23	19	0.2341	0.1922	0.2846	0.0000	0.1330	0.1614
4	14	31	10	0.2627	0.2385	0.3659	0.3113	0.1481	0.2961
5	10	31	9	0.1628	0.2233	0.4273	0.1868	0.0900	0.3060
6	9	24	10	0.1369	0.2519	0.3670	0.0000	0.1699	0.1719
7	10	23	15	0.1834	0.2281	0.3848	0.0816	0.2272	0.2660
8	10	27	11	0.3040	0.3358	0.3513	0.1972	0.2774	0.2622
9	20	68	37	0.1688	0.2021	0.3246	0.1295	0.1169	0.1421
10	10	20	11	0.2163	0.2768	0.4002	0.0797	0.1889	0.6420

Table 12.3 EBPs and direct estimates of subdomain poverty gaps

d	n_1	n_2	n_3	ebpg1	ebpg2	ebpg3	dirg1	dirg2	dirg3
1	9	36	15	0.0459	0.0371	0.0704	0.1017	0.0120	0.0710
2	9	13	15	0.0434	0.0343	0.0732	0.0070	0.0205	0.0849
3	5	23	19	0.0362	0.0297	0.0462	0.0000	0.0238	0.0252
4	14	31	10	0.0426	0.0382	0.0623	0.0196	0.0301	0.0279
5	10	31	9	0.0227	0.0339	0.0745	0.0105	0.0109	0.0779
6	9	24	10	0.0175	0.0392	0.0620	0.0000	0.0273	0.0414
7	10	23	15	0.0256	0.0361	0.0658	0.0160	0.0421	0.0296
8	10	27	11	0.0507	0.0569	0.0603	0.0322	0.0633	0.0446
9	20	68	37	0.0234	0.0311	0.0541	0.0243	0.0141	0.0155
10	10	20	11	0.0322	0.0454	0.0701	0.0182	0.0440	0.1315

Table 12.4 EBPs and direct estimates of subdomain income means

d	n_1	n_2	n_3	ebpi1	ebpi2	ebpi3	diri1	diri2	diri3
1	9	36	15	45,690	47,671	40,832	42,479	49,380	40,651
2	9	13	15	45,450	47,216	40,073	45,717	48,783	40,990
3	5	23	19	47,249	50,873	45,895	50,648	51,913	50,338
4	14	31	10	46,994	48,262	42,302	43,178	49,605	42,803
5	10	31	9	50,928	47,672	39,560	53,155	47,561	42,339
6	9	24	10	51,667	46,360	41,912	53,973	46,726	44,807
7	10	23	15	48,929	48,586	41,576	49,804	48,159	44,327
8	10	27	11	44,948	43,477	43,236	44,286	42,933	43,187
9	20	68	37	49,995	50,100	44,219	49,650	50,432	44,069
10	10	20	11	47,694	46,111	40,910	46,483	47,648	37,116

contain the direct estimates of poverty gaps by areas and age groups 1, 2, and 3, respectively.

Table 12.4 gives the EBPs of income means by subdomains. The columns labeled by ebpi1, ebpi2, and ebpi3 contain the EBPs of income means by areas and age groups 1, 2, and 3, respectively. The columns labeled by diri1, diri2, and diri3 contain the direct estimates of income means by areas and age groups 1, 2, and 3, respectively.

In general, the EBP method produces estimates that are, across domains, smoother than direct estimates. This in an interesting property when dealing with real data applications for doing poverty mapping.

References

Battese, G.E., Harter, R.M., Fuller, W.A.: An error-component model for prediction of county crop areas using survey and satellite data. J. Am. Stat. Assoc. **83**, 28–36 (1988)

Foster, J., Greer, J., Thorbecke, E.: A class of decomposable poverty measures. Econometrica **52**, 761–766 (1984)

González-Manteiga, W., Lombardía, M.J., Molina, I., Morales, D., Santamaría, L.: Bootstrap mean squared error of small-area EBLUP. J. Stat. Comput. Simul. **78**, 443–462 (2008a)

Marhuenda, Y., Molina, I., Morales, D., Rao, J.N.K.: Poverty mapping in small areas under a two-fold nested error regression model. J. R. Stat. Soc. A **180**(4), 1111–1136 (2017)

Molina, I., Rao, J.N.K.: Small area estimation of poverty indicators. Can. J. Stat. **38**, 369–385 (2010)

Rencher, A.C.: Multivariate statistical inference and applications. Wiley, New York (1998)

Chapter 13
Random Regression Coefficient Models

13.1 Introduction

Coefficients of auxiliary variables in the standard nested error regression (NER) model are not allowed to vary across sampling units or domains. This assumption is too rigid in many practical situations. In some small area estimation (SAE) problems, we can intuitively expect that the slope parameters of some explanatory variable are not constant and therefore they should take different values in different domains. The random regression coefficient (RRC) model gives a practical solution to this problem by assuming that the beta parameters are random and therefore they give a more flexible way of modeling.

This section describes a modification of the nested error regression (NER) model having random regression coefficients. In the framework of SAE, Prasad and Rao (1990) derived empirical best linear unbiased predictors (EBLUP) of domain linear parameters under a unit-level RRC model. They also derived a second order approximation of the mean squared error (MSE) of the EBLUP and gave an estimator of the MSE approximation. They considered a special case of the RRC model proposed by Dempster et al. (1981), with a single concomitant variable x and a null intercept parameter (regression through origin).

Moura and Hold (1999) used a class of models allowing for variation between areas because of: (1) differences in the distribution of unit-level or area-level variables between areas, and (2) area-specific components of variance which cannot be explained by covariates. Their family of models contains RRC models as particular cases. These authors derived EBLUPs of linear parameters, gave an approximation to the MSE of the EBLUP, and proposed MSE estimators.

Hobza and Morales (2013) applied a flexible class of RRC models to the prediction of domain linear parameters. They gave a Fisher-scoring algorithm to calculate the residual maximum likelihood estimators of the model parameters and they derived EBLUPs and MSEs estimators. They applied the introduced

D. Morales et al., *A Course on Small Area Estimation and Mixed Models*, Statistics for Social and Behavioral Sciences, https://doi.org/10.1007/978-3-030-63757-6_13

methodology to the estimation of household normalized net annual incomes in the Spanish Living Conditions Survey.

This chapter extends the results of Hobza and Morales (2013) by considering a model where the set of random effects has a multivariate normal distribution that includes all variances and covariances as unknown parameters. It also studies the more simple model without covariances and gives some R codes for the last model.

13.2 The RRC Model with Covariance Parameters

We start with a description of the more general model.

13.2.1 The Model

Let us consider the model

$$y_{dj} = \sum_{k=1}^{p} \beta_k x_{kdj} + \sum_{k=1}^{p} u_{kd} x_{kdj} + e_{dj}, \quad d = 1, \ldots, D, \; j = 1, \ldots, n_d, \tag{13.1}$$

where

$\boldsymbol{u}_d^* = (u_{1d}, \ldots, u_{pd})' \overset{iid}{\sim} N_p(\boldsymbol{0}, \boldsymbol{V}_{\boldsymbol{u}_d^*})$ and $e_{dj} \overset{ind}{\sim} N(0, w_{dj}^{-1}\sigma_e^2)$ are independent, $d = 1, \ldots, D$, $j = 1 \ldots, n_d$,

$\boldsymbol{V}_{\boldsymbol{u}_d^*}$ is a covariance matrix with components $\text{cov}(u_{k_1 d}, u_{k_2, d}) = \sigma_{k_1 k_2}$, $k_1, k_2 = 1, \ldots, p$.

In models with intercept, the first auxiliary variable is equal to one. The model (13.1) has p regression parameters and $1 + p + \frac{1}{2}p(p-1) = 1 + \frac{1}{2}p(p+1)$ variance component parameters. They are β_k, σ_e^2, $\sigma_{k_1 k_2}$, with $\sigma_{k_1 k_2} = \sigma_{k_2 k_1}$, $k, k_1, k_2 = 1, \ldots, p$.

An example of model (13.1) with intercept, $D = 2$, $n_1 = n_2 = 2$, $n = 4$, and $p = 2$ is

$$\begin{pmatrix} y_{11} \\ y_{12} \\ \hline y_{21} \\ y_{22} \end{pmatrix} = \begin{pmatrix} 1 \; x_{211} \\ 1 \; x_{212} \\ \hline 1 \; x_{221} \\ 1 \; x_{222} \end{pmatrix} \begin{pmatrix} \beta_1 \\ \beta_2 \end{pmatrix} + \begin{pmatrix} 1 \; 0 \\ 1 \; 0 \\ \hline 0 \; 1 \\ 0 \; 1 \end{pmatrix} \begin{pmatrix} u_{11} \\ u_{12} \end{pmatrix} + \begin{pmatrix} x_{211} & 0 \\ x_{212} & 0 \\ \hline 0 & x_{221} \\ 0 & x_{222} \end{pmatrix} \begin{pmatrix} u_{21} \\ u_{22} \end{pmatrix} + \begin{pmatrix} e_{11} \\ e_{12} \\ \hline e_{21} \\ e_{22} \end{pmatrix}.$$

In matrix notation model (13.1) is

$$\boldsymbol{y} = \boldsymbol{X}\boldsymbol{\beta} + \sum_{k=1}^{p} \boldsymbol{Z}_k \boldsymbol{u}_k + \boldsymbol{e}, \tag{13.2}$$

where $n = \sum_{d=1}^{D} n_d$, $\boldsymbol{\beta} = \boldsymbol{\beta}_{p\times 1}$, $\boldsymbol{y} = \underset{1\le d\le D}{\text{col}}(\boldsymbol{y}_d)$, $\boldsymbol{y}_d = \underset{1\le j\le n_d}{\text{col}}(y_{dj})$, $\boldsymbol{e} = \underset{1\le d\le D}{\text{col}}(\boldsymbol{e}_d)$, $\boldsymbol{e}_d = \underset{1\le j\le n_d}{\text{col}}(e_{dj})$, $\boldsymbol{u}_k = \underset{1\le d\le D}{\text{col}}(u_{kd})$, $\boldsymbol{X} = \underset{1\le d\le D}{\text{col}}(\boldsymbol{X}_d)$, $\boldsymbol{X}_d = \underset{1\le k\le p}{\text{col}'}(\boldsymbol{x}_{k,n_d}) = \underset{1\le j\le n_d}{\text{col}}(\boldsymbol{x}_{dj})$, $\boldsymbol{x}_{k,n_d} = \underset{1\le j\le n_d}{\text{col}}(x_{kdj})$, $\boldsymbol{x}_{dj} = \underset{1\le k\le p}{\text{col}'}(x_{kdj})$, $\boldsymbol{Z}_k = \underset{1\le d\le D}{\text{diag}}(\boldsymbol{x}_{k,n_d})$, $\boldsymbol{I}_a = \underset{1\le j\le a}{\text{diag}}(1)$, $\boldsymbol{W} = \underset{1\le d\le D}{\text{diag}}(\boldsymbol{W}_d)$, $\boldsymbol{W}_d = \underset{1\le j\le n_d}{\text{diag}}(w_{dj})$, with $w_{dj} > 0$ known, $d = 1,\dots,D$, $j = 1,\dots,n_d$. Covariance matrices are $\boldsymbol{V}_e = \text{var}(\boldsymbol{e}) = \sigma_e^2 \boldsymbol{W}^{-1}$, $\boldsymbol{V}_{k_1k_2} = \text{cov}(\boldsymbol{u}_{k_1}, \boldsymbol{u}_{k_2}) = \sigma_{k_1k_2}\boldsymbol{I}_D$, $k_1, k_2 = 1,\dots,p$, and

$$\boldsymbol{V} = \text{var}(\boldsymbol{y}) = \boldsymbol{V}_e + \sum_{k_1=1}^{p}\sum_{k_2=1}^{p} \boldsymbol{Z}_{k_1}\boldsymbol{V}_{k_1k_2}\boldsymbol{Z}'_{k_2} = \underset{1\le d\le D}{\text{diag}}(\boldsymbol{V}_d),$$

where

$$\boldsymbol{V}_d = \sigma_e^2 \boldsymbol{W}_d^{-1} + \sum_{k_1=1}^{p}\sum_{k_2=1}^{p} \sigma_{k_1k_2}\boldsymbol{x}_{k_1,n_d}\boldsymbol{x}'_{k_2,n_d}, \quad d = 1,\dots,D.$$

Let $\boldsymbol{u} = \underset{1\le k\le p}{\text{col}}(\boldsymbol{u}_k)$ and $\boldsymbol{Z} = \underset{1\le k\le p}{\text{col}'}(\boldsymbol{Z}_k)$. Under this notation, the variance of $\boldsymbol{u}$ is

$$\boldsymbol{V}_u = \text{var}(\boldsymbol{u}) = \big(\boldsymbol{V}_{k_1k_2}\big)_{k_1k_2=1,\dots,p}$$

and the model (13.2) can be written in the general form

$$\boldsymbol{y} = \boldsymbol{X}\boldsymbol{\beta} + \boldsymbol{Z}\boldsymbol{u} + \boldsymbol{e}.$$

If the variance components are known, then the BLUE of $\boldsymbol{\beta} = (\beta_1,\dots,\beta_p)'$ is (cf. (6.12))

$$\tilde{\boldsymbol{\beta}} = (\boldsymbol{X}'\boldsymbol{V}^{-1}\boldsymbol{X})^{-1}\boldsymbol{X}'\boldsymbol{V}^{-1}\boldsymbol{y} = \Big(\sum_{d=1}^{D}\boldsymbol{X}'_d\boldsymbol{V}_d^{-1}\boldsymbol{X}_d\Big)^{-1}\Big(\sum_{d=1}^{D}\boldsymbol{X}'_d\boldsymbol{V}_d^{-1}\boldsymbol{y}_d\Big)$$

and the BLUP of $\boldsymbol{u}$ is $\tilde{\boldsymbol{u}} = \boldsymbol{V}_u\boldsymbol{Z}'\boldsymbol{V}^{-1}(\boldsymbol{y} - \boldsymbol{X}\tilde{\boldsymbol{\beta}})$.

13.2.2 REML Estimators

In order to derive formulas for calculating the REML estimates of the unknown variance parameters we consider the alternative parametrization $\sigma^2 = \sigma_e^2$, $\varphi_{k_1k_2} = \sigma_{k_1k_2}/\sigma_e^2$, $k_1, k_2 = 1\dots,p$, and we define $\boldsymbol{\sigma} = (\sigma^2, \varphi_{k_1k_2}, k_1, k_2 = 1\dots,p)$ and $\boldsymbol{\Sigma} = \underset{1\le d\le D}{\text{diag}}(\boldsymbol{\Sigma}_d)$ where $\boldsymbol{\Sigma}_d = \sigma^{-2}\boldsymbol{V}_d$.

The REML log-likelihood is (cf. (6.32))

$$l_{REML}(\boldsymbol{\sigma}) = -\frac{1}{2}(n-p)\log 2\pi - \frac{1}{2}(n-p)\log\sigma^2 - \frac{1}{2}\log|\boldsymbol{K}'\boldsymbol{\Sigma}\boldsymbol{K}| - \frac{1}{2\sigma^2}\boldsymbol{y}'\boldsymbol{P}\boldsymbol{y},$$

where

$$\boldsymbol{P} = \boldsymbol{K}(\boldsymbol{K}'\boldsymbol{\Sigma}\boldsymbol{K})^{-1}\boldsymbol{K}' = \boldsymbol{\Sigma}^{-1} - \boldsymbol{\Sigma}^{-1}\boldsymbol{X}(\boldsymbol{X}'\boldsymbol{\Sigma}^{-1}\boldsymbol{X})^{-1}\boldsymbol{X}'\boldsymbol{\Sigma}^{-1},$$
$$\boldsymbol{K} = \boldsymbol{W} - \boldsymbol{W}\boldsymbol{X}(\boldsymbol{X}'\boldsymbol{W}\boldsymbol{X})^{-1}\boldsymbol{X}'\boldsymbol{W}$$

are such that $\boldsymbol{P}\boldsymbol{X} = \boldsymbol{0}$ and $\boldsymbol{P}\boldsymbol{\Sigma}\boldsymbol{P} = \boldsymbol{P}$. The matrix $\boldsymbol{\Sigma}$ can be written in the form

$$\boldsymbol{\Sigma} = \boldsymbol{W}^{-1} + \sum_{k_1=1}^{p}\sum_{k_2=1}^{p}\varphi_{k_1k_2}\boldsymbol{A}_{k_1k_2},$$

where $\boldsymbol{A}_{k_1k_2} = \boldsymbol{Z}_{k_1}\boldsymbol{Z}'_{k_2} = \underset{1\le d\le D}{\text{diag}}\,(\boldsymbol{x}_{k_1,n_d}\boldsymbol{x}'_{k_2,n_d})$, $k_1, k_2 = 1, \ldots, p$. In the same way as in (6.43) we obtain $\frac{\partial \boldsymbol{P}}{\partial\varphi_{k_1k_2}} = -\boldsymbol{P}\boldsymbol{A}_{k_1k_2}\boldsymbol{P}$. Thus, by taking partial derivatives of the REML log-likelihood with respect to σ^2 and $\varphi_{k_1k_2}$, $k_1, k_2 = 1, \ldots, p$, one gets the scores

$$S_{\sigma^2} = -\frac{n-p}{2\sigma^2} + \frac{1}{2\sigma^4}\boldsymbol{y}'\boldsymbol{P}\boldsymbol{y}, \quad S_{\varphi_{k_1k_2}} = -\frac{1}{2}\,\text{tr}\{\boldsymbol{P}\boldsymbol{A}_{k_1k_2}\} + \frac{1}{2\sigma^2}\boldsymbol{y}'\boldsymbol{P}\boldsymbol{A}_{k_1k_2}\boldsymbol{P}\boldsymbol{y},$$

and the second partial derivatives

$$H_{\sigma^2\sigma^2} = \frac{n-p}{2\sigma^4} - \frac{1}{\sigma^6}\boldsymbol{y}'\boldsymbol{P}\boldsymbol{y}, \quad H_{\sigma^2\varphi_{k_1k_2}} = -\frac{1}{2\sigma^4}\boldsymbol{y}'\boldsymbol{P}\boldsymbol{A}_{k_1k_2}\boldsymbol{P}\boldsymbol{y},$$
$$H_{\varphi_{k_1k_2}\varphi_{i_1i_2}} = \frac{1}{2}\,\text{tr}\{\boldsymbol{P}\boldsymbol{A}_{k_1k_2}\boldsymbol{P}\boldsymbol{A}_{i_1i_2}\} - \frac{1}{2\sigma^2}\boldsymbol{y}'\boldsymbol{P}\boldsymbol{A}_{k_1k_2}\boldsymbol{P}\boldsymbol{A}_{i_1i_2}\boldsymbol{P}\boldsymbol{y}$$
$$- \frac{1}{2\sigma^2}\boldsymbol{y}'\boldsymbol{P}\boldsymbol{A}_{i_1i_2}\boldsymbol{P}\boldsymbol{A}_{k_1k_2}\boldsymbol{P}\boldsymbol{y},$$

where $k_1, k_2, i_1, i_2 = 1, \ldots, p$. By taking expectations and multiplying by -1, we obtain the components of the Fisher information matrix. For $k_1, k_2, i_1, i_2 = 1, \ldots, p$, we have

$$F_{\sigma^2\sigma^2} = \frac{n-p}{2\sigma^4}, \quad F_{\sigma^2\varphi_{k_1k_2}} = \frac{1}{2\sigma^2}\,\text{tr}\{\boldsymbol{P}\boldsymbol{A}_{k_1k_2}\}, \quad F_{\varphi_{k_1k_2}\varphi_{i_1i_2}} = \frac{1}{2}\,\text{tr}\{\boldsymbol{P}\boldsymbol{A}_{k_1k_2}\boldsymbol{P}\boldsymbol{A}_{i_1i_2}\}.$$

To calculate the REML estimates, the Fisher-scoring updating formula, at iteration i, is

$$\boldsymbol{\sigma}^{(i+1)} = \boldsymbol{\sigma}^{(i)} + \boldsymbol{F}^{-1}(\boldsymbol{\sigma}^{(i)})\boldsymbol{S}(\boldsymbol{\sigma}^{(i)}),$$

where $S(\sigma)$ and $F(\sigma)$ are the vector of scores and the Fisher information matrix evaluated at σ. The following seeds can be used as starting values in the Fisher-scoring algorithm:

$$\sigma^{2(0)} = S^2/(p+2), \quad \varphi_{k_1 k_2} = \delta_{k_1 k_2}, \quad k_1, k_2 = 1, \ldots, p,$$

where $\delta_{k_1,k_2} = 1$ if $k_1 = k_2$, $\delta_{k_1,k_2} = 0$ if $k_1 \neq k_2$, $S^2 = \frac{1}{n-p}(\boldsymbol{y}-\boldsymbol{X}\hat{\boldsymbol{\beta}}_0)'\boldsymbol{W}(\boldsymbol{y}-\boldsymbol{X}\hat{\boldsymbol{\beta}}_0)$ and $\hat{\boldsymbol{\beta}}_0 = (\boldsymbol{X}'\boldsymbol{W}\boldsymbol{X})^{-1}\boldsymbol{X}'\boldsymbol{W}\boldsymbol{y}$.

13.2.3 EBLUP of the Domain Mean

Let us now consider a finite population U partitioned into D domains U_d, i.e. $U = \cup_{d=1}^D U_d$. Let N and N_d be the sizes of U and U_d, so that $N = \sum_{d=1}^D N_d$. We assume that the population target vector $\boldsymbol{y} = \boldsymbol{y}_{N\times 1}$ follows the RRC model (13.2) with the obvious size changes, i.e. with N and N_d in the place of n and n_d, respectively.

Let $s \subset U$ be a sample of $n \leq N$ units and let $r = U - s$ be the set of non-sampled units. The domain and subdomain subsets of s and r are denoted by s_d and r_d, respectively. The subindexes s and r in vectors or matrices are used to denote their sampled and the non-sampled parts. Without loss of generality, we renumber the population units and we write

$$\boldsymbol{y} = \begin{pmatrix} \boldsymbol{y}_s \\ \boldsymbol{y}_r \end{pmatrix}, \quad \boldsymbol{X} = \begin{pmatrix} \boldsymbol{X}_s \\ \boldsymbol{X}_r \end{pmatrix}, \quad \boldsymbol{e} = \begin{pmatrix} \boldsymbol{e}_s \\ \boldsymbol{e}_r \end{pmatrix}, \quad \boldsymbol{Z}_k = \begin{pmatrix} \boldsymbol{Z}_{sk} \\ \boldsymbol{Z}_{rk} \end{pmatrix}, \ k = 1, \ldots, p,$$

and

$$\boldsymbol{V} = \text{var}(\boldsymbol{y}) = \begin{pmatrix} \boldsymbol{V}_s & \boldsymbol{V}_{sr} \\ \boldsymbol{V}_{rs} & \boldsymbol{V}_r \end{pmatrix}.$$

Using Theorem 4.1, the EBLUP of the linear parameter $\eta = \boldsymbol{a}'\boldsymbol{y} = \boldsymbol{a}_s'\boldsymbol{y}_s + \boldsymbol{a}_r'\boldsymbol{y}_r$ is

$$\hat{\eta} = \boldsymbol{a}_s'\boldsymbol{y}_s + \boldsymbol{a}_r'\left[\boldsymbol{X}_r\hat{\boldsymbol{\beta}} + \hat{\boldsymbol{V}}_{rs}\hat{\boldsymbol{V}}_s^{-1}(\boldsymbol{y}_s - \boldsymbol{X}_s\hat{\boldsymbol{\beta}})\right],$$

where

$$\hat{\boldsymbol{\beta}} = (\boldsymbol{X}_s'\hat{\boldsymbol{V}}_s^{-1}\boldsymbol{X}_s)^{-1}\boldsymbol{X}_s'\hat{\boldsymbol{V}}_s^{-1}\boldsymbol{y}_s.$$

As $\boldsymbol{V}_{ers} = \boldsymbol{0}$, $\boldsymbol{V}_{rs} = \boldsymbol{Z}_r \boldsymbol{V}_u \boldsymbol{Z}_s' + \boldsymbol{V}_{ers} = \boldsymbol{Z}_r \boldsymbol{V}_u \boldsymbol{Z}_s'$ and $\hat{\boldsymbol{u}} = \hat{\boldsymbol{V}}_u \boldsymbol{Z}_s' \hat{\boldsymbol{V}}_s^{-1} (\boldsymbol{y}_s - \boldsymbol{X}_s \hat{\boldsymbol{\beta}})$, then

$$\hat{\eta} = \boldsymbol{a}_s' \boldsymbol{y}_s + \boldsymbol{a}_r' \left[\boldsymbol{X}_r \hat{\boldsymbol{\beta}} + \boldsymbol{Z}_r \hat{\boldsymbol{V}}_u \boldsymbol{Z}_s' \hat{\boldsymbol{V}}_s^{-1} (\boldsymbol{y}_s - \boldsymbol{X}_s \hat{\boldsymbol{\beta}}) \right] = \boldsymbol{a}_s' \boldsymbol{y}_s + \boldsymbol{a}_r' \left[\boldsymbol{X}_r \hat{\boldsymbol{\beta}} + \boldsymbol{Z}_r \hat{\boldsymbol{u}} \right]$$

$$= \boldsymbol{a}' \left[\boldsymbol{X} \hat{\boldsymbol{\beta}} + \sum_{k=1}^{p} \boldsymbol{Z}_k \hat{\boldsymbol{u}}_k \right] + \boldsymbol{a}_s' \left[\boldsymbol{y}_s - \boldsymbol{X}_s \hat{\boldsymbol{\beta}} - \sum_{k=1}^{p} \boldsymbol{Z}_{sk} \hat{\boldsymbol{u}}_k \right].$$

The domain mean is $\overline{Y}_d = \frac{1}{N_d} \sum_{j=1}^{N_d} y_{dj} = \mu_d + \overline{e}_d$, where $\overline{e}_d = \frac{1}{N_d} \sum_{j=1}^{N_d} e_{dj}$ and

$$\mu_d = \sum_{k=1}^{p} \overline{X}_{kd} \beta_k + \sum_{k=1}^{p} \overline{X}_{kd} u_{kd}, \qquad \overline{X}_{kd} = \frac{1}{N_d} \sum_{j=1}^{N_d} x_{kdj}.$$

The domain mean $\overline{Y}_d$ can be written in the form $\eta = \boldsymbol{a}' \boldsymbol{y}$, where

$$\boldsymbol{a}' = \frac{1}{N_d} (\boldsymbol{0}_{N_1}', \ldots, \boldsymbol{0}_{N_{d-1}}', \boldsymbol{1}_{N_d}', \boldsymbol{0}_{N_{d+1}}', \ldots, \boldsymbol{0}_{N_D}') = \frac{1}{N_d} \underset{1 \le \ell \le D}{\text{col}'} \{\delta_{d\ell} \boldsymbol{1}_{N_\ell}'\},$$

$\delta_{ab} = 1$ if $a = b$ and $\delta_{ab} = 0$ if $a \neq b$. It holds that $\boldsymbol{a}' \boldsymbol{X} = \overline{\boldsymbol{X}}_d = (\overline{X}_{1d}, \ldots, \overline{X}_{pd})$,

$$\boldsymbol{a}' \boldsymbol{Z}_k \hat{\boldsymbol{u}}_k = \frac{1}{N_d} \underset{1 \le \ell \le D}{\text{col}'} \{\delta_{d\ell} \boldsymbol{1}_{N_\ell}'\} \underset{1 \le \ell \le D}{\text{diag}} (\boldsymbol{x}_{k,N_\ell}) \hat{\boldsymbol{u}}_k = \underset{1 \le \ell \le D}{\text{col}'} \{\delta_{d\ell} \overline{X}_{k\ell}\} \hat{\boldsymbol{u}}_k = \overline{X}_{kd} \hat{u}_{kd}.$$

If $n_d > 0$, then the EBLUP of $\overline{Y}_d$ is

$$\widehat{\overline{Y}}_d^{eblup} = \sum_{k=1}^{p} \overline{X}_{kd} \hat{\beta}_k + \sum_{k=1}^{p} \overline{X}_{kd} \hat{u}_{kd} + f_d \left[\overline{y}_{s,d} - \sum_{k=1}^{p} \overline{X}_{s,kd} \hat{\beta}_k - \sum_{k=1}^{p} \overline{X}_{s,kd} \hat{u}_{kd} \right],$$

where $\overline{y}_{s,d} = \frac{1}{n_d} \sum_{j=1}^{n_d} y_{dj}$, $\overline{X}_{s,kd} = \frac{1}{n_d} \sum_{j=1}^{n_d} x_{kdj}$ and $f_d = \frac{n_d}{N_d}$. If $f_d \approx 0$, then the EBLUP of $\overline{Y}_d$ is approximately equal to

$$\hat{\mu}_d^{eblup} = \sum_{k=1}^{p} \overline{X}_{kd} \hat{\beta}_k + \sum_{k=1}^{p} \overline{X}_{kd} \hat{u}_{kd}.$$

The MSE of the EBLUP can be estimated by adapting the steps 1–6 of the parametric bootstrap procedure described in Sect. 8.5.

13.3 The RRC Model Without Covariance Parameters

For the ease of exposition, we consider now a slightly simpler model under which we derive more detailed formulas for the REML estimators and the formulas for the analytic approximation of the MSE of EBLUPs.

13.3.1 The Model

Let us consider the RRC model

$$y_{dj} = \sum_{k=1}^{p} \beta_k x_{kdj} + \sum_{k=1}^{p} u_{kd} x_{kdj} + e_{dj}, \quad d = 1, \ldots, D, \; j = 1, \ldots, n_d, \tag{13.3}$$

where $u_{kd} \overset{iid}{\sim} N(0, \sigma_k^2)$ and $e_{dj} \overset{iid}{\sim} N(0, w_{dj}^{-1} \sigma_e^2)$ are independent, $d = 1, \ldots, D$, $j = 1 \ldots, n_d, k = 1, \ldots, p$. The model variance and covariance parameters are σ_e^2, $\sigma_k^2, k = 1, \ldots, p$, $(p + 1$ parameters). In matrix notation model (13.3) is

$$\boldsymbol{y} = \boldsymbol{X}\boldsymbol{\beta} + \sum_{k=1}^{p} \boldsymbol{Z}_k \boldsymbol{u}_k + \boldsymbol{e}, \tag{13.4}$$

where the meaning of the used symbols is exactly the same as the one given below formula (13.2). The difference with respect to the model (13.2) is only in the variance matrices which are now simpler and have the form $\boldsymbol{V}_e = \text{var}(\boldsymbol{e}) = \sigma_e^2 \boldsymbol{W}^{-1}$, $\boldsymbol{V}_{u_k} = \text{var}(\boldsymbol{u}_k) = \sigma_k^2 \boldsymbol{I}_D$, $k = 1, \ldots, p$, and

$$\boldsymbol{V} = \text{var}(\boldsymbol{y}) = \boldsymbol{V}_e + \sum_{k=1}^{p} \boldsymbol{Z}_k \boldsymbol{V}_{u_k} \boldsymbol{Z}_k' = \underset{1 \le d \le D}{\text{diag}} (\boldsymbol{V}_d),$$

where

$$\boldsymbol{V}_d = \sigma_e^2 \boldsymbol{W}_d^{-1} + \sum_{k=1}^{p} \sigma_k^2 \boldsymbol{x}_{k,n_d} \boldsymbol{x}_{k,n_d}', \quad d = 1, \ldots, D.$$

We consider the alternative parameters $\sigma^2 = \sigma_e^2$, $\varphi_k = \sigma_k^2 / \sigma_e^2$, $k = 1, \ldots, p$, in such a way that $\boldsymbol{V} = \sigma^2 \boldsymbol{\Sigma}$ and $\boldsymbol{V}_d = \sigma^2 \boldsymbol{\Sigma}_d$, where

$$\boldsymbol{\Sigma}_d = \boldsymbol{W}_d^{-1} + \sum_{k=1}^{p} \varphi_k \boldsymbol{x}_{k,n_d} \boldsymbol{x}_{k,n_d}', \quad d = 1, \ldots, D.$$

Let $\boldsymbol{\theta} = (\sigma^2, \varphi_1, \ldots, \varphi_p)'$ be the vector of variance components, with $\sigma^2 > 0$, $\varphi_1 > 0, \ldots, \varphi_p > 0$. Let $\boldsymbol{u} = \underset{1\leq k\leq p}{\text{col}}(\boldsymbol{u}_k)$ and $\boldsymbol{Z} = \underset{1\leq k\leq p}{\text{col}'}(\boldsymbol{Z}_k)$. The variance of $\boldsymbol{u}$ is

$$\boldsymbol{V_u} = \text{var}(\boldsymbol{u}) = \underset{1\leq k\leq p}{\text{diag}}(\boldsymbol{V}_{\boldsymbol{u}_k}).$$

Using this notation, the model (13.4) can be written in the general form

$$\boldsymbol{y} = \boldsymbol{X\beta} + \boldsymbol{Zu} + \boldsymbol{e}.$$

If $\boldsymbol{\theta}$ is known, then the BLUE of $\boldsymbol{\beta} = (\beta_1, \ldots, \beta_p)'$ is (cf. (6.12))

$$\tilde{\boldsymbol{\beta}} = (\boldsymbol{X}'\boldsymbol{V}^{-1}\boldsymbol{X})^{-1}\boldsymbol{X}'\boldsymbol{V}^{-1}\boldsymbol{y} = \left(\sum_{d=1}^{D}\boldsymbol{X}_d'\boldsymbol{\Sigma}_d^{-1}\boldsymbol{X}_d\right)^{-1}\left(\sum_{d=1}^{D}\boldsymbol{X}_d'\boldsymbol{\Sigma}_d^{-1}\boldsymbol{y}_d\right)$$

and the BLUP of $\boldsymbol{u}$ is $\tilde{\boldsymbol{u}} = \boldsymbol{V}_u\boldsymbol{Z}'\boldsymbol{V}^{-1}(\boldsymbol{y} - \boldsymbol{X}\tilde{\boldsymbol{\beta}})$, i.e.

$$\tilde{\boldsymbol{u}} = \underset{1\leq k\leq p}{\text{diag}}(\boldsymbol{V}_{\boldsymbol{u}_k})\underset{1\leq k\leq p}{\text{col}}(\boldsymbol{Z}_k')\underset{1\leq d\leq D}{\text{diag}}(\boldsymbol{V}_d^{-1})\underset{1\leq d\leq D}{\text{col}}(\boldsymbol{y}_d - \boldsymbol{X}_d\tilde{\boldsymbol{\beta}})$$

$$= \begin{pmatrix} \varphi_1 \underset{1\leq d\leq D}{\text{diag}}(\boldsymbol{x}_{1,n_d}')\underset{1\leq d\leq D}{\text{col}}\left(\boldsymbol{\Sigma}_d^{-1}(\boldsymbol{y}_d - \boldsymbol{X}_d\tilde{\boldsymbol{\beta}})\right) \\ \varphi_2 \underset{1\leq d\leq D}{\text{diag}}(\boldsymbol{x}_{2,n_d}')\underset{1\leq d\leq D}{\text{col}}\left(\boldsymbol{\Sigma}_d^{-1}(\boldsymbol{y}_d - \boldsymbol{X}_d\tilde{\boldsymbol{\beta}})\right) \\ \vdots \\ \varphi_p \underset{1\leq d\leq D}{\text{diag}}(\boldsymbol{x}_{p,n_d}')\underset{1\leq d\leq D}{\text{col}}\left(\boldsymbol{\Sigma}_d^{-1}(\boldsymbol{y}_d - \boldsymbol{X}_d\tilde{\boldsymbol{\beta}})\right) \end{pmatrix}.$$

13.3.2 REML Estimators

In this section we follow the same steps as in the Sect. 13.2.2 and we derive the REML estimators under the model (13.4) with more details concerning the matrix calculations. The REML log-likelihood is

$$l_{REML}(\boldsymbol{\theta}) = -\frac{1}{2}(n-p)\log 2\pi - \frac{1}{2}(n-p)\log\sigma^2 - \frac{1}{2}\log|\boldsymbol{K}'\boldsymbol{\Sigma}\boldsymbol{K}| - \frac{1}{2\sigma^2}\boldsymbol{y}'\boldsymbol{P}\boldsymbol{y},$$

where $\boldsymbol{P} = \boldsymbol{K}(\boldsymbol{K}'\boldsymbol{\Sigma}\boldsymbol{K})^{-1}\boldsymbol{K}' = \boldsymbol{\Sigma}^{-1} - \boldsymbol{\Sigma}^{-1}\boldsymbol{X}(\boldsymbol{X}'\boldsymbol{\Sigma}^{-1}\boldsymbol{X})^{-1}\boldsymbol{X}'\boldsymbol{\Sigma}^{-1}$ and $\boldsymbol{K} = \boldsymbol{W} - \boldsymbol{W}\boldsymbol{X}(\boldsymbol{X}'\boldsymbol{W}\boldsymbol{X})^{-1}\boldsymbol{X}'\boldsymbol{W}$ are such that $\boldsymbol{P}\boldsymbol{X} = \boldsymbol{0}$ and $\boldsymbol{P}\boldsymbol{\Sigma}\boldsymbol{P} = \boldsymbol{P}$. The matrix $\boldsymbol{\Sigma}$ can be written in the form

$$\boldsymbol{\Sigma} = \boldsymbol{W}^{-1} + \sum_{k=1}^{p} \varphi_k \boldsymbol{A}_k,$$

where $\boldsymbol{A}_k = \boldsymbol{Z}_k \boldsymbol{Z}_k' = \underset{1 \le d \le D}{\text{diag}} (\boldsymbol{x}_{k,n_d} \boldsymbol{x}_{k,n_d}')$, $k = 1, \ldots, p$. As $\frac{\partial \boldsymbol{P}}{\partial \varphi_k} = -\boldsymbol{P}\boldsymbol{A}_k\boldsymbol{P}$, by taking partial derivatives with respect to σ^2 and φ_k, $k = 1, \ldots, p$, one gets

$$S_{\sigma^2} = -\frac{n-p}{2\sigma^2} + \frac{1}{2\sigma^4} \boldsymbol{y}'\boldsymbol{P}\boldsymbol{y}, \quad S_{\varphi_k} = -\frac{1}{2} \text{tr}\{\boldsymbol{P}\boldsymbol{A}_k\} + \frac{1}{2\sigma^2} \boldsymbol{y}'\boldsymbol{P}\boldsymbol{A}_k\boldsymbol{P}\boldsymbol{y}, \quad k = 1, \ldots, p.$$

The second partial derivatives are

$$H_{\sigma^2\sigma^2} = \frac{n-p}{2\sigma^4} - \frac{1}{\sigma^6} \boldsymbol{y}'\boldsymbol{P}\boldsymbol{y}, \quad H_{\sigma^2\varphi_k} = -\frac{1}{2\sigma^4} \boldsymbol{y}'\boldsymbol{P}\boldsymbol{A}_k\boldsymbol{P}\boldsymbol{y},$$
$$H_{\varphi_k\varphi_i} = \frac{1}{2} \text{tr}\{\boldsymbol{P}\boldsymbol{A}_k\boldsymbol{P}\boldsymbol{A}_i\} - \frac{1}{\sigma^2} \boldsymbol{y}'\boldsymbol{P}\boldsymbol{A}_k\boldsymbol{P}\boldsymbol{A}_i\boldsymbol{P}\boldsymbol{y}, \quad k, i = 1, \ldots, p.$$

By taking expectations and multiplying by -1, we obtain the components of the Fisher information matrix

$$F_{\sigma^2\sigma^2} = \frac{n-p}{2\sigma^4}, \; F_{\sigma^2\varphi_k} = \frac{1}{2\sigma^2} \text{tr}\{\boldsymbol{P}\boldsymbol{A}_k\}, \; F_{\varphi_k\varphi_i} = \frac{1}{2} \text{tr}\{\boldsymbol{P}\boldsymbol{A}_k\boldsymbol{P}\boldsymbol{A}_i\}, \; k, i = 1, \ldots, p.$$

To calculate the REML estimates, the Fisher-scoring updating formula, at iteration i, is

$$\boldsymbol{\theta}^{(i+1)} = \boldsymbol{\theta}^{(i)} + \boldsymbol{F}^{-1}(\boldsymbol{\theta}^{(i)})\boldsymbol{S}(\boldsymbol{\theta}^{(i)}).$$

The following seeds can be used as starting values in the Fisher-scoring algorithm:

$$\sigma^{2(0)} = \varphi_1^{(0)} = \ldots = \varphi_p^{(0)} = S^2/(p+2),$$

where $S^2 = \frac{1}{n-p}(\boldsymbol{y} - \boldsymbol{X}\hat{\boldsymbol{\beta}}_0)'\boldsymbol{W}(\boldsymbol{y} - \boldsymbol{X}\hat{\boldsymbol{\beta}}_0)$ and $\hat{\boldsymbol{\beta}}_0 = (\boldsymbol{X}'\boldsymbol{W}\boldsymbol{X})^{-1}\boldsymbol{X}'\boldsymbol{W}\boldsymbol{y}$.

13.3.2.1 Matrix Calculations for the RRC Model

In this section we show how to do the matrix calculations in the Fisher-scoring algorithm. We define

$$\boldsymbol{\Sigma} = \underset{1 \le d \le D}{\text{diag}} (\boldsymbol{\Sigma}_d), \; \boldsymbol{R} = (\boldsymbol{X}'\boldsymbol{\Sigma}^{-1}\boldsymbol{X})^{-1} = \left(\sum_{d=1}^{D} \boldsymbol{X}_d' \boldsymbol{\Sigma}_d^{-1} \boldsymbol{X}_d \right)^{-1}$$

such that

$$P = \Sigma^{-1} - \Sigma^{-1} X R X' \Sigma^{-1} = \underset{1\le d\le D}{\text{diag}} (\Sigma_d^{-1}) - \underset{1\le d\le D}{\text{col}} (\Sigma_d^{-1} X_d) R \underset{1\le d\le D}{\text{col}'} (X'_d \Sigma_d^{-1}).$$

For $k = 1, \ldots, p$, the components of the vector of scores are

$$S_{\sigma^2} = -\frac{n-p}{2\sigma^2} + \frac{1}{2\sigma^4} \sum_{d=1}^{D} y_d^t \Sigma_d^{-1} y_d - \frac{1}{2\sigma^4} \left(\sum_{d=1}^{D} y_d^t \Sigma_d^{-1} X_d \right) R \left(\sum_{d=1}^{D} X_d^t \Sigma_d^{-1} y_d \right),$$

$$\begin{aligned}
S_{\varphi_k} &= -\frac{1}{2} \text{tr}\{Z'_k P Z_k\} + \frac{1}{2\sigma^2} y' P Z_k Z'_k P y \\
&= -\frac{1}{2} \sum_{d=1}^{D} x'_{k,n_d} [\Sigma_d^{-1} - \Sigma_d^{-1} X_d R X'_d \Sigma_d^{-1}] x_{k,n_d} \\
&+ \frac{1}{2\sigma^2} \sum_{d=1}^{D} y'_d \Sigma_d^{-1} x_{k,n_d} x'_{k,n_d} \Sigma_d^{-1} y_d \\
&- \frac{1}{\sigma^2} \left(\sum_{d=1}^{D} y'_d \Sigma_d^{-1} x_{k,n_d} x'_{k,n_d} \Sigma_d^{-1} X_d \right) R \left(\sum_{d=1}^{D} X'_d \Sigma_d^{-1} y_d \right) \\
&+ \frac{1}{2\sigma^2} \left(\sum_{d=1}^{D} y'_d \Sigma_d^{-1} X_d \right) R \left(\sum_{d=1}^{D} X'_d \Sigma_d^{-1} x_{k,n_d} x'_{k,n_d} \Sigma_d^{-1} X_d \right) R \left(\sum_{d=1}^{D} X'_d \Sigma_d^{-1} y_d \right).
\end{aligned}$$

For $k, k_1, k_2 = 1, \ldots, p$, the components of the REML Fisher information matrix are

$$F_{\sigma^2\sigma^2} = \frac{n-p}{2\sigma^4},$$

$$F_{\sigma^2\varphi_k} = \frac{1}{2\sigma^2} \text{tr}\{Z'_k P Z_k\} = \frac{1}{2\sigma^2} \sum_{d=1}^{D} x'_{k,n_d} \left[\Sigma_d^{-1} - \Sigma_d^{-1} X_d R X'_d \Sigma_d^{-1} \right] x_{k,n_d},$$

$$\begin{aligned}
F_{\varphi_{k_1}\varphi_{k_2}} &= \frac{1}{2} \text{tr}\{Z'_{k_2} P Z_{k_1} Z'_{k_1} P Z_{k_2}\} = \frac{1}{2} \sum_{d=1}^{D} (x'_{k_2,n_d} \Sigma_d^{-1} x_{k_1,n_d})^2 \\
&- \sum_{d=1}^{D} x'_{k_2,n_d} \Sigma_d^{-1} x_{k_1,n_d} x'_{k_1,n_d} \Sigma_d^{-1} X_d R X'_d \Sigma_d^{-1} x_{k_2,n_d} \\
&+ \frac{1}{2} \sum_{d=1}^{D} x'_{k_2,n_d} \Sigma_d^{-1} X_d R \left(\sum_{d=1}^{D} X'_d \Sigma_d^{-1} x_{k_1,n_d} x'_{k_1,n_d} \Sigma_d^{-1} X_d \right) R X'_d \Sigma_d^{-1} x_{k_2,n_d}.
\end{aligned}$$

The inverse of matrix $\boldsymbol{\Sigma}_d$ can be calculated by applying iteratively the formula

$$(A+CBD)^{-1} = A^{-1} - A^{-1}C(B^{-1}+DA^{-1}C)^{-1}DA^{-1}.$$

STEP 1: Define $\boldsymbol{L}_{1d} = \boldsymbol{W}_d^{-1} + \varphi_1 \boldsymbol{x}_{1,n_d}\boldsymbol{x}'_{1,n_d}$. Take $A = \boldsymbol{W}_d^{-1}$, $C = \varphi_1\boldsymbol{x}_{1,n_d}$, $B = \boldsymbol{I}_1$, and $D = \boldsymbol{x}'_{1,n_d}$, so that

$$\boldsymbol{L}_{1d}^{-1} = \boldsymbol{W}_d - \boldsymbol{W}_d\varphi_1\boldsymbol{x}_{1,n_d}(1+\boldsymbol{x}'_{1,n_d}\boldsymbol{W}_d\varphi_1\boldsymbol{x}_{1,n_d})^{-1}\boldsymbol{x}'_{1,n_d}\boldsymbol{W}_d.$$

STEP 2: Define $\boldsymbol{L}_{2d} = \boldsymbol{L}_{1d} + \varphi_2\boldsymbol{x}_{2,n_d}\boldsymbol{x}'_{2,n_d}$. Take $A = \boldsymbol{L}_{1d}$, $C = \varphi_2\boldsymbol{x}_{2,n_d}$, $B = \boldsymbol{I}_1$ and $D = \boldsymbol{x}'_{2,n_d}$, so that

$$\boldsymbol{L}_{2d}^{-1} = \boldsymbol{L}_{1d}^{-1} - \boldsymbol{L}_{1d}^{-1}\varphi_2\boldsymbol{x}_{2,n_d}(1+\boldsymbol{x}'_{2,n_d}\boldsymbol{L}_{1d}^{-1}\varphi_2\boldsymbol{x}_{2,n_d})^{-1}\boldsymbol{x}'_{2,n_d}\boldsymbol{L}_{1d}^{-1}.$$

...

STEP p: Finally, $\boldsymbol{L}_{pd} = \boldsymbol{L}_{p-1d} + \varphi_p\boldsymbol{x}_{p,n_d}\boldsymbol{x}'_{p,n_d}$. Take $A = \boldsymbol{L}_{p-1d}$, $C = \varphi_p\boldsymbol{x}_{p,n_d}$, $B = \boldsymbol{I}_1$, and $D = \boldsymbol{x}'_{p,n_d}$, so that

$$\boldsymbol{L}_{pd}^{-1} = \boldsymbol{L}_{p-1d}^{-1} - \boldsymbol{L}_{p-1d}^{-1}\varphi_p\boldsymbol{x}_{p,n_d}(1+\boldsymbol{x}'_{p,n_d}\boldsymbol{L}_{p-1d}^{-1}\varphi_p\boldsymbol{x}_{p,n_d})^{-1}\boldsymbol{x}'_{p,n_d}\boldsymbol{L}_{p-1d}^{-1}.$$

13.3.3 EBLUP of a Domain Mean

The formulas for the EBLUP of the linear parameter $\overline{Y}_d$ under model (13.4) have exactly the same form as the ones derived under model (13.2) in Sect. 13.2.3. Namely, if $n_d > 0$, then the EBLUP of $\overline{Y}_d$ is

$$\widehat{\overline{Y}}_d^{eblup} = \sum_{k=1}^p \overline{X}_{kd}\hat{\beta}_k + \sum_{k=1}^p \overline{X}_{kd}\hat{u}_{kd} + f_d\left[\overline{y}_{s,d} - \sum_{k=1}^p \overline{X}_{s,kd}\hat{\beta}_k - \sum_{k=1}^p \overline{X}_{s,kd}\hat{u}_{kd}\right],$$

where $\overline{X}_{kd} = \frac{1}{N_d}\sum_{j=1}^{N_d} x_{kdj}$, $\overline{y}_{s,d} = \frac{1}{n_d}\sum_{j=1}^{n_d} y_{dj}$, $\overline{X}_{s,kd} = \frac{1}{n_d}\sum_{j=1}^{n_d} x_{kdj}$ and $f_d = \frac{n_d}{N_d}$. If $n_d = 0$, then the EBLUP of $\overline{Y}_d$ is the synthetic part

$$\hat{\mu}_d^{eblup} = \sum_{k=1}^p \overline{X}_{kd}\hat{\beta}_k + \sum_{k=1}^p \overline{X}_{kd}\hat{u}_{kd}.$$

13.3.4 MSE of the EBLUP

Let $\boldsymbol{\theta} = (\sigma^2, \varphi_1, \ldots, \varphi_p)'$ be the vector of variance components and $\hat{\boldsymbol{\theta}}$ be the corresponding REML estimate. The MSEs of the EBLUPs of $\overline{Y}_d$ and μ_d are (cf. pp. 220 and 217, respectively)

$$MSE(\hat{\overline{Y}}_d^{eblup}) = g_1(\boldsymbol{\theta})+g_2(\boldsymbol{\theta})+g_3(\boldsymbol{\theta})+g_4(\boldsymbol{\theta}), \quad MSE(\hat{\mu}_d^{eblup}) = g_1(\boldsymbol{\theta})+g_2(\boldsymbol{\theta})+g_3(\boldsymbol{\theta}),$$

where

$$g_1(\boldsymbol{\theta}) = \boldsymbol{a}_r'\boldsymbol{Z}_r\boldsymbol{T}_s\boldsymbol{Z}_r'\boldsymbol{a}_r,$$
$$g_2(\boldsymbol{\theta}) = [\boldsymbol{a}_r'\boldsymbol{X}_r - \boldsymbol{a}_r'\boldsymbol{Z}_r\boldsymbol{T}_s\boldsymbol{Z}_s'\boldsymbol{V}_{es}^{-1}\boldsymbol{X}_s]\boldsymbol{Q}_s[\boldsymbol{X}_r'\boldsymbol{a}_r - \boldsymbol{X}_s'\boldsymbol{V}_{es}^{-1}\boldsymbol{Z}_s\boldsymbol{T}_s\boldsymbol{Z}_r'\boldsymbol{a}_r],$$
$$g_3(\boldsymbol{\theta}) \approx \text{tr}\left\{(\nabla\boldsymbol{b}')\boldsymbol{V}_s(\nabla\boldsymbol{b}')'E\left[(\hat{\boldsymbol{\theta}}-\boldsymbol{\theta})(\hat{\boldsymbol{\theta}}-\boldsymbol{\theta})'\right]\right\},$$
$$g_4(\boldsymbol{\theta}) = \boldsymbol{a}_r'\boldsymbol{V}_{er}\boldsymbol{a}_r$$

and definitions of the symbols $\boldsymbol{T}_s$, $\boldsymbol{Q}_s$, and $\boldsymbol{b}'$ are given in Sect. 9.2 and will be revised on the following pages. The Prasad-Rao (PR) estimator of $MSE(\hat{\overline{Y}}_d^{eblup})$ is

$$mse_d = mse(\hat{\overline{Y}}_d^{eblup}) = g_1(\hat{\boldsymbol{\theta}}) + g_2(\hat{\boldsymbol{\theta}}) + 2g_3(\hat{\boldsymbol{\theta}}) + g_4(\hat{\boldsymbol{\theta}}).$$

Now, we present a detailed description of calculation of the terms $g_i(\boldsymbol{\theta})$, $i = 1, \ldots, 4$, under the present model.

Calculation of $g_1(\boldsymbol{\theta})$

To calculate $g_1(\boldsymbol{\theta}) = \boldsymbol{a}_r'\boldsymbol{Z}_r\boldsymbol{T}_s\boldsymbol{Z}_r'\boldsymbol{a}_r$, basic elements are

$$\boldsymbol{a}_r' = \frac{1}{N_d}\underset{1\le\ell\le D}{\text{col}'}(\delta_{d\ell}\boldsymbol{1}_{N_\ell-n_\ell}'), \quad \boldsymbol{V}_u = \sigma^2 \underset{1\le k\le p}{\text{diag}}(\varphi_k\boldsymbol{I}_D),$$
$$\boldsymbol{Z}_r = \underset{1\le k\le p}{\text{col}'}(\boldsymbol{Z}_{rk}), \quad \boldsymbol{Z}_{rk} = \underset{1\le d\le D}{\text{diag}}(\boldsymbol{x}_{k,N_d-n_d}),$$
$$\boldsymbol{T}_s = \boldsymbol{V}_u - \boldsymbol{V}_u\boldsymbol{Z}_s'\boldsymbol{V}_s^{-1}\boldsymbol{Z}_s\boldsymbol{V}_u = \sigma^2 \underset{1\le k\le p}{\text{diag}}(\varphi_k\boldsymbol{I}_D)$$
$$- \sigma^2 \underset{1\le k\le p}{\text{col}}(\varphi_k\boldsymbol{Z}_{sk}')\underset{1\le\ell\le D}{\text{diag}}(\boldsymbol{\Sigma}_{s\ell}^{-1})\underset{1\le k\le p}{\text{col}'}(\varphi_k\boldsymbol{Z}_{sk}) = (\boldsymbol{T}_{k_1k_2})_{k_1,k_2=1,\ldots,p},$$

where

$$\begin{aligned}
\boldsymbol{T}_{k_1k_2} &= \sigma^2\varphi_{k_1}\delta_{k_1k_2}\boldsymbol{I}_D - \sigma^2\varphi_{k_1}\varphi_{k_2}\boldsymbol{Z}'_{sk_1} \underset{1\le\ell\le D}{\text{diag}} (\boldsymbol{\Sigma}_{s\ell}^{-1})\boldsymbol{Z}_{sk_2} \\
&= \sigma^2\varphi_{k_1}\delta_{k_1k_2}\boldsymbol{I}_D - \sigma^2\varphi_{k_1}\varphi_{k_2} \underset{1\le\ell\le D}{\text{diag}} (\boldsymbol{x}'_{k_1,n_\ell}\boldsymbol{\Sigma}_{s\ell}^{-1}\boldsymbol{x}_{k_2,n_\ell})
\end{aligned}$$

and $\delta_{k_1k_2} = 0$ if $k_1 \neq k_2$, $\delta_{k_1k_2} = 1$ if $k_1 = k_2$. Therefore

$$\begin{aligned}
g_1(\theta) &= \boldsymbol{a}'_r\boldsymbol{Z}_r\boldsymbol{T}_s\boldsymbol{Z}'_r\boldsymbol{a}_r \\
&= \frac{1}{N_d^2} \underset{1\le\ell\le D}{\text{col}'} (\delta_{d\ell}\boldsymbol{1}'_{N_\ell-n_\ell}) \underset{1\le k\le p}{\text{col}'} (\boldsymbol{Z}_{rk})\boldsymbol{T}_s \underset{1\le k\le p}{\text{col}} (\boldsymbol{Z}'_{rk}) \underset{1\le\ell\le D}{\text{col}} (\delta_{d\ell}\boldsymbol{1}_{N_\ell-n_\ell}) \\
&= \frac{1}{N_d^2} \underset{1\le\ell\le D}{\text{col}'} (\delta_{d\ell}\boldsymbol{1}'_{N_\ell-n_\ell}) \sum_{k_1=1}^{p}\sum_{k_2=1}^{p} \boldsymbol{Z}_{rk_1}\boldsymbol{T}_{k_1k_2}\boldsymbol{Z}'_{rk_2} \underset{1\le\ell\le D}{\text{col}} (\delta_{d\ell}\boldsymbol{1}_{N_\ell-n_\ell}) \\
&= \frac{1}{N_d^2}\sum_{k_1=1}^{p}\sum_{k_2=1}^{p} \underset{1\le\ell\le D}{\text{col}'} (\delta_{d\ell}\boldsymbol{1}'_{N_\ell-n_\ell}\boldsymbol{x}_{k_1,N_\ell-n_\ell})\boldsymbol{T}_{k_1k_2} \underset{1\le\ell\le D}{\text{col}} (\delta_{d\ell}\boldsymbol{x}'_{k_2,N_\ell-n_\ell}\boldsymbol{1}_{N_\ell-n_\ell}) \\
&= (1-f_d)^2\sigma^2 \sum_{k_1=1}^{p}\sum_{k_2=1}^{p} \underset{1\le\ell\le D}{\text{col}'} (\delta_{d\ell}\overline{X}^*_{k_1\ell}) \\
&\quad \cdot \Big[\varphi_{k_1}\delta_{k_1k_2}\boldsymbol{I}_D - \varphi_{k_1}\varphi_{k_2} \underset{1\le\ell\le D}{\text{diag}} (\boldsymbol{x}'_{k_1,n_\ell}\boldsymbol{\Sigma}_{s\ell}^{-1}\boldsymbol{x}_{k_2,n_\ell})\Big] \underset{1\le\ell\le D}{\text{col}} (\delta_{d\ell}\overline{X}^*_{k_2\ell}) \\
&= (1-f_d)^2\sigma^2 \left\{ \sum_{k=1}^{p}\varphi_k\overline{X}^{*2}_{kd} - \sum_{k_1=1}^{p}\sum_{k_2=1}^{p}\varphi_{k_1}\varphi_{k_2}\overline{X}^*_{k_1d}\boldsymbol{x}'_{k_1,n_d}\boldsymbol{\Sigma}_{sd}^{-1}\boldsymbol{x}_{k_2,n_d}\overline{X}^*_{k_2d} \right\},
\end{aligned}$$

where $f_d = n_d/N_d$ and $\overline{X}^*_{kd} = \frac{1}{N_d-n_d}\sum_{j\in r_d} x_{kdj} = (1-f_d)^{-1}(\overline{X}_{kd} - f_d\overline{x}_{kd})$.

Calculation of $g_2(\boldsymbol{\theta})$

We recall that

$$\begin{aligned}
g_2(\boldsymbol{\theta}) &= [\boldsymbol{a}'_r\boldsymbol{X}_r - \boldsymbol{a}'_r\boldsymbol{Z}_r\boldsymbol{T}_s\boldsymbol{Z}'_s\boldsymbol{V}_{es}^{-1}\boldsymbol{X}_s]\boldsymbol{Q}_s[\boldsymbol{X}'_r\boldsymbol{a}_r - \boldsymbol{X}'_s\boldsymbol{V}_{es}^{-1}\boldsymbol{Z}_s\boldsymbol{T}_s\boldsymbol{Z}'_r\boldsymbol{a}_r] \\
&= [\boldsymbol{a}'_1 - \boldsymbol{a}'_2]\boldsymbol{Q}_s[\boldsymbol{a}_1 - \boldsymbol{a}_2],
\end{aligned}$$

where $\boldsymbol{Q}_s = (\boldsymbol{X}_s'\boldsymbol{V}^{-1}\boldsymbol{X}_s)^{-1} = \sigma^2\left(\sum_{d=1}^D \boldsymbol{X}_{sd}'\boldsymbol{\Sigma}_{sd}^{-1}\boldsymbol{X}_{sd}\right)^{-1}$ and $\boldsymbol{V}_{es}^{-1} = \sigma^{-2}\boldsymbol{W}_s$. The second vector is

$$\boldsymbol{a}_2' = \boldsymbol{a}_r' \underset{1\le k\le p}{\text{col}'}(\boldsymbol{Z}_{rk})\boldsymbol{T}_s \underset{1\le k\le p}{\text{col}}(\boldsymbol{Z}_{sk}')\sigma^{-2}\boldsymbol{W}_s\boldsymbol{X}_s = \sigma^{-2}\boldsymbol{a}_r' \sum_{k_1=1}^p\sum_{k_2=1}^p \boldsymbol{Z}_{rk_1}\boldsymbol{T}_{k_1k_2}\boldsymbol{Z}_{sk_2}'\boldsymbol{W}_s\boldsymbol{X}_s$$

$$= \frac{1}{N_d}\underset{1\le\ell\le D}{\text{col}'}(\delta_{d\ell}\boldsymbol{1}_{N_\ell-n_\ell}')\sum_{k_1=1}^p\sum_{k_2=1}^p \underset{1\le\ell\le D}{\text{diag}}(\boldsymbol{x}_{k_1,N_\ell-n_\ell})$$

$$\cdot\left[\varphi_{k_1}\delta_{k_1k_2}\boldsymbol{I}_D - \varphi_{k_1}\varphi_{k_2}\underset{1\le\ell\le D}{\text{diag}}(\boldsymbol{x}_{k_1,n_\ell}'\boldsymbol{\Sigma}_{s\ell}^{-1}\boldsymbol{x}_{k_2,n_\ell})\right]$$

$$\cdot\underset{1\le\ell\le D}{\text{diag}}(\boldsymbol{x}_{k_2,n_\ell}')\underset{1\le\ell\le D}{\text{diag}}(\boldsymbol{W}_{s\ell})\underset{1\le\ell\le D}{\text{col}}(\boldsymbol{X}_{s\ell})$$

$$= \frac{1}{N_d}\sum_{k_1=1}^p\sum_{k_2=1}^p \underset{1\le\ell\le D}{\text{col}'}(\delta_{d\ell}\boldsymbol{1}_{N_\ell-n_\ell}'\boldsymbol{x}_{k_1,N_\ell-n_\ell})$$

$$\cdot\left[\varphi_{k_1}\delta_{k_1k_2}\boldsymbol{I}_D - \varphi_{k_1}\varphi_{k_2}\underset{1\le\ell\le D}{\text{diag}}(\boldsymbol{x}_{k_1,n_\ell}'\boldsymbol{\Sigma}_{s\ell}^{-1}\boldsymbol{x}_{k_2,n_\ell})\right]$$

$$\cdot\underset{1\le\ell\le D}{\text{diag}}(\boldsymbol{x}_{k_2,n_\ell}')\underset{1\le\ell\le D}{\text{col}}(\boldsymbol{W}_{s\ell}\boldsymbol{X}_{s\ell})$$

$$=(1-f_d)\left\{\sum_{k=1}^p \varphi_k\overline{X}_{kd}^*\boldsymbol{x}_{k,n_d}' - \sum_{k_1=1}^p\sum_{k_2=1}^p \varphi_{k_1}\varphi_{k_2}\overline{X}_{k_1d}^*\boldsymbol{x}_{k_1,n_d}'\boldsymbol{\Sigma}_{sd}^{-1}\boldsymbol{x}_{k_2,n_d}\boldsymbol{x}_{k_2,n_d}'\right\}\boldsymbol{W}_{sd}\boldsymbol{X}_{sd}.$$

The first vector is

$$\boldsymbol{a}_1' = \boldsymbol{a}_r'\boldsymbol{X}_r = \frac{1}{N_d}\boldsymbol{1}_{N_d-n_d}'\boldsymbol{X}_{rd} = \frac{1}{N_d}\sum_{j\in r_d}\boldsymbol{x}_{dj} = (1-f_d)\overline{\boldsymbol{X}}_d^*, \quad \overline{\boldsymbol{X}}_d^* = (\overline{X}_{1d}^*,\ldots,\overline{X}_{pd}^*).$$

Calculation of $g_3(\boldsymbol{\theta})$

We recall that

$$g_3(\boldsymbol{\theta}) \approx \text{tr}\left\{(\nabla\boldsymbol{b}')\boldsymbol{V}_s(\nabla\boldsymbol{b}')'E\left[(\hat{\boldsymbol{\theta}}-\boldsymbol{\theta})(\hat{\boldsymbol{\theta}}-\boldsymbol{\theta})'\right]\right\},$$

where

$$\boldsymbol{b}' = \boldsymbol{a}_r'\boldsymbol{Z}_r\boldsymbol{V}_u\boldsymbol{Z}_s'\boldsymbol{V}_s^{-1} = \boldsymbol{a}_r'\underset{1\le k\le p}{\text{col}'}(\boldsymbol{Z}_{rk})\underset{1\le k\le p}{\text{diag}}(\varphi_k\boldsymbol{I}_D)\underset{1\le k\le p}{\text{col}}(\boldsymbol{Z}_{sk}')\boldsymbol{\Sigma}_s^{-1}$$

$$= \boldsymbol{a}_r'\sum_{k=1}^p\varphi_k\boldsymbol{Z}_{rk}\boldsymbol{Z}_{sk}'\underset{1\le\ell\le D}{\text{diag}}(\boldsymbol{\Sigma}_{s\ell}^{-1}).$$

The first derivative is $\frac{\partial \boldsymbol{b}'}{\partial \sigma^2} = 0$. As $\frac{\partial \boldsymbol{\Sigma}_{s\ell}}{\partial \varphi_k} = \boldsymbol{x}_{k,n_\ell}\boldsymbol{x}'_{k,n_\ell}$, the remaining derivatives are

$$\frac{\partial \boldsymbol{b}'}{\partial \varphi_k} = \boldsymbol{a}'_r \boldsymbol{Z}_{rk}\boldsymbol{Z}'_{sk} \underset{1\le\ell\le D}{\text{diag}} (\boldsymbol{\Sigma}_{s\ell}^{-1}) - \boldsymbol{a}'_r \left(\sum_{i=1}^{p} \varphi_i \boldsymbol{Z}_{ri}\boldsymbol{Z}'_{si}\right) \underset{1\le\ell\le D}{\text{diag}} (\boldsymbol{\Sigma}_{s\ell}^{-1}\boldsymbol{x}_{k,n_\ell}\boldsymbol{x}'_{k,n_\ell}\boldsymbol{\Sigma}_{s\ell}^{-1}),$$

$k = 1, \ldots, p$, where the formula for derivative of an inverse matrix given in Appendix A was used. As $\boldsymbol{Z}_{rk} = \underset{1\le\ell\le D}{\text{diag}} (\boldsymbol{x}_{k,N_\ell-n_\ell})$, then

$$\begin{aligned}
\frac{\partial \boldsymbol{b}'}{\partial \varphi_k} &= \frac{1}{N_d} \underset{1\le\ell\le D}{\text{col}'} (\delta_{d\ell}\boldsymbol{1}'_{N_\ell-n_\ell})\Big[\underset{1\le\ell\le D}{\text{diag}} (\boldsymbol{x}_{k,N_\ell-n_\ell}\boldsymbol{x}'_{k,n_\ell}\boldsymbol{\Sigma}_{s\ell}^{-1}) \\
&\quad - \sum_{i=1}^{p} \varphi_i \underset{1\le\ell\le D}{\text{diag}} (\boldsymbol{x}_{i,N_\ell-n_\ell}\boldsymbol{x}'_{i,n_\ell}) \underset{1\le\ell\le D}{\text{diag}} (\boldsymbol{\Sigma}_{s\ell}^{-1}\boldsymbol{x}_{k,n_\ell}\boldsymbol{x}'_{k,n_\ell}\boldsymbol{\Sigma}_{s\ell}^{-1})\Big] \\
&= \frac{1}{N_d} \underset{1\le\ell\le D}{\text{col}'} (\delta_{d\ell}\boldsymbol{1}'_{N_\ell-n_\ell}\boldsymbol{x}_{k,N_\ell-n_\ell}\boldsymbol{x}'_{k,n_\ell}\boldsymbol{\Sigma}_{s\ell}^{-1}) \\
&\quad - \frac{1}{N_d} \underset{1\le\ell\le D}{\text{col}'} \left(\delta_{d\ell}\left(\sum_{i=1}^{p} \varphi_i \boldsymbol{1}'_{N_\ell-n_\ell}\boldsymbol{x}_{i,N_\ell-n_\ell}\boldsymbol{x}'_{i,n_\ell}\right)\boldsymbol{\Sigma}_{s\ell}^{-1}\boldsymbol{x}_{k,n_\ell}\boldsymbol{x}'_{k,n_\ell}\boldsymbol{\Sigma}_{s\ell}^{-1}\right) \\
&= (1-f_d)\Big[\underset{1\le\ell\le D}{\text{col}'} (\delta_{d\ell}\overline{X}^*_{k\ell}\boldsymbol{x}'_{k,n_\ell}\boldsymbol{\Sigma}_{s\ell}^{-1}) \\
&\quad - \underset{1\le\ell\le D}{\text{col}'} \left(\delta_{d\ell}\left(\sum_{i=1}^{p} \varphi_i \overline{X}^*_{i\ell}\boldsymbol{x}'_{i,n_\ell}\right)\boldsymbol{\Sigma}_{s\ell}^{-1}\boldsymbol{x}_{k,n_\ell}\boldsymbol{x}'_{k,n_\ell}\boldsymbol{\Sigma}_{s\ell}^{-1}\right)\Big], \quad k = 1, \ldots, p.
\end{aligned}$$

Let us define $\boldsymbol{Q}(\boldsymbol{\theta}) = (q_{k_1,k_2})_{k_1,k_2=0,1,\ldots,p}$, where $q_{0,k} = q_{k,0} = 0$, $k = 0, 1, \ldots, p$ and

$$\begin{aligned}
q_{k_1,k_2} &= \frac{\partial \boldsymbol{b}'}{\partial \varphi_{k_1}} \boldsymbol{V}_s \left(\frac{\partial \boldsymbol{b}'}{\partial \varphi_{k_2}}\right)' = \sigma^2(1-f_d)^2\Big[\underset{1\le\ell\le D}{\text{col}'} (\delta_{d\ell}\overline{X}^*_{k_1\ell}\boldsymbol{x}'_{k_1,n_\ell}\boldsymbol{\Sigma}_{s\ell}^{-1}) \\
&\quad - \underset{1\le\ell\le D}{\text{col}'} \Big[\delta_{d\ell}\left(\sum_{i=1}^{p} \varphi_i \overline{X}^*_{i\ell}\boldsymbol{x}'_{i,n_\ell}\right)\boldsymbol{\Sigma}_{s\ell}^{-1}\boldsymbol{x}_{k_1,n_\ell}\boldsymbol{x}'_{k_1,n_\ell}\boldsymbol{\Sigma}_{s\ell}^{-1}\Big]\Big] \\
&\quad \cdot \underset{1\le\ell\le D}{\text{diag}} (\boldsymbol{\Sigma}_{s\ell})\Big[\underset{1\le\ell\le D}{\text{col}} (\delta_{d\ell}\boldsymbol{\Sigma}_{s\ell}^{-1}\boldsymbol{x}_{k_2,n_\ell}\overline{X}^*_{k_2\ell}) \\
&\quad - \underset{1\le\ell\le D}{\text{col}} \Big[\delta_{d\ell}\boldsymbol{\Sigma}_{s\ell}^{-1}\boldsymbol{x}_{k_2,n_\ell}\boldsymbol{x}'_{k_2,n_\ell}\boldsymbol{\Sigma}_{s\ell}^{-1}\left(\sum_{i=1}^{p} \varphi_i \boldsymbol{x}_{i,n_\ell}\overline{X}^*_{i\ell}\right)\Big]\Big]
\end{aligned}$$

for any $k_1, k_2 = 1, \ldots, p$. After further simplifications we obtain

$$
\begin{aligned}
q_{k_1,k_2} &= \sigma^2(1-f_d)^2 \overline{X}^*_{k_1 d} \boldsymbol{x}'_{k_1,n_d} \boldsymbol{\Sigma}^{-1}_{sd} \boldsymbol{x}_{k_2,n_d} \overline{X}^*_{k_2 d} \\
&- \sigma^2(1-f_d)^2 \overline{X}^*_{k_1 d} \boldsymbol{x}'_{k_1,n_d} \boldsymbol{\Sigma}^{-1}_{sd} \boldsymbol{x}_{k_2,n_d} \boldsymbol{x}'_{k_2,n_d} \boldsymbol{\Sigma}^{-1}_{sd} \left(\sum_{i=1}^{p} \varphi_i \boldsymbol{x}_{i,n_d} \overline{X}^*_{id} \right) \\
&- \sigma^2(1-f_d)^2 \left(\sum_{i=1}^{p} \varphi_i \overline{X}^*_{id} \boldsymbol{x}'_{i,n_d} \right) \boldsymbol{\Sigma}^{-1}_{sd} \boldsymbol{x}_{k_1,n_d} \boldsymbol{x}'_{k_1,n_d} \boldsymbol{\Sigma}^{-1}_{sd} \boldsymbol{x}_{k_2,n_d} \overline{X}^*_{k_2 d} \\
&+ \sigma^2(1-f_d)^2 \left(\sum_{i=1}^{p} \varphi_i \overline{X}^*_{id} \boldsymbol{x}'_{i,n_d} \right) \boldsymbol{\Sigma}^{-1}_{sd} \boldsymbol{x}_{k_1,n_d} \boldsymbol{x}'_{k_1,n_d} \boldsymbol{\Sigma}^{-1}_{sd} \boldsymbol{x}_{k_2,n_d} \boldsymbol{x}'_{k_2,n_d} \boldsymbol{\Sigma}^{-1}_{sd} \\
&\cdot \left(\sum_{i=1}^{p} \varphi_i \boldsymbol{x}_{i,n_d} \overline{X}^*_{id} \right), \quad k_1, k_2 = 1, \ldots, p.
\end{aligned}
$$

Then

$$
g_3(\boldsymbol{\theta}) \approx \operatorname{tr}\left\{ \boldsymbol{Q}(\boldsymbol{\theta}) \boldsymbol{F}^{-1}(\boldsymbol{\theta}) \right\},
$$

where $\boldsymbol{F}(\boldsymbol{\theta})$ is the REML Fisher information matrix.

Calculation of $g_4(\boldsymbol{\theta})$

We recall that $g_4(\boldsymbol{\theta}) = \boldsymbol{a}'_r \boldsymbol{V}_{er} \boldsymbol{a}_r$, where

$$
\boldsymbol{a}'_r = \frac{1}{N_d} \underset{1 \le \ell \le D}{\operatorname{col}'} (\delta_{d\ell} \boldsymbol{1}'_{N_\ell - n_\ell}), \quad \boldsymbol{V}^{-1}_{er} = \sigma^{-2} \boldsymbol{W}_r = \sigma^{-2} \underset{1 \le d \le D}{\operatorname{diag}} \{ \boldsymbol{W}_{rd} \}.
$$

Therefore

$$
\begin{aligned}
g_4(\boldsymbol{\theta}) &= \frac{1}{N_d} \underset{1 \le \ell \le D}{\operatorname{col}'} (\delta_{d\ell} \boldsymbol{1}'_{N_\ell - n_\ell}) \sigma^2 \underset{1 \le \ell \le D}{\operatorname{diag}} \{ \boldsymbol{W}^{-1}_{r\ell} \} \frac{1}{N_d} \underset{1 \le \ell \le D}{\operatorname{col}} (\delta_{d\ell} \boldsymbol{1}_{N_\ell - n_\ell}) \\
&= \frac{\sigma^2}{N_d^2} \boldsymbol{1}'_{N_d - n_d} \underset{j \in r_d}{\operatorname{diag}} \{ w^{-1}_{dj} \} \boldsymbol{1}_{N_d - n_d} = \frac{\sigma^2}{N_d^2} \sum_{j \in r_d} \frac{1}{w_{dj}}.
\end{aligned}
$$

13.4 R Codes for EBLUPs

This section gives R codes for fitting the RRC model to the survey data file `LFS20.txt`. The target variable y is the variable INCOME. As auxiliary variables we take REGISTERED and EDUCATION. The function `dir2` is employed for calculating direct estimators. The domains are defined by the variable AREA crossed by SEX. The parameters of interest are the income means by domains.

We install and/or load some R packages: `Matrix`, `lme4`, and `sae`.

```
if(!require(Matrix)){
  install.packages("Matrix")
  library(Matrix)
}
if(!require(lme4)){
  install.packages("lme4")
  library(lme4)
}
if(!require(sae)){
  install.packages("sae")
  library(sae)
}
```

The following code reads the data files and calculate some variables:

```
# Read unit-level data
dat <- read.table("LFS20.txt", header=TRUE, sep = "\t", dec = ".")
# Education level 2
edu2 <- as.numeric(dat$EDUCATION==2)
# Education level 3
edu3 <- as.numeric(dat$EDUCATION==3)
# Read domain-level data
aux <- read.table("Nds20.txt", header=TRUE, sep = "\t", dec = ".")
# Prop. of registered people
aux$mreg <- aux$reg/aux$N
# Proportion of edu2 people
aux$medu2 <- aux$edu2/aux$N
# Proportion of edu3 people
aux$medu3 <- aux$edu3/aux$N
```

We calculate direct estimators of domain average incomes and the population sizes by domain, by using `dir2` function described in Sect. 2.8.4. We also define some new variables.

```
income.dir <- dir2(data=dat$INCOME, w=dat$WEIGHT, domain=list(sex=dat$SEX,
                   area=dat$AREA))
diry <- income.dir$mean          # Direct estimates of domain means
hatNd <- income.dir$Nd           # Direct estimates of population sizes
nd <- income.dir$nd              # Sample sizes
fd <- nd/aux$N                   # Sample fractions
```

The following code calculates sample means by domains:

```
dat2 <- data.frame(income=dat$INCOME, edu2, edu3, reg=dat$REGISTERED)
smeans <- aggregate(dat2, by=list(sex=dat$SEX,area=dat$AREA), mean)
meany <- smeans$income              # Sample means of income
meanedu2 <- smeans$edu2             # Sample means of edu2
meanedu3 <- smeans$edu3             # Sample means of edu3
meanreg <- smeans$reg               # Sample means of registered
```

We fit a random regression coefficient model with INCOME as dependent variable and REGISTERED and EDUCATION as explanatory variables. The fitted model has a

Table 13.1 Estimated regression parameters of RRC model

Parameter	Estimate	Std. error	t-value	p-value
Intercept	40,187.0	485.6	82.75	0.00
Registered	−11,742.2	1124.9	−10.44	0.00
edu2	9704.1	674.8	14.38	0.00
edu3	20,013.6	957.0	20.91	0.00

random intercept and random slopes on the coefficients of the categories edu2 and edu3 of the variable EDUCATION. The employed R code is

```
dat$EDUCATION <- as.factor(dat$EDUCATION)
rrc <- lmer(formula=INCOME ~ REGISTERED + EDUCATION + (EDUCATION|AREA:SEX),
            data=dat, REML=FALSE)
summary(rrc)                                  # Summary of the fitting procedure
anova(rrc)                                    # Analysis of Variance Table
beta <- fixef(rrc); beta                      # Regression parameters
var <- as.data.frame(VarCorr(rrc))            # Variance parameters
ref <- ranef(rrc)[[1]]                        # Modes of the random effects
head(fitted(rrc))                             # Predicted values
residuals <- resid(rrc)                       # Residuals
p.values <- 2*pnorm(abs(coef(summary(rrc))[,3]), low=F)
p.values                                      # p values
```

Table 13.1 presents the estimated regression parameters and p-values.
We calculate the EBLUPs of income means by domain.

```
Xbeta <- beta[1] + beta[2]*aux$mreg + beta[3]*aux$medu2 + beta[4]*aux$medu3
Xubeta <- ref[,1] + aux$medu2*ref[,2] + aux$medu3*ref[,3]
mu <- Xbeta + Xubeta               # Projective estimates of income means
xbeta <- beta[1] + beta[2]*meanreg + beta[3]*meanedu2 + beta[4]*meanedu3
xubeta <- ref[,1] + meanedu2*ref[,2] + meanedu3*ref[,3]
mu.s <- meany - xbeta - xubeta
eb <- mu + fd*mu.s                 # EBLUPs of income means
```

Summary of results

```
output <- data.frame(Nd=aux$N[c(T,F)], hatNd=hatNd[c(T,F)] , nd=nd[c(T,F)],
          meany=round(meany[c(T,F)],0), dir=round(diry[c(T,F)],0),
          hatmu=round(mu[c(T,F)],0), eblup=round(eb[c(T,F)],0))
head(output, 10)
```

Figure 13.1 (left) plots the RRC model residuals $\hat{e}_d = y_d - \hat{y}_d$. The residuals are situated symmetrically around 0. Figure 13.1 (right) plots the EBLUPs and direct estimates of men income means by areas. The EBLUPs behave more smoothly than the direct estimators.

For the ten first areas, Table 13.2 gives a summary of results for men. The population sizes, the estimated population sizes, and the sample sizes are denoted by N_d, $\hat{N}_d$, and n_d, respectively. The columns meany and dir contain the sample means and the direct estimates of the population mean of the variable income. The projective predictors and the EBLUPs of the population means are labelled by hatmu and eblup, respectively.

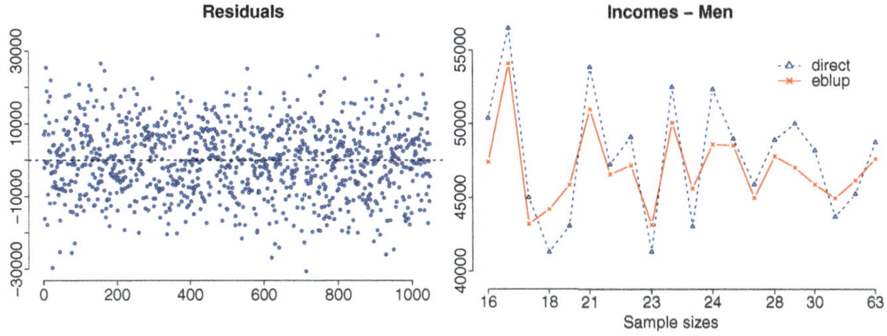

Fig. 13.1 Plots of residuals (left) and estimated men income means (right)

Table 13.2 Estimates of domain mean incomes for men

Area	N_d	$\hat{N}_d$	n_d	meany	dir	hatmu	eblup
1	8020	7950	29	49,729	50,005	47,013	47,021
2	3576	3522	18	43,847	45,019	43,191	43,199
3	4446	4359	24	51,968	52,482	50,053	50,065
4	4807	4740	27	45,748	45,889	44,974	44,977
5	3252	3375	22	46,247	47,250	46,575	46,574
6	5461	5316	30	48,562	48,193	45,863	45,879
7	4023	3984	23	48,082	49,103	47,167	47,177
8	3816	3796	23	42,005	41,310	43,140	43,130
9	10,785	10,873	63	48,202	48,746	47,607	47,612
10	3256	3195	19	43,684	43,068	45,878	45,865

References

Dempster, A.P., Rubin, D.B., Tsutakawa, R.K.: Estimation in covariance component models. J. Am. Stat. Assoc. **76**, 341–353 (1981)

Hobza, T., Morales, D.: Small area estimation under random regression coefficient models. J. Stat. Comput. Simul. **83**(11), 2160–2177 (2013)

Moura, F.A.S., Holt, D.: Small area estimation using multilevel models. Surv. Methodol. **25**(1), 73–80 (1999)

Prasad, N.G.N., Rao, J.N.K.: The estimation of the mean squared error of small-area estimators. J. Am. Stat. Assoc. **85**, 163–171 (1990)

Chapter 14
EBPs Under Unit-Level Logit Mixed Models

14.1 Introduction

The binomial-logit mixed models are generalized linear mixed models (GLMM) for dichotomous or counting variables that take into account variability between domains, which is not explained through auxiliary variables, by introducing random effects. The random effects are usually assumed to be normally distributed. Inferences based on GLMMs have some computational difficulties because the likelihood may involve high-dimensional integrals which cannot be evaluated analytically. This chapter describes the method of simulated moments (MSM), introduced by Jiang (1998), for fitting the GLMMs. This method approximates the method of moments (MM), is computationally attractive, and gives consistent estimators of model parameters. As alternative fitting methods, the EM and the ML-Laplace approximation algorithms are introduced.

Section 14.6 gives empirical best predictors (EBP) for estimating weighted sum of probabilities, where the weights are known quantities taken from administrative registers or census files. This is to say, the term "weighting" is distinct from sample survey weighting based on the survey design. The EBPs are sums of predicted values of probabilities (sometimes called soft estimates) rather than sums of target variable predictions that must be 0 or 1 (hard estimates). The estimation of domain averages of dichotomous variables (domain proportions) is also treated.

The statistical methodology given by Jiang and Lahiri (2001), Jiang (2003), and Hobza and Morales (2016) is summarized and EBPs are introduced under unit-level logit mixed models. Plug-in estimators are also considered.

The MSE is a standard accuracy measure for point estimators. Jiang and Lahiri (2001) and Jiang (2003) studied the approximation of the MSE of the EBP in the context of binary data and GLMM. Their approach is based on Taylor series expansions. They further gave a second order bias-corrected estimator of the MSE. As these MSE analytic estimators are computationally expensive in practice, the parametric bootstrap estimator introduced by González-Manteiga et al. (2007) in

D. Morales et al., *A Course on Small Area Estimation and Mixed Models*, Statistics for Social and Behavioral Sciences, https://doi.org/10.1007/978-3-030-63757-6_14

the context of logistic mixed models is proposed. This approach was later extended by González-Manteiga et al. (2008a,b) to nested error regression models and to multivariate area-level models, respectively.

The following sections introduce the unit-level logit mixed model, give three fitting algorithms, describe several model-based predictors of population-based and model-based parameters, and treat the problem of MSE estimation. The chapter also contains a section that illustrates the introduced methodology with R codes applied to synthetic data.

14.2 The Unit-Level Logit Mixed Model

Let us consider a set of domain random effects $\{v_d : d = 1, \dots, D\}$ that are i.i.d. $N(0, 1)$. In matrix notation, we write $\boldsymbol{v} = \underset{1 \le d \le D}{\text{col}}(v_d) \sim N_D(\boldsymbol{0}, \boldsymbol{I}_D)$, i.e.

$$f(\boldsymbol{v}) = (2\pi)^{-D/2} \exp\left\{-\frac{1}{2}\,\boldsymbol{v}'\boldsymbol{v}\right\}.$$

The unit-level logit mixed model assumes that the distribution of the target variable y_{dj}, conditioned to the random effect v_d, is

$$y_{dj}|_{v_d} \sim \text{Bin}(m_{dj}, p_{dj}), \quad d = 1, \dots, D, \ \ j = 1, \dots, n_d, \tag{14.1}$$

where m_{dj} is supposed to be a known size parameter. The binomial distribution is usually used for counting numbers of successes. If the sampling units are individuals we might be interested in the presence or absence of a certain characteristic and the corresponding size parameter is one. If the sampling units are e.g. households we can model the number of household members with a certain property. Then the size parameter is the number of household members. In both cases, it is possible to assume that the size parameters are known quantities for the sampled elements. For the natural parameter, the logit link is

$$\eta_{dj} = \log \frac{p_{dj}}{1 - p_{dj}} = \boldsymbol{x}_{dj}\boldsymbol{\beta} + \phi v_d, \quad d = 1, \dots, D, \ \ j = 1, \dots, n_d, \tag{14.2}$$

where $\phi > 0$ is a standard deviation parameter, $\boldsymbol{\beta} = \underset{1 \le k \le p}{\text{col}}(\beta_k)$ is the column vector of regression parameters, and $\boldsymbol{x}_{dj} = \underset{1 \le k \le p}{\text{col}'}(x_{djk})$ is the row vector of known auxiliary variables. Further, the model assumes that the y_{dj}'s are independent conditioned to the vector $\boldsymbol{v}$ of random effects. For the ease of exposition, this chapter uses the vector notation $\boldsymbol{y} = (\boldsymbol{y}_1', \dots, \boldsymbol{y}_D')'$ and $\boldsymbol{y}_d = (y_{d1}, \dots, y_{dn_d})'$,

$d = 1, \ldots, D$. It holds that

$$P(y_{dj}|\boldsymbol{v}) = \binom{m_{dj}}{y_{dj}} p_{dj}^{y_{dj}} (1 - p_{dj})^{m_{dj}-y_{dj}}, \quad P(\boldsymbol{y}|\boldsymbol{v}) = \prod_{d=1}^{D} \prod_{j=1}^{n_d} P(y_{dj}|\boldsymbol{v}),$$

where

$$p_{dj} = \frac{\exp\left\{\boldsymbol{x}_{dj}\boldsymbol{\beta} + \phi v_d\right\}}{1 + \exp\left\{\boldsymbol{x}_{dj}\boldsymbol{\beta} + \phi v_d\right\}} = \frac{\exp\{\eta_{dj}\}}{1 + \exp\{\eta_{dj}\}}, \quad 1 - p_{dj} = \frac{1}{1 + \exp\{\eta_{dj}\}}.$$

The marginal distribution of $\boldsymbol{y}$ is

$$P(\boldsymbol{y}) = \int_{R^D} P(\boldsymbol{y}|\boldsymbol{v}) f(\boldsymbol{v})\, d\boldsymbol{v} = \int_{R^D} \psi(\boldsymbol{y}, \boldsymbol{v})\, d\boldsymbol{v},$$

where

$$\begin{aligned}
\psi(\boldsymbol{y}, \boldsymbol{v}) &= (2\pi)^{-\frac{D}{2}} \exp\left\{\frac{-\boldsymbol{v}'\boldsymbol{v}}{2}\right\} \prod_{d=1}^{D} \prod_{j=1}^{n_d} \frac{\binom{m_{dj}}{y_{dj}} \exp\left\{y_{dj}(\boldsymbol{x}_{dj}\boldsymbol{\beta} + \phi v_d)\right\}}{\left[1 + \exp\left\{\boldsymbol{x}_{dj}\boldsymbol{\beta} + \phi v_d\right\}\right]^{m_{dj}}} \\
&= (2\pi)^{-\frac{D}{2}} \prod_{d=1}^{D} \prod_{j=1}^{n_d} \binom{m_{dj}}{y_{dj}} \exp\left\{\frac{-\boldsymbol{v}'\boldsymbol{v}}{2}\right\} \exp\left\{\sum_{k=1}^{p} \Big(\sum_{d=1}^{D} \sum_{j=1}^{n_d} y_{dj} x_{djk}\Big) \beta_k \right. \\
&\quad \left. + \phi \sum_{d=1}^{D} y_{d.} v_d - \sum_{d=1}^{D} \sum_{j=1}^{n_d} m_{dj} \log\left(1 + \exp\left\{\boldsymbol{x}_{dj}\boldsymbol{\beta} + \phi v_d\right\}\right)\right\}
\end{aligned}$$

and $y_{d.} = \sum_{j=1}^{n_d} y_{dj}$. A set of sufficient statistics for $\boldsymbol{\theta} = (\boldsymbol{\beta}', \phi)'$ is

$$\sum_{d=1}^{D} \sum_{j=1}^{n_d} y_{dj} x_{djk}, \quad k = 1, \ldots, p; \quad y_{d.}, \; d = 1, \ldots, D.$$

14.3 MSM Algorithm

This section derives an algorithm to fit the unit-level logit mixed model by using the method of simulated moments (MSM). Let $\boldsymbol{\theta} = (\theta_1, \ldots, \theta_{p+1})'$ be the vector of model parameters, where $\theta_1 = \beta_1, \ldots, \theta_p = \beta_p, \theta_{p+1} = \phi$. A set of equations for applying the method of moments (MM) is

$$0 = f_k(\boldsymbol{\theta}) = M_k(\boldsymbol{\theta}) - \hat{M}_k = \sum_{d=1}^{D}\sum_{j=1}^{n_d} E_\theta[y_{dj}]x_{djk} - \sum_{d=1}^{D}\sum_{j=1}^{n_d} y_{dj}x_{djk},\ k = 1, \dots p,$$

$$0 = f_{p+1}(\boldsymbol{\theta}) = M_{p+1}(\boldsymbol{\theta}) - \hat{M}_{p+1} = \sum_{d=1}^{D} E_\theta[y_{d.}^2] - \sum_{d=1}^{D} y_{d.}^2. \tag{14.3}$$

The updating formula of the Newton–Raphson algorithm for solving the system (14.3) of nonlinear equations is

$$\boldsymbol{\theta}^{(i+1)} = \boldsymbol{\theta}^{(i)} - \boldsymbol{H}^{-1}(\boldsymbol{\theta}^{(i)})\boldsymbol{f}(\boldsymbol{\theta}^{(i)}),$$

where

$$\boldsymbol{f}(\boldsymbol{\theta}) = \underset{1\le k\le p+1}{\text{col}}(f_k(\boldsymbol{\theta})), \quad \boldsymbol{H}(\boldsymbol{\theta}) = \left(\frac{\partial f_k(\boldsymbol{\theta})}{\partial \theta_\ell}\right)_{k,\ell=1,\dots,p+1}.$$

In what follows, we calculate the expectations appearing in $\boldsymbol{f}(\boldsymbol{\theta})$ and its partial derivatives. We start with the first p MM equations. The expectation of y_{dj} is

$$E_\theta[y_{dj}] = E_v\big[E_\theta[y_{dj}|\boldsymbol{v}]\big] = E_v[m_{dj}p_{dj}] = E_v\left[\frac{m_{dj}\exp\{\boldsymbol{x}_{dj}\boldsymbol{\beta} + \phi v_d\}}{1 + \exp\{\boldsymbol{x}_{dj}\boldsymbol{\beta} + \phi v_d\}}\right]$$
$$= \int_{-\infty}^{\infty} \frac{m_{dj}\exp\{\boldsymbol{x}_{dj}\boldsymbol{\beta} + \phi v_d\}}{1 + \exp\{\boldsymbol{x}_{dj}\boldsymbol{\beta} + \phi v_d\}} f(v_d)\,dv_d.$$

Under the assumption that it is possible to interchange the order of derivatives and expectation, the corresponding derivatives are

$$\frac{\partial E_\theta[y_{dj}]}{\partial \beta_k} = E_v\Big[m_{dj}\frac{\partial p_{dj}}{\partial \beta_k}\Big] = E_v\big[m_{dj}p_{dj}(1-p_{dj})x_{djk}\big],$$
$$\frac{\partial E_\theta[y_{dj}]}{\partial \phi} = E_v\Big[m_{dj}\frac{\partial p_{dj}}{\partial \phi}\Big] = E_v\big[m_{dj}p_{dj}(1-p_{dj})v_d\big].$$

The expectation of $y_{d.}^2$ is $E_\theta[y_{d.}^2] = E_v\big[E_\theta[y_{d.}^2|\boldsymbol{v}]\big]$, where

$$y_{d.}^2 = \sum_{j=1}^{n_d} y_{dj}^2 + \sum_{\substack{j_1=1 \\ j_1\neq j_2}}^{n_d}\sum_{j_2=1}^{n_d} y_{dj_1}y_{dj_2},$$
$$E_\theta[y_{dj}^2|\boldsymbol{v}] = \text{var}_\theta[y_{dj}|\boldsymbol{v}] + E_\theta^2[y_{dj}|\boldsymbol{v}] = m_{dj}p_{dj}(1-p_{dj}) + m_{dj}^2p_{dj}^2.$$

Therefore, we have

$$E_\theta[y_{d.}^2|\boldsymbol{v}] = \sum_{j=1}^{n_d} E_\theta[y_{dj}^2|\boldsymbol{v}] + \sum_{\substack{j_1=1 \\ j_1\neq j_2}}^{n_d}\sum_{j_2=1}^{n_d} E_\theta[y_{dj_1}|\boldsymbol{v}]\, E_\theta[y_{dj_2}|\boldsymbol{v}]$$

$$= \sum_{j=1}^{n_d} m_{dj}p_{dj}(1-p_{dj}) + \sum_{j=1}^{n_d} m_{dj}^2 p_{dj}^2 + \sum_{\substack{j_1=1 \\ j_1\neq j_2}}^{n_d}\sum_{j_2=1}^{n_d} m_{dj_1}p_{dj_1}m_{dj_2}p_{dj_2}$$

$$= \sum_{j=1}^{n_d} m_{dj}p_{dj}(1-p_{dj}) + \Big(\sum_{j=1}^{n_d} m_{dj}p_{dj}\Big)^2,$$

$$E_\theta[y_{d.}^2] = E_v\Big[\sum_{j=1}^{n_d} m_{dj}p_{dj}(1-p_{dj}) + \Big(\sum_{j=1}^{n_d} m_{dj}p_{dj}\Big)^2\Big].$$

Let us define $\xi_d = \sum_{j=1}^{n_d} m_{dj}p_{dj}$. The derivatives of $E_\theta[y_{d.}^2]$ are

$$\frac{\partial E_\theta[y_{d.}^2]}{\partial\beta_k} = E_v\Big[\frac{\partial}{\partial\beta_k}\Big\{\sum_{j=1}^{n_d} m_{dj}p_{dj}(1-p_{dj}) + \Big(\sum_{j=1}^{n_d} m_{dj}p_{dj}\Big)^2\Big\}\Big]$$

$$= E_v\Big[\sum_{j=1}^{n_d}\Big\{m_{dj}\frac{\partial p_{dj}}{\partial\beta_k}(1-p_{dj}) - m_{dj}p_{dj}\frac{\partial p_{dj}}{\partial\beta_k}\Big\}$$

$$+ 2\Big(\sum_{j=1}^{n_d} m_{dj}p_{dj}\Big)\Big(\sum_{j=1}^{n_d} m_{dj}\frac{\partial p_{dj}}{\partial\beta_k}\Big)\Big]$$

$$= \sum_{j=1}^{n_d} E_v\Big[\Big\{m_{dj}p_{dj}(1-p_{dj})^2 - m_{dj}p_{dj}^2(1-p_{dj})$$

$$+ 2m_{dj}p_{dj}(1-p_{dj})\xi_d\Big\}x_{djk}\Big]$$

$$= \sum_{j=1}^{n_d} E_v\left[m_{dj}p_{dj}(1-p_{dj})\left\{1-2(p_{dj}-\xi_d)\right\}x_{djk}\right],$$

$$\frac{\partial E_\theta[y_{d.}^2]}{\partial\phi} = \sum_{j=1}^{n_d} E_v\left[m_{dj}p_{dj}(1-p_{dj})\left\{1-2(p_{dj}-\xi_d)\right\}v_d\right].$$

The elements of the matrix of first partial derivatives are

$$H_{k\ell} = \frac{\partial f_k(\boldsymbol{\theta})}{\partial \theta_\ell} = \sum_{d=1}^{D} \sum_{j=1}^{n_d} \frac{\partial E_\theta[y_{dj}]}{\partial \theta_\ell} x_{djk}, \quad k = 1, \ldots, p, \; \ell = 1, \ldots, p+1,$$

$$H_{p+1\ell} = \frac{\partial f_{p+1}(\boldsymbol{\theta})}{\partial \theta_\ell} = \sum_{d=1}^{D} \frac{\partial E_\theta[y_{d.}^2]}{\partial \theta_\ell}, \quad \ell = 1, \ldots, p+1.$$

The expectations appearing in $\boldsymbol{f}(\boldsymbol{\theta})$ and $\boldsymbol{H}(\boldsymbol{\theta})$ can be approximated by Monte Carlo simulation. Assume that $\boldsymbol{\theta} = (\boldsymbol{\beta}', \phi)'$ is the current parameter in the iteration i of the Newton–Raphson algorithm. Then the expectations can be approximated as follows.

1. For $s = 1, \ldots, S$, $d = 1, \ldots, D$, generate $v_d^{(s)}$ i.i.d. $N(0, 1)$ and $v_d^{(S+s)} = -v_d^{(s)}$.
2. For $s = 1, \ldots, 2S$, $d = 1, \ldots, D$, $j = 1, \ldots, n_d$, calculate

$$p_{dj}^{(s)} = \frac{\exp\left\{\boldsymbol{x}_{dj}\boldsymbol{\beta} + \phi v_d^{(s)}\right\}}{1 + \exp\left\{\boldsymbol{x}_{dj}\boldsymbol{\beta} + \phi v_d^{(s)}\right\}}, \quad \xi_d^{(s)} = \sum_{j=1}^{n_d} m_{dj} p_{dj}^{(s)}.$$

3. Output 1: For $d = 1, \ldots, D$, $j = 1, \ldots, n_d$, do

$$\hat{E}_\theta[y_{dj}] = \frac{1}{2S} \sum_{s=1}^{2S} m_{dj} p_{dj}^{(s)},$$

$$\hat{E}_\theta[y_{d.}^2] = \frac{1}{2S} \sum_{s=1}^{2S} \left\{ \sum_{j=1}^{n_d} m_{dj} p_{dj}^{(s)} (1 - p_{dj}^{(s)}) + \left(\sum_{j=1}^{n_d} m_{dj} p_{dj}^{(s)} \right)^2 \right\}.$$

4. Output 2: For $k = 1, \ldots, p$, $d = 1, \ldots, D$, $j = 1, \ldots, n_d$, do

$$\frac{\partial \hat{E}_\theta[y_{dj}]}{\partial \beta_k} = \frac{1}{2S} \sum_{s=1}^{2S} m_{dj} p_{dj}^{(s)} (1 - p_{dj}^{(s)}) x_{djk},$$

$$\frac{\partial \hat{E}_\theta[y_{dj}]}{\partial \phi} = \frac{1}{2S} \sum_{s=1}^{2S} m_{dj} p_{dj}^{(s)} (1 - p_{dj}^{(s)}) v_d^{(s)}.$$

5. Output 3: For $k = 1, \ldots, p$, $d = 1, \ldots, D$, do

$$\frac{\partial \hat{E}_\theta[y_{d.}^2]}{\partial \beta_k} = \frac{1}{2S} \sum_{s=1}^{2S} \sum_{j=1}^{n_d} m_{dj} p_{dj}^{(s)} (1 - p_{dj}^{(s)}) \left\{1 - 2(p_{dj}^{(s)} - \xi_d^{(s)})\right\} x_{djk},$$

$$\frac{\partial \hat{E}_\theta[y_{d.}^2]}{\partial \phi} = \frac{1}{2S} \sum_{s=1}^{2S} \sum_{j=1}^{n_d} m_{dj} p_{dj}^{(s)} (1 - p_{dj}^{(s)}) \left\{1 - 2(p_{dj}^{(s)} - \xi_d^{(s)})\right\} v_d^{(s)}.$$

Let $\hat{\boldsymbol{f}}(\boldsymbol{\theta})$ and $\hat{\boldsymbol{H}}(\boldsymbol{\theta})$ be the approximations to $\boldsymbol{f}(\boldsymbol{\theta})$ and $\boldsymbol{H}(\boldsymbol{\theta})$, respectively. Then, the complete algorithm can be summarized as follows.

1. Set the initial values $i = 0$ and $\boldsymbol{\theta}^{(0)} = (\boldsymbol{\beta}^{(0)}, \phi^{(0)})$.
2. Repeat the following steps till convergence
   a. Run the Monte Carlo algorithm for $\boldsymbol{\theta}^{(i)}$ and calculate $\hat{\boldsymbol{f}}(\boldsymbol{\theta}^{(i)})$ and $\hat{\boldsymbol{H}}(\boldsymbol{\theta}^{(i)})$.
   b. Update $\boldsymbol{\theta}^{(i)}$ by using the equation
   $$\boldsymbol{\theta}^{(i+1)} = \boldsymbol{\theta}^{(i)} - \hat{\boldsymbol{H}}^{-1}(\boldsymbol{\theta}^{(i)})\hat{\boldsymbol{f}}(\boldsymbol{\theta}^{(i)}).$$
   c. Update the iteration index $i \leftarrow i + 1$.
3. Output: $\boldsymbol{\theta}^{(i)}$.

As algorithm seed for $\boldsymbol{\beta}$, we propose to use $\boldsymbol{\beta}^{(0)} = \tilde{\boldsymbol{\beta}}$, where $\tilde{\boldsymbol{\beta}}$ is the maximum likelihood estimator under the model without random effects. In that model the natural parameters are

$$\eta_{dj} = \boldsymbol{x}_{dj}\boldsymbol{\beta}, \quad d = 1, \ldots, D, \ \ j = 1, \ldots, n_d.$$

Concerning the variance parameters, we propose to use

$$\phi^{(0)} = \Big(\frac{1}{n}\sum_{d=1}^{D}\sum_{j=1}^{n_d}(\tilde{\eta}_{dj} - \hat{\eta}_{d.}^{dir})^2\Big)^{1/2}, \quad \hat{\eta}_{d.}^{dir} = \log\frac{\hat{p}_{d.}^{dir}}{1-\hat{p}_{d.}^{dir}}, \quad \hat{p}_{d.}^{dir} = \frac{1}{n_d}\sum_{j=1}^{n_d}\frac{y_{dj}}{m_{dj}},$$

where $\tilde{\eta}_{dj} = \boldsymbol{x}_{dj}\tilde{\boldsymbol{\beta}}$ and $n = \sum_{d=1}^{D} n_d$.

The asymptotic variance of the MSM estimators can be approximated by a Taylor expansion of $\boldsymbol{M}(\hat{\boldsymbol{\theta}}) = \underset{1\le k\le p+1}{\text{col}}(M_k(\hat{\boldsymbol{\theta}}))$ around $\boldsymbol{\theta}$. This is to say (cf. (14.3))

$$\hat{\boldsymbol{M}} = \boldsymbol{M}(\hat{\boldsymbol{\theta}}) \approx \boldsymbol{M}(\boldsymbol{\theta}) + \boldsymbol{H}(\boldsymbol{\theta})(\hat{\boldsymbol{\theta}} - \boldsymbol{\theta}), \quad \hat{\boldsymbol{\theta}} - \boldsymbol{\theta} \approx \boldsymbol{H}^{-1}(\boldsymbol{\theta})(\hat{\boldsymbol{M}} - \boldsymbol{M}(\boldsymbol{\theta})),$$

where $\hat{\boldsymbol{M}} = \underset{1\le k\le p+1}{\text{col}}(\hat{M}_k)$. Let us note that $\hat{\boldsymbol{M}}$ is an unbiased estimator of $\boldsymbol{M}(\boldsymbol{\theta})$ but the same need not hold for the parameter estimate $\hat{\boldsymbol{\theta}}$ since the system of equations is nonlinear. Nevertheless, under regularity conditions, $\hat{\boldsymbol{\theta}}$ is consistent estimator of $\boldsymbol{\theta}$, thus its variance can be approximated as

$$\text{var}(\hat{\boldsymbol{\theta}}) \approx E[(\hat{\boldsymbol{\theta}} - \boldsymbol{\theta})(\hat{\boldsymbol{\theta}} - \boldsymbol{\theta})'] \approx \boldsymbol{H}^{-1}(\boldsymbol{\theta})\text{var}(\hat{\boldsymbol{M}})\boldsymbol{H}^{-1}(\boldsymbol{\theta}).$$

An estimator of $\text{var}(\hat{\boldsymbol{\theta}})$ is

$$\widehat{\text{var}}(\hat{\boldsymbol{\theta}}) = \hat{\boldsymbol{H}}^{-1}(\hat{\boldsymbol{\theta}})\widehat{\text{var}}(\hat{\boldsymbol{M}})\hat{\boldsymbol{H}}^{-1}(\hat{\boldsymbol{\theta}}),$$

where $\hat{\boldsymbol{\theta}} = \boldsymbol{\theta}^{(i+1)}$ is taken from the output of the MSM algorithm and $\widehat{\text{var}}(\hat{\boldsymbol{M}})$ is an estimator of the covariance matrix of $\hat{\boldsymbol{M}}$.

A bootstrap algorithm to estimate $\text{var}(\hat{\boldsymbol{\theta}})$ is

1. Fit the model to the sample and calculate $\hat{\boldsymbol{\theta}}$.
2. From the fitted model, generate bootstrap samples

$$\{y_{dj}^{(b)} : d = 1, \ldots, D, j = 1, \ldots, n_d\}, \quad b = 1, \ldots, B.$$

3. Calculate $\hat{\boldsymbol{M}}^{(b)}$, $b = 1, \ldots, B$, and

$$\overline{\boldsymbol{M}} = \frac{1}{B}\sum_{b=1}^{B} \hat{\boldsymbol{M}}^{(b)}, \quad \widehat{\text{var}}_B(\hat{\boldsymbol{M}}) = \frac{1}{B}\sum_{b=1}^{B}(\hat{\boldsymbol{M}}^{(b)} - \overline{\boldsymbol{M}})(\hat{\boldsymbol{M}}^{(b)} - \overline{\boldsymbol{M}})'.$$

4. Output:

$$\widehat{\text{var}}_A(\hat{\boldsymbol{\theta}}) = \hat{\boldsymbol{H}}^{-1}(\hat{\boldsymbol{\theta}})\widehat{\text{var}}_B(\hat{\boldsymbol{M}})\hat{\boldsymbol{H}}^{-1}(\hat{\boldsymbol{\theta}}).$$

Alternatively, another estimator of $\text{var}(\hat{\theta})$ can be obtained by replacing steps 3 and 4 of the previous algorithm by the following steps

3. Fit the model to the bootstrap samples and calculate

$$\hat{\boldsymbol{\theta}}^{(b)}, \; b = 1, \ldots, B; \quad \overline{\boldsymbol{\theta}} = \frac{1}{B}\sum_{b=1}^{B} \hat{\boldsymbol{\theta}}^{(b)}.$$

4. Output:

$$\widehat{\text{var}}_B(\hat{\boldsymbol{\theta}}) = \frac{1}{B}\sum_{b=1}^{B}(\hat{\boldsymbol{\theta}}^{(b)} - \overline{\boldsymbol{\theta}})(\hat{\boldsymbol{\theta}}^{(b)} - \overline{\boldsymbol{\theta}})'.$$

14.4 EM Algorithm

14.4.1 Introduction

Although some earlier authors employed the EM algorithm in some special setups, Dempster et al. (1977) gave its name and its full description for the first time. The EM algorithm is used to find (local) maximum likelihood estimators of parameters in models involving random effects (whose realizations are not known) or missing values. Wu (1983) gave a rigorous proof of the convergence of the EM algorithm. This section gives a short description of the EM algorithm and develops the corresponding particularization to estimate the regression and variance parameters of the unit-level logit mixed model.

Let $\boldsymbol{y} \in R^n$ and $\boldsymbol{v} \in R^D$ be the observed and unobserved data, respectively; so that $(\boldsymbol{y}, \boldsymbol{v})$ is the hypothetical complete data and $\boldsymbol{y}$ is the incomplete observed data. Let $f_\theta(\boldsymbol{y}, \boldsymbol{v}) = f_\theta(\boldsymbol{y}|\boldsymbol{v}) f(\boldsymbol{v})$ be the joint probability density function (p.d.f.) of $(\boldsymbol{y}, \boldsymbol{v})$ and let $f_\theta(\boldsymbol{v}|\boldsymbol{y})$ be the conditional p.d.f. of $\boldsymbol{v}$ given $\boldsymbol{y}$. Here we note that if the variable $\boldsymbol{y}$ has discrete distribution, which is the case of the unit-level logit mixed model, we will use the notation $f_\theta(\boldsymbol{y}, \boldsymbol{v}) = P_\theta(\boldsymbol{y}|\boldsymbol{v}) f(\boldsymbol{v})$. The marginal log-likelihood of $\boldsymbol{y}$ is

$$\ell(\boldsymbol{\theta}) = \log \int_{R^D} f_\theta(\boldsymbol{y}, \boldsymbol{v})\, d\boldsymbol{v}$$

and the MLE of $\boldsymbol{\theta}$ is

$$\hat{\boldsymbol{\theta}} = \operatorname{argmax}_\theta \ell(\boldsymbol{\theta}).$$

Another expression of the marginal p.d.f. of $\boldsymbol{y}$ is

$$f_\theta(\boldsymbol{y}) = \frac{f_\theta(\boldsymbol{y}, \boldsymbol{v})}{f_\theta(\boldsymbol{v}|\boldsymbol{y})}.$$

Therefore

$$\ell(\boldsymbol{\theta}) = \log f_\theta(\boldsymbol{y}) = \log f_\theta(\boldsymbol{y}, \boldsymbol{v}) - \log f_\theta(\boldsymbol{v}|\boldsymbol{y}). \tag{14.4}$$

The problem of maximizing (14.4) is that $\boldsymbol{v}$ is unobservable. Let us then rewrite (14.4) using expectations with respect to $f_{\theta_0}(\boldsymbol{v}|\boldsymbol{y})$, where $\boldsymbol{\theta}_0 \in \Theta$ is the current value of the parameter (value at the current step of the EM algorithm). We obtain

$$\begin{aligned}
\ell(\boldsymbol{\theta}) &= \log f_\theta(\boldsymbol{y}) \cdot 1 = \log f_\theta(\boldsymbol{y}) \int_{R^D} f_{\theta_0}(\boldsymbol{v}|\boldsymbol{y})\, d\boldsymbol{v} = \int_{R^D} \log f_\theta(\boldsymbol{y})\, f_{\theta_0}(\boldsymbol{v}|\boldsymbol{y})\, d\boldsymbol{v} \\
&= \int_{R^D} \log f_\theta(\boldsymbol{y}, \boldsymbol{v})\, f_{\theta_0}(\boldsymbol{v}|\boldsymbol{y})\, d\boldsymbol{v} - \int_{R^D} \log f_\theta(\boldsymbol{v}|\boldsymbol{y})\, f_{\theta_0}(\boldsymbol{v}|\boldsymbol{y})\, d\boldsymbol{v} \\
&= M(\boldsymbol{\theta}|\boldsymbol{\theta}_0) - N(\boldsymbol{\theta}|\boldsymbol{\theta}_0).
\end{aligned} \tag{14.5}$$

In the formula (14.5), $\boldsymbol{\theta}_0$ and $\boldsymbol{\theta}$ can be interpreted as the parameter values at iterations i and $i+1$ of the EM algorithm. Proposition 14.1 says that when the EM algorithm moves from the actual value $\boldsymbol{\theta}_0$ to any other value $\boldsymbol{\theta}$, then $h(\boldsymbol{\theta}) = N(\boldsymbol{\theta}|\boldsymbol{\theta}_0)$ decreases. This is why the EM algorithm improves $\ell(\boldsymbol{\theta})$ in each step by only maximizing $M(\boldsymbol{\theta}|\boldsymbol{\theta}_0)$ in $\boldsymbol{\theta} \in \Theta$.

Proposition 14.1 *Under regularity assumptions, it holds that*

$$N(\boldsymbol{\theta}|\boldsymbol{\theta}_0) \leq N(\boldsymbol{\theta}_0|\boldsymbol{\theta}_0) \quad \forall \boldsymbol{\theta} \in \Theta.$$

Proof We give a sketch of the proof for the case $\Theta = R$. Let us define the function

$$h(\theta) = \int \log f_\theta(z) f_{\theta_0}(z)\, dz\,.$$

For the first derivative it holds (we assume that under regularity assumptions it is possible to interchange the order of integral and derivative)

$$\dot{h}(\theta) = \int \frac{\frac{\partial}{\partial\theta} f_\theta(z)}{f_\theta(z)} f_{\theta_0}(z)\, dz = 0 \quad \text{if} \quad \theta = \theta_0.$$

The second derivative is

$$\ddot{h}(\theta) = \int \frac{f_\theta(z)\frac{\partial^2}{\partial\theta^2} f_\theta(z) - \left(\frac{\partial}{\partial\theta} f_\theta(z)\right)^2}{f_\theta^2(z)} f_{\theta_0}(z)\, dz$$

and

$$\ddot{h}(\theta_0) = \int \frac{\partial^2 f_\theta(z)}{\partial\theta^2}\bigg|_{\theta=\theta_0} dz - \int \frac{1}{f_{\theta_0}(z)}\left(\frac{\partial f_\theta(z)}{\partial\theta}\bigg|_{\theta=\theta_0}\right)^2 dz$$
$$= 0 - \int \frac{1}{f_{\theta_0}(z)}\left(\frac{\partial f_\theta(z)}{\partial\theta}\bigg|_{\theta=\theta_0}\right)^2 dz < 0,$$

which means that the function $h(\theta)$ has maximum at θ_0. □

EM Algorithm The EM algorithm starts at an initial value $\boldsymbol{\theta}^{(0)}$. At stage $i+1$, the EM algorithm makes two steps

- Expectation step: Calculate the expectation

$$M(\boldsymbol{\theta}|\boldsymbol{\theta}^{(i)}) = \int_{R^D} \log f_{\boldsymbol{\theta}}(\boldsymbol{y}, \boldsymbol{v})\, f_{\boldsymbol{\theta}^{(i)}}(\boldsymbol{v}|\boldsymbol{y})\, d\boldsymbol{v}.$$

- Maximization step: Obtain $\theta^{(i+1)}$ by maximizing $M(\boldsymbol{\theta}|\boldsymbol{\theta}^{(i)})$, i.e.

$$\theta^{(i+1)} = \text{argmax}_{\theta\in\Theta}\, M(\boldsymbol{\theta}|\boldsymbol{\theta}^{(i)}).$$

14.4.2 EM Algorithm for the Logit Regression Model

Let us now return to the unit-level logit mixed model (14.1)–(14.2) and recall the notation $\boldsymbol{y}_d = (y_{d1}, \ldots, y_{dn_d})'$, $d = 1, \ldots, D$. Under the assumed model it holds that $\boldsymbol{y}_1, \ldots, \boldsymbol{y}_D$ are unconditionally independent and we have

$$\begin{aligned}
f_{\theta^{(i)}}(\boldsymbol{v}|\boldsymbol{y}) &= \frac{P_{\theta^{(i)}}(\boldsymbol{y}|\boldsymbol{v}) f(\boldsymbol{v})}{P_{\theta^{(i)}}(\boldsymbol{y})} = \frac{\left(\prod_{d=1}^D P_{\theta^{(i)}}(\boldsymbol{y}_d|v_d)\right)\left(\prod_{d=1}^D f(v_d)\right)}{\int_{R^D} P_{\theta^{(i)}}(\boldsymbol{y}|\boldsymbol{v}) f(\boldsymbol{v})\, d\boldsymbol{v}} \\
&= \frac{\prod_{d=1}^D f(v_d) P_{\theta^{(i)}}(\boldsymbol{y}_d|v_d)}{\prod_{d=1}^D \int_R P_{\theta^{(i)}}(\boldsymbol{y}_d|v_d) f(v_d)\, dv_d}, \\
M(\boldsymbol{\theta}|\boldsymbol{\theta}^{(i)}) &= \int_{R^D} (\log P_\theta(\boldsymbol{y}|\boldsymbol{v}) + \log f(\boldsymbol{v}))\, f_{\theta^{(i)}}(\boldsymbol{v}|\boldsymbol{y})\, d\boldsymbol{v} \\
&= \sum_{d=1}^D \int_{R^D} \log f(v_d) f_{\theta^{(i)}}(\boldsymbol{v}|\boldsymbol{y})\, d\boldsymbol{v} + \sum_{d=1}^D \int_{R^D} \log P_\theta(\boldsymbol{y}_d|v_d) f_{\theta^{(i)}}(\boldsymbol{v}|\boldsymbol{y})\, d\boldsymbol{v} \\
&= M_1(\boldsymbol{\theta}^{(i)}) + M_2(\boldsymbol{\theta}|\boldsymbol{\theta}^{(i)}),
\end{aligned}$$

where

$$\begin{aligned}
M_2(\boldsymbol{\theta}|\boldsymbol{\theta}^{(i)}) &= \sum_{d=1}^D \int_{R^D} \log P_\theta(\boldsymbol{y}_d|v_d) \frac{\prod_{\ell=1}^D P_{\theta^{(i)}}(\boldsymbol{y}_\ell|v_\ell) f(v_\ell)}{\prod_{\ell=1}^D \int_R P_{\theta^{(i)}}(\boldsymbol{y}_\ell|v_\ell) f(v_\ell)\, dv_\ell}\, d\boldsymbol{v} \\
&= \sum_{d=1}^D \Big\{ \int_R \log P_\theta(\boldsymbol{y}_d|v_d) \frac{P_{\theta^{(i)}}(\boldsymbol{y}_d|v_d) f(v_d)}{\int_R P_{\theta^{(i)}}(\boldsymbol{y}_d|v_d) f(v_d)\, dv_d}\, dv_d \\
&\quad \cdot \prod_{\ell \neq d} \int_R \frac{P_{\theta^{(i)}}(\boldsymbol{y}_\ell|v_\ell) f(v_\ell)}{\int_R P_{\theta^{(i)}}(\boldsymbol{y}_\ell|v_\ell) f(v_\ell)\, dv_\ell}\, dv_\ell \Big\} \\
&= \sum_{d=1}^D \frac{\int_R \log P_\theta(\boldsymbol{y}_d|v_d)\, P_{\theta^{(i)}}(\boldsymbol{y}_d|v_d) f(v_d)\, dv_d}{\int_R P_{\theta^{(i)}}(\boldsymbol{y}_d|v_d) f(v_d)\, dv_d}.
\end{aligned}$$

Let us define the Monte Carlo weights and functions

$$w_d^{(s)}(\boldsymbol{\theta}^{(i)}) = \frac{P_{\theta^{(i)}}(\boldsymbol{y}_d|v_d^{(s)})}{\sum_{s=1}^{2S} P_{\theta^{(i)}}(\boldsymbol{y}_d|v_d^{(s)})} = \frac{A_d^{(s)}(\boldsymbol{\theta}^{(i)})}{B_d(\boldsymbol{\theta}^{(i)})}, \quad m_d^{(s)}(\boldsymbol{\theta}) = \log P_\theta(\boldsymbol{y}_d|v_d^{(s)}),$$

where $v_d^{(s)}$, $s = 1, \ldots, 2S$, are realizations of $N(0, 1)$-distributed random variables and

$$\begin{aligned}
A_d^{(s)}(\boldsymbol{\theta}^{(i)}) &= \exp\Big\{ \sum_{j=1}^{n_d} y_{dj}(\boldsymbol{x}_{dj}\boldsymbol{\beta}^{(i)} + \phi^{(i)} v_d^{(s)}) \\
&\quad - \sum_{j=1}^{n_d} m_{dj} \log\Big[1 + \exp\{\boldsymbol{x}_{dj}\boldsymbol{\beta}^{(i)} + \phi^{(i)} v_d^{(s)}\}\Big]\Big\}, \\
B_d(\boldsymbol{\theta}^{(i)}) &= \sum_{s=1}^{2S} \exp\Big\{ \sum_{j=1}^{n_d} y_{dj}(\boldsymbol{x}_{dj}\boldsymbol{\beta}^{(i)} + \phi^{(i)} v_d^{(s)})
\end{aligned}$$

$$- \sum_{j=1}^{n_d} m_{dj} \log\Big[1 + \exp\{\boldsymbol{x}_{dj}\boldsymbol{\beta}^{(i)} + \phi^{(i)} v_d^{(s)}\}\Big]\Big\},$$

$$m_d^{(s)}(\boldsymbol{\theta}) = \sum_{j=1}^{n_d} \log\binom{m_{dj}}{y_{dj}} + \sum_{j=1}^{n_d} y_{dj}(\boldsymbol{x}_{dj}\boldsymbol{\beta} + \phi v_d^{(s)})$$

$$- \sum_{j=1}^{n_d} m_{dj} \log\Big[1 + \exp\{\boldsymbol{x}_{dj}\boldsymbol{\beta} + \phi v_d^{(s)}\}\Big].$$

The term $M_2(\boldsymbol{\theta}|\boldsymbol{\theta}^{(i)})$ can be estimated by means of the following antithetic Monte Carlo algorithm.

1. For $s = 1, \ldots, S$, generate $v_d^{(s)}$ i.i.d. $N(0, 1)$ and $v_d^{(S+s)} = -v_d^{(s)}$.
2. Calculate

$$\hat{M}_2(\boldsymbol{\theta}|\boldsymbol{\theta}^{(i)}) = \sum_{d=1}^{D} \sum_{s=1}^{2S} w_d^{(s)}(\boldsymbol{\theta}^{(i)})\, m_d^{(s)}(\boldsymbol{\theta}).$$

In each step i, the EM algorithm looks for a value of $\boldsymbol{\theta}$ improving $\hat{M}(\boldsymbol{\theta}|\boldsymbol{\theta}^{(i)}) = M_1(\boldsymbol{\theta}^{(i)}) + \hat{M}_2(\boldsymbol{\theta}|\boldsymbol{\theta}^{(i)})$. The first derivatives of $m_d^{(s)}(\boldsymbol{\theta})$ are

$$u_{d,k}^{(s)}(\boldsymbol{\theta}) = \frac{\partial m_d^{(s)}(\boldsymbol{\theta})}{\partial \beta_k} = \sum_{j=1}^{n_d} y_{dj}x_{djk} - \sum_{j=1}^{n_d} m_{dj} p_{dj}^{(s)}(\boldsymbol{\theta}) x_{djk}, \; k = 1, \ldots, p,$$

$$u_{d,p+1}^{(s)}(\boldsymbol{\theta}) = \frac{\partial m_d^{(s)}(\boldsymbol{\theta})}{\partial \phi} = \sum_{j=1}^{n_d} y_{dj} v_d^{(s)} - \sum_{j=1}^{n_d} m_{dj} p_{dj}^{(s)}(\boldsymbol{\theta}) v_d^{(s)},$$

where

$$p_{dj}^{(s)}(\boldsymbol{\theta}) = \frac{\exp\{\boldsymbol{x}_{dj}\boldsymbol{\beta} + \phi v_d^{(s)}\}}{1 + \exp\{\boldsymbol{x}_{dj}\boldsymbol{\beta} + \phi v_d^{(s)}\}}.$$

For $k, k_1, k_2 = 1, \ldots, p$, the second derivatives of $m_d^{(s)}(\boldsymbol{\theta})$ are

$$h_{d,k_1k_2}^{(s)}(\boldsymbol{\theta}) = \frac{\partial^2 m_d^{(s)}(\boldsymbol{\theta})}{\partial \beta_{k_1} \partial \beta_{k_2}} = - \sum_{j=1}^{n_d} m_{dj} x_{djk_1} x_{djk_2} p_{dj}^{(s)}(\boldsymbol{\theta})(1 - p_{dj}^{(s)}(\boldsymbol{\theta})),$$

$$h_{d,p+1k}^{(s)}(\boldsymbol{\theta}) = \frac{\partial^2 m_d^{(s)}(\boldsymbol{\theta})}{\partial \phi \partial \beta_k} = - \sum_{j=1}^{n_d} m_{dj} x_{djk} v_d^{(s)} p_{dj}^{(s)}(\boldsymbol{\theta})(1 - p_{dj}^{(s)}(\boldsymbol{\theta})),$$

$$h_{d,p+1p+1}^{(s)}(\boldsymbol{\theta}) = \frac{\partial^2 m_d^{(s)}(\boldsymbol{\theta})}{\partial \phi^2} = - \sum_{j=1}^{n_d} m_{dj} v_d^{(s)2} p_{dj}^{(s)}(\boldsymbol{\theta})(1 - p_{dj}^{(s)}(\boldsymbol{\theta})).$$

For $k, k_1, k_2 = 1, \ldots, p+1$, the derivatives of $\hat{M}(\boldsymbol{\theta}|\boldsymbol{\theta}^{(i)})$ are

$$u_k(\boldsymbol{\theta}|\boldsymbol{\theta}^{(i)}) = \frac{\partial \hat{M}(\boldsymbol{\theta}|\boldsymbol{\theta}^{(i)})}{\partial \theta_k} = \sum_{d=1}^{D}\sum_{s=1}^{2S} w_d^{(s)}(\boldsymbol{\theta}^{(i)}) u_{d,k}^{(s)}(\boldsymbol{\theta}),$$

$$h_{k_1k_2}(\boldsymbol{\theta}|\boldsymbol{\theta}^{(i)}) = \frac{\partial^2 \hat{M}(\boldsymbol{\theta}|\boldsymbol{\theta}^{(i)})}{\partial \theta_{k_1}\partial \theta_{k_2}} = \sum_{d=1}^{D}\sum_{s=1}^{2S} w_d^{(s)}(\boldsymbol{\theta}^{(i)}) h_{d,k_1k_2}^{(s)}(\boldsymbol{\theta}).$$

We define the vector of scores and the Hessian matrix as

$$\boldsymbol{u}(\boldsymbol{\theta}|\boldsymbol{\theta}^{(i)}) = \underset{1\le k\le p+1}{\text{col}} (u_k(\boldsymbol{\theta}|\boldsymbol{\theta}^{(i)})), \quad \boldsymbol{H}(\boldsymbol{\theta}|\boldsymbol{\theta}^{(i)}) = \left(h_{k_1k_2}(\boldsymbol{\theta}|\boldsymbol{\theta}^{(i)})\right)_{k_1,k_2=1,\ldots,p+1}.$$

Using this notation, the EM algorithm can be summarized in the following form.

1. Set the initial values $i = 0$ and $\boldsymbol{\theta}^{(0)} = (\boldsymbol{\beta}^{(0)}, \phi^{(0)})$.
2. Repeat the following steps till convergence
   a. E step: Calculate $\boldsymbol{u}(\boldsymbol{\theta}|\boldsymbol{\theta}^{(i)})$ and $\boldsymbol{H}(\boldsymbol{\theta}|\boldsymbol{\theta}^{(i)})$.
   b. M step: Update $\boldsymbol{\theta}^{(i)}$ by using the equation
   $$\boldsymbol{\theta}^{(i+1)} = \boldsymbol{\theta}^{(i)} - \boldsymbol{H}^{-1}(\boldsymbol{\theta}^{(i)}|\boldsymbol{\theta}^{(i)})\boldsymbol{u}(\boldsymbol{\theta}^{(i)}|\boldsymbol{\theta}^{(i)}).$$
   c. Update the iteration index $i \leftarrow i+1$.
3. Output: $\boldsymbol{\theta}^{(i)}$.

Step 2.b. can be substituted by the following more sophisticated steps.

1. Set the initial values $r = i$ and $\boldsymbol{\theta}^{(r)} = (\boldsymbol{\beta}^{(i)}, \phi^{(i)})$.
2. Repeat the following steps till convergence
   a. Update $\boldsymbol{\theta}^{(r)}$ by using the equation
   $$\boldsymbol{\theta}^{(r+1)} = \boldsymbol{\theta}^{(r)} - \boldsymbol{H}^{-1}(\boldsymbol{\theta}^{(r)}|\boldsymbol{\theta}^{(i)})\boldsymbol{u}(\boldsymbol{\theta}^{(r)}|\boldsymbol{\theta}^{(i)}).$$
   b. Update the iteration index $r \leftarrow r+1$.
3. Output: $\boldsymbol{\theta}^{(i+1)} \leftarrow \boldsymbol{\theta}^{(r+1)}$.

For approximating the asymptotic covariance matrix of $\hat{\boldsymbol{\theta}}$, we recall that

$$\ell(\boldsymbol{\theta}) = \log P_\theta(\boldsymbol{y}) = \sum_{d=1}^{D} \log \int_R P_\theta(\boldsymbol{y}_d|v_d) f(v_d)\, dv_d$$
$$\approx \sum_{d=1}^{D} \log\left[\frac{1}{2S}\sum_{s=1}^{2S} P_\theta(\boldsymbol{y}_d|v_d^{(s)})\right],$$

where $v_d^{(s)}$ are i.i.d. $N(0,1)$, $v_d^{(S+s)} = -v_d^{(s)}$, $s = 1, \ldots, S$, and

$$P_\theta(\boldsymbol{y}_d|v_d) = \prod_{j=1}^{n_d} \binom{m_{dj}}{y_{dj}} p_{dj}^{y_{dj}} (1-p_{dj})^{m_{dj}-y_{dj}} = \prod_{j=1}^{n_d} \frac{\binom{m_{dj}}{y_{dj}} \exp\left\{y_{dj}(\boldsymbol{x}_{dj}\boldsymbol{\beta} + \phi v_d)\right\}}{\left[1+\exp\left\{\boldsymbol{x}_{dj}\boldsymbol{\beta} + \phi v_d\right\}\right]^{m_{dj}}}$$

$$= \exp\left\{\sum_{j=1}^{n_d} \log \binom{m_{dj}}{y_{dj}} + \sum_{j=1}^{n_d} y_{dj}(\boldsymbol{x}_{dj}\boldsymbol{\beta} + \phi v_d)\right.$$

$$\left. - \sum_{j=1}^{n_d} m_{dj} \log\left[1+\exp\{\boldsymbol{x}_{dj}\boldsymbol{\beta} + \phi v_d\}\right]\right\}.$$

By taking derivatives with respect to β_k, $k = 1, \ldots, p$, and ϕ, we obtain

$$\frac{\partial \ell(\boldsymbol{\theta})}{\partial \beta_k} \approx \sum_{d=1}^{D} \frac{\sum_{s=1}^{2S} P_\theta(\boldsymbol{y}_d|v_d^{(s)}) \sum_{j=1}^{n_d} [y_{dj} - m_{dj} p_{dj}^{(s)}(\boldsymbol{\theta})] x_{djk}}{\sum_{s=1}^{2S} P_\theta(\boldsymbol{y}_d|v_d^{(s)})}$$

$$= \sum_{d=1}^{D} \sum_{s=1}^{2S} \frac{P_\theta(\boldsymbol{y}_d|v_d^{(s)}) u_{d,k}^{(s)}(\boldsymbol{\theta})}{\sum_{s=1}^{2S} P_\theta(\boldsymbol{y}_d|v_d^{(s)})},$$

$$\frac{\partial \ell(\boldsymbol{\theta})}{\partial \phi} \approx \sum_{d=1}^{D} \frac{\sum_{s=1}^{2S} P_\theta(\boldsymbol{y}_d|v_d^{(s)}) \sum_{j=1}^{n_d} [y_{dj} - m_{dj} p_{dj}^{(s)}(\boldsymbol{\theta})] v_d^{(s)}}{\sum_{s=1}^{2S} P_\theta(\boldsymbol{y}_d|v_d^{(s)})}$$

$$= \sum_{d=1}^{D} \sum_{s=1}^{2S} \frac{P_\theta(\boldsymbol{y}_d|v_d^{(s)}) u_{d,p+1}^{(s)}(\boldsymbol{\theta})}{\sum_{s=1}^{2S} P_\theta(\boldsymbol{y}_d|v_d^{(s)})}.$$

We recall that

$$\frac{\partial u_{d,k_1}^{(s)}(\boldsymbol{\theta})}{\partial \theta_{k_2}} = h_{d,k_1k_2}^{(s)}(\boldsymbol{\theta}), \quad k_1, k_2 = 1, \ldots, p+1.$$

Therefore, the second derivatives of $\ell(\boldsymbol{\theta})$ with respect to θ_{k_1} and θ_{k_2}, $k_1, k_2 = 1, \ldots, p+1$, are

$$g_{k_1k_2}(\boldsymbol{\theta}) = \frac{\partial^2 \ell(\boldsymbol{\theta})}{\partial \theta_{k_1} \partial \theta_{k_2}} = \sum_{d=1}^{D} \sum_{s=1}^{2S} \frac{P_\theta(\boldsymbol{y}_d|v_d^{(s)})}{\sum_{s=1}^{2S} P_\theta(\boldsymbol{y}_d|v_d^{(s)})}$$

$$\cdot \left\{ u_{d,k_1}^{(s)}(\boldsymbol{\theta}) u_{d,k_2}^{(s)}(\boldsymbol{\theta}) + h_{d,k_1k_2}^{(s)}(\boldsymbol{\theta}) - u_{d,k_1}^{(s)}(\boldsymbol{\theta}) \sum_{s=1}^{2S} \frac{P_\theta(\boldsymbol{y}_d|v_d^{(s)}) u_{d,k_2}^{(s)}(\boldsymbol{\theta})}{\sum_{s=1}^{2S} P_\theta(\boldsymbol{y}_d|v_d^{(s)})} \right\}$$

$$= \sum_{d=1}^{D}\sum_{s=1}^{2S} w_d^{(s)}(\boldsymbol{\theta})\Big\{u_{d,k_1}^{(s)}(\boldsymbol{\theta})u_{d,k_2}^{(s)}(\boldsymbol{\theta}) + h_{d,k_1k_2}^{(s)}(\boldsymbol{\theta})$$

$$- u_{d,k_1}^{(s)}(\boldsymbol{\theta})\sum_{s=1}^{2S} w_d^{(s)}(\boldsymbol{\theta})u_{d,k_2}^{(s)}(\boldsymbol{\theta})\Big\}.$$

The asymptotic covariance matrix of $\hat{\boldsymbol{\theta}}$ can be approximated by

$$\text{var}(\hat{\boldsymbol{\theta}}) \approx -\boldsymbol{G}^{-1}(\hat{\boldsymbol{\theta}}), \quad \boldsymbol{G}(\boldsymbol{\theta}) = \big(g_{k_1k_2}(\boldsymbol{\theta})\big)_{k_1,k_2=1,\dots,p+1}.$$

14.5 ML-Laplace Approximation Algorithm

14.5.1 Introduction

The marginal likelihood of the unit-level logit mixed model is intractable, but it can be approximated by several methods. The Laplace approximation is one of the most widely used in statistics. See the books of Evans and Swartz (2000) or Bleistein and Handelsman (1986) for more information on numerical approximation of integrals. The Laplace approximation is the default method in glmer function of the R package `lme4`. This approximation performs better for larger domain sizes n_d and its accuracy increases if higher order of Taylor expansion is used. However, this approximation is less accurate if the variance of random effects is large.

This section introduces Laplace's method to approximate a univariate integral and gives the Laplace approximation algorithm for estimating the parameters of the unit-level logit mixed model and for predicting (mode predictors) the values of the random effect v_d, $d = 1, \dots, D$.

Let $h : R \mapsto R$ be a twice continuously differentiable function with a global maximum at x_0. This is to say, let us assume that $\dot{h}(x_0) = 0$ and $\ddot{h}(x_0) < 0$. A Taylor series expansion of $h(x)$ around x_0 yields to

$$h(x) = h(x_0)+\dot{h}(x_0)(x-x_0)+\frac{1}{2}\ddot{h}(x_0)(x-x_0)^2+o(|x-x_0|^2) \approx h(x_0)+\frac{1}{2}\ddot{h}(x_0)(x-x_0)^2.$$

The univariate Laplace approximation of the integral of the function $\exp(h(x))$ is

$$\int_{-\infty}^{\infty} e^{h(x)}\,dx \approx \int_{-\infty}^{\infty} e^{h(x_0)} \exp\Big\{-\frac{1}{2}\big(-\ddot{h}(x_0)\big)(x-x_0)^2\Big\}\,dx$$

$$= (2\pi)^{1/2}\big(-\ddot{h}(x_0)\big)^{-1/2} e^{h(x_0)} \int_{-\infty}^{\infty} \frac{\exp\Big\{-\frac{1}{2}\Big(\frac{x-x_0}{(-\ddot{h}(x_0))^{-1/2}}\Big)^2\Big\}}{(2\pi)^{1/2}\big(-\ddot{h}(x_0)\big)^{-1/2}}\,dx$$

$$= (2\pi)^{1/2}\big(-\ddot{h}(x_0)\big)^{-1/2} e^{h(x_0)}, \tag{14.6}$$

where the property of the Gaussian integral was used in the last equality. For more details about the ML-Laplace approximation and algorithm, see e.g. Demidenko (2004).

14.5.2 The Laplace Approximation to the Likelihood

This section approximates the log-likelihood of the unit-level logit mixed model. The random effects $v_1, \ldots, v_D$ are i.i.d. $N(0, 1)$. Conditionally to v_d, the target variables $y_{d1}, \ldots, y_{dn_d}$ are independent with distributions

$$y_{dj}|_{v_d} \stackrel{ind}{\sim} \text{Bin}(m_{dj}, p_{dj}), \; p_{dj} = p_{dj}(v_d) = \frac{\exp\left\{\boldsymbol{x}_{dj}\boldsymbol{\beta} + \phi v_d\right\}}{1 + \exp\left\{\boldsymbol{x}_{dj}\boldsymbol{\beta} + \phi v_d\right\}},$$

$d = 1, \ldots, D$, $j = 1, \ldots, n_d$. Using the notation $\boldsymbol{y}_d = (y_{d1}, \ldots, y_{dn_d})'$, $d = 1, \ldots, D$, it holds that $\boldsymbol{y}_1, \ldots, \boldsymbol{y}_D$ are unconditionally independent with marginal p.d.f.

$$\begin{aligned}
P(\boldsymbol{y}_d) &= \int_{-\infty}^{\infty} \prod_{j=1}^{n_d} P(y_{dj}|v_d) f(v_d)\, dv_d \\
&= \int_{-\infty}^{\infty} \prod_{j=1}^{n_d} \left\{ \binom{m_{dj}}{y_{dj}} \exp\left\{ y_{dj}(\boldsymbol{x}_{dj}\boldsymbol{\beta} + \phi v_d) - m_{dj} \log\left(1 + \exp\{\boldsymbol{x}_{dj}\boldsymbol{\beta} + \phi v_d\}\right)\right\}\right\} \\
&\cdot (2\pi)^{-1/2} \exp\{-\frac{1}{2} v_d^2\}\, dv_d = (2\pi)^{-1/2} \prod_{j=1}^{n_d} \binom{m_{dj}}{y_{dj}} \int_{-\infty}^{\infty} \exp\Big\{ -\frac{v_d^2}{2} \\
&+ \sum_{j=1}^{n_d} \left\{ y_{dj}(\boldsymbol{x}_{dj}\boldsymbol{\beta} + \phi v_d) - m_{dj} \log\left(1 + \exp\{\boldsymbol{x}_{dj}\boldsymbol{\beta} + \phi v_d\}\right)\right\}\Big\}\, dv_d \\
&= (2\pi)^{-1/2} \prod_{j=1}^{n_d} \binom{m_{dj}}{y_{dj}} \int_{-\infty}^{\infty} \exp\left\{h_d(v_d)\right\} dv_d, \qquad (14.7)
\end{aligned}$$

where

$$h_d(v_d) = -\frac{v_d^2}{2} + \sum_{j=1}^{n_d} \left\{ y_{dj}(\boldsymbol{x}_{dj}\boldsymbol{\beta} + \phi v_d) - m_{dj} \log\left(1 + \exp\{\boldsymbol{x}_{dj}\boldsymbol{\beta} + \phi v_d\}\right)\right\}. \qquad (14.8)$$

In order to apply the Laplace approximation to the integral in (14.7), we need derivatives of the function h_d. These are

$$\dot{h}_d(v_d) = -v_d + \phi \sum_{j=1}^{n_d} \{y_{dj} - m_{dj} p_{dj}(v_d)\},$$

$$\ddot{h}_d(v_d) = -\Big(1 + \phi^2 \sum_{j=1}^{n_d} m_{dj} p_{dj}(v_d)(1 - p_{dj}(v_d))\Big).$$

For $\boldsymbol{\theta} = (\boldsymbol{\beta}', \phi)'$ fixed, the function $h_d(v_d) = h_d(v_d, \boldsymbol{\theta})$ (defined in (14.8)), can be maximized by using the Newton–Raphson algorithm. The updating equation is

$$v_d^{(i+1)} = v_d^{(i)} - \frac{\dot{h}_d(v_d^{(i)}, \boldsymbol{\theta})}{\ddot{h}_d(v_d^{(i)}, \boldsymbol{\theta})}, \quad d = 1, \ldots, D. \tag{14.9}$$

Let v_{0d} be the argument where the maximum of the function $h_d(v_d)$ is achieved. Thus, it holds $\dot{h}_d(v_{0d}) = 0$, $\ddot{h}_d(v_{0d}) < 0$ and by applying the approximation (14.6)–(14.7) in $v_d = v_{0d}$, we get

$$P(\boldsymbol{y}_d) \approx \prod_{j=1}^{n_d} \binom{m_{dj}}{y_{dj}} \cdot \Big(1 + \phi^2 \sum_{j=1}^{n_d} m_{dj} p_{dj}(v_{0d})(1 - p_{dj}(v_{0d}))\Big)^{-1/2}$$

$$\cdot \exp\Big\{ -\frac{v_{0d}^2}{2} + \sum_{j=1}^{n_{dj}} \{y_{dj}(\boldsymbol{x}_{dj}\boldsymbol{\beta} + \phi v_{0d}) - m_{dj} \log\big(1 + \exp\{\boldsymbol{x}_{dj}\boldsymbol{\beta} + \phi v_{0d}\}\big)\Big\}.$$

The log-likelihood is $\ell = \sum_{d=1}^{D} \ell_d$ and

$$\ell_d = \log P(\boldsymbol{y}_d) \approx \ell_{0d} = \sum_{j=1}^{n_d} \log \binom{m_{dj}}{y_{dj}} - \frac{1}{2} \log \xi_{0d} - \frac{v_{0d}^2}{2}$$

$$+ \sum_{j=1}^{n_d} \Big\{ y_{dj}(\boldsymbol{x}_{dj}\boldsymbol{\beta} + \phi v_{0d}) - m_{dj} \log\big(1 + \exp\{\boldsymbol{x}_{dj}\boldsymbol{\beta} + \phi v_{0d}\}\big)\Big\}, \tag{14.10}$$

where $p_{0dj} = p_{dj}(v_{0d})$ and $\xi_{0d} = 1 + \phi^2 \sum_{j=1}^{n_d} m_{dj} p_{0dj}(1 - p_{0dj})$. It holds that

$$\frac{\partial p_{0dj}}{\partial \beta_r} = x_{djr} p_{0dj}(1 - p_{0dj}) = x_{djr}(p_{0dj} - p_{0dj}^2),$$

$$\frac{\partial p_{0dj}}{\partial \phi} = v_{0d} p_{0dj}(1 - p_{0dj}) = v_{0d}(p_{0dj} - p_{0dj}^2),$$

$$\eta_{0dr} = \frac{\partial \xi_{0d}}{\partial \beta_r} = \phi^2 \sum_{j=1}^{n_d} m_{dj} x_{djr} [p_{0dj} - 3p_{0dj}^2 + 2p_{0dj}^3],$$

$$\eta_{0d} = \frac{\partial \xi_{0d}}{\partial \phi} = \phi^2 v_{0d} \sum_{j=1}^{n_d} m_{dj} [p_{0dj} - 3p_{0dj}^2 + 2p_{0dj}^3]$$

$$+2\phi \sum_{j=1}^{n_d} m_{dj} p_{0dj} \left(1 - p_{0dj}\right).$$

The first derivatives of ℓ_{0d}, with respect to β_r and ϕ, are

$$\frac{\partial \ell_{0d}}{\partial \beta_r} = -\frac{1}{2}\frac{\eta_{0dr}}{\xi_{0d}} + \sum_{j=1}^{n_d} (y_{dj} - m_{dj} p_{0dj}) x_{djr},$$

$$\frac{\partial \ell_{0d}}{\partial \phi} = -\frac{1}{2}\frac{\eta_{0d}}{\xi_{0d}} + v_{0d} \sum_{j=1}^{n_d} (y_{dj} - m_{dj} p_{0dj}).$$

It holds that

$$\frac{\partial \eta_{0dr}}{\partial \beta_s} = \phi^2 \sum_{j=1}^{n_d} m_{dj} x_{djr} x_{djs} [p_{0dj}(1 - p_{0dj}) - 6p_{0dj}^2(1 - p_{0dj}) + 6p_{0dj}^3(1 - p_{0dj})]$$

$$= \phi^2 \sum_{j=1}^{n_d} m_{dj} x_{djr} x_{djs} p_{0dj}(1 - p_{0dj})[1 - 6p_{0dj} + 6p_{0dj}^2],$$

$$\frac{\partial \eta_{0dr}}{\partial \phi} = \phi^2 v_{0d} \sum_{j=1}^{n_d} m_{dj} x_{djr} p_{0dj}(1 - p_{0dj})[1 - 6p_{0dj} + 6p_{0dj}^2]$$

$$+2\phi \sum_{j=1}^{n_d} m_{dj} x_{djr} [p_{0dj} - 3p_{0dj}^2 + 2p_{0dj}^3],$$

$$\frac{\partial \eta_{0d}}{\partial \beta_r} = \phi^2 v_{0d} \sum_{j=1}^{n_d} m_{dj} x_{djr} p_{0dj}(1 - p_{0dj})[1 - 6p_{0dj} + 6p_{0dj}^2]$$

$$+2\phi \sum_{j=1}^{n_d} m_{dj} x_{djr} [p_{0dj} - 3p_{0dj}^2 + 2p_{0dj}^3],$$

$$\frac{\partial \eta_{0d}}{\partial \phi} = \phi^2 v_{0d}^2 \sum_{j=1}^{n_d} m_{dj} p_{0dj}(1 - p_{0dj})[1 - 6p_{0dj} + 6p_{0dj}^2]$$

$$+4\phi v_{0d} \sum_{j=1}^{n_d} m_{dj} [p_{0dj} - 3p_{0dj}^2 + 2p_{0dj}^3] + 2\sum_{j=1}^{n_d} m_{dj} p_{0dj}(1 - p_{0dj}).$$

The second partial derivatives of ℓ_{0d} are

$$\frac{\partial^2 \ell_{0d}}{\partial \beta_s \partial \beta_r} = -\frac{1}{2} \frac{\frac{\partial \eta_{0dr}}{\partial \beta_s} \xi_{0d} - \eta_{0dr}\eta_{0ds}}{\xi_{0d}^2} - \sum_{j=1}^{n_d} m_{dj} x_{djr} x_{djs} p_{0dj}(1 - p_{0dj}),$$

$$\frac{\partial^2 \ell_{0d}}{\partial \phi \partial \beta_r} = -\frac{1}{2} \frac{\frac{\partial \eta_{0dr}}{\partial \phi} \xi_{0d} - \eta_{0dr}\eta_{0d}}{\xi_{0d}^2} - v_{0d} \sum_{j=1}^{n_d} m_{dj} x_{djr} p_{0dj}(1 - p_{0dj}),$$

$$\frac{\partial^2 \ell_{0d}}{\partial \phi^2} = -\frac{1}{2} \frac{\frac{\partial \eta_{0d}}{\partial \phi} \xi_{0d} - \eta_{0d}^2}{\xi_{0d}^2} - v_{0d}^2 \sum_{j=1}^{n_d} m_{dj} p_{0dj}(1 - p_{0dj}).$$

For $r, s = 1, \ldots, p$, the components of the score vector and the Hessian matrix are

$$U_{0r} = \sum_{d=1}^{D} \frac{\partial \ell_{0d}}{\partial \beta_r}, \; U_{0p+1} = \sum_{d=1}^{D} \frac{\partial \ell_{0d}}{\partial \phi},$$

$$H_{0rs} = H_{0sr} = \sum_{d=1}^{D} \frac{\partial^2 \ell_{0d}}{\partial \beta_s \partial \beta_r}, \; H_{rp+1} = H_{p+1r} = \sum_{d=1}^{D} \frac{\partial^2 \ell_{0d}}{\partial \phi \partial \beta_r}, \; H_{0p+1p+1} = \sum_{d=1}^{D} \frac{\partial^2 \ell_{0d}}{\partial \phi^2}.$$

In matrix form, we have $\boldsymbol{U}_0 = \boldsymbol{U}_0(\boldsymbol{\theta}) = \underset{1 \le r \le p+1}{\text{col}} (U_{0r})$ and $\boldsymbol{H}_0 = \boldsymbol{H}_0(\boldsymbol{\theta}) = (H_{0rs})_{r,s=1,\ldots,p+1}$. The Newton–Raphson algorithm maximizes $\ell_0(\boldsymbol{\theta})$ with fixed $v_d = v_{0d}$, $d = 1, \ldots, D$. The updating equation is

$$\boldsymbol{\theta}^{(i+1)} = \boldsymbol{\theta}^{(i)} - \boldsymbol{H}_0^{-1}(\boldsymbol{\theta}^{(i)}) \boldsymbol{U}_0(\boldsymbol{\theta}^{(i)}). \tag{14.11}$$

The final ML-Laplace approximation algorithm combines the two described Newton–Raphson algorithms and can be described by the following steps:

1. Set the initial values $i = 0, \varepsilon_1 > 0, \varepsilon_2 > 0, \varepsilon_3 > 0, \varepsilon_4 > 0, \boldsymbol{\theta}^{(0)}, \boldsymbol{\theta}^{(-1)} = \boldsymbol{\theta}^{(0)} + \mathbf{1}$, $v_d^{(0)} = 0, v_d^{(-1)} = 1, d = 1, \ldots, D$.
2. Until $\|\boldsymbol{\theta}^{(i)} - \boldsymbol{\theta}^{(i-1)}\|_2 < \varepsilon_1, |v_d^{(i)} - v_d^{(i-1)}| < \varepsilon_2, d = 1, \ldots, D$, do
   a. Apply algorithm (14.9) with seeds $v_d^{(i)}$, $d = 1, \ldots, D$, convergence tolerance ε_3 and $\boldsymbol{\theta} = \boldsymbol{\theta}^{(i)}$ fixed. Output: $v_d^{(i+1)}$, $d = 1, \ldots, D$.
   b. Apply algorithm (14.11) with seed $\boldsymbol{\theta}^{(i)}$, convergence tolerance ε_4 and $v_{0d} = v_d^{(i+1)}$ fixed, $d = 1, \ldots, D$. Output: $\boldsymbol{\theta}^{(i+1)}$.
   c. $i \leftarrow i + 1$.
3. Output: $\hat{\boldsymbol{\theta}} = \boldsymbol{\theta}^{(i)}$, $\hat{v}_d = v_d^{(i)}$, $d = 1, \ldots, D$.

Remark 14.1 The ML-Laplace approximation algorithm gives at convergence besides the ML estimators of the model parameters also the mode predictors, $\hat{v}_d$, of the random effects and the maximized marginal log-likelihood (logarithm of the joint p.d.f of the target vector $\boldsymbol{y}$). Since the ML estimators are consistent and asymptotically normal when $n_d \to \infty$, $d = 1, \ldots, D$, the algorithm can be also used to approximate the asymptotic covariance matrix (inverse of Fisher information matrix) which allows calculating Wald statistics for testing hypotheses on the model parameters. Namely, the asymptotic variance matrix of $\hat{\boldsymbol{\theta}}$ can be approximated by $\text{avar}(\hat{\boldsymbol{\theta}}) = -\boldsymbol{H}_0^{-1}(\hat{\boldsymbol{\theta}})$.

14.5.3 The AIC

Let $\hat{\boldsymbol{\beta}}$ and $\hat{\phi}$ be the ML estimators of $\boldsymbol{\beta}$ and ϕ, respectively. Let $\hat{v}_d$ be the likelihood mode of v_d under ML-Laplace estimation. The ML-Laplace non-corrected AIC is

$$AIC = 2(p+1) - 2\ell_L(\hat{\boldsymbol{\beta}}, \hat{\phi}, \hat{v}_1, \ldots, \hat{v}_D),$$

where the second term is the Laplace approximation (14.10) to the model log-likelihood, i.e.

$$\ell_L = \ell_L(\hat{\boldsymbol{\beta}}, \hat{\phi}, \hat{v}_1, \ldots, \hat{v}_D) = \sum_{d=1}^{D}\sum_{j=1}^{n_d} \log\binom{m_{dj}}{y_{dj}} + \sum_{d=1}^{D}\Big\{ -\frac{1}{2}\log\hat{\xi}_d - \frac{\hat{v}_d^2}{2}$$

$$+ \sum_{j=1}^{n_d}\Big\{ y_{dj}(\boldsymbol{x}_{dj}\hat{\boldsymbol{\beta}} + \hat{\phi}\hat{v}_d) - m_{dj}\log\big(1 + \exp\{\boldsymbol{x}_{dj}\hat{\boldsymbol{\beta}} + \hat{\phi}\hat{v}_d\}\big)\Big\}\Big\}$$

and

$$\hat{\xi}_d = 1 + \hat{\phi}^2 \sum_{j=1}^{n_d} m_{dj}\hat{p}_{dj}(1 - \hat{p}_{dj}), \quad \hat{p}_{dj} = \frac{\exp\big\{\boldsymbol{x}_{dj}\hat{\boldsymbol{\beta}} + \hat{\phi}\hat{v}_d\big\}}{1 + \exp\big\{\boldsymbol{x}_{dj}\hat{\boldsymbol{\beta}} + \hat{\phi}\hat{v}_d\big\}}.$$

14.6 Empirical Best Predictors

Let us now consider a finite population U of N elements partitioned into D domains U_d of sizes N_d, $d = 1, \ldots, D$. From the population, a sample s of size n is selected with subsamples s_d of sizes n_d from domains U_d. By $r = U - s$ and $r_d = U_d - s_d$ we denote the set of the non-sampled population units and the set of non-sampled population units from domain d, respectively. Let $\boldsymbol{y}_d = \underset{1 \le j \le N_d}{\text{col}}(y_{dj})$ be the random vector containing the values of a target variable on the N_d units of domain d. Let $\boldsymbol{y}_{ds}$ be the sub-vector of $\boldsymbol{y}_d$ corresponding to the units in the sample s_d and $\boldsymbol{y}_{dr}$ the sub-vector corresponding to the units in the non-sampled domain population

r_d. By reordering the domain units, we can write $\boldsymbol{y}_d = (\boldsymbol{y}'_{ds}, \boldsymbol{y}'_{dr})'$, where $\boldsymbol{y}_{ds} = \underset{1\le j\le n_d}{\text{col}}(y_{dj})$. We define $\boldsymbol{y}_s = \underset{1\le d\le D}{\text{col}}(\boldsymbol{y}_{ds})$ and $\boldsymbol{y}_r = \underset{1\le d\le D}{\text{col}}(\boldsymbol{y}_{dr})$ and we assume that $\boldsymbol{y} = (\boldsymbol{y}'_s, \boldsymbol{y}'_r)'$ follows the model (14.1)–(14.2) with population sizes N_d in the place of sample sizes n_d.

This section gives empirical best predictors (EBP) under the unit-level Bernoulli logit mixed model, i.e. we assume moreover that $m_{dj} = 1$, $d = 1, \ldots, D$, $j = 1, \ldots, N_d$.
Using the introduced notation, the conditional distribution of $\boldsymbol{y}_s$, given $\boldsymbol{v}$, is

$$P(\boldsymbol{y}_s|\boldsymbol{v}) = \prod_{d=1}^{D} P(\boldsymbol{y}_{ds}|v_d),$$

where

$$P(\boldsymbol{y}_{ds}|v_d) = \prod_{i=1}^{n_d} p_{di}^{y_{di}}(1-p_{di})^{1-y_{di}} = \prod_{i=1}^{n_d} \frac{\exp\{y_{di}(\boldsymbol{x}_{di}\boldsymbol{\beta} + \phi v_d)\}}{[1+\exp\{\boldsymbol{x}_{di}\boldsymbol{\beta} + \phi v_d\}]}$$

$$= \exp\left\{\sum_{i=1}^{n_d} y_{di}(\boldsymbol{x}_{di}\boldsymbol{\beta} + \phi v_d) - \sum_{i=1}^{n_d} \log\left[1+\exp\{\boldsymbol{x}_{di}\boldsymbol{\beta} + \phi v_d\}\right]\right\}$$

and the p.d.f. of $\boldsymbol{v}$ is

$$f(\boldsymbol{v}) = \prod_{d=1}^{D} f(v_d), \quad f(v_d) = (2\pi)^{-1/2}\exp\left\{-\frac{1}{2}v_d^2\right\}.$$

In the following sections we derive successively EBP for different quantities such as probabilities, sums of probabilities, and domain proportions.

14.6.1 EBP of p_{dj}

Let us start with the best predictor of $p_{dj} = p_{dj}(\boldsymbol{\theta}, v_d)$ which has the form $\hat{p}_{dj}(\boldsymbol{\theta}) = E_\theta[p_{dj}|\boldsymbol{y}_s]$ as can be seen by similar arguments as in the proof of Proposition 10.1. In this case, we have that $E_\theta[p_{dj}|\boldsymbol{y}_s] = E_\theta[p_{dj}|\boldsymbol{y}_{ds}]$ and

$$E_\theta[p_{dj}|\boldsymbol{y}_{ds}] = \frac{\int_R \frac{\exp\{\boldsymbol{x}_{dj}\boldsymbol{\beta}+\phi v_d\}}{1+\exp\{\boldsymbol{x}_{dj}\boldsymbol{\beta}+\phi v_d\}} P(\boldsymbol{y}_{ds}|v_d) f(v_d)\,dv_d}{\int_R P(\boldsymbol{y}_{ds}|v_d) f(v_d)\,dv_d}$$

$$= \frac{\mathcal{A}_{dj}(\boldsymbol{y}_{ds}, \boldsymbol{\theta})}{\mathcal{D}_d(\boldsymbol{y}_{ds}, \boldsymbol{\theta})} = \frac{A_{dj}(y_{d\cdot}, \boldsymbol{\theta})}{D_d(y_{d\cdot}, \boldsymbol{\theta})},$$

where $\mathcal{A}_{dj} = \mathcal{A}_{dj}(\boldsymbol{y}_{ds}, \boldsymbol{\theta})$, $\mathcal{D}_d = \mathcal{D}_d(\boldsymbol{y}_{ds}, \boldsymbol{\theta})$, $A_{dj} = A_{dj}(y_{d.}, \boldsymbol{\theta})$, and $D_d = D_d(y_{d.}, \boldsymbol{\theta})$ are

$$\mathcal{A}_{dj} = \int_R \frac{\exp\{\boldsymbol{x}_{dj}\boldsymbol{\beta} + \phi v_d\}}{1+\exp\{\boldsymbol{x}_{dj}\boldsymbol{\beta} + \phi v_d\}} \exp\left\{\sum_{i=1}^{n_d} y_{di}\boldsymbol{x}_{di}\boldsymbol{\beta} + \phi y_{d.} v_d \right.$$
$$\left. - \sum_{i=1}^{n_d} \log\left[1+\exp\{\boldsymbol{x}_{di}\boldsymbol{\beta} + \phi v_d\}\right]\right\} f(v_d)\, dv_d,$$

$$\mathcal{D}_d = \int_R \exp\left\{\sum_{i=1}^{n_d} y_{di}\boldsymbol{x}_{di}\boldsymbol{\beta} + \phi y_{d.} v_d - \sum_{i=1}^{n_d} \log\left[1+\exp\{\boldsymbol{x}_{di}\boldsymbol{\beta} + \phi v_d\}\right]\right\} f(v_d)\, dv_d,$$

$$A_{dj} = \int_R \frac{\exp\{\boldsymbol{x}_{dj}\boldsymbol{\beta}+\phi v_d\}}{1+\exp\{\boldsymbol{x}_{dj}\boldsymbol{\beta}+\phi v_d\}} \exp\left\{\phi y_{d.} v_d - \sum_{i=1}^{n_d} \log\left[1+\exp\{\boldsymbol{x}_{di}\boldsymbol{\beta}+\phi v_d\}\right]\right\} f(v_d)\, dv_d,$$

$$D_d = \int_R \exp\left\{\phi y_{d.} v_d - \sum_{i=1}^{n_d} \log\left[1+\exp\{\boldsymbol{x}_{di}\boldsymbol{\beta} + \phi v_d\}\right]\right\} f(v_d)\, dv_d \quad (14.12)$$

and $y_{d.} = \sum_{i=1}^{n_d} y_{di}$. The EBP of p_{dj} is $\hat{p}_{dj}^{ebp} = \hat{p}_{dj}(\hat{\boldsymbol{\theta}})$ and can be approximated as follows.

1. Estimate $\hat{\boldsymbol{\theta}} = (\hat{\boldsymbol{\beta}}', \hat{\phi})'$.
2. For $s = 1, \ldots, S$, generate $v_d^{(s)}$ i.i.d. $N(0, 1)$ and $v_d^{(S+s)} = -v_d^{(s)}$.
3. Calculate $\hat{p}_{dj}^{ebp} = \hat{A}_{dj}/\hat{D}_d$, where

$$\hat{A}_{dj} = \frac{1}{2S}\sum_{s=1}^{2S} \frac{\exp\{\boldsymbol{x}_{dj}\hat{\boldsymbol{\beta}} + \hat{\phi} v_d^{(s)}\}}{1+\exp\{\boldsymbol{x}_{dj}\hat{\boldsymbol{\beta}} + \hat{\phi} v_d^{(s)}\}}$$
$$\cdot \exp\left\{\hat{\phi} y_{d.} v_d^{(s)} - \sum_{i=1}^{n_d} \log\left[1+\exp\{\boldsymbol{x}_{di}\hat{\boldsymbol{\beta}} + \hat{\phi} v_d^{(s)}\}\right]\right\},$$

$$\hat{D}_d = \frac{1}{2S}\sum_{s=1}^{2S} \exp\left\{\hat{\phi} y_{d.} v_d^{(s)} - \sum_{i=1}^{n_d} \log\left[1+\exp\{\boldsymbol{x}_{di}\hat{\boldsymbol{\beta}} + \hat{\phi} v_d^{(s)}\}\right]\right\}. \quad (14.13)$$

If $n_d = 0$, then the EBP of p_{dj} is $\hat{p}_{dj}(\boldsymbol{\theta}) = E_\theta[p_{dj}]$, which is also called synthetic EBP (sEBP). The sEBP can be approximated as the EBP when $n_d > 0$, but running the following step 3.

3. Calculate

$$\hat{p}_{dj}^{sebp} = \frac{1}{2S}\sum_{s=1}^{2S} \frac{\exp\{\boldsymbol{x}_{dj}\hat{\boldsymbol{\beta}} + \hat{\phi} v_d^{(s)}\}}{1+\exp\{\boldsymbol{x}_{dj}\hat{\boldsymbol{\beta}} + \hat{\phi} v_d^{(s)}\}}.$$

The derived best predictors $\hat{p}_{dj}(\boldsymbol{\theta})$ have minimum mean squared error in the class of unbiased predictors. The same property, however, does not hold for the EBPs $\hat{p}_{dj}(\hat{\boldsymbol{\theta}})$ which are only asymptotically unbiased under the assumption that the estimates of the model parameters are consistent. But the domain sample sizes are typically small in SAE problems. Moreover, the EBPs are usually computationally demanding so it makes sense to consider also computationally simpler predictors. One such candidate is the plug-in predictor, whose calculation requires consistent estimates of the model parameters and good predictions of random effects. The plug-in predictor of p_{dj} is

$$\hat{p}_{dj}^{in} = \frac{\exp\{\boldsymbol{x}_{dj}\hat{\boldsymbol{\beta}} + \hat{\phi}\hat{v}_d\}}{1 + \exp\{\boldsymbol{x}_{dj}\hat{\boldsymbol{\beta}} + \hat{\phi}\hat{v}_d\}},$$

where $\hat{v}_d$ is a predictor (EBP or likelihood mode) of v_d, $d = 1, \ldots, D$. The predictors $\hat{v}_d$ may be taken as likelihood modes from the output of the ML-Laplace approximation algorithm. Other possibility is to take the EBP of v_d which is $\hat{v}_d^{ebp} = E_\theta[v_d|\boldsymbol{y}_{ds}]$ and it can be approximated by $\hat{v}_d^{ebp} = \hat{A}_{v,d}/\hat{D}_d$, where

$$\hat{A}_{v,d} = \frac{1}{2S}\sum_{s=1}^{2S} v_d^{(s)} \exp\left\{\hat{\phi} y_{d.} v_d^{(s)} - \sum_{i=1}^{n_d} \log\left[1 + \exp\{\boldsymbol{x}_{di}\hat{\boldsymbol{\beta}} + \hat{\phi} v_d^{(s)}\}\right]\right\}.$$

14.6.2 EBP of μ_d and $\overline{\mu}_d$

Let us now turn our attention to the population sum and the population mean

$$\mu_d = \sum_{j=1}^{N_d} p_{dj}, \quad \text{and} \quad \overline{\mu}_d = \frac{1}{N_d}\sum_{j=1}^{N_d} p_{dj}.$$

The EBP of μ_d is $\hat{\mu}_d^{ebp} = \hat{\mu}_d(\hat{\boldsymbol{\theta}}) = \sum_{j=1}^{N_d} \hat{p}_{dj}(\hat{\boldsymbol{\theta}})$ and the EBP of $\overline{\mu}_d$ is $\hat{\overline{\mu}}_d^{ebp} = \hat{\mu}_d(\hat{\boldsymbol{\theta}})/N_d$. The EBPs $\hat{\mu}_d^{ebp}$ and $\hat{\overline{\mu}}_d^{ebp}$ are functions of $\boldsymbol{x}_{dj}$, $d = 1, \ldots, D$, $j = 1, \ldots, N_d$. If at least one of the auxiliary variables is continuous, then for calculating the predictors it is necessary to have a data file containing the values of the explanatory variables in all the population units. But this kind of data file is not always available. The following remark presents a particular case where the calculation of $\hat{\mu}_d^{ebp}$ and $\hat{\overline{\mu}}_d^{ebp}$ requires less auxiliary data.

Remark 14.2 Suppose that the covariates are categorical and take finite number of values, i.e. $\boldsymbol{x}_{dj} \in \{z_1, \ldots, z_K\}$ for all d and j, and that the population sizes of categories crossed by domains are known. They can be obtained e.g. from some external data sources or administrative registers. Then

$$\mu_d = \mu_d(\boldsymbol{\theta}, v_d) = \sum_{j=1}^{N_d} p_{dj} = \sum_{k=1}^{K} N_{dk} q_{dk},$$

$$q_{dk} = q_{dk}(\boldsymbol{\theta}, v_d) = \frac{\exp\{\boldsymbol{z}_k \boldsymbol{\beta} + \phi v_d\}}{1 + \exp\{\boldsymbol{z}_k \boldsymbol{\beta} + \phi v_d\}},$$

where $N_{dk} = \#\{j \in U_d : \boldsymbol{x}_{dj} = \boldsymbol{z}_k\}$ is the known size of the covariate class $\boldsymbol{z}_k$ at domain d and U_d denotes the population of domain d.

Under this categorical setup, the plug-in predictor of μ_d can be written in the form

$$\mu_d(\hat{\boldsymbol{\theta}}, \hat{v}_d) = \sum_{k=1}^{K} N_{dk} q_{dk}(\hat{\boldsymbol{\theta}}, \hat{v}_d) = \sum_{k=1}^{K} N_{dk} \hat{q}_{dk}^{in}, \quad \hat{q}_{dk}^{in} = \frac{\exp\{\boldsymbol{z}_k \hat{\boldsymbol{\beta}} + \hat{\phi} \hat{v}_d\}}{1 + \exp\{\boldsymbol{z}_k \hat{\boldsymbol{\beta}} + \hat{\phi} \hat{v}_d\}}. \tag{14.14}$$

The best predictor of μ_d is

$$\hat{\mu}_d(\boldsymbol{\theta}) = E_\theta[\mu_d | \boldsymbol{y}_s] = E_\theta[\mu_d | \boldsymbol{y}_{ds}] = \sum_{k=1}^{K} N_{dk} E_\theta[q_{dk} | \boldsymbol{y}_{ds}],$$

where

$$E_\theta[q_{dk} | \boldsymbol{y}_{ds}] = \frac{\int_R \frac{\exp\{\boldsymbol{z}_k \boldsymbol{\beta} + \phi v_d\}}{1 + \exp\{\boldsymbol{z}_k \boldsymbol{\beta} + \phi v_d\}} P(\boldsymbol{y}_{ds} | v_d) f(v_d)\, dv_d}{\int_R P(\boldsymbol{y}_{ds} | v_d) f(v_d)\, dv_d} = \frac{A_{dk}^z(y_{d.}, \boldsymbol{\theta})}{D_d(y_{d.}, \boldsymbol{\theta})} = \frac{A_{dk}^z}{D_d},$$

the denominator D_d was defined in (14.12) and the numerator $A_{dk}^z = A_{dk}^z(y_{d.}, \boldsymbol{\theta})$ is

$$A_{dk}^z = \int_R \frac{\exp\{\boldsymbol{z}_k \boldsymbol{\beta} + \phi v_d\}}{1 + \exp\{\boldsymbol{z}_k \boldsymbol{\beta} + \phi v_d\}} \cdot \exp\left\{\phi y_{d.} v_d - \sum_{i=1}^{n_d} \log\left[1 + \exp\{\boldsymbol{x}_{di} \boldsymbol{\beta} + \phi v_d\}\right]\right\} f(v_d)\, dv_d.$$

The EBP of μ_d is $\hat{\mu}_d^{ebp} = \hat{\mu}_d(\hat{\boldsymbol{\theta}})$ and can be approximated as follows.

1. Estimate $\hat{\boldsymbol{\theta}} = (\hat{\boldsymbol{\beta}}', \hat{\phi})'$.
2. For $s = 1, \ldots, S$, generate $v_d^{(s)}$ i.i.d. $N(0, 1)$ and $v_d^{(S+s)} = -v_d^{(s)}$.
3. Calculate $\hat{\mu}_d^{ebp} = \sum_{k=1}^{K} N_{dk} \hat{q}_{dk}$, where $\hat{q}_{dk} = \hat{A}_{dk}^z / \hat{D}_d$ and

$$\hat{A}_{dk}^z = \frac{1}{2S} \sum_{s=1}^{2S} \frac{\exp\{\boldsymbol{z}_k \hat{\boldsymbol{\beta}} + \hat{\phi} v_d^{(s)}\}}{1 + \exp\{\boldsymbol{z}_k \hat{\boldsymbol{\beta}} + \hat{\phi} v_d^{(s)}\}} \tag{14.15}$$

$$\cdot \exp\left\{\hat{\phi} y_{d.} v_d^{(s)} - \sum_{i=1}^{n_d} \log\left[1+\exp\{\boldsymbol{x}_{di}\hat{\boldsymbol{\beta}} + \hat{\phi} v_d^{(s)}\}\right]\right\},$$

$$\hat{D}_d = \frac{1}{2S}\sum_{s=1}^{2S} \exp\left\{\hat{\phi} y_{d.} v_d^{(s)} - \sum_{i=1}^{n_d} \log\left[1+\exp\{\boldsymbol{x}_{di}\hat{\boldsymbol{\beta}} + \hat{\phi} v_d^{(s)}\}\right]\right\}. \quad (14.16)$$

In the case that $n_d = 0$ for a given domain d, the synthetic BP (sBP) of μ_d is $\hat{\mu}_d(\boldsymbol{\theta}) = E_\theta[\mu_d]$. We can approximate the sEBP of μ_d in the same way as the EBP when $n_d > 0$, but running the following step 3.

3. Calculate $\hat{\mu}_d^{sebp} = \sum_{k=1}^{K} N_{dk}\hat{q}_{dk}^{syn}$, where

$$\hat{q}_{dk}^{syn} = \frac{1}{2S}\sum_{s=1}^{2S}\left\{\frac{\exp\{z_k\hat{\boldsymbol{\beta}} + \hat{\phi} v_d^{(s)}\}}{1+\exp\{z_k\hat{\boldsymbol{\beta}} + \hat{\phi} v_d^{(s)}\}}\right\}.$$

14.6.3 EBP of y_{dj}

The best predictor of y_{dj} is $\hat{y}_{dj} = \hat{y}_{dj}(\boldsymbol{\theta}) = E_\theta[y_{dj}|\boldsymbol{y}_s]$. If $j \in s$, then $E_\theta[y_{dj}|\boldsymbol{y}_s] = y_{dj}$. If $j \in r$, then $E_\theta[y_{dj}|\boldsymbol{y}_s] = E_\theta[y_{dj}|\boldsymbol{y}_{ds}]$ and

$$\begin{aligned}\hat{y}_{dj}(\boldsymbol{\theta}) = E_\theta[y_{dj}|\boldsymbol{y}_{ds}] &= \frac{\int_R \int_{\{0,1\}} y_{dj} P(y_{dj}|v_d) P(\boldsymbol{y}_{ds}|v_d) f(v_d)\, dy_{dj}\, dv_d}{\int_R P(\boldsymbol{y}_{ds}|v_d) f(v_d)\, dv_d} \\ &= \frac{\int_R p_{dj} P(\boldsymbol{y}_{ds}|v_d) f(v_d)\, dv_d}{\int_R P(\boldsymbol{y}_{ds}|v_d) f(v_d)\, dv_d} = \frac{A_{dj}(y_{d.}, \boldsymbol{\theta})}{D_d(y_{d.}, \boldsymbol{\theta})} = E_\theta[p_{dj}|\boldsymbol{y}_{ds}] \\ &= \hat{p}_{dj}(\boldsymbol{\theta}), \end{aligned} \quad (14.17)$$

where $A_{dj} = A_{dj}(y_{d.}, \boldsymbol{\theta})$ and $D_d = D_d(y_{d.}, \boldsymbol{\theta})$ are defined in Sect. 14.6.1. The first line of Formula (14.17) expresses a sum as an integral with respect to the counting measure on {0, 1}, i.e.

$$\int_{\{0,1\}} y_{dj} P(y_{dj}|v_d)\, dy_{dj} = \sum_{y_{dj}=0}^{1} y_{dj} P(y_{dj}|v_d).$$

The EBP of y_{dj} is $\hat{y}_{dj} = \hat{y}_{dj}(\hat{\boldsymbol{\theta}})$. It holds that $\hat{y}_{dj} = y_{dj}$ if $j \in s$ and $\hat{y}_{dj} = \hat{p}_{dj}^{ebp}$ if $j \in r$, where $\hat{p}_{dj}^{ebp}$ is the EBP of p_{dj}, which is approximated in (14.13).

14.6.4 EBP of $\overline{Y}_d$

The aim of this section is to obtain EBP of the population means (proportions)

$$\overline{Y}_d = \frac{1}{N_d}\sum_{j=1}^{N_d} y_{dj}.$$

We consider separately models with at least one continuous auxiliary variable and models where all the auxiliary variables are categorical.

14.6.4.1 Predictors with Continuous Auxiliary Variables

This part considers two cases that are common in practice. In addition to the sample file S, we first suppose that a census file C with the following properties is available.

1. C contains the values $\boldsymbol{x}_{dj}$, $d = 1, \ldots, D$, $j = 1, \ldots, N_d$, of the auxiliary variables.
2. C has the same values $\boldsymbol{x}_{dj}$, $d = 1, \ldots, D$, $j \in s_d$, as the sample file S.
3. C uses the same unit identifier variable as the sample file S, so that each domain U_d of C can be partitioned in the sampled part s_d and the non-sampled part r_d.

The EBP and the plug-in predictor of $\overline{Y}_d$ are

$$\hat{\overline{Y}}_d^{ebp} = \hat{\overline{Y}}_d(\hat{\boldsymbol{\theta}}) = \frac{1}{N_d}\Big\{\sum_{j\in s_d} y_{dj} + \sum_{j\in r_d} \hat{p}_{dj}^{ebp}\Big\}, \qquad (14.18)$$

$$\hat{\overline{Y}}_d^{in} = \hat{\overline{Y}}_d^{in}(\hat{\boldsymbol{\theta}}) = \frac{1}{N_d}\Big\{\sum_{j\in s_d} y_{dj} + \sum_{j\in r_d} \hat{p}_{dj}^{in}\Big\}.$$

The empirical projective predictor (EPP) and the projective plug-in predictor (PIN) of $\overline{Y}_d$ are

$$\hat{\overline{Y}}_d^{epp} = \frac{1}{N_d}\sum_{j\in U_d} \hat{p}_{dj}^{ebp}, \quad \hat{\overline{Y}}_d^{pin} = \frac{1}{N_d}\sum_{j\in U_d} \hat{p}_{dj}^{in}. \qquad (14.19)$$

Design-based approximations of $\hat{\overline{Y}}_d^{ebp}$ and $\hat{\overline{Y}}_d^{in}$ are $\hat{\overline{Y}}_d^{dbp}$ and $\hat{\overline{Y}}_d^{din}$, respectively, where

$$\hat{\overline{Y}}_d^{dbp} = \frac{1}{N_d}\Big\{\sum_{j\in s_d} (y_{dj} - \hat{p}_{dj}^{ebp}) + \sum_{j\in s_d} \omega_{dj}\hat{p}_{dj}^{ebp}\Big\},$$

$$\hat{\overline{Y}}_d^{din} = \frac{1}{N_d}\Big\{\sum_{j\in s_d} (y_{dj} - \hat{p}_{dj}^{in}) + \sum_{j\in s_d} \omega_{dj}\hat{p}_{dj}^{in}\Big\},$$

and the ω_{dj}'s are the calibrated sampling weights. If $n_d = 0$, then the synthetic empirical projective predictor (sEPP) and the synthetic projective plug-in predictor (SPIN) of $\overline{Y}_d$ are

$$\hat{\overline{Y}}_d^{sepp} = \frac{1}{N_d}\sum_{j=1}^{N_d} \hat{p}_{dj}^{sebp}, \quad \hat{\overline{Y}}_d^{spin} = \frac{1}{N_d}\sum_{j=1}^{N_d} \tilde{p}_{dj}^{synth}, \quad \tilde{p}_{dj}^{synth} = \frac{\exp\{\boldsymbol{x}_{dj}\hat{\boldsymbol{\beta}}\}}{1+\exp\{\boldsymbol{x}_{dj}\hat{\boldsymbol{\beta}}\}}.$$

Second, we assume that C fulfills Property 1, but not Properties 2 and 3. As we cannot identify the sampled units in the census file C, we construct an enlarged file G by inserting n_d new registers of the sample file S within each domain of the census file C. The new registers will contain the sampled values of the auxiliary variables, i.e. $\boldsymbol{x}_{dj}$, $d = 1, \ldots, D$, $j \in s_d$. We also add a new column with a sample identifier taking the value one in the inserted registers and zero otherwise. The new enlarged census file G fulfills Properties 1–3 with domain sizes $N_d + n_d$, $d = 1, \ldots, D$. The approximated EBP (ABP) and the approximated plug-in predictor (AIN) of $\overline{Y}_d$ are

$$\hat{\overline{Y}}_d^{abp} = \frac{1}{N_d+n_d}\Big\{\sum_{j\in s_d} y_{dj} + \sum_{j\in U_d} \hat{p}_{dj}^{ebp}\Big\}, \quad \hat{\overline{Y}}_d^{ain} = \frac{1}{N_d+n_d}\Big\{\sum_{j\in s_d} y_{dj} + \sum_{j\in U_d} \hat{p}_{dj}^{in}\Big\}, \tag{14.20}$$

where the first and second sums are calculated by using data from G that correspond to the files S and C, respectively. This is to say, the data file of set r_d in the enlarged census file G is equal to the data file of set U_d of the original census file C.

14.6.4.2 Predictors with Categorical Auxiliary Variables

Under the categorical setup presented in Remark 14.2, the EBP and the plug-in predictor of $\overline{Y}_d = \frac{1}{N_d}\sum_{j=1}^{N_d} y_{dj}$ are

$$\hat{\overline{Y}}_d^{ebp} = \hat{\overline{Y}}_d(\hat{\boldsymbol{\theta}}) = \frac{1}{N_d}\Big\{\sum_{j\in s_d} y_{dj} + \sum_{k=1}^{K} N_{dk,r}\hat{q}_{dk}\Big\},$$

$$\hat{\overline{Y}}_d^{in} = \hat{\overline{Y}}_d^{in}(\hat{\boldsymbol{\theta}}) = \frac{1}{N_d}\Big\{\sum_{j\in s_d} y_{dj} + \sum_{k=1}^{K} N_{dk,r}\hat{q}_{dk}^{in}\Big\},$$

where $\hat{q}_{dk}$ and $\hat{q}_{dk}^{in}$ are defined in Sect. 14.6.2, $N_{dk,r} = \#\{j \in r_d : \boldsymbol{x}_{dj} = z_k\}$ is the size of the covariate class z_k at r_d and r_d denotes the non-sampled part of domain d.

14.7 MSE of Empirical Best Predictors

Jiang and Lahiri (2001) and Jiang (2003) gave approximations of the MSE of the EBPs of functions of fixed and random effects under models for binary data and GLMMs, respectively. Their approach is based on Taylor series expansions and was adapted to the unit-level logit mixed model by Hobza and Morales (2016). These authors gave full details of all the derivations needed for obtaining the formulas of the MSE approximations. They further implemented plug-in predictors of the MSE approximations. Due to the complexity of the algebraic calculations, this chapter does not introduce the MSE estimator based on the approximation given by Jiang and Lahiri (2001) and Jiang (2003) and developed by Hobza and Morales (2016). In what follows, the bootstrap-based estimators of the MSE are described.

14.7.1 Categorical Auxiliary Variables

This section assumes that the covariates are categorical such that $\boldsymbol{x}_{dj} \in \{z_1, \ldots, z_K\}$ and that $m_{dj} = 1, d = 1, \ldots, D, j = 1, \ldots, N_d$. There is no census file with the auxiliary variables. Let U_d, s_d, and $r_d = U_d - s_d$ denote the population, the sample, and the non-sample of domain d, respectively. The size of the covariate class z_k crossed by domain d is denoted by $N_{dk} = \#\{j \in U_d : \boldsymbol{x}_{dj} = z_k\}$. Similarly, the size of the covariate class z_k crossed by r_d is $N_{dk,r} = \#\{j \in r_d : \boldsymbol{x}_{dj} = z_k\}$.

Bootstrap Estimation of the MSE of a Predictor of μ_d

Let us consider the following functions of fixed and random effects

$$\mu_d = \mu_d(\boldsymbol{\theta}, v_d) = \sum_{j=1}^{N_d} p_{dj} = \sum_{k=1}^{K} N_{dk} q_{dk},$$

$$q_{dk} = q_{dk}(\boldsymbol{\theta}, v_d) = \frac{\exp\{z_k \boldsymbol{\beta} + \phi v_d\}}{1 + \exp\{z_k \boldsymbol{\beta} + \phi v_d\}}.$$

The following procedure calculates a parametric bootstrap estimator of $MSE(\hat{\mu}_d)$, where $\hat{\mu}_d$ is a predictor (EBP or plug-in) of μ_d.

1. Fit the model to the sample and calculate $\hat{\boldsymbol{\theta}} = (\hat{\boldsymbol{\beta}}', \hat{\phi})'$.
2. Repeat B times ($b = 1, \ldots, B$):

   a. *Bootstrap sample*: Generate a set of random effects $\{v_d^{*(b)} : d = 1, \ldots, D\}$ i.i.d. $N(0, 1)$. The bootstrap sample has the same units as the real data sample, i.e. $s_d^{*(b)} = s_d$, $b = 1, \ldots, B$. For $d = 1, \ldots, D$, $j \in s_d$, generate the elements of the bootstrap sample

$$y_{dj}^{*(b)} \sim \text{Bin}(1, p_{dj}^{*(b)}), \quad p_{dj}^{*(b)} = \frac{\exp\{\boldsymbol{x}_{dj}\hat{\boldsymbol{\beta}} + \hat{\phi} v_d^{*(b)}\}}{1 + \exp\{\boldsymbol{x}_{dj}\hat{\boldsymbol{\beta}} + \hat{\phi} v_d^{*(b)}\}}.$$

For $d = 1, \ldots, D$, $k = 1, \ldots, K$, calculate the bootstrap population quantities

$$\mu_d^{*(b)} = \sum_{j \in s_d} p_{dj}^{*(b)} + \sum_{k=1}^{K} N_{dk,r} q_{dk}^{*(b)}, \quad q_{dk}^{*(b)} = \frac{\exp\{z_k\hat{\boldsymbol{\beta}} + \hat{\phi} v_d^{*(b)}\}}{1 + \exp\{z_k\hat{\boldsymbol{\beta}} + \hat{\phi} v_d^{*(b)}\}}.$$

b. *Bootstrap model*: Fit a unit-level logit mixed model to the bootstrap sample $(y_{dj}^{*(b)}, \boldsymbol{x}_{dj})$, $d = 1, \ldots, D$, $j = 1, \ldots, n_d$. Calculate the parameter estimator $\hat{\boldsymbol{\theta}}^{*(b)} = (\hat{\boldsymbol{\beta}}^{*(b)\prime}, \hat{\phi}^{*(b)})'$ and the random effects predictors $\hat{v}_d^{*(b)}$, $d = 1, \ldots, D$. Calculate the predictor $\hat{\mu}_d^{*(b)}$ of the bootstrap population quantity $\mu_d^{*(b)}$. For example, the plug-in predictor of $\mu_d^{*(b)}$ is

$$\hat{\mu}_d^{*(b)} = \sum_{k=1}^{K} N_{dk} \hat{q}_{dk}^{in*(b)}, \quad \hat{q}_{dk}^{in*(b)} = \frac{\exp\{z_k\hat{\boldsymbol{\beta}}^{*(b)} + \hat{\phi}^{*(b)}\hat{v}_d^{*(b)}\}}{1 + \exp\{z_k\hat{\boldsymbol{\beta}}^{*(b)} + \hat{\phi}^{*(b)}\hat{v}_d^{*(b)}\}}.$$

3. Output: $mse^*(\hat{\mu}_d) = \frac{1}{B}\sum_{b=1}^{B}\big(\hat{\mu}_d^{*(b)} - \mu_d^{*(b)}\big)^2$.

Bootstrap Estimation of the MSE of a Predictor of $\overline{Y}_d$

For calculating a parametric bootstrap estimator of $MSE(\hat{\overline{Y}}_d)$, where $\hat{\overline{Y}}_d$ is a predictor (EBP or plug-in) of $\overline{Y}_d = \frac{1}{N_d}\sum_{j=1}^{N_d} y_{dj}$, the following procedure can be applied.

1. Fit the model to the sample and calculate $\hat{\boldsymbol{\theta}} = (\hat{\boldsymbol{\beta}}', \hat{\phi})'$.
2. Repeat B times ($b = 1, \ldots, B$):

a. *Bootstrap sample*: Generate a set of random effects $\{v_d^{*(b)} : d = 1, \ldots, D\}$ i.i.d. $N(0, 1)$. The bootstrap sample has the same units as the real data sample, i.e. $s_d^{*(b)} = s_d$, $b = 1, \ldots, B$. For $d = 1, \ldots, D$, $j \in s_d$, generate the bootstrap sample elements

$$y_{dj}^{*(b)} \sim \text{Bin}(1, p_{dj}^{*(b)}), \quad p_{dj}^{*(b)} = \frac{\exp\{\boldsymbol{x}_{dj}\hat{\boldsymbol{\beta}} + \hat{\phi} v_d^{*(b)}\}}{1 + \exp\{\boldsymbol{x}_{dj}\hat{\boldsymbol{\beta}} + \hat{\phi} v_d^{*(b)}\}}.$$

For $d = 1, \ldots, D$, $k = 1, \ldots, K$, calculate the bootstrap population quantities

$$\overline{Y}_d^{*(b)} = \frac{1}{N_d}\Big(\sum_{j\in s_d} y_{dj}^{*(b)} + \sum_{k=1}^{K} z_{dk}^{*(b)}\Big),$$

where

$$z_{dk}^{*(b)} \sim \text{Bin}(N_{dk,r}, q_{dk}^{*(b)}), \quad q_{dk}^{*(b)} = \frac{\exp\{z_k\hat{\boldsymbol{\beta}} + \hat{\phi}v_d^{*(b)}\}}{1+\exp\{z_k\hat{\boldsymbol{\beta}} + \hat{\phi}v_d^{*(b)}\}}.$$

b. *Bootstrap model*: Fit a unit-level logit mixed model to the bootstrap sample $(y_{dj}^{*(b)}, \boldsymbol{x}_{dj})$, $d = 1, \ldots, D$, $j = 1, \ldots, n_d$. Calculate the parameter estimator $\hat{\boldsymbol{\theta}}^{*(b)} = (\hat{\boldsymbol{\beta}}^{*(b)\prime}, \hat{\phi}^{*(b)})'$ and the random effects predictors $\hat{v}_d^{*(b)}$, $d = 1, \ldots, D$. Calculate the predictor $\hat{\overline{Y}}_d^{*(b)}$ of $\overline{Y}_d^{*(b)}$. For example, the plug-in predictor of $\overline{Y}_d^{*(b)} = \frac{1}{N_d}\sum_{j=1}^{N_d} y_{dj}^{*(b)}$ is

$$\hat{\overline{Y}}_d^{*(b)} = \frac{1}{N_d}\Big\{\sum_{j\in s_d} y_{dj}^{*(b)} + \sum_{k=1}^{K} N_{dk,r}\hat{q}_{dk}^{in*(b)}\Big\},$$

where

$$\hat{q}_{dk}^{in*(b)} = \frac{\exp\{z_k\hat{\boldsymbol{\beta}}^{*(b)} + \hat{\phi}^{*(b)}\hat{v}_d^{*(b)}\}}{1+\exp\{z_k\hat{\boldsymbol{\beta}}^{*(b)} + \hat{\phi}^{*(b)}\hat{v}_d^{*(b)}\}}.$$

3. Output: $mse^*(\hat{\overline{Y}}_d) = \frac{1}{B}\sum_{b=1}^{B}(\hat{\overline{Y}}_d^{*(b)} - \overline{Y}_d^{*(b)})^2$.

14.7.2 Continuous Auxiliary Variables

This section assumes that some auxiliary variables $\boldsymbol{x}_{dj}$ are continuous and available for the whole population and that $m_{dj} = 1$, $d = 1, \ldots, D$, $j = 1, \ldots, N_d$. It also assumes that all the $\boldsymbol{x}_{dj}$ are stored in census file. The census and sample files have the same set of identification variables, so that it is possible to divide the census file into the sample and non-sample parts. As before, $\boldsymbol{\theta} = (\boldsymbol{\beta}', \phi)'$ is the vector of parameters and U_d, s_d, and r_d denotes the population, the sample, and the non-sample of domain d. We finally assume that the same values $\boldsymbol{x}_{dj}$, $d = 1, \ldots, D$, $j \in s_d$, are stored in the sample and in the census files.

Bootstrap Estimation of the MSE of a Predictor of μ_d

Let us consider the following functions of fixed and random effects

$$\mu_d = \mu_d(\boldsymbol{\theta}, v_d) = \sum_{j\in U_d} p_{dj}, \quad p_{dj} = p_{dj}(\boldsymbol{\theta}, v_d) = \frac{\exp\{\boldsymbol{x}_{dj}\boldsymbol{\beta} + \phi v_d\}}{1+\exp\{\boldsymbol{x}_{dj}\boldsymbol{\beta} + \phi v_d\}}.$$

The following procedure calculates a parametric bootstrap estimator of $MSE(\hat{\mu}_d)$, where $\hat{\mu}_d = \hat{\mu}_d(\hat{\boldsymbol{\theta}})$ is a predictor (EBP or plug-in) of μ_d.

1. Fit the model to the sample and calculate $\hat{\boldsymbol{\theta}} = (\hat{\boldsymbol{\beta}}', \hat{\phi})'$.
2. Repeat B times ($b = 1, \ldots, B$):

   a. *Bootstrap population*: Generate random effects $\{v_d^{*(b)} : d = 1, \ldots, D\}$ i.i.d. $N(0, 1)$. For $d = 1, \ldots, D$, $j \in U_d$, generate

   $$y_{dj}^{*(b)} \sim \text{Bin}(1, p_{dj}^{*(b)}), \quad p_{dj}^{*(b)} = \frac{\exp\{\boldsymbol{x}_{dj}\hat{\boldsymbol{\beta}} + \hat{\phi} v_d^{*(b)}\}}{1+\exp\{\boldsymbol{x}_{dj}\hat{\boldsymbol{\beta}} + \hat{\phi} v_d^{*(b)}\}}$$

   and calculate the bootstrap population quantities

   $$\mu_d^{*(b)} = \sum_{j\in U_d} p_{dj}^{*(b)}.$$

   b. *Bootstrap sample*: The bootstrap sample has the same units as the real data sample, i.e. $s_d^{*(b)} = s_d$, $b = 1, \ldots, B$. It is not extracted at random. As the model is on the population, the source of randomness comes from the generation of the population.

   c. *Bootstrap model*: Fit a unit-level logit mixed model to the bootstrap sample $(y_{dj}^{*(b)}, \boldsymbol{x}_{dj})$, $d = 1, \ldots, D$, $j \in s_d$. Calculate the parameter estimator $\hat{\boldsymbol{\theta}}^{*(b)} = (\hat{\boldsymbol{\beta}}^{*(b)\prime}, \hat{\phi}^{*(b)})'$ and the random effects predictors $\hat{v}_d^{*(b)}$, $d = 1, \ldots, D$. Calculate the predictor $\hat{\mu}_d^{*(b)}$ of the bootstrap population quantity $\mu_d^{*(b)}$. For example, the plug-in predictor of $\mu_d^{*(b)}$ is

   $$\hat{\mu}_d^{*(b)} = \sum_{j\in U_d} \hat{p}_{dj}^{in*(b)},$$

   where

   $$\hat{p}_{dj}^{in*(b)} = \frac{\exp\{\boldsymbol{x}_{dj}\hat{\boldsymbol{\beta}}^{*(b)} + \hat{\phi}^{*(b)}\hat{v}_d^{*(b)}\}}{1+\exp\{\boldsymbol{x}_{dj}\hat{\boldsymbol{\beta}}^{*(b)} + \hat{\phi}^{*(b)}\hat{v}_d^{*(b)}\}}.$$

3. Output: $mse^*(\hat{\mu}_d) = \frac{1}{B}\sum_{b=1}^{B}\left(\hat{\mu}_d^{*(b)} - \mu_d^{*(b)}\right)^2$.

Bootstrap Estimation of the MSE of a Predictor of $\overline{Y}_d$

For calculating a parametric bootstrap estimator of $MSE(\hat{\overline{Y}}_d)$, where $\hat{\overline{Y}}_d$ is a predictor (EBP or plug-in) of

$$\overline{Y}_d = \frac{1}{N_d} \sum_{j \in U_d} y_{dj},$$

the following procedure can be applied.

1. Fit the model to the sample and calculate $\hat{\boldsymbol{\theta}} = (\hat{\boldsymbol{\beta}}', \hat{\phi})'$.
2. Repeat B times ($b = 1, \ldots, B$):

   a. *Bootstrap population*: Generate a set of random effects $\{v_d^{*(b)} : d = 1, \ldots, D\}$ i.i.d. $N(0, 1)$. For $d = 1, \ldots, D$, $j \in U_d$, generate

   $$y_{dj}^{*(b)} \sim \text{Bin}(1, p_{dj}^{*(b)}), \quad p_{dj}^{*(b)} = \frac{\exp\{\boldsymbol{x}_{dj}\hat{\boldsymbol{\beta}} + \hat{\phi} v_d^{*(b)}\}}{1 + \exp\{\boldsymbol{x}_{dj}\hat{\boldsymbol{\beta}} + \hat{\phi} v_d^{*(b)}\}}$$

   and calculate the bootstrap population quantities

   $$\overline{Y}_d^{*(b)} = \frac{1}{N_d} \sum_{j \in U_d} y_{dj}^{*(b)}.$$

   b. *Bootstrap sample*: The bootstrap sample has the same units as the real data sample. It is not extracted at random. This is to say, $s_d^{*(b)} = s_d$, $b = 1, \ldots, B$. The model is on the population, therefore the source of randomness comes from the generation of the population.
   c. *Bootstrap model*: Fit a unit-level logit mixed model to the bootstrap sample $(y_{dj}^{*(b)}, \boldsymbol{x}_{dj})$, $d = 1, \ldots, D$, $j \in s_d$. Calculate the parameter estimator $\hat{\boldsymbol{\theta}}^{*(b)} = (\hat{\boldsymbol{\beta}}^{*(b)\prime}, \hat{\phi}^{*(b)})'$ and the random effects predictors $\hat{v}_d^{*(b)}$, $d = 1, \ldots, D$. Calculate the predictor $\hat{\overline{Y}}_d^{*(b)}$ of the bootstrap population quantity $\overline{Y}_d^{*(b)}$. For example, the plug-in predictor of $\overline{Y}_d^{*(b)}$ is

   $$\hat{\overline{Y}}_d^{*(b)} = \frac{1}{N_d} \Big\{ \sum_{j \in s_d} y_{dj}^{*(b)} + \sum_{j \in r_d} \hat{p}_{dj}^{in*(b)} \Big\},$$

   where

   $$\hat{p}_{dj}^{in*(b)} = \frac{\exp\{\boldsymbol{x}_{dj}\hat{\boldsymbol{\beta}}^{*(b)} + \hat{\phi}^{*(b)} \hat{v}_d^{*(b)}\}}{1 + \exp\{\boldsymbol{x}_{dj}\hat{\boldsymbol{\beta}}^{*(b)} + \hat{\phi}^{*(b)} \hat{v}_d^{*(b)}\}}.$$

3. Output: $mse^*(\hat{\overline{Y}}_d) = \frac{1}{B} \sum_{b=1}^{B} (\hat{\overline{Y}}_d^{*(b)} - \overline{Y}_d^{*(b)})^2$.

Census File with Unidentified Sample Units

If the sample units are not identified in the census file, we can enlarge it by inserting n_d new registers within each domain. The new registers will contain the sampled values of the auxiliary variables, i.e. $\boldsymbol{x}_{dj}$, $d = 1, \ldots, D$, $j \in s_d$. We also add a new column with a sample identifier taking the value one in the inserted registers and zero otherwise. In this way, we construct a new census file with domain sizes $N_d + n_d$, $d = 1, \ldots, D$, and fulfilling the required conditions for programming a bootstrap resampling procedure with continuous auxiliary variables. Under this approach, we recall that the denominators in the formulas of $\hat{\mu}_d^{*(b)}$ and $\hat{\overline{Y}}_d^{*(b)}$ are $N_d + n_d$ instead of N_d and that the set r_d of the enlarged census file is equal to the set U_d of the original census file. The bootstrap procedure for estimating the MSE of the approximated plug-in predictor (or approximated EBP) of $\overline{Y}_d$ is

1. Fit the model to the sample and calculate $\hat{\boldsymbol{\theta}} = (\hat{\boldsymbol{\beta}}', \hat{\phi})'$.
2. Repeat B times ($b = 1, \ldots, B$):

    a. *Bootstrap population*: Generate a set of random effects $\{v_d^{*(b)} : d = 1, \ldots, D\}$ i.i.d. $N(0, 1)$. For $d = 1, \ldots, D$, $j = 1, \ldots, N_d + n_d$, generate the bootstrap enlarged population

    $$y_{dj}^{*(b)} \sim \text{Bin}(1, p_{dj}^{*(b)}), \quad p_{dj}^{*(b)} = \frac{\exp\{\boldsymbol{x}_{dj}\hat{\boldsymbol{\beta}} + \hat{\phi} v_d^{*(b)}\}}{1 + \exp\{\boldsymbol{x}_{dj}\hat{\boldsymbol{\beta}} + \hat{\phi} v_d^{*(b)}\}}, \quad j \in s_d \cup U_d,$$

    and calculate the bootstrap population quantities

    $$\overline{Y}_d^{*(b)} = \frac{1}{N_d + n_d}\Big(\sum_{j \in s_d} y_{dj}^{*(b)} + \sum_{j \in U_d} y_{dj}^{*(b)}\Big),$$

    where $j \in s_d$ and $j \in U_d$ denote domain summations over the whole sample and the whole census file, respectively.

    b. *Bootstrap sample*: The bootstrap sample, extracted from the enlarged census file, has the same units as the real data sample. It is not extracted at random. This is to say, $s_d^{*(b)} = s_d$, $b = 1, \ldots, B$. The model is on the population, therefore the source of randomness comes from the generation of the population.

    c. *Bootstrap model*: Fit a unit-level logit mixed model to the bootstrap sample $(y_{dj}^{*(b)}, \boldsymbol{x}_{dj})$, $d = 1, \ldots, D$, $j \in s_d$. Calculate the parameter estimator $\hat{\boldsymbol{\theta}}^{*(b)} = (\hat{\boldsymbol{\beta}}^{*(b)\prime}, \hat{\phi}^{*(b)})'$ and the random effects predictors $\hat{v}_d^{*(b)}$, $d = 1, \ldots, D$. Calculate the predictor $\hat{\overline{Y}}_d^{*(b)}$ of the bootstrap population quantity $\overline{Y}_d^{*(b)}$. For example, the plug-in predictor of $\overline{Y}_d^{*(b)}$ is

    $$\hat{\overline{Y}}_d^{ain*(b)} = \frac{1}{N_d + n_d}\Big\{\sum_{j \in s_d} y_{dj}^{*(b)} + \sum_{j \in U_d} \hat{p}_{dj}^{in*(b)}\Big\},$$

where

$$\hat{p}_{dj}^{in*(b)} = \frac{\exp\{\boldsymbol{x}_{dj}\hat{\boldsymbol{\beta}}^{*(b)} + \hat{\phi}^{*(b)}\hat{v}_d^{*(b)}\}}{1+\exp\{\boldsymbol{x}_{dj}\hat{\boldsymbol{\beta}}^{*(b)} + \hat{\phi}^{*(b)}\hat{v}_d^{*(b)}\}}.$$

3. Output: $mse^*(\hat{\overline{Y}}_d) = \frac{1}{B}\sum_{b=1}^{B}(\hat{\overline{Y}}_d^{ain*(b)} - \overline{Y}_d^{*(b)})^2$.

14.8 R Codes for EBPs

This section gives R codes for calculating the EBPs of domain means under the unit-level logit mixed model by using data from the survey data file `datLCS.txt`. The domains of interest are the counties. The target variable poor is derived from the variable *income* giving the individual equivalized annual net income (in euros). The variable poor takes the value 1 if income $< z_0$, where $z_0 = 7280$ is the poverty threshold. The auxiliary variables are the indicators of the categories *employed* and *unemployed* of the labor status variable. The following code reads the survey data file and calculates new variables.

```
dat <- read.table("datLCS.txt", header=TRUE, sep="\t", dec=",")
z0 <- 7280                               # Poverty threshold.
# Poverty variable: 1 if income<z0, 0 if income>z0
poor <- as.numeric(dat$income<z0)
# Labor situation: 0 if < 16 years, 1 if employed,
# 2 if unemployed, 3 if inactive.
work <- as.numeric(dat$lab=="1")        # Employed
nowork <- as.numeric(dat$lab=="2")      # Unemployed
inact <- as.numeric(dat$lab=="3")       # Inactive
```

We rename some variables

```
income <- dat$income; w <- dat$w; dom <- dat$dom
one <- rep(1,nrow(dat))                 # Auxiliary 1-variable
domains <- sort(unique(dat$dom))        # Domains sorted in ascending order
ndom <- length(domains)                 # Number of domains
```

The following code reads the file with the aggregated auxiliary variables and sort the file by domain.

```
aux <- read.table("auxLCS.txt", header=TRUE, sep="\t", dec=",")
# Totals of employed people
aux$Twork <- round(aux$TOT*aux$Mwork, 0)
# Totals of unemployed people
aux$Tnowork <- round(aux$TOT*aux$Mnowork, 0)
# Totals in the categories "inactive" or "<16 years"
aux$Tres <- aux$TOT - aux$Twork - aux$Tnowork
aux <- aux[order(aux$dom), ]    # Sort aux by dom (in ascending order)
```

We calculate sample sizes

```
# sample sizes by domains and others
nd <- tapply(X=one, INDEX=dom, FUN=sum)
ndworking <- tapply(one, list(dom,work), sum, default=0)
# sample sizes by domains and employment category
ndwork <- ndworking[,2]
ndnoworking <- tapply(one, list(dom,nowork), sum, default=0)
# sample sizes by domains and unemployment category
ndnowork <- ndnoworking[,2]
# sample sizes by domains and inactive or <16 categories
ndres <- nd - ndwork - ndnowork
```

We calculate direct estimators and estimated sizes by domains. We use `dir2` function described in Sect. 2.8.4.

```
# Direct estimates of poverty proportions
poor.dir <- dir2(data=poor, w=w, domain=list(dom))
dirp <- poor.dir$mean
# Estimated sizes
hatNd <- poor.dir$Nd.hat
# Direct estimates of totals of employed people
hatNwork <- tapply(work*w, dom, sum)
# Direct estimates of totals of unemployed people
hatNnowork <- tapply(nowork*w, dom, sum)
# Direct estimates of totals of remaining people
hatNdres <- hatNd - hatNwork - hatNnowork
```

We install and/or load the R package `lme4`.

```
if(!require(Matrix)){
  install.packages("Matrix")
  library(Matrix)
}
if(!require(lme4)){
  install.packages("lme4")
  library(lme4)
}
```

We fit a unit-level logit mixed model to the variable poor with the R library `lme4`

```
glmm <- glmer(formula=poor ~ work + nowork + (1|dom), data=dat,
              family=binomial)
summary(glmm)              # Summary of model results
pihat <- fitted(glmm)      # Predicted probabilities
beta <- fixef(glmm)        # Beta parameters
# Fixed effect for x0=1, x1=1, x2=0 (employed)
bwork <- beta[1] + beta[2]
# Fixed effect for x0=1, x1=0, x2=1 (unemployed)
bnowork <- beta[1] + beta[3]
# Fixed effect for x0=1, x1=0, x2=0 (remaining categories)
binact <- beta[1]
# Variance components
var <- as.data.frame(VarCorr(glmm))
# Standard deviation of the random effect
phi <- var$sdcor[1]
ud <- ranef(glmm)$dom      # Random effects
```

We calculate the plug-in predictors of poverty proportions by domains

```
# Poverty proportions by domains and employment category
pdworking <- tapply(pihat, list(dom,work), mean, default=0)
pdwork <- pdworking[,2]
# Poverty proportions by domains and unemployment category
pdnoworking <- tapply(pihat, list(dom,nowork), mean, default=0)
pdnowork <- pdnoworking[,2]
# Poverty proportions by domains and remaining categories
```

```
pdinactive <- tapply(pihat, list(dom,inact), mean, default=0)
pdinact <- pdinactive[,2]
# Poverty sample totals
totp <- tapply(poor, dom, sum)
# Plug-in estimates of poverty proportions by domains
plug.p <- (totp + (aux$Twork-ndwork)*pdwork +
          (aux$Tnowork-ndnowork)*pdnowork +
          (aux$Tres-ndres)*pdinact)/aux$TOT
```

We calculate the EBPs of poverty proportions by domains

```
S <- 1000                # Auxiliary terms
set.seed(123)            # Set seed of random number generator
# Generate the random values
v <- rnorm(n=S*ndom, mean=0, sd=1)
vv <- Map(split(v, rep(1:ndom,each=S)),
          split(-v, rep(1:ndom, each=S)), f=c)
vvphi <- lapply(vv, phi, FUN="*")
# Define some auxiliary variables
xbeta <- split(beta[1]+beta[2]*work+beta[3]*nowork, dat$dom)
# Calculate some terms of the EBP expression
a1 <- Map(phi*totp, vv, f="*")
a2 <- Dd <- vector()
term2 <- list()
for(d in 1:ndom){
  for(s in 1:(2*S)){
    a2[s] <- sum(log(1+exp(xbeta[[d]]+vvphi[[d]][s])))
  }
  term2[[d]] <- exp(a1[[d]] - a2)
  Dd[d] <- mean(term2[[d]])
}
expwork <- lapply(lapply(vvphi, bwork, FUN="+"), FUN=exp)
expnowork <- lapply(lapply(vvphi, bnowork, FUN="+"), FUN=exp)
expinact <- lapply(lapply(vvphi, binact, FUN="+"), FUN=exp)
expworkplus1 <- lapply(expwork, 1, FUN="+")
expnoworkplus1 <- lapply(expnowork, 1, FUN="+")
expinactplus1 <- lapply(expinact, 1, FUN="+")
Ads.work <- Map(Map(expwork, expworkplus1, f="/"), term2, f="*")
Ads.nowork <- Map(Map(expnowork, expnoworkplus1, f="/"), term2, f="*")
Ads.inact <- Map(Map(expinact, expinactplus1, f="/"), term2, f="*")
Ad.work <- sapply(Ads.work, FUN=mean)
Ad.nowork <- sapply(Ads.nowork, FUN=mean)
Ad.inact <- sapply(Ads.inact, FUN=mean)
# Calculating the EBPs by domains
ebp.work <- Ad.work/Dd
ebp.nowork <- Ad.nowork/Dd
ebp.inact <- Ad.inact/Dd
# Calculation of poverty proportion EBPs
ebp.p <- (totp + (aux$Twork-ndwork)*ebp.work +
         (aux$Tnowork-ndnowork)*ebp.nowork +
         (aux$Tres-ndres)*ebp.inact)/aux$TOT
```

The R code to save the results is

```
output <- data.frame(nd, dirp=round(dirp,3),
                     ebpp=round(ebp.p,3), plugp=round(plug.p,3))
head(output, 10)
```

For the ten first domains, Table 14.1 gives the sample sizes and the direct, EBP and plug-in estimates of the poverty proportions.

Table 14.1 Estimates of poverty proportions

dom	3	5	6	7	11	12	13	14	15	16
n_d	57	96	82	10	118	18	138	190	406	93
dirp	0.461	0.363	0.180	0.000	0.167	0.177	0.025	0.155	0.107	0.074
ebpp	0.355	0.256	0.176	0.087	0.151	0.172	0.046	0.178	0.088	0.091
plugp	0.352	0.254	0.175	0.056	0.151	0.168	0.046	0.178	0.088	0.090

References

Bleistein, N., Handelsman, R.: Asymptotic Expansions of Integrals. Dover, New York (1986). (eds.) Proceedings of Statistics Canada Symposium on Analysis of Data in Time, pp. 67–74 (1989)

Demidenko E.: Mixed Models, Theory and Applications. Wiley, New York (2004)

Dempster, A.P., Laird, N.M., Rubin, D.B.: Maximum likelihood from incomplete data via the EM algorithm. J. R. Stat. Soc. Ser. B. **39**(1), 1–38 (1977)

Evans, M., Swartz, T.: Approximating Integrals via Monte Carlo and Deterministic Methods. Oxford University Press, Oxford (2000)

González-Manteiga, W., Lombardía, M.J., Molina, I., Morales, D., Santamaría, L.: Estimation of the mean squared error of predictors of small area linear parameters under a logistic mixed model. Comput. Stat. Data Anal. **51**, 2720–2733 (2007)

González-Manteiga, W., Lombardía, M. J., Molina, I., Morales, D., Santamaría, L.: Bootstrap mean squared error of small-area EBLUP. J. Stat. Comput. Simul. **78**, 443–462 (2008a)

González-Manteiga, W., Lombardía, M. J., Molina, I., Morales, D., Santamaría, L.: Analytic and bootstrap approximations of prediction errors under a multivariate Fay-Herriot model. Comput. Stat. Data Anal. **52**, 5242–5252 (2008b)

Hobza, T., Morales, D.: Empirical best prediction under unit-level logit mixed models. J. Off. Stat. **32**(3), 661–69 (2016)

Jiang, J.: Consistent estimators in generalized linear models. J. Am. Stat. Assoc. **93**, 720–729 (1998)

Jiang, J.: Empirical best prediction for small-area inference based on generalized linear mixed models. J. Stat. Plan. Inference. **111**, 117–127 (2003)

Jiang, J., Lahiri, P.: Empirical best prediction for small area inference with binary data. Ann. Inst. Stat. Math. **53**, 217–243 (2001)

Wu, C.F.J.: On the convergence properties of the EM algorithm. Ann. Stat. **11**(1), 95–103 (1983)

Chapter 15
EBPs Under Unit-Level Two-Fold Logit Mixed Models

15.1 Introduction

Generalized linear mixed models (GLMMs) are typically employed to describe the behavior of non-normal variables in small area estimation problems. These models can be formulated at the area or unit level and can be applied to the prediction of counts and proportions.

The following contributions are examples of modeling with area-level GLMMs. Boubeta et al. (2016, 2017) introduced empirical best predictors (EBPs) of totals of poor people and poverty proportions based on Poisson mixed models. Molina et al. (2007) and López-Vizcaíno et al. (2013, 2015) introduced multinomial logit mixed models for predicting small area proportions and unemployment rates.

On the other hand, the unit-level GLMMs play also an important role in small area estimation. In particular, the binomial-logit mixed models have a prominent position. Under this kind of models, Ghosh et al. (2009) gave hierarchical and empirical Bayes predictors of proportions of people without health insurance for minority subpopulations. Erciulescu and Fuller (2016) studied the minimum mean squared error (MMSE) prediction method and compared plug-in and MMSE predictors of domain proportions. Militino et al. (2015) derived small area plug-in predictors of several indicators from information technology business. Hobza and Morales (2016) and Hobza et al. (2018) presented applications to estimating small area poverty indicators.

If the target variable takes the value of 1 on individuals having the studied property and takes the value of 0 otherwise, then their domain averages are proportions. The unit-level two-fold logit mixed models can be used to estimate proportions at different levels of aggregations or study the temporal behavior of domain proportions. These models have random effects taking into account the between-domains and the between-subdomain variability (within domains) that is not explained by the auxiliary variables. If the models are applied to temporal data,

D. Morales et al., *A Course on Small Area Estimation and Mixed Models*, Statistics for Social and Behavioral Sciences, https://doi.org/10.1007/978-3-030-63757-6_15

then the subdomains are substituted by time periods. The random effects are usually assumed to be normally distributed.

The maximum likelihood (ML) estimation of GLMM parameters has some computational difficulties because the likelihood may involve high-dimensional integrals that cannot be evaluated analytically. Section 15.3 maximizes the Laplace approximation to the log-likelihood for calculating ML estimators. By following Jiang and Lahiri (2001) and Jiang (2003), Sect. 15.4 introduces empirical best predictors (EBPs) and plug-in predictors for estimating domain proportions at different levels of aggregation. Section 15.6 presents a simulation experiment that empirically illustrates the consistency of the ML-Laplace estimators of model parameters. Finally, Sect. 15.7 gives some R codes for calculating EBPs of subdomain poverty proportions.

15.2 The Model

This section introduces the unit-level two-fold logit mixed model. Let D and T be the number of domains and time periods (or subdomains), respectively.

Let us consider two independent sets of random effects such that $\{v_{1,d} : d = 1, \dots, D\}$ are i.i.d. $N(0, 1)$ and $\{v_{2,dt} : d = 1, \dots, D, t = 1, \dots, T\}$ are i.i.d. $N(0, 1)$. In matrix notation, we have

$$\boldsymbol{v}_1 = \underset{1\leq d\leq D}{\operatorname{col}}(v_{1,d}) \sim N_D(\boldsymbol{0}, \boldsymbol{I}_D), \quad \boldsymbol{v}_{2,d} = \underset{1\leq t\leq T}{\operatorname{col}}(v_{2,dt}) \sim N(\boldsymbol{0}, \boldsymbol{I}_T),$$

$$\boldsymbol{v}_2 = \underset{1\leq d\leq D}{\operatorname{col}}(\boldsymbol{v}_{2,d}) \sim N(\boldsymbol{0}, \boldsymbol{I}_{DT}), \quad \boldsymbol{v} = (\boldsymbol{v}_1', \boldsymbol{v}_2')' \sim N(\boldsymbol{0}, \boldsymbol{I}_{D(T+1)}).$$

As the normal multivariate probability density function (p.d.f.) is

$$f_{N_k(\mu,\Sigma)}(\boldsymbol{x}) = \frac{1}{(2\pi)^{k/2}\sqrt{|\boldsymbol{\Sigma}|}} \exp\left\{-\frac{1}{2}(\boldsymbol{x}-\boldsymbol{\mu})'\boldsymbol{\Sigma}^{-1}(\boldsymbol{x}-\boldsymbol{\mu})\right\}, \tag{15.1}$$

we have

$$f_v(\boldsymbol{v}_1, \boldsymbol{v}_2) = (2\pi)^{-D(T+1)/2} \exp\left\{-\frac{1}{2}\boldsymbol{v}_1'\boldsymbol{v}_1 - \frac{1}{2}\boldsymbol{v}_2'\boldsymbol{v}_2\right\}.$$

Under the unit-level two-fold logit mixed model, the distribution of the target variable y_{dtj}, conditioned to the random effects $v_{1,d}$ and $v_{2,dt}$, is

$$y_{dtj}|_{v_{1,d},v_{2,dt}} \sim \operatorname{Bin}(m_{dtj}, p_{dtj}), \; d = 1, \dots, D, \; t = 1, \dots, T, \; j = 1, \dots, n_{dt}, \tag{15.2}$$

where m_{dtj} are known size parameters. For domains $d = 1, \dots, D$, subdomains or time periods $t = 1, \dots, T$, and sample units $j = 1, \dots, n_{dt}$, we assume

$$\eta_{dtj} = \log\frac{p_{dtj}}{1-p_{dtj}} = \boldsymbol{x}_{dtj}\boldsymbol{\beta} + \phi_1 v_{1,d} + \phi_2 v_{2,dt}, \tag{15.3}$$

where $\phi_1 > 0$ and $\phi_2 > 0$ are standard deviation parameters, $\boldsymbol{\beta} = \underset{1\le k\le p}{\text{col}}(\beta_k)$ is a vector of regression parameters, and $\boldsymbol{x}_{dtj} = \underset{1\le k\le p}{\text{col}'}(x_{dtjk})$ is a vector of auxiliary variables. The vector of unknown parameters of the assumed model is thus $\boldsymbol{\theta} = (\boldsymbol{\beta}', \phi_1, \phi_2)'$. Further, we assume that the y_{dtj}'s are independent conditioned to $\boldsymbol{v}_1$ and $\boldsymbol{v}_2$.

15.3 ML-Laplace Approximation Algorithm

Similarly as discussed in Sect. 14.5, the marginal likelihood of the unit-level two-fold logit mixed model (15.2)–(15.3) is intractable and must be approximated. This section describes the Laplace approximation of the model log-likelihood and the corresponding ML algorithm for estimating the model parameters and obtaining the mode predictors of the random effects.

Let us start with the Laplace approximation of a multiple integral of a general function $\exp(h(\boldsymbol{x}))$, where $h : R^k \mapsto R$ is a twice continuously differentiable function with a global maximum at the column vector $\boldsymbol{x}_0$. This is to say, let us assume that $\dot{h}(\boldsymbol{x}_0) = \left.\frac{dh}{d\boldsymbol{x}}\right|_{\boldsymbol{x}=\boldsymbol{x}_0} = \boldsymbol{0}$ and $\ddot{h}(\boldsymbol{x}_0) = \left.\frac{d^2h}{d\boldsymbol{x}^2}\right|_{\boldsymbol{x}=\boldsymbol{x}_0}$ is negative definite. A Taylor series expansion of $h(\boldsymbol{x})$ around $\boldsymbol{x}_0$ yields to

$$\begin{aligned} h(\boldsymbol{x}) &= h(\boldsymbol{x}_0) + \dot{h}'(\boldsymbol{x}_0)(\boldsymbol{x}-\boldsymbol{x}_0) + \frac{1}{2}(\boldsymbol{x}-\boldsymbol{x}_0)'\ddot{h}(x_0)(\boldsymbol{x}-\boldsymbol{x}_0) + o\big(\|\boldsymbol{x}-\boldsymbol{x}_0\|^2\big) \\ &\approx h(\boldsymbol{x}_0) + \frac{1}{2}(\boldsymbol{x}-\boldsymbol{x}_0)'\ddot{h}(x_0)(\boldsymbol{x}-\boldsymbol{x}_0). \end{aligned}$$

Using this expansion, the multivariate Laplace approximation of the integral of the function $\exp(h(\boldsymbol{x}))$ is

$$\begin{aligned} \int_{R^k} e^{h(x)}\,d\boldsymbol{x} &\approx \int_{R^k} e^{h(\boldsymbol{x}_0)} \exp\Big\{-\frac{1}{2}(\boldsymbol{x}-\boldsymbol{x}_0)'\big(-\ddot{h}(\boldsymbol{x}_0)\big)(\boldsymbol{x}-\boldsymbol{x}_0)\Big\}\,d\boldsymbol{x} \\ &= (2\pi)^{k/2}\big|-\ddot{h}(\boldsymbol{x}_0)\big|^{-1/2}e^{h(\boldsymbol{x}_0)}, \end{aligned} \tag{15.4}$$

where we used the fact that the integral of the multivariate normal p.d.f. $f(\boldsymbol{x})$, given in (15.1), is one.

15.3.1 The Laplace Approximation to the Likelihood

Let us now approximate the log-likelihood of the unit-level two-fold binomial mixed model. We recall that $\{v_{1,d} : d = 1, \ldots, D\}$ are i.i.d. $N(0, 1)$, $\{v_{2,dt} : d =$

$1,\ldots,D$, $t=1,\ldots,T\}$ are i.i.d. $N(0,1)$, and both sets of random effects are independent. Further, for $d=1,\ldots,D$, $t=1,\ldots,T$, and $j=1,\ldots,n_{dt}$, we recall that

$$y_{dtj}|_{v_{1,d},v_{2,dt}} \underset{ind}{\sim} \text{Bin}(m_{dtj}, p_{dtj}), \quad p_{dtj}=\frac{\exp\{\boldsymbol{x}_{dtj}\boldsymbol{\beta}+\phi_1 v_{1,d}+\phi_2 v_{2,dt}\}}{1+\exp\{\boldsymbol{x}_{dtj}\boldsymbol{\beta}+\phi_1 v_{1,d}+\phi_2 v_{2,dt}\}}.$$

Let us define $\boldsymbol{y}_d=(\boldsymbol{y}_{d1},\ldots,\boldsymbol{y}_{dT})'$, $\boldsymbol{y}_{dt}=(y_{dt1},\ldots,y_{dtn_{dt}})'$, $\boldsymbol{v}_d=(v_{1,d},\boldsymbol{v}_{2,d}')'$. The target vectors $\boldsymbol{y}_1,\ldots,\boldsymbol{y}_D$ are unconditionally independent, and the random vectors $\boldsymbol{v}_d$'s are independent and such that $\boldsymbol{v}_d\sim N_{T+1}(\boldsymbol{0},\boldsymbol{I}_{T+1})$, i.e.

$$f(v_{1,d},\boldsymbol{v}_{2,d})=(2\pi)^{-\frac{T+1}{2}}\exp\left\{-\frac{1}{2}v_{1,d}^2-\frac{1}{2}\boldsymbol{v}_{2,d}'\boldsymbol{v}_{2,d}\right\}.$$

Let us define $b_d=\prod_{t=1}^{T}\prod_{j=1}^{n_{dt}}\binom{m_{dtj}}{y_{dtj}}$. Using similar steps as in (14.7), we obtain the marginal distribution of $\boldsymbol{y}_d$ in the form

$$\begin{aligned}
P_d=P(\boldsymbol{y}_d)&=\int_{R^{T+1}}P(\boldsymbol{y}_d|v_{1,d},\boldsymbol{v}_{2,d})f(v_{1d})f(\boldsymbol{v}_{2,d})\,dv_{1,d}\,d\boldsymbol{v}_{2,d}\\
&=\int_{R^{T+1}}\Big[\prod_{t=1}^{T}\prod_{j=1}^{n_{dt}}P(y_{dtj}|v_{1,d},v_{2,dt})\Big]f(v_{1d})f(\boldsymbol{v}_{2,d})\,dv_{1,d}\,d\boldsymbol{v}_{2,d}\\
&=\int_{R^{T+1}}\Big[\prod_{t=1}^{T}\prod_{j=1}^{n_{dt}}\binom{m_{dtj}}{y_{dtj}}p_{dtj}^{y_{dtj}}(1-p_{dtj})^{m_{dtj}-y_{dtj}}\Big]f(v_{1d})f(\boldsymbol{v}_{2,d})\,dv_{1,d}\,d\boldsymbol{v}_{2,d}\\
&=b_d(2\pi)^{-\frac{T+1}{2}}\int_{R^{T+1}}\exp\Big\{-\sum_{t=1}^{T}\sum_{j=1}^{n_{dt}}m_{dtj}\log\big(1+\exp\{\boldsymbol{x}_{dtj}\boldsymbol{\beta}+\phi_1 v_{1,d}+\phi_2 v_{2,dt}\}\big)\\
&\quad+\sum_{t=1}^{T}\sum_{j=1}^{n_{dt}}y_{dtj}(\boldsymbol{x}_{dtj}\boldsymbol{\beta}+\phi_1 v_{1,d}+\phi_2 v_{2,dt})-\frac{1}{2}v_{1,d}^2-\frac{1}{2}\boldsymbol{v}_{2,d}'\boldsymbol{v}_{2,d}\Big\}\,dv_{1,d}\,d\boldsymbol{v}_{2,d}\\
&=b_d(2\pi)^{-\frac{T+1}{2}}\int_{R^{T+1}}\exp\{h(v_{1,d},v_{2,d1},\ldots,v_{2,dT})\}\,dv_{1,d}\,d\boldsymbol{v}_{2,d}, \qquad (15.5)
\end{aligned}$$

where

$$\begin{aligned}
h(\boldsymbol{v}_d)&=h(v_{1,d},v_{2,d1},\ldots,v_{2,dT})\\
&=-\sum_{t=1}^{T}\sum_{j=1}^{n_{dt}}m_{dtj}\log\big(1+\exp\{\boldsymbol{x}_{dtj}\boldsymbol{\beta}+\phi_1 v_{1,d}+\phi_2 v_{2,dt}\}\big)\\
&\quad+\sum_{t=1}^{T}\sum_{j=1}^{n_{dt}}y_{dtj}(\boldsymbol{x}_{dtj}\boldsymbol{\beta}+\phi_1 v_{1,d}+\phi_2 v_{2,dt})-\frac{1}{2}v_{1,d}^2-\frac{1}{2}\boldsymbol{v}_{2,d}'\boldsymbol{v}_{2,d}. \qquad (15.6)
\end{aligned}$$

In order to apply the Laplace approximation to the integral in (15.5), we need derivatives of the function h. The first derivatives of h, with respect to $v_{1,d}, v_{2,d1}, \ldots, v_{2,dT}$, are

$$\frac{\partial h(\boldsymbol{v}_d)}{\partial v_{1,d}} = \sum_{t=1}^{T}\sum_{j=1}^{n_{dt}} \{-m_{dtj}\phi_1 p_{dtj} + \phi_1 y_{dtj}\} - v_{1,d},$$

$$\frac{\partial h(\boldsymbol{v}_d)}{\partial v_{2,dt}} = \sum_{j=1}^{n_{dt}} \{-m_{dtj}\phi_2 p_{dtj} + \phi_2 y_{dtj}\} - v_{2,dt}.$$

The second derivatives of h are

$$\frac{\partial^2 h(\boldsymbol{v}_d)}{\partial v_{1,d}^2} = -\Big(1 + \sum_{t=1}^{T}\sum_{j=1}^{n_{dt}} m_{dtj}\phi_1^2 p_{dtj}(1-p_{dtj})\Big),$$

$$\frac{\partial^2 h(\boldsymbol{v}_d)}{\partial v_{1,d}\partial v_{2,dt}} = -\sum_{j=1}^{n_{dt}} m_{dtj}\phi_1\phi_2 p_{dtj}(1-p_{dtj}),$$

$$\frac{\partial^2 h(\boldsymbol{v}_d)}{\partial v_{2,dt}^2} = -\sum_{j=1}^{n_{dt}} m_{dtj}\phi_2^2 p_{dtj}(1-p_{dtj}) - 1, \quad \frac{\partial^2 h(\boldsymbol{v}_d)}{\partial v_{2,dt_1}\partial v_{2,dt_2}} = 0, \; t_1 \neq t_2.$$

Let us define the vector of first derivatives

$$\dot{\boldsymbol{h}}(\boldsymbol{v}_d) = \Big(\frac{\partial h(\boldsymbol{v}_d)}{\partial v_{1,d}}, \frac{\partial h(\boldsymbol{v}_d)}{\partial v_{2,d1}}, \ldots, \frac{\partial h(\boldsymbol{v}_d)}{\partial v_{2,dT}}\Big)'$$

and the Hessian matrix

$$\boldsymbol{H}(\boldsymbol{v}_d) = \begin{pmatrix} \frac{\partial^2 h(\boldsymbol{v}_d)}{\partial v_{1,d}^2} & \frac{\partial^2 h(\boldsymbol{v}_d)}{\partial v_{1,d}\partial v_{2,d1}} & \cdots & \frac{\partial^2 h(\boldsymbol{v}_d)}{\partial v_{1,d}\partial v_{2,dT}} \\ \frac{\partial^2 h(\boldsymbol{v}_d)}{\partial v_{2,d1}\partial v_{1,d}} & \frac{\partial^2 h(\boldsymbol{v}_d)}{\partial v_{2,d1}^2} & \cdots & \frac{\partial^2 h(\boldsymbol{v}_d)}{\partial v_{2,d1}\partial v_{2,dT}} \\ \vdots & \vdots & \ddots & \vdots \\ \frac{\partial^2 h(\boldsymbol{v}_d)}{\partial v_{2,dT}\partial v_{1,d}} & \frac{\partial^2 h(\boldsymbol{v}_d)}{\partial v_{2,dT}\partial v_{2,d1}} & \cdots & \frac{\partial^2 h(\boldsymbol{v}_d)}{\partial v_{2,dT}^2} \end{pmatrix}.$$

For fixed $\boldsymbol{\theta} = (\boldsymbol{\beta}', \phi_1, \phi_2)'$, the function $h(\boldsymbol{v}_d)$, defined for $d = 1, \ldots, D$ in (15.6), can be maximized using the Newton–Raphson algorithm. The updating equation is

$$\boldsymbol{v}_d^{(i+1)} = \boldsymbol{v}_d^{(i)} - \boldsymbol{H}^{-1}(\boldsymbol{v}_d^{(i)})\dot{\boldsymbol{h}}(\boldsymbol{v}_d^{(i)}). \tag{15.7}$$

Let us denote, for simplicity, by $\boldsymbol{v}_d$ the argument of maxima of the function $h(\boldsymbol{v}_d)$. Thus, it holds $\dot{\boldsymbol{h}}(\boldsymbol{v}_d) = 0$, and the matrix $\boldsymbol{H}(\boldsymbol{v}_d)$ is negative definite. The log-likelihood of the assumed model is $\ell = \ell(\boldsymbol{\theta}) = \sum_{d=1}^{D} \log P(\boldsymbol{y}_d) = \sum_{d=1}^{D} \ell_d$.

By applying (15.4) to (15.5) and defining $\boldsymbol{G}_d = -\boldsymbol{H}(\boldsymbol{v}_d)$, we get the Laplace approximation of ℓ_d for the argument maxima $\boldsymbol{v}_d$ of the function $h(\boldsymbol{v}_d)$, i.e.

$$\ell_d = \log P(\boldsymbol{y}_d) \approx \ell_{0d} = \log b_d + h(\boldsymbol{v}_d) - \frac{1}{2}\log|\boldsymbol{G}_d| = \log b_d \tag{15.8}$$
$$- \sum_{t=1}^{T}\sum_{j=1}^{n_{dt}} m_{dtj}\log\left(1+\exp\{\boldsymbol{x}_{dtj}\boldsymbol{\beta}+\phi_1 v_{1,d}+\phi_2 v_{2,dt}\}\right)$$
$$+ \sum_{t=1}^{T}\sum_{j=1}^{n_{dt}} y_{dtj}(\boldsymbol{x}_{dtj}\boldsymbol{\beta}+\phi_1 v_{1,d}+\phi_2 v_{2,dt}) - \frac{1}{2}v_{1,d}^2 - \frac{1}{2}\boldsymbol{v}_{2d}'\boldsymbol{v}_{2d} - \frac{1}{2}\log|\boldsymbol{G}_d|.$$

The Laplace approximation of the log-likelihood $\ell(\boldsymbol{\theta})$ then takes the form

$$\ell_0(\boldsymbol{\theta}) = \sum_{d=1}^{D}\ell_{0d}.$$

Using the formulas (A.2), we obtain the first partial derivatives of ℓ_{0d} ($r = 1, \ldots, p$)

$$\frac{\partial \ell_{0d}}{\partial \beta_r} = -\sum_{t=1}^{T}\sum_{j=1}^{n_{dt}} m_{dtj}x_{dtjr}p_{dtj} + \sum_{t=1}^{T}\sum_{j=1}^{n_{dt}} x_{dtjr}y_{dtj} - \frac{1}{2}\operatorname{tr}\Big(\boldsymbol{G}_d^{-1}\frac{\partial \boldsymbol{G}_d}{\partial \beta_r}\Big),$$

$$\frac{\partial \ell_{0d}}{\partial \phi_1} = -v_{1,d}\sum_{t=1}^{T}\sum_{j=1}^{n_{dt}} m_{dtj}p_{dtj} + v_{1,d}\sum_{t=1}^{T}\sum_{j=1}^{n_{dt}} y_{dtj} - \frac{1}{2}\operatorname{tr}\Big(\boldsymbol{G}_d^{-1}\frac{\partial \boldsymbol{G}_d}{\partial \phi_1}\Big),$$

$$\frac{\partial \ell_{0d}}{\partial \phi_2} = -\sum_{t=1}^{T} v_{2,dt}\sum_{j=1}^{n_{dt}} m_{dtj}p_{dtj} + \sum_{t=1}^{T} v_{2,dt}\sum_{j=1}^{n_{dt}} y_{dtj} - \frac{1}{2}\operatorname{tr}\Big(\boldsymbol{G}_d^{-1}\frac{\partial \boldsymbol{G}_d}{\partial \phi_2}\Big).$$

The second partial derivatives of ℓ_{0d} are ($r, s = 1, \ldots, p$)

$$\frac{\partial^2 \ell_{0d}}{\partial \beta_s \partial \beta_r} = -\sum_{t=1}^{T}\sum_{j=1}^{n_{dt}} m_{dtj}x_{dtjr}x_{dtjs}p_{dtj}(1-p_{dtj}) + \frac{1}{2}\operatorname{tr}\Big(\boldsymbol{G}_d^{-1}\frac{\partial \boldsymbol{G}_d}{\partial \beta_s}\boldsymbol{G}_d^{-1}\frac{\partial \boldsymbol{G}_d}{\partial \beta_r}\Big)$$
$$- \frac{1}{2}\operatorname{tr}\Big(\boldsymbol{G}_d^{-1}\frac{\partial^2 \boldsymbol{G}_d}{\partial \beta_s \partial \beta_r}\Big),$$

$$\frac{\partial^2 \ell_{0d}}{\partial \phi_1^2} = -v_{1,d}^2\sum_{t=1}^{T}\sum_{j=1}^{n_{dt}} m_{dtj}p_{dtj}(1-p_{dtj}) + \frac{1}{2}\operatorname{tr}\Big(\boldsymbol{G}_d^{-1}\frac{\partial \boldsymbol{G}_d}{\partial \phi_1}\boldsymbol{G}_d^{-1}\frac{\partial \boldsymbol{G}_d}{\partial \phi_1}\Big)$$
$$- \frac{1}{2}\operatorname{tr}\Big(\boldsymbol{G}_d^{-1}\frac{\partial^2 \boldsymbol{G}_d}{\partial \phi_1^2}\Big),$$

$$\frac{\partial^2 \ell_{0d}}{\partial \phi_2^2} = -\sum_{t=1}^{T} v_{2,dt}^2 \sum_{j=1}^{n_{dt}} m_{dtj} p_{dtj}(1-p_{dtj}) + \frac{1}{2}\operatorname{tr}\Big(\boldsymbol{G}_d^{-1}\frac{\partial \boldsymbol{G}_d}{\partial \phi_2}\boldsymbol{G}_d^{-1}\frac{\partial \boldsymbol{G}_d}{\partial \phi_2}\Big) - \frac{1}{2}\operatorname{tr}\Big(\boldsymbol{G}_d^{-1}\frac{\partial^2 \boldsymbol{G}_d}{\partial \phi_2^2}\Big).$$

The second cross-partial derivatives of ℓ_{0d} are ($r = 1, \ldots, p$)

$$\frac{\partial^2 \ell_{0d}}{\partial \phi_1 \partial \beta_r} = -v_{1,d} \sum_{t=1}^{T}\sum_{j=1}^{n_{dt}} m_{dtj} x_{dtjr} p_{dtj}(1-p_{dtj}) + \frac{1}{2}\operatorname{tr}\Big(\boldsymbol{G}_d^{-1}\frac{\partial \boldsymbol{G}_d}{\partial \phi_1}\boldsymbol{G}_d^{-1}\frac{\partial \boldsymbol{G}_d}{\partial \beta_r}\Big) - \frac{1}{2}\operatorname{tr}\Big(\boldsymbol{G}_d^{-1}\frac{\partial^2 \boldsymbol{G}_d}{\partial \phi_1 \partial \beta_r}\Big),$$

$$\frac{\partial^2 \ell_{0d}}{\partial \phi_2 \partial \beta_r} = -\sum_{t=1}^{T} v_{2,dt} \sum_{j=1}^{n_{dt}} m_{dtj} x_{dtjr} p_{dtj}(1-p_{dtj}) + \frac{1}{2}\operatorname{tr}\Big(\boldsymbol{G}_d^{-1}\frac{\partial \boldsymbol{G}_d}{\partial \phi_2}\boldsymbol{G}_d^{-1}\frac{\partial \boldsymbol{G}_d}{\partial \beta_r}\Big) - \frac{1}{2}\operatorname{tr}\Big(\boldsymbol{G}_d^{-1}\frac{\partial^2 \boldsymbol{G}_d}{\partial \phi_2 \partial \beta_r}\Big),$$

$$\frac{\partial^2 \ell_{0d}}{\partial \phi_2 \partial \phi_1} = -v_{1,d} \sum_{t=1}^{T} v_{2,dt} \sum_{j=1}^{n_{dt}} m_{dtj} p_{dtj}(1-p_{dtj}) + \frac{1}{2}\operatorname{tr}\Big(\boldsymbol{G}_d^{-1}\frac{\partial \boldsymbol{G}_d}{\partial \phi_2}\boldsymbol{G}_d^{-1}\frac{\partial \boldsymbol{G}_d}{\partial \phi_1}\Big) - \frac{1}{2}\operatorname{tr}\Big(\boldsymbol{G}_d^{-1}\frac{\partial^2 \boldsymbol{G}_d}{\partial \phi_2 \partial \phi_1}\Big).$$

Let us define the score vector and the Hessian matrix

$$\boldsymbol{U}_0(\boldsymbol{\theta}) = \Big(\frac{\partial \ell_0(\boldsymbol{\theta})}{\partial \theta_1}, \ldots, \frac{\partial \ell_0(\boldsymbol{\theta})}{\partial \theta_{p+2}}\Big)', \quad \boldsymbol{\Lambda}_0(\boldsymbol{\theta}) = \Big(\frac{\partial^2 \ell_0(\boldsymbol{\theta})}{\partial \theta_i \partial \theta_j}\Big)_{i,j=1,\ldots,p+2}.$$

For fixed $\boldsymbol{v}_d$, $d = 1, \ldots, D$, the Newton–Raphson algorithm maximizes $\ell = \ell(\boldsymbol{\theta})$. The updating equation is

$$\boldsymbol{\theta}^{(i+1)} = \boldsymbol{\theta}^{(i)} - \boldsymbol{\Lambda}_0^{-1}(\boldsymbol{\theta}^{(i)})\boldsymbol{U}_0(\boldsymbol{\theta}^{(i)}). \tag{15.9}$$

15.3.2 ML-Laplace Algorithm

The final ML-Laplace approximation algorithm combines the two described Newton–Raphson algorithms and can be described by the following steps:

1. Set the initial values $i = 0$, $\varepsilon_1 > 0$, $\varepsilon_2 > 0$, $\varepsilon_3 > 0$, $\varepsilon_4 > 0$, $\boldsymbol{\theta}^{(0)}$, $\boldsymbol{\theta}^{(-1)} = \boldsymbol{\theta}^{(0)} + \boldsymbol{1}$, $v_{1,d}^{(0)} = 0$, $v_{1,d}^{(-1)} = 1$, $v_{2,dt}^{(0)} = 0$, $v_{2,dt}^{(-1)} = 1$, $d = 1, \ldots, D$, and $t = 1, \ldots, T$.

2. Until $\|\boldsymbol{\theta}^{(i)} - \boldsymbol{\theta}^{(i-1)}\|_2 < \varepsilon_1$, $|v_{1,d}^{(i)} - v_{1,d}^{(i-1)}| < \varepsilon_2$, $|v_{2,dt}^{(i)} - v_{2,dt}^{(i-1)}| < \varepsilon_2$, $d = 1, \ldots, D$, and $t = 1, \ldots, T$, do
   (a) apply algorithm (15.7) with seeds $v_{1,d}^{(i)}, v_{2,dt}^{(i)}$, $d = 1, \ldots, D$, $t = 1, \ldots, T$, convergence tolerance ε_3, and $\boldsymbol{\theta} = \boldsymbol{\theta}^{(i)}$ fixed. Output: $v_{1,d}^{(i+1)}, v_{2,dt}^{(i+1)}$, $d = 1, \ldots, D$, $t = 1, \ldots, T$;
   (b) apply algorithm (15.9) with seed $\boldsymbol{\theta}^{(i)}$, convergence tolerance ε_4 and $v_{1,d} = v_{1,d}^{(i+1)}$, $v_{2,dt} = v_{2,dt}^{(i+1)}$ fixed, $d = 1, \ldots, D$, and $t = 1, \ldots, T$. Output: $\boldsymbol{\theta}^{(i+1)}$;
   (c) $i \leftarrow i + 1$.
3. Output: $\hat{\boldsymbol{\theta}} = \boldsymbol{\theta}^{(i)}$, $\hat{v}_{1,d} = v_{1,d}^{(i)}$, $\hat{v}_{2,dt} = v_{2,dt}^{(i)}$, $d = 1, \ldots, D$, $t = 1, \ldots, T$.

Remark 15.1 The ML-Laplace approximation algorithm gives at convergence besides the ML estimators of the model parameters also the mode predictors, $\hat{v}_{1,d}$, $\hat{v}_{2,dt}$ of the random effects and the maximized marginal log-likelihood (logarithm of the joint p.d.f. of the target vector $\boldsymbol{y}$). Since the ML estimators are consistent and asymptotically normal when $n_{dt} \to \infty$, $d = 1, \ldots, D$, $t = 1, \ldots, T$, the algorithm can also be used to approximate the asymptotic covariance matrix (inverse of Fisher information matrix) which allows calculating Wald statistics for testing hypotheses on the model parameters. Namely, the asymptotic variance matrix of $\hat{\boldsymbol{\theta}}$ can be approximated by $\text{avar}(\hat{\boldsymbol{\theta}}) = -\boldsymbol{\Lambda}_0^{-1}(\hat{\boldsymbol{\theta}})$.

15.3.3 Derivatives of $\boldsymbol{G}_d$

In this section we derive the formulas of the derivatives of the matrix $\boldsymbol{G}_d$ which are needed for calculation of the score vector $\boldsymbol{U}_0(\boldsymbol{\theta})$ and the Hessian matrix $\boldsymbol{\Lambda}_0(\boldsymbol{\theta})$. Let us define

$$P_2(x) = x - x^2, \quad P_3(x) = x - 3x^2 + 2x^3, \quad P_4(x) = x - 7x^2 + 12x^3 - 6x^4.$$

The derivatives of p_{dtj}, $P_2(p_{dtj})$, and $P_3(p_{dtj})$, with respect to β_r, are

$$\begin{aligned}
\frac{\partial p_{dtj}}{\partial \beta_r} &= x_{dtjr} p_{dtj}(1 - p_{dtj}) = x_{dtjr} P_2(p_{dtj}), \\
\frac{\partial P_2(p_{dtj})}{\partial \beta_r} &= x_{dtjr}\big[p_{dtj}(1 - p_{dtj}) - 2p_{dtj}^2(1 - p_{dtj})\big] \\
&= x_{dtjr}\big[p_{dtj} - 3p_{dtj}^2 + 2p_{dtj}^3\big] = x_{dtjr} P_3(p_{dtj}), \\
\frac{\partial P_3(p_{dtj})}{\partial \beta_r} &= x_{dtjr}\big[p_{dtj}(1 - p_{dtj}) - 6p_{dtj}^2(1 - p_{dtj}) + 6p_{dtj}^3(1 - p_{dtj})\big] \\
&= x_{dtjr}\big[p_{dtj} - 7p_{dtj}^2 + 12p_{dtj}^3 - 6p_{dtj}^4\big] = x_{dtjr} P_4(p_{dtj}).
\end{aligned}$$

Similarly, we obtain the derivatives of p_{dtj}, $P_2(p_{dtj})$, and $P_3(p_{dtj})$, with respect to ϕ_1 and ϕ_2.

Let us consider the matrix $\boldsymbol{G}_d = \left(g_{ij}\right)_{i,j=1,\ldots,T+1}$ as a function of $\boldsymbol{\theta}$. We recall that

$$g_{11} = 1 + \sum_{t=1}^{T}\sum_{j=1}^{n_{dt}} m_{dtj}\phi_1^2 P_2(p_{dtj}), \quad g_{1+t1+t} = \sum_{j=1}^{n_{dt}} m_{dtj}\phi_2^2 P_2(p_{dtj}) + 1,$$

$$g_{11+t} = g_{1+t1} = \sum_{j=1}^{n_{dt}} m_{dtj}\phi_1\phi_2 P_2(p_{dtj}), \qquad g_{1+t_1 1+t_2} = 0, \; t_1 \neq t_2.$$

The components of $\partial \boldsymbol{G}_d/\partial \beta_r$ are

$$\frac{\partial g_{11}}{\partial \beta_r} = \phi_1^2 \sum_{t=1}^{T}\sum_{j=1}^{n_{dt}} m_{dtj}x_{dtjr}P_3(p_{dtj}), \quad \frac{\partial g_{1+t1+t}}{\partial \beta_r} = \phi_2^2 \sum_{j=1}^{n_{dt}} m_{dtj}x_{dtjr}P_3(p_{dtj}),$$

$$\frac{\partial g_{11+t}}{\partial \beta_r} = \phi_1\phi_2 \sum_{j=1}^{n_{dt}} m_{dtj}x_{dtjr}P_3(p_{dtj}), \quad \frac{\partial g_{1+t_1 1+t_2}}{\partial \beta_r} = 0.$$

The components of $\partial \boldsymbol{G}_d/\partial \phi_1$ are

$$\frac{\partial g_{11}}{\partial \phi_1} = \sum_{t=1}^{T}\sum_{j=1}^{n_{dt}} m_{dtj}\left(2\phi_1 P_2(p_{dtj}) + v_{1,d}\phi_1^2 P_3(p_{dtj})\right),$$

$$\frac{\partial g_{1+t1+t}}{\partial \phi_1} = \phi_2^2 \sum_{j=1}^{n_{dt}} m_{dtj}v_{1,d}P_3(p_{dtj}),$$

$$\frac{\partial g_{11+t}}{\partial \phi_1} = \phi_2 \sum_{j=1}^{n_{dt}} m_{dtj}\left(P_2(p_{dtj}) + \phi_1 v_{1,d}P_3(p_{dtj})\right), \quad \frac{\partial g_{1+t_1 1+t_2}}{\partial \phi_1} = 0.$$

The components of $\partial \boldsymbol{G}_d/\partial \phi_2$ are

$$\frac{\partial g_{11}}{\partial \phi_2} = \phi_1^2 \sum_{t=1}^{T}\sum_{j=1}^{n_{dt}} m_{dtj}v_{2,dt}P_3(p_{dtj}),$$

$$\frac{\partial g_{1+t1+t}}{\partial \phi_2} = \sum_{j=1}^{n_{dt}} m_{dtj}\left(2\phi_2 P_2(p_{dtj}) + v_{2,dt}\phi_2^2 P_3(p_{dtj})\right),$$

$$\frac{\partial g_{11+t}}{\partial \phi_2} = \phi_1 \sum_{j=1}^{n_{dt}} m_{dtj}\left(P_2(p_{dtj}) + \phi_2 v_{2,dt}P_3(p_{dtj})\right), \quad \frac{\partial g_{1+t_1 1+t_2}}{\partial \phi_2} = 0.$$

The components of $\partial^2 \boldsymbol{G}_d/(\partial\beta_s\partial\beta_r)$ are

$$\frac{\partial^2 g_{11}}{\partial\beta_s\partial\beta_r} = \phi_1^2 \sum_{t=1}^{T}\sum_{j=1}^{n_{dt}} m_{dtj}x_{dtjs}x_{dtjr}P_4(p_{dtj}),$$

$$\frac{\partial^2 g_{1+t1+t}}{\partial\beta_s\partial\beta_r} = \phi_2^2 \sum_{j=1}^{n_{dt}} m_{dtj}x_{dtjs}x_{dtjr}P_4(p_{dtj}),$$

$$\frac{\partial^2 g_{11+t}}{\partial\beta_s\partial\beta_r} = \phi_1\phi_2 \sum_{j=1}^{n_{dt}} m_{dtj}x_{dtjs}x_{dtjr}P_4(p_{dtj}), \quad \frac{\partial^2 g_{1+t_1 1+t_2}}{\partial\beta_s\partial\beta_r} = 0.$$

The components of $\partial^2 \boldsymbol{G}_d/(\partial\phi_1\partial\beta_r)$ are

$$\frac{\partial^2 g_{11}}{\partial\phi_1\partial\beta_r} = 2\phi_1 \sum_{t=1}^{T}\sum_{j=1}^{n_{dt}} m_{dtj}x_{dtjr}P_3(p_{dtj}) + \phi_1^2 v_{1,d} \sum_{t=1}^{T} m_{dtj}x_{dtjr}P_4(p_{dtj}),$$

$$\frac{\partial^2 g_{1+t_1 1+t_2}}{\partial\phi_1\partial\beta_r} = 0, \quad \frac{\partial^2 g_{1+t1+t}}{\partial\phi_1\partial\beta_r} = \phi_2^2 v_{1,d} \sum_{j=1}^{n_{dt}} m_{dtj}x_{dtjr}P_4(p_{dtj}),$$

$$\frac{\partial^2 g_{11+t}}{\partial\phi_1\partial\beta_r} = \phi_2 \sum_{j=1}^{n_{dt}} m_{dtj}x_{dtjr}\big(P_3(p_{dtj}) + \phi_1 v_{1,d}P_4(p_{dtj})\big).$$

The components of $\partial^2 \boldsymbol{G}_d/(\partial\phi_2\partial\beta_r)$ are

$$\frac{\partial^2 g_{11}}{\partial\phi_2\partial\beta_r} = \phi_1^2 \sum_{t=1}^{T}\sum_{j=1}^{n_{dt}} m_{dtj}x_{dtjr}v_{2,dt}P_4(p_{dtj}),$$

$$\frac{\partial^2 g_{1+t1+t}}{\partial\phi_2\partial\beta_r} = \sum_{j=1}^{n_{dt}} m_{dtj}x_{dtjr}\big(2\phi_2 P_3(p_{dtj}) + \phi_2^2 v_{2,dt}P_4(p_{dtj})\big),$$

$$\frac{\partial^2 g_{11+t}}{\partial\phi_2\partial\beta_r} = \phi_1 \sum_{j=1}^{n_{dt}} m_{dtj}x_{dtjr}(P_3(p_{dtj}) + \phi_2 v_{2,dt}P_4(p_{dtj})), \quad \frac{\partial^2 g_{1+t_1 1+t_2}}{\partial\phi_2\partial\beta_r} = 0.$$

The components of $\partial^2 \boldsymbol{G}_d/\partial\phi_1^2$ are

$$\frac{\partial^2 g_{11}}{\partial\phi_1^2} = \sum_{t=1}^{T}\sum_{j=1}^{n_{dt}} m_{dtj}\big(2P_2(p_{dtj}) + 4\phi_1 v_{1,d} P_3(p_{dtj}) + \phi_1^2 v_{1,d}^2 P_4(p_{dtj})\big),$$

$$\frac{\partial^2 g_{1+t_1 1+t_2}}{\partial\phi_1^2} = 0, \quad \frac{\partial^2 g_{1+t1+t}}{\partial\phi_1^2} = \phi_2^2 v_{1,d}^2 \sum_{j=1}^{n_{dt}} m_{dtj} P_4(p_{dtj}),$$

$$\frac{\partial^2 g_{11+t}}{\partial\phi_1^2} = \phi_2 \sum_{j=1}^{n_{dt}} m_{dtj}\big(2v_{1,d} P_3(p_{dtj}) + \phi_1 v_{1,d}^2 P_4(p_{dtj})\big).$$

The components of $\partial^2 \boldsymbol{G}_d/(\partial\phi_2\partial\phi_1)$ are

$$\frac{\partial^2 g_{11}}{\partial\phi_2\partial\phi_1} = \sum_{t=1}^{T}\sum_{j=1}^{n_{dt}} m_{dtj}\big(2\phi_1 v_{2,dt} P_3(p_{dtj}) + \phi_1^2 v_{1,d} v_{2,dt} P_4(p_{dtj})\big),$$

$$\frac{\partial^2 g_{1+t1+t}}{\partial\phi_2\partial\phi_1} = \sum_{j=1}^{n_{dt}} m_{dtj} v_{1,d}\big(2\phi_2 P_3(p_{dtj}) + \phi_2^2 v_{2,dt} P_4(p_{dtj})\big), \quad \frac{\partial^2 g_{1+t_1 1+t_2}}{\partial\phi_2\partial\phi_1} = 0,$$

$$\frac{\partial^2 g_{11+t}}{\partial\phi_2\partial\phi_1} = \sum_{j=1}^{n_{dt}} m_{dtj}\Big(P_2(p_{dtj}) + \phi_1 v_{1,d} P_3(p_{dtj}) + \phi_2 v_{2,dt} P_3(p_{dtj})$$
$$+ \phi_1\phi_2 v_{1,d} v_{2,dt} P_4(p_{dtj})\Big).$$

The components of $\partial^2 \boldsymbol{G}_d/\partial\phi_2^2$ are

$$\frac{\partial^2 g_{11}}{\partial\phi_2^2} = \phi_1^2 \sum_{t=1}^{T}\sum_{j=1}^{n_{dt}} m_{dtj} v_{2,dt}^2 P_4(p_{dtj}),$$

$$\frac{\partial^2 g_{11+t}}{\partial\phi_2^2} = \phi_1 \sum_{j=1}^{n_{dt}} m_{dtj}(2v_{2,dt} P_3(p_{dtj}) + \phi_2 v_{2,dt}^2 P_4(p_{dtj})), \quad \frac{\partial^2 g_{1+t_1 1+t_2}}{\partial\phi_2^2} = 0,$$

$$\frac{\partial^2 g_{1+t1+t}}{\partial\phi_2^2} = \sum_{j=1}^{n_{dt}} m_{dtj}\big(2P_2(p_{dtj}) + 4\phi_2 v_{2,dt} P_3(p_{dtj}) + \phi_2^2 v_{2,dt}^2 P_4(p_{dtj})\big).$$

15.3.4 AIC

Let $\hat{\boldsymbol{\beta}}$, $\hat{\phi}_1$, and $\hat{\phi}_2$ be the ML estimators of $\boldsymbol{\beta}$, ϕ_1, and ϕ_2, respectively. Let $\hat{\boldsymbol{v}}$ be the ML mode predictor of $\boldsymbol{v}$ obtained in the output of the Laplace approximation algorithm. The Laplace non-corrected AIC is

$$AIC = 2(p+2) - 2\ell_L(\hat{\boldsymbol{\beta}}, \hat{\phi}_1, \hat{\phi}_2, \hat{\boldsymbol{v}}),$$

where the second term is the Laplace approximation (15.8) to the model log-likelihood, i.e.

$$\begin{aligned}\ell_L &= \ell_L(\hat{\boldsymbol{\beta}}, \hat{\phi}_1, \hat{\phi}_2, \hat{\boldsymbol{v}}) \\ &\approx \sum_{d=1}^{D} \Bigg\{ \log b_d - \sum_{t=1}^{T}\sum_{j=1}^{n_{dt}} m_{dtj} \log\big(1 + \exp\{\boldsymbol{x}_{dtj}\hat{\boldsymbol{\beta}} + \hat{\phi}_1\hat{v}_{1,d} + \hat{\phi}_2\hat{v}_{2,dt}\}\big) \\ &+ \sum_{t=1}^{T}\sum_{j=1}^{n_{dt}} y_{dtj}(\boldsymbol{x}_{dtj}\hat{\boldsymbol{\beta}} + \hat{\phi}_1\hat{v}_{1,d} + \hat{\phi}_2\hat{v}_{2,dt}) - \frac{1}{2}\hat{v}_{1,d}^2 - \frac{1}{2}\hat{\boldsymbol{v}}_{2d}'\hat{\boldsymbol{v}}_{2d} - \frac{1}{2}\log|\hat{\boldsymbol{G}}_d| \Bigg\},\end{aligned}$$

where $b_d = \prod_{t=1}^{T}\prod_{j=1}^{n_{dt}} \binom{m_{dtj}}{y_{dtj}}$ and $\hat{\boldsymbol{G}}_d = -\boldsymbol{H}(\hat{\boldsymbol{v}}_d)$.

15.4 Empirical Best Predictors

Let us now consider a finite population U of N elements partitioned into D domains U_d of sizes $N_d, d = 1, \ldots, D$. We further assume that each domain U_d is partitioned into T subdomains U_{dt} of sizes N_{dt}, $d = 1, \ldots, D$, $t = 1, \ldots, T$. Let us assume that the model (15.2)–(15.3), with population sizes N_{dt} in the place of sample sizes n_{dt}, holds for all the units of U. From the population, a sample s of size n is selected with subsamples s_{dt} of sizes n_{dt} from subdomains U_{dt}. By $r = U - s$ and $r_{dt} = U_{dt} - s_{dt}$, we denote the set of the non-sampled population units and the set of non-sampled population units from domain d and subdomain t, respectively. Let $\boldsymbol{y}$ be the random vector containing the values of a target variable on the N population units. Without loss of generality, we can reorder the elements of the population so that $\boldsymbol{y} = (\boldsymbol{y}_s', \boldsymbol{y}_r')'$, where $\boldsymbol{y}_s$ is the vector of n observed elements and $\boldsymbol{y}_r$ is the vector of $N - n$ unobserved elements, both sorted by domain and subdomain. In the following, the index s for the sample and the index r for the rest of the population will be used when appropriate. For example, the symbols $\boldsymbol{y}_{ds}$ and $\boldsymbol{y}_{dts}$ represent the sampled parts of the vectors $\boldsymbol{y}_d$ and $\boldsymbol{y}_{dt}$ after reordering, i.e.

$$\boldsymbol{y}_{dts} = \underset{1 \le j \le n_{dt}}{\text{col}}(y_{dtj}), \quad \boldsymbol{y}_{ds} = \underset{1 \le t \le T}{\text{col}}(\boldsymbol{y}_{dts}), \quad \boldsymbol{y}_s = \underset{1 \le d \le D}{\text{col}}(\boldsymbol{y}_{ds}).$$

This section derives empirical best predictors (EBPs) for the sums of probabilities and population averages

$$\mu_{dt} = \sum_{j=1}^{N_{dt}} p_{dtj} \quad \text{and} \quad \overline{Y}_{dt} = \frac{1}{N_{dt}} \sum_{j=1}^{N_{dt}} y_{dtj}, \quad d = 1, \ldots, D,\ t = 1, \ldots, T,$$

under the model (15.2)–(15.3), assuming that $m_{dtj} = 1$, $d = 1, \ldots, D$, $t = 1, \ldots, T$, $j = 1, \ldots, N_{dt}$.

Using the introduced notation, the conditional distribution of $\boldsymbol{y}_s$, given $\boldsymbol{v}_1$ and $\boldsymbol{v}_2$, is

$$P(\boldsymbol{y}_s|\boldsymbol{v}_1, \boldsymbol{v}_2) = \prod_{d=1}^{D} P(\boldsymbol{y}_{ds}|v_{1,d}, \boldsymbol{v}_{2,d}), \quad P(\boldsymbol{y}_{ds}|v_{1,d}, \boldsymbol{v}_{2,d}) = \prod_{t=1}^{T} P(\boldsymbol{y}_{dts}|v_{1,d}, v_{2,dt}),$$

where

$$P(\boldsymbol{y}_{dts}|v_{1,d}, v_{2,dt}) = \prod_{j=1}^{n_{dt}} p_{dtj}^{y_{dtj}} (1 - p_{dtj})^{1-y_{dtj}} = \prod_{j=1}^{n_{dt}} \frac{\exp\{y_{dtj}\eta_{dtj}\}}{[1 + \exp\{\eta_{dtj}\}]}$$

$$= \exp\Big\{\sum_{j=1}^{n_{dt}} y_{dtj}\eta_{dtj} - \sum_{j=1}^{n_{dt}} \log\big[1 + \exp\{\eta_{dtj}\}\big]\Big\}.$$

The p.d.f. of $\boldsymbol{v} = (\boldsymbol{v}_1, \boldsymbol{v}_2)$ is $f(\boldsymbol{v}_1, \boldsymbol{v}_2) = f(\boldsymbol{v}_1) f(\boldsymbol{v}_2)$, where

$$f(\boldsymbol{v}_1) = \prod_{d=1}^{D} f(v_{1,d}), \quad f(\boldsymbol{v}_2) = \prod_{d=1}^{D} f(\boldsymbol{v}_{2,d}), \quad f(\boldsymbol{v}_{2,d}) = \prod_{t=1}^{T} f(v_{2,dt}),$$

and

$$f(v_{1,d}) = (2\pi)^{-1/2} \exp\big\{ -\frac{1}{2} v_{1,d}^2 \big\}, \quad f(v_{2,dt}) = (2\pi)^{-1/2} \exp\big\{ -\frac{1}{2} v_{2,dt}^2 \big\}.$$

15.4.1 EBP of p_{dtj}

The best predictor of p_{dtj} is $\hat{p}_{dtj}(\boldsymbol{\theta}) = E_\theta[p_{dtj}|\boldsymbol{y}_s]$. In this case, we have that $E_\theta[p_{dtj}|\boldsymbol{y}_s] = E_\theta[p_{dtj}|\boldsymbol{y}_{ds}]$ and

$$E_\theta[p_{dtj}|\boldsymbol{y}_{ds}] = \frac{\int_{R^{T+1}} \frac{\exp\{\eta_{dtj}\}}{1+\exp\{\eta_{dtj}\}} \prod_{\tau=1}^{T} P(\boldsymbol{y}_{d\tau s}|v_{1,d}, v_{2,d\tau}) f(\boldsymbol{v}_{2,d}) f(v_{1,d}) d\boldsymbol{v}_{2,d} dv_{1,d}}{\int_{R^{T+1}} \prod_{\tau=1}^{T} P(\boldsymbol{y}_{d\tau s}|v_{1,d}, v_{2,d\tau}) f(\boldsymbol{v}_{2,d}) f(v_{1,d})\, d\boldsymbol{v}_{2,d}\, dv_{1,d}}.$$

Therefore $E_\theta[p_{dtj}|\boldsymbol{y}_d] = \mathcal{A}_{dtj}/\mathcal{D}_d = A_{dtj}/D_d$, where

$$\mathcal{A}_{dtj} = \int_R \prod_{\tau=1}^T \Bigg\{ \int_R \left(\frac{\exp\{\eta_{dtj}\}}{1+\exp\{\eta_{dtj}\}} \delta_{t\tau} + (1-\delta_{t\tau}) \right) \exp\Bigg\{ \sum_{i=1}^{n_{d\tau}} y_{d\tau i}\boldsymbol{x}_{d\tau i}\boldsymbol{\beta} + y_{d\tau.}(\phi_1 v_{1,d} + \phi_2 v_{2,d\tau}) - \sum_{i=1}^{n_{d\tau}} \log\left[1+\exp\{\eta_{d\tau i}\}\right] \Bigg\} f(v_{2,d\tau})\, dv_{2,d\tau} \Bigg\} \cdot f(v_{1,d})\, dv_{1,d},$$

$$\mathcal{D}_d = \int_R \prod_{\tau=1}^T \Bigg\{ \int_R \exp\Bigg\{ \sum_{i=1}^{n_{d\tau}} y_{d\tau i}\boldsymbol{x}_{d\tau i}\boldsymbol{\beta} + y_{d\tau.}(\phi_1 v_{1,d} + \phi_2 v_{2,d\tau}) - \sum_{i=1}^{n_{d\tau}} \log\left[1+\exp\{\eta_{d\tau i}\}\right] \Bigg\} f(v_{2,d\tau})\, dv_{2,d\tau} \Bigg\} f(v_{1,d})\, dv_{1,d},$$

$$A_{dtj} = \int_R \prod_{\tau=1}^T \Bigg\{ \int_R \exp\Big\{ \boldsymbol{x}_{dtj}\boldsymbol{\beta}\delta_{t\tau} + (y_{d\tau.} + \delta_{t\tau})(\phi_1 v_{1,d} + \phi_2 v_{2,d\tau}) - \sum_{i=1}^{n_{d\tau}} (1+\delta_{ij}\delta_{t\tau}) \log\left[1+\exp\{\eta_{d\tau i}\}\right] \Big\} f(v_{2,d\tau})\, dv_{2,d\tau} \Bigg\} f(v_{1,d})\, dv_{1,d},$$

$$D_d = \int_R \prod_{\tau=1}^T \Bigg\{ \int_R \exp\Big\{ y_{d\tau.}(\phi_1 v_{1,d} + \phi_2 v_{2,d\tau}) - \sum_{i=1}^{n_{d\tau}} \log\left[1+\exp\{\eta_{d\tau i}\}\right] \Big\} f(v_{2,d\tau})\, dv_{2,d\tau} \Bigg\} f(v_{1,d})\, dv_{1,d},$$

$y_{d\tau} = \sum_{j=1}^{n_{d\tau}} y_{d\tau j}$, and $\delta_{t\tau}$ is the Kronecker delta, i.e. $\delta_{t\tau} = 1$ if $t = \tau$ and $\delta_{t\tau} = 0$ otherwise.

The EBP of p_{dtj} is obtained by substituting $\boldsymbol{\theta}$ by its estimate $\hat{\boldsymbol{\theta}}$, i.e. it has the form $\hat{p}_{dtj}(\hat{\boldsymbol{\theta}})$, and can be approximated as follows:

1. Estimate $\hat{\boldsymbol{\theta}} = (\hat{\boldsymbol{\beta}}', \hat{\phi}_1, \hat{\phi}_2)'$.
2. For $s_1 = 1, \dots, S_1$, $s_2 = 1, \dots, S_2$, and $t = 1, \dots, T$, generate $v_{1,d}^{(s_1)}, v_{2,dt}^{(s_2)}$ i.i.d. $N(0, 1)$ and $v_{1,d}^{(S_1+s_1)} = -v_{1,d}^{(s_1)}$, $v_{2,dt}^{(S_2+s_2)} = -v_{2,dt}^{(s_2)}$.
3. Calculate $\hat{p}_{dtj}^{ebp} = \hat{p}_{dtj}(\hat{\boldsymbol{\theta}}) = \hat{A}_{dtj}/\hat{D}_d$, where

$$\hat{A}_{dtj} = \sum_{s_1=1}^{2S_1} \prod_{\tau=1}^T \sum_{s_2=1}^{2S_2} \exp\Bigg\{ \boldsymbol{x}_{dtj}\hat{\boldsymbol{\beta}}\delta_{t\tau} + (y_{d\tau.} + \delta_{t\tau})(\hat{\phi}_1 v_{1,d}^{(s_1)} + \hat{\phi}_2 v_{2,d\tau}^{(s_2)}) - \sum_{i=1}^{n_{d\tau}} (1+\delta_{ij}\delta_{t\tau}) \log\left[1+\exp\{\boldsymbol{x}_{d\tau i}\hat{\boldsymbol{\beta}} + \hat{\phi}_1 v_{1,d}^{(s_1)} + \hat{\phi}_2 v_{2,d\tau}^{(s_2)}\}\right] \Bigg\},$$

$$\hat{D}_d = \sum_{s_1=1}^{2S_1} \prod_{\tau=1}^{T} \sum_{s_2=1}^{2S_2} \exp\Big\{ y_{d\tau.}(\hat{\phi}_1 v_{1,d}^{(s_1)} + \hat{\phi}_2 v_{2,d\tau}^{(s_2)})$$

$$- \sum_{i=1}^{n_{d\tau}} \log\Big[1 + \exp\{\boldsymbol{x}_{d\tau i}\hat{\boldsymbol{\beta}} + \hat{\phi}_1 v_{1,d}^{(s_1)} + \hat{\phi}_2 v_{2,d\tau}^{(s_2)}\}\Big]\Big\}.$$

From the same reasons as described in Sect. 14.6.1, we also consider the computationally simpler plug-in predictor of p_{dtj}, which is given by the formula

$$\tilde{p}_{dtj} = \tilde{p}_{dtj}(\hat{\boldsymbol{\theta}}, \hat{v}_{1,d}, \hat{v}_{2,dt}) = \frac{\exp\{\boldsymbol{x}_{dtj}\hat{\boldsymbol{\beta}} + \hat{\phi}_1 \hat{v}_{1,d} + \hat{\phi}_2 \hat{v}_{2,dt}\}}{1 + \exp\{\boldsymbol{x}_{dtj}\hat{\boldsymbol{\beta}} + \hat{\phi}_1 \hat{v}_{1,d} + \hat{\phi}_2 \hat{v}_{2,dt}\}},$$

where the predictors $\hat{v}_{1,d}$ and $\hat{v}_{2,dt}$ may be taken as likelihood modes from the output of the ML-Laplace approximation algorithm.

15.4.2 EBP of μ_{dt} and $\overline{\mu}_{dt}$

This section deals with the estimation of the model parameters

$$\mu_{dt} = \mu_{dt}(\boldsymbol{\theta}) = \sum_{j=1}^{N_{dt}} p_{dtj}, \quad \overline{\mu}_{dt} = \mu_{dt}/N_{dt}.$$

The EBPs of μ_{dt} and $\overline{\mu}_{dt}$ are

$$\hat{\mu}_{dt}^{ebp} = \hat{\mu}_{dt}(\hat{\boldsymbol{\theta}}) = \sum_{j=1}^{N_{dt}} \hat{p}_{dtj}(\hat{\boldsymbol{\theta}}), \quad \hat{\overline{\mu}}_{dt} = \hat{\mu}_{dt}/N_{dt},$$

respectively, where $\hat{p}_{dtj}(\hat{\boldsymbol{\theta}})$ is the EBP of p_{dtj}. Similarly, the plug-in predictors of μ_{dt} and $\overline{\mu}_{dt}$ are

$$\tilde{\mu}_{dt} = \tilde{\mu}_{dt}(\hat{\boldsymbol{\theta}}) = \sum_{j=1}^{N_{dt}} \tilde{p}_{dtj}(\hat{\boldsymbol{\theta}}), \quad \tilde{\overline{\mu}}_{dt} = \tilde{\mu}_{dt}/N_{dt},$$

respectively, where $\tilde{p}_{dtj}(\hat{\boldsymbol{\theta}})$ is the plug-in predictor of p_{dtj}.

Since the presented EBPs are functions of $\boldsymbol{x}_{dtj}$, for calculating these predictors, it is necessary to have a data file containing the values of the explanatory variables in all the population units. But this kind of data file is rarely available. If in addition some of the auxiliary variables are continuous, the empirical best and the plug-in predictors are not applicable. The following remark presents a particular case where these predictors can be calculated.

Remark 15.2 Suppose that the covariates are categorical and take finite number of values, i.e. $\boldsymbol{x}_{dtj} \in \{z_1, \dots, z_K\}$ for all d, t, j. Then,

$$\mu_{dt} = \mu_{dt}(\boldsymbol{\theta}) = \sum_{j=1}^{N_{dt}} p_{dtj} = \sum_{k=1}^{K} N_{dtk} q_{dtk}, \tag{15.10}$$

$$q_{dtk} = q_{dtk}(\boldsymbol{\theta}, v_{1,d}, v_{2,dt}) = \frac{\exp\{z_k \boldsymbol{\beta} + \phi_1 v_{1,d} + \phi_2 v_{2,dt}\}}{1 + \exp\{z_k \boldsymbol{\beta} + \phi_1 v_{1,d} + \phi_2 v_{2,dt}\}},$$

where $N_{dtk} = \#\{j \in U_{dt} : \boldsymbol{x}_{dtj} = z_k\}$ is the size of the covariate class z_k in subdomain U_{dt} which must be available from external data sources (aggregated auxiliary information). Under this categorical setup, the plug-in predictor of μ_{dt} is

$$\tilde{\mu}_{dt} = \tilde{\mu}_{dt}(\hat{\boldsymbol{\theta}}) = \sum_{j=1}^{N_{dt}} \tilde{p}_{dtj} = \sum_{k=1}^{K} N_{dtk} \tilde{q}_{dtk}, \quad \tilde{q}_{dtk} = q_{dtk}(\hat{\boldsymbol{\theta}}, \hat{v}_{1,d}, \hat{v}_{2,dt}). \tag{15.11}$$

The best predictor of μ_{dt} is

$$\hat{\mu}_{dt}(\boldsymbol{\theta}) = E_\theta[\mu_{dt} | \boldsymbol{y}_s] = E_\theta[\mu_{dt} | \boldsymbol{y}_{ds}] = \sum_{k=1}^{K} N_{dtk} E_\theta[q_{dtk} | \boldsymbol{y}_{ds}], \tag{15.12}$$

where

$$E_\theta[q_{dtk} | \boldsymbol{y}_{ds}] = \frac{\int_{R^{T+1}} \frac{\exp\{z_k \boldsymbol{\beta} + \phi_1 v_{1,d} + \phi_2 v_{2,dt}\}}{1+\exp\{z_k \boldsymbol{\beta} + \phi_1 v_{1,d} + \phi_2 v_{2,dt}\}} h_{ds}(\boldsymbol{y}_{ds}, \boldsymbol{v}_{2,d}, v_{1,d}) \, d\boldsymbol{v}_{2,d} \, dv_{1,d}}{\int_{R^{T+1}} h_{ds}(\boldsymbol{y}_{ds}, \boldsymbol{v}_{2,d}, v_{1,d}) \, d\boldsymbol{v}_{2,d} \, dv_{1,d}},$$

$$h_{ds}(\boldsymbol{y}_{ds}, \boldsymbol{v}_{2,d}, v_{1,d}) = \prod_{\tau=1}^{T} P(\boldsymbol{y}_{d\tau s} | v_{1,d}, v_{2,d\tau}) f(\boldsymbol{v}_{2,d}) f(v_{1,d}).$$

Therefore $E_\theta[q_{dtk} | \boldsymbol{y}_{ds}] = A^z_{dtk} / D_d$, where D_d is defined above and

$$A^z_{dtk} = \int_R \prod_{\tau=1}^{T} \Big\{ \int_R \exp\Big\{ z_k \boldsymbol{\beta} \delta_{t\tau} + (y_{d\tau.} + \delta_{t\tau})(\phi_1 v_{1,d} + \phi_2 v_{2,d\tau})$$
$$- \delta_{t\tau} \log\big[1 + \exp\{z_k \boldsymbol{\beta} + \phi_1 v_{1,d} + \phi_2 v_{2,d\tau}\}\big]$$
$$- \sum_{i=1}^{n_{d\tau}} \log\big[1+\exp\{\boldsymbol{x}_{d\tau i} \boldsymbol{\beta} + \phi_1 v_{1,d} + \phi_2 v_{2,d\tau}\}\big]\Big\} f(v_{2,d\tau}) \, dv_{2,d\tau} \Big\} f(v_{1,d}) \, dv_{1,d}.$$

The EBP of μ_{dt} is $\hat{\mu}_{dt}^{ebp} = \hat{\mu}_{dt}(\hat{\boldsymbol{\theta}}) = E_{\hat{\theta}}[\mu_{dt}|\boldsymbol{y}_s]$, and it can be approximated as follows:

1. Estimate $\hat{\boldsymbol{\theta}} = (\hat{\boldsymbol{\beta}}', \hat{\phi}_1, \hat{\phi}_2)'$.
2. For $s_1 = 1, \ldots, S_1$ and $s_2 = 1, \ldots, S_2$, generate $v_{1,d}^{(s_1)}$, $v_{2,dt}^{(s_2)}$ i.i.d. $N(0, 1)$ and $v_{1,d}^{(S_1+s_1)} = -v_{1,d}^{(s_1)}$, $v_{2,dt}^{(S_2+s_2)} = -v_{2,dt}^{(s_2)}$.
3. Calculate $\hat{\mu}_{dt}(\hat{\boldsymbol{\theta}}) = \sum_{k=1}^{K} N_{dtk}\hat{q}_{dtk}$, where $\hat{q}_{dtk} = \hat{A}_{dtk}^{z}/\hat{D}_d$ and

$$\hat{A}_{dtk}^{z} = \sum_{s_1=1}^{2S_1}\prod_{\tau=1}^{T}\sum_{s_2=1}^{2S_2} \exp\Big\{z_k\hat{\boldsymbol{\beta}}\delta_{t\tau} + (y_{d\tau.} + \delta_{t\tau})(\hat{\phi}_1 v_{1,d}^{(s_1)} + \hat{\phi}_2 v_{2,d\tau}^{(s_2)})$$
$$- \delta_{t\tau}\log\Big[1 + \exp\{z_k\hat{\boldsymbol{\beta}} + \hat{\phi}_1 v_{1,d}^{(s_1)} + \hat{\phi}_2 v_{2,d\tau}^{(s_2)}\}\Big]$$
$$- \sum_{i=1}^{n_{d\tau}}\log\Big[1 + \exp\{\boldsymbol{x}_{d\tau i}\hat{\boldsymbol{\beta}} + \hat{\phi}_1 v_{1,d}^{(s_1)} + \hat{\phi}_2 v_{2,d\tau}^{(s_2)}\}\Big]\Big\},$$

$$\hat{D}_d = \sum_{s_1=1}^{2S_1}\prod_{\tau=1}^{T}\sum_{s_2=1}^{2S_2} \exp\big\{y_{d\tau.}(\hat{\phi}_1 v_{1,d}^{(s_1)} + \hat{\phi}_2 v_{2,d\tau}^{(s_2)})$$
$$- \sum_{i=1}^{n_{d\tau}}\log\Big[1 + \exp\{\boldsymbol{x}_{d\tau i}\hat{\boldsymbol{\beta}} + \hat{\phi}_1 v_{1,d}^{(s_1)} + \hat{\phi}_2 v_{2,d\tau}^{(s_2)}\}\Big]\big\}.$$

The EBP and the plug-in predictor of $\overline{\mu}_{dt}$ are $\hat{\overline{\mu}}_{dt}^{ebp} = \hat{\mu}_{dt}(\hat{\boldsymbol{\theta}})/N_{dt}$ and $\tilde{\overline{\mu}}_{dt} = \tilde{\mu}_{dt}(\hat{\boldsymbol{\theta}})/N_{dt}$, respectively.

15.4.3 EBP of y_{dtj}

The best predictor of y_{dtj} is $\hat{y}_{dtj}(\boldsymbol{\theta}) = E_\theta[y_{dtj}|\boldsymbol{y}_s]$. If $j \in s_{dt}$, then $E_\theta[y_{dtj}|\boldsymbol{y}_s] = y_{dtj}$. If $j \in r_{dt}$, then $E_\theta[y_{dtj}|\boldsymbol{y}_s] = E_\theta[y_{dtj}|\boldsymbol{y}_{ds}]$ and

$$E_\theta[y_{dtj}|\boldsymbol{y}_{ds}] = \frac{\int_{R^{T+1}}\int_{\{0,1\}} y_{dtj}g_{djs}(\boldsymbol{y}_{ds}, v_{1,d}, \boldsymbol{v}_{2,d})\,dy_{dtj}\,d\boldsymbol{v}_{2,d}\,dv_{1,d}}{\int_{R^{T+1}} g_{ds}(\boldsymbol{y}_{ds}, v_{1,d}, \boldsymbol{v}_{2,d})d\boldsymbol{v}_{2,d}dv_{1,d}}$$
$$= \frac{\int_{R^{T+1}} p_{dtj}g_{ds}(\boldsymbol{y}_{ds}, v_{1,d}, \boldsymbol{v}_{2,d})\,d\boldsymbol{v}_{2,d}\,dv_{1,d}}{\int_{R^{T+1}} g_{ds}(\boldsymbol{y}_{ds}, v_{1,d}, \boldsymbol{v}_{2,d})\,d\boldsymbol{v}_{2,d}\,dv_{1,d}} = \frac{A_{dtj}(\boldsymbol{y}_{ds}, \boldsymbol{\theta})}{D_d(\boldsymbol{y}_{ds}, \boldsymbol{\theta})}$$
$$= E_\theta[p_{dtj}|\boldsymbol{y}_{ds}] = \hat{p}_{dtj}(\boldsymbol{\theta}),$$

where $\int_{\{0,1\}} dy_{dtj}$ represents integral with respect to the counting measure, i.e. it is a sum over $y_{dtj} \in \{0, 1\}$,

$$g_{djs}(\boldsymbol{y}_{ds}, v_{1,d}, \boldsymbol{v}_{2,d}) = P(y_{dtj}|v_{1,d}, v_{2,dt})P(\boldsymbol{y}_{ds}|v_{1,d}, \boldsymbol{v}_{2,d})f(v_{1,d})f(\boldsymbol{v}_{2,d}),$$
$$g_{ds}(\boldsymbol{y}_{ds}, v_{1,d}, \boldsymbol{v}_{2,d}) = P(\boldsymbol{y}_{ds}|v_{1,d}, \boldsymbol{v}_{2,d})f(v_{1,d})f(\boldsymbol{v}_{2,d}),$$

and the numerator $A_{dtj} = A_{dtj}(\boldsymbol{y}_{ds}, \boldsymbol{\theta})$ and the denominator $D_d = D_d(\boldsymbol{y}_{ds}, \boldsymbol{\theta})$ are defined in Sect. 15.4.1. Therefore $\hat{y}_{dtj}(\boldsymbol{\theta}) = \hat{p}_{dtj}(\boldsymbol{\theta})$ if $j \in r_{dt}$.

The EBP of y_{dtj} is $\hat{y}_{dtj}^{ebp} = \hat{y}_{dtj}(\hat{\boldsymbol{\theta}})$. It holds that $\hat{y}_{dtj}^{ebp} = y_{dtj}$ if $j \in s_{dt}$ and $\hat{y}_{dtj}^{ebp} = \hat{p}_{dtj}^{ebp}$ if $j \in r_{dt}$, where $\hat{p}_{dtj}^{ebp}$ is the EBP of p_{dtj}, which is approximated in Sect. 15.4.1.

15.4.4 EBP of $\overline{Y}_{dt}$

Let us now turn our attention to the population averages (proportions if $m_{dtj} = 1$)

$$\overline{Y}_{dt} = \frac{1}{N_{dt}} \sum_{j=1}^{N_{dt}} y_{dtj}$$

and consider separately cases with at least one continuous auxiliary variable and all categorical auxiliary variables.

15.4.4.1 Predictors with Continuous Auxiliary Variables

In the case of at least one continuous auxiliary variable, we assume that a census file C with properties 1–3 given in Sect. 14.6.4.1 is available. In other words, the values of the auxiliary variables $\boldsymbol{x}_{dtj}$ are assumed to be known for all the population units. Under this setup, the EBP and the plug-in predictors of $\overline{Y}_{dt}$ are

$$\hat{\overline{Y}}_{dt}^{ebp} = \hat{\overline{Y}}_{dt}(\hat{\boldsymbol{\theta}}) = \frac{1}{N_{dt}} \Big[\sum_{j \in s_{dt}} y_{dtj} + \sum_{j \in r_{dt}} \hat{p}_{dtj}^{ebp} \Big],$$

$$\tilde{\overline{Y}}_{dt} = \tilde{\overline{Y}}_{dt}(\hat{\boldsymbol{\theta}}, \hat{v}_{1,d}, \hat{v}_{2,dt}) = \frac{1}{N_{dt}} \Big[\sum_{j \in s_{dt}} y_{dtj} + \sum_{j \in r_{dt}} \tilde{p}_{dtj} \Big],$$

where the empirical best predictor $\hat{p}_{dtj}^{ebp}$ and the plug-in predictor $\tilde{p}_{dtj}$ of p_{dtj} are given in Sect. 15.4.1.

15.4.4.2 Predictors with Categorical Auxiliary Variables

Under the categorical setup, the EBP and the plug-in predictors of $\overline{Y}_{dt}$ are

$$\hat{\overline{Y}}_{dt}^{ebp} = \hat{\overline{Y}}_{dt}(\hat{\boldsymbol{\theta}}) = \frac{1}{N_{dt}}\Big[\sum_{j\in s_{dt}} y_{dtj} + \sum_{k=1}^{K} N_{dtk,r}\hat{q}_{dtk}\Big],$$

$$\tilde{\overline{Y}}_{dt} = \tilde{\overline{Y}}_{dt}(\hat{\boldsymbol{\theta}}, \hat{v}_{1,d}, \hat{v}_{2,dt}) = \frac{1}{N_{dt}}\Big[\sum_{j\in s_{dt}} y_{dtj} + \sum_{k=1}^{K} N_{dtk,r}\tilde{q}_{dtk}\Big],$$

where $\hat{q}_{dtk}$ and $\tilde{q}_{dtk}$ are defined in Remark 15.2, $N_{dtk,r} = \#\{j \in r_{dt} : \boldsymbol{x}_{dtj} = z_k\}$ is the size of the covariate class z_k at r_{dt}, and r_{dt} denotes the non-sampled part of domain d at subdomain t.

15.5 MSE of Empirical Best Predictors

From the same reasons as stated at the beginning of Sect. 14.7, we describe here only bootstrap-based estimator of the MSE of the EBP. We treat the case $m_{dtj} = 1, d = 1, \ldots, D, t = 1, \ldots, T$, and $j = 1, \ldots, N_{dt}$, and we assume moreover that all the auxiliary variables are categorical, i.e. $\boldsymbol{x}_{dtj} \in \{z_1, \ldots, z_K\}$, and the corresponding sizes $N_{dtk,r} = \#\{j \in r_{dt} : \boldsymbol{x}_{dtj} = z_k\}$ of the covariate classes z_k at r_{dt} are available. Let us note that under this categorical setup, there is no need of generating the whole population in each bootstrap iteration.

For the case of continuous auxiliary variables, we refer to Sect. 14.7.2, which can be easily adapted to the assumed two-fold model.

15.5.1 Bootstrap Estimation of the MSE of the EBP of μ_{dt}

In this section we consider the function μ_{dt} given in (15.10). The following procedure calculates a parametric bootstrap estimator of $MSE(\hat{\mu}_{dt})$, where $\hat{\mu}_{dt} = \hat{\mu}_{dt}(\hat{\boldsymbol{\theta}})$ is the EBP of μ_{dt}:

1. Fit the model to the sample and calculate $\hat{\boldsymbol{\theta}} = (\hat{\boldsymbol{\beta}}', \hat{\phi}_1, \hat{\phi}_2)'$.
2. Repeat B times ($b = 1, \ldots, B$):

   (a) Generate two independent sets of random effects $\{v_{1,d}^{(b)} : d = 1, \ldots, D\}$ i.i.d. N(0,1) and $\{v_{2,dt}^{(b)} : d = 1, \ldots, D, t = 1, \ldots, T\}$ i.i.d. N(0,1). For $d = 1, \ldots, D$, $t = 1, \ldots, T$, and $j = 1, \ldots, n_{dt}$, generate the bootstrap sample

$$y_{dtj}^{(b)} \sim \text{Bin}(1, p_{dtj}^{(b)}), \quad p_{dtj}^{(b)} = \frac{\exp\{\boldsymbol{x}_{dtj}\hat{\boldsymbol{\beta}} + \hat{\phi}_1 v_{1,d}^{(b)} + \hat{\phi}_2 v_{2,dt}^{(b)}\}}{1 + \exp\{\boldsymbol{x}_{dtj}\hat{\boldsymbol{\beta}} + \hat{\phi}_1 v_{1,d}^{(b)} + \hat{\phi}_2 v_{2,dt}^{(b)}\}},$$

and calculate the true bootstrap population quantities

$$\mu_{dt}^{(b)} = \sum_{j=1}^{n_{dt}} p_{dtj}^{(b)} + \sum_{k=1}^{K} N_{dtk,r} q_{dtk}^{(b)}, \quad q_{dtk}^{(b)} = \frac{\exp\{z_k\hat{\boldsymbol{\beta}} + \hat{\phi}_1 v_{1,d}^{(b)} + \hat{\phi}_2 v_{2,dt}^{(b)}\}}{1 + \exp\{z_k\hat{\boldsymbol{\beta}} + \hat{\phi}_1 v_{1,d}^{(b)} + \hat{\phi}_2 v_{2,dt}^{(b)}\}}.$$

(b) The bootstrap sample has the same units as the real data sample. It is not extracted at random. The model is on the population, and therefore the source of randomness comes from the generation of the population. For each bootstrap sample, calculate $\hat{\boldsymbol{\theta}}^{(b)}$ and the EBP $\hat{\mu}_{dt}^{(b)} = \hat{\mu}_{dt}(\hat{\boldsymbol{\theta}}^{(b)})$.

3. Output: $mse^*(\hat{\mu}_{dt}) = \frac{1}{B}\sum_{b=1}^{B}\big(\hat{\mu}_{dt}^{(b)} - \mu_{dt}^{(b)}\big)^2$.

15.5.2 Bootstrap Estimation of the MSE of the EBP of $\overline{Y}_{dt}$

The following procedure calculates a parametric bootstrap estimator of $MSE(\hat{\overline{Y}}_{dt})$, where $\hat{\overline{Y}}_{dt} = \hat{\overline{Y}}_{dt}(\hat{\boldsymbol{\theta}})$ is the EBP of

$$\overline{Y}_{dt} = \frac{1}{N_{dt}} \sum_{j=1}^{N_{dt}} y_{dtj}.$$

1. Fit the model to the sample and calculate $\hat{\boldsymbol{\theta}} = (\hat{\boldsymbol{\beta}}', \hat{\phi}_1, \hat{\phi}_2)'$.
2. Repeat B times ($b = 1, \ldots, B$):

(a) Generate two independent sets of random effects $\{v_{1,d}^{(b)} : d = 1, \ldots, D\}$ i.i.d. $N(0,1)$ and $\{v_{2,dt}^{(b)} : d = 1, \ldots, D, t = 1, \ldots, T\}$ i.i.d. $N(0,1)$. For $d = 1, \ldots, D$, $t = 1, \ldots, T$, and $j = 1, \ldots, n_{dt}$, generate the bootstrap sample

$$y_{dtj}^{(b)} \sim \text{Bin}(1, p_{dtj}^{(b)}), \quad p_{dtj}^{(b)} = \frac{\exp\{\boldsymbol{x}_{dtj}\hat{\boldsymbol{\beta}} + \hat{\phi}_1 v_{1,d}^{(b)} + \hat{\phi}_2 v_{2,dt}^{(b)}\}}{1 + \exp\{\boldsymbol{x}_{dtj}\hat{\boldsymbol{\beta}} + \hat{\phi}_1 v_{1,d}^{(b)} + \hat{\phi}_2 v_{2,dt}^{(b)}\}},$$

and calculate the true bootstrap population quantities

$$\overline{Y}_{dt}^{(b)} = \frac{1}{N_{dt}}\left(\sum_{j=1}^{n_{dt}} y_{dtj}^{(b)} + \sum_{k=1}^{K} z_{dtk}^{(b)}\right), \quad \text{where} \quad z_{dtk}^{(b)} \sim \text{Bin}(N_{dtk,r}, q_{dtk}^{(b)})$$

and

$$q_{dtk}^{(b)} = \frac{\exp\{z_k\hat{\boldsymbol{\beta}} + \hat{\phi}_1 v_{1,d}^{(b)} + \hat{\phi}_2 v_{2,dt}^{(b)}\}}{1 + \exp\{z_k\hat{\boldsymbol{\beta}} + \hat{\phi}_1 v_{1,d}^{(b)} + \hat{\phi}_2 v_{2,dt}^{(b)}\}}.$$

(b) The bootstrap sample has the same units as the real data sample. It is not extracted at random. For each bootstrap sample, calculate $\hat{\boldsymbol{\theta}}^{(b)}$ and the EBP $\hat{\overline{Y}}_{dt}^{(b)} = \hat{\overline{Y}}_{dt}(\hat{\boldsymbol{\theta}}^{(b)})$.

3. Output: $mse^*(\hat{\overline{Y}}_{dt}^{ebp}) = \frac{1}{B}\sum_{b=1}^{B}\big(\hat{\overline{Y}}_{dt}^{(b)} - \overline{Y}_{dt}^{(b)}\big)^2$.

15.6 Simulation Experiment

This section presents a simulation experiment for empirical illustration of the consistency of the ML-Laplace estimators under the unit-level two-fold logit mixed model (15.2)–(15.3). We take $n_{dt} = 5, 10, 20$, $D = 50$, and $T = 5$. For $d = 1, \ldots, D$ and $j = 1, \ldots, n_{dt}$, we generate regressors x_{dtj1} and x_{dtj2}, classifying individuals into one of the three possible classes (e.g. unemployed—U, employed—E, and inactive—I), so that they take on values $(x_{dtj1}, x_{dtj2}) = (0, 0)$, $(0, 1)$, and $(1, 0)$ with probabilities equal to 0.2, 0.5, and 0.3, respectively, at time period (subdomain) 1. It means that the states U, E, and I are for the j-th individual in the area (domain) d at time period $t = 1$ generated with probabilities 0.2, 0.5, and 0.3, respectively. At the time periods $t = 2, 3, 4, 5$, the current states for each person are generated using the Markov process with probabilities of transition between states given by the following matrix:

	U	E	I
U	0.7	0.2	0.1
E	0.15	0.8	0.05
I	0.05	0.05	0.9

(let us note that for this matrix the stationary distribution of the process is $(7/29, 10/29, 12/29) \doteq (0.241, 0.345, 0.414)$).

We further generate two independent sets of random effects $\{v_{1,d} : d = 1, \ldots, D\}$ i.i.d. $N(0, 1)$ and $\{v_{2,dt} : d = 1, \ldots, D,\ t = 1, \ldots, T\}$ i.i.d. $N(0, 1)$

Table 15.1 RBIAS (left) and RRMSE (right) for $D = 50$

n_d	5	10	20	5	10	20
$\hat{\beta}_0$	0.0220	0.0073	0.0109	0.393	0.321	0.257
$\hat{\beta}_1$	−0.0134	−0.0058	−0.0071	0.333	0.241	0.161
$\hat{\beta}_2$	−0.0052	0.0007	−0.0052	0.226	0.158	0.114
$\hat{\phi}_1$	−0.2561	−0.1137	−0.0483	0.629	0.385	0.193
$\hat{\phi}_2$	−0.0691	−0.0358	−0.0239	0.273	0.185	0.148

and select $\boldsymbol{\beta} = (\beta_0, \beta_1, \beta_2) = (1/3, 1/2, -2/3)$, $\phi_1 = 0.4$, and $\phi_2 = 0.3$. For $d = 1, \ldots, D$, $t = 1, \ldots, T$, and $j = 1, \ldots, n_{dt}$, we take $m_{dtj} = 1$ and generate the target variables

$$y_{dtj} \sim Bin(1, p_{dtj}),\ p_{dtj} = \frac{\exp\{\beta_0 + x_{dtj1}\beta_1 + x_{dtj2}\beta_2 + \phi_1 v_{1,d} + \phi_2 v_{2,dt}\}}{1 + \exp\{\beta_0 + x_{dtj1}\beta_1 + x_{dtj2}\beta_2 + \phi_1 v_{1,d} + \phi_2 v_{2,dt}\}}.$$

The target of the simulation is to check the behavior of the fitting algorithms. We use the ML-Laplace approximation algorithm in the simulation experiment. The steps of the simulation experiment are:

1. Repeat $I = 1000$ times ($i = 1, \ldots, I$):

1.1. Generate a sample of size $n = \sum_{d=1}^{D} \sum_{t=1}^{T} n_{dt}$.

1.2. Calculate $\hat{\beta}_0^{(i)}, \hat{\beta}_1^{(i)}, \hat{\beta}_2^{(i)}, \hat{\phi}_1^{(i)}, \hat{\phi}_2^{(i)}$.

2. Output: For $\theta \in \{\beta_0, \beta_1, \beta_2, \phi_1, \phi_2\}$, calculate

$$RBIAS = \frac{\sum_{i=1}^{I}(\hat{\theta}^{(i)} - \theta)}{I|\theta|}, \qquad RRMSE = \frac{\sqrt{\frac{1}{I}\sum_{i=1}^{I}(\hat{\theta}^{(i)} - \theta)^2}}{|\theta|}.$$

Table 15.1 presents the obtained results. It can be seen that the relative bias and the relative root mean square error have a decreasing tendency when the sizes of samples increase, which is an expected behavior of ML estimators.

15.7 R Codes for EBPs

This section gives R codes for fitting the unit-level two-fold logit mixed model to the survey data file `LFS20.txt`. The domains are defined by the variable AREA, and the subdomains are obtained by crossing this variable with the age groups. The age groups (ageG) are defined by ageG = 1 if AGE < 25, ageG = 2 if $25 \leq$ AGE < 54, and ageG = 3 if AGE ≥ 54. The target parameters are the proportions of poor people by domains or subdomains. The variable poor is derived from the variable income giving the individual annual net income in euros. The variable poor takes the value of 1 if income $< z_0$, where $z_0 = 36500$ is the poverty threshold. As auxiliary variables, we take the dichotomic variables defining the three categories of the variable EDUCATION (primary or less, secondary, and superior). The categories are named edu1, edu2, and edu3, respectively. The following R code reads the aggregated auxiliary data in `Ndsa20.txt` (true population sizes) and calculates the corresponding sizes of covariate classes by area and age group.

```
aux <- read.table("Ndsa20.txt", header=TRUE, sep="\t", dec=".")
# Totals by subdomains
Ndt <- tapply(X=aux$N, INDEX=list(aux$area,aux$age), FUN=sum)
# Totals by subdomains-edu1
Ndt.edu1 <- tapply(aux$edu1, list(aux$area,aux$age), sum)
# Totals by subdomains-edu2
Ndt.edu2 <- tapply(aux$edu2, list(aux$area,aux$age), sum)
# Totals by subdomains-edu3
Ndt.edu3 <- tapply(aux$edu3, list(aux$area,aux$age), sum)
```

The following R code reads the unit-level data file and defines the age groups and variables indicating the EDUCATION categories:

```
dat <- read.table("LFS20.txt", header=TRUE, sep="\t", dec=".")
ns <- nrow(dat)                          # Global sample size
n.areas <- length(unique(dat$AREA))      # Number of areas
z0 <- 36500                              # Poverty threshold
# Poverty variable: 1 if INCOME<z0, 0 if INCOME>z0
poor <- as.numeric(dat$INCOME<z0)
Ga <- cut(dat$AGE, breaks=c(0,25,54,max(dat$AGE)), labels=1:3, right=TRUE)
ageG <- as.numeric(Ga)
edu2 <- as.numeric(dat$EDUCATION==2)
edu3 <- as.numeric(dat$EDUCATION==3)
one <- rep(1, ns)
```

We calculate sample sizes, counts, and means in subdomains.

```
# Sizes by subdomains
ndt <- tapply(X=one, INDEX=list(dat$AREA,ageG), FUN=sum)
# Sample counts of edu3
ndt.edu3 <- tapply(edu3, list(dat$AREA,ageG), sum)
# Sample counts of edu2
ndt.edu2 <- tapply(edu2, list(dat$AREA,ageG), sum)
# Sample counts of edu1
ndt.edu1 <- ndt - ndt.edu3 - ndt.edu2
# Sample counts of poor
ndtpoor <- tapply(poor, list(dat$AREA,ageG), sum)
# Sample means of poor
mdtpoor <- ndtpoor/ndt
```

We calculate direct estimators and estimated sizes by subdomains. We use `dir2` function described in Sect. 2.8.4.

```
dir.poor <- dir2(data=poor, w=dat$WEIGHT,
                 domain=list(area=dat$AREA, ageG=ageG))
hatNdt <- dir.poor$Nd.hat     # Estimated sizes
dir.p <- dir.poor$mean        # Direct estimates of poverty proportions
```

We install and/or load the R packages `lme4` and `Matrix`.

```
if(!require(Matrix)){
  install.packages("Matrix")
  library(Matrix)
}
if(!require(lme4)){
  install.packages("lme4")
  library(lme4)
}
```

We fit the unit-level two-fold logit mixed model (15.2)–(15.3) to the variable $y =$ POOR by applying the `glmer` function of the R library `lme4`.

```
glmm <- glmer(formula=poor ~ edu3 + edu2 + (1|AREA/ageG), data=dat,
              family=binomial)
summary(glmm)                          # Summary of model results
pihat <- fitted(glmm)                  # Predicted probabilities
beta <- fixef(glmm)                    # Beta parameters
bedu3 <- beta[1]+beta[2]               # Fixed effect for x1=1, x2=1, x3=0
bedu2 <- beta[1]+beta[3]               # Fixed effect for x1=1, x2=0, x3=1
bedu1 <- beta[1]                       # Fixed effect for x1=1, x2=0, x3=0
var <- as.data.frame(VarCorr(glmm))    # Variance parameters
# Standard deviation of area:age random effects
phi2.e <- var$sdcor[1]
# Standard deviation of area random effects
phi1.e <- var$sdcor[2]
r.effects <- ranef(glmm)               # Random effects
# Modes of the random effects for area
ud <- as.matrix(r.effects[[2]])
# Modes of the random effects for area:age
re.dt <- as.matrix(r.effects[[1]])
udt <- matrix(re.dt, nrow=n.areas)
```

Table 15.2 gives the estimates of the regression parameters. The estimated standard deviations of the random effects are $\phi_1 = 0.19186$ and $\phi_2 = 0.00007$.

Using the fitted model, we first calculate the plug-in estimators of poverty proportions by subdomains.

```
uud <- sweep(udt, 1, ud, "+")       # Random effects for subdomains
etadt.edu1 <- bedu1 + uud
# pihat by subdomains and education level 1
pd.edu1 <- exp(etadt.edu1)/(1+exp(etadt.edu1))
etadt.edu2 <- bedu2 + uud
# pihat by subdomains and education level 2
pd.edu2 <- exp(etadt.edu2)/(1+exp(etadt.edu2))
etadt.edu3 <- bedu3 + uud
# pihat by subdomains and education level 3
pd.edu3 <- exp(etadt.edu3)/(1+exp(etadt.edu3))
# Poverty proportion plug-in estimators
plug.p <- (ndtpoor + (Ndt.edu1-ndt.edu1)*pd.edu1 +
           (Ndt.edu2-ndt.edu2)*pd.edu2 +
           (Ndt.edu3-ndt.edu3)*pd.edu3)/Ndt
```

We prepare the data in the format used by the R function `calc.EBP.y` that calculates the EBPs of subdomain poverty proportions.

```
D <- n.areas
TT <- 3          # Number of levels of aggregation (or time instants)
K <- 3           # Number of covariate classes
# z_k for edu1, edu2, edu3
class.vec <- rbind(c(1,0,0), c(1,0,1), c(1,1,0))
# Calculating the unobserved covariate classes sizes
N.dtk.r <- array(0, dim=c(D,TT,K))
N.dtk.r[,,1] <- Ndt.edu1 - ndt.edu1
N.dtk.r[,,2] <- Ndt.edu2 - ndt.edu2
N.dtk.r[,,3] <- Ndt.edu3 - ndt.edu3
```

Table 15.2 Estimated parameters of the model (15.2)–(15.3)

Parameter	Estimate	Std. error	z-value	p-value
Intercept	−0.5117	0.1121	−4.565	0.00
edu3	−3.6006	0.5921	−6.081	0.00
edu2	−1.4526	0.1779	−8.167	0.00

```
S1 <- S2 <- 30                  # Number of points for Monte-Carlo approximation
beta.e <- as.matrix(beta)       # Estimated parameters beta in a column format
X <- cbind(one, edu3, edu2) # Matrix X of our model
y <- poor                       # Vector of the target variable y
```

We load the abind package and the function `calc.EBP.y`, we set the seed for the random number generator, and we calculate the EBPs of poverty proportions.

```
if(!require(abind)){
  install.packages("abind")
  library(abind)
}
set.seed(123)
ebp.p <- calc.EBP.y(y, X, beta.e, phi1.e, phi2.e,
                    S1, S2, D, TT, K, class.vec, N.dtk.r)
```

The R code to save the results is

```
dir.p <- round(data.frame(dir.p[1:20], dir.p[21:40], dir.p[41:60]),4)
output <- data.frame(ndt, round(ebp.p,4), round(plug.p,4), dir.p)
names(output) <- c("n1", "n2", "n3", "ebpp1", "ebpp2", "ebpp3",
                   "plugp1", "plugp2", "plugp3",
                   "dirp1", "dirp2", "dirp3")
head(output, 10)
```

Table 15.3 (left) gives the sample sizes by subdomains (AREA crossed by AGE). We note that sample sizes are very small. The columns labeled by ebpp1, ebpp2, and ebpp3 contain the EBPs of poverty proportions by areas and age groups 1, 2, and 3, respectively. The columns labeled by plugp1, plugp2, and plugp3 and dirp1, dirp2, and dirp3 contain the plug-in and direct estimates of poverty proportions by areas and age groups 1, 2, and 3, respectively.

The R code of the function `calc.EBP.y` is

```
calc.EBP.y <- function(y, X, beta.e, phi1.e, phi2.e,
                       S1, S2, D, TT, K, class.vec, w.dtk.r){
# Generate the random effects
v1 <- rnorm(D * S1,mean = 0,sd = 1)
v1 <- matrix(v1, nrow = D)
v1 <- cbind(v1, -v1)
v2 <- rnorm(D*TT*S2, mean=0, sd=1)
v2 <- array(v2, dim=c(D,TT,S2))
v2 <- abind(v2, -v2)
# Initialize variables
y.dtau <- matrix(0, D, TT)
```

Table 15.3 EBPs and direct estimates of subdomain poverty proportions

d	n_1	n_2	n_3	ebpp1	ebpp2	ebpp3	plugp1	plugp2	plugpp3	dirp1	dirp2	dirp3
1	9	36	15	0.2175	0.1923	0.3220	0.2183	0.1915	0.3221	0.4229	0.1098	0.2662
2	9	13	15	0.2254	0.1855	0.3543	0.2255	0.1855	0.3551	0.2049	0.2576	0.3303
3	5	23	19	0.1995	0.1607	0.2559	0.1875	0.1519	0.2431	0.0000	0.1330	0.1614
4	14	31	10	0.2085	0.1942	0.3082	0.2080	0.1931	0.3073	0.3113	0.1481	0.2961
5	10	31	9	0.1267	0.1821	0.3672	0.1199	0.1721	0.3526	0.1868	0.0900	0.3060
6	9	24	10	0.1090	0.2152	0.3236	0.0989	0.2002	0.3035	0.0000	0.1699	0.1719
7	10	23	15	0.1435	0.1806	0.3350	0.1444	0.1827	0.3384	0.0816	0.2272	0.2660
8	10	27	11	0.2445	0.2690	0.2876	0.2447	0.2695	0.2884	0.1972	0.2774	0.2622
9	20	68	37	0.1336	0.1768	0.2875	0.1118	0.1512	0.2490	0.1295	0.1169	0.1421
10	10	20	11	0.1624	0.2199	0.3144	0.1721	0.2324	0.3341	0.0797	0.1889	0.6420

```
q.dtk <- rep(0,K)  #dim=c(D, TT, K))
mu.dt <- matrix(0, D, TT)
  # Begin the loop for calculating the EBPs by domains
  for(d in 1:D) {
   # Begin the calculation of inner sum-products over s1, tau, s2
   sum.s1.D <- 0
   sum.s1.N <- matrix(0, TT, K)
   # Begin the loop s1
   for(s1 in 1:(2*S1)){
       prod.tau.D <- 1
       prod.tau.N <- matrix(1, TT, K)
       # Begin the loop tau
       for(tau in 1:TT){
          # Specify the matrix xx_dt and sample vector y.dt
          x.dt <- X[dat$AREA==d & dat$ageG==tau,]
          y.dt <- y[dat$AREA==d & dat$ageG==tau]
          # Calculate the sums of y.dt
          y.dtau[d,tau] <- sum(y.dt)
          sum.s2.D <- 0
          sum.s2.N <- matrix(0,TT,K)
          for(s2 in 1:(2*S2)){
            ran.term <- phi1.e*v1[d,s1] + phi2.e*v2[d,tau,s2]
            # Calculate the inner sum over i
            eta <- x.dt%*%beta.e + ran.term
            vec <- log(1+exp(eta))
            sum.i <- sum(vec)
            # End of the calculation of the inner sum over i
            exponent.D <- y.dtau[d,tau]*ran.term - sum.i
            sum.s2.D <- sum.s2.D + exp(exponent.D)
            # Calculate the sum over s2 for N.dtk
            for(t in 1:TT){
              for(k in 1:K){
                if(t==tau){
                  exponent <- class.vec[k,]%*%beta.e +
                              (y.dtau[d,tau]+1)*ran.term -
                              log(1+exp(class.vec[k,]%*%beta.e+ran.term)) -
                              sum.i
                }
                else{
                  exponent <- exponent.D
                }
                sum.s2.N[t,k] <- sum.s2.N[t,k] + exp(exponent)
              }
            }
          } # End of the loop in s2
          prod.tau.D <- prod.tau.D * sum.s2.D
          prod.tau.N <- prod.tau.N * sum.s2.N
       } # End of the loop in tau
       sum.s1.D <- sum.s1.D + prod.tau.D
       sum.s1.N <- sum.s1.N + prod.tau.N
       cat("d=", d, s1, "\n")
   } # End of the loop in s1
   # End of the calculation of the inner sum-products over s1, tau, s2
   D.term <- sum.s1.D
   N.dtk <- sum.s1.N
   q.dtk <- N.dtk/D.term
   for(t in 1:TT){
       mu.dt[d,t] <- as.numeric(w.dtk.r[d,t,] %*% q.dtk[t,] + y.dtau[d,t] )
   }
  } # End of the loop for calculating the EBPs by domains
  ebpp <- mu.dt/Ndt
  return(ebpp)
}
```

References

Boubeta, M., Lombardía, M.J., Morales, D.: Empirical best prediction under area-level Poisson mixed models. TEST **25**, 548–569 (2016)

Boubeta, M., Lombardía, M.J., Morales, D.: Poisson mixed models for studying the poverty in small areas. Comput. Stat. Data Anal. **107**, 32–47 (2017)

Erciulescu A.L., Fuller W.A.: Small area prediction under alternative model specifications. Stat. Trans. New Ser. **17**(1), 9–24 (2016)

Ghosh, M., Kim, D., Sinha, K., Maiti, T., Katzoff, M., and Parsons, V.L.: Hierarchical and Empirical Bayes small domain estimation and proportion of persons without health insurance for minority subpopulations. Surv. Methodol. **35**, 53–66 (2009)

Hobza, T., Morales, D.: Empirical best prediction under unit-level logit mixed models. J. Off. Stat. **32**(3), 661–69 (2016)

Hobza, T., Morales, D., Santamaría, L.: Small area estimation of poverty proportions under unit-level temporal binomial-logit mixed models. TEST **27**, 270–294 (2018)

Jiang, J.: Empirical best prediction for small-area inference based on generalized linear mixed models. J. Stat. Plan. Infer. **111**, 117–127 (2003)

Jiang, J., Lahiri, P.: Empirical best prediction for small area inference with binary data. Ann. Inst. Stat. Math. **53**, 217–243 (2001)

López-Vizcaíno, E., Lombardía, M.J., Morales, D.: Multinomial-based small area estimation of labour force indicators. Stat. Model. **13**(2), 153–178 (2013)

López-Vizcaíno, E., Lombardía, M.J., Morales, D.: Small area estimation of labour force indicators under a multinomial model with correlated time and area effects. J. R. Stat. Assoc. Ser. A **178**(3), 535–565 (2015)

Militino, A.F., Ugarte, M.D., Goicoa, T.: Deriving small area estimates from information technology business surveys. J. R. Stat. Soc. Ser. A **178**(Part 4), 1051–1067 (2015)

Molina, I., Saei, A., Lombardía, M.J.: Small area estimates of labour force participation under multinomial logit mixed model. J. R. Stat. Soc. Ser. A **170**, 975–1000 (2007)

Erciulescu, A.L., Fuller, W.A.: Small area prediction under alternative model specifications. Stat. Trans. New Ser. 17(1), 9–24 (2016)

Ghosh, M., Kim, D., Sinha, K., Maiti, T., Katzoff, M., and Parsons, V.L.: Hierarchical and empirical Bayes small domain estimation and prediction of proportions of persons without health insurance for minority subpopulations. Surv. Methodol. 35, 53–66 (2009)

Hobza, T., Morales, D.: Empirical best prediction under unit-level logit mixed models. J. Off. Stat. 32(3), 661–692 (2016)

Hobza, T., Morales, D., Santamaría, L.: Small area estimation of poverty proportions under unit-level temporal binomial-logit mixed models. TEST 27, 270–294 (2018)

Jiang, J.: Empirical best prediction for small-area inference based on generalized linear mixed models. J. Stat. Plan. Infer. 111, 117–127 (2003)

Jiang, J., Lahiri, P.: Empirical best prediction for small area inference with binary data. Ann. Inst. Stat. Math. 53, 217–243 (2001)

López-Vizcaíno, E., Lombardía, M.J., Morales, D.: Multinomial-based small area estimation of labour force indicators. Stat. Model. 13(2), 153–178 (2013)

López-Vizcaíno, E., Lombardía, M.J., Morales, D.: Small area estimation of labour force indicators under a multinomial model with correlated time and area effects. J. R. Stat. Soc. Ser. A 178(3), 535–565 (2015)

Militino, A.F., Ugarte, M.D., Goicoa, T.: Deriving small area estimates from information technology business surveys. J. R. Stat. Soc. Ser. A 178(Part 4), 1051–1067 (2015)

Molina, I., Saei, A., Lombardía, M.J.: Small area estimates of labour force participation under a multinomial logit mixed model. J. R. Stat. Soc. Ser. A 170(4), 975–1000 (2007)

Chapter 16
Fay–Herriot Models

16.1 Introduction

The unit-level models give a powerful tool for describing a target variable when they fit well to data. For calculating the empirical best linear unbiased predictors (EBLUPs) of domain linear parameters under a unit-level linear mixed model, we need an auxiliary data file with the domain averages of the selected auxiliary variables. This restriction reduces a lot the set of available auxiliary variables, and therefore it diminishes the prediction strength of the model. Further, the aggregated x-values are typically obtained from administrative registers, so they might not be calculated with exactly the same specifications as the corresponding ones in the survey sample. This problem appears from time to time when using data from different sources.

For calculating the empirical best predictors (EBPs) of domain nonlinear parameters under a unit-level linear mixed model, we need a census file with the values of the selected auxiliary variables. This is a big drawback for the Statistical Offices in most countries. Alternatively, if the vector of auxiliary variables takes a finite number of values (categories), the EBPs can be calculated if the population sizes of the domains crossed by categories are available. This solution helps, but models using only categorical auxiliary variables are not always good enough for predicting the target variable.

The abovementioned problems often appear when applying unit-level models to real data in small area estimation (SAE) problems. If good auxiliary individual information is not available, but data aggregated to the small areas can be found in administrative registers, then the model can be stated at the small area level. An area-level linear mixed model with random area effects was first proposed by Fay and Herriot (1979) to estimate average per capita income in small places of the USA. Since then, the EBLUP based on a Fay–Herriot model has been one of the most widely used estimators of small area parameters.

D. Morales et al., *A Course on Small Area Estimation and Mixed Models*, Statistics for Social and Behavioral Sciences, https://doi.org/10.1007/978-3-030-63757-6_16

The EBLUPs based on the Fay–Herriot model have the disadvantage of losing information because of aggregating unit-level data. However, they have the following advantages: (1) they can use more auxiliary variables, (2) they avoid the restriction of the unit-level model-based estimators of needing the same auxiliary variables in the survey file and the external administrative registers, and (3) they give good results (compared to unit-level models) when the number of small areas is large with small sample sizes.

The Fay–Herriot model has been the subject of numerous research works that have extended its applicability to many and different settings. Without intending to cover all the extensive existing literature, we present below some relevant contributions.

Many authors have investigated the estimation of the mean squared error (MSE) of the EBLUP. Some of them are Prasad and Rao (1990), Datta and Lahiri (2000), Das et al. (2004), Hall and Maiti (2006a,b), Slud and Maiti (2006), González-Manteiga et al. (2010), Datta et al. (2011), and Kubokawa (2011). These investigations were extended to the construction of confidence intervals and testing hypotheses by Diao et al. (2014), Molina et al. (2015), and Marhuenda et al. (2016). Related also with MSE estimation, Jiang and Tang (2011) studied the effect that estimators of model parameters have in the efficiency of the EBLUP.

Other authors have proposed the extension of the Fay–Herriot model. Sugasawa and Kubokawa (2015) and Polettini (2017) introduced parametric transformed and generalized semi-parametric Bayesian Fay–Herriot models, respectively. A different important problem appears when the auxiliary variables are measured with errors. Ybarra and Lohr (2008), Arima et al. (2015), Datta et al. (2018), and Burgard et al. (2020) proposed different approaches to treat this problem. Concerning the tools for model selection, Marhuenda et al. (2014) and Lombardía et al. (2017) investigated variants of the Akaike information criterion for the Fay–Herriot model. Other authors have investigated, for example, the extension of the basic model to non-normal distributions, the introduction of robust and semi-parametric procedures, the simultaneous inference, or the implementation of hierarchical Bayesian procedures.

This chapter does not attempt to review the published works on the Fay–Herriot model, but it rather presents an introduction to the basic aspects for its use in SAE problems. It is organized as follows. Section 16.2 extends the general prediction theorem to area-level linear mixed models. Section 16.3 introduces the Fay–Herriot model. Section 16.4 describes the generalized variance function method to introduce the error variances in the model. Section 16.5 studies several methods for estimating the model parameters. Section 16.6 approximates the MSE of the EBLUP. Section 16.7 derives hierarchical Bayes predictors. Section 16.8 introduces the transformed Fay–Herriot models and gives some comments about the selection of auxiliary variables in area-level models. Finally, Sect. 16.9 presents an application to living condition survey data and gives the employed R codes.

16.2 BLUPs Under Area-Level Linear Mixed Models

This section presents an adaptation of the general prediction theorem to the case of predicting linear combinations of fixed and random effects under area-level linear mixed models. The new theorem can be applied to obtain the best linear unbiased predictors (BLUPs) and the corresponding prediction errors.

Let us consider the area-level linear mixed model

$$\boldsymbol{y}_{D\times 1} = \boldsymbol{X}_{D\times p}\boldsymbol{\beta}_{p\times 1} + \boldsymbol{Z}_{D\times q}\boldsymbol{u}_{q\times 1} + \boldsymbol{e}_{D\times 1}, \tag{16.1}$$

where D is the number of domains, $\boldsymbol{X}$ and $\boldsymbol{Z}$ are design matrices, $\boldsymbol{\beta}$ is the vector of fixed effects, $\boldsymbol{u} \sim N(\boldsymbol{0}, \boldsymbol{V}_u)$ is the vector of random effects, $\boldsymbol{e} \sim N(\boldsymbol{0}, \boldsymbol{V}_e)$ is the vector of model errors, $\boldsymbol{u}$ and $\boldsymbol{e}$ are independent, and the variance matrices $\boldsymbol{V}_u$ and $\boldsymbol{V}_e$ are known. From (16.1), we have $E[\boldsymbol{y}] = \boldsymbol{X}\boldsymbol{\beta}$ and

$$\boldsymbol{V} = \text{var}(\boldsymbol{y}) = \boldsymbol{Z}\boldsymbol{V}_u\boldsymbol{Z}' + \boldsymbol{V}_e, \ \ \text{var}(\boldsymbol{u}) = \boldsymbol{V}_u, \ \ \text{cov}(\boldsymbol{y}, \boldsymbol{u}) = E\big[(\boldsymbol{y} - \boldsymbol{X}\boldsymbol{\beta})\boldsymbol{u}'\big] = \boldsymbol{Z}\boldsymbol{V}_u.$$

We are interested in predicting the linear combination of fixed and random effects

$$\tau = \boldsymbol{l}'\boldsymbol{\beta} + \boldsymbol{m}'\boldsymbol{u},$$

where $\boldsymbol{l}$ and $\boldsymbol{m}$ are known vectors of sizes $p \times 1$ and $q \times 1$, respectively. This section derives the BLUP of $\tau = \boldsymbol{l}'\boldsymbol{\beta} + \boldsymbol{m}'\boldsymbol{u}$. As the predictor is linear, it has the general form $\hat{\tau} = \boldsymbol{a}'\boldsymbol{y} + b$. As the predictor is unbiased, $E[\hat{\tau} - \tau] = 0$. From

$$E[\hat{\tau}] = \boldsymbol{a}'E[\boldsymbol{y}] + b = \boldsymbol{a}'\boldsymbol{X}\boldsymbol{\beta} + b, \quad E[\tau] = \boldsymbol{l}'\boldsymbol{\beta},$$

we have

$$0 = E[\hat{\tau} - \tau] = (\boldsymbol{a}'\boldsymbol{X} - \boldsymbol{l}')\boldsymbol{\beta} + b.$$

If $\boldsymbol{\beta}$ does not depend (functionally) on $\boldsymbol{X}$, then $b = 0$, $\hat{\tau} = \boldsymbol{a}'\boldsymbol{y}$ and $\boldsymbol{a}'\boldsymbol{X} = \boldsymbol{l}'$. Therefore, the best predictor $\hat{\tau}$ is a solution of the problem

$$\text{minimize var}(\hat{\tau} - \tau) \text{ on } \boldsymbol{a} \in R^D, \quad \text{subject to } \boldsymbol{a}'\boldsymbol{X} = \boldsymbol{l}'.$$

The following theorem summarizes the results of Henderson (1949; 1950; 1953; 1959; 1963) about BLUPs of linear combinations of fixed and random effects under linear mixed models.

Proposition 16.1 (Prediction Theorem) *The BLUP of* $\tau = \boldsymbol{l}'\boldsymbol{\beta} + \boldsymbol{m}'\boldsymbol{u}$ *is*

$$\hat{\tau}_B = \boldsymbol{l}'\hat{\boldsymbol{\beta}}_B + \boldsymbol{m}'\hat{\boldsymbol{u}}_B,$$

where

$$\hat{\boldsymbol{\beta}}_B = (\boldsymbol{X}'\boldsymbol{V}^{-1}\boldsymbol{X})^{-1}\boldsymbol{X}'\boldsymbol{V}^{-1}\boldsymbol{y} \quad \text{and} \quad \hat{\boldsymbol{u}}_B = \boldsymbol{V}_u\boldsymbol{Z}'\boldsymbol{V}^{-1}(\boldsymbol{y} - \boldsymbol{X}\hat{\boldsymbol{\beta}}_B).$$

The error variance of $\hat{\tau}_B$ *is*

$$var(\hat{\tau}_B - \tau) = \boldsymbol{m}'\boldsymbol{T}\boldsymbol{m} + (\boldsymbol{l}' - \boldsymbol{m}'\boldsymbol{T}\boldsymbol{Z}'\boldsymbol{V}_e^{-1}\boldsymbol{X})\boldsymbol{Q}(\boldsymbol{l}' - \boldsymbol{m}'\boldsymbol{T}\boldsymbol{Z}'\boldsymbol{V}_e^{-1}\boldsymbol{X})',$$

where $\boldsymbol{Q} = (\boldsymbol{X}'\boldsymbol{V}^{-1}\boldsymbol{X})^{-1}$ *and* $\boldsymbol{T} = \boldsymbol{V}_u - \boldsymbol{V}_u\boldsymbol{Z}'\boldsymbol{V}^{-1}\boldsymbol{Z}\boldsymbol{V}_u$.

Proof Note that

$$\begin{aligned}\mathrm{var}(\hat{\tau} - \tau) &= \mathrm{var}(\boldsymbol{a}'\boldsymbol{y} - \boldsymbol{l}'\boldsymbol{\beta} - \boldsymbol{m}'\boldsymbol{u}) = \mathrm{var}(\boldsymbol{a}'\boldsymbol{y} - \boldsymbol{m}'\boldsymbol{u}) = \mathrm{var}(\boldsymbol{a}'\boldsymbol{y}) + \mathrm{var}(\boldsymbol{m}'\boldsymbol{u}) \\ &- 2\mathrm{cov}(\boldsymbol{a}'\boldsymbol{y}, \boldsymbol{m}'\boldsymbol{u}) = \boldsymbol{a}'\boldsymbol{V}\boldsymbol{a} + \boldsymbol{m}'\boldsymbol{V}_u\boldsymbol{m} - 2\boldsymbol{a}'\boldsymbol{Z}\boldsymbol{V}_u\boldsymbol{m}.\end{aligned}$$

As $\boldsymbol{m}'\boldsymbol{V}_u\boldsymbol{m}$ does not depend on $\boldsymbol{a}$, we only have to solve the problem

$$\text{minimize } \boldsymbol{a}'\boldsymbol{V}\boldsymbol{a} - 2\boldsymbol{a}'\boldsymbol{Z}\boldsymbol{V}_u\boldsymbol{m} \text{ on } \boldsymbol{a} \in R^D, \qquad \text{subject to } \boldsymbol{a}'\boldsymbol{X} = \boldsymbol{l}'.$$

The Lagrangian function is

$$L(\boldsymbol{a}, \lambda) = \boldsymbol{a}'\boldsymbol{V}\boldsymbol{a} - 2\boldsymbol{a}'\boldsymbol{Z}\boldsymbol{V}_u\boldsymbol{m} + 2(\boldsymbol{a}'\boldsymbol{X} - \boldsymbol{l}')\lambda.$$

By taking partial derivatives with respect to $\boldsymbol{a}$, we have

$$0 = \frac{\partial L(\boldsymbol{a}, \lambda)}{\partial \boldsymbol{a}} = 2\boldsymbol{V}\boldsymbol{a} - 2\boldsymbol{Z}\boldsymbol{V}_u\boldsymbol{m} + 2\boldsymbol{X}\lambda \iff \boldsymbol{V}\boldsymbol{a} + \boldsymbol{X}\lambda = \boldsymbol{Z}\boldsymbol{V}_u\boldsymbol{m}.$$

By taking partial derivatives with respect to λ, we have

$$0 = \frac{\partial L(\boldsymbol{a}, \lambda)}{\partial \lambda} = \boldsymbol{a}'\boldsymbol{X} - \boldsymbol{l}' \iff \boldsymbol{a}'\boldsymbol{X} = \boldsymbol{l}' \iff \boldsymbol{X}'\boldsymbol{a} = \boldsymbol{l}.$$

In matrix notation, we have obtained the set of equations

$$\begin{pmatrix} \boldsymbol{V} & \boldsymbol{X} \\ \boldsymbol{X}' & 0 \end{pmatrix} \begin{pmatrix} \boldsymbol{a} \\ \lambda \end{pmatrix} = \begin{pmatrix} \boldsymbol{Z}\boldsymbol{V}_u\boldsymbol{m} \\ \boldsymbol{l} \end{pmatrix}.$$

In the same way as in (6.5) on page 115, we obtain the inverse matrix

$$\begin{pmatrix} \boldsymbol{V} & \boldsymbol{X} \\ \boldsymbol{X}' & 0 \end{pmatrix}^{-1} = \begin{pmatrix} \boldsymbol{V}^{-1} - \boldsymbol{V}^{-1}\boldsymbol{X}\boldsymbol{Q}\boldsymbol{X}'\boldsymbol{V}^{-1} & \boldsymbol{V}^{-1}\boldsymbol{X}\boldsymbol{Q} \\ \boldsymbol{Q}\boldsymbol{X}'\boldsymbol{V}^{-1} & -\boldsymbol{Q} \end{pmatrix},$$

where $\boldsymbol{Q} = (\boldsymbol{X}'\boldsymbol{V}^{-1}\boldsymbol{X})^{-1}$. Thus, the Lagrangian equations can be written in the form

$$\begin{pmatrix} \boldsymbol{a} \\ \lambda \end{pmatrix} = \begin{pmatrix} \boldsymbol{V}^{-1} - \boldsymbol{V}^{-1}\boldsymbol{X}\boldsymbol{Q}\boldsymbol{X}'\boldsymbol{V}^{-1} & \boldsymbol{V}^{-1}\boldsymbol{X}\boldsymbol{Q} \\ \boldsymbol{Q}\boldsymbol{X}'\boldsymbol{V}^{-1} & -\boldsymbol{Q} \end{pmatrix} \begin{pmatrix} \boldsymbol{Z}\boldsymbol{V}_u\boldsymbol{m} \\ \boldsymbol{l} \end{pmatrix}.$$

Therefore,

$$\begin{aligned} \boldsymbol{a} &= \boldsymbol{V}^{-1}\boldsymbol{Z}\boldsymbol{V}_u\boldsymbol{m} - \boldsymbol{V}^{-1}\boldsymbol{X}\boldsymbol{Q}\boldsymbol{X}'\boldsymbol{V}^{-1}\boldsymbol{Z}\boldsymbol{V}_u\boldsymbol{m} + \boldsymbol{V}^{-1}\boldsymbol{X}\boldsymbol{Q}\boldsymbol{l} \\ &= \boldsymbol{V}^{-1}\left[\boldsymbol{I} - \boldsymbol{X}\boldsymbol{Q}\boldsymbol{X}'\boldsymbol{V}^{-1}\right]\boldsymbol{Z}\boldsymbol{V}_u\boldsymbol{m} + \boldsymbol{V}^{-1}\boldsymbol{X}\boldsymbol{Q}\boldsymbol{l}. \end{aligned}$$

The BLUP of $\tau = \boldsymbol{l}'\boldsymbol{\beta} + \boldsymbol{m}'\boldsymbol{u}$ is

$$\begin{aligned} \hat{\tau}_B = \boldsymbol{a}'\boldsymbol{y} &= \boldsymbol{l}'\boldsymbol{Q}\boldsymbol{X}'\boldsymbol{V}^{-1}\boldsymbol{y} + \boldsymbol{m}'\boldsymbol{V}_u\boldsymbol{Z}'\left[\boldsymbol{I} - \boldsymbol{V}^{-1}\boldsymbol{X}\boldsymbol{Q}\boldsymbol{X}'\right]\boldsymbol{V}^{-1}\boldsymbol{y} \\ &= \boldsymbol{l}'\boldsymbol{Q}\boldsymbol{X}'\boldsymbol{V}^{-1}\boldsymbol{y} + \boldsymbol{m}'\boldsymbol{V}_u\boldsymbol{Z}'\boldsymbol{V}^{-1}\left[\boldsymbol{I} - \boldsymbol{X}\boldsymbol{Q}\boldsymbol{X}'\boldsymbol{V}^{-1}\right]\boldsymbol{y}. \end{aligned}$$

Let us define

$$\hat{\boldsymbol{\beta}}_B = \boldsymbol{Q}\boldsymbol{X}'\boldsymbol{V}^{-1}\boldsymbol{y} = (\boldsymbol{X}'\boldsymbol{V}^{-1}\boldsymbol{X})^{-1}\boldsymbol{X}'\boldsymbol{V}^{-1}\boldsymbol{y}.$$

Then, the BLUP of τ is

$$\hat{\tau}_B = \boldsymbol{l}'\hat{\boldsymbol{\beta}}_B + \boldsymbol{m}'\boldsymbol{V}_u\boldsymbol{Z}'\boldsymbol{V}^{-1}(\boldsymbol{y} - \boldsymbol{X}\hat{\boldsymbol{\beta}}_B).$$

If we further define

$$\hat{\boldsymbol{u}}_B = \boldsymbol{V}_u\boldsymbol{Z}'\boldsymbol{V}^{-1}(\boldsymbol{y} - \boldsymbol{X}\hat{\boldsymbol{\beta}}_B),$$

then

$$\hat{\tau}_B = \boldsymbol{l}'\hat{\boldsymbol{\beta}}_B + \boldsymbol{m}'\hat{\boldsymbol{u}}_B.$$

By substituting $\boldsymbol{a}' = \boldsymbol{m}'\boldsymbol{V}_u\boldsymbol{Z}'\boldsymbol{V}^{-1} - \boldsymbol{m}'\boldsymbol{V}_u\boldsymbol{Z}'\boldsymbol{V}^{-1}\boldsymbol{X}\boldsymbol{Q}\boldsymbol{X}'\boldsymbol{V}^{-1} + \boldsymbol{l}'\boldsymbol{Q}\boldsymbol{X}'\boldsymbol{V}^{-1}$ in the formula of $\mathrm{var}(\hat{\tau} - \tau)$, we obtain the error variance of $\hat{\tau}_B$. This is to say,

$$\begin{aligned} \mathrm{v}_B = \mathrm{var}(\hat{\tau}_B - \tau) &= \boldsymbol{a}'\boldsymbol{V}\boldsymbol{a} + \boldsymbol{m}'\boldsymbol{V}_u\boldsymbol{m} - 2\boldsymbol{a}'\boldsymbol{Z}\boldsymbol{V}_u\boldsymbol{m} \\ &= (\boldsymbol{m}'\boldsymbol{V}_u\boldsymbol{Z}'\boldsymbol{V}^{-1} - \boldsymbol{m}'\boldsymbol{V}_u\boldsymbol{Z}'\boldsymbol{V}^{-1}\boldsymbol{X}\boldsymbol{Q}\boldsymbol{X}'\boldsymbol{V}^{-1} + \boldsymbol{l}'\boldsymbol{Q}\boldsymbol{X}'\boldsymbol{V}^{-1})\boldsymbol{V} \\ &\cdot (\boldsymbol{V}^{-1}\boldsymbol{Z}\boldsymbol{V}_u\boldsymbol{m} - \boldsymbol{V}^{-1}\boldsymbol{X}\boldsymbol{Q}\boldsymbol{X}'\boldsymbol{V}^{-1}\boldsymbol{Z}\boldsymbol{V}_u\boldsymbol{m} + \boldsymbol{V}^{-1}\boldsymbol{X}\boldsymbol{Q}\boldsymbol{l}) + \boldsymbol{m}'\boldsymbol{V}_u\boldsymbol{m} \\ &- 2(\boldsymbol{m}'\boldsymbol{V}_u\boldsymbol{Z}'\boldsymbol{V}^{-1} - \boldsymbol{m}'\boldsymbol{V}_u\boldsymbol{Z}'\boldsymbol{V}^{-1}\boldsymbol{X}\boldsymbol{Q}\boldsymbol{X}'\boldsymbol{V}^{-1} + \boldsymbol{l}'\boldsymbol{Q}\boldsymbol{X}'\boldsymbol{V}^{-1})\boldsymbol{Z}\boldsymbol{V}_u\boldsymbol{m} \\ &= \boldsymbol{m}'\boldsymbol{V}_u\boldsymbol{Z}'\boldsymbol{V}^{-1}\boldsymbol{Z}\boldsymbol{V}_u\boldsymbol{m} - \boldsymbol{m}'\boldsymbol{V}_u\boldsymbol{Z}'\boldsymbol{V}^{-1}\boldsymbol{X}\boldsymbol{Q}\boldsymbol{X}'\boldsymbol{V}^{-1}\boldsymbol{Z}\boldsymbol{V}_u\boldsymbol{m} + \boldsymbol{m}'\boldsymbol{V}_u\boldsymbol{Z}'\boldsymbol{V}^{-1}\boldsymbol{X}\boldsymbol{Q}\boldsymbol{l} \end{aligned}$$

$$- \boldsymbol{m}'\boldsymbol{V}_u\boldsymbol{Z}'\boldsymbol{V}^{-1}\boldsymbol{X}\boldsymbol{Q}\boldsymbol{X}'\boldsymbol{V}^{-1}\boldsymbol{Z}\boldsymbol{V}_u\boldsymbol{m} + \boldsymbol{m}'\boldsymbol{V}_u\boldsymbol{Z}'\boldsymbol{V}^{-1}\boldsymbol{X}\underline{\boldsymbol{Q}\boldsymbol{X}'\boldsymbol{V}^{-1}\boldsymbol{X}}\boldsymbol{Q}\boldsymbol{X}'\boldsymbol{V}^{-1}\boldsymbol{Z}\boldsymbol{V}_u\boldsymbol{m}$$
$$- \boldsymbol{m}'\boldsymbol{V}_u\boldsymbol{Z}'\boldsymbol{V}^{-1}\boldsymbol{X}\underline{\boldsymbol{Q}\boldsymbol{X}'\boldsymbol{V}^{-1}\boldsymbol{X}}\boldsymbol{Q}\boldsymbol{l} + \boldsymbol{l}'\boldsymbol{Q}\boldsymbol{X}'\boldsymbol{V}^{-1}\boldsymbol{Z}\boldsymbol{V}_u\boldsymbol{m}$$
$$- \boldsymbol{l}'\underline{\boldsymbol{Q}\boldsymbol{X}'\boldsymbol{V}^{-1}\boldsymbol{X}}\boldsymbol{Q}\boldsymbol{X}'\boldsymbol{V}^{-1}\boldsymbol{Z}\boldsymbol{V}_u\boldsymbol{m} + \boldsymbol{l}'\underline{\boldsymbol{Q}\boldsymbol{X}'\boldsymbol{V}^{-1}\boldsymbol{X}}\boldsymbol{Q}\boldsymbol{l} + \boldsymbol{m}'\boldsymbol{V}_u\boldsymbol{m}$$
$$- 2\boldsymbol{m}'\boldsymbol{V}_u\boldsymbol{Z}'\boldsymbol{V}^{-1}\boldsymbol{Z}\boldsymbol{V}_u\boldsymbol{m} + 2\boldsymbol{m}'\boldsymbol{V}_u\boldsymbol{Z}'\boldsymbol{V}^{-1}\boldsymbol{X}\boldsymbol{Q}\boldsymbol{X}'\boldsymbol{V}^{-1}\boldsymbol{Z}\boldsymbol{V}_u\boldsymbol{m} - 2\boldsymbol{l}'\boldsymbol{Q}\boldsymbol{X}'\boldsymbol{V}^{-1}\boldsymbol{Z}\boldsymbol{V}_u\boldsymbol{m}.$$

As $\boldsymbol{Q}\boldsymbol{X}'\boldsymbol{V}^{-1}\boldsymbol{X} = \boldsymbol{I}_D$, we have

$$\begin{aligned} \mathrm{v}_B &= \boldsymbol{m}'\boldsymbol{V}_u\boldsymbol{m} - \boldsymbol{m}'\boldsymbol{V}_u\boldsymbol{Z}'\boldsymbol{V}^{-1}\boldsymbol{Z}\boldsymbol{V}_u\boldsymbol{m} + \boldsymbol{m}'\boldsymbol{V}_u\boldsymbol{Z}'\boldsymbol{V}^{-1}\boldsymbol{X}\boldsymbol{Q}\boldsymbol{X}'\boldsymbol{V}^{-1}\boldsymbol{Z}\boldsymbol{V}_u\boldsymbol{m} + \boldsymbol{l}'\boldsymbol{Q}\boldsymbol{l} \\ &\quad - 2\boldsymbol{l}'\boldsymbol{Q}\boldsymbol{X}'\boldsymbol{V}^{-1}\boldsymbol{Z}\boldsymbol{V}_u\boldsymbol{m} \\ &= \boldsymbol{m}'(\boldsymbol{V}_u - \boldsymbol{V}_u\boldsymbol{Z}'\boldsymbol{V}^{-1}\boldsymbol{Z}\boldsymbol{V}_u)\boldsymbol{m} + (\boldsymbol{l}' - \boldsymbol{m}'\boldsymbol{V}_u\boldsymbol{Z}'\boldsymbol{V}^{-1}\boldsymbol{X})\boldsymbol{Q}(\boldsymbol{l}' - \boldsymbol{m}'\boldsymbol{V}_u\boldsymbol{Z}'\boldsymbol{V}^{-1}\boldsymbol{X})'. \end{aligned}$$

For $\boldsymbol{T} = \boldsymbol{V}_u - \boldsymbol{V}_u\boldsymbol{Z}'\boldsymbol{V}^{-1}\boldsymbol{Z}\boldsymbol{V}_u$ and $\boldsymbol{V} = \boldsymbol{Z}\boldsymbol{V}_u\boldsymbol{Z}' + \boldsymbol{V}_e$, it holds that

$$\begin{aligned} \boldsymbol{m}'\boldsymbol{T}\boldsymbol{Z}'\boldsymbol{V}_e^{-1} &= \boldsymbol{m}'(\boldsymbol{V}_u - \boldsymbol{V}_u\boldsymbol{Z}'\boldsymbol{V}^{-1}\boldsymbol{Z}\boldsymbol{V}_u)\boldsymbol{Z}'\boldsymbol{V}_e^{-1} \\ &= \boldsymbol{m}'\boldsymbol{V}_u\boldsymbol{Z}'\boldsymbol{V}_e^{-1} - \boldsymbol{m}'\boldsymbol{V}_u\boldsymbol{Z}'\boldsymbol{V}^{-1}\boldsymbol{Z}\boldsymbol{V}_u\boldsymbol{Z}'\boldsymbol{V}_e^{-1} \\ &= \boldsymbol{m}'\boldsymbol{V}_u\boldsymbol{Z}'(\boldsymbol{V}_e^{-1} - \boldsymbol{V}^{-1}\boldsymbol{Z}\boldsymbol{V}_u\boldsymbol{Z}'\boldsymbol{V}_e^{-1}) \\ &= \boldsymbol{m}'\boldsymbol{V}_u\boldsymbol{Z}'\boldsymbol{V}^{-1}(\boldsymbol{V} - \boldsymbol{Z}\boldsymbol{V}_u\boldsymbol{Z}')\boldsymbol{V}_e^{-1} \\ &= \boldsymbol{m}'\boldsymbol{V}_u\boldsymbol{Z}'\boldsymbol{V}^{-1}\boldsymbol{V}_e\boldsymbol{V}_e^{-1} = \boldsymbol{m}'\boldsymbol{V}_u\boldsymbol{Z}'\boldsymbol{V}^{-1}. \end{aligned}$$

Therefore, we get

$$\mathrm{v}_B = \boldsymbol{m}'\boldsymbol{T}\boldsymbol{m} + (\boldsymbol{l}' - \boldsymbol{m}'\boldsymbol{T}\boldsymbol{Z}'\boldsymbol{V}_e^{-1}\boldsymbol{X})\boldsymbol{Q}(\boldsymbol{l}' - \boldsymbol{m}'\boldsymbol{T}\boldsymbol{Z}'\boldsymbol{V}_e^{-1}\boldsymbol{X})'. \qquad \square$$

If we are interested in predicting the domain mean $\mu_d = \boldsymbol{x}_d\boldsymbol{\beta} + z_d\boldsymbol{u}_d$, where $\boldsymbol{x}_d$ and z_d are the d-th row of matrices $\boldsymbol{X}$ and $\boldsymbol{Z}$, respectively, we can take $\boldsymbol{l}' = \boldsymbol{x}_d$ and $\boldsymbol{m}' = z_d$, and using the proposition, we get the BLUP of μ_d in the form $\hat{\mu}_d = \boldsymbol{x}_d\hat{\boldsymbol{\beta}}_B + z_d\hat{\boldsymbol{u}}_B$. The corresponding error variance of $\hat{\mu}_d$ is

$$\mathrm{var}(\hat{\mu}_d - \mu_d) = z_d\boldsymbol{T}z_d' + (\boldsymbol{x}_d - z_d\boldsymbol{T}\boldsymbol{Z}'\boldsymbol{V}_e^{-1}\boldsymbol{X})\boldsymbol{Q}(\boldsymbol{x}_d - z_d\boldsymbol{T}\boldsymbol{Z}'\boldsymbol{V}_e^{-1}\boldsymbol{X})'.$$

16.3 The Area-Level Fay–Herriot Model

The Fay–Herriot model is comprised of two stages. In the first stage, a model, called *sampling* model, is used to represent the sampling error of direct estimators. Let μ_d be the characteristic of interest in d-th area and y_d be a direct estimator of μ_d. In applications to real data, μ_d is typically the domain mean $\overline{Y}_d = \frac{1}{N_d}\sum_{i=1}^{N_d} y_{dj}$, where y_{dj} is the target variable measured at the unit level. The sampling model indicates that direct estimators y_d's are unbiased and can be expressed as

$$y_d = \mu_d + e_d, \quad d = 1, \ldots, D,$$

where D is the total number of areas or domains in which inference is made. Here, e_d's are sampling errors that are independent and normally distributed with known variances, that is, $e_d \sim N(0, \sigma_d^2)$, where σ_d^2 is the (assumed known) design-based variance of direct estimator y_d, $d = 1, \ldots, D$. In the second stage, the true area characteristics μ_d's are assumed to vary linearly with a number p of area-level auxiliary variables, that is,

$$\mu_d = \boldsymbol{x}_d \boldsymbol{\beta} + u_d, \quad d = 1, \ldots, D,$$

where $\boldsymbol{x}_d$ is a row vector containing the aggregated (population) values of p auxiliary variables for area d, $\boldsymbol{\beta}$ is the column vector of regression coefficients, and u_d's are model errors, typically assumed to be i.i.d. from $N(0, \sigma_u^2)$ with variance σ_u^2 unknown and independent of e_d's. Note that this model, called *linking* model, links the target quantities μ_d of all the areas through the common regression parameter $\boldsymbol{\beta}$. The Fay–Herriot model can be expressed as a single model in the form

$$y_d = \boldsymbol{x}_d \boldsymbol{\beta} + u_d + e_d, \quad d = 1, \ldots, D, \tag{16.2}$$

or in the matrix form $\boldsymbol{y} = \boldsymbol{X}\boldsymbol{\beta} + \boldsymbol{Z}\boldsymbol{u} + \boldsymbol{e}$, where $\boldsymbol{Z} = \boldsymbol{I}_D = \underset{1 \le d \le D}{\text{diag}}\,(1)$ and

$$\boldsymbol{y} = \underset{1 \le d \le D}{\text{col}}\,(y_d),\ \boldsymbol{X}_d = \underset{1 \le d \le D}{\text{col}}\,(\boldsymbol{x}_d),\ \boldsymbol{\beta} = \underset{1 \le k \le p}{\text{col}}\,(\beta_k),\ \boldsymbol{u} = \underset{1 \le d \le D}{\text{col}}\,(u_d),\ \boldsymbol{e} = \underset{1 \le d \le D}{\text{col}}\,(e_d).$$

It holds that $\boldsymbol{V}_u = \text{var}(\boldsymbol{u}) = \sigma_u^2 \boldsymbol{I}_D$, $\boldsymbol{V}_e = \text{var}(\boldsymbol{e}) = \text{diag}(\sigma_1^2, \ldots, \sigma_D^2)$ and $\boldsymbol{V} = \text{var}(\boldsymbol{y}) = \boldsymbol{Z}\boldsymbol{V}_u\boldsymbol{Z}' + \boldsymbol{V}_e = \boldsymbol{V}_u + \boldsymbol{V}_e = \text{diag}(\sigma_u^2 + \sigma_1^2, \ldots, \sigma_u^2 + \sigma_D^2)$, $\text{cov}(\boldsymbol{y}, \boldsymbol{u}) = \boldsymbol{V}_u$.

If σ_u^2 is known, the BLUP of $\mu_d = \boldsymbol{x}_d \boldsymbol{\beta} + u_d$ is $\tilde{\mu}_d = \boldsymbol{x}_d \tilde{\boldsymbol{\beta}} + \tilde{u}_d$, where

$$\tilde{\boldsymbol{\beta}} = (\boldsymbol{X}'\boldsymbol{V}^{-1}\boldsymbol{X})^{-1}\boldsymbol{X}'\boldsymbol{V}^{-1}\boldsymbol{y} \qquad \text{and} \qquad \tilde{\boldsymbol{u}} = \boldsymbol{V}_u \boldsymbol{V}^{-1}\left(\boldsymbol{y} - \boldsymbol{X}\tilde{\boldsymbol{\beta}}\right)$$

are the best linear unbiased estimator (BLUE) and predictor (BLUP) of $\boldsymbol{\beta} = (\beta_1, \ldots, \beta_p)'$ and $\boldsymbol{u} = (u_1, \ldots, u_D)'$. The components $\tilde{u}_d$ of the vector $\tilde{\boldsymbol{u}}$ can be obtained by applying Proposition 16.1 with $\boldsymbol{l} = \boldsymbol{0}_D$ and $\boldsymbol{m} = \underset{1 \le j \le D}{\text{col}}\,(\delta_{dj})$, where $\boldsymbol{0}_D$ is the $D \times 1$ zero vector and δ_{dj} is the Kronecker delta. It is easy to check that the components of $\tilde{\boldsymbol{u}}$ are

$$\tilde{u}_d = \frac{\sigma_u^2}{\sigma_u^2 + \sigma_d^2}\left(y_d - \boldsymbol{x}_d \tilde{\boldsymbol{\beta}}\right), \qquad d = 1, \ldots, D,$$

so that the BLUP of $\mu_d = \boldsymbol{x}_d\boldsymbol{\beta} + u_d$ can be written in the form

$$\tilde{\mu}_d = \boldsymbol{x}_d\tilde{\boldsymbol{\beta}} + \tilde{u}_d = \boldsymbol{x}_d\tilde{\boldsymbol{\beta}} + \frac{\sigma_u^2}{\sigma_u^2+\sigma_d^2}\left(y_d - \boldsymbol{x}_d\tilde{\boldsymbol{\beta}}\right) = \frac{\sigma_u^2}{\sigma_u^2+\sigma_d^2}\,y_d + \frac{\sigma_d^2}{\sigma_u^2+\sigma_d^2}\,\boldsymbol{x}_d\tilde{\boldsymbol{\beta}}. \tag{16.3}$$

Proposition 16.2 gives an alternative way of calculating the BLUPs. For a probability density function (p.d.f.), $f(x)$, we write $f(x) \propto g(x)$ if $g(x)$ is obtained from $f(x)$ by taking out multiplicative terms not depending on x. In that case, we say that g is proportional to f or g is a kernel of f.

Proposition 16.2 *If $\boldsymbol{\beta}$ and $\sigma_u^2 > 0$ are known, the best predictor (BP) of μ_d is*

$$E[\mu_d|y_d] = \frac{\sigma_u^2}{\sigma_u^2+\sigma_d^2}\,y_d + \frac{\sigma_d^2}{\sigma_u^2+\sigma_d^2}\,\boldsymbol{x}_d\boldsymbol{\beta},$$

so that the BLUP can be obtained from the BP by substituting $\boldsymbol{\beta}$ by $\tilde{\boldsymbol{\beta}}$.

Proof As $y_d \sim N(\boldsymbol{x}_d\boldsymbol{\beta}, \sigma_u^2+\sigma_d^2)$, $y_d|u_d \sim N(\boldsymbol{x}_d\boldsymbol{\beta}+u_d, \sigma_d^2)$, and $u_d \sim N(0, \sigma_u^2)$, it holds

$$\begin{aligned}
f(u_d|y_d) &\propto f(y_d|u_d)f(u_d)\\
&= \frac{1}{\sigma_d\sqrt{2\pi}}\exp\Big\{-\frac{1}{2\sigma_d^2}(y_d - \boldsymbol{x}_d\boldsymbol{\beta} - u_d)^2\Big\}\frac{1}{\sigma_u\sqrt{2\pi}}\exp\Big\{-\frac{1}{2\sigma_u^2}u_d^2\Big\}\\
&\propto \exp\left\{-\frac{u_d^2 - 2u_d(y_d - \boldsymbol{x}_d\boldsymbol{\beta})}{2\sigma_d^2}\right\}\exp\left\{-\frac{u_d^2}{2\sigma_u^2}\right\}\\
&= \exp\left\{-\frac{1}{2}\left[\left(\frac{1}{\sigma_d^2}+\frac{1}{\sigma_u^2}\right)u_d^2 - 2\frac{y_d - \boldsymbol{x}_d\boldsymbol{\beta}}{\sigma_d^2}u_d\right]\right\}\\
&= \exp\left\{-\frac{1}{2\frac{\sigma_d^2\sigma_u^2}{\sigma_d^2+\sigma_u^2}}\left[u_d^2 - 2\frac{\sigma_u^2}{\sigma_d^2+\sigma_u^2}(y_d - \boldsymbol{x}_d\boldsymbol{\beta})u_d\right]\right\},
\end{aligned}$$

which corresponds to a normal distribution with mean $E[u_d|y_d] = \frac{\sigma_u^2}{\sigma_d^2+\sigma_u^2}(y_d - \boldsymbol{x}_d\boldsymbol{\beta})$ and variance $\mathrm{var}(u_d|y_d) = \frac{\sigma_d^2\sigma_u^2}{\sigma_d^2+\sigma_u^2}$. Therefore,

$$\begin{aligned}
E[\mu_d|y_d] &= \boldsymbol{x}_d\boldsymbol{\beta} + E[u_d|y_d] = \boldsymbol{x}_d\boldsymbol{\beta} + \frac{\sigma_u^2}{\sigma_d^2+\sigma_u^2}(y_d - \boldsymbol{x}_d\boldsymbol{\beta})\\
&= \frac{\sigma_u^2}{\sigma_u^2+\sigma_d^2}\,y_d + \frac{\sigma_d^2}{\sigma_u^2+\sigma_d^2}\,\boldsymbol{x}_d\boldsymbol{\beta}.
\end{aligned}$$

□

The empirical BLUE (EBLUE) of $\boldsymbol{\beta}$ and the empirical BLUP (EBLUP) of $\boldsymbol{u}$ are obtained by plugging an estimator $\hat{\sigma}_u^2$ in the place of σ_u^2, i.e.

$$\hat{\boldsymbol{\beta}} = (\boldsymbol{X}'\hat{\boldsymbol{V}}^{-1}\boldsymbol{X})^{-1}\boldsymbol{X}'\hat{\boldsymbol{V}}^{-1}\boldsymbol{y}, \quad \hat{\boldsymbol{u}} = \hat{\boldsymbol{V}}_u\hat{\boldsymbol{V}}^{-1}\left(\boldsymbol{y} - \boldsymbol{X}\hat{\boldsymbol{\beta}}\right), \tag{16.4}$$

where $\hat{\boldsymbol{V}}_u = \hat{\sigma}_u^2\boldsymbol{I}_d$ and $\hat{\boldsymbol{V}} = \underset{1\leq d\leq D}{\text{diag}}\,(\hat{\sigma}_u^2 + \sigma_d^2)$. Under the model (16.2), the EBLUP of μ_d is obtained similarly from the BLUP (16.3), i.e.

$$\hat{\mu}_d = \frac{\hat{\sigma}_u^2}{\hat{\sigma}_u^2 + \sigma_d^2}\,y_d + \frac{\sigma_d^2}{\hat{\sigma}_u^2 + \sigma_d^2}\,\boldsymbol{x}_d\hat{\boldsymbol{\beta}}. \tag{16.5}$$

The EBLUP of μ_d is also employed to estimate the domain mean $\overline{Y}_d$, i.e. $\hat{\overline{Y}}_d^{eblup} = \hat{\mu}_d$. It is worth noting that:

1. If the direct estimator is precise (large n_d), then $\sigma_d^2 \approx 0$ and $\sigma_d^2 \ll \sigma_u^2$. Therefore,

$$\hat{\overline{Y}}_d^{eblup} \approx y_d = \hat{\overline{Y}}_d^{dir} \quad \text{(direct estimator)}.$$

2. If the direct estimator is not precise (small n_d), then $\sigma_d^2 \gg 0$ and $\sigma_d^2 \gg \sigma_u^2$. Therefore,

$$\hat{\overline{Y}}_d^{eblup} \approx \boldsymbol{x}_d\hat{\boldsymbol{\beta}} \quad \text{(synthetic estimator)}.$$

16.4 Sampling Error Variances

Let $\widehat{\text{var}}_\pi(y_d)$ be an estimator of the design-based variance, $\text{var}_\pi(y_d)$, of the direct estimator y_d. As the estimated variances, $\widehat{\text{var}}_\pi(y_d)$, $d = 1, \ldots, D$, are obtained in advance from the unit-level survey data, we equate $\sigma_d^2 = \widehat{\text{var}}_\pi(y_d)$ and treat them as known constants when dealing with the Fay–Herriot model. We can also use the *generalized variance function* (GVF) method to introduce the error variances, σ_d^2, in the Fay–Herriot model. For example, we may fit the log-linear model

$$\log \widehat{\text{var}}_\pi(y_d) = b_0 + b_1 y_d + b_2 n_d + \varepsilon_d, \tag{16.6}$$

where the ε_d's are i.i.d. $N(0, \sigma_A^2)$. The selected σ_d^2's are the predicted variance values under the GVF model (16.6), which are calculated by using the formula

$$\sigma_d^2 = \exp\{\hat{\sigma}_A^2/2\} \cdot \exp\{\hat{b}_0 + \hat{b}_1 y_d + \hat{b}_2 n_d\}.$$

The factor $\exp\{\sigma_A^2/2\}$ is the usual bias correction term in the log-linear regression analysis. Note that ignoring the correction term produces underestimation of the true variances when applying the GVF method.

In practice, it may occur that $\widehat{\text{var}}_\pi(y_d) = 0$ for some domains $d \in \mathcal{D}_0 \subset \mathcal{D} = \{1, \ldots, D\}$. This is quite common if the unit-level target variable, y_{dj}, is dichotomic. In those cases, $\log \widehat{\text{var}}_\pi(y_d) = -\infty$ and the GVF method is not applicable. A simple solution is to add a very small quantity (e.g. $\epsilon = 10^{-6}$) to the zero estimated variances. Alternatively, the GVF method can be applied to the subset of domains $\mathcal{D}_1 = \mathcal{D} - \mathcal{D}_0$ where $\widehat{\text{var}}_\pi(y_d) > 0$. For the domains $d \in \mathcal{D}_0$, the GVF estimates can be fixed to

$$\hat{\sigma}_{GVF,d}^2 = \varepsilon \cdot \min\{\hat{\sigma}_{GVF,d}^2 : d \in \mathcal{D}_1\}, \quad 0 < \varepsilon < 1, \ d \in \mathcal{D}_0.$$

The GVF method makes a smoothing of the estimates of the design-based variances. Applied statisticians should try not to do over- or under-smoothing. On the one hand, the GVF model (16.6) does not need to give accurate predictions of $\widehat{\text{var}}_\pi(y_d)$, $d = 1, \ldots, D$. We use this model when we do not trust the design-based estimates $\widehat{\text{var}}_\pi(y_d)$ of error variances σ_d^2, because the direct estimators of variances give bad estimates when domain sample sizes are small. In some problems of regression models, the dependent variable is measured without error. In the case of the GVF model, it is not like that. Therefore, we are not interested in obtaining a model with high predictive capacity. We only want to make a smoothing of the estimates of the design-based variances of the direct estimators of target quantities. We prefer a model having a reasonable interpretation.

On the other hand, a model with low predictive capacity may produce an over-smoothing of the variance estimates. To avoid this problem, it might be convenient to include in the GVF model (16.6) some new explanatory variables, for example, y_d^2, n_d^2, $n_d y_d$, or some component of $\boldsymbol{x}_d$ with p-values greater than 0.05. By doing so, the smoothing problem will be reduced or suppressed. This is a parsimonious solution if the new auxiliary variables are interpretable.

When determining the sampling error variances, σ_d^2, the unit-level dichotomic variables might produce computation problems. Consider a variable y_{dj} that takes the value of 1 if the unit j of domain d has a study property (e.g. "to be unemployed") and takes the value of 0 otherwise. Suppose that the target small area parameters are the domain averages (proportions) $\overline{Y}_d$, $d = 1, \ldots, D$. If the sample size n_{d_0}, of a domain d_0, is small, then it may happen that all the observed values, $y_{d_0 j}$, $j = 1, \ldots, n_{d_0}$, are zero (or one). In this case, the direct estimator is zero (one), and its estimated variance is zero. By applying the GVF method, the estimated final variance, $\sigma_{d_0}^2$, might be different from zero, but it can also be the case that it is zero. In the last situation, the EBLUP coincides with the direct estimator, and the Prasad–Rao estimator of the mean squared error of the EBLUP is zero. This kind of problems could be explained by the fact that the direct estimates $y_d = \hat{\overline{Y}}_d^{dir2}$ (cf. Sect. 2.5), $d = 1, \ldots, D$, do not follow the fitted Fay–Herriot model in some domains. We can take, among others, the following strategies for dealing with this problem:

1. not to publish the estimate "0" ("1"), because it is not credible,
2. to publish the synthetic estimate of the Fay–Herriot model,
3. redefine $\sigma^2_{d_0}$ as the smallest value of all the σ^2_d's divided by 2 (for example).

Using the last strategy, the EBLUP returns to be a linear convex combination between the direct and the synthetic estimators and with a value very close (but not equal) to the direct one.

16.5 Estimation of Model Parameters

We consider four procedures for estimating $\boldsymbol{\beta}$ and σ^2_u: (1) method of moments, (2) Henderson 3 method, (3) maximum likelihood method, and (4) the residual maximum likelihood method.

16.5.1 Prasad–Rao Estimator

By using the method of moments, Prasad and Rao (1990) introduced an unbiased estimator of σ^2_u. The Prasad–Rao estimator is

$$\hat{\sigma}^2_u = \frac{1}{D-p}\left[\sum_{d=1}^{D}\tilde{u}^2_d - \sum_{d=1}^{D}\sigma^2_d\Big(1 - \boldsymbol{x}_d\Big(\sum_{d=1}^{D}\boldsymbol{x}'_d\boldsymbol{x}_d\Big)^{-1}\boldsymbol{x}'_d\Big)\right], \tag{16.7}$$

where $\tilde{u}_d = y_d - \boldsymbol{x}_d\tilde{\boldsymbol{\beta}}$ and $\tilde{\boldsymbol{\beta}} = (\boldsymbol{X}'\boldsymbol{X})^{-1}\boldsymbol{X}'\boldsymbol{y} = \big(\sum_{d=1}^{D}\boldsymbol{x}'_d\boldsymbol{x}_d\big)^{-1}\big(\sum_{d=1}^{D}\boldsymbol{x}'_d y_d\big)$.

It may happen that $\hat{\sigma}^2_u$ takes a negative value, but $\mathrm{P}(\hat{\sigma}^2_u \le 0)$ tends to 0 when $D \to \infty$. If $\hat{\sigma}^2_u$ is negative, we equate it to zero, and we define

$$\tilde{\sigma}^2_u = \max\left\{\hat{\sigma}^2_u,\, 0\right\}. \tag{16.8}$$

16.5.2 Henderson 3 Estimator

For the linear mixed model

$$\boldsymbol{y} = \boldsymbol{X}\boldsymbol{\beta} + \boldsymbol{Z}\boldsymbol{u} + \boldsymbol{e},$$

where $\boldsymbol{u} \sim N(0, \sigma_u^2 \boldsymbol{I}_D)$ and $\boldsymbol{e} \sim N(0, \sigma_e^2 \boldsymbol{W}^{-1})$ are $D \times 1$ independent random vectors, the Henderson 3 method gives unbiased estimators of σ_e^2 and σ_u^2 from the expectations (cf. (6.69) and (6.70))

$$E[SSE(\boldsymbol{\beta}, \boldsymbol{u})] = \sigma_e^2[D - \mathrm{r}(\boldsymbol{X}, \boldsymbol{Z})],$$

$$E[SSR(\boldsymbol{u}|\boldsymbol{\beta})] = \mathrm{tr}\left\{\boldsymbol{Z}'\boldsymbol{W}[\boldsymbol{W}^{-1} - \boldsymbol{X}(\boldsymbol{X}'\boldsymbol{W}\boldsymbol{X})^{-1}\boldsymbol{X}']\boldsymbol{W}\boldsymbol{Z}\right\}\sigma_u^2 + \sigma_e^2[\mathrm{r}(\boldsymbol{X}, \boldsymbol{Z}) - \mathrm{r}(\boldsymbol{X})],$$

where $SSR(\boldsymbol{u}|\boldsymbol{\beta}) = SSE(\boldsymbol{\beta}) - SSE(\boldsymbol{\beta}, \boldsymbol{u})$ and $SSE(\boldsymbol{\beta})$ and $SSE(\boldsymbol{\beta}, \boldsymbol{u})$ are the sums of squared residuals from the models with fixed effects $\boldsymbol{y} = \boldsymbol{X}\boldsymbol{\beta} + \boldsymbol{e}$ and with random effects $\boldsymbol{y} = \boldsymbol{X}\boldsymbol{\beta} + \boldsymbol{Z}\boldsymbol{u} + \boldsymbol{e}$, respectively. It holds that

$$\begin{aligned} E[SSE(\boldsymbol{\beta})] &= E[SSR(\boldsymbol{u}|\boldsymbol{\beta})] + E[SSE(\boldsymbol{\beta}, \boldsymbol{u})] \\ &= \mathrm{tr}\left\{\boldsymbol{Z}'\boldsymbol{W}[\boldsymbol{W}^{-1} - \boldsymbol{X}(\boldsymbol{X}'\boldsymbol{W}\boldsymbol{X})^{-1}\boldsymbol{X}']\boldsymbol{W}\boldsymbol{Z}\right\}\sigma_u^2 + \sigma_e^2[D - \mathrm{r}(\boldsymbol{X})]. \end{aligned}$$

The Henderson 3 moment-based unbiased estimator of σ_{uH}^2 is (cf. (6.72) or (7.23))

$$\hat{\sigma}_{uH}^2 = \frac{SSE(\boldsymbol{\beta}) - \sigma_e^2[D - \mathrm{r}(\boldsymbol{X})]}{\mathrm{tr}\left\{\boldsymbol{Z}'\boldsymbol{W}[\boldsymbol{W}^{-1} - \boldsymbol{X}(\boldsymbol{X}'\boldsymbol{W}\boldsymbol{X})^{-1}\boldsymbol{X}']\boldsymbol{W}\boldsymbol{Z}\right\}}.$$

We have that $SSE(\boldsymbol{\beta}) = \boldsymbol{y}'\boldsymbol{P}_2\boldsymbol{y}$, where

$$\boldsymbol{P}_2 = [\boldsymbol{I} - \boldsymbol{X}(\boldsymbol{X}'\boldsymbol{W}\boldsymbol{X})^{-1}\boldsymbol{X}'\boldsymbol{W}]'\boldsymbol{W}[\boldsymbol{I} - \boldsymbol{X}(\boldsymbol{X}'\boldsymbol{W}\boldsymbol{X})^{-1}\boldsymbol{X}'\boldsymbol{W}] = \boldsymbol{W} - \boldsymbol{W}\boldsymbol{X}(\boldsymbol{X}'\boldsymbol{W}\boldsymbol{X})^{-1}\boldsymbol{X}'\boldsymbol{W}.$$

For the model (16.2), we have $\sigma_e^2 = 1$, $\boldsymbol{W} = \boldsymbol{V}_e^{-1}$, $\boldsymbol{Z} = \boldsymbol{I}_D$, and $\mathrm{r}(\boldsymbol{X}) = p$. Therefore,

$$\hat{\sigma}_{uH}^2 = \frac{\boldsymbol{y}'\boldsymbol{P}_2\boldsymbol{y} - (D - p)}{\mathrm{tr}\{\boldsymbol{P}_2\}},$$

where

$$\boldsymbol{P}_2 = \boldsymbol{V}_e^{-1} - \boldsymbol{V}_e^{-1}\boldsymbol{X}\boldsymbol{Q}\boldsymbol{X}'\boldsymbol{V}_e^{-1}, \quad \boldsymbol{Q} = (\boldsymbol{X}'\boldsymbol{V}_e^{-1}\boldsymbol{X})^{-1} = \Big(\sum_{d=1}^{D} \sigma_d^{-2}\boldsymbol{x}_d'\boldsymbol{x}_d\Big)^{-1}.$$

As $\boldsymbol{P}_2$ can be written in the form

$$\boldsymbol{P}_2 = \underset{1\le d\le D}{\mathrm{diag}}(\sigma_d^{-2}) - \underset{1\le d\le D}{\mathrm{diag}}(\sigma_d^{-2})\boldsymbol{X}\boldsymbol{Q}\boldsymbol{X}' \underset{1\le d\le D}{\mathrm{diag}}(\sigma_d^{-2}),$$

the denominator of $\hat{\sigma}_{uH}^2$ is

$$\mathrm{tr}\{\boldsymbol{P}_2\} = \sum_{d=1}^{D} \sigma_d^{-2} - \sum_{d=1}^{D} \sigma_d^{-4}\boldsymbol{x}_d\boldsymbol{Q}\boldsymbol{x}_d'.$$

This is to say,

$$\hat{\sigma}_{uH}^2 = \frac{\boldsymbol{y}'\boldsymbol{P}_2\boldsymbol{y} - (D-p)}{\sum_{d=1}^D \sigma_d^{-2} - \sum_{d=1}^D \sigma_d^{-4}\boldsymbol{x}_d\boldsymbol{Q}\boldsymbol{x}_d'}.$$

An estimator of $\boldsymbol{\beta}$ and a predictor of $\boldsymbol{u}$ can be obtained by applying the formulas (16.4).

For calculating the variance of $\hat{\sigma}_{uH}^2$, we need $\text{var}(\boldsymbol{y}'\boldsymbol{P}_2\boldsymbol{y})$. Using the formula (A.4), we obtain

$$\begin{aligned}\text{var}(\boldsymbol{y}'\boldsymbol{P}_2\boldsymbol{y}) &= 2\text{tr}\left\{(\boldsymbol{P}_2\text{var}(\boldsymbol{y}))^2\right\} + 4E[\boldsymbol{y}]'\boldsymbol{P}_2\text{var}(\boldsymbol{y})\boldsymbol{P}_2E[\boldsymbol{y}] \\ &= 2\text{tr}\left\{[\boldsymbol{P}_2(\sigma_u^2\boldsymbol{I}_D + \boldsymbol{V}_e)]^2\right\} + 4\boldsymbol{\beta}'\boldsymbol{X}'\boldsymbol{P}_2\text{var}(\boldsymbol{y})\boldsymbol{P}_2\boldsymbol{X}\boldsymbol{\beta}.\end{aligned}$$

As $\boldsymbol{P}_2\boldsymbol{X} = \boldsymbol{V}_e^{-1}\boldsymbol{X} - \boldsymbol{V}_e^{-1}\boldsymbol{X}\boldsymbol{Q}\boldsymbol{X}'\boldsymbol{V}_e^{-1}\boldsymbol{X} = 0$,

$$\text{var}(\boldsymbol{y}'\boldsymbol{P}_2\boldsymbol{y}) = 2\sigma_u^4\text{tr}\{\boldsymbol{P}_2^2\} + 4\sigma_u^2\text{tr}\{\boldsymbol{P}_2^2\boldsymbol{V}_e\} + 2\text{tr}\{\boldsymbol{P}_2\boldsymbol{V}_e\boldsymbol{P}_2\boldsymbol{V}_e\}.$$

Observe that

$$\begin{aligned}\boldsymbol{P}_2\boldsymbol{V}_e &= \boldsymbol{I}_D - \boldsymbol{V}_e^{-1}\boldsymbol{X}\boldsymbol{Q}\boldsymbol{X}', \\ (\boldsymbol{P}_2\boldsymbol{V}_e)^2 &= \boldsymbol{I}_D - \boldsymbol{V}_e^{-1}\boldsymbol{X}\boldsymbol{Q}\boldsymbol{X}' - \boldsymbol{V}_e^{-1}\boldsymbol{X}\boldsymbol{Q}\boldsymbol{X}' + \boldsymbol{V}_e^{-1}\boldsymbol{X}\boldsymbol{Q}\boldsymbol{X}'\boldsymbol{V}_e^{-1}\boldsymbol{X}\boldsymbol{Q}\boldsymbol{X}' \\ &= \boldsymbol{I}_D - \boldsymbol{V}_e^{-1}\boldsymbol{X}\boldsymbol{Q}\boldsymbol{X}' = \boldsymbol{P}_2\boldsymbol{V}_e.\end{aligned}$$

Therefore, $\boldsymbol{P}_2\boldsymbol{V}_e$ is idempotent and $\boldsymbol{X}'\boldsymbol{P}_2\boldsymbol{V}_e = 0$. Similarly, $\boldsymbol{V}_e\boldsymbol{P}_2$ is idempotent and $\boldsymbol{V}_e\boldsymbol{P}_2\boldsymbol{X} = 0$. In what follows, we calculate the traces appearing in the formula of $\text{var}(\boldsymbol{y}'\boldsymbol{P}_2\boldsymbol{y})$.

The trace in the third summand of $\text{var}(\boldsymbol{y}'\boldsymbol{P}_2\boldsymbol{y})$ is

$$\text{tr}\{(\boldsymbol{P}_2\boldsymbol{V}_e)^2\} = \text{tr}\{\boldsymbol{P}_2\boldsymbol{V}_e\} = D - \text{tr}\{\boldsymbol{V}_e^{-1}\boldsymbol{X}\boldsymbol{Q}\boldsymbol{X}'\} = D - \sum_{d=1}^D \sigma_d^{-2}\boldsymbol{x}_d\boldsymbol{Q}\boldsymbol{x}_d'. \quad (16.9)$$

As $\boldsymbol{P}_2^2$ takes the form

$$\begin{aligned}\boldsymbol{P}_2^2 = &\underset{1\le d\le D}{\text{diag}}(\sigma_d^{-4}) - \underset{1\le d\le D}{\text{diag}}(\sigma_d^{-4})\boldsymbol{X}\boldsymbol{Q}\boldsymbol{X}'\underset{1\le d\le D}{\text{diag}}(\sigma_d^{-2}) \\ &- \underset{1\le d\le D}{\text{diag}}(\sigma_d^{-2})\boldsymbol{X}\boldsymbol{Q}\boldsymbol{X}'\underset{1\le d\le D}{\text{diag}}(\sigma_d^{-4}) \\ &+ \underset{1\le d\le D}{\text{diag}}(\sigma_d^{-2})\boldsymbol{X}\boldsymbol{Q}\boldsymbol{X}'\underset{1\le d\le D}{\text{diag}}(\sigma_d^{-4})\boldsymbol{X}\boldsymbol{Q}\boldsymbol{X}'\underset{1\le d\le D}{\text{diag}}(\sigma_d^{-2}),\end{aligned}$$

the trace in the first summand of $\text{var}(\boldsymbol{y}'\boldsymbol{P}_2\boldsymbol{y})$ is

$$\text{tr}\{\boldsymbol{P}_2^2\} = \sum_{d=1}^{D}\sigma_d^{-4} - 2\sum_{d=1}^{D}\sigma_d^{-6}\boldsymbol{x}_d\boldsymbol{Q}\boldsymbol{x}_d' + \sum_{d=1}^{D}\sigma_d^{-4}\boldsymbol{x}_d\boldsymbol{Q}\left(\sum_{d=1}^{D}\sigma_d^{-4}\boldsymbol{x}_d'\boldsymbol{x}_d\right)\boldsymbol{Q}\boldsymbol{x}_d'.$$

Furthermore, we have

$$\begin{aligned}\boldsymbol{P}_2\boldsymbol{V}_e\boldsymbol{P}_2 &= \left(\boldsymbol{V}_e^{-1} - \boldsymbol{V}_e^{-1}\boldsymbol{X}\boldsymbol{Q}\boldsymbol{X}'\boldsymbol{V}_e^{-1}\right)\boldsymbol{V}_e\left(\boldsymbol{V}_e^{-1} - \boldsymbol{V}_e^{-1}\boldsymbol{X}\boldsymbol{Q}\boldsymbol{X}'\boldsymbol{V}_e^{-1}\right)\\ &= \left(\boldsymbol{I}_D - \boldsymbol{V}_e^{-1}\boldsymbol{X}\boldsymbol{Q}\boldsymbol{X}'\right)\left(\boldsymbol{V}_e^{-1} - \boldsymbol{V}_e^{-1}\boldsymbol{X}\boldsymbol{Q}\boldsymbol{X}'\boldsymbol{V}_e^{-1}\right)\\ &= \boldsymbol{V}_e^{-1} - \boldsymbol{V}_e^{-1}\boldsymbol{X}\boldsymbol{Q}\boldsymbol{X}'\boldsymbol{V}_e^{-1} - \boldsymbol{V}_e^{-1}\boldsymbol{X}\boldsymbol{Q}\boldsymbol{X}'\boldsymbol{V}_e^{-1} + \boldsymbol{V}_e^{-1}\boldsymbol{X}\boldsymbol{Q}\boldsymbol{X}'\boldsymbol{V}_e^{-1}\boldsymbol{X}\boldsymbol{Q}\boldsymbol{X}'\boldsymbol{V}_e^{-1}\\ &= \boldsymbol{V}_e^{-1} - \boldsymbol{V}_e^{-1}\boldsymbol{X}\boldsymbol{Q}\boldsymbol{X}'\boldsymbol{V}_e^{-1} = \boldsymbol{P}_2.\end{aligned}$$

The trace in the second summand is

$$\text{tr}\{\boldsymbol{P}_2\boldsymbol{V}_e\boldsymbol{P}_2\} = \text{tr}\{\boldsymbol{P}_2\} = \sum_{d=1}^{D}\sigma_d^{-2} - \sum_{d=1}^{D}\sigma_d^{-4}\boldsymbol{x}_d\boldsymbol{Q}\boldsymbol{x}_d'.$$

Therefore,

$$\begin{aligned}\text{var}(\boldsymbol{y}'\boldsymbol{P}_2\boldsymbol{y}) &= 2\sigma_u^4\left[\sum_{d=1}^{D}\sigma_d^{-4} - 2\sum_{d=1}^{D}\sigma_d^{-6}\boldsymbol{x}_d\boldsymbol{Q}\boldsymbol{x}_d' + \sum_{d=1}^{D}\sigma_d^{-4}\boldsymbol{x}_d\boldsymbol{Q}\left(\sum_{d=1}^{D}\sigma_d^{-4}\boldsymbol{x}_d'\boldsymbol{x}_d\right)\boldsymbol{Q}\boldsymbol{x}_d'\right]\\ &+ 4\sigma_u^2\left[\sum_{d=1}^{D}\sigma_d^{-2} - \sum_{d=1}^{D}\sigma_d^{-4}\boldsymbol{x}_d\boldsymbol{Q}\boldsymbol{x}_d'\right] + 2\left[D - \sum_{d=1}^{D}\sigma_d^{-2}\boldsymbol{x}_d\boldsymbol{Q}\boldsymbol{x}_d'\right].\end{aligned}$$

The variance of $\hat{\sigma}_{uH}^2$ is

$$\text{var}(\hat{\sigma}_{uH}^2) = \frac{\text{var}(\boldsymbol{y}'\boldsymbol{P}_2\boldsymbol{y})}{\left[\sum_{d=1}^{D}\sigma_d^{-2} - \sum_{d=1}^{D}\sigma_d^{-4}\boldsymbol{x}_d\boldsymbol{Q}\boldsymbol{x}_d'\right]^2}. \tag{16.10}$$

16.5.3 Maximum Likelihood Method

In what follows, we particularize the results of Sect. 6.4 to the case $m = 1, n = D$, $q_1 = D$, $\sigma_1^2 = \sigma_u^2$, and $\boldsymbol{\Sigma}_{u_1} = \boldsymbol{I}_D$. It holds that $\boldsymbol{y} \sim N(\boldsymbol{X}\boldsymbol{\beta}, \boldsymbol{V})$, with covariance matrix $\boldsymbol{V} = \text{diag}_{1\leq d\leq D}(\sigma_u^2 + \sigma_d^2)$. The log-likelihood is

$$\ell(\sigma_u^2, \boldsymbol{\beta}; \boldsymbol{y}) = -\frac{D}{2}\log 2\pi - \frac{1}{2}\log|\boldsymbol{V}| - \frac{1}{2}(\boldsymbol{y} - \boldsymbol{X\beta})'\boldsymbol{V}^{-1}(\boldsymbol{y} - \boldsymbol{X\beta}).$$

The partial derivatives of the log-likelihood are (cf. (6.15) and (6.17))

$$S_\beta = \boldsymbol{X}'\boldsymbol{V}^{-1}(\boldsymbol{y} - \boldsymbol{X\beta}) = \sum_{d=1}^{D} \boldsymbol{x}'_d \frac{1}{\sigma_u^2 + \sigma_d^2}(y_d - \boldsymbol{x}_d\boldsymbol{\beta}),$$

$$S_{\sigma_u^2} = -\frac{1}{2}\text{tr}(\boldsymbol{V}^{-1}\boldsymbol{G}_u) + \frac{1}{2}(\boldsymbol{y} - \boldsymbol{X\beta})'\boldsymbol{V}^{-1}\boldsymbol{G}_u\boldsymbol{V}^{-1}(\boldsymbol{y} - \boldsymbol{X\beta})$$

$$= -\frac{1}{2}\sum_{d=1}^{D}\frac{1}{\sigma_u^2 + \sigma_d^2} + \frac{1}{2}\sum_{d=1}^{D}\frac{1}{(\sigma_u^2 + \sigma_d^2)^2}(y_d - \boldsymbol{x}_d\boldsymbol{\beta})^2,$$

where $\boldsymbol{G}_u = \partial \boldsymbol{V}/\partial\sigma_u^2 = \boldsymbol{I}_D$. To calculate the second order partial derivatives, we use the formulas (6.18)–(6.20) to obtain

$$\boldsymbol{H}_{\beta\beta} = -\boldsymbol{X}'\boldsymbol{V}^{-1}\boldsymbol{X}, \quad \boldsymbol{H}_{\beta\sigma_u^2} = -\boldsymbol{X}'\boldsymbol{V}^{-2}(\boldsymbol{y} - \boldsymbol{X\beta}),$$

$$H_{\sigma_u^2\sigma_u^2} = \frac{1}{2}\text{tr}(\boldsymbol{V}^{-2}) - (\boldsymbol{y} - \boldsymbol{X\beta})'\boldsymbol{V}^{-3}(\boldsymbol{y} - \boldsymbol{X\beta}).$$

The components of the Fisher information matrix are (cf. (6.21))

$$\boldsymbol{F}_{\beta\beta} = \boldsymbol{X}'\boldsymbol{V}^{-1}\boldsymbol{X} = \sum_{d=1}^{D}\frac{1}{\sigma_u^2 + \sigma_d^2}\boldsymbol{x}'_d\boldsymbol{x}_d, \quad \boldsymbol{F}_{\beta\sigma_u^2} = \boldsymbol{F}_{\sigma_u^2\beta} = \boldsymbol{0},$$

$$F_{\sigma_u^2\sigma_u^2} = -\frac{1}{2}\text{tr}(\boldsymbol{V}^{-2}) + \text{tr}(\boldsymbol{V}^{-3}\boldsymbol{V}) = \frac{1}{2}\text{tr}(\boldsymbol{V}^{-2}) = \frac{1}{2}\sum_{d=1}^{D}\frac{1}{(\sigma_u^2 + \sigma_d^2)^2}.$$

At iteration i, the updating formulas of the Fisher-scoring algorithm are

$$\sigma_u^{2(i+1)} = \sigma_u^{2(i)} + F^{-1}_{\sigma_u^{2(i)}\sigma_u^{2(i)}} S_{\sigma_u^{2(i)}}, \quad \boldsymbol{\beta}^{(i+1)} = \boldsymbol{\beta}^{(i)} + \boldsymbol{F}^{-1}_{\beta^{(i)}\beta^{(i)}}\boldsymbol{S}_{\beta^{(i)}}.$$

Remark 16.1 Let us define

$$\boldsymbol{T} = (\boldsymbol{V}_e^{-1} + \sigma_u^{-2}\boldsymbol{I}_D)^{-1}.$$

By applying the inversion formula

$$(A + BCD)^{-1} = A^{-1} - A^{-1}B(C^{-1} + DA^{-1}B)^{-1}DA^{-1},$$

with $A = \sigma_u^{-2}\boldsymbol{I}_D$, $B = \boldsymbol{I}_D$, $C = \boldsymbol{V}_e^{-1} = \text{diag}_{1\le d\le D}(\sigma_d^{-2})$, and $D = \boldsymbol{I}_D$, we get

$$\boldsymbol{T} = \sigma_u^2\boldsymbol{I}_D - \sigma_u^4\boldsymbol{V}^{-1} \quad \text{and} \quad \boldsymbol{V}^{-1} = \frac{\sigma_u^2\boldsymbol{I}_D - \boldsymbol{T}}{\sigma_u^4}.$$

Therefore, an alternative formula for $F_{\sigma_u^2\sigma_u^2}$ is

$$F_{\sigma_u^2\sigma_u^2} = \frac{1}{2\sigma_u^8}\text{tr}\left((\sigma_u^2\boldsymbol{I}_d - \boldsymbol{T})^2\right) = \frac{1}{2\sigma_u^4}\left(D - \frac{2}{\sigma_u^2}\text{tr}(\boldsymbol{T}) + \frac{1}{\sigma_u^4}\text{tr}(\boldsymbol{T}^2)\right).$$

Remark 16.2 As mentioned in Sect. 6.4, under regularity assumptions, the ML estimators $\hat{\boldsymbol{\beta}}$ and $\hat{\sigma}_u^2$ are consistent and asymptotically normal with asymptotic distributions

$$N_p(\boldsymbol{\beta}, (\boldsymbol{X}'\boldsymbol{V}^{-1}\boldsymbol{X})^{-1}) \quad \text{and} \quad N(\sigma_u^2, F_{\sigma_u^2\sigma_u^2}^{-1}),$$

respectively.

16.5.4 Residual Maximum Likelihood Method

In what follows, we particularize the results of Sect. 6.5.3 to the case $m = 1, n = D$, $q_1 = D$, $\varphi_1 = \sigma_u^2, \sigma^2 = 1$, $\boldsymbol{\Sigma}_{u_1} = I_D$, and $\sigma^2\boldsymbol{\Sigma} = \boldsymbol{V}$. The REML log-likelihood is (cf. (6.41))

$$\ell_R(\sigma_u^2; \boldsymbol{y}) = -\frac{D-p}{2}\log 2\pi + \frac{1}{2}\log|\boldsymbol{X}'\boldsymbol{X}| - \frac{1}{2}\log|\boldsymbol{V}| - \frac{1}{2}\log|\boldsymbol{X}'\boldsymbol{V}^{-1}\boldsymbol{X}| - \frac{1}{2}\boldsymbol{y}'\boldsymbol{P}\boldsymbol{y},$$

where $\boldsymbol{P} = \boldsymbol{V}^{-1} - \boldsymbol{V}^{-1}\boldsymbol{X}(\boldsymbol{X}'\boldsymbol{V}^{-1}\boldsymbol{X})^{-1}\boldsymbol{X}'\boldsymbol{V}^{-1}$, $\boldsymbol{P}\boldsymbol{X} = \boldsymbol{0}$, and $\boldsymbol{P}\boldsymbol{V}\boldsymbol{P} = \boldsymbol{P}$. Further, we have

$$\frac{\partial \boldsymbol{V}}{\partial\sigma_u^2} = \boldsymbol{I}_D,\; \frac{\partial \boldsymbol{V}^{-1}}{\partial\sigma_u^2} = -\boldsymbol{V}^{-1}\frac{\partial \boldsymbol{V}}{\partial\sigma_u^2}\boldsymbol{V}^{-1} = -\boldsymbol{V}^{-2},\; \frac{\partial \log|\boldsymbol{V}|}{\partial\sigma_u^2} = \text{tr}\left\{\boldsymbol{V}^{-1}\frac{\partial \boldsymbol{V}}{\partial\sigma_u^2}\right\} = \text{tr}\{\boldsymbol{V}^{-1}\}.$$

By taking derivatives of ℓ_R with respect to σ_u^2, we get

$$\begin{aligned}\frac{\partial \ell_R}{\partial\sigma_u^2} &= -\frac{1}{2}\,\text{tr}\{\boldsymbol{V}^{-1}\} + \frac{1}{2}\,\text{tr}\{(\boldsymbol{X}'\boldsymbol{V}^{-1}\boldsymbol{X})^{-1}\boldsymbol{X}'\boldsymbol{V}^{-2}\boldsymbol{X}\} - \frac{1}{2}\,\boldsymbol{y}'\frac{\partial \boldsymbol{P}}{\partial\sigma_u^2}\boldsymbol{y}\\ &= -\frac{1}{2}\,\text{tr}\{\boldsymbol{V}^{-1}\} + \frac{1}{2}\,\text{tr}\{\boldsymbol{V}^{-1}\boldsymbol{X}(\boldsymbol{X}'\boldsymbol{V}^{-1}\boldsymbol{X})^{-1}\boldsymbol{X}'\boldsymbol{V}^{-1}\} - \frac{1}{2}\,\boldsymbol{y}'\frac{\partial \boldsymbol{P}}{\partial\sigma_u^2}\boldsymbol{y}\\ &= -\frac{1}{2}\,\text{tr}\{\boldsymbol{P}\} - \frac{1}{2}\,\boldsymbol{y}'\frac{\partial \boldsymbol{P}}{\partial\sigma_u^2}\boldsymbol{y}.\end{aligned}$$

Let us define $G = V^{-1}X(X'V^{-1}X)^{-1}$, so that $P = (I_D - GX')V^{-1}$. We have

$$\begin{aligned}\frac{\partial P}{\partial \sigma_u^2} &= -V^{-2} + V^{-2}X(X'V^{-1}X)^{-1}X'V^{-1} \\ &- V^{-1}X(X'V^{-1}X)^{-1}X'V^{-2}X(X'V^{-1}X)^{-1}X'V^{-1} + V^{-1}X(X'V^{-1}X)^{-1}X'V^{-2} \\ &= -V^{-2} + V^{-2}XG' - GX'V^{-2}XG' + GX'V^{-2} \\ &= -(I_D - GX')V^{-2}(I_D - GX')' = -PP' = -P^2.\end{aligned}$$

Therefore,

$$S_{\sigma_u^2} = \frac{\partial \ell_R}{\partial \sigma_u^2} = -\frac{1}{2}\operatorname{tr}\{P\} + \frac{1}{2}y'P^2y.$$

The second order derivative of the log-likelihood is

$$\frac{\partial^2 \ell_R}{\partial \sigma_u^2 \partial \sigma_u^2} = \frac{1}{2}\operatorname{tr}(P^2) - y'P^3y.$$

By changing the sign, taking expectations and using formula (A.3), we obtain the Fisher amount of information associated to σ_u^2. As $PVP = P$ and $PX = 0$, we have

$$F_{\sigma_u^2} = -\frac{1}{2}\operatorname{tr}(P^2) + \operatorname{tr}(P^3V) + \beta'X'P^3X\beta = \frac{1}{2}\operatorname{tr}(P^2).$$

We now describe three possibilities of how to estimate the model parameters β and σ_u^2.

Method 1 At iteration i, the updating formula of the Fisher-scoring algorithm is

$$\sigma_u^{2(i+1)} = \sigma_u^{2(i)} + F^{-1}_{\sigma_u^{2(i)}} S_{\sigma_u^{2(i)}}.$$

After convergence, we can estimate β with the REML estimator

$$\hat{\beta} = (X'\hat{V}^{-1}X)^{-1}X'\hat{V}^{-1}y, \quad \hat{V} = \operatorname{diag}(\hat{\sigma}_u^2 + \sigma_1^2, \ldots, \hat{\sigma}_u^2 + \sigma_D^2).$$

Method 2 We note that

$$Py = V^{-1}y - V^{-1}X(X'V^{-1}X)^{-1}X'V^{-1}y = V^{-1}(y - X\tilde{\beta}).$$

Therefore,

$$y'\frac{\partial P}{\partial \sigma_u^2}y = -y'P^2y = -(y - X\tilde{\beta})'V^{-2}(y - X\tilde{\beta}) = -\sum_{d=1}^{D} \frac{1}{(\sigma_u^2 + \sigma_d^2)^2}(y_d - x_d\tilde{\beta})^2.$$

We also have

$$\boldsymbol{P}=\frac{1}{\sigma_u^2}\Big(\boldsymbol{I}_D-\frac{1}{\sigma_u^2}\boldsymbol{R}\Big),\quad \operatorname{tr}(\boldsymbol{P})=\frac{1}{\sigma_u^2}\Big[D-\frac{1}{\sigma_u^2}\operatorname{tr}(\boldsymbol{R})\Big],$$

where

$$\boldsymbol{R}=\boldsymbol{T}+\boldsymbol{M},\quad \boldsymbol{M}=\boldsymbol{T}\boldsymbol{V}_e^{-1}\boldsymbol{X}(\boldsymbol{X}'\boldsymbol{V}^{-1}\boldsymbol{X})^{-1}\boldsymbol{X}'\boldsymbol{V}_e^{-1}\boldsymbol{T},$$

$$\boldsymbol{T}=\Big(\boldsymbol{V}_e^{-1}+\sigma_u^{-2}\boldsymbol{I}_D\Big)^{-1}=\operatorname{diag}_{1\le d\le D}\left(\frac{\sigma_u^2\sigma_d^2}{\sigma_u^2+\sigma_d^2}\right).$$

The first order derivative of the log-likelihood can be written in the form

$$\frac{\partial \ell_R}{\partial \sigma_u^2}=-\frac{1}{2}\operatorname{tr}(\boldsymbol{P})-\frac{1}{2}\boldsymbol{y}'\frac{\partial \boldsymbol{P}}{\partial \sigma_u^2}\boldsymbol{y}=-\frac{1}{2\sigma_u^2}\left[D-\frac{1}{\sigma_u^2}\operatorname{tr}(\boldsymbol{R})\right]+\frac{1}{2}\sum_{d=1}^{D}\frac{1}{(\sigma_u^2+\sigma_d^2)^2}(y_d-\boldsymbol{x}_d\tilde{\boldsymbol{\beta}})^2.$$

With the new notation, the Fisher amount of information associated to σ_u^2 is

$$F_{\sigma_u^2}=\frac{1}{2}\operatorname{tr}(\boldsymbol{P}^2)=\frac{1}{2\sigma_u^4}\left[D-\frac{2}{\sigma_u^2}\operatorname{tr}(\boldsymbol{R})+\frac{1}{\sigma_u^4}\operatorname{tr}(\boldsymbol{R}^2)\right].$$

The REML estimators may be obtained by applying the following Fisher-scoring algorithm that updates the variance and the regression parameters.

1. Set the seeds $\hat{\sigma}_{u,0}^2=\tilde{\sigma}_u^2=\max\{\hat{\sigma}_u^2,0\}$ and $\tilde{\boldsymbol{\beta}}_0=\tilde{\boldsymbol{\beta}}$, where $\hat{\sigma}_u^2$ and $\tilde{\boldsymbol{\beta}}$ are the moment estimators given by (16.7).
2. For $i=1,2,\ldots$, do

$$\tilde{\boldsymbol{\beta}}_i=\left(\sum_{d=1}^{D}\frac{\boldsymbol{x}_d'\boldsymbol{x}_d}{\hat{\sigma}_{u,i-1}^2+\sigma_d^2}\right)^{-1}\left(\sum_{d=1}^{D}\frac{\boldsymbol{x}_d'y_d}{\hat{\sigma}_{u,i-1}^2+\sigma_d^2}\right),\quad \hat{\sigma}_{u,i}^2=\hat{\sigma}_{u,i-1}^2+F_{i-1}^{-1}S_{i-1},$$

where

$$S_i=-\frac{1}{2\hat{\sigma}_{u,i}^2}\left(D-\frac{\operatorname{tr}(\hat{\boldsymbol{R}}_i)}{\hat{\sigma}_{u,i}^2}\right)+\frac{1}{2}\sum_{d=1}^{D}\frac{1}{(\hat{\sigma}_{u,i}^2+\sigma_d^2)^2}(y_d-\boldsymbol{x}_d\tilde{\boldsymbol{\beta}}_i)^2,$$

$$F_i=\frac{1}{2\hat{\sigma}_{u,i}^4}\left(D-\frac{2}{\hat{\sigma}_{u,i}^2}\operatorname{tr}\{\hat{\boldsymbol{R}}_i\}+\frac{1}{\hat{\sigma}_{u,i}^4}\operatorname{tr}\{\hat{\boldsymbol{R}}_i^2\}\right),$$

$$\operatorname{tr}\{\hat{\boldsymbol{R}}_i\}=\operatorname{tr}(\hat{\boldsymbol{T}}_i)+\operatorname{tr}(\hat{\boldsymbol{M}}_i),\quad \operatorname{tr}\{\hat{\boldsymbol{R}}_i^2\}=\operatorname{tr}(\hat{\boldsymbol{T}}_i^2)+2\operatorname{tr}(\hat{\boldsymbol{T}}_i\hat{\boldsymbol{M}}_i)+\operatorname{tr}(\hat{\boldsymbol{M}}_i^2),$$

$$\operatorname{tr}(\hat{\boldsymbol{T}}_i)=\sum_{d=1}^{D}\frac{\hat{\sigma}_{u,i}^2\sigma_d^2}{\hat{\sigma}_{u,i}^2+\sigma_d^2},\quad \operatorname{tr}(\hat{\boldsymbol{T}}_i^2)=\sum_{d=1}^{D}\frac{\hat{\sigma}_{u,i}^4\sigma_d^4}{(\hat{\sigma}_{u,i}^2+\sigma_d^2)^2},$$

$$\mathrm{tr}(\hat{\boldsymbol{M}}_i) = \mathrm{tr}\left[\left(\sum_{d=1}^{D} \frac{\hat{\sigma}_{u,i}^4 \boldsymbol{x}_d' \boldsymbol{x}_d}{(\hat{\sigma}_{u,i}^2 + \sigma_d^2)^2}\right)\left(\sum_{d=1}^{D} \frac{\boldsymbol{x}_d' \boldsymbol{x}_d}{\hat{\sigma}_{u,i}^2 + \sigma_d^2}\right)^{-1}\right],$$

$$\mathrm{tr}(\hat{\boldsymbol{T}}_i \hat{\boldsymbol{M}}_i) = \mathrm{tr}\left[\left(\sum_{d=1}^{D} \frac{\hat{\sigma}_{u,i}^6 \sigma_d^2 \boldsymbol{x}_d' \boldsymbol{x}_d}{(\hat{\sigma}_{u,i}^2 + \sigma_d^2)^3}\right)\left(\sum_{d=1}^{D} \frac{\boldsymbol{x}_d' \boldsymbol{x}_d}{\hat{\sigma}_{u,i}^2 + \sigma_d^2}\right)^{-1}\right],$$

$$\mathrm{tr}(\hat{\boldsymbol{M}}_i^2) = \mathrm{tr}\left[\left\{\left(\sum_{d=1}^{D} \frac{\hat{\sigma}_{u,i}^4 \boldsymbol{x}_d' \boldsymbol{x}_d}{(\hat{\sigma}_{u,i}^2 + \sigma_d^2)^2}\right)\left(\sum_{d=1}^{D} \frac{\boldsymbol{x}_d' \boldsymbol{x}_d}{\hat{\sigma}_{u,i}^2 + \sigma_d^2}\right)^{-1}\right\}^2\right].$$

3. Stop if $|\hat{\sigma}_{u,i}^2 - \hat{\sigma}_{u,i-1}^2| < \varepsilon_1$ and $\left[(\tilde{\boldsymbol{\beta}}_i - \tilde{\boldsymbol{\beta}}_{i-1})'(\tilde{\boldsymbol{\beta}}_i - \tilde{\boldsymbol{\beta}}_{i-1})\right]^{1/2} < \varepsilon_2$. Output: $\hat{\boldsymbol{\beta}}_{ML} = \tilde{\boldsymbol{\beta}}_i$ and $\hat{\sigma}_{u,ML}^2 = \hat{\sigma}_{u,i}^2$.

Method 3 Due to the derivations around formula (6.56), the following algorithm can also be used:

1. Set the seeds $\hat{\sigma}_{u,0}^2 = \tilde{\sigma}_u^2 = \max\{\hat{\sigma}_u^2, 0\}$ and $\hat{\boldsymbol{\beta}}_0 = \tilde{\boldsymbol{\beta}}$, where $\hat{\sigma}_u^2$ and $\tilde{\boldsymbol{\beta}}$ are the moment estimators given by (16.7).
2. For $i = 1, 2, \ldots$, do

$$\hat{\boldsymbol{\beta}}_i = \left(\sum_{d=1}^{D} \frac{\boldsymbol{x}_d' \boldsymbol{x}_d}{\hat{\sigma}_{u,i-1}^2 + \sigma_d^2}\right)^{-1} \left(\sum_{d=1}^{D} \frac{\boldsymbol{x}_d' y_d}{\hat{\sigma}_{u,i-1}^2 + \sigma_d^2}\right),$$

$$\hat{u}_{d,i} = \frac{\hat{\sigma}_{u,i-1}^2}{(\hat{\sigma}_{u,i-1}^2 + \sigma_d^2)}(y_d - \boldsymbol{x}_d \hat{\boldsymbol{\beta}}_i), \quad \hat{\sigma}_{u,i}^2 = \frac{\sum_{d=1}^{D} \hat{u}_{d,i}^2}{D - \frac{1}{\hat{\sigma}_{u,i-1}^2}\mathrm{tr}(\hat{\boldsymbol{R}}_{i-1})}.$$

3. Stop when $|\hat{\sigma}_{u,i}^2 - \hat{\sigma}_{u,i-1}^2| < \varepsilon_1$ and $\left[(\hat{\boldsymbol{\beta}}_i - \hat{\boldsymbol{\beta}}_{i-1})'(\hat{\boldsymbol{\beta}}_i - \hat{\boldsymbol{\beta}}_{i-1})\right]^{1/2} < \varepsilon_2$. Output: $\hat{\boldsymbol{\beta}}_{ML} = \hat{\boldsymbol{\beta}}_i$, $\hat{u}_d = \hat{u}_{d,i}$ and $\hat{\sigma}_{u,ML}^2 = \hat{\sigma}_{u,i}^2$

16.6 MSE of the EBLUP

This section derives an estimator of the mean squared error of the EBLUP of μ_d. Let us first assume that $\boldsymbol{\beta}$ and σ_u^2 are known and define $\gamma_d = \frac{\sigma_u^2}{\sigma_u^2 + \sigma_d^2}$. The BLUP of $\mu_d = \boldsymbol{x}_d \boldsymbol{\beta} + u_d$ is $\tilde{\mu}_d = \gamma_d y_d + (1 - \gamma_d)\boldsymbol{x}_d \boldsymbol{\beta}$. The mean squared error of $\tilde{\mu}_d$ is

$$\begin{aligned} MSE(\tilde{\mu}_d) &= E\big[(\tilde{\mu}_d - \mu_d)^2\big] = E\big[(\gamma_d y_d + (1 - \gamma_d)\boldsymbol{x}_d \boldsymbol{\beta} - \boldsymbol{x}_d \boldsymbol{\beta} - u_d)^2\big] \\ &= E\big[(\gamma_d (y_d - \boldsymbol{x}_d \boldsymbol{\beta}) - u_d)^2\big] = E\big[(\gamma_d (u_d + e_d) - u_d)^2\big] \\ &= E\big[(\gamma_d e_d - (1 - \gamma_d) u_d)^2\big] = \gamma_d^2 E[e_d^2] + (1 - \gamma_d)^2 E[u_d^2] \end{aligned}$$

$$- 2\gamma_d(1-\gamma_d)E[u_d]E[e_d] = \gamma_d^2\sigma_d^2 + (1-\gamma_d)^2\sigma_u^2$$
$$= \frac{\sigma_u^4\sigma_d^2}{(\sigma_u^2+\sigma_d^2)^2} + \frac{\sigma_u^2\sigma_d^4}{(\sigma_u^2+\sigma_d^2)^2} = \frac{\sigma_u^2\sigma_d^2(\sigma_u^2+\sigma_d^2)}{(\sigma_u^2+\sigma_d^2)^2} = \frac{\sigma_u^2\sigma_d^2}{\sigma_u^2+\sigma_d^2} = \gamma_d\sigma_d^2.$$

Let us now assume that σ_u^2 is known, but $\boldsymbol{\beta}$ is unknown. Define $\boldsymbol{Q} = (\boldsymbol{X}'\boldsymbol{V}^{-1}\boldsymbol{X})^{-1}$ and recall that $E[\boldsymbol{y}\boldsymbol{u}'] = \boldsymbol{V}_u$. The BLUP of μ_d is

$$\hat{\mu}_d^{blup} = \boldsymbol{a}_d'(\boldsymbol{X}\hat{\boldsymbol{\beta}} + \hat{\boldsymbol{u}}), \quad \boldsymbol{a}_d = \underset{1\leq i\leq D}{\text{col}}(\delta_{id}),$$

where δ_{id} is the Kronecker delta and

$$\hat{\boldsymbol{\beta}} = (\boldsymbol{X}'\boldsymbol{V}^{-1}\boldsymbol{X})^{-1}\boldsymbol{X}'\boldsymbol{V}^{-1}\boldsymbol{y} = \boldsymbol{Q}\boldsymbol{X}'\boldsymbol{V}^{-1}\boldsymbol{y},$$
$$\hat{\boldsymbol{u}} = \boldsymbol{V}_u\boldsymbol{V}^{-1}(\boldsymbol{y} - \boldsymbol{X}\hat{\boldsymbol{\beta}}) = \boldsymbol{V}_u\boldsymbol{V}^{-1}(\boldsymbol{I} - \boldsymbol{X}\boldsymbol{Q}\boldsymbol{X}'\boldsymbol{V}^{-1})\boldsymbol{y}.$$

Therefore,

$$\hat{\mu}_d^{blup} - \mu_d = \boldsymbol{a}_d'\boldsymbol{X}(\hat{\boldsymbol{\beta}} - \boldsymbol{\beta}) + \boldsymbol{a}_d'(\hat{\boldsymbol{u}} - \boldsymbol{u})$$

and

$$(\hat{\mu}_d^{blup} - \mu_d)^2 = \Big[\boldsymbol{a}_d'\boldsymbol{X}(\hat{\boldsymbol{\beta}} - \boldsymbol{\beta}) + \boldsymbol{a}_d'(\hat{\boldsymbol{u}} - \boldsymbol{u})\Big]\Big[(\hat{\boldsymbol{\beta}} - \boldsymbol{\beta})'\boldsymbol{X}'\boldsymbol{a}_d + (\hat{\boldsymbol{u}} - \boldsymbol{u})'\boldsymbol{a}_d\Big]$$
$$= \boldsymbol{a}_d'\boldsymbol{X}(\hat{\boldsymbol{\beta}} - \boldsymbol{\beta})(\hat{\boldsymbol{\beta}} - \boldsymbol{\beta})'\boldsymbol{X}'\boldsymbol{a}_d + \boldsymbol{a}_d'\boldsymbol{X}(\hat{\boldsymbol{\beta}} - \boldsymbol{\beta})(\hat{\boldsymbol{u}} - \boldsymbol{u})'\boldsymbol{a}_d$$
$$+ \boldsymbol{a}_d'(\hat{\boldsymbol{u}} - \boldsymbol{u})(\hat{\boldsymbol{\beta}} - \boldsymbol{\beta})'\boldsymbol{X}'\boldsymbol{a}_d + \boldsymbol{a}_d'(\hat{\boldsymbol{u}} - \boldsymbol{u})(\hat{\boldsymbol{u}} - \boldsymbol{u})'\boldsymbol{a}_d.$$

The mean squared error of $\hat{\mu}_d^{blup}$ is

$$MSE(\hat{\mu}_d^{blup}) = E\left[(\hat{\mu}_d^{blup} - \mu_d)^2\right]$$
$$= \boldsymbol{a}_d'\boldsymbol{X}E\left[(\hat{\boldsymbol{\beta}} - \boldsymbol{\beta})(\hat{\boldsymbol{\beta}} - \boldsymbol{\beta})'\right]\boldsymbol{X}'\boldsymbol{a}_d + \boldsymbol{a}_d'\boldsymbol{X}E\left[(\hat{\boldsymbol{\beta}} - \boldsymbol{\beta})(\hat{\boldsymbol{u}} - \boldsymbol{u})'\right]\boldsymbol{a}_d$$
$$+ \boldsymbol{a}_d'E\left[(\hat{\boldsymbol{u}} - \boldsymbol{u})(\hat{\boldsymbol{\beta}} - \boldsymbol{\beta})'\right]\boldsymbol{X}'\boldsymbol{a}_d + \boldsymbol{a}_d'E\left[(\hat{\boldsymbol{u}} - \boldsymbol{u})(\hat{\boldsymbol{u}} - \boldsymbol{u})'\right]\boldsymbol{a}_d.$$

We calculate each component of the previous formula,

$$\boldsymbol{R}_{11} = E\left[(\hat{\boldsymbol{\beta}} - \boldsymbol{\beta})(\hat{\boldsymbol{\beta}} - \boldsymbol{\beta})'\right] = \boldsymbol{Q}\boldsymbol{X}'\boldsymbol{V}^{-1}\boldsymbol{V}\boldsymbol{V}^{-1}\boldsymbol{X}\boldsymbol{Q} = \boldsymbol{Q}\boldsymbol{X}'\boldsymbol{V}^{-1}\boldsymbol{X}\boldsymbol{Q} = \boldsymbol{Q}\boldsymbol{Q}^{-1}\boldsymbol{Q} = \boldsymbol{Q},$$

$$\boldsymbol{R}_{12} = E[(\hat{\boldsymbol{\beta}} - \boldsymbol{\beta})(\hat{\boldsymbol{u}} - \boldsymbol{u})'] = E[\hat{\boldsymbol{\beta}}\hat{\boldsymbol{u}}'] - E[\hat{\boldsymbol{\beta}}\boldsymbol{u}']$$

$$
\begin{aligned}
&= E\big[\boldsymbol{Q}\boldsymbol{X}'\boldsymbol{V}^{-1}\boldsymbol{y}\boldsymbol{y}'(\boldsymbol{I}-\boldsymbol{V}^{-1}\boldsymbol{X}\boldsymbol{Q}\boldsymbol{X}')\boldsymbol{V}^{-1}\boldsymbol{V}_u\big] - E\big[\boldsymbol{Q}\boldsymbol{X}'\boldsymbol{V}^{-1}\boldsymbol{y}\boldsymbol{u}'\big]\\
&= \boldsymbol{Q}\boldsymbol{X}'\boldsymbol{V}^{-1}\boldsymbol{V}(\boldsymbol{I}-\boldsymbol{V}^{-1}\boldsymbol{X}\boldsymbol{Q}\boldsymbol{X}')\boldsymbol{V}^{-1}\boldsymbol{V}_u - \boldsymbol{Q}\boldsymbol{X}'\boldsymbol{V}^{-1}\boldsymbol{V}_u\\
&= -\boldsymbol{Q}(\boldsymbol{X}'\boldsymbol{V}^{-1}\boldsymbol{X})\boldsymbol{Q}\boldsymbol{X}'\boldsymbol{V}^{-1}\boldsymbol{V}_u = -\boldsymbol{Q}\boldsymbol{X}'\boldsymbol{V}^{-1}\boldsymbol{V}_u,\\
\boldsymbol{R}_{22} &= E\big[(\hat{\boldsymbol{u}}-\boldsymbol{u})(\hat{\boldsymbol{u}}-\boldsymbol{u})'\big] = E\big[\hat{\boldsymbol{u}}\hat{\boldsymbol{u}}'\big] - E\big[\hat{\boldsymbol{u}}\boldsymbol{u}'\big] - E\big[\boldsymbol{u}\hat{\boldsymbol{u}}'\big] + E\big[\boldsymbol{u}\boldsymbol{u}'\big]\\
&= \boldsymbol{V}_u\boldsymbol{V}^{-1}(\boldsymbol{I}-\boldsymbol{X}\boldsymbol{Q}\boldsymbol{X}'\boldsymbol{V}^{-1})\boldsymbol{V}(\boldsymbol{I}-\boldsymbol{V}^{-1}\boldsymbol{X}\boldsymbol{Q}\boldsymbol{X}')\boldsymbol{V}^{-1}\boldsymbol{V}_u\\
&- \boldsymbol{V}_u\boldsymbol{V}^{-1}(\boldsymbol{I}-\boldsymbol{X}\boldsymbol{Q}\boldsymbol{X}'\boldsymbol{V}^{-1})\boldsymbol{V}_u - \boldsymbol{V}_u(\boldsymbol{I}-\boldsymbol{V}^{-1}\boldsymbol{X}\boldsymbol{Q}\boldsymbol{X}')\boldsymbol{V}^{-1}\boldsymbol{V}_u + \boldsymbol{V}_u\\
&= \boldsymbol{V}_u\boldsymbol{V}^{-1}(\boldsymbol{I}-\boldsymbol{X}\boldsymbol{Q}\boldsymbol{X}'\boldsymbol{V}^{-1})\boldsymbol{V}\boldsymbol{V}^{-1}\boldsymbol{V}_u\\
&- \boldsymbol{V}_u\boldsymbol{V}^{-1}(\boldsymbol{I}-\boldsymbol{X}\boldsymbol{Q}\boldsymbol{X}'\boldsymbol{V}^{-1})\boldsymbol{V}\boldsymbol{V}^{-1}\boldsymbol{X}\boldsymbol{Q}\boldsymbol{X}'\boldsymbol{V}^{-1}\boldsymbol{V}_u\\
&- \boldsymbol{V}_u\boldsymbol{V}^{-1}(\boldsymbol{I}-\boldsymbol{X}\boldsymbol{Q}\boldsymbol{X}'\boldsymbol{V}^{-1})\boldsymbol{V}_u - \boldsymbol{V}_u(\boldsymbol{I}-\boldsymbol{V}^{-1}\boldsymbol{X}\boldsymbol{Q}\boldsymbol{X}')\boldsymbol{V}^{-1}\boldsymbol{V}_u + \boldsymbol{V}_u\\
&= -\boldsymbol{V}_u\boldsymbol{V}^{-1}\boldsymbol{X}\boldsymbol{Q}\boldsymbol{X}'\boldsymbol{V}^{-1}\boldsymbol{V}_u + \boldsymbol{V}_u\boldsymbol{V}^{-1}\boldsymbol{X}\boldsymbol{Q}(\boldsymbol{X}'\boldsymbol{V}^{-1}\boldsymbol{X})\boldsymbol{Q}\boldsymbol{X}'\boldsymbol{V}^{-1}\boldsymbol{V}_u\\
&- \boldsymbol{V}_u\boldsymbol{V}^{-1}\boldsymbol{V}_u + \boldsymbol{V}_u\boldsymbol{V}^{-1}\boldsymbol{X}\boldsymbol{Q}\boldsymbol{X}'\boldsymbol{V}^{-1}\boldsymbol{V}_u + \boldsymbol{V}_u\\
&= \boldsymbol{V}_u - \boldsymbol{V}_u\boldsymbol{V}^{-1}\boldsymbol{V}_u + \boldsymbol{V}_u\boldsymbol{V}^{-1}\boldsymbol{X}\boldsymbol{Q}\boldsymbol{X}'\boldsymbol{V}^{-1}\boldsymbol{V}_u.
\end{aligned}
$$

Note that $\boldsymbol{R}_{11} = \boldsymbol{Q}$ is a $p \times p$ matrix, whose diagonal elements might converge to zero as $D \to \infty$ under suitable regularity conditions on the auxiliary variables and the error variances σ_d^2, $d = 1, \ldots, D$. See e.g. Section 1.8 of Jiang (2007) or Section 4 of González-Manteiga et al. (2010). Therefore, $\hat{\boldsymbol{\beta}}$ is consistent under regularity assumptions. On the other hand, $\boldsymbol{R}_{22}$ is a $D \times D$ matrix, whose diagonal elements do not converge to zero as $D \to \infty$. Therefore, $\hat{u}_d - u_d$ and $\hat{\mu}_d - \mu_d$ do not converge to zero as $D \to \infty$. These facts are observed in simulations of area-level mixed models.

Coming back to the calculation of $MSE(\hat{\mu}_d^{blup})$, we have

$$
\begin{aligned}
MSE(\hat{\mu}_d^{blup}) &= \boldsymbol{a}_d'\boldsymbol{X}\boldsymbol{R}_{11}\boldsymbol{X}'\boldsymbol{a}_d + \boldsymbol{a}_d'\boldsymbol{X}\boldsymbol{R}_{12}\boldsymbol{a}_d + \boldsymbol{a}_d'\boldsymbol{R}_{12}'\boldsymbol{X}'\boldsymbol{a}_d + \boldsymbol{a}_d'\boldsymbol{R}_{22}\boldsymbol{a}_d\\
&= \boldsymbol{x}_d\boldsymbol{Q}\boldsymbol{x}_d' - 2\boldsymbol{x}_d\boldsymbol{Q}\boldsymbol{x}_d'\frac{\sigma_u^2}{\sigma_u^2+\sigma_d^2} + \sigma_u^2 - \frac{\sigma_u^4}{\sigma_u^2+\sigma_d^2} + \boldsymbol{x}_d\boldsymbol{Q}\boldsymbol{x}_d'\frac{\sigma_u^4}{(\sigma_u^2+\sigma_d^2)^2}\\
&= \frac{\sigma_u^2\sigma_d^2}{\sigma_u^2+\sigma_d^2} + \boldsymbol{x}_d\boldsymbol{Q}\boldsymbol{x}_d'\Big(1 - 2\frac{\sigma_u^2}{\sigma_u^2+\sigma_d^2} + \frac{\sigma_u^4}{(\sigma_u^2+\sigma_d^2)^2}\Big)\\
&= \frac{\sigma_u^2\sigma_d^2}{\sigma_u^2+\sigma_d^2} + \boldsymbol{x}_d\boldsymbol{Q}\boldsymbol{x}_d'\frac{\sigma_u^4+\sigma_d^4+2\sigma_u^2\sigma_d^2-2\sigma_u^4-2\sigma_u^2\sigma_d^2+\sigma_u^4}{(\sigma_u^2+\sigma_d^2)^2}\\
&= \frac{\sigma_u^2\sigma_d^2}{\sigma_u^2+\sigma_d^2} + \boldsymbol{x}_d\boldsymbol{Q}\boldsymbol{x}_d'\frac{\sigma_d^4}{(\sigma_u^2+\sigma_d^2)^2}.
\end{aligned}
$$

Given the variance component σ_u^2, we get

$$MSE(\hat{\mu}_d^{blup}) = g_1(\sigma_u^2) + g_2(\sigma_u^2),$$

$$g_1(\sigma_u^2) = \frac{\sigma_u^2\sigma_d^2}{\sigma_u^2+\sigma_d^2}, \quad g_2(\sigma_u^2) = \frac{\sigma_d^4}{(\sigma_u^2+\sigma_d^2)^2} \boldsymbol{x}_d \boldsymbol{Q} \boldsymbol{x}_d'.$$

For the case σ_u^2 unknown, Prasad and Rao (1990) gave an approximation to the MSE of the EBLUP under the Fay–Herriot model. The approximation is

$$MSE(\hat{\mu}_d) = MSE(\hat{\overline{Y}}_d^{eblup}) = g_1(\sigma_u^2) + g_2(\sigma_u^2) + g_3(\sigma_u^2) + o(D^{-1}),$$

where g_3 depends on the fitting method and g_1 and g_2 are given above.
For the moment-based estimator (16.7), Prasad and Rao (1990) obtained

$$g_3(\sigma_u^2) = \frac{\sigma_d^4}{(\sigma_u^2+\sigma_d^2)^3}\, \text{avar}(\hat{\sigma}_u^2), \quad \text{avar}(\hat{\sigma}_u^2) = \frac{2}{D}\left[\sigma_u^4 + \frac{2\sigma_u^2}{D}\sum_{d=1}^{D}\sigma_d^2 + \frac{1}{D}\sum_{d=1}^{D}\sigma_d^4\right].$$

For the Henderson 3 estimator, it holds that

$$g_3(\sigma_u^2) = \frac{\sigma_d^4}{(\sigma_u^2+\sigma_d^2)^3}\, \text{var}(\hat{\sigma}_{uH}^2),$$

where $\text{var}(\hat{\sigma}_{uH}^2)$ is given in (16.10).
For the ML or the REML estimators, Datta and Lahiri (2000) obtained

$$g_3(\sigma_u^2) = \frac{\sigma_d^4}{(\sigma_u^2+\sigma_d^2)^3}\, \text{avar}(\hat{\sigma}_u^2), \quad \text{avar}(\hat{\sigma}_u^2) = 2\Big(\sum_{d=1}^{D}(\sigma_u^2+\sigma_d^2)^{-2}\Big)^{-1}.$$

Note that for large n_d, we generally get $\sigma_d^2/\sigma_u^2 \approx 0$. Therefore, we have $g_1(\sigma_u^2) \approx \sigma_d^2$, $g_2(\sigma_u^2) \approx 0$ and $g_3(\sigma_u^2) \approx 0$ for moment-based, ML and REML estimators of model parameters. This is to say, in domains with large samples sizes, the model-based MSEs of the EBLUPs and the design-based variances of the direct estimators tend to be equal.

If we use the Prasad–Rao moment-based or the REML estimators, an MSE estimator is

$$mse(\hat{\mu}_d) = mse(\hat{\overline{Y}}_d^{eblup}) = g_1(\hat{\sigma}_u^2) + g_2(\hat{\sigma}_u^2) + 2g_3(\hat{\sigma}_u^2),$$

where the coefficient 2 stands for bias correction; for more details, see Sect. 9.4. If we use the ML estimators, an MSE estimator is

$$mse(\hat{\overline{Y}}_d^{eblup}) = g_1(\hat{\sigma}_u^2) + g_2(\hat{\sigma}_u^2) + 2g_3(\hat{\sigma}_u^2) - b(\hat{\sigma}_u^2)\nabla g_1(\hat{\sigma}_u^2),$$

where $\nabla g_1(\hat{\sigma}_u^2) = \sigma_d^4(\hat{\sigma}_u^2 + \sigma_d^2)^{-2}$ and

$$b(\hat{\sigma}_u^2) = -\text{tr}\Big\{\Big[\sum_{j=1}^{D}(\hat{\sigma}_u^2+\sigma_d^2)^{-1}\boldsymbol{x}_d'\boldsymbol{x}_d\Big]^{-1}\Big[\sum_{j=1}^{D}(\hat{\sigma}_u^2+\sigma_d^2)^{-2}\boldsymbol{x}_d'\boldsymbol{x}_d\Big]\Big\}\Big(\sum_{j=1}^{D}(\hat{\sigma}_u^2+\sigma_d^2)^{-2}\Big)^{-1}.$$

Remark 16.3 Let us assume that some of the y_d's are missing, while all the $\boldsymbol{x}_d$'s are recorded and known. More concretely, the set of domains is partitioned into two subsets $\mathcal{D}_0 = \{1, \ldots, D_0\}$ and $\mathcal{D}_1 = \{D_0+1, \ldots, D\}$, where $D_0 < D$. We assume that $n_d > 0$ and y_d is observed if $d \in \mathcal{D}_0$, and that y_d is not observed or $n_d = 0$ if $d \in \mathcal{D}_1$. Further, $\sigma_d^2 > 0$ is known for $d \in \mathcal{D}_0$, but it is missing for $d \in \mathcal{D}_1$. For the subset $\mathcal{D}_0$, we consider the vector and matrices

$$\boldsymbol{X}_0 = \underset{1\le d\le D_0}{\text{col}}(\boldsymbol{x}_d),\ \boldsymbol{V}_0 = \underset{1\le d\le D_0}{\text{diag}}(\sigma_u^2+\sigma_d^2),\ \boldsymbol{u}_0 = \underset{1\le d\le D_0}{\text{col}}(u_d),\ \boldsymbol{e}_0 = \underset{1\le d\le D_0}{\text{col}}(e_d).$$

For $d = 1, \ldots, D$, we define

$$\boldsymbol{a}_d' = \boldsymbol{x}_d(\boldsymbol{X}_0'\boldsymbol{V}_0^{-1}\boldsymbol{X}_0)^{-1}\boldsymbol{X}_0'\boldsymbol{V}_0^{-1},\ \boldsymbol{\delta}_d = \underset{1\le i\le D}{\text{col}}(\delta_{id}),\ a_{dd} = \boldsymbol{x}_d(\boldsymbol{X}_0'\boldsymbol{V}_0^{-1}\boldsymbol{X}_0)^{-1}\boldsymbol{X}_0'\boldsymbol{V}_0^{-1}\boldsymbol{\delta}_d.$$

The synthetic predictor is

$$\hat{\overline{Y}}_d^{syn} = \hat{\mu}_d^{syn} = \boldsymbol{x}_d\hat{\boldsymbol{\beta}},\quad d = 1, \ldots, D,$$

where $\hat{\boldsymbol{\beta}}$ and $\hat{\sigma}_u^2$ are calculated by using the data from $\mathcal{D}_0$. If $\hat{\boldsymbol{\beta}}$ and $\hat{\sigma}_u^2$ are asymptotically consistent and independent estimators of $\boldsymbol{\beta}$ and σ_u^2, as the ML and the REML estimators are, then the mean and variance of $\hat{\mu}_d^{syn}$ are

$$E[\hat{\mu}_d^{syn}] \approx \boldsymbol{x}_d\boldsymbol{\beta},\quad \text{var}(\hat{\mu}_d^{syn}) = \boldsymbol{x}_d\text{var}(\hat{\boldsymbol{\beta}})\boldsymbol{x}_d' \approx \boldsymbol{x}_d(\boldsymbol{X}_0'\boldsymbol{V}_0^{-1}\boldsymbol{X}_0)^{-1}\boldsymbol{x}_d'.$$

As $\mu_d = \boldsymbol{x}_d\boldsymbol{\beta} + u_d$, the MSE of $\hat{\mu}_d^{syn} = \boldsymbol{x}_d\hat{\boldsymbol{\beta}}$ is

$$\begin{aligned} MSE(\hat{\mu}_d^{syn}) &= E\big[(\hat{\mu}_d^{syn} - \mu_d)^2\big] = E\big[(\boldsymbol{x}_d\hat{\boldsymbol{\beta}} - \boldsymbol{x}_d\boldsymbol{\beta} - u_d)^2\big] = E\big[(\boldsymbol{x}_d(\hat{\boldsymbol{\beta}} - \boldsymbol{\beta}) - u_d)^2\big] \\ &= \boldsymbol{x}_d E\big[(\hat{\boldsymbol{\beta}} - \boldsymbol{\beta})(\hat{\boldsymbol{\beta}} - \boldsymbol{\beta})'\big]\boldsymbol{x}_d' + E[u_d^2] - 2E\big[\boldsymbol{x}_d(\hat{\boldsymbol{\beta}} - \boldsymbol{\beta})u_d\big] \\ &\approx \boldsymbol{x}_d(\boldsymbol{X}_0'\boldsymbol{V}_0^{-1}\boldsymbol{X}_0)^{-1}\boldsymbol{x}_d' + \sigma_u^2 - 2E\big[\boldsymbol{x}_d\hat{\boldsymbol{\beta}}u_d\big]. \end{aligned}$$

If $d \in \mathcal{D}_1$, then u_d and $\hat{\boldsymbol{\beta}}$ are independent, so that $E[\boldsymbol{x}_d\hat{\boldsymbol{\beta}}u_d] = \boldsymbol{x}_d E[\hat{\boldsymbol{\beta}}]E[u_d] = 0$ and

$$MSE(\hat{\mu}_d^{syn}) \approx \boldsymbol{x}_d(\boldsymbol{X}_0'\boldsymbol{V}_0^{-1}\boldsymbol{X}_0)^{-1}\boldsymbol{x}_d' + \sigma_u^2.$$

If $d \in \mathcal{D}_0$, then

$$E\big[\boldsymbol{x}_d\hat{\boldsymbol{\beta}}u_d\big] \approx E\big[\boldsymbol{x}_d(\boldsymbol{X}_0'\boldsymbol{V}_0^{-1}\boldsymbol{X}_0)^{-1}\boldsymbol{X}_0'\boldsymbol{V}_0^{-1}(\boldsymbol{X}_0\boldsymbol{\beta} + \boldsymbol{u}_0 + \boldsymbol{e}_0)u_d\big] = E\big[\boldsymbol{a}_d'\boldsymbol{u}_0u_d\big]$$

$$= E\left[u_d\sum_{i=1}^{D_0} a_{di}u_i\right] = a_{dd}\sigma_u^2,$$

where the a_{di}'s are the components of $\boldsymbol{a}_d$. Finally, the MSE of $\hat{\mu}_d^{syn}$ is

$$MSE(\hat{\mu}_d^{syn}) \approx \boldsymbol{x}_d(\boldsymbol{X}_0'\boldsymbol{V}_0^{-1}\boldsymbol{X}_0)^{-1}\boldsymbol{x}_d' + \sigma_u^2(1 - 2a_{dd}).$$

Remark 16.4 For calculating the Fay–Herriot EBLUP of a domain total, $\hat{Y}_d^{eblup}$, we can substitute the estimators $y_1, \ldots, y_D$ of the means $\overline{Y}_1, \ldots, \overline{Y}_D$ by the estimators $y_{1\cdot}, \ldots, y_{D\cdot}$ of the totals $Y_1, \ldots, Y_D$. All the theoretical developments given in this section remain applicable with small adaptations.

16.6.1 Parametric Bootstrap

As an alternative to the MSE approximation described above, a resampling method can be used. This section presents the parametric bootstrap method of González-Manteiga et al. (2010) for estimation of the MSE of the EBLUP $\hat{\mu}_d = \hat{\overline{Y}}_d^{eblup}$. This method provides consistent MSE estimates under the assumption that the estimates of the model parameters are consistent.

The parametric bootstrap procedure for estimating the MSE of $\hat{\mu}_d$ has the following steps:

1. Obtain the estimates $\hat{\sigma}_u^2$ and $\hat{\boldsymbol{\beta}}$ by fitting the Fay–Herriot model (16.2) to the initial data $\boldsymbol{y} = (y_1, \ldots, y_D)'$.
2. Repeat B times $(b = 1, \ldots, B)$:

   a. For $d = 1, \ldots, D$, generate $u_d^{*(b)}$ i.i.d. $N(0, \hat{\sigma}_u^2)$. Construct the vector $\boldsymbol{u}^{*(b)} = (u_1^{*(b)}, \ldots, u_D^{*(b)})'$, and calculate the true bootstrap quantities $\boldsymbol{\mu}^{*(b)} = \boldsymbol{X}\hat{\boldsymbol{\beta}} + \boldsymbol{u}^{*(b)}$ with elements $\mu_d^{*(b)}$.
   b. For $d = 1, \ldots, D$, generate independent variables $e_d^{*(b)} \sim N(0, \sigma_d^2)$. Construct the vector $\boldsymbol{e}^{*(b)} = (e_1^{*(b)}, \ldots, e_D^{*(b)})'$.

c. Calculate bootstrap data $\boldsymbol{y}^{*(b)}$ directly by applying the formula

$$\boldsymbol{y}^{*(b)} = \boldsymbol{X}\hat{\boldsymbol{\beta}} + \boldsymbol{u}^{*(b)} + \boldsymbol{e}^{*(b)}. \tag{16.11}$$

d. Fit the assumed model to the bootstrap vector $\boldsymbol{y}^{*(b)}$, and obtain the estimates $\hat{\sigma}_u^{2*(b)}$ and $\hat{\boldsymbol{\beta}}^{*(b)}$ of the bootstrap parameters $\hat{\sigma}_u^2$ and $\hat{\boldsymbol{\beta}}$, respectively.

e. For $d = 1, \ldots, D$, calculate the bootstrap EBLUP

$$\hat{\mu}_d^{*(b)} = \frac{\hat{\sigma}_u^{2*(b)}}{\hat{\sigma}_u^{2*(b)} + \sigma_d^2} y_d + \frac{\sigma_d^2}{\hat{\sigma}_u^{2*(b)} + \sigma_d^2} \boldsymbol{x}_d \hat{\boldsymbol{\beta}}^{*(b)}.$$

3. Output: parametric bootstrap estimator of $MSE(\hat{\mu}_d)$,

$$mse^*(\hat{\mu}_d) = \frac{1}{B}\sum_{b=1}^{B}\left(\hat{\mu}_d^{*(b)} - \mu_d^{*(b)}\right)^2, \qquad d = 1, \ldots, D.$$

16.7 Bayesian Prediction

This section introduces a hierarchical Bayes (HB) approach for predicting domain means and uses the same notation as in previous sections. For the Fay–Herriot model, Section 10.3 of Rao (2003) assumes the following hypotheses:

(1) $y_d = \mu_d + e_d$, $e_d \sim N(0, \sigma_d^2)$, $d = 1, \ldots, D$, where $e_1, \ldots, e_D$ are independent and $\sigma_1^2, \ldots, \sigma_D^2$ are known strictly positive constants.
(2) $\mu_d = \boldsymbol{x}_d\boldsymbol{\beta} + u_d$, $u_d \sim N(0, \sigma_u^2)$, $d = 1, \ldots, D$, where $u_1, \ldots, u_D$ are independent and $\sigma_u > 0$ is known.
(3) $\boldsymbol{\beta} \sim f_0$, where $f_0(\boldsymbol{\beta}) = 1$ in $\boldsymbol{\beta} \in R^p$. This is to say, f_0 is not a probability density function in R^p, but it is the density of the Lebesgue measure in R^p.
(4) $\boldsymbol{e} = \underset{1\le d\le D}{\text{col}}(e_d)$ and $\boldsymbol{u} = \underset{1\le d\le D}{\text{col}}(u_d)$ are mutually independent.

Under the assumptions (1)–(4), the conditional probability distributions of $\boldsymbol{y}$, given $(\boldsymbol{\beta}, \boldsymbol{u})$, and of $\boldsymbol{u}$ are

$$\boldsymbol{y}|\,(\boldsymbol{\beta}, \boldsymbol{u}) \sim N_D(\boldsymbol{X}\boldsymbol{\beta} + \boldsymbol{u}, \boldsymbol{V}_e), \quad \boldsymbol{u} \sim N_D(\boldsymbol{0}, \boldsymbol{V}_u).$$

We recall that the probability density function (p.d.f.) of a multivariate normal distribution in R^n is

$$f(\boldsymbol{y}|\,\boldsymbol{\mu}, \boldsymbol{\Sigma}) = \frac{1}{(2\pi)^{n/2}|\boldsymbol{\Sigma}|^{1/2}} \exp\Big\{-\frac{1}{2}\boldsymbol{y}'\boldsymbol{\Sigma}^{-1}\boldsymbol{y} + \boldsymbol{y}'\boldsymbol{\Sigma}^{-1}\boldsymbol{\mu} - \frac{1}{2}\boldsymbol{\mu}'\boldsymbol{\Sigma}^{-1}\boldsymbol{\mu}\Big\}. \tag{16.12}$$

By applying the inversion formula

$$(A+BCD)^{-1}=A^{-1}+A^{-1}B(C^{-1}+DA^{-1}B)^{-1}DA^{-1},$$

with $A=\boldsymbol{V}_e^{-1}$, $C=\boldsymbol{V}_u^{-1}$, $B=D=\boldsymbol{I}_D$, we obtain

$$\big(\boldsymbol{V}_e^{-1}+\boldsymbol{V}_u^{-1}\big)^{-1}=\boldsymbol{V}_e-\boldsymbol{V}_e(\boldsymbol{V}_u+\boldsymbol{V}_e)^{-1}\boldsymbol{V}_e \tag{16.13}$$

$$\boldsymbol{V}_e^{-1}-\boldsymbol{V}_e^{-1}\big(\boldsymbol{V}_e^{-1}+\boldsymbol{V}_u^{-1}\big)^{-1}\boldsymbol{V}_e^{-1}=\boldsymbol{V}_e^{-1}-\boldsymbol{V}_e^{-1}\big[\boldsymbol{V}_e-\boldsymbol{V}_e(\boldsymbol{V}_u+\boldsymbol{V}_e)^{-1}\boldsymbol{V}_e\big]\boldsymbol{V}_e^{-1}$$
$$=\boldsymbol{V}_e^{-1}-\boldsymbol{V}_e^{-1}+(\boldsymbol{V}_u+\boldsymbol{V}_e)^{-1}=(\boldsymbol{V}_u+\boldsymbol{V}_e)^{-1}.$$

Proposition 16.3 *Under (1)–(4), the conditional distribution of* $\boldsymbol{y}$*, given* $\boldsymbol{\beta}$*, is*

$$\boldsymbol{y}|\,\boldsymbol{\beta}\sim N_D(\boldsymbol{X}\boldsymbol{\beta},\,\boldsymbol{V}),\quad \boldsymbol{V}=\boldsymbol{V}_u+\boldsymbol{V}_e.$$

Proof Let us define $\boldsymbol{A}_{eu}=\boldsymbol{V}_e\boldsymbol{V}_u$ and $\boldsymbol{B}_{eu}=\boldsymbol{V}_e^{-1}+\boldsymbol{V}_u^{-1}$. It holds that

$$f(\boldsymbol{y}|\,\boldsymbol{\beta})=\int_{R^D}f(\boldsymbol{y}|\,\boldsymbol{\beta},\boldsymbol{u})f(\boldsymbol{u})\,d\boldsymbol{u}$$
$$=\frac{1}{(2\pi)^D|\boldsymbol{A}_{eu}|^{1/2}}\int_{R^D}\exp\Big\{-\frac{1}{2}(\boldsymbol{y}-\boldsymbol{X}\boldsymbol{\beta}-\boldsymbol{u})'\boldsymbol{V}_e^{-1}(\boldsymbol{y}-\boldsymbol{X}\boldsymbol{\beta}-\boldsymbol{u})-\frac{1}{2}\boldsymbol{u}'\boldsymbol{V}_u^{-1}\boldsymbol{u}\Big\}d\boldsymbol{u}$$
$$=\frac{|\boldsymbol{A}_{eu}|^{-1/2}}{(2\pi)^D}\int_{R^D}\exp\Big\{-\frac{1}{2}\boldsymbol{u}'\boldsymbol{B}_{eu}\boldsymbol{u}+\boldsymbol{u}'\boldsymbol{V}_e^{-1}(\boldsymbol{y}-\boldsymbol{X}\boldsymbol{\beta})-\frac{1}{2}(\boldsymbol{y}-\boldsymbol{X}\boldsymbol{\beta})'\boldsymbol{V}_e^{-1}(\boldsymbol{y}-\boldsymbol{X}\boldsymbol{\beta})\Big\}d\boldsymbol{u}$$
$$=\frac{|\boldsymbol{A}_{eu}|^{-1/2}}{(2\pi)^D}\frac{(2\pi)^{D/2}}{|\boldsymbol{B}_{eu}|^{1/2}}\exp\Big\{-\frac{1}{2}(\boldsymbol{y}-\boldsymbol{X}\boldsymbol{\beta})'\big[\boldsymbol{V}_e^{-1}-\boldsymbol{V}_e^{-1}\boldsymbol{B}_{eu}^{-1}\boldsymbol{V}_e^{-1}\big](\boldsymbol{y}-\boldsymbol{X}\boldsymbol{\beta})\Big\}$$
$$\cdot\int_{R^D}\frac{1}{(2\pi)^{D/2}|\boldsymbol{B}_{eu}|^{-1/2}}\exp\Big\{-\frac{1}{2}\boldsymbol{u}'\boldsymbol{B}_{eu}\boldsymbol{u}+\boldsymbol{u}'\boldsymbol{B}_{eu}\big[\boldsymbol{B}_{eu}^{-1}\boldsymbol{V}_e^{-1}(\boldsymbol{y}-\boldsymbol{X}\boldsymbol{\beta})\big]$$
$$-\frac{1}{2}\big[(\boldsymbol{y}-\boldsymbol{X}\boldsymbol{\beta})'\boldsymbol{V}_e^{-1}\boldsymbol{B}_{eu}^{-1}\big]\boldsymbol{B}_{eu}\big[\boldsymbol{B}_{eu}^{-1}\boldsymbol{V}_e^{-1}(\boldsymbol{y}-\boldsymbol{X}\boldsymbol{\beta})\big]\Big\}\,d\boldsymbol{u}.$$

We can do three simplifications in the last expression of $f(\boldsymbol{y}|\,\boldsymbol{\beta})$. First, we have the integral in R^D of the p.d.f. of a multivariate normal distribution with mean vector $\boldsymbol{\mu}=\boldsymbol{B}_{eu}^{-1}\boldsymbol{V}_e^{-1}(\boldsymbol{y}-\boldsymbol{X}\boldsymbol{\beta})$ and covariance matrix $\boldsymbol{\Sigma}=\boldsymbol{B}_{eu}^{-1}$, which is equal to one. Second, we apply (16.13), i.e.

$$\boldsymbol{V}_e^{-1}-\boldsymbol{V}_e^{-1}\boldsymbol{B}_{eu}^{-1}\boldsymbol{V}_e^{-1}=(\boldsymbol{V}_u+\boldsymbol{V}_e)^{-1}.$$

Third, we have

$$|\boldsymbol{A}_{eu}|^{1/2}|\boldsymbol{B}_{eu}|^{1/2}=|\boldsymbol{A}_{eu}\boldsymbol{B}_{eu}|^{1/2}=|\boldsymbol{V}_e\boldsymbol{V}_u(\boldsymbol{V}_e^{-1}+\boldsymbol{V}_u^{-1})|^{1/2}=|\boldsymbol{V}_u+\boldsymbol{V}_e|^{1/2}.$$

Therefore, we get

$$f(\mathbf{y}|\,\boldsymbol{\beta}) = \frac{1}{(2\pi)^{D/2}|\mathbf{V}_e+\mathbf{V}_u|^{1/2}}\exp\Big\{-\frac{1}{2}(\mathbf{y}-\mathbf{X}\boldsymbol{\beta})'(\mathbf{V}_u+\mathbf{V}_e)^{-1}(\mathbf{y}-\mathbf{X}\boldsymbol{\beta})\Big\},$$

which is the p.d.f. of a multivariate normal distribution with mean vector $\mu = \mathbf{X}\boldsymbol{\beta}$ and covariance matrix $\boldsymbol{\Sigma} = \mathbf{V}_u + \mathbf{V}_e = \mathbf{V}$. □

Proposition 16.4 *Under (1)–(4), the conditional distribution of* $\boldsymbol{\beta}$*, given* $\mathbf{y}$*, is*

$$\boldsymbol{\beta}|\,\mathbf{y} \sim N_p(\tilde{\boldsymbol{\beta}},\,\mathbf{V}_\beta),\quad \tilde{\boldsymbol{\beta}} = (\mathbf{X}'\mathbf{V}^{-1}\mathbf{X})^{-1}\mathbf{X}'\mathbf{V}^{-1}\mathbf{y},\ \ \mathbf{V}_\beta = (\mathbf{X}'\mathbf{V}^{-1}\mathbf{X})^{-1}.$$

Proof Let us first calculate the marginal p.d.f. of $\mathbf{y}$,

$$\begin{aligned} f(\mathbf{y}) &= \int_{R^p} f(\mathbf{y}|\,\boldsymbol{\beta})f_0(\boldsymbol{\beta})\,d\boldsymbol{\beta} = \int_{R^p} f(\mathbf{y}|\,\boldsymbol{\beta})\,d\boldsymbol{\beta} \\ &= \frac{1}{(2\pi)^{D/2}|\mathbf{V}|^{1/2}}\int_{R^p}\exp\Big\{-\frac{1}{2}(\mathbf{y}-\mathbf{X}\boldsymbol{\beta})'\mathbf{V}^{-1}(\mathbf{y}-\mathbf{X}\boldsymbol{\beta})\Big\}\,d\boldsymbol{\beta} \\ &= \frac{1}{(2\pi)^{D/2}|\mathbf{V}|^{1/2}}\int_{R^p}\exp\Big\{-\frac{1}{2}\boldsymbol{\beta}'\mathbf{X}'\mathbf{V}^{-1}\mathbf{X}\boldsymbol{\beta}+\boldsymbol{\beta}'\mathbf{X}'\mathbf{V}^{-1}\mathbf{y}-\frac{1}{2}\mathbf{y}'\mathbf{V}^{-1}\mathbf{y}\Big\}\,d\boldsymbol{\beta}. \end{aligned}$$

Therefore, we have

$$\begin{aligned} f(\mathbf{y}) &= \frac{(2\pi)^{p/2}}{(2\pi)^{D/2}|\mathbf{V}|^{1/2}|\mathbf{X}'\mathbf{V}^{-1}\mathbf{X}|^{1/2}}\exp\Big\{-\frac{1}{2}\mathbf{y}'\big[\mathbf{V}^{-1}-\mathbf{V}^{-1}\mathbf{X}(\mathbf{X}'\mathbf{V}^{-1}\mathbf{X})^{-1}\mathbf{X}'\mathbf{V}^{-1}\big]\mathbf{y}\Big\} \\ &\cdot\int_{R^p}\frac{|\mathbf{X}'\mathbf{V}^{-1}\mathbf{X}|^{1/2}}{(2\pi)^{p/2}}\exp\Big\{-\frac{1}{2}\boldsymbol{\beta}'(\mathbf{X}'\mathbf{V}^{-1}\mathbf{X})\boldsymbol{\beta}+\boldsymbol{\beta}'(\mathbf{X}'\mathbf{V}^{-1}\mathbf{X})\big[(\mathbf{X}'\mathbf{V}^{-1}\mathbf{X})^{-1}\mathbf{X}'\mathbf{V}^{-1}\mathbf{y}\big] \\ &-\frac{1}{2}\big[\mathbf{y}'\mathbf{V}^{-1}\mathbf{X}(\mathbf{X}'\mathbf{V}^{-1}\mathbf{X})^{-1}\big](\mathbf{X}'\mathbf{V}^{-1}\mathbf{X})\big[(\mathbf{X}'\mathbf{V}^{-1}\mathbf{X})^{-1}\mathbf{X}'\mathbf{V}^{-1}\mathbf{y}\big]\Big\}\,d\boldsymbol{\beta}. \end{aligned}$$

As the integral in R^p of the p.d.f. of a multivariate normal distribution with mean vector $\boldsymbol{\mu} = (\mathbf{X}'\mathbf{V}^{-1}\mathbf{X})^{-1}\mathbf{X}'\mathbf{V}^{-1}\mathbf{y}$ and covariance matrix $(\mathbf{X}'\mathbf{V}^{-1}\mathbf{X})^{-1}$ is one, we have

$$f(\mathbf{y}) = \frac{1}{(2\pi)^{(D-p)/2}|\mathbf{V}|^{1/2}|\mathbf{X}'\mathbf{V}^{-1}\mathbf{X}|^{1/2}}\exp\Big\{-\frac{1}{2}\mathbf{y}'\big[\mathbf{V}^{-1}-\mathbf{V}^{-1}\mathbf{X}(\mathbf{X}'\mathbf{V}^{-1}\mathbf{X})^{-1}\mathbf{X}'\mathbf{V}^{-1}\big]\mathbf{y}\Big\}.$$

The conditional distribution of $\boldsymbol{\beta}$, given $\mathbf{y}$, is

$$\begin{aligned} f(\boldsymbol{\beta}|\mathbf{y}) &= \frac{f(\mathbf{y}|\boldsymbol{\beta})f_0(\boldsymbol{\beta})}{f(\mathbf{y})} = \frac{|\mathbf{X}'\mathbf{V}^{-1}\mathbf{X}|^{1/2}}{(2\pi)^{p/2}}\frac{\exp\Big\{-\frac{1}{2}(\mathbf{y}-\mathbf{X}\boldsymbol{\beta})'\mathbf{V}^{-1}(\mathbf{y}-\mathbf{X}\boldsymbol{\beta})\Big\}}{\exp\Big\{-\frac{1}{2}\mathbf{y}'\big[\mathbf{V}^{-1}-\mathbf{V}^{-1}\mathbf{X}(\mathbf{X}'\mathbf{V}^{-1}\mathbf{X})^{-1}\mathbf{X}'\mathbf{V}^{-1}\big]\mathbf{y}\Big\}} \\ &= \frac{|\mathbf{X}'\mathbf{V}^{-1}\mathbf{X}|^{1/2}}{(2\pi)^{p/2}}\exp\Big\{-\frac{1}{2}\boldsymbol{\beta}'\mathbf{X}'\mathbf{V}^{-1}\mathbf{X}\boldsymbol{\beta}+\boldsymbol{\beta}'\mathbf{X}'\mathbf{V}^{-1}\mathbf{y}-\frac{1}{2}\mathbf{y}'\mathbf{V}^{-1}\mathbf{y}+\frac{1}{2}\mathbf{y}'\mathbf{V}^{-1}\mathbf{y} \\ &-\frac{1}{2}\mathbf{y}'\mathbf{V}^{-1}\mathbf{X}(\mathbf{X}'\mathbf{V}^{-1}\mathbf{X})^{-1}\mathbf{X}'\mathbf{V}^{-1}\mathbf{y}\Big\} \end{aligned}$$

$$= \frac{|X'V^{-1}X|^{1/2}}{(2\pi)^{p/2}} \exp\Big\{-\frac{1}{2}\beta' X' V^{-1} X\beta + \beta' X' V^{-1} y - \frac{1}{2} y' V^{-1} X (X' V^{-1} X)^{-1} X' V^{-1} y\Big\}.$$

We can write $f(\beta|y)$ in the form

$$f(\beta|y) = \frac{|X'V^{-1}X|^{1/2}}{(2\pi)^{p/2}} \exp\Big\{-\frac{1}{2}\beta'(X'V^{-1}X)\beta + \beta'(X'V^{-1}X)[(X'V^{-1}X)^{-1}X'V^{-1}y]$$

$$- \frac{1}{2}[y'V^{-1}X(X'V^{-1}X)^{-1}](X'V^{-1}X)[(X'V^{-1}X)^{-1}X'V^{-1}y]\Big\}$$

$$= \frac{1}{(2\pi)^{p/2}|V_\beta|^{1/2}} \exp\Big\{-\frac{1}{2}\beta' V_\beta^{-1}\beta + \beta' V_\beta^{-1}\tilde{\beta} - \frac{1}{2}\tilde{\beta}' V_\beta^{-1}\tilde{\beta}\Big\},$$

which is the p.d.f. of a multivariate normal distribution with mean vector $\mu = \tilde{\beta} = (X'V^{-1}X)^{-1}X'V^{-1}y$ and covariance matrix $V_\beta = (X'V^{-1}X)^{-1}$. □

Definition 16.1 The HB predictor of μ_d is the posterior mean $\tilde{\mu}_d^{HB} = \tilde{\mu}_d^{HB}(\sigma_u^2) = E[\mu_d|y]$.

Proposition 16.5 *Under (1)–(4), the HB predictor of* μ_d *is*

$$\tilde{\mu}_d^{HB} = E[\mu_d|y] = \frac{\sigma_u^2}{\sigma_u^2+\sigma_d^2}\, y_d + \frac{\sigma_d^2}{\sigma_u^2+\sigma_d^2}\, x_d\tilde{\beta},$$

and the posterior variance is

$$var(\mu_d|y) = \frac{\sigma_d^2\sigma_u^2}{\sigma_d^2+\sigma_u^2} + \frac{\sigma_d^4}{(\sigma_u^2+\sigma_d^2)^2}\, x_d V_\beta x_d' \triangleq g_1(\sigma_u^2) + g_2(\sigma_u^2).$$

This is to say, the HB predictor of μ_d *is equal to the BLUP, and the posterior variance is equal to the mean squared error of the BLUP.*

Proof Proposition 16.2 states that the conditioned distribution of μ_d, given y_d and β, is normal with mean and variance

$$E[\mu_d|y_d,\beta] = \frac{\sigma_u^2}{\sigma_u^2+\sigma_d^2}\, y_d + \frac{\sigma_d^2}{\sigma_u^2+\sigma_d^2}\, x_d\beta, \quad \text{var}(\mu_d|y_d,\beta) = \frac{\sigma_d^2\sigma_u^2}{\sigma_d^2+\sigma_u^2}.$$

Proposition 16.4 states that $\beta|\, y \sim N_p(\tilde{\beta},\, V_\beta)$. Therefore, the posterior mean is

$$E[\mu_d|y] = E\big[x_d\beta + u_d|y\big] = \int_{R^p}\int_R (x_d\beta + u_d) f(u_d,\beta|y)\, du_d\, d\beta$$

$$= \int_{R^p}\Big(\int_R (x_d\beta + u_d) f(u_d|\beta, y)\, du_d\Big) f(\beta|y)\, d\beta$$

$$= \int_{R^p} \left(\int_R (\mathbf{x}_d \boldsymbol{\beta} + u_d) f(u_d | \boldsymbol{\beta}, y_d)\, du_d \right) f(\boldsymbol{\beta} | \mathbf{y})\, d\boldsymbol{\beta}$$

$$= E\big[E[\mu_d | y_d, \boldsymbol{\beta}] \big| \, \mathbf{y}\big] = E\left[\frac{\sigma_u^2}{\sigma_u^2 + \sigma_d^2} y_d + \frac{\sigma_d^2}{\sigma_u^2 + \sigma_d^2} \mathbf{x}_d \boldsymbol{\beta} \,\middle|\, \mathbf{y} \right]$$

$$= \frac{\sigma_u^2}{\sigma_u^2 + \sigma_d^2} y_d + \frac{\sigma_d^2}{\sigma_u^2 + \sigma_d^2} \mathbf{x}_d \tilde{\boldsymbol{\beta}}.$$

Similarly, the posterior variance is

$$\mathrm{var}(\mu_d | \mathbf{y}) = E\big[\mathrm{var}(\mu_d | y_d, \boldsymbol{\beta}) \big| \, \mathbf{y}\big] + \mathrm{var}\big(E[\mu_d | y_d, \boldsymbol{\beta}] \big| \, \mathbf{y}\big)$$

$$= E\left[\frac{\sigma_d^2 \sigma_u^2}{\sigma_d^2 + \sigma_u^2} \,\middle|\, \mathbf{y} \right] + \mathrm{var}\left(\frac{\sigma_u^2}{\sigma_u^2 + \sigma_d^2} y_d + \frac{\sigma_d^2}{\sigma_u^2 + \sigma_d^2} \mathbf{x}_d \boldsymbol{\beta} \,\middle|\, \mathbf{y} \right)$$

$$= \frac{\sigma_d^2 \sigma_u^2}{\sigma_d^2 + \sigma_u^2} + \frac{\sigma_d^4}{(\sigma_u^2 + \sigma_d^2)^2} \mathbf{x}_d \mathbf{V}_\beta \mathbf{x}'_d,$$

which completes the proof. □

16.7.1 Unknown σ_u^2

In practice, σ_u^2 is not known, and we may replace it by an estimator $\hat{\sigma}_u^2$ to obtain the empirical Bayes predictor $\hat{\mu}_d^{EB} = \tilde{\mu}_d^{HB}(\hat{\sigma}_u^2)$, which coincides with the EBLUP $\hat{\mu}_d$ defined in (16.5). For applying the full Bayesian approach, a prior distribution for σ_u^2 is needed. In this case, we substitute assumptions (3) and (4) by

(3a) The joint prior density of $\boldsymbol{\beta}$ and σ_u^2 is $f(\boldsymbol{\beta}, \sigma_u^2) = f_0(\boldsymbol{\beta}) f_1(\sigma_u^2)$, where $f_0(\boldsymbol{\beta}) = 1$ in $\boldsymbol{\beta} \in R^p$ and $f_1(\sigma_u^2)$ is a p.d.f. in $(0, \infty)$.

(4a) Conditioned to σ_u^2, $\mathbf{e} = \underset{1 \le d \le D}{\mathrm{col}} (e_d)$ and $\mathbf{u} = \underset{1 \le d \le D}{\mathrm{col}} (u_d)$ are mutually independent.

The HB predictor of μ_d is the posterior mean $\hat{\mu}_d^{HB} = E[\mu_d | \mathbf{y}]$, i.e.

$$\hat{\mu}_d^{HB} = \int_0^\infty E[\mu_d | \mathbf{y}, \sigma_u^2]\, f_1(\sigma_u^2 | \mathbf{y})\, d\sigma_u^2 = E_{\sigma_u^2}[\tilde{\mu}_d^{HB}(\sigma_u^2)],$$

and the posterior variance is

$$\mathrm{var}(\mu_d | \mathbf{y}) = E_{\sigma_u^2}[g_1(\sigma_u^2) + g_2(\sigma_u^2) | \mathbf{y}] + \mathrm{var}_{\sigma_u^2}(\tilde{\mu}_d^{HB}(\sigma_u^2) | \mathbf{y}),$$

where $E_{\sigma_u^2}$ and $\mathrm{var}_{\sigma_u^2}$ denote expectation and variance with respect to the posterior p.d.f. of σ_u^2, which is written as $f_1(\sigma_u^2 | \mathbf{y})$.

Rao (2003) reviewed and discussed possible prior distributions for σ_u^2. In what follows, we give a short summary of his discussion. First, the flat improper prior p.d.f. $f_1(\sigma_u^2) = 1$ has a proper posterior p.d.f. $f_1(\sigma_u^2|\mathbf{y})$ if $D > p + 2$. Second, the inverted gamma prior distribution, with p.d.f.

$$f_1(\sigma_u^2) = \frac{b^a}{\Gamma(a)}(\sigma_u^2)^{-(a+1)} \exp\{-b/\sigma_u^2\}, \quad \sigma_u^2 > 0, \tag{16.14}$$

facilitates the calculation of the conditional p.d.f. required to program a Gibbs sampling algorithm for approximating the HB predictors $\hat{\mu}_d^{HB}$, $d = 1, \ldots, D$. Note that assuming (16.14) is equivalent to assuming $\sigma_u^{-2} \sim G(a, b)$, where $G(a, b)$ denotes the gamma distribution with shape parameter $a > 0$ and scale parameter $b > 0$, i.e.

$$f_1(\sigma_u^{-2}) = \frac{b^a}{\Gamma(a)}(\sigma_u^{-2})^{a-1} \exp\{-b\sigma_u^{-2}\}, \quad \sigma_u^{-2} > 0. \tag{16.15}$$

Third, the prior distribution $f_1(\sigma_u^2) = \kappa^{-1} h_1(\sigma_u^2)$, where

$$h_1(\sigma_u^2) = (\sigma_u^2 + \sigma_d^2)^2 \sum_{\ell=1}^{D} (\sigma_u^2 + \sigma_\ell^2)^{-2}, \quad \kappa = \int_0^\infty h_1(\sigma_u^2)\, d\sigma_u^2, \quad \sigma_u^2 > 0,$$

ensures that the posterior variance of μ_d is nearly unbiased for the frequentist mean squared error of the HB predictor. More concretely, it implies that

$$E_{\mathbf{y}}\big[\text{var}(\mu_d|\mathbf{y})\big] - MSE(\hat{\mu}_d^{HB}) = o(D^{-1}),$$

where $E_{\mathbf{y}}$ denotes expectation with respect to the marginal p.d.f. $f(\mathbf{y})$, and MSE is calculated with respect to the conditioned p.d.f. $f(\mathbf{y}|\boldsymbol{\beta}, \sigma_u^2)$.

If the posterior density $f_1(\sigma_u^2|\mathbf{y})$ can be easily calculated, then it is possible to approximate $\hat{\mu}_d^{HB}$ by generating random numbers from $f_1(\sigma_u^2|\mathbf{y})$ and applying the Monte Carlo method. The approximation algorithm is

1. For $i = 1, \ldots, I$, generate $\sigma_u^{2(i)} \sim f_1(\sigma_u^2|\mathbf{y})$.
2. The Monte Carlo approximation of $\hat{\mu}_d^{HB}$ is

$$\hat{\mu}_d^{HB} \approx \bar{\tilde{\mu}}_d^{HB} = \frac{1}{I} \sum_{i=1}^{I} \tilde{\mu}_d^{HB}(\sigma_u^{2(i)}).$$

3. The Monte Carlo approximation of $\text{var}(\mu_d|\mathbf{y})$ is

$$\text{var}(\mu_d|\mathbf{y}) \approx \frac{1}{I} \sum_{i=1}^{I} \{g_1(\sigma_u^{2(i)}) + g_2(\sigma_u^{2(i)})\} + \frac{1}{I-1} \sum_{i=1}^{I} \big(\tilde{\mu}_d^{HB}(\sigma_u^{2(i)}) - \bar{\tilde{\mu}}_d^{HB}\big)^2.$$

An alternative approach is to use a Markov chain Monte Carlo (MCMC) method, like the Gibbs sampling algorithm described in Section 10.3 of Rao (2003) or in Rao and Molina (2015). For this sake, three conditional p.d.f.s are needed. Proposition 16.6 gives the p.d.f. of $\boldsymbol{\mu}$ conditioned to $\boldsymbol{y}$, σ_u^2, and $\boldsymbol{\beta}$. Proposition 16.7 gives the p.d.f. of $\boldsymbol{\beta}$ conditioned to $\boldsymbol{\mu}$, σ_u^2, and $\boldsymbol{y}$. Finally, Proposition 16.8 gives the p.d.f. of σ_u^{-2} conditioned to $\boldsymbol{\beta}$, $\boldsymbol{\mu}$, and $\boldsymbol{y}$.

Proposition 16.6 *Under (1), (2), (3a), and (4a), it holds that*

$$\boldsymbol{\mu}|\boldsymbol{\beta}, \sigma_u^2, \boldsymbol{y} \sim N_D\big(\boldsymbol{V}_u\boldsymbol{V}^{-1}\boldsymbol{y} + \boldsymbol{V}_e\boldsymbol{V}^{-1}\boldsymbol{X}\boldsymbol{\beta},\ \boldsymbol{V}_u\boldsymbol{V}_e\boldsymbol{V}^{-1}\big). \tag{16.16}$$

In particular, it holds that

$$\mu_d|\boldsymbol{\beta}, \sigma_u^2, y_d \sim N\Big(\hat{\mu}_d^B, \frac{\sigma_d^2\sigma_u^2}{\sigma_d^2+\sigma_u^2}\Big),\ \hat{\mu}_d^B = \frac{\sigma_u^2}{\sigma_u^2+\sigma_d^2}\, y_d + \frac{\sigma_d^2}{\sigma_u^2+\sigma_d^2}\, \boldsymbol{x}_d\boldsymbol{\beta},\ d = 1, \ldots, D.$$

Proof The proof follows the same steps as the proof of Proposition 16.2. □

Proposition 16.7 *Under (1), (2), (3a), and (4a), it holds that*

$$\boldsymbol{\beta}|\boldsymbol{\mu}, \sigma_u^2, \boldsymbol{y} \sim N_p\big(\boldsymbol{\beta}^*, \sigma_u^2(\boldsymbol{X}'\boldsymbol{X})^{-1}\big), \quad \boldsymbol{\beta}^* = (\boldsymbol{X}'\boldsymbol{X})^{-1}\boldsymbol{X}'\boldsymbol{\mu}. \tag{16.17}$$

Proof As the conditional p.d.f.

$$f(\boldsymbol{\mu}, \sigma_u^2|\boldsymbol{y}) = \int_{R^p} f(\boldsymbol{\beta}, \boldsymbol{\mu}, \sigma_u^2|\boldsymbol{y})\, d\boldsymbol{\beta}$$

does not depend on $\boldsymbol{\beta}$, we get

$$f(\boldsymbol{\beta}|\boldsymbol{\mu}, \sigma_u^2, \boldsymbol{y}) = \frac{f(\boldsymbol{\beta}, \boldsymbol{\mu}, \sigma_u^2|\boldsymbol{y})}{f(\boldsymbol{\mu}, \sigma_u^2|\boldsymbol{y})} \propto f(\boldsymbol{\beta}, \boldsymbol{\mu}, \sigma_u^2|\boldsymbol{y}).$$

As $f(\sigma_u^2|\boldsymbol{y})$ does not depend on $\boldsymbol{\beta}$, it holds that

$$f(\boldsymbol{\beta}, \boldsymbol{\mu}, \sigma_u^2|\boldsymbol{y}) = f(\boldsymbol{\mu}|\boldsymbol{\beta}, \sigma_u^2, \boldsymbol{y}) f(\boldsymbol{\beta}|\sigma_u^2, \boldsymbol{y}) f(\sigma_u^2|\boldsymbol{y}) \propto f(\boldsymbol{\mu}|\boldsymbol{\beta}, \sigma_u^2, \boldsymbol{y}) f(\boldsymbol{\beta}|\sigma_u^2, \boldsymbol{y}).$$

As $\boldsymbol{V}_e\boldsymbol{V}_u^{-1} + \boldsymbol{I}_D = \boldsymbol{V}_u^{-1}\boldsymbol{V} = \sigma_u^{-2}\boldsymbol{V}$, Proposition 16.6 gives $f(\boldsymbol{\mu}|\boldsymbol{\beta}, \sigma_u^2, \boldsymbol{y})$. From Proposition 16.4, it follows that

$$\boldsymbol{\beta}|\sigma_u^2, \boldsymbol{y} \sim N_p\big((\boldsymbol{X}'\boldsymbol{V}^{-1}\boldsymbol{X})^{-1}\boldsymbol{X}'\boldsymbol{V}^{-1}\boldsymbol{y},\ (\boldsymbol{X}'\boldsymbol{V}^{-1}\boldsymbol{X})^{-1}\big)$$

and we have that $h(\boldsymbol{\beta}) = f(\boldsymbol{\beta}|\boldsymbol{\mu}, \sigma_u^2, \boldsymbol{y})$ is

$$
\begin{aligned}
h(\boldsymbol{\beta}) &\propto \exp\Big\{-\frac{1}{2}(\boldsymbol{\mu} - \boldsymbol{V}_u\boldsymbol{V}^{-1}\boldsymbol{y} - \boldsymbol{V}_e\boldsymbol{V}^{-1}\boldsymbol{X}\boldsymbol{\beta})'\boldsymbol{V}_u^{-1}\boldsymbol{V}_e^{-1}\boldsymbol{V}(\boldsymbol{\mu} - \boldsymbol{V}_u\boldsymbol{V}^{-1}\boldsymbol{y} - \boldsymbol{V}_e\boldsymbol{V}^{-1}\boldsymbol{X}\boldsymbol{\beta}) \\
&\quad - \frac{1}{2}(\boldsymbol{\beta} - (\boldsymbol{X}'\boldsymbol{V}^{-1}\boldsymbol{X})^{-1}\boldsymbol{X}'\boldsymbol{V}^{-1}\boldsymbol{y})'\boldsymbol{X}'\boldsymbol{V}^{-1}\boldsymbol{X}(\boldsymbol{\beta} - (\boldsymbol{X}'\boldsymbol{V}^{-1}\boldsymbol{X})^{-1}\boldsymbol{X}'\boldsymbol{V}^{-1}\boldsymbol{y})\Big\} \\
&\propto \exp\Big\{-\frac{1}{2}\boldsymbol{\beta}'\boldsymbol{X}'\boldsymbol{V}^{-1}\boldsymbol{V}_e\boldsymbol{V}_u^{-1}\boldsymbol{V}_e^{-1}\boldsymbol{V}\boldsymbol{V}_e\boldsymbol{V}^{-1}\boldsymbol{X}\boldsymbol{\beta} - \frac{1}{2}\boldsymbol{\beta}'\boldsymbol{X}'\boldsymbol{V}^{-1}\boldsymbol{X}\boldsymbol{\beta} \\
&\quad + \boldsymbol{\beta}'\boldsymbol{X}'\boldsymbol{V}^{-1}\boldsymbol{V}_e\boldsymbol{V}_u^{-1}\boldsymbol{V}_e^{-1}\boldsymbol{V}\boldsymbol{\mu} - \boldsymbol{\beta}'\boldsymbol{X}'\boldsymbol{V}^{-1}\boldsymbol{V}_e\boldsymbol{V}_u^{-1}\boldsymbol{V}_e^{-1}\boldsymbol{V}\boldsymbol{V}_u\boldsymbol{V}^{-1}\boldsymbol{y} \\
&\quad + \boldsymbol{\beta}'(\boldsymbol{X}'\boldsymbol{V}^{-1}\boldsymbol{X})(\boldsymbol{X}'\boldsymbol{V}^{-1}\boldsymbol{X})^{-1}\boldsymbol{X}'\boldsymbol{V}^{-1}\boldsymbol{y}\Big\} \\
&= \exp\Big\{-\frac{1}{2}\boldsymbol{\beta}'\boldsymbol{X}'\boldsymbol{V}^{-1}(\boldsymbol{V}_e\boldsymbol{V}_u^{-1} + \boldsymbol{I}_D)\boldsymbol{X}\boldsymbol{\beta} + \boldsymbol{\beta}'\boldsymbol{X}'\boldsymbol{V}_u^{-1}\boldsymbol{\mu} - \boldsymbol{\beta}'\boldsymbol{X}'\boldsymbol{V}^{-1}\boldsymbol{y} + \boldsymbol{\beta}'\boldsymbol{X}'\boldsymbol{V}^{-1}\boldsymbol{y}\Big\} \\
&= \exp\Big\{-\frac{1}{2\sigma_u^2}\boldsymbol{\beta}'\boldsymbol{X}'\boldsymbol{X}\boldsymbol{\beta} + \frac{1}{\sigma_u^2}\boldsymbol{\beta}'\boldsymbol{X}'\boldsymbol{X}[(\boldsymbol{X}'\boldsymbol{X})^{-1}\boldsymbol{X}'\boldsymbol{\mu}]\Big\},
\end{aligned}
$$

which is the kernel of the p.d.f. $N_p\big(\boldsymbol{\beta}^*, \sigma_u^2(\boldsymbol{X}'\boldsymbol{X})^{-1}\big)$ with $\boldsymbol{\beta}^* = (\boldsymbol{X}'\boldsymbol{X})^{-1}\boldsymbol{X}'\boldsymbol{\mu}$ (cf. (16.12)). □

Proposition 16.8 *Under (1), (2), (3a), (4a), and (16.15), it holds that*

$$
\sigma_u^{-2}|\boldsymbol{\beta}, \boldsymbol{\mu} \sim G(a^*, b^*), \quad a^* = a + \frac{D}{2}, \; b^* = b + \frac{1}{2}(\boldsymbol{\mu} - \boldsymbol{X}\boldsymbol{\beta})'(\boldsymbol{\mu} - \boldsymbol{X}\boldsymbol{\beta}). \tag{16.18}
$$

Proof As $\boldsymbol{\mu}|\sigma_u^{-2}, \boldsymbol{\beta} \sim N_D(\boldsymbol{X}\boldsymbol{\beta}, \boldsymbol{V}_u)$, it holds that

$$
\begin{aligned}
f(\sigma_u^{-2}|\boldsymbol{\mu}, \boldsymbol{\beta}) &\propto f(\boldsymbol{\mu}|\sigma_u^{-2}, \boldsymbol{\beta}) f(\sigma_u^{-2}|\boldsymbol{\beta}) = f(\boldsymbol{\mu}|\sigma_u^{-2}, \boldsymbol{\beta}) f(\sigma_u^{-2}) \\
&\propto (\sigma_u^{-2})^{D/2} \exp\big\{-\frac{1}{2}\sigma_u^{-2}(\boldsymbol{\mu} - \boldsymbol{X}\boldsymbol{\beta})'(\boldsymbol{\mu} - \boldsymbol{X}\boldsymbol{\beta}\big\}(\sigma_u^{-2})^{a-1}\exp\big\{-b\sigma_u^{-2}\big\} \\
&= (\sigma_u^{-2})^{a+\frac{D}{2}-1}\exp\Big\{-\sigma_u^{-2}\big(b + \frac{1}{2}(\boldsymbol{\mu} - \boldsymbol{X}\boldsymbol{\beta})'(\boldsymbol{\mu} - \boldsymbol{X}\boldsymbol{\beta})\big)\Big\},
\end{aligned}
$$

which is the kernel of the p.d.f. $G(a^*, b^*)$. □

The Gibbs sampling algorithm generates a large sequence of MCMC samples $\{\boldsymbol{\beta}^{(i)}, \boldsymbol{\mu}^{(i)}, \sigma_u^{2(i)}\}$, $i = i_0 + 1, \ldots, i_0 + I$, with the first i_0 burn-in iterations deleted. The three Gibbs conditional p.d.f.s have closed form, so the MCMC samples can be generated directly and iteratively from (16.16)–(16.18). The HB predictor of μ_d

and the posterior variance can be approximated by

$$\hat{\mu}_d^{HB} \approx \bar{\tilde{\mu}}_d^{HB} = \frac{1}{I}\sum_{i=i_0+1}^{i_0+I} \tilde{\mu}_d^{HB}(\sigma_u^{2(i)}),$$

$$\tilde{\mu}_d^{HB}(\sigma_u^{2(i)}) = \frac{\sigma_u^{2(i)}}{\sigma_u^{2(i)}+\sigma_d^2}\, y_d + \frac{\sigma_d^2}{\sigma_u^{2(i)}+\sigma_d^2}\, \boldsymbol{x}_d \tilde{\boldsymbol{\beta}}^{(i)},$$

where $\tilde{\boldsymbol{\beta}}^{(i)} = \left(\boldsymbol{X}'(\boldsymbol{V}^{(i)})^{-1}\boldsymbol{X}\right)^{-1}\boldsymbol{X}'(\boldsymbol{V}^{(i)})^{-1}\boldsymbol{y}$, $\boldsymbol{V}^{(i)} = \underset{1\le d\le D}{\text{diag}}\left(\sigma_u^{2(i)}+\sigma_d^2\right)$ and

$$\text{var}(\mu_d|\boldsymbol{y}) \approx \frac{1}{I}\sum_{i=i_0+1}^{i_0+I}\left\{g_1(\sigma_u^{2(i)})+g_2(\sigma_u^{2(i)})\right\}+\frac{1}{I-1}\sum_{i=i_0+1}^{i_0+I}\left(\tilde{\mu}_d^{HB}(\sigma_u^{2(i)})-\bar{\tilde{\mu}}_d^{HB}\right)^2.$$

For an introduction to MCMC algorithms, including the Gibbs sampling, see e.g. Chen et al. (2000) or Robert and Casella (2004).

16.8 Selection of Variables

This section introduces the transformed Fay–Herriot models and gives some comments about the selection of auxiliary variables in the area-level models.

16.8.1 Transformation of the Target Variable

In practical applications, it is sometimes convenient to apply the Fay–Herriot models to transformed data, e.g. if the data are nonlinear on the original scale. In the most common situations, the Fay–Herriot model is fitted on the logarithm of the direct estimates, and the parameter estimates are obtained under this model. Afterward, a back-transformation is done to obtain predictions in the original scale. We now describe this procedure for the log transformation more in detail.

Let us assume that z_d's, $d = 1,\ldots,D$, are direct estimators of quantities of interest with known variances and y_d is the log response, i.e. $y_d = \log(z_d)$. The log-transformed Fay–Herriot model has the form

$$y_d = \mu_d + e_d = \boldsymbol{x}'_d\boldsymbol{\beta} + u_d + e_d, \tag{16.19}$$

and the parameter of interest is defined as

$$\mu_d^* = \exp(\mu_d) = \exp(\boldsymbol{x}'_d\boldsymbol{\beta} + u_d).$$

In order to apply the log-transformed Fay–Herriot model, we need the variance estimators of the transformed variables. One possibility is to employ Taylor's approximation, which for general function h gives (cf. e.g. Neves et al. 2013)

$$\text{var}(h(z)) \approx h'[E(z)]^2 \cdot \text{var}(z),$$

where h' denotes the first derivative of the function h. For the log transformation, we thus obtain

$$\text{var}\big(\ln(z)\big) \approx \frac{\text{var}(z)}{[E(z)]^2} \approx [CV(z)]^2,$$

where CV denotes the coefficient of variation.

Applying the model (16.19), we obtain the EBLUP of μ_d,

$$\hat{\mu}_d = \boldsymbol{x}_d\hat{\boldsymbol{\beta}} + \hat{\gamma}_d(y_d - \boldsymbol{x}_d\hat{\boldsymbol{\beta}}) = \boldsymbol{x}_d\hat{\boldsymbol{\beta}} + \hat{u}_d, \tag{16.20}$$

where $\gamma_d = \hat{\sigma}_u^2(\hat{\sigma}_u^2 + \sigma_d^2)^{-1}$. A direct back-transformation of the EBLUP leads to

$$\hat{\mu}_d^* = \exp(\hat{\mu}_d) = \exp\{\boldsymbol{x}_d\hat{\boldsymbol{\beta}} + \hat{u}_d\},$$

which is a predictor of μ_d^* defined on the original scale. However, this predictor is biased. In fact, there are two types of bias. One comes from the nonlinearity of the back-transformation, and the other comes from the variability of parameter estimates. Slud and Maiti (2006) proposed a bias-corrected predictor of the form

$$\hat{\mu}_d^{*SM} = \exp\left\{\boldsymbol{x}_d\hat{\boldsymbol{\beta}} + \hat{u}_d + \frac{1}{2}\hat{\sigma}_u^2(1 - \hat{\gamma}_d)\right\}$$

and developed analytical estimator of its MSE. Although this predictor corrects bias due to the nonlinearity of the back-transformation, it ignores the bias due to the variability of the parameter estimates. While the first type of bias is of order $O(1)$, i.e. it is not negligible even for large D, the second type of bias is of order $o(1)$, i.e. it vanishes for large D. However, D need not to be large in some applications. In such cases, further bias corrections proposed by Chandra et al. (2018) can be used. Sugasawa and Kubokawa (2015) introduced a parametric transformed Fay–Herriot model for small area estimation. They considered the dual power transformation and provided consistent estimators of the transformation parameter, the regression coefficients and the variance component. For more details and for the MSE estimation under the transformed FH models, we refer readers to the cited papers.

16.8.2 Selection of Auxiliary Variables

In small area estimation, the models give the reference distributions to optimize the properties of the predictors of indicators or population parameters. These indicators are functions of the values that one or several dependent variables take in all the units of the population. To construct predictors, the unit-level models must predict the unobserved (out-of-sample) values of the dependent variables. On the other hand, the area-level models predict the observed (in-sample) values of functions of the sample values of the dependent variables (direct estimators). Having to predict unobserved or observed values is a substantial difference for the selection of the set of auxiliary variables and the fitting of the model to the data.

In the process of selecting a model, it is necessary to take into account what the final objective of the statistical analysis is. If we want to predict out-of-sample values, it is not convenient to over-parameterize. A perfect prediction of the observed values could lead to an absurd prediction of the unobserved values. It is recommended to use a model where all the auxiliary variables are significant, that is, with p-values lower than 0.05. This is what happens with the unit-level models in SAE. On the other hand, if we simply want to predict what is observed, then we can over-parameterize and include auxiliary variables with p-values greater than 0.05 in the model. One could even go to the extreme of selecting a saturated model with perfect prediction of what was observed. With some nuances, we could affirm that the area-level models in SAE are within this second typology.

On area-level models, the dependent variables are typically direct estimators of means or totals of population domains that, due to the small sample size, are not sufficiently reliable. For this reason, the area-level models describe the behavior of the direct estimators considering the relationships with aggregate auxiliary variables, the interdependencies of the data from different domains, and the possible temporal or spatial correlation structures. All this makes it possible to give predictions of the expected values (under the model) of the direct estimators. In a simplified way, the models draw a prediction line that passes through the point cloud formed by the direct estimates. It is therefore a smoothing process.

These intuitive ideas are important in establishing guidelines for the selection of auxiliary variables. An over-parameterized model will tend to reproduce direct (unreliable) estimates with excess variability across domains (poor spatial stability). That is not recommended. On the other hand, a model where all the auxiliary variables are significant could produce an over-smoothing effect and would not collect variability between domains well enough. A point of consensus between both extremes may be to include the significant auxiliary variables that are detected plus some other, with p-value greater than 0.05, that has a reasonable interpretation.

The auxiliary variables that enter a unit-level or area-level model must have a relationship with the dependent variable that can be reasonably interpreted in accordance with the laws of science (sociology, demographic economics, epidemiology, etc.) that govern the data. Interpretability is a factor to be taken into account in the selection of the model, apart from being a tool for its justification before a general public. Finally, it should be noted that the interpretation of the relationship between auxiliary and dependent variables varies with the level of aggregation. This must be taken into account when modeling to avoid making selection errors.

In the example of Sect. 16.9, a Fay–Herriot model is fitted to data from the `datLCS.txt` file, where one of the explanatory variables (proportion of unemployed population) has a p-value greater than 0.05. This variable is included in the model since it has a reasonable interpretation. In those territories where the proportion of unemployed population is big, the average annual household income tends to be small and vice versa.

16.9 R Codes for EBLUPs

This section gives R codes for fitting the Fay–Herriot model to the survey data file `datLCS.txt`. The target variable y is the direct estimator of the domain mean of the variable income. As auxiliary variables, x_1, x_2, and x_3, we take the domain means of the labor status situations, employed, unemployed, and inactive.
We install and/or load the R package `sae`.

```
if(!require(sae)){
   install.packages("sae")
   library(sae)
   }
```

The following code reads the data files and renames some variables:

```
# Read data
dat <- read.table("datLCS.txt",header=TRUE,sep="\t",dec=",")
# Read aux
aux <- read.table("auxLCS.txt",header=TRUE,sep="\t",dec=",")
# Sort aux by dom
aux <- aux[order(aux$dom),]
# Rename variables
income <- dat$income; w <- dat$w; dom <- dat$dom; D <- nrow(aux)
Mwork <- aux$Mwork; Mnowork <- aux$Mnowork
Minact <- aux$Minact; Nd <- aux$TOT
```

We calculate direct estimators of domain average incomes, their variance estimates, and the population sizes by domain. We also define some new variables.

```
income.dir <- dir2(data=income, w, domain=list(dom=dom))
diry <- income.dir$mean; vardiry <- income.dir$var.mean
CVdir <- round(100*sqrt(vardiry)/diry, 2)
hatNd <- income.dir$Nd.hat; nd <- income.dir$nd
log.var <- log(vardiry)
```

We apply the generalized variance function (GVF) approach. We fit the model

$$\log \widehat{\text{var}}_\pi(y_d) = b_0 + b_1 y_d + b_2 n_d + b_3 y_d n_d + \varepsilon_d \quad \text{(Model A)},$$

where the ε_d's are i.i.d. $N(0, \sigma_3^2)$.

The following code calculates the estimated parameters and p-values of Model A. It also calculates GVF variance estimators:

```
model.A <- lm(log.var ~ diry + nd + diry:nd)      # GVF Model A
summary(model.A)                                   # Model coefficients
res <- resid(model.A)                              # Residuals
p <- predict(model.A)                              # Fitted values
# Error variance of GVF model
v <- deviance(model.A)/df.residual(model.A)
Vgvf <- exp(p)*exp(v/2)                            # GVF variance estimates
CVgvf <- round(100*sqrt(Vgvf)/diry, 2)             # GVF CV estimates
```

Table 16.1 presents the estimated parameters and p-values.

Figure 16.1 (left) presents the residuals of the fitted GVF Model A. The residuals are randomly situated above and below the line $y = 0$ with no apparent pattern. Figure 16.1 (right) plots the predicted variances σ_d^2 versus the observed (design-based estimated) variances $\sigma_{\pi,d}^2$. The plotted points are situated close to the straight line $y = x$ and symmetrically around it, which shows the high prediction power of the fitted Model A.

Figure 16.2 (left) plots the observed variances $\sigma_{\pi,d}^2$ and the predicted variances σ_d^2 versus the direct estimates y_d of poverty proportions. The dispersion graph shows that predicted variances follow the same pattern as the observed variances, but they have a smoother behavior. Figure 16.2 (right) plots the design-based estimated log-variances $\log \sigma_{\pi,d}^2$ versus the direct estimates y_d of poverty proportions but with dots colored by the quartile sample size levels of 1, 2, 3, and 4.

We select a Fay–Herriot model. First, we fit the model with the three auxiliary variables, but employed is not significant. Because of interpretability, we select the

Table 16.1 Estimated parameters of GVF Model A

Parameter	Estimate		Std. error	t-value	p-value
b_0	9.78		8.61×10^{-1}	11.367	0.0000
b_1	2.96×10^{-4}		5.72×10^{-5}	5.175	0.0000
b_2	1.39×10^{-2}		1.03×10^{-2}	1.345	0.1922
b_3	-1.20×10^{-6}		6.71×10^{-7}	−1.794	0.0866

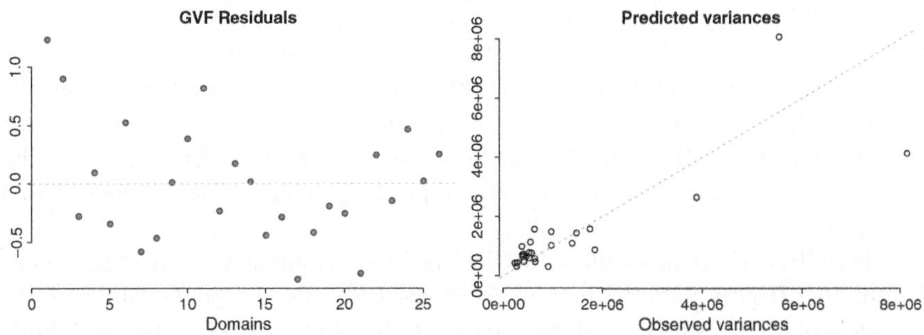

Fig. 16.1 Dispersion graphs of residuals and predicted variances

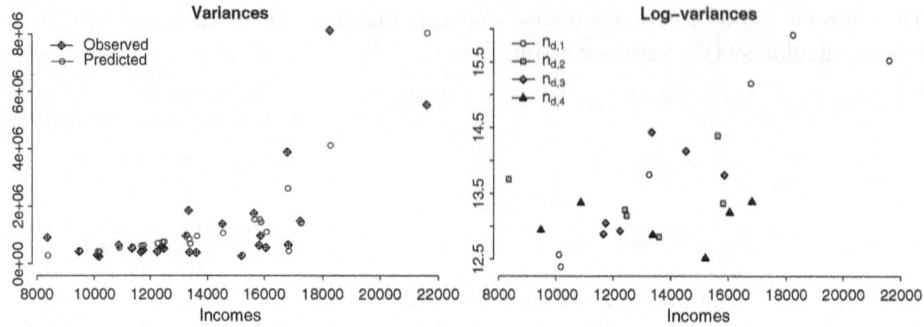

Fig. 16.2 Dispersion graphs of variances and log-variances versus income estimates

Table 16.2 Estimated parameters of FH model

Parameter	Estimate	Std. error	z-value	p-value
Intercept	26690.30	6774.74	3.9397	0.0001
Mnowork	−31385.60	21347.94	−1.4702	0.1415
Minact	−25340.52	13965.29	−1.8145	0.0696

model with unemployed (Mnowork) and inactive (Minact) as regressors. Table 16.2 presents the estimated parameters and p-values.

```
# Model 123
M123 <- eblupFH(diry ~ Mwork + Mnowork + Minact, Vgvf)
M123$fit$estcoef; M123$fit$goodness[2]
# Take Mwork out. It has the highest p-value
M23 <- eblupFH(diry ~ Mnowork + Minact, Vgvf)
M23$fit$estcoef; M23$fit$goodness[2]
# Select model M23
```

We calculate the EBLUPs of income means by domain and the corresponding MSE estimators.

```
# EBLUPs
eblup <- M23$eblup
mseM23 <- mseFH(diry ~ Mnowork + Minact, Vgvf)
MSEeblup <- mseM23$mse
# CV with respect to direct estimator
CVeblup <- round(100*sqrt(MSEeblup)/diry, 2)
```

Figure 16.3 plots the FH model residuals $\hat{e}_d = y_d - \hat{y}_d^{eblup}$ versus the direct estimates y_d. For small values of y_d, the residuals tend to be negative, and therefore the EBLUPs $\hat{y}_d^{eblup}$ tend to be higher than the direct estimates. On the other hand, for large values of y_d, the residuals tend to be positive and the EBLUPs tend to be smaller than the direct estimates. This is a typical smoothing effect of model-based estimators.

Figure 16.4 (left) plots the EBLUPs and direct estimates of income means. Figure 16.4 (right) plots the root-MSEs of the EBLUPs and predicted root-GVF variances of direct estimators of income means. We observe that EBLUPs and direct

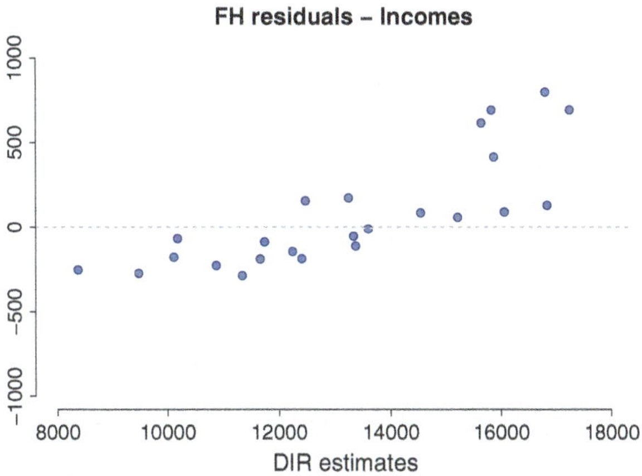

Fig. 16.3 Dispersion graphs of residuals versus direct estimates of incomes

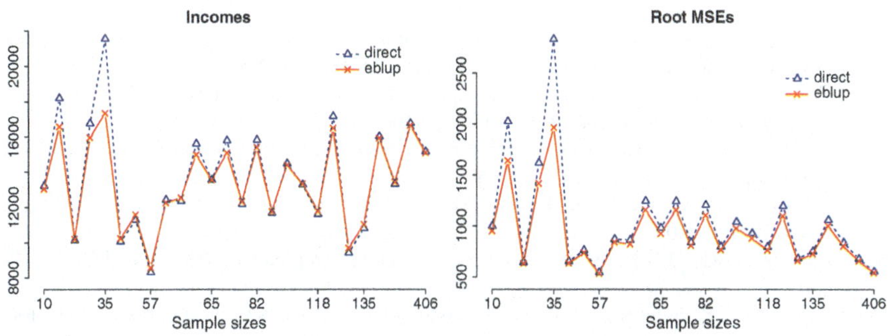

Fig. 16.4 Plots of EBLUPs of average incomes (left) and root-MSEs (right)

estimators tend to coincide as soon as the sample size increases.
The R code to save the results is

```
output <- data.frame(nd, DIR=round(diry), Vdir=round(vardiry),
                     Vgvf=round(Vgvf), CVgvf,
                     EB=round(eblup), MSEeb=round(MSEeblup), CVeb=CVeblup)
head(output, 10)
```

For the ten first areas, Table 16.3 gives the direct estimators (DIR) and the EBLUPs (EB) of the domain income means. The design-based estimates of the variances and coefficients of variations of direct estimators are labeled by Vdir and CVdir, respectively. The labels of the corresponding GVF-based estimates are Vgvf and CVgvf, respectively. The labels of the estimated MSEs and coefficients of variations of EBLUPs are MSEeb and CVeb (with respect to direct estimator), respectively. Table 16.3 shows that estimated MSEs of EBLUPs are smaller than estimated variances of direct estimators.

Table 16.3 Direct and EBLUP estimates of domain mean incomes

dom	n_d	DIR	Vdir	CVdir	Vgvf	CVgvf	EB	MSEeb	CVeb
3	57	8361	905,785	11.38	304,238	6.60	8611	293,413	6.48
5	96	13,334	1,850,153	10.20	867,992	6.99	13,384	783,290	6.64
6	82	15,869	968,480	6.20	1,476,571	7.66	15,449	1,244,240	7.03
7	10	13,245	969,944	7.44	1,018,458	7.62	13,067	924,141	7.26
11	118	11,662	393,578	5.38	639,527	6.86	11,847	593,605	6.61
12	18	16,785	3,884,039	11.74	2,649,280	9.70	15,982	2,031,123	8.49
13	138	16,057	547,935	4.61	1,130,715	6.62	15,961	1,012,397	6.27
14	190	13,370	386,865	4.65	710,698	6.31	13,478	655,385	6.05
15	406	15,211	271,868	3.43	310,468	3.66	15,147	299,409	3.60
16	93	14,531	1,389,995	8.11	1,089,531	7.18	14,441	965,069	6.76

References

Arima, S., Datta, G.S., Liseo, B.: Bayesian estimators for small area models when auxiliary information is measured with error. Scand. J. Stat. **42**, 518–529 (2015)

Burgard, J.P., Esteban, M.D., Morales, D., Pérez, A.: A Fay-Herriot model when auxiliary variables are measured with error. TEST **29**, 166–195 (2020)

Chandra, H., Aditya, K., Kumar, S.: Small-area estimation under a log-transformed area-level model. J. Stat. Theory Pract. **12**(3), 497–505 (2018)

Chen, M.-H., Shao, Q.-M., Ibrahim, J.G.: Monte Carlo Methods in Bayesian Computation. Springer, Berlin (2000)

Das, K., Jiang, J., Rao, J.N.K.: Mean squared error of empirical predictor. Ann. Stat. **32**, 818–840 (2004)

Datta, G.S., Lahiri, P.: A unified measure of uncertainty of estimated best linear unbiased predictors in small area estimation problems. Stat. Sin. **10**, 613–627 (2000)

Datta, G., Kubokawa, T., Molina, I., Rao, J.N.K.: Estimation of mean squared error of model-based small area estimators. TEST **20**(2), 367–388 (2011)

Datta, G., Delaigle, A., Hall, P., Wang, L.: Semi-parametric prediction intervals in small areas when auxiliary data are measured with error. Stat. Sin. **28**, 2309–2335 (2018)

Diao, L., Smith, D.D., Datta, G.S., Maiti, T., Opsomer, J.D.: Accurate confidence interval estimation of small area parameters under the Fay-Herriot model. Scand. J. Stat. **41**(2), 497–515 (2014)

Fay, R.E., Herriot, R.A.: Estimates of income for small places: an application of James-Stein procedures to census data. J. Am. Stat. Assoc. **74**, 269–277 (1979)

González-Manteiga, W., Lombardía, M.J., Molina, I., Morales, D., Santamaría, L.: Small area estimation under Fay-Herriot models with nonparametric estimation of heteroscedasticity. Stat. Model. **10**(2), 215–239 (2010)

Hall, P., Maiti, T.: Nonparametric estimation of mean-squared prediction error in nested-error regression models. Ann. Stat. **34**(4), 1733–1750 (2006a)

Hall, P., Maiti, T.: On parametric bootstrap methods for small-area prediction. J. Royal Stat. Soc. B **68**, 221–238 (2006b)

Henderson, C.R.: Estimation of changes in herd environment. J. Dairy Sci. **32**, 706 (1949)

Henderson, C.R.: Estimation of genetic parameters. Ann. Math. Stat. **21**, 309–310 (1950)

Henderson, C.R.: Estimation of variance and covariance components. Biometrics **9**, 226–252 (1953)

Henderson, C.R., Kempthorne, O., Searle, S.R., Von Krosigk, C.M.: The estimation of environmental and genetic trends from records subject to culling. Biometrics **15**, 192–218 (1959)

Henderson, C.R.: Selection index and expected genetic advance. Stat. Gen. Plant Breed. **NAS-NRC 982**, 141–163 (1963)
Jiang, J.: Linear and generalized linear mixed models and their applications. Springer (2007)
Jiang, J., Tang, E.: The best EBLUP in the Fay-Herriot model. Ann. Inst. Stat. Math. **63**, 1123–1140 (2011)
Kubokawa, T.: On measuring uncertainty of small area estimators. J. Jpn. Stat. Soc. **41**(2), 93–119 (2011)
Lombardía, M.J., López-Vizcaíno, E., Rueda, C.: Mixed generalized Akaike information criterion for small area models. J. R. Stat. Assoc. A **180**(4), 1229–1252 (2017)
Marhuenda, Y., Morales, D., Pardo, M.C.: Information criteria for Fay-Herriot model selection. Comput. Stat. Data Anal. **70**, 268–280 (2014)
Marhuenda, Y., Morales, D., Pardo, M.C.: Tests for the variance parameter in the Fay-Herriot model. Statistics **50**(1), 27–42 (2016)
Molina, I., Rao, J.N.K., Datta, G.S.: Small area estimation under a Fay-Herriot model with preliminary testing for the presence of random area effects. Surv. Methodol. **41**(1), 1–19 (2015)
Neves, A., Silva, D., Correa, S.: Small domain estimation for the Brazilian service sector survey. Estadistica **65**(185), 13–37 (2013)
Polettini, S.: A generalised semiparametric Bayesian Fay-Herriot Model for small area estimation shrinking both means and variances. Bayesian Anal. **12**(3), 729–752 (2017)
Prasad, N.G.N., Rao, J.N.K.: The estimation of the mean squared error of small-area estimators. J. Am. Stat. Assoc. **85**, 163–171 (1990)
Rao, J.N.K.: Small Area Estimation. Wiley, Hoboken (2003)
Rao, J.N.K., Molina, I.: Small Area Estimation, 2nd edn. Wiley, Hoboken (2015)
Robert, C., Casella, G.: Monte Carlo Statistical Methods. Springer, Berlin (2004)
Slud, E.V., Maiti, T.: Mean-squared error estimation in transformed Fay-Herriot models. J. R. Stat. Soc B **68**(2), 239–257 (2006)
Sugasawa, S., Kubokawa, T.: Parametric transformed Fay-Herriot model for small area estimation. J. Multivar. Anal. **139**, 295–311 (2015)
Ybarra, L.M.R., Lohr, S.L.: Small area estimation when auxiliary information is measured with error. Biometrika **95**(4), 919–931 (2008)

Chapter 17
Area-Level Temporal Linear Mixed Models

17.1 Introduction

Temporal data offer information that can be used to obtain more efficient estimates at the current instant, that is, borrowing strength from time is a good strategy. In the small area estimation setup, Choudry and Rao (1989) extended the basic Fay–Herriot model including several time instants and considering an autocorrelated structure for sampling errors. Rao and Yu (1994) proposed a model that borrows information across areas and over time and that includes unexplained area–time variation. Esteban et al. (2012a,b) applied temporal area-level linear mixed models to derive empirical best linear unbiased predictors (EBLUP) of poverty indicators in the Spanish Living Condition Survey.

Other models with temporal correlation have been introduced in the literature. In what follows, some of them are commented. Ghosh et al. (1996) proposed a slightly more complicated time correlated area-level model to estimate the median income of four-person families for the fifty American states and the District of Columbia. You and Rao (2000) and Datta et al. (2002) used the Rao–Yu model, but replacing the AR(1) process by a random walk. Datta et al. (1999) considered a similar model but added extra terms to the linking models to reflect seasonal variation in their application. They applied their model to estimate monthly unemployment rates for nine American states and the District of Columbia. You et al. (2001) applied the Rao–Yu model to estimate monthly unemployment rates for census metropolitan areas in Canada. Pfeffermann and Burck (1990) and Singh et al. (1991) considered a model with time-varying random slopes obeying an autoregressive process.

This chapter describes two temporal extensions of the basic Fay–Herriot area-level model. The first model assumes that the domain-time random effects are independent. The second model assumes an AR(1) correlation structure across time within each domain. For the two models, the residual maximum likelihood (REML) fitting algorithm is given, the EBLUPs of domain means are derived, and

D. Morales et al., *A Course on Small Area Estimation and Mixed Models*, Statistics for Social and Behavioral Sciences, https://doi.org/10.1007/978-3-030-63757-6_17

the problem of estimating the mean squared errors (MSE) is addressed. Further, some simulation results and R codes are given.

17.2 Area-Level Model with Independent Time Effects

In this section we present a temporal generalization of the Fay–Herriot model defined in (16.2).

17.2.1 The Model

Let $u_{1,d} \sim N(0, \sigma_1^2)$, $u_{2,dt} \sim N(0, \sigma_2^2)$ and $e_{dt} \sim N(0, \sigma_{dt}^2)$, $d = 1, \ldots, D$, $t = 1, \ldots, T$, be independent domain and domain-time random effects and sampling errors. The subindex d represents the domain and the subindex t denotes the time period. Alternatively, d may represent domains and t subdomains in a nested population. The area-level linear mixed model with independent time effects is defined by the equation

$$y_{dt} = \boldsymbol{x}_{dt}\boldsymbol{\beta} + u_{1,d} + u_{2,dt} + e_{dt}, \quad d = 1, \ldots, D, \ t = 1, \ldots, T, \tag{17.1}$$

where y_{dt} is a direct estimator of the characteristic of interest and $\boldsymbol{x}_{dt}$ is a vector containing the aggregated values of p auxiliary variables. The error variances, σ_{dt}^2, are assumed to be known, and they are typically taken as the estimated design-based variances of the direct estimators y_{dt}. In the matrix form, the model (17.1) can be written alternatively as

$$\boldsymbol{y} = \boldsymbol{X}\boldsymbol{\beta} + \boldsymbol{Z}_1\boldsymbol{u}_1 + \boldsymbol{Z}_2\boldsymbol{u}_2 + \boldsymbol{e},$$

where $\boldsymbol{y} = \underset{1 \le d \le D}{\text{col}}(\boldsymbol{y}_d)$, $\boldsymbol{y}_d = \underset{1 \le t \le T}{\text{col}}(y_{dt})$, $\boldsymbol{u}_1 = \underset{1 \le d \le D}{\text{col}}(u_{1,d})$, $\boldsymbol{u}_2 = \underset{1 \le d \le D}{\text{col}}(\boldsymbol{u}_{2,d})$, $\boldsymbol{u}_{2,d} = \underset{1 \le t \le T}{\text{col}}(u_{2,dt})$, $\boldsymbol{e} = \underset{1 \le d \le D}{\text{col}}(\boldsymbol{e}_d)$, $\boldsymbol{e}_d = \underset{1 \le t \le T}{\text{col}}(e_{dt})$, $\boldsymbol{X} = \underset{1 \le d \le D}{\text{col}}(\boldsymbol{X}_d)$, $\boldsymbol{X}_d = \underset{1 \le t \le T}{\text{col}}(\boldsymbol{x}_{dt})$, $\boldsymbol{x}_{dt} = \underset{1 \le j \le p}{\text{col}'}(x_{dtj})$, $\boldsymbol{\beta} = \underset{1 \le j \le p}{\text{col}}(\beta_j)$, $\boldsymbol{Z}_1 = \underset{1 \le d \le D}{\text{diag}}(\boldsymbol{1}_T)$, $\boldsymbol{Z}_2 = \boldsymbol{I}_M$, and $M = DT$. The multivariate distributions of the independent vectors of random effects and errors are $\boldsymbol{u}_1 \sim N(\boldsymbol{0}, \boldsymbol{V}_{u_1})$, $\boldsymbol{u}_2 \sim N(\boldsymbol{0}, \boldsymbol{V}_{u_2})$, and $\boldsymbol{e} \sim N(\boldsymbol{0}, \boldsymbol{V}_e)$, with covariance matrices

$$\boldsymbol{V}_{u_1} = \sigma_1^2\boldsymbol{I}_D, \ \boldsymbol{V}_{u_2} = \sigma_2^2\boldsymbol{I}_M, \ \boldsymbol{V}_e = \underset{1 \le d \le D}{\text{diag}}(\boldsymbol{V}_{ed}), \ \boldsymbol{V}_{ed} = \underset{1 \le t \le T}{\text{diag}}(\sigma_{dt}^2).$$

The model (17.1) can also be written in the form $\boldsymbol{y} = \boldsymbol{X}\boldsymbol{\beta} + \boldsymbol{Z}\boldsymbol{u} + \boldsymbol{e}$, where $\boldsymbol{Z} = [\boldsymbol{Z}_1, \boldsymbol{Z}_2]$ and $\boldsymbol{u} = (\boldsymbol{u}_1', \boldsymbol{u}_2')'$. For σ_1^2, σ_2^2 known, the estimators and predictors

BLU of $\boldsymbol{\beta}$ and $\boldsymbol{u}$ are (cf. Proposition 16.1)

$$\tilde{\boldsymbol{\beta}} = (\boldsymbol{X}'\boldsymbol{V}^{-1}\boldsymbol{X})^{-1}\boldsymbol{X}'\boldsymbol{V}^{-1}\boldsymbol{y} = \Big(\sum_{d=1}^{D}\boldsymbol{X}'_d\boldsymbol{V}_d^{-1}\boldsymbol{X}_d\Big)^{-1}\Big(\sum_{d=1}^{D}\boldsymbol{X}'_d\boldsymbol{V}_d^{-1}\boldsymbol{y}_d\Big)$$

and

$$\begin{aligned}
\tilde{\boldsymbol{u}} &= \boldsymbol{V}_u\boldsymbol{Z}'\boldsymbol{V}^{-1}(\boldsymbol{y}-\boldsymbol{X}\tilde{\boldsymbol{\beta}}) \\
&= \begin{pmatrix} \sigma_1^2\boldsymbol{I}_D & \boldsymbol{0} \\ \boldsymbol{0} & \sigma_2^2\boldsymbol{I}_M \end{pmatrix} \begin{pmatrix} \underset{1\le d\le D}{\text{diag}}\,(\boldsymbol{1}'_T) \\ \boldsymbol{I}_M \end{pmatrix} \underset{1\le d\le D}{\text{diag}}\,(\boldsymbol{V}_d^{-1}) \underset{1\le d\le D}{\text{col}}\,(\boldsymbol{y}_d-\boldsymbol{X}_d\tilde{\boldsymbol{\beta}}) \\
&= \begin{pmatrix} \sigma_1^2 \underset{1\le d\le D}{\text{col}}\big(\boldsymbol{1}'_T\boldsymbol{V}_d^{-1}(\boldsymbol{y}_d-\boldsymbol{X}_d\tilde{\boldsymbol{\beta}})\big) \\ \sigma_2^2 \underset{1\le d\le D}{\text{col}}\big(\boldsymbol{V}_d^{-1}(\boldsymbol{y}_d-\boldsymbol{X}_d\tilde{\boldsymbol{\beta}})\big) \end{pmatrix},
\end{aligned}$$

where $\boldsymbol{V}_u = \text{diag}(\sigma_1^2\boldsymbol{I}_D, \sigma_2^2\boldsymbol{I}_M)$, $\boldsymbol{V}_d = \text{var}(\boldsymbol{y}_d) = \sigma_1^2\boldsymbol{1}_T\boldsymbol{1}'_T + \sigma_2^2\boldsymbol{I}_T + \boldsymbol{V}_{ed}$, and

$$\boldsymbol{V} = \text{var}(\boldsymbol{y}) = \sigma_1^2\boldsymbol{Z}_1\boldsymbol{Z}'_1 + \sigma_2^2 \underset{1\le d\le D}{\text{diag}}\,(\boldsymbol{I}_T) + \boldsymbol{V}_e = \underset{1\le d\le D}{\text{diag}}\,(\boldsymbol{V}_d).$$

The empirical versions, EBLUE of $\boldsymbol{\beta}$ and EBLUP of $\boldsymbol{u}$, are obtained by plugging estimators $\hat{\sigma}_1^2, \hat{\sigma}_2^2$ in the place of σ_1^2, σ_2^2, i.e.

$$\hat{\boldsymbol{\beta}} = (\boldsymbol{X}'\hat{\boldsymbol{V}}^{-1}\boldsymbol{X})^{-1}\boldsymbol{X}'\hat{\boldsymbol{V}}^{-1}\boldsymbol{y}, \quad \hat{\boldsymbol{u}} = \hat{\boldsymbol{V}}_u\boldsymbol{Z}'\hat{\boldsymbol{V}}^{-1}\left(\boldsymbol{y}-\boldsymbol{X}\hat{\boldsymbol{\beta}}\right), \tag{17.2}$$

where $\hat{\boldsymbol{V}}_u = \text{diag}(\hat{\sigma}_1^2\boldsymbol{I}_D, \hat{\sigma}_2^2\boldsymbol{I}_M)$ and $\hat{\boldsymbol{V}} = \hat{\sigma}_1^2\boldsymbol{Z}_1\boldsymbol{Z}'_1 + \hat{\sigma}_2^2 \underset{1\le d\le D}{\text{diag}}\,(\boldsymbol{I}_T) + \boldsymbol{V}_e$.

17.2.2 Residual Maximum Likelihood Estimation

In what follows, we present the REML method for parameter estimation. In fact, we particularize the results of Sect. 6.5.3 to the case $m = 2$, $n = M$, $q_1 = D$, $q_2 = M$, $\varphi_1 = \sigma_1^2$, $\varphi_2 = \sigma_2^2$, $\sigma^2 = 1$, $\boldsymbol{\Sigma}_{u_1} = I_D$, and $\boldsymbol{\Sigma}_{u_2} = I_M$. The REML log-likelihood of the temporal model (17.1) is (cf. (6.41))

$$l_{reml}(\sigma_1^2, \sigma_2^2) = -\frac{M-p}{2}\log 2\pi + \frac{1}{2}\log|\boldsymbol{X}'\boldsymbol{X}| - \frac{1}{2}\log|\boldsymbol{V}| - \frac{1}{2}\log|\boldsymbol{X}'\boldsymbol{V}^{-1}\boldsymbol{X}| - \frac{1}{2}\boldsymbol{y}'\boldsymbol{P}\boldsymbol{y},$$

where

$$\boldsymbol{P} = \boldsymbol{V}^{-1} - \boldsymbol{V}^{-1}\boldsymbol{X}(\boldsymbol{X}'\boldsymbol{V}^{-1}\boldsymbol{X})^{-1}\boldsymbol{X}'\boldsymbol{V}^{-1}, \quad \boldsymbol{P}\boldsymbol{V}\boldsymbol{P} = \boldsymbol{P}, \quad \boldsymbol{P}\boldsymbol{X} = \boldsymbol{0}.$$

Let $\boldsymbol{\theta} = (\theta_1, \theta_2) = (\sigma_1^2, \sigma_2^2)$, $\boldsymbol{V}_1 = \frac{\partial \boldsymbol{V}}{\partial \sigma_1^2} = \underset{1\leq d\leq D}{\text{diag}} (\boldsymbol{1}_T \boldsymbol{1}_T')$, and $\boldsymbol{V}_2 = \frac{\partial \boldsymbol{V}}{\partial \sigma_2^2} = \underset{1\leq d\leq D}{\text{diag}} (\boldsymbol{I}_T)$, then (cf. (6.43))

$$\boldsymbol{P}_a = \frac{\partial \boldsymbol{P}}{\partial \theta_a} = -\boldsymbol{P}\frac{\partial \boldsymbol{V}}{\partial \theta_a}\boldsymbol{P} = -\boldsymbol{P}\boldsymbol{V}_a\boldsymbol{P}, \quad a = 1, 2.$$

By taking derivatives of l_{reml} with respect to θ_a, we get (cf. (6.42)) the elements of the score vector $\boldsymbol{S} = \boldsymbol{S}(\boldsymbol{\theta}) = (S_1, S_2)'$,

$$S_a = \frac{\partial l_{reml}}{\partial \theta_a} = -\frac{1}{2}\text{tr}(\boldsymbol{P}\boldsymbol{V}_a) + \frac{1}{2}\boldsymbol{y}'\boldsymbol{P}\boldsymbol{V}_a\boldsymbol{P}\boldsymbol{y}, \quad a = 1, 2.$$

By taking again derivatives with respect to θ_a and θ_b, taking expectations, and changing the sign, we get, in the same way as on page 128, the elements of the Fisher information matrix $\boldsymbol{F} = \boldsymbol{F}(\boldsymbol{\theta}) = (F_{ab})_{a,b=1,2}$, where

$$F_{ab} = \frac{1}{2}\text{tr}(\boldsymbol{P}\boldsymbol{V}_a\boldsymbol{P}\boldsymbol{V}_b), \quad a, b = 1, 2. \tag{17.3}$$

For calculating the REML estimators, the Fisher-scoring updating formula is

$$\boldsymbol{\theta}^{(i+1)} = \boldsymbol{\theta}^{(i)} + \boldsymbol{F}^{-1}(\boldsymbol{\theta}^{(i)})\boldsymbol{S}(\boldsymbol{\theta}^{(i)}).$$

As algorithm seeds it is possible to take $\theta_0^{(0)} = \theta_1^{(0)} = S^2/2$, where $S^2 = \frac{1}{M-p}(\boldsymbol{y} - \boldsymbol{X}\tilde{\boldsymbol{\beta}})'\boldsymbol{V}_e^{-1}(\boldsymbol{y} - \boldsymbol{X}\tilde{\boldsymbol{\beta}})$ and $\tilde{\boldsymbol{\beta}} = (\boldsymbol{X}'\boldsymbol{V}_e^{-1}\boldsymbol{X})^{-1}\boldsymbol{X}'\boldsymbol{V}_e^{-1}\boldsymbol{y}$. The REML estimator of $\boldsymbol{\beta}$ is

$$\hat{\boldsymbol{\beta}} = (\boldsymbol{X}'\hat{\boldsymbol{V}}^{-1}\boldsymbol{X})^{-1}\boldsymbol{X}'\hat{\boldsymbol{V}}^{-1}\boldsymbol{y}.$$

The asymptotic distributions of the REML estimators of $\boldsymbol{\theta}$ and $\boldsymbol{\beta}$ are (see, e.g., Section 1.3.2 in Jiang 2007)

$$\hat{\boldsymbol{\theta}} \sim N_2(\boldsymbol{\theta}, \boldsymbol{F}^{-1}(\boldsymbol{\theta})), \quad \hat{\boldsymbol{\beta}} \sim N_p(\boldsymbol{\beta}, (\boldsymbol{X}'\boldsymbol{V}^{-1}\boldsymbol{X})^{-1}).$$

Asymptotic confidence intervals at the level $1 - \alpha$ for θ_a and β_j are

$$\hat{\theta}_a \pm z_{1-\alpha/2}\, \nu_{aa}^{1/2}, \ a = 1, 2, \quad \hat{\beta}_j \pm z_{1-\alpha/2}\, q_{jj}^{1/2}, \ j = 1, \ldots, p,$$

where $\hat{\boldsymbol{\theta}} = \boldsymbol{\theta}^{\kappa}$, $\boldsymbol{F}^{-1}(\boldsymbol{\theta}^{\kappa}) = (\nu_{ab})_{a,b=1,2}$, $(\boldsymbol{X}'\boldsymbol{V}^{-1}(\boldsymbol{\theta}^{\kappa})\boldsymbol{X})^{-1} = (q_{ij})_{i,j=1,\ldots,p}$, κ is the last iteration of the Fisher-scoring algorithm and z_α is the α-quantile of the $N(0, 1)$) distribution. If $\hat{\beta}_j = \beta_0$, then the p-value for testing $H_0 : \beta_j = 0$ is

$$p = 2P_{H_0}(\hat{\beta}_j > |\beta_0|) = 2P(N(0, 1) > |\beta_0|/\sqrt{q_{jj}}).$$

Let us define the matrices

$$Q = (X'V^{-1}X)^{-1} = \Big(\sum_{d=1}^{D} X_d' V_d^{-1} X_d\Big)^{-1},$$

$$P = \underset{1\le d\le D}{\operatorname{diag}}(V_d^{-1}) - \underset{1\le d\le D}{\operatorname{col}}(V_d^{-1}X_d)Q\underset{1\le d\le D}{\operatorname{col}'}(X_d'V_d^{-1}).$$

Some matrix calculations for applying the REML Fisher-scoring algorithm are

$$PV_a = \underset{1\le d\le D}{\operatorname{diag}}(V_d^{-1}V_{ad}) - \underset{1\le d\le D}{\operatorname{col}}(V_d^{-1}X_d)Q\underset{1\le d\le D}{\operatorname{col}'}(X_d'V_d^{-1}V_{ad}),$$

$$\operatorname{tr}(PV_a) = \sum_{d=1}^{D}\operatorname{tr}(V_d^{-1}V_{ad}) - \sum_{d=1}^{D}\operatorname{tr}(X_d'V_d^{-1}V_{ad}V_d^{-1}X_dQ),$$

and

$$\operatorname{tr}(PV_aPV_b) = \sum_{d=1}^{D}\operatorname{tr}(V_d^{-1}V_{ad}V_d^{-1}V_{bd}) - 2\sum_{d=1}^{D}\operatorname{tr}(X_d'V_d^{-1}V_{ad}V_d^{-1}V_{bd}V_d^{-1}X_dQ)$$
$$+\operatorname{tr}\left\{\Big(\sum_{d=1}^{D}X_d'V_d^{-1}V_{ad}V_d^{-1}X_d\Big)Q\Big(\sum_{d=1}^{D}X_d'V_d^{-1}V_{bd}V_d^{-1}X_d\Big)Q\right\},$$

$$y'PV_aPy = \sum_{d=1}^{D}y_d'V_d^{-1}V_{ad}V_d^{-1}y_d - \Big(\sum_{d=1}^{D}y_d'V_d^{-1}V_{ad}V_d^{-1}X_d\Big)Q\Big(\sum_{d=1}^{D}y_d'V_d^{-1}X_d\Big)'$$
$$-\Big(\sum_{d=1}^{D}y_d'V_d^{-1}X_d\Big)Q\Big(\sum_{d=1}^{D}X_d'V_d^{-1}V_{ad}V_d^{-1}y_d\Big)$$
$$+\Big(\sum_{d=1}^{D}y_d'V_d^{-1}X_d\Big)Q\Big(\sum_{d=1}^{D}X_d'V_d^{-1}V_{ad}V_d^{-1}X_d\Big)Q\Big(\sum_{d=1}^{D}y_d'V_d^{-1}X_d\Big)',$$

where $a = 1, 2$ and $V_{1d} = \mathbf{1}_T\mathbf{1}_T'$, $V_{2d} = I_T$.

17.2.3 EBLUP and Mean Squared Error

This section considers the problem of predicting the linear combination of fixed and random effects $\mu_{dt} = x_{dt}\beta + u_{1,d} + u_{2,dt}$. If the variance components are known, the BLUP of μ_{dt} is $\tilde{\mu}_{dt} = x_{dt}\tilde{\beta} + \tilde{u}_{1,d} + \tilde{u}_{2,dt}$. The corresponding EBLUP of μ_{dt} is obtained by substituting the estimators of the variance components σ_1^2, σ_2^2, and it has the form

$$\hat{\mu}_{dt} = x_{dt}\hat{\beta} + \hat{u}_{1,d} + \hat{u}_{2,dt},$$

where $\hat{\boldsymbol{\beta}}$ and $\hat{\boldsymbol{u}}$ are given in (17.2). The EBLUP of μ_{dt} can also be employed for estimating the domain-time mean $\overline{Y}_{dt}$, in this case the notation $\hat{\overline{Y}}_{dt}^{eblup} = \hat{\mu}_{dt}$ is used.

Notice that the parameter of interest can be written in the form $\mu_{dt} = \boldsymbol{a}'(\boldsymbol{X}\boldsymbol{\beta} + \boldsymbol{Z}\boldsymbol{u})$, where $\boldsymbol{a} = \underset{1\leq\ell\leq D}{\text{col}}\,(\underset{1\leq k\leq T}{\text{col}}\,(\delta_{d\ell}\delta_{tk}))$ is a vector having a one in the cell $t + (d-1)T$ and having zeros in the remaining cells. Thus, for approximating the MSE of $\hat{\overline{Y}}_{dt}^{eblup}$, it is possible to follow the same steps as in Sect. 9.2.3 and to obtain the formulas

$$
\begin{aligned}
MSE(\hat{\overline{Y}}_{dt}^{eblup}) &= g_1(\boldsymbol{\theta}) + g_2(\boldsymbol{\theta}) + g_3(\boldsymbol{\theta}), \qquad (17.4)\\
g_1(\boldsymbol{\theta}) &= \boldsymbol{a}'\boldsymbol{Z}\boldsymbol{T}\boldsymbol{Z}'\boldsymbol{a},\\
g_2(\boldsymbol{\theta}) &= [\boldsymbol{a}'\boldsymbol{X} - \boldsymbol{a}'\boldsymbol{Z}\boldsymbol{T}\boldsymbol{Z}'\boldsymbol{V}_e^{-1}\boldsymbol{X}]\boldsymbol{Q}[\boldsymbol{X}'\boldsymbol{a} - \boldsymbol{X}'\boldsymbol{V}_e^{-1}\boldsymbol{Z}\boldsymbol{T}\boldsymbol{Z}'\boldsymbol{a}],\\
g_3(\boldsymbol{\theta}) &\approx \text{tr}\left\{(\nabla\boldsymbol{b}')\boldsymbol{V}(\nabla\boldsymbol{b}')'E\left[(\hat{\boldsymbol{\theta}} - \boldsymbol{\theta})(\hat{\boldsymbol{\theta}} - \boldsymbol{\theta})'\right]\right\},
\end{aligned}
$$

where $\boldsymbol{T} = \boldsymbol{V}_u - \boldsymbol{V}_u\boldsymbol{Z}'\boldsymbol{V}^{-1}\boldsymbol{Z}\boldsymbol{V}_u$ and $\boldsymbol{b}' = \boldsymbol{a}'\boldsymbol{Z}\boldsymbol{V}_u\boldsymbol{Z}'\boldsymbol{V}^{-1}$. The MSE of $\hat{\overline{Y}}_{dt}^{eblup}$ can be estimated by parametric bootstrap or by using the analytic estimator

$$
mse(\hat{\overline{Y}}_{dt}^{eblup}) = g_1(\hat{\boldsymbol{\theta}}) + g_2(\hat{\boldsymbol{\theta}}) + 2g_3(\hat{\boldsymbol{\theta}}),
$$

where $\hat{\boldsymbol{\theta}}$ is the REML estimator. For more details, see Sect. 9.4. The following sections give the calculations of g_1, g_2, and g_3, and as an alternative, a parametric bootstrap procedure for estimating the mean squared error is described.

Calculation of $g_1(\boldsymbol{\theta})$

It holds that $g_1(\boldsymbol{\theta}) = \boldsymbol{a}'\boldsymbol{Z}\boldsymbol{T}\boldsymbol{Z}'\boldsymbol{a}$, where

$$
\boldsymbol{T} = \boldsymbol{V}_u - \boldsymbol{V}_u\boldsymbol{Z}'\boldsymbol{V}^{-1}\boldsymbol{Z}\boldsymbol{V}_u = \begin{pmatrix} \sigma_1^2\boldsymbol{I}_D & \boldsymbol{0} \\ \boldsymbol{0} & \sigma_2^2\boldsymbol{I}_M \end{pmatrix} - \begin{pmatrix} \sigma_1^2\boldsymbol{I}_D & \boldsymbol{0} \\ \boldsymbol{0} & \sigma_2^2\boldsymbol{I}_M \end{pmatrix} \begin{pmatrix} \underset{1\leq\ell\leq D}{\text{diag}}\,(\boldsymbol{1}_T') \\ \boldsymbol{I}_M \end{pmatrix}
$$

$$
\cdot \underset{1\leq\ell\leq D}{\text{diag}}\,(\boldsymbol{V}_\ell^{-1})(\underset{1\leq\ell\leq D}{\text{diag}}\,(\boldsymbol{1}_T), \boldsymbol{I}_M)\begin{pmatrix} \sigma_1^2\boldsymbol{I}_D & \boldsymbol{0} \\ \boldsymbol{0} & \sigma_2^2\boldsymbol{I}_M \end{pmatrix} = \begin{pmatrix} \boldsymbol{T}_{11} & \boldsymbol{T}_{12} \\ \boldsymbol{T}_{21} & \boldsymbol{T}_{22} \end{pmatrix}
$$

and

$$\boldsymbol{T}_{11} = \sigma_1^2 \boldsymbol{I}_D - \sigma_1^4 \underset{1\leq \ell \leq D}{\text{diag}} (\mathbf{1}_T' \boldsymbol{V}_\ell^{-1} \mathbf{1}_T)) = \underset{1\leq \ell \leq D}{\text{diag}} (T_{11\ell}),$$

$$\boldsymbol{T}_{12} = -\sigma_1^2\sigma_2^2 \underset{1\leq \ell \leq D}{\text{diag}} (\mathbf{1}_T' \boldsymbol{V}_\ell^{-1}) = \underset{1\leq \ell \leq D}{\text{diag}} (\boldsymbol{T}_{12\ell}) = \boldsymbol{T}_{21}',$$

$$\boldsymbol{T}_{22} = \sigma_2^2 \boldsymbol{I}_M - \sigma_2^4 \underset{1\leq \ell \leq D}{\text{diag}} (\boldsymbol{V}_\ell^{-1}) = \underset{1\leq \ell \leq D}{\text{diag}} (\boldsymbol{T}_{22\ell}).$$

Let $\boldsymbol{a}_t = \underset{1\leq k \leq T}{\text{col}} (\delta_{tk})$, then $\boldsymbol{a} = \underset{1\leq \ell \leq D}{\text{col}} (\delta_{d\ell}\boldsymbol{a}_t)$ and

$$g_1(\boldsymbol{\theta}) = \boldsymbol{a}'\boldsymbol{Z}\boldsymbol{T}\boldsymbol{Z}'\boldsymbol{a} = \boldsymbol{a}'\Big(\underset{1\leq \ell \leq D}{\text{diag}} (\mathbf{1}_T), \boldsymbol{I}_M\Big) \begin{pmatrix} \boldsymbol{T}_{11} & \boldsymbol{T}_{12} \\ \boldsymbol{T}_{21} & \boldsymbol{T}_{22} \end{pmatrix} \begin{pmatrix} \underset{1\leq \ell \leq D}{\text{diag}} (\mathbf{1}_T') \\ \boldsymbol{I}_M \end{pmatrix} \boldsymbol{a}$$

$$= \boldsymbol{a}' \underset{1\leq \ell \leq D}{\text{diag}} (\mathbf{1}_T T_{11\ell} \mathbf{1}_T' + \mathbf{1}_T \boldsymbol{T}_{12\ell} + \boldsymbol{T}_{21\ell}\mathbf{1}_T' + \boldsymbol{T}_{22\ell})\boldsymbol{a}$$

$$= \boldsymbol{a}_t'(\mathbf{1}_T T_{11d} \mathbf{1}_T' + \mathbf{1}_T \boldsymbol{T}_{12d} + \boldsymbol{T}_{21d}\mathbf{1}_T' + \boldsymbol{T}_{22d})\boldsymbol{a}_t.$$

Calculation of $g_2(\boldsymbol{\theta})$

It holds that

$$g_2(\boldsymbol{\theta}) = [\boldsymbol{a}'\boldsymbol{X} - \boldsymbol{a}'\boldsymbol{Z}\boldsymbol{T}\boldsymbol{Z}'\boldsymbol{V}_e^{-1}\boldsymbol{X}]\boldsymbol{Q}[\boldsymbol{X}'\boldsymbol{a} - \boldsymbol{X}'\boldsymbol{V}_e^{-1}\boldsymbol{Z}\boldsymbol{T}\boldsymbol{Z}'\boldsymbol{a}] = (\boldsymbol{a}_1' - \boldsymbol{a}_2')\boldsymbol{Q}(\boldsymbol{a}_1 - \boldsymbol{a}_2),$$

where $\boldsymbol{a}_1' = \boldsymbol{a}'\boldsymbol{X} = \boldsymbol{a}_t'\boldsymbol{X}_d = \boldsymbol{x}_{dt}$ and

$$\boldsymbol{a}_2' = \boldsymbol{a}'\boldsymbol{Z}\boldsymbol{T}\boldsymbol{Z}'\boldsymbol{V}_e^{-1}\boldsymbol{X}$$

$$= \boldsymbol{a}'\Big(\underset{1\leq \ell \leq D}{\text{diag}} (\mathbf{1}_T), \boldsymbol{I}_M\Big) \begin{pmatrix} \boldsymbol{T}_{11} & \boldsymbol{T}_{12} \\ \boldsymbol{T}_{21} & \boldsymbol{T}_{22} \end{pmatrix} \begin{pmatrix} \underset{1\leq \ell \leq D}{\text{diag}} (\mathbf{1}_T') \\ \boldsymbol{I}_M \end{pmatrix} \underset{1\leq \ell \leq D}{\text{diag}} (\boldsymbol{V}_{e\ell}^{-1}) \underset{1\leq \ell \leq D}{\text{col}} (\boldsymbol{X}_\ell)$$

$$= \boldsymbol{a}_t' \left\{\mathbf{1}_T T_{11d} \mathbf{1}_T' + \mathbf{1}_T \boldsymbol{T}_{12d} + \boldsymbol{T}_{21d}\mathbf{1}_T' + \boldsymbol{T}_{22d}\right\} \boldsymbol{V}_{ed}^{-1} \boldsymbol{X}_d.$$

Calculation of $g_3(\boldsymbol{\theta})$

It holds that

$$g_3(\boldsymbol{\theta}) \approx \text{tr} \left\{(\nabla \boldsymbol{b}')\boldsymbol{V}(\nabla \boldsymbol{b}')' E\left[(\hat{\boldsymbol{\theta}} - \boldsymbol{\theta})(\hat{\boldsymbol{\theta}} - \boldsymbol{\theta})'\right]\right\},$$

where

$$\begin{aligned} b' &= a'ZV_uZ'V^{-1} \\ &= a'\big(\operatorname*{diag}_{1\le\ell\le D}(\mathbf{1}_T), I_M\big)\begin{pmatrix} \sigma_1^2 I_D & \mathbf{0} \\ \mathbf{0} & \sigma_2^2 I_M \end{pmatrix}\begin{pmatrix} \operatorname*{diag}_{1\le\ell\le D}(\mathbf{1}_T') \\ I_M \end{pmatrix}\operatorname*{diag}_{1\le\ell\le D}(V_\ell^{-1}) \\ &= a'\operatorname*{diag}_{1\le\ell\le D}\big([\sigma_1^2\mathbf{1}_T\mathbf{1}_T' + \sigma_2^2 I_T]V_\ell^{-1}\big). \end{aligned}$$

The derivatives of b' are

$$\begin{aligned} \frac{\partial b'}{\partial\sigma_1^2} &= \operatorname*{col'}_{1\le\ell\le D}\big(\delta_{d\ell}a_t'\mathbf{1}_T\mathbf{1}_T'V_\ell^{-1}\big) - \operatorname*{col'}_{1\le\ell\le D}\big(\delta_{d\ell}a_t'[\sigma_1^2\mathbf{1}_T\mathbf{1}_T' + \sigma_2^2 I_T]V_\ell^{-1}V_{\ell 1}V_\ell^{-1}\big), \\ \frac{\partial b'}{\partial\sigma_2^2} &= \operatorname*{col'}_{1\le\ell\le D}(\delta_{d\ell}a_t'I_TV_\ell^{-1}) - \operatorname*{col'}_{1\le\ell\le D}(\delta_{d\ell}a_t'[\sigma_1^2\mathbf{1}_T\mathbf{1}_T' + \sigma_2^2 I_T]V_\ell^{-1}V_{\ell 2}V_\ell^{-1}), \end{aligned}$$

where $V_{\ell 1} = \frac{\partial V_\ell}{\partial\sigma_1^2} = \mathbf{1}_T\mathbf{1}_T'$ and $V_{\ell 2} = \frac{\partial V_\ell}{\partial\sigma_2^2} = I_T$. Let us define

$$\begin{aligned} \frac{\partial b_d'}{\partial\sigma_1^2} &= a_t'\mathbf{1}_T\mathbf{1}_T'V_d^{-1} - a_t'[\sigma_1^2\mathbf{1}_T\mathbf{1}_T' + \sigma_2^2 I_T]V_d^{-1}\mathbf{1}_T\mathbf{1}_T'V_d^{-1}, \\ \frac{\partial b_d'}{\partial\sigma_2^2} &= a_t'I_TV_d^{-1} - a_t'[\sigma_1^2\mathbf{1}_T\mathbf{1}_T' + \sigma_2^2 I_T]V_d^{-1}I_TV_d^{-1}. \end{aligned}$$

The components of $(\nabla b')V(\nabla b')'$ are

$$\begin{aligned} q_{11} &= \frac{\partial b'}{\partial\sigma_1^2}\operatorname*{diag}_{1\le\ell\le D}(V_\ell)\left(\frac{\partial b'}{\partial\sigma_1^2}\right)' = \frac{\partial b_d'}{\partial\sigma_1^2}V_d\left(\frac{\partial b_d'}{\partial\sigma_1^2}\right)' \\ &= a_t'\mathbf{1}_T\mathbf{1}_T'V_d^{-1}\mathbf{1}_T\mathbf{1}_T'a_t - 2a_t'\mathbf{1}_T\mathbf{1}_T'V_d^{-1}\mathbf{1}_T\mathbf{1}_T'V_d^{-1}[\sigma_1^2\mathbf{1}_T\mathbf{1}_T' + \sigma_2^2 I_T]a_t \\ &+ a_t'[\sigma_1^2\mathbf{1}_T\mathbf{1}_T' + \sigma_2^2 I_T]V_d^{-1}\mathbf{1}_T\mathbf{1}_T'V_d^{-1}\mathbf{1}_T\mathbf{1}_T'V_d^{-1}[\sigma_1^2\mathbf{1}_T\mathbf{1}_T' + \sigma_2^2 I_T]a_t, \end{aligned}$$

$$\begin{aligned} q_{22} &= \frac{\partial b'}{\partial\sigma_2^2}\operatorname*{diag}_{1\le\ell\le D}(V_\ell)\left(\frac{\partial b'}{\partial\sigma_2^2}\right)' = \frac{\partial b_d'}{\partial\sigma_2^2}V_d\left(\frac{\partial b_d'}{\partial\sigma_2^2}\right)' \\ &= a_t'V_d^{-1}a_t - 2a_t'V_d^{-2}[\sigma_1^2\mathbf{1}_T\mathbf{1}_T' + \sigma_2^2 I_T]a_t \\ &+ a_t'[\sigma_1^2\mathbf{1}_T\mathbf{1}_T' + \sigma_2^2 I_T]V_d^{-3}[\sigma_1^2\mathbf{1}_T\mathbf{1}_T' + \sigma_2^2 I_T]a_t, \end{aligned}$$

$$
\begin{aligned}
q_{12} &= \frac{\partial \boldsymbol{b}'}{\partial \sigma_1^2} \operatorname*{diag}_{1\le \ell \le D}(\boldsymbol{V}_\ell) \left(\frac{\partial \boldsymbol{b}'}{\partial \sigma_2^2}\right)' = \frac{\partial \boldsymbol{b}_d'}{\partial \sigma_1^2} \boldsymbol{V}_d \left(\frac{\partial \boldsymbol{b}_d'}{\partial \sigma_2^2}\right)' \\
&= \boldsymbol{a}_t' \boldsymbol{1}_T \boldsymbol{1}_T' \boldsymbol{V}_d^{-1} \boldsymbol{a}_t - \boldsymbol{a}_t' \boldsymbol{1}_T \boldsymbol{1}_T' \boldsymbol{V}_d^{-2} [\sigma_1^2 \boldsymbol{1}_T \boldsymbol{1}_T' + \sigma_2^2 \boldsymbol{I}_T] \boldsymbol{a}_t \\
&\quad - \boldsymbol{a}_t' [\sigma_1^2 \boldsymbol{1}_T \boldsymbol{1}_T' + \sigma_2^2 \boldsymbol{I}_T] \boldsymbol{V}_d^{-1} \boldsymbol{1}_T \boldsymbol{1}_T' \boldsymbol{V}_d^{-1} \boldsymbol{a}_t \\
&\quad + \boldsymbol{a}_t' [\sigma_1^2 \boldsymbol{1}_T \boldsymbol{1}_T' + \sigma_2^2 \boldsymbol{I}_T] \boldsymbol{V}_d^{-1} \boldsymbol{1}_T \boldsymbol{1}_T' \boldsymbol{V}_d^{-2} [\sigma_1^2 \boldsymbol{1}_T \boldsymbol{1}_T' + \sigma_2^2 \boldsymbol{I}_T] \boldsymbol{a}_t .
\end{aligned}
$$

Finally, we have

$$
g_3(\boldsymbol{\theta}) \approx \operatorname{tr} \left\{ \begin{pmatrix} q_{11} & q_{12} \\ q_{21} & q_{22} \end{pmatrix} \begin{pmatrix} F_{11} & F_{12} \\ F_{21} & F_{22} \end{pmatrix}^{-1} \right\},
$$

where F_{ab} is the generic element of the REML Fisher information matrix that was calculated for deriving the updating formula of the Fisher-scoring algorithm, and it is given in (17.3).

Parametric Bootstrap

The parametric bootstrap procedure for estimating the MSE of $\hat{\overline{Y}}_{dt}^{eblup}$ has the following steps.

1. Calculate the REML estimators $\hat{\boldsymbol{\theta}} = (\hat{\sigma}_1^2, \hat{\sigma}_2^2)$ and $\hat{\boldsymbol{\beta}}$.
2. Repeat B times ($b = 1, \ldots, B$).
   a. For $d = 1, \ldots, D$, generate $u_{1,d}^{*(b)}$ i.i.d. $N(0, \hat{\sigma}_1^2)$. Construct the vector $\boldsymbol{u}_1^{*(b)} = (u_{1,1}^{*(b)}, \ldots, u_{1,D}^{*(b)})'$.
   b. For $d = 1, \ldots, D$, $t = 1, \ldots, T$, generate $u_{2,dt}^{*(b)}$ i.i.d. $N(0, \hat{\sigma}_2^2)$. Construct the vector $\boldsymbol{u}_2^{*(b)} = \operatorname*{col}_{1\le d\le D}(\operatorname*{col}_{1\le t\le T}(u_{2,dt}^{*(b)}))$.
   c. For $d = 1, \ldots, D$, $t = 1, \ldots, T$, generate independent variables $e_{dt}^{*(b)} \sim N(0, \sigma_{dt}^2)$. Construct the vector $\boldsymbol{e}^{*(b)} = \operatorname*{col}_{1\le d\le D}(\operatorname*{col}_{1\le t\le T}(e_{dt}^{*(b)}))$.
   d. Calculate the bootstrap vector
   $$
   \boldsymbol{y}^{*(b)} = \boldsymbol{X}\hat{\boldsymbol{\beta}} + \boldsymbol{Z}_1 \boldsymbol{u}_1^{*(b)} + \boldsymbol{Z}_2 \boldsymbol{u}_2^{*(b)} + \boldsymbol{e}^{*(b)}. \tag{17.5}
   $$
   e. Fit the assumed model to the bootstrap vector $\boldsymbol{y}^{*(b)}$, calculate the estimators $\hat{\boldsymbol{\theta}}^{*(b)}$, $\hat{\boldsymbol{\beta}}^{*(b)}$ of the model parameters, the true value of the mixed parameter $\boldsymbol{\mu}^{*(b)} = \boldsymbol{X}\hat{\boldsymbol{\beta}} + \boldsymbol{Z}_1 \boldsymbol{u}_1^{*(b)} + \boldsymbol{Z}_2 \boldsymbol{u}_2^{*(b)}$, and the EBLUP $\hat{\boldsymbol{\mu}}^{*(b)}$ with components $\mu_{dt}^{*(b)}$ and $\hat{\mu}_{dt}^{*(b)}$, respectively.

3. Output: for $d = 1, \ldots, D, t = 1, \ldots, T$, calculate

$$mse^*(\hat{\overline{Y}}_{dt}^{eblup}) = \frac{1}{B}\sum_{b=1}^{B}\left(\hat{\mu}_{dt}^{*(b)} - \mu_{dt}^{*(b)}\right)^2.$$

17.2.4 Simulations

Two simulation experiments are carried out under the area-level model with independent time effects. They investigate the behavior of the REML estimators of model parameters, the EBLUPs of domain means, and the bootstrap MSE estimators.

For $d = 1, \ldots, D, t = 1, \ldots, T$, the explanatory and dependent variables are

$$x_{dt} = \frac{1}{5}(b_{dt} - a_{dt})U_{dt} + a_{dt}, \; U_{dt} = \frac{t}{T+1}, \; a_{dt} = 1, \; b_{dt} = 1 + \frac{1}{D}(T(d-1)+t),$$

$$y_{dt} = \beta_1 + \beta_2 x_{dt} + u_{1,d} + u_{2,dt} + e_{dt}, \; \beta_1 = 0, \; \beta_2 = 1,$$

where $u_{1,d} \sim N(0, \sigma_1^2)$ with $\sigma_1^2 = 0.75$, $u_{2,dt} \sim N(0, \sigma_2^2)$ with $\sigma_2^2 = 0.75$, $e_{dt} \sim N(0, \sigma_{dt}^2)$ and

$$\sigma_{dt}^2 = \frac{2}{25}\frac{T(d-1)+t-1}{M-1} + 0.8.$$

Simulation 1 has the following steps.

1. Repeat $I = 10^4$ times $(i = 1, \ldots, I)$.
   1.1. Generate a sample of size $M = DT$ and calculate $\mu_{dt}^{(i)} = \beta_1 + \beta_2 x_{dt} + u_{1,d}^{(i)} + u_{2,dt}^{(i)}$.
   1.2. Calculate the REML estimator $\hat{\tau}^{(i)} \in \{\hat{\beta}_1^{(i)}, \hat{\beta}_2^{(i)}, \hat{\sigma}_1^{2(i)}, \hat{\sigma}_2^{2(i)}\}$ and the EBLUP $\hat{\mu}_{dt}^{(i)}$.
2. For $\tau \in \{\beta_1, \beta_2, \sigma_1^2, \sigma_2^2\}$ and μ_{dt}, $d = 1, \ldots, D, t = 1, \ldots, T$, calculate

$$BIAS(\hat{\tau}) = \frac{1}{I}\sum_{i=1}^{I}(\hat{\tau}^{(i)} - \tau), \quad MSE(\hat{\tau}) = \frac{1}{I}\sum_{i=1}^{I}(\hat{\tau}^{(i)} - \tau)^2,$$

$$BIAS_{dt} = \frac{1}{I}\sum_{i=1}^{I}(\hat{\mu}_{dt}^{(i)} - \mu_{dt}^{(i)}), \quad MSE_{dt} = \frac{1}{I}\sum_{i=1}^{I}(\hat{\mu}_{dt}^{(i)} - \mu_{dt}^{(i)})^2,$$

$$BIAS = \frac{1}{M}\sum_{d=1}^{D}\sum_{t=1}^{T} BIAS_{dt}, \quad MSE = \frac{1}{M}\sum_{d=1}^{D}\sum_{t=1}^{T} MSE_{dt}.$$

Table 17.1 Results of Simulation 1 for $T = 20$

D	50	100	200	300
$BIAS(\hat{\beta}_1)$	0.00043	0.00042	0.00084	0.00124
$BIAS(\hat{\beta}_2)$	0.00002	0.00000	−0.00045	−0.00001
$BIAS(\hat{\sigma}_1^2)$	−0.00073	0.00047	0.00025	0.00094
$BIAS(\hat{\sigma}_2^2)$	0.00014	0.00007	−0.00019	0.00002
$BIAS$	0.00025	0.00006	−0.00003	0.00015
$MSE(\hat{\beta}_1)$	0.02490	0.01245	0.00621	0.00410
$MSE(\hat{\beta}_2)$	0.00223	0.00114	0.00056	0.00038
$MSE(\hat{\sigma}_1^2)$	0.02591	0.01294	0.00630	0.00431
$MSE(\hat{\sigma}_2^2)$	0.00192	0.00095	0.00047	0.00032
MSE	0.15962	0.15960	0.15958	0.15960

Table 17.1 presents the results of Simulation 1 for $T = 20$. It shows that the bias is always close to zero and that the MSE decreases when the number of domains D grows, so that the REML estimators seem to be consistent. From the last line, we observe that the MSE of the EBLUP is almost constant with increasing D. This behavior is expected since with increasing number of small areas it also grows the number of terms we have to predict. This is to say, the ratio of the sample size to the number of quantities to be predicted is constant.

Simulation 2 studies the behavior of the estimators of the MSE of the EBLUP. Both the analytic and parametric bootstrap estimators are investigated. More concretely, we investigate the behavior of the estimators $mse(\hat{Y}_{dt}^{eblup})$ and $mse^*(\hat{Y}_{dt}^{eblup})$. To do this, such estimators are compared with the empirical MSE of $\hat{\mu}_{dt}$ obtained at the output of Simulation 1. The procedure is as follows.

1. For $D = 50, 100, 200, 300$, take the values of MSE_{dt} from the output of Simulation 1.
2. Repeat $I = 500$ times ($i = 1, \ldots, I$).

   2.1. Generate a sample $(y_{dt}^{(i)}, x_{dt})$, $d = 1, \ldots, D$, $t = 1, \ldots, T$.
   2.2. Calculate the estimators $\hat{\beta}_1^{(i)}$, $\hat{\beta}_2^{(i)}$, $\hat{\sigma}_1^{2(i)}$, and $\hat{\sigma}_2^{2(i)}$.
   2.3. For $d = 1, \ldots, D$, $t = 1, \ldots, T$, calculate $mse_{dt}^{0(i)} = mse(\hat{Y}_{dt}^{eblup(i)})$.
   2.4. Repeat $B = 500$ times ($b = 1, \ldots, B$).

      2.4.1. Generate $u_{1,d}^{*(ib)}, u_{2,dt}^{*(ib)}, e_{dt}^{*(ib)}$, $d = 1, \ldots, D$, $t = 1, \ldots, T$, by using $\hat{\sigma}_1^{2(i)}$, $\hat{\sigma}_2^{2(i)}$ instead of σ_1^2, σ_2^2.
      2.4.2. Generate a bootstrap sample $\{y_{dt}^{*(ib)}; d = 1, \ldots, D, t = 1, \ldots, T\}$ from the model

$$y_{dt}^{*(ib)} = \hat{\beta}_1^{(i)} + \hat{\beta}_2^{(i)} x_{dt} + u_{1,d}^{*(ib)} + u_{2,dt}^{*(ib)} + e_{dt}^{*(ib)}.$$

      2.4.3. Calculate $\mu_{dt}^{*(ib)} = \hat{\beta}_1^{(i)} + \hat{\beta}_2^{(i)} x_{dt} + u_{1,d}^{*(ib)} + u_{2,dt}^{*(ib)}$.

Table 17.2 B^a and E^a, $a = 0, 1$, with $T = 20$

D	50	100	200	300
B^0	0.000444	0.000267	−0.000159	−0.000087
E^0	0.000009	0.000002	0.000006	0.000006
B^1	0.000367	0.000189	−0.000184	−0.000068
E^1	0.000177	0.000174	0.000174	0.000176

2.4.4. Calculate the estimators of the parameters of the bootstrap model, $\hat{\beta}_1^{*(ib)}$, $\hat{\beta}_2^{*(ib)}$, $\hat{\sigma}_1^{2*(ib)}$, $\hat{\sigma}_2^{2*(ib)}$.

2.4.5. For $d = 1, \ldots, D$, $t = 1, \ldots, T$, calculate the EBLUP $\hat{\mu}_{dt}^{*(ib)} = \hat{\beta}_1^{*(ib)} + \hat{\beta}_2^{*(ib)} x_{dt} + \hat{u}_{1,d}^{*(ib)} + \hat{u}_{2,dt}^{*(ib)}$.

2.5. For $d = 1, \ldots, D$, $t = 1, \ldots, T$, calculate

$$mse_{dt}^{1(i)} = \frac{1}{B} \sum_{b=1}^{B} \left(\hat{\mu}_{dt}^{*(ib)} - \mu_{dt}^{*(ib)} \right)^2 .$$

3. Output: for $d = 1, \ldots, D$, $t = 1, \ldots, T$, calculate

$$B_{dt}^a = \frac{1}{I} \sum_{i=1}^{I} \left(mse_{dt}^{a(i)} - MSE_{dt} \right),\; E_{dt}^a = \frac{1}{I} \sum_{i=1}^{I} \left(mse_{dt}^{a(i)} - MSE_{dt} \right)^2,\; a = 0, 1,$$

$$B^a = \frac{1}{M} \sum_{d=1}^{D} \sum_{t=1}^{T} B_{dt}^a,\; E^a = \frac{1}{M} \sum_{d=1}^{D} \sum_{t=1}^{T} E_{dt}^a,\; a = 0, 1.$$

Table 17.2 presents the results of Simulation 2. It shows that B^a and E^a, $a = 0, 1$, tend to zero as D grows. The analytical estimator mse_{dt}^0 has a very good behavior. It is basically unbiased, and its average quadratic error is much smaller than that of the basic bootstrap estimator mse_{dt}^1.

17.3 Area-Level Model with Correlated Time Effects

In this section we present a generalization of the previous model by introducing a correlation structure across time within each domain.

17.3.1 The Model

Let $u_{1,d} \sim N(0, \sigma_1^2)$ and $e_{dt} \sim N(0, \sigma_{dt}^2)$, $d = 1, \ldots, D$, $t = 1, \ldots, T$, be independent domain random effects and domain-time errors. Let $u_{2,dt}$, $d = 1, \ldots, D$,

$t = 1, \ldots, T$, be domain-time random effects such that $(u_{2,d1}, \ldots, u_{2,dT})'$, $d = 1, \ldots, D$, are i.i.d. AR(1)-distributed random vectors with autocorrelation and variance parameters ϕ and σ_2^2, respectively. Let us also assume that the $u_{1,d}$'s, the $u_{2,dt}$'s, and the e_{dt}'s are mutually independent. The subindex d represents the domain and the subindex t denotes the time period. The area-level linear mixed model with AR(1)-correlated time effects is a temporal generalization of the Fay–Herriot model introduced by Rao and Yu (1994) and applied to the estimation of poverty proportions by Esteban et al. (2012b). It is defined by the equation

$$y_{dt} = \boldsymbol{x}_{dt}\boldsymbol{\beta} + u_{1,d} + u_{2,dt} + e_{dt}, \quad d = 1, \ldots, D, \ t = 1, \ldots, T, \tag{17.6}$$

where y_{dt} is a direct estimator of the characteristic of interest, $\boldsymbol{x}_{dt}$ is a row vector containing the aggregated values of p auxiliary variables, and $\boldsymbol{\beta}$ is a vector of regression parameters. The error variances, σ_{dt}^2, are assumed to be known, and they are typically taken as the estimated design-based variances of the direct estimators y_{dt}. In the matrix form, the model (17.6) can be written alternatively as

$$\boldsymbol{y} = \boldsymbol{X}\boldsymbol{\beta} + \boldsymbol{Z}_1\boldsymbol{u}_1 + \boldsymbol{Z}_2\boldsymbol{u}_2 + \boldsymbol{e},$$

where $\boldsymbol{y} = \underset{1\le d\le D}{\text{col}}(\boldsymbol{y}_d)$, $\boldsymbol{y}_d = \underset{1\le t\le T}{\text{col}}(y_{dt})$, $\boldsymbol{u}_1 = \underset{1\le d\le D}{\text{col}}(u_{1,d})$, $\boldsymbol{u}_2 = \underset{1\le d\le D}{\text{col}}(\boldsymbol{u}_{2,d})$, $\boldsymbol{u}_{2,d} = \underset{1\le t\le T}{\text{col}}(u_{2,dt})$, $\boldsymbol{e} = \underset{1\le d\le D}{\text{col}}(\boldsymbol{e}_d)$, $\boldsymbol{e}_d = \underset{1\le t\le T}{\text{col}}(e_{dt})$, $\boldsymbol{X} = \underset{1\le d\le D}{\text{col}}(\boldsymbol{X}_d)$, $\boldsymbol{X}_d = \underset{1\le t\le T}{\text{col}}(\boldsymbol{x}_{dt})$, $\boldsymbol{x}_{dt} = \underset{1\le j\le p}{\text{col}'}(x_{dtj})$, $\boldsymbol{\beta} = \underset{1\le j\le p}{\text{col}}(\beta_j)$, $\boldsymbol{Z}_1 = \underset{1\le d\le D}{\text{diag}}(\boldsymbol{1}_T)$, $\boldsymbol{Z}_2 = \boldsymbol{I}_M$, and $M = DT$. The model (17.6) assumes that $\boldsymbol{u}_1 \sim N(\boldsymbol{0}, \boldsymbol{V}_{u_1})$, $\boldsymbol{u}_2 \sim N(\boldsymbol{0}, \boldsymbol{V}_{u_2})$, and $\boldsymbol{e} \sim N(\boldsymbol{0}, \boldsymbol{V}_e)$ are independent with covariance matrices

$$\boldsymbol{V}_{u_1} = \sigma_1^2\boldsymbol{I}_D, \quad \boldsymbol{V}_{u_2} = \sigma_2^2\Omega(\phi), \quad \boldsymbol{V}_e = \underset{1\le d\le D}{\text{diag}}(\boldsymbol{V}_{ed}),$$

$$\Omega(\phi) = \underset{1\le d\le D}{\text{diag}}(\Omega_d(\phi)), \quad \boldsymbol{V}_{ed} = \underset{1\le t\le T}{\text{diag}}(\sigma_{dt}^2),$$

where the σ_{dt}^2's are known and

$$\Omega_d = \Omega_d(\phi) = \frac{1}{1-\phi^2}\begin{pmatrix} 1 & \phi & \ldots & \phi^{T-2} & \phi^{T-1} \\ \phi & 1 & \ddots & & \phi^{T-2} \\ \vdots & \ddots & \ddots & \ddots & \vdots \\ \phi^{T-2} & & \ddots & 1 & \phi \\ \phi^{T-1} & \phi^{T-2} & \ldots & \phi & 1 \end{pmatrix}_{T\times T}.$$

The model (17.6) can also be written as $\boldsymbol{y} = \boldsymbol{X}\boldsymbol{\beta} + \boldsymbol{Z}\boldsymbol{u} + \boldsymbol{e}$, where $\boldsymbol{Z} = [\boldsymbol{Z}_1, \boldsymbol{Z}_2]$ and $\boldsymbol{u} = (\boldsymbol{u}_1', \boldsymbol{u}_2')'$. For $\sigma_1^2, \sigma_2^2, \phi$ known, the BLU estimator and predictor of $\boldsymbol{\beta}$ and

$\boldsymbol{u}$ are (cf. Proposition 16.1)

$$\tilde{\boldsymbol{\beta}} = (\boldsymbol{X}'\boldsymbol{V}^{-1}\boldsymbol{X})^{-1}\boldsymbol{X}'\boldsymbol{V}^{-1}\boldsymbol{y} \quad \text{and} \quad \tilde{\boldsymbol{u}} = \boldsymbol{V}_u\boldsymbol{Z}'\boldsymbol{V}^{-1}(\boldsymbol{y} - \boldsymbol{X}\tilde{\boldsymbol{\beta}}),$$

where $\boldsymbol{V}_u = \text{diag}(\boldsymbol{V}_{u_1}, \boldsymbol{V}_{u_2})$ and

$$\begin{aligned} \boldsymbol{V} = \text{var}(\boldsymbol{y}) &= \sigma_1^2 \boldsymbol{Z}_1 \boldsymbol{Z}_1' + \sigma_2^2 \underset{1\le d\le D}{\text{diag}} (\Omega_d(\phi)) + \boldsymbol{V}_e \\ &= \underset{1\le d\le D}{\text{diag}} (\sigma_1^2 \boldsymbol{1}_T \boldsymbol{1}_T' + \sigma_2^2 \Omega_d(\phi) + \boldsymbol{V}_{ed}) = \underset{1\le d\le D}{\text{diag}} (\boldsymbol{V}_d). \end{aligned}$$

For calculating $\tilde{\boldsymbol{\beta}}$ and $\tilde{\boldsymbol{u}}$, it is possible to apply the formulas

$$\tilde{\boldsymbol{\beta}} = \Big(\sum_{d=1}^{D} \boldsymbol{X}_d'\boldsymbol{V}_d^{-1}\boldsymbol{X}_d\Big)^{-1}\Big(\sum_{d=1}^{D} \boldsymbol{X}_d'\boldsymbol{V}_d^{-1}\boldsymbol{y}_d\Big),$$

and

$$\begin{aligned} \tilde{\boldsymbol{u}} &= \boldsymbol{V}_u\boldsymbol{Z}'\boldsymbol{V}^{-1}(\boldsymbol{y} - \boldsymbol{X}\tilde{\boldsymbol{\beta}}) \\ &= \begin{pmatrix} \sigma_1^2 \boldsymbol{I}_D & \boldsymbol{0} \\ \boldsymbol{0} & \sigma_2^2\Omega(\phi) \end{pmatrix} \begin{pmatrix} \underset{1\le d\le D}{\text{diag}} (\boldsymbol{1}_T') \\ \boldsymbol{I}_M \end{pmatrix} \underset{1\le d\le D}{\text{diag}} (\boldsymbol{V}_d^{-1}) \underset{1\le d\le D}{\text{col}} (\boldsymbol{y}_d - \boldsymbol{X}_d\tilde{\boldsymbol{\beta}}) \\ &= \begin{pmatrix} \sigma_1^2 \underset{1\le d\le D}{\text{col}} (\boldsymbol{1}_T'\boldsymbol{V}_d^{-1}(\boldsymbol{y}_d - \boldsymbol{X}_d\tilde{\boldsymbol{\beta}})) \\ \sigma_2^2 \underset{1\le d\le D}{\text{col}} (\Omega_d(\phi)\boldsymbol{V}_d^{-1}(\boldsymbol{y}_d - \boldsymbol{X}_d\tilde{\boldsymbol{\beta}})) \end{pmatrix}. \end{aligned}$$

The corresponding empirical versions, EBLUE $\hat{\boldsymbol{\beta}}$ of $\boldsymbol{\beta}$ and EBLUP $\hat{\boldsymbol{u}}$ of $\boldsymbol{u}$, are obtained by plugging estimators $\hat{\sigma}_1^2, \hat{\sigma}_2^2, \hat{\phi}$ in the place of σ_1^2, σ_2^2 and ϕ.

17.3.2 *Residual Maximum Likelihood Estimation*

This section follows the same steps as Sect. 17.2.2 to derive a Fisher-scoring algorithm for calculating the REML estimators of the variance components.

The REML log-likelihood of the model (17.6), $l_{reml} = l_{reml}(\sigma_1^2, \sigma_2^2, \phi)$, is

$$l_{reml} = -\frac{M-p}{2}\log 2\pi + \frac{1}{2}\log|\boldsymbol{X}'\boldsymbol{X}| - \frac{1}{2}\log|\boldsymbol{V}| - \frac{1}{2}\log|\boldsymbol{X}'\boldsymbol{V}^{-1}\boldsymbol{X}| - \frac{1}{2}\boldsymbol{y}'\boldsymbol{P}\boldsymbol{y},$$

where

$$\boldsymbol{P} = \boldsymbol{V}^{-1} - \boldsymbol{V}^{-1}\boldsymbol{X}(\boldsymbol{X}'\boldsymbol{V}^{-1}\boldsymbol{X})^{-1}\boldsymbol{X}'\boldsymbol{V}^{-1}, \quad \boldsymbol{P}\boldsymbol{V}\boldsymbol{P} = \boldsymbol{P}, \quad \boldsymbol{P}\boldsymbol{X} = \boldsymbol{0}.$$

Let $\boldsymbol{\theta} = (\theta_1, \theta_2, \theta_3) = (\sigma_1^2, \sigma_2^2, \phi)$ be the vector of parameters and define

$$\boldsymbol{V}_1 = \frac{\partial \boldsymbol{V}}{\partial \sigma_1^2} = \underset{1\le d\le D}{\text{diag}} \; (\mathbf{1}_T\mathbf{1}_T'), \quad \boldsymbol{V}_2 = \frac{\partial \boldsymbol{V}}{\partial \sigma_2^2} = \underset{1\le d\le D}{\text{diag}} \; (\Omega_d(\phi)),$$

$$\boldsymbol{V}_3 = \frac{\partial \boldsymbol{V}}{\partial \phi} = \sigma_2^2 \underset{1\le d\le D}{\text{diag}} \; (\dot{\Omega}_d(\phi)),$$

where the derivative of matrix $\Omega_d(\phi)$ with respect to ϕ is

$$\dot{\Omega}_d(\phi) = \frac{1}{1-\phi^2}\begin{pmatrix} 0 & 1 & \ldots\ldots & (T-1)\phi^{T-2} \\ 1 & 0 & \ddots & (T-2)\phi^{T-3} \\ \vdots & \ddots & \ddots \quad \ddots & \vdots \\ (T-2)\phi^{T-3} & & \ddots \quad 0 & 1 \\ (T-1)\phi^{T-2} & \ldots\ldots & 1 & 0 \end{pmatrix} + \frac{2\phi\Omega_d(\phi)}{1-\phi^2}.$$

Further, we have

$$\boldsymbol{P}_a = \frac{\partial \boldsymbol{P}}{\partial \theta_a} = -\boldsymbol{P}\frac{\partial \boldsymbol{V}}{\partial \theta_a}\boldsymbol{P} = -\boldsymbol{P}\boldsymbol{V}_a\boldsymbol{P}, \quad a = 1, 2, 3.$$

By taking derivatives of l_{reml} with respect to θ_a, we get elements of the score vector $\boldsymbol{S} = \boldsymbol{S}(\boldsymbol{\theta}) = (S_1, S_2, S_3)'$,

$$S_a = \frac{\partial l_{reml}}{\partial \theta_a} = -\frac{1}{2}\text{tr}(\boldsymbol{P}\boldsymbol{V}_a) + \frac{1}{2}\boldsymbol{y}'\boldsymbol{P}\boldsymbol{V}_a\boldsymbol{P}\boldsymbol{y}, \quad a = 1, 2, 3.$$

By taking again derivatives with respect to θ_a and θ_b, taking expectations, and changing the sign, we get the elements of the Fisher information matrix $\boldsymbol{F} = \boldsymbol{F}(\boldsymbol{\theta}) = (F_{ab})_{a,b=1,2,3}$, where

$$F_{ab} = \frac{1}{2}\text{tr}(\boldsymbol{P}\boldsymbol{V}_a\boldsymbol{P}\boldsymbol{V}_b), \quad a, b = 1, 2, 3.$$

The updating formula of the REML Fisher-scoring algorithm is

$$\boldsymbol{\theta}^{(i+1)} = \boldsymbol{\theta}^{(i)} + \boldsymbol{F}^{-1}(\boldsymbol{\theta}^{(i)})\boldsymbol{S}(\boldsymbol{\theta}^{(i)}).$$

Some possible algorithm seeds are $\theta_0^{(0)} = \theta_1^{(0)} = S^2/2, \theta_3^{(0)} = 0$, where

$$S^2 = \frac{1}{M-p}(\boldsymbol{y} - \boldsymbol{X}\tilde{\boldsymbol{\beta}})'\boldsymbol{V}_e^{-1}(\boldsymbol{y} - \boldsymbol{X}\tilde{\boldsymbol{\beta}}), \quad \tilde{\boldsymbol{\beta}} = (\boldsymbol{X}'\boldsymbol{V}_e^{-1}\boldsymbol{X})^{-1}\boldsymbol{X}'\boldsymbol{V}_e^{-1}\boldsymbol{y}.$$

The REML estimator of $\boldsymbol{\beta}$ is

$$\hat{\boldsymbol{\beta}} = (\boldsymbol{X}'\hat{\boldsymbol{V}}^{-1}\boldsymbol{X})^{-1}\boldsymbol{X}'\hat{\boldsymbol{V}}^{-1}\boldsymbol{y},$$

where

$$\hat{\boldsymbol{V}} = \hat{\sigma}_1^2\boldsymbol{Z}_1\boldsymbol{Z}_1' + \hat{\sigma}_2^2 \underset{1\leq d\leq D}{\text{diag}}\,(\boldsymbol{\Omega}_d(\hat{\phi})) + \boldsymbol{V}_e.$$

For comments concerning the asymptotic distributions and confidence intervals of the REML estimators of $\boldsymbol{\theta}$ and $\boldsymbol{\beta}$, we refer the reader to Sect. 17.2.2.

17.3.3 EBLUP and Mean Squared Error

This section deals with the prediction of the linear combination of fixed and random effects, $\mu_{dt} = \boldsymbol{x}_{dt}\boldsymbol{\beta} + u_{1,d} + u_{2,dt}$, with the EBLUP $\hat{\mu}_{dt} = \boldsymbol{x}_{dt}\hat{\boldsymbol{\beta}} + \hat{u}_{1,d} + \hat{u}_{2,dt}$. For estimating $\overline{Y}_{dt}$, it is possible to employ the EBLUP of μ_{dt} in which case the notation $\hat{\overline{Y}}_{dt}^{eblup} = \hat{\mu}_{dt}$ is used.

The mixed parameter of interest can be written in the form $\mu_{dt} = \boldsymbol{a}'\,(\boldsymbol{X\beta} + \boldsymbol{Zu})$, where $\boldsymbol{a} = \underset{1\leq \ell\leq D}{\text{col}}\,(\underset{1\leq k\leq T}{\text{col}}\,(\delta_{d\ell}\delta_{tk}))$ is a vector having a one in the cell $t + (d-1)T$ and having zeros in the remaining cells. So in order to obtain an approximation of the MSE of $\hat{\overline{Y}}_{dt}^{eblup}$, one can follow the procedure discussed in Sect. 17.2.3 and obtain for $\boldsymbol{\theta} = (\sigma_1^2, \sigma_2^2, \phi)$ the formula

$$MSE(\hat{\overline{Y}}_{dt}^{eblup}) = g_1(\boldsymbol{\theta}) + g_2(\boldsymbol{\theta}) + g_3(\boldsymbol{\theta}),$$

where the functions g_1, g_2, and g_3 are given in (17.4). The MSE of $\hat{\overline{Y}}_{dt}^{eblup}$ can be estimated by parametric bootstrap or by using the analytic estimator

$$mse(\hat{\overline{Y}}_{dt}^{eblup}) = g_1(\hat{\boldsymbol{\theta}}) + g_2(\hat{\boldsymbol{\theta}}) + 2g_3(\hat{\boldsymbol{\theta}}),$$

where $\hat{\boldsymbol{\theta}} = (\hat{\sigma}_1^2, \hat{\sigma}_2^2, \hat{\phi})$ is the corresponding REML estimator.

The term g_3 of MSE was derived in Sect. 9.2.3. Its expression can be obtained under the hypotheses and conditions of regularity established by Kackar and Harville (1984), Prasad and Rao (1990), Datta and Lahiri (2000), and Das et al. (2004). Esteban et al. (2012b) used the term g_3 without checking the corresponding hypotheses. These authors empirically justified the use of g_3 by simulation experiments. This chapter follows the same approach and presents additional simulations to justify the use of the term g_3.

The following sections give the calculations of g_1, g_2, and g_3 and describe a parametric bootstrap procedure for estimating the mean squared error.

Calculation of $g_1(\boldsymbol{\theta})$

It holds that $g_1(\boldsymbol{\theta}) = \boldsymbol{a}'\boldsymbol{Z}\boldsymbol{T}\boldsymbol{Z}'\boldsymbol{a}$, where

$$\boldsymbol{T} = \boldsymbol{V}_u - \boldsymbol{V}_u\boldsymbol{Z}'\boldsymbol{V}^{-1}\boldsymbol{Z}\boldsymbol{V}_u = \begin{pmatrix} \sigma_1^2\boldsymbol{I}_D & \boldsymbol{0} \\ \boldsymbol{0} & \sigma_2^2\Omega(\phi) \end{pmatrix} - \begin{pmatrix} \sigma_1^2\boldsymbol{I}_D & \boldsymbol{0} \\ \boldsymbol{0} & \sigma_2^2\Omega(\phi) \end{pmatrix} \begin{pmatrix} \underset{1\le\ell\le D}{\text{diag}}\,(\boldsymbol{1}_T') & \\ & \boldsymbol{I}_M \end{pmatrix}$$

$$\cdot \underset{1\le\ell\le D}{\text{diag}}\,(\boldsymbol{V}_\ell^{-1})(\underset{1\le\ell\le D}{\text{diag}}\,(\boldsymbol{1}_T), \boldsymbol{I}_M)\begin{pmatrix} \sigma_1^2\boldsymbol{I}_D & \boldsymbol{0} \\ \boldsymbol{0} & \sigma_2^2\Omega(\phi) \end{pmatrix} = \begin{pmatrix} \boldsymbol{T}_{11} & \boldsymbol{T}_{12} \\ \boldsymbol{T}_{21} & \boldsymbol{T}_{22} \end{pmatrix}$$

and

$$\boldsymbol{T}_{11} = \sigma_1^2\boldsymbol{I}_D - \sigma_1^4 \underset{1\le\ell\le D}{\text{diag}}\,(\boldsymbol{1}_T'\boldsymbol{V}_\ell^{-1}\boldsymbol{1}_T)) = \underset{1\le\ell\le D}{\text{diag}}\,(T_{11\ell}),$$

$$\boldsymbol{T}_{12} = -\sigma_1^2\sigma_2^2 \underset{1\le\ell\le D}{\text{diag}}\,(\boldsymbol{1}_T'\boldsymbol{V}_\ell^{-1}\Omega_\ell(\phi)) = \underset{1\le\ell\le D}{\text{diag}}\,(\boldsymbol{T}_{12\ell}) = \boldsymbol{T}_{21}',$$

$$\boldsymbol{T}_{22} = \sigma_2^2 \underset{1\le\ell\le D}{\text{diag}}\,(\Omega_\ell(\phi)) - \sigma_2^4 \underset{1\le\ell\le D}{\text{diag}}\,(\Omega_\ell(\phi)\boldsymbol{V}_\ell^{-1}\Omega_\ell(\phi)) = \underset{1\le\ell\le D}{\text{diag}}\,(\boldsymbol{T}_{22\ell}).$$

Let $\boldsymbol{a}_t = \underset{1\le k\le T}{\text{col}}\,(\delta_{tk})$, then $\boldsymbol{a} = \underset{1\le\ell\le D}{\text{col}}\,(\delta_{d\ell}\boldsymbol{a}_t)$, and using the same steps as on page 467, we obtain

$$g_1(\boldsymbol{\theta}) = \boldsymbol{a}'\boldsymbol{Z}\boldsymbol{T}\boldsymbol{Z}'\boldsymbol{a} = \boldsymbol{a}_t'(\boldsymbol{1}_T T_{11d}\boldsymbol{1}_T' + \boldsymbol{1}_T\boldsymbol{T}_{12d} + \boldsymbol{T}_{21d}\boldsymbol{1}_T' + \boldsymbol{T}_{22d})\boldsymbol{a}_t.$$

Calculation of $g_2(\boldsymbol{\theta})$

The formula for $g_2(\boldsymbol{\theta})$ has exactly the same form as the one given on page 467 with the current version of the matrix $\boldsymbol{T}$.

Calculation of $g_3(\boldsymbol{\theta})$

It holds that

$$g_3(\boldsymbol{\theta}) \approx \text{tr}\left\{(\nabla\boldsymbol{b}')\boldsymbol{V}(\nabla\boldsymbol{b}')'E\left[(\hat{\boldsymbol{\theta}} - \boldsymbol{\theta})(\hat{\boldsymbol{\theta}} - \boldsymbol{\theta})'\right]\right\},$$

where

$$
\begin{aligned}
\boldsymbol{b}' &= \boldsymbol{a}'\boldsymbol{Z}\boldsymbol{V}_u\boldsymbol{Z}'\boldsymbol{V}^{-1} \\
&= \boldsymbol{a}'\big(\underset{1\le\ell\le D}{\text{diag}}\,(\boldsymbol{1}_T), \boldsymbol{I}_M\big)\begin{pmatrix}\sigma_1^2\boldsymbol{I}_D & \boldsymbol{0}\\ \boldsymbol{0} & \sigma_2^2\Omega(\phi)\end{pmatrix}\begin{pmatrix}\underset{1\le\ell\le D}{\text{diag}}\,(\boldsymbol{1}_T') \\ \boldsymbol{I}_M\end{pmatrix}\underset{1\le\ell\le D}{\text{diag}}\,(\boldsymbol{V}_\ell^{-1}) \\
&= \boldsymbol{a}'\underset{1\le\ell\le D}{\text{diag}}\,\big([\sigma_1^2\boldsymbol{1}_T\boldsymbol{1}_T' + \sigma_2^2\Omega_\ell(\phi)]\boldsymbol{V}_\ell^{-1}\big) \\
&= \underset{1\le\ell\le D}{\text{col}'}\big(\delta_{d\ell}\boldsymbol{a}_t'[\sigma_1^2\boldsymbol{1}_T\boldsymbol{1}_T' + \sigma_2^2\Omega_\ell(\phi)]\boldsymbol{V}_\ell^{-1}\big).
\end{aligned}
$$

The partial derivatives of $\boldsymbol{b}'$, with respect to σ_1^2, σ_2^2, and ϕ, are

$$
\begin{aligned}
\frac{\partial \boldsymbol{b}'}{\partial\sigma_1^2} &= \underset{1\le\ell\le D}{\text{col}'}\big(\delta_{d\ell}\boldsymbol{a}_t'\boldsymbol{1}_T\boldsymbol{1}_T'\boldsymbol{V}_\ell^{-1}\big) - \underset{1\le\ell\le D}{\text{col}'}\big(\delta_{d\ell}\boldsymbol{a}_t'[\sigma_1^2\boldsymbol{1}_T\boldsymbol{1}_T' + \sigma_2^2\Omega_\ell(\phi)]\boldsymbol{V}_\ell^{-1}\boldsymbol{V}_{\ell 1}\boldsymbol{V}_\ell^{-1}\big), \\
\frac{\partial \boldsymbol{b}'}{\partial\sigma_2^2} &= \underset{1\le\ell\le D}{\text{col}'}\big(\delta_{d\ell}\boldsymbol{a}_t'\Omega_\ell(\phi)\boldsymbol{V}_\ell^{-1}\big) - \underset{1\le\ell\le D}{\text{col}'}\big(\delta_{d\ell}\boldsymbol{a}_t'[\sigma_1^2\boldsymbol{1}_T\boldsymbol{1}_T' + \sigma_2^2\Omega_\ell(\phi)]\boldsymbol{V}_\ell^{-1}\boldsymbol{V}_{\ell 2}\boldsymbol{V}_\ell^{-1}\big), \\
\frac{\partial \boldsymbol{b}'}{\partial\phi} &= \sigma_2^2\underset{1\le\ell\le D}{\text{col}'}\big(\delta_{d\ell}\boldsymbol{a}_t'\dot{\Omega}_\ell(\phi)\boldsymbol{V}_\ell^{-1}\big) - \underset{1\le\ell\le D}{\text{col}'}\big(\delta_{d\ell}\boldsymbol{a}_t'[\sigma_1^2\boldsymbol{1}_T\boldsymbol{1}_T' + \sigma_2^2\Omega_\ell(\phi)]\boldsymbol{V}_\ell^{-1}\boldsymbol{V}_{\ell 3}\boldsymbol{V}_\ell^{-1}\big),
\end{aligned}
$$

where $\boldsymbol{V}_{\ell 1} = \frac{\partial \boldsymbol{V}_\ell}{\partial\sigma_1^2} = \boldsymbol{1}_T\boldsymbol{1}_T'$, $\boldsymbol{V}_{\ell 2} = \frac{\partial \boldsymbol{V}_\ell}{\partial\sigma_2^2} = \Omega_\ell(\phi)$, and $\boldsymbol{V}_{\ell 3} = \frac{\partial \boldsymbol{V}_\ell}{\partial\phi} = \sigma_2^2\dot{\Omega}_\ell(\phi)$.

Associated to each partial derivative, we define

$$
\begin{aligned}
\frac{\partial \boldsymbol{b}_d'}{\partial\sigma_1^2} &= \boldsymbol{a}_t'\boldsymbol{1}_T\boldsymbol{1}_T'\boldsymbol{V}_d^{-1} - \boldsymbol{a}_t'[\sigma_1^2\boldsymbol{1}_T\boldsymbol{1}_T' + \sigma_2^2\Omega_d(\phi)]\boldsymbol{V}_d^{-1}\boldsymbol{1}_T\boldsymbol{1}_T'\boldsymbol{V}_d^{-1}, \\
\frac{\partial \boldsymbol{b}_d'}{\partial\sigma_2^2} &= \boldsymbol{a}_t'\Omega_d(\phi)\boldsymbol{V}_d^{-1} - \boldsymbol{a}_t'[\sigma_1^2\boldsymbol{1}_T\boldsymbol{1}_T' + \sigma_2^2\Omega_d(\phi)]\boldsymbol{V}_d^{-1}\Omega_d(\phi)\boldsymbol{V}_d^{-1}, \\
\frac{\partial \boldsymbol{b}_d'}{\partial\phi} &= \sigma_2^2\boldsymbol{a}_t'\dot{\Omega}_d(\phi)\boldsymbol{V}_d^{-1} - \sigma_2^2\boldsymbol{a}_t'[\sigma_1^2\boldsymbol{1}_T\boldsymbol{1}_T' + \sigma_2^2\Omega_d(\phi)]\boldsymbol{V}_d^{-1}\dot{\Omega}_d(\phi)\boldsymbol{V}_d^{-1}.
\end{aligned}
$$

Using this notation, for the components of $(\nabla\boldsymbol{b}')\boldsymbol{V}(\nabla\boldsymbol{b}')'$, it holds that

$$
\begin{aligned}
q_{11} &= \frac{\partial \boldsymbol{b}'}{\partial\sigma_1^2}\underset{1\le\ell\le D}{\text{diag}}\,(\boldsymbol{V}_\ell)\left(\frac{\partial \boldsymbol{b}'}{\partial\sigma_1^2}\right)' = \frac{\partial \boldsymbol{b}_d'}{\partial\sigma_1^2}\boldsymbol{V}_d\left(\frac{\partial \boldsymbol{b}_d'}{\partial\sigma_1^2}\right)' \\
&= \boldsymbol{a}_t'\boldsymbol{1}_T\boldsymbol{1}_T'\boldsymbol{V}_d^{-1}\boldsymbol{1}_T\boldsymbol{1}_T'\boldsymbol{a}_t - 2\boldsymbol{a}_t'\boldsymbol{1}_T\boldsymbol{1}_T'\boldsymbol{V}_d^{-1}\boldsymbol{1}_T\boldsymbol{1}_T'\boldsymbol{V}_d^{-1}[\sigma_1^2\boldsymbol{1}_T\boldsymbol{1}_T' + \sigma_2^2\Omega_d(\phi)]\boldsymbol{a}_t \\
&\quad + \boldsymbol{a}_t'[\sigma_1^2\boldsymbol{1}_T\boldsymbol{1}_T' + \sigma_2^2\Omega_d(\phi)]\boldsymbol{V}_d^{-1}\boldsymbol{1}_T\boldsymbol{1}_T'\boldsymbol{V}_d^{-1}\boldsymbol{1}_T\boldsymbol{1}_T'\boldsymbol{V}_d^{-1}[\sigma_1^2\boldsymbol{1}_T\boldsymbol{1}_T' + \sigma_2^2\Omega_d(\phi)]\boldsymbol{a}_t,
\end{aligned}
$$

$$
\begin{aligned}
q_{22} &= \frac{\partial \boldsymbol{b}'}{\partial \sigma_2^2} \operatorname*{diag}_{1\le \ell \le D} (\boldsymbol{V}_\ell) \left(\frac{\partial \boldsymbol{b}'}{\partial \sigma_2^2}\right)' = \frac{\partial \boldsymbol{b}_d'}{\partial \sigma_2^2} \boldsymbol{V}_d \left(\frac{\partial \boldsymbol{b}_d'}{\partial \sigma_2^2}\right)' \\
&= \boldsymbol{a}_t' \Omega_d(\phi) \boldsymbol{V}_d^{-1} \Omega_d(\phi) \boldsymbol{a}_t - 2\boldsymbol{a}_t' \Omega_d(\phi) \boldsymbol{V}_d^{-1} \Omega_d(\phi) \boldsymbol{V}_d^{-1} [\sigma_1^2 \boldsymbol{1}_T \boldsymbol{1}_T' + \sigma_2^2 \Omega_d(\phi)] \boldsymbol{a}_t \\
&+ \boldsymbol{a}_t' [\sigma_1^2 \boldsymbol{1}_T \boldsymbol{1}_T' + \sigma_2^2 \Omega_d(\phi)] \boldsymbol{V}_d^{-1} \Omega_d(\phi) \boldsymbol{V}_d^{-1} \Omega_d(\phi) \boldsymbol{V}_d^{-1} [\sigma_1^2 \boldsymbol{1}_T \boldsymbol{1}_T' + \sigma_2^2 \Omega_d(\phi)] \boldsymbol{a}_t, \\
q_{33} &= \frac{\partial \boldsymbol{b}'}{\partial \phi} \operatorname*{diag}_{1\le \ell \le D} (\boldsymbol{V}_\ell) \left(\frac{\partial \boldsymbol{b}'}{\partial \phi}\right)' = \frac{\partial \boldsymbol{b}_d'}{\partial \phi} \boldsymbol{V}_d \left(\frac{\partial \boldsymbol{b}_d'}{\partial \phi}\right)' = \sigma_2^4 \boldsymbol{a}_t' \dot{\Omega}_d(\phi) \boldsymbol{V}_d^{-1} \dot{\Omega}_d(\phi) \boldsymbol{a}_t \\
&- 2\sigma_2^4 \boldsymbol{a}_t' \dot{\Omega}_d(\phi) \boldsymbol{V}_d^{-1} \dot{\Omega}_d(\phi) \boldsymbol{V}_d^{-1} [\sigma_1^2 \boldsymbol{1}_T \boldsymbol{1}_T' + \sigma_2^2 \Omega_d(\phi)] \boldsymbol{a}_t + \sigma_2^4 \boldsymbol{a}_t' [\sigma_1^2 \boldsymbol{1}_T \boldsymbol{1}_T' \\
&+ \sigma_2^2 \Omega_d(\phi)] \boldsymbol{V}_d^{-1} \dot{\Omega}_d(\phi) \boldsymbol{V}_d^{-1} \dot{\Omega}_d(\phi) \boldsymbol{V}_d^{-1} [\sigma_1^2 \boldsymbol{1}_T \boldsymbol{1}_T' + \sigma_2^2 \Omega_d(\phi)] \boldsymbol{a}_t, \\
q_{12} &= \frac{\partial \boldsymbol{b}'}{\partial \sigma_1^2} \operatorname*{diag}_{1\le \ell \le D} (\boldsymbol{V}_\ell) \left(\frac{\partial \boldsymbol{b}'}{\partial \sigma_2^2}\right)' = \frac{\partial \boldsymbol{b}_d'}{\partial \sigma_1^2} \boldsymbol{V}_d \left(\frac{\partial \boldsymbol{b}_d'}{\partial \sigma_2^2}\right)' \\
&= \boldsymbol{a}_t' \boldsymbol{1}_T \boldsymbol{1}_T' \boldsymbol{V}_d^{-1} \Omega_d(\phi) \boldsymbol{a}_t - \boldsymbol{a}_t' \boldsymbol{1}_T \boldsymbol{1}_T' \boldsymbol{V}_d^{-1} \Omega_d(\phi) \boldsymbol{V}_d^{-1} [\sigma_1^2 \boldsymbol{1}_T \boldsymbol{1}_T' + \sigma_2^2 \Omega_d(\phi)] \boldsymbol{a}_t \\
&- \boldsymbol{a}_t' [\sigma_1^2 \boldsymbol{1}_T \boldsymbol{1}_T' + \sigma_2^2 \Omega_d(\phi)] \boldsymbol{V}_d^{-1} \boldsymbol{1}_T \boldsymbol{1}_T' \boldsymbol{V}_d^{-1} \Omega_d(\phi) \boldsymbol{a}_t \\
&+ \boldsymbol{a}_t' [\sigma_1^2 \boldsymbol{1}_T \boldsymbol{1}_T' + \sigma_2^2 \Omega_d(\phi)] \boldsymbol{V}_d^{-1} \boldsymbol{1}_T \boldsymbol{1}_T' \boldsymbol{V}_d^{-1} \Omega_d(\phi) \boldsymbol{V}_d^{-1} [\sigma_1^2 \boldsymbol{1}_T \boldsymbol{1}_T' + \sigma_2^2 \Omega_d(\phi)] \boldsymbol{a}_t, \\
q_{13} &= \frac{\partial \boldsymbol{b}'}{\partial \sigma_1^2} \operatorname*{diag}_{1\le \ell \le D} (\boldsymbol{V}_\ell) \left(\frac{\partial \boldsymbol{b}'}{\partial \phi}\right)' = \frac{\partial \boldsymbol{b}_d'}{\partial \sigma_1^2} \boldsymbol{V}_d \left(\frac{\partial \boldsymbol{b}_d'}{\partial \phi}\right)' \\
&= \sigma_2^2 \boldsymbol{a}_t' \boldsymbol{1}_T \boldsymbol{1}_T' \boldsymbol{V}_d^{-1} \dot{\Omega}_d(\phi) \boldsymbol{a}_t - \sigma_2^2 \boldsymbol{a}_t' \boldsymbol{1}_T \boldsymbol{1}_T' \boldsymbol{V}_d^{-1} \dot{\Omega}_d(\phi) \boldsymbol{V}_d^{-1} [\sigma_1^2 \boldsymbol{1}_T \boldsymbol{1}_T' + \sigma_2^2 \Omega_d(\phi)] \boldsymbol{a}_t \\
&- \sigma_2^2 \boldsymbol{a}_t' [\sigma_1^2 \boldsymbol{1}_T \boldsymbol{1}_T' + \sigma_2^2 \Omega_d(\phi)] \boldsymbol{V}_d^{-1} \boldsymbol{1}_T \boldsymbol{1}_T' \boldsymbol{V}_d^{-1} \dot{\Omega}_d(\phi) \boldsymbol{a}_t \\
&+ \sigma_2^2 \boldsymbol{a}_t' [\sigma_1^2 \boldsymbol{1}_T \boldsymbol{1}_T' + \sigma_2^2 \Omega_d(\phi)] \boldsymbol{V}_d^{-1} \boldsymbol{1}_T \boldsymbol{1}_T' \boldsymbol{V}_d^{-1} \dot{\Omega}_d(\phi) \boldsymbol{V}_d^{-1} [\sigma_1^2 \boldsymbol{1}_T \boldsymbol{1}_T' + \sigma_2^2 \Omega_d(\phi)] \boldsymbol{a}_t, \\
q_{23} &= \frac{\partial \boldsymbol{b}'}{\partial \sigma_2^2} \operatorname*{diag}_{1\le \ell \le D} (\boldsymbol{V}_\ell) \left(\frac{\partial \boldsymbol{b}'}{\partial \phi}\right)' = \frac{\partial \boldsymbol{b}_d'}{\partial \sigma_2^2} \boldsymbol{V}_d \left(\frac{\partial \boldsymbol{b}_d'}{\partial \phi}\right)' = \sigma_2^2 \boldsymbol{a}_t' \Omega_d(\phi) \boldsymbol{V}_d^{-1} \dot{\Omega}_d(\phi) \boldsymbol{a}_t \\
&- \sigma_2^2 \boldsymbol{a}_t' \Omega_d(\phi) \boldsymbol{V}_d^{-1} \dot{\Omega}_d(\phi) \boldsymbol{V}_d^{-1} [\sigma_1^2 \boldsymbol{1}_T \boldsymbol{1}_T' + \sigma_2^2 \Omega_d(\phi)] \boldsymbol{a}_t \\
&- \sigma_2^2 \boldsymbol{a}_t' [\sigma_1^2 \boldsymbol{1}_T \boldsymbol{1}_T' + \sigma_2^2 \Omega_d(\phi)] \boldsymbol{V}_d^{-1} \Omega_d(\phi) \boldsymbol{V}_d^{-1} \dot{\Omega}_d(\phi) \boldsymbol{a}_t \\
&+ \sigma_2^2 \boldsymbol{a}_t' [\sigma_1^2 \boldsymbol{1}_T \boldsymbol{1}_T' + \sigma_2^2 \Omega_d(\phi)] \boldsymbol{V}_d^{-1} \Omega_d(\phi) \boldsymbol{V}_d^{-1} \dot{\Omega}_d(\phi) \boldsymbol{V}_d^{-1} [\sigma_1^2 \boldsymbol{1}_T \boldsymbol{1}_T' + \sigma_2^2 \Omega_d(\phi)] \boldsymbol{a}_t.
\end{aligned}
$$

Finally, we have

$$
g_3(\boldsymbol{\theta}) \approx \operatorname{tr} \left\{ \begin{pmatrix} q_{11} & q_{12} & q_{13} \\ q_{21} & q_{22} & q_{23} \\ q_{31} & q_{32} & q_{33} \end{pmatrix} \begin{pmatrix} F_{11} & F_{12} & F_{13} \\ F_{21} & F_{22} & F_{23} \\ F_{31} & F_{32} & F_{33} \end{pmatrix}^{-1} \right\},
$$

where F_{ab} is the generic element of the REML Fisher information matrix that was calculated for deriving the updating formula of the Fisher-scoring algorithm.

Parametric Bootstrap

The following algorithm calculates the basic parametric bootstrap estimator, mse^*_{dt}, of the MSE of $\hat{\overline{Y}}^{eblup}_{dt}$.

1. Calculate the REML estimators $\hat{\boldsymbol{\theta}} = (\hat{\sigma}_1^2, \hat{\sigma}_2^2, \hat{\phi})$ and $\hat{\boldsymbol{\beta}}$.
2. Repeat B times ($b = 1, \ldots, B$).
   a. For $d = 1, \ldots, D$, generate $u^{*(b)}_{1,d}$ i.i.d. $N(0, \hat{\sigma}_1^2)$. Construct the vector $\boldsymbol{u}^{*(b)}_1 = (u^{*(b)}_{1,1}, \ldots, u^{*(b)}_{1,D})'$.
   b. For $d = 1, \ldots, D$, generate independent vectors $\boldsymbol{u}^{*(b)}_{2,d} = (u^{*(b)}_{2,d1}, \ldots, u^{*(b)}_{2,dT})'$ with AR(1) correlation structure and distribution $N_T(\mathbf{0}, \hat{\sigma}_2 \Omega_d(\hat{\phi}))$. Construct the vector $\boldsymbol{u}^{*(b)}_2 = \underset{1 \leq d \leq D}{\text{col}} (\boldsymbol{u}^{*(b)}_{2,d})$.
   c. For $d = 1, \ldots, D$, $t = 1, \ldots, T$, generate independent variables $e^{*(b)}_{dt} \sim N(0, \sigma^2_{dt})$. Construct the vector $\boldsymbol{e}^{*(b)} = \underset{1 \leq d \leq D}{\text{col}} (\underset{1 \leq t \leq T}{\text{col}} (e^{*(b)}_{dt}))$.
   d. Calculate the bootstrap vector

$$\boldsymbol{y}^{*(b)} = \boldsymbol{X}\hat{\boldsymbol{\beta}} + \boldsymbol{Z}_1 \boldsymbol{u}^{*(b)}_1 + \boldsymbol{Z}_2 \boldsymbol{u}^{*(b)}_2 + \boldsymbol{e}^{*(b)}. \tag{17.7}$$

   e. Fit the assumed model to the bootstrap vector $\boldsymbol{y}^{*(b)}$, calculate the estimators $\hat{\boldsymbol{\theta}}^{*(b)}$, $\hat{\boldsymbol{\beta}}^{*(b)}$ of the model parameters, the true value of the mixed parameter $\boldsymbol{\mu}^{*(b)} = \boldsymbol{X}\hat{\boldsymbol{\beta}} + \boldsymbol{Z}_1 \boldsymbol{u}^{*(b)}_1 + \boldsymbol{Z}_2 \boldsymbol{u}^{*(b)}_2$, and the EBLUP $\hat{\boldsymbol{\mu}}^{*(b)}$ with components $\mu^{*(b)}_{dt}$ and $\hat{\mu}^{*(b)}_{dt}$, respectively.
3. Output: for $d = 1, \ldots, D$, $t = 1, \ldots, T$, calculate

$$mse^*_{dt} = \frac{1}{B} \sum_{b=1}^{B} \left(\hat{\mu}^{*(b)}_{dt} - \mu^{*(b)}_{dt} \right)^2.$$

17.3.4 Simulations

This section presents two simulation experiments for investigating the behavior of the REML estimators of model parameters, the EBLUPs of domain means, and the bootstrap MSE estimators. The explanatory and target variables are generated in the same way as described at the beginning of Sect. 17.2.4. The only difference is that the AR(1) random effects u_{dt} are, for $d = 1, \dots, D$, generated as follows:

$$u_{2,d1} = (1-\phi^2)^{-1/2}\varepsilon_{d1}, \quad u_{2,dt} = \phi u_{2,dt-1} + \varepsilon_{dt}, \quad t = 2, \dots, T,$$

where $\varepsilon_{dt} \sim N(0, \sigma_2^2)$, $t = 1, \dots, T$, with $\sigma_2^2 = 0.75$ and $\phi = 0.5$.

Simulation 1 has the following steps.

1. Repeat $I = 10^4$ times ($i = 1, \dots, I$).
   1.1. Generate a sample of size $M = DT$ and calculate $\mu_{dt}^{(i)} = \beta_1 + \beta_2 x_{dt} + u_{1,d}^{(i)} + u_{2,dt}^{(i)}$.
   1.2. Calculate $\hat{\tau}^{(i)} \in \{\hat{\beta}_1^{(i)}, \hat{\beta}_2^{(i)}, \hat{\sigma}_1^{2(i)}, \hat{\sigma}_2^{2(i)}, \hat{\phi}^{(i)}\}$ by using the REML method and the EBLUP $\hat{\mu}_{dt}^{(i)}$.
2. For each $\hat{\tau} \in \{\beta_1, \beta_2, \sigma_1^2, \sigma_2^2, \phi\}$ and for $\hat{\mu}_{dt}$, $d = 1, \dots, D$, $t = 1, \dots, T$, calculate

$$BIAS(\hat{\tau}) = \frac{1}{I}\sum_{i=1}^{I}(\hat{\tau}^{(i)} - \tau), \quad MSE(\hat{\tau}) = \frac{1}{I}\sum_{i=1}^{I}(\hat{\tau}^{(i)} - \tau)^2,$$

$$BIAS_{dt} = \frac{1}{I}\sum_{i=1}^{I}(\hat{\mu}_{dt}^{(i)} - \mu_{dt}^{(i)}), \quad MSE_{dt} = \frac{1}{I}\sum_{i=1}^{I}(\hat{\mu}_{dt}^{(i)} - \mu_{dt}^{(i)})^2,$$

$$BIAS = \frac{1}{M}\sum_{d=1}^{D}\sum_{t=1}^{T} BIAS_{dt}, \quad MSE = \frac{1}{M}\sum_{d=1}^{D}\sum_{t=1}^{T} MSE_{dt}.$$

Table 17.3 presents the results of Simulation 1. It shows that the bias is always close to zero and that the MSE decreases when the number of domains grows, so that the REML estimators are consistent. In the case of MSE of EBLUP (the last line of the table), we again do not observe any differences with increasing number of small areas D.

Simulation 2 studies the behavior of the MSE estimators of the EBLUP. The analytic and bootstrap estimators, mse_{dt} and mse_{dt}^*, of MSE_{dt} are investigated. For this, the considered estimators are compared with the Monte Carlo MSE of $\hat{\mu}_{dt}$ obtained in Simulation 1. The procedure is as follows:

1. For $D = 50, 100, 200, 300$, take the values MSE_{dt} from the output of Simulation 1.
2. Repeat $I = 500$ times ($i = 1, \dots, I$).
   2.1. Generate a sample $(y_{dt}^{(i)}, x_{dt})$, $d = 1, \dots, D$, $t = 1, \dots, T$.
   2.2. Calculate the estimators $\hat{\beta}_1^{(i)}, \hat{\beta}_2^{(i)}, \hat{\boldsymbol{\theta}}^{(i)} = (\hat{\sigma}_1^{2(i)}, \hat{\sigma}_2^{2(i)}, \hat{\phi}^{(i)})$.

Table 17.3 Result of Simulation 1 with $T = 20$

D	50	100	200	300	400	500
$BIAS(\hat{\beta}_1)$	0.0002	0.0002	0.0009	0.0014	−0.0001	−0.0002
$BIAS(\hat{\beta}_2)$	0.0001	−0.0001	−0.0001	0.0000	0.0001	0.0000
$BIAS(\hat{\sigma}_1^2)$	−0.0020	−0.0004	−0.0001	0.0008	−0.0004	0.0000
$BIAS(\hat{\sigma}_2^2)$	−0.0007	−0.0004	−0.0005	−0.0002	−0.0001	−0.0002
$BIAS(\hat{\phi})$	−0.0012	−0.0008	−0.0001	−0.0002	−0.0003	−0.0001
$BIAS$	0.0003	0.0000	0.0000	0.0002	0.0001	−0.0001
$MSE(\hat{\beta}_1)$	0.0259	0.0131	0.0064	0.0042	0.0033	0.0026
$MSE(\hat{\beta}_2)$	0.0002	0.0001	0.0001	0.0000	0.0000	0.0000
$MSE(\hat{\sigma}_1^2)$	0.0335	0.0163	0.0081	0.0055	0.0041	0.0032
$MSE(\hat{\sigma}_2^2)$	0.0021	0.0010	0.0005	0.0004	0.0003	0.0002
$MSE(\hat{\phi})$	0.0015	0.0008	0.0004	0.0003	0.0002	0.0002
MSE	0.1541	0.1541	0.1541	0.1541	0.1541	0.1541

2.3. For $d = 1, \ldots, D$, $t = 1, \ldots, T$, calculate

$$mse_{dt}^{0(i)} = mse(\hat{\overline{Y}}_{dt}^{eblup(i)}) = g_1(\hat{\boldsymbol{\theta}}^{(i)}) + g_2(\hat{\boldsymbol{\theta}}^{(i)}) + 2g_3(\hat{\boldsymbol{\theta}}^{(i)}).$$

2.4. Repeat $B = 500$ times ($b = 1, \ldots, B$).

2.4.1. Generate $u_{1,d}^{*(ib)}$, $u_{2,dt}^{*(ib)}$, $e_{dt}^{*(ib)}$, $d = 1, \ldots, D$, $t = 1, \ldots, T$, by using $\hat{\sigma}_1^{2(i)}$, $\hat{\sigma}_2^{2(i)}$, $\hat{\phi}^{(i)}$ instead of σ_1^2, σ_2^2, ϕ.

2.4.2. Generate a bootstrap sample $\{y_{dt}^{*(ib)}; d = 1, \ldots, D, t = 1, \ldots, T\}$, from the model

$$y_{dt}^{*(ib)} = \hat{\beta}_1^{(i)} + \hat{\beta}_2^{(i)} x_{dt} + u_{1,d}^{*(ib)} + u_{2,dt}^{*(ib)} + e_{dt}^{*(ib)}.$$

2.4.3. Calculate $\mu_{dt}^{*(ib)} = \hat{\beta}_1^{(i)} + \hat{\beta}_2^{(i)} x_{dt} + u_{1,d}^{*(ib)} + u_{2,dt}^{*(ib)}$.

2.4.4. Calculate the estimators of the bootstrap model, $\hat{\beta}_1^{*(ib)}$, $\hat{\beta}_2^{*(ib)}$, $\hat{\sigma}_1^{2*(ib)}$, $\hat{\sigma}_2^{2*(ib)}$, and $\hat{\phi}^{*(ib)}$.

2.4.5. For $d = 1, \ldots, D$, $t = 1, \ldots, T$, calculate the EBLUP $\hat{\mu}_{dt}^{*(ib)} = \hat{\beta}_1^{*(ib)} + \hat{\beta}_2^{*(ib)} x_{dt} + \hat{u}_{1,d}^{*(ib)} + \hat{u}_{2,dt}^{*(ib)}$.

2.5. For $d = 1, \ldots, D$, $t = 1, \ldots, T$, calculate

$$mse_{dt}^{1(i)} = \frac{1}{B} \sum_{b=1}^{B} \left(\hat{\mu}_{dt}^{*(ib)} - \mu_{dt}^{*(ib)} \right)^2.$$

Table 17.4 B^a y E^a, $a = 0, 1$, with $T = 20$

D	50	100	200	300
B^0	0.003526	0.001745	0.000476	0.000345
E^0	0.000018	0.000006	0.000006	0.000006
B^1	0.000301	0.000249	−0.000283	−0.000124
E^1	0.000164	0.000164	0.000163	0.000165

3. Output: for $d = 1, \ldots, D$, $t = 1, \ldots, T$, calculate

$$B_{dt}^a = \frac{1}{I}\sum_{i=1}^{I}(mse_{dt}^{a(i)} - MSE_{dt}), \quad E_{dt}^a = \frac{1}{I}\sum_{i=1}^{I}(mse_{dt}^{a(i)} - MSE_{dt})^2, \quad a = 0, 1,$$

$$B^a = \frac{1}{M}\sum_{d=1}^{D}\sum_{t=1}^{T} B_{dt}^a, \quad E^a = \frac{1}{M}\sum_{d=1}^{D}\sum_{t=1}^{T} E_{dt}^a, \quad a = 0, 1.$$

Table 17.4 presents the results of Simulation 2. It shows that B^a and E^a, $a = 0, 1$, tend to zero as D increases. The analytical estimator mse_{dt}^0 has good behavior. It has a low bias and its average quadratic error is smaller than that of the basic bootstrap estimator mse_{dt}^1.

17.4 R Codes for EBLUPs

This section gives R codes for fitting area-level temporal linear mixed models to the aggregated data. The data is taken from the R package `saery`. First, install and/or load the package and load the data set.

```
if(!require(saery)){
  install.packages("saery")
  library(saery)
}
data(datos)
```

Redefine variables and calculate auxiliary information

```
sigma2edi <- datos[,6]                  # Sampling variance
X <- as.matrix(datos[,5])               # Auxiliary variable
ydi <- datos[,3]                        # Target variable
D <- length(unique(datos[,1]))          # Number of domains
md <- table(datos[,1])                  # Number of time instants per domain
```

Fit the non-correlated area-level temporal linear mixed model. The function `fit.saery` fits three types of models with `model` argument (non-correlated by default), and it uses `conf.level` argument to calculate confidence intervals for regression parameters and variance components.

```
output.fit.indep <- fit.saery(X, ydi, D, md, sigma2edi, model="INDEP",
                              conf.level=0.9)
output.fit.indep                        # Summary of results
```

Table 17.5 Direct and EBLUP estimates of domain means under the model with independent random effects

Domain	Period	dir	eblup	var.dir	mse.eblup	resid
1	1	1.2055	1.2253	0.1600	0.1434	−0.0197
1	2	0.7105	0.7844	0.1600	0.1434	−0.0739
1	3	2.2769	2.1801	0.1601	0.1435	0.0968
1	4	0.4804	0.5801	0.1601	0.1435	−0.0997
1	5	0.8867	0.9425	0.1602	0.1435	−0.0558
1	6	1.4257	1.4232	0.1602	0.1436	0.0025
1	7	2.6338	2.4999	0.1602	0.1436	0.1338
1	8	2.1265	2.0488	0.1603	0.1436	0.0777
1	9	1.7019	1.6715	0.1603	0.1437	0.0304
1	10	1.0213	1.0663	0.1604	0.1437	−0.0450

```
output.fit.indep$Regression         # Regression parameters
output.fit.indep$SIGMA              # Variance components
```

Calculate the EBLUP, the explicit MSE estimator, and some plots. The package `saery` does not give the bootstrap MSE estimation under the model with independent random effects.

```
eblup.output.indep <- eblup.saery(X, ydi, D, md, sigma2edi,
                                  model="INDEP", plot=TRUE)
head(round(eblup.output.indep, 4), 10)       # Summary of results
```

For the first domain and the ten first periods, Table 17.5 gives direct estimators, the EBLUPs, and their corresponding MSEs under the model with independent random effects.

The `eblup.saery` function of the package `saery` can return some plots with the argument `plot=TRUE`. Figure 17.1 plots the residuals sorted by direct estimator, and Fig. 17.2 plots the root-MSE estimates. The EBLUP outperforms the direct estimator.

Fit the AR(1)-correlated area-level temporal linear mixed model.

```
output.fit.ar1 <- fit.saery(X, ydi, D, md, sigma2edi, model="AR1",
                            conf.level=0.9)
output.fit.ar1                          # Summary of results
output.fit.ar1$Regression               # Regression parameters
output.fit.ar1$SIGMA                    # Variance components
```

Calculate the EBLUP and the explicit MSE estimator under the AR(1)-correlated area-level temporal linear mixed model.

```
eblup.output.ar1 <- eblup.saery(X, ydi, D, md, sigma2edi, "AR1", type="III")
head(round(eblup.output.ar1, 4), 10)          # Summary of results
```

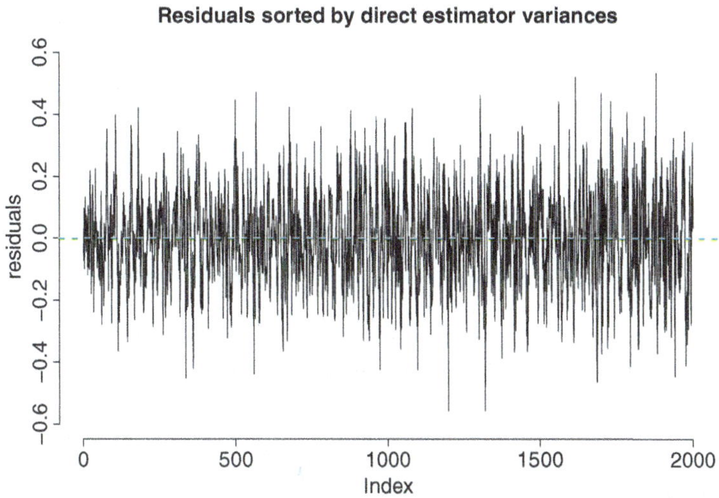

Fig. 17.1 Dispersion graphs of residuals

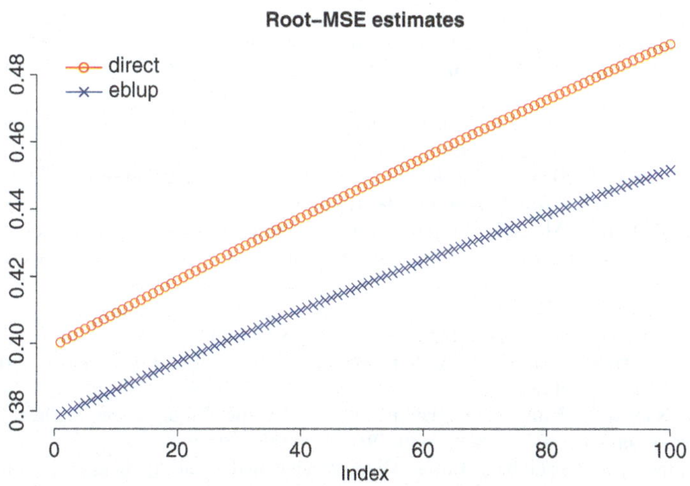

Fig. 17.2 Root-MSE estimates

For the first domain and the ten first periods, Table 17.6 gives the direct estimators, the EBLUPs, and their corresponding MSEs under the AR(1)-correlated area-level temporal linear mixed model. The EBLUP outperforms the direct estimator.

Table 17.6 Direct and EBLUP estimates of domain means under the model with AR(1)-correlated random effects

Domain	Period	dir	eblup	var.dir	mse.eblup	resid
1	1	1.2055	1.1632	0.1600	0.1556	0.0423
1	2	0.7105	0.9165	0.1600	0.1315	−0.2060
1	3	2.2769	1.9090	0.1601	0.1264	0.3678
1	4	0.4804	0.7177	0.1601	0.1233	−0.2373
1	5	0.8867	0.9414	0.1602	0.1210	−0.0548
1	6	1.4257	1.4852	0.1602	0.1193	−0.0595
1	7	2.6338	2.4146	0.1602	0.1181	0.2192
1	8	2.1265	2.0989	0.1603	0.1172	0.0276
1	9	1.7019	1.6805	0.1603	0.1167	0.0214
1	10	1.0213	1.1681	0.1604	0.1165	−0.1468

References

Choudry, G.H., Rao, J.N.K.: Small area estimation using models that combine time series and cross sectional data. In: Singh, A.C., Whitridge, P. (eds.) Proceedings of Statistics Canada Symposium on Analysis of Data in Time, pp. 67–74 (1989)

Das, K., Jiang, J., Rao, J.N.K.: Mean squared error of empirical predictor. Ann. Stat. **32**, 818–840 (2004)

Datta, G.S., Lahiri, P.: A unified measure of uncertainty of estimated best linear unbiased predictors in small area estimation problems. Stat. Sin. **10**, 613–627 (2000)

Datta, G.S., Lahiri, P., Maiti, T., Lu, K.L.: Hierarchical Bayes estimation of unemployment rates for the U.S. States. J. Am. Stat. Assoc. **94**, 1074–1082 (1999)

Datta, G.S., Lahiri, P., Maiti, T.: Empirical Bayes estimation of median income of four-person families by state using time series and cross-sectional data. J. Stat. Plan. Inference **102**, 83–97 (2002)

Esteban, M.D., Morales, D., Pérez, A., Santamaría, L.: Two area-level time models for estimating small area poverty indicators. J. Indian Soc. Agric. Stat. **66**(1), 75–89 (2012a)

Esteban, M.D., Morales, D., Pérez, A., Santamaría, L.: Small area estimation of poverty proportions under area-level time models. Comput. Stat. Data Anal. **56**, 2840–2855 (2012b)

Ghosh, M., Nangia, N., Kim, D.: Estimation of median income of four-person families: a Bayesian time series approach. J. Am. Stat. Assoc. **91**, 1423–1431 (1996)

Jiang, J.: Linear and Generalized Linear Mixed Models and Their Applications. Springer, New York (2007)

Kackar, R.N., Harville, D.A.: Approximations for standard errors of estimators of fixed and random effects in mixed linear models. J. Am. Stat. Assoc. **79**, 853–862 (1984)

Pfeffermann, D., Burck, L.: Robust small area estimation combining time series and cross-sectional data. Surv. Methodol. **16**, 217–237 (1990)

Prasad, N.G.N., Rao, J.N.K.: The estimation of the mean squared error of small-area estimators. J. Am. Stat. Assoc. **85**, 163–171 (1990)

Rao, J.N.K., Yu, M.: Small area estimation by combining time series and cross sectional data. Can. J. Stat. **22**, 511–528 (1994)

Singh, A.C., Mantel, H., Thomas, B.W.: Time series generalizations of Fay-Herriot estimator for small areas. In: Proceedings of Survey Research Methods Section, pp. 455–460. American Statistical Association, Alexandria (1991)

You, Y., Rao, J.N.K.: Hierarchical Bayes estimation of small area means using multi-level models. Surv. Methodol. **26**, 173–181 (2000)

You, Y., Rao, J.N.K., Gambino, J.: Model-based unemployment rate estimation for the Canadian Labour Force Survey: a hierarchical approach. Technical report, Household Survey Method Division. Statistics Canada (2001)

Chapter 18
Area-Level Spatio-Temporal Linear Mixed Models

18.1 Introduction

Many different extensions of the Fay–Herriot model have been proposed in the literature, and the generalizations to models including spatial correlation have played a prominent role. Cressie (1993) pointed out that when there is unexplained spatial correlation in the data, not considering it in the model will lead to erroneous inferences. In small area estimation, if areas are properly delimited regions of the population, closer areas tend to have more similar socioeconomic characteristics. When data from neighbor areas are correlated, considering spatial correlation in the model leads to more efficient small area estimators. This was taken into account in the basic Fay–Herriot model by Singh et al. (2005), Petrucci and Salvati (2006), Pratesi and Salvati (2008), and Molina et al. (2009), who considered an extension of the Fay–Herriot model by assuming that area effects $\{u_d\}$ follow a spatial autoregressive process of order 1 or SAR(1). See Anselin (1988) and Cressie (1993) for details on the SAR(1) process. Bayesian spatial models have been considered by Moura and Migon (2002), Banerjee et al. (2004), and You and Zhou (2011).

Marhuenda et al. (2013) introduced some spatio-temporal extensions of the Fay–Herriot model. Their models incorporate historical data similarly as the model proposed by Rao and Yu (1994) and at the same time include spatial correlation among data from neighboring areas.

This chapter describes the basic area-level linear mixed model with a SAR(1) vector of area random effects. The paradigm is that taking into account the spatial correlation among data from different areas allows to borrow even more strength from the areas. This chapter also presents the spatio-temporal models of Marhuenda et al. (2013) that take into account the temporal and spatial correlations for improving the predictions. The chapter gives REML Fisher-scoring fitting algorithms, derives the EBLUPs of linear parameters, presents simulation results, and gives some R codes.

D. Morales et al., *A Course on Small Area Estimation and Mixed Models*, Statistics for Social and Behavioral Sciences, https://doi.org/10.1007/978-3-030-63757-6_18

18.2 Area-Level Spatial Linear Mixed Model

The first model treated in this chapter is the most simple area-level linear mixed model with spatially correlated area random effects.

18.2.1 The Model

Consider a finite population partitioned into D small areas. The basic FH model relates linearly the quantity of inferential interest for d-th small area, μ_d, to a vector of p area-level auxiliary covariates $\boldsymbol{x}_d = (x_{d1}, \ldots, x_{dp})$ and includes a random effect u_d associated to the area, that is,

$$\mu_d = \boldsymbol{x}_d \boldsymbol{\beta} + u_d, \quad d = 1, \ldots, D. \tag{18.1}$$

Here $\boldsymbol{\beta}$ is the $p \times 1$ vector of regression parameters, and the random effects $\{u_d;\, d = 1, \ldots, D\}$ are independent and identically distributed, each with mean 0 and variance σ_u^2. Model (18.1) is called linking model since all small areas are linked by the common parameter $\boldsymbol{\beta}$. Moreover, the FH model assumes that a design-unbiased direct estimator y_d of μ_d is available for each small area $d = 1, \ldots, D$ and that these direct estimators can be expressed as

$$y_d = \mu_d + e_d, \quad d = 1, \ldots, D, \tag{18.2}$$

where $\{e_d;\, d = 1, \ldots, D\}$ are independent sampling errors, independent of the random effects u_d. It is further assumed that e_d has mean 0 and variance σ_d^2, which is known, $d = 1, \ldots, D$. For more details, see e.g. Ghosh an Rao (1994). Model (18.2) is called sampling model. Combining both, the linking model (18.1) and the sampling model (18.2), we obtain the linear mixed model

$$y_d = \boldsymbol{x}_d \boldsymbol{\beta} + u_d + e_d, \quad d = 1, \ldots, D. \tag{18.3}$$

Let us define vectors $\boldsymbol{y} = (y_1, \ldots, y_D)'$, $\boldsymbol{u} = (u_1, \ldots, u_D)'$ and $\boldsymbol{e} = (e_1, \ldots, e_D)'$ and matrices $\boldsymbol{X} = \underset{1 \le d \le D}{\text{col}}(\boldsymbol{x}_d)$ and $\boldsymbol{V}_e = \text{diag}(\sigma_1^2, \ldots, \sigma_D^2)$. Then the model in matrix notation is

$$\boldsymbol{y} = \boldsymbol{X}\boldsymbol{\beta} + \boldsymbol{u} + \boldsymbol{e}. \tag{18.4}$$

Model (18.4) can be extended to allow for spatially correlated area effects as follows. Let $\boldsymbol{u}$ be the result of a SAR process with unknown autoregression parameter ρ and proximity matrix $\boldsymbol{W}$ (see Anselin (1988) and Cressie (1993)), i.e.

$$\boldsymbol{u} = \rho \boldsymbol{W} \boldsymbol{u} + \boldsymbol{v}, \tag{18.5}$$

where $\boldsymbol{v} = (v_1, \ldots, v_D)'$ is a vector with mean $\mathbf{0}$ and covariance matrix $\sigma_u^2 \boldsymbol{I}_D$, where $\boldsymbol{I}_D$ denotes the $D \times D$ identity matrix and σ_u^2 is an unknown parameter. We assume that the matrix $(\boldsymbol{I}_D - \rho \boldsymbol{W})$ is non-singular. Then $\boldsymbol{u}$ can be expressed as

$$\boldsymbol{u} = (\boldsymbol{I}_D - \rho \boldsymbol{W})^{-1} \boldsymbol{v}. \tag{18.6}$$

The elements w_{ij} of the proximity matrix $\boldsymbol{W}$ are weights, which define spatial relation among each pair of domains. The two main approaches used for weights definition are based on contiguity of domains or on distance between domains. By convention, $w_{ij} = 0$ for the diagonal elements. We further assume that the proximity matrix $\boldsymbol{W}$ is defined in row standardized form, that is, $\boldsymbol{W}$ is row stochastic. Then, $\rho \in (-1, 1)$ is called spatial autocorrelation parameter. See e.g. Banerjee et al. (2004). Hereafter, the vector of variance components will be denoted as $\boldsymbol{\theta} = (\theta_1, \theta_2)' = (\sigma_u^2, \rho)'$. Equation (18.6) implies that $\boldsymbol{u}$ has mean vector $\mathbf{0}$ and covariance matrix equal to

$$\mathbf{G}(\boldsymbol{\theta}) = \sigma_u^2 [(\boldsymbol{I}_D - \rho \mathbf{W})'(\boldsymbol{I}_D - \rho \mathbf{W})]^{-1}. \tag{18.7}$$

Since $\boldsymbol{e}$ is independent of $\boldsymbol{u}$, the covariance matrix of $\boldsymbol{y}$ is equal to

$$\boldsymbol{V}(\boldsymbol{\theta}) = \boldsymbol{G}(\boldsymbol{\theta}) + \boldsymbol{V}_e.$$

Combining (18.4) and (18.6), the model is

$$\boldsymbol{y} = \mathbf{X}\boldsymbol{\beta} + (\boldsymbol{I}_D - \rho \mathbf{W})^{-1} \boldsymbol{v} + \boldsymbol{e}. \tag{18.8}$$

Under model (18.8), the spatial BLUP of the quantity of interest $\mu_d = \boldsymbol{x}_d \boldsymbol{\beta} + u_d$ is (cf. Proposition 16.1)

$$\tilde{\mu}_d(\boldsymbol{\theta}) = \boldsymbol{x}_d \tilde{\boldsymbol{\beta}}(\boldsymbol{\theta}) + \mathbf{b}_d' \mathbf{G}(\boldsymbol{\theta}) \boldsymbol{V}^{-1}(\boldsymbol{\theta})[\boldsymbol{y} - \boldsymbol{X}\tilde{\boldsymbol{\beta}}(\boldsymbol{\theta})], \tag{18.9}$$

where $\tilde{\boldsymbol{\beta}}(\boldsymbol{\theta}) = [\boldsymbol{X}'\boldsymbol{V}^{-1}(\boldsymbol{\theta})\boldsymbol{X}]^{-1}\boldsymbol{X}'\boldsymbol{V}^{-1}(\boldsymbol{\theta})\boldsymbol{y}$ is the generalized least squares estimator of the regression parameter $\boldsymbol{\beta}$ and $\mathbf{b}_d'$ is the $1 \times D$ vector $(0, \ldots, 0, 1, 0, \ldots, 0)$ with 1 in the d-th position. The spatial BLUP $\tilde{\mu}_d(\boldsymbol{\theta})$ depends on the unknown vector of variance components $\boldsymbol{\theta} = (\sigma_u^2, \rho)'$. The predictor $\hat{\mu}_d = \tilde{\mu}_d(\hat{\boldsymbol{\theta}})$, obtained by replacing $\boldsymbol{\theta}$ in expression (18.9) by a consistent estimator $\hat{\boldsymbol{\theta}} = (\hat{\sigma}_u^2, \hat{\rho})'$, is the spatial EBLUP (see Singh et al. (2005) and Petrucci and Salvati (2006)).

18.2.2 Fitting Methods Based on the Likelihood

Assuming normality of the random effects and the errors, the variance components $\boldsymbol{\theta} = (\sigma_u^2, \rho)'$ can be estimated by ML or REML procedures. In fact, under regularity conditions, the estimators derived from these two methods (and using the normal likelihood) remain consistent at order $O_p(D^{-1/2})$ even without the normality assumption, for details see Jiang (2007).

A maximum likelihood estimator (MLE) of $\boldsymbol{\theta} = (\sigma_u^2, \rho)'$ is obtained by maximizing the log-likelihood of $\boldsymbol{\theta}$ given the data vector $\boldsymbol{y}$,

$$\ell(\boldsymbol{\theta}; \boldsymbol{y}) = c - \frac{1}{2}\log|\boldsymbol{V}(\boldsymbol{\theta})| - \frac{1}{2}(\boldsymbol{y} - \boldsymbol{X}\boldsymbol{\beta})'\boldsymbol{V}^{-1}(\boldsymbol{\theta})(\boldsymbol{y} - \boldsymbol{X}\boldsymbol{\beta}),$$

where c denotes a constant. In practice, an iterative algorithm such as the Fisher-scoring algorithm can be applied to maximize the likelihood. Let $\boldsymbol{S}(\boldsymbol{\theta}) = (S_{\sigma_u^2}, S_\rho)'$ be the scores or derivatives of the log-likelihood with respect to σ_u^2 and ρ, and let $\boldsymbol{F}(\boldsymbol{\theta})$ be the Fisher information matrix obtained from $\ell(\boldsymbol{\theta}; \boldsymbol{y})$, with elements

$$\boldsymbol{F}(\boldsymbol{\theta}) = \begin{pmatrix} F_{\sigma_u^2,\sigma_u^2} & F_{\sigma_u^2,\rho} \\ F_{\rho,\sigma_u^2} & F_{\rho,\rho} \end{pmatrix},$$

where e.g.

$$F_{\sigma_u^2,\rho} = -E\left[\frac{\partial^2 \ell(\boldsymbol{\theta}; \boldsymbol{y})}{\partial\sigma_u^2 \partial\rho}\right].$$

Then the Fisher-scoring algorithm starts with an initial estimate $\boldsymbol{\theta}^{(0)} = (\sigma_u^{2(0)}, \rho^{(0)})'$, and at each iteration i, this estimate is updated with the equation

$$\boldsymbol{\theta}^{(i+1)} = \boldsymbol{\theta}^{(i)} + \boldsymbol{F}^{-1}(\boldsymbol{\theta}^{(i)})\boldsymbol{S}(\boldsymbol{\theta}^{(i)}). \tag{18.10}$$

The ML equation for $\boldsymbol{\beta}$ obtained by equating the corresponding score to zero yields

$$\tilde{\boldsymbol{\beta}}(\boldsymbol{\theta}) = [\boldsymbol{X}'\boldsymbol{V}^{-1}(\boldsymbol{\theta})\boldsymbol{X}]^{-1}\boldsymbol{X}'\boldsymbol{V}^{-1}(\boldsymbol{\theta})\boldsymbol{y}. \tag{18.11}$$

Let us denote

$$\boldsymbol{C}(\rho) = (\boldsymbol{I}_D - \rho\boldsymbol{W})'(\boldsymbol{I}_D - \rho\boldsymbol{W})$$

and

$$\boldsymbol{P}(\boldsymbol{\theta}) = \boldsymbol{V}^{-1}(\boldsymbol{\theta}) - \boldsymbol{V}^{-1}(\boldsymbol{\theta})\boldsymbol{X}\left[\boldsymbol{X}'\boldsymbol{V}^{-1}(\boldsymbol{\theta})\boldsymbol{X}\right]^{-1}\boldsymbol{X}'\boldsymbol{V}^{-1}(\boldsymbol{\theta}). \tag{18.12}$$

Then the derivative of $\boldsymbol{C}(\rho)$ with respect to ρ is

$$\frac{\partial \boldsymbol{C}(\rho)}{\partial\rho} = -\boldsymbol{W} - \boldsymbol{W}' + 2\rho\boldsymbol{W}'\boldsymbol{W},$$

and the derivatives of $\boldsymbol{V}(\boldsymbol{\theta})$ with respect to σ_u^2 and ρ are

$$\frac{\partial \boldsymbol{V}(\boldsymbol{\theta})}{\partial \sigma_u^2} = \boldsymbol{C}^{-1}(\rho), \quad \frac{\partial \boldsymbol{V}(\boldsymbol{\theta})}{\partial \rho} = -\sigma_u^2 \boldsymbol{C}^{-1}(\rho)\frac{\partial \boldsymbol{C}(\rho)}{\partial \rho}\boldsymbol{C}^{-1}(\rho) \triangleq \boldsymbol{A}(\boldsymbol{\theta}).$$

The scores associated to σ_u^2 and ρ, after replacing (18.11) and using (A.2), are (cf. (6.17))

$$S_{\sigma_u^2} = -\frac{1}{2}\operatorname{tr}\left\{\boldsymbol{V}^{-1}(\boldsymbol{\theta})\boldsymbol{C}^{-1}(\rho)\right\} + \frac{1}{2}\boldsymbol{y}'\boldsymbol{P}(\boldsymbol{\theta})\boldsymbol{C}^{-1}(\rho)\boldsymbol{P}(\boldsymbol{\theta})\boldsymbol{y},$$
$$S_{\rho} = -\frac{1}{2}\operatorname{tr}\left\{\boldsymbol{V}^{-1}(\boldsymbol{\theta})\boldsymbol{A}(\boldsymbol{\theta})\right\} + \frac{1}{2}\boldsymbol{y}'\boldsymbol{P}(\boldsymbol{\theta})\boldsymbol{A}(\boldsymbol{\theta})\boldsymbol{P}(\boldsymbol{\theta})\boldsymbol{y}.$$

The elements of the Fisher information matrix are (cf. (6.21))

$$F_{\sigma_u^2,\sigma_u^2} = \frac{1}{2}\operatorname{tr}\left\{\boldsymbol{V}^{-1}(\boldsymbol{\theta})\boldsymbol{C}^{-1}(\rho)\boldsymbol{V}^{-1}(\boldsymbol{\theta})\boldsymbol{C}^{-1}(\rho)\right\},$$
$$F_{\sigma_u^2,\rho} = F_{\rho,\sigma_u^2} = \frac{1}{2}\operatorname{tr}\left\{\boldsymbol{V}^{-1}(\boldsymbol{\theta})\boldsymbol{A}(\boldsymbol{\theta})\boldsymbol{V}^{-1}(\boldsymbol{\theta})\boldsymbol{C}^{-1}(\rho)\right\},$$
$$F_{\rho,\rho} = \frac{1}{2}\operatorname{tr}\left\{\boldsymbol{V}^{-1}(\boldsymbol{\theta})\boldsymbol{A}(\boldsymbol{\theta})\boldsymbol{V}^{-1}(\boldsymbol{\theta})\boldsymbol{A}(\boldsymbol{\theta})\right\}.$$

A restricted maximum likelihood estimator (REML) of $\boldsymbol{\theta}$ is obtained by maximizing the restricted likelihood, which is the likelihood of $\boldsymbol{\theta}$ after eliminating the vector of coefficients $\boldsymbol{\beta}$. The restricted log-likelihood is given by (cf. (6.41))

$$\ell_R(\boldsymbol{\theta};\boldsymbol{y}) = c - -\frac{1}{2}\log|\boldsymbol{V}(\boldsymbol{\theta})| - \frac{1}{2}\log|\boldsymbol{X}'\boldsymbol{V}^{-1}(\boldsymbol{\theta})\boldsymbol{X}| - \frac{1}{2}\boldsymbol{y}'\boldsymbol{P}(\boldsymbol{\theta})\boldsymbol{y},$$

where $\boldsymbol{P}(\boldsymbol{\theta})$ is defined in (18.12). Using the following properties of the matrix $\boldsymbol{P}(\boldsymbol{\theta})$,

$$\boldsymbol{P}(\boldsymbol{\theta})\boldsymbol{V}(\boldsymbol{\theta})\boldsymbol{P}(\boldsymbol{\theta}) = \boldsymbol{P}(\boldsymbol{\theta}), \quad \frac{\partial \boldsymbol{P}(\boldsymbol{\theta})}{\partial \theta_j} = -\boldsymbol{P}(\boldsymbol{\theta})\frac{\partial \boldsymbol{V}(\boldsymbol{\theta})}{\partial \theta_j}\boldsymbol{P}(\boldsymbol{\theta}),$$

we obtain the scores corresponding to this restricted log-likelihood (cf. (6.42)),

$$S_{\sigma_u^2}^R = -\frac{1}{2}\operatorname{tr}\left\{\boldsymbol{P}(\boldsymbol{\theta})\boldsymbol{C}^{-1}(\rho)\right\} + \frac{1}{2}\boldsymbol{y}'\boldsymbol{P}(\boldsymbol{\theta})\boldsymbol{C}^{-1}(\rho)\boldsymbol{P}(\boldsymbol{\theta})\boldsymbol{y},$$
$$S_{\rho}^R = -\frac{1}{2}\operatorname{tr}\{\boldsymbol{P}\boldsymbol{\theta})\boldsymbol{A}(\boldsymbol{\theta})\} + \frac{1}{2}\boldsymbol{y}'\boldsymbol{P}(\boldsymbol{\theta})\boldsymbol{A}(\boldsymbol{\theta})\boldsymbol{P}(\boldsymbol{\theta})\boldsymbol{y}.$$

Finally, the elements of the Fisher information obtained from ℓ_R are

$$F^R_{\sigma_u^2,\sigma_u^2} = \frac{1}{2}\,\mathrm{tr}\{\boldsymbol{P}(\boldsymbol{\theta})\boldsymbol{C}^{-1}(\rho)\boldsymbol{P}(\boldsymbol{\theta})\boldsymbol{C}^{-1}(\rho)\},\quad F^R_{\rho,\rho} = \frac{1}{2}\,\mathrm{tr}\{\boldsymbol{P}(\boldsymbol{\theta})\boldsymbol{A}(\boldsymbol{\theta})\boldsymbol{P}(\boldsymbol{\theta})\boldsymbol{A}(\boldsymbol{\theta})\},$$

$$F^R_{\sigma_u^2,\rho} = F^R_{\rho,\sigma_u^2} = \frac{1}{2}\,\mathrm{tr}\{\boldsymbol{P}(\boldsymbol{\theta})\boldsymbol{A}(\boldsymbol{\theta})\boldsymbol{P}(\boldsymbol{\theta})\boldsymbol{C}^{-1}(\rho)\}.$$

For more details on calculating the scores and Fisher information matrix corresponding to restricted log-likelihood, we refer the reader to Sect. 6.5. A REML estimator of $\boldsymbol{\theta}$ can be obtained by applying the iterative Fisher-scoring formula (18.10) with the Fisher information matrix and the score vector replaced by their corresponding REML versions.

18.2.3 Parametric Bootstrap Estimation of the MSE

This section extends the parametric bootstrap method of González-Manteiga et al. (2010) to the area-level linear mixed model with spatial correlation (18.4)–(18.5), to derive an estimator for the MSE of the EBLUP $\hat{\mu}_d = \tilde{\mu}_d(\hat{\boldsymbol{\theta}})$, which is consistent if the estimators of the model parameters are consistent. In order to check the consistency of the estimators of the MSE, we can use, as in González-Manteiga et al. (2010), the asymptotic formula of the MSE obtained by Singh et al. (2005). The extended parametric bootstrap is composed of the following steps:

1. Obtain the estimates $\hat{\boldsymbol{\theta}} = (\hat{\sigma}_u^2, \hat{\rho})'$ and $\hat{\boldsymbol{\beta}} = \tilde{\boldsymbol{\beta}}(\hat{\boldsymbol{\theta}})$ by fitting the model (18.8) to the initial data $\boldsymbol{y} = (y_1, \ldots, y_D)'$.
2. Repeat B times ($b = 1, \ldots, B$).
   a. Generate a vector $\boldsymbol{t}_1^{*(b)}$ whose D elements are independent $N(0, 1)$. Build bootstrap vectors $\boldsymbol{v}^{*(b)} = \hat{\sigma}_u \boldsymbol{t}_1^{*(b)}$ and $\boldsymbol{u}^{*(b)} = (\boldsymbol{I}_D - \hat{\rho}\boldsymbol{W})^{-1}\boldsymbol{v}^{*(b)}$, and calculate $\boldsymbol{\mu}^{*(b)} = \boldsymbol{X}\hat{\boldsymbol{\beta}} + \boldsymbol{u}^{*(b)}$ with elements $\mu_d^{*(b)}$.
   b. Generate a vector $\boldsymbol{t}_2^{*(b)}$ with D independent $N(0, 1)$ elements, which is independent of $\boldsymbol{t}_1^{*(b)}$. Then, construct the vector of random errors as $\boldsymbol{e}^{*(b)} = \boldsymbol{V}_e^{1/2}\boldsymbol{t}_2^{*(b)}$.
   c. Obtain bootstrap data $\boldsymbol{y}^{*(b)}$ directly applying the formula

$$\boldsymbol{y}^{*(b)} = \boldsymbol{\mu}^{*(b)} + \boldsymbol{e}^{*(b)} = \boldsymbol{X}\hat{\boldsymbol{\beta}} + \boldsymbol{u}^{*(b)} + \boldsymbol{e}^{*(b)}. \tag{18.13}$$

   d. Fit the model (18.8) to the bootstrap data $\boldsymbol{y}^{*(b)}$ and obtain the estimates $\hat{\boldsymbol{\theta}}^{*(b)}$ and $\hat{\boldsymbol{\beta}}^{*(b)}$ of the bootstrap parameters $\hat{\boldsymbol{\theta}}$ and $\hat{\boldsymbol{\beta}}$, respectively.
   e. Calculate the bootstrap spatial EBLUP

$$\hat{\mu}_d^{*(b)} = \boldsymbol{x}_d\hat{\boldsymbol{\beta}}^{*(b)} + \boldsymbol{b}_d'\boldsymbol{G}(\hat{\boldsymbol{\theta}}^{*(b)})\boldsymbol{V}^{-1}(\hat{\boldsymbol{\theta}}^{*(b)})\big(\boldsymbol{y}^{*(b)} - \boldsymbol{X}\hat{\boldsymbol{\beta}}^{*(b)}\big).$$

3. A parametric bootstrap estimator of $MSE(\hat{\mu}_d)$ is

$$mse_d^* = \frac{1}{B}\sum_{b=1}^{B}\left(\hat{\mu}_d^{*(b)} - \mu_d^{*(b)}\right)^2. \tag{18.14}$$

18.3 Area-Level Spatio-Temporal Linear Mixed Model 1

In this section we generalize the basic spatial model by introducing independent domain-time random effects.

18.3.1 The Model

Let y_{dt} be a direct estimator of the target population parameter, and let $\boldsymbol{x}_{dt}$ be a vector containing the aggregated values of p auxiliary variables. Subindexes d and t are used for domains and time instants, respectively. Let us consider the model

$$y_{dt} = \boldsymbol{x}_{dt}\boldsymbol{\beta} + u_{1d} + u_{2dt} + e_{dt}, \quad d = 1, \dots, D, \quad t = 1, \dots, T. \tag{18.15}$$

Model (18.15) can alternatively written as

$$\boldsymbol{y} = \boldsymbol{X}\boldsymbol{\beta} + \boldsymbol{Z}_1\boldsymbol{u}_1 + \boldsymbol{Z}_2\boldsymbol{u}_2 + \boldsymbol{e}, \tag{18.16}$$

where

$$\boldsymbol{y} = \underset{1\le d\le D}{\text{col}}(\underset{1\le t\le T}{\text{col}}(y_{dt})), \quad \boldsymbol{e} = \underset{1\le d\le D}{\text{col}}(\underset{1\le t\le T}{\text{col}}(e_{dt})),$$

$$\boldsymbol{u}_1 = \underset{1\le d\le D}{\text{col}}(u_{1d}), \quad \boldsymbol{u}_2 = \underset{1\le d\le D}{\text{col}}(\boldsymbol{u}_{2d}), \quad \boldsymbol{u}_{2d} = \underset{1\le t\le T}{\text{col}}(u_{2dt}),$$

$$\boldsymbol{X} = \underset{1\le d\le D}{\text{col}}(\underset{1\le t\le T}{\text{col}}(\boldsymbol{x}_{dt})), \quad \boldsymbol{x}_{dt} = \underset{1\le j\le p}{\text{col}'}(x_{dtj}), \quad \boldsymbol{\beta} = \underset{1\le j\le p}{\text{col}}(\beta_j),$$

$$\boldsymbol{Z}_1 = \underset{1\le d\le D}{\text{diag}}(\boldsymbol{1}_T), \quad \boldsymbol{Z}_2 = \boldsymbol{I}_M, \quad M = DT.$$

We assume that $\boldsymbol{u}_1 \sim N(\boldsymbol{0}, \boldsymbol{V}_{u_1})$, $\boldsymbol{u}_2 \sim N(\boldsymbol{0}, \boldsymbol{V}_{u_2})$, and $\boldsymbol{e} \sim N(\boldsymbol{0}, \boldsymbol{V}_e)$ are independent with covariance matrices

$$\boldsymbol{V}_{u_1} = \sigma_1^2\Omega_1(\rho_1), \quad \Omega_1(\rho_1) = \left[(\boldsymbol{I}_D - \rho_1\boldsymbol{W})'(\boldsymbol{I}_D - \rho_1\boldsymbol{W})\right]^{-1} \triangleq \boldsymbol{C}^{-1}(\rho_1),$$

$$\boldsymbol{V}_{u_2} = \sigma_2^2\boldsymbol{I}_M, \quad \boldsymbol{V}_e = \underset{1\le d\le D}{\text{diag}}(\boldsymbol{V}_{ed}), \quad \boldsymbol{V}_{ed} = \underset{1\le t\le T}{\text{diag}}(\sigma_{dt}^2),$$

and known variances σ_{dt}^2's. We assume that the rows of the proximity matrix $\boldsymbol{W}$ are stochastic vectors, i.e. with components summing up to one. The vector $\boldsymbol{u}_1$ is distributed according to a SAR(1) stochastic process, and the variables u_{2dt} are i.i.d. normal. The variance of $\boldsymbol{y}$ is

$$\text{var}(\boldsymbol{y}) = \boldsymbol{V} = \boldsymbol{V}(\boldsymbol{\theta}) = \boldsymbol{Z}_1\boldsymbol{V}_{u_1}\boldsymbol{Z}_1' + \boldsymbol{Z}_2\boldsymbol{V}_{u_2}\boldsymbol{Z}_2' + \boldsymbol{V}_e = \boldsymbol{Z}_1\boldsymbol{V}_{u_1}\boldsymbol{Z}_1' + \underset{1\le d\le D}{\text{diag}}\,(\sigma_2^2\boldsymbol{I}_T + \boldsymbol{V}_{ed}).$$

Its inverse can be calculated with the formula

$$(A + CBD)^{-1} = A^{-1} - A^{-1}C(B^{-1} + DA^{-1}C)^{-1}DA^{-1},$$

where $A = \underset{1\le d\le D}{\text{diag}}\,(\sigma_2^2\boldsymbol{I}_T + \boldsymbol{V}_{ed})$, $B = \boldsymbol{V}_{u_1}$, $C = \boldsymbol{Z}_1$ and $D = \boldsymbol{Z}_1'$. Then

$$\boldsymbol{V}^{-1} = A^{-1} - A^{-1}\boldsymbol{Z}_1(\boldsymbol{V}_{u_1}^{-1} + \boldsymbol{Z}_1'A^{-1}\boldsymbol{Z}_1)^{-1}\boldsymbol{Z}_1'A^{-1},$$

where $A^{-1} = \underset{1\le d\le D}{\text{diag}}\,(A_d^{-1})$ and $A_d = \sigma_2^2\boldsymbol{I}_T + \boldsymbol{V}_{ed}$. Observe that applying the above formula avoids inverting an $M \times M$ matrix, instead we only have to invert D matrices of order $T \times T$ and one matrix of order $D \times D$.

Let us define the parameter $\boldsymbol{\theta} = (\sigma_1^2, \rho_1, \sigma_2^2)$. The formula

$$\frac{\partial \boldsymbol{C}^{-1}(\rho_1)}{\partial \rho_1} = -\boldsymbol{C}^{-1}(\rho_1)\frac{\partial \boldsymbol{C}(\rho_1)}{\partial \rho_1}\boldsymbol{C}^{-1}(\rho_1)$$

is used for calculating the partial derivatives of $\boldsymbol{V} = \boldsymbol{V}(\boldsymbol{\theta})$ with respect to the components of $\boldsymbol{\theta}$, i.e.

$$\begin{aligned}
\boldsymbol{V}_1 &= \frac{\partial \boldsymbol{V}}{\partial \sigma_1^2} = \boldsymbol{Z}_1\frac{\partial \boldsymbol{V}_{u_1}}{\partial \sigma_1^2}\boldsymbol{Z}_1' = \boldsymbol{Z}_1\boldsymbol{\Omega}_1(\rho_1)\boldsymbol{Z}_1',\\
\boldsymbol{V}_2 &= \frac{\partial \boldsymbol{V}}{\partial \rho_1} = \boldsymbol{Z}_1\frac{\partial \boldsymbol{V}_{u_1}}{\partial \rho_1}\boldsymbol{Z}_1' = -\sigma_1^2\boldsymbol{Z}_1\boldsymbol{C}^{-1}(\rho_1)\frac{\partial \boldsymbol{C}(\rho_1)}{\partial \rho_1}\boldsymbol{C}^{-1}(\rho_1)\boldsymbol{Z}_1',\\
\boldsymbol{V}_3 &= \frac{\partial \boldsymbol{V}}{\partial \sigma_2^2} = \boldsymbol{I}_M,
\end{aligned}$$

where

$$\begin{aligned}
\frac{\partial \boldsymbol{C}(\rho_1)}{\partial \rho_1} &= \frac{\partial}{\partial \rho_1}\{(\boldsymbol{I}_D - \rho_1\boldsymbol{W})'(\boldsymbol{I}_D - \rho_1\boldsymbol{W})\}\\
&= -\boldsymbol{W}' + \rho_1\boldsymbol{W}'\boldsymbol{W} - \boldsymbol{W} + \rho_1\boldsymbol{W}'\boldsymbol{W} = -\boldsymbol{W} - \boldsymbol{W}' + 2\rho_1\boldsymbol{W}'\boldsymbol{W}.
\end{aligned}$$

For $\boldsymbol{\theta}$ known, the BLU estimator and predictor of $\boldsymbol{\beta}$ and $\boldsymbol{u}$ are (cf. Proposition 16.1)

$$\tilde{\boldsymbol{\beta}} = \tilde{\boldsymbol{\beta}}(\boldsymbol{\theta}) = (\boldsymbol{X}'\boldsymbol{V}^{-1}\boldsymbol{X})^{-1}\boldsymbol{X}'\boldsymbol{V}^{-1}\boldsymbol{y} \quad \text{and} \quad \tilde{\boldsymbol{u}} = \tilde{\boldsymbol{u}}(\boldsymbol{\theta}) = \boldsymbol{V}_u\boldsymbol{Z}'\boldsymbol{V}^{-1}(\boldsymbol{y} - \boldsymbol{X}\tilde{\boldsymbol{\beta}}),$$

where $\boldsymbol{V}_u = \text{diag}(\boldsymbol{V}_{u_1}, \boldsymbol{V}_{u_2})$, $\boldsymbol{Z} = [\boldsymbol{Z}_1, \boldsymbol{Z}_2]$, and $\boldsymbol{u} = (\boldsymbol{u}_1', \boldsymbol{u}_2')'$. The vector $\tilde{\boldsymbol{u}}$ can be calculated by applying the formula

$$\tilde{\boldsymbol{u}} = \begin{pmatrix} \boldsymbol{V}_{u_1}\boldsymbol{Z}_1' \\ \boldsymbol{V}_{u_2}\boldsymbol{Z}_2' \end{pmatrix} \boldsymbol{V}^{-1}(\boldsymbol{y} - \boldsymbol{X}\tilde{\boldsymbol{\beta}}) = \begin{pmatrix} \sigma_1^2\Omega_1(\rho_1)\boldsymbol{Z}_1'\boldsymbol{V}^{-1}(\boldsymbol{y} - \boldsymbol{X}\tilde{\boldsymbol{\beta}}) \\ \sigma_2^2\boldsymbol{Z}_2'\boldsymbol{V}^{-1}(\boldsymbol{y} - \boldsymbol{X}\tilde{\boldsymbol{\beta}}) \end{pmatrix}.$$

The BLUP predictor of μ_{dt} is

$$\tilde{\mu}_{dt} = \tilde{\mu}_{dt}(\boldsymbol{\theta}) = \boldsymbol{x}_{dt}\tilde{\boldsymbol{\beta}} + \tilde{u}_{1d} + \tilde{u}_{2dt}.$$

The corresponding EBLUP of μ_{dt} is obtained by substituting the estimators of the variance components $\boldsymbol{\theta}$, and it has the form

$$\hat{\mu}_{dt} = \tilde{\mu}_{dt}(\hat{\boldsymbol{\theta}}) = \boldsymbol{x}_{dt}\hat{\boldsymbol{\beta}} + \hat{u}_{1,d} + \hat{u}_{2,dt}.$$

18.3.2 Residual Maximum Likelihood Estimation

Following steps of Sect. 17.2.2, the Fisher-scoring algorithm for the REML estimate of the parameter $\boldsymbol{\theta}$ can be derived. The updating equation is

$$\boldsymbol{\theta}^{(i+1)} = \boldsymbol{\theta}^{(i)} + \boldsymbol{F}^{-1}(\boldsymbol{\theta}^{(i)})\boldsymbol{S}(\boldsymbol{\theta}^{(i)}),$$

where $\boldsymbol{\theta} = (\theta_1, \theta_2, \theta_3) = (\sigma_1^2, \rho_1, \sigma_2^2)$, the components of the score vector are

$$S_a = -\frac{1}{2}\text{tr}(\boldsymbol{P}\boldsymbol{V}_a) + \frac{1}{2}\boldsymbol{y}'\boldsymbol{P}\boldsymbol{V}_a\boldsymbol{P}\boldsymbol{y}, \quad a = 1, 2, 3,$$

the components of the Fisher information matrix are

$$F_{ab} = \frac{1}{2}\text{tr}(\boldsymbol{P}\boldsymbol{V}_a\boldsymbol{P}\boldsymbol{V}_b), \quad a, b = 1, 2, 3,$$

matrices $\boldsymbol{V}_a$, $a = 1, 2, 3$, are defined on the previous page, and

$$\boldsymbol{P} = \boldsymbol{V}^{-1} - \boldsymbol{V}^{-1}\boldsymbol{X}(\boldsymbol{X}'\boldsymbol{V}^{-1}\boldsymbol{X})^{-1}\boldsymbol{X}'\boldsymbol{V}^{-1}.$$

This algorithm requires starting values of $\boldsymbol{\theta}$ (seeds). We may obtain seeds by considering the model without $\boldsymbol{u}_1$. For this last model, we might consider the

Henderson 3 estimator $\hat{\sigma}^2_{u_2H}$ of the only remaining variance σ_2^2. Therefore, we might propose the following seeds: $\sigma_1^{2(0)} = \sigma_2^{2(0)} = \frac{1}{2}\hat{\sigma}^2_{u_2H}$, $\rho_1^{(0)} = 0.3$.

The REML estimator of $\boldsymbol{\beta}$ is

$$\hat{\boldsymbol{\beta}} = (\boldsymbol{X}'\hat{\boldsymbol{V}}^{-1}\boldsymbol{X})^{-1}\boldsymbol{X}'\hat{\boldsymbol{V}}^{-1}\boldsymbol{y},$$

where $\hat{\boldsymbol{V}} = \boldsymbol{V}(\hat{\boldsymbol{\theta}})$. For comments concerning the asymptotic distributions and confidence intervals of the REML estimators of $\boldsymbol{\theta}$ and $\boldsymbol{\beta}$, we refer the reader to Sect. 17.2.2.

18.3.3 Simulations

This section presents two simulation experiments that investigate the behavior of the REML estimators of model parameters, the EBLUPs of domain means, and the bootstrap MSE estimators.

For $d = 1, \ldots, D$, $t = 1, \ldots, T$, the explanatory and target variables are

$$x_{dt} = \frac{1}{5}(b_{dt} - a_{dt})U_{dt} + a_{dt},\ U_{dt} = \frac{t}{T+1},\ a_{dt} = 1,\ b_{dt} = 1 + \frac{1}{D}(T(d-1)+t),$$

$$y_{dt} = \beta_1 + \beta_2 x_{dt} + u_{1d} + u_{2dt} + e_{dt},\ \beta_1 = 0,\ \beta_2 = 1,$$

where $u_{2dt} \sim N(0, \sigma_2^2)$ and $e_{dt} \sim N(0, \sigma_{dt}^2)$ are independent with $\sigma_2^2 = 1$ and

$$\sigma_{dt}^2 = \frac{2}{25}\frac{T(d-1)+t-1}{M-1} + 0.8.$$

The vector $\boldsymbol{u}_1 = \underset{1\le d\le D}{\text{col}}(u_{1d})$ is generated from the distribution $N_D(0, \sigma_1^2\Omega_1(\rho_1))$, using the proximity matrix

$$\boldsymbol{W} = \begin{pmatrix} 0 & 1 & 0 & \ldots & 0 \\ 1/2 & 0 & 1/2 & \ldots & 0 \\ \vdots & \ddots & \ddots & \ddots & \vdots \\ 0 & \ldots & 1/2 & 0 & 1/2 \\ 0 & \ldots & 0 & 1 & 0 \end{pmatrix}_{D\times D}, \quad \sigma_1^2 = 1, \rho_1 = 0.5. \tag{18.17}$$

The steps of Simulation 1 are:

1. Repeat $I = 4000$ times ($i = 1, \ldots, I$).
   1.1. Generate $y_{dt}^{(i)}$ and calculate $\mu_{dt}^{(i)} = \beta_1 + \beta_2 x_{dt} + u_{1d}^{(i)} + u_{2dt}^{(i)}$, $d = 1, \ldots, D$, $t = 1, \ldots, T$.

1.2. Calculate $\hat{\tau}^{(i)} \in \{\hat{\beta}_1^{(i)}, \hat{\beta}_2^{(i)}, \hat{\sigma}_1^{2(i)}, \hat{\rho}_1^{(i)}, \hat{\sigma}_2^{2(i)}\}$ by using the REML method and the EBLUP $\hat{\mu}_{dt}^{(i)} = \hat{\beta}_1^{(i)} + \hat{\beta}_2^{(i)} x_{dt} + \hat{u}_{1d}^{(i)} + \hat{u}_{2dt}^{(i)}$.

2. For each $\tau \in \{\beta_1, \beta_2, \sigma_1^2, \rho_1, \sigma_2^2\}$ and for each $\hat{\mu}_{dt}$, $d = 1, \ldots, D$, $t = 1, \ldots, T$, calculate

$$BIAS(\hat{\tau}) = \frac{1}{I}\sum_{i=1}^{I}(\hat{\tau}^{(i)} - \tau), \quad MSE(\hat{\tau}) = \frac{1}{I}\sum_{i=1}^{I}(\hat{\tau}^{(i)} - \tau)^2,$$

$$BIAS_{dt} = \frac{1}{I}\sum_{i=1}^{I}(\hat{\mu}_{dt}^{(i)} - \mu_{dt}^{(i)}), \quad MSE_{dt} = \frac{1}{I}\sum_{i=1}^{I}(\hat{\mu}_{dt}^{(i)} - \mu_{dt}^{(i)})^2,$$

$$BIAS = \frac{1}{M}\sum_{d=1}^{D}\sum_{t=1}^{T} BIAS_{dt}, \quad MSE = \frac{1}{M}\sum_{d=1}^{D}\sum_{t=1}^{T} MSE_{dt}.$$

The simulation experiment is repeated for each of the six combinations of sample sizes appearing in Table 18.1. The results of the simulation experiments are presented in Table 18.2. Table 18.2 shows that bias is always close to zero and that MSE decreases as the number of domains increases, so that the estimators are empirically consistent. Let us note that the MSE of the EBLUP (presented on the last line of the table) is almost constant with increasing D since with increasing number of areas we also augment the number of parameters to predict.

Simulation 2 studies the behavior of the bootstrap MSE estimator of the EBLUP. The bootstrap estimator can be obtained in the same way as described in Sect. 18.2.3 with the only difference that the bootstrap samples have to be generated from the actual model. The steps of Simulation 2 are:

1. For $D = 50, 100, 200, 400$, take the values MSE_{dt} from the output of Simulation 1.
2. Repeat $I = 200$ times ($i = 1, \ldots, I$).

2.1. Generate a sample $y_{dt}^{(i)}$, $d = 1, \ldots, D$, $t = 1, \ldots, T$.
2.2. Calculate the parameter estimates $\hat{\beta}_1^{(i)}, \hat{\beta}_2^{(i)}, \hat{\sigma}_1^{2(i)}, \hat{\rho}_1^{(i)}, \hat{\sigma}_2^{2(i)}$ by using the REML method.
2.3. Repeat $B = 100$ times ($b = 1, \ldots, B$).

Table 18.1 Sample sizes

D	50	100	200	300	400	500
T	5	5	5	5	5	5
M	250	500	1000	1500	2000	2500

Table 18.2 Results of Simulation 1

D	50	100	200	300	400	500
$BIAS(\hat{\beta}_1)$	−0.0046	0.0047	0.0014	−0.0012	−0.0002	−0.0009
$BIAS(\hat{\beta}_2)$	0.0012	−0.0005	−0.0011	0.0003	−0.0003	−0.0002
$MSE(\hat{\beta}_1)$	0.1595	0.0851	0.0435	0.0278	0.0219	0.0169
$BIAS(\hat{\sigma}_1^2)$	−0.0350	−0.0138	−0.0064	−0.0065	−0.0036	−0.0053
$BIAS(\hat{\rho}_1)$	−0.0169	−0.0073	−0.0034	−0.0029	−0.0014	−0.0009
$BIAS(\hat{\sigma}_2^2)$	0.0006	−0.0012	0.0010	−0.0005	−0.0009	0.0001
$BIAS$	−0.0003	0.0005	−0.0002	−0.0003	0.0001	0.0001
$MSE(\hat{\beta}_2)$	0.0159	0.0083	0.0042	0.0028	0.0020	0.0016
$MSE(\hat{\sigma}_1^2)$	0.1101	0.0547	0.0285	0.0192	0.0141	0.0116
$MSE(\hat{\rho}_1)$	0.0267	0.0115	0.0059	0.0038	0.0030	0.0023
$MSE(\hat{\sigma}_2^2)$	0.0388	0.0191	0.0100	0.0064	0.0049	0.0040
MSE	0.5784	0.5739	0.5715	0.5716	0.5712	0.5710

2.3.1. Generate a bootstrap sample $y_{dt}^{*(ib)}$ with the parameters $\hat{\beta}_1^{(i)}$, $\hat{\beta}_2^{(i)}$, $\hat{\sigma}_1^{2(i)}$, $\hat{\rho}_1^{(i)}$, and $\hat{\sigma}_2^{2(i)}$ from step 2.2. Calculate the true bootstrap quantities $\mu_{dt}^{*(ib)} = \hat{\beta}_1^{(i)} + \hat{\beta}_2^{(i)} x_{dt} + u_{1d}^{*(ib)} + u_{2dt}^{*(ib)}$.

2.3.2. Calculate the REML estimators of the parameters of the bootstrap model $\hat{\beta}_1^{*(ib)}, \hat{\beta}_2^{*(ib)}, \hat{\sigma}_1^{2*(ib)}, \hat{\rho}_1^{*(ib)}, \hat{\sigma}_2^{2*(ib)}$ and the EBLUP $\hat{\mu}_{dt}^{*(ib)} = \hat{\beta}_1^{*(ib)} + \hat{\beta}_2^{*(ib)} x_{dt} + \hat{u}_{1d}^{*(ib)} + \hat{u}_{2dt}^{*(ib)}$.

2.4. Calculate

$$mse_{dt}^{(i)} = \frac{1}{B}\sum_{b=1}^{B}(\hat{\mu}_{dt}^{*(ib)} - \mu_{dt}^{*(ib)})^2.$$

3. For $d = 1, \ldots, D$, $t = 1, \ldots, T$, calculate

$$B_{dt} = \frac{1}{I}\sum_{i=1}^{I}(mse_{dt}^{(i)} - MSE_{dt}), \quad E_{dt} = \frac{1}{I}\sum_{i=1}^{I}(mse_{dt}^{(i)} - MSE_{dt})^2,$$

$$B = \frac{1}{M}\sum_{d=1}^{D}\sum_{t=1}^{T} B_{dt}, \quad E = \frac{1}{M}\sum_{d=1}^{D}\sum_{t=1}^{T} E_{dt}.$$

The simulation experiment is repeated for four combinations of sample sizes appearing in Table 18.1. The simulation results are presented in Table 18.3 that shows that bias B is always close to zero and that the MSE E slightly decreases as the number of domains increases. Intuitively, we can say that it is caused by increasing precision of the estimates of the variance parameters.

Table 18.3 Results of Simulation 2

D	50	100	200	400
B	−0.0032	−0.0052	−0.0020	−0.0025
E	0.0082	0.0075	0.0071	0.0069

18.4 Area-Level Spatio-Temporal Linear Mixed Model 2

In this section we present briefly a model that is based on the previous one but introduces AR(1)-correlation structure across time within domains.

18.4.1 The Model

Let y_{dt} be a direct estimator of the characteristic of interest and let $\boldsymbol{x}_{dt}$ be a vector containing the aggregated values of p of auxiliary variables. The subindex d is used for domains, and the subindex t for time instants. Let us consider the model

$$y_{dt} = \boldsymbol{x}_{dt}\boldsymbol{\beta} + u_{1d} + u_{2dt} + e_{dt}, \quad d = 1, \dots, D, \quad t = 1, \dots, T, \tag{18.18}$$

which can be alternatively written in the form

$$\boldsymbol{y} = \boldsymbol{X}\boldsymbol{\beta} + \boldsymbol{Z}_1\boldsymbol{u}_1 + \boldsymbol{Z}_2\boldsymbol{u}_2 + \boldsymbol{e}, \tag{18.19}$$

where definition of the symbols is exactly the same as the one given on page 495. We assume that $\boldsymbol{u}_1 \sim N(\boldsymbol{0}, \boldsymbol{V}_{u_1})$, $\boldsymbol{u}_2 \sim N(\boldsymbol{0}, \boldsymbol{V}_{u_2})$, and $\boldsymbol{e} \sim N(\boldsymbol{0}, \boldsymbol{V}_e)$ are independent with covariance matrices

$$\boldsymbol{V}_{u_1} = \sigma_1^2\Omega_1(\rho_1), \quad \Omega_1(\rho_1) = \left[(\boldsymbol{I}_D - \rho_1\boldsymbol{W})'(\boldsymbol{I}_D - \rho_1\boldsymbol{W})\right]^{-1} \triangleq \boldsymbol{C}^{-1}(\rho_1),$$

$$\boldsymbol{V}_{u_2} = \sigma_2^2\Omega_2(\rho_2), \quad \Omega_2(\rho_2) = \underset{1\le d\le D}{\text{diag}}\,(\Omega_{2d}(\rho_2)),$$

$$\Omega_{2d} = \Omega_{2d}(\rho_2) = \frac{1}{1-\rho_2^2}\begin{pmatrix} 1 & \rho_2 & \dots & \rho_2^{T-2} & \rho_2^{T-1} \\ \rho_2 & 1 & \ddots & & \rho_2^{T-2} \\ \vdots & \ddots & \ddots & \ddots & \vdots \\ \rho_2^{T-2} & & \ddots & 1 & \rho_2 \\ \rho_2^{T-1} & \rho_2^{T-2} & \dots & \rho_2 & 1 \end{pmatrix}_{T\times T},$$

$$\boldsymbol{V}_e = \underset{1\le d\le D}{\text{diag}}\,(\boldsymbol{V}_{ed}), \quad \boldsymbol{V}_{ed} = \underset{1\le t\le T}{\text{diag}}\,(\sigma_{dt}^2),$$

where the variance parameters σ_{dt}^2's are known. The rows of the proximity matrix $\boldsymbol{W}$ are assumed to be stochastic vectors, i.e. their components sum up to one. The vector $\boldsymbol{u}_1$ is distributed as a SAR(1) stochastic process, and the vectors $\boldsymbol{u}_{2d}$ are independent with homogeneous AR(1) distributions (they all have the same variance and autocorrelation parameters).

The covariance matrix $\text{var}(\boldsymbol{y}) = \boldsymbol{V} = \boldsymbol{V}(\boldsymbol{\theta})$ of the vector $\boldsymbol{y}$ is

$$\boldsymbol{V} = \boldsymbol{Z}_1 \boldsymbol{V}_{u_1} \boldsymbol{Z}_1' + \boldsymbol{Z}_2 \boldsymbol{V}_{u_2} \boldsymbol{Z}_2' + \boldsymbol{V}_e = \boldsymbol{Z}_1 \boldsymbol{V}_{u_1} \boldsymbol{Z}_1' + \underset{1 \le d \le D}{\text{diag}} (\sigma_2^2 \Omega_{2d}(\rho_2) + \boldsymbol{V}_{ed}).$$

Its inverse can be calculated in the same way as described on page 496, and it has the form

$$\boldsymbol{V}^{-1} = A^{-1} - A^{-1} \boldsymbol{Z}_1 (\boldsymbol{V}_{u_1}^{-1} + \boldsymbol{Z}_1' A^{-1} \boldsymbol{Z}_1)^{-1} \boldsymbol{Z}_1' A^{-1},$$

where $A^{-1} = \underset{1 \le d \le D}{\text{diag}} (A_d^{-1})$ and $A_d = \sigma_2^2 \Omega_{2d}(\rho_2) + \boldsymbol{V}_{ed}$. Observe that after applying the inversion formula we substitute the inversion of one matrix of order $M \times M$ by the inversion of D matrices of order $T \times T$ and one matrix of order $D \times D$.

Let us define the parameter $\boldsymbol{\theta} = (\sigma_1^2, \rho_1, \sigma_2^2, \rho_2)$. We apply the formula

$$\frac{\partial \boldsymbol{C}^{-1}(\rho_1)}{\partial \rho_1} = -\boldsymbol{C}^{-1}(\rho_1) \frac{\partial \boldsymbol{C}(\rho_1)}{\partial \rho_1} \boldsymbol{C}^{-1}(\rho_1),$$

to obtain the partial derivatives of $\boldsymbol{V}$ with respect to $\boldsymbol{\theta}$. Namely,

$$\boldsymbol{V}_1 = \frac{\partial \boldsymbol{V}}{\partial \sigma_1^2} = \boldsymbol{Z}_1 \frac{\partial \boldsymbol{V}_{u_1}}{\partial \sigma_1^2} \boldsymbol{Z}_1' = \boldsymbol{Z}_1 \Omega_1(\rho_1) \boldsymbol{Z}_1',$$

$$\boldsymbol{V}_2 = \frac{\partial \boldsymbol{V}}{\partial \rho_1} = \boldsymbol{Z}_1 \frac{\partial \boldsymbol{V}_{u_1}}{\partial \rho_1} \boldsymbol{Z}_1' = -\sigma_1^2 \boldsymbol{Z}_1 \boldsymbol{C}^{-1}(\rho_1) \frac{\partial \boldsymbol{C}(\rho_1)}{\partial \rho_1} \boldsymbol{C}^{-1}(\rho_1) \boldsymbol{Z}_1',$$

$$\boldsymbol{V}_3 = \frac{\partial \boldsymbol{V}}{\partial \sigma_2^2} = \underset{1 \le d \le D}{\text{diag}} (\Omega_{2d}(\rho_2)),$$

$$\boldsymbol{V}_4 = \frac{\partial \boldsymbol{V}}{\partial \rho_2} = \sigma_2^2 \underset{1 \le d \le D}{\text{diag}} \left(\frac{\partial \Omega_{2d}(\rho_2)}{\partial \rho_2} \right),$$

where

$$\begin{aligned} \frac{\partial \boldsymbol{C}(\rho_1)}{\partial \rho_1} &= \frac{\partial}{\partial \rho_1} \{ (\boldsymbol{I}_D - \rho_1 \boldsymbol{W})' (\boldsymbol{I}_D - \rho_1 \boldsymbol{W}) \} \\ &= -\boldsymbol{W}' + \rho_1 \boldsymbol{W}' \boldsymbol{W} - \boldsymbol{W} + \rho_1 \boldsymbol{W}' \boldsymbol{W} = -\boldsymbol{W} - \boldsymbol{W}' + 2\rho_1 \boldsymbol{W}' \boldsymbol{W}, \end{aligned}$$

and

$$\frac{\partial \Omega_{2d}(\rho_2)}{\partial \rho_2} = \frac{1}{1-\rho_2^2}\begin{pmatrix} 0 & 1 & \dots\dots & (T-1)\rho_2^{T-2} \\ 1 & 0 & \ddots & (T-2)\rho_2^{T-3} \\ \vdots & \ddots & \ddots & \ddots & \vdots \\ (T-2)\rho_2^{T-3} & & \ddots & 0 & 1 \\ (T-1)\rho_2^{T-2} & \dots\dots & & 1 & 0 \end{pmatrix} + \frac{2\rho_2\Omega_{2d}(\rho_2)}{1-\rho_2^2}.$$

For $\boldsymbol{\theta}$ known, the BLU estimator and predictor of $\boldsymbol{\beta}$ and $\boldsymbol{u}$ are (cf. Proposition 16.1)

$$\tilde{\boldsymbol{\beta}} = \tilde{\boldsymbol{\beta}}(\boldsymbol{\theta}) = (\boldsymbol{X}'\boldsymbol{V}^{-1}\boldsymbol{X})^{-1}\boldsymbol{X}'\boldsymbol{V}^{-1}\boldsymbol{y} \quad \text{and} \quad \tilde{\boldsymbol{u}} = \tilde{\boldsymbol{u}}(\boldsymbol{\theta}) = \boldsymbol{V}_u\boldsymbol{Z}'\boldsymbol{V}^{-1}(\boldsymbol{y} - \boldsymbol{X}\tilde{\boldsymbol{\beta}}),$$

where $\boldsymbol{V}_u = \text{diag}(\boldsymbol{V}_{u_1}, \boldsymbol{V}_{u_2})$, $\boldsymbol{Z} = [\boldsymbol{Z}_1, \boldsymbol{Z}_2]$, and $\boldsymbol{u} = (\boldsymbol{u}_1', \boldsymbol{u}_2')'$. To calculate $\tilde{\boldsymbol{u}}$, it is possible to apply the formula

$$\tilde{\boldsymbol{u}} = \begin{pmatrix} \boldsymbol{V}_{u_1}\boldsymbol{Z}_1' \\ \boldsymbol{V}_{u_2}\boldsymbol{Z}_2' \end{pmatrix} \boldsymbol{V}^{-1}(\boldsymbol{y} - \boldsymbol{X}\tilde{\boldsymbol{\beta}}) = \begin{pmatrix} \sigma_1^2\Omega_1(\rho_1)\boldsymbol{Z}_1'\boldsymbol{V}^{-1}(\boldsymbol{y} - \boldsymbol{X}\tilde{\boldsymbol{\beta}}) \\ \sigma_2^2\Omega_2(\rho_2)\boldsymbol{Z}_2'\boldsymbol{V}^{-1}(\boldsymbol{y} - \boldsymbol{X}\tilde{\boldsymbol{\beta}}) \end{pmatrix}.$$

The BLUP predictor of μ_{dt} is

$$\tilde{\mu}_{dt} = \tilde{\mu}_{dt}(\boldsymbol{\theta}) = \boldsymbol{x}_{dt}\tilde{\boldsymbol{\beta}} + \tilde{u}_{1d} + \tilde{u}_{2dt}.$$

The corresponding EBLUP of μ_{dt} is obtained by substituting the estimate $\hat{\boldsymbol{\theta}}$ of the variance components $\boldsymbol{\theta}$, and it has the form

$$\hat{\mu}_{dt} = \tilde{\mu}_{dt}(\hat{\boldsymbol{\theta}}) = \boldsymbol{x}_{dt}\hat{\boldsymbol{\beta}} + \hat{u}_{1,d} + \hat{u}_{2,dt}.$$

18.4.2 Residual Maximum Likelihood Estimation

The REML estimates of the model parameters can be obtained in the same way as described in Sect. 18.3.2 with the actual version of the covariance matrix $\boldsymbol{V}$ and the parameter $\boldsymbol{\theta} = (\sigma_1^2, \rho_1, \sigma_2^2, \rho_2)$. For obtaining seeds for the Fisher-scoring algorithm, we can take the reduced model without $\boldsymbol{u}_1$ and with $\rho_2 = 0$ as a reference. For this reduced model, it is easy to calculate the Henderson 3 estimator $\hat{\sigma}_{u_2H}^2$ of the only remaining variance σ_2^2. Therefore, a possible set of algorithm seeds is $\sigma_1^{2(0)} = \sigma_2^{2(0)} = \frac{1}{2}\hat{\sigma}_{u_2H}^2$, $\rho_1^{(0)} = \rho_2^{(0)} = 0.3$.

18.4.3 Simulations

Two simulation experiments are carried out under the area-level spatio-temporal linear mixed model 2. Simulation 1 investigates the behavior of the REML estimators of model parameters and the EBLUPs of domain means. It follows exactly the same steps as Simulation 1 described in Sect. 18.3.3 with $\boldsymbol{\theta} = (\sigma_1^2, \rho_1, \sigma_2^2, \rho_2)$ and the domain-time random effects u_{2dt} generated as follows:

$$u_{2d1} = (1-\rho_2^2)^{-1/2}\varepsilon_{d1}, \qquad u_{2dt} = \rho_2 u_{2dt-1} + \varepsilon_{dt},\ t = 2, \ldots, T,$$

where $\varepsilon_{dt} \overset{i.i.d.}{\sim} N(0, \sigma_2^2)$, $d = 1, \ldots, D$, $t = 1, \ldots, T$, and $\rho_2 = 0.5$.

The simulation experiment is repeated for the six combinations of sample sizes appearing in Table 18.1. Table 18.4 presents the simulation results. Table 18.4 shows that the bias is always close to zero and that the MSE decreases as the number of domains increases, so that the REML estimators are empirically consistent. Concerning the MSE of EBLUP and its bootstrap estimates presented below, we observe the same pattern as discussed in Sect. 18.3.3.

Simulation 2 studies the behavior of the bootstrap MSE estimators. Again, it follows the same steps as Simulation 2 in Sect. 18.3.3 with $\boldsymbol{\theta} = (\sigma_1^2, \rho_1, \sigma_2^2, \rho_2)$ and the AR(1)-correlated domain-time random effects u_{2dt} generated as described above.

The simulation experiment is repeated for the four combinations of sample sizes appearing in Table 18.1. Table 18.5 presents the simulation results, and it shows that the bias B is always close to zero and that the MSE E decreases as the number of domains increases.

Table 18.4 Results of Simulation 1

D	50	100	200	300	400	500
$BIAS(\hat{\beta}_1)$	0.0021	0.0059	0.0015	0.0023	0.0027	−0.0001
$BIAS(\hat{\beta}_2)$	0.0004	−0.0016	−0.0019	−0.0012	−0.0025	−0.0011
$BIAS(\hat{\sigma}_1^2)$	0.0413	−0.0248	−0.0308	−0.0337	−0.0269	−0.0226
$BIAS(\hat{\rho}_1)$	−0.0245	0.0061	0.0083	0.0108	0.0075	0.0053
$BIAS(\hat{\sigma}_2^2)$	−0.0174	−0.0082	−0.0039	−0.0050	−0.0013	−0.0017
$BIAS(\hat{\rho}_2)$	−0.0896	−0.0293	−0.0113	−0.0030	−0.0023	−0.0011
$BIAS$	−0.0016	0.0000	−0.0006	−0.0008	−0.0002	−0.0001
$MSE(\hat{\beta}_1)$	0.1966	0.1050	0.0515	0.0344	0.0257	0.0199
$MSE(\hat{\beta}_2)$	0.0212	0.0109	0.0053	0.0036	0.0025	0.0020
$MSE(\hat{\sigma}_1^2)$	0.2519	0.1603	0.1025	0.0706	0.0550	0.0449
$MSE(\hat{\rho}_1)$	0.0455	0.0241	0.0139	0.0086	0.0069	0.0054
$MSE(\hat{\sigma}_2^2)$	0.0430	0.0228	0.0109	0.0072	0.0056	0.0044
$MSE(\hat{\rho}_2)$	0.0412	0.0191	0.0112	0.0074	0.0056	0.0044
MSE	0.5500	0.5446	0.5424	0.5418	0.5413	0.5409

Table 18.5 Results of Simulation 2

D	50	100	200	400
B	0.0009	−0.0032	0.0010	−0.0007
E	0.0086	0.0072	0.0067	0.0065

18.5 R Codes for EBLUPs

This section gives R codes for fitting area-level spatial and spatio-temporal linear mixed models to the aggregated data. The data is taken from the R package `sae`. First, install and/or load the package and load the data set `grapes`.

```
if(!require(sae)){
  install.packages("sae")
  library(sae)
}
data(grapes); dim(grapes)              # Load data set
data(grapesprox); dim(grapesprox)      # Load proximity matrix
apply(grapesprox, 1, sum)              # Check that rows sum up to one
```

Fit the spatial Fay–Herriot model by using the ML method.

```
resultML <- eblupSFH(grapehect ~ area + workdays - 1, var, grapesprox,
                     method="ML", data=grapes)
head(resultML$eblup, 10)               # EBPLUPs of first 10 domains
resultML$fit$estcoef                   # Regression coefficients
resultML$fit$refvar                    # Variance of random effect
resultML$fit$spatialcorr               # Spatial correlation parameters
resultML$fit$goodness                  # AIC
```

Fit the spatial Fay–Herriot model using the REML method

```
resultREML <- eblupSFH(grapehect ~ area + workdays - 1, var, grapesprox,
                       data=grapes)
head(resultREML$eblup, 10)             # EBPLUPs of first 10 domains
resultREML$fit$estcoef                 # Regression coefficients
resultREML$fit$refvar                  # Variance of random effect
resultREML$fit$spatialcorr             # Spatial correlation parameters
resultREML$fit$goodness                # AIC
```

Load the data set `spacetime`.

```
data(spacetime); dim(spacetime)               # Load data set
data(spacetimeprox); dim(spacetimeprox)       # Load proximity matrix
D <- nrow(spacetimeprox)                      # number of domains
T <- length(unique(spacetime$Time))           # number of time instant
```

Fit the spatio-temporal Fay–Herriot model with uncorrelated time effects for each domain

```
resultS <- eblupSTFH(Y ~ X1 + X2, D, T, Var, spacetimeprox,
                     model="S", data=spacetime)
resultS$fit$estcoef                          # Regression coefficients
resultS$fit$estvarcomp                       # Variance components
# EPLUPs for the last time domain
subset(resultS$eblup, spacetime$Time==T)
```

Table 18.6 EBLUPs estimates of spatio-temporal FH model with uncorrelated (Model Indep) and AR(1)-correlated (Model AR1) time effects for the last time domain

Area	Model Indep	Model AR1
2	0.2749	0.2734
3	0.1774	0.1772
8	0.0964	0.0965
12	0.1385	0.1374
13	0.2911	0.2913
16	0.3195	0.3189
17	0.0700	0.0691
25	0.1722	0.1738
43	0.1435	0.1440
45	0.2261	0.2281
46	0.1436	0.1435

Fit the spatio-temporal Fay–Herriot model with AR(1)-correlated time effects for each domain

```
resultST <- eblupSTFH(Y ~ X1 + X2, D, T, Var, spacetimeprox,
                      model="ST", data=spacetime)
resultST$fit$estcoef                          # Regression coefficients
resultST$fit$estvarcomp                       # Variance components
# EPLUPs for the last time domain
subset(resultST$eblup, spacetime$Time==T)
```

The R code to save the results is

```
output <- data.frame(Area=subset(spacetime$Area, spacetime$Time==T),
                     Indep=subset(resultS$eblup, spacetime$Time==T),
                     AR1=subset(resultST$eblup, spacetime$Time==T))
head(output, 10)
```

Results obtained by the two spatio-temporal models are presented in Table 18.6.

References

Anselin, L.: Spatial Econometrics. Methods and Models. Kluwer, Boston (1988)

Banerjee, S., Carlin, B., Gelfand, A.: Hierarchical Modeling and Analysis for Spatial Data. Chapman and Hall, New York (2004)

Cressie, N.: Statistics for Spatial Data. Wiley, New York (1993)

Ghosh, M., Rao, J.N.K.: Small area estimation: an appraisal (with discussion). Stat. Sci. **9**, 55–93 (1994)

González-Manteiga, W., Lombardía, M. J., Molina, I., Morales, D., Santamaría, L.: Small area estimation under Fay-Herriot models with nonparametric estimation of heteroscedasticity. Stat. Model. **10**(2), 215–239 (2010)

Jiang, J.: Linear and Generalized Linear Mixed Models and Their Applications. Springer, Berlin (2007)

Marhuenda, Y., Molina, I., Morales, D.: Small area estimation with spatio-temporal Fay-Herriot models. Comput. Stat. Data Anal. **58**, 308–325 (2013)

Molina, I., Salvati, N., Pratesi, M.: Bootstrap for estimating the MSE of the spatial EBLUP. Comput. Stat. **24**, 441–458 (2009)

Moura, F.A.S., Migon, H.S.: Bayesian spatial models for small area estimation of proportions. Stat. Model. **2**(3), 183–201 (2002)

Petrucci, A., Salvati, N.: Small area estimation for spatial correlation in watershed erosion assessment. J. Agric. Biol. Environ. Stat. **11**, 169–182 (2006)

Pratesi, M., Salvati, N.: Small area estimation: the EBLUP estimator based on spatially correlated random area effects. Stat. Methods Appl. **17**, 113–141 (2008)

Rao, J.N.K., Yu, M.: Small area estimation by combining time series and cross sectional data. Can. J. Stat. **22**, 511–528 (1994)

Singh, B., Shukla, G., Kundu, D.: Spatio-temporal models in small area estimation. Surv. Methodol. **31**, 183–195 (2005)

You, Y., Zhou, Q.M.: Hierarchical Bayes small area estimation under a spatial model with application to health survey data. Surv. Methodol. **37**, 25–37 (2011)

Chapter 19
Area-Level Bivariate Linear Mixed Models

19.1 Introduction

Fay and Herriot (1979) proposed an area-level linear mixed model with random area effects to estimate average per capita income in small places of the United States. The Fay–Herriot model is widely employed to estimate univariate small area parameters. However, statisticians are often required to estimate correlated descriptive measures. Multivariate models include the correlation of several variables, and they are useful tools to treat this kind of problems.

Some papers can be found in the literature of small area estimation where bivariate linear mixed models are employed. Fay (1987) and Datta et al. (1991) compared the precision of small area estimators obtained from univariate models for each response variable with the ones obtained from a bivariate model. Datta et al. (1996) also used a bivariate Fay–Herriot model for obtaining hierarchical Bayes estimates of median income of four-person families for the US states. González-Manteiga et al. (2008b) studied a class of bivariate Fay–Herriot models with a common random effect for all the components of the target vector. They further introduced bootstrap approximations to prediction errors.

Benavent and Morales (2016) introduced a class of multivariate Fay–Herriot (MFH) models with one random effect per component of the target vector and allowing for different covariance patterns between the components of the vector of random effects. They applied the developed methodology to data from the Spanish living conditions surveys of 2005 and 2006 and presented two applications. The target of the first application was the estimation of 2006 poverty proportions and gaps. The second application jointly estimated 2005 and 2006 poverty proportions.

Ubaidillah et al. (2019) studied the application of MFH models for small area estimation of household consumption per capita expenditure on food and non-food. Esteban et al. (2020) made adaptations of MFH models to the estimation of small

D. Morales et al., *A Course on Small Area Estimation and Mixed Models*, Statistics for Social and Behavioral Sciences, https://doi.org/10.1007/978-3-030-63757-6_19

area compositional parameters, like proportions of people in the four categories of the variable labor status: under 16 years, employed, unemployed, and inactive. Arima et al. (2017) and Burgard et al. (2020) introduced variants of the bivariate Fay–Herriot model that take into account the measurement error of the auxiliary variables. Benavent and Morales (2020) extended the bivariate Fay–Herriot model to the temporal setup. These works, and others not cited here, demonstrate the utility of multivariate area models for estimation problems in small areas.

This chapter describes the bivariate Fay–Herriot model under complete parametrization, and it gives the Fisher-scoring algorithms to calculate the maximum likelihood (ML) and the residual maximum likelihood (REML) estimators of the model parameters and approximates the matrices of mean squared crossed errors of the empirical best linear unbiased predictors of population linear parameters. Some simulations and R codes illustrate the theoretical developments. Contrary to what is usual in this book, this chapter does not use bold characters to denote vectors and matrices.

19.2 The Bivariate Fay–Herriot Model

Let U be a finite population partitioned into D domains $U_1, \ldots, U_D$. Let $\mu_d = (\mu_{d1}, \mu_{d2})'$ be a vector of characteristics of interest in the domain d, and let $y_d = (y_{d1}, y_{d2})'$ be a vector of direct estimators of μ_d calculated by using the data of the target survey sample. Let us note again that we do not use the bold notation of vectors and matrices in this chapter, since the majority of the symbols would be printed in bold.

The bivariate Fay–Herriot model is defined in two stages. The first stage indicates that direct estimators $y_d, d = 1, \ldots, D$, are unbiased and follow the *sampling model*

$$y_d = \mu_d + e_d, \quad d = 1, \ldots, D, \tag{19.1}$$

where the vectors $e_d = (e_{d1}, e_{d2})' \sim N(0, V_{ed})$ are independent and the 2×2 covariance matrices V_{ed} are known. In most cases, V_{ed} is taken to be the design-based covariance matrix of direct estimator $y_d, d = 1, \ldots, D$.

In the second stage the true area characteristic μ_{dk} is assumed to be linearly related to p_k explanatory variables, $k = 1, 2$, $d = 1, \ldots, D$. Let $x'_{dk} = (x_{dk1}, \ldots, x_{dkp_k})$ be a row vector containing the true aggregated (population) values of p_k explanatory variables for μ_{dk}, and let $X_d = \text{diag}(x'_{d1}, x'_{d2})$ be a $2 \times p$ block-diagonal matrix with $p = p_1 + p_2$. Let $\beta_k = (\beta_{k1}, \ldots, \beta_{kp_k})'$ be a column vector of size p_k containing the regression parameters for μ_{dk}, and let $\beta = \left(\beta'_1, \beta'_2\right)'_{p\times 1}$. The *linking model* is

$$\mu_d = X_d\beta + u_d, \quad u_d = (u_{d1}, u_{d2})' \sim N_2(0, V_{ud}), \quad d = 1, \ldots, D. \tag{19.2}$$

The 2×2 covariance matrices V_{ud} depend on $\theta = (\theta_1, \theta_2, \theta_3)'$ with 3 unknown parameters, $\theta_1 = \sigma_{u1}^2$, $\theta_2 = \sigma_{u2}^2$, and $\theta_3 = \rho$, i.e.

$$V_{ud} = \begin{pmatrix} \sigma_{u1}^2 & \rho\sigma_{u1}\sigma_{u2} \\ \rho\sigma_{u1}\sigma_{u2} & \sigma_{u2}^2 \end{pmatrix}.$$

The bivariate Fay–Herriot (BFH) model can be expressed as a single model in the form

$$y_d = X_d\beta + u_d + e_d, \quad d = 1, \dots, D, \tag{19.3}$$

or in the linear mixed model matrix form

$$y = X\beta + Zu + e,$$

where

$$y = \underset{1\le d\le D}{\text{col}}(y_d),\ u = \underset{1\le d\le D}{\text{col}}(u_d),\ e = \underset{1\le d\le D}{\text{col}}(e_d),\ X = \underset{1\le d\le D}{\text{col}}(X_d),\ Z = I_{2D}.$$

We finally assume that u_d, e_d, $d = 1, \dots, D$, are independent. The BFH model (19.3) is a reparametrization of Model 3 introduced by Benavent and Morales (2016).

Under model (19.3), it holds that $y \sim N(X\beta, V)$ with

$$E[y] = X\beta \quad \text{and} \quad V = \text{var}(y) = Z'V_uZ + V_e = V_u + V_e = \underset{1\le d\le D}{\text{diag}}(V_d),$$

where

$$V_u = V_u(\theta) = \underset{1\le d\le D}{\text{diag}}(V_{ud}),\ V_e = \underset{1\le d\le D}{\text{diag}}(V_{ed}),\ V_d = V_{ud} + V_{ed}, \quad d = 1, \dots, D.$$

It also holds that

$$E[yy'] = E\big[(X\beta+Zu+e)[(X\beta+Zu+e)']\big] = X\beta\beta'X'+ZV_uZ'+V_e = X\beta\beta'X'+V.$$

As the BFH model is a linear mixed model, the best linear unbiased estimator (BLUE) of β and the best linear unbiased predictors (BLUP) of u and μ are (cf. Proposition 16.1)

$$\hat\beta_B = (X'V^{-1}X)^{-1}X'V^{-1}y,\ \hat u_B = V_uV^{-1}(y - X\hat\beta_B),\ \hat\mu_B = X\hat\beta_B + \hat u_B. \tag{19.4}$$

Let us define $Q = (X'V^{-1}X)^{-1} = \left(\sum_{d=1}^{D} X'_d V_d^{-1} X_d\right)^{-1}$. Alternative formulas are

$$\hat{\beta}_B = Q \sum_{d=1}^{D} X'_d V_d^{-1} y_d, \quad \hat{u}_B = \underset{1\le d\le D}{\text{col}}(\hat{u}_{Bd}), \quad \hat{u}_{Bd} = V_{ud} V_d^{-1}(y_d - X_d\hat{\beta}_B).$$

By substituting θ in the formulas (19.4) by an estimator $\hat{\theta}$, we obtain the empirical BLUE (EBLUE) of β and the empirical BLUP (EBLUP) of u and $\mu = X\beta + Zu$, i.e.

$$\hat{\beta}_E = (X'\hat{V}^{-1}X)^{-1}X'\hat{V}^{-1}y, \; \hat{u}_E = \hat{V}_u\hat{V}^{-1}(y - X\hat{\beta}_E), \; \hat{u}_{Ed} = \hat{V}_{ud}\hat{V}_d^{-1}(y_d - X_d\hat{\beta}_E),$$
$$\hat{\mu}_E = X\hat{\beta}_E + Z\hat{u}_E, \quad \hat{\mu}_{Ed} = X_d\hat{\beta}_E + \hat{u}_{Ed}, \tag{19.5}$$

where $\hat{V}_u = V_u(\hat{\theta})$ and $\hat{V} = \hat{V}_u + V_e$ are obtained by plugging $\hat{\theta}$ in the place of θ.

19.3 Properties of the BLUPs

By BLUP$_0$, we will denote the BLUP when all the model parameters, β and θ, are known. Proposition 19.1 derives the BLUP$_0$ of u_d under the assumed normality. It calculates the best predictor (BP) of u_d, which is the expectation of u_d given y_d. The BLUP $\hat{u}_{Bd}$ can be obtained by substituting β by $\hat{\beta}_B$ in the expression of $\hat{u}_d^{bp}$.

Proposition 19.1 *The BP of u_d under model (19.3) is the BLUP$_0$, i.e.*

$$\hat{u}_d^{bp} = E\big[u_d|y_d; \beta, \theta\big] = \Phi_d V_{ed}^{-1}(y_d - X_d\beta) = V_{ud}V_d^{-1}(y_d - X_d\beta),$$

where $\Phi_d = \left(V_{ed}^{-1} + V_{ud}^{-1}\right)^{-1}$.

Proof We recall that the n-variate normal probability density function is

$$f(y|\mu, \Sigma) = \frac{1}{(2\pi)^{n/2}|\Sigma|^{1/2}} \exp\{-\frac{1}{2}(y-\mu)'\Sigma^{-1}(y-\mu)\} \propto \exp\{-\frac{1}{2}y'\Sigma^{-1}y + \mu'\Sigma^{-1}y\},$$

with density kernel on the right hand side of the symbol $\propto$ (proportional to). The conditional distribution of u_d, given y_d, is

$$\begin{aligned} f(u_d|y_d) \propto f(y_d|u_d)f(u_d) &= \\ &= \frac{1}{2\pi|V_{ed}|^{1/2}} \exp\Big\{-\frac{1}{2}(y_d - X_d\beta - u_d)'V_{ed}^{-1}(y_d - X_d\beta - u_d)\Big\} \\ &\cdot \frac{1}{2\pi|V_{ud}|^{1/2}} \exp\Big\{-\frac{1}{2}u'_d V_{ud}^{-1} u_d\Big\} \end{aligned}$$

$$\propto \exp\Big\{ -\frac{1}{2}u_d' V_{ed}^{-1} u_d + u_d' V_{ed}^{-1}(y_d - X_d\beta)\Big\} \exp\Big\{ -\frac{1}{2}u_d' V_{ud}^{-1} u_d\Big\}$$
$$= \exp\Big\{ -\frac{1}{2}u_d'\Big(V_{ed}^{-1} + V_{ud}^{-1}\Big)u_d + u_d'\Phi_d^{-1}\Big[\Phi_d V_{ed}^{-1}(y_d - X_d\beta)\Big]\Big\}.$$

Therefore $f(u_d|y_d)$ is a bivariate normal distribution with parameters

$$\text{var}(u_d|y_d;\beta,\theta) = \Big(V_{ed}^{-1} + V_{ud}^{-1}\Big)^{-1} = \Phi_d,\;\; E\big[u_d|y_d;\beta,\theta\big] = \Phi_d V_{ed}^{-1}(y_d - X_d\beta).$$

By applying the inversion formula,

$$(A+BCD)^{-1} = A^{-1} - A^{-1}B(C^{-1} + DA^{-1}B)^{-1}DA^{-1},$$

with $A = V_{ed}^{-1}$, $C = V_{ud}^{-1}$, and $B = D = I_2$, we have

$$\Phi_d = \Big(V_{ed}^{-1} + V_{ud}^{-1}\Big)^{-1} = V_{ed} - V_{ed}(V_{ud} + V_{ed})^{-1}V_{ed}.$$

Finally, we get

$$\hat{u}_d^{bp} = E\big[u_d|y_d;\beta,\theta\big] = \big[V_{ed} - V_{ed}(V_{ud}+V_{ed})^{-1}V_{ed}\big]\, V_{ed}^{-1}(y_d - X_d\beta)$$
$$= \big[I_2 - V_{ed}V_d^{-1}\big](y_d - X_d\beta) = \big[(V_d - V_{ed})V_d^{-1}\big](y_d - X_d\beta)$$
$$= V_{ud}V_d^{-1}(y_d - X_d\beta).$$

□

We now derive the mean squared error of the BLUP of μ. For this sake, we need to calculate several moments first. The BLUE of β has the following moments:

$$E[\hat{\beta}_B] = QX'V^{-1}E[y] = QX'V^{-1}X\beta = QQ^{-1}\beta = \beta,$$
$$\text{var}(\hat{\beta}_B) = QX'V^{-1}\text{var}(y)V^{-1}XQ = QX'V^{-1}VV^{-1}XQ = QX'V^{-1}XQ = QQ^{-1}Q = Q,$$
$$E[\hat{\beta}_B\hat{\beta}_B'] = \text{var}(\hat{\beta}_B) + E[\hat{\beta}_B]E[\hat{\beta}_B'] = Q + \beta\beta',$$
$$E[\hat{\beta}_B y'] = QX'V^{-1}E[yy'] = QX'V^{-1}(X\beta\beta'X' + V) = \beta\beta'X' + QX',$$
$$E[\hat{\beta}_B u'] = QX'V^{-1}E[yu'] = QX'V^{-1}V_u.$$

The moments of $y - X\hat{\beta}_B$ are

$$A_1 = E\big[y - X\hat{\beta}_B\big] = X\beta - XE\big[\hat{\beta}_B\big] = X\beta - X\beta = 0,$$
$$A_2 = E\big[(y - X\hat{\beta}_B)(y - X\hat{\beta}_B)'\big] = E\big[yy'\big] + XE\big[\hat{\beta}_B\hat{\beta}_B'\big]X' - XE\big[\hat{\beta}_B y'\big] - E\big[y\hat{\beta}_B'\big]X'$$
$$= X\beta\beta'X' + V + X(Q + \beta\beta')X' - X(\beta\beta'X' + QX') - (X\beta\beta' + XQ)X'$$
$$= X\beta\beta'X' + V + XQX' + X\beta\beta'X' - 2X\beta\beta'X' - 2XQX' = V - XQX'.$$

The BLUP of u has the following moments:

$$
\begin{aligned}
E\big[\hat{u}_B\big] &= V_u V^{-1} E[y - X\hat{\beta}_B] = 0, \\
\operatorname{var}(\hat{u}_B) &= E\big[\hat{u}_B \hat{u}_B'\big] = V_u V^{-1} E[(y - X\hat{\beta}_B)(y - X\hat{\beta}_B)']V^{-1} V_u \\
&= V_u V^{-1}(V - XQX')V^{-1} V_u = V_u V^{-1} V_u - V_u V^{-1} XQX' V^{-1} V_u, \\
E\big[\hat{u}_B u'\big] &= V_u V^{-1} E[(y - X\hat{\beta}_B)u'] = V_u V^{-1}\{E\big[yu'\big] - XE\big[\hat{\beta}_B u'\big]\} \\
&= V_u V^{-1}\{V_u - XQX' V^{-1} V_u\} = V_u V^{-1} V_u - V_u V^{-1} XQX' V^{-1} V_u, \\
\operatorname{cov}(\hat{\beta}_B, \hat{u}_B) &= E[(\hat{\beta}_B - \beta)\hat{u}_B'] = E\big[\hat{\beta}_B \hat{u}_B'\big] = QX'V^{-1} E\big[y(y - X\hat{\beta}_B)'\big]V^{-1} V_u \\
&= QX'V^{-1}\{E\big[yy'\big] - E\big[y\hat{\beta}_B'\big]X'\}V^{-1} V_u \\
&= QX'V^{-1}\{X\beta\beta'X' + V - (X\beta\beta' + XQ)X'\}V^{-1} V_u \\
&= QX'V^{-1}\{V - XQX'\}V^{-1} V_u = QX'V^{-1} V_u - QQ^{-1} QX'V^{-1} V_u = 0.
\end{aligned}
$$

As $E\big[\hat{u}_B \hat{u}_B'\big] = E\big[\hat{u}_B u'\big] = E\big[u\hat{u}_B'\big]$ and $E\big[\hat{\beta}_B \hat{u}_B'\big] = 0$, the moments of $\hat{u}_B - u$ are

$$
\begin{aligned}
B_1 &= E\big[\hat{u}_B - u\big] = E\big[\hat{u}_B\big] - E[u] = 0, \\
B_2 &= E[(\hat{u}_B - u)(\hat{u}_B - u)'] = E\big[\hat{u}_B \hat{u}_B'\big] + E\big[uu'\big] - E\big[\hat{u}_B u'\big] - E\big[u\hat{u}_B'\big] \\
&= E\big[uu'\big] - E\big[\hat{u}_B u'\big] = V_u - V_u V^{-1} V_u + V_u V^{-1} XQX' V^{-1} V_u, \\
B_3 &= E[(\hat{\beta}_B - \beta)(\hat{u}_B - u)'] = E\big[\hat{\beta}_B \hat{u}_B'\big] + E\big[\beta u'\big] - E\big[\hat{\beta}_B u'\big] - E\big[\beta\hat{u}_B'\big] \\
&= 0 + 0 - QX'V^{-1} V_u + 0 = -QX'V^{-1} V_u.
\end{aligned}
$$

The BLUP of μ has the following moments and mean square error:

$$
\begin{aligned}
E\big[\hat{\mu}_B\big] &= E\big[X\hat{\beta}_B + \hat{u}_B\big] = X\beta, \\
\operatorname{var}(\hat{\mu}_B) &= X\operatorname{var}(\hat{\beta}_B)X' + \operatorname{var}(\hat{u}_B) + X\operatorname{cov}(\hat{\beta}_B, \hat{u}_B) + \operatorname{cov}(\hat{u}_B, \hat{\beta}_B)X' \\
&= XQX' + V_u V^{-1} V_u - V_u V^{-1} XQX' V^{-1} V_u, \\
MSE(\hat{\mu}_B) &= E[(\hat{\mu}_B - \mu)(\hat{\mu}_B - \mu)'] \\
&= E\big[(X(\hat{\beta}_B - \beta) + (\hat{u}_B - u))(X(\hat{\beta}_B - \beta) + (\hat{u}_B - u))'\big] \\
&= X\operatorname{var}(\hat{\beta}_B)X' + E\big[(\hat{u}_B - u)(\hat{u}_B - u)'\big] + XE\big[(\hat{\beta}_B - \beta)(\hat{u}_B - u)'\big] \\
&+ E\big[(\hat{u}_B - u)(\hat{\beta}_B - \beta)'\big]X' = XQX' + V_u - V_u V^{-1} V_u \\
&+ V_u V^{-1} XQX' V^{-1} V_u - XQX' V^{-1} V_u - V_u V^{-1} XQX' \\
&= V_u - V_u V^{-1} V_u + (X - V_u V^{-1} X)Q(X - V_u V^{-1} X)'.
\end{aligned}
$$

As the matrix $T = V_u - V_u V^{-1} V_u$ fulfills the relation

$$TV_e^{-1} = \big(V_u - V_u V^{-1} V_u\big) V_e^{-1} = V_u\big(I - V^{-1} V_u\big) V_e^{-1}$$
$$= V_u V^{-1}(V - V_u) V_e^{-1} = V_u V^{-1} V_e V_e^{-1} = V_u V^{-1},$$

the mean squared error of $\hat{\mu}_B$ can also be written in the form

$$MSE\left(\hat{\mu}_B\right) = G_1(\theta) + G_2(\theta),\ G_1(\theta) = T,\ G_2(\theta) = \big(X - TV_e^{-1}X\big)Q\big(X' - X'V_e^{-1}T\big). \tag{19.6}$$

19.4 Maximum Likelihood Estimation

This section derives the Fisher-scoring algorithm for calculating the ML estimator of the vector of model parameters $\psi = (\beta', \theta')'$, where $\theta = (\theta_1, \theta_2, \theta_3)' = (\sigma_{u1}^2, \sigma_{u2}^2, \rho)'$. The log-likelihood of the model (19.3) is $l = \sum_{d=1}^{D} l_d$, where

$$l_d = -\log 2\pi - \frac{1}{2}\log|V_d| - \frac{1}{2}(y_d - X_d\beta)' V_d^{-1}(y_d - X_d\beta), \quad d = 1\ldots, D.$$

For the Fisher-scoring algorithm, we need derivatives of the log-likelihood l to the second order. We start with calculating the derivatives of the matrix V_d. Let us remind that $V_d = V_{ud} + V_{ed}$ and V_{ed} is assumed to be known and it does not depend on θ. The first derivatives of V_{ud} with respect to $\theta_1 = \sigma_{u1}^2$, $\theta_2 = \sigma_{u2}^2$ and $\theta_3 = \rho$ are

$$V_{ud1} = \frac{\partial V_{ud}}{\partial \sigma_{u1}^2} = \begin{pmatrix} 1 & \frac{\rho\sigma_{u2}}{2\sigma_{u1}} \\ \frac{\rho\sigma_{u2}}{2\sigma_{u1}} & 0 \end{pmatrix}, \quad V_{ud2} = \frac{\partial V_{ud}}{\partial \sigma_{u2}^2} = \begin{pmatrix} 0 & \frac{\rho\sigma_{u1}}{2\sigma_{u2}} \\ \frac{\rho\sigma_{u1}}{2\sigma_{u2}} & 1 \end{pmatrix},$$
$$V_{ud3} = \frac{\partial V_{ud}}{\partial \rho} = \begin{pmatrix} 0 & \sigma_{u1}\sigma_{u2} \\ \sigma_{u1}\sigma_{u2} & 0 \end{pmatrix}. \tag{19.7}$$

The second derivatives of V_{ud} are

$$V_{ud11} = \frac{\partial^2 V_{ud}}{\partial \sigma_{u1}^2 \partial \sigma_{u1}^2} = \begin{pmatrix} 0 & \frac{-\rho\sigma_{u2}}{4\sigma_{u1}^3} \\ \frac{-\rho\sigma_{u2}}{4\sigma_{u1}^3} & 0 \end{pmatrix}, \quad V_{ud22} = \frac{\partial^2 V_{ud}}{\partial \sigma_{u2}^2 \partial \sigma_{u2}^2} = \begin{pmatrix} \frac{-\rho\sigma_{u1}}{4\sigma_{u2}^3} & 0 \\ 0 & \frac{-\rho\sigma_{u1}}{4\sigma_{u2}^3} \end{pmatrix},$$
$$V_{ud33} = \frac{\partial^2 V_{ud}}{\partial \rho \partial \rho} = \begin{pmatrix} 0 & 0 \\ 0 & 0 \end{pmatrix}, \quad V_{ud12} = \frac{\partial^2 V_{ud}}{\partial \sigma_{u1}^2 \partial \sigma_{u2}^2} = \begin{pmatrix} 0 & \frac{\rho}{4\sigma_{u1}\sigma_{u2}} \\ \frac{\rho}{4\sigma_{u1}\sigma_{u2}} & 0 \end{pmatrix},$$
$$V_{ud13} = \frac{\partial^2 V_{ud}}{\partial \sigma_{u1}^2 \partial \rho} = \begin{pmatrix} 0 & \frac{\sigma_{u2}}{2\sigma_{u1}} \\ \frac{\sigma_{u2}}{2\sigma_{u1}} & 0 \end{pmatrix}, \quad V_{ud23} = \frac{\partial^2 V_{ud}}{\partial \sigma_{u2}^2 \partial \rho} = \begin{pmatrix} 0 & \frac{\sigma_{u1}}{2\sigma_{u2}} \\ \frac{\sigma_{u1}}{2\sigma_{u2}} & 0 \end{pmatrix}. \tag{19.8}$$

By using the formulas given in (A.2), we calculate the first partial derivatives of l_d, with respect to β_{1j_1}, β_{2j_2}, and θ_a, $j_k = 1, \ldots, p_k$, $k = 1, 2$, $a, b = 1, 2, 3$, i.e.

$$\frac{\partial l_d}{\partial \beta_{1j_1}} = (x_{d1j_1}, 0)V_d^{-1}(y_d - X_d\beta), \quad \frac{\partial l_d}{\partial \beta_{2j_2}} = (0, x_{d2j_2})V_d^{-1}(y_d - X_d\beta),$$

$$\frac{\partial l_d}{\partial \theta_a} = -\frac{1}{2}\operatorname{tr}\big(V_d^{-1}V_{uda}\big) + \frac{1}{2}(y_d - X_d\beta)'V_d^{-1}V_{uda}V_d^{-1}(y_d - X_d\beta).$$

For $j_k = 1, \ldots, p_k$, $k = 1, 2$, $a, b = 1, 2, 3$, the second partial derivatives of l_d are

$$\frac{\partial^2 l_d}{\partial \beta_{1j_1}\partial \beta_{1j_2}} = -(x_{d1j_1}, 0)V_d^{-1}(x_{d1j_2}, 0)', \quad \frac{\partial^2 l_d}{\partial \beta_{2j_1}\partial \beta_{2j_2}} = -(0, x_{d2j_1})V_d^{-1}(0, x_{d2j_2})',$$

$$\frac{\partial^2 l_d}{\partial \beta_{1j_1}\partial \beta_{2j_2}} = -(x_{d1j_1}, 0)V_d^{-1}(0, x_{d2j_2})',$$

$$\frac{\partial^2 l_d}{\partial \beta_{1j_1}\partial \theta_a} = -(x_{d1j}, 0)V_d^{-1}V_{uda}V_d^{-1}(y_d - X_d\beta),$$

$$\frac{\partial^2 l_d}{\partial \beta_{2j_2}\partial \theta_a} = -(0, x_{d2j})V_d^{-1}V_{uda}V_d^{-1}(y_d - X_d\beta),$$

$$\begin{aligned}\frac{\partial^2 l_d}{\partial \theta_a\partial \theta_b} &= \frac{1}{2}\operatorname{tr}\big(V_d^{-1}V_{udb}V_d^{-1}V_{uda}\big) - \frac{1}{2}\operatorname{tr}\big(V_d^{-1}V_{udab}\big)\\ &- (y_d - X_d\beta)'V_d^{-1}V_{udb}V_d^{-1}V_{uda}V_d^{-1}(y_d - X_d\beta)\\ &+ \frac{1}{2}(y_d - X_d\beta)'V_d^{-1}V_{udab}V_d^{-1}(y_d - X_d\beta).\end{aligned}$$

By changing the sign and taking expectations, we have (cf. (A.3))

$$F_{d\beta_{1j_1}\beta_{1j_2}} = -E\Big[\frac{\partial^2 l_d}{\partial \beta_{1j_1}\partial \beta_{1j_2}}\Big] = (x_{d1j_1}, 0)V_d^{-1}(x_{d1j_2}, 0)',$$

$$F_{d\beta_{2j_1}\beta_{2j_2}} = (0, x_{d2j_1})V_d^{-1}(0, x_{d2j_2})',$$

$$F_{d\beta_{1j_1}\beta_{2j_2}} = (x_{d1j_1}, 0)V_d^{-1}(0, x_{d2j_2})', \; F_{d\beta_{1j_1}\theta_a} = F_{d\beta_{2j_2}\theta_a} = 0,$$

$$F_{d\theta_a\theta_b} = \frac{1}{2}\operatorname{tr}\big(V_d^{-1}V_{udb}V_d^{-1}V_{uda}\big).$$

For $j_k = 1, \ldots, p_k$, $k = 1, 2$, $a, b = 1, 2, 3$, the components of the score vector are

$$U_{\beta_{1j_1}} = \sum_{d=1}^{D}\frac{\partial l_d}{\partial \beta_{1j_1}}, \quad U_{\beta_{2j_2}} = \sum_{d=1}^{D}\frac{\partial l_d}{\partial \beta_{2j_2}}, \quad U_{\theta_a} = \sum_{d=1}^{D}\frac{\partial l_d}{\partial \theta_a}.$$

The $(p+3)\times 1$ score vector is $U(\psi)=(U'_\beta(\psi), U'_\theta(\psi))'$, where

$$U_\beta(\psi)=\left(U_{\beta_1 1},\ldots,U_{\beta_1 p_1},U_{\beta_2 1},\ldots,U_{\beta_2 p_2}\right)', \quad U_\theta(\psi)'=\left(U_{\theta_1},U_{\theta_2},U_{\theta_3}\right)'.$$

The components of the Fisher information matrix are

$$F_{\beta_{1j_1}\beta_{1j_2}}=\sum_{d=1}^{D}F_{d\beta_{1j_1}\beta_{1j_2}},\ F_{\beta_{2j_1}\beta_{2j_2}}=\sum_{d=1}^{D}F_{d\beta_{2j_1}\beta_{2j_2}},\ F_{\beta_{1j_1}\beta_{2j_2}}=\sum_{d=1}^{D}F_{d\beta_{1j_1}\beta_{2j_2}},$$

$$F_{\beta_{1j_1}\theta_a}=\sum_{d=1}^{D}F_{d\beta_{1j_1}\theta_a},\ \ F_{\beta_{2j_2}\theta_a}=\sum_{d=1}^{D}F_{d\beta_{2j_2}\theta_a},\ \ F_{\theta_a\theta_b}=\sum_{d=1}^{D}F_{d\theta_a\theta_b}.$$

The blocks of the Fisher information matrix are

$$F_{\beta_1\beta_1}=\left(F_{\beta_{1j_1}\beta_{1j_2}}\right)_{j_1,j_2=1,\ldots,p_1},\quad F_{\beta_2\beta_2}=\left(F_{\beta_{2j_1}\beta_{2j_2}}\right)_{j_1,j_2=1,\ldots,p_2},$$
$$F_{\beta_1\beta_2}=\left(F_{\beta_{1j_1}\beta_{2j_2}}\right)_{j_1=1,\ldots,p_1;j_2=1,\ldots,p_2},\quad F_{\beta_1\theta}=\left(F_{\beta_{1j_1}\theta_a}\right)_{j=1,\ldots,p_1;a=1,2,3},$$
$$F_{\beta_2\theta}=\left(F_{\beta_{2j_2}\theta_a}\right)_{j=1,\ldots,p_2;a=1,2,3},\quad F_{\theta\theta}=\left(F_{\theta_a\theta_b}\right)_{a=1,2,3;b=1,2,3}.$$

The $(p+3)\times(p+3)$ Fisher information matrix is

$$F(\theta)=\begin{pmatrix}F_{\beta_1\beta_1} & F_{\beta_1\beta_2} & F_{\beta_1\theta}\\ F_{\beta_2\beta_1} & F_{\beta_2\beta_2} & F_{\beta_2\theta}\\ F_{\theta\beta_1} & F_{\theta\beta_2} & F_{\theta\theta}\end{pmatrix}=\begin{pmatrix}F_{\beta\beta} & 0\\ 0 & F_{\theta\theta}\end{pmatrix},\quad F_{\beta\beta}(\theta)=\begin{pmatrix}F_{\beta_1\beta_1} & F_{\beta_1\beta_2}\\ F_{\beta_2\beta_1} & F_{\beta_2\beta_2}\end{pmatrix}_{p\times p}.$$

The ML Fisher-scoring algorithm is as follows:

1. Set the initial values $\beta^{(0)}=(\beta_1^{(0)\prime},\beta_2^{(0)\prime})'$, $\theta^{(0)}=(\theta_1^{(0)},\theta_2^{(0)},\theta_3^{(0)})'$, $\psi^{(0)}=(\beta^{(0)\prime},\theta^{(0)\prime})'$, $\varepsilon>0$, $i=0$.
2. Repeat the following steps until the tolerance or the boundary conditions are fulfilled.

   (a) Updating equations: Do

   $$\theta^{(i+1)}=\theta^{(i)}+F_{\theta\theta}^{-1}(\theta^{(i)})U_\theta(\psi^{(i)}),\quad \beta^{(i+1)}=\beta^{(i)}+F_{\beta\beta}^{-1}(\theta^{(i)})U_\beta(\psi^{(i)})$$

   and $\psi^{(i+1)}=(\beta^{(i+1)\prime},\theta^{(i+1)\prime})'$.

   (b) Boundary condition: If $\theta_1^{(i+1)}>0$, $\theta_2^{(i+1)}>0$, and $\left|\theta_3^{(i+1)}\right|<1$, continue. Otherwise, put $\hat{\psi}=\psi^{(i)}$ and stop.

(c) Tolerance condition: If $|\psi_j^{(i+1)} - \psi_j^{(i)}| < \varepsilon$, $j = 1, \ldots, p+3$, put $\hat{\psi} = \psi^{(i+1)}$ and stop. Otherwise, continue.
(d) Updating iteration index: put $i = i + 1$.

3. Output: $\hat{\psi}$, $F^{-1}(\hat{\theta})$.

The output $\hat{\psi}$ of the Fisher-scoring algorithm is the ML estimate of ψ. Possible starting parameter values are $\beta_1^{(0)} = \hat{\beta}_1^{(0)}$, $\beta_2^{(0)} = \hat{\beta}_2^{(0)}$, $\hat{\theta}_{3,0} = 0$, and $\hat{\theta}_{k,0} = \hat{\sigma}^2_{uk,0}$, $k = 1, 2$, where $(\hat{\beta}_1^{(0)}, \hat{\sigma}^2_{u1,0})$, and $(\hat{\beta}_2^{(0)}, \hat{\sigma}^2_{u2,0})$ are the REML or the ML estimators of the corresponding marginal univariate Fay–Herriot models.

19.5 Residual Maximum Likelihood Estimation

The REML method maximizes the joint probability density function of a vector of $2D - p$ independent contrasts $\omega = W'y$, where W is a $2D \times (2D - p)$ matrix with linearly independent columns and such that $W'W = I_{2D-p}$ and $W'X = 0$. It holds that ω is independent of the BLUE $\hat{\beta}_B$ given in (19.4). The joint probability density function of ω is the REML likelihood. The REML log-likelihood of model (19.3) is (cf. (6.41))

$$l_{reml}(\theta) = -\frac{2D-p}{2}\log 2\pi + \frac{1}{2}\log|X'X| - \frac{1}{2}\log|V| - \frac{1}{2}\log|X'V^{-1}X| - \frac{1}{2}y'Py,$$

where $\theta = (\theta_1, \theta_2, \theta_3)'$, $P = V^{-1} - V^{-1}X(X'V^{-1}X)^{-1}X'V^{-1}$, $PVP = P$ and $PX = 0$. By applying the formulas

$$\frac{\partial \log|V|}{\partial\theta} = \operatorname{tr}\Big(V^{-1}\frac{\partial V}{\partial\theta}\Big), \quad \frac{\partial V^{-1}}{\partial\theta} = -V^{-1}\frac{\partial V}{\partial\theta}V^{-1},$$

we calculate the first partial derivatives of l_{reml} with respect to θ_a, i.e.

$$\begin{aligned}
\frac{\partial l_{reml}(\theta)}{\partial\theta_a} &= -\frac{1}{2}\operatorname{tr}\Big(V^{-1}\frac{\partial V}{\partial\theta_a}\Big) + \frac{1}{2}\operatorname{tr}\Big((X'V^{-1}X)^{-1}X'V^{-1}\frac{\partial V}{\partial\theta_a}V^{-1}X\Big) - \frac{1}{2}y'\frac{\partial P}{\partial\theta_a}y \\
&= -\frac{1}{2}\operatorname{tr}\Big(V^{-1}\frac{\partial V}{\partial\theta_a}\Big) + \frac{1}{2}\operatorname{tr}\Big(V^{-1}X(X'V^{-1}X)^{-1}X'V^{-1}\frac{\partial V}{\partial\theta_a}\Big) - \frac{1}{2}y'\frac{\partial P}{\partial\theta_a}y \\
&= -\frac{1}{2}\operatorname{tr}\Big([V^{-1} - V^{-1}X(X'V^{-1}X)^{-1}X'V^{-1}]\frac{\partial V}{\partial\theta_a}\Big) - \frac{1}{2}y'\frac{\partial P}{\partial\theta_a}y \\
&= -\frac{1}{2}\operatorname{tr}\Big(P\frac{\partial V}{\partial\theta_a}\Big) - \frac{1}{2}y'\frac{\partial P}{\partial\theta_a}y, \quad a = 1, 2, 3.
\end{aligned}$$

Let us define $G = V^{-1}X(X'V^{-1}X)^{-1}$, so that $P = (I - GX')V^{-1} = V^{-1}(I - XG')$. The first partial derivatives of $P = V^{-1} - V^{-1}X(X'V^{-1}X)^{-1}X'V^{-1}$ with respect to θ_a are

$$\begin{aligned}
\frac{\partial P}{\partial \theta_a} &= -V^{-1}\frac{\partial V}{\partial \theta_a}V^{-1} + V^{-1}\frac{\partial V}{\partial \theta_a}V^{-1}X(X'V^{-1}X)^{-1}X'V^{-1} \\
&\quad + V^{-1}X(X'V^{-1}X)^{-1}X'V^{-1}\frac{\partial V}{\partial \theta_a}V^{-1} \\
&\quad - V^{-1}X(X'V^{-1}X)^{-1}X'V^{-1}\frac{\partial V}{\partial \theta_a}V^{-1}X(X'V^{-1}X)^{-1}X'V^{-1} \\
&= -V^{-1}\frac{\partial V}{\partial \theta_a}V^{-1} + V^{-1}\frac{\partial V}{\partial \theta_a}V^{-1}XG' + GX'V^{-1}\frac{\partial V}{\partial \theta_a}V^{-1} - GX'V^{-1}\frac{\partial V}{\partial \theta_a}V^{-1}XG' \\
&= -(I - GX')V^{-1}\frac{\partial V}{\partial \theta_a}V^{-1}(I - GX')' = -P\frac{\partial V}{\partial \theta_a}P, \quad a = 1, 2, 3.
\end{aligned}$$

Therefore, the score vector is $S(\theta) = (S_1, S_2, S_3)'$, where

$$S_a = S_a(\theta) = \frac{\partial l_{reml}}{\partial \theta_a} = -\frac{1}{2}\,\text{tr}(PV_a) + \frac{1}{2}\,y'PV_aPy, \quad a = 1, 2, 3,$$

where $P = \left(P_{d_1 d_2}\right)_{d_1, d_2 = 1, \ldots, D}$, $V_a = \frac{\partial V}{\partial \theta_a} = \underset{1 \le d \le D}{\text{diag}}\,(V_{uda})$, and the matrices of first derivatives V_{ud1}, V_{ud2}, and V_{ud3} were calculated in (19.7), and

$$P_{dd} = V_d^{-1} - V_d^{-1}X_d Q X_d' V_d^{-1}, \quad P_{d_1 d_2} = -V_{d_1}^{-1}X_{d_1} Q X_{d_2}' V_{d_2}^{-1}.$$

For $a = 1, 2, 3$, we have

$$\begin{aligned}
PV_a &= \left[\underset{1 \le d \le D}{\text{diag}}\,(V_d^{-1}) - \underset{1 \le d \le D}{\text{col}}\,(V_d^{-1}X_d)Q\,\underset{1 \le d \le D}{\text{col}'}\,(X_d'V_d^{-1})\right]\underset{1 \le d \le D}{\text{diag}}\,(V_{uda}) \\
&= \underset{1 \le d \le D}{\text{diag}}\,(V_d^{-1}V_{uda}) - \underset{1 \le d \le D}{\text{col}}\,(V_d^{-1}X_d)Q\,\underset{1 \le d \le D}{\text{col}'}\,(X_d'V_d^{-1}V_{uda}).
\end{aligned}$$

For $a, b = 1, 2, 3$, we have

$$\begin{aligned}
PV_aPV_b &= \left[\underset{1 \le d \le D}{\text{diag}}\,(V_d^{-1}V_{uda}) - \underset{1 \le d \le D}{\text{col}}\,(V_d^{-1}X_d)Q\,\underset{1 \le d \le D}{\text{col}'}\,(X_d'V_d^{-1}V_{uda})\right] \\
&\quad \cdot \left[\underset{1 \le d \le D}{\text{diag}}\,(V_d^{-1}V_{udb}) - \underset{1 \le d \le D}{\text{col}}\,(V_d^{-1}X_d)Q\,\underset{1 \le d \le D}{\text{col}'}\,(X_d'V_d^{-1}V_{udb})\right] \\
&= \underset{1 \le d \le D}{\text{diag}}\,(V_d^{-1}V_{uda}V_d^{-1}V_{udb}) - \underset{1 \le d \le D}{\text{col}}\,(V_d^{-1}V_{uda}V_d^{-1}X_d)Q
\end{aligned}$$

$$\cdot \underset{1\le d\le D}{\text{col}'}(X_d'V_d^{-1}V_{udb}) - \underset{1\le d\le D}{\text{col}}(V_d^{-1}X_d)Q\underset{1\le d\le D}{\text{col}'}(X_d'V_d^{-1}V_{uda}V_d^{-1}V_{udb})$$

$$+ \underset{1\le d\le D}{\text{col}}(V_d^{-1}X_d)Q\Big(\sum_{d=1}^{D}X_d'V_d^{-1}V_{uda}V_d^{-1}X_d\Big)Q\underset{1\le d\le D}{\text{col}'}(X_d'V_d^{-1}V_{udb}).$$

For $a = 1, 2, 3$, we have

$$\text{tr}(PV_a) = \sum_{d=1}^{D}\text{tr}(V_d^{-1}V_{uda}) - \sum_{d=1}^{D}\text{tr}(V_d^{-1}X_dQX_d'V_d^{-1}V_{uda}) = \sum_{d=1}^{D}\text{tr}(P_{dd}V_{uda}),$$

$$y'PV_aPy = \sum_{d=1}^{D}y_d'V_d^{-1}V_{uda}V_d^{-1}y_d - 2\sum_{d_1=1}^{D}\sum_{d_2=1}^{D}y_{d_1}'V_{d_1}^{-1}V_{ud_1a}V_{d_1}^{-1}X_{d_1}QX_{d_2}'V_{d_2}^{-1}y_{d_2}$$

$$+ \sum_{d_1=1}^{D}\sum_{d_2=1}^{D}y_{d_1}'V_{d_1}^{-1}X_{d_1}Q\Big(\sum_{d=1}^{D}X_d'V_d^{-1}V_{uda}V_d^{-1}X_d\Big)QX_{d_2}'V_{d_2}^{-1}y_{d_2}.$$

For $a, b = 1, 2, 3$, the second partial derivatives of the REML log-likelihood function are

$$\frac{\partial l_{reml}^2(\theta)}{\partial\theta_a\partial\theta_b} = \frac{1}{2}\text{tr}\big(PV_bPV_a\big) - \frac{1}{2}\text{tr}\big(PV_{ab}\big) - \frac{1}{2}y'(PV_bPV_aP + PV_aPV_bP)y$$

$$+ \frac{1}{2}y'PV_{ab}Py$$

$$= \frac{1}{2}\text{tr}\big(PV_aPV_b\big) - \frac{1}{2}\text{tr}\big(PV_{ab}\big) - y'PV_aPV_bPy + \frac{1}{2}y'PV_{ab}Py,$$

where the last equality follows from the fact that V_a is symmetric, $a = 1, 2, 3$, $V_{ab} = \frac{\partial V_a}{\partial\theta_b} = \underset{1\le d\le D}{\text{diag}}(V_{udab})$, and the matrices V_{udab} were calculated in (19.8). By changing the sign, taking expectations, and applying $PX = 0$, $PV = I - V^{-1}XQX'$, and the formula

$$E\big[y'Ay\big] = \text{tr}\big(A\text{var}(y)\big) + E[y]'AE[y],$$

we get for $a, b = 1, 2, 3$ the components of the Fisher information matrix, i.e.

$$F_{ab} = F_{ab}(\theta) = -\frac{1}{2}\text{tr}\big(PV_aPV_b\big) + \text{tr}\big(PV_aPV_bPV\big) + \beta'X'PV_aPV_bPX\beta$$

$$+ \frac{1}{2}\text{tr}\big(PV_{ab}\big) - \frac{1}{2}\text{tr}\big(PV_{ab}PV\big) - \frac{1}{2}\beta'X'PV_{ab}PX\beta$$

$$= -\frac{1}{2}\text{tr}\big(PV_aPV_b\big) + \text{tr}\big(PV_aPV_b[I - V^{-1}XQX']\big)$$

$$+\frac{1}{2}\operatorname{tr}\big(PV_{ab}\big)-\frac{1}{2}\operatorname{tr}\big(PV_{ab}[I-V^{-1}XQX']\big)$$
$$=\frac{1}{2}\operatorname{tr}\big(PV_aPV_b\big)-\operatorname{tr}\big(PV_aPV_bV^{-1}XQX'\big)+\frac{1}{2}\operatorname{tr}\big(PV_{ab}V^{-1}XQX'\big)$$
$$=\frac{1}{2}\operatorname{tr}\big(PV_aPV_b\big).$$

Therefore, the Fisher information matrix is

$$F(\theta)=\big(F_{a,b}\big)_{a,b=1,2,3},\quad F_{ab}=F_{ab}(\theta)=\frac{1}{2}\operatorname{tr}(PV_aPV_b),\quad a,b=1,2,3.$$

The trace of PV_aPV_b can be calculated as

$$\operatorname{tr}(PV_aPV_b)=\sum_{d=1}^{D}\operatorname{tr}\big(V_d^{-1}V_{uda}V_d^{-1}V_{udb}\big)-\sum_{d=1}^{D}\operatorname{tr}\big(V_{udb}V_d^{-1}V_{uda}V_d^{-1}X_dQX_d'V_d^{-1}\big)$$
$$-\sum_{d=1}^{D}\operatorname{tr}\big(V_{uda}V_d^{-1}V_{udb}V_d^{-1}X_dQX_d'V_d^{-1}\big)$$
$$+\operatorname{tr}\Big(\Big\{\sum_{d=1}^{D}X_d'V_d^{-1}V_{udb}V_d^{-1}X_d\Big\}Q\Big\{\sum_{d=1}^{D}X_d'V_d^{-1}V_{uda}V_d^{-1}X_d\Big\}Q\Big).$$

The REML Fisher-scoring algorithm is as follows:

1. Set the initial values $\theta^{(0)}$, $\varepsilon>0$, and $i=0$.
2. Repeat the following steps until the tolerance or the boundary conditions are fulfilled.
   (a) Updating equation: Put $\theta^{(i+1)}=\theta^{(i)}+F^{-1}(\theta^{(i)})S(\theta^{(i)})$.
   (b) Boundary condition: If $\theta_1^{(i+1)}>0$, $\theta_2^{(i+1)}>0$, and $|\theta_3^{(i+1)}|<1$, continue. Otherwise, put $\hat{\theta}=\theta^{(i)}$ and stop.
   (c) Tolerance condition: If $|\theta_a^{(i+1)}-\theta_a^{(i)}|<\varepsilon$, $a=1,2,3$, put $\hat{\theta}=\theta^{(i+1)}$ and stop. Otherwise, continue.
   (d) Updating iteration index: put $i=i+1$.
3. Output: $\hat{\theta}$, $F^{-1}(\hat{\theta})$.

The output $\hat{\theta}$ of the Fisher-scoring algorithm is the REML estimate of θ. As starting parameter values, it is possible to take $\hat{\theta}_{3,0}=0$, $\hat{\theta}_{k,0}=\hat{\sigma}_{uk,0}^2$, $k=1,2$, where $\hat{\sigma}_{uk,0}^2$ is the REML or the ML or the Prasad and Rao (1990) moment-based estimator of σ_{uk}^2 in the k-th marginal Fay–Herriot model. They can be calculated with the R library `sae`.

By plugging $\hat{\theta}$ in V_u, we get estimated variance matrices $\hat{V}_u=V_u(\hat{\theta})$ and $\hat{V}=\hat{V}_u+V_e$. By substituting $\hat{V}_u$ in (19.4), we obtain the EBLUP of $\mu=X\beta+Zu$

defined in (19.5), i.e.

$$\hat{\beta}_E = (X'\hat{V}^{-1}X)^{-1}X'\hat{V}^{-1}y, \quad \hat{u}_E = \hat{V}_u Z'\hat{V}^{-1}(y - X\hat{\beta}_E), \quad \hat{\mu}_E = X\hat{\beta}_E + Z\hat{u}_E.$$

Alternative formulas for $\hat{\beta}_E = \underset{1\le i\le p}{\text{col}}\,(\hat{\beta}_{Ei})$ and $\hat{u}_E = \underset{1\le d\le D}{\text{col}}\,(\hat{u}_d)$ are

$$\hat{\beta}_E = \Big(\sum_{d=1}^{D} X_d'\hat{V}_d^{-1}X_d\Big)^{-1}\sum_{d=1}^{D} X_d'\hat{V}_d^{-1}y_d, \; \hat{u}_{Ed} = \hat{V}_{ud}\hat{V}_d^{-1}\big(y_d - X_d\hat{\beta}_E\big), \; d = 1,\dots,D.$$

The asymptotic distributions of the REML estimators $\hat{\theta}$ and $\hat{\beta}_E$ (see e.g. Section 1.3.2 in Jiang (2007)),

$$\hat{\theta} \sim N_3(\theta, F^{-1}(\theta)), \quad \hat{\beta}_E \sim N_p(\beta, (X'V^{-1}X)^{-1}),$$

can be used to construct $(1-\alpha)$-level asymptotic confidence intervals for the components θ_a of θ and β_i of β, i.e.

$$\hat{\theta}_a \pm z_{\alpha/2}\, \nu_{aa}^{1/2}, \; a = 1, 2, 3, \quad \hat{\beta}_{Ei} \pm z_{\alpha/2}\, q_{ii}^{1/2}, \; i = 1, \dots, p, \tag{19.9}$$

where $F^{-1}(\hat{\theta}) = (\nu_{ab})_{a,b=1,2,3}$, $(X'V^{-1}(\hat{\theta})X)^{-1} = (q_{ij})_{i,j=1,\dots,p}$ and z_α is the α-quantile of the $N(0,1)$ distribution. For $\hat{\beta}_{Ei} = \beta_0$, the p-value for testing the hypothesis $H_0: \beta_i = 0$ is

$$p\text{-value} = 2P_{H_0}(\hat{\beta}_{Ei} > |\beta_0|) = 2P(N(0,1) > |\beta_0|/\sqrt{q_{ii}}). \tag{19.10}$$

We remark that we have changed the notation in (19.9) and (19.10), where β_i denotes the i-th component of the vector β and not the vector of regression parameters of the i-th category.

19.6 The Matrix of Mean Squared Crossed Errors

Prasad and Rao (1990) gave an approximation to the MSE of the EBLUP of μ_d under the univariate Fay–Herriot model when their proposed moment-based estimator of the variance σ_u^2 is employed. Datta and Lahiri (2000) extended the results of Prasad and Rao (1990) to the case of the general longitudinal model. They further considered ML and REML estimators of the variance components. For the general linear model, Das, Jiang and Rao (2004) derived the MSE of the EBLUP under REML and ML. Their proof contains the general longitudinal model considered by Datta and Lahiri (2000) as a special case. However, none of the three papers study the approximation of the matrix of mean squared crossed errors of the EBLUP vector $\hat{\mu}_E$. They deal with the approximation of the MSEs of the components of $\hat{\mu}_E$ under the univariate model. Although the bivariate Fay–Herriot model (19.3) can be written in the form of the general linear mixed model considered by Das, Jiang and Rao (2004), the approximation of the matrix of mean squared crossed errors was not covered by this paper. This matrix was approximated by Benavent and Morales (2016) under some classes of multivariate Fay–Herriot models.

This section uses the notation $f(D) = O(D)_{m\times m}$ and $f(D) = o(D)_{m\times m}$ for matrix-valued functions such that $f(D)/D$ is element-wise uniformly bounded and converges to a zero $m \times m$ matrix, as $D \to \infty$, respectively. Similarly, the notations $f(D) = O_p(D)$ and $f(D) = o_p(D)$ are used for convergence in probability when f and g are matrix-valued stochastic functions. Further, the following hypotheses H1–H5 are assumed.

H1 $0 < p < \infty$.

H2 The determinant $|V_u| > 0$. The matrices V_{ed}, $d = 1, \dots D$, are positive definite with uniformly bounded elements and determinants bounded away from zero. This is to say, there exist $\kappa > 0$ such that $|V_{ed}| > \kappa$, $d = 1, \dots, D$.

H3 $|x_{dkj}| \leq x < \infty$, $X'X = O(D)_{p\times p}$.

H4 $(X'V^{-1}X)^{-1} = O(D^{-1})_{p\times p}$ and $XQX' = O(D^{-1})_{2D\times 2D}$.

H5 $\hat{\theta}_a = k_a + y'C_a y$ is an unbiased, consistent, and translation invariant estimator of θ_a, $a = 1, 2, 3$, where $k_a = O(1)$ and

$$C_a = \operatorname*{diag}_{1\leq d\leq D}\left\{O(D^{-1})_{2\times 2}, \cdots, O(D^{-1})_{2\times 2}\right\} + O(D^{-2})_{2D\times 2D}.$$

Assumption H2 implies that $V = O(1)$ and $V^{-1} = O(1)$. Assumption H5 and formula (A.3) on covariances of quadratic forms imply that $\mathrm{cov}(\hat{\theta}_a, \hat{\theta}_b) = o(1)$.

This section approximates the matrix of mean squared crossed errors of the EBLUP, i.e.

$$MSE\left(\hat{\mu}_E\right) = E\left(\left(\hat{\mu}_E - \mu\right)\left(\hat{\mu}_E - \mu\right)'\right).$$

By adding and subtracting $\hat{\mu}_B$, we have $\hat{\mu}_E - \mu = \hat{\mu}_B - \mu + \hat{\mu}_E - \hat{\mu}_B$. Therefore,

$$\begin{aligned}\left(\hat{\mu}_E - \mu\right)\left(\hat{\mu}_E - \mu\right)' &= \left(\hat{\mu}_B - \mu\right)\left(\hat{\mu}_B - \mu\right)' + \left(\hat{\mu}_B - \mu\right)\left(\hat{\mu}_E - \hat{\mu}_B\right)' \\ &\quad + \left(\hat{\mu}_E - \hat{\mu}_B\right)\left(\hat{\mu}_B - \mu\right)' + (\hat{\mu}_E - \hat{\mu}_B)(\hat{\mu}_E - \hat{\mu}_B)'.\end{aligned} \tag{19.11}$$

Under the assumption of normality on u and e and for unbiased and translation invariant estimators of θ, Kackar and Harville (1984) proved that the expectations of the second and third terms in (19.11) are null. Therefore, by taking expectations, we get

$$MSE\left(\hat{\mu}_E\right) = MSE\left(\hat{\mu}_B\right) + E\left[\left(\hat{\mu}_E - \hat{\mu}_B\right)\left(\hat{\mu}_E - \hat{\mu}_B\right)'\right]. \tag{19.12}$$

From (19.6), we have

$$MSE\left(\hat{\mu}_B\right) = G_1(\theta) + G_2(\theta),$$

where $T = V_u - V_u V^{-1} V_u$, $Q = (X'V^{-1}X)^{-1}$, and

$$G_1(\theta) = T, \quad G_2(\theta) = \left(X - TV_e^{-1}X\right) Q \left(X' - X'V_e^{-1}T\right). \tag{19.13}$$

For calculating the second summand in (19.12), we write $\hat{\mu}_B = \hat{\mu}(\theta)$ and $\hat{\mu}_E = \hat{\mu}(\hat{\theta})$. A Taylor series expansion $\hat{\mu}(\hat{\theta})$ around θ yields to

$$(\hat{\mu}_E - \hat{\mu}_B)(\hat{\mu}_E - \hat{\mu}_B)' \approx S(\hat{\theta} - \theta)(\hat{\theta} - \theta)'S' + o_p(D^{-1}),$$

where $S = \left(\frac{\partial \hat{\mu}_{dk}(\hat{\theta})}{\partial \hat{\theta}_a}\Big|_{\hat{\theta}=\theta} : d = 1, \ldots, D, k = 1, 2;\ a = 1, 2, 3\right)$ is a $2D \times 3$ matrix, i.e.

$$S = \frac{\partial \hat{\mu}(\hat{\theta})}{\partial \hat{\theta}}\Bigg|_{\hat{\theta}=\theta} = \underset{1\leq a\leq 3}{\text{col}'}\,(S^{(a)}), \quad S^{(a)} = \frac{\partial \hat{\mu}(\hat{\theta})}{\partial \hat{\theta}_a}\Bigg|_{\hat{\theta}=\theta} = \underset{1\leq d\leq D}{\text{col}}\,(\underset{1\leq k\leq 2}{\text{col}}\,(s_{dk}^{(a)})),$$

$$s_{dk}^{(a)} = \frac{\partial \hat{\mu}_{dk}(\hat{\theta})}{\partial \hat{\theta}_a}\Bigg|_{\hat{\theta}=\theta}.$$

The row index of matrix S takes the values $(d, k) = (1, 1), (1, 2), \ldots, (D, 1), (D, 2)$, and the column index is $a = 1, 2, 3$. In the new notation, we have

$$\begin{aligned} S(\hat{\theta} - \theta)(\hat{\theta} - \theta)'S' &= \underset{1\leq a\leq 3}{\text{col}'}\,(S^{(a)})\ \underset{1\leq a\leq 3}{\text{col}}\,(\hat{\theta}_a - \theta_a)\ \underset{1\leq a\leq 3}{\text{col}'}\,(\hat{\theta}_a - \theta_a)\ \underset{1\leq a\leq 3}{\text{col}}\,(S^{(a)\prime}) \\ &= \sum_{a=1}^{3} S^{(a)}(\hat{\theta}_a - \theta_a) \sum_{b=1}^{3} (\hat{\theta}_b - \theta_b) S^{(b)\prime} \\ &= \sum_{a=1}^{3}\sum_{b=1}^{3} (\hat{\theta}_a - \theta_a)(\hat{\theta}_b - \theta_b) S^{(a)} S^{(b)\prime}. \end{aligned}$$

By taking expectations, we get

$$E[S(\hat{\theta} - \theta)(\hat{\theta} - \theta)'S'] = \sum_{a=1}^{3}\sum_{b=1}^{3} E\left[(\hat{\theta}_a - \theta_a)(\hat{\theta}_b - \theta_b) S^{(a)} S^{(b)\prime}\right].$$

Theorem 19.1 *Let us assume that H1–H5 holds, then*

$$E[(\hat{\theta}_a-\theta_a)(\hat{\theta}_b-\theta_b)S^{(a)}S^{(b)\prime}]=cov(\hat{\theta}_a,\hat{\theta}_b)L^{(a)}VL^{(b)\prime}+o(D^{-1})_{2D\times 2D},$$

where $L^{(a)}=(I-R)V_{ua}V^{-1}$, $R=V_uV^{-1}$ *and* $V_{ua}=\frac{\partial V_u}{\partial\theta_a}$.

Proof By Lemma 19.1, the components of the vectors $S^{(a)}=\left(s_{11}^{(a)},\ldots,s_{D2}^{(a)}\right)$ are linear functions of $v=Zu+e$, i.e.

$$s_{dk}^{(a)}=(F_{dk}^{(a)}+L_{dk}^{(a)})'v,\quad k=1,2,\ d=1,\ldots,D,$$

where $F_{dk}^{(a)\prime}$ and $L_{dk}^{(a)\prime}$ are the (d,k)-th rows of the $2D\times 2D$ matrices

$$F^{(a)}=\underset{1\le d\le D}{\text{col}}\left(\underset{1\le k\le 2}{\text{col}}(F_{dk}^{(a)\prime})\right)\quad\text{and}\quad L^{(a)}=\underset{1\le d\le D}{\text{col}}\left(\underset{1\le k\le 2}{\text{col}}(L_{dk}^{(a)\prime})\right),$$

respectively, which are defined in (19.15) within Sect. 19.7. By H5, we have $\hat{\theta}_a=k+y'C_ay$, with $E[\hat{\theta}_a]=\theta_a$. As $\hat{\theta}_a$ is translation invariant and $v=Zu+e=y-X\beta$, we have $\hat{\theta}_a(y)=\hat{\theta}_a(y-X\beta)=\hat{\theta}_a(v)$ and $\hat{\theta}_a(v)=k+v'C_av$, with $\theta_a=k+E[v'C_av]$. By subtracting, we get $\hat{\theta}_a-\theta_a=v'C_av-E[v'C_av]$. By defining $q_a=v'C_av$, we have $\hat{\theta}_a-\theta_a=q_a-E[q_a]$.

As $v\sim N(0,V)$, we apply Lemma 19.3 with $\lambda_1=F_{dk_1}^{(a)}+L_{dk_1}^{(a)}$, $\lambda_2=F_{dk_2}^{(b)}+L_{dk_2}^{(b)}$, $s_1=s_{dk_1}^{(a)}$, $s_2=s_{dk_2}^{(b)}$, and $q_1=v'C_av$, $q_2=v'C_bv$, $k_1,k_2=1,2$, $d=1,\ldots,D$. We obtain

$$\begin{aligned}E[s_{dk_1}^{(a)}s_{dk_2}^{(b)}(\hat{\theta}_a-\theta_a)(\hat{\theta}_b-\theta_b)]&=\text{cov}(s_{dk_1}^{(a)},s_{dk_2}^{(b)})\text{cov}(\hat{\theta}_a,\hat{\theta}_b)\\&\quad+8(F_{dk_1}^{(a)}+L_{dk_1}^{(a)})'VC_aVC_bV(F_{dk_2}^{(b)}+L_{dk_2}^{(b)}).\end{aligned}$$

In matrix notation, we have

$$\begin{aligned}E[(\hat{\theta}_a-\theta_a)(\hat{\theta}_b-\theta_b)S^{(a)}S^{(b)\prime}]&=\text{cov}(S^{(a)},S^{(b)})\text{cov}(\hat{\theta}_a,\hat{\theta}_b)\\&\quad+8(F^{(a)}+L^{(a)})VC_aVC_bV(F^{(b)}+L^{(b)})'.\end{aligned}$$

From Lemma 19.4, we get

$$\text{cov}(S^{(a)},S^{(b)})=L^{(a)}VL^{(b)\prime}+O(D^{-1})_{2D\times 2D}.$$

By applying Lemma 19.2 and an adapted version of Lemma 19.7 of Prasad and Rao (1990), we finally obtain

$$(F^{(a)}+L^{(a)})VC_aVC_bV(F^{(b)}+L^{(b)})'=O(D^{-2})_{2D\times 2D}$$

and

$$E[(\hat{\theta}_a-\theta_a)(\hat{\theta}_b-\theta_b)S^{(a)}S^{(b)\prime}] = \text{cov}(\hat{\theta}_a,\hat{\theta}_b)L^{(a)}VL^{(b)\prime} + \text{cov}(\hat{\theta}_a,\hat{\theta}_b)\,O(D^{-1}) + o(D^{-1})_{2D\times 2D} = \text{cov}(\hat{\theta}_a,\hat{\theta}_b)L^{(a)}VL^{(b)\prime} + o(D^{-1})_{2D\times 2D}. \quad \square$$

Corollary 19.1 *If H1-H5 holds, then*

$$E\left[\left(\hat{\mu}_E-\hat{\mu}_B\right)\left(\hat{\mu}_E-\hat{\mu}_B\right)'\right] = G_3(\theta) + o(D^{-1})_{2D\times 2D},$$
$$MSE(\hat{\mu}_E) = G_1(\theta)+G_2(\theta)+G_3(\theta)+o(D^{-1})_{2D\times 2D},$$

where

$$G_3(\theta) = \sum_{a=1}^{3}\sum_{b=1}^{3} cov(\hat{\theta}_a,\hat{\theta}_b)L^{(a)}VL^{(b)\prime}.$$

Similarly as in Prasad and Rao (1990), Datta and Lahiri (2000), and Das, Jiang and Rao (2004), $MSE(\hat{\mu}_E)$ is estimated with

$$mse(\hat{\mu}_E) = G_1(\hat{\theta}) + G_2(\hat{\theta}) + 2G_3(\hat{\theta}). \tag{19.14}$$

In the section of simulations, we further consider a bootstrap-based MSE estimator. In fact, it is similar to one of the bootstrap alternatives considered by González-Manteiga et al. (2008b).

19.7 Auxiliary Results

This section gives some lemmas and auxiliary results for approximating the matrix of mean squared crossed errors.

Lemma 19.1 *Let* $v = Zu + e$ *be the vector containing the random part of model (19.3). Under H1-H2, it holds that*

$$S^{(a)} = \frac{\partial\hat{\mu}(\theta)}{\partial\theta_a} = (F^{(a)}+L^{(a)})v, \quad a=1,2,3,$$

where

$$F^{(a)} = -(I-R)XQX'\frac{\partial V^{-1}}{\partial\theta_a}A - \frac{\partial R}{\partial\theta_a}XQX'V^{-1},\ L^{(a)} = \frac{\partial R}{\partial\theta_a} = (I-R)\frac{\partial V}{\partial\theta_a}V^{-1},$$

$$A = I - XQX'V^{-1}, \quad R = V_u V^{-1}, \quad Q = (X'V^{-1}X)^{-1}. \tag{19.15}$$

Proof The BLUP of μ is

$$\hat{\mu}_B = X\hat{\beta}_B + R(y - X\hat{\beta}_B) = X\hat{\beta}_B - RX\hat{\beta}_B + Ry = XQX'V^{-1}y - RXQX'V^{-1}y + Ry.$$

By substituting y by $X\beta + v$, we have

$$\begin{aligned}\hat{\mu}_B &= XQX'V^{-1}X\beta - RXQX'V^{-1}X\beta + RX\beta + XQX'V^{-1}v \\ &\quad - RXQX'V^{-1}v + Rv = X\beta + XQX'V^{-1}v + RAv.\end{aligned}$$

By taking partial derivatives with respect to θ_a, we get

$$\begin{aligned}S^{(a)} = \frac{\partial\hat{\mu}_B}{\partial\theta_a} &= -XQX'\frac{\partial V^{-1}}{\partial\theta_a}XQX'V^{-1}v + XQX'\frac{\partial V^{-1}}{\partial\theta_a}v + \frac{\partial R}{\partial\theta_a}Av + R\frac{\partial A}{\partial\theta_a}v \\ &= XQX'\frac{\partial V^{-1}}{\partial\theta_a}(I - XQX'V^{-1})v + \frac{\partial R}{\partial\theta_a}Av + R\frac{\partial A}{\partial\theta_a}v \\ &= XQX'\frac{\partial V^{-1}}{\partial\theta_a}Av + \frac{\partial R}{\partial\theta_a}Av + R\frac{\partial A}{\partial\theta_a}v.\end{aligned} \tag{19.16}$$

The partial derivative of A with respect to θ_a is

$$\frac{\partial A}{\partial\theta_a} = XQX'\frac{\partial V^{-1}}{\partial\theta_a}XQX'V^{-1} - XQX'\frac{\partial V^{-1}}{\partial\theta_a} = -XQX'\frac{\partial V^{-1}}{\partial\theta_a}A.$$

Therefore,

$$\begin{aligned}S^{(a)} &= XQX'\frac{\partial V^{-1}}{\partial\theta_a}Av - RXQX'\frac{\partial V^{-1}}{\partial\theta_a}Av + \frac{\partial R}{\partial\theta_a}Av \\ &= \left[(I-R)XQX'\frac{\partial V^{-1}}{\partial\theta_a}A + \frac{\partial R}{\partial\theta_a}A\right]v \\ &= \left[(I-R)XQX'\frac{\partial V^{-1}}{\partial\theta_a}A - \frac{\partial R}{\partial\theta_a}XQX'V^{-1} + \frac{\partial R}{\partial\theta_a}\right]v = [F^{(a)} + L^{(a)}]v,\end{aligned}$$

where

$$F^{(a)} = (I-R)XQX'\frac{\partial V^{-1}}{\partial\theta_a}A - \frac{\partial R}{\partial\theta_a}XQX'V^{-1}, \quad L^{(a)} = \frac{\partial R}{\partial\theta_a}.$$

As the partial derivatives of V^{-1} and $R = V_u V^{-1}$ are

$$\frac{\partial V^{-1}}{\partial \theta_a} = -V^{-1}\frac{\partial V}{\partial \theta_a}V^{-1},$$
$$\frac{\partial R}{\partial \theta_a} = \frac{\partial V_u}{\partial \theta_a}V^{-1} + V_u\frac{\partial V^{-1}}{\partial \theta_a} = \frac{\partial V}{\partial \theta_a}V^{-1} - V_u V^{-1}\frac{\partial V}{\partial \theta_a}V^{-1} = (I-R)\frac{\partial V}{\partial \theta_a}V^{-1},$$

we finally get

$$F^{(a)} = -(I-R)XQX'V^{-1}\frac{\partial V}{\partial \theta_a}V^{-1}A - (I-R)\frac{\partial V}{\partial \theta_a}V^{-1}XQX'V^{-1},$$
$$L^{(a)} = (I-R)\frac{\partial V}{\partial \theta_a}V^{-1}.$$
□

Lemma 19.2 *Under H1–H4, it holds that:*

(i) $L^{(a)} = \underset{1\le d\le D}{\text{diag}}\,(L_d^{(a)})$, *with* $L_d^{(a)} = O(1)_{2\times 2}$, $d = 1, \ldots, D$.

(ii) $F^{(a)} = O(D^{-1})_{2D\times 2D}$.

Proof As $L_d^{(a)} = \frac{\partial V_d}{\partial \theta_a}V_d^{-1} - V_{ud}V_d^{-1}\frac{\partial V_d}{\partial \theta_a}V_d^{-1}$, (i) follows from H1–H4. On the other hand,

$$F^{(a)} = (I-R)XQX'\frac{\partial V^{-1}}{\partial \theta_a}A - L^{(a)}XQX'V^{-1}.$$

From H4, we have $Q = (X'V^{-1}X)^{-1} = O(D^{-1})_{p\times p}$ and $XQX' = O(D^{-1})_{2D\times 2D}$. From $L^{(a)} = O(1)_{2D\times 2D}$ and $V^{-1} = O(1)_{2D\times 2D}$, the second summand of $F^{(a)}$ is

$$L^{(a)}XQX'V^{-1} = O(D^{-1})_{2D\times 2D}.$$

Concerning the first summand of $F^{(a)}$, we have

$$I - R = O(1)_{2D\times 2D},\ \frac{\partial V^{-1}}{\partial \theta_a} = O(1)_{2D\times 2D},\ (I-R)XQX'\frac{\partial V^{-1}}{\partial \theta_a} = O(D^{-1})_{2D\times 2D}.$$

If we post-multiply by $A = I - XQX'V^{-1}$, we get

$$(I-R)XQX'\frac{\partial V^{-1}}{\partial \theta_a}A = (I-R)XQX'\frac{\partial V^{-1}}{\partial \theta_a} - (I-R)XQX'\frac{\partial V^{-1}}{\partial \theta_a}XQX'V^{-1},$$

where the first term is $O(D^{-1})$ and

$$XQX'\frac{\partial V^{-1}}{\partial \theta_a}XQX' = O(D^{-1})_{2D\times 2D}O(1)_{2D\times 2D}O(D^{-1})_{2D\times 2D} = O(D^{-2})_{2D\times 2D}.$$

Therefore, the first summand of $F^{(a)}$ is

$$(I-R)XQX'\frac{\partial V^{-1}}{\partial \theta_a}A = O(D^{-1})_{2D\times 2D}.$$ □

Lemma 19.3 *Let $v \sim N(0, V)$, $s_1 = \lambda_1' v$, $s_2 = \lambda_2' v$, $q_1 = v'A_1 v$, and $q_2 = v'A_2 v$, where λ_i and A_i, $i = 1, 2$, are nonstochastic vectors and matrices, respectively. Then,*

$$E[s_1 s_2 (q_1 - E[q_1])(q_2 - E[q_2])] = cov(q_1, q_2)cov(s_1, s_2) + 8\lambda_1' V A_1 V A_2 \lambda_2.$$

Proof By applying Lemma 19.6 of Prasad and Rao (1990), we have

$$\begin{aligned} E &= E[s_1 s_2(q_1 - E[q_1])(q_2 - E[q_2])] = \lambda_1' E[v(v'A_1 v v'A_2 v)v']\lambda_2 \\ &- E[q_1]\lambda_1' E[v(v'A_2 v)v']\lambda_2 - E[q_2]\lambda_1' E[v(v'A_1 v)v']\lambda_2 + E[q_1]E[q_2]\lambda_1' V\lambda_2. \end{aligned}$$

By applying Lemma 19.5(c) of Prasad and Rao (1990), we have

$$\begin{aligned} E[v(v'A_1 v v'A_2 v)v'] &= \mathrm{tr}(A_1 V)\mathrm{tr}(A_2 V)V + 2\mathrm{tr}(A_1 V)VA_2 V + 2\mathrm{tr}(A_2 V)VA_1 V \\ &+ 2\mathrm{tr}(A_1 V A_2 V)V + 4VA_1 V A_2 V + 4VA_2 V A_1 V. \end{aligned}$$

By applying Lemma 19.5(a) of Prasad and Rao (1990), we have

$$E[v(v'A_a v)v'] = \mathrm{tr}(A_a V)V + 2VA_a V, \quad i = 1, 2.$$

Further, $E[q_a] = \mathrm{tr}(A_a V)$, $\mathrm{cov}(q_1, q_2) = 2\mathrm{tr}(A_1 V A_2 V)$, and $\mathrm{cov}(s_1, s_2) = \lambda_1' V\lambda_2$. By substitution, we get

$$\begin{aligned} E &= E[q_1]E[q_2]\lambda_1' V\lambda_2 + 2E[q_1]\lambda_1' VA_2 V\lambda_2 + 2E[q_2]\lambda_1' VA_1 V\lambda_2 \\ &+ 2\mathrm{tr}(A_1 V A_2 V)\lambda_1' V\lambda_2 + 8\lambda_1' VA_1 V A_2 V\lambda_2 - E[q_1]E[q_2]\lambda_1' V\lambda_2 \\ &- 2E[q_1]\lambda_1' VA_2 V\lambda_2 - E[q_1]E[q_2]\lambda_1' V\lambda_2 - 2E[q_2]\lambda_1' VA_1 V\lambda_2 \\ &+ E[q_1]E[q_2]\lambda_1' V\lambda_2 = 2\mathrm{tr}(A_1 V A_2 V)\lambda_1' V\lambda_2 + 8\lambda_1' VA_1 V A_2 V\lambda_2 \\ &= \mathrm{cov}(q_1, q_2)\mathrm{cov}(s_1, s_2) + 8\lambda_1' VA_1 V A_2 \lambda_2. \end{aligned}$$ □

Lemma 19.4 *Under H1–H4, it holds*

$$cov(S^{(a)}, S^{(b)}) = L^{(a)} V L^{(b)\prime} + O(D^{-1})_{2D\times 2D}.$$

Proof From Lemma 19.1, we get $S^{(a)} = (L^{(a)} + F^{(a)})v$, $a = 1, 2, 3$, where $v \sim N(0, V)$, and from Lemma 19.2 it follows

$$F^{(a)} = O(D^{-1})_{2D\times 2D},\ \ L^{(a)} = \underset{1\le d\le D}{\text{diag}}\ (L_d^{(a)}),\ L_d^{(a)} = O(1)_{2\times 2},\ d = 1, \ldots, D. \tag{19.17}$$

On the other hand, we have

$$\begin{aligned}\text{cov}(S^{(a)}, S^{(b)}) &= (L^{(a)} + F^{(a)})V(L^{(b)} + F^{(b)})' \\ &= L^{(a)}VL^{(b)\prime} + L^{(a)}VF^{(b)\prime} + F^{(a)}VL^{(b)\prime} + F^{(a)}VF^{(b)\prime}.\end{aligned}$$

From (19.17), we get $F^{(a)}VF^{(b)\prime} = O(D^{-1})_{2D\times 2D}$ and

$$L^{(a)}VF^{(b)\prime} = O(D^{-1})_{2D\times 2D},\quad F^{(a)}VL^{(b)\prime} = O(D^{-1})_{2D\times 2D}.$$

Therefore, $\text{cov}(S^{(a)}, S^{(b)}) = L^{(a)}VL^{(b)\prime} + O(D^{-1})_{2D\times 2D}$. □

Lemma 19.5 (Prasad and Rao 1990) *Let A_1 and A_2 be nonstochastic matrices of order n and $y \sim N_n(\mathbf{0}, V)$, where V is positive definite. Then,*

(a) $E[y(y'A_sy)y'] = tr(A_sV)V + 2VA_sV, \quad s = 1, 2,$

(b) $E[(y'A_1y)(y'A_2y)] = 2tr(A_1VA_2V) + tr(A_1V)tr(A_2V),$

(c) $E[y(y'A_1y)(y'A_2y)y'] = tr(A_1V)tr(A_2V)V + 2tr(A_1V)VA_2V$
$+2tr(A_2V)VA_1V + 2tr(A_1VA_2V)V + 4VA_1VA_2V + 4VA_2VA_1V.$

Lemma 19.6 (Prasad and Rao 1990) *Let $y \sim N_n(0, V)$, $z_j = \lambda_j' y$, and $q_j = y'A_jy$, $j = 1, \ldots, p$, where λ_j and A_j are nonstochastic of order $n \times 1$ and $n \times n$, respectively. Let $z = (z_1, \ldots, z_p)'$, $q = (q_1, \ldots, q_p)'$ have covariance matrices V_z and V_q, respectively. Then,*

$$E[(z'(q - E[q]))^2] = tr(V_zV_q) + 4\sum_{j=1}^{p}\sum_{i=1}^{p}\{\lambda_j'VA_jVA_iV\lambda_i + \lambda_j'VA_iVA_jV\lambda_i\},$$

$$\begin{aligned}&E[z_iz_j(q_i - E[q_i])(q_j - E[q_j])] = \lambda_i'E[y(y'A_iy)(y'A_jy)y']\lambda_j \\ &- E[q_i]\lambda_i'E[y(y'A_jy)y']\lambda_j - E[q_i]\lambda_i'E[y(y'A_iy)y']\lambda_j + E[q_i]E[q_j]\lambda_i'V\lambda_j.\end{aligned}$$

Lemma 19.7 (Prasad and Rao 1990) *Let us assume: (a) $V = \underset{1\le d\le D}{diag}\,(V_d)$, (b) $C = \underset{1\le d\le D}{diag}\,\big(O(D^{-1})_{2\times 2}\big) + O(D^{-2})_{2D\times 2D}$, (c) $r = \underset{1\le d\le D}{col}\,\underset{1\le j\le R}{col}\,\big(O(D^{-1})\big)$, and (d) $s_i = \underset{1\le d\le D}{col}\,\underset{1\le j\le R}{col}\,\big(\delta_{id}O(1)\big)$, where V_d is an 2×2 matrix with bounded elements.*

Then the following results hold: (e) $VCVCV = O(D^{-2})_{2D\times 2D}$, *(f)* $s_i' \sum s_i = O(1)$, *and (g)* $(r+s_i)'VCVCV(r+s_i) = O(D^{-2})$.

19.8 Simulations

This section presents some simulation experiments to analyze the behavior of the REML fitting algorithm, the EBLUPs, and the parametric bootstrap estimator of the MSE. Let us consider the particular BFH model (19.3)

$$y_d = X_d\beta + u_d + e_d, \quad d = 1, \ldots, D, \tag{19.18}$$

with $p_1 = p_2 = 2$, $p = 4$, $\beta_1 = (\beta_{11}, \beta_{12})' = (1,1)'$, $\beta_2 = (\beta_{21}, \beta_{22})' = (1,1)'$, $\mu_{x1} = \mu_{x2} = 1$, $\sigma_{x11} = 1$ and $\sigma_{x22} = 3/2$. For $k = 1, 2$, $d = 1, \ldots, D$, generate $X_d = \text{diag}(x_{d1}, x_{d2})_{2\times 4}$, where $x_{d1} = (x_{d11}, x_{d12})$, $x_{d2} = (x_{d21}, x_{d22})$, $x_{d11} = x_{d21} = 1$,

$$x_{d12} = \mu_{x1} + \sigma_{x11}^{1/2} U_{d1}, \quad x_{d22} = \mu_{x2} + \sigma_{x22}^{1/2} U_{d2}, \quad U_{dk} \overset{\text{ind}}{\sim} U(0,1).$$

For $d = 1, \ldots, D$, $u_d \sim N_2(0, V_{ud})$ and $e_d \sim N_2(0, V_{ed})$, where

$$V_{ud} = \begin{pmatrix} \theta_1 & \theta_3\sqrt{\theta_1}\sqrt{\theta_2} \\ \theta_3\sqrt{\theta_1}\sqrt{\theta_2} & \theta_2 \end{pmatrix}, \quad V_{ed} = \begin{pmatrix} 1 & c \\ c & 1 \end{pmatrix},$$

and $\theta_1 = 1, \theta_2 = 3/2, \theta_3 = 1/2, c = 1/4$.

Simulation 1

The target of Simulation 1 is to check the behavior of the REML algorithm for fitting the BFH model (19.18). The steps of Simulation 1 are:

1. Generate x_{dk}, $d = 1, \ldots, D$, $k = 1, 2$.
2. Repeat $I = 1000$ times ($i = 1, \ldots, 1000$).
   - 2.1. Generate $u_d^{(i)} \sim N_2(0, V_{ud})$, $e_d^{(i)} \sim N_2(0, V_{ed})$, $y_d^{(i)} = X_d\beta + u_d^{(i)} + e_d^{(i)}$, $d = 1, \ldots, D$.
   - 2.2. For every $\eta \in \{\beta_{11}, \beta_{12}, \beta_{21}, \beta_{22}, \theta_1, \theta_2, \theta_3\}$, calculate the REML estimator $\hat{\eta}^{(i)} \in \{\hat{\beta}_{11}^{(i)}, \hat{\beta}_{12}^{(i)}, \hat{\beta}_{21}^{(i)}, \hat{\beta}_{22}^{(i)}, \hat{\theta}_1^{(i)}, \hat{\theta}_2^{(i)}, \hat{\theta}_3^{(i)}\}$.

Table 19.1 Biases $BIAS(\hat{\eta})$ (left) and root-MSEs $RMSE(\hat{\eta})$ (right)

	η	$D=50$	$D=100$	$D=200$	$D=50$	$D=100$	$D=200$
β_{11}	1	−0.0419	0.0262	−0.0033	0.8750	0.6331	0.4681
β_{12}	1	0.0310	−0.0155	0.0044	0.5727	0.4157	0.3010
β_{21}	1	0.0015	0.0055	−0.0087	0.9826	0.6688	0.4807
β_{22}	1	0.0037	0.0004	0.0067	0.5778	0.3995	0.2902
θ_1	1	0.0126	−0.0023	−0.0002	0.4174	0.2987	0.2086
θ_2	3/2	0.0233	0.0051	0.0107	0.4971	0.3606	0.2477
θ_3	1/2	0.0027	0.0006	0.0007	0.2242	0.1621	0.1086

3. Output:

$$RMSE(\hat{\eta}) = \left(\frac{1}{I}\sum_{i=1}^{I}(\hat{\eta}^{(i)} - \eta)^2\right)^{1/2}, \quad BIAS(\hat{\eta}) = \frac{1}{I}\sum_{i=1}^{I}(\hat{\eta}^{(i)} - \eta).$$

Table 19.1 presents the simulation results. The column labeled by η contains the values of the true model parameters. Simulation 1 shows that the REML Fisher-scoring algorithm works properly because BIAS (left) and RMSE (right) decrease as D increases.

Simulation 2

Simulation 2 investigates the performance of the EBLUPs of the mean parameters μ_{dk}. The steps of Simulation 2 are:

1. Generate x_{dk}, $d=1,\ldots,D$, $k=1,2$.
2. Repeat $I=10^4$ times ($i=1,\ldots,I$).
   2.1. Generate $\{(e_d^{(i)}, u_d^{(i)}, y_d^{(i)}) : d=1,\ldots,D\}$ from the BFH model (19.18).
   2.2. Calculate the true means $\mu_d^{(i)} = X_d\beta + u_d^{(i)}$, $d=1,\ldots,D$.
   2.3. Fit the BFH model (19.18) to the simulated data $(y_d^{(i)}, X_d)$, $d=1,\ldots,D$.
   2.4. Calculate the EBLUP $\hat{\mu}_d^{(i)}$ under the BFH model (19.18).
3. For $d=1,\ldots,D$, $k=1,2$, calculate the performance measures
   3.1. $E_{\mu,dk} = \left(\frac{1}{I}\sum_{i=1}^{I}(\hat{\mu}_{dk}^{(i)} - \mu_{dk}^{(i)})^2\right)^{1/2}$, $B_{\mu,dk} = \frac{1}{I}\sum_{i=1}^{I}(\hat{\mu}_{dk}^{(i)} - \mu_{dk}^{(i)})$.
   3.2. $E_{\mu,k} = \frac{1}{D}\sum_{d=1}^{D}E_{\mu,dk}$, $B_{\mu,k} = \frac{1}{D}\sum_{d=1}^{D}B_{\mu,dk}$.

Table 19.2 presents the results of Simulation 2. From the table, one can observe that the EBLUPs are basically unbiased and their MSE is almost constant with increasing D, since with increasing number of areas we also augment the number of parameters to predict.

Table 19.2 Biases $B_{\mu,k}$ (left) and root-MSEs $E_{\mu,k}$ (right)

k	$D = 50$	$D = 100$	$D = 200$	$D = 50$	$D = 100$	$D = 200$
1	0.00441	0.00128	0.00232	0.72923	0.71524	0.70516
2	0.00422	0.00300	−0.00017	0.79611	0.78282	0.77669

Simulation 3

Simulation 3 investigates the performance of the parametric bootstrap estimator of the MSE of the EBLUP. The steps of Simulation 3 are:

1. Take $mse_{\mu,dk} = E^2_{\mu,dk}$ from the output of Simulation 2.
2. Generate x_{dk}, $d = 1, \dots, D$, $k = 1, 2$.
3. Repeat $I = 10^2$ times ($i = 1, \dots, I$).

   3.1. Generate $\{(e_d^{(i)}, u_d^{(i)}, y_d^{(i)}) : d = 1, \dots, D\}$ from model (19.18).
   3.2. Calculate the REML estimators $\hat{\beta}_{11}^{(i)}, \hat{\beta}_{12}^{(i)}, \hat{\beta}_{21}^{(i)}, \hat{\beta}_{22}^{(i)}, \hat{\theta}_1^{(i)}, \hat{\theta}_2^{(i)}, \hat{\theta}_3^{(i)}$ by using the sample data $(y_d^{(i)}, X_d)$, $d = 1, \dots, D$.
   3.3. Repeat B times.

   (a) Generate $u_d^{*(ib)} \sim N_2(0, \hat{V}_{ud}^{(i)})$, $e_d^{*(ib)} \sim N_2(0, V_{ed})$ and calculate

$$y_d^{*(ib)} = \mu_d^{*(ib)} + e_d^{*(ib)}, \quad \mu_d^{*(ib)} = X_d \hat{\beta}^{(i)} + u_d^{*(ib)}, \quad d = 1, \dots, D. \tag{19.19}$$

   (b) Calculate the REML estimators $\hat{\beta}_{11}^{*(ib)}, \hat{\beta}_{12}^{*(ib)}, \hat{\beta}_{21}^{*(ib)}, \hat{\beta}_{22}^{*(ib)}, \hat{\theta}_1^{*(ib)}, \hat{\theta}_2^{*(ib)}$, $\hat{\theta}_3^{*(ib)}$ by using the bootstrap data $(y_d^{*(ib)}, X_d)$, $d = 1, \dots, D$.
   (c) Calculate $\hat{\mu}_d^{*(ib)}$ based on the BFH model (19.19).

   3.5. For $d = 1, \dots, D$, $k = 1, 2$, calculate $mse_{\mu,dk}^{*(i)} = \frac{1}{B} \sum_{b=1}^{B} \big(\hat{\mu}_{dk}^{*(ib)} - \mu_{dk}^{*(ib)}\big)^2$.

4. For $d = 1, \dots, D$, $k = 1, 2$, calculate

$$E_{\mu,dk}^{*} = \Big(\frac{1}{I} \sum_{i=1}^{I} (mse_{\mu,dk}^{*(i)} - mse_{\mu,dk})^2\Big)^{1/2}, \quad E_{\mu,k}^{*} = \frac{1}{D} \sum_{d=1}^{D} E_{\mu,dk}^{*}.$$

Table 19.3 presents the averages of across domains root-MSEs $E_{\mu,k}^{*}$. Figure 19.1 presents the boxplots of $E_{\mu,dk}^{*}$ for $D = 100$. Simulation 3 suggests that running the parametric bootstrap algorithm with $B = 400$ gives a good approximation to the MSE.

Table 19.3 Average across domains of root-MSEs of MSE estimators

$D = 100$	k	$B = 50$	$B = 100$	$B = 200$	$B = 400$	$B = 600$
$E^*_{\mu,k}$	1	0.12975	0.11192	0.09670	0.07341	0.08276
	2	0.13755	0.10665	0.08785	0.07700	0.08354

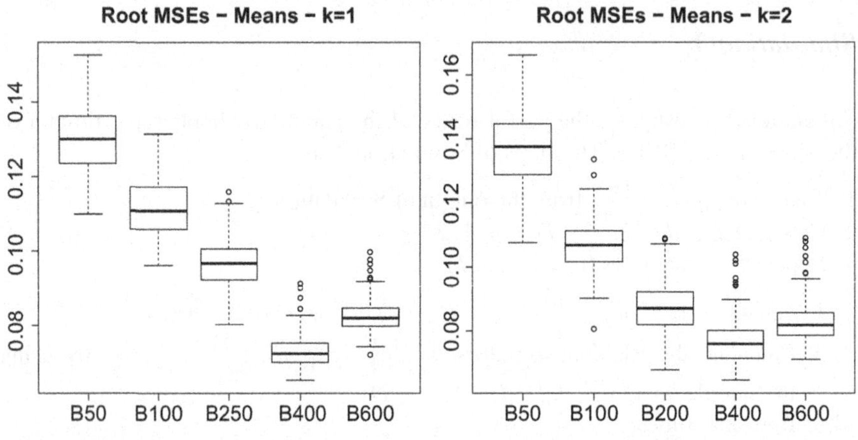

Fig. 19.1 Root-MSEs of parametric bootstrap estimators of EBLUPs ($D = 100$)

19.9 R Codes for EBLUPs

19.9.1 Main Program

This section gives R codes for fitting the bivariate Fay–Herriot model to the survey data file `datLCS.txt`. The target variables y_1 and y_2 are the direct estimators of the domain poverty proportion and gap. As auxiliary variables x, we take the domain means of the variable ss giving the national social security status (1 if registered as worker and 0 otherwise). The main code fits a bivariate Fay–Herriot model by the REML method and calculates the EBLUPs and the corresponding MSEs. Section 19.9.2 gives the employed R functions.

We install the R packages `sae` and `Matrix`.

```
if(!require(sae)){
  install.packages("sae")
  library(sae)
}
if(!require(Matrix)){
  install.packages("Matrix")
  library(Matrix)
}
```

The following code reads the data files and renames some variables. The poverty threshold z0 is fixed at 7280 euros. The variable poor is defined as 1 if income<z0 and 0 otherwise. The variable gap is defined by the formula (z0-income)·poor/z0.

```
dat <- read.table("datLCS.txt", header=TRUE, sep="\t", dec=",")    # Read dat
```

```
# poverty threshold fixed at 7280
z0 <- 7280
# variable poor
poor <- as.numeric(dat$income<z0)
# variable gap
gap <- (z0-dat$income)*poor/z0
# Read auxiliary data
aux <- read.table("auxLCS2.txt", header=TRUE, sep="\t", dec=",")
aux <- aux[order(aux$dom),]      # Sort aux by dom
D <- nrow(aux)                   # Number of domains
ss <- aux$ss                     # Select ss as aux variable
```

We calculate the direct estimators of domain poverty proportions and gaps, their variance and covariance estimates, and the population sizes by domain. For this sake, we apply the R function `direct.bfh` and we look for those cases where the estimates of the variances of the direct estimators are equal to zero.

```
poor.dir <- direct.bfh(data1=poor, data2=gap, w=dat$w,
                       domain=list(domain=dat$dom))
y <- lapply(1:D,
            function(d) matrix(c(poor.dir$y1.mean[d], poor.dir$y2.mean[d])))
sel.pprop <- poor.dir$var.y1.mean>0
sel.pgap <- poor.dir$var.y2.mean>0
poor.dir$var.y1.mean[!sel.pprop] <- min(poor.dir$var.y1.mean[sel.pprop])
poor.dir$var.y2.mean[!sel.pgap] <- min(poor.dir$var.y2.mean[sel.pgap])
```

We define the covariance matrices V_{ed} of the vectors of sampling errors.

```
Ved <- list()
for(d in 1:D){
  Ved[[d]] <- matrix(NA, nrow=2, ncol=2)
  sup <- as.numeric(poor.dir$cov.y12.mean[d])
  Ved[[d]][upper.tri(Ved[[d]])] <- sup
  Ved[[d]][lower.tri(Ved[[d]])] <- sup
  diag(Ved[[d]]) <- as.numeric(c(poor.dir$var.y1.mean[d],
                                 poor.dir$var.y2.mean[d]))
}
```

The REML iterative algorithm for fitting the BFH model requires starting values (seeds). For setting the seeds, we fit an univariate FH model to each coordinate (y_1=poor and y_2=gap). We use the function `mseFH` of the R package `sae` for fitting the FH models.

```
# Target variables
directy1 <- poor.dir$y1.mean; directy2 <- poor.dir$y2.mean
# Sampling error variances
vardiry1 <- poor.dir$var.y1.mean; vardiry2 <- poor.dir$var.y2.mean
fit.FH.sae1 <- mseFH(directy1 ~ ss, vardiry1)    # FH model for y1
fit.FH.sae2 <- mseFH(directy2 ~ ss, vardiry2)    # FH model for y2
```

For fitting a bivariate Fay–Herriot model to the data (y_1, y_2), with auxiliary variable ss in both coordinates, the R function `REML.BFH` employs the REML method. The following code calculates the function inputs. The matrix X contains the auxiliary variables, i.e. the intercepts (ones) and ss for each coordinate.

```
# Construct the matrix of auxiliary variables.
X <- lapply(1:D,
            function(d) as.matrix(t(bdiag(as.numeric(c(1,ss[d])),
                                          as.numeric(c(1,ss[d]))))))
# Number of regression parameters
p <- ncol(X[[1]])
```

```
# Seeds for variance component parameters
thetas.0 <- c(fit.FH.sae1$est$fit$refvar, fit.FH.sae2$est$fit$refvar, 0)
Vud <- UveU(thetas.0)
```

We apply the R function `REML.BFH` to fit a bivariate Fay–Herriot model.

```
fit <- try(REML.BFH(X, y, D, Ved, Vud), TRUE)
cat("BFH model parameters", fit[[1]], "\n")
cat("Number of iterations", fit[[3]], "\n")
cat("Errors: ", fit[[4]], "\n")
```

The R function `BETA.U.BFH` calculates the random effect predictions and the estimates of the regression parameters. The R function `pvalue` calculates the corresponding standard errors, z-values, and p-values. The R function `CI` calculates the asymptotic confidence intervals for the regression parameters.

```
beta.u.hat <- BETA.U.BFH(X, y, D, Ved , fit[[1]])
beta.hat <- beta.u.hat[[1]]          # Estimates of regression parameters
theta.hat <- fit[[1]]                # Estimates of variance components
u <- beta.u.hat[[2]]                 # Predictions of random effects
pv <- pvalue(beta.hat, fit)          # Apply function pvalue
# Anova table with standard errors, z-values and p-values
betas <- data.frame(beta.hat, pv, Sig=pv[[3]]<0.05)
row.names(betas) <- c("Intercept.p", "ss.p", "Intercept.g",  "ss.g")
# Apply function CI for confidence intervals
C.I <- CI(beta.hat, fit)
```

The estimates of the variance component parameters are $\hat{\sigma}_{u1}^2 = 0.00861$, $\hat{\sigma}_{u2}^2 = 0.00192$, and $\hat{\rho} = 0.88006$. The corresponding 95% confidence intervals are $(0.00277, 0.01445)$, $(0.00064, 0.00320)$, and $(0.76145, 0.99867)$. Table 19.4 presents the estimated regression parameters and p-values. It also contains their 95% confidence intervals.

Figure 19.2 (left) plots the residuals of the first coordinate versus the direct estimates of the poverty proportions. Figure 19.2 (right) plots the residuals of the second coordinate versus the direct estimates of the poverty gaps. We observe that the divergence between EBLUPs and direct estimates is greater when the direct estimates are big. The model is smoothing the behavior of the direct estimates.

In the variable `eblup.bfh` we calculate the EBLUPs of poverty proportions and gaps.

```
eblup.bfh <- mapply(FUN="+", lapply(X, FUN="%*%", beta.hat), u)
eblup.bfh.1 <- eblup.bfh[1,]     # EBLUPs of poverty proportions
eblup.bfh.2 <- eblup.bfh[2,]     # EBLUPs of poverty gaps
```

The R function `MSE.BFH` calculates analytic estimators of the MSEs. The following code also calculates the coefficients of variation of direct estimators and EBLUPs.

Table 19.4 Estimated parameters and CIs of BFH model

Parameter	Estimate	Std. error	z-value	p-value	Inf.lim	Sup.lim
Intercept.p	0.4720	0.1496	3.1539	0.0016	0.1787	0.7653
ss.p	−0.5495	0.2377	2.3121	0.0208	−1.0153	−0.0837
Intercept.g	0.1858	0.0698	2.6610	0.0078	0.0489	0.3226
ss.g	−0.2193	0.1106	1.9815	0.0475	−0.4361	−0.0024

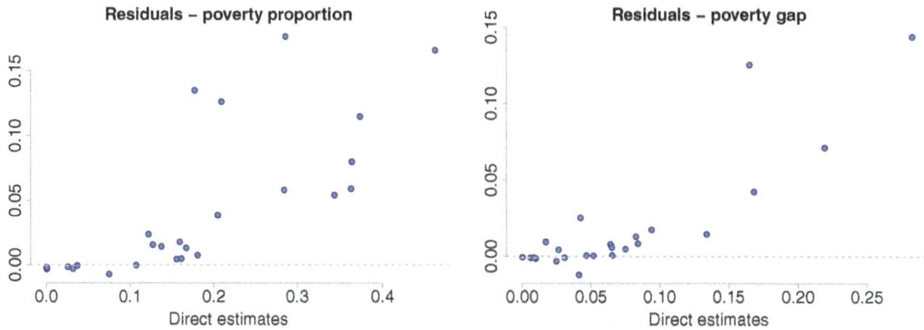

Fig. 19.2 Dispersion graphs of residuals of poverty proportions (left) and gaps (right) versus direct estimates

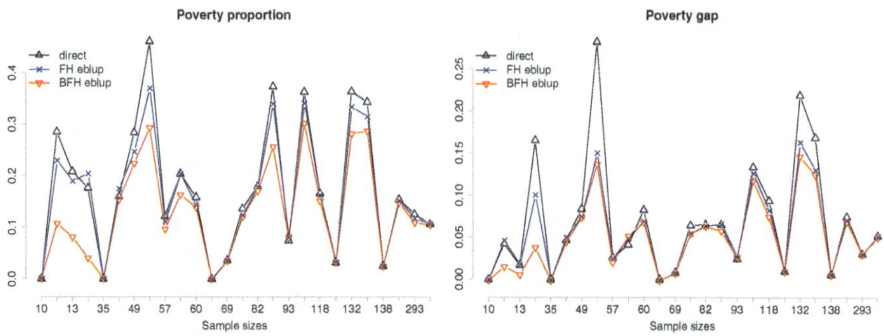

Fig. 19.3 EBLUP and direct estimates of poverty proportions (left) and gaps (right)

```
mse.bfh <- MSE.BFH(X, D, Ved, fit)     # Apply function mse.bfh
# MSE estimates
mse.bfh.1 <- mse.bfh[[1]]; mse.bfh.2 <- mse.bfh[[2]]
# coefficients of variation
CVdir1 <- round(100*sqrt(poor.dir$var.y1.mean/abs(poor.dir$y1.mean)), 2)
CVdir2 <- round(100*sqrt(poor.dir$var.y2.mean/abs(poor.dir$y2.mean)), 2)
CV.bfh.1 <- round(100*sqrt(mse.bfh.1/abs(eblup.bfh.1)), 2)
CV.bfh.2 <- round(100*sqrt(mse.bfh.2/abs(eblup.bfh.2)), 2)
```

The R code to save the results is

```
output1 <- data.frame(nd=poor.dir$nd, DIR=round(poor.dir$y1.mean,5),
                      Vdir=round(poor.dir$var.y1.mean,5), CVdir=CVdir1,
                      EB=round(eblup.bfh.1,5), MSEeb=round(mse.bfh.1,5),
                      CVeb=CV.bfh.1)
head(output1, 10)
output2 <- data.frame(nd=poor.dir$nd, DIR=round(poor.dir$y2.mean,5),
                      Vdir=round(poor.dir$var.y2.mean,5), CVdir=CVdir2,
                      EB=round(eblup.bfh.2,5), MSEeb=round(mse.bfh.2,5),
                      CVeb=CV.bfh.2)
head(output2, 10)
```

Figure 19.3 plots the direct estimates (direct), the EBLUPs based on the Fay–Herriot model (FH eblup), and the EBLUPs based on the bivariate Fay–Herriot model (BFH eblup) of poverty proportions (left) and poverty gaps (right). The three types of estimators tend to coincide as the sample size increases.

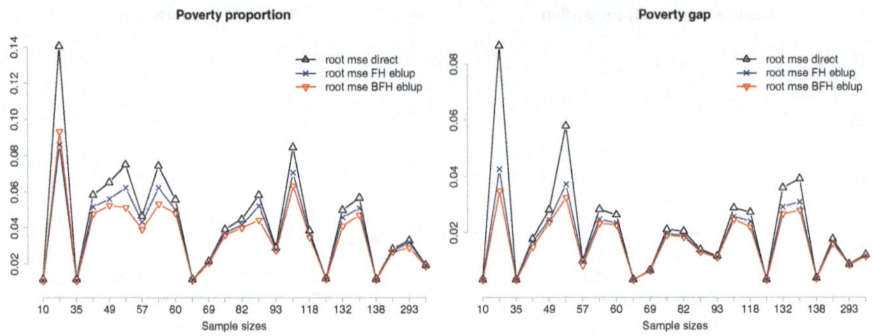

Fig. 19.4 Root-MSEs of estimators of poverty proportions (left) and gaps (right)

Table 19.5 Direct and EBLUP estimates of poverty proportions

dom	n_d	DIR	Vdir	CVdir	EB	MSEeb	CVeb
3	57	0.46122	0.00561	11.03	0.29430	0.00269	9.55
5	96	0.36285	0.00714	14.03	0.30309	0.00403	11.52
6	82	0.18046	0.00200	10.53	0.17238	0.00161	9.67
7	10	0.00000	0.00014	-	0.00270	0.00013	21.78
11	118	0.16681	0.00148	9.41	0.15310	0.00122	8.93
12	18	0.17700	0.00860	22.04	0.04157	0.00233	23.65
13	138	0.02533	0.00014	7.34	0.02637	0.00013	7.12
14	190	0.15525	0.00080	7.19	0.15023	0.00073	6.97
15	406	0.10684	0.00038	5.94	0.10649	0.00036	5.79
16	93	0.07426	0.00084	10.61	0.08081	0.00076	9.71

Figure 19.4 plots the root-MSEs of the direct estimates (root mse direct), of the EBLUPs based on the Fay–Herriot model (root mse FH eblup) and of the EBLUPs based on the bivariate Fay–Herriot model (root mse BFH eblup) for poverty proportions (left) and poverty gaps (right). The EBLUPs based on the bivariate Fay–Herriot model have the smallest root-MSEs.

For the ten first areas, Table 19.5 gives the direct estimators (DIR) and the EBLUPs (EB), based on the fitted BFH model, of the domain poverty proportions. The design-based estimates of the variances and coefficients of variations of direct estimators are labeled by Vdir and CVdir, respectively. The labels of the estimated MSEs and coefficients of variations of EBLUPs are MSEeb and CVeb, respectively. Table 19.5 shows that estimated MSEs of EBLUPs are smaller than estimated design-based variances of direct estimators. But we have to note that in the case of EBLUP the MSE estimators were calculated using only the terms $G_1(\hat{\theta})$ and $G_2(\hat{\theta})$.

For the ten first areas, Table 19.6 gives the direct estimators (DIR) and the EBLUPs (EB), based on the fitted BFH model, of the domain poverty gaps. This table shows that estimated MSEs of EBLUPs (calculated using only terms $G_1(\hat{\theta})$, $G_2(\hat{\theta})$) are smaller than estimated variances of direct estimators.

Table 19.6 Direct and EBLUP estimates of poverty gaps

dom	n_d	DIR	Vdir	CVdir	EB	MSEeb	CVeb
3	57	0.28223	0.00336	10.91	0.13786	0.00107	8.80
5	96	0.13363	0.00084	7.91	0.11835	0.00062	7.22
6	82	0.06521	0.00042	8.04	0.06385	0.00035	7.36
7	10	0.00000	0.00001	-	0.00000	0.00001	33.96
11	118	0.09395	0.00075	8.92	0.07574	0.00049	8.03
12	18	0.16513	0.00748	21.29	0.03879	0.00123	17.81
13	138	0.00582	0.00002	5.16	0.00611	0.00002	5.01
14	190	0.07491	0.00032	6.50	0.06932	0.00026	6.16
15	406	0.05169	0.00015	5.39	0.05065	0.00014	5.22
16	93	0.02449	0.00014	7.50	0.02696	0.00013	6.89

19.9.2 R Functions for the BFH Model

This section contains the R function employed in the main program of Sect. 19.9.1. The R function `direct.bfh` calculates the direct estimators of two domain totals and means and the corresponding direct estimators of their variances and covariances.

```
direct.bfh <- function(data1, data2, w, domain) {
 if(!inherits(data1,"data.frame")){
   last <- length(domain) + 1
# Population and sample size
   Nd <- aggregate(w, by=domain, sum)[,last]
   nd <- aggregate(rep(1, length(data1)), by=domain, sum)[,last]
# y1 direct
   y1 <- aggregate(w*data1, by=domain, sum)
   names(y1) <- c(names(domain), "y1")
   y1.mean <- y1[,last]/Nd
# y2 direct
   y2 <- aggregate(w*data2, by=domain, sum)
   names(y2) <- c(names(domain), "y2")
   y2.mean <- y2[,last]/Nd
# Variances and covariances
   difference1 <- difference2 <- difference12 <- list()
   for(d in 1:length(y1.mean)){
     condition <- domain[[1]]==y1[d,1]
     difference1[[d]] <- w[condition]*(w[condition]-1) *
                         (data1[condition]-y1.mean[d])^2
     difference2[[d]] <- w[condition]*(w[condition]-1) *
                         (data2[condition]-y2.mean[d])^2
     difference12[[d]] <- w[condition]*(w[condition]-1) *
                          (data1[condition]-y1.mean[d]) *
                          (data2[condition]-y2.mean[d])
   }
   var.y1 <- unlist(lapply(difference1, sum))
   var.y2 <- unlist(lapply(difference2, sum))
   cov.y12 <- unlist(lapply(difference12, sum))
   var.y1.mean <- var.y1/Nd^2
   var.y2.mean <- var.y2/Nd^2
   cov.y12.mean <- cov.y12/Nd^2
   return(cbind(y1, var.y1, y1.mean, var.y1.mean,
                y2, var.y2, y2.mean, var.y2.mean,
                cov.y12, cov.y12.mean, nd, Nd))
  }
```

```
    else{
      warning("Only a numeric or integer vector must be called as data",
              call. = FALSE)
    }
  }
```

The R function `UveU` calculates the variance matrix V_{ud} of the random effect u_d.

```
UveU <- function(thetas) {
  Vuu <- matrix(0, nrow=2, ncol=2)
  sup <- c(thetas[3]*sqrt(thetas[1])*sqrt(thetas[2]))
  Vuu[upper.tri(Vuu)] <- sup
  Vuu <- Vuu + t(Vuu)
  diag(Vuu) <- c(thetas[1], thetas[2])
  return(Vuu)
}
```

The R function `Thetas` calculates the vector of parameters $\boldsymbol{\theta} = (\theta_1, \theta_2, \theta_3)' = (\sigma_{u1}^2, \sigma_{u2}^2, \rho)'$.

```
Thetas <- function(Vus) {
  Zethas <- vector()
  Zethas <- c(Vus[1,1], Vus[2,2], Vus[1,2]/(sqrt(Vus[1,1]*Vus[2,2])))
  return(Zethas)
}
```

The R function `FirstDer` calculates the matrices of first derivatives V_{ud1}, V_{ud2}, and V_{ud3} of V_{ud} with respect to θ_1, θ_2, and θ_3, respectively.

```
FirstDer <- function(thetas){
  Vud1 <- matrix(0, nrow=2, ncol=2)
  Vud1[1,1] <- 1
  Vud1[1,2] <- Vud1[2,1] <- thetas[3]*sqrt(thetas[2])/(2*sqrt(thetas[1]))
  Vud2 <- matrix(0, nrow=2, ncol=2)
  Vud2[2,2] <- 1
  Vud2[1,2] <- Vud2[2,1] <- thetas[3]*sqrt(thetas[1])/(2*sqrt(thetas[2]))
  Vud3 <- matrix(0, nrow=2, ncol=2)
  Vud3[1,2] <- Vud3[2,1] <- sqrt(thetas[1])*sqrt(thetas[2])
  return(list(Vud1, Vud2, Vud3))
}
```

The R function `REML.BFH` calculates the REML estimators of the parameters of the bivariate Fay–Herriot model.

```
REML.BFH <- function(X, y, D, Ved, Vud, MAXITER = 100) {
  Vud.f  <- Vud
  Xd <- X
  yd <- y
  theta.f <- Thetas(Vud)
  Vuds <- FirstDer(theta.f)
  Bad <- 0
  FLAG <- 0
  p <- ncol(X[[1]])
  # Iteration loop of Fisher-scoring algorithm
  for(ITER in 1:MAXITER){
    Vd.inv <- Vinvyd <- VinvXd <- list()
    VinvVud1 <- XtVinvVud1VinvX <- VinvVud1VinvVud1 <-
    XtVinvVud1VinvVud1VinvX <- list()
    tr.VinvVud1 <- ytVinvX <- ytVinvVud1Vinvy <- ytVinvVud1VinvX <-
    SumXtVinvVud1VinvX <- tr.VinvVud1VinvVud1 <- 0
    VinvVud2 <- XtVinvVud2VinvX <- VinvVud2VinvVud2 <-
    XtVinvVud2VinvVud2VinvX <- list()
    tr.VinvVud2 <- ytVinvVud2Vinvy <- ytVinvVud2VinvX <-
    SumXtVinvVud2VinvX <- tr.VinvVud2VinvVud2 <- 0
```

```
VinvVud3 <- XtVinvVud3VinvX <- VinvVud3VinvVud3 <-
XtVinvVud3VinvVud3VinvX <- list()
tr.VinvVud3 <- ytVinvVud3Vinvy <- ytVinvVud3VinvX <-
SumXtVinvVud3VinvX <- tr.VinvVud3VinvVud3 <- 0
VinvVud1VinvVud2 <- XtVinvVud1VinvVud2VinvX <- list()
tr.VinvVud1VinvVud2 <- 0
VinvVud1VinvVud3 <- XtVinvVud1VinvVud3VinvX <- list()
tr.VinvVud1VinvVud3 <- 0
VinvVud2VinvVud3 <- XtVinvVud2VinvVud3VinvX <- list()
tr.VinvVud2VinvVud3 <- 0
Q.inv <- matrix(0, nrow=p, ncol=p)
# Domain loop 1
for(d in 1:D) {
  Vd <- Vud.f+Ved[[d]]
  if(abs(det(Vd))<1e-09 || abs(det(Vd))>1e+11){
    FLAG <- 1
    Bad <- Bad+1
    break
  }
  Vd.inv[[d]] <- solve(Vd)
  Vinvyd[[d]] <- Vd.inv[[d]]%*%yd[[d]]
  VinvXd[[d]] <- Vd.inv[[d]]%*%Xd[[d]]
  Q.inv <- Q.inv + t(Xd[[d]])%*%VinvXd[[d]]
  # Score S1
  VinvVud1[[d]] <- Vd.inv[[d]]%*%Vuds[[1]]
  tr.VinvVud1 <- tr.VinvVud1 + sum(diag(VinvVud1[[d]]))
  XtVinvVud1VinvX[[d]] <- t(VinvXd[[d]])%*%Vuds[[1]]%*%VinvXd[[d]]
  ytVinvX <- ytVinvX + t(yd[[d]])%*%VinvXd[[d]]
  ytVinvVud1Vinvy <- ytVinvVud1Vinvy +
                     t(Vinvyd[[d]])%*%Vuds[[1]]%*%Vinvyd[[d]]
  ytVinvVud1VinvX <- ytVinvVud1VinvX +
                     t(Vinvyd[[d]])%*%Vuds[[1]]%*%VinvXd[[d]]
  SumXtVinvVud1VinvX <- SumXtVinvVud1VinvX + XtVinvVud1VinvX[[d]]
  # Score S2
  VinvVud2[[d]] <- Vd.inv[[d]]%*%Vuds[[2]]
  tr.VinvVud2 <-  tr.VinvVud2 + sum(diag(VinvVud2[[d]]))
  XtVinvVud2VinvX[[d]] <- t(VinvXd[[d]])%*%Vuds[[2]]%*%VinvXd[[d]]
  ytVinvVud2Vinvy <- ytVinvVud2Vinvy +
                     t(Vinvyd[[d]])%*%Vuds[[2]]%*%Vinvyd[[d]]
  ytVinvVud2VinvX <- ytVinvVud2VinvX +
                     t(Vinvyd[[d]])%*%Vuds[[2]]%*%VinvXd[[d]]
  SumXtVinvVud2VinvX <- SumXtVinvVud2VinvX + XtVinvVud2VinvX[[d]]
  # Score S3
  VinvVud3[[d]] <- Vd.inv[[d]]%*%Vuds[[3]]
  tr.VinvVud3 <- tr.VinvVud3 + sum(diag(VinvVud3[[d]]))
  XtVinvVud3VinvX[[d]] <- t(VinvXd[[d]])%*%Vuds[[3]]%*%VinvXd[[d]]
  ytVinvVud3Vinvy <- ytVinvVud3Vinvy +
                     t(Vinvyd[[d]])%*%Vuds[[3]]%*%Vinvyd[[d]]
  ytVinvVud3VinvX <- ytVinvVud3VinvX +
                     t(Vinvyd[[d]])%*%Vuds[[3]]%*%VinvXd[[d]]
  SumXtVinvVud3VinvX <- SumXtVinvVud3VinvX + XtVinvVud3VinvX[[d]]
  # Fisher information element F11
  VinvVud1VinvVud1[[d]] <- VinvVud1[[d]]%*%VinvVud1[[d]]
  tr.VinvVud1VinvVud1 <- tr.VinvVud1VinvVud1 +
                         sum(diag(VinvVud1VinvVud1[[d]]))
  XtVinvVud1VinvVud1VinvX[[d]] <-
             t(Xd[[d]])%*%VinvVud1VinvVud1[[d]]%*%VinvXd[[d]]
  # Fisher information element F22
  VinvVud2VinvVud2[[d]] <- VinvVud2[[d]]%*%VinvVud2[[d]]
  tr.VinvVud2VinvVud2 <- tr.VinvVud2VinvVud2 +
                         sum(diag(VinvVud2VinvVud2[[d]]))
  XtVinvVud2VinvVud2VinvX[[d]] <-
             t(Xd[[d]])%*%VinvVud2VinvVud2[[d]]%*%VinvXd[[d]]
  # Fisher information element F33
  VinvVud3VinvVud3[[d]] <- VinvVud3[[d]]%*%VinvVud3[[d]]
  tr.VinvVud3VinvVud3 <- tr.VinvVud3VinvVud3 +
                         sum(diag(VinvVud3VinvVud3[[d]]))
```

```
    XtVinvVud3VinvVud3VinvX[[d]] <-
              t(Xd[[d]])%*%VinvVud3VinvVud3[[d]]%*%VinvXd[[d]]
    # Fisher information element F12
    VinvVud1VinvVud2[[d]] <- VinvVud1[[d]]%*%VinvVud2[[d]]
    tr.VinvVud1VinvVud2 <- tr.VinvVud1VinvVud2 +
                           sum(diag(VinvVud1VinvVud2[[d]]))
    XtVinvVud1VinvVud2VinvX[[d]] <-
              t(Xd[[d]])%*%VinvVud1VinvVud2[[d]]%*%VinvXd[[d]]
    # Fisher information element F13
    VinvVud1VinvVud3[[d]] <- VinvVud1[[d]]%*%VinvVud3[[d]]
    tr.VinvVud1VinvVud3 <- tr.VinvVud1VinvVud3 +
                           sum(diag(VinvVud1VinvVud3[[d]]))
    XtVinvVud1VinvVud3VinvX[[d]] <-
              t(Xd[[d]])%*%VinvVud1VinvVud3[[d]]%*%VinvXd[[d]]
    # Fisher information element F23
    VinvVud2VinvVud3[[d]] <- VinvVud2[[d]]%*%VinvVud3[[d]]
    tr.VinvVud2VinvVud3 <- tr.VinvVud2VinvVud3 +
                           sum(diag(VinvVud2VinvVud3[[d]]))
    XtVinvVud2VinvVud3VinvX[[d]] <-
              t(Xd[[d]])%*%VinvVud2VinvVud3[[d]]%*%VinvXd[[d]]
    #
    if(FLAG==1){
      FLAG <- 0
      ITER <- MAXITER
      break
    }
  } # End of the domain loop 1
  Q <- solve(Q.inv)
  tr.XtVinvVud1VinvXQ <- tr.XtVinvVud2VinvXQ <- tr.XtVinvVud3VinvXQ <-
  tr.XtVinvVud4VinvXQ <- tr.XtVinvVud5VinvXQ <- tr.XtVinvVud6VinvXQ <- 0
  tr.XtVinvVud1VinvVud1VinvXQ <- tr.XtVinvVud2VinvVud2VinvXQ <-
  tr.XtVinvVud3VinvVud3VinvXQ <- 0
  tr.XtVinvVud1VinvVud2VinvXQ <- tr.XtVinvVud1VinvVud3VinvXQ <-
  tr.XtVinvVud2VinvVud3VinvXQ <- 0
  # Domain loop 2
  for(d in 1:D){
    tr.XtVinvVud1VinvXQ <- tr.XtVinvVud1VinvXQ +
                           sum(diag(XtVinvVud1VinvX[[d]]%*%Q))
    tr.XtVinvVud2VinvXQ <- tr.XtVinvVud2VinvXQ +
                           sum(diag(XtVinvVud2VinvX[[d]]%*%Q))
    tr.XtVinvVud3VinvXQ <- tr.XtVinvVud3VinvXQ +
                           sum(diag(XtVinvVud3VinvX[[d]]%*%Q))
    tr.XtVinvVud1VinvVud1VinvXQ <- tr.XtVinvVud1VinvVud1VinvXQ +
    sum(diag(XtVinvVud1VinvVud1VinvX[[d]]%*%Q))
    tr.XtVinvVud2VinvVud2VinvXQ <- tr.XtVinvVud2VinvVud2VinvXQ +
    sum(diag(XtVinvVud2VinvVud2VinvX[[d]]%*%Q))
    tr.XtVinvVud3VinvVud3VinvXQ <- tr.XtVinvVud3VinvVud3VinvXQ +
    sum(diag(XtVinvVud3VinvVud3VinvX[[d]]%*%Q))
    tr.XtVinvVud1VinvVud2VinvXQ <- tr.XtVinvVud1VinvVud2VinvXQ +
    sum(diag(XtVinvVud1VinvVud2VinvX[[d]]%*%Q))
    tr.XtVinvVud1VinvVud3VinvXQ <- tr.XtVinvVud1VinvVud3VinvXQ +
    sum(diag(XtVinvVud1VinvVud3VinvX[[d]]%*%Q))
    tr.XtVinvVud2VinvVud3VinvXQ <- tr.XtVinvVud2VinvVud3VinvXQ +
    sum(diag(XtVinvVud2VinvVud3VinvX[[d]]%*%Q))
  } # End of the domain loop 2
  tr.PV1 <- tr.VinvVud1 - tr.XtVinvVud1VinvXQ
  tr.PV2 <- tr.VinvVud2 - tr.XtVinvVud2VinvXQ
  tr.PV3 <- tr.VinvVud3 - tr.XtVinvVud3VinvXQ
  SumXtVinvVud1VinvXQ <- SumXtVinvVud1VinvX%*%Q
  SumXtVinvVud2VinvXQ <- SumXtVinvVud2VinvX%*%Q
  SumXtVinvVud3VinvXQ <- SumXtVinvVud3VinvX%*%Q
  tr.PV1PV1  <- tr.VinvVud1VinvVud1 - 2*tr.XtVinvVud1VinvVud1VinvXQ +
  sum(diag(SumXtVinvVud1VinvXQ%*%SumXtVinvVud1VinvXQ))
  tr.PV1PV2  <- tr.VinvVud1VinvVud2 - 2*tr.XtVinvVud1VinvVud2VinvXQ +
  sum(diag(SumXtVinvVud1VinvXQ%*%SumXtVinvVud2VinvXQ))
  tr.PV1PV3  <- tr.VinvVud1VinvVud3 - 2*tr.XtVinvVud1VinvVud3VinvXQ +
  sum(diag(SumXtVinvVud1VinvXQ%*%SumXtVinvVud3VinvXQ))
```

```
    tr.PV2PV2  <- tr.VinvVud2VinvVud2 - 2*tr.XtVinvVud2VinvVud2VinvXQ +
    sum(diag(SumXtVinvVud2VinvXQ%*%SumXtVinvVud2VinvXQ))
    tr.PV2PV3  <- tr.VinvVud2VinvVud3 - 2*tr.XtVinvVud2VinvVud3VinvXQ +
    sum(diag(SumXtVinvVud2VinvXQ%*%SumXtVinvVud3VinvXQ))
    tr.PV3PV3  <- tr.VinvVud3VinvVud3 - 2*tr.XtVinvVud3VinvVud3VinvXQ +
    sum(diag(SumXtVinvVud3VinvXQ%*%SumXtVinvVud3VinvXQ))
    ytVinvXQ <- ytVinvX%*%Q
    ytPV1Py  <- ytVinvVud1Vinvy - 2*ytVinvVud1VinvX%*%t(ytVinvXQ) +
    ytVinvXQ%*%SumXtVinvVud1VinvX%*%t(ytVinvXQ)
    ytPV2Py  <- ytVinvVud2Vinvy - 2*ytVinvVud2VinvX%*%t(ytVinvXQ) +
    ytVinvXQ%*%SumXtVinvVud2VinvX%*%t(ytVinvXQ)
    ytPV3Py  <- ytVinvVud3Vinvy - 2*ytVinvVud3VinvX%*%t(ytVinvXQ) +
    ytVinvXQ%*%SumXtVinvVud3VinvX%*%t(ytVinvXQ)
    # Scores
    S1 <- -0.5*tr.PV1 + 0.5*ytPV1Py
    S2 <- -0.5*tr.PV2 + 0.5*ytPV2Py
    S3 <- -0.5*tr.PV3 + 0.5*ytPV3Py
    # Fisher information matrix
    F11  <- 0.5*tr.PV1PV1
    F12  <- 0.5*tr.PV1PV2
    F13  <- 0.5*tr.PV1PV3
    F22  <- 0.5*tr.PV2PV2
    F23  <- 0.5*tr.PV2PV3
    F33  <- 0.5*tr.PV3PV3
    #
    if(ITER>1){
      F.sig.prev <- Fsig
      Q.prev <- Q
      theta.f.prev <- theta.f
    }
    Ssig <- rbind(S1, S2, S3)
    Fsig <- matrix(c(F11, F12, F13, F12, F22, F23, F13, F23, F33),
                   nrow=3, byrow=TRUE)
    #
    Fsig.inv <- solve(Fsig)
    dif <- Fsig.inv%*%Ssig
    theta.rml <- theta.f
    theta.f <- theta.rml + dif
    Vud.f <- UveU(theta.f)
    Vuds <- FirstDer(theta.f)
    # Stop because of warning
    if( theta.f[1]<=0 || theta.f[2]<=0 || abs(theta.f[3])>1){
      if(ITER==1){
        return(list(theta.rml, ITER, ITER))
        cat("WARNING: Stop at the first iteration. Out of parametric space")
      }
      return(list(theta.f.prev, F.sig.prev, ITER, Bad, Q.prev))
      cat("WARNING: Out of parametric space")
    }
    # Stop because of warning
    if(det(Vud.f)<=0){
      return(list(theta.f.prev, F.sig.prev, ITER, Bad, Q.prev))
      cat("WARNING: Singularity of Vud matrix")
    }
    # Stopping rule for iterative algorithm
    if(identical(as.numeric(abs(dif)<1e-06), rep(1,3))){
      break
    }
  }
  # End of Iteration loop (Fisher-scoring algorithm)
  return(list(theta.f, Fsig, ITER, Bad, Q))
}
```

By using the REML method with a Fisher-scoring algorithm, the R function BETA.U.BFH calculates the mode predictors $\hat{u}_d$ of the random effects and the estimators $\hat{\beta}$ of the regression parameters of a bivariate Fay–Herriot model.

```
BETA.U.BFH <- function(X, y, D, Ved, theta.hat) {
  Vd.inv <- Vd.hat <- Xd <- yd <- u <- list()
  p <- ncol(X[[1]]); Xd <- X; yd <- y
  Vudbeta <- UveU(theta.hat)
  Q.inv <- matrix(0, nrow=p, ncol=p)
  XVy <- rep(0,p)
  for(d in 1:D){
    Vd.hat[[d]] <- Vudbeta + Ved[[d]]
    Vd.inv[[d]] <- solve(Vd.hat[[d]])
    Q.inv <- Q.inv + t(Xd[[d]])%*%Vd.inv[[d]]%*%Xd[[d]]
    XVy <- XVy + t(Xd[[d]])%*%Vd.inv[[d]]%*%yd[[d]]
  }
  Q <- solve(Q.inv); betax <- Q%*%XVy
  for(d in 1:D){
    u[[d]] <- Vudbeta%*%Vd.inv[[d]]%*%(yd[[d]]-X[[d]]%*%betax)
  }
  return(list(betax, u))
}
```

The R function pvalue calculates the standard error, the z-value, and the p-value for testing if the regression parameters are equal to zero.

```
pvalue <- function(beta.fitted, fit) {
  if(!inherits(fit,"try-error")){
    stderror <- sqrt(diag(fit[[5]]))
    z <- abs(as.vector(beta.fitted))/stderror
    p <- pnorm(z, lower.tail=F)
    return(data.frame(stderror, z, "p-value"=2*p))
  }
  else{
    warning("Only a converging algorithm is allowed", call. = FALSE)
  }
}
```

The R function CI calculates asymptotic confidence intervals for the regression and the variance component parameters.

```
CI <- function(beta.fitted, fit, conf.level = 0.95) {
  if(!inherits(fit,"try-error")){
    alpha <- 1-conf.level; k <- 1-alpha/2; z <- qnorm(k)
    Finv <- solve(fit[[2]]); Q <- fit[[5]]
    lower.beta <- beta.fitted - z*sqrt(diag(Q))
    lower.theta <- fit[[1]] - z*sqrt(diag(Finv))
    upper.beta <- beta.fitted + z*sqrt(diag(Q))
    upper.theta <- fit[[1]] + z*sqrt(diag(Finv))
    return(list(data.frame(Inf.beta=lower.beta, Sup.beta=upper.beta),
                data.frame(Inf.theta=lower.theta, Sup.theta=upper.theta)))
  }
  else{
    warning("Only a converging algorithm is allowed", call. = FALSE)
  }
}
```

For each domain d and coordinate k, $d = 1, \ldots, D$, $k = 1, 2$, the R function MSE.BFH calculates the analytic estimator of the MSE of the BLUP. It calculates the summands G_1 and G_2.

```
MSE.BFH <- function(X, D, Ved, fit){
  p <- ncol(X[[1]])
  Vud <- UveU(fit[[1]])
  Vu <- bdiag(lapply(1:D, function(d) Vud ))
  Ve <- bdiag(Ved); V <- Vu+Ve; V.inv <- solve(V);  Z <- diag(2*D)
  # G1
  Te <- Vu - tcrossprod(Vu,Z)%*%V.inv%*%crossprod(Z,Vu)
```

```
    G1 <- Z%*%tcrossprod(Te,Z)
    # G2
    equis.matrix <- sapply(1:p,
                           function(j) sapply(1:D, function(i) X[[i]][,j]))
    Q <- fit[[5]]
    M <- equis.matrix - G1%*%crossprod(solve(Ve), equis.matrix)
    G2 <- M%*%tcrossprod(Q,M)
    # MSEs
    mse.k1 <- diag(G1+G2)[c(T,F)]
    mse.k2 <- diag(G1+G2)[c(F,T)]
    return(list(mse.k1, mse.k2))
}
```

References

Arima, S., Bell, W.R., Datta, G.S., Franco, C., Liseo, B.: Multivariate Fay-Herriot Bayesian estimation of small area means under functional measurement error. J. R. Stat. Soc. Ser. A **180**(4), 1191–1209 (2017)

Benavent, R., Morales, D.: Multivariate Fay-Herriot models for small area estimation. Comput. Stat. Data Anal. **94**, 372–390 (2016)

Benavent, R., Morales, D.: Small area estimation under a temporal bivariate area-level linear mixed model with independent time effects. Stat. Method. Appl. (2020). https://doi.org/10.1007/s10260-020-00521-x

Burgard, J.P., Esteban, M.D., Morales, D., Pérez, A.: Small area estimation under a measurement error bivariate Fay-Herriot model. Stat. Method. Appl. (2020) https://doi.org/10.1007/s10260-020-00515-9

Das, K., Jiang, J., Rao, J.N.K.: Mean squared error of empirical predictor. Ann. Stat. **32**, 818–840 (2004)

Datta, G.S., Fay, R.E., Ghosh, M.: Hierarchical and empirical Bayes multivariate analysis in small area estimation. In: Proceedings of Bureau of the Census 1991 Annual Research Conference, pp. 63–79. U. S. Bureau of the Census, Washington (1991)

Datta, G.S., Ghosh, M., Nangia, N., Natarajan, K.: Estimation of median income of four-person families: a Bayesian approach. In: Berry, D.A., Chaloner, K.M., Geweke, J.M. (eds.) Bayesian Analysis in Statistics and Econometrics, pp. 129–140. Wiley, New York (1996)

Datta, G.S., Lahiri, P.: A unified measure of uncertainty of estimated best linear unbiased predictors in small area estimation problems. Stat. Sin. **10**, 613–627 (2000)

Esteban, M.D., Lombardía, M.J., López-Vizcaíno, E., Morales, D., Pérez, A.: Small area estimation of proportions under area-level compositional mixed models. TEST (2020). https://doi.org/10.1007/s11749-019-00688-w

Fay, R. E.: Application of multivariate regression of small domain estimation. In: Platek, R., Rao, J. N.K., Särndal, C.E., Singh, M.P. (eds.) Small Area Statistics, pp. 91–102. Wiley, New York (1987)

Fay, R.E., Herriot, R.A.: Estimates of income for small places: an application of James-Stein procedures to census data. J. Am. Stat. Assoc. **74**, 269–277 (1979)

González-Manteiga, W., Lombardía, M.J., Molina, I., Morales, D., Santamaría, L.: Analytic and bootstrap approximations of prediction errors under a multivariate Fay-Herriot model. Comput. Stat. Data Anal. **52**, 5242–5252 (2008b)

Jiang, J.: Linear and Generalized Linear Mixed Models and Their Applications. Springer, Berlin (2007)

Kackar, R.N., Harville, D.A.: Approximations for standard errors of estimators of fixed and random effects in mixed linear models. J. Am. Stat. Assoc. **79**, 853–862 (1984)

Prasad, N.G.N., Rao, J.N.K.: The estimation of the mean squared error of small-area estimators. J. Am. Stat. Assoc. **85**, 163–171 (1990)
Ubaidillah, A., Notodiputro, K.A., Kurnia, A., Mangku, I.W.: Multivariate Fay-Herriot models for small area estimation with application to household consumption per capita expenditure in Indonesia. J. Appl. Stat. **46**(15), 2845–2861 (2019)

Chapter 20
Area-Level Poisson Mixed Models

20.1 Introduction

Poisson regression models are generalized linear models (GLMs) that are used for counts, i.e. for target variables counting some event of interest. In these models, the hypothesis of linearity is not assumed. Instead, the GLMs assume that a link function of the mean of the target variable is linearly related to some covariates. The GLMs also relax the hypothesis of normality and assume that the distribution belongs to the exponential family.

Sometimes, the GLMs cannot explain the variability of the target variable through the selected auxiliary variables. It may happen that observations from different domains are independent, but observations within the same domain are dependent because they share common properties. The generalized linear mixed models (GLMMs) are extensions of the GLMs that capture the variability between domains by introducing random effects. The random effects are usually assumed to be normally distributed. Despite the usefulness of the GLMMs, inferences based on these models have some computational difficulties because the likelihood may involve high-dimensional integrals that cannot be evaluated analytically. Several methods have been proposed to overcome this problem, most of them relying on the Taylor linearization and/or the Laplace's method for integral approximations. See the review of Jiang and Lahiri (2006). EM-type algorithms assisted by Monte Carlo methods are also applied. The penalized quasi-likelihood (PQL) algorithm is used in combination with a Gaussian approximation of the marginal density that provides approximate maximum likelihood estimators of variance components. Nevertheless, the PQL method may lead to inconsistent and biased estimators. This is why Jiang (1998) proposed the method of simulated moments (MSM) for fitting GLMMs. This method is computationally attractive and gives consistent estimators of model parameters. See Breslow and Clayton (1993), Lin (2007), and MacNab and Lin (2009) for more details of fitting algorithms.

D. Morales et al., *A Course on Small Area Estimation and Mixed Models*, Statistics for Social and Behavioral Sciences, https://doi.org/10.1007/978-3-030-63757-6_20

This chapter describes the area-level Poisson mixed model studied by Boubeta et al. (2016) in the framework of small area estimation. It also presents several fitting methods, namely method of moments (MM), EM algorithm, ML-Laplace approximation algorithm, and PQL algorithm, derives empirical best and plug-in predictors of domain counts, and gives some R codes.

20.2 The Model

Let us consider a set of area random effects such that $\{v_d : d = 1, \ldots, D\}$ are i.i.d. $N(0, 1)$. In vector notation, we have $\boldsymbol{v} = \underset{1\leq d\leq D}{\text{col}}(v_d) \sim N_D(\boldsymbol{0}, \boldsymbol{I}_D)$ with probability density function

$$f_v(\boldsymbol{v}) = (2\pi)^{-D/2} \exp\Big\{-\frac{1}{2}\,\boldsymbol{v}'\boldsymbol{v}\Big\}.$$

The distribution of the target variable y_d, conditioned to the random effect v_d, is

$$y_d|_{v_d} \sim \text{Poiss}(m_d p_d), \quad d = 1, \ldots, D,$$

where the offset (or size) parameters $m_d > 0$ are known. For the natural parameter, we assume

$$\eta_d = \log \mu_d = \log m_d + \log p_d = \log m_d + \boldsymbol{x}_d\boldsymbol{\beta} + \phi v_d, \quad d = 1, \ldots, D,$$

where $\boldsymbol{\beta} = \underset{1\leq k\leq p}{\text{col}}(\beta_k)$ and $\boldsymbol{x}_d = \underset{1\leq k\leq p}{\text{col}'}(x_{dk})$. Further, we assume that the y_d's are independent conditioned to $\boldsymbol{v}$, and we use the notation $\boldsymbol{y} = \underset{1\leq d\leq D}{\text{col}}(y_d)$. It holds that

$$P(y_d|\boldsymbol{v}) = P(y_d|v_d) = \frac{1}{y_d!}\exp\{-m_d p_d\} m_d^{y_d} p_d^{y_d}, \quad p_d = \exp\{\boldsymbol{x}_d\boldsymbol{\beta} + \phi v_d\},$$

$$P(\boldsymbol{y}|\boldsymbol{v}) = \prod_{d=1}^{D} P(y_d|v_d), \quad P(\boldsymbol{y}) = \int_{R^D} P(\boldsymbol{y}|\boldsymbol{v}) f_v(\boldsymbol{v})\, d\boldsymbol{v} = \int_{R^D} \psi(\boldsymbol{y}, \boldsymbol{v})\, d\boldsymbol{v},$$

where

$$\psi(\boldsymbol{y}, \boldsymbol{v}) = (2\pi)^{-\frac{D}{2}} \exp\left\{\frac{-\boldsymbol{v}'\boldsymbol{v}}{2}\right\} \prod_{d=1}^{D} \frac{\exp\{-m_d p_d\} m_d^{y_d} \exp\{y_d(\boldsymbol{x}_d\boldsymbol{\beta} + \phi v_d)\}}{y_d!}$$
$$= (2\pi)^{-\frac{D}{2}} \exp\left\{\frac{-\boldsymbol{v}'\boldsymbol{v}}{2}\right\} \left(\prod_{d=1}^{D} y_d!\right)^{-1}$$

$$\cdot \exp\Big\{\sum_{d=1}^{D}\{-m_d\exp\{\boldsymbol{x}_d\boldsymbol{\beta}+\phi v_d\}+y_d\log m_d\}\Big\}$$

$$\cdot \exp\Big\{\sum_{k=1}^{p}\Big(\sum_{d=1}^{D}y_dx_{dk}\Big)\beta_k+\phi\sum_{d=1}^{D}y_dv_d\Big\}.$$

20.3 MM Algorithm

This section derives an algorithm to fit the area-level Poisson mixed model by using the method of moments (MM). Boubeta et al. (2016) considered the following set of equations for estimating the unknown parameter $\boldsymbol{\theta}=(\boldsymbol{\beta}',\phi)'$ by the method of moments:

$$0=f_k(\boldsymbol{\theta})=M_k(\boldsymbol{\theta})-\hat{M}_k=\sum_{d=1}^{D}E_\theta[y_d]x_{dk}-\sum_{d=1}^{D}y_dx_{dk},\quad k=1,\dots p,$$

$$0=f_{p+1}(\boldsymbol{\theta})=M_{p+1}(\boldsymbol{\theta})-\hat{M}_{p+1}=\sum_{d=1}^{D}E_\theta[y_d^2]-\sum_{d=1}^{D}y_d^2.$$

The updating formula of the Newton–Raphson algorithm to solve this system of nonlinear equations is

$$\boldsymbol{\theta}^{(i+1)}=\boldsymbol{\theta}^{(i)}-\boldsymbol{H}^{-1}(\boldsymbol{\theta}^{(i)})\boldsymbol{f}(\boldsymbol{\theta}^{(i)}),$$

where $\theta_1=\beta_1,\dots,\theta_p=\beta_p,\theta_{p+1}=\phi$ and

$$\boldsymbol{\theta}=\underset{1\le k\le p+1}{\text{col}}(\theta_k),\quad \boldsymbol{f}(\boldsymbol{\theta})=\underset{1\le k\le p+1}{\text{col}}(f_k(\boldsymbol{\theta})),\quad \boldsymbol{H}(\boldsymbol{\theta})=\left(\frac{\partial f_k(\boldsymbol{\theta})}{\partial\theta_r}\right)_{k,r=1,\dots,p+1}.$$

Let us now calculate the expectations appearing in $\boldsymbol{f}(\boldsymbol{\theta})$ and its partial derivatives. We start with the first p MM equations. Note that

$$-\frac{1}{2}(v_d-\phi)^2=-\frac{1}{2}(v_d^2-2\phi v_d+\phi^2)=-\frac{1}{2}v_d^2+\phi v_d-\frac{1}{2}\phi^2.$$

The expectation of y_d is

$$E_\theta[y_d]=E_v\big[E_\theta[y_d|\boldsymbol{v}]\big]=E_v[m_dp_d]=E_v\left[m_d\exp\{\boldsymbol{x}_d\boldsymbol{\beta}+\phi v_d\}\right]$$
$$=\int_{-\infty}^{\infty}m_d\exp\{\boldsymbol{x}_d\boldsymbol{\beta}+\phi v_d\}\,(2\pi)^{-1/2}\exp\Big\{-\frac{1}{2}v_d^2\Big\}\,dv_d$$

$$= m_d \exp\big\{\boldsymbol{x}_d\boldsymbol{\beta} + \frac{1}{2}\phi^2\big\} \int_{-\infty}^{\infty} (2\pi)^{-1/2} \exp\big\{-\frac{1}{2}(v_d - \phi)^2\big\}\, dv_d$$

$$= m_d \exp\big\{\boldsymbol{x}_d\boldsymbol{\beta} + \frac{1}{2}\phi^2\big\}.$$

Therefore,

$$f_k(\boldsymbol{\theta}) = \sum_{d=1}^{D} m_d \exp\big\{\boldsymbol{x}_d\boldsymbol{\beta} + \frac{1}{2}\phi^2\big\} x_{dk} - \sum_{d=1}^{D} y_d x_{dk}, \quad k = 1, \ldots p.$$

The derivatives of $E_\theta[y_d]$ are

$$\frac{\partial E_\theta[y_d]}{\partial \beta_k} = m_d \exp\big\{\boldsymbol{x}_d\boldsymbol{\beta} + \frac{1}{2}\phi^2\big\} x_{dk}, \quad \frac{\partial E_\theta[y_d]}{\partial \phi} = m_d \exp\big\{\boldsymbol{x}_d\boldsymbol{\beta} + \frac{1}{2}\phi^2\big\}\phi.$$

The expectation of y_d^2 is $E_\theta[y_d^2] = E_v\big[E_\theta[y_d^2|\boldsymbol{v}]\big]$, where

$$E_\theta[y_d^2|\boldsymbol{v}] = \mathrm{var}_\theta[y_d|\boldsymbol{v}] + E_\theta^2[y_d|\boldsymbol{v}] = m_d p_d + m_d^2 p_d^2.$$

Therefore,

$$E_\theta[y_d^2] = E_v\big[E_\theta[y_d^2|\boldsymbol{v}]\big] = \int_{-\infty}^{\infty} m_d p_d f_v(v_d)\, dv_d + \int_{-\infty}^{\infty} m_d^2 p_d^2 f_v(v_d)\, dv_d.$$

Note that

$$-\frac{1}{2}(v_d - 2\phi)^2 = -\frac{1}{2}(v_d^2 - 4\phi v_d + 4\phi^2) = -\frac{1}{2}v_d^2 + 2\phi v_d - 2\phi^2.$$

We have

$$\int_{-\infty}^{\infty} p_d^2 f_v(v_d)\, dv_d = \int_{-\infty}^{\infty} \exp\big\{2\boldsymbol{x}_d\boldsymbol{\beta} + 2\phi v_d\big\}(2\pi)^{-1/2} \exp\big\{-\frac{1}{2}v_d^2\big\}\, dv_d$$

$$= \exp\big\{2\boldsymbol{x}_d\boldsymbol{\beta} + 2\phi^2\big\} \int_{-\infty}^{\infty} (2\pi)^{-1/2} \exp\big\{-\frac{1}{2}(v_d - 2\phi)^2\big\}\, dv_d$$

$$= \exp\big\{2\boldsymbol{x}_d\boldsymbol{\beta} + 2\phi^2\big\}$$

and

$$E_\theta[y_d^2] = m_d \exp\big\{\boldsymbol{x}_d\boldsymbol{\beta} + \frac{1}{2}\phi^2\big\} + m_d^2 \exp\big\{2\boldsymbol{x}_d\boldsymbol{\beta} + 2\phi^2\big\}.$$

Therefore,

$$f_{p+1}(\boldsymbol{\theta}) = \sum_{d=1}^{D} \left\{ m_d \exp\left\{ \boldsymbol{x}_d \boldsymbol{\beta} + \frac{1}{2}\phi^2 \right\} + m_d^2 \exp\left\{ 2\boldsymbol{x}_d \boldsymbol{\beta} + 2\phi^2 \right\} \right\} - \sum_{d=1}^{D} y_d^2.$$

The derivatives of $E_\theta[y_d^2]$ are

$$\frac{\partial E_\theta[y_d^2]}{\partial \beta_k} = m_d \exp\left\{ \boldsymbol{x}_d \boldsymbol{\beta} + \frac{1}{2}\phi^2 \right\} x_{dk} + 2m_d^2 \exp\left\{ 2\boldsymbol{x}_d \boldsymbol{\beta} + 2\phi^2 \right\} x_{dk},$$

$$\frac{\partial E_\theta[y_d^2]}{\partial \phi} = m_d \exp\left\{ \boldsymbol{x}_d \boldsymbol{\beta} + \frac{1}{2}\phi^2 \right\} \phi + 4m_d^2 \exp\left\{ 2\boldsymbol{x}_d \boldsymbol{\beta} + 2\phi^2 \right\} \phi.$$

The elements of the Jacobian matrix are

$$H_{kr} = \frac{\partial f_k(\boldsymbol{\theta})}{\partial \theta_r} = \sum_{d=1}^{D} \frac{\partial E_\theta[y_d]}{\partial \theta_r} x_{dk}, \quad k = 1, \ldots, p,\ r = 1, \ldots, p+1,$$

$$H_{p+1r} = \frac{\partial f_{p+1}(\boldsymbol{\theta})}{\partial \theta_r} = \sum_{d=1}^{D} \frac{\partial E_\theta[y_d^2]}{\partial \theta_r}, \quad r = 1, \ldots, p+1.$$

Let us note that since the expectations and its derivatives can be expressed in an explicit form under the assumed model, it is not necessary to employ the method of simulated moments, and the classic method of moments can be used.

The MM algorithm can be summarized as follows:

1. Set the initial values $i = 0$ and $\boldsymbol{\theta}^{(0)} = (\boldsymbol{\beta}^{(0)}, \phi^{(0)})$.
2. Repeat the following steps until convergence:
   a. Update $\boldsymbol{\theta}^{(i)}$ by using the equation
   $$\boldsymbol{\theta}^{(i+1)} = \boldsymbol{\theta}^{(i)} - \boldsymbol{H}^{-1}(\boldsymbol{\theta}^{(i)}) \boldsymbol{f}(\boldsymbol{\theta}^{(i)}).$$
   b. Update the iteration index $i \leftarrow i + 1$.
3. Output: $\boldsymbol{\theta}^{(i)}$.

As algorithm seed for $\boldsymbol{\beta}$, we propose to use $\boldsymbol{\beta}^{(0)} = \tilde{\boldsymbol{\beta}}$, where $\tilde{\boldsymbol{\beta}}$ is the maximum likelihood estimator under the model with no random effects. In that model, the linear parameters are

$$\eta_d = \boldsymbol{x}_d \boldsymbol{\beta}, \quad d = 1, \ldots, D.$$

Concerning the variance parameters, one possible choice is

$$\phi^{(0)} = \Big(\frac{1}{D}\sum_{d=1}^{D}(\tilde{\eta}_d - \hat{\eta}_d^{dir})^2\Big)^{1/2}, \quad \tilde{\eta}_d = \boldsymbol{x}_d\tilde{\boldsymbol{\beta}}, \quad \hat{\eta}_d^{dir} = \log \hat{p}_d^{dir}, \quad \hat{p}_d^{dir} = \frac{y_d + 1}{m_d + 1}.$$

The variance–covariance matrix of the MM estimators can be estimated by parametric bootstrap, i.e.

1. Fit the model to the sample and calculate $\hat{\boldsymbol{\theta}}$.
2. Generate bootstrap samples $\{y_d^{(b)} : d = 1, \ldots, D\}$, $b = 1, \ldots, B$, from the fitted model.
3. Fit the model to the bootstrap samples and calculate $\hat{\boldsymbol{\theta}}^{(b)}$, $b = 1, \ldots, B$, $\overline{\boldsymbol{\theta}} = \frac{1}{B}\sum_{b=1}^{B}\hat{\boldsymbol{\theta}}^{(b)}$.
4. Output: $\widehat{\text{var}}_B(\hat{\boldsymbol{\theta}}) = \frac{1}{B}\sum_{b=1}^{B}(\hat{\boldsymbol{\theta}}^{(b)} - \overline{\boldsymbol{\theta}})(\hat{\boldsymbol{\theta}}^{(b)} - \overline{\boldsymbol{\theta}})'$.

20.4 EM Algorithm

We remind that the EM algorithm starts in an initial value $\boldsymbol{\theta}^{(0)}$ and, at stage $i + 1$, it makes two steps (cf. Sect. 14.4).
Expectation step: Calculate the expectation

$$M(\boldsymbol{\theta}|\boldsymbol{\theta}^{(i)}) = \int_{R^D} \log f_\theta(\boldsymbol{y}, \boldsymbol{v})\, f_{\theta^{(i)}}(\boldsymbol{v}|\boldsymbol{y})\, d\boldsymbol{v}\,.$$

Maximization step: Obtain $\boldsymbol{\theta}^{(i+1)}$ by maximizing $M(\boldsymbol{\theta}|\boldsymbol{\theta}^{(i)})$, i.e.

$$\boldsymbol{\theta}^{(i+1)} = \text{argmax}_{\theta\in\Theta}\, M(\boldsymbol{\theta}|\boldsymbol{\theta}^{(i)}).$$

For the area-level Poisson mixed model, we have

$$f_{\theta^{(i)}}(\boldsymbol{v}|\boldsymbol{y}) = \frac{P_{\theta^{(i)}}(\boldsymbol{y}|\boldsymbol{v})f(\boldsymbol{v})}{P_{\theta^{(i)}}(\boldsymbol{y})} = \frac{\prod_{d=1}^{D} P_{\theta^{(i)}}(y_d|v_d)f(v_d)}{\prod_{d=1}^{D}\int_R P_{\theta^{(i)}}(y_d|v_d)f(v_d)\,dv_d},$$

$$\begin{aligned}
M(\boldsymbol{\theta}|\boldsymbol{\theta}^{(i)}) &= \int_{R^D} (\log P_\theta(\boldsymbol{y}|\boldsymbol{v}) + \log f(\boldsymbol{v}))\, f_{\theta^{(i)}}(\boldsymbol{v}|\boldsymbol{y})\, d\boldsymbol{v} \\
&= \int_{R^D} \sum_{d=1}^{D}\big\{\log P_\theta(y_d|v_d) + \log f(v_d)\big\}\, f_{\theta^{(i)}}(\boldsymbol{v}|\boldsymbol{y})\, d\boldsymbol{v} \\
&= \sum_{d=1}^{D}\int_{R^D}\big\{\log P_\theta(y_d|v_d) + \log f(v_d)\big\}\, \frac{\prod_{\ell=1}^{D} P_{\theta^{(i)}}(y_\ell|v_\ell)f(v_\ell)}{\prod_{\ell=1}^{D}\int_R P_{\theta^{(i)}}(y_\ell|v_\ell)f(v_\ell)\,dv_\ell}
\end{aligned}$$

$$
\cdot\, dv_1 \cdots dv_D
$$

$$
= \sum_{d=1}^{D} \left\{ \int_R \{ \log P_\theta(y_d|v_d) + \log f(v_d) \} \frac{P_{\theta^{(i)}}(y_d|v_d) f(v_d)}{\int_R P_{\theta^{(i)}}(y_d|v_d) f(v_d)\, dv_d} dv_d \right.
$$

$$
\left. \cdot \prod_{\ell \neq d} \int_R \frac{P_{\theta^{(i)}}(y_\ell|v_\ell) f(v_\ell)}{\int_R P_{\theta^{(i)}}(y_\ell|v_\ell) f(v_\ell)\, dv_\ell} dv_\ell \right\}
$$

$$
= \sum_{d=1}^{D} \frac{\int_R \{ \log P_\theta(y_d|v_d) + \log f(v_d) \} P_{\theta^{(i)}}(y_d|v_d) f(v_d)\, dv_d}{\int_R P_{\theta^{(i)}}(y_d|v_d) f(v_d)\, dv_d}
$$

$$
= \sum_{d=1}^{D} \frac{N_d^{(i)}}{D_d^{(i)}},
$$

where

$$
N_d^{(i)} = \int_R \Big(- m_d \exp\{\boldsymbol{x}_d \boldsymbol{\beta} + \phi v_d\} + y_d(\boldsymbol{x}_d \boldsymbol{\beta} + \phi v_d) + \log m_d^{y_d} - \log y_d! - \frac{1}{2}(v_d^2 + \log 2\pi) \Big)
$$

$$
\cdot \exp\left\{ -m_d \exp\{\boldsymbol{x}_d \boldsymbol{\beta}^{(i)} + \phi^{(i)} v_d\} + y_d(\boldsymbol{x}_d \boldsymbol{\beta}^{(i)} + \phi^{(i)} v_d) + y_d \log m_d - \log y_d! \right\}
$$

$$
\cdot (2\pi)^{-1/2} \exp\left\{ -\frac{1}{2} v_d^2 \right\} dv_d ,
$$

$$
D_d^{(i)} = \int_R \exp\left\{ -m_d \exp\{\boldsymbol{x}_d \boldsymbol{\beta}^{(i)} + \phi^{(i)} v_d\} + y_d(\boldsymbol{x}_d \boldsymbol{\beta}^{(i)} + \phi^{(i)} v_d) + y_d \log m_d - \log y_d! \right\}
$$

$$
\cdot (2\pi)^{-1/2} \exp\left\{ -\frac{1}{2} v_d^2 \right\} dv_d .
$$

The terms $N_d^{(i)}$ and $D_d^{(i)}$ can be approximated by the following Monte Carlo procedure:

1. For $s = 1, \ldots, S$, generate $v_d^{(s)}$ i.i.d. $N(0, 1)$ and $v_d^{(S+s)} = -v_d^{(s)}$.
2. Calculate

$$
\hat{N}_d^{(i)} = \frac{1}{2S} \sum_{s=1}^{2S} \left\{ \Big(- m_d \exp\{\boldsymbol{x}_d \boldsymbol{\beta} + \phi v_d^{(s)}\} + y_d(\boldsymbol{x}_d \boldsymbol{\beta} + \phi v_d^{(s)}) + \log m_d^{y_d} \right.
$$

$$
- \log y_d! - \frac{1}{2}(v_d^{(s)2} + \log 2\pi) \Big) \exp\left\{ - m_d \exp\{\boldsymbol{x}_d \boldsymbol{\beta}^{(i)} + \phi^{(i)} v_d^{(s)}\} \right.
$$

$$
\left. \left. + y_d(\boldsymbol{x}_d \boldsymbol{\beta}^{(i)} + \phi^{(i)} v_d^{(s)}) + y_d \log m_d - \log y_d! \right\} \right\},
$$

$$
\hat{D}_d^{(i)} = \frac{1}{2S} \sum_{s=1}^{2S} \exp\left\{ - m_d \exp\{\boldsymbol{x}_d \boldsymbol{\beta}^{(i)} + \phi^{(i)} v_d^{(s)}\} + y_d(\boldsymbol{x}_d \boldsymbol{\beta}^{(i)} + \phi^{(i)} v_d^{(s)}) \right.
$$

$$
\left. + y_d \log m_d - \log y_d! \right\}.
$$

The Monte Carlo approximation of $M(\boldsymbol{\theta}|\boldsymbol{\theta}^{(i)})$ is then

$$\hat{M}(\boldsymbol{\theta}|\boldsymbol{\theta}^{(i)}) = \sum_{d=1}^{D} \frac{\hat{N}_d^{(i)}}{\hat{D}_d^{(i)}}.$$

By defining the Monte Carlo weights and functions

$$w_d^{(s)}(\boldsymbol{\theta}^{(i)}) = \frac{\exp\left\{-m_d \exp\{\boldsymbol{x}_d\boldsymbol{\beta}^{(i)} + \phi^{(i)} v_d^{(s)}\} + y_d(\boldsymbol{x}_d\boldsymbol{\beta}^{(i)} + \phi^{(i)} v_d^{(s)})\right\}}{\sum_{s=1}^{2S} \exp\left\{-m_d \exp\{\boldsymbol{x}_d\boldsymbol{\beta}^{(i)} + \phi^{(i)} v_d^{(s)}\} + y_d(\boldsymbol{x}_d\boldsymbol{\beta}^{(i)} + \phi^{(i)} v_d^{(s)})\right\}},$$

$$m_d^{(s)}(\boldsymbol{\theta}) = -m_d \exp\{\boldsymbol{x}_d\boldsymbol{\beta} + \phi v_d^{(s)}\} + y_d(\boldsymbol{x}_d\boldsymbol{\beta} + \phi v_d^{(s)}) + \log m_d^{y_d} - \log y_d! \\ - \frac{1}{2}\log 2\pi - \frac{1}{2} v_d^{(s)2},$$

we can write the approximation in a different form, which is more suitable for the subsequent maximization. The alternative algorithm for approximating $M(\boldsymbol{\theta}|\boldsymbol{\theta}^{(i)})$ is

1. For $s = 1, \ldots, S$, generate $v_d^{(s)}$ i.i.d. $N(0, 1)$ and $v_d^{(S+s)} = -v_d^{(s)}$.
2. Calculate

$$\hat{M}(\boldsymbol{\theta}|\boldsymbol{\theta}^{(i)}) = \sum_{d=1}^{D}\sum_{s=1}^{2S} w_d^{(s)}(\boldsymbol{\theta}^{(i)})\, m_d^{(s)}(\boldsymbol{\theta}).$$

In each step i, the EM algorithm looks for a value of $\boldsymbol{\theta}$ improving $\hat{M}(\boldsymbol{\theta}|\boldsymbol{\theta}^{(i)})$.

The first derivatives of $m_d^{(s)}(\boldsymbol{\theta})$ are

$$u_{d,k}^{(s)}(\boldsymbol{\theta}) = \frac{\partial m_d^{(s)}(\boldsymbol{\theta})}{\partial \beta_k} = [y_d - m_d p_d^{(s)}(\boldsymbol{\theta})]x_{dk}, \quad k = 1, \ldots, p,$$

$$u_{d,p+1}^{(s)}(\boldsymbol{\theta}) = \frac{\partial m_d^{(s)}(\boldsymbol{\theta})}{\partial \phi} = [y_d - m_d p_d^{(s)}(\boldsymbol{\theta})]v_d^{(s)},$$

where $p_d^{(s)}(\boldsymbol{\theta}) = \exp\{\boldsymbol{x}_d\boldsymbol{\beta} + \phi v_d^{(s)}\}$. The second derivatives of $m_d^{(s)}(\boldsymbol{\theta})$ are

$$h_{d,k_1k_2}^{(s)}(\boldsymbol{\theta}) = \frac{\partial^2 m_d^{(s)}(\boldsymbol{\theta})}{\partial \beta_{k_1}\partial \beta_{k_2}} = -m_d p_d^{(s)}(\boldsymbol{\theta}) x_{dk_1} x_{dk_2}, \quad k_1, k_2 = 1, \ldots, p,$$

$$h_{d,p+1k}^{(s)}(\boldsymbol{\theta}) = \frac{\partial^2 m_d^{(s)}(\boldsymbol{\theta})}{\partial \phi \partial \beta_k} = -m_d p_d^{(s)}(\boldsymbol{\theta}) x_{dk} v_d^{(s)}, \quad k = 1, \ldots, p,$$

$$h_{d,p+1p+1}^{(s)}(\boldsymbol{\theta}) = \frac{\partial^2 m_d^{(s)}(\boldsymbol{\theta})}{\partial \phi^2} = -m_d p_d^{(s)}(\boldsymbol{\theta}) v_d^{(s)2}.$$

The derivatives of $\hat{M}(\boldsymbol{\theta}|\boldsymbol{\theta}^{(i)})$ are

$$u_k(\boldsymbol{\theta}|\boldsymbol{\theta}^{(i)}) = \frac{\partial \hat{M}(\boldsymbol{\theta}|\boldsymbol{\theta}^{(i)})}{\partial \theta_k} = \sum_{d=1}^{D}\sum_{s=1}^{2S} w_d^{(s)}(\boldsymbol{\theta}^{(i)}) u_{d,k}^{(s)}(\boldsymbol{\theta}),\ k = 1, \ldots, p+1,$$

$$h_{k_1k_2}(\boldsymbol{\theta}|\boldsymbol{\theta}^{(i)}) = \frac{\partial^2 \hat{M}(\boldsymbol{\theta}|\boldsymbol{\theta}^{(i)})}{\partial \theta_{k_1}\partial \theta_{k_2}} = \sum_{d=1}^{D}\sum_{s=1}^{2S} w_d^{(s)}(\boldsymbol{\theta}^{(i)}) h_{d,k_1k_2}^{(s)}(\boldsymbol{\theta}),\ k_1, k_2 = 1, \ldots, p+1.$$

We define the vector of scores and the Hessian matrix in the form

$$\boldsymbol{u}(\boldsymbol{\theta}|\boldsymbol{\theta}^{(i)}) = \underset{1\le k\le p+1}{\text{col}}(u_k(\boldsymbol{\theta}|\boldsymbol{\theta}^{(i)})), \quad \boldsymbol{H}(\boldsymbol{\theta}|\boldsymbol{\theta}^{(i)}) = \big(h_{k_1k_2}(\boldsymbol{\theta}|\boldsymbol{\theta}^{(i)})\big)_{k_1,k_2=1,\ldots,p+1}.$$

The EM algorithm can be summarized as follows:

1. For $s = 1, \ldots, S$, generate $v_d^{(s)}$ i.i.d. $N(0, 1)$ and $v_d^{(S+s)} = -v_d^{(s)}$.
2. Set the initial values $i = 0$ and $\boldsymbol{\theta}^{(0)} = (\boldsymbol{\beta}^{(0)}, \phi^{(0)})$.
3. Repeat the following steps until convergence:
   a. E step: Calculate $\boldsymbol{u}(\boldsymbol{\theta}|\boldsymbol{\theta}^{(i)})$ and $\boldsymbol{H}(\boldsymbol{\theta}|\boldsymbol{\theta}^{(i)})$ in $\boldsymbol{\theta} = \boldsymbol{\theta}^{(i)}$.
   b. M step: Update $\boldsymbol{\theta}^{(i)}$ by using the equation
   $$\boldsymbol{\theta}^{(i+1)} = \boldsymbol{\theta}^{(i)} - \boldsymbol{H}^{-1}(\boldsymbol{\theta}^{(i)}|\boldsymbol{\theta}^{(i)})\boldsymbol{u}(\boldsymbol{\theta}^{(i)}|\boldsymbol{\theta}^{(i)}).$$
   c. Update the iteration index $i \leftarrow i + 1$.
4. Output: $\boldsymbol{\theta}^{(i)}$.

The step 3.b can be substituted by the more sophisticated following steps:

1. Set the initial values $r = i$ and $\boldsymbol{\theta}^{(r)} = (\boldsymbol{\beta}^{(i)}, \phi^{(i)})$.
2. Repeat the following steps until convergence:
   a. Update $\boldsymbol{\theta}^{(r)}$ by using the equation
   $$\boldsymbol{\theta}^{(r+1)} = \boldsymbol{\theta}^{(r)} - \boldsymbol{H}^{-1}(\boldsymbol{\theta}^{(r)}|\boldsymbol{\theta}^{(i)})\boldsymbol{u}(\boldsymbol{\theta}^{(r)}|\boldsymbol{\theta}^{(i)}).$$
   b. Update the iteration index $r \leftarrow r + 1$.
3. Output: $\boldsymbol{\theta}^{(i+1)} \leftarrow \boldsymbol{\theta}^{(r)}$.

20.5 ML-Laplace Approximation Algorithm

Let $h : R \mapsto R$ be a twice continuously differentiable function with a global maximum at x_0. This is to say, let us assume that $\dot{h}(x_0) = 0$ and $\ddot{h}(x_0) < 0$. The univariate Laplace approximation of the integral $\int_R \exp(h(x))dx$ is (cf. (14.6))

$$\int_{-\infty}^{\infty} e^{h(x)}\, dx \approx (2\pi)^{1/2}\big(-\ddot{h}(x_0)\big)^{-1/2} e^{h(x_0)}. \tag{20.1}$$

Let us now approximate the log-likelihood of the area-level Poisson mixed model. We recall that $v_1, \ldots, v_D$ are i.i.d. $N(0,1)$ and that

$$y_d|v_d \underset{ind}{\sim} \text{Poiss}(m_d p_d), \quad p_d = \exp\{\boldsymbol{x}_d\boldsymbol{\beta} + \phi v_d\}, \quad d = 1, \ldots, D.$$

It holds that $y_1, \ldots, y_D$ are unconditionally independent with marginal p.d.f.

$$\begin{aligned}
P(y_d) &= \int_{-\infty}^{\infty} P(y_d|v_d) f(v_d)\, dv_d \\
&= \int_{-\infty}^{\infty} \frac{1}{y_d!} \exp\{-m_d p_d\} m_d^{y_d} p_d^{y_d} (2\pi)^{-1/2} \exp\{-\frac{1}{2}v_d^2\}\, dv_d = \frac{m_d^{y_d}}{(2\pi)^{1/2} y_d!} \\
&\cdot \int_{-\infty}^{\infty} \exp\big\{-m_d \exp\{\boldsymbol{x}_d\boldsymbol{\beta} + \phi v_d\} + y_d(\boldsymbol{x}_d\boldsymbol{\beta} + \phi v_d) - \frac{1}{2}v_d^2\big\}\, dv_d \\
&= \frac{m_d^{y_d}}{(2\pi)^{1/2} y_d!} \exp\{y_d \boldsymbol{x}_d\boldsymbol{\beta}\} \qquad (20.2) \\
&\cdot \int_{-\infty}^{\infty} \exp\big\{-m_d \exp\{\boldsymbol{x}_d\boldsymbol{\beta} + \phi v_d\} + \phi y_d v_d - \frac{1}{2}v_d^2\big\}\, dv_d \\
&= \frac{m_d^{y_d}}{(2\pi)^{1/2} y_d!} \exp\{y_d \boldsymbol{x}_d\boldsymbol{\beta}\} \int_{-\infty}^{\infty} \exp\big\{h(v_d)\big\}\, dv_d, \qquad (20.3)
\end{aligned}$$

where

$$h(v_d) = -m_d \exp\{\boldsymbol{x}_d\boldsymbol{\beta} + \phi v_d\} + \phi y_d v_d - \frac{1}{2}v_d^2. \tag{20.4}$$

In order to apply the Laplace approximation to the integral in (20.3), we need the first and second derivatives of the function h. These are

$$\begin{aligned}
\dot{h}(v_d) &= -m_d \phi \exp\{\boldsymbol{x}_d\boldsymbol{\beta} + \phi v_d\} + \phi y_d - v_d, \\
\ddot{h}(v_d) &= -(1 + m_d \phi^2 \exp\{\boldsymbol{x}_d\boldsymbol{\beta} + \phi v_d\}).
\end{aligned}$$

For $\boldsymbol{\theta} = (\boldsymbol{\beta}', \phi)' = \boldsymbol{\theta}_0$ fixed, the function $h(v_d) = h(v_d, \boldsymbol{\theta})$, defined in (20.4), can be maximized by the Newton–Raphson algorithm. The updating equation is

$$v_d^{(i+1)} = v_d^{(i)} - \frac{\dot{h}(v_d^{(i)}, \boldsymbol{\theta}_0)}{\ddot{h}(v_d^{(i)}, \boldsymbol{\theta}_0)}. \tag{20.5}$$

Let us denote by v_{0d} the argument of maxima of the function $h(v_d)$. Thus, it holds $\dot{h}(v_{0d}) = 0$, $\ddot{h}(v_{0d}) < 0$, and by applying the approximation (20.1) in $v_d = v_{0d}$ to (20.3), we get

$$P(y_d) \approx \frac{m_d^{y_d}}{y_d!} \exp\{y_d \boldsymbol{x}_d \boldsymbol{\beta}\}(1 + m_d \phi^2 \exp\{\boldsymbol{x}_d \boldsymbol{\beta} + \phi v_{0d}\})^{-1/2}$$
$$\cdot \exp\Big\{ - m_d \exp\{\boldsymbol{x}_d \boldsymbol{\beta} + \phi v_{0d}\} + \phi y_d v_{0d} - \frac{1}{2} v_{0d}^2 \Big\}.$$

The log-likelihood is

$$\ell = \sum_{d=1}^{D} \log P(y_d)$$

and can be approximated as

$$\ell \approx \ell_0 = \sum_{d=1}^{D} \Big\{ y_d \log m_d - \log y_d! + y_d \boldsymbol{x}_d \boldsymbol{\beta} - \frac{1}{2} \log(1 + m_d \phi^2 \exp\{\boldsymbol{x}_d \boldsymbol{\beta} + \phi v_{0d}\})$$
$$- m_d \exp\{\boldsymbol{x}_d \boldsymbol{\beta} + \phi v_{0d}\} + \phi y_d v_{0d} - \frac{1}{2} v_{0d}^2 \Big\}. \tag{20.6}$$

For ease of exposition, let us define $p_{0d} = \exp\{\boldsymbol{x}_d \boldsymbol{\beta} + \phi v_{0d}\}$ and $\xi_{0d} = 1 + m_d \phi^2 p_{0d}$. The approximated log-likelihood ℓ_0 takes the form

$$\ell_0 = \sum_{d=1}^{D} \Big\{ y_d \log m_d - \log y_d! + y_d \boldsymbol{x}_d \boldsymbol{\beta} - \frac{1}{2} \log \xi_{0d} - m_d p_{0d} + \phi y_d v_{0d} - \frac{1}{2} v_{0d}^2 \Big\}. \tag{20.7}$$

It holds that

$$\frac{\partial p_{0d}}{\partial \beta_k} = x_{dk} p_{0d}, \quad \frac{\partial p_{0d}}{\partial \phi} = v_{0d} p_{0d}, \quad \frac{\partial \xi_{0d}}{\partial \beta_k} = m_d \phi^2 x_{dk} p_{0d}, \quad \frac{\partial \xi_{0d}}{\partial \phi} = (2 + \phi v_{0d}) m_d \phi p_{0d}.$$

The first partial derivatives of ℓ_0 with respect to β_k and ϕ are

$$\frac{\partial \ell_0}{\partial \beta_k} = \sum_{d=1}^{D} \Big\{ y_d x_{dk} - \frac{m_d x_{dk}}{2} \frac{\phi^2 p_{0d}}{\xi_{0d}} - m_d x_{dk} p_{0d} \Big\}, \quad k = 1, \ldots, p,$$

$$\frac{\partial \ell_0}{\partial \phi} = \sum_{d=1}^{D} \Big\{ y_d v_{0d} - \frac{m_d}{2} \frac{2\phi p_{0d} + \phi^2 v_{0d} p_{0d}}{\xi_{0d}} - m_d v_{0d} p_{0d} \Big\}.$$

The second partial derivatives of ℓ_0 are

$$\frac{\partial^2 \ell_0}{\partial \beta_r \partial \beta_k} = -\sum_{d=1}^{D} \left\{ \frac{m_d x_{dk} \phi^2}{2} \frac{p_{0d} x_{dr}(1+m_d\phi^2 p_{0d}) - p_{0d} m_d \phi^2 x_{dr} p_{0d}}{\xi_{0d}^2} + m_d x_{dk} x_{dr} p_{0d} \right\}$$

$$= -\sum_{d=1}^{D} \left\{ \frac{m_d x_{dk} x_{dr} \phi^2}{2} \frac{p_{0d}}{\xi_{0d}^2} + m_d x_{dk} x_{dr} p_{0d} \right\} = -\sum_{d=1}^{D} m_d x_{dk} x_{dr} p_{0d} \left(\frac{\phi^2}{2\xi_{0d}^2} + 1 \right),$$

$$\frac{\partial^2 \ell_0}{\partial \phi \partial \beta_k} = -\sum_{d=1}^{D} \left\{ \frac{m_d x_{dk}}{2} \frac{(2\phi p_{0d} + \phi^2 p_{0d} v_{0d})(1 + m_d \phi^2 p_{0d})}{\xi_{0d}^2} \right.$$

$$\left. - \frac{m_d x_{dk}}{2} \frac{\phi^3 p_{0d}^2 m_d (2 + \phi v_{0d})}{\xi_{0d}^2} + m_d x_{dk} v_{0d} p_{0d} \right\}$$

$$= -\sum_{d=1}^{D} \left\{ \frac{m_d x_{dk} \phi p_{0d}}{2} \frac{(2+\phi v_{0d})[1 + m_d \phi^2 p_{0d} - m_d \phi^2 p_{0d}]}{\xi_{0d}^2} + m_d x_{dk} v_{0d} p_{0d} \right\}$$

$$= -\sum_{d=1}^{D} \left\{ \frac{m_d x_{dk} \phi p_{0d}}{2} \frac{(2 + \phi v_{0d})}{\xi_{0d}^2} + m_d x_{dk} v_{0d} p_{0d} \right\}$$

$$= -\sum_{d=1}^{D} m_d x_{dk} v_{0d} p_{0d} \left(\frac{\phi(2 + \phi v_{0d})}{2 v_{0d} \xi_{0d}^2} + 1 \right),$$

$$\frac{\partial^2 \ell_0}{\partial \phi^2} = -\sum_{d=1}^{D} \left\{ \frac{m_d}{2} \frac{p_{0d}[2 + 4\phi v_{0d} + \phi^2 v_{0d}^2](1 + m_d \phi^2 p_{0d})}{\xi_{0d}^2} \right.$$

$$\left. - \frac{m_d}{2} \frac{(2 + \phi v_{0d})(2 + \phi v_{0d}) m_d \phi^2 p_{0d}^2}{\xi_{0d}^2} + m_d v_{0d}^2 p_{0d} \right\}$$

$$= -\sum_{d=1}^{D} \left\{ \frac{m_d p_{0d}}{2} \frac{[-2 + (2 + \phi v_{0d})^2](1 + m_d \phi^2 p_{0d})}{\xi_{0d}^2} \right.$$

$$\left. - \frac{m_d p_{0d}}{2} \frac{(2 + \phi v_{0d})^2 m_d \phi^2 p_{0d}}{\xi_{0d}^2} + m_d v_{0d}^2 p_{0d} \right\}$$

$$= -\sum_{d=1}^{D} \left\{ \frac{m_d p_{0d}}{2} \frac{-2(1 + m_d \phi^2 p_{0d})}{\xi_{0d}^2} \right.$$

$$\left. + \frac{m_d p_{0d}}{2} \frac{(2 + \phi v_{0d})^2 (1 + m_d \phi^2 p_{0d} - m_d \phi^2 p_{0d})}{\xi_{0d}^2} + m_d v_{0d}^2 p_{0d} \right\}$$

$$= -\sum_{d=1}^{D} \left\{ \frac{m_d p_{0d}}{2} \frac{(2+\phi v_{0d})^2 - 2\xi_{0d}}{\xi_{0d}^2} + m_d v_{0d}^2 p_{0d} \right\}$$

$$= -\sum_{d=1}^{D} m_d v_{0d}^2 p_{0d} \Big(\frac{(2+\phi v_{0d})^2 - 2\xi_{0d}}{2 v_{0d}^2 \xi_{0d}^2} + 1 \Big).$$

For $k, r = 1, \ldots, p$, the components of the score vector and the Hessian matrix are

$$U_{0k} = \frac{\partial \ell_0}{\partial \beta_k}, \; U_{0p+1} = \frac{\partial \ell_0}{\partial \phi}, \quad H_{0kr} = H_{0rk} = \frac{\partial^2 \ell_0}{\partial \beta_r \partial \beta_k},$$

$$H_{0kp+1} = H_{0p+1k} = \frac{\partial^2 \ell_0}{\partial \phi \partial \beta_k}, \quad H_{0p+1p+1} = \frac{\partial^2 \ell_0}{\partial \phi^2}.$$

In matrix form, we have $\boldsymbol{U}_0 = \boldsymbol{U}_0(\boldsymbol{\theta}) = \underset{1\le k\le p+1}{\text{col}}(U_{0k})$, and $\boldsymbol{H}_0 = \boldsymbol{H}_0(\boldsymbol{\theta}) = (H_{0kr})_{k,r=1,\ldots,p+1}$.

The Newton–Raphson algorithm maximizes $\ell_0 = \ell_0(\boldsymbol{\theta})$, with fixed $v_d = v_{0d}$, $d = 1, \ldots, D$. The updating equation is

$$\boldsymbol{\theta}^{(i+1)} = \boldsymbol{\theta}^{(i)} - \boldsymbol{H}_0^{-1}(\boldsymbol{\theta}^{(i)}) \boldsymbol{U}_0(\boldsymbol{\theta}^{(i)}). \tag{20.8}$$

The final ML-Laplace approximation algorithm combines the two described Newton–Raphson algorithms and can be described by the following steps:

1. Set the initial values $i = 0, \epsilon_1 > 0, \epsilon_2 > 0, \epsilon_3 > 0, \epsilon_4 > 0, \boldsymbol{\theta}^{(0)}, \boldsymbol{\theta}^{(-1)} = \boldsymbol{\theta}^{(0)} + \mathbf{1}$, $v_d^{(0)} = 0, v_d^{(-1)} = 1, d = 1, \ldots, D$.
2. Until $\|\boldsymbol{\theta}^{(i)} - \boldsymbol{\theta}^{(i-1)}\|_2 < \varepsilon_1, |v_d^{(i)} - v_d^{(i-1)}| < \varepsilon_2, d = 1, \ldots, D$, do
   a. apply algorithm (20.5) with seeds $v_d^{(i)}$, $d = 1, \ldots, D$, convergence tolerance ε_3, and $\boldsymbol{\theta} = \boldsymbol{\theta}^{(i)}$ fixed. Output: $v_d^{(i+1)}$, $d = 1, \ldots, D$;
   b. apply algorithm (20.8) with seed $\boldsymbol{\theta}^{(i)}$, convergence tolerance ε_4, and $v_{0d} = v_d^{(i+1)}$ fixed, $d = 1, \ldots, D$. Output: $\boldsymbol{\theta}^{(i+1)}$;
   c. $i \leftarrow i + 1$.
3. Output: $\hat{\boldsymbol{\theta}} = \boldsymbol{\theta}^{(i)}$, $\hat{v}_d = v_d^{(i)}$, $d = 1, \ldots, D$.

For more details about the output of the ML-Laplace algorithm, see Remark 14.1. Let $\hat{\boldsymbol{\beta}}$ and $\hat{\phi}$ be the ML estimators of $\boldsymbol{\beta}$ and ϕ, respectively. Let $\hat{v}_d$ be the mode predictor of v_d under ML estimation. The Laplace-approximated AIC is

$$AIC = 2(p+1) - 2\ell_0(\hat{\boldsymbol{\beta}}, \hat{\phi}, \hat{v}_1, \ldots, \hat{v}_D),$$

where the second term is the Laplace approximation (20.7) to the model log-likelihood, i.e.

$$\ell_0 = \ell_0(\hat{\boldsymbol{\beta}}, \hat{\phi}, \hat{v}_1, \ldots, \hat{v}_D)$$
$$= \sum_{d=1}^{D} \left\{ y_d \log m_d - \log y_d! + y_d \boldsymbol{x}_d \hat{\boldsymbol{\beta}} - \frac{1}{2} \log \hat{\xi}_d - m_d \hat{p}_d + \hat{\phi} y_d \hat{v}_d - \frac{1}{2} \hat{v}_d^2 \right\},$$

where $\hat{p}_d = \exp\left\{\boldsymbol{x}_d \hat{\boldsymbol{\beta}} + \hat{\phi}\hat{v}_d\right\}$ and $\hat{\xi}_d = 1 + m_d \hat{\phi}^2 \hat{p}_d$.

20.6 PQL Algorithm

The PQL estimator of $\boldsymbol{\beta}$ and predictor of $\boldsymbol{v}$ (see Breslow and Clayton 1993) maximize the joint log-likelihood

$$\ell = \log \psi(\boldsymbol{y}, \boldsymbol{v})$$
$$= -\frac{D}{2} \log 2\pi - \frac{1}{2} \sum_{d=1}^{D} v_d^2 - \sum_{d=1}^{D} \log y_d! + \sum_{d=1}^{D} \{y_d \log m_d - m_d \exp\{\boldsymbol{x}_d \boldsymbol{\beta} + \phi v_d\}\}$$
$$+ \sum_{k=1}^{p} \Big(\sum_{d=1}^{D} y_d x_{dk}\Big) \beta_k + \phi \sum_{d=1}^{D} y_d v_d.$$

The first partial derivatives of ℓ with respect to $\boldsymbol{\beta}$ and $\boldsymbol{v}$ are

$$U_k = \frac{\partial \ell}{\partial \beta_k} = -\sum_{d=1}^{D} m_d \exp\{\boldsymbol{x}_d \boldsymbol{\beta} + \phi v_d\} x_{dk} + \sum_{d=1}^{D} y_d x_{dk}, \quad k = 1, \ldots, p,$$

$$U_{p+d} = \frac{\partial \ell}{\partial v_d} = -v_d - m_d \exp\{\boldsymbol{x}_d \boldsymbol{\beta} + \phi v_d\}\phi + y_d \phi, \quad d = 1, \ldots, D.$$

The second partial derivatives of ℓ with respect to $\boldsymbol{\beta}$ and $\boldsymbol{v}$ are

$$H_{k_1 k_2} = \frac{\partial^2 \ell}{\partial \beta_{k_1} \partial \beta_{k_2}} = -\sum_{d=1}^{D} m_d \exp\{\boldsymbol{x}_d \boldsymbol{\beta} + \phi v_d\} x_{dk_1} x_{dk_2}, \quad k_1, k_2 = 1, \ldots, p,$$

$$H_{k p+d} = \frac{\partial^2 \ell}{\partial \beta_k \partial v_d} = -m_d \exp\{\boldsymbol{x}_d \boldsymbol{\beta} + \phi v_d\} x_{dk} \phi, \quad k = 1, \ldots, p, \; d = 1, \ldots, D,$$

$$H_{p+d p+d} = \frac{\partial^2 \ell}{\partial v_d^2} = -1 - m_d \exp\{\boldsymbol{x}_d \boldsymbol{\beta} + \phi v_d\} \phi^2, \quad d = 1, \ldots, D,$$

$$H_{p+d_1 p+d_2} = \frac{\partial^2 \ell}{\partial v_{d_1} \partial v_{d_2}} = 0, \quad d_1, d_2 = 1, \ldots, D, \; d_1 \neq d_2.$$

In matrix form, we have $\boldsymbol{\xi} = (\boldsymbol{\beta}', \boldsymbol{v}')'$, $\boldsymbol{U} = \boldsymbol{U}(\boldsymbol{\xi}) = \underset{1 \le k \le p+D}{\text{col}}(U_k)$ and $\boldsymbol{H} = \boldsymbol{H}(\boldsymbol{\xi}) = (H_{kr})_{k,r=1,\ldots,p+D}$. The Newton–Raphson algorithm maximizes the log-likelihood ℓ, with fixed ϕ. The updating equation is

$$\boldsymbol{\xi}^{(i+1)} = \boldsymbol{\xi}^{(i)} - \boldsymbol{H}^{-1}(\boldsymbol{\xi}^{(i)})\boldsymbol{U}(\boldsymbol{\xi}^{(i)}). \tag{20.9}$$

At the step i of the algorithm, the penalized maximum likelihood estimation of ϕ maximizes the joint likelihood of $\eta_1^{(i)}, \ldots, \eta_D^{(i)}$, where $\eta_d^{(i)} = \log m_d + \boldsymbol{x}_d\boldsymbol{\beta}^{(i)} + \phi v_d^{(i)}$ and

$$\eta_d^{(i)} \sim N(\log m_d + \boldsymbol{x}_d\boldsymbol{\beta}^{(i)},\ \phi^2), \quad d = 1, \ldots, D.$$

The joint log-likelihood of $\eta_1^{(i)}, \ldots, \eta_D^{(i)}$ is

$$\ell^{(i)}(\phi) = -\frac{D}{2}\log 2\pi - D\log\phi - \frac{1}{2}\frac{1}{\phi^2}\sum_{d=1}^{D}(\eta_d^{(i)} - \log m_d - \boldsymbol{x}_d\boldsymbol{\beta}^{(i)})^2.$$

By taking the first derivative of $\ell^{(i)}(\phi)$ with respect to ϕ, ignoring the dependence of $\eta_d^{(i)}$ on ϕ and equating to zero, we get

$$0 = U^{(i)} = \frac{\partial \ell^{(i)}(\phi)}{\partial \phi} = -\frac{D}{\phi} + \frac{1}{\phi^3}\sum_{d=1}^{D}(\eta_d^{(i)} - \log m_d - \boldsymbol{x}_d\boldsymbol{\beta}^{(i)})^2,$$

which means

$$\phi^2 = \frac{1}{D}\sum_{d=1}^{D}(\eta_d^{(i)} - \log m_d - \boldsymbol{x}_d\boldsymbol{\beta}^{(i)})^2.$$

As $\eta_d^{(i)} - \log m_d - \boldsymbol{x}_d\boldsymbol{\beta}^{(i)} = \phi v_d^{(i)}$, we propose the following PQL updating equation for ϕ:

$$\phi^{(i+1)2} = \phi^{(i)2}\frac{1}{D}\sum_{d=1}^{D} v_d^{(i)2}. \tag{20.10}$$

The PQL algorithm calculates predictors of $\boldsymbol{v}$ and estimators of $\boldsymbol{\beta}$ and ϕ. The algorithm is

1. Set $i = 1$, where i counts the external iterations. Set the values of $\boldsymbol{\beta}^{(0)}$, $\boldsymbol{v}^{(0)}$, and $\phi^{(0)}$.
2. Run algorithm (20.9). Use $\phi^{(i-1)}$ as a known value and $\boldsymbol{\beta}^{(i-1)}$ and $\boldsymbol{v}^{(i-1)}$ as algorithm seeds. Let $\boldsymbol{\beta}^{(i)}$ and $\boldsymbol{v}^{(i)}$ be the output of Algorithm (20.9).

3. Update ϕ by using the updating Eq. (20.10), i.e.

$$\phi^{(i)2} = \phi^{(i-1)2} \frac{1}{D} \sum_{d=1}^{D} v_d^{(i)2}$$

and increase the iteration index $i \leftarrow i + 1$.
4. Repeat steps (2) and (3) until the convergence of $\boldsymbol{\beta}^{(i)}$, $\boldsymbol{v}^{(i)}$, and $\phi^{(i)}$.

20.7 Empirical Best Predictors

This section gives best predictors (BPs) and empirical best predictors (EBPs) for the area-level Poisson mixed model. Assume that the parameters $\boldsymbol{\theta} = (\boldsymbol{\beta}', \phi)'$ are known. Under the assumed model, the BP is an unbiased predictor $\hat{p}_d = \hat{p}_d(\boldsymbol{\theta}) = \hat{p}_d(\boldsymbol{\theta}, y_d)$ of $p_d = p_d(\boldsymbol{\theta}, v_d) = \exp\{\boldsymbol{x}_d\boldsymbol{\beta} + \phi v_d\}$ that minimizes the MSE. To obtain the BP of p_d, we have to solve the problem

$$\min_{\hat{p}_d} E\big[(\hat{p}_d - p_d)^2\big] \quad \text{subject to} \quad E[\hat{p}_d - p_d] = 0,$$

where the expectation is taken with respect to the joint p.d.f. of y_d and v_d.

Proposition 20.1 *The BP of p_d is $\hat{p}_d = E[p_d|y_d]$.*

Proof Let $\hat{p}_d$ be a predictor of p_d. For ease of exposition, we use this simplified notation. Let us just recall that $\hat{p}_d$ is a function of y_d and p_d is a function of v_d. We have

$$\begin{aligned}
E\big[(\hat{p}_d - p_d)^2\big] &= E\Big[E\big[(\hat{p}_d - p_d)^2 |\, y_d\big]\Big] \\
&= E\Big[E\big[\{(\hat{p}_d - E[p_d|y_d]) + (E[p_d|y_d] - p_d)\}^2 |\, y_d\big]\Big] \\
&= E\Big[E\big[(\hat{p}_d - E[p_d|y_d])^2 |\, y_d\big]\Big] + E\Big[E\big[(E[p_d|y_d] - p_d)^2 |\, y_d\big]\Big] \\
&+ 2E\Big[(\hat{p}_d - E[p_d|y_d]) E\big[E[p_d|y_d] - p_d |\, y_d\big]\Big].
\end{aligned}$$

The last summand is zero because

$$E\big[E[p_d|y_d] - p_d |\, y_d\big] = E[p_d|y_d] - E[p_d|y_d] = 0.$$

By taking $\hat{p}_d = E[p_d|y_d]$, the first summand is zero and $E\big[(\hat{p}_d - p_d)^2\big]$ is minimized. Further, it holds that

$$E\big[\hat{p}_d - p_d\big] = E\big[E[p_d|y_d] - p_d\big] = E\Big[E\big[E[p_d|y_d] - p_d|\, y_d\big]\Big]$$
$$= E\Big[E[p_d|y_d] - E[p_d|y_d]\Big] = 0.$$

This is to say, $\hat{p}_d = E[p_d|y_d]$ is unbiased for p_d. □

Since the conditional distribution of y_d, given v_d, is

$$P(y_d|v_d) = \frac{m_d^{y_d}}{y_d!} \exp\left\{y_d(\boldsymbol{x}_d\boldsymbol{\beta} + \phi v_d) - m_d \exp\{\boldsymbol{x}_d\boldsymbol{\beta} + \phi v_d\}\right\}$$

and the p.d.f. of v_d is

$$f(v_d) = (2\pi)^{-1/2} \exp\Big\{-\frac{1}{2}\, v_d^2\Big\},$$

the best predictor $\hat{p}_d(\boldsymbol{\theta})$ of p_d can be written in the form

$$\hat{p}_d(\boldsymbol{\theta}) = E[p_d|y_d] = \frac{\int_R \exp\{\boldsymbol{x}_d\boldsymbol{\beta} + \phi v_d\} P(y_d|v_d) f(v_d)\, dv_d}{\int_R P(y_d|v_d) f(v_d)\, dv_d} = \frac{A_d(y_d, \boldsymbol{\theta})}{D_d(y_d, \boldsymbol{\theta})},$$

where

$$A_d(y_d, \boldsymbol{\theta}) = \int_R \exp\left\{(y_d + 1)(\boldsymbol{x}_d\boldsymbol{\beta} + \phi v_d) - m_d \exp\left\{\boldsymbol{x}_d\boldsymbol{\beta} + \phi v_d\right\}\right\} f(v_d)\, dv_d,$$

$$D_d(y_d, \boldsymbol{\theta}) = \int_R \exp\left\{y_d(\boldsymbol{x}_d\boldsymbol{\beta} + \phi v_d) - m_d \exp\{\boldsymbol{x}_d\boldsymbol{\beta} + \phi v_d\}\right\} f(v_d)\, dv_d.$$

The EBP of p_d is $\hat{p}_d(\hat{\boldsymbol{\theta}})$ and can be approximated by estimating the integrals with a Monte Carlo method that uses antithetic variables to reduce the variability. The approximation algorithm is

1. Estimate $\hat{\boldsymbol{\theta}} = (\hat{\boldsymbol{\beta}}', \hat{\phi})'$.
2. For $s = 1, \ldots, S$, generate $v_d^{(s)}$ i.i.d. $N(0, 1)$ and $v_d^{(S+s)} = -v_d^{(s)}$.
3. Calculate $\hat{p}_d(\hat{\boldsymbol{\theta}}) = \hat{A}_d / \hat{D}_d$, where

$$\hat{A}_d = \frac{1}{2S} \sum_{s=1}^{2S} \exp\left\{(y_d + 1)(\boldsymbol{x}_d\hat{\boldsymbol{\beta}} + \hat{\phi} v_d^{(s)}) - m_d \exp\{\boldsymbol{x}_d\hat{\boldsymbol{\beta}} + \hat{\phi} v_d^{(s)}\}\right\},$$

$$\hat{D}_d = \frac{1}{2S} \sum_{s=1}^{2S} \exp\left\{y_d(\boldsymbol{x}_d\hat{\boldsymbol{\beta}} + \hat{\phi} v_d^{(s)}) - m_d \exp\{\boldsymbol{x}_d\hat{\boldsymbol{\beta}} + \hat{\phi} v_d^{(s)}\}\right\}.$$

The EBP of $\mu_d = m_d p_d$ is $\hat{\mu}_d(\hat{\boldsymbol{\theta}}) = m_d \hat{p}_d(\hat{\boldsymbol{\theta}})$.

Computationally simpler predictor of p_d is the plug-in predictor $\tilde{p}_d = \exp\{\boldsymbol{x}_d\hat{\boldsymbol{\beta}} + \hat{\phi}\hat{v}_d\}$, where $\hat{v}_d$ is a predictor (EBP or likelihood mode) of v_d, $d = 1, \ldots, D$. Let us note that not all of the presented fitting algorithms give a prediction of v_d, see e.g. the MM algorithm. Alternatively, the EBPs of the random effects can be used. The BP of v_d is

$$\hat{v}_d(\boldsymbol{\theta}) = E_\theta[v_d|y_d] = \frac{\int_R v_d P(y_d|v_d) f(v_d)\, dv_d}{\int_R P(y_d|v_d) f(v_d)\, dv_d} = \frac{A_{v,d}(y_d, \boldsymbol{\theta})}{D_d(y_d, \boldsymbol{\theta})},$$

where

$$A_{v,d}(y_d, \boldsymbol{\theta}) = \int_R v_d \exp\{y_d(\boldsymbol{x}_d\boldsymbol{\beta} + \phi v_d) - m_d \exp\{\boldsymbol{x}_d\boldsymbol{\beta} + \phi v_d\}\} f(v_d)\, dv_d.$$

The EBP of v_d is $\hat{v}_d = \hat{v}_d(\hat{\boldsymbol{\theta}})$ and can be approximated as follows:

1. Estimate $\hat{\boldsymbol{\theta}} = (\hat{\boldsymbol{\beta}}', \hat{\phi})'$.
2. For $s = 1, \ldots, S$, generate $v_d^{(s)}$ i.i.d. $N(0, 1)$ and $v_d^{(S+s)} = -v_d^{(s)}$.
3. Calculate $\hat{v}_d(\hat{\boldsymbol{\theta}}) = \hat{A}_{v,d}/\hat{D}_d$, where $\hat{D}_d$ is defined above and

$$\hat{A}_{v,d} = \frac{1}{2S}\sum_{s=1}^{2S} v_d^{(s)} \exp\left\{y_d(\boldsymbol{x}_d\hat{\boldsymbol{\beta}} + \hat{\phi}v_d^{(s)}) - m_d \exp\{\boldsymbol{x}_d\hat{\boldsymbol{\beta}} + \hat{\phi}v_d^{(s)}\}\right\}.$$

20.8 MSE of the EBP

This section presents an approximation and gives analytic and bootstrap estimators of the MSE of the EBP of

$$p_d(\boldsymbol{\theta}, v_d) = \exp\{\boldsymbol{x}_d\boldsymbol{\beta} + \phi v_d\}.$$

Let us recall that the BP of $p_d(\boldsymbol{\theta}, v_d)$ is

$$\hat{p}_d(\boldsymbol{\theta}) = E[p_d(\boldsymbol{\theta}, v_d)|y_d] = \frac{A_d(y_d, \boldsymbol{\theta})}{D_d(y_d, \boldsymbol{\theta})} \triangleq \psi_d(y_d, \boldsymbol{\theta}), \tag{20.11}$$

and the EBP is $\hat{p}_d = \hat{p}_d(\hat{\boldsymbol{\theta}}) = \psi_d(y_d, \hat{\boldsymbol{\theta}})$.

20.8.1 Approximation of the MSE

The MSE of the EBP can be decomposed in the form

$$
\begin{aligned}
MSE(\hat{p}_d) &= E\big[(\hat{p}_d(\hat{\boldsymbol{\theta}}) - p_d(\boldsymbol{\theta}, v_d))^2\big] \\
&= E\big[\big(\{\hat{p}_d(\hat{\boldsymbol{\theta}}) - \hat{p}_d(\boldsymbol{\theta})\} + \{\hat{p}_d(\boldsymbol{\theta}) - p_d(\boldsymbol{\theta}, v_d)\}\big)^2\big] \\
&= E\big[\big(\hat{p}_d(\hat{\boldsymbol{\theta}}) - \hat{p}_d(\boldsymbol{\theta})\big)^2\big] + E\big[\big(\hat{p}_d(\boldsymbol{\theta}) - p_d(\boldsymbol{\theta}, v_d)\big)^2\big],
\end{aligned}
$$

because

$$
\begin{aligned}
&E\big[\big(\hat{p}_d(\hat{\boldsymbol{\theta}}) - \hat{p}_d(\boldsymbol{\theta})\big)\big(\hat{p}_d(\boldsymbol{\theta}) - p_d(\boldsymbol{\theta}, v_d)\big)\big] \\
&= E\big[\big(\hat{p}_d(\hat{\boldsymbol{\theta}}) - \hat{p}_d(\boldsymbol{\theta})\big)E\big[\hat{p}_d(\boldsymbol{\theta}) - p_d(\boldsymbol{\theta}, v_d)|y_d\big]\big] = 0.
\end{aligned}
$$

The second term of $MSE(\hat{p}_d)$ is the MSE of the BP $\hat{p}_d(\boldsymbol{\theta})$, i.e.

$$
\begin{aligned}
g_d(\boldsymbol{\theta}) &= E\big[\big(\hat{p}_d(\boldsymbol{\theta}) - p_d(\boldsymbol{\theta}, v_d)\big)^2\big] = E\big[\hat{p}_d^2(\boldsymbol{\theta})\big] + E\big[p_d^2(\boldsymbol{\theta}, v_d)\big] \\
&- 2E\big[\hat{p}_d(\boldsymbol{\theta})E\big[p_d(\boldsymbol{\theta}, v_d)|y_d\big]\big] = E\big[p_d^2(\boldsymbol{\theta}, v_d)\big] - E\big[\hat{p}_d^2(\boldsymbol{\theta})\big].
\end{aligned}
$$

The first term of $g_d(\boldsymbol{\theta})$ is

$$
E\big[p_d^2(\boldsymbol{\theta}, v_d)\big] = \int_R \exp\{2\boldsymbol{x}_d\boldsymbol{\beta} + 2\phi v_d\}\, f(v_d)\, dv_d = \exp\{2\boldsymbol{x}_d\boldsymbol{\beta} + 2\phi^2\}.
$$

The second term of $g_d(\boldsymbol{\theta})$ is

$$
E\big[\hat{p}_d^2(\boldsymbol{\theta})\big] = E[\psi_d^2(y_d, \boldsymbol{\theta})] = \sum_{j=0}^{\infty} \psi_d^2(j, \boldsymbol{\theta}) p_d(j, \boldsymbol{\theta}),
$$

where

$$
\begin{aligned}
p_d(j, \boldsymbol{\theta}) &= P(y_d = j) = \int_R P(y_d = j|v_d) f(v_d)\, dv_d \\
&= \frac{m_d^j}{j!} \int_R \exp\{j(\boldsymbol{x}_d\boldsymbol{\beta} + \phi v_d) - m_d \exp\{\boldsymbol{x}_d\boldsymbol{\beta} + \phi v_d\}\}\, f(v_d) dv_d \\
&= \frac{m_d^j}{j!} D_d(j, \boldsymbol{\theta}).
\end{aligned}
$$

Concerning the first term of $MSE(\hat{p}_d)$, using the Taylor expansion we obtain

$$\hat{p}_d(\hat{\boldsymbol{\theta}}) - \hat{p}_d(\boldsymbol{\theta}) = \psi_d(y_d, \hat{\boldsymbol{\theta}}) - \psi_d(y_d, \boldsymbol{\theta}) = \left(\frac{\partial}{\partial \boldsymbol{\theta}} \psi_d(y_d, \boldsymbol{\theta})\right)' (\hat{\boldsymbol{\theta}} - \boldsymbol{\theta}) + o(\|\hat{\boldsymbol{\theta}} - \boldsymbol{\theta}\|).$$

Hereafter, the symbols $o(.)$ and $O(.)$ are understood in an appropriate sense, for example, in probability. Under the assumption $\|\hat{\boldsymbol{\theta}} - \boldsymbol{\theta}\| = O(D^{-\frac{1}{2}})$, we have

$$E\left[\left(\hat{p}_d(\hat{\boldsymbol{\theta}}) - \hat{p}_d(\boldsymbol{\theta})\right)^2\right] = \frac{1}{D} E\left[\left(\left(\frac{\partial}{\partial \boldsymbol{\theta}} \psi_d(y_d, \boldsymbol{\theta})\right)' \sqrt{D}(\hat{\boldsymbol{\theta}} - \boldsymbol{\theta})\right)^2\right] + o(1/D).$$

Now, it is necessary to consider the following trick. Let $\hat{\boldsymbol{\theta}}_{d-}$ be an estimator based on $\boldsymbol{y}_{d-} = \underset{d' \neq d}{\text{col}}(\boldsymbol{y}_{d'})$, and denote $\hat{p}_{d-} = \psi_d(y_d, \hat{\boldsymbol{\theta}}_{d-})$. Then, by the independence of y_d and $\boldsymbol{y}_{d-}$, we have

$$\begin{aligned}
a_d(\boldsymbol{\theta}) &= E\left[\left(\left(\frac{\partial}{\partial \boldsymbol{\theta}} \psi_d(y_d, \boldsymbol{\theta})\right)' \sqrt{D}(\hat{\boldsymbol{\theta}}_{d-} - \boldsymbol{\theta})\right)^2\right] \\
&= \sum_{j=0}^{\infty} E\left[\left(\left(\frac{\partial}{\partial \boldsymbol{\theta}} \psi_d(y_d, \boldsymbol{\theta})\right)' \sqrt{D}(\hat{\boldsymbol{\theta}}_{d-} - \boldsymbol{\theta})\right)^2 \middle| y_d = j\right] p_d(j, \boldsymbol{\theta}) \\
&= \sum_{j=0}^{\infty} \left(\frac{\partial}{\partial \boldsymbol{\theta}} \psi_d(j, \boldsymbol{\theta})\right)' \boldsymbol{V}_d(\boldsymbol{\theta}) \left(\frac{\partial}{\partial \boldsymbol{\theta}} \psi_d(j, \boldsymbol{\theta})\right) p_d(j, \boldsymbol{\theta}),
\end{aligned}$$

where $\boldsymbol{V}_d(\boldsymbol{\theta}) = DE\left[(\hat{\boldsymbol{\theta}}_{d-} - \boldsymbol{\theta})(\hat{\boldsymbol{\theta}}_{d-} - \boldsymbol{\theta})' | y_d{=}j\right] {=} DE\left[(\hat{\boldsymbol{\theta}}_{d-}{-}\boldsymbol{\theta})(\hat{\boldsymbol{\theta}}_{d-}{-}\boldsymbol{\theta})'\right]$. Therefore,

$$MSE(\hat{p}_{d-}) = g_d(\boldsymbol{\theta}) + \frac{1}{D} a_d(\boldsymbol{\theta}) + o(1/D).$$

If moreover the conditions

$$\|\hat{\boldsymbol{\theta}}_{d-} - \boldsymbol{\theta}\| = O(D^{-\frac{1}{2}}) \quad \text{and} \quad \|\hat{\boldsymbol{\theta}} - \hat{\boldsymbol{\theta}}_{d-}\| = o(D^{-\frac{1}{2}}) \tag{20.12}$$

are fulfilled, one may replace $\hat{\boldsymbol{\theta}}_{d-}$ by $\hat{\boldsymbol{\theta}}$, an estimator of $\boldsymbol{\theta}$ based on all data, and may still obtain (for more details, see Jiang 2003)

$$MSE(\hat{p}_d) = g_d(\boldsymbol{\theta}) + \frac{1}{D} c_d(\boldsymbol{\theta}) + o(1/D),$$

where

$$c_d(\boldsymbol{\theta}) = \sum_{j=0}^{\infty} \Big(\frac{\partial}{\partial \boldsymbol{\theta}} \psi_d(j, \boldsymbol{\theta}) \Big)' \boldsymbol{V}(\boldsymbol{\theta}) \Big(\frac{\partial}{\partial \boldsymbol{\theta}} \psi_d(j, \boldsymbol{\theta}) \Big) p_d(j, \boldsymbol{\theta}),$$

$$\boldsymbol{V}(\boldsymbol{\theta}) = DE\Big[(\hat{\boldsymbol{\theta}} - \boldsymbol{\theta})(\hat{\boldsymbol{\theta}} - \boldsymbol{\theta})'\Big].$$

20.8.2 Analytic Estimation of the MSE for MM Estimators

Let us now consider the estimation of the MSE of EBP. A plug-in estimator of $MSE(\hat{p}_d)$ is

$$mse^P(\hat{p}_d) = g_d(\hat{\boldsymbol{\theta}}) + \frac{1}{D} c_d(\hat{\boldsymbol{\theta}}).$$

By the Taylor expansion of $c_d(\hat{\boldsymbol{\theta}})$ around $\boldsymbol{\theta}$ and the consistency of $\hat{\boldsymbol{\theta}}$, we have that $E[c_d(\hat{\boldsymbol{\theta}}) - c_d(\boldsymbol{\theta})] = o(1)$. However, the bias $E[g_d(\hat{\boldsymbol{\theta}}) - g_d(\boldsymbol{\theta})]$ may not be of the order $o(D^{-1})$. But by the Taylor expansion of $g_d(\hat{\boldsymbol{\theta}})$ around $\boldsymbol{\theta}$, we have

$$g_d(\hat{\boldsymbol{\theta}}) = g_d(\boldsymbol{\theta}) + \Big(\frac{\partial}{\partial \boldsymbol{\theta}} g_d(\boldsymbol{\theta}) \Big)' (\hat{\boldsymbol{\theta}} - \boldsymbol{\theta}) + \frac{1}{2} (\hat{\boldsymbol{\theta}} - \boldsymbol{\theta})' \Big(\frac{\partial^2}{\partial \boldsymbol{\theta}^2} g_d(\boldsymbol{\theta}) \Big) (\hat{\boldsymbol{\theta}} - \boldsymbol{\theta}) + o(\|\hat{\boldsymbol{\theta}} - \boldsymbol{\theta}\|^2).$$

Thus, under the assumptions

$$\|\hat{\boldsymbol{\theta}} - \boldsymbol{\theta}\| = O(D^{-\frac{1}{2}}) \quad \text{and} \quad E(\hat{\boldsymbol{\theta}} - \boldsymbol{\theta}) = O(D^{-1}), \tag{20.13}$$

we obtain

$$E[g_d(\hat{\boldsymbol{\theta}})] = g_d(\boldsymbol{\theta}) + \frac{1}{D} b_d(\boldsymbol{\theta}) + o(D^{-1}),$$

where

$$b_d(\boldsymbol{\theta}) = \Big(\frac{\partial}{\partial \boldsymbol{\theta}} g_d(\boldsymbol{\theta}) \Big)' DE[\hat{\boldsymbol{\theta}} - \boldsymbol{\theta}] + \frac{1}{2} E\Big[\sqrt{D}(\hat{\boldsymbol{\theta}} - \boldsymbol{\theta})' \Big(\frac{\partial^2}{\partial \boldsymbol{\theta}^2} g_d(\boldsymbol{\theta}) \Big) \sqrt{D}(\hat{\boldsymbol{\theta}} - \boldsymbol{\theta}) \Big]. \tag{20.14}$$

To illustrate that there are some estimators satisfying the above given assumptions, we can mention the following statement of Jiang (2003). If his regularity conditions (23)–(25) hold and the x_{dk}'s are bounded, then the MM estimators $\hat{\boldsymbol{\theta}}$ and $\hat{\boldsymbol{\theta}}_{d-}$ fulfill both the assumptions (20.12) and (20.13).

Boubeta et al. (2016) adapted the MSE calculations given by Jiang and Lahiri (2001) and Jiang (2003) to the case of area-level Poisson mixed models and proved the following proposition.

Proposition 20.2 *Let us assume that the area-level Poisson mixed model is fitted with the MM algorithm. Under regularity conditions, it holds that* $b_d(\boldsymbol{\theta}) = B_d(\boldsymbol{\theta}) + o(1)$, *where*

$$B_d(\boldsymbol{\theta}) = \frac{1}{2}\left\{E[r_{D,d}] - \left(\frac{\partial}{\partial\boldsymbol{\theta}} g_d(\boldsymbol{\theta})\right)'\left(\frac{\partial}{\partial\boldsymbol{\theta}} \boldsymbol{M}(\boldsymbol{\theta})\right)^{-1} E[\boldsymbol{q}_D]\right\},$$

$$r_{D,d} = \Delta_D' \boldsymbol{R}_d(\boldsymbol{\theta})\Delta_D, \quad \boldsymbol{R}_d(\boldsymbol{\theta}) = \left(\left(\frac{\partial}{\partial\boldsymbol{\theta}} \boldsymbol{M}(\boldsymbol{\theta})\right)^{-1}\right)'\left(\frac{\partial^2}{\partial\boldsymbol{\theta}^2} g_d(\boldsymbol{\theta})\right)\left(\frac{\partial}{\partial\boldsymbol{\theta}} \boldsymbol{M}(\boldsymbol{\theta})\right)^{-1},$$

$$\boldsymbol{q}_D = \underset{1\le k\le p+1}{col}(q_{Dk}), \quad \boldsymbol{M}(\boldsymbol{\theta}) = \underset{1\le k\le p+1}{col}(M_k(\boldsymbol{\theta})), \quad \hat{\boldsymbol{M}} = \underset{1\le k\le p+1}{col}(\hat{M}_k),$$

$$q_{Dk} = \Delta_D' \boldsymbol{Q}_k(\boldsymbol{\theta})\Delta_D, \quad \boldsymbol{Q}_k(\boldsymbol{\theta}) = \left(\left(\frac{\partial}{\partial\boldsymbol{\theta}} \boldsymbol{M}(\boldsymbol{\theta})\right)^{-1}\right)'\left(\frac{\partial^2}{\partial\boldsymbol{\theta}^2} M_k(\boldsymbol{\theta})\right)\left(\frac{\partial}{\partial\boldsymbol{\theta}} \boldsymbol{M}(\boldsymbol{\theta})\right)^{-1},$$

$$\Delta_D = \sqrt{D}(\hat{\boldsymbol{M}} - \boldsymbol{M}(\boldsymbol{\theta})), \quad \frac{\partial}{\partial\boldsymbol{\theta}} \boldsymbol{M}(\boldsymbol{\theta}) = \left(\frac{\partial}{\partial\theta_{k_2}} M_{k_1}(\boldsymbol{\theta})\right)_{k_1,k_2=1,\ldots,p+1}.$$

Combining the assertion of the proposition with the previous formulas, we obtain a theoretical estimator of $MSE(\hat{p}_d)$, with bias of order $O(D^{-1})$, in the form

$$\widehat{MSE}(\hat{p}_d) = g_d(\hat{\boldsymbol{\theta}}) + \frac{1}{D} c_d(\hat{\boldsymbol{\theta}}) - \frac{1}{D} B_d(\boldsymbol{\theta}) = mse^P(\hat{p}_d) - \frac{1}{D} B_d(\boldsymbol{\theta}).$$

The corresponding practical estimator is

$$mse(\hat{p}_d) = \hat{g}_d(\hat{\boldsymbol{\theta}}) + \frac{1}{D} \hat{c}_d(\hat{\boldsymbol{\theta}}) - \frac{1}{D} \hat{B}_d(\hat{\boldsymbol{\theta}}).$$

For the case of MM estimators, Boubeta et al. (2016) gave Monte Carlo approximations $\hat{g}_d(\hat{\boldsymbol{\theta}})$ and $\hat{c}_d(\hat{\boldsymbol{\theta}})$ of $g_d(\hat{\boldsymbol{\theta}})$ and $c_d(\hat{\boldsymbol{\theta}})$, respectively, and bootstrap estimators $\hat{B}_d(\hat{\boldsymbol{\theta}})$ of $B_d(\boldsymbol{\theta})$. This is to say, they gave a practical way of calculating $mse(\hat{p}_d)$.

20.8.3 Bootstrap Estimation of the MSE

Alternatively, the MSE can be estimated by a resampling method. The following procedure calculates a parametric bootstrap estimator of $MSE(\hat{p}_d)$:

1. Fit the model to the sample and calculate the estimate $\hat{\boldsymbol{\theta}} = (\hat{\boldsymbol{\beta}}', \hat{\phi})'$.
2. Repeat B times ($b = 1, \ldots, B$):

a. Generate $v_d^{*(b)}$ as i.i.d. N(0,1), $d = 1, \ldots, D$. Calculate $p_d^{*(b)} = \exp\{\boldsymbol{x}_d\hat{\boldsymbol{\beta}} + \hat{\phi}v_d^{*(b)}\}$ and $y_d^{*(b)} \sim \text{Poiss}(m_d p_d^{*(b)})$.
b. For each bootstrap sample, calculate the estimator $\hat{\boldsymbol{\theta}}^{*(b)}$ and the EBP $\hat{p}_d^{*(b)} = \hat{p}_d(\hat{\boldsymbol{\theta}}^{*(b)})$ following the previous expressions.

3. Output: $mse^*(\hat{p}_d) = \frac{1}{B}\sum_{b=1}^{B}\left(\hat{p}_d^{*(b)} - p_d^{*(b)}\right)^2$.

20.9 R Codes for EBPs

This section gives R codes for fitting the area-level Poisson mixed model to the survey data file `datLCS.txt`. We install the R package `lme4`.

```
if(!require(Matrix)){
  install.packages("Matrix")
  library(Matrix)
}
if(!require(lme4)){
  install.packages("lme4")
  library(lme4)
}
```

The target variable y gives the counts of poor people by domains. As auxiliary variable x_1, we take the domain means of the labor status situation inactive (Minact). The following R codes read the unit-level data file and define the variable poor:

```
dat <- read.table("datLCS.txt", header=TRUE, sep="\t", dec=",")
# Poverty variable: 1 if yes (income < z0), 0 if no (income > z0).
z0 <- 7280                            # poverty threshold.
poor <- as.numeric(dat$income < z0)   # variable poor
one <- rep(1, nrow(dat))              # variable one
domains <- sort(unique(dat$dom))      # domains
ndom <- length(domains)               # number of domains
```

The following code reads the auxiliary data file.

```
aux <- read.table("auxLCS.txt", header=TRUE, sep="\t", dec=",")
aux <- aux[order(aux$dom),]           # sort aux by dom
```

We calculate sample sizes and sample counts.

```
# sample sizes by domains
nd <- tapply(X=one, INDEX=dat$dom, FUN=sum)
# sample counts of poor people by domains
ndpoor <- tapply(poor, dat$dom, sum)
```

We calculate direct estimators of sizes and poverty proportions by domains, by using `dir2` function described in Sect. 2.8.4. We also obtain the design-based estimates of the root-MSEs of the plug-in predictors of poverty proportions.

```
# Direct estimates of poverty proportions
poor.dir <- dir2(data=poor, w=dat$w, domain=list(dat$dom))
dirp <- poor.dir$mean                           # poverty proportions
hatNd <- poor.dir$Nd.hat                        # estimated sizes
# root-MSE estimates of direct estimators
rmsedirp <- sqrt(poor.dir$var.mean)
```

Table 20.1 Estimated parameters of Poisson model

Parameter	Estimate	Std. error	z-value	p-value
Intercept	−4.139	1.239	−3.340	0.0008
Minact	6.181	3.600	1.717	0.0860

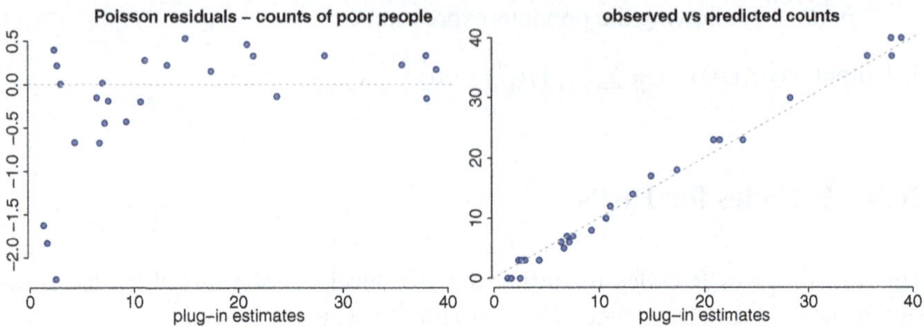

Fig. 20.1 Plots of residuals (left) and observed values (right) versus plug-in estimates

We fit an area-level Poisson mixed model to the sample counts of poor people. We apply the ML-Laplace approximation algorithm. Table 20.1 presents the estimated parameters and p-values. The estimated random effect standard deviation is $\hat{\phi} = 0.6525$.

```
glmm <- glmer(formula=ndpoor ~ Minact + (1|dom), family=poisson,
              offset=log(nd), data=aux)
summary(glmm)
beta <- fixef(glmm)                               # regression parameters
var <- as.data.frame(VarCorr(glmm))               # variance parameters
phi <- var$sdcor                                  # standard deviation parameter
# ML-Laplace predictions of random effects
ud <- ranef(glmm)$dom
```

We calculate the plug-in predictors of sample counts of poor people and poverty proportions.

```
# plug-in estimators of sample counts of poor people by domains
plugmu <- fitted(glmm)
# plug-in estimators of poverty proportions by domains
plugp <- plugmu/nd
```

Figure 20.1 (left) plots the residuals of the fitted Poisson mixed model versus the plug-in predictions. Figure 20.1 (right) plots the observed counts of poor people versus the plug-in predictions of counts. We observe that the selected Poisson mixed model has a good fit to the data.

We calculate the EBPs of sample counts of poor people and poverty proportions.

```
S <- 1000
set.seed(123)              # Set seed of random number generator
# Generate the random values
v <- rnorm(n=S*ndom, mean=0, sd=1)
vv <- Map(split(v, rep(1:ndom,each=S)), split(-v, rep(1:ndom, each=S)), f=c)
vvphi <- lapply(vv, phi, FUN="*")
Minact <- aux$Minact
# Define some auxiliary variables
xbeta.vvphi <- Map(beta[1]+beta[2]*aux$Minact, vvphi, f="+")
sum2 <- mapply(lapply(xbeta.vvphi, FUN=exp), nd, FUN="*")
Ad <- apply(exp(mapply(xbeta.vvphi, ndpoor+1, FUN="*") - sum2), 2, FUN=mean)
```

```
Dd <- apply(exp(mapply(xbeta.vvphi, ndpoor, FUN="*") - sum2), 2, FUN=mean)
ebp.p <- Ad/Dd          # EBPs of poverty proportions
ebp.mu <- ebp.p*nd      # EBPs of sample counts of poor people
```

We calculate the parametric bootstrap estimates of the root-MSEs of the plug-in predictors and EBPs of poverty proportions.

```
set.seed(123); B <- 200
mseplugpcum.star <- mseebppcum.star <- 0
for (b in 1:B){
  cat("iteration = ", b, "\n")
  v.star <- rnorm(n=ndom, mean=0, sd=1)
  # True proportions of bootsrap model
  p.star <- exp(beta[1] + beta[2]*Minact + phi*v.star)
  # Bootstrap counts of poor people
  y.star <- rpois(n=ndom, lambda=nd*p.star)
  mod.star <- glmer(formula=y.star ~ Minact + (1|dom), family=poisson,
                    offset=log(nd), data=aux)
  # regression parameters of bootstrap model
  b.star <- fixef(mod.star)
  # standard deviation of bootstrap model
  phi.star <- as.data.frame(VarCorr(mod.star))$sdcor
  # Estimated proportions of bootstrap model
  plugp.star <- fitted(mod.star)/nd
  # Calculation of EBP with bootstrap data
  v.star <- rnorm(n=S*ndom, mean=0, sd=1)   # Standard normal random numbers
  vv.star <- Map(split(v.star, rep(1:ndom,each=S)),
                 split(-v.star, rep(1:ndom, each=S)), f=c)
  vvphi.star <- lapply(vv.star, phi, FUN="*")
  # Define some auxiliary variables
  xbeta.vvphi.star <- Map(beta[1]+beta[2]*Minact, vvphi.star, f="+")
  sum2.star <- mapply(lapply(xbeta.vvphi.star, FUN=exp), nd, FUN="*")
  Ad.star <- apply(exp(mapply(xbeta.vvphi.star, y.star+1, FUN="*") -
                   sum2.star), 2, FUN=mean)
  Dd.star <- apply(exp(mapply(xbeta.vvphi.star, y.star, FUN="*") -
                   sum2.star), 2, FUN=mean)
  ebp.p.star <- Ad.star/Dd.star       # EBPs of poverty proportions
  mseplugpcum.star <- mseplugpcum.star + (plugp.star - p.star)^2
  mseebppcum.star <- mseebppcum.star + (ebp.p.star - p.star)^2
}
# root-MSE bootstrap estimates of plug-in
rmseplugp.star <- sqrt(mseplugpcum.star/B)
# root-MSE bootstrap estimates of EBPs
rmseebpp.star <- sqrt(mseebppcum.star/B)
```

The R code to save the results is

```
output <- data.frame(nd, ndmu=ndpoor, ebpmu=round(ebp.mu,2),
                     plugmu=round(plugmu,2), dirp, ebp.p, plugp,
                     rdirp=rmsedirp, rplugp=rmseplugp.star,
                     rebpp=rmseebpp.star)
head(round(output,4), 10)
```

For the ten first domains, Table 20.2 gives the sample sizes (n_d), the sample counts of poor people (ndmu), and the corresponding plug-in (plugmu) and EB (ebpmu) predictors. It also gives the direct estimators (dirp) and the plug-in (plugp) and EB (ebpp) predictors of poverty proportions. The right part of the table presents the estimates of the root-MSEs of the direct estimators, plug-in predictors, and EBPs. The first ones are calculated with the R function `dir2`, so that they are design-based estimates. The second and third ones are calculated by parametric bootstrap, and they are model based.

If a domain sample size is zero, then the direct estimator cannot be calculated, and Table 20.2 gives 0.0000. From the table, one can see that in most domains, the root-MSEs of the plug-in and EB predictors are smaller than those of direct estimators. Moreover, the plug-in predictor and the EBP show very similar behavior.

Table 20.2 Estimates of sample counts, poverty proportions, and root-MSEs

dom	n_d	ndmu	ebpmu	plugmu	dirp	ebpp	plugp	rdirp	rplugp	rebpp
3	57	23	21.00	20.83	0.4612	0.3685	0.3655	0.0749	0.0553	0.0538
5	96	23	21.60	21.43	0.3629	0.2250	0.2233	0.0845	0.0357	0.0352
6	82	14	13.18	13.16	0.1805	0.1607	0.1605	0.0447	0.0390	0.0380
7	10	0	1.42	1.30	0.0000	0.1418	0.1297	0.0000	0.1295	0.1211
11	118	18	17.44	17.33	0.1668	0.1478	0.1469	0.0384	0.0314	0.0321
12	18	3	3.09	2.97	0.1770	0.1718	0.1649	0.0927	0.0877	0.0849
13	138	5	6.71	6.64	0.0253	0.0486	0.0481	0.0117	0.0274	0.0397
14	190	37	35.91	35.61	0.1552	0.1890	0.1874	0.0283	0.0235	0.0236
15	406	37	37.95	37.98	0.1068	0.0935	0.0936	0.0194	0.0200	0.0198
16	93	7	7.60	7.49	0.0743	0.0817	0.0806	0.0289	0.0368	0.0357

References

Boubeta, M., Lombardía, M.J., Morales, D.: Empirical best prediction under area-level Poisson mixed models. TEST **25**, 548–569 (2016)

Breslow, N.E., Clayton, D.G.: Approximate inference in generalized linear mixed models. J. Am. Stat. Assoc. **88**, 9–25 (1993)

Jiang, J.: Consistent estimators in generalized linear models. J. Am. Stat. Assoc. **93**, 720–729 (1998)

Jiang, J., P. Lahiri.: Empirical best prediction for small area inference with binary data. Ann. Inst. Stat. Math. **53**, 217–243 (2001)

Jiang, J.: Empirical best prediction for small-area inference based on generalized linear mixed models. J. Stat. Plann. Inference **111**, 117–127 (2003)

Jiang J., Lahiri P.: Mixed model prediction and small area estimation. TEST **15**, 1–96 (2006)

Lin, X.: Estimation using penalized quasi-likelihood and quasi-pseudo-likelihood in Poisson mixed models. Lifetime Data Anal. **13**, 533–544 (2007)

MacNab, Y.C., Lin, X.: On empirical Bayes penalized quasi-likelihood inference in GLMMs and in Bayesian disease mapping and ecological modelling. Comput. Stat. Data Anal. **53**, 2950–2967 (2009)

Chapter 21
Area-Level Temporal Poisson Mixed Models

21.1 Introduction

For estimating counts and proportions, the small area estimation (SAE) model-based approach employs models at the unit or at the area level. In both cases linear mixed models (LMM) and generalized linear mixed models (GLMM) are used. Nevertheless, the literature shows more applications based on area-level than on unit-level models. This is because the former applications are in general easier to carry out because of the simplicity of the procedure and the availability of aggregated data from administrative registers. Under the area-level approach, Esteban et al. (2012a,b), Marhuenda et al. (2013, 2014), and Morales et al. (2015) derived poverty proportion estimators based on linear mixed models. They reported poverty indicators for Spanish provinces. Saei and Chambers (2003), Johnson et al. (2010), Chandra et al. (2011), Chambers et al. (2012), and López-Vizcaíno et al. (2013, 2015) applied area-level logit regression models to the estimation of domain counts or proportions. Similarly, Tzavidis et al. (2015) and Boubeta et al. (2016, 2017) applied Poisson regression models for estimating the same type of parameters.

This chapter presents the temporal Poisson mixed models employed by Boubeta et al. (2017) for estimating domain counts and proportions. The first Poisson model uses independent time random effects. The second model assumes that time random effects follow an autoregressive process of order one. For the sake of brevity, the mathematical development is only presented for the first model. Further information about the derivations for the second model and method of simulated moments (MSM) can be found in Boubeta et al. (2017).

The chapter presents the algebraic calculations required for programming the ML-Laplace approximation algorithm. It also gives empirical best predictors (EBP) of domain proportions and counts. The statistical methodology is taken and adapted from Jiang and Lahiri (2001) and Jiang (2003). In addition to the EBP, a plug-in predictor is also given. For estimating the EBP mean squared error (MSE), the parametric bootstrap MSE estimator introduced by González-Manteiga et al. (2007,

D. Morales et al., *A Course on Small Area Estimation and Mixed Models*, Statistics for Social and Behavioral Sciences, https://doi.org/10.1007/978-3-030-63757-6_21

2008) in the context of logistic and normal mixed models, is adapted to the current context. Finally, some R codes are given and applied to simulated data.

21.2 The Model with Independent Time Effects

Let us consider two independent sets of domain and domain-time random effects such that $\{v_{1,d} : d = 1, \ldots, D\}$ are i.i.d. $N(0, 1)$ and $\{v_{2,dt} : d = 1, \ldots, D, t = 1, \ldots, T\}$ are i.i.d. $N(0, 1)$. In matrix notation, we have

$$\boldsymbol{v}_1 = \underset{1 \le d \le D}{\text{col}}(v_{1,d}) \sim N_D(\boldsymbol{0}, \boldsymbol{I}_D), \quad \boldsymbol{v}_{2d} = \underset{1 \le t \le T}{\text{col}}(v_{2,dt}) \sim N(\boldsymbol{0}, \boldsymbol{I}_T),$$

$$\boldsymbol{v}_2 = \underset{1 \le d \le D}{\text{col}}(\boldsymbol{v}_{2d}) \sim N(\boldsymbol{0}, \boldsymbol{I}_{DT}), \quad \boldsymbol{v} = (\boldsymbol{v}_1', \boldsymbol{v}_2')' \sim N(\boldsymbol{0}, \boldsymbol{I}_{D(T+1)}).$$

For the joint density function of vector $\boldsymbol{v}$ it holds

$$f_v(\boldsymbol{v}_1, \boldsymbol{v}_2) = (2\pi)^{-D(T+1)/2} \exp\left\{-\frac{1}{2}\,\boldsymbol{v}_1'\boldsymbol{v}_1 - \frac{1}{2}\,\boldsymbol{v}_2'\boldsymbol{v}_2\right\}.$$

The distribution of the target variable y_{dt}, conditioned to the random effects $v_{1,d}$ and $v_{2,dt}$, is

$$y_{dt}|_{v_{1,d}, v_{2,dt}} \sim \text{Poisson}(m_{dt} p_{dt}), \quad d = 1, \ldots, D, \ t = 1, \ldots, T,$$

where the offset (or size) parameters $m_{dt} > 0$ are known. For the natural parameters, $d = 1, \ldots, D, t = 1, \ldots, T$, we assume

$$\eta_{dt} = \log \mu_{dt} = \log m_{dt} + \log p_{dt} = \log m_{dt} + \boldsymbol{x}_{dt}\boldsymbol{\beta} + \phi_1 v_{1,d} + \phi_2 v_{2,dt},$$

where $\mu_{dt} = E[y_{dt}|v_{1,d}, v_{2,dt}]$, $\boldsymbol{\beta} = \underset{1 \le k \le p}{\text{col}}(\beta_k)$ is the column vector of regression parameters and $\boldsymbol{x}_{dt} = \underset{1 \le k \le p}{\text{col}'}(x_{dtk})$ is the row vector of known auxiliary variables. Further, we assume that the y_{dt}'s are independent conditioned to $\boldsymbol{v}_1$ and $\boldsymbol{v}_2$. Let us denote

$$\boldsymbol{y}_d = \underset{1 \le t \le T}{\text{col}}(y_{dt}), \quad \boldsymbol{y} = \underset{1 \le d \le D}{\text{col}}(\boldsymbol{y}_d).$$

It holds that

$$P(y_{dt}|\boldsymbol{v}) = P(y_{dt}|v_{1,d}, v_{2,dt}) = \frac{1}{y_{dt}!} \exp\{-m_{dt} p_{dt}\} m_{dt}^{y_{dt}} p_{dt}^{y_{dt}},$$

$$p_{dt} = \exp\{\boldsymbol{x}_{dt}\boldsymbol{\beta} + \phi_1 v_{1,d} + \phi_2 v_{2,dt}\},$$

$$P(\boldsymbol{y}|\boldsymbol{v}) = \prod_{d=1}^{D}\prod_{t=1}^{T} P(y_{dt}|\boldsymbol{v}),$$

$$P(\boldsymbol{y}) = \int_{R^{D(T+1)}} P(\boldsymbol{y}|\boldsymbol{v}) f_v(\boldsymbol{v}_1, \boldsymbol{v}_2)\, d\boldsymbol{v}_1 d\boldsymbol{v}_2 = \int_{R^{D(T+1)}} \psi(\boldsymbol{y}, \boldsymbol{v})\, d\boldsymbol{v},$$

where

$$\psi(\boldsymbol{y}, \boldsymbol{v}) = (2\pi)^{-\frac{D(T+1)}{2}} \exp\left\{\frac{-\boldsymbol{v}_1'\boldsymbol{v}_1 - \boldsymbol{v}_2'\boldsymbol{v}_2}{2}\right\} \prod_{d=1}^{D}\prod_{t=1}^{T} \frac{\exp\{-m_{dt}p_{dt}\} m_{dt}^{y_{dt}} p_{dt}^{y_{dt}}}{y_{dt}!}$$

$$= c(\boldsymbol{y}) \exp\left\{\frac{-\boldsymbol{v}_1'\boldsymbol{v}_1 - \boldsymbol{v}_2'\boldsymbol{v}_2}{2}\right\}$$

$$\cdot \exp\left\{\sum_{d=1}^{D}\sum_{t=1}^{T}\{-m_{dt}\exp\{\boldsymbol{x}_{dt}\boldsymbol{\beta} + \phi_1 v_{1,d} + \phi_2 v_{2,dt}\}\}\right\}$$

$$\cdot \exp\left\{\sum_{k=1}^{p}\Big(\sum_{d=1}^{D}\sum_{t=1}^{T} y_{dt}x_{dtk}\Big)\beta_k + \phi_1 \sum_{d=1}^{D} y_{d.}v_{1,d} + \phi_2 \sum_{d=1}^{D}\sum_{t=1}^{T} y_{dt}v_{2,dt}\right\},$$

$c(\boldsymbol{y}) = (2\pi)^{-\frac{D(T+1)}{2}} \prod_{d=1}^{D}\prod_{t=1}^{T}\left(m_{dt}^{y_{dt}}/y_{dt}!\right)$ and $y_{d.} = \sum_{t=1}^{T} y_{dt}$.

Remark 21.1 The presented model can be easily generalized to a model with correlated time effects. The only difference would be in the definition of the domain-time random effects $v_{2,dt}$. Let us, for example, consider that $\{v_{2,dt} : d = 1, \ldots, D,\ t = 1, \ldots, T\}$ are AR(1)-correlated within each domain d and independent between domains. For $d = 1, \ldots, D$, we assume

$$\boldsymbol{v}_{2d} = \underset{1\leq t\leq T}{\text{col}}\,(v_{2,dt}) \sim N(\boldsymbol{0}, \Omega_d(\rho)),\ \boldsymbol{v}_2 = \underset{1\leq d\leq D}{\text{col}}\,(\boldsymbol{v}_{2d}) \sim N(\boldsymbol{0}, \Omega(\rho)),$$

$$\Omega(\rho) = \underset{1\leq d\leq D}{\text{diag}}\,(\Omega_d(\rho)),\quad \Omega_d = \Omega_d(\rho) = \frac{\boldsymbol{A}_d(\rho)}{1-\rho^2},$$

$$\boldsymbol{A}_d(\rho) = \begin{pmatrix} 1 & \rho & \ldots & \rho^{T-2} & \rho^{T-1} \\ \rho & 1 & \ddots & & \rho^{T-2} \\ \vdots & \ddots & \ddots & \ddots & \vdots \\ \rho^{T-2} & & \ddots & 1 & \rho \\ \rho^{T-1} & \rho^{T-2} & \ldots & \rho & 1 \end{pmatrix}_{T\times T},$$

where $\boldsymbol{v}_{21}, \ldots, \boldsymbol{v}_{2D}$ are independent and $\boldsymbol{v}_1$ and $\boldsymbol{v}_2$ are independent. As the normal multivariate p.d.f of $\boldsymbol{X} \sim N_n(\boldsymbol{\mu}, \boldsymbol{\Sigma})$ is

$$f_{N_n(\mu,\Sigma)}(\boldsymbol{x}) = \frac{1}{(2\pi)^{n/2}\sqrt{|\boldsymbol{\Sigma}|}} \exp\left\{-\frac{1}{2}(\boldsymbol{x}-\boldsymbol{\mu})'\boldsymbol{\Sigma}^{-1}(\boldsymbol{x}-\boldsymbol{\mu})\right\},$$

we have

$$f_v(\boldsymbol{v}_1, \boldsymbol{v}_2) = (2\pi)^{-D(T+1)/2} |\Omega_d(\rho)|^{-D/2} \exp\left\{-\frac{1}{2}\boldsymbol{v}_1'\boldsymbol{v}_1 - \frac{1}{2}\sum_{d=1}^{D} \boldsymbol{v}_{2d}'\Omega_d^{-1}(\rho)\boldsymbol{v}_{2d}\right\}.$$

Under this setup, we obtain the marginal p.d.f. of $\boldsymbol{y}$ in the form

$$P(\boldsymbol{y}) = \int_{R^{D(T+1)}} \psi(\boldsymbol{y}, \boldsymbol{v})\, d\boldsymbol{v},$$

where

$$\psi(\boldsymbol{y}, \boldsymbol{v}) = (2\pi)^{-\frac{D(T+1)}{2}} |\Omega_d(\rho)|^{-D/2} \exp\left\{-\frac{1}{2}\boldsymbol{v}_1'\boldsymbol{v}_1 - \frac{1}{2}\sum_{d=1}^{D} \boldsymbol{v}_{2d}'\Omega_d^{-1}(\rho)\boldsymbol{v}_{2d}\right\}$$
$$\cdot \prod_{d=1}^{D}\prod_{t=1}^{T} \frac{\exp\{-m_{dt}p_{dt}\} m_{dt}^{y_{dt}} p_{dt}^{y_{dt}}}{y_{dt}!}$$
$$= c(\boldsymbol{y})|\Omega_d(\rho)|^{-D/2} \exp\left\{-\frac{1}{2}\boldsymbol{v}_1'\boldsymbol{v}_1 - \frac{1}{2}\sum_{d=1}^{D} \boldsymbol{v}_{2d}'\Omega_d^{-1}(\rho)\boldsymbol{v}_{2d}\right\}$$
$$\cdot \exp\left\{-\sum_{d=1}^{D}\sum_{t=1}^{T} m_{dt} \exp\{\boldsymbol{x}_{dt}\boldsymbol{\beta} + \phi_1 v_{1,d} + \phi_2 v_{2,dt}\}\right\}$$
$$\cdot \exp\left\{\sum_{k=1}^{p}\Big(\sum_{d=1}^{D}\sum_{t=1}^{T} y_{dt}x_{dtk}\Big)\beta_k + \phi_1 \sum_{d=1}^{D} y_{d.}v_{1,d} + \phi_2 \sum_{d=1}^{D}\sum_{t=1}^{T} y_{dt}v_{2,dt}\right\},$$

$c(\boldsymbol{y}) = (2\pi)^{-\frac{D(T+1)}{2}} \prod_{d=1}^{D}\prod_{t=1}^{T}\left(m_{dt}^{y_{dt}}/y_{dt}!\right)$ and $y_{d.} = \sum_{t=1}^{T} y_{dt}$.

This model with AR(1)-correlated time random effect was largely described by Boubeta et al. (2017). These authors fitted the model to data of the Spanish living conditions survey and applied the developed methodology to the estimation of the proportion of people under the poverty line by counties and sex in Galicia (a region in north-west of Spain). This chapter does not present their mathematical derivations.

21.3 ML-Laplace Approximation Algorithm

Let $h : R \mapsto R$ be a twice continuously differentiable function with a global maximum at x_0. This is to say, let us assume that $\dot{h}(x_0) = 0$ and $\ddot{h}(x_0) < 0$. As shown in Sect. 14.5, the univariate Laplace approximation of the integral of the function $\exp(h(x))$ is

$$\int_{-\infty}^{\infty} e^{h(x)}\,dx \approx (2\pi)^{1/2}\big(-\ddot{h}(x_0)\big)^{-1/2} e^{h(x_0)}. \tag{21.1}$$

Let us now approximate the log-likelihood of the area-level Poisson mixed model with independent time effects. We recall that the sets of random effects $\{v_{1,d}\}$ and $\{v_{2,dt}\}$ are independent and i.i.d N(0,1). For $d = 1, \dots, D$, $t = 1, \dots, T$, we recall that

$$y_{dt}|_{v_{1,d},v_{2,dt}} \underset{ind}{\sim} \text{Poiss}(m_{dt}p_{dt}), \quad p_{dt} = \exp\left\{\boldsymbol{x}_d\boldsymbol{\beta} + \phi_1 v_{1,d} + \phi_2 v_{2,dt}\right\}.$$

Under the introduced notation, it holds that $\boldsymbol{y}_1, \dots, \boldsymbol{y}_D$ are unconditionally independent with marginal p.d.f.

$$P(\boldsymbol{y}_d) = \int_{-\infty}^{\infty} P(\boldsymbol{y}_d|v_{1,d}) f(v_{1,d})\,dv_{1,d}, \quad d = 1, \dots, D,$$

and that $y_{d1}, \dots, y_{dT}$ are conditionally independent given $v_{1,d}$. We have

$$P(\boldsymbol{y}_d|v_{1,d}) = \prod_{t=1}^{T} P(y_{dt}|v_{1,d}),$$

$$P_{dt} = P(y_{dt}|v_{1,d}) = \int_{-\infty}^{\infty} P(y_{dt}|v_{1,d}, v_{2,dt}) f(v_{2,dt})\,dv_{2,dt}$$

and

$$\begin{aligned}
P_{dt} &= \int_{-\infty}^{\infty} \frac{m_{dt}^{y_{dt}}}{y_{dt}!} \exp\{-m_{dt}p_{dt}\} p_{dt}^{y_{dt}} (2\pi)^{-1/2} \exp\{-\frac{1}{2}v_{2,dt}^2\}\,dv_{2,dt} \\
&= \frac{m_{dt}^{y_{dt}}}{(2\pi)^{1/2} y_{dt}!} \int_{-\infty}^{\infty} \exp\Big\{ - m_{dt}\exp\{\boldsymbol{x}_{dt}\boldsymbol{\beta} + \phi_1 v_{1,d} + \phi_2 v_{2,dt}\} \\
&\quad + y_{dt}(\boldsymbol{x}_{dt}\boldsymbol{\beta} + \phi_1 v_{1,d} + \phi_2 v_{2,dt}) - \frac{1}{2}v_{2,dt}^2\Big\}\,dv_{2,dt} \\
&= \frac{m_{dt}^{y_{dt}}}{(2\pi)^{1/2} y_{dt}!} \exp\{y_{dt}(\boldsymbol{x}_{dt}\boldsymbol{\beta} + \phi_1 v_{1,d})\} \\
&\quad \cdot \int_{-\infty}^{\infty} \exp\{-m_{dt}\exp\{\boldsymbol{x}_{dt}\boldsymbol{\beta} + \phi_1 v_{1,d} + \phi_2 v_{2,dt}\} + \phi_2 y_{dt} v_{2,dt} - \frac{1}{2}v_{2,dt}^2\}\,dv_{2,dt} \\
&= \frac{m_{dt}^{y_{dt}}}{(2\pi)^{1/2} y_{dt}!} \exp\{y_{dt}(\boldsymbol{x}_{dt}\boldsymbol{\beta} + \phi_1 v_{1,d})\} \int_{-\infty}^{\infty} \exp\left\{h_{dt}(v_{2,dt})\right\}\,dv_{2,dt},
\end{aligned} \tag{21.2}$$

where

$$h_{dt}(v_{2,dt}) = -m_{dt}\exp\{\boldsymbol{x}_{dt}\boldsymbol{\beta}+\phi_1 v_{1,d}+\phi_2 v_{2,dt}\}+\phi_2 y_{dt} v_{2,dt} - \frac{1}{2}v_{2,dt}^2 \,. \tag{21.3}$$

The first two derivatives of the function $h_{dt}(v_{2,dt})$ are

$$\dot{h}_{dt}(v_{2,dt}) = -m_{dt}\phi_2\exp\{\boldsymbol{x}_{dt}\boldsymbol{\beta}+\phi_1 v_{1,d}+\phi_2 v_{2,dt}\}+\phi_2 y_{dt} - v_{2,dt},$$
$$\ddot{h}_{dt}(v_{2,dt}) = -(1+m_{dt}\phi_2^2\exp\{\boldsymbol{x}_{dt}\boldsymbol{\beta}+\phi_1 v_{1,d}+\phi_2 v_{2,dt}\}).$$

For $d = 1,\ldots,D$, $t = 1,\ldots,T$, the Newton–Raphson algorithm maximizes $h_{dt}(v_{2,dt}) = h_{dt}(v_{1,d}, v_{2,dt}, \boldsymbol{\theta})$, defined in (21.3), with $\boldsymbol{\theta} = (\boldsymbol{\beta}', \phi_1, \phi_2)' = \boldsymbol{\theta}_0$ and $v_{1,d}$ fixed. The updating equation is

$$v_{2,dt}^{(i+1)} = v_{2,dt}^{(i)} - \frac{\dot{h}_{dt}(v_{1,d}, v_{2,dt}^{(i)}, \boldsymbol{\theta}_0)}{\ddot{h}_{dt}(v_{1,d}, v_{2,dt}^{(i)}, \boldsymbol{\theta}_0)}. \tag{21.4}$$

Let us denote by $v_{2,dt0}$ the argument of maxima of the function $h_{dt}(v_{2,dt})$. Thus, it holds $\dot{h}_{dt}(v_{2,dt0}) = 0$, $\ddot{h}_{dt}(v_{2,dt0}) < 0$ and by applying (21.1) with $v_{2,dt} = v_{2,dt0}$ to (21.2), we get an approximation of $P_{dt} = P(y_{dt}|v_{1,d})$, i.e.

$$\begin{aligned}
P_{dt} &\approx \frac{m_{dt}^{y_{dt}}}{y_{dt}!}\exp\{y_{dt}(\boldsymbol{x}_{dt}\boldsymbol{\beta}+\phi_1 v_{1,d})\}(1+m_{dt}\phi_2^2\exp\{\boldsymbol{x}_{dt}\boldsymbol{\beta}+\phi_1 v_{1,d}+\phi_2 v_{2,dt0}\})^{-1/2} \\
&\cdot \exp\Big\{-m_{dt}\exp\{\boldsymbol{x}_{dt}\boldsymbol{\beta}+\phi_1 v_{1,d}+\phi_2 v_{2,dt0}\}+\phi_2 y_{dt} v_{2,dt0} - \frac{1}{2}v_{2,dt0}^2\Big\} \\
&= \frac{m_{dt}^{y_{dt}}}{y_{dt}!}(1+m_{dt}\phi_2^2\exp\{\boldsymbol{x}_{dt}\boldsymbol{\beta}+\phi_1 v_{1,d}+\phi_2 v_{2,dt0}\})^{-1/2} \\
&\cdot \exp\Big\{-m_{dt}\exp\{\boldsymbol{x}_{dt}\boldsymbol{\beta}+\phi_1 v_{1,d}+\phi_2 v_{2,dt0}\}+y_{dt}(\boldsymbol{x}_{dt}\boldsymbol{\beta}+\phi_1 v_{1,d}+\phi_2 v_{2,dt0}) \\
&- \frac{1}{2}v_{2,dt0}^2\Big\}.
\end{aligned}$$

The marginal p.d.f. of $\boldsymbol{y}_d$ is

$$\begin{aligned}
P(\boldsymbol{y}_d) &= \int_{-\infty}^{\infty} P(\boldsymbol{y}_d|v_{1,d})f(v_{1,d})\,dv_{1,d} = \int_{-\infty}^{\infty}\prod_{t=1}^{T} P(y_{dt}|v_{1,d})f(v_{1,d})\,dv_{1,d} \\
&\approx \frac{\left(\prod_{t=1}^{T}\frac{m_{dt}^{y_{dt}}}{y_{dt}!}\right)}{(2\pi)^{1/2}}\int_{-\infty}^{\infty}\exp\Big\{-\frac{1}{2}\sum_{t=1}^{T}\log(1+m_{dt}\phi_2^2\exp\{\boldsymbol{x}_{dt}\boldsymbol{\beta}+\phi_1 v_{1,d}+\phi_2 v_{2,dt0}\})\Big\} \\
&\cdot \exp\Big\{\sum_{t=1}^{T}\{-m_{dt}\exp\{\boldsymbol{x}_{dt}\boldsymbol{\beta}+\phi_1 v_{1,d}+\phi_2 v_{2,dt0}\}+y_{dt}(\boldsymbol{x}_{dt}\boldsymbol{\beta}+\phi_1 v_{1,d}+\phi_2 v_{2,dt0})\}\Big\} \\
&\cdot \exp\Big\{-\sum_{t=1}^{T}\frac{1}{2}v_{2,dt0}^2\Big\}\exp\Big\{-\frac{1}{2}v_{1,d}^2\Big\}\,dv_{1,d}\,.
\end{aligned}$$

Therefore, we can write

$$P(y_d) \approx \Big(\prod_{t=1}^{T} \frac{m_{dt}^{y_{dt}}}{y_{dt}!}\Big)\frac{1}{(2\pi)^{1/2}}\int_{-\infty}^{\infty} \exp\{h_d(v_{1,d})\}\, dv_{1,d}, \tag{21.5}$$

where

$$h_d(v_{1,d}) = \sum_{t=1}^{T}\Big\{-\frac{1}{2}\log(1+m_{dt}\phi_2^2\exp\{\boldsymbol{x}_{dt}\boldsymbol{\beta}+\phi_1 v_{1,d}+\phi_2 v_{2,dt0}\})$$
$$-m_{dt}\exp\{\boldsymbol{x}_{dt}\boldsymbol{\beta}+\phi_1 v_{1,d}+\phi_2 v_{2,dt0}\}+y_{dt}(\boldsymbol{x}_{dt}\boldsymbol{\beta}+\phi_1 v_{1,d}+\phi_2 v_{2,dt0})-\frac{1}{2}v_{2,dt0}^2\Big\}$$
$$-\frac{1}{2}v_{1,d}^2 = \sum_{t=1}^{T}\Big\{-\frac{1}{2}\log(1+m_{dt}\phi_2^2 p_{0dt})-m_{dt}p_{0dt}+y_{dt}\log p_{0dt}-\frac{1}{2}v_{2,dt0}^2\Big\}$$
$$-\frac{1}{2}v_{1,d}^2, \tag{21.6}$$

and $p_{0dt} = \exp\{\boldsymbol{x}_{dt}\boldsymbol{\beta}+\phi_1 v_{1,d}+\phi_2 v_{2,dt0}\}$. The derivatives of p_{0dt} and $\log p_{0dt}$, with respect to $v_{1,d}$, are

$$\frac{\partial p_{0dt}}{\partial v_{1,d}} = p_{0dt}\phi_1, \quad \frac{\partial \log p_{0dt}}{\partial v_{1,d}} = \phi_1.$$

The derivatives of $h_d(v_{1,d})$, with respect to $v_{1,d}$, are

$$\dot h_d(v_{1,d}) = \sum_{t=1}^{T}\Big\{-\frac{1}{2}\frac{m_{dt}\phi_1\phi_2^2 p_{0dt}}{1+m_{dt}\phi_2^2 p_{0dt}}-m_{dt}\phi_1 p_{0dt}+y_{dt}\phi_1\Big\}-v_{1,d},$$

$$\ddot h_d(v_{1,d}) = \sum_{t=1}^{T}\Big\{-\frac{1}{2}m_{dt}\phi_1\phi_2^2\frac{p_{0dt}\phi_1[1+m_{dt}\phi_2^2 p_{0dt}]-p_{0dt}m_{dt}\phi_1\phi_2^2 p_{0dt}}{[1+m_{dt}\phi_2^2 p_{0dt}]^2}$$
$$-m_{dt}\phi_1^2 p_{0dt}\Big\}-1 = -\Big(1+\sum_{t=1}^{T}\Big\{\frac{1}{2}\frac{m_{dt}\phi_1^2\phi_2^2 p_{0dt}}{[1+m_{dt}\phi_2^2 p_{0dt}]^2}+m_{dt}\phi_1^2 p_{0dt}\Big\}\Big).$$

For $\boldsymbol{\theta} = (\boldsymbol{\beta}', \phi_1, \phi_2) = \boldsymbol{\theta}_0$ and $v_{2,dt0}$ fixed, the function $h_d(v_{1,d}) = h_d(v_{1,d}, v_{2,dt0}, \boldsymbol{\theta})$, defined in (21.6), can be maximized using the Newton–Raphson algorithm. The updating equation is

$$v_{1,d}^{(i+1)} = v_{1,d}^{(i)} - \frac{\dot h_d(v_{1,d}^{(i)}, v_{2,dt0}, \boldsymbol{\theta}_0)}{\ddot h_d(v_{1,d}^{(i)}, v_{2,dt0}, \boldsymbol{\theta}_0)}, \quad d = 1, \ldots, D. \tag{21.7}$$

Let us denote by $v_{1,d0}$ the argument of maxima of the function $h_d(v_{1,d})$. Thus, it holds $\dot{h}_d(v_{1,d0}) = 0$, $\ddot{h}_d(v_{1,d0}) < 0$ and by applying (21.1) with $v_{1,d} = v_{1,d0}$ to (21.5), we get

$$P(\boldsymbol{y}_d) \approx \Big(\prod_{t=1}^{T} \frac{m_{dt}^{y_{dt}}}{y_{dt}!}\Big) \Big(1+\sum_{t=1}^{T}\Big\{\frac{1}{2}\frac{m_{dt}\phi_1^2\phi_2^2 p_{dt0}}{[1+m_{dt}\phi_2^2 p_{dt0}]^2} + m_{dt}\phi_1^2 p_{dt0}\Big\}\Big)^{-1/2}$$
$$\cdot \exp\Big\{\sum_{t=1}^{T}\Big\{y_{dt}\log p_{dt0} - \frac{1}{2}\log(1+m_{dt}\phi_2^2 p_{dt0}) - m_{dt}p_{dt0} - \frac{1}{2}v_{2,dt0}^2\Big\}$$
$$-\frac{1}{2}v_{1,d0}^2\Big\},$$

where $p_{dt0} = \exp\{\boldsymbol{x}_{dt}\boldsymbol{\beta} + \phi_1 v_{1,d0} + \phi_2 v_{2,dt0}\}$. The approximated log-likelihood is

$$\ell = \sum_{d=1}^{D} \log P(\boldsymbol{y}_d) \approx \ell_0 = \sum_{d=1}^{D}\sum_{t=1}^{T} \ell_{dt} - \frac{1}{2}\sum_{d=1}^{D}\ell_d,$$

where

$$\ell_{dt} = y_{dt}\log m_{dt} - \log y_{dt}! + y_{dt}\log p_{dt0} - \frac{1}{2}\log(1+m_{dt}\phi_2^2 p_{dt0}) - m_{dt}p_{dt0} - \frac{1}{2}v_{2,dt0}^2$$

and

$$\ell_d = \log\Big(1+\sum_{t=1}^{T}\Big\{\frac{1}{2}\frac{m_{dt}\phi_1^2\phi_2^2 p_{dt0}}{[1+m_{dt}\phi_2^2 p_{dt0}]^2} + m_{dt}\phi_1^2 p_{dt0}\Big\}\Big) + v_{1,d0}^2.$$

For ease of exposition, let us define $\xi_{dt0} = 1 + m_{dt}\phi_2^2 p_{dt0}$. It holds that

$$\frac{\partial p_{dt0}}{\partial \beta_k} = x_{dtk}p_{dt0}, \quad \frac{\partial p_{dt0}}{\partial \phi_1} = v_{1,d0}p_{dt0}, \quad \frac{\partial p_{dt0}}{\partial \phi_2} = v_{2,dt0}p_{dt0},$$

$$\frac{\partial \xi_{dt0}}{\partial \beta_k} = m_{dt}\phi_2^2 x_{dtk}p_{dt0}, \quad \frac{\partial \xi_{dt0}}{\partial \phi_1} = m_{dt}\phi_2^2 v_{1,d0}p_{dt0}, \quad \frac{\partial \xi_{dt0}}{\partial \phi_2} = (2+\phi_2 v_{2,dt0})m_{dt}\phi_2 p_{dt0}.$$

The first derivatives of ℓ_{dt} with respect to β_k, ϕ_1 and ϕ_2 are

$$\frac{\partial \ell_{dt}}{\partial \beta_k} = y_{dt}x_{dtk} - \frac{m_{dt}x_{dtk}}{2}\frac{\phi_2^2 p_{dt0}}{\xi_{dt0}} - m_{dt}x_{dtk}p_{dt0},$$
$$\frac{\partial \ell_{dt}}{\partial \phi_1} = y_{dt}v_{1,d0} - \frac{m_{dt}}{2}\frac{\phi_2^2 v_{1,d0}p_{dt0}}{\xi_{dt0}} - m_{dt}v_{1,d0}p_{dt0},$$
$$\frac{\partial \ell_{dt}}{\partial \phi_2} = y_{dt}v_{2,dt0} - \frac{m_{dt}}{2}\frac{(2+\phi_2 v_{2,dt0})\phi_2 p_{dt0}}{\xi_{dt0}} - m_{dt}v_{2,dt0}p_{dt0}.$$

The second partial derivatives of ℓ_{dt}, with respect to β_r and β_k, are

$$\frac{\partial^2 \ell_{dt}}{\partial \beta_r \partial \beta_k} = -\frac{m_{dt}x_{dtk}\phi_2^2}{2} \frac{p_{dt0}x_{dtr}(1+m_{dt}\phi_2^2 p_{dt0}) - p_{dt0}m_{dt}\phi_2^2 x_{dtr}p_{dt0}}{\xi_{dt0}^2}$$

$$- m_{dt}x_{dtk}x_{dtr}p_{dt0} = -\frac{m_{dt}x_{dtk}x_{dtr}\phi_2^2}{2} \frac{p_{dt0}}{\xi_{dt0}^2} - m_{dt}x_{dtk}x_{dtr}p_{dt0}$$

$$= -m_{dt}x_{dtk}x_{dtr}p_{dt0}\left(\frac{\phi_2^2}{2\xi_{dt0}^2} + 1\right).$$

The second partial derivatives of ℓ_{dt}, with respect to ϕ_1 and β_k, are

$$\frac{\partial^2 \ell_{dt}}{\partial \phi_1 \partial \beta_k} = -\frac{m_{dt}x_{dtk}\phi_2^2}{2} \frac{p_{dt0}v_{1,d0}(1+m_{dt}\phi_2^2 p_{dt0}) - p_{dt0}m_{dt}\phi_2^2 v_{1,d0}p_{dt0}}{\xi_{dt0}^2}$$

$$- m_{dt}x_{dtk}v_{1,d0}p_{dt0} = -\frac{m_{dt}x_{dtk}v_{1,d0}\phi_2^2}{2} \frac{p_{dt0}}{\xi_{dt0}^2} - m_{dt}x_{dtk}v_{1,d0}p_{dt0}$$

$$= -m_{dt}x_{dtk}v_{1,d0}p_{dt0}\left(\frac{\phi_2^2}{2\xi_{dt0}^2} + 1\right).$$

The second partial derivatives of ℓ_{dt}, with respect to ϕ_2 and β_k, are

$$\frac{\partial^2 \ell_{dt}}{\partial \phi_2 \partial \beta_k} = -\frac{m_{dt}x_{dtk}}{2} \frac{(2\phi_2 p_{dt0} + \phi_2^2 p_{dt0}v_{2,dt0})(1+m_{dt}\phi_2^2 p_{dt0})}{\xi_{dt0}^2}$$

$$- \frac{\phi_2^3 p_{dt0}^2 m_{dt}(2+\phi_2 v_{2,dt0})}{\xi_{dt0}^2} - m_{dt}x_{dtk}v_{2,dt0}p_{dt0}$$

$$= -\frac{m_{dt}x_{dtk}\phi_2 p_{dt0}}{2} \frac{(2+\phi_2 v_{2,dt0})[1+m_{dt}\phi_2^2 p_{dt0} - m_{dt}\phi_2^2 p_{dt0}]}{\xi_{dt0}^2}$$

$$- m_{dt}x_{dtk}v_{2,dt0}p_{dt0} = -\frac{m_{dt}x_{dtk}\phi_2 p_{dt0}}{2} \frac{(2+\phi_2 v_{2,dt0})}{\xi_{dt0}^2}$$

$$- m_{dt}x_{dtk}v_{2,dt0}p_{dt0} = -m_{dt}x_{dtk}v_{2,dt0}p_{dt0}\left(\frac{\phi_2(2+\phi_2 v_{2,dt0})}{2v_{2,dt0}\xi_{dt0}^2} + 1\right).$$

The second partial derivatives of ℓ_{dt}, with respect to ϕ_1 and ϕ_1, are

$$\frac{\partial^2 \ell_{dt}}{\partial \phi_1^2} = -\frac{m_{dt} v_{1,d0} \phi_2^2}{2} \frac{p_{dt0} v_{1,d0}(1 + m_{dt}\phi_2^2 p_{dt0}) - p_{dt0} m_{dt} \phi_2^2 v_{1,d0} p_{dt0}}{\xi_{dt0}^2}$$

$$- m_{dt} v_{1,d0}^2 p_{dt0} = -\frac{m_{dt} v_{1,d0}^2 \phi_2^2}{2} \frac{p_{dt0}}{\xi_{dt0}^2} - m_{dt} v_{1,d0}^2 p_{dt0}$$

$$= -m_{dt} v_{1,d0}^2 p_{dt0} \left(\frac{\phi_2^2}{2\xi_{dt0}^2} + 1 \right).$$

The second partial derivatives of ℓ_{dt}, with respect to ϕ_2 and ϕ_1, are

$$\frac{\partial^2 \ell_{dt}}{\partial \phi_2 \partial \phi_1} = -\frac{m_{dt} v_{1,d0}}{2} \frac{(2\phi_2 p_{dt0} + \phi_2^2 p_{dt0} v_{2,dt0})(1 + m_{dt}\phi_2^2 p_{dt0})}{\xi_{dt0}^2}$$

$$- \frac{\phi_2^3 p_{dt0}^2 m_{dt}(2 + \phi_2 v_{2,dt0})}{\xi_{dt0}^2} - m_{dt} v_{1,d0} v_{2,dt0} p_{dt0}$$

$$= -\frac{m_{dt} v_{1,d0} \phi_2 p_{dt0}}{2} \frac{(2 + \phi_2 v_{2,dt0})[1 + m_{dt}\phi_2^2 p_{dt0} - m_{dt}\phi_2^2 p_{dt0}]}{\xi_{dt0}^2}$$

$$- m_{dt} v_{1,d0} v_{2,dt0} p_{dt0} = -\frac{m_{dt} v_{1,d0} \phi_2 p_{dt0}}{2} \frac{(2 + \phi_2 v_{2,dt0})}{\xi_{dt0}^2}$$

$$- m_{dt} v_{1,d0} v_{2,dt0} p_{dt0} = -m_{dt} v_{1,d0} v_{2,dt0} p_{dt0} \left(\frac{\phi_2(2 + \phi_2 v_{2,dt0})}{2 v_{2,dt0} \xi_{dt0}^2} + 1 \right).$$

The second partial derivatives of ℓ_{dt}, with respect to ϕ_2 and ϕ_2, are

$$\frac{\partial^2 \ell_{dt}}{\partial \phi_2^2} = -\frac{m_{dt}}{2} \frac{p_{dt0}[2 + 4\phi_2 v_{2,dt0} + \phi_2^2 v_{2,dt0}^2](1 + m_{dt}\phi_2^2 p_{dt0})}{\xi_{dt0}^2}$$

$$- \frac{(2 + \phi_2 v_{2,dt0})(2 + \phi_2 v_{2,dt0}) m_{dt} \phi_2^2 p_{dt0}^2}{\xi_{dt0}^2} - m_{dt} v_{2,dt0}^2 p_{dt0}$$

$$= -\frac{m_{dt} p_{dt0}}{2} \frac{[-2 + (2 + \phi_2 v_{2,dt0})^2](1 + m_{dt}\phi_2^2 p_{dt0})}{\xi_{dt0}^2}$$

$$- \frac{(2 + \phi_2 v_{2,dt0})^2 m_{dt} \phi_2^2 p_{dt0}}{\xi_{dt0}^2} - m_{dt} v_{2,dt0}^2 p_{dt0}$$

$$= -\frac{m_{dt} p_{dt0}}{2} \frac{-2(1 + m_{dt}\phi_2^2 p_{dt0}) + (2 + \phi_2 v_{2,dt0})^2 (1 + m_{dt}\phi_2^2 p_{dt0} - m_{dt}\phi_2^2 p_{dt0})}{\xi_{dt0}^2}$$

$$- m_{dt}v_{2,dt0}^2 p_{dt0} = -\frac{m_{dt}p_{dt0}}{2}\frac{(2+\phi_2 v_{2,dt0})^2 - 2\xi_{dt0}}{\xi_{dt0}^2} - m_{dt}v_{2,dt0}^2 p_{dt0}$$

$$= -m_{dt}v_{2,dt0}^2 p_{dt0}\left(\frac{(2+\phi_2 v_{2,dt0})^2 - 2\xi_{dt0}}{2v_{2,dt0}^2\xi_{dt0}^2} + 1\right).$$

Let us denote $\boldsymbol{\theta} = (\beta_1, \ldots, \beta_p, \phi_1, \phi_2) = (\theta_1, \ldots, \theta_{p+2})$ and $\ell_d = \ell_d(\boldsymbol{\theta})$, where

$$\ell_d(\boldsymbol{\theta}) = \log\left(1 + \sum_{t=1}^{T}\left\{\frac{1}{2}\frac{m_{dt}\phi_1^2\phi_2^2 p_{dt0}}{[1+m_{dt}\phi_2^2 p_{dt0}]^2} + m_{dt}\phi_1^2 p_{dt0}\right\}\right) + v_{1,d0}^2$$

$$= \log\left(1 + \sum_{t=1}^{T} s_{dt}\right) + v_{1,d0}^2,$$

and

$$s_{dt} = s_{dt}(\boldsymbol{\theta}) = \frac{1}{2}\frac{m_{dt}\phi_1^2\phi_2^2 p_{dt0}}{[1+m_{dt}\phi_2^2 p_{dt0}]^2} + m_{dt}\phi_1^2 p_{dt0},\ d = 1, \ldots, D,\ t = 1, \ldots, T.$$

The first order partial derivatives of ℓ_d are

$$\frac{\partial \ell_d}{\partial \theta_k} = \frac{\sum_{t=1}^{T}\frac{\partial s_{dt}}{\partial \theta_k}}{1 + \sum_{t=1}^{T} s_{dt}}, \qquad k = 1, \ldots, p+2.$$

The second order partial derivatives of ℓ_d, with respect to θ_r and θ_k, $r = 1, \ldots, p+2$, $k = 1, \ldots, p+2$, are

$$\frac{\partial^2 \ell_d}{\partial \theta_r \partial \theta_k} = \frac{\left(\sum_{t=1}^{T}\frac{\partial^2 s_{dt}}{\partial \theta_r \partial \theta_k}\right)\left(1 + \sum_{t=1}^{T} s_{dt}\right) - \left(\sum_{t=1}^{T}\frac{\partial s_{dt}}{\partial \theta_k}\right)\left(\sum_{t=1}^{T}\frac{\partial s_{dt}}{\partial \theta_r}\right)}{\left(1 + \sum_{t=1}^{T} s_{dt}\right)^2}.$$

For $k, r = 1, \ldots, p+2$ the components of the score vector and the Hessian matrix are

$$U_{0k} = \frac{\partial \ell_0}{\partial \theta_k} = \sum_{d=1}^{D}\sum_{t=1}^{T}\frac{\partial \ell_{dt}}{\partial \theta_k} - \frac{1}{2}\sum_{d=1}^{D}\frac{\partial \ell_d}{\partial \theta_k},$$

$$H_{0kr} = \frac{\partial^2 \ell_0}{\partial \theta_r \partial \theta_k} = \sum_{d=1}^{D}\sum_{t=1}^{T}\frac{\partial \ell_{dt}^2}{\partial \theta_r \partial \theta_k} - \frac{1}{2}\sum_{d=1}^{D}\frac{\partial \ell_d^2}{\partial \theta_r \partial \theta_k}.$$

In matrix form, we have $\boldsymbol{U}_0 = \boldsymbol{U}_0(\boldsymbol{\theta}) = \underset{1\leq k\leq p+2}{\text{col}}(U_{0k})$ and $\boldsymbol{H}_0 = \boldsymbol{H}_0(\boldsymbol{\theta}) = (H_{0kr})_{k,r=1,\dots,p+2}$.

The Newton–Raphson algorithm maximizes $\ell_0 = \ell_0(\boldsymbol{\theta})$, with fixed $v_{1,d} = v_{1,d0}$ and $v_{2,dt} = v_{2,dt0}$, $d = 1, \dots, D$, $t = 1, \dots, T$. The updating equation is

$$\boldsymbol{\theta}^{(i+1)} = \boldsymbol{\theta}^{(i)} - \boldsymbol{H}_0^{-1}(\boldsymbol{\theta}^{(i)})\boldsymbol{U}_0(\boldsymbol{\theta}^{(i)}). \tag{21.8}$$

Combining the three described Newton–Raphson algorithms we get the final ML-Laplace approximation algorithm in the form:

1. Set the initial values $i = 0$, $\epsilon_1 > 0$, $\epsilon_2 > 0$, $\epsilon_3 > 0$, $\epsilon_4 > 0$, $\boldsymbol{\theta}^{(0)}$, $\boldsymbol{\theta}^{(-1)} = \boldsymbol{\theta}^{(0)} + \mathbf{1}$, $v_{1,d}^{(0)} = 0$, $v_{1,d}^{(-1)} = 1$, $v_{2,dt}^{(0)} = 0$, $v_{2,dt}^{(-1)} = 1$, $d = 1, \dots, D$, $t = 1, \dots, T$.
2. Until $\|\boldsymbol{\theta}^{(i)} - \boldsymbol{\theta}^{(i-1)}\|_2 < \varepsilon_1$, $|v_{1,d}^{(i)} - v_{1,d}^{(i-1)}| < \varepsilon_2$, $|v_{2,dt}^{(i)} - v_{2,dt}^{(i-1)}| < \varepsilon_2$, $d = 1, \dots, D$, $t = 1, \dots, T$, do

   a. Apply algorithm (21.4) with seeds $v_{2,dt}^{(i)}$, $d = 1, \dots, D$, $t = 1, \dots, T$, convergence tolerance ε_3 and $v_{1,d} = v_{1,d}^{(i)}$, $\boldsymbol{\theta}_0 = \boldsymbol{\theta}^{(i)}$ fixed. Output: $v_{2,dt}^{(i+1)}$, $d = 1, \dots, D$, $t = 1, \dots, T$.
   b. Apply algorithm (21.7) with seeds $v_{1,d}^{(i)}$, $d = 1, \dots, D$, convergence tolerance ε_3 and $v_{2,dt0} = v_{2,dt}^{(i+1)}$, $\boldsymbol{\theta}_0 = \boldsymbol{\theta}^{(i)}$ fixed. Output: $v_{1,d}^{(i+1)}$, $d = 1, \dots, D$.
   c. Apply algorithm (21.8) with seed $\boldsymbol{\theta}^{(i)}$, convergence tolerance ε_4 and $v_{1,d0} = v_{1,d}^{(i+1)}$, $v_{2,dt0} = v_{2,dt}^{(i+1)}$ fixed, $d = 1, \dots, D$, $t = 1, \dots, T$. Output: $\boldsymbol{\theta}^{(i+1)}$.
   d. $i \leftarrow i + 1$.

3. Output: $\hat{\boldsymbol{\theta}} = \boldsymbol{\theta}^{(i)}$, $\hat{v}_{1,d} = v_{1,d}^{(i)}$, $\hat{v}_{2,dt} = v_{2,dt}^{(i)}$, $d = 1, \dots, D$, $t = 1, \dots, T$.

21.4 Empirical Best Predictors

This section derives the empirical best predictors (EBP) of p_{dt}, $v_{1,d}$, and $v_{2,dt}$ under the area-level Poisson mixed model with independent time effects.

The conditional distribution of $\boldsymbol{y}$, given $\boldsymbol{v}_1$ and $\boldsymbol{v}_2$, is

$$P(\boldsymbol{y}|\boldsymbol{v}_1, \boldsymbol{v}_2) = \prod_{d=1}^{D} P(\boldsymbol{y}_d|v_{1,d}, \boldsymbol{v}_{2,d}), \quad P(\boldsymbol{y}_d|v_{1,d}, \boldsymbol{v}_{2,d}) = \prod_{t=1}^{T} P(y_{dt}|v_{1,d}, v_{2,dt}),$$

where

$$P(y_{dt}|v_{1,d}, v_{2,dt}) = \frac{1}{y_{dt}!} \exp\{-m_{dt}p_{dt}\} m_{dt}^{y_{dt}} p_{dt}^{y_{dt}}$$
$$= c_{dt} \exp\Big\{ y_{dt}(\boldsymbol{x}_{dt}\boldsymbol{\beta} + \phi_1 v_{1,d} + \phi_2 v_{2,dt})$$

$$- m_{dt} \exp\{\boldsymbol{x}_{dt}\boldsymbol{\beta} + \phi_1 v_{1,d} + \phi_2 v_{2,dt}\}\Big\}$$

and $c_{dt} = 1/(y_{dt}!)\, m_{dt}^{y_{dt}}$. The p.d.f. of $\boldsymbol{v} = (\boldsymbol{v}_1, \boldsymbol{v}_2)$ is

$$f(\boldsymbol{v}_1, \boldsymbol{v}_2) = f(\boldsymbol{v}_1) f(\boldsymbol{v}_2), \quad f(\boldsymbol{v}_1) = \prod_{d=1}^{D} f(v_{1,d}),$$

$$f(\boldsymbol{v}_2) = \prod_{d=1}^{D} f(\boldsymbol{v}_{2,d}), \quad f(\boldsymbol{v}_{2,d}) = \prod_{t=1}^{T} f(v_{2,dt}),$$

where

$$f(v_{1,d}) = (2\pi)^{-1/2} \exp\Big\{-\frac{1}{2} v_{1,d}^2\Big\}, \quad f(v_{2,dt}) = (2\pi)^{-1/2} \exp\Big\{-\frac{1}{2} v_{2,dt}^2\Big\}.$$

The best predictor of p_{dt} is $\hat{p}_{dt}(\boldsymbol{\theta}) = E_\theta[p_{dt}|\boldsymbol{y}]$ (cf. Sect. 20.7). In this case, we have that $E_\theta[p_{dt}|\boldsymbol{y}] = E_\theta[p_{dt}|\boldsymbol{y}_d]$ and

$$E_\theta[p_{dt}|\boldsymbol{y}_d] = \frac{A_{dt}(\boldsymbol{y}_d, \boldsymbol{\theta})}{D_d(\boldsymbol{y}_d, \boldsymbol{\theta})} = \frac{A_{dt}}{D_d},$$

where

$$A_{dt} = \int_{R^{T+1}} \exp\{\boldsymbol{x}_{dt}\boldsymbol{\beta} + \phi_1 v_{1,d} + \phi_2 v_{2,dt}\} P(\boldsymbol{y}_d|v_{1,d}, \boldsymbol{v}_{2,d}) f(v_{1,d}) f(\boldsymbol{v}_{2,d})\, dv_{1,d} d\boldsymbol{v}_{2,d},$$

$$D_d = \int_{R^{T+1}} P(\boldsymbol{y}_d|v_{1,d}, \boldsymbol{v}_{2,d}) f(v_{1,d}) f(\boldsymbol{v}_{2,d})\, dv_{1,d} d\boldsymbol{v}_{2,d}.$$

It holds that

$$A_{dt} = \int_{R^{T+1}} \prod_{\tau=1}^{T} \exp\Big\{(y_{d\tau} + \delta_{t\tau})(\boldsymbol{x}_{d\tau}\boldsymbol{\beta} + \phi_1 v_{1,d} + \phi_2 v_{2,d\tau})$$
$$- \nu_{d\tau} \exp\{\boldsymbol{x}_{d\tau}\boldsymbol{\beta} + \phi_1 v_{1,d} + \phi_2 v_{2,d\tau}\}\Big\} f(v_{1,d}) f(\boldsymbol{v}_{2,d})\, dv_{1,d} d\boldsymbol{v}_{2,d}$$
$$= \int_R \prod_{\tau=1}^{T} \Big[\int_R \exp\{(y_{d\tau} + \delta_{t\tau})(\boldsymbol{x}_{d\tau}\boldsymbol{\beta} + \phi_1 v_{1,d} + \phi_2 v_{2,d\tau})$$
$$- \nu_{d\tau} \exp\{\boldsymbol{x}_{d\tau}\boldsymbol{\beta} + \phi_1 v_{1,d} + \phi_2 v_{2,d\tau}\}\} f(v_{2,d\tau})\, dv_{2,d\tau}\Big] f(v_{1,d})\, dv_{1,d},$$

$$D_d = \int_{R^{T+1}} \prod_{\tau=1}^{T} \exp\Big\{y_{d\tau}(\boldsymbol{x}_{d\tau}\boldsymbol{\beta} + \phi_1 v_{1,d} + \phi_2 v_{2,d\tau})$$

$$- \nu_{d\tau} \exp\{\boldsymbol{x}_{d\tau}\boldsymbol{\beta} + \phi_1 v_{1,d} + \phi_2 v_{2,d\tau}\}\Big\} f(v_{1,d}) f(\boldsymbol{v}_{2,d})\, dv_{1,d} d\boldsymbol{v}_{2,d}$$

$$= \int_R \prod_{\tau=1}^{T} \Big[\int_R \exp\big\{ y_{d\tau}(\boldsymbol{x}_{d\tau}\boldsymbol{\beta} + \phi_1 v_{1,d} + \phi_2 v_{2,d\tau})$$

$$- \nu_{d\tau} \exp\{\boldsymbol{x}_{d\tau}\boldsymbol{\beta} + \phi_1 v_{1,d} + \phi_2 v_{2,d\tau}\}\big\} f(v_{2,d\tau})\, dv_{2,d\tau} \Big] f(v_{1,d})\, dv_{1,d},$$

where $\delta_{t\tau}$ is the Kronecker delta, i.e. $\delta_{t\tau} = 1$ if $t = \tau$ and $\delta_{t\tau} = 0$ otherwise.

The EBP of p_{dt} is $\hat{p}_{dt}(\hat{\boldsymbol{\theta}})$ and it can be approximated by a Monte Carlo method as follows.

1. Estimate $\hat{\boldsymbol{\theta}} = (\hat{\boldsymbol{\beta}}', \hat{\phi}_1, \hat{\phi}_2)'$.
2. For $s_1 = 1, \ldots, S_1$, $s_2 = 1, \ldots, S_2$, $\tau = 1, \ldots, T$, generate $v_{1,d}^{(s_1)}$, $v_{2,d\tau}^{(s_2)}$ i.i.d. $N(0, 1)$ and $v_{1,d}^{(S_1+s_1)} = -v_{1,d}^{(s_1)}$, $v_{2,d\tau}^{(S_2+s_2)} = -v_{2,d\tau}^{(s_2)}$.
3. Calculate $\hat{p}_{dt}(\hat{\boldsymbol{\theta}}) = \hat{A}_{dt}/\hat{D}_d$, where

$$\hat{A}_{dt} = \sum_{s_1=1}^{2S_1} \prod_{\tau=1}^{T} \sum_{s_2=1}^{2S_2} \exp\Big\{ (y_{d\tau} + \delta_{t\tau})(\boldsymbol{x}_{d\tau}\hat{\boldsymbol{\beta}} + \hat{\phi}_1 v_{1,d}^{(s_1)} + \hat{\phi}_2 v_{2,d\tau}^{(s_2)})$$

$$- \nu_{d\tau} \exp\{\boldsymbol{x}_{d\tau}\hat{\boldsymbol{\beta}} + \hat{\phi}_1 v_{1,d}^{(s_1)} + \hat{\phi}_2 v_{2,d\tau}^{(s_2)}\}\Big\},$$

$$\hat{D}_d = \sum_{s_1=1}^{2S_1} \prod_{\tau=1}^{T} \sum_{s_2=1}^{2S_2} \exp\Big\{ y_{d\tau}(\boldsymbol{x}_{d\tau}\hat{\boldsymbol{\beta}} + \hat{\phi}_1 v_{1,d}^{(s_1)} + \hat{\phi}_2 v_{2,d\tau}^{(s_2)})$$

$$- \nu_{d\tau} \exp\{\boldsymbol{x}_{d\tau}\hat{\boldsymbol{\beta}} + \hat{\phi}_1 v_{1,d}^{(s_1)} + \hat{\phi}_2 v_{2,d\tau}^{(s_2)}\}\Big\}.$$

The EBP of $\mu_{dt} = m_{dt} p_{dt}$ is $\hat{\mu}_{dt}(\hat{\boldsymbol{\theta}}) = m_{dt}\hat{p}_{dt}(\hat{\boldsymbol{\theta}})$.

Let us note that EBP's $\hat{p}_{dt}(\hat{\boldsymbol{\theta}})$ are usually computationally demanding and need not to be unbiased, for more details see Sect. 14.6.1. Other option is to consider computationally simpler plug-in predictor of p_{dt} which has the form

$$\tilde{p}_{dt} = \exp\{\boldsymbol{x}_{dt}\hat{\boldsymbol{\beta}} + \hat{\phi}_1 \hat{v}_{1,d} + \hat{\phi}_2 \hat{v}_{2,dt}\},$$

where $\hat{v}_{1,d}$, $\hat{v}_{2,dt}$ can be taken from the output of the ML-Laplace algorithm. Alternatively, the EBP of the random effects can be used to calculate the plug-in estimator. The BP of $v_{1,d}$ is

$$\hat{v}_{1,d}(\boldsymbol{\theta}) = E_\theta[v_{1,d}|\boldsymbol{y}_d] = \frac{\int_{R^{T+1}} v_{1,d} P(\boldsymbol{y}_d|v_{1,d}, \boldsymbol{v}_{2,d}) f(v_{1,d}) f(\boldsymbol{v}_{2,d})\, dv_{1,d} d\boldsymbol{v}_{2,d}}{\int_{R^{T+1}} P(\boldsymbol{y}_d|v_{1,d}, \boldsymbol{v}_{2,d}) f(v_{1,d}) f(\boldsymbol{v}_{2,d})\, dv_{1,d} d\boldsymbol{v}_{2,d}}$$

$$= \frac{A_{1,d}(\boldsymbol{y}_d, \boldsymbol{\theta})}{D_d(\boldsymbol{y}_d, \boldsymbol{\theta})},$$

where

$$A_{1,d}(y_d,\theta) = \int_R \prod_{\tau=1}^{T} \Big[\int_R \exp\big\{ y_{d\tau}(x_{d\tau}\beta + \phi_1 v_{1,d} + \phi_2 v_{2,d\tau}) \\ - \nu_{d\tau} \exp\{x_{d\tau}\beta + \phi_1 v_{1,d} + \phi_2 v_{2,d\tau}\}\big\} f(v_{2,d\tau})\, dv_{2,d\tau}\Big] v_{1,d} f(v_{1,d})\, dv_{1,d}.$$

The EBP of $v_{1,d}$ is $\hat{v}_{1,d} = \hat{v}_{1,d}(\hat{\theta})$ and it can be approximated as follows.

1. Estimate $\hat{\theta} = (\hat{\beta}', \hat{\phi}_1, \hat{\phi}_2)'$.
2. For $s_1 = 1, \ldots, S_1$, $s_2 = 1, \ldots, S_2$, $\tau = 1, \ldots, T$, generate $v_{1,d}^{(s_1)}$, $v_{2,d\tau}^{(s_2)}$ i.i.d. $N(0,1)$ and $v_{1,d}^{(S_1+s_1)} = -v_{1,d}^{(s_1)}$, $v_{2,d\tau}^{(S_2+s_2)} = -v_{2,d\tau}^{(s_2)}$.
3. Calculate $\hat{v}_{1,d}(\hat{\theta}) = \hat{A}_{1,d}/\hat{D}_d$, where

$$\hat{A}_{1,d} = \sum_{s_1=1}^{2S_1} \prod_{\tau=1}^{T} \sum_{s_2=1}^{2S_2} \exp\Big\{ y_{d\tau}(x_{d\tau}\hat{\beta} + \hat{\phi}_1 v_{1,d}^{(s_1)} + \hat{\phi}_2 v_{2,d\tau}^{(s_2)}) \\ - \nu_{d\tau} \exp\{x_{d\tau}\hat{\beta} + \hat{\phi}_1 v_{1,d}^{(s_1)} + \hat{\phi}_2 v_{2,d\tau}^{(s_2)}\}\Big\} v_{1,d}^{(s_1)},$$

$$\hat{D}_d = \sum_{s_1=1}^{2S_1} \prod_{\tau=1}^{T} \sum_{s_2=1}^{2S_2} \exp\Big\{ y_{d\tau}(x_{d\tau}\hat{\beta} + \hat{\phi}_1 v_{1,d}^{(s_1)} + \hat{\phi}_2 v_{2,d\tau}^{(s_2)}) \\ - \nu_{d\tau} \exp\{x_{d\tau}\hat{\beta} + \hat{\phi}_1 v_{1,d}^{(s_1)} + \hat{\phi}_2 v_{2,d\tau}^{(s_2)}\}\Big\}.$$

The BP of $v_{2,dt}$ is

$$\hat{v}_{2,dt}(\theta) = E_\theta[v_{2,dt}|y_{dt}] = \frac{\int_{R^2} v_{2,dt} P(y_{dt}|v_{1,d}, v_{2,dt}) f(v_{1,d}) f(v_{2,dt})\, dv_{1,d} dv_{2,dt}}{\int_{R^2} P(y_{dt}|v_{1,d}, v_{2,dt}) f(v_{1,d}) f(v_{2,dt})\, dv_{1,d} dv_{2,dt}} \\ = \frac{A_{2,dt}(y_{dt},\theta)}{D_{dt}(y_{dt},\theta)},$$

where

$$A_{2,dt}(y_{dt},\theta) = \int_{R^2} \exp\Big\{ y_{dt}(x_{dt}\beta + \phi_1 v_{1,d} + \phi_2 v_{2,dt}) \\ - m_{dt} \exp\{x_{dt}\beta + \phi_1 v_{1,d} + \phi_2 v_{2,dt}\}\Big\} v_{2,dt} f(v_{2,dt}) f(v_{1,d})\, dv_{2,dt} dv_{1,d},$$

and

$$D_{dt}(y_{dt},\theta) = \int_{R^2} \exp\Big\{ y_{dt}(x_{dt}\beta + \phi_1 v_{1,d} + \phi_2 v_{2,dt}) \\ - m_{dt} \exp\{x_{dt}\beta + \phi_1 v_{1,d} + \phi_2 v_{2,dt}\}\Big\} f(v_{2,dt}) f(v_{1,d})\, dv_{2,dt} dv_{1,d}.$$

The EBP of $v_{2,dt}$ is $\hat{v}_{2,dt} = \hat{v}_{2,dt}(\hat{\boldsymbol{\theta}})$ and it can be approximated as follows.

1. Estimate $\hat{\boldsymbol{\theta}} = (\hat{\boldsymbol{\beta}}', \hat{\phi}_1, \hat{\phi}_2)'$.
2. For $s_1 = 1, \ldots, S_1$, $s_2 = 1, \ldots, S_2$, generate $v_{1,d}^{(s_1)}$, $v_{2,dt}^{(s_2)}$ i.i.d. $N(0, 1)$ and $v_{1,d}^{(S_1+s_1)} = -v_{1,d}^{(s_1)}$, $v_{2,dt}^{(S_2+s_2)} = -v_{2,dt}^{(s_2)}$.
3. Calculate $\hat{v}_{2,dt}(\hat{\boldsymbol{\theta}}) = \hat{A}_{2,dt}/\hat{D}_{dt}$, where

$$\hat{A}_{2,dt} = \sum_{s_1=1}^{2S_1}\sum_{s_2=1}^{2S_2} \exp\Big\{y_{dt}(\boldsymbol{x}_{dt}\hat{\boldsymbol{\beta}} + \hat{\phi}_1 v_{1,d}^{(s_1)} + \hat{\phi}_2 v_{2,dt}^{(s_2)}) - m_{dt}\exp\{\boldsymbol{x}_{dt}\hat{\boldsymbol{\beta}} + \hat{\phi}_1 v_{1,d}^{(s_1)} + \hat{\phi}_2 v_{2,dt}^{(s_2)}\}\Big\} v_{2,dt}^{(s_2)},$$

$$\hat{D}_{dt} = \sum_{s_1=1}^{2S_1}\sum_{s_2=1}^{2S_2} \exp\Big\{y_{dt}(\boldsymbol{x}_{dt}\hat{\boldsymbol{\beta}} + \hat{\phi}_1 v_{1,d}^{(s_1)} + \hat{\phi}_2 v_{2,dt}^{(s_2)}) - m_{dt}\exp\{\boldsymbol{x}_{dt}\hat{\boldsymbol{\beta}} + \hat{\phi}_1 v_{1,d}^{(s_1)} + \hat{\phi}_2 v_{2,dt}^{(s_2)}\}\Big\}.$$

21.4.1 Bootstrap Estimation of the MSE

One possibility how to estimate the mean squared error of the EBP is to use a resampling method. The following procedure calculates a parametric bootstrap estimator of $MSE(\hat{p}_{dt})$.

1. Fit the model to the sample and calculate the estimate $\hat{\boldsymbol{\theta}} = (\hat{\boldsymbol{\beta}}', \hat{\phi}_1, \hat{\phi}_2)'$.
2. Repeat B times ($b = 1, \ldots, B$):
   a. Generate $v_{1,d}^{*(b)}$ and $v_{2,dt}^{*(b)}$ i.i.d. N(0,1), $d = 1, \ldots, D$, $t = 1, \ldots, T$. Calculate $p_{dt}^{*(b)} = \exp\{\boldsymbol{x}_{dt}\hat{\boldsymbol{\beta}} + \hat{\phi}_1 v_{1,d}^{*(b)} + \hat{\phi}_2 v_{2,dt}^{*(b)}\}$ and $y_{dt}^{*(b)} \sim \text{Poiss}(m_{dt}p_{dt}^{*(b)})$.
   b. On the basis of the bootstrap sample, calculate the estimate $\hat{\boldsymbol{\theta}}^{*(b)}$ and the EBP $\hat{p}_{dt}^{*(b)} = \hat{p}_{dt}(\hat{\boldsymbol{\theta}}^{*(b)})$ following the previous expressions.
3. Output: $mse^*(\hat{p}_{dt}) = \frac{1}{B}\sum_{b=1}^{B}\big(\hat{p}_{dt}^{*(b)} - p_{dt}^{*(b)}\big)^2$.

21.5 Simulation Experiment

The target of the simulation experiment is to empirically investigate the behavior of the bias and the MSE of the ML-Laplace estimators under the area-level Poisson mixed model with independent time effects. We take $D = 50, 100, 150$, $T = 4, 5, 9, 12$, $\beta_0 = -3$, $\beta_1 = 0.8$ and $\phi_1 = \phi_2 = 0.5$. For $d = 1, \ldots, D$, $t =$

Table 21.1 RBIAS (left) and RRMSE (right) in percentage

D		RBIAS in %				RRMSE in %			
		$T=4$	$T=5$	$T=9$	$T=12$	$T=4$	$T=5$	$T=9$	$T=12$
50	$\hat{\beta}_0$	0.5788	−0.2069	−0.0745	0.2091	5.7378	5.6174	5.0059	5.1732
	$\hat{\beta}_1$	−3.6057	0.6501	−0.2496	−1.4040	33.9986	34.4622	31.2976	31.4854
	$\hat{\phi}_1$	−5.0801	−4.6058	−4.1311	−3.6423	15.5854	14.1816	12.4302	12.3847
	$\hat{\phi}_2$	−0.7538	−1.0005	−0.8336	−0.6912	8.6294	7.8262	5.4636	4.8226
100	$\hat{\beta}_0$	0.5541	0.0237	−0.0751	0.2246	4.2332	3.8625	3.5918	3.5573
	$\hat{\beta}_1$	−3.7502	−0.8501	−1.1080	−1.6497	26.5024	23.4685	23.2016	23.3184
	$\hat{\phi}_1$	−3.9064	−2.2826	−1.7960	−3.2090	11.4934	9.1613	8.6055	8.3768
	$\hat{\phi}_2$	−1.2457	−0.7535	−1.0636	−1.0177	6.0855	5.6455	4.0863	3.6071
150	$\hat{\beta}_0$	0.1253	0.2350	0.3817	−0.0730	3.3286	3.3555	3.2290	2.6620
	$\hat{\beta}_1$	−1.0388	−1.3094	−2.5767	0.0374	21.7230	22.1789	20.0607	18.1947
	$\hat{\phi}_1$	−1.2860	−2.1321	−2.2118	−2.1803	8.6248	7.7478	7.4273	7.2987
	$\hat{\phi}_2$	−1.0496	−0.9248	−1.0696	−1.0034	4.9742	4.7510	3.3362	2.8830

$1, \ldots, T$, we put $m_{dt} = 100$, $x_{dt} = D^{-1}(d + tT^{-1})$ and generate $v_{1,d} \sim N(0,1)$, $v_{2,dt} \sim N(0,1)$ and

$$y_{dt} \sim \text{Poisson}(m_{dt} p_{dt}), \quad p_{dt} = \exp\{\beta_0 + x_{dt}\beta_1 + \phi_1 v_{1,d} + \phi_2 v_{2,dt}\}. \tag{21.9}$$

The steps of the simulation for checking the behavior of the ML-Laplace fitting algorithm are

1. Repeat K=1000 times ($k = 1, \ldots, K$).

1.1. Generate a sample $\{y_{dt}^{(k)} : d = 1, \ldots, D,\ t = 1, \ldots, T\}$, according to (21.9).
1.2. Calculate $\hat{\beta}_0^{(k)}, \hat{\beta}_1^{(k)}, \hat{\phi}_1^{(k)}, \hat{\phi}_2^{(k)}$.

2. Output: For $\theta \in \{\beta_0, \beta_1, \phi_1, \phi_2\}$, calculate

$$RBIAS = \frac{\sum_{k=1}^{K}(\hat{\theta}^{(k)} - \theta)}{K|\theta|}, \quad RRMSE = \frac{\sqrt{\frac{1}{K}\sum_{k=1}^{K}(\hat{\theta}^{(k)} - \theta)^2}}{|\theta|}.$$

Table 21.1 presents the RBIAS (left) and the RRMSE (right) in percentages for the parameter estimators calculated by the ML-Laplace approximation algorithm. Table 21.1 shows that both performance measures decrease as D or T increases.

21.6 R Codes for EBPs

This section gives R codes for fitting the area-level Poisson mixed model with independent time effects to data simulated under model (21.9). We install the R package lme4

```
if(!require(lme4)){
  install.packages("lme4")
  library(lme4)
}
```

We take $D = 50$ domains and $T = 4$ time instants. The following R codes simulates the area-level temporal count data.

```
# Number of domains and time instants
D <- 50; T <- 4
d1 <- seq(1:D); dd <- rep(d1,T)                    # Domain index
# Regression parameters
beta0 <- -3; beta1 <- 0.8
beta <- c(beta0, beta1)
phi1 <- phi2 <- 0.5                                # Standard deviation parameters
set.seed(123)                                      # Set the simulation seed
vv1 <- rnorm(D, 0, 1); v1 <- rep(vv1, T)           # Domain random effects
v2 <- rnorm(D*T, mean=0, sd=1)                     # Domain-time random effects
m <- rep(100, D*T)                                 # Offset parameter
# Time index and x-variables
x <- tt <- vector()
for (t in 1:T){
  for (d in 1:D){
    x[(t-1)*D+d] <- (d + t/T)/D
    tt[(t-1)*D+d] <- t
  }
}
# Generation of y
eta <- beta0 + beta1*x + phi1*v1 + phi2*v2         # Natural parameter
p <- exp(eta)                                      # Proportion parameter
mu <- m*p                                          # Mean parameter
y <- rpois(D*T, mu)                                # Poisson counts
```

We fit an area-level Poisson mixed model with independent time effects to the simulated sample counts. We apply the ML-Laplace approximation algorithm. Table 21.2 presents the estimated parameters and p-values. The estimated random effect standard deviations are $\hat{\phi}_1 = 0.40428$ and $\hat{\phi}_2 = 0.52069$.

```
glmm <- glmer(formula=y ~ x + (1|dd/tt), family=poisson, offset=log(m))
summary(glmm)
hbeta <- fixef(glmm)                         # Estimated regression parameters
# Estimated variances and standard deviations
var <- as.data.frame(VarCorr(glmm))
hphi2 <- var$sdcor[1]                        # Estimated standard deviation of udt
hphi1 <- var$sdcor[2]                        # Estimated standard deviation of ud
```

We calculate the modes of the random effects, the plug-in predictors of the sample counts and the model residuals.

```
r.effects <- ranef(glmm)          # Modes of random effects
udt <- r.effects[[1]]             # Modes of udt
```

Table 21.2 Estimated parameters of Poisson model

Parameter	Estimate	Std. error	z-value	p-value
β_0	−2.9847	0.1530	−19.51	0.0000
β_1	0.7638	0.2520	3.03	0.0024

```
ud <- r.effects[[2]]          # Modes of ud
mutilde <- fitted(glmm)       # Plug-in predictions of counts
residuals <- resid(glmm)      # Model residuals
# Summary of results
result <- data.frame(d=dd, t=tt, m, x, mu, y, mutilde)
```

Figure 21.1 (left) plots the raw residuals $\hat{r}_{dt} = y_{dt} - \tilde{\mu}_{dt}$ of the fitted Poisson mixed model versus the plug-in predictions of counts $\tilde{\mu}_{dt} = m_{dt}\tilde{p}_{dt}$. Figure 21.1 (right) plots the observed counts y_{dt} versus the plug-in predictions $\tilde{\mu}_{dt}$ of counts.

Figure 21.1 (left) shows that most of the residuals $\hat{r}_{dt}$'s are in the interval $(-2, 1)$. By observing Fig. 21.1 (right), we may conclude that the selected Poisson mixed model has a relatively good fit to the data.

We calculate the EBPs of sample counts.

```
set.seed(123)
S1 <- S2 <- 100
for (t in 1:T){                                    # Loop for time instants
  for (d in 1:D){                                  # Loop for domains
    numdt <- dendt <- 0
    for (s1 in 1:S1) {                             # Loop for outer sum
      v1 <- rnorm(n=1, mean=0, sd=1)
      v1 <- c(v1, -v1)
      numdtau <- dendtau <- 1
      for (tau in 1:T){                            # Loop for product
        condition <- tt==tau & dd==d
        vv2 <- rnorm(n=S2, mean=0, sd=1)
        vv2 <- c(vv2, -vv2)
        ydtau <- y[condition]
        deltau <- as.numeric(t==tau)
        xbetau <- hbeta[1] + hbeta[2]*x[condition] + hphi1*v1 + hphi2*vv2
        sum1num <- (ydtau + deltau)*xbetau
        sum2 <- m[condition] * exp(xbetau)
        # Numerator term
        numdtau <- numdtau * sum(exp(sum1num-sum2))
        sum1den <- ydtau * xbetau
        # Denominator term
        dendtau <- dendtau * sum(exp(sum1den-sum2))
      }
      # EBP numerator
      numdt <- numdt + numdtau
      # EBP denominator
      dendt <- dendt + dendtau
    }
    condition2 <- tt==t&dd==d
    # EBP
    result$ebp[condition2] <- m[condition2]*numdt/dendt
  }
}
```

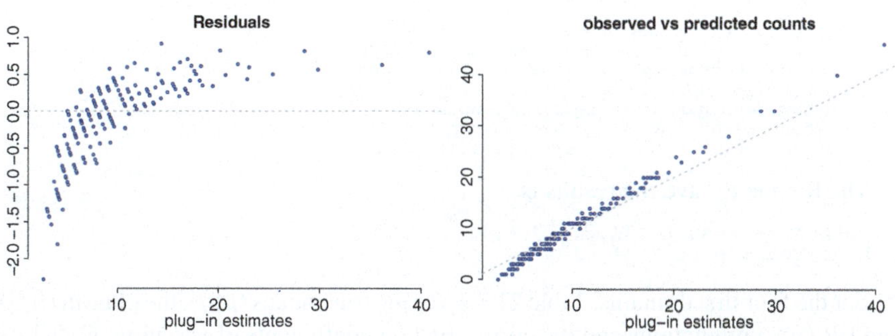

Fig. 21.1 Plots of residuals (left) and observed values (right) versus plug-in estimates

We calculate the parametric bootstrap estimates of the root-MSEs of the EBPs of sample counts.

```
set.seed(123); B <- 200
mseebppcum.star <- 0
dif.ebp.star <- dif.plugin.star <- list()
ebp.star <- plugin.star <- list()
for (b in 1:B){
  cat("iteration = ", b, "\n")
  # Generation of y
  v1.star <- rnorm(n=D, mean=0, sd=1)
  v2.star <- rnorm(n=D*T, mean=0, sd=1)
  p.star <- exp(hbeta[1] + hbeta[2]*x + hphi1*v1.star + hphi2*v2.star)
  y.star <- rpois(D*T, m*p.star)
  # Fitting boostrap model
  glmm.star <- glmer(formula=y.star ~ x + (1|dd/tt), family=poisson,
                     offset=log(m))
  hbeta.star <- fixef(glmm.star)
  var.star <- as.data.frame(VarCorr(glmm.star))
  hphi2.star <- var.star$sdcor[1]
  hphi1.star <- var.star$sdcor[2]
  # Plug-in bootstrap predictions of counts
  plugin.star[[b]] <- as.numeric(fitted(glmm.star))
  dif.plugin.star[[b]] <- (plugin.star[[b]] - m*p.star)^2
  # Loops to calculate EBPs
  ebp.star[[b]] <- vector()
  for (t in 1:T){
    for (d in 1:D){
      numdt.star <- dendt.star <- 0
      for (s1 in 1:S1) {
        vv1.star <- rnorm(n=1, mean=0, sd=1)
        vv1.star <- c(vv1.star, -vv1.star)
        numdtau.star <- dendtau.star <- 1
        for (tau in 1:T){
          condition <- tt==tau & dd==d
          vv2.star <- rnorm(n=S2, mean=0, sd=1)
          vv2.star <- c(vv2.star, -vv2.star)
          ydtau.star <- y.star[condition]
          deltau.star <- as.numeric(t==tau)
          xbetau.star <- hbeta.star[1] + hbeta.star[2]*x[condition] +
          hphi1.star*vv1.star + hphi2.star*vv2.star
          sum1num.star <- (ydtau.star + deltau.star)*xbetau.star
          sum2.star <- m[condition] * exp(xbetau.star)
          numdtau.star <- numdtau.star * sum(exp(sum1num.star-sum2.star))
          sum1den.star <- ydtau.star * xbetau.star
          dendtau.star <- dendtau.star * sum(exp(sum1den.star-sum2.star))
        }
        numdt.star <- numdt.star + numdtau.star
        dendt.star <- dendt.star + dendtau.star
      }
      condition2 <- tt==t&dd==d
      # EBP
      ebp.star[[b]][condition2] <- m[condition2]*numdt.star/dendt.star
    }
  }
  dif.ebp.star[[b]] <- (ebp.star[[b]] - m*p.star)^2
}
mse.plugin.star <- Reduce(f="+", dif.plugin.star)/B
mse.ebp.star <- Reduce(f="+", dif.ebp.star)/B
# root-MSE bootstrap estimates of plug-in and EBPs
result$rmseplugp.star <- sqrt(mse.plugin.star)
result$rmseebpp.star <- sqrt(mse.ebp.star)
result$cv.plugp.star <- 100*rmseplugp.star/mu
result$cv.eb.star <- 100*rmseebpp.star/mu
```

The R code to save the results is

```
output <- result[c(1,2,5,7,8,11,12)]
head(round(output,3), 10)
```

For the four first domains, Table 21.3 gives the true means (μ_{dt}), the plug-in ($\tilde{\mu}_{dt}$) and EB ($\hat{\mu}_{dt}$) predictions and the corresponding coefficients of variation, $CV(\tilde{\mu}_{dt})$ and $CV(\hat{\mu}_{dt})$, with common denominator μ_{dt}.

Table 21.3 Plug-in and EB predictions of sample counts

d	t	μ_{dt}	$\tilde{\mu}_{dt}$	$\hat{\mu}_{dt}$	$CV(\tilde{\mu}_{dt})$	$CV(\hat{\mu}_{dt})$
1	1	4.356	4.073	3.892	50.13	51.40
1	2	2.701	4.621	4.506	71.34	72.45
1	3	5.736	4.629	4.521	33.608	35.15
1	4	11.662	11.338	11.845	18.07	18.80
2	1	4.535	4.826	4.563	47.87	49.59
2	2	5.252	7.309	7.226	37.31	37.92
2	3	6.811	6.664	6.539	29.14	29.41
2	4	8.974	7.327	7.274	22.60	22.77
3	1	11.191	8.918	8.266	19.37	20.14
3	2	10.147	8.928	8.268	20.01	20.32
3	3	13.608	14.224	13.586	15.81	16.20
3	4	10.135	11.897	11.423	21.30	21.67
4	1	10.943	6.363	6.636	22.21	22.94
4	2	4.658	3.540	3.451	46.47	48.18
4	3	3.361	4.054	4.009	66.83	68.92
4	4	7.330	9.090	9.633	30.36	31.27

References

Boubeta, M., Lombardía, M.J., Morales, D.: Empirical best prediction under area-level Poisson mixed models. TEST, **25**, 548–569 (2016)

Boubeta, M., Lombardía, M.J., Morales, D.: Poisson mixed models for studying the poverty in small areas. Comput. Stat. Data Anal. **107**, 32–47 (2017)

Chambers R., Salvati, N., Tzavidis, N.: M-Quantile regression for binary data with application to small area estimation. Centre for Statistical and Survey Methodology, University of Wollongong (Working Paper 12–12, 2012, 24). http://ro.uow.edu.au/cssmwp/101. Accessed Jan 2016

Chandra, H., Salvati, N., Sud, U.C.: Disaggregate-level estimates of indebtedness in the state of Uttar Pradesh in India - an application of small area estimation technique. J. Appl. Stat. **38**(11), 2413–2432 (2011)

Esteban, M.D., Morales, D., Pérez, A., Santamaría, L.: Two area-level time models for estimating small area poverty indicators. J. Indian Soc. Agric. Stat. **66**(1), 75–89 (2012a)

Esteban, M.D., Morales, D., Pérez, A., Santamaría, L.: Small area estimation of poverty proportions under area-level time models. Comput. Stat. Data Anal. **56**, 2840–2855 (2012b)

González-Manteiga, W., Lombardía, M.J., Molina, I., Morales, D., Santamaría, L.: Estimation of the mean squared error of predictors of small area linear parameters under a logistic mixed model. Comput. Stat. Data Anal. **51**, 2720–2733 (2007)

González-Manteiga, W., Lombardía, M.J., Molina, I., Morales, D., Santamaría, L.: Bootstrap mean squared error of small-area EBLUP. J. Stat. Comput. Simul. **78**, 443–462 (2008)

Jiang, J.: Empirical best prediction for small-area inference based on generalized linear mixed models. J. Stat. Plan. Inference **111**, 117–127 (2003)

Jiang, J., Lahiri, P.: Empirical best prediction for small area inference with binary data. Ann. Inst. Stat. Math. **53**, 217–243 (2001)

Johnson, F.A., Chandra, H., Brown, J., Padmadas, S.: Estimating district-level births attended by skilled attendants in Ghana using demographic health survey and census data: an application of small area estimation technique. J. Off. Stat. **26**(2), 341–359 (2010)

López-Vizcaíno, E., Lombardía, M.J., Morales, D.: Multinomial-based small area estimation of labour force indicators. Stat. Model. **13**(2), 153–178 (2013)

López-Vizcaíno, E., Lombardía, M.J., Morales, D.: Small area estimation of labour force indicators under a multinomial model with correlated time and area effects. J. R. Stat. Assoc. Ser. A. **178**(3), 535–565 (2015)

Marhuenda, Y., Molina, I., Morales, D.: Small area estimation with spatio-temporal Fay-Herriot models. Comput. Stat. Data Anal. **58**, 308–325 (2013)

Marhuenda, Y., Morales, D., Pardo, M.C.: Information criteria for Fay-Herriot model selection. Comput. Stat. Data Anal. **70**, 268–280 (2014)

Morales, D., Pagliarella, M.C., Salvatore, R.: Small area estimation of poverty indicators under partitioned area-level time models. Stat. Oper. Res. Trans. **39**(1), 19–34 (2015)

Saei, A., Chambers, R.: Small area estimation under linear and generalized linear mixed models with time and area effects. Methodology Working Paper No. M03/15. Southampton Statistical Sciences Research Institute, University of Southampton (2003)

Tzavidis, N., Ranalli, M. G., Salvati, N., Dreassi, E., Chambers, R.: Robust small area prediction for counts. Stat. Methods Med. Res. **24**(3), 373–395 (2015)

Appendix A
Some Useful Formulas

Some formulas for calculating inverses of block matrices are

$$\begin{pmatrix} A_{11} & A_{12} \\ A_{21} & A_{22} \end{pmatrix}^{-1} = \begin{pmatrix} A^{11} & A^{12} \\ A^{21} & A^{22} \end{pmatrix}, \tag{A.1}$$

where

$$A^{11} = \left(A_{11} - A_{12}A_{22}^{-1}A_{21}\right)^{-1}, \quad A^{12} = -A^{11}A_{12}A_{22}^{-1}, \quad A^{21} = -A_{22}^{-1}A_{21}A^{11},$$

$$A^{22} = \left(A_{22} - A_{21}A_{11}^{-1}A_{12}\right)^{-1} = A_{22}^{-1} + A_{22}^{-1}A_{21}A^{11}A_{12}A_{22}^{-1}.$$

Further,

$$\begin{pmatrix} A & B \\ B' & C \end{pmatrix}^{-1} = \begin{pmatrix} A^{-1} & 0 \\ 0 & 0 \end{pmatrix} + \begin{pmatrix} -A^{-1}B \\ I \end{pmatrix}\left(C - B'A^{-1}B\right)^{-1}\left(-B'A^{-1}, I\right),$$

$$(A + BCD)^{-1} = A^{-1} - A^{-1}B(C^{-1} + DA^{-1}B)^{-1}DA^{-1},$$

$$(A + uv')^{-1} = A^{-1} - \frac{A^{-1}uv'A^{-1}}{1 + v'A^{-1}u}.$$

The following formulas may be useful for calculating determinants of matrices in block or other form,

$$\begin{vmatrix} A_{11} & A_{12} \\ A_{21} & A_{22} \end{vmatrix} = |A_{11}|\left|A_{22} - A_{21}A_{11}^{-1}A_{12}\right| = |A_{22}|\left|A_{11} - A_{12}A_{22}^{-1}A_{21}\right|,$$

$$|A + yy'| = |A|(1 + y'A^{-1}y) \quad \text{if } A \text{ is not singular},$$

$$|a1_n1_n' + bI_n| = b^{n-1}(na + b), \quad a > 0,\ b > 0.$$

D. Morales et al., *A Course on Small Area Estimation and Mixed Models*, Statistics for Social and Behavioral Sciences, https://doi.org/10.1007/978-3-030-63757-6

Some formulas for matrix derivatives are

$$\frac{\partial \log |\boldsymbol{V}|}{\partial \sigma_i^2} = \mathrm{tr}\left\{\boldsymbol{V}^{-1}\frac{\partial \boldsymbol{V}}{\partial \sigma_i^2}\right\}, \quad \frac{\partial \boldsymbol{V}^{-1}}{\partial \sigma_i^2} = -\boldsymbol{V}^{-1}\frac{\partial \boldsymbol{V}}{\partial \sigma_i^2}\boldsymbol{V}^{-1}. \tag{A.2}$$

Let us now consider formulas for expectation of quadratic forms, and let $\boldsymbol{y}$ be a random vector and $\boldsymbol{Q}$ a symmetric matrix. It holds

$$E[\boldsymbol{y}'\boldsymbol{Q}\boldsymbol{y}] = \mathrm{tr}\{\boldsymbol{Q}\,\mathrm{var}(\boldsymbol{y})\} + E[\boldsymbol{y}]'\boldsymbol{Q}E[\boldsymbol{y}]. \tag{A.3}$$

If, moreover, $\boldsymbol{y}$ has multivariate normal distribution, then

$$\mathrm{var}(\boldsymbol{y}'\boldsymbol{Q}\boldsymbol{y}) = 2\mathrm{tr}\left\{(\boldsymbol{Q}\,\mathrm{var}(\boldsymbol{y}))^2\right\} + 4E[\boldsymbol{y}]'\boldsymbol{Q}\,\mathrm{var}(\boldsymbol{y})\,\boldsymbol{Q}E[\boldsymbol{y}], \tag{A.4}$$

$$\mathrm{cov}(\boldsymbol{y}'\boldsymbol{A}\boldsymbol{y},\, \boldsymbol{y}'\boldsymbol{B}\boldsymbol{y}) = 2\mathrm{tr}\{\boldsymbol{A}\mathrm{var}(\boldsymbol{y})\boldsymbol{B}\mathrm{var}(\boldsymbol{y})\} + 4E[\boldsymbol{y}]'\boldsymbol{A}\mathrm{var}(\boldsymbol{y})\boldsymbol{B}E[\boldsymbol{y}]. \tag{A.5}$$

Lemma A.1 (Prasad-Rao 1990) *Let $\boldsymbol{A}_1$ and $\boldsymbol{A}_2$ be $n \times n$ matrices, and let $\boldsymbol{y} \sim N_n(\boldsymbol{0}, \boldsymbol{V})$, where $\boldsymbol{V}$ is positive definite. Then,*

(a) $E[\boldsymbol{y}(\boldsymbol{y}'\boldsymbol{A}_s\boldsymbol{y})\boldsymbol{y}'] = tr(\boldsymbol{A}_s\boldsymbol{V})\boldsymbol{V} + 2\boldsymbol{V}\boldsymbol{A}_s\boldsymbol{V}, \quad s = 1, 2,$

(b) $E[(\boldsymbol{y}'\boldsymbol{A}_1\boldsymbol{y})(\boldsymbol{y}'\boldsymbol{A}_2\boldsymbol{y})] = 2tr(\boldsymbol{A}_1\boldsymbol{V}\boldsymbol{A}_2\boldsymbol{V}) + tr(\boldsymbol{A}_1\boldsymbol{V})tr(\boldsymbol{A}_2\boldsymbol{V}),$

(c) $E[\boldsymbol{y}(\boldsymbol{y}'\boldsymbol{A}_1\boldsymbol{y})(\boldsymbol{y}'\boldsymbol{A}_2\boldsymbol{y})\boldsymbol{y}'] = tr(\boldsymbol{A}_1\boldsymbol{V})tr(\boldsymbol{A}_2\boldsymbol{V})\boldsymbol{V} + 2tr(\boldsymbol{A}_1\boldsymbol{V})\boldsymbol{V}\boldsymbol{A}_2\boldsymbol{V}$

$+2tr(\boldsymbol{A}_2\boldsymbol{V})\boldsymbol{V}\boldsymbol{A}_1\boldsymbol{V} + 2tr(\boldsymbol{A}_1\boldsymbol{V}\boldsymbol{A}_2\boldsymbol{V})\boldsymbol{V} + 4\boldsymbol{V}\boldsymbol{A}_1\boldsymbol{V}\boldsymbol{A}_2\boldsymbol{V} + 4\boldsymbol{V}\boldsymbol{A}_2\boldsymbol{V}\boldsymbol{A}_1\boldsymbol{V}.$

Lemma A.2 (Prasad-Rao 1990) *Let $\boldsymbol{y} \sim N_n(\boldsymbol{0}, \boldsymbol{V})$, $z_j = \boldsymbol{\lambda}_j'\boldsymbol{y}$ and $q_j = \boldsymbol{y}'\boldsymbol{A}_j\boldsymbol{y}$, $j = 1, \ldots, p$, where $\boldsymbol{\lambda}_j$ is $n \times 1$ and $\boldsymbol{A}_j$ is $n \times n$. Let $\boldsymbol{z} = (z_1, \ldots, z_p)'$ and $\boldsymbol{q} = (q_1, \ldots, q_p)'$ with covariance matrices $\boldsymbol{V}_z$ and $\boldsymbol{V}_q$, respectively. Then,*

$$E[(\boldsymbol{z}'(\boldsymbol{q} - E[\boldsymbol{q}]))^2] = tr(\boldsymbol{V}_z\boldsymbol{V}_q) + 4\sum_{j=1}^{p}\sum_{i=1}^{p}\{\boldsymbol{\lambda}_j'\boldsymbol{V}\boldsymbol{A}_j\boldsymbol{V}\boldsymbol{A}_i\boldsymbol{V}\boldsymbol{\lambda}_i + \boldsymbol{\lambda}_j'\boldsymbol{V}\boldsymbol{A}_i\boldsymbol{V}\boldsymbol{A}_j\boldsymbol{V}\boldsymbol{\lambda}_i\},$$

and

$$E[z_iz_j(q_i - E[q_i])(q_j - E[q_j])] = \boldsymbol{\lambda}_i'E[\boldsymbol{y}(\boldsymbol{y}'\boldsymbol{A}_i\boldsymbol{y})(\boldsymbol{y}'\boldsymbol{A}_j\boldsymbol{y})\boldsymbol{y}']\boldsymbol{\lambda}_j - E[q_i]\boldsymbol{\lambda}_i'E[\boldsymbol{y}(\boldsymbol{y}'\boldsymbol{A}_j\boldsymbol{y})\boldsymbol{y}']\boldsymbol{\lambda}_j - E[q_i]\boldsymbol{\lambda}_i'E[\boldsymbol{y}(\boldsymbol{y}'\boldsymbol{A}_i\boldsymbol{y})\boldsymbol{y}']\boldsymbol{\lambda}_j + E[q_i]E[q_j]\boldsymbol{\lambda}_i'\boldsymbol{V}\boldsymbol{\lambda}_j.$$

Index

A
Additive parameter, 245, 317
Akaike information criterion (AIC), 372, 402, 559
ANCOVA model, 92
ANOVA estimator, 99
Area-level models, 3
Asymptotic distribution, 122, 127, 272, 359, 464, 522
Average income, 252, 255, 258, 323

B
Basic synthetic estimator, 42
Bayesian prediction, 443
Benchmarking, 60
Best linear unbiased estimator (BLUE), 75, 79, 157, 269, 335, 340
Best linear unbiased predictor (BLUP), 78, 79, 83, 85, 97, 99, 113, 157, 192, 269, 335, 340, 421, 426, 512
Best predictor (BP), 244, 426, 562
Bivariate Fay-Herriot model, 510
Bootstrap, 409

C
Census, 13
Conditional distribution, 240, 243, 306

D
Design-based bias, 15
Design-based bootstrap, 26, 60
Design-based jackknife, 27, 61
Design-based variance, 15
Dir1 estimator, 17, 195
Dir2 estimator, 23, 196
Domain mean, 191, 245, 338, 343, 378, 408

E
ELL predictor, 256, 258
EM algorithm, 360, 362, 552
Empirical best linear unbiased estimator (EBLUE), 158, 512
Empirical best linear unbiased predictor (EBLUP), 158, 192, 210, 286, 337, 343, 427, 463, 474, 497, 503, 512
Empirical best predictors (EBP), 245, 246, 249, 251, 253–255, 317, 373, 403, 563, 586
Error variance, 77, 80, 85, 87, 100, 427
Estimation error, 76
Expansive estimator, 85

F
Fay–Herriot model, 424
Finite population, 14, 16
Fisher information matrix, 121, 123, 160, 164, 433, 493, 517
Fisher-scoring algorithm, 121, 123, 161, 165, 433, 517
Fixed effects, 97, 101, 231

G
Generalized inverse, 93
Generalized linear mixed models, 4

D. Morales et al., *A Course on Small Area Estimation and Mixed Models*, Statistics for Social and Behavioral Sciences, https://doi.org/10.1007/978-3-030-63757-6

Generalized regression estimator, 2, 49, 106
Generalized variance function, 428
Gibbs sampling, 450

H
Henderson 3 (H3) estimators, 137, 143, 148, 171, 174, 228, 275, 429
Hierarchical Bayes predictor, 446, 448

I
Inclusion probabilities, 15

L
Labor force survey, 16, 55
Lagrangian function, 58, 79, 81, 114, 422
Laplace approximation, 393
Least squares estimator, 83, 102, 104, 105, 112
Likelihood equations, 93
Linear estimator, 76
Linear mixed models, 4, 112, 117
Linear models, 4, 92
Linear regression estimator, 85
Logistic regression estimator, 54
Logit mixed model, 354
Log-likelihood, 93, 119, 122, 132, 134, 159, 433, 492, 515

M
Maximum likelihood estimators, 99, 119, 122, 159, 163, 229, 432, 492, 515
Mean squared error, 210, 220, 227, 232, 259, 287, 380, 437, 466, 476, 565
ML-Laplace algorithm, 367, 371, 397, 555, 576
MM estimators, 355, 429, 549
Model-assisted estimator, 195
Moments, 144, 175, 179
MSE estimator, 221, 223, 228, 229, 344, 380, 526, 567
MSE matrix, 522

N
Nested error regression model, 156, 190, 223, 241
Newton-Raphson algorithm, 371, 397, 584

P
Parametric bootstrap, 194, 259, 261, 324, 380, 409, 442, 469, 480, 494, 568, 588
Penalized quasi-likelihood (PQL), 560
Poisson mixed model, 548
Population, 13
Population size, 16
Post-stratified estimator, 45
Poverty gap, 255, 257, 322
Poverty proportion, 249, 250, 254, 257, 321
Prediction theorem, 80, 421
Prediction theory, 73, 76
Predictively unbiased, 77, 80
Projective estimator, 99, 104, 106

R
Ratio estimator, 21, 76
R codes, 30, 62, 88, 107, 148, 183, 201, 235, 263, 297, 325, 349, 386, 412, 454, 483, 505, 534, 569, 590
Regression synthetic estimator, 102
REML estimators, 123, 127, 135, 166, 169, 229, 434, 463, 474, 497, 503, 518
REML Fisher information matrix, 125, 128, 167, 169, 271, 336, 341, 435, 464, 475, 521
REML Fisher-scoring algorithm, 126, 128, 168, 170, 272, 436, 464, 475, 497, 521
REML log-likelihood, 126, 128, 167, 271, 335, 340, 434, 463, 474, 518
R functions, 185, 539
RRC model, 334, 339

S
Sample indicator, 15
Sample size dependent estimator, 47
Sampling weights, 16, 101
Simple random sampling, 15
Simulations, 196, 232, 294, 295, 411, 470, 480, 498, 504, 531, 588
Small area estimation, 1
Spatial Fay–Herriot model, 490
Spatio-temporal Fay–Herriot model, 495, 501
Subdomain-level NER model, 253
Subdomain mean, 286
Sum of squared residuals, 82, 138, 171
Survey sample, 1, 14, 16
Synthetic predictor, 441

T
Taylor linearization, 54, 215, 221
Temporal Fay–Herriot model, 462, 472
Temporal Poisson mixed model, 574
Two-fold logit mixed model, 392
Two-fold NER model, 268, 302

U
Unemployment rate, 54
Unit-level models, 3

W
Weight calibration, 58

The manufacturer's authorised representative in the EU is Springer Nature Customer Service Centre GmbH, Europaplatz 3, 69115 Heidelberg, Germany. If you have any concerns regarding our products, please contact ProductSafety@springernature.com

Printed and bound by CPI Group (UK) Ltd, Croydon, CR0 4YY
22/09/2026
02221048-0014